# Beginning & Intermediate Algebra

# Beginning & Intermediate Algebra

## Fourth Edition

### *Elayn Martin-Gay*

*University of New Orleans*

PEARSON

Prentice Hall

Upper Saddle River, New Jersey 07458

**Library of Congress Cataloging-in-Publication Data**

Martin-Gay, K. Elayn
  Beginning & intermediate algebra/K. Elayn Martin-Gay—4th ed.
    p. cm.
  Includes index.
  ISBN 0-13-600731-7
  1. Algebra  I. Title

**President:** *Greg Tobin*
**Editor in Chief:** *Paul Murphy*
**Editorial Director, Mathematics:** *Christine Hoag*
**Sponsoring Editor:** *Mary Beckwith*
**Assistant Editor:** *Christine Whitlock*
**Editorial Assistant:** *Georgina Brown*
**Production Management:** *Elm Street Publishing Services*
**Senior Managing Editor:** *Linda Mihatov Behrens*
**Operations Specialist:** *Ilene Kahn*
**Senior Operations Supervisor:** *Diane Peirano*
**Vice President and Executive Director of Development:** *Carol Trueheart*
**Development Editor:** *Lisa Collette*
**Media Producer:** *Audra J. Walsh*
**Lead Media Project Manager:** *Richard Bretan*
**Software Development:** *MyMathLab: Jennifer Sparkes, Media Producer*
*MathXL: Janet Szykolony, Software Editor*
*TestGen: Ted Hartman, Software Editor*
**Vice President and Director of Marketing:** *Amy Cronin*
**Executive Marketing Manager:** *Kate Valentine*
**Senior Marketing Manager:** *Michelle Renda*
**Marketing Manager:** *Marlana Voerster*
**Marketing Assistants:** *Jill Kapinus and Nathaniel Koven*
**Senior Art Director:** *Juan R. López*
**Interior Designer:** *Mike Fruhbeis*
**Cover Art Direction, Design, and Illustration:** *Kenny Beck*
**AV Project Manager:** *Thomas Benfatti*
**Director, Image Resource Center:** *Melinda Patelli*
**Manager, Rights and Permissions:** *Zina Arabia*
**Manager, Visual Research:** *Beth Brenzel*
**Image Permission Coordinator:** *Craig Jones*
**Photo Researcher:** *Kathy Ringrose*
**Art Studios:** *Scientific Illustrators/Laserwords*
**Compositor:** *ICC Macmillan Inc.*

**PEARSON**
Prentice
Hall

© 2009, 2005, 2001, 1996 Pearson Education, Inc.
Pearson Prentice Hall
Pearson Education, Inc.
Upper Saddle River, New Jersey 07458

10 9 8 7 6 5 4 3

ISBN-10     0-13-600731-7
ISBN-13 978-0-13-600731-9

Pearson Education LTD., *London*
Pearson Education Australia PTY. Limited, *Sydney*
Pearson Education Singapore PTE. LTD.
Pearson Education North Asia, LTD., *Hong Kong*
Pearson Education Canada, LTD., *Toronto*
Pearson Educacíon de Mexico, S.A., de C.V.
Pearson Education, Japan, *Tokyo*
Pearson Education Malaysia, PTE. LTD.

This book is dedicated in memory of **Karin Elizabeth Wagner**

*I find the great thing in this world is not so much where we stand, as in what direction we are moving: To reach the port of heaven, we must sail sometimes with the wind and sometimes against it— but we must sail, and not drift, nor lie at anchor.*

—Oliver Wendell Holmes

# Contents

CHAPTER

# 4

## SOLVING SYSTEMS OF LINEAR EQUATIONS  245

CHAPTER

# 5

## EXPONENTS AND POLYNOMIALS  300

CHAPTER

# 6

## FACTORING POLYNOMIALS  366

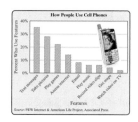

# Tools to Help Students Succeed

Your textbook includes a number of features designed to help you succeed in this math course—as well as the next math course you take. These features include:

| Feature | Benefit | Page |
|---|---|---|
| **Well-crafted Exercise Sets:** We learn math by doing math | The exercise sets in your text offer an ample number of exercises carefully ordered so you can master basic mathematical skills and concepts while developing all-important problem solving skills.<br><br>Exercise sets include Mixed Practice exercises to help you master multiple key concepts, as well as Vocabulary and Readiness Check, Writing, Applications, Concept Check, Concept Extension, and Review and Preview Exercises. | 208–212 |
| **Study Skills Builders:** Maximize your chances for success | Study Skills Builders reinforce the material in *Section 1.1—Tips for Success in Mathematics.*<br><br>Study Skills Builders are a great resource for study ideas and self-assessment to maximize your opportunity for success in this course. Take your new study skills with you to help you succeed in your next math course. | 92 |
| **The Bigger Picture:** Succeed in this math course and the next one you take | The Bigger Picture focuses on the key concepts of this course—simplifying expressions and solving equations and inequalities—and asks you to keep an ongoing study guide so you can simplify expressions and solve equations and inequalities, and recognize the difference between them.<br><br>A strong foundation in simplifying expressions and solving equations and inequalities will help you succeed in this algebra course, as well as the next math course you take. | 153 |
| **Examples:** Step-by-step instruction for you | Examples in the text provide you with clear, concise step-by-step instructions to help you learn. Annotations in the examples provide additional instruction. | 77 |
| **Helpful Hints:** Help where you'll need it most | Helpful Hints provide tips and advice at exact locations where students need it most. Strategically placed where you might have the most difficulty, Helpful Hints will help you work through common trouble spots. | 77 |
| **Practice Exercises:** Immediate reinforcement | New Practice exercises offer immediate reinforcement after every example. Try each Practice exercise after studying the corresponding example to make sure you have a good working knowledge of the concept. | 77 |
| **Integrated Review:** Mid-chapter progress check | To ensure you understand the key concepts covered in the first sections of the chapter, work the exercises in the Integrated Review before you continue with the rest of the chapter. | 213 |
| **Vocabulary and Readiness Check, Vocabulary Check:** Key terms and vocabulary | Use the Vocabulary and Readiness Checks to build your vocabulary and warm-up on concepts in the section. Make sure you understand key terms and vocabulary in each chapter with the end-of-chapter Vocabulary Check. | 208, 234 |
| **Chapter Highlights:** Study smart | Chapter Highlights outline the key concepts of the chapter along with examples to help you focus your studying efforts as you prepare for your test. | 234–238 |
| **Chapter Test:** Take a practice test | In preparation for your classroom test, take this practice test to make sure you understand the key topics in the chapter. Be sure to use the **Chapter Test Prep Video** included with this text to see the author present a fully worked-out solution to each exercise in the Chapter Test. | 242 |
| **Practice Final Exam:** Take a practice final | In preparation for your final, take the practice final exam found in Appendix A.2. **Martin-Gay's Interactive DVD/CD Lecture Series** includes the Practice Final Exam and provides you with full video solutions to each exercise. Overview clips provide a brief overview on how to approach different problem types. | 841 |

# Martin-Gay's VIDEO RESOURCES
## Help Students Succeed

### MARTIN-GAY'S CHAPTER TEST PREP VIDEO (AVAILABLE WITH THIS TEXT) TEST PREP VIDEO

- Provides students with help during their most "teachable moment"—while they are studying for a test.
- Text author Elayn Martin-Gay presents step-by-step solutions to the exact exercises found in each Chapter Test in the book.
- Easy video navigation allows students to instantly access the worked-out solutions to the exercises they want to review.
- Close captioned in English and Spanish.

### NEW MARTIN-GAY'S INTERACTIVE DVD/CD LECTURE SERIES

Martin-Gay's video series has been comprehensively updated to address the way today's students study and learn. The new videos offer students active learning at their pace, with the following resources and more:

- **A complete lecture** for each section of the text, presented by Elayn Martin-Gay. Students can easily review a section or a specific topic before a homework assignment, quiz, or test. Exercises in the text marked with the 🎥 are worked on the video.
- A **new interface** with menu and navigation features helps students quickly find and focus on the examples and exercises they need to review.
- Martin-Gay's "pop-ups" reinforce key terms and definitions and are a great support for multiple learning styles.
- A new **Practice Final Exam Video** helps students prepare for the final exam. This Practice Final Exam is included in the text in Appendix A.2. At the click of a button, students can watch the full solutions to each exercise on the exam when they need help. Overview clips provide a brief overview on how to approach different problem types—just as they will need to do on a Final Exam.
- **Interactive Concept Checks** allow students to check their understanding of essential concepts. Like the concept checks in the text, these multiple choice exercises focus on common misunderstandings. After making their answer selection, students are told whether they're correct or not, and why! Elayn also presents the full solution.
- **Study Skills Builders** help students develop effective study habits and reinforce the advice provided in Section 1.1, Tips for Success in Mathematics, found in the text and video.
- **Close-captioned in Spanish and English**
- Ask your bookstore for information about Martin-Gay's *Beginning & Intermediate Algebra*, Fourth Edition Interactive DVD/CD Lecture Series or visit www.mypearsonstore.com.

You will find Interactive Concept Checks and Study Skills Builders in the following sections on the Interactive DVD/CD Lecture Series:

| **Interactive Concept Checks Section** | | **Study Skills Builders Section** |
|---|---|---|
| 1.3 | 10.1 | 1.2 Time Management |
| 1.4 | 10.2 | 2.4 Are You Familiar with the Resources Available with Your Textbook? |
| 1.5 | 10.4 | |
| 1.6 | 10.5 | 2.5 Have You Decided to Complete This Course Successfully? |
| 1.7 | 10.6 | |
| 1.8 | 11.1 | 2.6 How Well Do You Know Your Textbook? |
| 2.1 | 11.2 | |
| 2.2 | 11.3 | 2.7 Tips for Studying for an Exam |
| 2.3 | 11.4 | 3.1 How Are Your Homework Assignments Going? |
| 2.8 | 11.6 | |
| 3.3 | 12.1 | 3.2 Doing Your Homework Online |
| 3.4 | 12.2 | 4.5 Tips for Studying for an Exam |
| 3.5 | 12.3 | 5.1 Are You Satisfied with Your Performance on a Particular Quiz? |
| 3.6 | 12.5 | |
| 4.1 | 12.6 | 5.2 Are You Organized? |
| 4.2 | 13.1 | |
| 4.3 | 13.2 | 5.5 Are You Familiar with the Resources Available with Your Textbook? |
| 4.4 | 13.3 | |
| 5.3 | 14.1 | 5.7 What to Do the Day of an Exam |
| 5.4 | 14.2 | 6.4 Have You Decided to Complete This Course Successfully? |
| 5.6 | 14.3 | |
| 6.1 | | 6.7 How Well Do You Know Your Textbook? |
| 6.2 | | |
| 6.3 | | 7.4 Tips for Studying for an Exam |
| 6.5 | | 8.2 How Are Your Homework Assignments Going? |
| 6.6 | | 9.4 What to Do the Day of an Exam |
| Ch. 6 Integrated Review | | 10.3 Are You Familiar with the Resources Available with Your Textbook? |
| 7.1 | | |
| 7.2 | | 10.7 Have You Decided to Complete This Course Successfully? |
| 7.3 | | |
| 7.5 | | 11.5 How Well Do You Know Your Textbook? |
| 7.6 | | |
| 7.7 | | 12.4 Tips for Studying for an Exam |
| 8.1 | | 12.7 Are You Satisfied with Your Performance on a Particular Quiz or Exam? |
| 8.3 | | |
| 8.4 | | 13.4 What to Do the Day of an Exam |
| 9.1 | | 14.4 How Are Your Homework Assignments Going? |
| 9.2 | | |
| 9.3 | | 14.5 Preparing for Your Final Exam |

# Additional Resources to Help You Succeed

| Student Study Pack |
|---|
| A single, easy-to-use package—available bundled with your textbook or by itself—for purchase through your bookstore. This package contains the following resources to help you succeed: |

**Student Solutions Manual**
- Contains worked-out solutions to odd-numbered exercises from each section exercise set, all Practice exercises, Vocabulary and Readiness Check Exercises, and all exercises found in the Chapter Review, Chapter Tests, and Integrated Reviews.

**Martin-Gay's Interactive Video Lectures**
- Text author Elayn Martin-Gay presents the key concepts from every section of the text with 15–20 minute mini-lectures. Students can easily review a section or a specific topic before a homework assignment, quiz, or test.
- A new interface allows easy navigation through the lesson.
- Includes fully worked-out solutions to exercises marked with an icon in each section. Also includes *Section 1.1, Tips for Success in Mathematics.*
- Close-captioned in English and Spanish.

**Pearson Tutor Center**

| Online Homework and Tutorial Resources |
|---|

**MyMathLab®** *MyMathLab*

MyMathLab is a series of text-specific, easily customizable online courses for Pearson Education's textbooks in mathematics and statistics. Powered by CourseCompass™ (our online teaching and learning environment) and MathXL® (our online homework, tutorial, and assessment system), MyMathLab gives you the tools you need to deliver all or a portion of your course online, whether your students are in a lab setting or working from home. MyMathLab provides a rich and flexible set of course materials, featuring free-response exercises that are algorithmically generated for unlimited practice and mastery. Students can also use online tools, such as video lectures, animations, and a multimedia textbook, to independently improve their understanding and performance. Instructors can use MyMathLab's homework and test managers to select and assign online exercises correlated directly to the textbook, and they can also create and assign their own online exercises and import TestGen tests for added flexibility. MyMathLab's online gradebook—designed specifically for mathematics and statistics—automatically tracks students' homework and test results and gives the instructor control over how to calculate final grades. Instructors can also add offline (paper-and-pencil) grades to the gradebook. Includes access to the Pearson Tutor Center. MyMathLab is available to qualified adopters. For more information, visit our website at www.mymathlab.com or contact your sales representative.

**MathXL®** PRACTICE

MathXL® is a powerful online homework, tutorial, and assessment system that accompanies Pearson Education's textbooks in mathematics or statistics. With MathXL, instructors can create, edit, and assign online homework and tests using algorithmically generated exercises correlated at the objective level to the textbook. They can also create and assign their own online exercises and import TestGen tests for added flexibility. All student work is tracked in MathXL's online gradebook. Students can take chapter tests in MathXL and receive personalized study plans based on their test results. The study plan diagnoses weaknesses and links students directly to tutorial exercises for the objectives they need to study and retest. Students can also access supplemental animations and video clips directly from selected exercises. MathXL is available to qualified adopters. For more information, visit our website at www.mathxl.com, or contact your sales representative.

# Preface

## ABOUT THE BOOK

*Beginning & Intermediate Algebra, Fourth Edition* was written to provide a **solid foundation in algebra** for students who might not have had previous experience in algebra. Specific care has been taken to ensure that students have the most **up-to-date and relevant** text preparation for their next mathematics course, as well as to help students succeed in nonmathematical courses that require a grasp of algebraic fundamentals. I have tried to achieve this by writing a user-friendly text that is keyed to objectives and contains many worked-out examples. The basic concepts of graphing and functions are introduced early. Problem solving techniques, real-life and real-data applications, data interpretation, appropriate use of technology, mental mathematics, number sense, critical thinking, decision-making, and geometric concepts are emphasized and integrated throughout the book.

The many factors that contributed to the success of the previous editions have been retained. In preparing this edition, I considered the comments and suggestions of colleagues throughout the country, students, and many users of the prior editions. The AMATYC Crossroads in Mathematics: Standards for Introductory College Mathematics before Calculus and the MAA and NCTM standards (plus Addenda), together with advances in technology, also influenced the writing of this text.

Throughout the series, pedagogical features are designed to develop student proficiency in algebra and problem solving, and to prepare students for future courses.

## WHAT'S NEW IN THE FOURTH EDITION

**New Martin-Gay's Interactive DVD/CD Lecture Series,** featuring Elayn Martin-Gay, provides students with active learning at their pace. The new videos offer the following resources and more:

- A complete lecture for each section of the text. A new interface with menu and navigation features allows students to quickly find and focus on the examples and exercises they need to review.
- The new **Practice Final Exam** helps students prepare for an end of course final. Students can watch full video solutions to each exercise. Overview clips give students a brief overview on how to approach different problem types.
- **Interactive Concept Check exercises** that allow students to check their understanding of essential concepts.

**A New Three-Point System** to support students and instructors:

- **Examples** in the text prepare students for class and homework.
- New **Practice Exercises** are paired with each text example and are ready-made for use in the classroom and can be assigned as homework.
- **Classroom Examples,** found only in the Annotated Instructor's Edition, are also paired with each example to give instructors a convenient way to further illustrate skills and concepts.

### ENHANCED EXERCISE SETS

- **New Vocabulary and Readiness exercises** appear at the beginning of exercise sets. The **Vocabulary** exercises reinforce a student's understanding of new terms so that forthcoming instructions in the exercise set are clear. The **Readiness** exercises serve as a warm-up on the skills and concepts necessary to complete the exercise set.

- **NEW! Concept Check exercises** have been added to the section exercise sets. These exercises are related to the Concept Check(s) found within the section. They help students measure their understanding of key concepts by focusing on common trouble areas. These exercises may ask students to identify a common error, and/or provide an explanation.

- **New Mixed Review** exercises are included at the end of the Chapter Review. These exercises require students to determine the problem type and strategy needed in order to solve it.

- **New Practice Final Exam—included in Appendix A.2** helps students prepare for an end of course final. Martin-Gay's Interactive DVD/CD Lecture Series provides students with solutions to all exercises on the final. Overview clips give a brief introduction on how to approach different types of problems.

### INCREASED EMPHASIS ON STUDY SKILLS AND STUDENT SUCCESS

- **NEW! Study Skills Builders** (formerly Study Skill Reminders) Found at the end of many exercise sets, Study Skills Builders allow instructors to assign exercises that will help students improve their study skills and take responsibility for their part of the learning process. Study Skills Builders reinforce the material found in Section 1.1, "Tips for Success in Mathematics" and serve as an excellent tool for self-assessment.

- **NEW! The Bigger Picture** is a recurring feature, starting in Section 1.7, that focuses on the key concepts of the course—simplifying expressions, and solving equations and inequalities. Students develop an ongoing study guide to help them be able to simplify expressions and solve equations and inequalities, and to know the difference between them. By working the exercises and developing this study guide throughout the text, students can begin to transition from thinking "section by section" to thinking about how the mathematics in this course is part of the "bigger picture" of mathematics in general. A completed outline is provided in Appendix A so students have a model for their work. It's great preparation for the Practice Final in A.2.

## CONTINUING SUPPORT FOR TEST PREPARATION

- **TEST PREP VIDEO** **Chapter Test Prep Video** provides students with help during their most "teachable moment"—while they are studying for a test. Included with every copy of the student edition of the text, this video provides fully worked-out solutions by the author to every exercise from each Chapter Test in the text. The easy video navigation allows students to instantly access the solutions to the exercises they want to review. The problems are solved by the author in the same manner as in the text.

- **Chapter Test Files in TestGen®** provide algorithms specific to each exercise from each Chapter Test in the text. Allows for easy replication of Chapter Tests with consistent, algorithmically generated problem types for additional assignments or assessments purposes.

## CONTENT CHANGES IN THE FOURTH EDITION

The sections on problem solving in Chapter 2 have been revised to make them more accessible to students. The changes include:

- Section 2.4 has been reorganized into three objectives: solving problems involving direct translation; solving problems involving relationships among unknown quantities; and finding consecutive integers.

- A new Section 2.6, Percent and Mixture Problem Solving, covers percent applications, discount and mark up, percent increase and decrease, and mixture applications.

- Section 2.7, now covers distance, money, and interest applications.

- There is now one section on Equations of Lines, Section 3.5, that reviews the slope-intercept form and introduces the point-slope form. This section now contains mixed exercises that help students decide what equation form to use.

- Section 6.4, Factoring Trinomials of the Form $ax^2 + bx + c$ by Grouping, is new for those instructors who prefer this method of factoring trinomials.

- Renewed Progression of writing equations of parallel and perpendicular lines in Sections 3.4, 3.5, and 8.1. Section 3.4—compare the slopes of parallel and perpendicular lines. Section 3.5—write equations of vertical and horizontal parallel and perpendicular lines. Section 8.1 full coverage of writing equations of parallel and perpendicular lines.

- A new Section 8.3 devoted entirely to graphing piecewise-defined functions and shifting and reflecting common graphs.

- Appendix A.2 is a Practice Final Exam that is an excellent review for students. The entire exam is worked by the author on the Interactive Video Lectures.

## KEY PEDAGOGICAL FEATURES

The following key features have been retained and/or updated for the Fourth Edition of the text:

**Problem Solving Process**   This is formally introduced in Chapter 2 with a four-step process that is integrated throughout the text. The four steps are **Understand, Translate, Solve,** and **Interpret.** The repeated use of these steps in a variety of examples shows their wide applicability. Reinforcing the steps can increase students' comfort level and confidence in tackling problems.

**Exercise Sets Revised and Updated**   The exercise sets have been carefully examined and extensively revised. Special focus was placed on making sure that even- and odd-numbered exercises are paired.

Each text section ends with an exercise set, usually divided into two parts. Both parts contain graded exercises. The **first part is carefully keyed** to at least one worked example in the text. Once a student has gained confidence in a skill, **the second part contains exercises not keyed to examples.** Exercises and examples marked with a (📹) have been worked out step-by-step by the author in the interactive videos that accompany this text.

Throughout the text exercises there is an emphasis on data and graphical interpretation via tables, charts, and graphs. The ability to interpret data and read and create a variety of types of graphs is developed gradually so students become comfortable with it. Similarly, throughout the text there is integration of geometric concepts, such as perimeter and area. Exercises and examples marked with a geometry icon ($\triangle$) have been identified for convenience.

**Examples**   Detailed step-by-step examples were added, deleted, replaced, or updated as needed. Many of these reflect real life. Additional instructional support is provided in the annotated examples.

**Helpful Hints**   Helpful Hints contain practical advice on applying mathematical concepts. Strategically placed where students are most likely to need immediate reinforcement, Helpful Hints help students avoid common trouble areas and mistakes.

**Concept Checks**   This feature allows students to gauge their grasp of an idea as it is being presented in the text. Concept Checks stress conceptual understanding at the point-of-use and help suppress misconceived notions before they start. Answers appear at the bottom of the page. Exercises related to Concept Checks are now included in the exercise sets.

**Mixed Practice Exercises** These exercises combine objectives within a section. They require students to determine the problem type and strategy needed in order to solve it. In doing so, students need to think about key concepts to proceed with a correct method of solving—just as they would need to do on a test.

**Concept Extensions** These exercises require students to combine several skills of concepts to solve exercises in the section.

**Integrated Reviews** A unique, mid-chapter exercise set that helps students assimilate new skills and concepts that they have learned separately over several sections. These reviews provide yet another opportunity for students to work with "mixed" exercises as they master the topics.

**Vocabulary Check** Provides an opportunity for students to become more familiar with the use of mathematical terms as they strengthen their verbal skills. These appear at the end of each chapter before the Chapter Highlights.

**Chapter Highlights** Found at the end of every chapter, these contain key definitions and concepts with examples to help students understand and retain what they have learned and help them organize their notes and study for tests.

**Chapter Review** The end of every chapter contains a comprehensive review of topics introduced in the chapter. The Chapter Review offers exercises keyed to every section in the chapter, as well as **Mixed Review (NEW!)** exercises that are not keyed to sections.

**Cumulative Review** Follows every chapter in the text except Chapter 1. Each odd-numbered exercise contained in the Cumulative Review is an earlier worked example in the text that is referenced in the back of the book along with the answer.

**Writing Exercises** \ These exercises occur in almost every exercise set and require students to provide a written response to explain concepts or justify their thinking.

**Applications** Real-world and real-data applications have been thoroughly updated and many new applications are included. These exercises occur in almost every exercise set and show the relevance of mathematics and help students gradually and continuously develop their problem solving skills.

**Review and Preview Exercises** These exercises occur in each exercise set (except Chapter 1) and are keyed to earlier sections. They review concepts learned earlier in the text that will be needed in the next section or chapter.

**Exercise Set Resource Icons** at the opening of each exercise set remind students of the resources available for extra practice and support:

See Student Resource descriptions on pages xxiii–xxiv for details on the individual resources available.

**Exercise Icons** These icons facilitate the assignment of specialized exercises and let students know what resources can support them.

- Video icon: exercise worked on Martin-Gay's Interactive DVD/CD Lectures.
- △ Triangle icon: identifies exercises involving geometric concepts.
- \ Pencil icon: indicates a written response is needed.
- Calculator icons: optional exercises intended to be solved using a scientific or graphing calculator.

**Group Activities**   Found at the end of each chapter, these activities are for individual or group completion, and are usually hands-on or data-based activities that extend the concepts found in the chapter allowing students to make decisions and interpretations and to think and write about algebra.

**Optional: Calculator Exploration Boxes and Calculator Exercises**   The optional Calculator Explorations provide key strokes and exercises at appropriate points to provide an opportunity for students to become familiar with these tools. Section exercises that are best completed by using a calculator are identified by 🖩 or 🖩 for ease of assignment.

---

## A Word about Textbook Design and Student Success

The design of developmental mathematics textbooks has become increasingly important. As students and instructors have told Pearson in focus groups and market research surveys, these textbooks cannot look "cluttered" or "busy." A "busy" design can distract a student from what is most important in the text. It can also heighten math anxiety.

As a result of the conversations and meetings we have had with students and instructors, we concluded the design of this text should be understated and focused on the most important pedagogical elements. Students and instructors helped us to identify the primary elements that are central to student success. These primary elements include:

- Exercise Sets
- Examples and Practice Problems
- Helpful Hints
- Rules, Property, and Definition boxes

As you will notice in this text, these primary features are the most prominent elements in the design. We have made every attempt to make sure these elements are the features the eye is drawn to. The remaining features, the secondary elements in the design, blend into the "fabric" or "grain" of the overall design. These secondary elements complement the primary elements without becoming distractions.

Pearson's thanks goes to all of the students and instructors (as noted by the author in Acknowledgments) who helped us develop the design of this text. At every step in the design process, their feedback proved valuable in helping us to make the right decisions. Thanks to your input, we're confident the design of this text will be both practical and engaging as it serves its educational and learning purposes.

Sincerely,

**Paul Murphy**

Editor-in-Chief
Developmental Mathematics
Pearson Arts & Sciences

---

## INSTRUCTOR AND STUDENT RESOURCES

The following resources are available to help instructors and students use this text more effectively.

### INSTRUCTOR RESOURCES

*Annotated Instructor's Edition*

- Answers to all exercises printed on the same text page or in the Graphing Answer Section.
- Teaching Tips throughout the text placed at key points.

- Classroom Examples paired with each text example for use as an additional resource during class.
- General tips and suggestions for classroom or group activities.
- Graphing Answer Section includes graphical answers and answers to the Group Activities.

*Instructor's Solutions Manual*

- Solutions to the even-numbered exercises
- Solutions to every Vocabulary Check exercise
- Solutions to every Practice exercise
- Solutions to every exercise in the Integrated Reviews, Chapter Reviews, Chapter Tests, and Cumulative Reviews

*Instructor's Resource Manual with Tests and Mini-Lectures*

- **NEW!** Includes Mini-Lectures for every section from the text
- Additional Exercises now 3 forms per section, to help instructors support students at different skill and ability levels.
- Free Response Test Forms, Multiple Choice Test Forms, Cumulative Tests, Group Activities
- Answers to all items

*Martin-Gay's Instructor to Instructor Videos*

- Text author Elayn Martin-Gay presents tips, hints, and suggestions for engaging students and presenting key topics.
- Available as part of the Instructor Resource Kit.
- Contact your sales representative for more information.

*Instructor Resource Kit*

The Martin-Gay Instructor Resource Kit contains tools and resources to help instructors succeed in the classroom. The kit includes:

- Instructor-to-Instructor CD Videos that offer tips, suggestions, and strategies for engaging students and presenting key topics
- PDF files of the Instructor's Solutions Manual and the Instructor's Resource Manual
- Powerpoint Lecture Slides and Active Learning Questions

## MYMATHLAB®

*(Instructor Version 0-13-147898-2)*

MyMathLab is a series of text-specific, easily customizable, online courses for Pearson Education's textbooks in mathematics and statistics. Powered by CourseCompass™ (our online teaching and learning environment) and MathXL® (our online homework, tutorial, and assessment system), MyMathLab gives you the tools you need to deliver all or a portion of your course online, whether your students are in a lab setting or working from home. MyMathLab provides a rich and flexible set of course materials, featuring free-response exercises that are algorithmically generated for unlimited practice and mastery. Students can also use online tools, such as video lectures, animations, and a multimedia textbook, to independently improve their understanding and performance. Instructors can use MyMathLab's homework and test managers to select and assign online exercises correlated directly to the textbook, and they can also create and assign their own online exercises and import TestGen tests for added flexibility. MyMathLab's online gradebook—designed specifically for mathematics and statistics—automatically tracks students' homework and test results and gives the instructor control over how to calculate final grades. Instructors can also add offline (paper-and-pencil) grades to the gradebook. **Includes access to the Pearson Tutor Center. Students**

**can receive tutoring via toll free phone, fax, email, and Internet.** MyMathLab is available to qualified adopters. For more information, visit our website at www.mymathlab.com or contact your sales representative.

## MATHXL®

*(Instructor version 0-13-147895-8)*

MathXL® is a powerful online homework, tutorial, and assessment system that accompanies Pearson Education's textbooks in mathematics or statistics. With MathXL, instructors can create, edit, and assign online homework and tests using algorithmically generated exercises correlated at the objective level to the textbook. They can also create and assign their own online exercises and import TestGen tests for added flexibility. All student work is tracked in MathXL's online gradebook. Students can take chapter tests in MathXL and receive personalized study plans based on their test results. The study plan diagnoses weaknesses and links students directly to tutorial exercises for the objectives they need to study and retest. Students can also access supplemental animations and video clips directly from selected exercises. MathXL is available to qualified adopters. For more information, visit our website at www.mathxl.com, or contact your sales representative.

## INTERACT MATH TUTORIAL WEBSITE: WWW.INTERACTMATH.COM

Get practice and tutorial help online! This interactive tutorial website provides algorithmically generated practice exercises that correlate directly to the exercises in the textbook. Students can retry an exercise as many times as they like with new values each time for unlimited practice and mastery. Every exercise is accompanied by an interactive guided solution that provides helpful feedback for incorrect answers, and students can also view a worked-out sample problem that steps them through an exercise similar to the one they're working on.

## TESTGEN®

TestGen enables instructors to build, edit, print, and administer tests using a computerized bank of questions developed to cover all the objectives of the text. TestGen is algorithmically based, allowing instructors to create multiple but equivalent versions of the same question or test with the click of a button. Instructors can also modify test bank questions or add new questions. Tests can be printed or administered online. The software and testbank are available for download from Pearson Education's online catalog.

## STUDENT RESOURCES

*Student Solutions Manual*

- Solutions to the odd-numbered section exercises
- Solutions to the Practice exercises
- Solutions to the Vocabulary and Readiness Check
- Solutions to every exercise found in the Chapter Reviews and Chapter Tests, Cumulative Reviews, and Integrated Reviews

## MARTIN-GAY'S INTERACTIVE DVD/CD LECTURE SERIES

Martin-Gay's video series has been comprehensively updated to address the way today's students study and learn. The new videos offer students active learning at their pace, with the following resources and more:

- **A complete lecture** for each section of the text, presented by Elayn Martin-Gay. Students can easily review a section or a specific topic before a homework assignment, quiz, or test. Exercises in the text marked with the 🕮 are worked on the video.

- A **new interface** with menu and navigation features allows students to quickly find and focus on examples and exercises they need to review.
- Pop-ups reinforce key terms and definitions.
- A new **Practice Final Exam Video** helps students prepare for the final exam. This Practice Final Exam is included in the text in Appendix A.2. At the click of a button, students can watch the full solutions to each exercise on the exam that they need help with. Overview clips provide a brief overview on how to approach different problem types.
- **Interactive Concept Checks** allow students to check their understanding of essential concepts. These multiple choice exercises focus on common misunderstandings. After making their answer selection, students are told whether they're correct or not, and why! Elayn also presents the full solution.
- **Study Skills Builders** help students develop effective study habits and reinforce the advice provided in Section 1.1, Tips for Success in Mathematics, found in the text and video.
- **Close-captioned in Spanish and English**

*Beginning and Intermediate Algebra* Fourth Edition *Student Study Pack*
The Student Study Pack includes:

- Martin-Gay's Interactive DVD/CD Lecture Series
- Student Solutions Manual
- Pearson Tutor Center

*Chapter Test Prep Video CD—Standalone* **TEST PREP VIDEO**

- Includes fully worked-out solutions to every problem from each Chapter Test in the text.

### MATHXL® TUTORIALS ON CD

This interactive tutorial CD-ROM provides algorithmically generated practice exercises that are correlated at the objective level to the exercises in the textbook. Every practice exercise is accompanied by an example and a guided solution designed to involve students in the solution process. Selected exercises may also include a video clip to help students visualize concepts. The software provides helpful feedback for incorrect answers and can generate printed summaries of students' progress.

### INTERACT MATH TUTORIAL WEBSITE: WWW.INTERACTMATH.COM

Get practice and tutorial help online! This interactive tutorial website provides algorithmically generated practice exercises that correlate directly to the exercises in the textbook. Students can retry an exercise as many times as they like with new values each time for unlimited practice and mastery. Every exercise is accompanied by an interactive guided solution that provides helpful feedback for incorrect answers, and students can also view a worked-out sample problem that steps them through an exercise similar to the one they're working on.

## ACKNOWLEDGMENTS

There are many people who helped me develop this text, and I will attempt to thank some of them here. Carrie Green and Edutorial Services were *invaluable* for contributing to the overall accuracy of the text. Suellen Robinson, Lisa Collette, and Kim Lane were *invaluable* for their many suggestions and contributions during the development and writing of this Fourth Edition. Karin Kipp provided guidance throughout the production process.

A special thanks to my Editor-in-Chief, Paul Murphy, for all of his assistance, support, and contributions to this project. A very special thank you goes to my Sponsoring Editor, Mary Beckwith, for being there 24/7/365, as my students say. Last, my thanks to the staff at Pearson Education for all their support: Linda Behrens, Ilene Kahn, Juan López, Mike Fruhbeis, Kenny Beck, Richard Bretan, Tom Benfatti, Kate Valentine, Michelle Renda, Chris Hoag, and Greg Tobin.

I would like to thank the following reviewers for their input and suggestions:

Sandi Athanassiou, *University of Missouri—Columbia*
Michelle Beerman, *Pasco Hernandez Community College*
Monika Bender, *Central Texas College*
Bob Hervey, *Hillsborough Community College*
Susan Poss, *Spartanburg Community College*
Jorge Romero, *Hillsborough Community College*
Joseph Wakim, *Brevard Community College*
Flo Wilson, *Central Texas College*
Marie Caruso and students, *Middlesex Community College*

I would also like to thank the following dedicated group of instructors who participated in our focus groups, Martin-Gay Summits, and our design review. Their feedback and insights have helped to strengthen this text. These instructors include:

Cedric Atkins, *Mott Community College*
Michelle Beermann, *Pasco Hernandez Community College*
Laurel Berry, *Bryant & Stratton*
John Beyers, *University of Maryland University College*
Lisa and Bob Brown, *Community College Baltimore County–Essex*
Gail Burkett, *Palm Beach Community College*
Cheryl Cantwell, *Seminole Community College*
Jackie Cohen, *Augusta State*
Julie Dewan, *Mohawk Community College*
Janice Ervin, *Central Piedmont Community College*
Karen Estes, *St. Petersburg College*
Cindy Gaddis, *Tyler Junior College*
Pauline Hall, *Iowa State*
Sonya Johnson, *Central Piedmont Community College*
Irene Jones, *Fullerton College*
Paul Jones, *University of Cincinnati*
Nancy Lange, *Inver Hills Community College*
Sandy Lofstock, *St. Petersburg College*
Jean McArthur, *Joliet Junior College*
Marcia Molle, *Metropolitan Community College*
Greg Nguyen, *Fullerton College*
Linda Padilla, *Joliet Junior College*
Rena Petrello, *Moorpark College*
Ena Salter, *Manatee Community College*
Carole Shapero, *Oakton Community College*
Ann Smallen, *Mohawk Community College*
Jennifer Strehler, *Oakton Community College*
Tanomo Taguchi, *Fullerton College*
Sam Tinsley, *Richland College*
Linda Tucker, *Rose State College*
Leigh Ann Wheeler, *Greenville Technical Community College*
Jenny Wilson, *Tyler Junior College*
Valerie Wright, *Central Piedmont Community College*

A special thank you to those students who participated in our design review: Katherine Browne, Mike Bulfin, Nancy Canipe, Ashley Carpenter, Jeff Chojnachi, Roxanne Davis, Mike Dieter, Amy Dombrowski, Kay Herring, Todd Jaycox, Kaleena Levan, Matt Montgomery, Tony Plese, Abigail Polkinghorn, Harley Price, Eli Robinson, Avery Rosen, Robyn Schott, Cynthia Thomas, and Sherry Ward.

## ADDITIONAL ACKNOWLEDGMENTS

As usual, I would like to thank my husband, Clayton, for his constant encouragement. I would also like to thank my children, Eric and Bryan. They are now both attending college and I miss them dearly. I would also like to thank my extended family for their help and wonderful sense of humor. Their contributions are too numerous to list. They are Rod and Karen Pasch; Peter, Michael, Christopher, and Matthew Callac; Jessica and Matt Chavez; Perry and Melissa Landrum; Stuart and Earline Martin; Josh, Mandy, Bailey, Ethan, Avery, and Mia Barnes; Mark, Sabrina, and Madison Martin; Leo and Barbara Miller; and Jewett Gay.

*Elayn Martin-Gay*

## ABOUT THE AUTHOR

Elayn Martin-Gay has taught mathematics at the University of New Orleans for more than 25 years. Her numerous teaching awards include the local University Alumni Association's Award for Excellence in Teaching, and Outstanding Developmental Educator at University of New Orleans, presented by the Louisiana Association of Developmental Educators.

Prior to writing textbooks, Elayn Martin-Gay developed an acclaimed series of lecture videos to support developmental mathematics students in their quest for success. These highly successful videos originally served as the foundation material for her texts. Today, the videos are specific to each book in the Martin-Gay series. The author has also created Chapter Test Prep Videos to help students during their most "teachable moment"—as they prepare for a test, along with Instructor-to-Instructor videos that provide teaching tips, hints, and suggestions for each developmental mathematics course, including basic mathematics, prealgebra, beginning algebra, and intermediate algebra.

Elayn is the author of 12 published textbooks as well as multimedia interactive mathematics, all specializing in developmental mathematics courses. She has participated as an author across the broadest range of educational materials: textbooks, videos, tutorial software, and courseware. This offers an opportunity of various combinations for an integrated teaching and learning package offering great consistency for the student.

# Application Index

# Beginning & Intermediate Algebra

CHAPTER

# 1

# Review of Real Numbers

The apparent magnitude of a star is the measure of its brightness as seen by someone on Earth. The smaller the apparent magnitude, the brighter the star. Below, the apparent magnitudes of some stars are listed.

Around 150 B.C., a Greek astronomer, Hipparchus, devised a system of classifying the brightness of stars. Hipparchus's system is the basis of the apparent magnitude scale used by modern astronomers. In Exercises 77 through 82, Section 1.2, we shall see how this scale is used to describe the brightness of objects such as the sun, the moon, and some planets.

The power of mathematics is its flexibility. We apply numbers to almost every aspect of our lives, from an ordinary trip to the grocery store to a rocket launched into space. The power of algebra is its generality. Using letters to represent numbers, we tie together the trip to the grocery store and the launched rocket.

In this chapter we review the basic symbols and words—the language—of arithmetic and introduce using variables in place of numbers. This is our starting place in the study of algebra.

| Star | Apparent Magnitude | Star | Apparent Magnitude |
|------|------|------|------|
| Arcturus | −0.04 | Spica | 0.98 |
| Sirius | −1.46 | Rigel | 0.12 |
| Vega | 0.03 | Regulus | 1.35 |
| Antares | 0.96 | Canopus | −0.72 |
| Sun | −26.7 | Hadar | 0.61 |

(*Source: Norton's 2000.0: Star Atlas and Reference Handbook*, 18th ed., Longman Group, UK, 1989)

# 1.1 TIPS FOR SUCCESS IN MATHEMATICS

Before reading this section, remember that your instructor is your best source for information. Please see your instructor for any additional help or information.

**OBJECTIVE 1 ▶ Getting ready for this course.** Now that you have decided to take this course, remember that a *positive attitude* will make all the difference in the world. Your belief that you can succeed is just as important as your commitment to this course. Make sure you are ready for this course by having the time and positive attitude that it takes to succeed.

Next, make sure you have scheduled your math course at a time that will give you the best chance for success. For example, if you are also working, you may want to check with your employer to make sure that your work hours will not conflict with your course schedule. Also, schedule your class during a time of day when you are more attentive and do your best work.

On the day of your first class period, double-check your schedule and allow yourself extra time to arrive in case of traffic problems or difficulty locating your classroom. Make sure that you bring at least your textbook, paper, and a writing instrument. Are you required to have a lab manual, graph paper, calculator, or some other supply besides this text? If so, also bring this material with you.

**OBJECTIVE 2 ▶ General tips for success.** Below are some general tips that will increase your chance for success in a mathematics class. Many of these tips will also help you in other courses you may be taking.

*Exchange names and phone numbers* with at least one other person in class. This contact person can be a great help if you miss an assignment or want to discuss math concepts or exercises that you find difficult.

*Choose to attend all class periods and be on time.* If possible, sit near the front of the classroom. This way, you will see and hear the presentation better. It may also be easier for you to participate in classroom activities.

*Do your homework.* You've probably heard the phrase "practice makes perfect" in relation to music and sports. It also applies to mathematics. You will find that the more time you spend solving mathematics problems, the easier the process becomes. Be sure to schedule enough time to complete your assignments before the next class period.

*Check your work.* Review the steps you made while working a problem. Learn to check your answers in the original problems. You may also compare your answers with the answers to selected exercises section in the back of the book. If you have made a mistake, try to figure out what went wrong. Then correct your mistake. If you can't find what went wrong, don't erase your work or throw it away. Bring your work to your instructor, a tutor in a math lab, or a classmate. It is easier for someone to find where you had trouble if they look at your original work.

*Learn from your mistakes and be patient with yourself.* Everyone, even your instructor, makes mistakes. (That definitely includes me—Elayn Martin-Gay.) Use your errors to learn and to become a better math student. The key is finding and understanding your errors.

Was your mistake a careless one, or did you make it because you can't read your own math writing? If so, try to work more slowly or write more neatly and make a conscious effort to carefully check your work.

Did you make a mistake because you don't understand a concept? Take the time to review the concept or ask questions to better understand it.

Did you skip too many steps? Skipping steps or trying to do too many steps mentally may lead to preventable mistakes.

*Know how to get help if you need it.* It's all right to ask for help. In fact, it's a good idea to ask for help whenever there is something that you don't understand. Make sure you know when your instructor has office hours and how to find his or her office. Find out whether math tutoring services are available on your campus. Check out the hours, location, and requirements of the tutoring service. Videotapes and software are available with this text. Learn how to access these resources.

***Organize your class materials,*** including homework assignments, graded quizzes and tests, and notes from your class or lab. All of these items will be valuable references throughout your course especially when studying for upcoming tests and the final exam. Make sure that you can locate these materials when you need them.

***Read your textbook before class.*** Reading a mathematics textbook is unlike leisure reading such as reading a novel or newspaper. Your pace will be much slower. It is helpful to have paper and a pencil with you when you read. Try to work out examples on your own as you encounter them in your text. You should also write down any questions that you want to ask in class. When you read a mathematics textbook, some of the information in a section may be unclear. But when you hear a lecture or watch a videotape on that section, you will understand it much more easily than if you had not read your text beforehand.

***Don't be afraid to ask questions.*** Instructors are not mind readers. Many times we do not know a concept is unclear until a student asks a question. You are not the only person in class with questions. Other students are normally grateful that someone has spoken up.

***Hand in assignments on time.*** This way you can be sure that you will not lose points for being late. Show every step of a problem and be neat and organized. Also be sure that you understand which problems are assigned for homework. You can always double-check this assignment with another student in your class.

**OBJECTIVE 3 ▶ Using this text.** There are many helpful resources that are available to you in this text. It is important that you become familiar with and use these resources. They should increase your chances for success in this course.

- The main section of exercises in each exercise set is referenced by an example(s). Use this referencing if you have trouble completing an assignment from the exercise set.

- If you need extra help in a particular section, look at the beginning of the section to see what videotapes and software are available.

- Make sure that you understand the meaning of the icons that are beside many exercises. The video icon 📽 tells you that the corresponding exercise may be viewed on the videotape that corresponds to that section. The pencil icon ✎ tells you that this exercise is a writing exercise in which you should answer in complete sentences. The △ icon tells you that the exercise involves geometry.

- Practice exercises are located immediately after a worked-out Example in the text. This exercise is similar to the previous Example and is a great way for you to immediately reinforce a learned concept. All answers to the practice exercises are found in the back of the text.

- Integrated Reviews in each chapter offer you a chance to practice—in one place—the many concepts that you have learned separately over several sections.

- There are many opportunities at the end of each chapter to help you understand the concepts of the chapter.

   **Chapter Highlights** contain chapter summaries and examples.

   **Chapter Reviews** contain review problems organized by section.

   **Chapter Tests** are sample tests to help you prepare for an exam.

   **Cumulative Reviews** are reviews consisting of material from the beginning of the book to the end of that particular chapter.

- *The Bigger Picture.* This feature contains the directions for building an outline to be used throughout the course. The purpose of this outline is to help you make the transition from thinking "section by section" to thinking about how the mathematics in this course is part of a bigger picture.

- *Study Skills Builder.* This feature is found at the end of many exercise sets. In order to increase your chance of success in this course, please read and answer the questions in the Study Skills Builder. For your convenience, the table below contains selected Study Skills Builder titles and their location.

| *Study Skills Builder Title* | *Page of First Occurrence* |
| --- | --- |
| Learning New Terms | Page 57 |
| What to Do the Day of an Exam | Page 141 |
| Have You Decided to Complete This Course Successfully? | Page 81 |
| Organizing a Notebook | Page 122 |
| How Are Your Homework Assignments Going? | Page 92 |
| Are You Familiar with Your Textbook Supplements? | Page 198 |
| How Well Do You Know Your Textbook? | Page 311 |
| Are You Organized? | Page 329 |
| Are You Satisfied with Your Performance on a Particular Quiz or Exam? | Page 390 |
| Tips for Studying for an Exam | Page 525 |
| Are You Getting All the Mathematics Help That You Need? | Page 402 |
| How Are You Doing? | Page 457 |
| Are You Preparing for Your Final Exam? | Page 790 |

See the Preface at the beginning of this text for a more thorough explanation of the features of this text.

**OBJECTIVE 4 ▶ Getting help.** If you have trouble completing assignments or understanding the mathematics, get help as soon as you need it! This tip is presented as an objective on its own because it is so important. In mathematics, usually the material presented in one section builds on your understanding of the previous section. This means that if you don't understand the concepts covered during a class period, there is a good chance that you will not understand the concepts covered during the next class period. If this happens to you, get help as soon as you can.

Where can you get help? Many suggestions have been made in the section on where to get help, and now it is up to you to do it. Try your instructor, a tutoring center, or a math lab, or you may want to form a study group with fellow classmates. If you do decide to see your instructor or go to a tutoring center, make sure that you have a neat notebook and are ready with your questions.

**OBJECTIVE 5 ▶ Preparing for and taking an exam.** Make sure that you allow yourself plenty of time to prepare for a test. If you think that you are a little "math anxious," it may be that you are not preparing for a test in a way that will ensure success. The way that you prepare for a test in mathematics is important. To prepare for a test,

1. Review your previous homework assignments. You may also want to rework some of them.
2. Review any notes from class and section-level quizzes you have taken. (If this is a final exam, also review chapter tests you have taken.)
3. Review concepts and definitions by reading the Highlights at the end of each chapter.
4. Practice working out exercises by completing the Chapter Review found at the end of each chapter. (If this is a final exam, go through a Cumulative Review. There is

one found at the end of each chapter except Chapter 1. Choose the review found at the end of the latest chapter that you have covered in your course.) *Don't stop here!*

5. It is important that you place yourself in conditions similar to test conditions to find out how you will perform. In other words, as soon as you feel that you know the material, get a few blank sheets of paper and take a sample test. There is a Chapter Test available at the end of each chapter. During this sample test, do not use your notes or your textbook. Once you complete the Chapter Test, check your answers in the back of the book. If any answer is incorrect, there is a CD available with each exercise of each chapter test worked. Use this CD or your instructor to correct your sample test. Your instructor may also provide you with a review sheet. If you are not satisfied with the results, study the areas that you are weak in and try again.

6. Get a good night's sleep before the exam.

7. On the day of the actual test, allow yourself plenty of time to arrive at where you will be taking your exam.

When taking your test,

1. Read the directions on the test carefully.

2. Read each problem carefully as you take the test. Make sure that you answer the question asked.

3. Pace yourself by first completing the problems you are most confident with. Then work toward the problems you are least confident with. Watch your time so you do not spend too much time on one particular problem.

4. If you have time, check your work and answers.

5. Do not turn your test in early. If you have extra time, spend it double-checking your work.

**OBJECTIVE 6 ▶ Managing your time.** As a college student, you know the demands that classes, homework, work, and family place on your time. Some days you probably wonder how you'll ever get everything done. One key to managing your time is developing a schedule. Here are some hints for making a schedule:

1. Make a list of all of your weekly commitments for the term. Include classes, work, regular meetings, extracurricular activities, etc. You may also find it helpful to list such things as laundry, regular workouts, grocery shopping, etc.

2. Next, estimate the time needed for each item on the list. Also make a note of how often you will need to do each item. Don't forget to include time estimates for reading, studying, and homework you do outside of your classes. You may want to ask your instructor for help estimating the time needed.

3. In the following exercise set, you are asked to block out a typical week on the schedule grid given. Start with items with fixed time slots like classes and work.

4. Next, include the items on your list with flexible time slots. Think carefully about how best to schedule some items such as study time.

5. Don't fill up every time slot on the schedule. Remember that you need to allow time for eating, sleeping, and relaxing! You should also allow a little extra time in case some items take longer than planned.

6. If you find that your weekly schedule is too full for you to handle, you may need to make some changes in your workload, classload, or in other areas of your life. You may want to talk to your advisor, manager or supervisor at work, or someone in your college's academic counseling center for help with such decisions.

**Note:** Don't forget in this chapter we begin a feature called Study Skills Builder. The purpose of this feature is to remind you of some of the information given in this section and to further expand on some topics in this section.

## 1.1 EXERCISE SET

| | | | | |
|---|---|---|---|---|
| PRACTICE | WATCH | DOWNLOAD | READ | REVIEW |

1. What is your instructor's name?

2. What are your instructor's office location and office hours?

3. What is the best way to contact your instructor?

4. What does the ⬥ icon mean?

5. What does the ⬥ icon mean?

6. What does the △ icon mean?

7. Where are answers located in this text?

8. What Exercise Set answers are available to you in the answers section?

9. What Chapter Review, Chapter Test, and Cumulative Review answers are available to you in the answer section?

10. Search after the solution of any worked example in this text. What are Practice exercises?

11. When might be the best time to work a Practice exercise?

12. Where are the answers to Practice exercises?

13. List any similarities between a Practice exercise and the worked example before it.

14. Go to the Highlights section at the end of this chapter. Describe how this section may be helpful to you when preparing for a test.

15. Do you have the name and contact information of at least one other student in class?

16. Will your instructor allow you to use a calculator in this class?

17. Are videotapes, CDs, and/or tutorial software available to you? If so, where?

18. Is there a tutoring service available? If so, what are its hours?

19. Have you attempted this course before? If so, write down ways that you might improve your chances of success during this second attempt.

20. List some steps that you can take if you begin having trouble understanding the material or completing an assignment.

21. Read or reread objective 6 and fill out the schedule grid below.

22. Study your filled-out grid from Exercise 21. Decide whether you have the time necessary to successfully complete this course and any other courses you may be registered for.

| | Monday | Tuesday | Wednesday | Thursday | Friday | Saturday | Sunday |
|---|---|---|---|---|---|---|---|
| 4:00 A.M. | | | | | | | |
| 5:00 A.M. | | | | | | | |
| 6:00 A.M. | | | | | | | |
| 7:00 A.M. | | | | | | | |
| 8:00 A.M. | | | | | | | |
| 9:00 A.M. | | | | | | | |
| 10:00 A.M. | | | | | | | |
| 11:00 A.M. | | | | | | | |
| 12:00 P.M. | | | | | | | |
| 1:00 P.M. | | | | | | | |
| 2:00 P.M. | | | | | | | |
| 3:00 P.M. | | | | | | | |
| 4:00 P.M. | | | | | | | |
| 5:00 P.M. | | | | | | | |
| 6:00 P.M. | | | | | | | |
| 7:00 P.M. | | | | | | | |
| 8:00 P.M. | | | | | | | |
| 9:00 P.M. | | | | | | | |
| 10:00 P.M. | | | | | | | |
| 11:00 P.M. | | | | | | | |
| Midnight | | | | | | | |
| 1:00 A.M. | | | | | | | |
| 2:00 A.M. | | | | | | | |
| 3:00 A.M. | | | | | | | |

# 1.2 SYMBOLS AND SETS OF NUMBERS

## OBJECTIVES

1 Use a number line to order numbers.

2 Translate sentences into mathematical statements.

3 Identify natural numbers, whole numbers, integers, rational numbers, irrational numbers, and real numbers.

4 Find the absolute value of a real number.

**OBJECTIVE 1 ▶ Using a number line to order numbers.** We begin with a review of the set of natural numbers and the set of whole numbers and how we use symbols to compare these numbers. A **set** is a collection of objects, each of which is called a **member** or **element** of the set. A pair of brace symbols { } encloses the list of elements and is translated as "the set of" or "the set containing."

> **Natural Numbers**
> The set of **natural numbers** is $\{1, 2, 3, 4, 5, 6, \ldots\}$.

> **Whole Numbers**
> The set of **whole numbers** is $\{0, 1, 2, 3, 4, \ldots\}$.

The three dots (an ellipsis) at the end of the list of elements of a set means that the list continues in the same manner indefinitely.

These numbers can be pictured on a **number line.** We will use number lines often to help us visualize distance and relationships between numbers. Visualizing mathematical concepts is an important skill and tool, and later we will develop and explore other visualizing tools.

To draw a number line, first draw a line. Choose a point on the line and label it 0. To the right of 0, label any other point 1. Being careful to use the same distance as from 0 to 1, mark off equally spaced distances. Label these points 2, 3, 4, 5, and so on. Since the whole numbers continue indefinitely, it is not possible to show every whole number on this number line. The arrow at the right end of the line indicates that the pattern continues indefinitely.

Picturing whole numbers on a number line helps us to see the order of the numbers. Symbols can be used to describe concisely in writing the order that we see.

The **equal symbol** $=$ means "is equal to."

The symbol $\neq$ means "is not equal to."

These symbols may be used to form a **mathematical statement.** The statement might be true or it might be false. The two statements below are both true.

$2 = 2$    states that "two is equal to two"

$2 \neq 6$    states that "two is not equal to six"

If two numbers are not equal, then one number is larger than the other. The symbol $>$ means "is greater than." The symbol $<$ means "is less than." For example,

$2 > 0$    states that "two is greater than zero"

$3 < 5$    states that "three is less than five"

On a number line, we see that a number **to the right of** another number is **larger.** Similarly, a number **to the left of** another number is smaller. For example, 3 is to the left of 5 on a number line, which means that 3 is less than 5, or $3 < 5$. Similarly, 2 is to the right of 0 on a number line, which means 2 is greater than 0, or $2 > 0$. Since 0 is to the left of 2, we can also say that 0 is less than 2, or $0 < 2$.

The symbols $\neq$, $<$, and $>$ are called **inequality symbols.**

$2 > 0$ or $0 < 2$

$3 < 5$

> ▶ **Helpful Hint**
>
> Notice that $2 > 0$ has exactly the same meaning as $0 < 2$. Switching the order of the numbers and reversing the "direction of the inequality symbol" does not change the meaning of the statement.
>
> $$5 > 3 \text{ has the same meaning as } 3 < 5.$$
>
> Also notice that, when the statement is true, the inequality arrow points to the smaller number.

**EXAMPLE 1** Insert $<$, $>$, or $=$ in the space between each pair of numbers to make each statement true.

**a.** 2    3          **b.** 7    4          **c.** 72    27

*Solution*

**a.** $2 < 3$ since 2 is to the left of 3 on the number line.

**b.** $7 > 4$ since 7 is to the right of 4 on the number line.

**c.** $72 > 27$ since 72 is to the right of 27 on the number line.          ☐

**PRACTICE**

**1** Insert $<$, $>$, or $=$ in the space between each pair of numbers to make each statement true.

**a.** 5    8          **b.** 6    4          **c.** 16    82

Two other symbols are used to compare numbers. The symbol $\leq$ means "is less than or equal to." The symbol $\geq$ means "is greater than or equal to." For example,

$$7 \leq 10 \text{ states that "seven is less than or equal to ten"}$$

This statement is true since $7 < 10$ is true. If either $7 < 10$ or $7 = 10$ is true, then $7 \leq 10$ is true.

$$3 \geq 3 \text{ states that "three is greater than or equal to three"}$$

This statement is true since $3 = 3$ is true. If either $3 > 3$ or $3 = 3$ is true, then $3 \geq 3$ is true.

The statement $6 \geq 10$ is false since neither $6 > 10$ nor $6 = 10$ is true. The symbols $\leq$ and $\geq$ are also called **inequality symbols.**

**EXAMPLE 2** Tell whether each statement is true or false.

**a.** $8 \geq 8$          **b.** $8 \leq 8$          **c.** $23 \leq 0$          **d.** $23 \geq 0$

*Solution*

**a.** True, since $8 = 8$ is true.

**b.** True, since $8 = 8$ is true.

**c.** False, since neither $23 < 0$ nor $23 = 0$ is true.

**d.** True, since $23 > 0$ is true.          ☐

**PRACTICE**

**2** Tell whether each statement is true or false.

**a.** $9 \geq 3$          **b.** $3 \geq 8$          **c.** $25 \leq 25$          **d.** $4 \leq 14$

**OBJECTIVE 2 ▶ Translating sentences.** Now, let's use the symbols discussed above to translate sentences into mathematical statements.

**EXAMPLE 3**    Translate each sentence into a mathematical statement.

**a.** Nine is less than or equal to eleven.

**b.** Eight is greater than one.

**c.** Three is not equal to four.

*Solution*

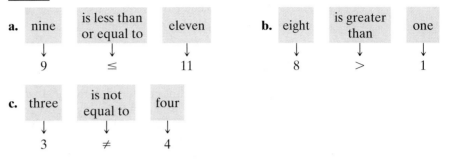

**c.**

| three | is not equal to | four |
| :---: | :---: | :---: |
| ↓ | ↓ | ↓ |
| 3 | ≠ | 4 |

**PRACTICE**
**3**    Translate each sentence into a mathematical statement.

**a.** Three is less than eight.

**b.** Fifteen is greater than or equal to nine.

**c.** Six is not equal to seven.

**OBJECTIVE 3 ▶ Identifying common sets of numbers.** Whole numbers are not sufficient to describe many situations in the real world. For example, quantities smaller than zero must sometimes be represented, such as temperatures less than 0 degrees.

We can picture numbers less than zero on a number line as follows:

$$\overset{\longleftarrow}{\underset{-5\ -4\ -3\ -2\ -1\ \ 0\ \ 1\ \ 2\ \ 3\ \ 4\ \ 5}{+\!+\!+\!+\!+\!+\!+\!+\!+\!+\!+}}\overset{\longrightarrow}{}$$

Numbers less than 0 are to the left of 0 and are labeled −1, −2, −3, and so on. A − sign, such as the one in −1, tells us that the number is to the left of 0 on a number line. In words, −1 is read "negative one." A + sign or no sign tells us that a number lies to the right of 0 on the number line. For example, 3 and +3 both mean positive three.

The numbers we have pictured are called the set of **integers.** Integers to the left of 0 are called **negative integers;** integers to the right of 0 are called **positive integers.** The integer **0 is neither positive nor negative.**

> **Integers**
> The set of **integers** is $\{ \ldots, -3, -2, -1, 0, 1, 2, 3, \ldots \}$.

Notice the ellipses (three dots) to the left and to the right of the list for the integers. This indicates that the positive integers and the negative integers continue indefinitely.

**EXAMPLE 4** Use an integer to express the number in the following. "Pole of Inaccessibility, Antarctica, is the coldest location in the world, with an average annual temperature of 72 degrees below zero." (*Source: The Guinness Book of Records*)

<u>Solution</u>  The integer $-72$ represents 72 degrees below zero.  ☐

**PRACTICE**
**4**  Use an integer to express the number in the following: Fred overdrew his checking account and now owes his bank 52 dollars.

A problem with integers in real-life settings arises when quantities are smaller than some integer but greater than the next smallest integer. On a number line, these quantities may be visualized by points between integers. Some of these quantities between integers can be represented as a quotient of integers. For example,

The point on a number line halfway between 0 and 1 can be represented by $\frac{1}{2}$, a quotient of integers.

The point on a number line halfway between 0 and $-1$ can be represented by $-\frac{1}{2}$. Other quotients of integers and their graphs are shown.

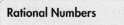

These numbers, each of which can be represented as a quotient of integers, are examples of **rational numbers.** It's not possible to list the set of rational numbers using the notation that we have been using. For this reason, we will use a different notation.

---

**Rational Numbers**

$$\left\{ \frac{a}{b} \middle| a \text{ and } b \text{ are integers and } b \neq 0 \right\}$$

---

We read this set as "the set of all numbers $\frac{a}{b}$ such that $a$ and $b$ are integers and **$b$ is not equal to 0.**" Notice that every integer is also a rational number since each integer can be expressed as a quotient of integers. For example, the integer 5 is also a rational number since $5 = \frac{5}{1}$.

The number line also contains points that cannot be expressed as quotients of integers. These numbers are called **irrational numbers** because they cannot be represented by rational numbers. For example, $\sqrt{2}$ and $\pi$ are irrational numbers.

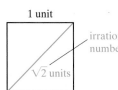

1 unit

irrational number

$\sqrt{2}$ units

---

**Irrational Numbers**

The set of **irrational numbers** is

{Nonrational numbers that correspond to points on the number line}.

That is, an irrational number is a number that cannot be expressed as a quotient of integers.

---

Both rational numbers and irrational numbers can be written as decimal numbers. The decimal equivalent of a rational number will either terminate or repeat in a pattern. For example, upon dividing we find that

$$\frac{3}{4} = 0.75 \text{ (decimal number terminates or ends) and}$$

$$\frac{2}{3} = 0.66666\ldots \text{ (decimal number repeats in a pattern)}$$

The decimal representation of an irrational number will neither terminate nor repeat. For example, the decimal representations of irrational numbers $\sqrt{2}$ and $\pi$ are

$$\sqrt{2} = 1.414213562\ldots \text{ (decimal number does not terminate or repeat in a pattern)}$$

$$\pi = 3.141592653\ldots \text{ (decimal number does not terminate or repeat in a pattern)}$$

(For further review of decimals, see the Appendix.)

Combining the natural numbers with the irrational numbers gives the set of **real numbers.** One and only one point on a number line corresponds to each real number.

---

**Real Numbers**

The set of **real numbers** is

{All numbers that correspond to points on the number line}

---

▶ **Helpful Hint**

From our previous definitions, we have that

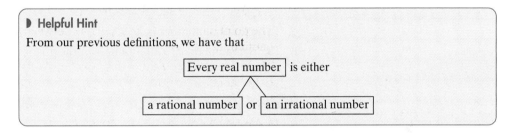

---

On the following number line, we see that real numbers can be positive, negative, or 0. Numbers to the left of 0 are called **negative numbers;** numbers to the right of 0 are called **positive numbers.** Positive and negative numbers are also called **signed numbers.**

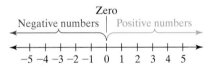

Several different sets of numbers have been discussed in this section. The following diagram shows the relationships among these sets of real numbers.

**Common Sets of Numbers**

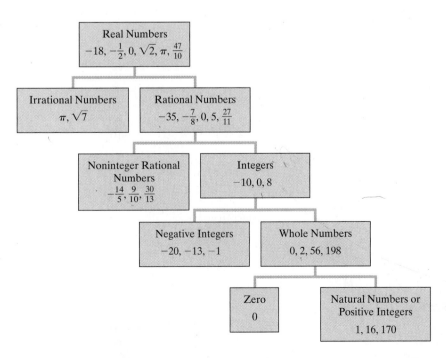

**EXAMPLE 5**  Given the set $\left\{-2, 0, \frac{1}{4}, -1.5, 112, -3, 11, \sqrt{2}\right\}$, list the numbers in this set that belong to the set of:

**a.** Natural numbers

**b.** Whole numbers

**c.** Integers

**d.** Rational numbers

**e.** Irrational numbers

**f.** Real numbers

_Solution_

**a.** The natural numbers are 11 and 112.

**b.** The whole numbers are 0, 11, and 112.

**c.** The integers are $-3, -2, 0, 11,$ and 112.

**d.** Recall that integers are rational numbers also. The rational numbers are $-3, -2,$ $-1.5, 0, \frac{1}{4}, 11,$ and 112.

**e.** The irrational number is $\sqrt{2}$.

**f.** The real numbers are all numbers in the given set. □

**PRACTICE**

**5**  Given the set $\left\{25, \frac{7}{3}, -15, \frac{-3}{4}, \sqrt{5}, -3.7, 8.8, -99\right\}$, list the numbers in this set that belong to the set of:

**a.** Natural numbers

**b.** Whole numbers

**c.** Integers

**d.** Rational numbers

**e.** Irrational numbers

**f.** Real numbers

We can now extend the meaning and use of inequality symbols such as $<$ and $>$ to apply to all real numbers.

---

**Order Property for Real Numbers**

Given any two real numbers $a$ and $b$, $a < b$ if $a$ is to the left of $b$ on a number line. Similarly, $a > b$ if $a$ is to the right of $b$ on a number line.

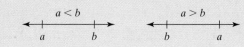

---

**EXAMPLE 6**  Insert $<$, $>$, or $=$ in the appropriate space to make each statement true.

**a.** $-1 \quad 0$

**b.** $7 \quad \dfrac{14}{2}$

**c.** $-5 \quad -6$

_Solution_

**a.** $-1 < 0$ since $-1$ is to the left of 0 on a number line.

$$-1 < 0$$

**b.** $7 = \dfrac{14}{2}$ since $\dfrac{14}{2}$ simplifies to 7.

**c.** $-5 > -6$ since $-5$ is to the right of $-6$ on the number line.

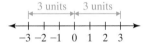

$$-5 > -6$$

☐

**PRACTICE**
**6**   Insert $<, >,$ or $=$ in the appropriate space to make each statement true.

**a.** 0    3          **b.** 15    $-5$          **c.** 3    $\dfrac{12}{4}$

---

**OBJECTIVE 4 ▶ Finding the absolute value of a real number.** A number line not only gives us a picture of the real numbers, it also helps us visualize the distance between numbers. The distance between a real number $a$ and 0 is given a special name called the **absolute value** of $a$. "The absolute value of $a$" is written in symbols as $|a|$.

> **Absolute Value**
> The absolute value of a real number $a$, denoted by $|a|$, is the distance between $a$ and 0 on a number line.

For example, $|3| = 3$ and $|-3| = 3$ since both 3 and $-3$ are a distance of 3 units from 0 on a number line.

> **▶ Helpful Hint**
> Since $|a|$ is a distance, $|a|$ is always either positive or 0, never negative. That is, **for any real number $a$, $|a| \geq 0$.**

**EXAMPLE 7**   Find the absolute value of each number.

**a.** $|4|$          **b.** $|-5|$          **c.** $|0|$          **d.** $\left|-\dfrac{1}{2}\right|$          **e.** $|5.6|$

*Solution*

**a.** $|4| = 4$ since 4 is 4 units from 0 on a number line.

**b.** $|-5| = 5$ since $-5$ is 5 units from 0 on a number line.

**c.** $|0| = 0$ since 0 is 0 units from 0 on a number line.

**d.** $\left|-\dfrac{1}{2}\right| = \dfrac{1}{2}$ since $-\dfrac{1}{2}$ is $\dfrac{1}{2}$ unit from 0 on a number line.

**e.** $|5.6| = 5.6$ since 5.6 is 5.6 units from 0 on a number line.

☐

**PRACTICE**
**7**   Find the absolute value of each number.

**a.** $|-8|$          **b.** $|9|$          **c.** $|-2.5|$          **d.** $\left|\dfrac{5}{11}\right|$          **e.** $|\sqrt{3}|$

**EXAMPLE 8** Insert $<$, $>$, or $=$ in the appropriate space to make each statement true.

**a.** $|0|$ ___ 2 **b.** $|-5|$ ___ 5 **c.** $|-3|$ ___ $|-2|$ **d.** $|5|$ ___ $|6|$ **e.** $|-7|$ ___ $|6|$

*Solution*

**a.** $|0| < 2$ since $|0| = 0$ and $0 < 2$. **b.** $|-5| = 5$ since $5 = 5$.
**c.** $|-3| > |-2|$ since $3 > 2$. **d.** $|5| < |6|$ since $5 < 6$.
**e.** $|-7| > |6|$ since $7 > 6$.

**PRACTICE**
**8** Insert $<$, $>$, or $=$ in the appropriate space to make each statement true.

**a.** $|8|$ ___ $|-8|$ **b.** $|-3|$ ___ 0 **c.** $|-7|$ ___ $|-11|$ **d.** $|3|$ ___ $|2|$ **e.** $|0|$ ___ $|-4|$

## VOCABULARY & READINESS CHECK

*Use the choices below to fill in each blank.*

| real | natural | whole | irrational |
| $|b|$ | inequality | integers | rational |

**1.** The _____ numbers are $\{0, 1, 2, 3, 4, \ldots\}$.
**2.** The _____ numbers are $\{1, 2, 3, 4, 5, \ldots\}$.
**3.** The symbols $\neq$, $\leq$, and $>$ are called _____ symbols.
**4.** The _____ are $\{\ldots, -3, -2, -1, 0, 1, 2, 3, \ldots\}$.
**5.** The _____ numbers are {all numbers that correspond to points on the number line}.
**6.** The _____ numbers are $\left\{\dfrac{a}{b} \middle| a \text{ and } b \text{ are integers, } b \neq 0\right\}$.
**7.** The _____ numbers are {nonrational numbers that correspond to points on the number line}.
**8.** The distance between a number $b$ and 0 on a number line is _____.

## 1.2 EXERCISE SET

*Insert $<$, $>$, or $=$ in the appropriate space to make the statement true. See Example 1.*

**1.** 7 ___ 3
**2.** 9 ___ 15
**3.** 6.26 ___ 6.26
**4.** 2.13 ___ 1.13
**5.** 0 ___ 7
**6.** 20 ___ 0
**7.** $-2$ ___ 2
**8.** $-4$ ___ $-6$

**9.** The freezing point of water is 32° Fahrenheit. The boiling point of water is 212° Fahrenheit. Write an inequality statement using $<$ or $>$ comparing the numbers 32 and 212.

**10.** The freezing point of water is 0° Celsius. The boiling point of water is 100° Celsius. Write an inequality statement using $<$ or $>$ comparing the numbers 0 and 100.

**11.** The spring 2007 tuition and fees for a Texas resident undergraduate student at University of Texas at El Paso were approximately $2631 for a 15-credit load. At the same time the tuition and fees for a Florida resident attending University of Florida were approximately $2456. Write an inequality statement using $<$ or $>$ comparing the numbers 2631 and 2456. (*Source:* UTEP and UF)

**12.** The average salary in the San Jose, California, area for a chemical engineer is $67,841. The average salary for a database administrator in the same area is $75,657. Write an inequality statement using $<$ or $>$ comparing the numbers 67,841 and 75,657. (*Source: The Wall Street Journal*)

*Are the following statements true or false? See Example 2.*

**13.** $11 \leq 11$
**14.** $4 \geq 7$
**15.** $10 > 11$
**16.** $17 > 16$
**17.** $3 + 8 \geq 3(8)$
**18.** $8 \cdot 8 \leq 8 \cdot 7$
**19.** $7 > 0$
**20.** $4 < 7$

△ **21.** An angle measuring 30° is shown and an angle measuring 45° is shown. Use the inequality symbol ≤ or ≥ to write a statement comparing the numbers 30 and 45.

△ **22.** The sum of the measures of the angles of a triangle is 180°. The sum of the measures of the angles of a parallelogram is 360°. Use the inequality symbol ≤ or ≥ to write a statement comparing the numbers 360 and 180.

*Write each sentence as a mathematical statement. See Example 3.*

**23.** Eight is less than twelve.

**24.** Fifteen is greater than five.

**25.** Five is greater than or equal to four.

**26.** Negative ten is less than or equal to thirty-seven.

**27.** Fifteen is not equal to negative two.

**28.** Negative seven is not equal to seven.

*Use integers to represent the values in each statement. See Example 4.*

**29.** Driskill Mountain, in Louisiana, has an altitude of 535 feet. New Orleans, Louisiana, lies 8 feet below sea level. (*Source:* U.S. Geological Survey)

**30.** During a Green Bay Packers football game, the team gained 23 yards and then lost 12 yards on consecutive plays.

**31.** From 2005 to 2010, the population of Washington, D.C., is expected to decrease by approximately 21,350. (*Source:* U.S. Census Bureau)

**32.** From 2005 to 2010, the population of Alaska is expected to grow by about 33,000 people. (*Source:* U.S. Census Bureau)

**33.** Aaron Miller deposited $350 in his savings account. He later withdrew $126.

**34.** Aris Peña was deep-sea diving. During her dive, she ascended 30 feet and later descended 50 feet.

*The graph below is called a bar graph. This particular graph shows the annual numbers of recreational visitors to U.S. National Parks. Each bar represents a different year, and the height of the bar represents the number of visitors (in millions) in that year.*

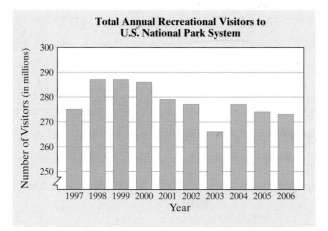

**35.** In which year(s) was the number of visitors the greatest?

**36.** What was the greatest number of visitors shown?

**37.** In what year(s) was the number of visitors greater than 280 million?

**38.** In what year(s) was the number of visitors less than 270 million?

**39.** Write an inequality statement comparing the number of annual visitors in 2001 and 2006.

**40.** Do you notice any trends shown by this bar graph?

*Tell which set or sets each number belongs to: natural numbers, whole numbers, integers, rational numbers, irrational numbers, and real numbers. See Example 5.*

**41.** 0

**42.** $\frac{1}{4}$

**43.** −2

**44.** $-\frac{1}{2}$

**45.** 6

**46.** 5

**47.** $\frac{2}{3}$

**48.** $\sqrt{3}$

**49.** $-\sqrt{5}$

**50.** $-1\frac{5}{9}$

*Tell whether each statement is true or false.*

**51.** Every rational number is also an integer.

**52.** Every negative number is also a rational number.

**53.** Every natural number is positive.

**54.** Every rational number is also a real number.

**55.** 0 is a real number.

**56.** Every real number is also a rational number.

**57.** Every whole number is an integer.

**58.** $\frac{1}{2}$ is an integer.

**59.** A number can be both rational and irrational.

**60.** Every whole number is positive.

*Insert <, >, or = in the appropriate space to make a true statement. See Examples 6 through 8.*

**61.** −10    −100

**62.** −200    −20

**63.** 32    5.2

**64.** 7.1    −7

**65.** $\frac{18}{3}$    $\frac{24}{3}$

**66.** $\frac{8}{2}$    $\frac{12}{3}$

**67.** −51    −50

**68.** $|-20|$    −200

**69.** $|-5|$    −4

**70.** 0    $|0|$

**71.** $|-1|$    $|1|$

**72.** $\left|\frac{2}{5}\right|$    $\left|-\frac{2}{5}\right|$

**73.** $|-2|$    $|-3|$

**74.** −500    $|-50|$

**75.** $|0|$    $|-8|$

**76.** $|-12|$    $\frac{24}{2}$

## CONCEPT EXTENSIONS

*The apparent magnitude of a star is the measure of its brightness as seen by someone on Earth. The smaller the apparent magnitude, the brighter the star. Use the apparent magnitudes in the table to answer Exercises 77 through 82.*

| Star | Apparent Magnitude | Star | Apparent Magnitude |
|------|--------------------|------|--------------------|
| Arcturus | −0.04 | Spica | 0.98 |
| Sirius | −1.46 | Rigel | 0.12 |
| Vega | 0.03 | Regulus | 1.35 |
| Antares | 0.96 | Canopus | −0.72 |
| Sun | −26.7 | Hadar | 0.61 |

(*Source: Norton's 2000: Star Atlas and Reference Handbook,* 18th ed., Longman Group, UK, 1989)

**77.** The apparent magnitude of the sun is −26.7. The apparent magnitude of the star Arcturus is −0.04. Write an inequality statement comparing the numbers −0.04 and −26.7.

**78.** The apparent magnitude of Antares is 0.96. The apparent magnitude of Spica is 0.98. Write an inequality statement comparing the numbers 0.96 and 0.98.

**79.** Which is brighter, the sun or Arcturus?

**80.** Which is dimmer, Antares or Spica?

**81.** Which star listed is the brightest?

**82.** Which star listed is the dimmest?

*Rewrite the following inequalities so that the inequality symbol points in the opposite direction and the resulting statement has the same meaning as the given one.*

**83.** $25 \geq 20$

**84.** $-13 \leq 13$

**85.** $0 < 6$

**86.** $5 > 3$

**87.** $-10 > -12$

**88.** $-4 < -2$

**89.** In your own words, explain how to find the absolute value of a number.

**90.** Give an example of a real-life situation that can be described with integers but not with whole numbers.

## 1.3 FRACTIONS

### OBJECTIVES

1 Write fractions in simplest form.

2 Multiply and divide fractions.

3 Add and subtract fractions.

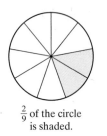

$\frac{2}{9}$ of the circle is shaded.

**OBJECTIVE 1 ▶ Writing fractions in simplest form.** A quotient of two numbers such as $\frac{2}{9}$ is called a **fraction.** In the fraction $\frac{2}{9}$, the top number, 2, is called the **numerator** and the bottom number, 9, is called the **denominator.**

A fraction may be used to refer to part of a whole. For example, $\frac{2}{9}$ of the circle to the left is shaded. The denominator 9 tells us how many equal parts the whole circle is divided into and the numerator 2 tells us how many equal parts are shaded.

To simplify fractions, we can factor the numerator and the denominator. In the statement $3 \cdot 5 = 15$, 3 and 5 are called **factors** and 15 is the **product.** (The raised dot symbol indicates multiplication.)

$$
\begin{array}{ccccc}
3 & \cdot & 5 & = & 15 \\
\uparrow & & \uparrow & & \uparrow \\
\text{factor} & & \text{factor} & & \text{product}
\end{array}
$$

To **factor** 15 means to write it as a product. The number 15 can be factored as $3 \cdot 5$ or as $1 \cdot 15$.

A fraction is said to be **simplified** or in **lowest terms** when the numerator and the denominator have no factors in common other than 1. For example, the fraction $\frac{5}{11}$ is in lowest terms since 5 and 11 have no common factors other than 1.

To help us simplify fractions, we write the numerator and the denominator as a product of **prime numbers.**

> **Prime Number**
>
> A prime number is a natural number, other than 1, whose only factors are 1 and itself. The first few prime numbers are
>
> $$2, 3, 5, 7, 11, 13, 17, 19, 23, 29, \text{ and so on.}$$

A natural number, other than 1, that is not a prime number is called a **composite number.** Every composite number can be written as a product of prime numbers. We call this product of prime numbers the prime factorization of the composite number.

**EXAMPLE 1**   Write each of the following numbers as a product of primes.

**a.** 40          **b.** 63

*Solution*

**a.** First, write 40 as the product of any two whole numbers, other than 1.

$$40 = 4 \cdot 10$$

Next, factor each of these numbers. Continue this process until all of the factors are prime numbers.

$$40 = 4 \quad \cdot \quad 10$$
$$= 2 \cdot 2 \cdot 2 \cdot 5$$

All the factors are now prime numbers. Then 40 written as a product of primes is

$$40 = 2 \cdot 2 \cdot 2 \cdot 5$$

**b.** $63 = 9 \quad \cdot \quad 7$
$$= 3 \cdot 3 \cdot 7$$

**PRACTICE**

**1**   Write each of the following numbers as a product of primes.

**a.** 36                    **b.** 75

To use prime factors to write a fraction in lowest terms, apply the fundamental principle of fractions.

> **Fundamental Principle of Fractions**
>
> If $\dfrac{a}{b}$ is a fraction and $c$ is a nonzero real number, then
>
> $$\frac{a \cdot c}{b \cdot c} = \frac{a}{b}$$

To understand why this is true, we use the fact that since $c$ is not zero, then $\dfrac{c}{c} = 1$.

$$\frac{a \cdot c}{b \cdot c} = \frac{a}{b} \cdot \frac{c}{c} = \frac{a}{b} \cdot 1 = \frac{a}{b}$$

We will call this process dividing out the common factor of $c$.

**EXAMPLE 2** Write each fraction in lowest terms.

**a.** $\dfrac{42}{49}$     **b.** $\dfrac{11}{27}$     **c.** $\dfrac{88}{20}$

*Solution*

**a.** Write the numerator and the denominator as products of primes; then apply the fundamental principle to the common factor 7.

$$\frac{42}{49} = \frac{2 \cdot 3 \cdot 7}{7 \cdot 7} = \frac{2 \cdot 3}{7} = \frac{6}{7}$$

**b.** $\dfrac{11}{27} = \dfrac{11}{3 \cdot 3 \cdot 3}$

There are no common factors other than 1, so $\dfrac{11}{27}$ is already in lowest terms.

**c.** $\dfrac{88}{20} = \dfrac{2 \cdot 2 \cdot 2 \cdot 11}{2 \cdot 2 \cdot 5} = \dfrac{22}{5}$

**PRACTICE**
**2** Write each fraction in lowest terms.

**a.** $\dfrac{63}{72}$     **b.** $\dfrac{64}{12}$     **c.** $\dfrac{7}{25}$

**Concept Check** ✓

Explain the error in the following steps.

**a.** $\dfrac{15}{55} = \dfrac{1\,\cancel{5}}{5\,\cancel{5}} = \dfrac{1}{5}$     **b.** $\dfrac{6}{7} = \dfrac{5+1}{5+2} = \dfrac{1}{2}$

**OBJECTIVE 2 ▶ Multiplying and dividing fractions.** To multiply two fractions, multiply numerator times numerator to obtain the numerator of the product; multiply denominator times denominator to obtain the denominator of the product.

---
**Multiplying Fractions**

$$\frac{a}{b} \cdot \frac{c}{d} = \frac{a \cdot c}{b \cdot d}, \qquad \text{if } b \neq 0 \text{ and } d \neq 0$$
---

**EXAMPLE 3** Multiply $\dfrac{2}{15}$ and $\dfrac{5}{13}$. Write the product in lowest terms.

*Solution*     $\dfrac{2}{15} \cdot \dfrac{5}{13} = \dfrac{2 \cdot 5}{15 \cdot 13}$    Multiply numerators.
                              Multiply denominators.

Next, simplify the product by dividing the numerator and the denominator by any common factors.

$$= \frac{2 \cdot 5}{3 \cdot 5 \cdot 13}$$
$$= \frac{2}{39}$$

**PRACTICE**
**3** Multiply $\dfrac{3}{8}$ and $\dfrac{7}{9}$. Write the product in lowest terms.

Before dividing fractions, we first define **reciprocals.** Two fractions are reciprocals of each other if their product is 1. For example $\dfrac{2}{3}$ and $\dfrac{3}{2}$ are reciprocals since $\dfrac{2}{3} \cdot \dfrac{3}{2} = 1$. Also, the reciprocal of 5 is $\dfrac{1}{5}$ since $5 \cdot \dfrac{1}{5} = \dfrac{5}{1} \cdot \dfrac{1}{5} = 1$.

To divide fractions, multiply the first fraction by the reciprocal of the second fraction.

---

**Dividing Fractions**

$$\frac{a}{b} \div \frac{c}{d} = \frac{a}{b} \cdot \frac{d}{c}, \qquad \text{if } b \neq 0, d \neq 0, \text{ and } c \neq 0$$

---

**EXAMPLE 4**  Divide. Write all quotients in lowest terms.

**a.** $\dfrac{4}{5} \div \dfrac{5}{16}$  **b.** $\dfrac{7}{10} \div 14$  **c.** $\dfrac{3}{8} \div \dfrac{3}{10}$

*Solution*

**a.** $\dfrac{4}{5} \div \dfrac{5}{16} = \dfrac{4}{5} \cdot \dfrac{16}{5} = \dfrac{4 \cdot 16}{5 \cdot 5} = \dfrac{64}{25}$

**b.** $\dfrac{7}{10} \div 14 = \dfrac{7}{10} \div \dfrac{14}{1} = \dfrac{7}{10} \cdot \dfrac{1}{14} = \dfrac{7 \cdot 1}{2 \cdot 5 \cdot 2 \cdot 7} = \dfrac{1}{20}$.

**c.** $\dfrac{3}{8} \div \dfrac{3}{10} = \dfrac{3}{8} \cdot \dfrac{10}{3} = \dfrac{3 \cdot 2 \cdot 5}{2 \cdot 2 \cdot 2 \cdot 3} = \dfrac{5}{4}$

**PRACTICE**
**4**  Divide. Write all quotients in lowest terms.

**a.** $\dfrac{3}{4} \div \dfrac{4}{9}$  **b.** $\dfrac{5}{12} \div 15$  **c.** $\dfrac{7}{6} \div \dfrac{7}{15}$

---

**OBJECTIVE 3 ▶ Adding and subtracting fractions.**  To add or subtract fractions with the same denominator, combine numerators and place the sum or difference over the common denominator.

---

**Adding and Subtracting Fractions with the Same Denominator**

$$\frac{a}{b} + \frac{c}{b} = \frac{a + c}{b}, \qquad \text{if } b \neq 0$$

$$\frac{a}{b} - \frac{c}{b} = \frac{a - c}{b}, \qquad \text{if } b \neq 0$$

---

**EXAMPLE 5**  Add or subtract as indicated. Write each result in lowest terms.

**a.** $\dfrac{2}{7} + \dfrac{4}{7}$  **b.** $\dfrac{3}{10} + \dfrac{2}{10}$  **c.** $\dfrac{9}{7} - \dfrac{2}{7}$  **d.** $\dfrac{5}{3} - \dfrac{1}{3}$

*Solution*

**a.** $\dfrac{2}{7} + \dfrac{4}{7} = \dfrac{2 + 4}{7} = \dfrac{6}{7}$

**b.** $\dfrac{3}{10} + \dfrac{2}{10} = \dfrac{3 + 2}{10} = \dfrac{5}{10} = \dfrac{5}{2 \cdot 5} = \dfrac{1}{2}$

**c.** $\dfrac{9}{7} - \dfrac{2}{7} = \dfrac{9 - 2}{7} = \dfrac{7}{7} = 1$

**d.** $\dfrac{5}{3} - \dfrac{1}{3} = \dfrac{5 - 1}{3} = \dfrac{4}{3}$

**PRACTICE**
**5**  Add or subtract as indicated. Write each result in lowest terms.

**a.** $\dfrac{8}{5} - \dfrac{3}{5}$  **b.** $\dfrac{8}{5} - \dfrac{2}{5}$  **c.** $\dfrac{3}{5} + \dfrac{1}{5}$  **d.** $\dfrac{5}{12} + \dfrac{1}{12}$

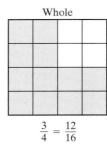

Whole

$$\frac{3}{4} = \frac{12}{16}$$

To add or subtract fractions without the same denominator, first write the fractions as **equivalent fractions** with a common denominator. Equivalent fractions are fractions that represent the same quantity. For example, $\frac{3}{4}$ and $\frac{12}{16}$ are equivalent fractions since they represent the same portion of a whole, as the diagram shows. Count the larger squares and the shaded portion is $\frac{3}{4}$. Count the smaller squares and the shaded portion is $\frac{12}{16}$. Thus, $\frac{3}{4} = \frac{12}{16}$.

We can write equivalent fractions by multiplying a given fraction by 1, as shown in the next example. Multiplying a fraction by 1 does not change the value of the fraction.

**EXAMPLE 6**  Write $\frac{2}{5}$ as an equivalent fraction with a denominator of 20.

**Solution**  Since $5 \cdot 4 = 20$, multiply the fraction by $\frac{4}{4}$. Multiplying by $\frac{4}{4} = 1$ does not change the value of the fraction.

Multiply by $\frac{4}{4}$ or 1.

$$\frac{2}{5} = \frac{2}{5} \cdot \frac{4}{4} = \frac{2 \cdot 4}{5 \cdot 4} = \frac{8}{20}$$

**PRACTICE**
**6**  Write $\frac{2}{3}$ as an equivalent fraction with a denominator of 21.

**EXAMPLE 7**  Add or subtract as indicated. Write each answer in lowest terms.

**a.** $\frac{2}{5} + \frac{1}{4}$      **b.** $\frac{1}{2} + \frac{17}{22} - \frac{2}{11}$      **c.** $3\frac{1}{6} - 1\frac{11}{12}$

**Solution**

**a.** Fractions must have a common denominator before they can be added or subtracted. Since 20 is the smallest number that both 5 and 4 divide into evenly, 20 is the **least common denominator.** Write both fractions as equivalent fractions with denominators of 20. Since

$$\frac{2}{5} \cdot \frac{4}{4} = \frac{2 \cdot 4}{5 \cdot 4} = \frac{8}{20} \qquad \text{and} \qquad \frac{1}{4} \cdot \frac{5}{5} = \frac{1 \cdot 5}{4 \cdot 5} = \frac{5}{20}$$

then

$$\frac{2}{5} + \frac{1}{4} = \frac{8}{20} + \frac{5}{20} = \frac{13}{20}$$

**b.** The least common denominator for denominators 2, 22, and 11 is 22. First, write each fraction as an equivalent fraction with a denominator of 22. Then add or subtract from left to right.

$$\frac{1}{2} = \frac{1}{2} \cdot \frac{11}{11} = \frac{11}{22}, \qquad \frac{17}{22} = \frac{17}{22}, \qquad \text{and} \qquad \frac{2}{11} = \frac{2}{11} \cdot \frac{2}{2} = \frac{4}{22}$$

Then

$$\frac{1}{2} + \frac{17}{22} - \frac{2}{11} = \frac{11}{22} + \frac{17}{22} - \frac{4}{22} = \frac{24}{22} = \frac{12}{11}$$

**c.** To find $3\frac{1}{6} - 1\frac{11}{12}$, let's use a vertical format.

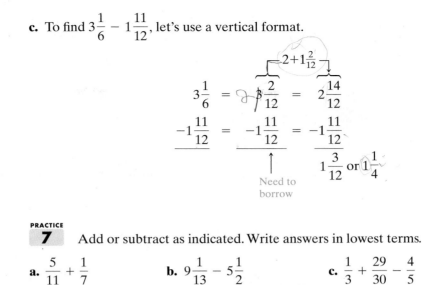

$$
\begin{array}{rcccc}
3\frac{1}{6} & = & 3\frac{2}{12} & = & 2\frac{14}{12} \\[2mm]
-1\frac{11}{12} & = & -1\frac{11}{12} & = & -1\frac{11}{12} \\
\hline
& & & & 1\frac{3}{12} \ \text{or}\ 1\frac{1}{4}
\end{array}
$$

Need to borrow

**PRACTICE**
**7**   Add or subtract as indicated. Write answers in lowest terms.

**a.** $\dfrac{5}{11} + \dfrac{1}{7}$
**b.** $9\dfrac{1}{13} - 5\dfrac{1}{2}$
**c.** $\dfrac{1}{3} + \dfrac{29}{30} - \dfrac{4}{5}$

## VOCABULARY & READINESS CHECK

*Use the choices below to fill in each blank. Some choices may be used more than once.*

| | | | |
|---|---|---|---|
| simplified | reciprocals | equivalent | denominator |
| product | factors | fraction | numerator |

**1.** A quotient of two numbers, such as $\dfrac{5}{8}$, is called a ___fraction___ .

**2.** In the fraction $\dfrac{3}{11}$, the number 3 is called the _____ and the number 11 is called the _____ .

**3.** To factor a number means to write it as a ___product___

**4.** A fraction is said to be _____ when the numerator and the denominator have no common factors other than 1.

**5.** In $7 \cdot 3 = 21$, the numbers 7 and 3 are called _____ and the number 21 is called the _____ .

**6.** The fractions $\dfrac{2}{9}$ and $\dfrac{9}{2}$ are called _____ .

**7.** Fractions that represent the same quantity are called ___equival___ fractions.

*Represent the shaded part of each geometric figure by a fraction*

**8.**   3/8

**9.**   1/4

**10.**   2/7

**11.**

## 1.3 | EXERCISE SET

**MyMathLab** — Powered by CourseCompass™ and MathXL®

Math XL  PRACTICE   WATCH   DOWNLOAD   READ   REVIEW

*Write each number as a product of primes. See Example 1.*

**1.** 33
**2.** 60
**3.** 98
**4.** 27
**5.** 20
**6.** 56
**7.** 75
**8.** 32
**9.** 45
**10.** 24

*Write the fraction in lowest terms. See Example 2.*

**11.** $\dfrac{2}{4}$
**12.** $\dfrac{3}{6}$
**13.** $\dfrac{10}{15}$
**14.** $\dfrac{15}{20}$
**15.** $\dfrac{3}{7}$
**16.** $\dfrac{5}{9}$
**17.** $\dfrac{18}{30}$
**18.** $\dfrac{42}{45}$

*Multiply or divide as indicated. Write the answer in lowest terms. See Examples 3 and 4.*

**19.** $\frac{1}{2} \cdot \frac{3}{4}$

**20.** $\frac{1}{8} \cdot \frac{3}{5}$

**21.** $\frac{2}{3} \cdot \frac{3}{4}$

**22.** $\frac{7}{8} \cdot \frac{3}{21}$

**23.** $\frac{1}{2} \div \frac{7}{12}$

**24.** $\frac{7}{12} \div \frac{1}{2}$

**25.** $\frac{3}{4} \div \frac{1}{20}$

**26.** $\frac{3}{5} \div \frac{9}{10}$

**27.** $\frac{7}{10} \cdot \frac{5}{21}$

**28.** $\frac{3}{35} \cdot \frac{10}{63}$

**29.** $2\frac{7}{9} \cdot \frac{1}{3}$

**30.** $\frac{1}{4} \cdot 5\frac{5}{6}$

*The area of a plane figure is a measure of the amount of surface of the figure. Find the area of each figure below. (The area of a rectangle is the product of its length and width. The area of a triangle is $\frac{1}{2}$ the product of its base and height.)*

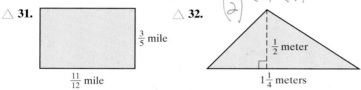

△ **31.**

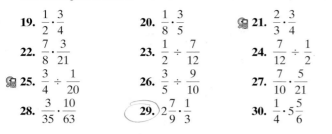

$\frac{3}{5}$ mile

$\frac{11}{12}$ mile

△ **32.**

$\frac{1}{2}$ meter

$1\frac{1}{4}$ meters

*Add or subtract as indicated. Write the answer in lowest terms. See Example 5.*

**33.** $\frac{4}{5} - \frac{1}{5}$

**34.** $\frac{6}{7} - \frac{1}{7}$

**35.** $\frac{4}{5} + \frac{1}{5}$

**36.** $\frac{6}{7} + \frac{1}{7}$

**37.** $\frac{17}{21} - \frac{10}{21}$

**38.** $\frac{18}{35} - \frac{11}{35}$

**39.** $\frac{23}{105} + \frac{4}{105}$

**40.** $\frac{13}{132} + \frac{35}{132}$

*Write each fraction as an equivalent fraction with the given denominator. See Example 6.*

**41.** $\frac{7}{10}$ with a denominator of 30

**42.** $\frac{2}{3}$ with a denominator of 9

**43.** $\frac{2}{9}$ with a denominator of 18

**44.** $\frac{8}{7}$ with a denominator of 56

**45.** $\frac{4}{5}$ with a denominator of 20

**46.** $\frac{4}{5}$ with a denominator of 25

*Add or subtract as indicated. Write the answer in lowest terms. See Example 7.*

**47.** $\frac{2}{3} + \frac{3}{7}$

**48.** $\frac{3}{4} + \frac{1}{6}$

**49.** $2\frac{13}{15} - 1\frac{1}{5}$

**50.** $5\frac{2}{9} - 3\frac{1}{6}$

**51.** $\frac{5}{22} - \frac{5}{33}$

**52.** $\frac{7}{10} - \frac{8}{15}$

**53.** $\frac{12}{5} - 1$

**54.** $2 - \frac{3}{8}$

*Each circle below represents a whole, or 1. Use subtraction to determine the unknown part of the circle.*

**55.**

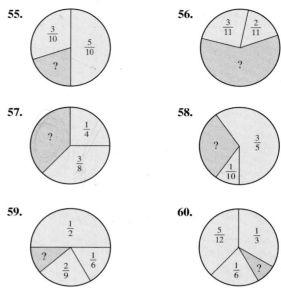

$\frac{3}{10}$  $\frac{5}{10}$  ?

**56.**

$\frac{3}{11}$  $\frac{2}{11}$  ?

**57.**

?  $\frac{1}{4}$  $\frac{3}{8}$

**58.**

?  $\frac{3}{5}$  $\frac{1}{10}$

**59.**

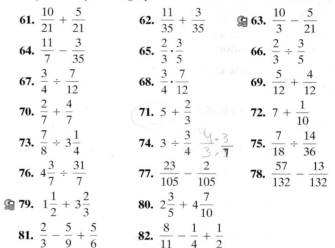

$\frac{1}{2}$  ?  $\frac{2}{9}$  $\frac{1}{6}$

**60.**

$\frac{5}{12}$  $\frac{1}{3}$  $\frac{1}{6}$  ?

**MIXED PRACTICE**

*Perform the following operations. Write answers in lowest terms.*

**61.** $\frac{10}{21} + \frac{5}{21}$

**62.** $\frac{11}{35} + \frac{3}{35}$

**63.** $\frac{10}{3} - \frac{5}{21}$

**64.** $\frac{11}{7} - \frac{3}{35}$

**65.** $\frac{2}{3} \cdot \frac{3}{5}$

**66.** $\frac{2}{3} \div \frac{3}{5}$

**67.** $\frac{3}{4} \div \frac{7}{12}$

**68.** $\frac{3}{4} \cdot \frac{7}{12}$

**69.** $\frac{5}{12} + \frac{4}{12}$

**70.** $\frac{2}{7} + \frac{4}{7}$

**71.** $5 + \frac{2}{3}$

**72.** $7 + \frac{1}{10}$

**73.** $\frac{7}{8} \div 3\frac{1}{4}$

**74.** $3 \div \frac{3}{4}$

**75.** $\frac{7}{18} \div \frac{14}{36}$

**76.** $4\frac{3}{7} \div \frac{31}{7}$

**77.** $\frac{23}{105} - \frac{2}{105}$

**78.** $\frac{57}{132} - \frac{13}{132}$

**79.** $1\frac{1}{2} + 3\frac{2}{3}$

**80.** $2\frac{3}{5} + 4\frac{7}{10}$

**81.** $\frac{2}{3} - \frac{5}{9} + \frac{5}{6}$

**82.** $\frac{8}{11} - \frac{1}{4} + \frac{1}{2}$

*The perimeter of a plane figure is the total distance around the figure. Find the perimeter of each figure in Exercises 83 and 84.*

△ **83.**

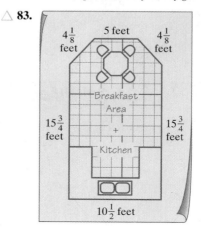

$4\frac{1}{8}$ feet   5 feet   $4\frac{1}{8}$ feet

Breakfast Area

$15\frac{3}{4}$ feet   $15\frac{3}{4}$ feet

Kitchen

$10\frac{1}{2}$ feet

△ **84.**

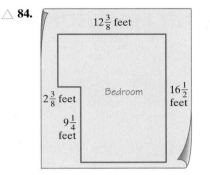

**85.** Yelena Isinbaeva currently holds the women's pole vault world record at $5\frac{1}{50}$ meters. The men's pole vault world record is currently held by Sergei Bubka, at $1\frac{3}{25}$ meters higher than the women's record. What is the current men's pole vault record? (*Source:* International Association of Athletics Federations)

**86.** The Preakness, one of the horse races in the Triple Crown, is a $1\frac{3}{16}$-mile race. The Belmont, another of the three races, is $\frac{5}{16}$ of a mile longer than the Preakness. How long is the Belmont? (*Source: Sports Illustrated*)

**87.** In your own words, explain how to add two fractions with different denominators.

**88.** In your own words, explain how to multiply two fractions.

*The following trail chart is given to visitors at the Lakeview Forest Preserve.*

| Trail Name | Distance (miles) |
|---|---|
| Robin Path | $3\frac{1}{2}$ |
| Red Falls | $5\frac{1}{2}$ |
| Green Way | $2\frac{1}{8}$ |
| Autumn Walk | $1\frac{3}{4}$ |

**89.** How much longer is Red Falls Trail than Green Way Trail?

**90.** Find the total distance traveled by someone who hiked along all four trails.

**CONCEPT EXTENSIONS**

*The breakdown of science and engineering doctorate degrees awarded in the United States is summarized in the graph on the next column, called a circle graph or a pie chart. Use the graph to answer the questions. (Source: National Science Foundation)*

**91.** What fraction of science and engineering doctorates are awarded in the physical sciences?

**92.** Engineering doctorates make up what fraction of all science and engineering doctorates awarded in the United States?

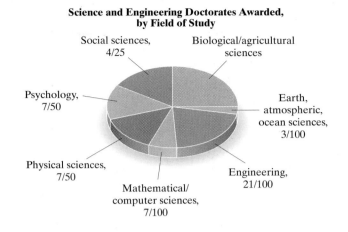

**Science and Engineering Doctorates Awarded, by Field of Study**

**93.** What fraction of all science and engineering doctorates are awarded in the biological and agricultural sciences?

**94.** Social sciences and psychology doctorates together make up what fraction of all science and engineering doctorates awarded in the United States?

*In 2006, Gap Inc. operated a total of 3054 stores worldwide. The following chart shows the store breakdown by brand. (Source: Gap Inc.)*

| Brand | Number of Stores |
|---|---|
| Gap (Domestic) | 1335 |
| Gap (International) | 256 |
| Banana Republic | 498 |
| Old Navy | 960 |
| Forth & Towne | 5 |
| Total | 3054 |

**95.** What fraction of Gap-brand stores were Old Navy stores? Simplify this fraction.

**96.** What fraction of Gap-brand stores were either domestic or international Gap stores or Forth & Towne stores? Simplify this fraction.

*The area of a plane figure is a measure of the amount of surface of the figure. Find the area of each figure. (The area of a triangle is $\frac{1}{2}$ the product of its base and height. The area of a rectangle is the product of its length and width. Recall that area is measured in square units.)*

△ **97.**    △ **98.**

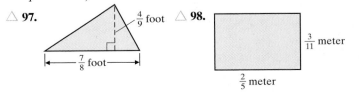

# 1.4 INTRODUCTION TO VARIABLE EXPRESSIONS AND EQUATIONS

## OBJECTIVES

**1** Define and use exponents and the order of operations.

**2** Evaluate algebraic expressions, given replacement values for variables.

**3** Determine whether a number is a solution of a given equation.

**4** Translate phrases into expressions and sentences into equations.

**OBJECTIVE 1 ▶ Using exponents and the order of operations.** Frequently in algebra, products occur that contain repeated multiplication of the same factor. For example, the volume of a cube whose sides each measure 2 centimeters is $(2 \cdot 2 \cdot 2)$ cubic centimeters. We may use **exponential notation** to write such products in a more compact form. For example,

$$2 \cdot 2 \cdot 2 \quad \textit{may be written as} \quad 2^3.$$

The 2 in $2^3$ is called the **base**; it is the repeated factor. The 3 in $2^3$ is called the **exponent** and is the number of times the base is used as a factor. The expression $2^3$ is called an **exponential expression.**

$$\underset{\text{base}}{\nearrow} 2^{\overset{\text{exponent}}{\nwarrow 3}} = 2 \cdot 2 \cdot 2 = 8$$

2 is a factor 3 times

2 cm

Volume is $(2 \cdot 2 \cdot 2)$ cubic centimeters.

### EXAMPLE 1 Evaluate the following:

**a.** $3^2$ [read as "3 squared" or as "3 to the second power"]

**b.** $5^3$ [read as "5 cubed" or as "5 to the third power"]

**c.** $2^4$ [read as "2 to the fourth power"]

**d.** $7^1$     **e.** $\left(\dfrac{3}{7}\right)^2$

*Solution*

**a.** $3^2 = 3 \cdot 3 = 9$      **b.** $5^3 = 5 \cdot 5 \cdot 5 = 125$

**c.** $2^4 = 2 \cdot 2 \cdot 2 \cdot 2 = 16$      **d.** $7^1 = 7$

**e.** $\left(\dfrac{3}{7}\right)^2 = \left(\dfrac{3}{7}\right)\left(\dfrac{3}{7}\right) = \dfrac{9}{49}$

**PRACTICE**
**1** Evaluate:

**a.** $1^3$      **b.** $5^2$      **c.** $\left(\dfrac{1}{10}\right)^2$      **d.** $9^1$      **e.** $\left(\dfrac{2}{5}\right)^3$

> ▶ **Helpful Hint**
>
> $2^3 \neq 2 \cdot 3$ since $2^3$ indicates repeated **multiplication** of the same factor.
>
> $$2^3 = 2 \cdot 2 \cdot 2 = 8, \text{ whereas } 2 \cdot 3 = 6.$$

Using symbols for mathematical operations is a great convenience. However, the more operation symbols presented in an expression, the more careful we must be when performing the indicated operation. For example, in the expression $2 + 3 \cdot 7$, do we add first or multiply first? To eliminate confusion, **grouping symbols** are used. Examples of grouping symbols are parentheses ( ), brackets [ ], braces { }, and the fraction bar. If we wish $2 + 3 \cdot 7$ to be simplified by adding first, we enclose $2 + 3$ in parentheses.

$$(2 + 3) \cdot 7 = 5 \cdot 7 = 35$$

If we wish to multiply first, $3 \cdot 7$ may be enclosed in parentheses.

$$2 + (3 \cdot 7) = 2 + 21 = 23$$

To eliminate confusion when no grouping symbols are present, use the following agreed upon order of operations.

PEMDAS

---

**Order of Operations**

Simplify expressions using the order below. If grouping symbols such as parentheses are present, simplify expressions within those first, starting with the innermost set. If fraction bars are present, simplify the numerator and the denominator separately.

1. Evaluate exponential expressions.
2. Perform multiplications or divisions in order from left to right.
3. Perform additions or subtractions in order from left to right.

---

Now simplify $2 + 3 \cdot 7$. There are no grouping symbols and no exponents, so we multiply and then add.

$$2 + 3 \cdot 7 = 2 + 21 \quad \text{Multiply.}$$
$$= 23 \quad \text{Add.}$$

**EXAMPLE 2**   Simplify each expression.

**a.** $6 \div 3 + 5^2$  **b.** $\dfrac{2(12 + 3)}{|-15|}$  **c.** $3 \cdot 10 - 7 \div 7$  **d.** $3 \cdot 4^2$  **e.** $\dfrac{3}{2} \cdot \dfrac{1}{2} - \dfrac{1}{2}$

*Solution*

**a.** Evaluate $5^2$ first.

$$6 \div 3 + 5^2 = 6 \div 3 + 25$$

Next divide, then add.

$$= 2 + 25 \quad \text{Divide.}$$
$$= 27 \quad \text{Add.}$$

**b.** First, simplify the numerator and the denominator separately.

$$\frac{2(12 + 3)}{|-15|} = \frac{2(15)}{15} \quad \begin{array}{l}\text{Simplify numerator and} \\ \text{denominator separately.}\end{array}$$
$$= \frac{30}{15}$$
$$= 2 \quad \text{Simplify.}$$

**c.** Multiply and divide from left to right. Then subtract.

$$3 \cdot 10 - 7 \div 7 = 30 - 1$$
$$= 29 \quad \text{Subtract.}$$

**d.** In this example, only the 4 is squared. The factor of 3 is not part of the base because no grouping symbol includes it as part of the base.

$$3 \cdot 4^2 = 3 \cdot 16 \quad \text{Evaluate the exponential expression.}$$
$$= 48 \quad \text{Multiply.}$$

**e.** The order of operations applies to operations with fractions in exactly the same way as it applies to operations with whole numbers.

$$\frac{3}{2} \cdot \frac{1}{2} - \frac{1}{2} = \frac{3}{4} - \frac{1}{2} \qquad \text{Multiply.}$$

$$= \frac{3}{4} - \frac{2}{4} \qquad \text{The least common denominator is 4.}$$

$$= \frac{1}{4} \qquad \text{Subtract.} \qquad \square$$

**PRACTICE**
**2** Simplify each expression

**a.** $6 + 3 \cdot 9$

**b.** $4^3 \div 8 + 3$

**c.** $\left(\frac{2}{3}\right)^2 \cdot |-8|$

**d.** $\dfrac{9(14 - 6)}{|-2|}$

**e.** $\dfrac{7}{4} \cdot \dfrac{1}{4} - \dfrac{1}{4}$

> ▶ **Helpful Hint**
> Be careful when evaluating an exponential expression. In $3 \cdot 4^2$, the exponent 2 applies only to the base 4. In $(3 \cdot 4)^2$, we multiply first because of parentheses, so the exponent 2 applies to the product $3 \cdot 4$.
>
> $$3 \cdot 4^2 = 3 \cdot 16 = 48 \qquad (3 \cdot 4)^2 = (12)^2 = 144$$

Expressions that include many grouping symbols can be confusing. When simplifying these expressions, keep in mind that grouping symbols separate the expression into distinct parts. Each is then simplified separately.

**EXAMPLE 3**    Simplify $\dfrac{3 + |4 - 3| + 2^2}{6 - 3}$.

**Solution**    The fraction bar serves as a grouping symbol and separates the numerator and denominator. Simplify each separately. Also, the absolute value bars here serve as a grouping symbol. We begin in the numerator by simplifying within the absolute value bars.

$$\frac{3 + |4 - 3| + 2^2}{6 - 3} = \frac{3 + |1| + 2^2}{6 - 3} \qquad \text{Simplify the expression inside the absolute value bars.}$$

$$= \frac{3 + 1 + 2^2}{3} \qquad \text{Find the absolute value and simplify the denominator.}$$

$$= \frac{3 + 1 + 4}{3} \qquad \text{Evaluate the exponential expression.}$$

$$= \frac{8}{3} \qquad \text{Simplify the numerator.} \qquad \square$$

**PRACTICE**
**3** Simplify $\dfrac{6^2 - 5}{3 + |6 - 5| \cdot 8}$

**EXAMPLE 4**   Simplify $3[4 + 2(10 - 1)]$.

*Solution*   Notice that both parentheses and brackets are used as grouping symbols. Start with the innermost set of grouping symbols.

$$3[4 + 2(10 - 1)] = 3[4 + 2(9)] \quad \text{Simplify the expression in parentheses.}$$
$$= 3[4 + 18] \quad \text{Multiply.}$$
$$= 3[22] \quad \text{Add.}$$
$$= 66 \quad \text{Multiply.}$$

> ▶ **Helpful Hint**
>
> Be sure to follow order of operations and resist the temptation to incorrectly add 4 and 2 first.

**PRACTICE**
**4**   Simplify   $4[25 - 3(5 + 3)]$.

**EXAMPLE 5**   Simplify $\dfrac{8 + 2 \cdot 3}{2^2 - 1}$.

*Solution*

$$\frac{8 + 2 \cdot 3}{2^2 - 1} = \frac{8 + 6}{4 - 1} = \frac{14}{3}$$

**PRACTICE**
**5**   Simplify   $\dfrac{36 \div 9 + 5}{5^2 - 3}$.

**OBJECTIVE 2** ▶ **Evaluating algebraic expressions.** In algebra, we use symbols, usually letters such as $x$, $y$, or $z$, to represent unknown numbers. A symbol that is used to represent a number is called a **variable**. An **algebraic expression** is a collection of numbers, variables, operation symbols, and grouping symbols. For example,

$$2x, \quad -3, \quad 2x + 10, \quad 5(p^2 + 1), \quad \text{and} \quad \frac{3y^2 - 6y + 1}{5}$$

are algebraic expressions. The expression $2x$ means $2 \cdot x$. Also, $5(p^2 + 1)$ means $5 \cdot (p^2 + 1)$ and $3y^2$ means $3 \cdot y^2$. If we give a specific value to a variable, we can **evaluate an algebraic expression.** To evaluate an algebraic expression means to find its numerical value once we know the values of the variables.

Algebraic expressions often occur during problem solving. For example, the expression

$$16t^2$$

gives the distance in feet (neglecting air resistance) that an object will fall in $t$ seconds. (See Exercise 63 in this section.)

**EXAMPLE 6** Evaluate each expression if $x = 3$ and $y = 2$.

**a.** $2x - y$      **b.** $\dfrac{3x}{2y}$      **c.** $\dfrac{x}{y} + \dfrac{y}{2}$      **d.** $x^2 - y^2$

*Solution*

**a.** Replace $x$ with $3$ and $y$ with $2$.

$$2x - y = 2(3) - 2 \quad \text{Let } x = 3 \text{ and } y = 2.$$
$$= 6 - 2 \qquad \text{Multiply.}$$
$$= 4 \qquad\quad \text{Subtract.}$$

**b.** $\dfrac{3x}{2y} = \dfrac{3 \cdot 3}{2 \cdot 2} = \dfrac{9}{4} \quad$ Let $x = 3$ and $y = 2$.

**c.** Replace $x$ with $3$ and $y$ with $2$. Then simplify.

$$\frac{x}{y} + \frac{y}{2} = \frac{3}{2} + \frac{2}{2} = \frac{5}{2}$$

**d.** Replace $x$ with $3$ and $y$ with $2$.

$$x^2 - y^2 = 3^2 - 2^2 = 9 - 4 = 5$$

**PRACTICE**

**6** Evaluate each expression if $x = 2$ and $y = 5$.

**a.** $2x + y$      **b.** $\dfrac{4x}{3y}$      **c.** $\dfrac{3}{x} + \dfrac{x}{y}$      **d.** $x^3 + y^2$

**OBJECTIVE 3 ▶ Determining whether a number is a solution of an equation.** Many times a problem-solving situation is modeled by an equation. An **equation** is a mathematical statement that two expressions have equal value. The equal symbol "=" is used to equate the two expressions. For example, $3 + 2 = 5$, $7x = 35$, $\dfrac{2(x - 1)}{3} = 0$, and $I = PRT$ are all equations.

> ▶ **Helpful Hint**
> An equation contains the equal symbol "=". An algebraic expression does not.

**Concept Check** ☑

Which of the following are equations? Which are expressions?

**a.** $5x = 8$      **b.** $5x - 8$      **c.** $12y + 3x$      **d.** $12y = 3x$

When an equation contains a variable, deciding which values of the variable make an equation a true statement is called **solving** an equation for the variable. A **solution** of an equation is a value for the variable that makes the equation true. For example, 3 is a solution of the equation $x + 4 = 7$, because if $x$ is replaced with 3 the statement is true.

$$x + 4 = 7$$
$$\downarrow$$
$$3 + 4 = 7 \quad \text{Replace } x \text{ with 3.}$$
$$7 = 7 \quad \text{True}$$

Similarly, 1 is not a solution of the equation $x + 4 = 7$, because $1 + 4 = 7$ is **not** a true statement.

**EXAMPLE 7**   Decide whether 2 is a solution of $3x + 10 = 8x$.

_Solution_   Replace $x$ with 2 and see if a true statement results.

$$3x + 10 = 8x \qquad \text{Original equation}$$
$$3(2) + 10 \stackrel{?}{=} 8(2) \qquad \text{Replace } x \text{ with 2.}$$
$$6 + 10 \stackrel{?}{=} 16 \qquad \text{Simplify each side.}$$
$$16 = 16 \qquad \text{True}$$

Since we arrived at a true statement after replacing $x$ with 2 and simplifying both sides of the equation, 2 is a solution of the equation.  □

**PRACTICE**
**7**   Decide whether 4 is a solution of $9x - 6 = 7x$.

**OBJECTIVE 4 ▶ Translating phrases to expressions and sentences to equations.** Now that we know how to represent an unknown number by a variable, let's practice translating phrases into algebraic expressions and sentences into equations. Oftentimes solving problems requires the ability to translate word phrases and sentences into symbols. Below is a list of some key words and phrases to help us translate.

> ▶ **Helpful Hint**
> Order matters when subtracting and also dividing, so be especially careful with these translations.

| Addition (+) | Subtraction (−) | Multiplication (·) | Division (÷) | Equality (=) |
|---|---|---|---|---|
| Sum | Difference of | Product | Quotient | Equals |
| Plus | Minus | Times | Divide | Gives |
| Added to | Subtracted from | Multiply | Into | Is/was/should be |
| More than | Less than | Twice | Ratio | Yields |
| Increased by | Decreased by | Of | Divided by | Amounts to |
| Total | Less | | | Represents/ Is the same as |

**EXAMPLE 8**   Write an algebraic expression that represents each phrase. Let the variable $x$ represent the unknown number.

a. The sum of a number and 3
b. The product of 3 and a number
c. Twice a number
d. 10 decreased by a number
e. 5 times a number, increased by 7

_Solution_

a. $x + 3$ since "sum" means to add
b. $3 \cdot x$ and $3x$ are both ways to denote the product of 3 and $x$
c. $2 \cdot x$ or $2x$
d. $10 - x$ because "decreased by" means to subtract
e. $\underbrace{5x}_{5 \text{ times a number}} + 7$   □

**PRACTICE**
**8** Write an algebraic expression that represents each phase. Let the variable $x$ represent the unknown number.

**a.** Six times a number

**b.** A number decreased by 8

**c.** The product of a number and 9

**d.** Two times a number, plus 3

**e.** The sum of 7 and a number

---

▶ **Helpful Hint**

Make sure you understand the difference when translating phrases containing "decreased by," "subtracted from," and "less than."

| *Phrase* | *Translation* |
|---|---|
| A number decreased by 10 | $x - 10$ |
| A number subtracted from 10 | $10 - x$ |
| 10 less than a number | $x - 10$ |
| A number less 10 | $x - 10$ |

} Notice the order.

Now let's practice translating sentences into equations.

**EXAMPLE 9** Write each sentence as an equation or inequality. Let $x$ represent the unknown number.

**a.** The quotient of 15 and a number is 4.

**b.** Three subtracted from 12 is a number.

**c.** Four times a number, added to 17, is not equal to 21.

**d.** Triple a number is less than 48.

*Solution*

**a.** In words:　the quotient of 15 and a number　　is　　4

　　Translate:　　$\dfrac{15}{x}$　　　$=$　　4

**b.** In words:　three subtracted **from** 12　　is　　a number

　　Translate:　　$12 - 3$　　　$=$　　$x$

Care must be taken when the operation is subtraction. The expression $3 - 12$ would be incorrect. Notice that $3 - 12 \neq 12 - 3$.

**c.** In words:　four times a number　added to　17　is not equal to　21

　　Translate:　　$4x$　　　$+$　　17　　$\neq$　　21

**d.** In words:　triple a number　is less than　48

　　Translate:　　$3x$　　　$<$　　48

**PRACTICE**
**9**   Write each sentence as an equation or inequality. Let $x$ represent the unknown number.

**a.** A number increased by 7 is equal to 13.

**b.** Two less than a number is 11.

**c.** Double a number, added to 9, is not equal to 25.

**d.** Five times 11 is greater than or equal to an unknown number.

---

## Calculator Explorations

### Exponents

To evaluate exponential expressions on a scientific calculator, find the key marked $\boxed{y^x}$ or $\boxed{\wedge}$. To evaluate, for example, $3^5$, press the following keys: $\boxed{3}\ \boxed{y^x}\ \boxed{5}\ \boxed{=}$

The display should read $\boxed{\phantom{xxxx}243}$ or $\boxed{\begin{array}{l}3\wedge 5\\ \phantom{xxx}243\end{array}}$

### Order of Operations

Some calculators follow the order of operations, and others do not. To see whether or not your calculator has the order of operations built in, use your calculator to find $2 + 3 \cdot 4$. To do this, press the following sequence of keys:

$\boxed{2}\ \boxed{+}\ \boxed{3}\ \boxed{\times}\ \boxed{4}\ \boxed{=}$

$\updownarrow$ or

$\boxed{\text{ENTER}}$

The correct answer is 14 because the order of operations is to multiply before we add. If the calculator displays $\boxed{\phantom{xxx}14}$, then it has the order of operations built in.

Even if the order of operations is built in, parentheses must sometimes be inserted. For example, to simplify $\dfrac{5}{12 - 7}$, press the keys

$\boxed{5}\ \boxed{\div}\ \boxed{(}\ \boxed{1}\ \boxed{2}\ \boxed{-}\ \boxed{7}\ \boxed{)}\ \boxed{=}$

$\updownarrow$ or

$\boxed{\text{ENTER}}$

The display should read $\boxed{\phantom{xxx}1}$ or $\boxed{\begin{array}{l}5/(12 - 7)\\ \phantom{xxxxxx}1\end{array}}$

*Use a calculator to evaluate each expression.*

**1.** $5^4$                           **2.** $7^4$

**3.** $9^5$                           **4.** $8^6$

**5.** $2(20 - 5)$                      **6.** $3(14 - 7) + 21$

**7.** $24(862 - 455) + 89$            **8.** $99 + (401 + 962)$

**9.** $\dfrac{4623 + 129}{36 - 34}$   **10.** $\dfrac{956 - 452}{89 - 86}$

# VOCABULARY & READINESS CHECK

*Use the choices below to fill in each blank.*

add ✓       ✓multiply       ʳequation       ✓variable       ✓base       ✓grouping

subtract ✓       ✓divide       ✓expression       solution       ✓solving       ✓exponent

**1.** In the expression $5^2$, the 5 is called the ___base___ and the 2 is called the ___expon___.

**2.** The symbols ( ), [ ], and { } are examples of ___group___ symbols.

**3.** A symbol that is used to represent a number is called a(n) ___variable___

**4.** A collection of numbers, variables, operation symbols, and grouping symbols is called a(n) ___expression___

**5.** A mathematical statement that two expressions are equal is called a(n) ___equation___

**6.** A value for the variable that makes an equation a true statement is called a(n) ___solution___

**7.** Deciding what values of a variable make an equation a true statement is called ___solving___ the equation.

**8.** To simplify the expression $1 + 3 \cdot 6$, first ___m___.

**9.** To simplify the expression $(1 + 3) \cdot 6$, first ___a___.

**10.** To simplify the expression $(20 - 4) \cdot 2$, first ___s___.

**11.** To simplify the expression $20 - 4 \div 2$, first ___✓___.

## 1.4 EXERCISE SET

*Evaluate. See Example 1.*

**1.** $3^5$

**2.** $2^5$

**3.** $3^3$

**4.** $4^4$

**5.** $1^5$

**6.** $1^8$

**7.** $5^1$

**8.** $8^1$

**9.** $\left(\dfrac{1}{5}\right)^3$

**10.** $\left(\dfrac{6}{11}\right)^2$

**11.** $\left(\dfrac{2}{3}\right)^4$

**12.** $\left(\dfrac{1}{2}\right)^5$

**13.** $7^2$

**14.** $9^2$

**15.** $4^2$

**16.** $4^3$

**17.** $(1.2)^2$

**18.** $(0.07)^2$

### MIXED PRACTICE

*Simplify each expression. See Examples 2 through 5.*

**19.** $5 + 6 \cdot 2$

**20.** $8 + 5 \cdot 3$

**21.** $4 \cdot 8 - 6 \cdot 2$

**22.** $12 \cdot 5 - 3 \cdot 6$

**23.** $2(8 - 3)$

**24.** $5(6 - 2)$

**25.** $2 + (5 - 2) + 4^2$

**26.** $6 - 2 \cdot 2 + 2^5$

**27.** $5 \cdot 3^2$

**28.** $2 \cdot 5^2$

**29.** $\dfrac{1}{4} \cdot \dfrac{2}{3} - \dfrac{1}{6}$

**30.** $\dfrac{3}{4} \cdot \dfrac{1}{2} + \dfrac{2}{3}$

**31.** $\dfrac{6 - 4}{9 - 2}$

**32.** $\dfrac{8 - 5}{24 - 20}$

**33.** $2[5 + 2(8 - 3)]$

**34.** $3[4 + 3(6 - 4)]$

**35.** $\dfrac{19 - 3 \cdot 5}{6 - 4}$

**36.** $\dfrac{4 \cdot 3 + 2}{4 + 3 \cdot 2}$

**37.** $\dfrac{|6 - 2| + 3}{8 + 2 \cdot 5}$

**38.** $\dfrac{15 - |3 - 1|}{12 - 3 \cdot 2}$

**39.** $\dfrac{3 + 3(5 + 3)}{3^2 + 1}$

**40.** $\dfrac{3 + 6(8 - 5)}{4^2 + 2}$

**41.** $\dfrac{6 + |8 - 2| + 3^2}{18 - 3}$

**42.** $\dfrac{16 + |13 - 5| + 4^2}{17 - 5}$

**43.** Are parentheses necessary in the expression $2 + (3 \cdot 5)$? Explain your answer.

**44.** Are parentheses necessary in the expression $(2 + 3) \cdot 5$? Explain your answer.

*For Exercises 45 and 46, match each expression in the first column with its value in the second column.*

**45.**
| | | |
|---|---|---|
| **a.** $(6 + 2) \cdot (5 + 3)$ | | 19 |
| **b.** $(6 + 2) \cdot 5 + 3$ | | 22 |
| **c.** $6 + 2 \cdot 5 + 3$ | | 64 |
| **d.** $6 + 2 \cdot (5 + 3)$ | | 43 |

**46.**
| | | |
|---|---|---|
| **a.** $(1 + 4) \cdot 6 - 3$ | | 15 |
| **b.** $1 + 4 \cdot (6 - 3)$ | | 13 |
| **c.** $1 + 4 \cdot 6 - 3$ | | 27 |
| **d.** $(1 + 4) \cdot (6 - 3)$ | | 22 |

*Evaluate each expression when $x = 1$, $y = 3$, and $z = 5$. See Example 6.*

**47.** $3y$

**48.** $4x$

**49.** $\dfrac{z}{5x}$

**50.** $\dfrac{y}{2z}$

**51.** $3x - 2$

**52.** $6y - 8$

**53.** $|2x + 3y|$

**54.** $|5z - 2y|$

**55.** $5y^2$

**56.** $2z^2$

*Evaluate each expression if $x = 12$, $y = 8$, and $z = 4$. See Example 6.*

**57.** $\dfrac{x}{z} + 3y$

**58.** $\dfrac{y}{z} + 8x$

**59.** $x^2 - 3y + x$

**60.** $y^2 - 3x + y$

**61.** $\dfrac{x^2 + z}{y^2 + 2z}$

**62.** $\dfrac{y^2 + x}{x^2 + 3y}$

*Neglecting air resistance, the expression $16t^2$ gives the distance in feet an object will fall in t seconds.*

**63.** Complete the chart below. To evaluate $16t^2$, remember to first find $t^2$, then multiply by 16.

| Time t (in seconds) | Distance $16t^2$ (in feet) |
|---|---|
| 1 | |
| 2 | |
| 3 | |
| 4 | |

**64.** Does an object fall the same distance *during* each second? Why or why not? (See Exercise 63.)

*Decide whether the given number is a solution of the given equation. See Example 7.*

**65.** Is 5 a solution of $3x + 30 = 9x$?

**66.** Is 6 a solution of $2x + 7 = 3x$?

**67.** Is 0 a solution of $2x + 6 = 5x - 1$?

**68.** Is 2 a solution of $4x + 2 = x + 8$?

**69.** Is 8 a solution of $2x - 5 = 5$?

**70.** Is 6 a solution of $3x - 10 = 8$?

**71.** Is 2 a solution of $x + 6 = x + 6$?

**72.** Is 10 a solution of $x + 6 = x + 6$?

**73.** Is 0 a solution of $x = 5x + 15$?

**74.** Is 1 a solution of $4 = 1 - x$?

*Write each phrase as an algebraic expression. Let x represent the unknown number. See Example 8.*

**75.** Fifteen more than a number

**76.** One-half times a number

**77.** Five subtracted from a number

**78.** The quotient of a number and 9

**79.** Three times a number, increased by 22

**80.** The product of 8 and a number, decreased by 10

*Write each sentence as an equation or inequality. Use x to represent any unknown number. See Example 9.*

**81.** One increased by two equals the quotient of nine and three.

**82.** Four subtracted from eight is equal to two squared.

**83.** Three is not equal to four divided by two.

**84.** The difference of sixteen and four is greater than ten.

**85.** The sum of 5 and a number is 20.

**86.** Twice a number is 17.

**87.** Thirteen minus three times a number is 13.

**88.** Seven subtracted from a number is 0.

**89.** The quotient of 12 and a number is $\dfrac{1}{2}$.

**90.** The sum of 8 and twice a number is 42.

**91.** In your own words, explain the difference between an expression and an equation.

**92.** Determine whether each is an expression or an equation.

    **a.** $3x^2 - 26$     **b.** $3x^2 - 26 = 1$

    **c.** $2x - 5 = 7x - 5$     **d.** $9y + x - 8$

## CONCEPT EXTENSIONS

**93.** Insert parentheses so that the following expression simplifies to 32.

$$20 - 4 \cdot 4 \div 2$$

**94.** Insert parentheses so that the following expression simplifies to 28.

$$2 \cdot 5 + 3^2$$

*Solve the following.*

**95.** The perimeter of a figure is the distance around the figure. The expression $2l + 2w$ represents the perimeter of a rectangle when $l$ is its length and $w$ is its width. Find the perimeter of the following rectangle by substituting 8 for $l$ and 6 for $w$.

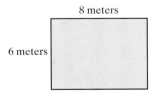

8 meters

6 meters

**96.** The expression $a + b + c$ represents the perimeter of a triangle when $a$, $b$, and $c$ are the lengths of its sides. Find the perimeter of the following triangle.

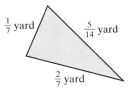

$\dfrac{1}{7}$ yard      $\dfrac{5}{14}$ yard

$\dfrac{2}{7}$ yard

△ **97.** The area of a figure is the total enclosed surface of the figure. Area is measured in square units. The expression $lw$ represents the area of a rectangle when $l$ is its length and $w$ is its width. Find the area of the following rectangular-shaped lot.

△ **98.** A trapezoid is a four-sided figure with exactly one pair of parallel sides. The expression $\frac{1}{2}h(B + b)$ represents its area, when $B$ and $b$ are the lengths of the two parallel sides and $h$ is the height between these sides. Find the area if $B = 15$ inches, $b = 7$ inches, and $h = 5$ inches.

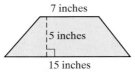

7 inches

5 inches

15 inches

**99.** The expression $\frac{I}{PT}$ represents the rate of interest being charged if a loan of $P$ dollars for $T$ years required $I$ dollars in interest to be paid. Find the interest rate if a \$650 loan for 3 years to buy a used IBM personal computer requires \$126.75 in interest to be paid.

**100.** The expression $\frac{d}{t}$ represents the average speed $r$ in miles per hour if a distance of $d$ miles is traveled in $t$ hours. Find the rate to the nearest whole number if the distance between Dallas, Texas, and Kaw City, Oklahoma, is 432 miles, and it takes Peter Callac 8.5 hours to drive the distance.

**101.** Verizon long-distance service offers a "Talk to the World Plan," which offers lower rates on international phone calls to Verizon subscribers. This plan charges \$3.00 per month, plus \$0.12 per minute for calls to Japan. The expression $3.00 + 0.12m$ represents the long-distance charge for a call to Japan last month by a Verizon customer enrolled in this plan. Find the monthly bill for the customer whose only call to Japan lasted 84 minutes.

**102.** In forensics, the density of a substance is used to help identify it. The expression $\frac{M}{V}$ represents the density of an object with a mass of $M$ grams and a volume of $V$ milliliters. Find the density of an object having a mass of 29.76 grams and a volume of 12 milliliters.

---

## 1.5 ADDING REAL NUMBERS

**OBJECTIVES**

1. Add real numbers with the same sign.
2. Add real numbers with unlike signs.
3. Solve problems that involve addition of real numbers.
4. Find the opposite of a number.

**OBJECTIVE 1 ▶ Adding real numbers with the same sign.** Real numbers can be added, subtracted, multiplied, divided, and raised to powers, just as whole numbers can. We use a number line to help picture the addition of real numbers.

**EXAMPLE 1** Add: $3 + 2$

*Solution* Recall that 3 and 2 are called addends. We start at 0 on a number line, and draw an arrow representing the addend 3. This arrow is three units long and points to the right since 3 is positive. From the tip of this arrow, we draw another arrow representing the addend 2. The number below the tip of this arrow is the sum, 5.

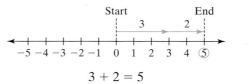

$$3 + 2 = 5$$

**PRACTICE**

**1** Add using a number line: $2 + 4$.

**EXAMPLE 2** Add: $-1 + (-2)$

*Solution* Here, $-1$ and $-2$ are addends. We start at 0 on a number line, and draw an arrow representing $-1$. This arrow is one unit long and points to the left since $-1$ is negative. From the tip of this arrow, we draw another arrow representing $-2$. The number below the tip of this arrow is the sum, $-3$.

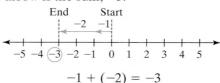

$$-1 + (-2) = -3$$

**PRACTICE**
**2**    Add using a number line:   $-2 + (-3)$.

---

Thinking of signed numbers as money earned or lost might help make addition more meaningful. Earnings can be thought of as positive numbers. If $1 is earned and later another $3 is earned, the total amount earned is $4. In other words, $1 + 3 = 4$.

On the other hand, losses can be thought of as negative numbers. If $1 is lost and later another $3 is lost, a total of $4 is lost. In other words, $(-1) + (-3) = -4$.

Using a number line each time we add two numbers can be time consuming. Instead, we can notice patterns in the previous examples and write rules for adding signed numbers. When adding two numbers with the same sign, notice that the sign of the sum is the same as the sign of the addends.

> **Adding Two Numbers with the Same Sign**
> Add their absolute values. Use their common sign as the sign of the sum.

**EXAMPLE 3**    Add.

**a.** $-3 + (-7)$          **b.** $-1 + (-20)$          **c.** $-2 + (-10)$

*Solution*    Notice that each time, we are adding numbers with the same sign.

**a.** $-3 + (-7) = -10$ ⟵ Add their absolute values: $3 + 7 = 10$.
          ⟶ Use their common sign.

**b.** $-1 + (-20) = -21$ ⟵ Add their absolute values: $1 + 20 = 21$.
          ⟶ Common sign.

**c.** $-2 + (-10) = -12$ ⟵ Add their absolute values.
          ⟶ Common sign.

**PRACTICE**
**3**    Add.   **a.** $-5 + (-8)$          **b.** $-31 + (-1)$

---

**OBJECTIVE 2 ▶ Adding real numbers with unlike signs.**  Adding numbers whose signs are not the same can also be pictured on a number line.

**EXAMPLE 4**    Add:   $-4 + 6$

*Solution*

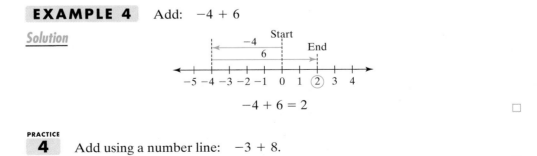

$$-4 + 6 = 2$$

**PRACTICE**
**4**    Add using a number line:   $-3 + 8$.

---

Using temperature as an example, if the thermometer registers 4 degrees below 0 degrees and then rises 6 degrees, the new temperature is 2 degrees above 0 degrees. Thus, it is reasonable that $-4 + 6 = 2$.

Once again, we can observe a pattern: When adding two numbers with different signs, the sign of the sum is the same as the sign of the addend whose absolute value is larger.

> **Adding Two Numbers with Different Signs**
> Subtract the smaller absolute value from the larger absolute value. Use the sign of the number whose absolute value is larger as the sign of the sum.

**EXAMPLE 5**  Add.

a. $3 + (-7)$        b. $-2 + 10$        c. $0.2 + (-0.5)$

*Solution*  Notice that each time, we are adding numbers with different signs.

a. $3 + (-7) = -4$ ◄—— Subtract their absolute values: $7 - 3 = 4$.
     └————— The negative number, $-7$, has the larger absolute value so the sum is negative.

b. $-2 + 10 = 8$ ◄——— Subtract their absolute values: $10 - 2 = 8$.
     └————— The positive number, $10$, has the larger absolute value so the sum is positive.

c. $0.2 + (-0.5) = -0.3$ ◄—— Subtract their absolute values: $0.5 - 0.2 = 0.3$.
     └————— The negative number, $-0.5$, has the larger absolute value so the sum is negative.

**PRACTICE**
**5**  Add.

a. $15 + (-18)$        b. $-19 + 20$        c. $-0.6 + 0.4$

**EXAMPLE 6**  Add.

a. $-8 + (-11)$        b. $-5 + 35$        c. $0.6 + (-1.1)$

d. $-\dfrac{7}{10} + \left(-\dfrac{1}{10}\right)$        e. $11.4 + (-4.7)$        f. $-\dfrac{3}{8} + \dfrac{2}{5}$

*Solution*

a. $-8 + (-11) = -19$        Same sign. Add absolute values and use the common sign.

b. $-5 + 35 = 30$        Different signs. Subtract absolute values and use the sign of the number with the larger absolute value.

c. $0.6 + (-1.1) = -0.5$        Different signs.

d. $-\dfrac{7}{10} + \left(-\dfrac{1}{10}\right) = -\dfrac{8}{10} = -\dfrac{4}{5}$        Same sign.

e. $11.4 + (-4.7) = 6.7$

f. $-\dfrac{3}{8} + \dfrac{2}{5} = -\dfrac{15}{40} + \dfrac{16}{40} = \dfrac{1}{40}$

▶ **Helpful Hint**
Don't forget that a common denominator is needed when adding or subtracting fractions. The common denominator here is 40.

**PRACTICE**
**6**  Add.

a. $-\dfrac{3}{5} + \left(-\dfrac{2}{5}\right)$        b. $3 + (-9)$

c. $2.2 + (-1.7)$        d. $-\dfrac{2}{7} + \dfrac{3}{10}$

**EXAMPLE 7**  Add.

**a.** $3 + (-7) + (-8)$

**b.** $[7 + (-10)] + [-2 + |-4|]$

*Solution*

**a.** Perform the additions from left to right.

$$3 + (-7) + (-8) = -4 + (-8) \quad \text{Adding numbers with different signs.}$$
$$= -12 \quad \text{Adding numbers with like signs.}$$

**b.** Simplify inside brackets first.

> ▶ **Helpful Hint**
> Don't forget that brackets are grouping symbols. We simplify within them first.

$$[7 + (-10)] + [-2 + |-4|] = [-3] + [-2 + 4]$$
$$= [-3] + [2]$$
$$= -1 \quad \text{Add.} \qquad \square$$

**PRACTICE**
**7**  Add.

**a.** $8 + (-5) + (-9)$

**b.** $[-8 + 5] + [-5 + |-2|]$

**OBJECTIVE 3 ▶ Solving problems by adding real numbers.** Positive and negative numbers are often used in everyday life. Stock market returns show gains and losses as positive and negative numbers. Temperatures in cold climates often dip into the negative range, commonly referred to as "below zero" temperatures. Bank statements report deposits and withdrawals as positive and negative numbers.

**EXAMPLE 8**  **Finding the Gain or Loss of a Stock**

During a three-day period, a share of Fremont General Corporation stock recorded the following gains and losses:

|  Monday  |  Tuesday  |  Wednesday  |
|----------|-----------|-------------|
| a gain of $2 | a loss of $1 | a loss of $3 |

Find the overall gain or loss for the stock for the three days.

*Solution*  Gains can be represented by positive numbers. Losses can be represented by negative numbers. The overall gain or loss is the sum of the gains and losses.

In words:  gain  plus  loss  plus  loss
           ↓     ↓     ↓     ↓     ↓
Translate:  2   +   (−1)   +   (−3) = −2

The overall loss is $2. $\qquad \square$

**PRACTICE**
**8**  During a three-day period, a share of McDonald's stock recorded the following gain and losses:

|  Monday  |  Tuesday  |  Wednesday  |
|----------|-----------|-------------|
| a loss of $5 | a gain of $8 | a loss of $2 |

Find the overall gain or loss for the stock for the three days.

**OBJECTIVE 4 ▶ Finding the opposite of a number.** To help us subtract real numbers in the next section, we first review the concept of opposites. The graphs of 4 and −4 are shown on a number line below.

Notice that 4 and −4 lie on opposite sides of 0, and each is 4 units away from 0.

This relationship between −4 and +4 is an important one. Such numbers are known as **opposites** or **additive inverses** of each other.

> **Opposites or Additive Inverses**
> Two numbers that are the same distance from 0 but lie on opposite sides of 0 are called opposites or additive inverses of each other.

Let's discover another characteristic about opposites. Notice that the sum of a number and its opposite is 0.

$$10 + (-10) = 0$$
$$-3 + 3 = 0$$
$$\frac{1}{2} + \left(-\frac{1}{2}\right) = 0$$

In general, we can write the following:

> The sum of a number $a$ and its opposite $-a$ is 0.
> $$a + (-a) = 0$$

This is why opposites are also called additive inverses. Notice that this also means that the opposite of 0 is then 0 since $0 + 0 = 0$.

**EXAMPLE 9** Find the opposite or additive inverse of each number.

**a.** 5 **b.** −6 **c.** $\frac{1}{2}$ **d.** −4.5

*Solution*

**a.** The opposite of 5 is −5. Notice that 5 and −5 are on opposite sides of 0 when plotted on a number line and are equal distances away.

**b.** The opposite of −6 is 6.

**c.** The opposite of $\frac{1}{2}$ is $-\frac{1}{2}$.

**d.** The opposite of −4.5 is 4.5.

**PRACTICE**
**9** Find the opposite or additive inverse of each number.

**a.** $-\frac{5}{9}$ **b.** 8 **c.** 6.2 **d.** −3

We use the symbol "−" to represent the phrase "the opposite of" or "the additive inverse of." In general, if $a$ is a number, we write the opposite or additive inverse of $a$ as $-a$. We know that the opposite of $-3$ is 3. Notice that this translates as

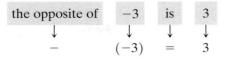

This is true in general.

> If $a$ is a number, then $-(-a) = a$.

**EXAMPLE 10**   Simplify each expression.

**a.** $-(-10)$   **b.** $-\left(-\dfrac{1}{2}\right)$   **c.** $-(-2x)$   **d.** $-|-6|$

*Solution*

**a.** $-(-10) = 10$   **b.** $-\left(-\dfrac{1}{2}\right) = \dfrac{1}{2}$   **c.** $-(-2x) = 2x$

**d.** Since $|-6| = 6$, then $-|-6| = -6$.

**PRACTICE**
**10**   Simplify each expression.

**a.** $-|-15|$   **b.** $-\left(-\dfrac{3}{5}\right)$

**c.** $-(-5y)$   **d.** $-(-8)$

# VOCABULARY & READINESS CHECK

*Use the choices below to fill in each blank.*

positive number       $n$       opposites
negative number       0       $-n$

**1.** Two numbers that are the same distance from 0 but lie on opposite sides of 0 are called _____.

**2.** The sum of a number and its opposite is always _____.

**3.** If $n$ is a number, then $-(-n) =$ _____.

*Tell whether the sum is a positive number, a negative number, or 0. Do not actually find the sum.*

**4.** $-80 + (-127) =$ _____.

**5.** $-162 + 164 =$ _____.

**6.** $-162 + 162 =$ _____.

**7.** $-1.26 + (-8.3) =$ _____.

**8.** $-3.68 + 0.27 =$ _____.

**9.** $-\dfrac{2}{3} + \dfrac{2}{3} =$ _____.

## 1.5 | EXERCISE SET

MyMathLab   PRACTICE   WATCH   DOWNLOAD   READ   REVIEW

**MIXED PRACTICE**

*Add. See Examples 1 through 7.*

**1.** $6 + 3$

**2.** $9 + (-12)$

**3.** $-6 + (-8)$

**4.** $-6 + (-14)$

**5.** $8 + (-7)$

**6.** $6 + (-4)$

**7.** $-14 + 2$

**8.** $-10 + 5$

**9.** $-2 + (-3)$

**10.** $-7 + (-4)$

**11.** $-9 + (-3)$

**12.** $7 + (-5)$

**13.** $-7 + 3$

**14.** $-5 + 9$

**15.** $10 + (-3)$

**16.** $8 + (-6)$

**17.** $5 + (-7)$

**18.** $3 + (-6)$

**19.** $-16 + 16$

**20.** $23 + (-23)$

**21.** $27 + (-46)$

**22.** $53 + (-37)$

**23.** $-18 + 49$

**24.** $-26 + 14$

**25.** $-33 + (-14)$

**26.** $-18 + (-26)$

**27.** $6.3 + (-8.4)$

**28.** $9.2 + (-11.4)$

**29.** $|-8| + (-16)$

**30.** $|-6| + (-61)$

**31.** $117 + (-79)$

**32.** $144 + (-88)$

**33.** $-9.6 + (-3.5)$

**34.** $-6.7 + (-7.6)$

**35.** $-\dfrac{3}{8} + \dfrac{5}{8}$

**36.** $-\dfrac{5}{12} + \dfrac{7}{12}$

**37.** $-\dfrac{7}{16} + \dfrac{1}{4}$

**38.** $-\dfrac{5}{9} + \dfrac{1}{3}$

**39.** $-\dfrac{7}{10} + \left(-\dfrac{3}{5}\right)$

**40.** $-\dfrac{5}{6} + \left(-\dfrac{2}{3}\right)$

**41.** $-15 + 9 + (-2)$

**42.** $-9 + 15 + (-5)$

**43.** $-21 + (-16) + (-22)$

**44.** $-18 + (-6) + (-40)$

**45.** $-23 + 16 + (-2)$

**46.** $-14 + (-3) + 11$

**47.** $|5 + (-10)|$

**48.** $|7 + (-17)|$

**49.** $6 + (-4) + 9$

**50.** $8 + (-2) + 7$

**51.** $[-17 + (-4)] + [-12 + 15]$

**52.** $[-2 + (-7)] + [-11 + 22]$

**53.** $|9 + (-12)| + |-16|$

**54.** $|43 + (-73)| + |-20|$

**55.** $-1.3 + [0.5 + (-0.3) + 0.4]$

**56.** $-3.7 + [0.1 + (-0.6) + 8.1]$

*Solve. See Example 8.*

**57.** The low temperature in Anoka, Minnesota, was $-15°$ last night. During the day it rose only $9°$. Find the high temperature for the day.

**58.** On January 2, 1943, the temperature was $-4°$ at 7:30 a.m. in Spearfish, South Dakota. Incredibly, it got $49°$ warmer in the next 2 minutes. To what temperature did it rise by 7:32?

**59.** The deepest canyon in the world is the Great Canyon of the Yarlung Tsangpo in Tibet. The bottom of the canyon is 17,657 feet below the surrounding terrain, called the rim. If you are standing 1230 feet above the bottom of the canyon, how far from the rim are you?

**60.** The lowest point in Africa is $-512$ feet at Lake Assal in Djibouti. If you are standing at a point 658 feet above Lake Assal, what is your elevation? (*Source:* Microsoft Encarta)

*A negative net income results when a company's expenses are more than the money brought in.*

**61.** The table below shows net incomes for Ford Motor Company's Automotive sector for the years 2004, 2005, and 2006. Find the total net income for three years.

| Year | Net Income (in millions) |
|------|--------------------------|
| 2004 | $-\$155$ |
| 2005 | $-\$3895$ |
| 2006 | $-\$5200$ |

(*Source:* Ford Motor Company)

**62.** The table below shows net incomes for Continental Airlines for the years 2004, 2005, and 2006. Find the total net income for these years.

| Year | Net Income (in millions) |
|------|--------------------------|
| 2004 | $-\$409$ |
| 2005 | $-\$68$ |
| 2006 | $\$343$ |

(*Source:* Continental Airlines)

*In golf, scores that are under par for the entire round are shown as negative scores; positive scores are shown for scores that are over par, and 0 is par.*

**63.** Paula Creamer was the winner of the 2007 LPGA SBS Open at Turtle Bay. Her scores were $-5, -2,$ and $-2$. What was her overall score? (*Source:* Ladies Professional Golf Association)

**64.** During the 2007 PGA Buick Invitational Golf Tournament, Tiger Woods won with scores of $-6, 0, -3,$ and $-6$. What was his overall score? (*Source:* Professional Golf Association)

*Find each additive inverse or opposite. See Example 9.*

**65.** 6          **66.** 4

**67.** $-2$         **68.** $-8$

**69.** 0          **70.** $-\dfrac{1}{4}$

**71.** $|-6|$       **72.** $|-11|$

**73.** In your own words, explain how to find the opposite of a number.

**74.** In your own words, explain why 0 is the only number that is its own opposite.

*Simplify each of the following. See Example 10.*

**75.** $-|-2|$         **76.** $-(-3)$

**77.** $-|0|$          **78.** $\left|-\dfrac{2}{3}\right|$

**79.** $-\left|-\dfrac{2}{3}\right|$     **80.** $-(-7)$

**81.** Explain why adding a negative number to another negative number always gives a negative sum.

**82.** When a positive and a negative number are added, sometimes the sum is positive, sometimes it is zero, and sometimes it is negative. Explain why and when this happens.

*Decide whether the given number is a solution of the given equation.*

**83.** Is $-4$ a solution of $x + 9 = 5$?

**84.** Is 10 a solution of $7 = -x + 3$?

**85.** Is $-1$ a solution of $y + (-3) = -7$?

**86.** Is $-6$ a solution of $1 = y + 7$?

**CONCEPT EXTENSIONS**

*The following bar graph shows each month's average daily low temperature in degrees Fahrenheit for Barrow, Alaska. Use this graph to answer Exercises 87 through 92.*

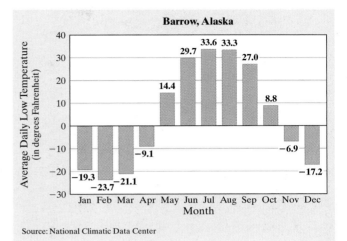

Source: National Climatic Data Center

**87.** For what month is the graphed temperature the highest?

**88.** For what month is the graphed temperature the lowest?

**89.** For what month is the graphed temperature positive *and* closest to 0°?

**90.** For what month is the graphed temperature negative *and* closest to 0°?

**91.** Find the average of the temperatures shown for the months of April, May, and October. (To find the average of three temperatures, find their sum and divide by 3.)

**92.** Find the average of the temperatures shown for the months of January, September, and October.

*If a is a positive number and b is a negative number, fill in the blanks with the words positive or negative.*

**93.** $-a$ is _____.      **94.** $-b$ is _____.

**95.** $a + a$ is _____.     **96.** $b + b$ is _____.

---

## 1.6 SUBTRACTING REAL NUMBERS

**OBJECTIVES**

1 Subtract real numbers.

2 Add and subtract real numbers.

3 Evaluate algebraic expressions using real numbers.

4 Solve problems that involve subtraction of real numbers.

**OBJECTIVE 1 ▶ Subtracting real numbers.** Now that addition of signed numbers has been discussed, we can explore subtraction. We know that $9 - 7 = 2$. Notice that $9 + (-7) = 2$, also. This means that

$$9 - 7 = 9 + (-7)$$

Notice that the difference of 9 and 7 is the same as the sum of 9 and the opposite of 7. In general, we have the following.

> **Subtracting Two Real Numbers**
> If $a$ and $b$ are real numbers, then $a - b = a + (-b)$.

In other words, to find the difference of two numbers, add the first number to the opposite of the second number.

**EXAMPLE 1**   Subtract.

**a.** $-13 - 4$     **b.** $5 - (-6)$     **c.** $3 - 6$     **d.** $-1 - (-7)$

*Solution*

**a.** $-13 - 4 = -13 + (-4)$   Add $-13$ to the opposite of $+4$, which is $-4$.

$\qquad\qquad = -17$

**b.** $5 - (-6) = 5 + (6)$     Add $5$ to the opposite of $-6$, which is $6$.

$\qquad\qquad = 11$

**c.** $3 - 6 = 3 + (-6)$     Add $3$ to the opposite of $6$, which is $-6$.

$\qquad\quad = -3$

**d.** $-1 - (-7) = -1 + (7) = 6$

**PRACTICE**

**1**   Subtract.

**a.** $-7 - 6$     **b.** $-8 - (-1)$     **c.** $9 - (-3)$     **d.** $5 - 7$

---

▶ **Helpful Hint**

Study the patterns indicated.

No change ————————— Change to addition.
———————————— Change to opposite.

$$5 - 11 = 5 + (-11) = -6$$
$$-3 - 4 = -3 + (-4) = -7$$
$$7 - (-1) = 7 + (1) = 8$$

---

**EXAMPLE 2**   Subtract.

**a.** $5.3 - (-4.6)$     **b.** $-\dfrac{3}{10} - \dfrac{5}{10}$     **c.** $-\dfrac{2}{3} - \left(-\dfrac{4}{5}\right)$

*Solution*

**a.** $5.3 - (-4.6) = 5.3 + (4.6) = 9.9$

**b.** $-\dfrac{3}{10} - \dfrac{5}{10} = -\dfrac{3}{10} + \left(-\dfrac{5}{10}\right) = -\dfrac{8}{10} = -\dfrac{4}{5}$

**c.** $-\dfrac{2}{3} - \left(-\dfrac{4}{5}\right) = -\dfrac{2}{3} + \left(\dfrac{4}{5}\right) = -\dfrac{10}{15} + \dfrac{12}{15} = \dfrac{2}{15}$   The common denominator is $15$.

**PRACTICE**

**2**   Subtract.

**a.** $8.4 - (-2.5)$     **b.** $-\dfrac{5}{8} - \left(-\dfrac{1}{8}\right)$     **c.** $-\dfrac{3}{4} - \dfrac{1}{5}$

**EXAMPLE 3**   Subtract 8 from −4.

*Solution*   Be careful when interpreting this: The order of numbers in subtraction is important. 8 is to be subtracted **from** −4.

$$-4 - 8 = -4 + (-8) = -12$$   □

**PRACTICE**
**3**   Subtract 5 from −2.

**OBJECTIVE 2 ▶ Adding and subtracting real numbers.** If an expression contains additions and subtractions, just write the subtractions as equivalent additions. Then simplify from left to right.

**EXAMPLE 4**   Simplify each expression.

**a.** $-14 - 8 + 10 - (-6)$     **b.** $1.6 - (-10.3) + (-5.6)$

*Solution*

**a.** $-14 - 8 + 10 - (-6) = -14 + (-8) + 10 + 6$
$$= -6$$

**b.** $1.6 - (-10.3) + (-5.6) = 1.6 + 10.3 + (-5.6)$
$$= 6.3$$   □

**PRACTICE**
**4**   Simplify each expression.

**a.** $-15 - 2 - (-4) + 7$     **b.** $3.5 + (-4.1) - (-6.7)$

When an expression contains parentheses and brackets, remember the order of operations. Start with the innermost set of parentheses or brackets and work your way outward.

**EXAMPLE 5**   Simplify each expression.

**a.** $-3 + [(-2 - 5) - 2]$     **b.** $2^3 - |10| + [-6 - (-5)]$

*Solution*

**a.** Start with the innermost sets of parentheses. Rewrite $-2 - 5$ as a sum.

$$
\begin{aligned}
-3 + [(-2 - 5) - 2] &= -3 + [(-2 + (-5)) - 2] \\
&= -3 + [(-7) - 2] &&\text{Add: } -2 + (-5). \\
&= -3 + [-7 + (-2)] &&\text{Write } -7 - 2 \text{ as a sum.} \\
&= -3 + [-9] &&\text{Add.} \\
&= -12 &&\text{Add.}
\end{aligned}
$$

**b.** Start simplifying the expression inside the brackets by writing $-6 - (-5)$ as a sum.

$$
\begin{aligned}
2^3 - |10| + [-6 - (-5)] &= 2^3 - |10| + [-6 + 5] \\
&= 2^3 - |10| + [-1] &&\text{Add.} \\
&= 8 - 10 + (-1) &&\text{Evaluate } 2^3 \text{ and } |10|. \\
&= 8 + (-10) + (-1) &&\text{Write } 8 - 10 \text{ as a sum.} \\
&= -2 + (-1) &&\text{Add.} \\
&= -3 &&\text{Add.}
\end{aligned}
$$   □

**PRACTICE**
**5**   Simplify each expression.

**a.** $-4 + [(-8 - 3) - 5]$     **b.** $|-13| - 3^2 + [2 - (-7)]$

**OBJECTIVE 3 ▶ Evaluating algebraic expressions.** Knowing how to evaluate expressions for given replacement values is helpful when checking solutions of equations and when solving problems whose unknowns satisfy given expressions. The next example illustrates this.

**EXAMPLE 6**   Find the value of each expression when $x = 2$ and $y = -5$.

**a.** $\dfrac{x - y}{12 + x}$    **b.** $x^2 - 3y$

*Solution*

**a.** Replace $x$ with 2 and $y$ with $-5$. Be sure to put parentheses around $-5$ to separate signs. Then simplify the resulting expression.

$$\frac{x - y}{12 + x} = \frac{2 - (-5)}{12 + 2}$$
$$= \frac{2 + 5}{14}$$
$$= \frac{7}{14} = \frac{1}{2}$$

**b.** Replace the $x$ with 2 and $y$ with $-5$ and simplify.

$$x^2 - 3y = 2^2 - 3(-5)$$
$$= 4 - 3(-5)$$
$$= 4 - (-15)$$
$$= 4 + 15$$
$$= 19$$

**PRACTICE**
**6**   Find the value of each expression when $x = -3$ and $y = 4$.

**a.** $\dfrac{7 - x}{2y + x}$    **b.** $y^2 + x$

**OBJECTIVE 4 ▶ Solving problems by subtracting real numbers.** One use of positive and negative numbers is in recording altitudes above and below sea level, as shown in the next example.

**EXAMPLE 7**   **Finding the Difference in Elevations**

The lowest point on the surface of the Earth is the Dead Sea, at an elevation of 1349 feet below sea level. The highest point is Mt. Everest, at an elevation of 29,035 feet. How much of a variation in elevation is there between these two world extremes? (*Source:* National Geographic Society)

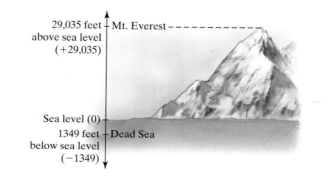

*Solution*    To find the variation in elevation between the two heights, find the difference of the high point and the low point.

In words:    high point    minus    low point

Translate:    29,035    −    (−1349)    = 29,035 + 1349

= 30,384 feet

Thus, the variation in elevation is 30,384 feet.    □

**PRACTICE**

**7**    On Tuesday morning, a bank account balance was $282. On Thursday the account balance had dropped to −$75. Find the overall change in this account balance.

A knowledge of geometric concepts is needed by many professionals, such as doctors, carpenters, electronic technicians, gardeners, machinists, and pilots, just to name a few. With this in mind, we review the geometric concepts of **complementary** and **supplementary angles.**

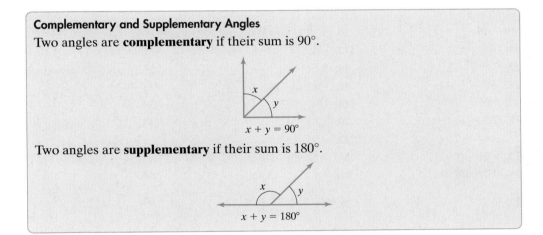

**Complementary and Supplementary Angles**

Two angles are **complementary** if their sum is 90°.

$x + y = 90°$

Two angles are **supplementary** if their sum is 180°.

$x + y = 180°$

**EXAMPLE 8**    Find each unknown complementary or supplementary angle.

a.

b.

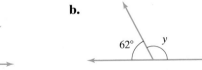

*Solution*

**a.** These angles are complementary, so their sum is 90°. This means that $x$ is 90° − 38°.

$$x = 90° − 38° = 52°$$

**b.** These angles are supplementary, so their sum is 180°. This means that $y$ is 180° − 62°.

$$y = 180° − 62° = 118°$$    □

**PRACTICE**

**8**    Find each unknown complementary or supplementary angle.

a.

b.

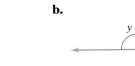

## VOCABULARY & READINESS CHECK

*Translate each phrase. Let x represent "a number." Use the choices below to fill in each blank.*

$$7 - x \qquad x - 7$$

**1.** 7 minus a number _____

**2.** 7 subtracted from a number _____

**3.** A number decreased by 7 _____

**4.** 7 less a number _____

**5.** A number less than 7 _____

**6.** A number subtracted from 7 _____

## 1.6 EXERCISE SET

### MIXED PRACTICE

*Subtract. See Examples 1 through 5.*

**1.** $-6 - 4$

**2.** $-12 - 8$

**3.** $4 - 9$

**4.** $8 - 11$

**5.** $16 - (-3)$

**6.** $12 - (-5)$

**7.** $\dfrac{1}{2} - \dfrac{1}{3}$

**8.** $\dfrac{3}{4} - \dfrac{7}{8}$

**9.** $-16 - (-18)$

**10.** $-20 - (-48)$

**11.** $-6 - 5$

**12.** $-8 - 4$

**13.** $7 - (-4)$

**14.** $3 - (-6)$

**15.** $-6 - (-11)$

**16.** $-4 - (-16)$

**17.** $16 - (-21)$

**18.** $15 - (-33)$

**19.** $9.7 - 16.1$

**20.** $8.3 - 11.2$

**21.** $-44 - 27$

**22.** $-36 - 51$

**23.** $-21 - (-21)$

**24.** $-17 - (-17)$

**25.** $-2.6 - (-6.7)$

**26.** $-6.1 - (-5.3)$

**27.** $-\dfrac{3}{11} - \left(-\dfrac{5}{11}\right)$

**28.** $-\dfrac{4}{7} - \left(-\dfrac{1}{7}\right)$

**29.** $-\dfrac{1}{6} - \dfrac{3}{4}$

**30.** $-\dfrac{1}{10} - \dfrac{7}{8}$

**31.** $8.3 - (-0.62)$

**32.** $4.3 - (-0.87)$

*Perform the operation. See Example 3.*

**33.** Subtract $-5$ from 8.

**34.** Subtract 3 from $-2$.

**35.** Subtract $-1$ from $-6$.

**36.** Subtract 17 from 1.

**37.** Subtract 8 from 7.

**38.** Subtract 9 from $-4$.

**39.** Decrease $-8$ by 15.

**40.** Decrease 11 by $-14$.

**41.** In your own words, explain why $5 - 8$ simplifies to a negative number.

**42.** Explain why $6 - 11$ is the same as $6 + (-11)$.

*Simplify each expression. (Remember the order of operations.) See Examples 4 and 5.*

**43.** $-10 - (-8) + (-4) - 20$

**44.** $-16 - (-3) + (-11) - 14$

**45.** $5 - 9 + (-4) - 8 - 8$

**46.** $7 - 12 + (-5) - 2 + (-2)$

**47.** $-6 - (2 - 11)$

**48.** $-9 - (3 - 8)$

**49.** $3^3 - 8 \cdot 9$

**50.** $2^3 - 6 \cdot 3$

**51.** $2 - 3(8 - 6)$

**52.** $4 - 6(7 - 3)$

**53.** $(3 - 6) + 4^2$

**54.** $(2 - 3) + 5^2$

**55.** $-2 + [(8 - 11) - (-2 - 9)]$

**56.** $-5 + [(4 - 15) - (-6) - 8]$

**57.** $|-3| + 2^2 + [-4 - (-6)]$

**58.** $|-2| + 6^2 + (-3 - 8)$

*Evaluate each expression when $x = -5$, $y = 4$, and $t = 10$. See Example 6.*

**59.** $x - y$

**60.** $y - x$

**61.** $|x| + 2t - 8y$

**62.** $|x + t - 7y|$

**63.** $\dfrac{9 - x}{y + 6}$

**64.** $\dfrac{15 - x}{y + 2}$

**65.** $y^2 - x$

**66.** $t^2 - x$

**67.** $\dfrac{|x - (-10)|}{2t}$

**68.** $\dfrac{|5y - x|}{6t}$

*Solve. See Example 7.*

**69.** Within 24 hours in 1916, the temperature in Browning, Montana, fell from 44 degrees to $-56$ degrees. How large a drop in temperature was this?

**70.** Much of New Orleans is below sea level. If George descends 12 feet from an elevation of 5 feet above sea level, what is his new elevation?

**71.** In a series of plays, the San Francisco 49ers gain 2 yards, lose 5 yards, and then lose another 20 yards. What is their total gain or loss of yardage?

**72.** In some card games, it is possible to have a negative score. Lavonne Schultz currently has a score of 15 points. She then loses 24 points. What is her new score?

**73.** Pythagoras died in the year −475 (or 475 B.C.). When was he born, if he was 94 years old when he died?

**74.** The Greek astronomer and mathematician Geminus died in 60 A.D. at the age of 70. When was he born?

**75.** A commercial jet liner hits an air pocket and drops 250 feet. After climbing 120 feet, it drops another 178 feet. What is its overall vertical change?

**76.** Tyson Industries stock posted a loss of 1.625 points yesterday. If it drops another 0.75 point today, find its overall change for the two days.

**77.** The highest point in Africa is Mt. Kilimanjaro, Tanzania, at an elevation of 19,340 feet. The lowest point is Lake Assal, Djibouti, at 512 feet below sea level. How much higher is Mt. Kilimanjaro than Lake Assal? (*Source:* National Geographic Society)

**78.** The airport in Bishop, California, is at an elevation of 4101 feet above sea level. The nearby Furnace Creek Airport in Death Valley, California, is at an elevation of 226 feet below sea level. How much higher in elevation is the Bishop Airport than the Furnace Creek Airport? (*Source:* National Climatic Data Center)

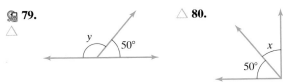

*Find each unknown complementary or supplementary angle. See Example 8.*

**79.**

**80.**

**81.**

**82.**

*Decide whether the given number is a solution of the given equation.*

**83.** Is −4 a solution of $x - 9 = 5$?

**84.** Is 3 a solution of $x - 10 = -7$?

**85.** Is −2 a solution of $-x + 6 = -x - 1$?

**86.** Is −10 a solution of $-x - 6 = -x - 1$?

**87.** Is 2 a solution of $-x - 13 = -15$?

**88.** Is 5 a solution of $4 = 1 - x$?

**CONCEPT EXTENSIONS**

*Recall from the last section the bar graph below that shows each month's average daily low temperature in degrees Fahrenheit for Barrow, Alaska. Use this graph to answer Exercises 89 through 91.*

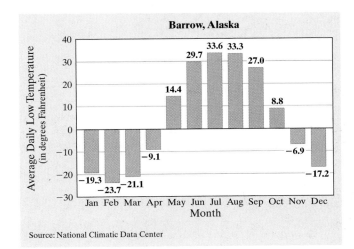

**89.** Record the monthly increases and decreases in the low temperature from the previous month.

| Month | Monthly Increase or Decrease |
|---|---|
| February | |
| March | |
| April | |
| May | |
| June | |
| July | |
| August | |
| September | |
| October | |
| November | |
| December | |

**90.** Which month had the greatest increase in temperature?

**91.** Which month had the greatest decrease in temperature?

*If a is a positive number and b is a negative number, determine whether each statement is true or false.*

**92.** $a - b$ is always a positive number.

**93.** $b - a$ is always a negative number.

**94.** $|b| - |a|$ is always a positive number.

**95.** $|b - a|$ is always a positive number.

*Without calculating, determine whether each answer is positive or negative. Then use a calculator to find the exact difference.*

**96.** $56{,}875 - 87{,}262$

**97.** $4.362 - 7.0086$

# INTEGRATED REVIEW OPERATIONS ON REAL NUMBERS

Sections 1.1–1.6

*Answer the following with positive, negative, or 0.*

**1.** The opposite of a positive number is a _____ number.

**2.** The sum of two negative numbers is a _____ number.

**3.** The absolute value of a negative number is a _____ number.

**4.** The absolute value of zero is _____.

**5.** The reciprocal of a positive number is a _____ number.

**6.** The sum of a number and its opposite is _____.

**7.** The absolute value of a positive number is a _____ number.

**8.** The opposite of a negative number is a _____ number.

*Fill in the chart:*

| | Number | Opposite | Absolute Value |
|---|---|---|---|
| **9.** | $\frac{1}{7}$ | | |
| **10.** | $-\frac{12}{5}$ | | |
| **11.** | | $-3$ | |
| **12.** | | $\frac{9}{11}$ | |

*Perform each indicated operation and simplify.*

**13.** $-19 + (-23)$

**14.** $7 - (-3)$

**15.** $-15 + 17$

**16.** $-8 - 10$

**17.** $18 + (-25)$

**18.** $-2 + (-37)$

**19.** $-14 - (-12)$

**20.** $5 - 14$

**21.** $4.5 - 7.9$

**22.** $-8.6 - 1.2$

**23.** $-\frac{3}{4} - \frac{1}{7}$

**24.** $\frac{2}{3} - \frac{7}{8}$

**25.** $-9 - (-7) + 4 - 6$

**26.** $11 - 20 + (-3) - 12$

**27.** $24 - 6(14 - 11)$

**28.** $30 - 5(10 - 8)$

**29.** $(7 - 17) + 4^2$

**30.** $9^2 + (10 - 30)$

**31.** $|-9| + 3^2 + (-4 - 20)$

**32.** $|-4 - 5| + 5^2 + (-50)$

**33.** $-7 + [(1 - 2) + (-2 - 9)]$

**34.** $-6 + [(-3 + 7) + (4 - 15)]$

**35.** Subtract 5 from 1.

**36.** Subtract $-2$ from $-3$.

**37.** Subtract $-\frac{2}{5}$ from $\frac{1}{4}$.

**38.** Subtract $\frac{1}{10}$ from $-\frac{5}{8}$.

**39.** $2(19 - 17)^3 - 3(-7 + 9)^2$

**40.** $3(10 - 9)^2 + 6(20 - 19)^3$

*Evaluate each expression when $x = -2$, $y = -1$, and $z = 9$.*

**41.** $x - y$

**42.** $x + y$

**43.** $y + z$

**44.** $z - y$

**45.** $\dfrac{|5z - x|}{y - x}$

**46.** $\dfrac{|-x - y + z|}{2z}$

# 1.7 MULTIPLYING AND DIVIDING REAL NUMBERS

**OBJECTIVES**

1  Multiply and divide real numbers.

2  Evaluate algebraic expressions using real numbers.

**OBJECTIVE 1 ▶ Multiplying and dividing real numbers.** In this section, we discover patterns for multiplying and dividing real numbers. To discover sign rules for multiplication, recall that multiplication is repeated addition. Thus $3 \cdot 2$ means that 2 is an addend 3 times. That is,

$$2 + 2 + 2 = 3 \cdot 2$$

which equals 6. Similarly, $3 \cdot (-2)$ means $-2$ is an addend 3 times. That is,

$$(-2) + (-2) + (-2) = 3 \cdot (-2)$$

Since $(-2) + (-2) + (-2) = -6$, then $3 \cdot (-2) = -6$. This suggests that the product of a positive number and a negative number is a negative number.

What about the product of two negative numbers? To find out, consider the following pattern.

Factor decreases by 1 each time

$$\left. \begin{array}{l} -3 \cdot 2 = -6 \\ -3 \cdot 1 = -3 \\ -3 \cdot 0 = 0 \end{array} \right\} \text{Product increases by 3 each time.}$$

This pattern continues as

Factor decreases by 1 each time

$$\left. \begin{array}{l} -3 \cdot -1 = 3 \\ -3 \cdot -2 = 6 \end{array} \right\} \text{Product increases by 3 each time.}$$

This suggests that the product of two negative numbers is a positive number.

> **Multiplying Real Numbers**
> **1.** The product of two numbers with the *same* sign is a positive number.
> **2.** The product of two numbers with *different* signs is a negative number.

**EXAMPLE 1**   Multiply.

**a.** $(-8)(4)$          **b.** $14(-1)$          **c.** $-9(-10)$

*Solution*

**a.** $-8(4) = -32$          **b.** $14(-1) = -14$          **c.** $-9(-10) = 90$          □

**PRACTICE**
**1**   Multiply.

**a.** $8(-5)$          **b.** $(-3)(-4)$          **c.** $(-6)(9)$

We know that every whole number multiplied by zero equals zero. This remains true for real numbers.

> **Zero as a Factor**
> If $b$ is a real number, then $b \cdot 0 = 0$. Also, $0 \cdot b = 0$.

**EXAMPLE 2**    Perform the indicated operations.

**a.** $(7)(0)(-6)$     **b.** $(-2)(-3)(-4)$     **c.** $(-1)(5)(-9)$     **d.** $(-4)(-11) - (5)(-2)$

*Solution*

**a.** By the order of operations, we multiply from left to right. Notice that, because one of the factors is 0, the product is 0.

$$(7)(0)(-6) = 0(-6) = 0$$

**b.** Multiply two factors at a time, from left to right.

$$(-2)(-3)(-4) = (6)(-4) \quad \text{Multiply } (-2)(-3).$$
$$= -24$$

**c.** Multiply from left to right.

$$(-1)(5)(-9) = (-5)(-9) \quad \text{Multiply } (-1)(5).$$
$$= 45$$

**d.** Follow the rules for order of operation.

$$(-4)(-11) - (5)(-2) = 44 - (-10) \quad \text{Find each product.}$$
$$= 44 + 10 \quad \quad \text{Add 44 to the opposite of } -10.$$
$$= 54 \quad \quad \quad \text{Add.} \qquad \square$$

**PRACTICE**
**2**    Perform the indicated operations.

**a.** $(-1)(-5)(-6)$                    **b.** $(-3)(-2)(4)$

**c.** $(-4)(0)(5)$                      **d.** $(-2)(-3) - (-4)(5)$

> ▶ **Helpful Hint**
> You may have noticed from the example that if we multiply:
>
> - an *even* number of negative numbers, the product is *positive*.
> - an *odd* number of negative numbers, the product is *negative*.

Multiplying signed decimals or fractions is carried out exactly the same way as multiplying by integers.

**EXAMPLE 3**    Multiply.

**a.** $(-1.2)(0.05)$       **b.** $\dfrac{2}{3} \cdot \left(-\dfrac{7}{10}\right)$       **c.** $\left(-\dfrac{4}{5}\right)(-20)$

*Solution*

**a.** The product of two numbers with different signs is negative.

$$(-1.2)(0.05) = -[(1.2)(0.05)]$$
$$= -0.06$$

**b.** $\dfrac{2}{3} \cdot \left(-\dfrac{7}{10}\right) = -\dfrac{2 \cdot 7}{3 \cdot 10} = -\dfrac{2 \cdot 7}{3 \cdot 2 \cdot 5} = -\dfrac{7}{15}$

**c.** $\left(-\dfrac{4}{5}\right)(-20) = \dfrac{4 \cdot 20}{5 \cdot 1} = \dfrac{4 \cdot 4 \cdot 5}{5 \cdot 1} = \dfrac{16}{1}$ or $16$       $\square$

**PRACTICE**
**3**   Multiply.

**a.** $(0.23)(-0.2)$         **b.** $\left(-\dfrac{3}{5}\right) \cdot \left(\dfrac{4}{9}\right)$         **c.** $\left(-\dfrac{7}{12}\right)(-24)$

Now that we know how to multiply positive and negative numbers, let's see how we find the values of $(-4)^2$ and $-4^2$, for example. Although these two expressions look similar, the difference between the two is the parentheses. In $(-4)^2$, the parentheses tell us that the base, or repeated factor, is $-4$. In $-4^2$, only 4 is the base. Thus,

$$(-4)^2 = (-4)(-4) = 16 \quad \text{The base is } -4.$$

$$-4^2 = -(4 \cdot 4) = -16 \quad \text{The base is } 4.$$

**EXAMPLE 4**   Evaluate.

**a.** $(-2)^3$      **b.** $-2^3$      **c.** $(-3)^2$      **d.** $-3^2$

*Solution*

**a.** $(-2)^3 = (-2)(-2)(-2) = -8$      The base is $-2$.
**b.** $-2^3 = -(2 \cdot 2 \cdot 2) = -8$      The base is 2.
**c.** $(-3)^2 = (-3)(-3) = 9$      The base is $-3$.
**d.** $-3^2 = -(3 \cdot 3) = -9$      The base is 3.

**PRACTICE**
**4**   Evaluate.

**a.** $(-6)^2$         **b.** $-6^2$         **c.** $(-4)^3$         **d.** $-4^3$

> ▶ **Helpful Hint**
> Be careful when identifying the base of an exponential expression.
>
> $$(-3)^2 \qquad\qquad\qquad -3^2$$
> $$\text{Base is } -3 \qquad\qquad \text{Base is } 3$$
> $$(-3)^2 = (-3)(-3) = 9 \quad -3^2 = -(3 \cdot 3) = -9$$

Just as every difference of two numbers $a - b$ can be written as the sum $a + (-b)$, so too every quotient of two numbers can be written as a product. For example, the quotient $6 \div 3$ can be written as $6 \cdot \dfrac{1}{3}$. Recall that the pair of numbers 3 and $\dfrac{1}{3}$ has a special relationship. Their product is 1 and they are called reciprocals or **multiplicative inverses** of each other.

> **Reciprocals or Multiplicative Inverses**
> Two numbers whose product is 1 are called reciprocals or multiplicative inverses of each other.

Notice that **0 has no multiplicative inverse** since 0 multiplied by any number is never 1 but always 0.

**EXAMPLE 5** Find the reciprocal of each number.

**a.** 22        **b.** $\dfrac{3}{16}$        **c.** $-10$        **d.** $-\dfrac{9}{13}$

*Solution*

**a.** The reciprocal of 22 is $\dfrac{1}{22}$ since $22 \cdot \dfrac{1}{22} = 1$.

**b.** The reciprocal of $\dfrac{3}{16}$ is $\dfrac{16}{3}$ since $\dfrac{3}{16} \cdot \dfrac{16}{3} = 1$.

**c.** The reciprocal of $-10$ is $-\dfrac{1}{10}$.

**d.** The reciprocal of $-\dfrac{9}{13}$ is $-\dfrac{13}{9}$.

**PRACTICE**

**5** Find the reciprocal of each number.

**a.** $\dfrac{8}{3}$        **b.** 15        **c.** $-\dfrac{2}{7}$        **d.** $-5$

We may now write a quotient as an equivalent product.

> **Quotient of Two Real Numbers**
> If $a$ and $b$ are real numbers and $b$ is not 0, then
> $$a \div b = \frac{a}{b} = a \cdot \frac{1}{b}$$

In other words, the quotient of two real numbers is the product of the first number and the multiplicative inverse or reciprocal of the second number.

**EXAMPLE 6** Use the definition of the quotient of two numbers to divide.

**a.** $-18 \div 3$        **b.** $\dfrac{-14}{-2}$        **c.** $\dfrac{20}{-4}$

*Solution*

**a.** $-18 \div 3 = -18 \cdot \dfrac{1}{3} = -6$        **b.** $\dfrac{-14}{-2} = -14 \cdot -\dfrac{1}{2} = 7$

**c.** $\dfrac{20}{-4} = 20 \cdot -\dfrac{1}{4} = -5$

**PRACTICE**

**6** Use the definition of the quotient of two numbers to divide.

**a.** $\dfrac{16}{-2}$        **b.** $24 \div (-6)$        **c.** $\dfrac{-35}{-7}$

Since the quotient $a \div b$ can be written as the product $a \cdot \dfrac{1}{b}$, it follows that sign patterns for dividing two real numbers are the same as sign patterns for multiplying two real numbers.

> **Multiplying and Dividing Real Numbers**
> **1.** The product or quotient of two numbers with the *same* sign is a positive number.
> **2.** The product or quotient of two numbers with *different* signs is a negative number.

**EXAMPLE 7**   Divide.

**a.** $\dfrac{-24}{-4}$    **b.** $\dfrac{-36}{3}$    **c.** $\dfrac{2}{3} \div \left( -\dfrac{5}{4} \right)$    **d.** $-\dfrac{3}{2} \div 9$

*Solution*

**a.** $\dfrac{-24}{-4} = 6$    **b.** $\dfrac{-36}{3} = -12$    **c.** $\dfrac{2}{3} \div \left( -\dfrac{5}{4} \right) = \dfrac{2}{3} \cdot \left( -\dfrac{4}{5} \right) = -\dfrac{8}{15}$

**d.** $-\dfrac{3}{2} \div 9 = -\dfrac{3}{2} \cdot \dfrac{1}{9} = -\dfrac{3 \cdot 1}{2 \cdot 9} = -\dfrac{3 \cdot 1}{2 \cdot 3 \cdot 3} = -\dfrac{1}{6}$

**PRACTICE**
**7**   Divide.

**a.** $\dfrac{-18}{-6}$    **b.** $\dfrac{-48}{3}$    **c.** $\dfrac{3}{5} \div \left( -\dfrac{1}{2} \right)$    **d.** $-\dfrac{4}{9} \div 8$

The definition of the quotient of two real numbers does not allow for division by 0 because 0 does not have a multiplicative inverse. There is no number we can multiply 0 by to get 1. How then do we interpret $\dfrac{3}{0}$? We say that division by 0 is not allowed or not defined and that $\dfrac{3}{0}$ does not represent a real number. The denominator of a fraction can never be 0.

Can the numerator of a fraction be 0? Can we divide 0 by a number? Yes. For example,

$$\frac{0}{3} = 0 \cdot \frac{1}{3} = 0$$

In general, the quotient of 0 and any nonzero number is 0.

---

**Zero as a Divisor or Dividend**

**1.** The quotient of any nonzero real number and 0 is undefined. In symbols, if $a \neq 0, \dfrac{a}{0}$ is **undefined.**

**2.** The quotient of 0 and any real number except 0 is 0. In symbols, if $a \neq 0, \dfrac{0}{a} = 0$.

---

**EXAMPLE 8**   Perform the indicated operations.

**a.** $\dfrac{1}{0}$    **b.** $\dfrac{0}{-3}$    **c.** $\dfrac{0(-8)}{2}$

*Solution*

**a.** $\dfrac{1}{0}$ is undefined    **b.** $\dfrac{0}{-3} = 0$    **c.** $\dfrac{0(-8)}{2} = \dfrac{0}{2} = 0$

**PRACTICE**
**8**   Perform the indicated operations.

**a.** $\dfrac{0}{-2}$    **b.** $\dfrac{-4}{0}$    **c.** $\dfrac{-5}{6(0)}$

Notice that $\dfrac{12}{-2} = -6$, $-\dfrac{12}{2} = -6$, and $\dfrac{-12}{2} = -6$. This means that

$$\frac{12}{-2} = -\frac{12}{2} = \frac{-12}{2}$$

In words, a single negative sign in a fraction can be written in the denominator, in the numerator, or in front of the fraction without changing the value of the fraction. Thus,

$$\frac{1}{-7} = \frac{-1}{7} = -\frac{1}{7}$$

In general, if $a$ and $b$ are real numbers, $b \neq 0$, $\dfrac{a}{-b} = \dfrac{-a}{b} = -\dfrac{a}{b}$.

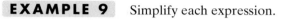

Examples combining basic arithmetic operations along with the principles of order of operations help us to review these concepts.

**EXAMPLE 9**  Simplify each expression.

**a.** $\dfrac{(-12)(-3) + 3}{-7 - (-2)}$ 　　　　 **b.** $\dfrac{2(-3)^2 - 20}{-5 + 4}$

*Solution*

**a.** First, simplify the numerator and denominator separately, then divide.

$$\frac{(-12)(-3) + 3}{-7 - (-2)} = \frac{36 + 3}{-7 + 2}$$

$$= \frac{39}{-5} \text{ or } -\frac{39}{5}$$

**b.** Simplify the numerator and denominator separately, then divide.

$$\frac{2(-3)^2 - 20}{-5 + 4} = \frac{2 \cdot 9 - 20}{-5 + 4} = \frac{18 - 20}{-5 + 4} = \frac{-2}{-1} = 2$$

**PRACTICE**

**9**  Simplify each expression.

**a.** $\dfrac{(-8)(-11) - 4}{-9 - (-4)}$ 　　　　 **b.** $\dfrac{3(-2)^3 - 9}{-6 + 3}$

**OBJECTIVE 2 ▶ Evaluating algebraic expressions using real numbers.** Using what we have learned about multiplying and dividing real numbers, we continue to practice evaluating algebraic expressions.

**EXAMPLE 10**  If $x = -2$ and $y = -4$, evaluate each expression.

**a.** $5x - y$ 　　　 **b.** $x^4 - y^2$ 　　　 **c.** $\dfrac{3x}{2y}$

*Solution*

**a.** Replace $x$ with $-2$ and $y$ with $-4$ and simplify.

$$5x - y = 5(-2) - (-4) = -10 - (-4) = -10 + 4 = -6$$

**b.** Replace $x$ with $-2$ and $y$ with $-4$.

$$x^4 - y^2 = (-2)^4 - (-4)^2 \quad \text{Substitute the given values for the variables.}$$

$$= 16 - (16) \quad \text{Evaluate exponential expressions.}$$

$$= 0 \quad \text{Subtract.}$$

**c.** Replace $x$ with $-2$ and $y$ with $-4$ and simplify.

$$\frac{3x}{2y} = \frac{3(-2)}{2(-4)} = \frac{-6}{-8} = \frac{3}{4}$$

**PRACTICE**
**10**   If $x = -5$ and $y = -2$, evaluate each expression.

**a.** $7y - x$ 　　　　 **b.** $x^2 - y^3$ 　　　　 **c.** $\dfrac{2x}{3y}$

---

## Calculator Explorations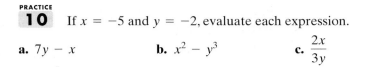

### Entering Negative Numbers on a Scientific Calculator

To enter a negative number on a scientific calculator, find a key marked $\boxed{+/-}$. (On some calculators, this key is marked $\boxed{\text{CHS}}$ for "change sign.") To enter $-8$, for example, press the keys $\boxed{8}$ $\boxed{+/-}$. The display will read $\boxed{-8}$.

### Entering Negative Numbers on a Graphing Calculator

To enter a negative number on a graphing calculator, find a key marked $\boxed{(-)}$. Do not confuse this key with the key $\boxed{-}$, which is used for subtraction. To enter $-8$, for example, press the keys $\boxed{(-)}$ $\boxed{8}$. The display will read $\boxed{-8}$.

### Operations with Real Numbers

To evaluate $-2(7 - 9) - 20$ on a calculator, press the keys
$\boxed{2}$ $\boxed{+/-}$ $\boxed{\times}$ $\boxed{(}$ $\boxed{7}$ $\boxed{-}$ $\boxed{9}$ $\boxed{)}$ $\boxed{-}$ $\boxed{2}$ $\boxed{0}$ $\boxed{=}$ , or
$\boxed{(-)}$ $\boxed{2}$ $\boxed{(}$ $\boxed{7}$ $\boxed{-}$ $\boxed{9}$ $\boxed{)}$ $\boxed{-}$ $\boxed{2}$ $\boxed{0}$ $\boxed{\text{ENTER}}$ .

The display will read $\boxed{-16}$ or $\boxed{\begin{array}{r} -2(7 - 9) - 20 \\ -16 \end{array}}$ .

*Use a calculator to simplify each expression.*

**1.** $-38(26 - 27)$ 　　　　 **2.** $-59(-8) + 1726$

**3.** $134 + 25(68 - 91)$ 　　　　 **4.** $45(32) - 8(218)$

**5.** $\dfrac{-50(294)}{175 - 265}$ 　　　　 **6.** $\dfrac{-444 - 444.8}{-181 - 324}$

**7.** $9^5 - 4550$ 　　　　 **8.** $5^8 - 6259$

**9.** $(-125)^2$ (Be careful.) 　　　　 **10.** $-125^2$ (Be careful.)

---

## VOCABULARY & READINESS CHECK

*Use the choices below to fill in each blank.*

　　positive 　　　　 0 　　　　 negative 　　　　 undefined

**1.** If $n$ is a real number, then $n \cdot 0 = \underline{\quad}$ and $0 \cdot n = \underline{\quad}$.

**2.** If $n$ is a real number, but not 0, then $\dfrac{0}{n} = \underline{\quad}$ and we say $\dfrac{n}{0}$ is _____.

**3.** The product of two negative numbers is a _____number.

**4.** The quotient of two negative numbers is a _____ number.

**5.** The quotient of a positive number and a negative number is a _____ number.

**6.** The product of a positive number and a negative number is a _____ number.

**7.** The reciprocal of a positive number is a _____ number.

**8.** The opposite of a positive number is a _____ number.

## 1.7 | EXERCISE SET

**Multiply. See Examples 1 through 3.**

**1.** $-6(4)$    **2.** $-8(5)$
**3.** $2(-1)$    **4.** $7(-4)$
**5.** $-5(-10)$    **6.** $-6(-11)$
**7.** $-3 \cdot 4$    **8.** $-2 \cdot 8$
**9.** $-7 \cdot 0$    **10.** $-6 \cdot 0$
**11.** $2(-9)$    **12.** $3(-5)$
**13.** $-\dfrac{1}{2}\left(-\dfrac{3}{5}\right)$    **14.** $-\dfrac{1}{8}\left(-\dfrac{1}{3}\right)$
**15.** $-\dfrac{3}{4}\left(-\dfrac{8}{9}\right)$    **16.** $-\dfrac{5}{6}\left(-\dfrac{3}{10}\right)$
**17.** $5(-1.4)$    **18.** $6(-2.5)$
**19.** $-0.2(-0.7)$    **20.** $-0.5(-0.3)$
**21.** $-10(80)$    **22.** $-20(60)$
**23.** $4(-7)$    **24.** $5(-9)$
**25.** $(-5)(-5)$    **26.** $(-7)(-7)$
**27.** $\dfrac{2}{3}\left(-\dfrac{4}{9}\right)$    **28.** $\dfrac{2}{7}\left(-\dfrac{2}{11}\right)$
**29.** $-11(11)$    **30.** $-12(12)$
**31.** $-\dfrac{20}{25}\left(\dfrac{5}{16}\right)$    **32.** $-\dfrac{25}{36}\left(\dfrac{6}{15}\right)$
**33.** $(-1)(2)(-3)(-5)$    **34.** $(-2)(-3)(-4)(-2)$

**Perform the indicated operations. See Example 2.**

**35.** $(-2)(5) - (-11)(3)$    **36.** $8(-3) - 4(-5)$
**37.** $(-6)(-1)(-2) - (-5)$    **38.** $20 - (-4)(3)(-2)$

**Decide whether each statement is true or false.**

**39.** The product of three negative integers is negative.
**40.** The product of three positive integers is positive.
**41.** The product of four negative integers is negative.
**42.** The product of four positive integers is positive.

**Evaluate. See Example 4.**

**43.** $(-2)^4$    **44.** $-2^4$
**45.** $-1^5$    **46.** $(-1)^5$
**47.** $(-5)^2$    **48.** $-5^2$
**49.** $-7^2$    **50.** $(-7)^2$

**Find each reciprocal or multiplicative inverse. See Example 5.**

**51.** $9$    **52.** $100$    **53.** $\dfrac{2}{3}$
**54.** $\dfrac{1}{7}$    **55.** $-14$    **56.** $-8$
**57.** $-\dfrac{3}{11}$    **58.** $-\dfrac{6}{13}$    **59.** $0.2$
**60.** $1.5$    **61.** $\dfrac{1}{-6.3}$    **62.** $\dfrac{1}{-8.9}$

**Divide. See Examples 6 through 8.**

**63.** $\dfrac{18}{-2}$    **64.** $\dfrac{20}{-10}$    **65.** $\dfrac{-16}{-4}$
**66.** $\dfrac{-18}{-6}$    **67.** $\dfrac{-48}{12}$    **68.** $\dfrac{-60}{5}$
**69.** $\dfrac{0}{-4}$    **70.** $\dfrac{0}{-9}$    **71.** $-\dfrac{15}{3}$
**72.** $-\dfrac{24}{8}$    **73.** $\dfrac{5}{0}$    **74.** $\dfrac{3}{0}$
**75.** $\dfrac{-12}{-4}$    **76.** $\dfrac{-45}{-9}$    **77.** $\dfrac{30}{-2}$
**78.** $\dfrac{14}{-2}$    **79.** $\dfrac{6}{7} \div \left(-\dfrac{1}{3}\right)$    **80.** $\dfrac{4}{5} \div \left(-\dfrac{1}{2}\right)$
**81.** $-\dfrac{5}{9} \div \left(-\dfrac{3}{4}\right)$    **82.** $-\dfrac{1}{10} \div \left(-\dfrac{8}{11}\right)$
**83.** $-\dfrac{4}{9} \div \dfrac{4}{9}$    **84.** $-\dfrac{5}{12} \div \dfrac{5}{12}$

### MIXED PRACTICE

**Simplify. See Example 9.**

**85.** $\dfrac{-9(-3)}{-6}$    **86.** $\dfrac{-6(-3)}{-4}$
**87.** $\dfrac{12}{9 - 12}$    **88.** $\dfrac{-15}{1 - 4}$
**89.** $\dfrac{-6^2 + 4}{-2}$    **90.** $\dfrac{3^2 + 4}{5}$
**91.** $\dfrac{8 + (-4)^2}{4 - 12}$    **92.** $\dfrac{6 + (-2)^2}{4 - 9}$
**93.** $\dfrac{22 + (3)(-2)}{-5 - 2}$    **94.** $\dfrac{-20 + (-4)(3)}{1 - 5}$
**95.** $\dfrac{-3 - 5^2}{2(-7)}$    **96.** $\dfrac{-2 - 4^2}{3(-6)}$
**97.** $\dfrac{6 - 2(-3)}{4 - 3(-2)}$    **98.** $\dfrac{8 - 3(-2)}{2 - 5(-4)}$
**99.** $\dfrac{-3 - 2(-9)}{-15 - 3(-4)}$    **100.** $\dfrac{-4 - 8(-2)}{-9 - 2(-3)}$
**101.** $\dfrac{|5 - 9| + |10 - 15|}{|2(-3)|}$    **102.** $\dfrac{|-3 + 6| + |-2 + 7|}{|-2 \cdot 2|}$

**If $x = -5$ and $y = -3$, evaluate each expression. See Example 10.**

**103.** $3x + 2y$    **104.** $4x + 5y$
**105.** $2x^2 - y^2$    **106.** $x^2 - 2y^2$
**107.** $x^3 + 3y$    **108.** $y^3 + 3x$
**109.** $\dfrac{2x - 5}{y - 2}$    **110.** $\dfrac{2y - 12}{x - 4}$
**111.** $\dfrac{-3 - y}{x - 4}$    **112.** $\dfrac{4 - 2x}{y + 3}$

**113.** At the end of 2006, Delta Airlines posted a net loss of $6203 million, which we will write as $-\$6203$ million. If this continues, what will Delta's income be after four years? (*Source:* Delta Airlines)

**114.** At the end of the third quarter of 2006, General Motors reported a net loss of $115 million. If this continued, what would General Motor's income be after four more quarters? (*Source:* General Motors)

*Decide whether the given number is a solution of the given equation.*

**115.** Is 7 a solution of $-5x = -35$?

**116.** Is $-4$ a solution of $2x = x - 1$?

**117.** Is $-20$ a solution of $\dfrac{x}{10} = 2$?

**118.** Is $-3$ a solution of $\dfrac{45}{x} = -15$?

**119.** Is 5 a solution of $-3x - 5 = -20$?

**120.** Is $-4$ a solution of $2x + 4 = x + 8$?

**CONCEPT EXTENSIONS**

**121.** Explain why the product of an even number of negative numbers is a positive number.

**122.** If $a$ and $b$ are any real numbers, is the statement $a \cdot b = b \cdot a$ always true? Why or why not?

**123.** Find any real numbers that are their own reciprocal.

**124.** Explain why 0 has no reciprocal.

*If q is a negative number, r is a negative number, and t is a positive number, determine whether each expression simplifies to a positive or negative number. If it is not possible to determine, state so.*

**125.** $\dfrac{q}{r \cdot t}$

**126.** $q^2 \cdot r \cdot t$

**127.** $q + t$

**128.** $t + r$

**129.** $t(q + r)$

**130.** $r(q - t)$

*Write each of the following as an expression and evaluate.*

**131.** The sum of $-2$ and the quotient of $-15$ and 3

**132.** The sum of 1 and the product of $-8$ and $-5$

**133.** Twice the sum of $-5$ and $-3$

**134.** 7 subtracted from the quotient of 0 and 5

## STUDY SKILLS BUILDER

### Learning New Terms

Many of the terms used in this text may be new to you. It will be helpful to make a list of new mathematical terms and symbols as you encounter them and to review them frequently. Placing these new terms (including page references) on $3 \times 5$ index cards might help you later when you're preparing for a quiz.

*Answer the following.*

**1.** Name one way you might place a word and its definition on a $3 \times 5$ card.

**2.** How do new terms stand out in this text so that they can be found?

## THE BIGGER PICTURE   SIMPLIFYING EXPRESSIONS

This is a special feature that we introduce in this section. Among other concepts introduced later in this text, it is very important for you to be able to simplify expressions and solve equations and to know the difference between the two. To help with this, we began an outline below and expand this outline throughout the text. Although suggestions are given, this outline should be in your own words. Once you complete the new portion of your outline, try the exercises to the right. Remember: Study your outline often as you proceed through this text.

**I. Simplifying Expressions**

  **A. Real Numbers**

    **1. Add:**

    $-1.7 + (-0.21) = -1.91$  Adding like signs. Add absolute values. Attach the common sign.

    $-7 + 3 = -4$  Adding unlike signs. Subtract absolute values. Attach the sign of the number with the larger absolute value.

    **2. Subtract:** Add the first number to the opposite of the second number.

    $\dfrac{1}{7} - \dfrac{1}{3} = \dfrac{3}{21} + \left(-\dfrac{7}{21}\right) = -\dfrac{4}{21}$

**3. Multiply or Divide:** Multiply or divide as usual. If the signs of the two numbers are the same, the answer is positive. If the signs of the two numbers are different, the answer is negative.

$$-\dfrac{3}{8} \cdot \dfrac{7}{11} = -\dfrac{21}{88}, \quad -42 \div (-10) = 4.2$$

*Perform the indicated operations.*

**1.** $-0.2(25)$

**2.** $86 - 100$

**3.** $-\dfrac{1}{7} + \left(-\dfrac{3}{5}\right)$

**4.** $\dfrac{-40}{-5}$

**5.** $(-7)^2$

**6.** $-7^2$

**7.** $\dfrac{|-42|}{-|-2|}$

**8.** $\dfrac{8.6}{0}$

**9.** $\dfrac{0}{8.6}$

**10.** $-25 - (-13)$

**11.** $-8.3 - 8.3$

**12.** $-\dfrac{8}{9}\left(-\dfrac{3}{16}\right)$

**13.** $2 + 3(8 - 11)^3$

**14.** $-2\dfrac{1}{2} \div \left(-3\dfrac{1}{4}\right)$

**15.** $20 \div 2 \cdot 5$

**16.** $-2[(1 - 5) - (7 - 17)]$

# 1.8 PROPERTIES OF REAL NUMBERS

**OBJECTIVE 1 ▶ Using the commutative and associative properties.** In this section we give names to properties of real numbers with which we are already familiar. Throughout this section, the variables $a$, $b$, and $c$ represent real numbers.

We know that order does not matter when adding numbers. For example, we know that $7 + 5$ is the same as $5 + 7$. This property is given a special name—the **commutative property of addition.** We also know that order does not matter when multiplying numbers. For example, we know that $-5(6) = 6(-5)$. This property means that multiplication is commutative also and is called the **commutative property of multiplication.**

**Commutative Properties**

*Addition:*            $a + b = b + a$

*Multiplication:*        $a \cdot b = b \cdot a$

These properties state that the *order* in which any two real numbers are added or multiplied does not change their sum or product. For example, if we let $a = 3$ and $b = 5$, then the commutative properties guarantee that

$$3 + 5 = 5 + 3 \quad \text{and} \quad 3 \cdot 5 = 5 \cdot 3$$

▶ **Helpful Hint**

Is subtraction also commutative? Try an example. Does $3 - 2 = 2 - 3$? **No!** The left side of this statement equals 1; the right side equals $-1$. There is no commutative property of subtraction. Similarly, there is no commutative property for division. For example, $10 \div 2$ does not equal $2 \div 10$.

**EXAMPLE 1**   Use a commutative property to complete each statement.

**a.** $x + 5 = $ _____          **b.** $3 \cdot x = $ _____

*Solution*

**a.** $x + 5 = 5 + x$   By the commutative property of addition

**b.** $3 \cdot x = x \cdot 3$    By the commutative property of multiplication         ☐

**PRACTICE**
**1**   Use a commutative property to complete each statement.

**a.** $x \cdot 8 = $ ____          **b.** $x + 17 = $ _____

**Concept Check** ☑

Which of the following pairs of actions are commutative?

**a.** "raking the leaves" and "bagging the leaves"

**b.** "putting on your left glove" and "putting on your right glove"

**c.** "putting on your coat" and "putting on your shirt"

**d.** "reading a novel" and "reading a newspaper"

Let's now discuss grouping numbers. We know that when we add three numbers, the way in which they are grouped or associated does not change their sum. For example, we know that $2 + (3 + 4) = 2 + 7 = 9$. This result is the same if we group the numbers differently. In other words, $(2 + 3) + 4 = 5 + 4 = 9$, also. Thus, $2 + (3 + 4) = (2 + 3) + 4$. This property is called the **associative property of addition.**

We also know that changing the grouping of numbers when multiplying does not change their product. For example, $2 \cdot (3 \cdot 4) = (2 \cdot 3) \cdot 4$ (check it). This is the **associative property of multiplication.**

---

**Associative Properties**

*Addition:*                           $(a + b) + c = a + (b + c)$

*Multiplication:*                    $(a \cdot b) \cdot c = a \cdot (b \cdot c)$

---

These properties state that the way in which three numbers are *grouped* does not change their sum or their product.

**EXAMPLE 2**   Use an associative property to complete each statement.

**a.** $5 + (4 + 6) = $ _____          **b.** $(-1 \cdot 2) \cdot 5 = $ _____

*Solution*

**a.** $5 + (4 + 6) = (5 + 4) + 6$   By the associative property of addition
**b.** $(-1 \cdot 2) \cdot 5 = -1 \cdot (2 \cdot 5)$   By the associative property of multiplication   ☐

**PRACTICE**
**2**   Use an associative property to complete each statement.

**a.** $(2 + 9) + 7 = $ _____          **b.** $-4 \cdot (2 \cdot 7) = $ _____

---

▶ **Helpful Hint**
Remember the difference between the commutative properties and the associative properties. The commutative properties have to do with the *order* of numbers, and the associative properties have to do with the *grouping* of numbers.

---

Let's now illustrate how these properties can help us simplify expressions.

**EXAMPLE 3**   Simplify each expression.

**a.** $10 + (x + 12)$          **b.** $-3(7x)$

*Solution*

**a.** $10 + (x + 12) = 10 + (12 + x)$   By the commutative property of addition
$= (10 + 12) + x$   By the associative property of addition
$= 22 + x$   Add.
**b.** $-3(7x) = (-3 \cdot 7)x$   By the associative property of multiplication
$= -21x$   Multiply.   ☐

**PRACTICE**
**3**   Simplify each expression.

**a.** $(5 + x) + 9$          **b.** $5(-6x)$

**OBJECTIVE 2 ▶ Using the distributive property.** The **distributive property of multiplication over addition** is used repeatedly throughout algebra. It is useful because it allows us to write a product as a sum or a sum as a product.

We know that $7(2 + 4) = 7(6) = 42$. Compare that with $7(2) + 7(4) = 14 + 28 = 42$. Since both original expressions equal 42, they must equal each other, or

$$7(2 + 4) = 7(2) + 7(4)$$

This is an example of the distributive property. The product on the left side of the equal sign is equal to the sum on the right side. We can think of the 7 as being distributed to each number inside the parentheses.

---

**Distributive Property of Multiplication Over Addition**

$$a(b + c) = ab + ac$$

---

Since multiplication is commutative, this property can also be written as

$$(b + c)a = ba + ca$$

The distributive property can also be extended to more than two numbers inside the parentheses. For example,

$$3(x + y + z) = 3(x) + 3(y) + 3(z)$$
$$= 3x + 3y + 3z$$

Since we define subtraction in terms of addition, the distributive property is also true for subtraction. For example

$$2(x - y) = 2(x) - 2(y)$$
$$= 2x - 2y$$

**EXAMPLE 4**   Use the distributive property to write each expression without parentheses. Then simplify if possible.

**a.** $2(x + y)$          **b.** $-5(-3 + 2z)$          **c.** $5(x + 3y - z)$
**d.** $-1(2 - y)$          **e.** $-(3 + x - w)$          **f.** $4(3x + 7) + 10$

*Solution*

**a.** $2(x + y) = 2 \cdot x + 2 \cdot y$
$$= 2x + 2y$$

**b.** $-5(-3 + 2z) = -5(-3) + (-5)(2z)$
$$= 15 - 10z$$

**c.** $5(x + 3y - z) = 5(x) + 5(3y) - 5(z)$
$$= 5x + 15y - 5y$$

**d.** $-1(2 - y) = (-1)(2) - (-1)(y)$

$\qquad\qquad = -2 + y$

> **▶ Helpful Hint**
> Notice in part **(e)** that $-(3 + x - w)$ is first rewritten as $-1(3 + x - w)$.

**e.** $-(3 + x - w) = -1(3 + x - w)$

$\qquad\qquad = (-1)(3) + (-1)(x) - (-1)(w)$

$\qquad\qquad = -3 - x + w$

**f.** $4(3x + 7) + 10 = 4(3x) + 4(7) + 10$   Apply the distributive property.

$\qquad\qquad\qquad = 12x + 28 + 10$   Multiply.

$\qquad\qquad\qquad = 12x + 38$   Add.   □

**PRACTICE**

**4**  Use the distributive property to write each expression without parentheses. Then simplify, if possible.

**a.** $5(x - y)$

**b.** $-6(4 + 2t)$

**c.** $2(3x - 4y - z)$

**d.** $(3 - y) \cdot (-1)$

**e.** $-(x - 7 + 2s)$

**f.** $2(7x + 4) + 6$

We can use the distributive property in reverse to write a sum as a product.

**EXAMPLE 5**   Use the distributive property to write each sum as a product.

**a.** $8 \cdot 2 + 8 \cdot x$

**b.** $7s + 7t$

_Solution_

**a.** $8 \cdot 2 + 8 \cdot x = 8(2 + x)$

**b.** $7s + 7t = 7(s + t)$   □

**PRACTICE**

**5**  Use the distributive property to write each sum as a product.

**a.** $5 \cdot w + 5 \cdot 3$

**b.** $9w + 9z$

**OBJECTIVE 3 ▶ Using the identity and inverse properties.** Next, we look at the **identity properties.**

The number 0 is called the identity for addition because when 0 is added to any real number, the result is the same real number. In other words, the _identity_ of the real number is not changed.

The number 1 is called the identity for multiplication because when a real number is multiplied by 1, the result is the same real number. In other words, the _identity_ of the real number is not changed.

> **Identities for Addition and Multiplication**
> 0 is the identity element for addition.
> $$a + 0 = a \qquad \text{and} \qquad 0 + a = a$$
> 1 is the identity element for multiplication.
> $$a \cdot 1 = a \qquad \text{and} \qquad 1 \cdot a = a$$

Notice that 0 is the _only_ number that can be added to any real number with the result that the sum is the same real number. Also, 1 is the _only_ number that can be multiplied by any real number with the result that the product is the same real number.

**Additive inverses** or **opposites** were introduced in Section 1.5. Two numbers are called additive inverses or opposites if their sum is 0. The additive inverse or opposite of 6 is $-6$ because $6 + (-6) = 0$. The additive inverse or opposite of $-5$ is 5 because $-5 + 5 = 0$.

**Reciprocals** or **multiplicative inverses** were introduced in Section 1.3. Two nonzero numbers are called reciprocals or multiplicative inverses if their product is 1. The reciprocal or multiplicative inverse of $\frac{2}{3}$ is $\frac{3}{2}$ because $\frac{2}{3} \cdot \frac{3}{2} = 1$. Likewise, the reciprocal of $-5$ is $-\frac{1}{5}$ because $-5\left(-\frac{1}{5}\right) = 1$.

**Concept Check** ☑

Which of the following, $1, -\frac{10}{3}, \frac{3}{10}, 0, \frac{10}{3}, -\frac{3}{10}$, is the

**a.** opposite of $-\frac{3}{10}$?  **b.** reciprocal of $-\frac{3}{10}$?

---

**Additive or Multiplicative Inverses**

The numbers $a$ and $-a$ are additive inverses or opposites of each other because their sum is 0; that is,

$$a + (-a) = 0$$

The numbers $b$ and $\frac{1}{b}$ (for $b \neq 0$) are reciprocals or multiplicative inverses of each other because their product is 1; that is,

$$b \cdot \frac{1}{b} = 1$$

---

**EXAMPLE 6** Name the property or properties illustrated by each true statement.

*Solution*

**a.** $3 \cdot y = y \cdot 3$ — Commutative property of multiplication (order changed)
**b.** $(x + 7) + 9 = x + (7 + 9)$ — Associative property of addition (grouping changed)
**c.** $(b + 0) + 3 = b + 3$ — Identity element for addition
**d.** $0.2 \cdot (z \cdot 5) = 0.2 \cdot (5 \cdot z)$ — Commutative property of multiplication (order changed)
**e.** $-2 \cdot \left(-\frac{1}{2}\right) = 1$ — Multiplicative inverse property
**f.** $-2 + 2 = 0$ — Additive inverse property
**g.** $-6 \cdot (y \cdot 2) = (-6 \cdot 2) \cdot y$ — Commutative and associative properties of multiplication (order and grouping changed)

**PRACTICE 6** Name the property or properties illustrated by each true statement.

**a.** $(7 \cdot 3x) \cdot 4 = (3x \cdot 7) \cdot 4$
**b.** $6 + (3 + y) = (6 + 3) + y$
**c.** $8 + (t + 0) = 8 + t$
**d.** $-\frac{3}{4} \cdot \left(-\frac{4}{3}\right) = 1$
**e.** $(2 + x) + 5 = 5 + (2 + x)$
**f.** $3 + (-3) = 0$
**g.** $(-3b) \cdot 7 = (-3 \cdot 7) \cdot b$

**Answers to Concept Check:**
**a.** $\frac{3}{10}$  **b.** $-\frac{10}{3}$

# VOCABULARY & READINESS CHECK

*Use the choices below to fill in each blank.*

distributive property

opposites or additive inverses

reciprocals or multiplicative inverses

associative property of multiplication

associative property of addition

commutative property of multiplication

commutative property of addition

**1.** $x + 5 = 5 + x$ is a true statement by the _____.

**2.** $x \cdot 5 = 5 \cdot x$ is a true statement by the _____.

**3.** $3(y + 6) = 3 \cdot y + 3 \cdot 6$ is true by the _____.

**4.** $2 \cdot (x \cdot y) = (2 \cdot x) \cdot y$ is a true statement by the _____.

**5.** $x + (7 + y) = (x + 7) + y$ is a true statement by the _____.

**6.** The numbers $-\dfrac{2}{3}$ and $-\dfrac{3}{2}$ are called _____.

**7.** The numbers $-\dfrac{2}{3}$ and $\dfrac{2}{3}$ are called _____.

## 1.8 | EXERCISE SET

**MyMathLab** Powered by CourseCompass™ and MathXL™   Math XL PRACTICE   WATCH   DOWNLOAD   READ   REVIEW

*Use a commutative property to complete each statement. See Example 1.*

**1.** $x + 16 =$ _____   **2.** $4 + y =$ _____

**3.** $-4 \cdot y =$ _____   **4.** $-2 \cdot x =$ _____

**5.** $xy =$ ____   **6.** $ab =$ ____

**7.** $2x + 13 =$ _____   **8.** $19 + 3y =$ _____

*Use an associative property to complete each statement. See Example 2.*

**9.** $(xy) \cdot z =$ _____   **10.** $3 \cdot (xy) =$ _____

**11.** $2 + (a + b) =$ _____   **12.** $(y + 4) + z =$ _____

**13.** $4 \cdot (ab) =$ _____   **14.** $(-3y) \cdot z =$ _____

**15.** $(a + b) + c =$ _____

**16.** $6 + (r + s) =$ _____

*Use the commutative and associative properties to simplify each expression. See Example 3.*

**17.** $8 + (9 + b)$   **18.** $(r + 3) + 11$

**19.** $4(6y)$   **20.** $2(42x)$

**21.** $\dfrac{1}{5}(5y)$   **22.** $\dfrac{1}{8}(8z)$

**23.** $(13 + a) + 13$   **24.** $7 + (x + 4)$

**25.** $-9(8x)$   **26.** $-3(12y)$

**27.** $\dfrac{3}{4}\left(\dfrac{4}{3}s\right)$   **28.** $\dfrac{2}{7}\left(\dfrac{7}{2}r\right)$

**29.** Write an example that shows that division is not commutative.

**30.** Write an example that shows that subtraction is not commutative.

*Use the distributive property to write each expression without parentheses. Then simplify the result. See Example 4.*

**31.** $4(x + y)$   **32.** $7(a + b)$

**33.** $9(x - 6)$   **34.** $11(y - 4)$

**35.** $2(3x + 5)$   **36.** $5(7 + 8y)$

**37.** $7(4x - 3)$   **38.** $3(8x - 1)$

**39.** $3(6 + x)$   **40.** $2(x + 5)$

**41.** $-2(y - z)$   **42.** $-3(z - y)$

**43.** $-7(3y + 5)$   **44.** $-5(2r + 11)$

**45.** $5(x + 4m + 2)$

**46.** $8(3y + z - 6)$

**47.** $-4(1 - 2m + n)$

**48.** $-4(4 + 2p + 5)$

**49.** $-(5x + 2)$

**50.** $-(9r + 5)$

**51.** $-(r - 3 - 7p)$

**52.** $-(q - 2 + 6r)$

**53.** $\dfrac{1}{2}(6x + 8)$

**54.** $\dfrac{1}{4}(4x - 2)$

**55.** $-\dfrac{1}{3}(3x - 9y)$

**56.** $-\dfrac{1}{5}(10a - 25b)$

**57.** $3(2r + 5) - 7$

**58.** $10(4s + 6) - 40$

**59.** $-9(4x + 8) + 2$

**60.** $-11(5x + 3) + 10$

**61.** $-4(4x + 5) - 5$

**62.** $-6(2x + 1) - 1$

*Use the distributive property to write each sum as a product. See Example 5.*

**63.** $4\cdot1 + 4\cdot y$

**64.** $14\cdot z + 14\cdot5$

**65.** $11x + 11y$

**66.** $9a + 9b$

**67.** $(-1)\cdot5 + (-1)\cdot x$

**68.** $(-3)a + (-3)b$

**69.** $30a + 30b$

**70.** $25x + 25y$

*Name the properties illustrated by each true statement. See Example 6.*

**71.** $3\cdot5 = 5\cdot3$

**72.** $4(3 + 8) = 4\cdot3 + 4\cdot8$

**73.** $2 + (x + 5) = (2 + x) + 5$

**74.** $(x + 9) + 3 = (9 + x) + 3$

**75.** $9(3 + 7) = 9\cdot3 + 9\cdot7$

**76.** $1\cdot9 = 9$

**77.** $(4\cdot y)\cdot9 = 4\cdot(y\cdot9)$

**78.** $6\cdot\dfrac{1}{6} = 1$

**79.** $0 + 6 = 6$

**80.** $(a + 9) + 6 = a + (9 + 6)$

**81.** $-4(y + 7) = -4\cdot y + (-4)\cdot7$

**82.** $(11 + r) + 8 = (r + 11) + 8$

**83.** $-4\cdot(8\cdot3) = (8\cdot-4)\cdot3$

**84.** $r + 0 = r$

## CONCEPT EXTENSIONS

*Fill in the table with the opposite (additive inverse), and the reciprocal (multiplicative inverse). Assume that the value of each expression is not 0.*

| | Expression | Opposite | Reciprocal |
|---|---|---|---|
| **85.** | 8 | | |
| **86.** | $-\dfrac{2}{3}$ | | |
| **87.** | $x$ | | |
| **88.** | $4y$ | | |
| **89.** | | | $\dfrac{1}{2x}$ |
| **90.** | | $7x$ | |

*Determine which pairs of actions are commutative.*

**91.** "taking a test" and "studying for the test"

**92.** "putting on your shoes" and "putting on your socks"

**93.** "putting on your left shoe" and "putting on your right shoe"

**94.** "reading the sports section" and "reading the comics section"

**95.** Explain why 0 is called the identity element for addition.

**96.** Explain why 1 is called the identity element for multiplication.

# CHAPTER 1 GROUP ACTIVITY

## Sections 1.3, 1.4, 1.5

### Magic Squares

A magic square is a set of numbers arranged in a square table so that the sum of the numbers in each column, row, and diagonal is the same. For instance, in the magic square below, the sum of each column, row, and diagonal is 15. Notice that no number is used more than once in the magic square.

| 2 | 9 | 4 |
|---|---|---|
| 7 | 5 | 3 |
| 6 | 1 | 8 |

The properties of magic squares have been known for a very long time and once were thought to be good luck charms. The ancient Egyptians and Greeks understood their patterns. A magic square even made it into a famous work of art. The engraving titled *Melencolia I*, created by German artist Albrecht Dürer in 1514, features the following four-by-four magic square on the building behind the central figure.

| 16 | 3 | 2 | 13 |
|---|---|---|---|
| 5 | 10 | 11 | 8 |
| 9 | 6 | 7 | 12 |
| 4 | 15 | 14 | 1 |

### Group Exercises

**1.** Verify that what is shown in the Dürer engraving is, in fact, a magic square. What is the common sum of the columns, rows, and diagonals?

**2.** Negative numbers can also be used in magic squares. Complete the following magic square:

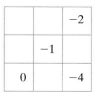

**3.** Use the numbers −12, −9, −6, −3, 0, 3, 6, 9, and 12 to form a magic square.

## CHAPTER 1  VOCABULARY CHECK

*Fill in each blank with one of the words or phrases listed below.*

| | | | | |
|---|---|---|---|---|
| set | inequality symbols | opposites | absolute value | numerator |
| denominator | grouping symbols | exponent | base | reciprocals |
| variable | equation | solution | | |

**1.** The symbols ≠, <, and > are called ___inequality___.

**2.** A mathematical statement that two expressions are equal is called an _____.

**3.** The ___absolute___ of a number is the distance between that number and 0 on the number line.

**4.** A symbol used to represent a number is called a ___variable___.

**5.** Two numbers that are the same distance from 0 but lie on opposite sides of 0 are called ___abs value___.

**6.** The number in a fraction above the fraction bar is called the ___n_____.

**7.** A _____ of an equation is a value for the variable that makes the equation a true statement.

**8.** Two numbers whose product is 1 are called _____.

**9.** In $2^3$, the 2 is called the _____ and the 3 is called the _____.

**10.** The number in a fraction below the fraction bar is called the _____.

**11.** Parentheses and brackets are examples of _____.

**12.** A _____ is a collection of objects.

> ▶ **Helpful Hint**
>
> Are you preparing for your test? Don't forget to take the Chapter 1 Test on page 72. Then check your answers at the back of the text and use the Chapter Test Prep Video CD to see the fully worked-out solutions to any of the exercises you want to review.

## CHAPTER 1  HIGHLIGHTS

| DEFINITIONS AND CONCEPTS | EXAMPLES |
|---|---|
| **SECTION 1.2   SYMBOLS AND SETS OF NUMBERS** | |

| DEFINITIONS AND CONCEPTS | EXAMPLES |
|---|---|
| A **set** is a collection of objects, called **elements,** enclosed in braces. | $\{a, c, e\}$ |
| **Natural Numbers:** $\{1, 2, 3, 4, \dots\}$ <br> **Whole Numbers:** $\{0, 1, 2, 3, 4, \dots\}$ | Given the set $\left\{-3.4, \sqrt{3}, 0, \frac{2}{3}, 5, -4\right\}$, list the numbers that belong to the set of |
| **Integers:** $\{\dots, -3, -2, -1, 0, 1, 2, 3, \dots\}$ | Natural numbers: 5 |
| **Rational Numbers:** {real numbers that can be expressed as a quotient of integers} | Whole numbers: 0, 5 <br> Integers: −4, 0, 5 |
| **Irrational Numbers:** {real numbers that cannot be expressed as a quotient of integers} | Rational numbers: −4, −3.4, 0, $\frac{2}{3}$, 5 <br><br> Irrational Numbers: $\sqrt{3}$                    *(continued)* |

| DEFINITIONS AND CONCEPTS | EXAMPLES |
|---|---|

SECTION 1.2   SYMBOLS AND SETS OF NUMBERS (continued)

**Real Numbers:** {all numbers that correspond to a point on the number line}

Real numbers: $-4, -3.4, 0, \frac{2}{3}, \sqrt{3}, 5$

A line used to picture numbers is called a **number line.**

The **absolute value** of a real number $a$, denoted by $|a|$, is the distance between $a$ and 0 on the number line.

$|5| = 5 \qquad |0| = 0 \qquad |-2| = 2$

**Symbols:** = is equal to

$\qquad \neq$ is not equal to

$\qquad >$ is greater than

$\qquad <$ is less than

$\qquad \leq$ is less than or equal to

$\qquad \geq$ is greater than or equal to

$-7 = -7$

$3 \neq -3$

$4 > 1$

$1 < 4$

$6 \leq 6$

$18 \geq -\frac{1}{3}$

*Order Property for Real Numbers*

For any two real numbers $a$ and $b$, $a$ is less than $b$ if $a$ is to the left of $b$ on a number line.

$-3 < 0 \qquad 0 > -3 \qquad 0 < 2.5 \qquad 2.5 > 0$

SECTION 1.3   FRACTIONS

A quotient of two integers is called a **fraction.** The **numerator** of a fraction is the top number. The **denominator** of a fraction is the bottom number.

$\dfrac{13}{17} \leftarrow \text{numerator}$
$\phantom{\dfrac{13}{17}} \leftarrow \text{denominator}$

If $a \cdot b = c$, then $a$ and $b$ are **factors** and $c$ is the **product.**

$$\underset{\text{factor}}{7} \quad \cdot \quad \underset{\text{factor}}{9} \quad = \quad \underset{\text{product}}{63}$$

A fraction is in **lowest terms** when the numerator and the denominator have no factors in common other than 1.

$\dfrac{13}{17}$ is in lowest terms.

**To write a fraction in lowest terms,** factor the numerator and the denominator; then apply the fundamental principle.

Write in lowest terms.

$$\frac{6}{14} = \frac{2 \cdot 3}{2 \cdot 7} = \frac{3}{7}$$

Two fractions are **reciprocals** if their product is 1.
The reciprocal of $\dfrac{a}{b}$ is $\dfrac{b}{a}$.

The reciprocal of $\dfrac{6}{25}$ is $\dfrac{25}{6}$.

**To multiply fractions,** numerator times numerator is the numerator of the product and denominator times denominator is the denominator of the product.

Perform the indicated operations.

$$\frac{2}{5} \cdot \frac{3}{7} = \frac{6}{35}$$

**To divide fractions,** multiply the first fraction by the reciprocal of the second fraction.

$$\frac{5}{9} \div \frac{2}{7} = \frac{5}{9} \cdot \frac{7}{2} = \frac{35}{18}$$

**To add fractions with the same denominator,** add the numerators and place the sum over the common denominator.

$$\frac{5}{11} + \frac{3}{11} = \frac{8}{11}$$

**To subtract fractions with the same denominator,** subtract the numerators and place the difference over the common denominator.
Fractions that represent the same quantity are called **equivalent fractions.**

$$\frac{13}{15} - \frac{3}{15} = \frac{10}{15} = \frac{2}{3}$$

$$\frac{1}{5} = \frac{1 \cdot 4}{5 \cdot 4} = \frac{4}{20}$$

$\dfrac{1}{5}$ and $\dfrac{4}{20}$ are equivalent fractions.

| DEFINITIONS AND CONCEPTS | EXAMPLES |
|---|---|

The expression $a^n$ is an **exponential expression.** The number $a$ is called the **base;** it is the repeated factor. The number $n$ is called the **exponent;** it is the number of times that the base is a factor.

$$4^3 = 4 \cdot 4 \cdot 4 = 64$$
$$7^2 = 7 \cdot 7 = 49$$

*Order of Operations*

Simplify expressions in the following order. If grouping symbols are present, simplify expressions within those first, starting with the innermost set. Also, simplify the numerator and the denominator of a fraction separately.

**1.** Simplify exponential expressions.

**2.** Multiply or divide in order from left to right.

**3.** Add or subtract in order from left to right.

$$\frac{8^2 + 5(7 - 3)}{3 \cdot 7} = \frac{8^2 + 5(4)}{21}$$
$$= \frac{64 + 5(4)}{21}$$
$$= \frac{64 + 20}{21}$$
$$= \frac{84}{21}$$
$$= 4$$

A symbol used to represent a number is called a **variable.**

Examples of variables are:

$$q, x, z$$

An **algebraic expression** is a collection of numbers, variables, operation symbols, and grouping symbols.

Examples of algebraic expressions are:

$$5x, 2(y - 6), \frac{q^2 - 3q + 1}{6}$$

**To evaluate an algebraic expression** containing a variable, substitute a given number for the variable and simplify.

Evaluate $x^2 - y^2$ if $x = 5$ and $y = 3$.
$$x^2 - y^2 = (5)^2 - 3^2$$
$$= 25 - 9$$
$$= 16$$

A mathematical statement that two expressions are equal is called an **equation.**

Equations:

$$3x - 9 = 20$$
$$A = \pi r^2$$

A **solution** of an equation is a value for the variable that makes the equation a true statement.

Determine whether 4 is a solution of $5x + 7 = 27$.
$$5x + 7 = 27$$
$$5(4) + 7 \stackrel{?}{=} 27$$
$$20 + 7 \stackrel{?}{=} 27$$
$$27 = 27 \quad \text{True}$$

4 is a solution.

*To Add Two Numbers with the Same Sign*

Add.

**1.** Add their absolute values.

**2.** Use their common sign as the sign of the sum.

$$10 + 7 = 17$$
$$-3 + (-8) = -11$$

*To Add Two Numbers with Different Signs*

**1.** Subtract their absolute values.

**2.** Use the sign of the number whose absolute value is larger as the sign of the sum.

$$-25 + 5 = -20$$
$$14 + (-9) = 5$$

*(continued)*

| **DEFINITIONS AND CONCEPTS** | **EXAMPLES** |
|---|---|

| | |
|---|---|
| Two numbers that are the same distance from 0 but lie on opposite sides of 0 are called **opposites** or **additive inverses**. The opposite of a number $a$ is denoted by $-a$. | The opposite of $-7$ is 7.<br>The opposite of 123 is $-123$. |
| The sum of a number $a$ and its opposite, $-a$, is 0.<br><br>$$a + (-a) = 0$$<br><br>If $a$ is a number, then $-(-a) = a$. | $$-4 + 4 = 0$$<br>$$12 + (-12) = 0$$<br>$$-(-8) = 8$$<br>$$-(-14) = 14$$ |

| | |
|---|---|
| **To subtract two numbers** $a$ and $b$, add the first number $a$ to the opposite of the second number $b$.<br><br>$$a - b = a + (-b)$$ | Subtract.<br><br>$$3 - (-44) = 3 + 44 = 47$$<br>$$-5 - 22 = -5 + (-22) = -27$$<br>$$-30 - (-30) = -30 + 30 = 0$$ |

| | |
|---|---|
| *Quotient of two real numbers*<br><br>$$\frac{a}{b} = a \cdot \frac{1}{b}$$ | Multiply or divide.<br><br>$$\frac{42}{2} = 42 \cdot \frac{1}{2} = 21$$ |
| *Multiplying and Dividing Real Numbers*<br><br>The product or quotient of two numbers with the same sign is a positive number. The product or quotient of two numbers with different signs is a negative number. | $$7 \cdot 8 = 56 \quad -7 \cdot (-8) = 56$$<br>$$-2 \cdot 4 = -8 \qquad 2 \cdot (-4) = -8$$<br>$$\frac{90}{10} = 9 \qquad \frac{-90}{-10} = 9$$<br>$$\frac{42}{-6} = -7 \qquad \frac{-42}{6} = -7$$ |
| *Products and Quotients Involving Zero*<br><br>The product of 0 and any number is 0.<br><br>$$b \cdot 0 = 0 \quad \text{and} \quad 0 \cdot b = 0$$ | $$-4 \cdot 0 = 0 \qquad 0 \cdot \left(-\frac{3}{4}\right) = 0$$ |
| The quotient of a nonzero number and 0 is undefined.<br><br>$$\frac{b}{0} \text{ is undefined.}$$ | $$\frac{-85}{0} \text{ is undefined.}$$ |
| The quotient of 0 and any nonzero number is 0.<br><br>$$\frac{0}{b} = 0$$ | $$\frac{0}{18} = 0 \qquad \frac{0}{-47} = 0$$ |

| | |
|---|---|
| *Commutative Properties*<br>Addition: $a + b = b + a$<br>Multiplication: $a \cdot b = b \cdot a$ | $$3 + (-7) = -7 + 3$$<br>$$-8 \cdot 5 = 5 \cdot (-8)$$ |
| *Associative Properties*<br>Addition: $(a + b) + c = a + (b + c)$<br>Multiplication: $(a \cdot b) \cdot c = a \cdot (b \cdot c)$ | $$(5 + 10) + 20 = 5 + (10 + 20)$$<br>$$(-3 \cdot 2) \cdot 11 = -3 \cdot (2 \cdot 11)$$ |

| DEFINITIONS AND CONCEPTS | EXAMPLES |
|---|---|

SECTION 1.8   PROPERTIES OF REAL NUMBERS (continued)

Two numbers whose product is 1 are called **multiplicative inverses** or **reciprocals.** The reciprocal of a nonzero number $a$ is $\frac{1}{a}$ because $a \cdot \frac{1}{a} = 1$.

The reciprocal of 3 is $\frac{1}{3}$.

The reciprocal of $-\frac{2}{5}$ is $-\frac{5}{2}$.

**Distributive Property**   $a(b + c) = a \cdot b + a \cdot c$

$$5(6 + 10) = 5 \cdot 6 + 5 \cdot 10$$
$$-2(3 + x) = -2 \cdot 3 + (-2)(x)$$

**Identities**   $a + 0 = a$    $0 + a = a$
$a \cdot 1 = a$    $1 \cdot a = a$

$$5 + 0 = 5 \qquad 0 + (-2) = -2$$
$$-14 \cdot 1 = -14 \qquad 1 \cdot 27 = 27$$

**Inverses**
Addition or opposite:   $a + (-a) = 0$

$$7 + (-7) = 0$$

Multiplication or reciprocal:   $b \cdot \frac{1}{b} = 1$

$$3 \cdot \frac{1}{3} = 1$$

---

### 📖 STUDY SKILLS BUILDER

**Are You Preparing for a Test on Chapter 1?**

Below I have listed some *common trouble areas* for topics covered in Chapter 1. After studying for your test—but before taking your test—read these.

- Do you know the difference between $|-3|$, $-|-3|$, and $-(-3)$?

  $|-3| = 3$;   $-|-3| = -3$;   and   $-(-3) = 3$   (Section 1.2)

- Evaluate $x - y$ if $x = 7$ and $y = -3$.

  $x - y = 7 - (-3) = 10$   (Sections 1.4 and 1.6)

- Make sure you are familiar with order of operations. Sometimes the simplest-looking expressions can give you the most trouble.

  $1 + 2(3 + 6) = 1 + 2(9) = 1 + 18 = 19$   (Section 1.4)

- Do you know the difference between $(-3)^2$ and $-3^2$?

  $(-3)^2 = 9$   and   $-3^2 = -9$   (Section 1.7)

- Do you know that these fractions are equivalent?

  $-\frac{1}{3} = \frac{-1}{3} = \frac{1}{-3}$   (Section 1.7)

- Do you know the difference between opposite and reciprocal? If not, study the table below.

| Number | Opposite | Reciprocal |
|---|---|---|
| 5 | $-5$ | $\frac{1}{5}$ |
| $-\frac{4}{7}$ | $\frac{4}{7}$ | $-\frac{7}{4}$ |
| $-\frac{1}{3}$ | $\frac{1}{3}$ | $-3$ |

(Sections 1.5 and 1.7)

Remember: This is simply a checklist of selected topics given to check your understanding. For a review of Chapter 1 in the text, see the material at the end of Chapter 1.

---

## CHAPTER 1  REVIEW

*(1.2)* *Insert* $<, >,$ *or* $=$ *in the appropriate space to make the following statements true.*

**1.** 8    10

**2.** 7    2

**3.** $-4$    $-5$

**4.** $\frac{12}{2}$    $-8$

**5.** $|-7|$    $|-8|$

**6.** $|-9|$    $-9$

**7.** $-|-1|$    $-1$

**8.** $|-14|$    $-(-14)$

**9.** 1.2    1.02

**10.** $-\frac{3}{2}$    $-\frac{3}{4}$

*Translate each statement into symbols.*

**11.** Four is greater than or equal to negative three.

**12.** Six is not equal to five.

**13.** 0.03 is less than 0.3.

**14.** New York City has 155 museums and 400 art galleries. Write an inequality comparing the numbers 155 and 400. (*Source:* Absolute Trivia.com)

*Given the following sets of numbers, list the numbers in each set that also belong to the set of:*

**a.** Natural numbers      **b.** Whole numbers

**c.** Integers           **d.** Rational numbers

**e.** Irrational numbers     **f.** Real numbers

**15.** $\left\{-6, 0, 1, 1\frac{1}{2}, 3, \pi, 9.62\right\}$

**16.** $\left\{-3, -1.6, 2, 5, \frac{11}{2}, 15.1, \sqrt{5}, 2\pi\right\}$

*The following chart shows the gains and losses in dollars of Density Oil and Gas stock for a particular week.*

| Day | Gain or Loss in Dollars |
|---|---|
| Monday | +1 |
| Tuesday | −2 |
| Wednesday | +5 |
| Thursday | +1 |
| Friday | −4 |

**17.** Which day showed the greatest loss?

**18.** Which day showed the greatest gain?

**(1.3)** *Write the number as a product of prime factors.*

**19.** 36                   **20.** 120

*Perform the indicated operations. Write results in lowest terms.*

**21.** $\frac{8}{15} \cdot \frac{27}{30}$          **22.** $\frac{7}{8} \div \frac{21}{32}$

**23.** $\frac{7}{15} + \frac{5}{6}$          **24.** $\frac{3}{4} - \frac{3}{20}$

**25.** $2\frac{3}{4} + 6\frac{5}{8}$          **26.** $7\frac{1}{6} - 2\frac{2}{3}$

**27.** $5 \div \frac{1}{3}$            **28.** $2 \cdot 8\frac{3}{4}$

**29.** Determine the unknown part of the given circle.

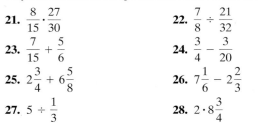

*Find the area and the perimeter of each figure.*

△ **30.**                     △ **31.**

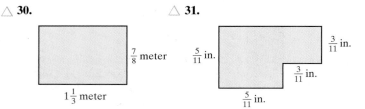

$\frac{7}{8}$ meter     $\frac{5}{11}$ in.           $\frac{3}{11}$ in.

$\frac{3}{11}$ in.

$1\frac{1}{3}$ meter              $\frac{5}{11}$ in.

△ **32.** A trim carpenter needs a piece of quarter round molding $6\frac{1}{8}$ feet long for a bathroom. She finds a piece $7\frac{1}{2}$ feet long. How long a piece does she need to cut from the $7\frac{1}{2}$-foot-long molding in order to use it in the bathroom?

*In December 1998, Nkem Chukwu gave birth to the world's first surviving octuplets in Houston, Texas. The following chart gives the octuplets' birthweights. The babies are listed in order of birth.*

| Baby's Name | Gender | Birthweight (pounds) |
|---|---|---|
| Ebuka | girl | $1\frac{1}{2}$ |
| Chidi | girl | $1\frac{11}{16}$ |
| Echerem | girl | $1\frac{3}{4}$ |
| Chima | girl | $1\frac{5}{8}$ |
| Odera | girl | $\frac{11}{16}$ |
| Ikem | boy | $1\frac{1}{8}$ |
| Jioke | boy | $1\frac{13}{16}$ |
| Gorom | girl | $1\frac{1}{8}$ |

(*Source:* Texas Children's Hospital, Houston, Texas)

**33.** What was the total weight of the boy octuplets?

**34.** What was the total weight of the girl octuplets?

**35.** Find the combined weight of all eight octuplets.

**36.** Which baby weighed the most?

**37.** Which baby weighed the least?

**38.** How much more did the heaviest baby weigh than the lightest baby?

**39.** By March 1999, Chima weighed $5\frac{1}{2}$ pounds. How much weight had she gained since birth?

**40.** By March 1999, Ikem weighed $4\frac{5}{32}$ pounds. How much weight had he gained since birth?

**(1.4)** *Simplify each expression.*

**41.** $2^4$                    **42.** $5^2$

**43.** $\left(\frac{2}{7}\right)^2$              **44.** $\left(\frac{3}{4}\right)^3$

**45.** $6 \cdot 3^2 + 2 \cdot 8$        **46.** $68 - 5 \cdot 2^3$

**47.** $3(1 + 2 \cdot 5) + 4$     **48.** $8 + 3(2 \cdot 6 - 1)$

**49.** $\frac{4 + |6 - 2| + 8^2}{4 + 6 \cdot 4}$       **50.** $5[3(2 + 5) - 5]$

*Translate each word statement to symbols.*

**51.** The difference of twenty and twelve is equal to the product of two and four.

**52.** The quotient of nine and two is greater than negative five.

*Evaluate each expression if $x = 6$, $y = 2$, and $z = 8$.*

**53.** $2x + 3y$

**54.** $x(y + 2z)$

**55.** $\dfrac{x}{y} + \dfrac{z}{2y}$

**56.** $x^2 - 3y^2$

△ **57.** The expression $180 - a - b$ represents the measure of the unknown angle of the given triangle. Replace $a$ with 37 and $b$ with 80 to find the measure of the unknown angle.

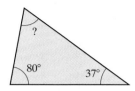

*Decide whether the given number is a solution to the given equation.*

**58.** Is $x = 3$ a solution of $7x - 3 = 18$?

**59.** Is $x = 1$ a solution of $3x^2 + 4 = x - 1$?

**(1.5)** *Find the additive inverse or the opposite.*

**60.** $-9$

**61.** $\dfrac{2}{3}$

**62.** $|-2|$

**63.** $-|-7|$

*Find the following sums.*

**64.** $-15 + 4$

**65.** $-6 + (-11)$

**66.** $\dfrac{1}{16} + \left(-\dfrac{1}{4}\right)$

**67.** $-8 + |-3|$

**68.** $-4.6 + (-9.3)$

**69.** $-2.8 + 6.7$

**70.** The lowest elevation in North America is $-282$ feet at Death Valley in California. If you are standing at a point 728 feet above Death Valley, what is your elevation? (*Source:* National Geographic Society)

**(1.6)** *Perform the indicated operations.*

**71.** $6 - 20$

**72.** $-3.1 - 8.4$

**73.** $-6 - (-11)$

**74.** $4 - 15$

**75.** $-21 - 16 + 3(8 - 2)$

**76.** $\dfrac{11 - (-9) + 6(8 - 2)}{2 + 3 \cdot 4}$

*If $x = 3$, $y = -6$, and $z = -9$, evaluate each expression.*

**77.** $2x^2 - y + z$

**78.** $\dfrac{y - x + 5x}{2x}$

**(1.7)** *Find the multiplicative inverse or reciprocal.*

**79.** $-6$

**80.** $\dfrac{3}{5}$

*Simplify each expression.*

**81.** $6(-8)$

**82.** $(-2)(-14)$

**83.** $\dfrac{-18}{-6}$

**84.** $\dfrac{42}{-3}$

**85.** $\dfrac{4(-3) + (-8)}{2 + (-2)}$

**86.** $\dfrac{3(-2)^2 - 5}{-14}$

**87.** $\dfrac{-6}{0}$

**88.** $\dfrac{0}{-2}$

**89.** $-4^2 - (-3 + 5) \div (-1) \cdot 2$

**90.** $-5^2 - (2 - 20) \div (-3) \cdot 3$

*If $x = -5$ and $y = -2$, evaluate each expression.*

**91.** $x^2 - y^4$

**92.** $x^2 - y^3$

**93.** During the 1999 LPGA Sara Lee Classic, Michelle McGann had scores of $-9$, $-7$, and $+1$ in three rounds of golf. Find her average score per round. (*Source:* Ladies Professional Golf Association)

**94.** During the 1999 PGA Masters Tournament, Bob Estes had scores of $-1, 0, -3$, and $0$ in four rounds of golf. Find his average score per round. (*Source:* Professional Golf Association)

**(1.8)** *Name the property illustrated.*

**95.** $-6 + 5 = 5 + (-6)$

**96.** $6 \cdot 1 = 6$

**97.** $3(8 - 5) = 3 \cdot 8 + 3 \cdot (-5)$

**98.** $4 + (-4) = 0$

**99.** $2 + (3 + 9) = (2 + 3) + 9$

**100.** $2 \cdot 8 = 8 \cdot 2$

**101.** $6(8 + 5) = 6 \cdot 8 + 6 \cdot 5$

**102.** $(3 \cdot 8) \cdot 4 = 3 \cdot (8 \cdot 4)$

**103.** $4 \cdot \dfrac{1}{4} = 1$

**104.** $8 + 0 = 8$

*Use the distributive property to write each expression without parentheses.*

**105.** $5(y - 2)$

**106.** $-3(z + y)$

**107.** $-(7 - x + 4z)$

**108.** $\dfrac{1}{2}(6z - 10)$

**109.** $-4(3x + 5) - 7$

**110.** $-8(2y + 9) - 1$

## MIXED REVIEW

*Insert $<$, $>$, or $=$ in the space between each pair of numbers.*

**111.** $-|-11|$ ____ $|11.4|$

**112.** $-1\dfrac{1}{2}$ ____ $-2\dfrac{1}{2}$

*Perform the indicated operations.*

**113.** $-7.2 + (-8.1)$

**114.** $14 - 20$

**115.** $4(-20)$

**116.** $\dfrac{-20}{4}$

**117.** $-\dfrac{4}{5}\left(\dfrac{5}{16}\right)$

**118.** $-0.5(-0.3)$

**119.** $8 \div 2 \cdot 4$

**120.** $(-2)^4$

**121.** $\dfrac{-3 - 2(-9)}{-15 - 3(-4)}$

**122.** $5 + 2[(7 - 5)^2 + (1 - 3)]$

**123.** $-\dfrac{5}{8} \div \dfrac{3}{4}$

**124.** $\dfrac{-15 + (-4)^2 + |-9|}{10 - 2 \cdot 5}$

# CHAPTER 1 TEST

Remember to use the Chapter Test Prep Video CD to see the fully worked-out solutions to any of the exercises you want to review.

*Translate the statement into symbols.*

**1.** The absolute value of negative seven is greater than five.

**2.** The sum of nine and five is greater than or equal to four.

*Simplify the expression.*

**3.** $-13 + 8$

**4.** $-13 - (-2)$

**5.** $12 \div 4 \cdot 3 - 6 \cdot 2$

**6.** $(13)(-3)$

**7.** $(-6)(-2)$

**8.** $\dfrac{|-16|}{-8}$

**9.** $\dfrac{-8}{0}$

**10.** $\dfrac{|-6| + 2}{5 - 6}$

**11.** $\dfrac{1}{2} - \dfrac{5}{6}$

**12.** $-1\dfrac{1}{8} + 5\dfrac{3}{4}$

**13.** $(2 - 6) \div \dfrac{-2 - 6}{-3 - 1} - \dfrac{1}{2}$

**14.** $3(-4)^2 - 80$

**15.** $6[5 + 2(3 - 8) - 3]$

**16.** $\dfrac{-12 + 3 \cdot 8}{4}$

**17.** $\dfrac{(-2)(0)(-3)}{-6}$

*Insert* $<, >,$ *or* $=$ *in the appropriate space to make each of the following statements true.*

**18.** $-3 \quad -7$

**19.** $4 \quad -8$

**20.** $2 \quad |-3|$

**21.** $|-2| \quad -1 - (-3)$

**22.** In the state of Massachusetts, there are 2221 licensed child care centers and 10,993 licensed home-based child care providers. Write an inequality statement comparing the numbers 2221 and 10,993. (*Source:* Children's Foundation)

**23.** Given $\left\{-5, -1, 0, \dfrac{1}{4}, 1, 7, 11.6, \sqrt{7}, 3\pi\right\}$, list the numbers in this set that also belong to the set of:

   **a.** Natural numbers

   **b.** Whole numbers

   **c.** Integers

   **d.** Rational numbers

   **e.** Irrational numbers

   **f.** Real numbers

*If* $x = 6, y = -2,$ *and* $z = -3,$ *evaluate each expression.*

**24.** $x^2 + y^2$

**25.** $x + yz$

**26.** $2 + 3x - y$

**27.** $\dfrac{y + z - 1}{x}$

*Identify the property illustrated by each expression.*

**28.** $8 + (9 + 3) = (8 + 9) + 3$

**29.** $6 \cdot 8 = 8 \cdot 6$

**30.** $-6(2 + 4) = -6 \cdot 2 + (-6) \cdot 4$

**31.** $\dfrac{1}{6}(6) = 1$

**32.** Find the opposite of $-9$.

**33.** Find the reciprocal of $-\dfrac{1}{3}$.

*The New Orleans Saints were 22 yards from the goal when the following series of gains and losses occurred.*

| Gains and Losses in Yards | |
| --- | --- |
| First Down | 5 |
| Second Down | $-10$ |
| Third Down | $-2$ |
| Fourth Down | 29 |

**34.** During which down did the greatest loss of yardage occur?

**35.** Was a touchdown scored?

**36.** The temperature at the Winter Olympics was a frigid 14 degrees below zero in the morning, but by noon it had risen 31 degrees. What was the temperature at noon?

**37.** United HealthCare is a health insurance provider. In 3 consecutive recent years, it had net incomes of $356 million, $460 million, and $-$166 million. What was United HealthCare's total net income for these three years? (*Source:* United HealthCare Corp.)

**38.** Jean Avarez decided to sell 280 shares of stock, which decreased in value by $1.50 per share yesterday. How much money did she lose?

# 2

# Equations, Inequalities, and Problem Solving

There is an expanding market for high-speed trains as more countries, such as China, turn to bullet trains. In April 2007, a French TGV, short for "train à grande vitesse" or "very fast train," broke the previous world speed record on rails. (It should be noted that the current record for overall train speed is held by a magnetically levitated train called the Maglev, from Japan. This train hovers above the rails.)

The bar graph below shows a history of train speed records. In Section 2.4, Exercise 49, you will have the opportunity to calculate the speeds of the Maglev and the TGV.

Much of mathematics relates to deciding which statements are true and which are false. For example, the statement $x + 7 = 15$ is an equation stating that the sum $x + 7$ has the same value as 15. Is this statement true or false? It is false for some values of $x$ and true for just one value of $x$, namely 8. Our purpose in this chapter is to learn ways of deciding which values make an equation or an inequality true.

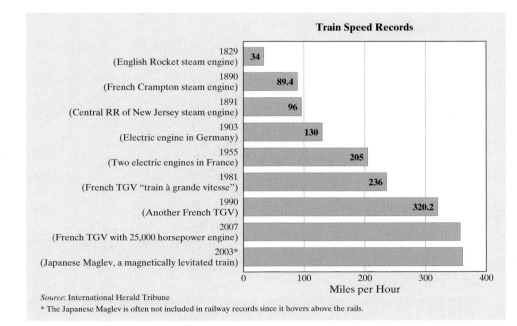

**Train Speed Records**

*Source*: International Herald Tribune

\* The Japanese Maglev is often not included in railway records since it hovers above the rails.

# 2.1 SIMPLIFYING ALGEBRAIC EXPRESSIONS

As we explore in this section, an expression such as $3x + 2x$ is not as simple as possible, because—even without replacing $x$ by a value—we can perform the indicated addition.

**OBJECTIVE 1 ▶ Identifying terms, like terms, and unlike terms.** Before we practice simplifying expressions, some new language of algebra is presented. A **term** is a number or the product of a number and variables raised to powers.

*Terms*

$$-y, \quad 2x^3, \quad -5, \quad 3xz^2, \quad \frac{2}{y}, \quad 0.8z$$

The **numerical coefficient** (sometimes also simply called the **coefficient**) of a term is the numerical factor. The numerical coefficient of $3x$ is 3. Recall that $3x$ means $3 \cdot x$.

| Term | Numerical Coefficient |
|---|---|
| $3x$ | 3 |
| $\dfrac{y^3}{5}$ | $\dfrac{1}{5}$  since $\dfrac{y^3}{5}$ means $\dfrac{1}{5} \cdot y^3$ |
| $0.7ab^3c^5$ | 0.7 |
| $z$ | 1 |
| $-y$ | $-1$ |
| $-5$ | $-5$ |

▶ **Helpful Hint**

The term $-y$ means $-1y$ and thus has a numerical coefficient of $-1$. The term $z$ means $1z$ and thus has a numerical coefficient of 1.

**EXAMPLE 1**   Identify the numerical coefficient in each term.

**a.** $-3y$     **b.** $22z^4$     **c.** $y$     **d.** $-x$     **e.** $\dfrac{x}{7}$

*Solution*

**a.** The numerical coefficient of $-3y$ is $-3$.

**b.** The numerical coefficient of $22z^4$ is $22$.

**c.** The numerical coefficient of $y$ is 1, since $y$ is $1y$.

**d.** The numerical coefficient of $-x$ is $-1$, since $-x$ is $-1x$.

**e.** The numerical coefficient of $\dfrac{x}{7}$ is $\dfrac{1}{7}$, since $\dfrac{x}{7}$ means $\dfrac{1}{7} \cdot x$.

**PRACTICE**

**1**   Identify the numerical coefficients in each term.

**a.** $t$     **b.** $-7x$     **c.** $-\dfrac{w}{5}$     **d.** $43x^4$     **e.** $-b$

Terms with the same variables raised to exactly the same powers are called **like terms.** Terms that aren't like terms are called **unlike terms.**

| Like Terms | Unlike Terms | |
|---|---|---|
| $3x, 2x$ | $5x, 5x^2$ | Why? Same variable $x$, but different powers $x$ and $x^2$ |
| $-6x^2y, 2x^2y, 4x^2y$ | $7y, 3z, 8x^2$ | Why? Different variables |
| $2ab^2c^3, ac^3b^2$ | $6abc^3, 6ab^2$ | Why? Different variables and different powers |

> ▶ **Helpful Hint**
>
> In like terms, each variable and its exponent must match exactly, but these factors don't need to be in the same order.
>
> $$2x^2y \text{ and } 3yx^2 \text{ are like terms.}$$

**EXAMPLE 2**   Determine whether the terms are like or unlike.

**a.** $2x, 3x^2$      **b.** $4x^2y, x^2y, -2x^2y$        **c.** $-2yz, -3zy$       **d.** $-x^4, x^4$

*Solution*

**a.** Unlike terms, since the exponents on $x$ are not the same.

**b.** Like terms, since each variable and its exponent match.

**c.** Like terms, since $zy = yz$ by the commutative property.

**d.** Like terms.                                                                  □

**PRACTICE**
**2**    Determine whether the terms are like or unlike.

**a.** $-4xy, 5yx$                              **b.** $5q, -3q^2$

**c.** $3ab^2, -2ab^2, 43ab^2$                 **d.** $y^5, \dfrac{y^5}{2}$

**OBJECTIVE 2 ▶ Combining like terms.**  An algebraic expression containing the sum or difference of like terms can be simplified by applying the distributive property. For example, by the distributive property, we rewrite the sum of the like terms $3x + 2x$ as

$$3x + 2x = (3 + 2)x = 5x$$

Also,

$$-y^2 + 5y^2 = (-1 + 5)y^2 = 4y^2$$

Simplifying the sum or difference of like terms is called **combining like terms.**

**EXAMPLE 3**   Simplify each expression by combining like terms.

**a.** $7x - 3x$              **b.** $10y^2 + y^2$            **c.** $8x^2 + 2x - 3x$

*Solution*

**a.** $7x - 3x = (7 - 3)x = 4x$

**b.** $10y^2 + y^2 = 10y^2 + 1y^2 = (10 + 1)y^2 = 11y^2$

**c.** $8x^2 + 2x - 3x = 8x^2 + (2 - 3)x = 8x^2 - x$                              □

**PRACTICE**
**3**    Simplify each expression by combining like terms.

**a.** $4x^2 + 3x^2$            **b.** $-3y + y$            **c.** $5x - 3x^2 + 8x^2$

**EXAMPLE 4** Simplify each expression by combining like terms.

**a.** $2x + 3x + 5 + 2$      **b.** $-5a - 3 + a + 2$      **c.** $4y - 3y^2$

**d.** $2.3x + 5x - 6$      **e.** $-\dfrac{1}{2}b + b$

_Solution_ Use the distributive property to combine like terms.

**a.** $2x + 3x + 5 + 2 = (2 + 3)x + (5 + 2)$
$$= 5x + 7$$

**b.** $-5a - 3 + a + 2 = -5a + 1a + (-3 + 2)$
$$= (-5 + 1)a + (-3 + 2)$$
$$= -4a - 1$$

**c.** $4y - 3y^2$ These two terms cannot be combined because they are unlike terms.

**d.** $2.3x + 5x - 6 = (2.3 + 5)x - 6$
$$= 7.3x - 6$$

**e.** $-\dfrac{1}{2}b + b = -\dfrac{1}{2}b + 1b = \left(-\dfrac{1}{2} + 1\right)b = \dfrac{1}{2}b$     □

**PRACTICE**

**4** Use the distributive property to combine like terms.

**a.** $3y + 8y - 7 + 2$      **b.** $6x - 3 - x - 3$      **c.** $\dfrac{3}{4}t - t$

**d.** $9y + 3.2y + 10 + 3$      **e.** $5z - 3z^4$

The examples above suggest the following:

> **Combining Like Terms**
>
> To **combine like terms,** add the numerical coefficients and multiply the result by the common variable factors.

**OBJECTIVE 3 ▶ Using the distributive property.** Simplifying expressions makes frequent use of the distributive property to also remove parentheses.

**EXAMPLE 5** Find each product by using the distributive property to remove parentheses.

**a.** $5(x + 2)$      **b.** $-2(y + 0.3z - 1)$      **c.** $-(x + y - 2z + 6)$

_Solution_

**a.** $5(x + 2) = 5 \cdot x + 5 \cdot 2$      Apply the distributive property.
$$= 5x + 10 \quad\quad\quad\text{Multiply.}$$

**b.** $-2(y + 0.3z - 1) = -2(y) + (-2)(0.3z) + (-2)(-1)$      Apply the distributive property.
$$= -2y - 0.6z + 2 \quad\quad\text{Multiply.}$$

**c.** $-(x + y - 2z + 6) = -1(x + y - 2z + 6)$      Distribute $-1$ over each term.
$$= -1(x) - 1(y) - 1(-2z) - 1(6)$$
$$= -x - y + 2z - 6 \quad\quad\quad □$$

**PRACTICE**

**5** Find each product by using the distributive property to remove parentheses.

**a.** $3(2x - 7)$      **b.** $-5(3x - 4z - 5)$      **c.** $-(2x - y + z - 2)$

> ▶ **Helpful Hint**
>
> If a "−" sign precedes parentheses, the sign of each term inside the parentheses is changed when the distributive property is applied to remove parentheses.
>
> **Examples:**
>
> $$-(2x + 1) = -2x - 1 \qquad -(-5x + y - z) = 5x - y + z$$
> $$-(x - 2y) = -x + 2y \qquad -(-3x - 4y - 1) = 3x + 4y + 1$$

When simplifying an expression containing parentheses, we often use the distributive property in both directions—first to remove parentheses and then again to combine any like terms.

**EXAMPLE 6**    Simplify the following expressions.

**a.** $3(2x - 5) + 1$   **b.** $-2(4x + 7) - (3x - 1)$   **c.** $9 - 3(4x + 10)$

*Solution*

**a.** $3(2x - 5) + 1 = 6x - 15 + 1$      Apply the distributive property.
               $= 6x - 14$      Combine like terms.

**b.** $-2(4x + 7) - (3x - 1) = -8x - 14 - 3x + 1$      Apply the distributive property.
                 $= -11x - 13$      Combine like terms.

> ▶ **Helpful Hint**
>
> Don't forget to use the distributive property to multiply before adding or subtracting like terms.

**c.** $9 - 3(4x + 10) = 9 - 12x - 30$      Apply the distributive property.
              $= -21 - 12x$      Combine like terms.   ☐

**PRACTICE**
**6**    Simplify the following expressions.

**a.** $4(9x + 1) + 6$   **b.** $-7(2x - 1) - (6 - 3x)$   **c.** $8 - 5(6x + 5)$

**OBJECTIVE 4 ▶ Writing word phrases as algebraic expressions.** Next, we practice writing word phrases as algebraic expressions.

**EXAMPLE 7**    Write the phrase below as an algebraic expression. Then simplify if possible.

"Subtract $4x - 2$ from $2x - 3$."

*Solution*    "Subtract $4x - 2$ **from** $2x - 3$" translates to $(2x - 3) - (4x - 2)$. Next, simplify the algebraic expression.

$$(2x - 3) - (4x - 2) = 2x - 3 - 4x + 2 \quad \text{Apply the distributive property.}$$
$$= -2x - 1 \quad \text{Combine like terms.} \quad ☐$$

**PRACTICE**
**7**    Write the phrase below as an algebraic expression. Then simplify if possible.

"Subtract $7x - 1$ from $2x + 3$."

**EXAMPLE 8** Write the following phrases as algebraic expressions and simplify if possible. Let $x$ represent the unknown number.

**a.** Twice a number, added to 6

**b.** The difference of a number and 4, divided by 7

**c.** Five added to 3 times the sum of a number and 1

**d.** The sum of twice a number, 3 times the number, and 5 times the number

*Solution*

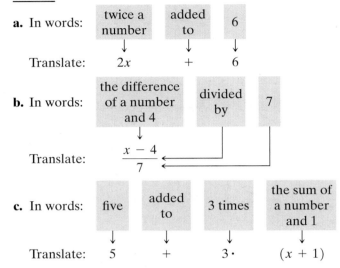

**a.** In words:  twice a number  added to  6

Translate:  $2x$  $+$  6

**b.** In words:  the difference of a number and 4  divided by  7

Translate:  $\dfrac{x-4}{7}$

**c.** In words:  five  added to  3 times  the sum of a number and 1

Translate:  5  $+$  3·  $(x+1)$

Next, we simplify this expression.

$$5 + 3(x + 1) = 5 + 3x + 3 \qquad \text{Use the distributive property.}$$
$$= 8 + 3x \qquad \text{Combine like terms.}$$

**d.** The phrase "the sum of" means that we add.

In words:  twice a number  added to  3 times the number  added to  5 times the number

Translate:  $2x$  $+$  $3x$  $+$  $5x$

Now let's simplify.

$$2x + 3x + 5x = 10x \qquad \text{Combine like terms.} \qquad \square$$

**PRACTICE**

**8** Write the following phrases as algebraic expressions and simplify if possible. Let $x$ represent the unknown number.

**a.** Three added to double a number

**b.** Six subtracted from the sum of 5 and a number

**c.** Two times the sum of 3 and a number, increased by 4

**d.** The sum of a number, half the number, and 5 times the number

## VOCABULARY & READINESS CHECK

*Use the choices below to fill in each blank. Some choices may be used more than once.*

| like | numerical coefficient | term | distributive |
| unlike | combine like terms | expression | |

1. $23y^2 + 10y - 6$ is called a(n) _____ while $23y^2$, $10y$, and $-6$ are each called a(n) _____ .
2. To simplify $x + 4x$, we _____ .
3. The term $y$ has an understood _____ of 1.
4. The terms $7z$ and $7y$ are _____ terms and the terms $7z$ and $-z$ are _____ terms.
5. For the term $-\frac{1}{2}xy^2$, the number $-\frac{1}{2}$ is the _____ .
6. $5(3x - y)$ equals $15x - 5y$ by the _____ property.

*Fill in the blank with the numerical coefficient of each term. See Example 1.*

7. $-7y$ _____    8. $3x$ _____    9. $x$ _____    10. $-y$ _____    11. $-\dfrac{5y}{3}$ _____    12. $-\dfrac{2}{3}z$ _____

*Indicate whether the following lists of terms are like or unlike. See Example 2.*

13. $5y, -y$ _____    14. $-2x^2y, 6xy$ _____    15. $2z, 3z^2$ _____    16. $b^2a, -\dfrac{7}{8}ab^2$ _____

## 2.1 | EXERCISE SET

*MyMathLab*  Powered by CourseCompass™ and MathXL®   |   Math XP PRACTICE   WATCH   DOWNLOAD   READ   REVIEW

*Simplify each expression by combining any like terms. See Examples 3 and 4.*

1. $7y + 8y$
2. $3x + 2x$
3. $8w - w + 6w$
4. $c - 7c + 2c$
5. $3b - 5 - 10b - 4$
6. $6g + 5 - 3g - 7$
7. $m - 4m + 2m - 6$
8. $a + 3a - 2 - 7a$
9. $5g - 3 - 5 - 5g$
10. $8p + 4 - 8p - 15$
11. $6.2x - 4 + x - 1.2$
12. $7.9y - 0.7 - y + 0.2$
13. $6x - 5x + x - 3 + 2x$
14. $8h + 13h - 6 + 7h - h$
15. $7x^2 + 8x^2 - 10x^2$
16. $8x^3 + x^3 - 11x^3$
17. $6x + 0.5 - 4.3x - 0.4x + 3$
18. $0.4y - 6.7 + y - 0.3 - 2.6y$
19. In your own words, explain how to combine like terms.
20. Do like terms contain the same numerical coefficients? Explain your answer.

*Simplify each expression. First use the distributive property to remove any parentheses. See Examples 5 and 6.*

21. $5(y - 4)$
22. $7(r - 3)$
23. $-2(x + 2)$
24. $-4(y + 6)$
25. $7(d - 3) + 10$
26. $9(z + 7) - 15$
27. $-5(2x - 3y + 6)$
28. $-2(4x - 3z - 1)$
29. $-(3x - 2y + 1)$
30. $-(y + 5z - 7)$
31. $5(x + 2) - (3x - 4)$
32. $4(2x - 3) - 2(x + 1)$

*Write each of the following as an algebraic expression. Simplify if possible. See Example 7.*

33. Add $6x + 7$ to $4x - 10$.
34. Add $3y - 5$ to $y + 16$.
35. Subtract $7x + 1$ from $3x - 8$.
36. Subtract $4x - 7$ from $12 + x$.
37. Subtract $5m - 6$ from $m - 9$.
38. Subtract $m - 3$ from $2m - 6$.

## MIXED PRACTICE

*Simplify each expression. See Examples 3 through 7.*

**39.** $2k - k - 6$

**40.** $7c - 8 - c$

**41.** $-9x + 4x + 18 - 10x$

**42.** $5y - 14 + 7y - 20y$

**43.** $-4(3y - 4) + 12y$

**44.** $-3(2x + 5) - 6x$

**45.** $3(2x - 5) - 5(x - 4)$

**46.** $2(6x - 1) - (x - 7)$

**47.** $-2(3x - 4) + 7x - 6$

**48.** $8y - 2 - 3(y + 4)$

**49.** $5k - (3k - 10)$

**50.** $-11c - (4 - 2c)$

**51.** Subtract $6x - 1$ from $3x + 4$

**52.** Subtract $4 + 3y$ from $8 - 5y$

**53.** $3.4m - 4 - 3.4m - 7$

**54.** $2.8w - 0.9 - 0.5 - 2.8w$

**55.** $\frac{1}{3}(7y - 1) + \frac{1}{6}(4y + 7)$

**56.** $\frac{1}{5}(9y + 2) + \frac{1}{10}(2y - 1)$

**57.** $2 + 4(6x - 6)$

**58.** $8 + 4(3x - 4)$

**59.** $0.5(m + 2) + 0.4m$

**60.** $0.2(k + 8) - 0.1k$

**61.** $10 - 3(2x + 3y)$

**62.** $14 - 11(5m + 3n)$

**63.** $6(3x - 6) - 2(x + 1) - 17x$

**64.** $7(2x + 5) - 4(x + 2) - 20x$

**65.** $\frac{1}{2}(12x - 4) - (x + 5)$

**66.** $\frac{1}{3}(9x - 6) - (x - 2)$

*Write each phrase as an algebraic expression and simplify if possible. Let x represent the unknown number. See Examples 7 and 8.*

**67.** Twice a number, decreased by four

**68.** The difference of a number and two, divided by five

**69.** Seven added to double a number

**70.** Eight more than triple a number

**71.** Three-fourths of a number, increased by twelve

**72.** Eleven, increased by two-thirds of a number

**73.** The sum of 5 times a number and $-2$, added to 7 times the number

**74.** The sum of 3 times a number and 10, **subtracted from** 9 times the number

**75.** Eight times the sum of a number and six

**76.** Six times the difference of a number and five

**77.** Double a number, minus the sum of the number and ten

**78.** Half a number, minus the product of the number and eight

**79.** Seven, multiplied by the quotient of a number and six

**80.** The product of a number and ten, less twenty

**81.** The sum of 2, three times a number, $-9$, and four times the number

**82.** The sum of twice a number, $-1$, five times the number, and $-12$

## REVIEW AND PREVIEW

*Evaluate the following expressions for the given values. See Section 1.7.*

**83.** If $x = -1$ and $y = 3$, find $y - x^2$.

**84.** If $g = 0$ and $h = -4$, find $gh - h^2$.

**85.** If $a = 2$ and $b = -5$, find $a - b^2$.

**86.** If $x = -3$, find $x^3 - x^2 + 4$.

**87.** If $y = -5$ and $z = 0$, find $yz - y^2$.

**88.** If $x = -2$, find $x^3 - x^2 - x$.

## CONCEPT EXTENSIONS

△ **89.** Recall that the perimeter of a figure is the total distance around the figure. Given the following rectangle, express the perimeter as an algebraic expression containing the variable $x$.

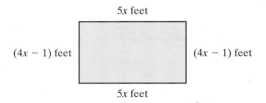

5x feet

(4x − 1) feet       (4x − 1) feet

5x feet

△ **90.** Given the following triangle, express its perimeter as an algebraic expression containing the variable $x$.

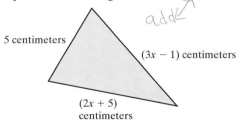

add

5 centimeters

(3x − 1) centimeters

(2x + 5) centimeters

*Given the following two rules, determine whether each scale in Exercises 91 through 94 is balanced or not.*

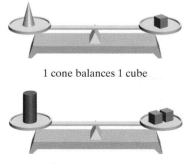

1 cone balances 1 cube

1 cylinder balances 2 cubes

**91.**
**92.**
**93.**
**94.**

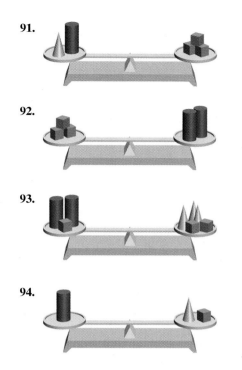

*Write each algebraic expression described.*

**95.** Write an expression with 4 terms that simplifies to $3x - 4$.

**96.** Write an expression of the form _____ (_____ + _____) whose product is $6x + 24$.

**97.** To convert from feet to inches, we multiply by 12. For example, the number of inches in 2 feet is $12 \cdot 2$ inches. If one board has a length of $(x + 2)$ *feet* and a second board has a length of $(3x - 1)$ *inches*, express their total length in inches as an algebraic expression.

**98.** The value of 7 nickels is $5 \cdot 7$ cents. Likewise, the value of $x$ nickels is $5x$ cents. If the money box in a drink machine contains $x$ *nickels*, $3x$ *dimes*, and $(30x - 1)$ *quarters*, express their total value in cents as an algebraic expression.

*For Exercises 99 through 104, see the example below.*

**Example**

Simplify $-3xy + 2x^2y - (2xy - 1)$.

*Solution*

$$-3xy + 2x^2y - (2xy - 1)$$
$$= -3xy + 2x^2y - 2xy + 1 = -5xy + 2x^2y + 1$$

*Simplify each expression.*

**99.** $5b^2c^3 + 8b^3c^2 - 7b^3c^2$

**100.** $4m^4p^2 + m^4p^2 - 5m^2p^4$

**101.** $3x - (2x^2 - 6x) + 7x^2$

**102.** $9y^2 - (6xy^2 - 5y^2) - 8xy^2$

**103.** $-(2x^2y + 3z) + 3z - 5x^2y$

**104.** $-(7c^3d - 8c) - 5c - 4c^3d$

---

📖 **STUDY SKILLS BUILDER**

**Have You Decided to Complete This Course Successfully?**

Ask yourself if one of your current goals is to complete this course successfully.

If it is not a goal of yours, ask yourself why? One common reason is fear of failure. Amazingly enough, fear of failure alone can be strong enough to keep many of us from doing our best in any endeavor.

Another common reason is that you simply haven't taken the time to think about or write down your goals for this course. To help accomplish this, answer the questions below.

Self-Check

**1.** Write down your goal(s) for this course.

**2.** Now list steps you will take to make sure your goal(s) in Question 1 are accomplished.

**3.** Rate your commitment to this course with a number between 1 and 5. Use the diagram below to help.

| High Commitment | | Average Commitment | Not committed at all | |
|---|---|---|---|---|
| 5 | 4 | 3 | 2 | 1 |

**4.** If you have rated your personal commitment level (from the exercise above) as a 1, 2, or 3, list the reasons why this is so. Then determine whether it is possible to increase your commitment level to a 4 or 5.

Good luck, and don't forget that a positive attitude will make a big difference.

# 2.2 THE ADDITION AND MULTIPLICATION PROPERTIES OF EQUALITY

**OBJECTIVES**

1 Define linear equations and use the addition property of equality to solve linear equations.

2 Use the multiplication property of equality to solve linear equations.

3 Use both properties of equality to solve linear equations.

4 Write word phrases as algebraic expressions.

**OBJECTIVE 1 ▶ Defining linear equations and using the addition property.** Recall from Section 1.4 that an equation is a statement that two expressions have the same value. Also, a value of the variable that makes an equation a true statement is called a solution or root of the equation. The process of finding the solution of an equation is called **solving** the equation for the variable. In this section we concentrate on solving **linear equations** in one variable.

---

**Linear Equation in One Variable**

**A linear equation in one variable** can be written in the form

$$ax + b = c$$

where $a$, $b$, and $c$ are real numbers and $a \neq 0$.

---

Evaluating a linear equation for a given value of the variable, as we did in Section 1.4, can tell us whether that value is a solution, but we can't rely on evaluating an equation as our method of solving it.

Instead, to solve a linear equation in $x$, we write a series of simpler equations, all *equivalent* to the original equation, so that the final equation has the form

$$x = \textbf{number} \qquad \textbf{or} \qquad \textbf{number} = x$$

**Equivalent equations** are equations that have the same solution. This means that the "number" above is the solution to the original equation.

The first property of equality that helps us write simpler equivalent equations is the **addition property of equality.**

---

**Addition Property of Equality**

If $a$, $b$, and $c$ are real numbers, then

$$a = b \qquad \text{and} \qquad a + c = b + c$$

are equivalent equations.

---

This property guarantees that adding the same number to both sides of an equation does not change the solution of the equation. Since subtraction is defined in terms of addition, we may also **subtract the same number from both sides** without changing the solution.

A good way to picture a true equation is as a balanced scale. Since it is balanced, each side of the scale weighs the same amount.

If the same weight is added to or subtracted from each side, the scale remains balanced.

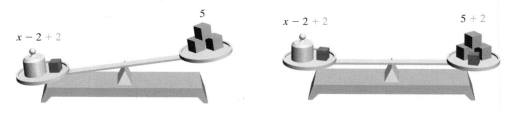

We use the addition property of equality to write equivalent equations until the variable is by itself on one side of the equation, and the equation looks like "$x$ = number" or "number = $x$."

**EXAMPLE 1**   Solve $x - 7 = 10$ for $x$.

*Solution*   To solve for $x$, we want $x$ alone on one side of the equation. To do this, we add 7 to both sides of the equation.

$$x - 7 = 10$$
$$x - 7 + 7 = 10 + 7 \quad \text{Add 7 to both sides.}$$
$$x = 17 \quad \text{Simplify.}$$

The solution of the equation $x = 17$ is obviously 17. Since we are writing equivalent equations, the solution of the equation $x - 7 = 10$ is also 17.

**Check:**   To check, replace $x$ with 17 in the original equation.

$$x - 7 = 10$$
$$17 - 7 \stackrel{?}{=} 10 \quad \text{Replace } x \text{ with 17 in the original equation.}$$
$$10 = 10 \quad \text{True}$$

Since the statement is true, 17 is the solution.   □

**PRACTICE**
**1**   Solve: $x + 3 = -5$ for $x$.

**Concept Check** ✓

Use the addition property to fill in the blank so that the middle equation simplifies to the last equation.

$$x - 5 = 3$$
$$x - 5 + \underline{\quad} = 3 + \underline{\quad}$$
$$x = 8$$

**EXAMPLE 2**   Solve $y + 0.6 = -1.0$ for $y$.

*Solution*   To get $y$ alone on one side of the equation, subtract 0.6 from both sides of the equation.

$$y + 0.6 = -1.0$$
$$y + 0.6 - 0.6 = -1.0 - 0.6 \quad \text{Subtract 0.6 from both sides.}$$
$$y = -1.6 \quad \text{Combine like terms.}$$

**Check:**   To check the proposed solution, $-1.6$, replace $y$ with $-1.6$ in the original equation.

$$y + 0.6 = -1.0$$
$$-1.6 + 0.6 \stackrel{?}{=} -1.0 \quad \text{Replace } y \text{ with } -1.6 \text{ in the original equation.}$$
$$-1.0 = -1.0 \quad \text{True}$$

The solution is $-1.6$.   □

**PRACTICE**
**2**   Solve: $y - 0.3 = -2.1$ for $y$.

Many times, it is best to simplify one or both sides of an equation before applying the addition property of equality.

**Answer to Concept Check:**   5

**EXAMPLE 3**  Solve: $2x + 3x - 5 + 7 = 10x + 3 - 6x - 4$

*Solution*  First we simplify both sides of the equation.

$$2x + 3x - 5 + 7 = 10x + 3 - 6x - 4$$
$$5x + 2 = 4x - 1 \qquad \text{Combine like terms on each side of the equation.}$$

Next, we want all terms with a variable on one side of the equation and all numbers on the other side.

$$5x + 2 - 4x = 4x - 1 - 4x \qquad \text{Subtract } 4x \text{ from both sides.}$$
$$x + 2 = -1 \qquad \text{Combine like terms.}$$
$$x + 2 - 2 = -1 - 2 \qquad \text{Subtract 2 from both sides to get } x \text{ alone.}$$
$$x = -3 \qquad \text{Combine like terms.}$$

**Check:**
$$2x + 3x - 5 + 7 = 10x + 3 - 6x - 4 \qquad \text{Original equation}$$
$$2(-3) + 3(-3) - 5 + 7 \stackrel{?}{=} 10(-3) + 3 - 6(-3) - 4 \qquad \text{Replace } x \text{ with } -3.$$
$$-6 - 9 - 5 + 7 \stackrel{?}{=} -30 + 3 + 18 - 4 \qquad \text{Multiply.}$$
$$-13 = -13 \qquad \text{True}$$

The solution is $-3$.  □

**PRACTICE**
**3**  Solve: $8x - 5x - 3 + 9 = x + x + 3 - 7$

If an equation contains parentheses, we use the distributive property to remove them.

**EXAMPLE 4**  Solve: $7 = -5(2a - 1) - (-11a + 6)$.

*Solution*
$$7 = -5(2a - 1) - (-11a + 6)$$
$$7 = -10a + 5 + 11a - 6 \qquad \text{Apply the distributive property.}$$
$$7 = a - 1 \qquad \text{Combine like terms.}$$
$$7 + 1 = a - 1 + 1 \qquad \text{Add 1 to both sides to get } a \text{ alone.}$$
$$8 = a \qquad \text{Combine like terms.}$$

Check to see that 8 is the solution.  □

**PRACTICE**
**4**  Solve: $2 = 4(2a - 3) - (7a + 4)$

> ▶ **Helpful Hint**
> We may solve an equation so that the variable is alone on either side of the equation. For example, $8 = a$ is equivalent to $a = 8$.

When solving equations, we may sometimes encounter an equation such as

$$-x = 5.$$

This equation is not solved for $x$ because $x$ is not isolated. One way to solve this equation for $x$ is to recall that

"$-$" can be read as "the opposite of."

We can read the equation $-x = 5$ then as "the opposite of $x = 5$." If the opposite of $x$ is 5, this means that $x$ is the opposite of 5 or $-5$.

In summary,

$$-x = 5 \quad \text{and} \quad x = -5$$

are equivalent equations and $x = -5$ is solved for $x$.

**OBJECTIVE 2 ▶ Using the multiplication property.** As useful as the addition property of equality is, it cannot help us solve every type of linear equation in one variable. For example, adding or subtracting a value on both sides of the equation does not help solve

$$\frac{5}{2}x = 15.$$

Instead, we apply another important property of equality, the **multiplication property of equality.**

---

**Multiplication Property of Equality**

If $a$, $b$, and $c$ are real numbers and $c \neq 0$, then

$$a = b \quad \text{and} \quad ac = bc$$

are equivalent equations.

---

This property guarantees that multiplying both sides of an equation by the same nonzero number does not change the solution of the equation. Since division is defined in terms of multiplication, we may also **divide both sides of the equation by the same nonzero number** without changing the solution.

**EXAMPLE 5** Solve: $\dfrac{5}{2}x = 15$.

_Solution_ To get $x$ alone, multiply both sides of the equation by the reciprocal of $\dfrac{5}{2}$, which is $\dfrac{2}{5}$.

$$\frac{5}{2}x = 15$$

$$\frac{2}{5} \cdot \frac{5}{2}x = \frac{2}{5} \cdot 15 \quad \text{Multiply both sides by } \frac{2}{5}.$$

$$\left(\frac{2}{5} \cdot \frac{5}{2}\right)x = \frac{2}{5} \cdot 15 \quad \text{Apply the associative property.}$$

$$1x = 6 \quad \text{Simplify.}$$

or

$$x = 6$$

**Check:** Replace $x$ with 6 in the original equation.

$$\frac{5}{2}x = 15 \quad \text{Original equation}$$

$$\frac{5}{2}(6) \stackrel{?}{=} 15 \quad \text{Replace } x \text{ with 6.}$$

$$15 = 15 \quad \text{True}$$

The solution is 6. □

> ▶ **Helpful Hint**
>
> Don't forget to multiply *both* sides by $\dfrac{2}{5}$.

**PRACTICE**
**5** Solve: $\dfrac{4}{5}x = 16$

In the equation $\frac{5}{2}x = 15$, $\frac{5}{2}$ is the coefficient of $x$. When the coefficient of $x$ is a *fraction,* we will get $x$ alone by multiplying by the reciprocal. When the coefficient of $x$ is an integer or a decimal, it is usually more convenient to divide both sides by the coefficient. (Dividing by a number is, of course, the same as multiplying by the reciprocal of the number.)

**EXAMPLE 6** Solve: $-3x = 33$

_Solution_ Recall that $-3x$ means $-3 \cdot x$. To get $x$ alone, we divide both sides by the coefficient of $x$, that is, $-3$.

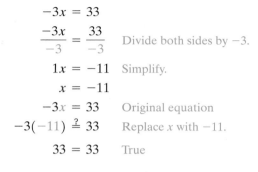

$$-3x = 33$$
$$\frac{-3x}{-3} = \frac{33}{-3} \qquad \text{Divide both sides by } -3.$$
$$1x = -11 \qquad \text{Simplify.}$$
$$x = -11$$

**Check:**
$$-3x = 33 \qquad \text{Original equation}$$
$$-3(-11) \stackrel{?}{=} 33 \qquad \text{Replace } x \text{ with } -11.$$
$$33 = 33 \qquad \text{True}$$

The solution is $-11$. □

**PRACTICE**
**6** Solve: $8x = -96$

---

**EXAMPLE 7** Solve: $\frac{y}{7} = 20$

_Solution_ Recall that $\frac{y}{7} = \frac{1}{7}y$. To get $y$ alone, we multiply both sides of the equation by 7, the reciprocal of $\frac{1}{7}$.

$$\frac{y}{7} = 20$$
$$\frac{1}{7}y = 20$$
$$7 \cdot \frac{1}{7}y = 7 \cdot 20 \qquad \text{Multiply both sides by 7.}$$
$$1y = 140 \qquad \text{Simplify.}$$
$$y = 140$$

**Check:**
$$\frac{y}{7} = 20 \qquad \text{Original equation}$$
$$\frac{140}{7} \stackrel{?}{=} 20 \qquad \text{Replace } y \text{ with 140.}$$
$$20 = 20 \qquad \text{True}$$

The solution is 140. □

**PRACTICE**
**7** Solve: $\frac{x}{5} = 13$

---

**OBJECTIVE 3 ▶ Using both the addition and multiplication properties.** Next, we practice solving equations using both properties.

**EXAMPLE 8**   Solve: $12a - 8a = 10 + 2a - 13 - 7$

*Solution*   First, simplify both sides of the equation by combining like terms.

$$12a - 8a = 10 + 2a - 13 - 7$$
$$4a = 2a - 10 \qquad \text{Combine like terms.}$$

To get all terms containing a variable on one side, subtract $2a$ from both sides.

$$4a - 2a = 2a - 10 - 2a \qquad \text{Subtract } 2a \text{ from both sides.}$$
$$2a = -10 \qquad \text{Simplify.}$$
$$\frac{2a}{2} = \frac{-10}{2} \qquad \text{Divide both sides by 2.}$$
$$a = -5 \qquad \text{Simplify.}$$

**Check:**   Check by replacing $a$ with $-5$ in the original equation. The solution is $-5$. ☐

**PRACTICE**
**8**   Solve: $6b - 11b = 18 + 2b - 6 + 9$

**OBJECTIVE 4 ▶ Writing word phrases as algebraic expressions.** Next, we practice writing word phrases as algebraic expressions.

**EXAMPLE 9**

**a.** The sum of two numbers is 8. If one number is 3, find the other number.

**b.** The sum of two numbers is 8. If one number is $x$, write an expression representing the other number.

**c.** An 8-foot board is cut into two pieces. If one piece is $x$ feet, express the length of the other piece in terms of $x$.   ✱ $(8 - x)$

*Solution*

**a.** If the sum of two numbers is 8 and one number is 3, we find the other number by subtracting 3 from 8. The other number is $8 - 3$ or 5.

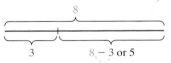

**b.** If the sum of two numbers is 8 and one number is $x$, we find the other number by subtracting $x$ from 8. The other number is represented by $8 - x$.

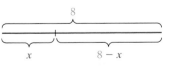

**c.** If an 8-foot board is cut into two pieces and one piece is $x$ feet, we find the other length by subtracting $x$ from 8. The other piece is $(8 - x)$ feet.

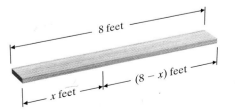

☐

PRACTICE
**9**

**a.** The sum of two numbers is 9. If one number is 2, find the other number.

**b.** The sum of two numbers is 9. If one number is $x$, write an expression representing the other number.

**c.** A 9-foot rope is cut into two pieces. If one piece is $x$ feet, express the length of the other piece in terms of $x$.

---

**EXAMPLE 10**   If $x$ is the first of three consecutive integers, express the sum of the three integers in terms of $x$. Simplify if possible.

*Solution*   An example of three consecutive integers is

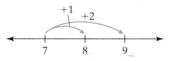

The second consecutive integer is always 1 more than the first, and the third consecutive integer is 2 more than the first. If $x$ is the first of three consecutive integers, the three consecutive integers are

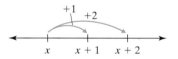

Their sum is

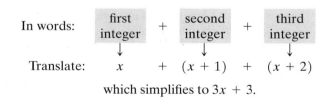

which simplifies to $3x + 3$.

PRACTICE
**10**   If $x$ is the first of three consecutive *even* integers, express their sum in terms of $x$.

---

Below are examples of consecutive even and odd integers.

*Consecutive Even integers:*

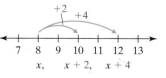

*Consecutive Odd integers:*

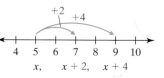

> ▶ **Helpful Hint**
> If $x$ is an odd integer, then $x + 2$ is the next odd integer. This 2 simply means that odd integers are always 2 units from each other. (The same is true for even integers. They are always 2 units from each other.)

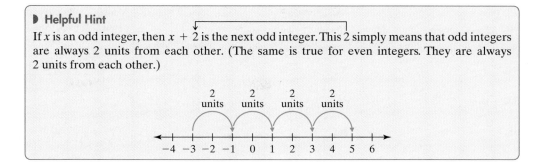

## VOCABULARY & READINESS CHECK

*Use the choices below to fill in each blank. Some choices will be used more than once.*

| | | | | |
|---|---|---|---|---|
| addition | solving | expression | true | multiplication |
| equivalent | equation | solution | false | |

1. The difference between an equation and an expression is that a(n) _____ contains an equal sign, whereas an _____ does not.

2. _____ equations are equations that have the same solution.

3. A value of the variable that makes the equation a true statement is called a(n) _____ of the equation.

4. The process of finding the solution of an equation is called _____ the equation for the variable.

5. By the _____ property of equality, $x = -2$ and $x + 10 = -2 + 10$ are equivalent equations.

6. True or false: The equations $x = \dfrac{1}{2}$ and $\dfrac{1}{2} = x$ are equivalent equations. _____

7. By the _____ property of equality, $y = \dfrac{1}{2}$ and $5 \cdot y = 5 \cdot \dfrac{1}{2}$ are equivalent equations.

8. True or false: The equations $\dfrac{z}{4} = 10$ and $4 \cdot \dfrac{z}{4} = 10$ are equivalent equations. _____

9. True or false: The equations $-7x = 30$ and $\dfrac{-7x}{-7} = \dfrac{30}{7}$ are equivalent equations. _____

10. By the _____ property of equality, $9x = -63$ and $\dfrac{9x}{9} = \dfrac{-63}{9}$ are equivalent equations.

*Solve each equation mentally.*

**11.** $3a = 27$   **12.** $9c = 54$   **13.** $5b = 10$   **14.** $7t = 14$

## 2.2 | EXERCISE SET

*Solve each equation. Check each solution. See Examples 1 and 2.*

**1.** $x + 7 = 10$    **2.** $x + 14 = 25$

**3.** $x - 2 = -4$    **4.** $y - 9 = 1$

**5.** $3 + x = -11$    **6.** $8 + z = -8$

**7.** $r - 8.6 = -8.1$    **8.** $t - 9.2 = -6.8$

**9.** $8x = 7x - 3$    **10.** $2x = x - 5$

**11.** $5b - 0.7 = 6b$    **12.** $9x + 5.5 = 10x$

**13.** $7x - 3 = 6x$    **14.** $18x - 9 = 19x$

*Solve each equation. See Examples 3 and 4.*

**15.** $3x - 6 = 2x + 5$    **16.** $7y + 2 = 6y + 2$

**17.** $3t - t - 7 = t - 7$    **18.** $4c + 8 - c = 8 + 2c$

**19.** $7x + 2x = 8x - 3$    **20.** $3n + 2n = 7 + 4n$

**21.** $-2(x + 1) + 3x = 14$

**22.** $10 = 8(3y - 4) - 23y + 20$

*Solve each equation. See Example 6.*

**23.** $-5x = 20$    **24.** $-7x = -49$

**25.** $3x = 0$    **26.** $-2x = 0$

**27.** $-x = -12$

**28.** $-y = 8$

**29.** $3x + 2x = 50$

**30.** $-y + 4y = 33$

*Solve each equation. See Examples 5 and 7.*

**31.** $\frac{2}{3}x = -8$

**32.** $\frac{3}{4}n = -15$

**33.** $\frac{1}{6}d = \frac{1}{2}$

**34.** $\frac{1}{8}v = \frac{1}{4}$

**35.** $\frac{a}{-2} = 1$

**36.** $\frac{d}{15} = 2$

**37.** $\frac{k}{7} = 0$

**38.** $\frac{f}{-5} = 0$

**39.** In your own words, explain the addition property of equality.

**40.** In your own words, explain the multiplication property of equality.

**MIXED PRACTICE**

*Solve each equation. Check each solution. See Examples 1 through 8.*

**41.** $2x - 4 = 16$

**42.** $3x - 1 = 26$

**43.** $-x + 2 = 22$

**44.** $-x + 4 = -24$

**45.** $6a + 3 = 3$

**46.** $8t + 5 = 5$

**47.** $6x + 10 = -20$

**48.** $-10y + 15 = 5$

**49.** $5 - 0.3k = 5$

**50.** $2 + 0.4p = 2$

**51.** $-2x + \frac{1}{2} = \frac{7}{2}$

**52.** $-3n - \frac{1}{3} = \frac{8}{3}$

**53.** $\frac{x}{3} + 2 = -5$

**54.** $\frac{b}{4} - 1 = -7$

**55.** $10 = 2x - 1$

**56.** $12 = 3j - 4$

**57.** $6z - 8 - z + 3 = 0$

**58.** $4a + 1 + a - 11 = 0$

**59.** $10 - 3x - 6 - 9x = 7$

**60.** $12x + 30 + 8x - 6 = 10$

**61.** $\frac{5}{6}x = 10$

**62.** $-\frac{3}{4}x = 9$

**63.** $1 = 0.4x - 0.6x - 5$

**64.** $19 = 0.4x - 0.9x - 6$

**65.** $z - 5z = 7z - 9 - z$

**66.** $t - 6t = -13 + t - 3t$

**67.** $0.4x - 0.6x - 5 = 1$

**68.** $0.4x - 0.9x - 6 = 19$

**69.** $6 - 2x + 8 = 10$

**70.** $-5 - 6y + 6 = 19$

**71.** $-3a + 6 + 5a = 7a - 8a$

**72.** $4b - 8 - b = 10b - 3b$

**73.** $20 = -3(2x + 1) + 7x$

**74.** $-3 = -5(4x + 3) + 21x$

*See Example 9.*

**75.** Two numbers have a sum of 20. If one number is $p$, express the other number in terms of $p$.

**76.** Two numbers have a sum of 13. If one number is $y$, express the other number in terms of $y$.

**77.** A 10-foot board is cut into two pieces. If one piece is $x$ feet long, express the other length in terms of $x$.

**78.** A 5-foot piece of string is cut into two pieces. If one piece is $x$ feet long, express the other length in terms of $x$.

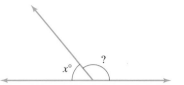

**79.** Two angles are *supplementary* if their sum is 180°. If one angle measures $x°$, express the measure of its supplement in terms of $x$.

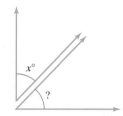

**80.** Two angles are *complementary* if their sum is 90°. If one angle measures $x°$, express the measure of its complement in terms of $x$.

**81.** In a mayoral election, April Catarella received 284 more votes than Charles Pecot. If Charles received $n$ votes, how many votes did April receive?

**82.** The length of the top of a computer desk is $1\frac{1}{2}$ feet longer than its width. If its width measures $m$ feet, express its length as an algebraic expression in $m$.

**83.** The Verrazano-Narrows Bridge in New York City is the longest suspension bridge in North America. The Golden Gate Bridge in San Francisco is 60 feet shorter than the Verrazano-Narrows Bridge. If the length of the Verrazano-Narrows Bridge is $m$ feet, express the length of the Golden Gate Bridge as an algebraic expression in $m$. (*Source: World Almanac*, 2000).

**84.** The longest interstate highway in the U.S. is I-90, which connects Seattle, Washington, and Boston, Massachusetts. The second longest interstate highway, I-80 (connecting San Francisco, California, and Teaneck, New Jersey), is 178.5 miles shorter than I-90. If the length of I-80 is $m$ miles, express the length of I-90 as an algebraic expression in $m$.

*(Source:* U.S. Department of Transportation–Federal Highway Administration)

**85.** In a recent election, Pat Ahumada ran against Solomon P. Ortiz for one of Texas's seats in the U.S. House of Representatives. Ahumada received 47,628 fewer votes than Ortiz. If Ahumada received $n$ votes, how many did Ortiz receive? *(Source:* Voter News Service)

**86.** In a recent U.S. Senate race in Maine, Susan M. Collins received 30,898 more votes than Joseph E. Brennan. If Joseph received $n$ votes, how many did Susan receive? *(Source:* Voter News Service)

**87.** The area of the Sahara Desert in Africa is 7 times the area of the Gobi Desert in Asia. If the area of the Gobi Desert is $x$ square miles, express the area of the Sahara Desert as an algebraic expression in $x$.

**88.** The largest meteorite in the world is the Hoba West located in Namibia. Its weight is 3 times the weight of the Armanty meteorite located in Outer Mongolia. If the weight of the Armanty meteorite is $y$ kilograms, express the weight of the Hoba West meteorite as an algebraic expression in $y$.

*Write each algebraic expression described. Simplify if possible. See Example 10.*

**89.** If $x$ represents the first of two consecutive odd integers, express the sum of the two integers in terms of $x$.

**90.** If $x$ is the first of four consecutive even integers, write their sum as an algebraic expression in $x$.

**91.** If $x$ is the first of four consecutive integers, express the sum of the first integer and the third integer as an algebraic expression containing the variable $x$.

**92.** If $x$ is the first of two consecutive integers, express the sum of 20 and the second consecutive integer as an algebraic expression containing the variable $x$.

**93.** Classrooms on one side of the science building are all numbered with consecutive even integers. If the first room on this side of the building is numbered $x$, write an expression in $x$

for the sum of five classroom numbers in a row. Then simplify this expression.

**94.** Two sides of a quadrilateral have the same length, $x$, while the other two sides have the same length, both being the next consecutive odd integer. Write the sum of these lengths. Then simplify this expression.

### REVIEW AND PREVIEW

*Simplify each expression. See Section 2.1.*

**95.** $5x + 2(x - 6)$

**96.** $-7y + 2y - 3(y + 1)$

**97.** $-(x - 1) + x$

**98.** $-(3a - 3) + 2a - 6$

*Insert* $<, >,$ *or* $=$ *in the appropriate space to make each statement true. See Sections 1.2 and 1.7.*

**99.** $(-3)^2 > -3^2$      **100.** $(-2)^4 \quad -2^4$

**101.** $(-2)^3 \quad -2^3$      **102.** $(-4)^3 \quad -4^3$

### CONCEPT EXTENSIONS

△ **103.** The sum of the angles of a triangle is $180°$. If one angle of a triangle measures $x°$ and a second angle measures $(2x + 7)°$, express the measure of the third angle in terms of $x$. Simplify the expression.

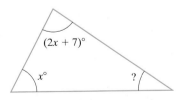

△ **104.** A quadrilateral is a four-sided figure like the one shown below whose angle sum is $360°$. If one angle measures $x°$, a second angle measures $3x°$, and a third angle measures $5x°$, express the measure of the fourth angle in terms of $x$. Simplify the expression.

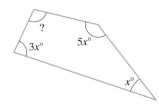

**105.** Write two terms whose sum is $-3x$.

**106.** Write four terms whose sum is $2y - 6$.

*Use the addition property to fill in the blank so that the middle equation simplifies to the last equation. See the Concept Check in this section.*

**107.**
$$x - 4 = -9$$
$$x - 4 + (\quad) = -9 + (\quad)$$
$$x = -5$$

**108.**
$$a + 9 = 15$$
$$a + 9 + (\quad) = 15 + (\quad)$$
$$a = 6$$

*Fill in the blanks with numbers of your choice so that each equation has the given solution.* Note: *Each blank may be replaced with a different number.*

**109** _____ $+ x =$ _____ ; Solution: $-3$

**110.** $x -$ _____ $=$ _____ ; Solution: $-10$

**111.** Let $x = 1$ and then $x = 2$ in the equation $x + 5 = x + 6$. Is either number a solution? How many solutions do you think this equation has? Explain your answer.

**112.** Let $x = 1$ and then $x = 2$ in the equation $x + 3 = x + 3$. Is either number a solution? How many solutions do you think this equation has? Explain your answer.

*Fill in the blank with a number so that each equation has the given solution.*

**113.** $6x =$ _____ ; solution: $-8$

**114.** _____ $x = 10$; solution: $\dfrac{1}{2}$

**115.** A licensed nurse practitioner is instructed to give a patient 2100 milligrams of an antibiotic over a period of 36 hours. If the antibiotic is to be given every 4 hours starting immediately, how much antibiotic should be given in each dose? To answer this question, solve the equation $9x = 2100$.

**116.** Suppose you are a pharmacist and a customer asks you the following question. His child is to receive 13.5 milliliters of a nausea medicine over a period of 54 hours. If the nausea medicine is to be administered every 6 hours starting immediately, how much medicine should be given in each dose?

*Use a calculator to determine whether the given value is a solution of the given equation.*

**117.** $8.13 + 5.85y = 20.05y - 8.91$; $y = 1.2$

**118.** $3(a + 4.6) = 5a + 2.5$; $a = 6.3$

*Solve each equation.*

**119.** $-3.6x = 10.62$

**120.** $4.95y = -31.185$

**121.** $7x - 5.06 = -4.92$

**122.** $0.06y + 2.63 = 2.5562$

---

### 📖 STUDY SKILLS BUILDER

**How Are Your Homework Assignments Going?**

It is very important in mathematics to keep up with homework. Why? Many concepts build on each other. Often your understanding of a day's concepts depends on an understanding of the previous day's material.

Remember that completing your homework assignment involves a lot more than attempting a few of the problems assigned.

To complete a homework assignment, remember these four things:

- Attempt all of it.
- Check it.
- Correct it.
- If needed, ask questions about it.

**Self-Check**

*Take a moment and review your completed homework assignments. Answer the questions below based on this review.*

**1.** Approximate the fraction of your homework you have attempted.

**2.** Approximate the fraction of your homework you have checked (if possible).

**3.** If you are able to check your homework, have you corrected it when errors have been found?

**4.** When working homework, if you do not understand a concept, what do you do?

## 2.3  SOLVING LINEAR EQUATIONS

### OBJECTIVES

1  Apply a general strategy for solving a linear equation.

2  Solve equations containing fractions.

3  Solve equations containing decimals.

4  Recognize identities and equations with no solution.

**OBJECTIVE 1 ▶ Applying a general strategy for solving a linear equation.** We now present a general strategy for solving linear equations. One new piece of strategy is a suggestion to "clear an equation of fractions" as a first step. Doing so makes the equation more manageable, since operating on integers is more convenient than operating on fractions.

> **Solving Linear Equations in One Variable**
>
> **STEP 1.** Multiply on both sides by the LCD to clear the equation of fractions if they occur.
>
> **STEP 2.** Use the distributive property to remove parentheses if they occur.
>
> **STEP 3.** Simplify each side of the equation by combining like terms.
>
> **STEP 4.** Get all variable terms on one side and all numbers on the other side by using the addition property of equality.
>
> **STEP 5.** Get the variable alone by using the multiplication property of equality.
>
> **STEP 6.** Check the solution by substituting it into the original equation.

**EXAMPLE 1**   Solve: $4(2x - 3) + 7 = 3x + 5$

*Solution*   There are no fractions, so we begin with Step 2.

$$4(2x - 3) + 7 = 3x + 5$$

**STEP 2.**  $8x - 12 + 7 = 3x + 5$   Apply the distributive property.

**STEP 3.**  $8x - 5 = 3x + 5$   Combine like terms.

**STEP 4.** Get all variable terms on the same side of the equation by subtracting $3x$ from both sides, then adding 5 to both sides.

$$8x - 5 - 3x = 3x + 5 - 3x \quad \text{Subtract } 3x \text{ from both sides.}$$
$$5x - 5 = 5 \quad \text{Simplify.}$$
$$5x - 5 + 5 = 5 + 5 \quad \text{Add 5 to both sides.}$$
$$5x = 10 \quad \text{Simplify.}$$

**STEP 5.** Use the multiplication property of equality to get $x$ alone.

$$\frac{5x}{5} = \frac{10}{5} \quad \text{Divide both sides by 5.}$$
$$x = 2 \quad \text{Simplify.}$$

**STEP 6.** Check.

> **▶ Helpful Hint**
>
> When checking solutions, remember to use the original written equation.

$$4(2x - 3) + 7 = 3x + 5 \quad \text{Original equation}$$
$$4[2(2) - 3] + 7 \stackrel{?}{=} 3(2) + 5 \quad \text{Replace } x \text{ with 2.}$$
$$4(4 - 3) + 7 \stackrel{?}{=} 6 + 5$$
$$4(1) + 7 \stackrel{?}{=} 11$$
$$4 + 7 \stackrel{?}{=} 11$$
$$11 = 11 \quad \text{True}$$

The solution is 2 or the solution set is $\{2\}$.

**PRACTICE**
**1**   Solve: $2(4a - 9) + 3 = 5a - 6$

**EXAMPLE 2**    Solve: $8(2 - t) = -5t$

*Solution*    First, we apply the distributive property.

$$\overset{\frown}{8(2 - t)} = -5t$$

**STEP 2.**      $16 - 8t = -5t$      Use the distributive property.

**STEP 4.**    $16 - 8t + 8t = -5t + 8t$    To get variable terms on one side, add $8t$ to both sides.

                  $16 = 3t$      Combine like terms.

**STEP 5.**         $\dfrac{16}{3} = \dfrac{3t}{3}$      Divide both sides by 3.

           $\dfrac{16}{3} = t$      Simplify.

**STEP 6.**   Check.

$$8(2 - t) = -5t \qquad \text{Original equation}$$

$$8\left(2 - \frac{16}{3}\right) \overset{?}{=} -5\left(\frac{16}{3}\right) \qquad \text{Replace } t \text{ with } \frac{16}{3}.$$

$$8\left(\frac{6}{3} - \frac{16}{3}\right) \overset{?}{=} -\frac{80}{3} \qquad \text{The LCD is 3.}$$

$$8\left(-\frac{10}{3}\right) \overset{?}{=} -\frac{80}{3} \qquad \text{Subtract fractions.}$$

$$-\frac{80}{3} = -\frac{80}{3} \qquad \text{True}$$

The solution is $\dfrac{16}{3}$.

**PRACTICE**
**2**    Solve: $7(x - 3) = -6x$

**OBJECTIVE 2** ▶ **Solving equations containing fractions.** If an equation contains fractions, we can clear the equation of fractions by multiplying both sides by the LCD of all denominators. By doing this, we avoid working with time-consuming fractions.

**EXAMPLE 3**    Solve: $\dfrac{x}{2} - 1 = \dfrac{2}{3}x - 3$

*Solution*    We begin by clearing fractions. To do this, we multiply both sides of the equation by the LCD of 2 and 3, which is 6.

$$\frac{x}{2} - 1 = \frac{2}{3}x - 3$$

**STEP 1.**      $6\left(\dfrac{x}{2} - 1\right) = 6\left(\dfrac{2}{3}x - 3\right)$    Multiply both sides by the LCD, 6.

**STEP 2.**      $6\left(\dfrac{x}{2}\right) - 6(1) = 6\left(\dfrac{2}{3}x\right) - 6(3)$    Apply the distributive property.

                $3x - 6 = 4x - 18$    Simplify.

> ▶ **Helpful Hint**
> Don't forget to multiply *each* term by the LCD.

There are no longer grouping symbols and no like terms on either side of the equation, so we continue with Step 4.

$$3x - 6 = 4x - 18$$

**STEP 4.** $3x - 6 - 3x = 4x - 18 - 3x$    To get variable terms on one side, subtract $3x$ from both sides.

$$-6 = x - 18$$    Simplify.

$$-6 + 18 = x - 18 + 18$$    Add 18 to both sides.

$$12 = x$$    Simplify.

**STEP 5.** The variable is now alone, so there is no need to apply the multiplication property of equality.

**STEP 6.** Check.

$$\frac{x}{2} - 1 = \frac{2}{3}x - 3$$    Original equation

$$\frac{12}{2} - 1 \overset{?}{=} \frac{2}{3} \cdot 12 - 3$$    Replace $x$ with 12.

$$6 - 1 \overset{?}{=} 8 - 3$$    Simplify.

$$5 = 5$$    True

The solution is 12. □

**PRACTICE**
**3**   Solve: $\dfrac{3}{5}x - 2 = \dfrac{2}{3}x - 1$

 *study*

**EXAMPLE 4**   Solve: $\dfrac{2(a + 3)}{3} = 6a + 2$

*Solution*   We clear the equation of fractions first.

$$\frac{2(a + 3)}{3} = 6a + 2$$

**STEP 1.** $3 \cdot \dfrac{2(a + 3)}{3} = 3(6a + 2)$    Clear the fraction by multiplying both sides by the LCD, 3.

$$2(a + 3) = 3(6a + 2)$$

**STEP 2.** Next, we use the distributive property and remove parentheses.

$$2a + 6 = 18a + 6$$    Apply the distributive property.

**STEP 4.** $2a + 6 - 6 = 18a + 6 - 6$    Subtract 6 from both sides.

$$2a = 18a$$

$$2a - 18a = 18a - 18a$$    Subtract $18a$ from both sides.

$$-16a = 0$$

**STEP 5.** $\dfrac{-16a}{-16} = \dfrac{0}{-16}$    Divide both sides by $-16$.

$$a = 0$$    Write the fraction in simplest form.

**STEP 6.** To check, replace $a$ with 0 in the original equation. The solution is 0. □

**PRACTICE**
**4**   Solve: $\dfrac{4(y + 3)}{3} = 5y - 7$

**OBJECTIVE 3 ▶ Solving equations containing decimals.** When solving a problem about money, you may need to solve an equation containing decimals. If you choose, you may multiply to clear the equation of decimals.

**EXAMPLE 5** Solve: $0.25x + 0.10(x - 3) = 0.05(22)$

*Solution* First we clear this equation of decimals by multiplying both sides of the equation by 100. Recall that multiplying a decimal number by 100 has the effect of moving the decimal point 2 places to the right.

$$0.25x + 0.10(x - 3) = 0.05(22)$$

> **Helpful Hint**
>
> By the distributive property, 0.10 is multiplied by $x$ and $-3$. Thus to multiply each term here by 100, we only need to multiply 0.10 by 100.

**STEP 1.** $0.25x + 0.10(x - 3) = 0.05(22)$    Multiply both sides by 100.

$$25x + 10(x - 3) = 5(22)$$

**STEP 2.** $25x + 10x - 30 = 110$    Apply the distributive property.

**STEP 3.** $35x - 30 = 110$    Combine like terms.

**STEP 4.** $35x - 30 + 30 = 110 + 30$    Add 30 to both sides.

$$35x = 140$$    Combine like terms.

**STEP 5.** $\dfrac{35x}{35} = \dfrac{140}{35}$    Divide both sides by 35.

$$x = 4$$

**STEP 6.** To check, replace $x$ with 4 in the original equation. The solution is 4.   ☐

**PRACTICE**
**5**   Solve: $0.35x + 0.09(x + 4) = 0.03(12)$

---

**OBJECTIVE 4 ▶ Recognizing identities and equations with no solution.** So far, each equation that we have solved has had a single solution. However, not every equation in one variable has a single solution. Some equations have no solution, while others have an infinite number of solutions. For example,

$$x + 5 = x + 7$$

has no solution since no matter which **real number** we replace $x$ with, the equation is false.

     **real number** $+ 5 =$ same **real number** $+ 7$    **FALSE**

On the other hand,

$$x + 6 = x + 6$$

has infinitely many solutions since $x$ can be replaced by any real number and the equation is always true.

     **real number** $+ 6 =$ same **real number** $+ 6$    **TRUE**

The equation $x + 6 = x + 6$ is called an **identity.** The next few examples illustrate special equations like these.

**EXAMPLE 6** Solve: $-2(x - 5) + 10 = -3(x + 2) + x$

*Solution*

$$-2(x - 5) + 10 = -3(x + 2) + x$$
$$-2x + 10 + 10 = -3x - 6 + x$$    Apply the distributive property on both sides.
$$-2x + 20 = -2x - 6$$    Combine like terms.
$$-2x + 20 + 2x = -2x - 6 + 2x$$    Add $2x$ to both sides.
$$20 = -6$$    Combine like terms.

The final equation contains no variable terms, and there is no value for $x$ that makes $20 = -6$ a true equation. We conclude that there is **no solution** to this equation. In set notation, we can indicate that there is no solution with the empty set, { }, or use the empty set or null set symbol, $\varnothing$. In this chapter, we will simply write *no solution*.   ☐

**PRACTICE**
**6**   Solve: $4(x + 4) - x = 2(x + 11) + x$

**EXAMPLE 7**   Solve: $3(x - 4) = 3x - 12$

*Solution*

$$3(x - 4) = 3x - 12$$
$$3x - 12 = 3x - 12 \quad \text{Apply the distributive property.}$$

The left side of the equation is now identical to the right side. Every real number may be substituted for $x$ and a true statement will result. We arrive at the same conclusion if we continue.

$$3x - 12 = 3x - 12$$
$$3x - 12 + 12 = 3x - 12 + 12 \quad \text{Add 12 to both sides.}$$
$$3x = 3x \quad\quad\quad\quad\quad \text{Combine like terms.}$$
$$3x - 3x = 3x - 3x \quad\quad\quad \text{Subtract } 3x \text{ from both sides.}$$
$$0 = 0$$

Again, one side of the equation is identical to the other side. Thus, $3(x - 4) = 3x - 12$ is an **identity** and **all real numbers** are solutions. In set notation, this is {all real numbers}.

□

**PRACTICE**
**7**   Solve: $12x - 18 = 9(x - 2) + 3x$

**Answers to Concept Check:**
**a.** Every real number is a solution.
**b.** The solution is 0.
**c.** There is no solution.

**Concept Check** ☑

Suppose you have simplified several equations and obtain the following results. What can you conclude about the solutions to the original equation?

**a.** $7 = 7$        **b.** $x = 0$        **c.** $7 = -4$

---

**Calculator Explorations**

**Checking Equations**

We can use a calculator to check possible solutions of equations. To do this, replace the variable by the possible solution and evaluate both sides of the equation separately.

*Equation:*   $3x - 4 = 2(x + 6)$        *Solution: $x = 16$*

$$3x - 4 = 2(x + 6) \quad \text{Original equation}$$
$$3(16) - 4 \stackrel{?}{=} 2(16 + 6) \quad \text{Replace } x \text{ with 16.}$$

Now evaluate each side with your calculator.

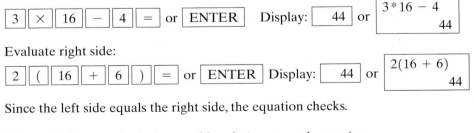

Since the left side equals the right side, the equation checks.

*Use a calculator to check the possible solutions to each equation.*

**1.** $2x = 48 + 6x; \quad x = -12$        **2.** $-3x - 7 = 3x - 1; \quad x = -1$

**3.** $5x - 2.6 = 2(x + 0.8); \quad x = 4.4$        **4.** $-1.6x - 3.9 = -6.9x - 25.6; \quad x = 5$

**5.** $\dfrac{564x}{4} = 200x - 11(649); \quad x = 121$        **6.** $20(x - 39) = 5x - 432; \quad x = 23.2$

## VOCABULARY & READINESS CHECK

*Throughout algebra, it is important to be able to identify equations and expressions.*

Remember,
- an equation contains an equals sign and
- an expression does not.

Among other things,
- we solve equations and
- we simplify or perform operations on expressions.

*Identify each as an equation or an expression.*

**1.** $x = -7$ _____

**2.** $x - 7$ _____

**3.** $4y - 6 + 9y + 1$ _____

**4.** $4y - 6 = 9y + 1$ _____

**5.** $\dfrac{1}{x} - \dfrac{x-1}{8}$ _____

**6.** $\dfrac{1}{x} - \dfrac{x-1}{8} = 6$ _____

**7.** $0.1x + 9 = 0.2x$ _____

**8.** $0.1x^2 + 9y - 0.2x^2$ _____

## 2.3 | EXERCISE SET

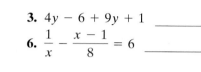

*Solve each equation. See Examples 1 and 2.*

**1.** $-4y + 10 = -2(3y + 1)$

**2.** $-3x + 1 = -2(4x + 2)$

**3.** $15x - 8 = 10 + 9x$

**4.** $15x - 5 = 7 + 12x$

**5.** $-2(3x - 4) = 2x$

**6.** $-(5x - 10) = 5x$

**7.** $5(2x - 1) - 2(3x) = 1$

**8.** $3(2 - 5x) + 4(6x) = 12$

**9.** $-6(x - 3) - 26 = -8$

**10.** $-4(n - 4) - 23 = -7$

**11.** $8 - 2(a + 1) = 9 + a$

**12.** $5 - 6(2 + b) = b - 14$

**13.** $4x + 3 = -3 + 2x + 14$

**14.** $6y - 8 = -6 + 3y + 13$

**15.** $-2y - 10 = 5y + 18$

**16.** $-7n + 5 = 8n - 10$

*Solve each equation. See Examples 3 through 5.*

**17.** $\dfrac{2}{3}x + \dfrac{4}{3} = -\dfrac{2}{3}$

**18.** $\dfrac{4}{5}x - \dfrac{8}{5} = -\dfrac{16}{5}$

**19.** $\dfrac{3}{4}x - \dfrac{1}{2} = 1$

**20.** $\dfrac{2}{9}x - \dfrac{1}{3} = 1$

**21.** $0.50x + 0.15(70) = 35.5$

**22.** $0.40x + 0.06(30) = 9.8$

**23.** $\dfrac{2(x + 1)}{4} = 3x - 2$

**24.** $\dfrac{3(y + 3)}{5} = 2y + 6$

**25.** $x + \dfrac{7}{6} = 2x - \dfrac{7}{6}$

**26.** $\dfrac{5}{2}x - 1 = x + \dfrac{1}{4}$

**27.** $0.12(y - 6) + 0.06y = 0.08y - 0.7$

**28.** $0.60(z - 300) + 0.05z = 0.70z - 205$

*Solve each equation. See Examples 6 and 7.*

**29.** $4(3x + 2) = 12x + 8$

**30.** $14x + 7 = 7(2x + 1)$

**31.** $\dfrac{x}{4} + 1 = \dfrac{x}{4}$

**32.** $\dfrac{x}{3} - 2 = \dfrac{x}{3}$

**33.** $3x - 7 = 3(x + 1)$

**34.** $2(x - 5) = 2x + 10$

**35.** $-2(6x - 5) + 4 = -12x + 14$

**36.** $-5(4y - 3) + 2 = -20y + 17$

### MIXED PRACTICE

*Solve. See Examples 1 through 7.*

**37.** $\dfrac{6(3 - z)}{5} = -z$

**38.** $\dfrac{4(5 - w)}{3} = -w$

**39.** $-3(2t - 5) + 2t = 5t - 4$

**40.** $-(4a - 7) - 5a = 10 + a$

**41.** $5y + 2(y - 6) = 4(y + 1) - 2$

**42.** $9x + 3(x - 4) = 10(x - 5) + 7$

**43.** $\dfrac{3(x - 5)}{2} = \dfrac{2(x + 5)}{3}$

**44.** $\dfrac{5(x - 1)}{4} = \dfrac{3(x + 1)}{2}$

**45.** $0.7x - 2.3 = 0.5$

**46.** $0.9x - 4.1 = 0.4$

**47.** $5x - 5 = 2(x + 1) + 3x - 7$

**48.** $3(2x - 1) + 5 = 6x + 2$

**49.** $4(2n + 1) = 3(6n + 3) + 1$

**50.** $4(4y + 2) = 2(1 + 6y) + 8$

**51.** $x + \dfrac{5}{4} = \dfrac{3}{4}x$

**52.** $\dfrac{7}{8}x + \dfrac{1}{4} = \dfrac{3}{4}x$

**53.** $\dfrac{x}{2} - 1 = \dfrac{x}{5} + 2$

**54.** $\dfrac{x}{5} - 7 = \dfrac{x}{3} - 5$

**55.** $2(x + 3) - 5 = 5x - 3(1 + x)$

**56.** $4(2 + x) + 1 = 7x - 3(x - 2)$

**57.** $0.06 - 0.01(x + 1) = -0.02(2 - x)$

**58.** $-0.01(5x + 4) = 0.04 - 0.01(x + 4)$

**59.** $\dfrac{9}{2} + \dfrac{5}{2}y = 2y - 4$

**60.** $3 - \dfrac{1}{2}x = 5x - 8$

**61.** $-2y - 10 = 5y + 18$

**62.** $7n + 5 = 10n - 10$

**63.** $0.6x - 0.1 = 0.5x + 0.2$

**64.** $0.2x - 0.1 = 0.6x - 2.1$

**65.** $0.02(6t - 3) = 0.12(t - 2) + 0.18$

**66.** $0.03(2m + 7) = 0.06(5 + m) - 0.09$

**REVIEW AND PREVIEW**

*Write each phrase as an algebraic expression. Use x for the unknown number. See Section 2.1.*

**67.** A number subtracted from $-8$

**68.** Three times a number

**69.** The sum of $-3$ and twice a number

**70.** The difference of 8 and twice a number

**71.** The product of 9 and the sum of a number and 20

**72.** The quotient of $-12$ and the difference of a number and 3

*See Section 2.1.*

**73.** A plot of land is in the shape of a triangle. If one side is $x$ meters, a second side is $(2x - 3)$ meters and a third side is $(3x - 5)$ meters, express the perimeter of the lot as a simplified expression in $x$.

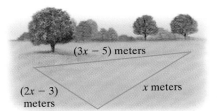

(3x − 5) meters

(2x − 3) meters    x meters

**74.** A portion of a board has length $x$ feet. The other part has length $(7x - 9)$ feet. Express the total length of the board as a simplified expression in $x$.

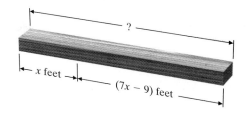

? 

x feet     (7x − 9) feet

**CONCEPT EXTENSIONS**

*See the Concept Check in this section.*

**75. a.** Solve: $x + 3 = x + 3$

   **b.** If you simplify an equation and get $0 = 0$, what can you conclude about the solution(s) of the original equation?

   **c.** On your own, construct an equation for which every real number is a solution.

**76. a.** Solve: $x + 3 = x + 5$

   **b.** If you simplify an equation and get $3 = 5$, what can you conclude about the solution(s) of the original equation?

   **c.** On your own, construct an equation that has no solution.

*Match each equation in the first column with its solution in the second column. Items in the second column may be used more than once.*

**77.** $5x + 1 = 5x + 1$

**78.** $3x + 1 = 3x + 2$

**79.** $2x - 6x - 10 = -4x + 3 - 10$

**80.** $x - 11x - 3 = -10x - 1 - 2$

**81.** $9x - 20 = 8x - 20$

**82.** $-x + 15 = x + 15$

**a.** all real numbers

**b.** no solution

**c.** 0

**83.** Explain the difference between simplifying an expression and solving an equation.

**84.** On your own, write an expression and then an equation. Label each.

*For Exercises 85 and 86, **a.** Write an equation for perimeter. **b.** Solve the equation in part (a). **c.** Find the length of each side.*

**85.** The perimeter of a geometric figure is the sum of the lengths of its sides. The perimeter of the following pentagon (five-sided figure) is 28 centimeters.

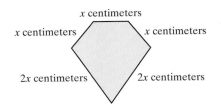

x centimeters

x centimeters          x centimeters

2x centimeters        2x centimeters

△ **86.** The perimeter of the following triangle is 35 meters.

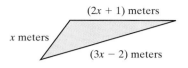

(2x + 1) meters

x meters

(3x − 2) meters

*Fill in the blanks with numbers of your choice so that each equation has the given solution. Note: Each blank may be replaced by a different number.*

**87.** x + ___ = 2x − ___; solution: 9

**88.** −5x − ___ = ___; solution: 2

*Solve.*

**89.** 1000(7x − 10) = 50(412 + 100x)

**90.** 1000(x + 40) = 100(16 + 7x)

**91.** 0.035x + 5.112 = 0.010x + 5.107

**92.** 0.127x − 2.685 = 0.027x − 2.38

*For Exercises 93 through 96, see the example below.*

**Example**

Solve: $t(t + 4) = t^2 − 2t + 6$.

*Solution*

$$t(t + 4) = t^2 − 2t + 6$$
$$t^2 + 4t = t^2 − 2t + 6$$
$$t^2 + 4t − t^2 = t^2 − 2t + 6 − t^2$$
$$4t = −2t + 6$$
$$4t + 2t = −2t + 6 + 2t$$
$$6t = 6$$
$$t = 1$$

*Solve each equation.*

**93.** $x(x − 3) = x^2 + 5x + 7$

**94.** $t^2 − 6t = t(8 + t)$

**95.** $2z(z + 6) = 2z^2 + 12z − 8$

**96.** $y^2 − 4y + 10 = y(y − 5)$

---

## THE BIGGER PICTURE  SIMPLIFYING EXPRESSIONS AND SOLVING EQUATIONS

Now we continue our outline started in Section 1.7. Although suggestions are given, this outline should be in your own words. Once you complete this new portion, try the exercises to the right.

**I.** Simplifying Expressions

   **A.** Real Numbers

      **1.** Add (Section 1.5)

      **2.** Subtract (Section 1.6)

      **3.** Multiply or Divide (Section 1.7)

**II.** **Solving Equations**

   **A.** **Linear Equations:** power on variable is 1 and there are no variables in the denominator

$$7(x − 3) = 4x + 6$$    Linear equation. Simplify both sides, then get variable terms on one side, numbers on the other side.

$$7x − 21 = 4x + 6$$    Use the distributive property.

$$7x = 4x + 27$$    Add 21 to both sides.

$$3x = 27$$    Subtract 4x from both sides.

$$x = 9$$    Divide both sides by 3.

*Solve.*

**1.** $3x − 4 = 3(2x − 1) + 7$

**2.** $5 + 2x = 5(x + 1)$

**3.** $\dfrac{x + 3}{2} = 1$

**4.** $\dfrac{x − 2}{2} − \dfrac{x − 4}{3} = \dfrac{5}{6}$

**5.** $\dfrac{7}{5} + \dfrac{y}{10} = 2$

**6.** $5 + 2x = 2(x + 1)$

**7.** $4(x − 2) + 3x = 9(x − 1) − 2$

**8.** $6(x + 1) − 2 = 6x + 4$

# INTEGRATED REVIEW  SOLVING LINEAR EQUATIONS

Sections 2.1–2.3

*Solve. Feel free to use the steps given in Section 2.3.*

**1.** $x - 10 = -4$

**2.** $y + 14 = -3$

**3.** $9y = 108$

**4.** $-3x = 78$

**5.** $-6x + 7 = 25$

**6.** $5y - 42 = -47$

**7.** $\dfrac{2}{3}x = 9$

**8.** $\dfrac{4}{5}z = 10$

**9.** $\dfrac{r}{-4} = -2$

**10.** $\dfrac{y}{-8} = 8$

**11.** $6 - 2x + 8 = 10$

**12.** $-5 - 6y + 6 = 19$

**13.** $2x - 7 = 2x - 27$

**14.** $3 + 8y = 8y - 2$

**15.** $-3a + 6 + 5a = 7a - 8a$

**16.** $4b - 8 - b = 10b - 3b$

**17.** $-\dfrac{2}{3}x = \dfrac{5}{9}$

**18.** $-\dfrac{3}{8}y = -\dfrac{1}{16}$

**19.** $10 = -6n + 16$

**20.** $-5 = -2m + 7$

**21.** $3(5c - 1) - 2 = 13c + 3$

**22.** $4(3t + 4) - 20 = 3 + 5t$

**23.** $\dfrac{2(z + 3)}{3} = 5 - z$

**24.** $\dfrac{3(w + 2)}{4} = 2w + 3$

**25.** $-2(2x - 5) = -3x + 7 - x + 3$

**26.** $-4(5x - 2) = -12x + 4 - 8x + 4$

**27.** $0.02(6t - 3) = 0.04(t - 2) + 0.02$

**28.** $0.03(m + 7) = 0.02(5 - m) + 0.03$

**29.** $-3y = \dfrac{4(y - 1)}{5}$

**30.** $-4x = \dfrac{5(1 - x)}{6}$

**31.** $\dfrac{5}{3}x - \dfrac{7}{3} = x$

**32.** $\dfrac{7}{5}n + \dfrac{3}{5} = -n$

**33.** $\dfrac{1}{10}(3x - 7) = \dfrac{3}{10}x + 5$

**34.** $\dfrac{1}{7}(2x - 5) = \dfrac{2}{7}x + 1$

**35.** $5 + 2(3x - 6) = -4(6x - 7)$

**36.** $3 + 5(2x - 4) = -7(5x + 2)$

---

## 2.4  AN INTRODUCTION TO PROBLEM SOLVING

### OBJECTIVES

Apply the steps for problem solving as we

**1** Solve problems involving direct translations.

**2** Solve problems involving relationships among unknown quantities.

**3** Solve problems involving consecutive integers.

**OBJECTIVE 1 ▶ Solving direct translation problems.** In previous sections, you practiced writing word phrases and sentences as algebraic expressions and equations to help prepare for problem solving. We now use these translations to help write equations that model a problem. The problem-solving steps given next may be helpful.

**General Strategy for Problem Solving**

**1.** UNDERSTAND the problem. During this step, become comfortable with the problem. Some ways of doing this are:

Read and reread the problem.

Choose a variable to represent the unknown.

Construct a drawing, whenever possible.

Propose a solution and check. Pay careful attention to how you check your proposed solution. This will help when writing an equation to model the problem.

**2.** TRANSLATE the problem into an equation.

**3.** SOLVE the equation.

**4.** INTERPRET the results: *Check* the proposed solution in the stated problem and state your conclusion.

Much of problem solving involves a direct translation from a sentence to an equation.

**EXAMPLE 1** **Finding an Unknown Number**

Twice a number, added to seven, is the same as three subtracted from the number. Find the number.

_Solution_ Translate the sentence into an equation and solve.

| In words: | twice a number | added to | seven | is the same as | three subtracted from the number |
|---|---|---|---|---|---|
| | ↓ | ↓ | ↓ | ↓ | ↓ |
| Translate: | $2x$ | $+$ | $7$ | $=$ | $x - 3$ |

To solve, begin by subtracting $x$ from both sides to isolate the variable term.

$$2x + 7 = x - 3$$
$$2x + 7 - x = x - 3 - x \quad \text{Subtract } x \text{ from both sides.}$$
$$x + 7 = -3 \quad \text{Combine like terms.}$$
$$x + 7 - 7 = -3 - 7 \quad \text{Subtract 7 from both sides.}$$
$$x = -10 \quad \text{Combine like terms.}$$

Check the solution in the problem as it was originally stated. To do so, replace "number" in the sentence with $-10$. Twice "$-10$" added to 7 is the same as 3 subtracted from "$-10$."

$$2(-10) + 7 = -10 - 3$$
$$-13 = -13$$

The unknown number is $-10$. ☐

**PRACTICE**
**1** Three times a number, minus 6, is the same as two times a number, plus 3. Find the number.

> ▶ **Helpful Hint**
> When checking solutions, go back to the original stated problem, rather than to your equation in case errors have been made in translating to an equation.

**EXAMPLE 2** **Finding an Unknown Number**

Twice the sum of a number and 4 is the same as four times the number, decreased by 12. Find the number.

_Solution_

1. UNDERSTAND. Read and reread the problem. If we let

$$x = \text{the unknown number, then}$$

"the sum of a number and 4" translates to "$x + 4$" and "four times the number" translates to "$4x$."

2. TRANSLATE.

| twice | sum of a number and 4 | is the same as | four times the number | decreased by | 12 |
|---|---|---|---|---|---|
| ↓ | ↓ | ↓ | ↓ | ↓ | ↓ |
| $2$ | $(x + 4)$ | $=$ | $4x$ | $-$ | $12$ |

**3.** SOLVE.

$$2(x + 4) = 4x - 12$$

$$2x + 8 = 4x - 12 \qquad \text{Apply the distributive property.}$$

$$2x + 8 - 4x = 4x - 12 - 4x \quad \text{Subtract } 4x \text{ from both sides.}$$

$$-2x + 8 = -12$$

$$-2x + 8 - 8 = -12 - 8 \qquad \text{Subtract 8 from both sides.}$$

$$-2x = -20$$

$$\frac{-2x}{-2} = \frac{-20}{-2} \qquad \text{Divide both sides by } -2.$$

$$x = 10$$

**4.** INTERPRET.

**Check:**  Check this solution in the problem as it was originally stated. To do so, replace "number" with 10. Twice the sum of "10" and 4 is 28, which is the same as 4 times "10" decreased by 12.

**State:**  The number is 10. ☐

**PRACTICE**

**2**   Three times a number, decreased by 4, is the same as double the difference of the number and 1.

**OBJECTIVE 2 ▶ Solving problems involving relationships among unknown quantities.**
The next three examples have to do with relationships among unknown quantities.

**EXAMPLE 3**   **Finding the Length of a Board**

Balsa wood sticks are commonly used for building models (for example, bridge models). A 48-inch Balsa wood stick is to be cut into two pieces so that the longer piece is 3 times the shorter. Find the length of each piece.

*Solution*

**1.** UNDERSTAND the problem. To do so, read and reread the problem. You may also want to propose a solution. For example, if 10 inches represents the length of the shorter piece, then $3(10) = 30$ inches is the length of the longer piece, since it is 3 times the length of the shorter piece. This guess gives a total board length of 10 inches + 30 inches = 40 inches, too short. However, the purpose of proposing a solution is not to guess correctly, but to help better understand the problem and how to model it.

   Since the length of the longer piece is given in terms of the length of the shorter piece, let's let

$$x = \text{length of shorter piece, then}$$
$$3x = \text{length of longer piece}$$

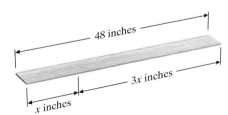

**2.** TRANSLATE the problem. First, we write the equation in words.

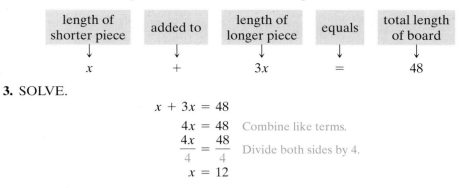

| length of shorter piece | added to | length of longer piece | equals | total length of board |
|:---:|:---:|:---:|:---:|:---:|
| ↓ | ↓ | ↓ | ↓ | ↓ |
| $x$ | $+$ | $3x$ | $=$ | $48$ |

**3.** SOLVE.

$$x + 3x = 48$$
$$4x = 48 \quad \text{Combine like terms.}$$
$$\frac{4x}{4} = \frac{48}{4} \quad \text{Divide both sides by 4.}$$
$$x = 12$$

**4.** INTERPRET.

**Check:** Check the solution in the stated problem. If the shorter piece of board is 12 inches, the longer piece is $3 \cdot (12 \text{ inches}) = 36$ inches and the sum of the two pieces is 12 inches + 36 inches = 48 inches.

**State:** The shorter piece of Balsa wood is 12 inches and the longer piece of Balsa wood is 36 inches.  □

> ▶ **Helpful Hint**
>
> Make sure that units are included in your answer, if appropriate.

**PRACTICE**

**3** A 45-inch board is to be cut into two pieces so that the longer piece is 4 times the shorter. Find the length of each piece.

---

**EXAMPLE 4** **Finding the Number of Democratic and Republican Representatives**

In a recent year, the U.S. House of Representatives had a total of 435 Democrats and Republicans. There were 31 more Democratic representatives than Republican representatives. Find the number of representatives from each party. (*Source:* Office of the Clerk of the U.S. House of Representatives)

*Solution*

**1.** UNDERSTAND. Read and reread the problem. Let's suppose that there were 200 Republican representatives. Since there were 31 more Democrats than Republicans, there must have been $200 + 31 = 231$ Democrats. The total number of Democrats and Republicans was then $200 + 231 = 431$. This is incorrect since the total should be 435, but now we have a better understanding of the problem.

In general, if we let

$$x = \text{number of Republicans, then}$$
$$x + 31 = \text{number of Democrats}$$

**2.** TRANSLATE. First we write the equation in words.

| Number of Republicans | added to | number of Democrats | equals | 435 |
|:---:|:---:|:---:|:---:|:---:|
| ↓ | ↓ | ↓ | ↓ | ↓ |
| $x$ | $+$ | $(x + 31)$ | $=$ | $435$ |

**3.** SOLVE.

$$x + (x + 31) = 435$$
$$2x + 31 = 435$$
$$2x + 31 - 31 = 435 - 31$$
$$2x = 404$$
$$\frac{2x}{2} = \frac{404}{2}$$
$$x = 202$$

**4.** INTERPRET.

**Check:** If there were 202 Republican representatives, then there were $202 + 31 = 233$ Democratic representatives. The total number of Democratic or Republican representatives is $202 + 233 = 435$. The results check.

**State:** There were 202 Republican and 233 Democratic representatives in Congress.

☐

**PRACTICE**
**4**    In a recent year, there were 6 more Democratic State Governors than Republican State Governors. Find the number of State Governors from each party. (We are only counting the 50 states.) (*Source:* National Conference of State Legislatures).

---

△ **EXAMPLE 5**   **Finding Angle Measures**

If the two walls of the Vietnam Veterans Memorial in Washington, D.C., were connected, an isosceles triangle would be formed. The measure of the third angle is 97.5° more than the measure of either of the other two equal angles. Find the measure of the third angle. (*Source:* National Park Service)

*Solution*

**1.** UNDERSTAND. Read and reread the problem. We then draw a diagram (recall that an isosceles triangle has two angles with the same measure) and let

$$x = \text{degree measure of one angle}$$
$$x = \text{degree measure of the second equal angle}$$
$$x + 97.5 = \text{degree measure of the third angle}$$

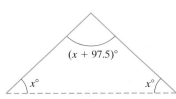

**2.** TRANSLATE. Recall that the sum of the measures of the angles of a triangle equals 180.

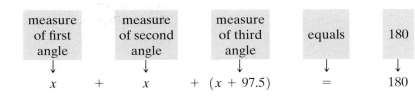

| measure of first angle | | measure of second angle | | measure of third angle | equals | 180 |
|---|---|---|---|---|---|---|
| ↓ | | ↓ | | ↓ | ↓ | ↓ |
| $x$ | $+$ | $x$ | $+$ | $(x + 97.5)$ | $=$ | 180 |

**3.** SOLVE.

$$x + x + (x + 97.5) = 180$$
$$3x + 97.5 = 180 \qquad \text{Combine like terms.}$$
$$3x + 97.5 - 97.5 = 180 - 97.5 \quad \text{Subtract 97.5 from both sides.}$$
$$3x = 82.5$$
$$\frac{3x}{3} = \frac{82.5}{3} \qquad \text{Divide both sides by 3.}$$
$$x = 27.5$$

**4.** INTERPRET.

**Check:** If $x = 27.5$, then the measure of the third angle is $x + 97.5 = 125$. The sum of the angles is then $27.5 + 27.5 + 125 = 180$, the correct sum.

**State:** The third angle measures 125°.*

☐

**PRACTICE**
**5**    The second angle of a triangle measures three times as large as the first. If the third angle measures 55° more than the first, find the measures of all three angles.

---

*The two walls actually meet at an angle of 125 degrees 12 minutes. The measurement of 97.5° given in the problem is an approximation.

**OBJECTIVE 3 ▶ Solving consecutive integer problems.** The next example has to do with consecutive integers. Recall what we have learned thus far about these integers.

|  | *Example* | *General Representation* |
|---|---|---|
| *Consecutive Integers* | 11,　12,　13 $\underset{+1\quad+1}{}$ | Let $x$ be an integer.　$x,\ x+1,\ x+2$ $\underset{+1\quad+1}{}$ |
| *Consecutive Even Integers* | 38,　40,　42 $\underset{+2\quad+2}{}$ | Let $x$ be an even integer.　$x,\ x+2,\ x+4$ $\underset{+2\quad+2}{}$ |
| *Consecutive Odd Integers* | 57,　59,　61 $\underset{+2\quad+2}{}$ | Let $x$ be an odd integer.　$x,\ x+2,\ x+4$ $\underset{+2\quad+2}{}$ |

**EXAMPLE 6**　Some states have a single area code for the entire state. Two such states have area codes that are consecutive odd integers. If the sum of these integers is 1208, find the two area codes. (*Source:* North American Numbering Plan Administration)

*Solution:*

**1.** UNDERSTAND. Read and reread the problem. If we let

$$x = \text{the first odd integer, then}$$
$$x + 2 = \text{the next odd integer}$$

> ▶ **Helpful Hint**
> Remember, the 2 here means that odd integers are 2 units apart, for example, the odd integers 13 and 13 + 2 = 15.

**2.** TRANSLATE.

| first odd integer | the sum of | next odd integer | is | 1208 |
|---|---|---|---|---|
| ↓ | ↓ | ↓ | ↓ | ↓ |
| $x$ | $+$ | $(x+2)$ | $=$ | 1208 |

**3.** SOLVE.

$$x + x + 2 = 1208$$
$$2x + 2 = 1208$$
$$2x + 2 - 2 = 1208 - 2$$
$$2x = 1206$$
$$\frac{2x}{2} = \frac{1206}{2}$$
$$x = 603$$

**4.** INTERPRET.

**Check:**　If $x = 603$, then the next odd integer $x + 2 = 603 + 2 = 605$. Notice their sum, $603 + 605 = 1208$, as needed.

**State:**　The area codes are 603 and 605.

**Note:**　New Hampshire's area code is 603 and South Dakota's area code is 605.　☐

**PRACTICE**
**6**　The sum of three consecutive even integers is 144. Find the integers.

# VOCABULARY & READINESS CHECK

*Fill in the table.*

| | | | | | |
|---|---|---|---|---|---|
| **1.** | A number: $x$ | → | Double the number: $2x$ | → | Double the number, decreased by 31: $2x-31$ |
| **2.** | A number: $x$ | → | Three times the number: $3x$ | → | Three times the number, increased by 17: $3x+17$ |
| **3.** | A number: $x$ | → | The sum of the number and 5: $x+5$ | → | Twice the sum of the number and 5: $2(x+5)$ |
| **4.** | A number: $x$ | → | The difference of the number and 11: $x-11$ | → | Seven times the difference of a number and 11: $7x(x-11)$ |
| **5.** | A number: $y$ | → | The difference of 20 and the number: $20-y$ | → | The difference of 20 and the number, divided by 3: $(20-y)\div 3$ |
| **6.** | A number: $y$ | → | The sum of $-10$ and the number: $(-10+x)$ | → | The sum of $-10$ and the number, divided by 9: |

## 2.4 EXERCISE SET

*Write each of the following as equations. Then solve. See Examples 1 and 2.*

1. The sum of twice a number, and 7, is equal to the sum of a number and 6. Find the number.

2. The difference of three times a number, and 1, is the same as twice a number. Find the number.

3. Three times a number, minus 6, is equal to two times a number, plus 8. Find the number.

4. The sum of 4 times a number, and $-2$, is equal to the sum of 5 times a number, and $-2$. Find the number.

5. Twice the difference of a number and 8 is equal to three times the sum of the number and 3. Find the number.

6. Five times the sum of a number and $-1$ is the same as 6 times the number. Find the number.

7. Four times the sum of $-2$ and a number is the same as five times the number increased by $\frac{1}{2}$. Find the number.

8. If the difference of a number and four is doubled, the result is $\frac{1}{4}$ less than the number. Find the number.

*Solve. See Examples 3 through 5.*

9. A 17-foot piece of string is cut into two pieces so that the longer piece is 2 feet longer than twice the shorter piece. Find the lengths of both pieces.

```
|←———————————— 17 feet ————————————→|
```

10. A 25-foot wire is to be cut so that the longer piece is one foot longer than 5 times the shorter piece. Find the length of each piece.

11. The largest meteorite in the world is the Hoba West located in Namibia. Its weight is 3 times the weight of the Armanty meteorite located in Outer Mongolia. If the sum of their weights is 88 tons, find the weight of each.

12. The area of the Sahara Desert is 7 times the area of the Gobi Desert. If the sum of their areas is 4,000,000 square miles, find the area of each desert.

13. The countries with the most cinema screens in the world are China and the United States. China has 5806 more cinema screens than the United States whereas the total screens for both countries is 78,994. Find the number of cinema screens for both countries. (*Source:* Film Distributor's Association)

14. The countries with the most television stations in the world are Russia and China. Russia has 4066 more television stations than China whereas the total stations for both countries is 10,546. Find the number of television stations for both countries. (*Source:* Central Intelligence Agency, *The World Factbook 2006*)

15. The flag of Equatorial Guinea contains an isosceles triangle. (Recall that an isosceles triangle contains two angles with the

same measure.) If the measure of the third angle of the triangle is 30° more than twice the measure of either of the other two angles, find the measure of each angle of the triangle. (*Hint:* Recall that the sum of the measures of the angles of a triangle is 180°.)

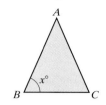

16. Recall that the sum of the measures of the angles of a triangle is 180°. In the triangle below, angle *C* has the same measure as angle *B*, and angle *A* measures 42° less than angle *B*. Find the measure of each angle.

*Solve. See Example 6. Fill in the table. Most of the first row has been completed for you.*

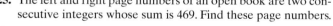

|  | First Integer : | Next Integers | : | Indicated Sum |
|---|---|---|---|---|
| 17. Three consecutive integers: | Integer: $x$ | $x+1$ $\quad$ $x+2$ | | Sum of the three consecutive integers, simplified: $(x+1)+(x+2)$ |
| 18. Three consecutive integers: | Integer: $x$ | $x+1$ $\quad$ $x+2$ | | Sum of the second and third consecutive integers, simplified: $3x+3$ |
| 19. Three consecutive even integers: | Even integer: $x$ | $x+2$ $\quad$ $x+4$ | | Sum of the first and third even consecutive integers, simplified: $x+(x+4)$ |
| 20. Three consecutive odd integers: | Odd integer: $x$ | $x+2$ $\quad$ $x+4$ | | Sum of the three consecutive odd integers, simplified: $x+(x+2)+(x+4)$ |
| 21. Four consecutive integers: | Integer: $x$ | $x+1$ $\quad$ $x+2$ $\quad$ $x+3$ | | Sum of the four consecutive integers, simplified: $x+(x+1)+(x+2)(x+3)$ |
| 22. Four consecutive integers: | Integer: $x$ | $x+1$ $\quad$ $x+2$ $\quad$ $x+3$ | | Sum of the first and fourth consecutive integers, simplified: $4x+6$ |
| 23. Three consecutive odd integers: | Odd integer: $x$ | $x+2$ $\quad$ $x+4$ | | Sum of the second and third consecutive odd integers, simplified: $3x+6$ |
| 24. Three consecutive even integers: | Even integer: $x$ | $x+2$ $\quad$ $x+4$ | | Sum of the three consecutive even integers, simplified: $4x+6$ |

25. The left and right page numbers of an open book are two consecutive integers whose sum is 469. Find these page numbers.

26. The room numbers of two adjacent classrooms are two consecutive even numbers. If their sum is 654, find the classroom numbers.

$2x+2=654$
$x=326$

27. To make an international telephone call, you need the code for the country you are calling. The codes for Belgium, France, and Spain are three consecutive integers whose sum is 99. Find the code for each country. (*Source: The World Almanac and Book of Facts, 2007*)

28. To make an international telephone call, you need the code for the country you are calling. The codes for Mali Republic, Côte d'Ivoire, and Niger are three consecutive odd integers whose sum is 675. Find the code for each country.

**MIXED PRACTICE**

*Solve. See Examples 1 through 6.*

29. A 25-inch piece of steel is cut into three pieces so that the second piece is twice as long as the first piece, and the third piece is one inch more than five times the length of the first piece. Find the lengths of the pieces.

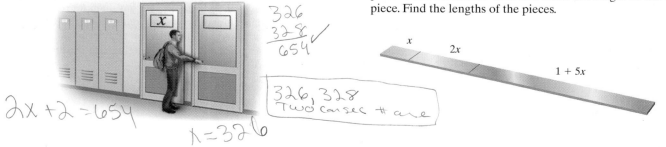

**30.** A 46-foot piece of rope is cut into three pieces so that the second piece is three times as long as the first piece, and the third piece is two feet more than seven times the length of the first piece. Find the lengths of the pieces.

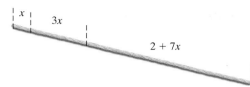

**31.** Five times a number, subtracted from ten, is triple the number. Find the number.

**32.** Nine is equal to ten subtracted from double a number. Find the number. $9 = 2x - 10$

**33.** The greatest producer of diamonds in carats is Botswana. This country produces about four times the amount produced in Angola. If the total produced in both countries is 40,000,000 carats, find the amount produced in each country. (*Source: Diamond Facts 2006.*)

ANGOLA

BOTSWANA

**34.** Beetles have the greatest number of different species. There are twenty times the number of beetle species as grasshopper species, and the total number of species for both is 420,000. Find the number of species for each type of insect.

**35.** The measures of the angles of a triangle are 3 consecutive even integers. Find the measure of each angle.

**36.** A quadrilateral is a polygon with 4 sides. The sum of the measures of the 4 angles in a quadrilateral is 360°. If the measures of the angles of a quadrilateral are consecutive odd integers, find the measures.

**37.** For the 2006 Winter Olympics, the total number of medals won by athletes in each of the countries of Russia, Austria, Canada, and the United States are four consecutive integers whose sum is 94. Find the number of medals for each country.

**38.** The code to unlock a student's combination lock happens to be three consecutive odd integers whose sum is 51. Find the integers.

**39.** If the sum of a number and five is tripled, the result is one less than twice the number. Find the number.

**40.** Twice the sum of a number and six equals three times the sum of the number and four. Find the number.

**41.** In a recent election in Illinois for a seat in the United States House of Representatives, Jerry Weller received 20,196 more votes than opponent John Pavich. If the total number of votes was 196,554, find the number of votes for each candidate. (*Source: The Washington Post*)

**42.** In a recent election in New York for a seat in the United States House of Representatives, Timothy Bishop received 35,650 more votes than opponent Italo Zanzi. If the total number of votes was 158,192, find the number of votes for each candidate. (*Source: The New York Times*)

**43.** Two angles are supplementary if their sum is 180°. The larger angle measures eight degree more than three times the measure of a smaller angle. If $x$ represents the measure of the smaller angle and these two angles are supplementary, find the measure of each angle.

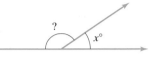

**44.** Two angles are complementary if their sum is 90°. The larger angle measures three degrees less than twice the measure of a smaller angle. If $x$ represents the measure of the smaller angle and these two angles are complementary, find the measure of each angle.

$= 59°$ $\qquad x + 2x - 3 = 90$

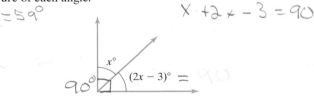

$90°$ $(2x - 3)° = 90$

**45.** If the quotient of a number and 4 is added to $\frac{1}{2}$, the result is $\frac{3}{4}$. Find the number.

**46.** If $\frac{3}{4}$ is added to three times a number, the result is $\frac{1}{2}$ subtracted from twice the number. Find the number.

**47.** The flag of Brazil contains a parallelogram. One angle of the parallelogram is 15° less than twice the measure of the angle next to it. Find the measure of each angle of the parallelogram. (*Hint:* Recall that opposite angles of a parallelogram have the same measure and that the sum of the measures of the angles is 360°.)

$2(x - 15)$
$15° - 2$
$360° = 2(x - 15°)$

**48.** The sum of the measures of the angles of a parallelogram is 360°. In the parallelogram below, angles $A$ and $D$ have the same measure as well as angles $C$ and $B$. If the measure of angle $C$ is twice the measure of angle $A$, find the measure of each angle.

**49.** Currently, the two fastest trains are the Japanese Maglev and the French TGV. The sum of their fastest speeds is 718.2 miles per hour. If the speed of the Maglev is 3.8 mph faster than the speed of the TGV, find the speeds of each.

**50.** The Pentagon is the world's largest office building in terms of floor space. It has three times the amount of floor space as the Empire State Building. If the total floor space for these two buildings is approximately 8700 thousand square feet, find the floor space of each building.

**51.** One-third of a number is five-sixths. Find the number.

**52.** Seven-eighths of a number is one-half. Find the number.

**53.** The number of counties in California and the number of counties in Montana are consecutive even integers whose sum is 114. If California has more counties than Montana, how many counties does each state have? (*Source: The World Almanac and Book of Facts 2007*)

**54.** A student is building a bookcase with stepped shelves for her dorm room. She buys a 48-inch board and wants to cut the board into three pieces with lengths equal to three consecutive even integers. Find the three board lengths.

**55.** In Super Bowl XLI in Miami, Florida, the Indianapolis Colts won over the Chicago Bears with a 12-point lead. If the total of the two scores was 46, find the individual team scores. (*Source:* National Football League)

**56.** During the 2007 Rose Bowl, University of Southern California beat Michigan by 14 points. If their combined scores total 50, find the individual team scores. (*Source: ESPN Sports Almanac*)

**57.** A geodesic dome, based on the design by Buckminster Fuller, is composed of two different types of triangular panels. One of these is an isosceles triangle. In one geodesic dome, the measure of the third angle is 76.5° more than the measure of either of the two equal angles. Find the measure of the third angle. (*Source:* Buckminster Fuller Institute)

**58.** The measures of the angles of a particular triangle are such that the second and third angles are each four times larger than the smallest angle. Find the measures of the angles of this triangle.

**59.** A 40-inch board is to be cut into three pieces so that the second piece is twice as long as the first piece and the third piece is 5 times as long as the first piece. If $x$ represents the length of the first piece, find the lengths of all three pieces.

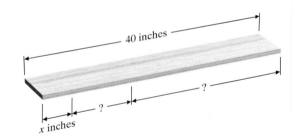

**60.** A 30-foot piece of siding is cut into three pieces so that the second piece is four times as long as the first piece and the third piece is five times as long as the first piece. If $x$ represents the length of the first piece, find the lengths of all three pieces.

*The graph below shows the states with the highest tourism budgets. Use the graph for Exercises 61 through 66.*

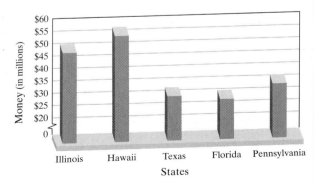

Source: Travel Industry Association

**61.** Which state spends the most money on tourism?

**62.** Which states spend between $30 and $40 million on tourism?

**63.** The states of Texas and Florida spend a total of $60.5 million for tourism. The state of Texas spends $1.7 million more than the state of Florida. Find the amount that each state spends on tourism.

**64.** The states of Hawaii and Pennsylvania spend a total of $91.1 million for tourism. The state of Hawaii spends $14.2 million less than twice the amount of money that the state of Pennsylvania spends. Find the amount that each state spends on tourism.

*Compare the heights of the bars in the graph with your results of the exercises below. Are your answers reasonable?*

**65.** Exercise 63

**66.** Exercise 64

**REVIEW AND PREVIEW**

*Evaluate each expression for the given values. See Section 1.4.*

**67.** $2W + 2L$;   $W = 7$ and $L = 10$

**68.** $\frac{1}{2} Bh$;   $B = 14$ and $h = 22$

**69.** $\pi r^2$;   $r = 15$

**70.** $r \cdot t$;   $r = 15$ and $t = 2$

**CONCEPT EXTENSIONS**

**71.** In your own words, explain why a solution of a word problem should be checked using the original wording of the problem and not the equation written from the wording.

**72.** Give an example of how you recently solved a problem using mathematics.

# 2.5 FORMULAS AND PROBLEM SOLVING

**OBJECTIVES**

**1** Use formulas to solve problems.

**2** Solve a formula or equation for one of its variables.

**OBJECTIVE 1 ▶ Using formulas to solve problems.** An equation that describes a known relationship among quantities, such as distance, time, volume, weight, and money is called a **formula.** These quantities are represented by letters and are thus variables of the formula. Here are some common formulas and their meanings.

| *Formulas and Their Meanings* |
| --- |
| $A = lw$ <br> Area of a rectangle = length · width |
| $I = PRT$ <br> Simple interest = principal · rate · time |
| $P = a + b + c$ <br> Perimeter of a triangle = side $a$ + side $b$ + side $c$ |
| $d = rt$ <br> distance = rate · time |
| $V = lwh$ <br> Volume of a rectangular solid = length · width · height |
| $F = \left(\dfrac{9}{5}\right)C + 32$   or   $F = 1.8C + 32$ <br><br> degrees Fahrenheit = $\left(\dfrac{9}{5}\right)$ · degrees Celsius + 32 |

Formulas are valuable tools because they allow us to calculate measurements as long as we know certain other measurements. For example, if we know we traveled a distance of 100 miles at a rate of 40 miles per hour, we can replace the variables $d$ and $r$ in the formula $d = rt$ and find our time, $t$.

$$d = rt \qquad \text{Formula.}$$
$$100 = 40t \qquad \text{Replace } d \text{ with 100 and } r \text{ with 40.}$$

This is a linear equation in one variable, $t$. To solve for $t$, divide both sides of the equation by 40.

$$\frac{100}{40} = \frac{40t}{40} \quad \text{Divide both sides by 40.}$$

$$\frac{5}{2} = t \quad \text{Simplify.}$$

The time traveled is $\frac{5}{2}$ hours or $2\frac{1}{2}$ hours.

In this section we solve problems that can be modeled by known formulas. We use the same problem-solving steps that were introduced in the previous section. These steps have been slightly revised to include formulas.

**EXAMPLE 1**   **Finding Time Given Rate and Distance**

A glacier is a giant mass of rocks and ice that flows downhill like a river. Portage Glacier in Alaska is about 6 miles, or 31,680 *feet*, long and moves 400 *feet* per year. Icebergs are created when the front end of the glacier flows into Portage Lake. How long does it take for ice at the head (beginning) of the glacier to reach the lake?

_Solution_

1. UNDERSTAND. Read and reread the problem. The appropriate formula needed to solve this problem is the distance formula, $d = rt$. To become familiar with this formula, let's find the distance that ice traveling at a rate of 400 feet per year travels in 100 years. To do so, we let time $t$ be 100 years and rate $r$ be the given 400 feet per year, and substitute these values into the formula $d = rt$. We then have that distance $d = 400(100) = 40,000$ feet. Since we are interested in finding how long it takes ice to travel 31,680 feet, we now know that it is less than 100 years.

   Since we are using the formula $d = rt$, we let

   $t = $ the time in years for ice to reach the lake

   $r = $ rate or speed of ice

   $d = $ distance from beginning of glacier to lake

2. TRANSLATE. To translate to an equation, we use the formula $d = rt$ and let distance $d = 31,680$ feet and rate $r = 400$ feet per year.

$$d = r \cdot t$$

$$31,680 = 400 \cdot t \quad \text{Let } d = 31,680 \text{ and } r = 400.$$

**3. SOLVE.** Solve the equation for $t$. To solve for $t$, divide both sides by 400.

$$\frac{31,680}{400} = \frac{400 \cdot t}{400} \quad \text{Divide both sides by 400.}$$

$$79.2 = t \qquad \text{Simplify.}$$

**4. INTERPRET.**

> ▶ **Helpful Hint**
>
> Don't forget to include units, if appropriate.

**Check:** To check, substitute 79.2 for $t$ and 400 for $r$ in the distance formula and check to see that the distance is 31,680 feet.

**State:** It takes 79.2 years for the ice at the head of Portage Glacier to reach the lake. □

PRACTICE

**1** The Stromboli Volcano, in Italy, began erupting in 2002, after a dormant period of over 17 years. In 2007, a vulcanologist measured the lava flow to be moving at 5 meters/ second. If the path the lava followed to the sea is 580 meters long, how long does it take the lava to reach the sea? (*Source:* Thorsten Boeckel and CNN)

⚠ **EXAMPLE 2**  **Calculating the Length of a Garden**

Charles Pecot can afford enough fencing to enclose a rectangular garden with a perimeter of 140 feet. If the width of his garden must be 30 feet, find the length.

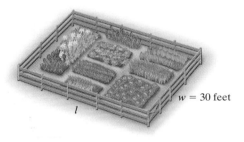

$w = 30$ feet

$l$

*Solution*

**1. UNDERSTAND.** Read and reread the problem. The formula needed to solve this problem is the formula for the perimeter of a rectangle, $P = 2l + 2w$. Before continuing, let's become familiar with this formula.

$l$ = the length of the rectangular garden

$w$ = the width of the rectangular garden

$P$ = perimeter of the garden

**2. TRANSLATE.** To translate to an equation, we use the formula $P = 2l + 2w$ and let perimeter $P = 140$ feet and width $w = 30$ feet.

$$P = 2l + 2w$$

$$140 = 2l + 2(30) \quad \text{Let } P = 140 \text{ and } w = 30.$$

**3. SOLVE.**

$$140 = 2l + 2(30)$$

$$140 = 2l + 60 \qquad\qquad \text{Multiply } 2(30).$$

$$140 - 60 = 2l + 60 - 60 \quad \text{Subtract 60 from both sides.}$$

$$80 = 2l \qquad\qquad\qquad \text{Combine like terms.}$$

$$40 = l \qquad\qquad\qquad \text{Divide both sides by 2.}$$

**4. INTERPRET.**

**Check:**   Substitute 40 for $l$ and 30 for $w$ in the perimeter formula and check to see that the perimeter is 140 feet.

**State:**   The length of the rectangular garden is 40 feet.   □

**PRACTICE**
**2**   Evelyn Gryk fenced in part of her back yard for a dog run. The dog run was 40 feet in length and used 98 feet of fencing. Find the width of the dog run.

---

**EXAMPLE 3**   **Finding an Equivalent Temperature**

The average minimum temperature for July in Shanghai, China, is 77° Fahrenheit. Find the equivalent temperature in degrees Celsius.

*Solution*

1. **UNDERSTAND.**  Read and reread the problem. A formula that can be used to solve this problem is the formula for converting degrees Celsius to degrees Fahrenheit, $F = \frac{9}{5}C + 32$. Before continuing, become familiar with this formula. Using this formula, we let

$$C = \text{temperature in degrees Celsius, and}$$
$$F = \text{temperature in degrees Fahrenheit.}$$

2. **TRANSLATE.**  To translate to an equation, we use the formula $F = \frac{9}{5}C + 32$ and let degrees Fahrenheit $F = 77$.

$$\text{Formula:} \quad F = \frac{9}{5}C + 32$$
$$\text{Substitute:} \quad 77 = \frac{9}{5}C + 32 \quad \text{Let } F = 77.$$

3. **SOLVE.**

$$77 = \frac{9}{5}C + 32$$
$$77 - 32 = \frac{9}{5}C + 32 - 32 \quad \text{Subtract 32 from both sides.}$$
$$45 = \frac{9}{5}C \quad\quad\quad\quad\quad \text{Combine like terms.}$$
$$\frac{5}{9} \cdot 45 = \frac{5}{9} \cdot \frac{9}{5}C \quad\quad\quad \text{Multiply both sides by } \frac{5}{9}.$$
$$25 = C \quad\quad\quad\quad\quad\quad \text{Simplify.}$$

4. **INTERPRET.**

**Check:**   To check, replace $C$ with 25 and $F$ with 77 in the formula and see that a true statement results.

**State:**   Thus, 77° Fahrenheit is equivalent to 25° Celsius.

**Note:**   There is a formula for directly converting degrees Fahrenheit to degrees Celsius. It is $C = \frac{5}{9}(F - 32)$, as we shall see in Example 8.   □

**PRACTICE**
**3**   The average minimum temperature for July in Sydney, Australia, is 8° Celsius. Find the equivalent temperature in degrees Fahrenheit.

In the next example, we again use the formula for perimeter of a rectangle as in Example 2. In Example 2, we knew the width of the rectangle. In this example, both the length and width are unknown.

### EXAMPLE 4   Finding Road Sign Dimensions

The length of a rectangular road sign is 2 feet less than three times its width. Find the dimensions if the perimeter is 28 feet.

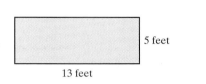

5 feet

13 feet

GARDEN CITY 15
Exit 90 and 91

#### Solution

**1.** UNDERSTAND. Read and reread the problem. Recall that the formula for the perimeter of a rectangle is $P = 2l + 2w$. Draw a rectangle and guess the solution. If the width of the rectangular sign is 5 feet, its length is 2 feet less than 3 times the width or $3(5 \text{ feet}) - 2 \text{ feet} = 13 \text{ feet}$. The perimeter $P$ of this rectangle, drawn above, is then $2(13 \text{ feet}) + 2(5 \text{ feet}) = 36 \text{ feet}$, too much. We now know that the width is less than 5 feet.

Let

$w = $ the width of the rectangular sign; then

$3w - 2 = $ the length of the sign.

$w$

$3w - 2$

Draw a rectangle and label it with the assigned variables, as shown in the left margin.

**2.** TRANSLATE.

Formula:     $P = 2l + 2w$

Substitute:   $28 = 2(3w - 2) + 2w.$

**3.** SOLVE.

$$28 = 2(3w - 2) + 2w$$
$$28 = 6w - 4 + 2w \qquad \text{Apply the distributive property.}$$
$$28 = 8w - 4$$
$$28 + 4 = 8w - 4 + 4 \qquad \text{Add 4 to both sides.}$$
$$32 = 8w$$
$$\frac{32}{8} = \frac{8w}{8} \qquad\qquad \text{Divide both sides by 8.}$$
$$4 = w$$

**4.** INTERPRET.

**Check:**   If the width of the sign is 4 feet, the length of the sign is $3(4 \text{ feet}) - 2 \text{ feet} = 10 \text{ feet}$. This gives a perimeter of $P = 2(4 \text{ feet}) + 2(10 \text{ feet}) = 28 \text{ feet}$, the correct perimeter.

**State:**   The width of the sign is 4 feet and the length of the sign is 10 feet.   □

**PRACTICE**

**4**     The new street signs along Route 114 have a length that is 3 inches more than 5 times the width. Find the dimensions of the signs if the perimeter of the signs is 66 inches.

**OBJECTIVE 2 ▶ Solving a formula for one of its variables.** We say that the formula $F = \frac{9}{5}C + 32$ is solved for $F$ because $F$ is alone on one side of the equation and the other side of the equation contains no $F$'s. Suppose that we need to convert many Fahrenheit temperatures to equivalent degrees Celsius. In this case, it is easier to perform this task by solving the formula $F = \frac{9}{5}C + 32$ for $C$. (See Example 8.) For this reason, it is important to be able to solve an equation for any one of its specified variables. For example, the formula $d = rt$ is solved for $d$ in terms of $r$ and $t$. We can also solve $d = rt$ for $t$ in terms of $d$ and $r$. To solve for $t$, divide both sides of the equation by $r$.

$$d = rt$$
$$\frac{d}{r} = \frac{rt}{r} \quad \text{Divide both sides by } r.$$
$$\frac{d}{r} = t \quad \text{Simplify.}$$

To solve a formula or an equation for a specified variable, we use the same steps as for solving a linear equation. These steps are listed next.

---

**Solving Equations for a Specified Variable**

**STEP 1.** Multiply on both sides to clear the equation of fractions if they occur.

**STEP 2.** Use the distributive property to remove parentheses if they occur.

**STEP 3.** Simplify each side of the equation by combining like terms.

**STEP 4.** Get all terms containing the specified variable on one side and all other terms on the other side by using the addition property of equality.

**STEP 5.** Get the specified variable alone by using the multiplication property of equality.

---

△ **EXAMPLE 5** Solve $V = lwh$ for $l$.

**Solution** This formula is used to find the volume of a box. To solve for $l$, divide both sides by $wh$.

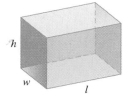

$$V = lwh$$
$$\frac{V}{wh} = \frac{lwh}{wh} \quad \text{Divide both sides by } wh.$$
$$\frac{V}{wh} = l \quad \text{Simplify.}$$

Since we have $l$ alone on one side of the equation, we have solved for $l$ in terms of $V$, $w$, and $h$. Remember that it does not matter on which side of the equation we isolate the variable. ☐

**PRACTICE**
**5** Solve $I = Prt$ for $r$.

---

**EXAMPLE 6** Solve $y = mx + b$ for $x$.

**Solution** The term containing the variable we are solving for, $mx$, is on the right side of the equation. Get $mx$ alone by subtracting $b$ from both sides.

$$y = mx + b$$
$$y - b = mx + b - b \quad \text{Subtract } b \text{ from both sides.}$$
$$y - b = mx \quad \text{Combine like terms.}$$

Next, solve for $x$ by dividing both sides by $m$.

$$\frac{y-b}{m} = \frac{mx}{m}$$

$$\frac{y-b}{m} = x \qquad \text{Simplify.} \qquad \square$$

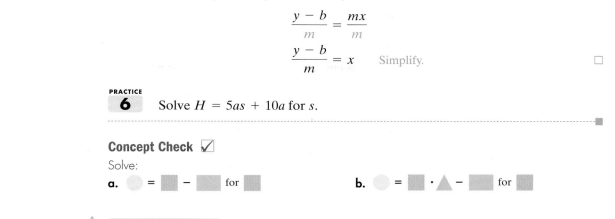

**PRACTICE**
**6**   Solve $H = 5as + 10a$ for $s$.

**Concept Check** ☑

Solve:

**a.**  ● = ■ − ▨  for ▨

**b.**  ● = ■ · ▲ − ▨  for ▨

△   **EXAMPLE 7**   Solve $P = 2l + 2w$ for $w$.

*Solution*   This formula relates the perimeter of a rectangle to its length and width. Find the term containing the variable $w$. To get this term, $2w$, alone subtract $2l$ from both sides.

▶ **Helpful Hint**

The 2's may *not* be divided out here. Although 2 is a factor of the denominator, 2 is *not* a factor of the numerator since it is not a factor of both terms in the numerator.

$$P = 2l + 2w$$
$$P - 2l = 2l + 2w - 2l \qquad \text{Subtract } 2l \text{ from both sides.}$$
$$P - 2l = 2w \qquad \text{Combine like terms.}$$
$$\frac{P - 2l}{2} = \frac{2w}{2} \qquad \text{Divide both sides by 2.}$$
$$\frac{P - 2l}{2} = w \qquad \text{Simplify.} \qquad \square$$

**PRACTICE**
**7**   Solve $N = F + d(n - 1)$ for $d$.

The next example has an equation containing a fraction. We will first clear the equation of fractions and then solve for the specified variable.

**EXAMPLE 8**   Solve $F = \frac{9}{5}C + 32$ for $C$.

*Solution*
$$F = \frac{9}{5}C + 32$$

$$5(F) = 5\left(\frac{9}{5}C + 32\right) \qquad \begin{array}{l}\text{Clear the fraction by multiplying}\\ \text{both sides by the LCD.}\end{array}$$

$$5F = 9C + 160 \qquad \text{Distribute the 5.}$$

$$5F - 160 = 9C + 160 - 160 \qquad \begin{array}{l}\text{To get the term containing the variable}\\ C \text{ alone, subtract 160 from both sides.}\end{array}$$

$$5F - 160 = 9C \qquad \text{Combine like terms.}$$

$$\frac{5F - 160}{9} = \frac{9C}{9} \qquad \text{Divide both sides by 9.}$$

$$\frac{5F - 160}{9} = C \qquad \text{Simplify.}$$

**Note:**   Another equivalent way to write this formula is $C = \frac{5}{9}(F - 32)$.   $\square$

**Answers to Concept Check:**

**a.**  ● + ■     **b.**  $\dfrac{● + ■}{▲}$

**PRACTICE**
**8**   Solve $A = \frac{1}{2}a(b + B)$ for B.

## 2.5 | EXERCISE SET

*MyMathLab*  PRACTICE  WATCH  DOWNLOAD  READ  REVIEW

*Substitute the given values into each given formula and solve for the unknown variable. If necessary, round to one decimal place. See Examples 1 through 3.*

△ **1.** $A = bh$;  $A = 45, b = 15$ (Area of a parallelogram)

**2.** $d = rt$;  $d = 195, t = 3$ (Distance formula)

△ **3.** $S = 4lw + 2wh$;  $S = 102, l = 7, w = 3$ (Surface area of a special rectangular box)

△ **4.** $V = lwh$;  $l = 14, w = 8, h = 3$ (Volume of a rectangular box)

△ **5.** $A = \frac{1}{2}h(B + b)$;  $A = 180, B = 11, b = 7$ (Area of a trapezoid)

△ **6.** $A = \frac{1}{2}h(B + b)$;  $A = 60, B = 7, b = 3$ (Area of a trapezoid)

△ **7.** $P = a + b + c$;  $P = 30, a = 8, b = 10$ (Perimeter of a triangle)

△ **8.** $V = \frac{1}{3}Ah$;  $V = 45, h = 5$ (Volume of a pyramid)

**9.** $C = 2\pi r$;  $C = 15.7$ (use the approximation 3.14 or a calculator approximation for $\pi$) (Circumference of a circle)

**10.** $A = \pi r^2$;  $r = 4.5$ (use the approximation 3.14 or a calculator approximation for $\pi$) (Area of a circle)

**11.** $I = PRT$;  $I = 3750, P = 25,000, R = 0.05$ (Simple interest formula)

**12.** $I = PRT$;  $I = 1,056,000, R = 0.055, T = 6$ (Simple interest formula)

**13.** $V = \frac{1}{3}\pi r^2 h$;  $V = 565.2, r = 6$ (use a calculator approximation for $\pi$) (Volume of a cone)

**14.** $V = \frac{4}{3}\pi r^3$;  $r = 3$ (use a calculator approximation for $\pi$) (Volume of a sphere)

*Solve each formula for the specified variable. See Examples 5 through 8.*

**15.** $f = 5gh$ for $h$

△ **16.** $A = \pi ab$ for $b$

**17.** $V = lwh$ for $w$

**18.** $T = mnr$ for $n$

**19.** $3x + y = 7$ for $y$

**20.** $-x + y = 13$ for $y$

**21.** $A = P + PRT$ for $R$

**22.** $A = P + PRT$ for $T$

**23.** $V = \frac{1}{3}Ah$ for $A$

**24.** $D = \frac{1}{4}fk$ for $k$

**25.** $P = a + b + c$ for $a$

**26.** $PR = x + y + z + w$ for $z$

**27.** $S = 2\pi rh + 2\pi r^2$ for $h$

△ **28.** $S = 4lw + 2wh$ for $h$

*Solve. See Examples 1 through 4.*

**29.** For the purpose of purchasing new baseboard and carpet,
   **a.** Find the area and perimeter of the room below (neglecting doors).
   **b.** Identify whether baseboard has to do with area or perimeter and the same with carpet.

11.5 ft    9 ft

**30.** For the purpose of purchasing lumber for a new fence and seed to plant grass,
   **a.** Find the area and perimeter of the yard below.
   **b.** Identify whether a fence has to do with area or perimeter and the same with grass seed.  $P = a + b + c$

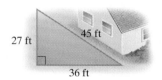

27 ft    45 ft    36 ft

**31.** A frame shop charges according to both the amount of framing needed to surround the picture and the amount of glass needed to cover the picture.
   **a.** Find the area and perimeter of the trapezoid-shaped framed picture below.
   **b.** Identify whether the amount of framing has to do with perimeter or area and the same with the amount of glass.

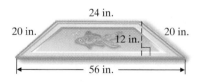

24 in.
20 in.    20 in.    12 in.
56 in.

**32.** A decorator is painting and placing a border completely around the parallelogram-shaped wall.
   **a.** Find the area and perimeter of the wall below.
   **b.** Identify whether the border has to do with perimeter or area and the same with paint.

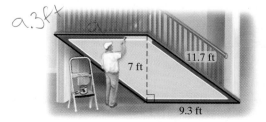

a.3ft

7 ft    11.7 ft    9.3 ft

**33.** The world's largest pink ribbon, the sign of the fight against breast cancer, was erected out of pink post-it notes on a billboard in New York City in October, 2004. If the area of the rectangular billboard covered by the ribbon is approximately 3990 square feet, and the width of the billboard was approximately 57 feet, what was the height of this billboard?

**34.** The world's largest sign for Coca-Cola is located in Arica, Chile. The rectangular sign has a length of 400 feet and has an area of 52,400 square feet. Find the width of the sign. (*Source: Fabulous Facts about Coca-Cola, Atlanta, GA*)

**35.** Convert Nome, Alaska's 14°F high temperature to Celsius.

**36.** Convert Paris, France's low temperature of −5°C to Fahrenheit.

**37.** The X-30 is a "space plane" that skims the edge of space at 4000 miles per hour. Neglecting altitude, if the circumference of the Earth is approximately 25,000 miles, how long will it take for the X-30 to travel around the Earth?

**38.** In the United States, a notable hang glider flight was a 303-mile, $8\frac{1}{2}$ hour flight from New Mexico to Kansas. What was the average rate during this flight?

**39.** An architect designs a rectangular flower garden such that the width is exactly two-thirds of the length. If 260 feet of antique picket fencing are to be used to enclose the garden, find the dimensions of the garden.

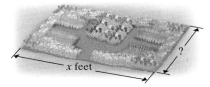

**40.** If the length of a rectangular parking lot is 10 meters less than twice its width, and the perimeter is 400 meters, find the length of the parking lot.

**41.** A flower bed is in the shape of a triangle with one side twice the length of the shortest side, and the third side is 30 feet more than the length of the shortest side. Find the dimensions if the perimeter is 102 feet.

**42.** The perimeter of a yield sign in the shape of an isosceles triangle is 22 feet. If the shortest side is 2 feet less than the other two sides, find the length of the shortest side. (*Hint:* An isosceles triangle has two sides the same length.)

**43.** The Cat is a high-speed catamaran auto ferry that operates between Bar Harbor, Maine, and Yarmouth, Nova Scotia. The Cat can make the 138-mile trip in about $2\frac{1}{2}$ hours. Find the catamaran speed for this trip. (*Source:* Bay Ferries)

**44.** A family is planning their vacation to Disney World. They will drive from a small town outside New Orleans, Louisiana, to Orlando, Florida, a distance of 700 miles. They plan to average a rate of 55 mph. How long will this trip take?

**45.** Piranha fish require 1.5 cubic feet of water per fish to maintain a healthy environment. Find the maximum number of piranhas you could put in a tank measuring 8 feet by 3 feet by 6 feet.

**46.** Find the maximum number of goldfish you can put in a cylindrical tank whose diameter is 8 meters and whose height is 3 meters if each goldfish needs 2 cubic meters of water.

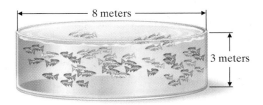

**47.** A lawn is in the shape of a trapezoid with a height of 60 feet and bases of 70 feet and 130 feet. How many whole bags of fertilizer must be purchased to cover the lawn if each bag covers 4000 square feet?

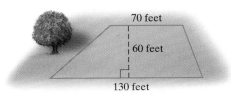

△ **48.** If the area of a right-triangularly shaped sail is 20 square feet and its base is 5 feet, find the height of the sail.

△ **49.** Maria's Pizza sells one 16-inch cheese pizza or two 10-inch cheese pizzas for $9.99. Determine which size gives more pizza.

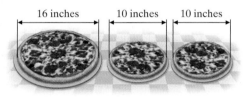

△ **50.** Find how much rope is needed to wrap around the Earth at the equator, if the radius of the Earth is 4000 miles. (*Hint:* Use 3.14 for $\pi$ and the formula for circumference.)

△ **51.** The perimeter of a geometric figure is the sum of the lengths of its sides. If the perimeter of the following pentagon (five-sided figure) is 48 meters, find the length of each side.

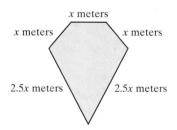

△ **52.** The perimeter of the following triangle is 82 feet. Find the length of each side.

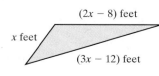

**53.** A Japanese "bullet" train set a new world record for train speed at 361 miles per hour during a manned test run on the Yamanashi Maglev Test Line in 2003. How long does it take this train to travel 72.2 miles at this speed? Give the result in hours; then convert to minutes.

**54.** In 1983, the Hawaiian volcano Kilauea began erupting in a series of episodes still occurring at the time of this writing. At times, the lava flows advanced at speeds of up to 0.5 kilometer per hour. In 1983 and 1984 lava flows destroyed 16 homes in the Royal Gardens subdivision, about 6 km away from the eruption site. Roughly how long did it take the lava to reach Royal Gardens? Assume that the lava traveled at its fastest rate, 0.5 kph. (*Source:* U.S. Geological Survey Hawaiian Volcano Observatory)

△ **55.** The perimeter of an equilateral triangle is 7 inches more than the perimeter of a square, and the side of the triangle is 5 inches longer than the side of the square. Find the side of the triangle. (*Hint:* An equilateral triangle has three sides the same length.)

△ **56.** A square animal pen and a pen shaped like an equilateral triangle have equal perimeters. Find the length of the sides of each pen if the sides of the triangular pen are fifteen less than twice a side of the square pen.

**57.** Find how long it takes a person to drive 135 miles on I-10 if she merges onto I-10 at 10 a.m. and drives nonstop with her cruise control set on 60 mph.

**58.** Beaumont, Texas, is about 150 miles from Toledo Bend. If Leo Miller leaves Beaumont at 4 a.m. and averages 45 mph, when should he arrive at Toledo Bend?

△ **59.** The longest runway at Los Angeles International Airport has the shape of a rectangle and an area of 1,813,500 square feet. This runway is 150 feet wide. How long is the runway? (*Source:* Los Angeles World Airports)

**60.** Normal room temperature is about 78°F. Convert this temperature to Celsius.

**61.** The highest temperature ever recorded in Europe was 122°F in Seville, Spain, in August of 1881. Convert this record high temperature to Celsius. (*Source:* National Climatic Data Center)

**62.** The lowest temperature ever recorded in Oceania was −10°C at the Haleakala Summit in Maui, Hawaii, in January 1961. Convert this record low temperature to Fahrenheit. (*Source:* National Climatic Data Center)

△ **63.** The CART FedEx Championship Series is an open-wheeled race car competition based in the United States. A CART car has a maximum length of 199 inches, a maximum width of 78.5 inches, and a maximum height of 33 inches. When the CART series travels to another country for a grand prix, teams must ship their cars. Find the volume of the smallest shipping crate needed to ship a CART car of maximum dimensions. (*Source:* Championship Auto Racing Teams, Inc.)

**CART Racing Car**

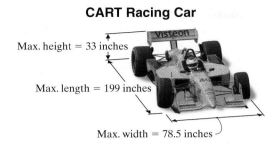

Max. height = 33 inches

Max. length = 199 inches

Max. width = 78.5 inches

**64.** On a road course, a CART car's speed can average up to around 105 mph. Based on this speed, how long would it take a CART driver to travel from Los Angeles to New York City, a distance of about 2810 miles by road, without stopping? Round to the nearest tenth of an hour.

**65.** The Hoberman Sphere is a toy ball that expands and contracts. When it is completely closed, it has a diameter of 9.5 inches. Find the volume of the Hoberman Sphere when it is completely closed. Use 3.14 for $\pi$. Round to the nearest whole cubic inch. (*Source:* Hoberman Designs, Inc.)

**66.** When the Hoberman Sphere (see Exercise 65) is completely expanded, its diameter is 30 inches. Find the volume of the Hoberman Sphere when it is completely expanded. Use 3.14 for $\pi$. Round to the nearest whole cubic inch. (*Source:* Hoberman Designs, Inc.)

**67.** The average temperature on the planet Mercury is 167°C. Convert this temperature to degrees Fahrenheit. (*Source:* National Space Science Data Center)

**68.** The average temperature on the planet Jupiter is −227°F. Convert this temperature to degrees Celsius. Round to the nearest degree. (*Source:* National Space Science Data Center)

**REVIEW AND PREVIEW**

*Write the following phrases as algebraic expressions. See Section 2.1.*

**69.** Nine divided by the sum of a number and 5

**70.** Half the product of a number and five

**71.** Three times the sum of a number and four

**72.** Double the sum of ten and four times a number

**73.** Triple the difference of a number and twelve

**74.** A number minus the sum of the number and six

**CONCEPT EXTENSIONS**

*Solve. See the Concept Check in this section.*

**75.** ■ − ● · ▮ = ▲  for ●

**76.** ⬣ · ■ + ▲ = ●  for ■

**77.** Dry ice is a name given to solidified carbon dioxide. At −78.5° Celsius it changes directly from a solid to a gas. Convert this temperature to Fahrenheit.

**78.** Lightning bolts can reach a temperature of 50,000° Fahrenheit. Convert this temperature to Celsius.

**79.** The distance from the sun to the Earth is approximately 93,000,000 miles. If light travels at a rate of 186,000 miles per second, how long does it take light from the sun to reach us?

**80.** Light travels at a rate of 186,000 miles per second. If our moon is 238,860 miles from the Earth, how long does it take light from the moon to reach us? (Round to the nearest tenth of a second.)

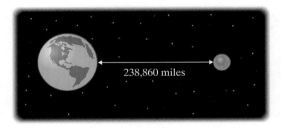

238,860 miles

**81.** A glacier is a giant mass of rocks and ice that flows downhill like a river. Exit Glacier, near Seward, Alaska, moves at a rate of 20 inches a day. Find the distance in feet the glacier moves in a year. (Assume 365 days a year. Round to 2 decimal places.)

**82.** Flying fish do not *actually* fly, but glide. They have been known to travel a distance of 1300 feet at a rate of 20 miles per hour. How many seconds did it take to travel this distance? (*Hint:* First convert miles per hour to feet per second. Recall that 1 mile = 5280 feet. Round to the nearest tenth of a second.)

**83.** Stalactites join stalagmites to form columns. A column found at Natural Bridge Caverns near San Antonio, Texas, rises 15 feet and has a *diameter* of only 2 inches. Find the volume of this column in cubic inches. (*Hint:* Use the formula for volume of a cylinder and use a calculator approximation for $\pi$. Round to the nearest tenth of an inch.)

**84.** Find the temperature at which the Celsius measurement and Fahrenheit measurement are the same number.

**85.** The formula $A = bh$ is used to find the area of a parallelogram. If the base of a parallelogram is doubled and its height is doubled, how does this affect the area?

**86.** The formula $V = LWH$ is used to find the volume of a box. If the length of a box is doubled, the width is doubled, and the height is doubled, how does this affect the volume?

---

## STUDY SKILLS BUILDER

### Organizing a Notebook

It's never too late to get organized. If you need ideas about organizing a notebook for your mathematics course, try some of these:

- Use a spiral or ring binder notebook with pockets and use it for mathematics only.

- Start each page by writing the book's section number you are working on at the top.

- When your instructor is lecturing, take notes. *Always* include any examples your instructor works for you.

- Place your worked-out homework exercises in your notebook immediately after the lecture notes from that section. This way, a section's worth of material is together.

- Homework exercises: Attempt and check all assigned homework.

- Place graded quizzes in the pockets of your notebook or a special section of your binder.

### Self-Check

*Check your notebook organization by answering the following questions.*

**1.** Do you have a spiral or ring binder notebook for your mathematics course only?

**2.** Have you ever had to flip through several sheets of notes and work in your mathematics notebook to determine what section's work you are in?

**3.** Are you now writing the textbook's section number at the top of each notebook page?

**4.** Have you ever lost or had trouble finding a graded quiz or test?

**5.** Are you now placing all your graded work in a dedicated place in your notebook?

**6.** Are you attempting all of your homework and placing all of your work in your notebook?

**7.** Are you checking and correcting your homework in your notebook? If not, why not?

**8.** Are you writing in your notebook the examples your instructor works for you in class?

# 2.6 PERCENT AND MIXTURE PROBLEM SOLVING

This section is devoted to solving problems in the categories listed. The same problem-solving steps used in previous sections are also followed in this section. They are listed below for review.

---

**General Strategy for Problem Solving**

1. UNDERSTAND the problem. During this step, become comfortable with the problem. Some ways of doing this are as follows:

   Read and reread the problem.

   Choose a variable to represent the unknown.

   Construct a drawing, whenever possible.

   Propose a solution and check. Pay careful attention to how you check your proposed solution. This will help writing an equation to model the problem.

2. TRANSLATE the problem into an equation.

3. SOLVE the equation.

4. INTERPRET the results: *Check* the proposed solution in the stated problem and *state* your conclusion.

---

**OBJECTIVE 1 ▶ Solving percent equations.** Many of today's statistics are given in terms of percent: a basketball player's free throw percent, current interest rates, stock market trends, and nutrition labeling, just to name a few. In this section, we first explore percent, percent equations, and applications involving percents. See Appendix B.2 if a further review of percents is needed.

**EXAMPLE 1**   The number 63 is what percent of 72?

*Solution*

1. UNDERSTAND. Read and reread the problem. Next, let's suppose that the percent is 80%. To check, we find 80% of 72.

$$80\% \text{ of } 72 = 0.80(72) = 57.6$$

This is close, but not 63. At this point, though, we have a better understanding of the problem, we know the correct answer is close to and greater than 80%, and we know how to check our proposed solution later.

Let $x$ = the unknown percent.

2. TRANSLATE. Recall that "is" means "equals" and "of" signifies multiplying. Let's translate the sentence directly.

| the number 63 | is | what percent | of | 72 |
|:---:|:---:|:---:|:---:|:---:|
| ↓ | ↓ | ↓ | ↓ | ↓ |
| 63 | = | $x$ | · | 72 |

3. SOLVE.

$$63 = 72x$$
$$0.875 = x \qquad \text{Divide both sides by 72.}$$
$$87.5\% = x \qquad \text{Write as a percent.}$$

**4.** INTERPRET.

**Check:** Verify that 87.5% of 72 is 63.

**State:** The number 63 is 87.5% of 72. □

**PRACTICE**
**1** The number 35 is what percent of 56?

---

**EXAMPLE 2** The number 120 is 15% of what number?

*Solution*

**1.** UNDERSTAND. Read and reread the problem.

Let $x$ = the unknown number.

**2.** TRANSLATE.

| the number 120 | is | 15% | of | what number |
|:---:|:---:|:---:|:---:|:---:|
| ↓ | ↓ | ↓ | ↓ | ↓ |
| 120 | = | 15% | · | $x$ |

**3.** SOLVE.

$$120 = 0.15x \quad \text{Write 15\% as 0.15.}$$
$$800 = x \quad \text{Divide both sides by 0.15.}$$

**4.** INTERPRET.

**Check:** Check the proposed solution by finding 15% of 800 and verifying that the result is 120.

**State:** Thus, 120 is 15% of 800. □

**PRACTICE**
**2** The number 198 is 55% of what number?

---

The next example contains a circle graph. This particular circle graph shows percents of American travelers in certain categories. Since the circle graph represents all American travelers, the percents should add to 100%.

> ▶ **Helpful Hint**
> The percents in a circle graph should have a sum of 100%.

**EXAMPLE 3** The circle graph below shows the purpose of trips made by American travelers. Use this graph to answer the following questions.

**Purpose of American Travelers**

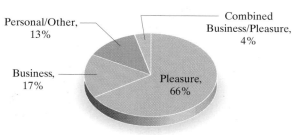

Personal/Other, 13%
Combined Business/Pleasure, 4%
Business, 17%
Pleasure, 66%

*Source:* Travel Industry Association of America

**a.** What percent of trips made by American travelers are solely for the purpose of business?

**b.** What percent of trips made by American travelers are for the purpose of business or combined business/pleasure?

**c.** On an airplane flight of 253 Americans, how many of these people might we expect to be traveling solely for business?

### Solution

**a.** From the circle graph, we see that 17% of trips made by American travelers are solely for the purpose of business.

**b.** From the circle graph, we know that 17% of trips are solely for business and 4% of trips are for combined business/pleasure. The sum 17% + 4% or 21% of trips made by American travelers are for the purpose of business or combined business/pleasure.

**c.** Since 17% of trips made by American travelers are for business, we find 17% of 253. Remember that "of" translates to "multiplication."

$$17\% \text{ of } 253 = 0.17(253) \quad \text{Replace "of" with the operation of multiplication.}$$
$$= 43.01$$

We might then expect that about 43 American travelers on the flight are traveling solely for business. □

**PRACTICE**

**3** Use the Example 3 circle graph to answer each question.

**a.** What percent of trips made by American travelers are for combined business/pleasure?

**b.** What percent of trips made by American travelers are for the purpose of business, pleasure, or combined business/pleasure?

**c.** On a flight of 325 Americans, how many of these people might we expect to be traveling for business/pleasure?

-------

**OBJECTIVE 2 ▶ Solving discount and mark-up problems.** The next example has to do with discounting the price of a cell phone.

**EXAMPLE 4** Cell Phones Unlimited recently reduced the price of a $140 phone by 20%. What is the discount and the new price?

### Solution

**1.** UNDERSTAND. Read and reread the problem. Make sure you understand the meaning of the word "discount." Discount is the amount of money by which the cost of an item has been decreased. To find the discount, we simply find 20% of $140. In other words, we have the formulas,

$$\text{discount} = \text{percent} \cdot \text{original price} \quad \text{Then}$$

$$\text{new price} = \text{original price} - \text{discount}$$

**2, 3. TRANSLATE and SOLVE.**

$$
\boxed{\text{discount}} \;=\; \boxed{\text{percent}} \cdot \boxed{\text{original price}}
$$

$$
\begin{aligned}
&= \quad 20\% \quad \cdot \quad \$140\\
&= \quad 0.20 \quad \cdot \quad \$140\\
&= \$28
\end{aligned}
$$

Thus, the discount in price is $28.

$$
\boxed{\text{new price}} \;=\; \boxed{\text{original price}} \;-\; \boxed{\text{discount}}
$$

$$
\begin{aligned}
&= \quad \$140 \quad - \quad \$28\\
&= \$112
\end{aligned}
$$

**4. INTERPRET.**

**Check:**  Check your calculations in the formulas, and also see if our results are reasonable. They are.

**State:**  The discount in price is $28 and the new price is $112.

> **PRACTICE**
> **4**   A used treadmill, originally purchased for $480, was sold at a garage sale at a discount of 85% of the original price. What was the discount and the new price?

A concept similar to discount is mark-up. What is the difference between the two? A discount is subtracted from the original price while a mark-up is added to the original price. For mark-ups,

$$
\boxed{\text{mark-up} = \text{percent} \cdot \text{original price}}
$$

$$
\boxed{\text{new price} = \text{original price} + \text{mark-up}}
$$

> ▶ **Helpful Hint**
> Discounts are subtracted from the original price while mark-ups are added.

Mark-up exercises can be found in Exercise Set 2.7.

**OBJECTIVE 3 ▶ Solving percent increase and percent decrease problems.** Percent increase or percent decrease is a common way to describe how some measurement has increased or decreased. For example, crime increased by 8%, teachers received a 5.5% increase in salary, or a company decreased its employees by 10%. The next example is a review of percent increase.

**EXAMPLE 5**   The cost of attending a public college rose from $9258 in 1996 to $12,796 in 2006. Find the percent increase, rounded to the nearest tenth of a percent. (*Source:* The College Board)

*Solution*

**1.** UNDERSTAND. Read and reread the problem. Let's guess that the percent increase is 20%. To see if this is the case, we find 20% of $9258 to find the *increase* in cost. Then we add this increase to $9258 to find the *new cost*. In other words, 20% ($9258) = 0.20($9258) = $1851.60, the *increase* in cost. The new cost then would be $9258 + $1851.60 = $11,109.60, less than the actual new cost of $12,976. We now know that the increase is greater than 20% and we know how to check our proposed solution.

Let $x$ = the percent increase.

**2.** TRANSLATE. First, find the **increase,** and then the **percent increase.** The increase in cost is found by

In words:   increase   =   new cost   −   old cost   or

Translate:   increase   =   $12,796   −   $9258
                        =   $3538

Next, find the percent increase. <u>The percent increase or percent decrease is always a</u> percent of the original number or, in this case, the old cost.

In words:   increase   is   what percent increase   of   old cost

Translate:   $3538   =   $x$   ·   $9258

**3.** SOLVE.

$$3538 = x \cdot 9258 \qquad \text{Divide both sides by 9258.}$$
$$0.382 \approx x \qquad \text{Round to 3 decimal places.}$$
$$38.2\% \approx x \qquad \text{Write as a percent.}$$

**4.** INTERPRET.

**Check:**   Check the proposed solution, as shown in Step 1.

**State:**   The percent increase in cost is approximately 38.2%.    □

**PRACTICE**
**5**   The average price of a single family home in the United States rose from $198,900 in 2000 to $299,800 in 2005. Find the percent increase. Round to the nearest tenth of a percent. (*Source:* Federal Housing Finance Board)

Percent decrease is found using a similar method. First find the decrease, then determine what percent of the original or first amount is that decrease.

Read the next example carefully. For Example 5, we were asked to find percent increase. In Example 6, we are given the percent increase and asked to find the number before the increase.

**EXAMPLE 6**   Most of the movie screens globally project analog film, but the number of cinemas using digital is increasing. Find the number of digital screens worldwide last year if, after a 153% increase the number this year is 849. Round to the nearest whole number. (*Source:* Motion Picture Association of America)

*Solution*

**1.** UNDERSTAND. Read and reread the problem. Let's guess a solution and see how we would check our guess. If the number of digital screens worldwide last year was 400, we would see if 400 plus the increase is 849; that is,

$$400 + 153\%(400) = 400 + 1.53(400) = 2.53(400) = 1012$$

Since 1012 is too large, we know that our guess of 400 is too large. We also have a better understanding of the problem. Let

$$x = \text{number of digital screens last year}$$

**2. TRANSLATE.** To translate an equation, we remember that

In words:

| number of digital screens last year | plus | increase | equals | number of digital screens this year |
|---|---|---|---|---|
| ↓ | ↓ | ↓ | ↓ | ↓ |

Translate:  $x$  $+$  $1.53x$  $=$  $849$

**3. SOLVE.**

$$2.53x = 849 \quad \text{Add like terms.}$$
$$x = \frac{849}{2.53}$$
$$x \approx 336$$

**4. INTERPRET.**

**Check:** Recall that $x$ represents the number of digital screens worldwide last year. If this number is approximately 336, let's see if 336 plus the increase is close to 849. (We use the word "close" since 336 is rounded.)

$$336 + 153\%(336) = 336 + 1.53(336) = 2.53(336) = 850.08$$

which is close to 849.

**State:** There were approximately 336 digital screens worldwide last year. ☐

**PRACTICE**
**6**  In 2005, 535 new feature films were released in the United States. This was an increase of 2.8% over the number of new feature films released in 2004. Find the number of new feature films released in 2004. (*Source:* Motion Picture Association of America)

**OBJECTIVE 4 ▶ Solving mixture problems.** Mixture problems involve two or more different quantities being combined to form a new mixture. These applications range from Dow Chemical's need to form a chemical mixture of a required strength to Planter's Peanut Company's need to find the correct mixture of peanuts and cashews, given taste and price constraints.

**EXAMPLE 7**  **Calculating Percent for a Lab Experiment**

A chemist working on his doctoral degree at Massachusetts Institute of Technology needs 12 liters of a 50% acid solution for a lab experiment. The stockroom has only 40% and 70% solutions. How much of each solution should be mixed together to form 12 liters of a 50% solution?

*Solution:*

**1. UNDERSTAND.** First, read and reread the problem a few times. Next, guess a solution. Suppose that we need 7 liters of the 40% solution. Then we need $12 - 7 = 5$ liters of the 70% solution. To see if this is indeed the solution, find the amount of pure acid in 7 liters of the 40% solution, in 5 liters of the 70% solution, and in 12 liters of a 50% solution, the required amount and strength.

| number of liters | × | acid strength | = | amount of pure acid |
|---|---|---|---|---|
| ↓ | | ↓ | | ↓ |
| 7 liters | × | 40% | = | 7(0.40) or 2.8 liters |
| 5 liters | × | 70% | = | 5(0.70) or 3.5 liters |
| 12 liters | × | 50% | = | 12(0.50) or 6 liters |

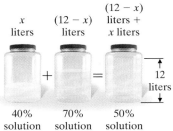

x
liters

(12 − x)
liters

(12 − x)
liters +
x liters

40%
solution

70%
solution

50%
solution

12
liters

Since 2.8 liters + 3.5 liters = 6.3 liters and not 6, our guess is incorrect, but we have gained some valuable insight into how to model and check this problem.

Let

$$x = \text{number of liters of 40\% solution; then}$$
$$12 - x = \text{number of liters of 70\% solution.}$$

**2.** TRANSLATE. To help us translate to an equation, the following table summarizes the information given. Recall that the amount of acid in each solution is found by multiplying the acid strength of each solution by the number of liters.

| | *No. of Liters* · | *Acid Strength* = | *Amount of Acid* |
|---|---|---|---|
| **40% Solution** | $x$ | 40% | $0.40x$ |
| **70% Solution** | $12 - x$ | 70% | $0.70(12 - x)$ |
| **50% Solution Needed** | 12 | 50% | $0.50(12)$ |

*(handwritten note: Why isn't 12−X in the 40% column too?)*

The amount of acid in the final solution is the sum of the amounts of acid in the two beginning solutions.

In words:  | acid in 40% solution | + | acid in 70% solution | = | acid in 50% mixture |

Translate:  $0.40x$   +   $0.70(12 - x)$   =   $0.50(12)$

**3.** SOLVE.

$$0.40x + 0.70(12 - x) = 0.50(12)$$
$$0.4x + 8.4 - 0.7x = 6 \qquad \text{Apply the distributive property.}$$
$$-0.3x + 8.4 = 6 \qquad \text{Combine like terms.}$$
$$-0.3x = -2.4 \qquad \text{Subtract 8.4 from both sides.}$$
$$x = 8 \qquad \text{Divide both sides by } -0.3.$$

**4.** INTERPRET.

**Check:**   To check, recall how we checked our guess.

**State:**   If 8 liters of the 40% solution are mixed with 12 − 8 or 4 liters of the 70% solution, the result is 12 liters of a 50% solution.   □

**PRACTICE**
**7**   Hamida Barash was responsible for refilling the eye wash stations in the chemistry lab. She needed 6 liters of 3% strength eyewash to refill the dispensers. The supply room only had 2% and 5% eyewash in stock. How much of each solution should she mix to produce the needed 3% strength eyewash?

*(handwritten: 6 · 0.3)*
*(handwritten: X, 6 − X)*

# VOCABULARY & READINESS CHECK

*Tell whether the percent labels in the circle graphs are correct.*

**1.**
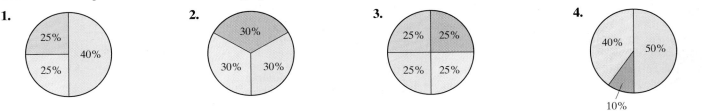
25%  40%  25%

**2.**
30%  30%  30%

**3.**
25%  25%  25%  25%

**4.**
40%  50%  10%

## 2.6 EXERCISE SET

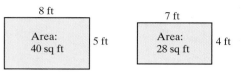

*Find each number described. See Examples 1 and 2.*

1. What number is 16% of 70?
2. What number is 88% of 1000?
3. The number 28.6 is what percent of 52?
4. The number 87.2 is what percent of 436?
5. The number 45 is 25% of what number?
6. The number 126 is 35% of what number?

*The circle graph below shows the uses of U.S. corn production. Use this graph for Exercises 7 through 10. See Example 3.*

**U.S. Corn Production Use**

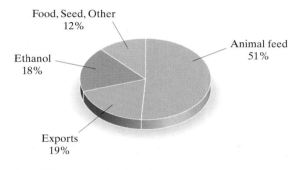

*Source*: USDA, American Farm Bureau Federation

7. What percent of corn production is used for animal feed or ethanol?
8. What percent of corn production is *not* used for exports?
9. The U.S. corn production in 2006–2007 was 10,535 million bushels. How many bushels were used to make ethanol?
10. How many bushels of the 2006–2007 corn production was used for food, seed, or other? (See Exercise 9.)

*Solve. If needed, round answers to the nearest cent. See Example 4.*

11. A used automobile dealership recently reduced the price of a used compact car by 8%. If the price of the car before discount was $18,500, find the discount and the new price.
12. A music store is advertising a 25%-off sale on all new releases. Find the discount and the sale price of a newly released CD that regularly sells for $12.50.
13. A birthday celebration meal is $40.50 including tax. Find the total cost if a 15% tip is added to the cost.
14. A retirement dinner for two is $65.40 including tax. Find the total cost if a 20% tip is added to the cost.

*Solve. See Example 5.*

15. The number of different cars sold in the United States rose from 208 in 1997 to 280 in 2007. Find the percent increase. Round to the nearest whole percent. (*Source: New York Times*, May 2007)
16. The cost of attending a private college rose from $19,000 in 2000 to $22,200 in 2006. Find the percent increase. Round to the nearest whole percent.

17. By decreasing each dimension by 1 unit, the area of a rectangle decreased from 40 square feet (on the left) to 28 square feet (on the right). Find the percent decrease in area.

| | | |
|---|---|---|
| 8 ft | | 7 ft |
| Area: 40 sq ft | 5 ft | Area: 28 sq ft   4 ft |

18. By decreasing the length of the side by one unit, the area of a square decreased from 100 square meters to 81 square meters. Find the percent decrease in area.

| 10 m | 9 m |
|---|---|
| Area: 100 sq m | Area: 81 sq m |

*Solve. See Example 6.*

19. Find the original price of a pair of shoes if the sale price is $78 after a 25% discount.
20. Find the original price of a popular pair of shoes if the increased price is $80 after a 25% increase.
21. Find last year's salary if after a 4% pay raise, this year's salary is $44,200.
22. Find last year's salary if after a 3% pay raise, this year's salary is $55,620.

*Solve. For each exercise, a table is given for you to complete and use to write an equation that models the situation. See Example 7.*

23. How much pure acid should be mixed with 2 gallons of a 40% acid solution in order to get a 70% acid solution?

| | Number of Gallons | · | Acid Strength | = | Amount of Acid |
|---|---|---|---|---|---|
| **Pure Acid** | X | | 100% | | X |
| **40% Acid Solution** | 2−x | | .40 | | .40(2−x) |
| **70% Acid Solution Needed** | 2 | | .70 | | 2(.70) |

24. How many cubic centimeters (cc) of a 25% antibiotic solution should be added to 10 cubic centimeters of a 60% antibiotic solution in order to get a 30% antibiotic solution?

| | Number of Cubic cm | · | Antibiotic Strength | = | Amount of Antibiotic |
|---|---|---|---|---|---|
| **25% Antibiotic Solution** | | | | | |
| **60% Antibiotic Solution** | | | | | |
| **30% Antibiotic Solution Needed** | | | | | |

**25.** Community Coffee Company wants a new flavor of Cajun coffee. How many pounds of coffee worth $7 a pound should be added to 14 pounds of coffee worth $4 a pound to get a mixture worth $5 a pound?

|  | Number of Pounds | · | Cost per Pound | = | Value |
|---|---|---|---|---|---|
| **$7 per lb Coffee** | X |  | 7 |  | 7x |
| **$4 per lb Coffee** | 14 - x |  | 4 |  | 4(14-x) |
| **$5 per lb Coffee Wanted** | x + 14 |  | 5 |  |  |

**26.** Planter's Peanut Company wants to mix 20 pounds of peanuts worth $3 a pound with cashews worth $5 a pound in order to make an experimental mix worth $3.50 a pound. How many pounds of cashews should be added to the peanuts?

$$60 + 5x = 3.50(20 + x)$$

|  | Number of Pounds | · | Cost per Pound | = | Value |
|---|---|---|---|---|---|
| **$3 per lb Peanuts** | 20 |  | 3 |  | 60 |
| **$5 per lb Cashews** | X |  | 5 |  |  |
| **$3.50 per lb Mixture Wanted** | 20+x |  | 3.50 |  | 3.50(20+x) |

## MIXED PRACTICE

*Solve. If needed, round money amounts to two decimal places and all other amounts to one decimal place. See Examples 1 through 7.*

**27.** Find 23% of 20.

**28.** Find 140% of 86.

**29.** The number 40 is 80% of what number?

**30.** The number 56.25 is 45% of what number?

**31.** The number 144 is what percent of 480?

**32.** The number 42 is what percent of 35?

*The graph shows the communities in the United States that have the highest percents of citizens that shop by catalog. Use the graph to answer Exercises 33 through 36.*

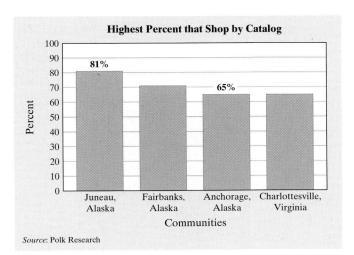

**Highest Percent that Shop by Catalog**

Source: Polk Research

**33.** Estimate the percent of the population in Fairbanks, Alaska, who shops by catalog.

**34.** Estimate the percent of the population in Charlottesville, Virginia, who shops by catalog.

**35.** According to the *World Almanac*, Anchorage has a population of 275,043. How many catalog shoppers might we predict live in Anchorage? Round to the nearest whole number.

**36.** According to the *World Almanac*, Juneau has a population of 30,987. How many catalog shoppers might we predict live in Juneau? Round to the nearest whole number.

*For Exercises 37 and 38, fill in the percent column in each table. Each table contains a worked-out example.*

**37.**

| Ford Motor Company Model Year 2006 Vehicle Sales Worldwide | | |
|---|---|---|
|  | Thousands of Vehicles | Percent of Total (Rounded to Nearest Percent) |
| **North America** | 3051 |  |
| **Europe** | 1846 |  |
| **Asia-Pacific-Africa** | 589 |  |
| **South America** | 381 |  |
| **Rest of the World** | 730 | Example: $\frac{730}{6597} \approx 11\%$ |
| **Total** | 6597 |  |

Source: Ford Motor Company

**38.**

| Kraft Foods North America Volume Food Produced in a year | | |
|---|---|---|
| **Food Group** | Volume (in pounds) | Percent (Round to Nearest Percent) |
| **Cheese, Meals, and Enhancers** | 6183 |  |
| **Biscuits, Snacks, and Confectionaries** | 2083 | Example: $\frac{2083}{13,741} \approx 15\%$ |
| **Beverages, Desserts, and Cereals** | 3905 |  |
| **Oscar Mayer and Pizza** | 1570 |  |
| **Total** | 13,741 |  |

Source: Kraft Foods, North America

**39.** Nordstrom advertised a 25%-off sale. If a London Fog coat originally sold for $256, find the decrease in price and the sale price.

**40.** A gasoline station decreased the price of a $0.95 cola by 15%. Find the decrease in price and the new price.

**41.** Iceberg lettuce is grown and shipped to stores for about 40 cents a head, and consumers purchase it for about 86 cents a head. Find the percent increase. (*Source: Statistical Abstract of the United States*)

**42.** The lettuce consumption per capita in 1980 was about 25.6 pounds, and in 2005 the consumption dropped to about 22.4 pounds. Find the percent decrease. (*Source: Statistical Abstract of the United States.*)

**43.** Smart Cards (cards with an embedded computer chip) have been growing in popularity in recent years. In 2006, about 1900 million Smart Cards were expected to be issued. This represents a 726% increase from the number of cards that were issued in 2001. How many Smart Cards were issued in 2001? Round to the nearest million. (*Source:* The Freedonia Group)

**44.** Fuel ethanol production is projected to be 10,800 million gallons in 2009. This represents a 44% increase from the number of gallons produced in 2007. How many millions of gallons were produced in 2007? (*Source:* Renewable Fuels Association)

**45.** How much of an alloy that is 20% copper should be mixed with 200 ounces of an alloy that is 50% copper in order to get an alloy that is 30% copper?

**46.** How much water should be added to 30 gallons of a solution that is 70% antifreeze in order to get a mixture that is 60% antifreeze?

**47.** A junior one-day admission to Hershey Park amusement park in Hershey, Pennsylvania, is $27. This price is increased by 70% for (nonsenior) adults. Find the mark-up and the adult price. (*Note:* Prices given are approximations.)

**48.** The price of a biology book recently increased by 10%. If this book originally cost $99.90, find the mark-up and the new price.

**49.** By doubling each dimension, the area of a parallelogram increased from 36 square centimeters to 144 square centimeters. Find the percent increase in area.

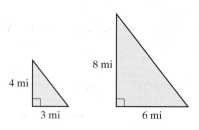

**50.** By doubling each dimension, the area of a triangle increased from 6 square miles to 24 square miles. Find the percent increase in area.

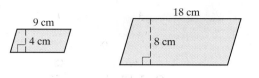

**51.** A company recently downsized its number of employees by 35%. If there are still 78 employees, how many employees were there prior to the layoffs?

**52.** The average number of children born to each U.S. woman has decreased by 44% since 1920. If this average is now 1.9, find the average in 1920. Round to the nearest tenth.

**53.** The owner of a local chocolate shop wants to develop a new trail mix. How many pounds of chocolate-covered peanuts worth $5 a pound should be mixed with 10 pounds of granola bites worth $2 a pound to get a mixture worth $3 per pound?

**54.** A new self-tanning lotion for everyday use is to be sold. First, an experimental lotion mixture is made by mixing 800 ounces of everyday moisturizing lotion worth $0.30 an ounce with self-tanning lotion worth $3 per ounce. If the experimental lotion is to cost $1.20 per ounce, how many ounces of the self-tanning lotion should be in the mixture?

**55.** The number of farms in the United States was 2.19 million in 2000. By 2006, the number had dropped to 2.09 million. What was the percent of decrease? Round to the nearest tenth of a percent. (*Source:* USDA: National Agricultural Statistical Service)

**56.** The average size of farms in the United States was 436 acres in 2000. By 2005, the average size had increased to 444 acres. What was the percent increase? Round to the nearest tenth of a percent. (*Source:* USDA: National Agricultural Statistical Service)

**57.** The number of Supreme Court decisions has been decreasing in recent years. During the 2005–2006 term, 182 decisions were announced. This is a 45.7% decrease from the number of decisions announced during the 1982–1983 term. How many decisions were announced during 1982–1983? Round to the nearest whole. (*Source: World Almanac*)

**58.** The total number of movie screens in the United States has been increasing in recent years. In 2005, there were 37,092 indoor movie screens. This is a 4.3% increase from the number of indoor movie screens in 2000. How many movie screens were operating in 2000? Round to the nearest whole. (*Source:* National Association of Theater Owners)

**59.** Scoville units are used to measure the hotness of a pepper. Measuring 577 thousand Scoville units, the "Red Savina" habañero pepper was known as the hottest chili pepper. That has recently changed with the discovery of Naga Jolokia pepper from India. It measures 48% hotter than the habañero. Find the measure of the Naga Jolokia pepper. Round to the nearest thousand units.

**60.** At this writing, the women's world record for throwing a disc (like a heavy Frisbee) was set by Jennifer Griffin of the United States in 2000. Her throw was 138.56 meters. The men's world record was set by Christian Sandstrom of Sweden in 2002. His throw was 80.4% farther than Jennifer's. Find the distance of his throw. Round to the nearest meter. (*Source:* World Flying Disc Federation)

**61.** A recent survey showed that 42% of recent college graduates named flexible hours as their most desired employment benefit. In a graduating class of 860 college students, how many would you expect to rank flexible hours as their top priority in job benefits? (Round to the nearest whole.) (*Source:* JobTrak.com)

**62.** A recent survey showed that 64% of U.S. colleges have Internet access in their classrooms. There are approximately 9800 post-secondary institutions in the United States. How many of these would you expect to have Internet access in their classrooms? (*Source:* Market Data Retrieval, National Center for Education Statistics)

**REVIEW AND PREVIEW**

*Place* $<, >,$ *or* $=$ *in the appropriate space to make each a true statement. See Sections 1.2, 1.4, and 1.7.*

**63.** $-5$   $-7$

**64.** $\dfrac{12}{3}$   $2^2$

**65.** $|-5|$   $-(-5)$

**66.** $-3^3$   $(-3)^3$

**67.** $(-3)^2$   $-3^2$

**68.** $|-2|$   $-|-2|$

**CONCEPT EXTENSIONS**

**69.** Is it possible to mix a 10% acid solution and a 40% acid solution to obtain a 60% acid solution? Why or why not?

**70.** Must the percents in a circle graph have a sum of 100%? Why or why not?

**71.** A trail mix is made by combining peanuts worth $3 a pound, raisins worth $2 a pound, and M & M's worth $4 a pound. Would it make good business sense to sell the trail mix for $1.98 a pound? Why or why not?

**72. a.** Can an item be marked-up by more than 100%? Why or why not?

   **b.** Can an item be discounted by more than 100%? Why or why not?

*Standardized nutrition labels like the one below have been displayed on food items since 1994. The percent column on the right shows the percent of daily values (based on a 2000-calorie diet) shown at the bottom of the label. For example, a serving of this food contains 4 grams of total fat, where the recommended daily fat based on a 2000-calorie diet is less than 65 grams of fat. This means that* $\dfrac{4}{65}$ *or approximately 6% (as shown) of your daily recommended fat is taken in by eating a serving of this food. Use this nutrition label to answer Exercises 73 through 75.*

**Nutrition Facts**

Serving Size 18 Crackers (31g)
Servings Per Container About 9

Amount Per Serving

Calories 130          Calories from Fat 35

% Daily Value*

| | |
|---|---|
| Total Fat 4g | 6% |
| Saturated Fat 0.5g | 3% |
| Polyunsaturated Fat 0g | |
| Monounsaturated Fat 1.5g | |
| Cholesterol 0mg | 0% |
| Sodium 230mg | $x$ |
| Total Carbohydrate 23g | $y$ |
| Dietary Fiber 2g | 8% |
| Sugars 3g | |
| Protein 2g | |

Vitamin A 0%    •    Vitamin C 0%
Calcium 2%    •    Iron 6%

* Percent Daily Values are based on a 2,000 calorie diet. Your daily values may be higher or lower depending on your calorie needs.

| | | Calories | 2,000 | 2,500 |
|---|---|---|---|---|
| Total Fat | Less than | | 65g | 80g |
| Sat. Fat | Less than | | 20g | 25g |
| Cholesterol | Less than | | 300mg | 300mg |
| Sodium | Less than | | 2400mg | 2400mg |
| Total Carbohydrate | | | 300g | 375g |
| Dietary Fiber | | | 25g | 30g |

**73.** Based on a 2000-calorie diet, what percent of daily value of sodium is contained in a serving of this food? In other words, find $x$ in the label. (Round to the nearest tenth of a percent.)

**74.** Based on a 2000-calorie diet, what percent of daily value of total carbohydrate is contained in a serving of this food? In other words, find $y$ in the label. (Round to the nearest tenth of a percent.)

**75.** Notice on the nutrition label that one serving of this food contains 130 calories and 35 of these calories are from fat. Find the percent of calories from fat. (Round to the nearest tenth of a percent.) It is recommended that no more than 30% of calorie intake come from fat. Does this food satisfy this recommendation?

*Use the nutrition label below to answer Exercises 76 through 78.*

**NUTRITIONAL INFORMATION PER SERVING**

| Serving Size: 9.8 oz. | Servings Per Container: 1 |
|---|---|
| Calories . . . . . . . . . . . . . . . .280 | Polyunsaturated Fat . . . . .1g |
| Protein . . . . . . . . . . . . . . . . .12g | Saturated Fat . . . . . . . . . . 3g |
| Carbohydrate . . . . . . . . . . . 45g | Cholesterol . . . . . . . . . 20mg |
| Fat . . . . . . . . . . . . . . . . . . . . .6g | Sodium . . . . . . . . . . . 520mg |
| Percent of Calories from Fat....? | Potassium . . . . . . . . . 220mg |

**76.** If fat contains approximately 9 calories per gram, find the percent of calories from fat in one serving of this food. (Round to the nearest tenth of a percent.)

**77.** If protein contains approximately 4 calories per gram, find the percent of calories from protein from one serving of this food. (Round to the nearest tenth of a percent.)

**78.** Find a food that contains more than 30% of its calories per serving from fat. Analyze the nutrition label and verify that the percents shown are correct.

# 2.7 FURTHER PROBLEM SOLVING

**OBJECTIVES**

**1** Solve problems involving distance.

**2** Solve problems involving money.

**3** Solve problems involving interest.

This section is devoted to solving problems in the categories listed. The same problem-solving steps used in previous sections are also followed in this section. They are listed below for review.

**General Strategy for Problem Solving**

**1.** UNDERSTAND the problem. During this step, become comfortable with the problem. Some ways of doing this are:

Read and reread the problem.

Choose a variable to represent the unknown.

Construct a drawing, whenever possible.

Propose a solution and check. Pay careful attention to how you check your proposed solution. This will help writing an equation to model the problem.

**2.** TRANSLATE the problem into an equation.

**3.** SOLVE the equation.

**4.** INTERPRET the results: *Check* the proposed solution in the stated problem and *state* your conclusion.

**OBJECTIVE 1 ▶ Solving distance problems.** Our first example involves distance. For a review of the distance formula, $d = r \cdot t$, see Section 2.5, Example 1 and the table before the example.

**EXAMPLE 1**    **Finding Time Given Rate and Distance**

Marie Antonio, a bicycling enthusiast, rode her 21-speed at an average speed of 18 miles per hour on level roads and then slowed down to an average of 10 mph on the hilly roads of the trip. If she covered a distance of 98 miles, how long did the entire trip take if traveling the level roads took the same time as traveling the hilly roads?

*Solution*

**1.** UNDERSTAND the problem. To do so, read and reread the problem. The formula $d = r \cdot t$ is needed. At this time, let's guess a solution. Suppose that she spent 2 hours traveling on the level roads. This means that she also spent 2 hours traveling on the hilly roads, since the times spent were the same. What is her total distance? Her distance on the level road is rate $\cdot$ time = 18(2) = 36 miles. Her distance on the hilly roads is rate $\cdot$ time = 10(2) = 20 miles. This gives a total distance of 36 miles + 20 miles = 56 miles, not the correct distance of 98 miles. Remember that the purpose of guessing a solution is not to guess correctly (although this may happen) but to help better understand the problem and how to model it with an equation. We are looking for the length of the entire trip, so we begin by letting

$$x = \text{the time spent on level roads.}$$

Because the same amount of time is spent on hilly roads, then also

$$x = \text{the time spent on hilly roads.}$$

2. TRANSLATE. To help us translate to an equation, we now summarize the information from the problem on the following chart. Fill in the rates given, the variables used to represent the times, and use the formula $d = r \cdot t$ to fill in the distance column.

|  | *Rate · Time = Distance* | | |
|---|---|---|---|
| *Level* | 18 | $x$ | $18x$ |
| *Hilly* | 10 | $x$ | $10x$ |

Since the entire trip covered 98 miles, we have that

In words:   total distance   =   level distance   +   hilly distance

Translate:         98          =         $18x$          +         $10x$

3. SOLVE.

$$98 = 28x \quad \text{Add like terms.}$$
$$\frac{98}{28} = \frac{28x}{28} \quad \text{Divide both sides by 28.}$$
$$3.5 = x$$

4. INTERPRET the results.

**Check:**   Recall that $x$ represents the time spent on the level portion of the trip and also the time spent on the hilly portion. If Marie rides for 3.5 hours at 18 mph, her distance is $18(3.5) = 63$ miles. If Marie rides for 3.5 hours at 10 mph, her distance is $10(3.5) = 35$ miles. The total distance is 63 miles + 35 miles = 98 miles, the required distance.

**State:**   The time of the entire trip is then 3.5 hours + 3.5 hours or 7 hours.   ☐

**PRACTICE**

**1**   Sat Tranh took a short hike with his friends up Mt. Wachusett. They hiked uphill at a steady pace of 1.5 miles per hour, and downhill at a rate of 4 miles per hour. If the time to climb the mountain took an hour more than the time to hike down, how long was the entire hike?

**EXAMPLE 2**   **Finding Train Speeds**

The Kansas City Southern Railway operates in 10 states and Mexico. Suppose two trains leave Neosho, Missouri, at the same time. One travels north and the other travels south at a speed that is 15 miles per hour faster. In 2 hours, the trains are 230 miles apart. Find the speed of each train.

*Kansas City Southern Railway*

*Solution*

1. UNDERSTAND the problem. Read and reread the problem. Guess a solution and check. Let's let

$$x = \text{speed of train traveling north}$$

Because the train traveling south is 15 mph faster, we have

$$x + 15 = \text{speed of train traveling south}$$

2. TRANSLATE. Just as for Example 1, let's summarize our information on a chart. Use the formula $d = r \cdot t$ to fill in the distance column.

|  | $r$ | $\cdot$ | $t$ | $=$ | $d$ |
|---|---|---|---|---|---|
| **North Train** | $x$ | | 2 | | $2x$ |
| **South Train** | $x + 15$ | | 2 | | $2(x + 15)$ |

Since the total distance between the trains is 230 miles, we have

| In words: | north train distance | $+$ | south train distance | $=$ | total distance |
|---|---|---|---|---|---|
|  | $\downarrow$ | | $\downarrow$ | | $\downarrow$ |
| Translate: | $2x$ | $+$ | $2(x + 15)$ | $=$ | 230 |

3. SOLVE.

$$
\begin{aligned}
2x + 2x + 30 &= 230 && \text{Use the distributive property.}\\
4x + 30 &= 230 && \text{Combine like terms.}\\
4x &= 200 && \text{Subtract 30 from both sides.}\\
\frac{4x}{4} &= \frac{200}{4} && \text{Divide both sides by 4.}\\
x &= 50 && \text{Simplify.}
\end{aligned}
$$

4. INTERPRET the results.

**Check:** Recall that $x$ is the speed of the train traveling north, or 50 mph. In 2 hours, this train travels a distance of $2(50) = 100$ miles. The speed of the train traveling south is $x + 15$ or $50 + 15 = 65$ mph. In 2 hours, this train travels $2(65) = 130$ miles. The total distance of the trains is $100$ miles $+ 130$ miles $= 230$ miles, the required distance.

**State:** The northbound train's speed is 50 mph and the southbound train's speed is 65 mph. □

**PRACTICE**

**2**  The Kansas City Southern Railway has a station in Mexico City, Mexico. Suppose two trains leave Mexico City at the same time. One travels east and the other west at a speed that is 10 mph slower. In 1.5 hours, the trains are 171 miles apart. Find the speed of each train.

**OBJECTIVE 2 ▶ Solving money problems.** The next example has to do with finding an unknown number of a certain denomination of coin or bill. These problems are extremely useful in that they help you understand the difference between the number of coins or bills and the total value of the money.

For example, suppose there are seven $5-bills. The *number* of $5-bills is 7 and the *total value* of the money is $5(7) = $35.

Study the table below for more examples.

| Denomination of Coin or Bill | Number of Coins or Bills | Value of Coins or Bills |
|---|---|---|
| 20-dollar bills | 17 | $20(17) = \$340$ |
| nickels | 31 | $\$0.05(31) = \$1.55$ |
| quarters | $x$ | $\$0.25(x) = \$0.25x$ |

### EXAMPLE 3   Finding Numbers of Denominations

Part of the proceeds from a local talent show was $2420 worth of $10 and $20 bills. If there were 37 more $20 bills than $10 bills, find the number of each denomination.

*Solution*

**1.** UNDERSTAND the problem. To do so, read and reread the problem. If you'd like, let's guess a solution. Suppose that there are 25 $10 bills. Since there are 37 more $20 bills, we have $25 + 37 = 62$ $20 bills. The total amount of money is $\$10(25) + \$20(62) = \$1490$, below the given amount of $2420. Remember that our purpose for guessing is to help us better understand the problem.

We are looking for the number of each denomination, so we let

$$x = \text{number of } \$10 \text{ bills}$$

There are 37 more $20 bills, so

$$x + 37 = \text{number of } \$20 \text{ bills}$$

**2.** TRANSLATE. To help us translate to an equation, study the table below

| Denomination | Number of Bills | Value of Bills (in dollars) |
|---|---|---|
| $10 bills | $x$ | $10x$ |
| $20 bills | $x + 37$ | $20(x + 37)$ |

Since the total value of these bills is $2420, we have

In words:   value of $10 bills   plus   value of $20 bills   is   2420

Translate:   $10x$   $+$   $20(x + 37)$   $=$   $2420$

**3.** SOLVE:
$$10x + 20x + 740 = 2420 \quad \text{Use the distributive property.}$$
$$30x + 740 = 2420 \quad \text{Add like terms.}$$
$$30x = 1680 \quad \text{Subtract 740 from both sides.}$$
$$\frac{30x}{30} = \frac{1680}{30} \quad \text{Divide both sides by 30.}$$
$$x = 56$$

**4.** INTERPRET the results.

**Check:**   Since $x$ represents the number of $10 bills, we have 56 $10 bills and $56 + 37$, or 93 $20 bills. The total amount of these bills is $\$10(56) + \$20(93) = \$2420$, the correct total.

**State:**   There are 56 $10 bills and 93 $20 bills.   □

PRACTICE

**3**  A stack of $5 and $20 bills was counted by the treasurer of an organization. The total value of the money was $1710 and there were 47 more $5 bills than $20 bills. Find the number of each type of bill.

**OBJECTIVE 3 ▶ Solving interest problems.** The next example is an investment problem. For a review of the simple interest formula, $I = PRT$, see the table at the beginning of Section 2.5 and also Exercises 11 and 12 in that exercise set.

**EXAMPLE 4**  Finding the Investment Amount  ~~total~~

Rajiv Puri invested part of his $20,000 inheritance in a mutual funds account that pays 7% simple interest yearly and the rest in a certificate of deposit that pays 9% simple interest yearly. At the end of one year, Rajiv's investments earned $1550. Find the amount he invested at each rate.

*Solution*

1. UNDERSTAND. Read and reread the problem. Next, guess a solution. Suppose that Rajiv invested $8000 in the 7% fund and the rest, $12,000, in the fund paying 9%. To check, find his interest after one year. Recall the formula, $I = PRT$, so the interest from the 7% fund = $8000(0.07)(1) = $560. The interest from the 9% fund = $12,000(0.09)(1) = $1080. The sum of the interests is $560 + $1080 = $1640. Our guess is incorrect, since the sum of the interests is not $1550, but we now have a better understanding of the problem.

    Let

$$x = \text{amount of money in the account paying 7\%.}$$

    The rest of the money is $20,000 less $x$ or

$$20,000 - x = \text{amount of money in the account paying 9\%.}$$

2. TRANSLATE. We apply the simple interest formula $I = PRT$ and organize our information in the following chart. Since there are two different rates of interest and two different amounts invested, we apply the formula twice.

| | **Principal** · | **Rate** · | **Time** = | **Interest** |
|---|---|---|---|---|
| **7% Fund** | $x$ | 0.07 | 1 | $x(0.07)(1)$ or $0.07x$ |
| **9% Fund** | $20,000 - x$ | 0.09 | 1 | $(20,000 - x)(0.09)(1)$ or $0.09(20,000 - x)$ |
| **Total** | 20,000 | | | 1550 |

*how do u know to put 9% and inheritance together*

The total interest earned, $1550, is the sum of the interest earned at 7% and the interest earned at 9%.

In words:  | interest at 7% | + | interest at 9% | = | total interest |

Translate:  $0.07x$  +  $0.09(20,000 - x)$  =  1550

3. SOLVE.  *↳ what is x?*

$$0.07x + 0.09(20,000 - x) = 1550$$
$$0.07x + 1800 - 0.09x = 1550 \quad \text{Apply the distributive property.}$$
$$1800 - 0.02x = 1550 \quad \text{Combine like terms.}$$
$$-0.02x = -250 \quad \text{Subtract 1800 from both sides.}$$
$$x = 12,500 \quad \text{Divide both sides by } -0.02.$$

**4. INTERPRET.**

**Check:** If $x = 12{,}500$, then $20{,}000 - x = 20{,}000 - 12{,}500$ or 7500. These solutions are reasonable, since their sum is $20,000 as required. The annual interest on $12,500 at 7% is $875; the annual interest on $7500 at 9% is $675, and $875 + $675 = $1550.

**State:** The amount invested at 7% is $12,500. The amount invested at 9% is $7500. □

**PRACTICE**
**4**   Suzanne Scarpulla invested $30,000, part of it in a high-risk venture that yielded 11.5% per year, and the rest in a secure mutual fund paying interest of 6% per year. At the end of one year, Suzanne's investments earned $2790. Find the amount she invested at each rate.

## 2.7 EXERCISE SET

**MyMathLab**  *Powered by CourseCompass and MathXL*

Math XP — PRACTICE   WATCH   DOWNLOAD   READ   REVIEW

*Solve. See Examples 1 and 2.*

1. A jet plane traveling at 500 mph overtakes a propeller plane traveling at 200 mph that had a 2-hour head start. How far from the starting point are the planes?

2. How long will it take a bus traveling at 60 miles per hour to overtake a car traveling at 40 mph if the car had a 1.5-hour head start?

3. A bus traveled on a level road for 3 hours at an average speed 20 miles per hour faster than it traveled on a winding road. The time spent on the winding road was 4 hours. Find the average speed on the level road if the entire trip was 305 miles.

4. The Jones family drove to Disneyland at 50 miles per hour and returned on the same route at 40 mph. Find the distance to Disneyland if the total driving time was 7.2 hours.

*Complete the table. The first and sixth rows have been completed for you. See Example 3.*

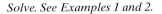

*Tells me*

| | Number of Coins or Bills | Value of Coins or Bills (in dollars) |
|---|---|---|
| **pennies** | $x$ | $0.01x$ |
| 5. **dimes** | $y$ | |
| 6. **quarters** | $z$ | |
| 7. **nickels** | $(x + 7)$ | |
| 8. **half-dollars** | $(20 - z)$ | |
| **$5 bills** | $9x$ | $5(9x)$ |
| 9. **$20 bills** | $4y$ | |
| 10. **$100 bills** | $97z$ | |
| 11. **$50 bills** | $(35 - x)$ | |
| 12. **$10 bills** | $(15 - y)$ | |

*expression for Value*

*.25z   → 5 Times   how many I have*

13. Part of the proceeds from a garage sale was $280 worth of $5 and $10 bills. If there were 20 more $5 bills than $10 bills, find the number of each denomination.

| | Number of Bills | Value of Bills |
|---|---|---|
| **$5 bills** | | |
| **$10 bills** | | |
| **Total** | | |

14. A bank teller is counting $20 and $50-dollar bills. If there are six times as many $20 bills as $50 bills and the total amount of money is $3910, find the number of each denomination.

*Value of one · amt = $*

| | Number of Bills | Value of Bills |
|---|---|---|
| 1st **$20 bills** | $6x$ | $120 \cdot x$ |
| 2nd **$50 bills** | $x$ | $50 \cdot x$ |
| **Total** | $3910$ | $3910$ |

$120x + 50x = 3910$

*Solve. See Example 4.*

15. Zoya Lon invested part of her $25,000 advance at 8% annual simple interest and the rest at 9% annual simple interest. If her total yearly interest from both accounts was $2135, find the amount invested at each rate.

16. Karen Waugtal invested some money at 9% annual simple interest and $250 more than that amount at 10% annual simple interest. If her total yearly interest was $101, how much was invested at each rate?

17. Sam Mathius invested part of his $10,000 bonus in a fund that paid an 11% profit and invested the rest in stock that suffered a 4% loss. Find the amount of each investment if his overall net profit was $650.

18. Bruce Blossum invested a sum of money at 10% annual simple interest and invested twice that amount at 12% annual

simple interest. If his total yearly income from both investments was $2890, how much was invested at each rate?

**19.** The Concordia Theatre contains 500 seats and the ticket prices for a recent play were $43 for adults and $28 for children. For one matinee, if the total proceeds were $16,805, how many of each type of ticket were sold?

**20.** A zoo in Oklahoma charged $22 for adults and $15 for children. During a summer day, 732 zoo tickets were sold and the total receipts were $12,912. How many children and how many adult tickets were sold?

## MIXED PRACTICE

**21.** How can $54,000 be invested, part at 8% annual simple interest and the remainder at 10% annual simple interest, so that the interest earned by the two accounts will be equal?

**22.** Ms. Mills invested her $20,000 bonus in two accounts. She took a 4% loss on one investment and made a 12% profit on another investment, but ended up breaking even. How much was invested in each account?

**23.** Alan and Dave Schaferkötter leave from the same point driving in opposite directions, Alan driving at 55 miles per hour and Dave at 65 mph. Alan has a one-hour head start. How long will they be able to talk on their car phones if the phones have a 250-mile range?

**24.** Kathleen and Cade Williams leave simultaneously from the same point hiking in opposite directions, Kathleen walking at 4 miles per hour and Cade at 5 mph. How long can they talk on their walkie-talkies if the walkie-talkies have a 20-mile radius?

**25.** A youth organization collected nickels and dimes for a charity drive. By the end of the 1-day drive, the youth had collected $56.35. If there were three times as many dimes as nickels, how many of each type of coin was collected?

**26.** A collection of dimes and quarters are retrieved from a soft drink machine. There are five times as many dimes as quarters and the total value of the coins is $27.75. Find the number of dimes and the number of quarters.

**27.** If $3000 is invested at 6% annual simple interest, how much should be invested at 9% annual simple interest so that the total yearly income from both investments is $585?

**28.** Trudy Waterbury, a financial planner, invested a certain amount of money at 9% annual simple interest, twice that amount at 10% annual simple interest, and three times that amount at 11% annual simple interest. Find the amount invested at each rate if her total yearly income from the investments was $2790.

**29.** Two hikers are 11 miles apart and walking toward each other. They meet in 2 hours. Find the rate of each hiker if one hiker walks 1.1 mph faster than the other.

**30.** Nedra and Latonya Dominguez are 12 miles apart hiking toward each other. How long will it take them to meet if Nedra walks at 3 mph and Latonya walks 1 mph faster?

**31.** Mark Martin can row upstream at 5 mph and downstream at 11 mph. If Mark starts rowing upstream until he gets tired

and then rows downstream to his starting point, how far did Mark row if the entire trip took 4 hours?

**32.** On a 255-mile trip, Gary Alessandrini traveled at an average speed of 70 mph, got a speeding ticket, and then traveled at 60 mph for the remainder of the trip. If the entire trip took 4.5 hours and the speeding ticket stop took 30 minutes, how long did Gary speed before getting stopped?

## REVIEW AND PREVIEW

*Perform the indicated operations. See Sections 1.5 and 1.6.*

**33.** $3 + (-7)$

**34.** $(-2) + (-8)$

**35.** $\dfrac{3}{4} - \dfrac{3}{16}$

**36.** $-11 + 2.9$

**37.** $-5 - (-1)$

**38.** $-12 - 3$

## CONCEPT EXTENSIONS

**39.** A stack of $20, $50, and $100 bills was retrieved as part of an FBI investigation. There were 46 more $50 bills than $100 bills. Also, the number of $20 bills was 7 times the number of $100 bills. If the total value of the money was $9550, find the number of each type of bill.

**40.** A man places his pocket change in a jar every day. The jar is full and his children have counted the change. The total value is $44.86. Let $x$ represent the number of quarters and use the information below to find the number of each type of coin.

There are: 136 more dimes than quarters
         8 times as many nickels as quarters
         32 more than 16 times as many pennies as quarters

*To "break even" in a manufacturing business, revenue R (income)* **must equal** *the cost C of production, or R = C.*

**41.** The cost $C$ to produce $x$ number of skateboards is given by $C = 100 + 20x$. The skateboards are sold wholesale for $24 each, so revenue $R$ is given by $R = 24x$. Find how many skateboards the manufacturer needs to produce and sell to break even. (*Hint:* Set the expression for $R$ equal to the expression for $C$, then solve for $x$.)

**42.** The revenue $R$ from selling $x$ number of computer boards is given by $R = 60x$, and the cost $C$ of producing them is given by $C = 50x + 5000$. Find how many boards must be sold to break even. Find how much money is needed to produce the break-even number of boards.

**43.** The cost $C$ of producing $x$ number of paperback books is given by $C = 4.50x + 2400$. Income $R$ from these books is given by $R = 7.50x$. Find how many books should be produced and sold to break even.

**44.** Find the break-even quantity for a company that makes $x$ number of computer monitors at a cost $C$ given by $C = 870 + 70x$ and receives revenue $R$ given by $R = 105x$.

**45.** Exercises 41 through 44 involve finding the break-even point for manufacturing. Discuss what happens if a company makes and sells fewer products than the break-even point. Discuss what happens if more products than the break-even point are made and sold.

---

## STUDY SKILLS BUILDER

**What to Do the Day of an Exam**

Your first exam may be soon. On the day of an exam, don't forget to try the following:

- Allow yourself plenty of time to arrive.
- Read the directions on the test carefully.
- Read each problem carefully as you take your test. Make sure that you answer the question asked.
- Watch your time and pace yourself so that you may attempt each problem on your test.
- Check your work and answers.
- ***Do not turn your test in early.*** If you have extra time, spend it double-checking your work.

Good luck!

*Answer the following questions based on your most recent mathematics exam, whenever that was.*

1. How soon before class did you arrive?
2. Did you read the directions on the test carefully?
3. Did you make sure you answered the question asked for each problem on the exam?
4. Were you able to attempt each problem on your exam?
5. If your answer to question 4 is no, list reasons why.
6. Did you have extra time on your exam?
7. If your answer to question 6 is yes, describe how you spent that extra time.

---

# 2.8  SOLVING LINEAR INEQUALITIES

## OBJECTIVES

1 Define linear inequality in one variable, graph solution sets on a number line, and use interval notation.

2 Solve linear inequalities.

3 Solve compound inequalities.

4 Solve inequality applications.

**OBJECTIVE 1 ▶ Graphing solution sets to linear inequalities and using interval notation.**
In Chapter 1, we reviewed these inequality symbols and their meanings:

$<$ means "is less than"        $\leq$ means "is less than or equal to"

$>$ means "is greater than"    $\geq$ means "is greater than or equal to"

| *Equations* | *Inequalities* |
|---|---|
| $x = 3$ | $x \leq 3$ |
| $5n - 6 = 14$ | $5n - 6 > 14$ |
| $12 = 7 - 3y$ | $12 \leq 7 - 3y$ |
| $\dfrac{x}{4} - 6 = 1$ | $\dfrac{x}{4} - 6 > 1$ |

A linear inequality is similar to a linear equation except that the equality symbol is replaced with an inequality symbol.

> **Linear Inequality in One Variable**
>
> A **linear inequality in one variable** is an inequality that can be written in the form
>
> $$ax + b < c$$
>
> where $a$, $b$, and $c$ are real numbers and $a$ is not 0.

This definition and all other definitions, properties, and steps in this section also hold true for the inequality symbols, $>$, $\geq$, and $\leq$.

A **solution of an inequality** is a value of the variable that makes the inequality a true statement. The solution set is the set of all solutions. For the inequality $x < 3$, replacing $x$ with any number less than 3, that is, to the left of 3 on a number line, makes the resulting inequality true. This means that any number less than 3 is a solution of the inequality $x < 3$.

Since there are infinitely many such numbers, we cannot list all the solutions of the inequality. We *can* use set notation and write

$\{x \qquad | \qquad x < 3\}$. Recall that this is read

$\uparrow \qquad \uparrow \qquad \underbrace{\qquad}$

the    such    $\uparrow$

set of    that    $x$ is less than 3.

all $x$

We can also picture the solutions on a number line. If we use open/closed-circle notation, the graph of $\{x|x < 3\}$ looks like the following.

In this text, a convenient notation, called **interval notation,** will be used to write solution sets of inequalities. To help us understand this notation, a different graphing notation will be used. Instead of an open circle, we use a parenthesis; instead of a closed circle, we use a bracket. With this new notation, the graph of $\{x|x < 3\}$ now looks like

and can be represented in interval notation as $(-\infty, 3)$. The symbol $-\infty$, read as "negative infinity," does not indicate a number, but does indicate that the shaded arrow to the left never ends. In other words, the interval $(-\infty, 3)$ includes *all* numbers less than 3.

Picturing the solutions of an inequality on a number line is called **graphing** the solutions or graphing the inequality, and the picture is called the **graph** of the inequality.

To graph $\{x|x \le 3\}$ or simply $x \le 3$, shade the numbers to the left of 3 and place a bracket at 3 on the number line. The bracket indicates that 3 **is** a solution: 3 **is** less than or equal to 3. In interval notation, we write $(-\infty, 3]$.

> ▶ **Helpful Hint**
>
> When writing an inequality in interval notation, it may be easier to first graph the inequality, then write it in interval notation. To help, think of the number line as approaching $-\infty$ to the left and $+\infty$ or $\infty$ to the right. Then simply write the interval notation by following your shading from left to right.
>
>

**EXAMPLE 1**    Graph $x \ge -1$. Then write the solutions in interval notation.

**Solution**    We place a bracket at $-1$ since the inequality symbol is $\ge$ and $-1$ is greater than or equal to $-1$. Then we shade to the right of $-1$.

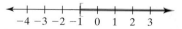

In interval notation, this is $[-1, \infty)$.          ☐

**PRACTICE**

**1**    Graph $x < 5$. Then write the solutions in interval notation.

**OBJECTIVE 2 ▶ Solving linear inequalities.** When solutions of a linear inequality are not immediately obvious, they are found through a process similar to the one used to solve a linear equation. Our goal is to get the variable alone, and we use properties of inequality similar to properties of equality.

---

**Addition Property of Inequality**

If $a$, $b$, and $c$ are real numbers, then

$$a < b \qquad \text{and} \qquad a + c < b + c$$

are equivalent inequalities.

---

This property also holds true for subtracting values, since subtraction is defined in terms of addition. In other words, adding or subtracting the same quantity from both sides of an inequality does not change the solution of the inequality.

**EXAMPLE 2**   Solve $x + 4 \le -6$ for $x$. Graph the solution set and write it in interval notation.

*Solution*   To solve for $x$, subtract 4 from both sides of the inequality.

$$x + 4 \le -6 \qquad \text{Original inequality}$$
$$x + 4 - 4 \le -6 - 4 \qquad \text{Subtract 4 from both sides.}$$
$$x \le -10 \qquad \text{Simplify.}$$

The solution set is $(-\infty, -10]$.

**PRACTICE**
**2**   Solve $x + 11 \ge 6$. Graph the solution set and write it in interval notation.

---

▶ **Helpful Hint**

Notice that any number less than or equal to $-10$ is a solution to $x \le -10$. For example, solutions include

$$-10, \ -200, \ -11\frac{1}{2}, \ -7\pi, \ -\sqrt{130}, \ -50.3$$

---

An important difference between linear equations and linear inequalities is shown when we multiply or divide both sides of an inequality by a nonzero real number. For example, start with the true statement $6 < 8$ and multiply both sides by 2. As we see below, the resulting inequality is also true.

$$6 < 8 \qquad \text{True}$$
$$2(6) < 2(8) \qquad \text{Multiply both sides by 2.}$$
$$12 < 16 \qquad \text{True}$$

But if we start with the same true statement $6 < 8$ and multiply both sides by $-2$, the resulting inequality is not a true statement.

$$6 < 8 \qquad \text{True}$$
$$-2(6) < -2(8) \qquad \text{Multiply both sides by } -2.$$
$$-12 < -16 \qquad \text{False}$$

Notice, however, that if we reverse the direction of the inequality symbol, the resulting inequality is true.

$$-12 < -16 \qquad \text{False}$$
$$-12 > -16 \qquad \text{True}$$

This demonstrates the multiplication property of inequality.

> **Multiplication Property of Inequality**
> **1.** If $a$, $b$, and $c$ are real numbers, and $c$ is **positive,** then
> $$a < b \quad \text{and} \quad ac < bc$$
> are equivalent inequalities.
> **2.** If $a$, $b$, and $c$ are real numbers, and $c$ is **negative,** then
> $$a < b \quad \text{and} \quad ac > bc$$
> are equivalent inequalities.

Because division is defined in terms of multiplication, this property also holds true when dividing both sides of an inequality by a nonzero number. If we multiply or divide both sides of an inequality by a negative number, **the direction of the inequality sign must be reversed for the inequalities to remain equivalent.**

> ▶ **Helpful Hint**
> Whenever both sides of an inequality are multiplied or divided by a negative number, the direction of the inequality symbol **must be** reversed to form an equivalent inequality.

**EXAMPLE 3** Solve $-2x \le -4$. Graph the solution set and write it in interval notation.

*Solution* Remember to reverse the direction of the inequality symbol when dividing by a negative number.

$$-2x \le -4$$

$$\frac{-2x}{-2} \ge \frac{-4}{-2} \quad \text{Divide both sides by } -2 \text{ and reverse the direction of the inequality sign.}$$

$$x \ge 2 \quad \text{Simplify.}$$

> ▶ **Helpful Hint**
> Don't forget to reverse the direction of the inequality sign.

The solution set $[2, \infty)$ is graphed as shown.

**PRACTICE**
**3** Solve $-5x \ge -15$. Graph the solution set and write it in interval notation.

**EXAMPLE 4** Solve $2x < -4$. Graph the solution set and write it in interval notation.

*Solution*

$$2x < -4$$

> ▶ **Helpful Hint**
> Do not reverse the inequality sign.

$$\frac{2x}{2} < \frac{-4}{2} \quad \text{Divide both sides by 2.}$$
$$\quad \text{Do not reverse the direction of the inequality sign.}$$

$$x < -2 \quad \text{Simplify.}$$

The solution set $(-\infty, -2)$ is graphed as shown.

**PRACTICE**
**4**   Solve $3x > -9$. Graph the solution set and write it in interval notation.

## Concept Check ✓

Fill in the blank with $<$, $>$, $\leq$, or $\geq$.

**a.** Since $-8 < -4$, then $3(-8)$_____$3(-4)$.

**b.** Since $5 \geq -2$, then $\dfrac{5}{-7}$ _____ $\dfrac{-2}{-7}$.

**c.** If $a < b$, then $2a$_____$2b$.

**d.** If $a \geq b$, then $\dfrac{a}{-3}$ _____ $\dfrac{b}{-3}$.

The following steps may be helpful when solving inequalities. Notice that these steps are similar to the ones given in Section 2.3 for solving equations.

---

**Solving Linear Inequalities in One Variable**

**STEP 1.** Clear the inequality of fractions by multiplying both sides of the inequality by the lowest common denominator (LCD) of all fractions in the inequality.

**STEP 2.** Remove grouping symbols such as parentheses by using the distributive property.

**STEP 3.** Simplify each side of the inequality by combining like terms.

**STEP 4.** Write the inequality with variable terms on one side and numbers on the other side by using the addition property of inequality.

**STEP 5.** Get the variable alone by using the multiplication property of inequality.

---

▶ **Helpful Hint**
Don't forget that if both sides of an inequality are multiplied or divided by a negative number, the direction of the inequality sign must be reversed.

---

**EXAMPLE 5**   Solve $-4x + 7 \geq -9$. Graph the solution set and write it in interval notation.

*Solution*
$$-4x + 7 \geq -9$$
$$-4x + 7 - 7 \geq -9 - 7 \quad \text{Subtract 7 from both sides.}$$
$$-4x \geq -16 \quad \text{Simplify.}$$
$$\frac{-4x}{-4} \leq \frac{-16}{-4} \quad \text{Divide both sides by } -4 \text{ and reverse the direction of the inequality sign.}$$
$$x \leq 4 \quad \text{Simplify.}$$

The solution set $(-\infty, 4]$ is graphed as shown.

**PRACTICE**
**5**   Solve $45 - 7x \leq -4$. Graph the solution set and write it in interval notation.

**EXAMPLE 6** Solve $2x + 7 \leq x - 11$. Graph the solution set and write it in interval notation.

*Solution*

$$2x + 7 \leq x - 11$$

$2x + 7 - x \leq x - 11 - x$    Subtract $x$ from both sides.

$x + 7 \leq -11$    Combine like terms.

$x + 7 - 7 \leq -11 - 7$    Subtract 7 from both sides.

$x \leq -18$    Combine like terms.

The graph of the solution set $(-\infty, -18]$ is shown.

$$\xleftarrow{\hspace{3cm}} \underset{-20\;-19\;-18\;-17\;-16\;-15\;-14}{+\;+\;]\;+\;+\;+\;+} \xrightarrow{\hspace{1cm}}$$

**PRACTICE**

**6** Solve $3x + 20 \leq 2x + 13$. Graph the solution set and write it in interval notation.

---

**EXAMPLE 7** Solve $-5x + 7 < 2(x - 3)$. Graph the solution set and write it in interval notation.

*Solution*

$$-5x + 7 < 2(x - 3)$$

$-5x + 7 < 2x - 6$    Apply the distributive property.

$-5x + 7 - 2x < 2x - 6 - 2x$    Subtract $2x$ from both sides.

$-7x + 7 < -6$    Combine like terms.

$-7x + 7 - 7 < -6 - 7$    Subtract 7 from both sides.

$-7x < -13$    Combine like terms.

$\dfrac{-7x}{-7} > \dfrac{-13}{-7}$    Divide both sides by $-7$ and reverse the direction of the inequality sign.

$x > \dfrac{13}{7}$    Simplify.

The graph of the solution set $\left(\dfrac{13}{7}, \infty\right)$ is shown.

$$\overset{\frac{13}{7}}{\xleftarrow{\hspace{3cm}} \underset{-2\;-1\;\;0\;\;1\;\;2\;\;3\;\;4}{+\;+\;+\;+\;(\;+\;+} \xrightarrow{\hspace{1cm}}}$$

**PRACTICE**

**7** Solve $6 - 5x > 3(x - 4)$. Graph the solution set and write it in interval notation.

---

**EXAMPLE 8** Solve $2(x - 3) - 5 \leq 3(x + 2) - 18$. Graph the solution set and write it in interval notation.

*Solution*

$$2(x - 3) - 5 \leq 3(x + 2) - 18$$

$2x - 6 - 5 \leq 3x + 6 - 18$    Apply the distributive property.

$2x - 11 \leq 3x - 12$    Combine like terms.

$-x - 11 \leq -12$    Subtract $3x$ from both sides.

$-x \leq -1$    Add 11 to both sides.

$$\frac{-x}{-1} \ge \frac{-1}{-1}$$   Divide both sides by $-1$ and reverse the direction of the inequality sign.

$$x \ge 1$$   Simplify.

The graph of the solution set $[1, \infty)$ is shown.

**8**   Solve $3(x - 4) - 5 \le 5(x - 1) - 12$. Graph the solution set and write it in interval notation.

**OBJECTIVE 3 ▶ Solving compound inequalities.** Inequalities containing one inequality symbol are called **simple inequalities,** while inequalities containing two inequality symbols are called **compound inequalities.** A compound inequality is really two simple inequalities in one. The compound inequality

$$3 < x < 5 \quad \text{means} \quad 3 < x \text{ and } x < 5$$

This can be read "$x$ is greater than 3 and less than 5."

A solution of a compound inequality is a value that is a solution of both of the simple inequalities that make up the compound inequality. For example,

$$4\frac{1}{2} \text{ is a solution of } 3 < x < 5 \text{ since } 3 < 4\frac{1}{2} \text{ and } 4\frac{1}{2} < 5.$$

To graph $3 < x < 5$, place parentheses at both 3 and 5 and shade between.

**EXAMPLE 9**   Graph $2 < x \le 4$. Write the solutions in interval notation.

*Solution*   Graph all numbers greater than 2 and less than or equal to 4. Place a parenthesis at 2, a bracket at 4, and shade between.

In interval notation, this is $(2, 4]$.

**9**   Graph $-3 \le x < 1$. Write the solutions in interval notation.

When we solve a simple inequality, we isolate the variable on one side of the inequality. When we solve a compound inequality, we isolate the variable in the middle part of the inequality. Also, when solving a compound inequality, we must perform the same operation to all **three** parts of the inequality: left, middle, and right.

**EXAMPLE 10**   Solve $-1 \le 2x - 3 < 5$. Graph the solution set and write it in interval notation.

*Solution*
$$-1 \le 2x - 3 < 5$$

$$-1 + 3 \le 2x - 3 + 3 < 5 + 3 \quad \text{Add 3 to all three parts.}$$

$$2 \le 2x < 8 \quad \text{Combine like terms.}$$

$$\frac{2}{2} \le \frac{2x}{2} < \frac{8}{2} \quad \text{Divide all three parts by 2.}$$

$$1 \le x < 4 \quad \text{Simplify.}$$

The graph of the solution set $[1, 4)$ is shown.

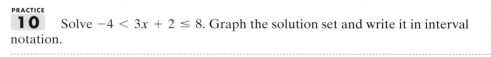

**PRACTICE**
**10** Solve $-4 < 3x + 2 \leq 8$. Graph the solution set and write it in interval notation.

---

**EXAMPLE 11** Solve $3 \leq \dfrac{3x}{2} + 4 \leq 5$. Graph the solution set and write it in interval notation.

*Solution*

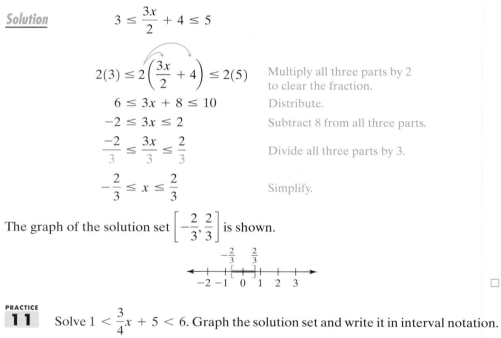

$$3 \leq \dfrac{3x}{2} + 4 \leq 5$$

$$2(3) \leq 2\left(\dfrac{3x}{2} + 4\right) \leq 2(5) \qquad \text{Multiply all three parts by 2 to clear the fraction.}$$

$$6 \leq 3x + 8 \leq 10 \qquad \text{Distribute.}$$

$$-2 \leq 3x \leq 2 \qquad \text{Subtract 8 from all three parts.}$$

$$\dfrac{-2}{3} \leq \dfrac{3x}{3} \leq \dfrac{2}{3} \qquad \text{Divide all three parts by 3.}$$

$$-\dfrac{2}{3} \leq x \leq \dfrac{2}{3} \qquad \text{Simplify.}$$

The graph of the solution set $\left[-\dfrac{2}{3}, \dfrac{2}{3}\right]$ is shown.

**PRACTICE**
**11** Solve $1 < \dfrac{3}{4}x + 5 < 6$. Graph the solution set and write it in interval notation.

---

**OBJECTIVE 4 ▶ Solving inequality applications.** Problems containing words such as "at least," "at most," "between," "no more than," and "no less than" usually indicate that an inequality should be solved instead of an equation. In solving applications involving linear inequalities, use the same procedure you use to solve applications involving linear equations.

**EXAMPLE 12** **Staying within Budget**

Marie Chase and Jonathan Edwards are having their wedding reception at the Gallery Reception Hall. They may spend at most $2000 for the reception. If the reception hall charges a $100 cleanup fee plus $36 per person, find the greatest number of people that they can invite and still stay within their budget.

*Solution*

1. **UNDERSTAND.** Read and reread the problem. Next, guess a solution. If 40 people attend the reception, the cost is $100 + \$36(40) = \$100 + \$1440 = \$1540$. Let $x = $ the number of people who attend the reception.

**2.** TRANSLATE.

| In words: | cleanup fee | + | cost per person | must be less than or equal to | $2000 |
|---|---|---|---|---|---|
| | ↓ | | ↓ | ↓ | ↓ |
| Translate: | 100 | + | 36x | ≤ | 2000 |

**3.** SOLVE.

$$100 + 36x \leq 2000$$
$$36x \leq 1900 \quad \text{Subtract 100 from both sides.}$$
$$x \leq 52\frac{7}{9} \quad \text{Divide both sides by 36.}$$

**4.** INTERPRET.

**Check:** Since $x$ represents the number of people, we round down to the nearest whole, or 52. Notice that if 52 people attend, the cost is

$$\$100 + \$36(52) = \$1972. \text{ If 53 people attend, the cost is}$$
$$\$100 + \$36(53) = \$2008, \text{ which is more than the given } \$2000.$$

**State:** Marie Chase and Jonathan Edwards can invite at most 52 people to the reception. □

**PRACTICE**
**12** Kasonga is eager to begin his education at his local community college. He has budgeted $1500 for college this semester. His local college charges a $300 matriculation fee and costs an average of $375 for tuition, fees, and books for each three-credit course. Find the greatest number of classes Kasonga can afford to take this semester.

## VOCABULARY & READINESS CHECK

*Use the choices below to fill in each blank.*

   expression          inequality          equation

**1.** $6x - 7(x + 9)$ _____

**2.** $6x = 7(x + 9)$ _____

**3.** $6x < 7(x + 9)$ _____

**4.** $5y - 2 \geq -38$ _____

**5.** $\dfrac{9}{7} = \dfrac{x + 2}{14}$ _____

**6.** $\dfrac{9}{7} - \dfrac{x + 2}{14}$ _____

*Decide which number listed is not a solution to each given inequality.*

**7.** $x \geq -3$; $-3, 0, -5, \pi$ _____

**8.** $x < 6$; $-6, |-6|, 0, -3.2$ _____

**9.** $x < 4.01$; $4, -4.01, 4.1, -4.1$ _____

**10.** $x \geq -3$; $-4, -3, -2, -(-2)$ _____

**2.8** | **EXERCISE SET**

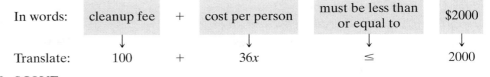

*Graph each set of numbers given in interval notation. Then write an inequality statement in x describing the numbers graphed.*

**1.** $[2, \infty)$

**2.** $(-3, \infty)$

**3.** $(-\infty, -5)$

**4.** $(-\infty, 4]$

*Graph each inequality on a number line. Then write the solutions in interval notation. See Example 1.*

**5.** $x \leq -1$

**6.** $y < 0$

**7.** $x < \dfrac{1}{2}$

**8.** $z < -\dfrac{2}{3}$

**9.** $y \geq 5$

**10.** $x > 3$

*Solve each inequality. Graph the solution set and write it in interval notation. See Examples 2 through 4.*

**11.** $2x < -6$

**12.** $3x > -9$

**13.** $x - 2 \geq -7$

**14.** $x + 4 \leq 1$

**15.** $-8x \leq 16$

**16.** $-5x < 20$

*Solve each inequality. Graph the solution set and write it in interval notation. See Examples 5 and 6.*

**17.** $3x - 5 > 2x - 8$

**18.** $3 - 7x \geq 10 - 8x$

**19.** $4x - 1 \leq 5x - 2x$

**20.** $7x + 3 < 9x - 3x$

*Solve each inequality. Graph the solution set and write it in interval notation. See Examples 7 and 8.*

**21.** $x - 7 < 3(x + 1)$

**22.** $3x + 9 \leq 5(x - 1)$

**23.** $-6x + 2 \geq 2(5 - x)$

**24.** $-7x + 4 > 3(4 - x)$

**25.** $4(3x - 1) \leq 5(2x - 4)$

**26.** $3(5x - 4) \leq 4(3x - 2)$

**27.** $3(x + 2) - 6 > -2(x - 3) + 14$

**28.** $7(x - 2) + x \leq -4(5 - x) - 12$

**MIXED PRACTICE**

*Solve the following inequalities. Graph each solution set and write it in interval notation.*

**29.** $-2x \leq -40$

**30.** $-7x > 21$

**31.** $-9 + x > 7$

**32.** $y - 4 \leq 1$

**33.** $3x - 7 < 6x + 2$

**34.** $2x - 1 \geq 4x - 5$

**35.** $5x - 7x \geq x + 2$

**36.** $4 - x < 8x + 2x$

**37.** $\frac{3}{4}x > 2$

**38.** $\frac{5}{6}x \geq -8$

**39.** $3(x - 5) < 2(2x - 1)$

**40.** $5(x + 4) < 4(2x + 3)$

**41.** $4(2x + 1) < 4$

**42.** $6(2 - x) \geq 12$

**43.** $-5x + 4 \geq -4(x - 1)$

**44.** $-6x + 2 < -3(x + 4)$

**45.** $-2(x - 4) - 3x < -(4x + 1) + 2x$

**46.** $-5(1 - x) + x \leq -(6 - 2x) + 6$

**47.** $-3x + 6 \geq 2x + 6$

**48.** $-(x - 4) < 4$

**49.** Explain how solving a linear inequality is similar to solving a linear equation.

**50.** Explain how solving a linear inequality is different from solving a linear equation.

*Graph each inequality. Then write the solutions in interval notation. See Example 9.*

**51.** $-1 < x < 3$

**52.** $2 \leq y \leq 3$

**53.** $0 \leq y < 2$

**54.** $-1 \leq x \leq 4$

*Solve each inequality. Graph the solution set and write it in interval notation. See Examples 10 and 11.*

**55.** $-3 < 3x < 6$

**56.** $-5 < 2x < -2$

**57.** $2 \leq 3x - 10 \leq 5$

**58.** $4 \leq 5x - 6 \leq 19$

**59.** $-4 < 2(x - 3) \leq 4$

**60.** $0 < 4(x + 5) \leq 8$

**61.** $-2 < 3x - 5 < 7$

**62.** $1 < 4 + 2x \leq 7$

**63.** $-6 < 3(x - 2) \leq 8$

**64.** $-5 \leq 2(x + 4) < 8$

**65.** Explain how solving a linear inequality is different from solving a compound inequality.

**66.** Explain how solving a linear inequality is similar to solving a compound inequality.

*Solve. See Example 12.*

**67.** Six more than twice a number is greater than negative fourteen. Find all numbers that make this statement true.

**68.** Five times a number, increased by one, is less than or equal to ten. Find all such numbers.

**69.** Dennis and Nancy Wood are celebrating their 30th wedding anniversary by having a reception at Tiffany Oaks reception hall. They have budgeted $3000 for their reception. If the reception hall charges a $50.00 cleanup fee plus $34 per person, find the greatest number of people that they may invite and still stay within their budget.

**70.** A surprise retirement party is being planned for Pratep Puri. A total of $860 has been collected for the event, which is to be held at a local reception hall. This reception hall charges a cleanup fee of $40 and $15 per person for drinks and light snacks. Find the greatest number of people that may be invited and still stay within $860.

**71.** Find the values for $x$ so that the perimeter of this rectangle is no greater than 100 centimeters.

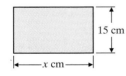

15 cm

$x$ cm

**72.** Find the values for $x$ so that the perimeter of this triangle is no longer than 87 inches.

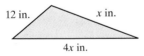

12 in.    $x$ in.

4$x$ in.

**73.** A financial planner has a client with $15,000 to invest. If he invests $10,000 in a certificate of deposit paying 11% annual simple interest, at what rate does the remainder of the money need to be invested so that the two investments together yield at least $1600 in yearly interest?

**74.** Alex earns $600 per month plus 4% of all his sales over $1000. Find the minimum sales that will allow Alex to earn at least $3000 per month.

**75.** Ben Holladay bowled 146 and 201 in his first two games. What must he bowl in his third game to have an average of at least 180?

**76.** On an NBA team the two forwards measure 6′8″ and 6′6″ and the two guards measure 6′0″ and 5′9″ tall. How tall a center should they hire if they wish to have a starting team average height of at least 6′5″?

**77.** High blood cholesterol levels increase the risk of heart disease in adults. Doctors recommend that total blood cholesterol be less than 200 milligrams per deciliter. Total cholesterol levels from 200 up to 240 milligrams per deciliter are considered borderline. Any total cholesterol reading above 240 milligrams per deciliter is considered high.

Letting $x$ represent a patient's total blood cholesterol level, write a series of three inequalities that describe the ranges corresponding to recommended, borderline, and high levels of total blood cholesterol.

**78.** In 1971, T. Theodore Fujita created the Fujita Scale (or F Scale), which uses ratings from 0 to 5 to classify tornadoes, based on the damage that wind intensity causes to structures as the tornado passes through. This scale was updated by meteorologists and wind engineers working for the National Oceanic and Atmospheric Administration (NOAA) and implemented in February 2007. This new scale is called the Enhanced F Scale. An EF-0 tornado has wind speeds between 65 and 85 mph, inclusive. The winds in an EF-1 tornado range from 86 to 110 mph. In an EF-2 tornado, winds are from 111 to 135 mph. An EF-3 tornado has wind speeds ranging from 136 to 165 mph. Wind speeds in an EF-4 tornado are clocked at 166 to 200 mph. The most violent tornadoes are ranked at EF-5, with wind speeds of at least 201 mph. (*Source:* Storm Prediction Center, NOAA)

Letting $y$ represent a tornado's wind speed, write a series of six inequalities that describe the wind speed ranges corresponding to each Enhanced Fujita Scale rank.

**79.** Twice a number, increased by one, is between negative five and seven. Find all such numbers.

**80.** Half a number, decreased by four, is between two and three. Find all such numbers.

**81.** The temperatures in Ohio range from $-39°$C to $45°$C. Use a compound inequality to convert these temperatures to Fahrenheit temperatures. (*Hint:* Use $C = \frac{5}{9}(F - 32)$.)

**82.** Mario Lipco has scores of 85, 95, and 92 on his algebra tests. Use a compound inequality to find the range of scores he can make on his final exam in order to receive an A in the course. The final exam counts as three tests, and an A is received if the final course average is from 90 to 100. (*Hint:* The average of a list of numbers is their sum divided by the number of numbers in the list.)

**REVIEW AND PREVIEW**

*Evaluate the following. See Section 1.4.*

**83.** $(2)^3$    **84.** $(3)^3$

**85.** $(1)^{12}$    **86.** $0^5$

**87.** $\left(\frac{4}{7}\right)^2$    **88.** $\left(\frac{2}{3}\right)^3$

*This broken line graph shows the average annual per person expenditure on newspapers for the given years. Use this graph for Exercises 89 through 92. (Source: Veronis Suhler Stevenson)*

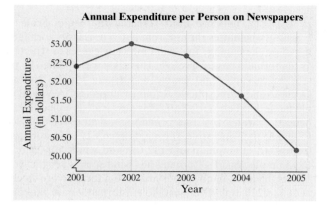

**Annual Expenditure per Person on Newspapers**

**89.** What was the average per person expenditure on newspapers in 2003?

**90.** What was the average per person expenditure on newspapers in 2005?

**91.** What year had the greatest drop in newspaper expenditures?

**92.** What years had per person newspaper expenditures over $52?

## CONCEPT EXTENSIONS

**93.** The formula $C = 3.14d$ can be used to approximate the circumference of a circle given its diameter. Waldo Manufacturing manufactures and sells a certain washer with an outside circumference of 3 centimeters. The company has decided that a washer whose actual circumference is in the interval $2.9 \le C \le 3.1$ centimeters is acceptable. Use a compound inequality and find the corresponding interval for diameters of these washers. (Round to 3 decimal places.)

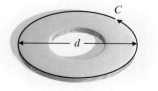

**94.** Bunnie Supplies manufactures plastic Easter eggs that open. The company has determined that if the circumference of the opening of each part of the egg is in the interval $118 \le C \le 122$ millimeters, the eggs will open and close comfortably. Use a compound inequality and find the corresponding interval for diameters of these openings. (Round to 2 decimal places.)

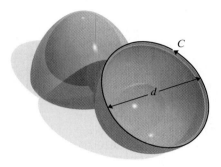

*For Exercises 95 through 98, see the example below.*

Solve $x(x - 6) > x^2 - 5x + 6$. Graph the solution set and write it in interval notation.

*Solution*

$$x(x - 6) > x^2 - 5x + 6$$
$$x^2 - 6x > x^2 - 5x + 6$$
$$x^2 - 6x - x^2 > x^2 - 5x + 6 - x^2$$
$$-6x > -5x + 6$$
$$-x > 6$$
$$\frac{-x}{-1} < \frac{6}{-1}$$
$$x < -6$$

The solution set $(-\infty, -6)$ is graphed as shown.

*Solve each inequality. Graph the solution set and write it in interval notation.*

**95.** $x(x + 4) > x^2 - 2x + 6$

**96.** $x(x - 3) \ge x^2 - 5x - 8$

**97.** $x^2 + 6x - 10 < x(x - 10)$

**98.** $x^2 - 4x + 8 < x(x + 8)$

## THE BIGGER PICTURE    SIMPLIFYING EXPRESSIONS AND SOLVING EQUATIONS AND INEQUALITIES

Now we continue our outline from Sections 1.7 and 2.3. Although suggestions are given, this outline should be in your own words. Once you complete this new portion, try the exercises to the right.

   **I.** Simplifying Expressions

      **A.** Real Numbers

         **1.** Add (Section 1.5)

         **2.** Subtract (Section 1.6)

         **3.** Multiply or Divide (Section 1.7)

  **II.** Solving Equations

      **A.** Linear Equations (Section 2.3)

**III. Solving Inequalities**

      **A. Linear Inequalities:** Same as linear equations, except there are inequality symbols, $\le, <, \ge, >$ Remember, if you multiply or divide by a negative number, then reverse the direction of the inequality symbol.

$$-4x - 11 \le 1 \qquad \text{Linear inequality.}$$
$$-4x \le 12 \qquad \text{Add 11 to both sides.}$$
$$\frac{-4x}{-4} \ge \frac{12}{-4} \qquad \begin{array}{l}\text{Divide both sides by } -4 \text{ and}\\\text{reverse the direction of the}\\\text{inequality symbol.}\end{array}$$
$$x \ge -3 \qquad \text{Simplify.}$$

*Solve each equation or inequality. Write inequality solutions in interval notation.*

  **1.** $-5x = 15$

  **2.** $-5x > 15$

  **3.** $9y - 14 = -12$

  **4.** $9x - 3 = 5x - 4$

  **5.** $4(x - 2) \le 5x + 7$

  **6.** $5(4x - 1) = 2(10x - 1)$

  **7.** $-5.4 = 0.6x - 9.6$

  **8.** $\dfrac{1}{3}(x - 4) < \dfrac{1}{4}(x + 7)$

  **9.** $3y - 5(y - 4) = -2(y - 10)$

**10.** $\dfrac{7(x - 1)}{3} = \dfrac{2(x + 1)}{5}$

# CHAPTER 2 GROUP ACTIVITY

## Investigating Averages

Sections 2.1–2.8

*Materials:*

- small rubber ball or crumpled paper ball
- bucket or waste can

This activity may be completed by working in groups or individually.

**1.** Try shooting the ball into the bucket or waste can 5 times. Record your results below.

  **Shots Made**        **Shots Missed**

**2.** Find your shooting percent for the 5 shots (that is, the percent of the shots you actually made out of the number you tried).

**3.** Suppose you are going to try an additional 5 shots. How many of the next 5 shots will you have to make to have a 50% shooting percent for all 10 shots? An 80% shooting percent?

**4.** Did you solve an equation in Question 3? If so, explain what you did. If not, explain how you could use an equation to find the answers.

**5.** Now suppose you are going to try an additional 22 shots. How many of the next 22 shots will you have to make to have at least a 50% shooting percent for all 27 shots? At least a 70% shooting percent?

**6.** Choose one of the sports played at your college that is currently in season. How many regular-season games are scheduled? What is the team's current percent of games won?

**7.** Suppose the team has a goal of finishing the season with a winning percent better than 110% of their current wins. At least how many of the remaining games must they win to achieve their goal?

# CHAPTER 2 VOCABULARY CHECK

*Fill in each blank with one of the words or phrases listed below.*

like terms          numerical coefficient          linear inequality in one variable

equivalent equations          formula          compound inequalities

linear equation in one variable

1. Terms with the same variables raised to exactly the same powers are called _____.
2. A _____ can be written in the form $ax + b = c$.
3. Equations that have the same solution are called _____.
4. Inequalities containing two inequality symbols are called _____.
5. An equation that describes a known relationship among quantities is called a _____.
6. A _____ can be written in the form $ax + b < c$, (or $>$, $\leq$, $\geq$).
7. The _____ of a term is its numerical factor.

> ▶ **Helpful Hint**
>
> Are you preparing for your test? Don't forget to take the Chapter 2 Test on page 162. Then check your answers at the back of the text and use the Chapter Test Prep Video CD to see the fully worked-out solutions to any of the exercises you want to review.

# CHAPTER 2 HIGHLIGHTS

| DEFINITIONS AND CONCEPTS | EXAMPLES |
| --- | --- |

### SECTION 2.1 SIMPLIFYING ALGEBRAIC EXPRESSIONS

| DEFINITIONS AND CONCEPTS | EXAMPLES |
| --- | --- |
| The **numerical coefficient** of a **term** is its numerical factor. | **Term**     **Numerical Coefficient**<br><br>$-7y$        $-7$<br>$x$        $1$<br>$\frac{1}{5}a^2b$        $\frac{1}{5}$ |
| Terms with the same variables raised to exactly the same powers are **like terms.** | **Like Terms**     **Unlike Terms**<br>$12x, -x$        $3y, 3y^2$<br>$-2xy, 5yx$        $7a^2b, -2ab^2$ |
| **To combine like terms,** add the numerical coefficients and multiply the result by the common variable factor. | $9y + 3y = 12y$<br>$-4z^2 + 5z^2 - 6z^2 = -5z^2$ |
| To remove parentheses, apply the distributive property. | $-4(x + 7) + 10(3x - 1)$<br>$= -4x - 28 + 30x - 10$<br>$= 26x - 38$ |

| DEFINITIONS AND CONCEPTS | EXAMPLES |
|---|---|

SECTION 2.2   THE ADDITION AND MULTIPLICATION PROPERTIES OF EQUALITY

A **linear equation in one variable** can be written in the form $ax + b = c$ where $a, b,$ and $c$ are real numbers and $a \neq 0$.

*Linear Equations*

$$-3x + 7 = 2$$
$$3(x - 1) = -8(x + 5) + 4$$

**Equivalent equations** are equations that have the same solution.

$x - 7 = 10$ and $x = 17$
are equivalent equations.

*Addition Property of Equality*

Adding the same number to or subtracting the same number from both sides of an equation does not change its solution.

$$y + 9 = 3$$
$$y + 9 - 9 = 3 - 9$$
$$y = -6$$

*Multiplication Property of Equality*

Multiplying both sides or dividing both sides of an equation by the same nonzero number does not change its solution.

$$\frac{2}{3}a = 18$$
$$\frac{3}{2}\left(\frac{2}{3}a\right) = \frac{3}{2}(18)$$
$$a = 27$$

SECTION 2.3   SOLVING LINEAR EQUATIONS

*To Solve Linear Equations*

Solve: $\dfrac{5(-2x + 9)}{6} + 3 = \dfrac{1}{2}$

**1.** Clear the equation of fractions.

**1.** $6 \cdot \dfrac{5(-2x + 9)}{6} + 6 \cdot 3 = 6 \cdot \dfrac{1}{2}$

$5(-2x + 9) + 18 = 3$

**2.** Remove any grouping symbols such as parentheses.

**2.** $-10x + 45 + 18 = 3$    Distributive property

**3.** Simplify each side by combining like terms.

**3.** $-10x + 63 = 3$    Combine like terms.

**4.** Write variable terms on one side and numbers on the other side using the addition property of equality.

**4.** $-10x + 63 - 63 = 3 - 63$   Subtract 63.
$-10x = -60$

**5.** Get the variable alone using the multiplication property of equality.

**5.** $\dfrac{-10x}{-10} = \dfrac{-60}{-10}$    Divide by $-10$.
$x = 6$

**6.** Check by substituting in the original equation.

**6.** $\dfrac{5(-2x + 9)}{6} + 3 = \dfrac{1}{2}$

$\dfrac{5(-2 \cdot 6 + 9)}{6} + 3 \stackrel{?}{=} \dfrac{1}{2}$

$\dfrac{5(-3)}{6} + 3 \stackrel{?}{=} \dfrac{1}{2}$

$-\dfrac{5}{2} + \dfrac{6}{2} \stackrel{?}{=} \dfrac{1}{2}$

$\dfrac{1}{2} = \dfrac{1}{2}$    True

| **DEFINITIONS AND CONCEPTS** | **EXAMPLES** |
| --- | --- |

*Problem-Solving Steps*

The height of the Hudson volcano in Chili is twice the height of the Kiska volcano in the Aleutian Islands. If the sum of their heights is 12,870 feet, find the height of each.

**1.** UNDERSTAND the problem.

**1.** Read and reread the problem. Guess a solution and check your guess.

Let $x$ be the height of the Kiska volcano. Then $2x$ is the height of the Hudson volcano.

$$x\rceil \qquad 2x\rceil$$
Kiska   Hudson

**2.** TRANSLATE the problem.

**2.** In words:

| height of Kiska | added to | height of Hudson | is | 12,870 |
| --- | --- | --- | --- | --- |
| ↓ | ↓ | ↓ | ↓ | ↓ |

Translate:   $x \qquad + \qquad 2x \qquad = \qquad 12{,}870$

**3.**

$$x + 2x = 12{,}870$$
$$3x = 12{,}870$$
$$x = 4290$$

**3.** SOLVE.

**4.** *Check:* If $x$ is 4290 then $2x$ is 2(4290) or 8580. Their sum is $4290 + 8580$ or 12,870, the required amount.

**4.** INTERPRET the results.

*State:* Kiska volcano is 4290 feet high and Hudson volcano is 8580 feet high.

*Formulas*

An equation that describes a known relationship among quantities is called a **formula.**

$A = lw$ (area of a rectangle)
$I = PRT$ (simple interest)

**To solve a formula for a specified variable,** use the same steps as for solving a linear equation. Treat the specified variable as the only variable of the equation.

Solve: $P = 2l + 2w$ for $l$.
$$P = 2l + 2w$$
$$P - 2w = 2l + 2w - 2w \quad \text{Subtract } 2w.$$
$$P - 2w = 2l$$
$$\frac{P - 2w}{2} = \frac{2l}{2} \qquad \text{Divide by 2.}$$
$$\frac{P - 2w}{2} = l \qquad \text{Simplify.}$$

If all values for the variables in a formula are known except for one, this unknown value may be found by substituting in the known values and solving.

If $d = 182$ miles and $r = 52$ miles per hour in the formula $d = r \cdot t$, find $t$.

$$d = r \cdot t$$
$$182 = 52 \cdot t \quad \text{Let } d = 182 \text{ and } r = 52.$$
$$3.5 = t$$

The time is 3.5 hours.

| DEFINITIONS AND CONCEPTS | EXAMPLES |
|---|---|

Use the same problem-solving steps to solve a problem containing percents.

**1.** UNDERSTAND.

**2.** TRANSLATE.

**3.** SOLVE.

**4.** INTERPRET.

32% of what number is 36.8?

**1.** Read and reread. Propose a solution and check.
  Let $x$ = the unknown number.

**2.**

| 32% | of | what number | is | 36.8 |
|---|---|---|---|---|
| ↓ | ↓ | ↓ | ↓ | ↓ |
| 32% | · | $x$ | = | 36.8 |

**3.** *Solve:*
$$32\% \cdot x = 36.8$$
$$0.32x = 36.8$$
$$\frac{0.32x}{0.32} = \frac{36.8}{0.32} \quad \text{Divide by 0.32.}$$
$$x = 115 \quad \text{Simplify.}$$

**4.** *Check, then state:* 32% of 115 is 36.8.

---

**1.** UNDERSTAND.

**2.** TRANSLATE.

How many liters of a 20% acid solution must be mixed with a 50% acid solution in order to obtain 12 liters of a 30% solution?

**1.** Read and reread. Guess a solution and check.
  Let $x$ = number of liters of 20% solution.
  Then $12 - x$ = number of liters of 50% solution.

**2.**

|  | No. of Liters | · | Acid Strength | = | Amount of Acid |
|---|---|---|---|---|---|
| **20% Solution** | $x$ | | 20% | | $0.20x$ |
| **50% Solution** | $12 - x$ | | 50% | | $0.50(12 - x)$ |
| **30% Solution Needed** | 12 | | 30% | | $0.30(12)$ |

In words:

| acid in 20% solution | + | acid in 50% solution | = | acid in 30% solution |
|---|---|---|---|---|
| ↓ | | ↓ | | ↓ |

Translate:  $0.20x + 0.50(12 - x) = 0.30(12)$

**3.** SOLVE.

**3.** Solve: $0.20x + 0.50(12 - x) = 0.30(12)$
$$0.20x + 6 - 0.50x = 3.6 \quad \text{Apply the distributive}$$
$$-0.30x + 6 = 3.6 \quad \text{property.}$$
$$-0.30x = -2.4 \quad \text{Subtract 6.}$$
$$x = 8 \quad \text{Divide by } -0.30.$$

**4.** INTERPRET.

**4.** *Check, then state:*
If 8 liters of a 20% acid solution are mixed with $12 - 8$ or 4 liters of a 50% acid solution, the result is 12 liters of a 30% solution.

| DEFINITIONS AND CONCEPTS | EXAMPLES |
|---|---|

**Problem-Solving Steps**

A collection of dimes and quarters has a total value of $19.55. If there are three times as many quarters as dimes, find the number of quarters.

**1.** UNDERSTAND.

**1.** Read and reread. Propose a solution and check.

$$\text{Let } x = \text{number of dimes and}$$
$$3x = \text{number of quarters.}$$

**2.** TRANSLATE.

**2.** In words:

| value of dimes | + | value of quarters | = | 19.55 |
|---|---|---|---|---|
| ↓ | | ↓ | | ↓ |
| Translate:  $0.10x$ | + | $0.25(3x)$ | = | 19.55 |

**3.** SOLVE.

**3.** Solve: $0.10x + 0.75x = 19.55$    Multiply.

$$0.85x = 19.55 \qquad \text{Add like terms.}$$
$$x = 23 \qquad \text{Divide by 0.85.}$$

**4.** INTERPRET.

**4.** Check, then state.

The number of dimes is 23 and the number of quarters is 3(23) or 69. The total value of this money is

$$0.10(23) + 0.25(69) = 19.55, \text{ so our result checks.}$$

The number of quarters is 69.

A **linear inequality in one variable** is an inequality that can be written in one of the forms:

$$ax + b < c \qquad ax + b \le c$$
$$ax + b > c \qquad ax + b \ge c$$

where $a, b,$ and $c$ are real numbers and $a$ is not 0.

*Linear Inequalities*

$$2x + 3 < 6 \qquad\qquad 5(x - 6) \ge 10$$
$$\frac{x - 2}{5} > \frac{5x + 7}{2} \qquad \frac{-(x + 8)}{9} \le \frac{-2x}{11}$$

*Addition Property of Inequality*

Adding the same number to or subtracting the same number from both sides of an inequality does not change the solutions.

$$y + 4 \le -1$$
$$y + 4 - 4 \le -1 - 4 \qquad \text{Subtract 4.}$$
$$y \le -5$$

*Multiplication Property of Inequality*

Multiplying or dividing both sides of an inequality by the same positive number does not change its solutions.

$$\frac{1}{3}x > -2$$
$$3\left(\frac{1}{3}x\right) > 3 \cdot -2 \qquad \text{Multiply by 3.}$$
$$x > -6$$

Multiplying or dividing both sides of an inequality by the same **negative number and reversing the direction of the inequality sign** does not change its solutions.

$$-2x \le 4$$
$$\frac{-2x}{-2} \ge \frac{4}{-2} \qquad \text{Divide by } -2, \text{ reverse inequality sign.}$$
$$x \ge -2$$

| DEFINITIONS AND CONCEPTS | EXAMPLES |
|---|---|

SECTION 2.8   SOLVING LINEAR INEQUALITIES (continued)

**To Solve Linear Inequalities**

**1.** Clear the equation of fractions.

**2.** Remove grouping symbols.

**3.** Simplify each side by combining like terms.

**4.** Write variable terms on one side and numbers on the other side using the addition property of inequality.

**5.** Get the variable alone using the multiplication property of inequality.

Solve: $3(x + 2) \leq -2 + 8$

**1.** No fractions to clear. $3(x + 2) \leq -2 + 8$

**2.** $3x + 6 \leq -2 + 8$     Distributive property

**3.** $3x + 6 \leq 6$     Combine like terms.

**4.** $3x + 6 - 6 \leq 6 - 6$     Subtract 6.

$3x \leq 0$

**5.** $\dfrac{3x}{3} \leq \dfrac{0}{3}$     Divide by 3.

$x \leq 0$

Inequalities containing two inequality symbols are called **compound inequalities.**

**Compound Inequalities**

$$-2 < x < 6$$

$$5 \leq 3(x - 6) < \frac{20}{3}$$

**To solve a compound inequality,** isolate the variable in the middle part of the inequality. Perform the same operation to all three parts of the inequality: left, middle, right.

Solve: $-2 < 3x + 1 < 7$

$-2 - 1 < 3x + 1 - 1 < 7 - 1$     Subtract 1.

$-3 < 3x < 6$

$\dfrac{-3}{3} < \dfrac{3x}{3} < \dfrac{6}{3}$     Divide by 3.

$-1 < x < 2$

# CHAPTER 2 REVIEW

*(2.1)* *Simplify the following expressions.*

**1.** $5x - x + 2x$

**2.** $0.2z - 4.6x - 7.4z$

**3.** $\dfrac{1}{2}x + 3 + \dfrac{7}{2}x - 5$

**4.** $\dfrac{4}{5}y + 1 + \dfrac{6}{5}y + 2$

**5.** $2(n - 4) + n - 10$

**6.** $3(w + 2) - (12 - w)$

**7.** Subtract $7x - 2$ from $x + 5$.

**8.** Subtract $1.4y - 3$ from $y - 0.7$.

*Write each of the following as algebraic expressions.*

**9.** Three times a number decreased by 7

**10.** Twice the sum of a number and 2.8 added to 3 times the number

*(2.2)* *Solve each equation.*

**11.** $8x + 4 = 9x$

**12.** $5y - 3 = 6y$

**13.** $\dfrac{2}{7}x + \dfrac{5}{7}x = 6$

**14.** $3x - 5 = 4x + 1$

**15.** $2x - 6 = x - 6$

**16.** $4(x + 3) = 3(1 + x)$

**17.** $6(3 + n) = 5(n - 1)$

**18.** $5(2 + x) - 3(3x + 2) = -5(x - 6) + 2$

*Use the addition property to fill in the blank so that the middle equation simplifies to the last equation.*

**19.**       $x - 5 = 3$

    $x - 5 + \underline{\quad} = 3 + \underline{\quad}$

        $x = 8$

**20.**       $x + 9 = -2$

    $x + 9 - \underline{\quad} = -2 - \underline{\quad}$

        $x = -11$

*Choose the correct algebraic expression.*

**21.** The sum of two numbers is 10. If one number is $x$, express the other number in terms of $x$.

    **a.** $x - 10$          **b.** $10 - x$

    **c.** $10 + x$         **d.** $10x$

**22.** Mandy is 5 inches taller than Melissa. If $x$ inches represents the height of Mandy, express Melissa's height in terms of $x$.

    **a.** $x - 5$          **b.** $5 - x$

    **c.** $5 + x$         **d.** $5x$

△ **23.** If one angle measures $x°$, express the measure of its complement in terms of $x$.

   **a.** $(180 - x)°$         **b.** $(90 - x)°$

   **c.** $(x - 180)°$        **d.** $(x - 90)°$

△ **24.** If one angle measures $(x + 5)°$, express the measure of its supplement in terms of $x$.

   **a.** $(185 + x)°$

   **b.** $(95 + x)°$

   **c.** $(175 - x)°$

   **d.** $(x - 170)°$

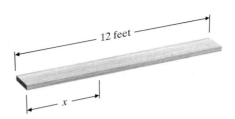

*Solve each equation.*

**25.** $\dfrac{3}{4}x = -9$           **26.** $\dfrac{x}{6} = \dfrac{2}{3}$

**27.** $-5x = 0$             **28.** $-y = 7$

**29.** $0.2x = 0.15$        **30.** $\dfrac{-x}{3} = 1$

**31.** $-3x + 1 = 19$      **32.** $5x + 25 = 20$

**33.** $7(x - 1) + 9 = 5x$    **34.** $7x - 6 = 5x - 3$

**35.** $-5x + \dfrac{3}{7} = \dfrac{10}{7}$      **36.** $5x + x = 9 + 4x - 1 + 6$

**37.** Write the sum of three consecutive integers as an expression in $x$. Let $x$ be the first integer.

**38.** Write the sum of the first and fourth of four consecutive even integers. Let $x$ be the first even integer.

**(2.3)** *Solve each equation.*

**39.** $\dfrac{5}{3}x + 4 = \dfrac{2}{3}x$      **40.** $\dfrac{7}{8}x + 1 = \dfrac{5}{8}x$

**41.** $-(5x + 1) = -7x + 3$   **42.** $-4(2x + 1) = -5x + 5$

**43.** $-6(2x - 5) = -3(9 + 4x)$

**44.** $3(8y - 1) = 6(5 + 4y)$

**45.** $\dfrac{3(2 - z)}{5} = z$       **46.** $\dfrac{4(n + 2)}{5} = -n$

**47.** $0.5(2n - 3) - 0.1 = 0.4(6 + 2n)$

**48.** $-9 - 5a = 3(6a - 1)$    **49.** $\dfrac{5(c + 1)}{6} = 2c - 3$

**50.** $\dfrac{2(8 - a)}{3} = 4 - 4a$

▨ **51.** $200(70x - 3560) = -179(150x - 19{,}300)$

**52.** $1.72y - 0.04y = 0.42$

**(2.4)** *Solve each of the following.*

**53.** The height of the Washington Monument is 50.5 inches more than 10 times the length of a side of its square base. If the sum of these two dimensions is 7327 inches, find the height of the Washington Monument. (*Source:* National Park Service)

**54.** A 12-foot board is to be divided into two pieces so that one piece is twice as long as the other. If $x$ represents the length of the shorter piece, find the length of each piece.

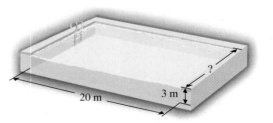

**55.** In a recent year, Kellogg Company acquired Keebler Foods Company. After the merger, the total number of Kellogg and Keebler manufacturing plants was 53. The number of Kellogg plants was one less than twice the number of Keebler plants. How many of each type of plant were there? (*Source: Kellogg Company 2000 Annual Report*)

**56.** Find three consecutive integers whose sum is $-114$.

**57.** The quotient of a number and 3 is the same as the difference of the number and two. Find the number.

**58.** Double the sum of a number and 6 is the opposite of the number. Find the number.

**(2.5)** *Substitute the given values into the given formulas and solve for the unknown variable.*

**59.** $P = 2l + 2w$;  $P = 46, l = 14$

**60.** $V = lwh$;  $V = 192, l = 8, w = 6$

*Solve each equation for the indicated variable.*

**61.** $y = mx + b$ for $m$

**62.** $r = vst - 5$ for $s$

**63.** $2y - 5x = 7$ for $x$

**64.** $3x - 6y = -2$ for $y$

△ **65.** $C = \pi D$ for $\pi$

△ **66.** $C = 2\pi r$ for $\pi$

△ **67.** A swimming pool holds 900 cubic meters of water. If its length is 20 meters and its height is 3 meters, find its width.

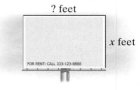

**68.** The perimeter of a rectangular billboard is 60 feet and has a length 6 feet longer than its width. Find the dimensions of the billboard.

**69.** A charity 10K race is given annually to benefit a local hospice organization. How long will it take to run/walk a 10K race (10 kilometers or 10,000 meters) if your average pace is 125 **meters** per minute? Give your time in hours and minutes.

**70.** On April 28, 2001, the highest temperature recorded in the United States was 104°F, which occurred in Death Valley, California. Convert this temperature to degrees Celsius. (*Source:* National Weather Service)

*(2.6)* *Find each of the following.*

**71.** The number 9 is what percent of 45?

**72.** The number 59.5 is what percent of 85?

**73.** The number 137.5 is 125% of what number?

**74.** The number 768 is 60% of what number?

**75.** The price of a small diamond ring was recently increased by 11%. If the ring originally cost $1900, find the mark-up and the new price of the ring.

**76.** A recent survey found that 66.9% of Americans use the Internet. If a city has a population of 76,000 how many people in that city would you expect to use the Internet? (*Source:* UCLA Center for Communication Policy)

**77.** Thirty gallons of a 20% acid solution is needed for an experiment. Only 40% and 10% acid solutions are available. How much of each should be mixed to form the needed solution?

**78.** The ACT Assessment is a college entrance exam taken by about 60% of college-bound students. The national average score was 20.7 in 1993 and rose to 21.0 in 2001. Find the percent increase. (Round to the nearest hundredth of a percent.)

*The graph below shows the percent(s) of cell phone users who have engaged in various behaviors while driving and talking on their cell phones. Use this graph to answer Exercises 79 through 82.*

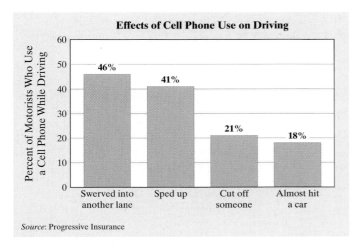

Source: Progressive Insurance

**79.** What percent of motorists who use a cell phone while driving have almost hit another car?

**80.** What is the most common effect of cell phone use on driving?

**81.** If a cell-phone service has an estimated 4600 customers who use their cell phones while driving, how many of these customers would you expect to have cut someone off while driving and talking on their cell phones?

**82.** Do the percents in the graph have a sum of 100%? Why or why not?

**83.** In 2005, Lance Armstrong incredibly won his seventh Tour de France, the first man in history to win more than five Tour de France Championships. Suppose he rides a bicycle up a category 2 climb at 10 km/hr and rides down the same distance at a speed of 50 km/hr. Find the distance traveled if the total time on the mountain was 3 hours.

*(2.7)* *Solve.*

**84.** A $50,000 retirement pension is to be invested into two accounts: a money market fund that pays 8.5% and a certificate of deposit that pays 10.5%. How much should be invested at each rate in order to provide a yearly interest income of $4550?

**85.** A pay phone is holding its maximum number of 500 coins consisting of nickels, dimes, and quarters. The number of quarters is twice the number of dimes. If the value of all the coins is $88.00, how many nickels were in the pay phone?

**86.** How long will it take an Amtrak passenger train to catch up to a freight train if their speeds are 60 and 45 mph and the freight train had an hour and a half head start?

*(2.8)* *Solve and graph the solution of each of the following inequalities.*

**87.** $x > 0$

**88.** $x \le -2$

**89.** $0.5 \le y < 1.5$

**90.** $-1 < x < 1$

**91.** $-3x > 12$

**92.** $-2x \ge -20$

**93.** $x + 4 \ge 6x - 16$

**94.** $5x - 7 > 8x + 5$

**95.** $-3 < 4x - 1 < 2$

**96.** $2 \le 3x - 4 < 6$

**97.** $4(2x - 5) \le 5x - 1$

**98.** $-2(x - 5) > 2(3x - 2)$

**99.** Tina earns $175 per week plus a 5% commission on all her sales. Find the minimum amount of sales to ensure that she earns at least $300 per week.

**100.** Ellen Catarella shot rounds of 76, 82, and 79 golfing. What must she shoot on her next round so that her average will be below 80?

## MIXED REVIEW

*Solve each equation.*

**101.** $6x + 2x - 1 = 5x + 11$

**102.** $2(3y - 4) = 6 + 7y$

**103.** $4(3 - a) - (6a + 9) = -12a$

**104.** $\frac{x}{3} - 2 = 5$

**105.** $2(y + 5) = 2y + 10$

**106.** $7x - 3x + 2 = 2(2x - 1)$

*Solve.*

**107.** The sum of six and twice a number is equal to seven less than the number. Find the number.

**108.** A 23-inch piece of string is to be cut into two pieces so that the length of the longer piece is three more than four times the shorter piece. If $x$ represents the length of the shorter piece, find the lengths of both pieces.

*Solve for the specified variable.*

**109.** $V = \frac{1}{3} Ah$ for $h$

**110.** What number is 26% of 85?

**111.** The number 72 is 45% of what number?

**112.** A company recently increased their number of employees from 235 to 282. Find the percent increase.

*Solve each inequality. Graph the solution set.*

**113.** $4x - 7 > 3x + 2$

**114.** $-5x < 20$

**115.** $-3(1 + 2x) + x \geq -(3 - x)$

# CHAPTER 2 TEST TEST PREP VIDEO

Remember to use the Chapter Test Prep Video CD to see the fully worked-out solutions to any of the exercises you want to review.

*Simplify each of the following expressions.*

**1.** $2y - 6 - y - 4$

**2.** $2.7x + 6.1 + 3.2x - 4.9$

**3.** $4(x - 2) - 3(2x - 6)$

**4.** $7 + 2(5y - 3)$

*Solve each of the following equations.*

**5.** $-\frac{4}{5}x = 4$

**6.** $4(n - 5) = -(4 - 2n)$

**7.** $5y - 7 + y = -(y + 3y)$

**8.** $4z + 1 - z = 1 + z$

**9.** $\frac{2(x + 6)}{3} = x - 5$

**10.** $\frac{1}{2} - x + \frac{3}{2} = x - 4$

**11.** $-0.3(x - 4) + x = 0.5(3 - x)$

**12.** $-4(a + 1) - 3a = -7(2a - 3)$

**13.** $-2(x - 3) = x + 5 - 3x$

*Solve each of the following applications.*

**14.** A number increased by two-thirds of the number is 35. Find the number.

△ **15.** A gallon of water seal covers 200 square feet. How many gallons are needed to paint two coats of water seal on a deck that measures 20 feet by 35 feet?

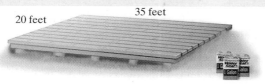

**16.** Some states have a single area code for the entire state. Two such states have area codes where one is double the other. If the sum of these integers is 1203, find the two area codes. (*Source:* North American Numbering Plan Administration)

**17.** Sedric Angell invested an amount of money in Amoxil stock that earned an annual 10% return, and then he invested twice the original amount in IBM stock that earned an annual 12% return. If his total return from both investments was $2890, find how much he invested in each stock.

**18.** Two trains leave Los Angeles simultaneously traveling on the same track in opposite directions at speeds of 50 and 64 mph. How long will it take before they are 285 miles apart?

**19.** Find the value of $x$ if $y = -14$, $m = -2$, and $b = -2$ in the formula $y = mx + b$.

*Solve each of the following equations for the indicated variable.*

△ **20.** $V = \pi r^2 h$ for $h$

**21.** $3x - 4y = 10$ for $y$

*Solve and graph each of the following inequalities.*

**22.** $3x - 5 \geq 7x + 3$

**23.** $x + 6 > 4x - 6$

**24.** $-2 < 3x + 1 < 8$

**25.** $\frac{2(5x + 1)}{3} > 2$

# CHAPTER 2 CUMULATIVE REVIEW

1. Given the set $\left\{-2, 0, \frac{1}{4}, -1.5, 112, -3, 11, \sqrt{2}\right\}$, list the numbers in this set that belong to the set of:
   a. Natural numbers
   b. Whole numbers
   c. Integers
   d. Rational numbers
   e. Irrational numbers
   f. Real numbers

2. Given the set $\left\{7, 2, -\frac{1}{5}, 0, \sqrt{3}, -185, 8\right\}$, list the numbers in this set that belong to the set of:
   a. Natural numbers
   b. Whole Numbers
   c. Integers
   d. Rational Numbers
   e. Irrational numbers
   f. Real numbers

3. Find the absolute value of each number.
   a. $|4|$        b. $|-5|$
   c. $|0|$        d. $\left|-\frac{1}{2}\right|$
   e. $|5.6|$

4. Find the absolute value of each number.
   a. $|5|$        b. $|-8|$        c. $\left|-\frac{2}{3}\right|$

5. Write each of the following numbers as a product of primes.
   a. 40        b. 63

6. Write each number as a product of primes.
   a. 44        b. 90

7. Write $\frac{2}{5}$ as an equivalent fraction with a denominator of 20.

8. Write $\frac{2}{3}$ as an equivalent fraction with a denominator of 24.

9. Simplify $3[4 + 2(10 - 1)]$.

10. Simplify $5[16 - 4(2 + 1)]$.

11. Decide whether 2 is a solution of $3x + 10 = 8x$.

12. Decide whether 3 is a solution of $5x - 2 = 4x$.

*Add.*

13. $-1 + (-2)$        14. $(-2) + (-8)$
15. $-4 + 6$           16. $-3 + 10$
17. Simplify each expression.
   a. $-(-10)$        b. $-\left(-\frac{1}{2}\right)$
   c. $-(-2x)$        d. $-|-6|$
18. Simplify each expression.
   a. $-(-5)$        b. $-\left(-\frac{2}{3}\right)$
   c. $-(-a)$        d. $-|-3|$

19. Subtract.
   a. $5.3 - (-4.6)$        b. $-\frac{3}{10} - \frac{5}{10}$
   c. $-\frac{2}{3} - \left(-\frac{4}{5}\right)$

20. Subtract
   a. $-2.7 - 8.4$        b. $-\frac{4}{5} - \left(-\frac{3}{5}\right)$
   c. $\frac{1}{4} - \left(-\frac{1}{2}\right)$

21. Find each unknown complementary or supplementary angle.
   a.        b.

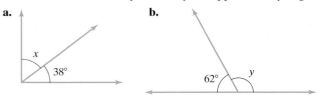

22. Find each unknown complementary or supplementary angle.
   a.        b.

23. Find each product.
   a. $(-1.2)(0.05)$        b. $\frac{2}{3} \cdot \left(-\frac{7}{10}\right)$
   c. $\left(-\frac{4}{5}\right)(-20)$

24. Find each product.
   a. $(4.5)(-0.08)$        b. $-\frac{3}{4} \cdot \left(-\frac{8}{17}\right)$

25. Find each quotient.
   a. $\frac{-24}{-4}$        b. $\frac{-36}{3}$
   c. $\frac{2}{3} \div \left(-\frac{5}{4}\right)$
   d. $-\frac{3}{2} \div 9$

26. Find each quotient.
   a. $\frac{-32}{8}$        b. $\frac{-108}{-12}$
   c. $\frac{-5}{7} \div \left(\frac{-9}{2}\right)$

27. Use a commutative property to complete each statement.
   a. $x + 5 = \rule{1.5cm}{0.4pt}$
   b. $3 \cdot x = \rule{1.5cm}{0.4pt}$

28. Use a commutative property to complete each statement.
   a. $y + 1 = \rule{1.5cm}{0.4pt}$
   b. $y \cdot 4 = \rule{1.5cm}{0.4pt}$

**29.** Use the distributive property to write each sum as a product.

   **a.** $8 \cdot 2 + 8 \cdot x$

   **b.** $7s + 7t$

**30.** Use the distributive property to write each sum as a product.

   **a.** $4 \cdot y + 4 \cdot \dfrac{1}{3}$

   **b.** $0.10x + 0.10y$

**31.** Subtract $4x - 2$ from $2x - 3$.

**32.** Subtract $10x + 3$ from $-5x + 1$.

*Solve.*

**33.** $y + 0.6 = -1.0$

**34.** $\dfrac{5}{6} + x = \dfrac{2}{3}$

**35.** $7 = -5(2a - 1) - (-11a + 6)$

**36.** $-3x + 1 - (-4x - 6) = 10$

**37.** $\dfrac{y}{7} = 20$

**38.** $\dfrac{x}{4} = 18$

**39.** $4(2x - 3) + 7 = 3x + 5$

**40.** $6x + 5 = 4(x + 4) - 1$

**41.** Twice the sum of a number and 4 is the same as four times the number, decreased by 12. Find the number.

**42.** A number increased by 4 is the same as 3 times the number decreased by 8. Find the number.

**43.** Solve $V = lwh$ for $l$.

**44.** Solve $C = 2\pi r$ for $r$.

**45.** Solve $x + 4 \le -6$ for $x$. Graph the solution set and write it in interval notation.

**46.** Solve $x - 3 > 2$ for $x$. Graph the solution set and write it in interval notation.

# 3

# Graphs and Introduction to Functions

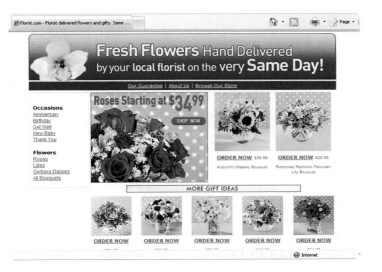

O nline advertising is a way to promote services and products via the Internet. This fairly new way of advertising is quickly growing in popularity as more of the world is looking to the Internet for news and information. The broken-line graph below shows the yearly revenue generated by online advertising.

In Chapter 3's Integrated Review, Exercise 16, you will have the opportunity to use a linear equation, generated by the years 2003–2010, to predict online advertising revenue.

In the previous chapter we learned to solve and graph the solutions of linear equations and inequalities in one variable. Now we define and present techniques for solving and graphing linear equations and inequalities in two variables.

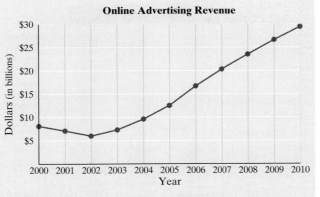

**Online Advertising Revenue**

*Source*: PriceWaterHouse Cooper's IAB Internet Advertising Revenue Report

# 3.1 READING GRAPHS AND THE RECTANGULAR COORDINATE SYSTEM

**OBJECTIVES**

1 Read bar and line graphs.

2 Define the rectangular coordinate system and plot ordered pairs of numbers.

3 Graph paired data to create a scatter diagram.

4 Determine whether an ordered pair is a solution of an equation in two variables.

5 Find the missing coordinate of an ordered pair solution, given one coordinate of the pair.

In today's world, where the exchange of information must be fast and entertaining, graphs are becoming increasingly popular. They provide a quick way of making comparisons, drawing conclusions, and approximating quantities.

**OBJECTIVE 1 ▶ Reading bar and line graphs.** A **bar graph** consists of a series of bars arranged vertically or horizontally. The bar graph in Example 1 shows a comparison of worldwide Internet users by country. The names of the countries are listed vertically and a bar is shown for each country. Corresponding to the length of the bar for each country is a number along a horizontal axis. These horizontal numbers are number of Internet users in millions.

### EXAMPLE 1

The following bar graph shows the estimated number of Internet users worldwide by country, as of a recent year.

**a.** Find the country that has the most Internet users and approximate the number of users.

**b.** How many more users are in the United States than in China?

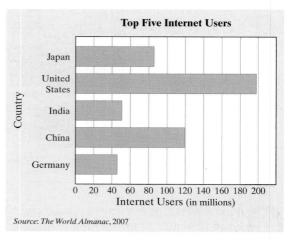

*Source: The World Almanac, 2007*

#### Solution

**a.** Since these bars are arranged horizontally, we look for the longest bar, which is the bar representing the United States. To approximate the number associated with this country, we move from the right edge of this bar vertically downward to the Internet user axis. This country has approximately 198 million Internet users.

**b.** The United States has approximately 198 million Internet users. China has approximately 120 million Internet users. To find how many more users are in the United States, we subtract 198 − 120 = 78 or 78 million more Internet users.

**PRACTICE**

**1** Use the graph from Example 1 to answer the following.

**a.** Find the country shown with the fewest Internet users and approximate the number of users.

**b.** How many more users are in India than in Germany?

A **line graph** consists of a series of points connected by a line. The next graph is an example of a line graph. It is also sometimes called a **broken line graph.**

**EXAMPLE 2**    The line graph shows the relationship between time spent smoking a cigarette and pulse rate. Time is recorded along the horizontal axis in minutes, with 0 minutes being the moment a smoker lights a cigarette. Pulse is recorded along the vertical axis in heartbeats per minute.

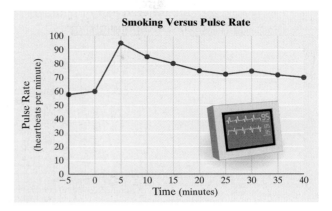

**a.** What is the pulse rate 15 minutes after a cigarette is lit?

**b.** When is the pulse rate the lowest?

**c.** When does the pulse rate show the greatest change?

*Solution*

**a.** We locate the number 15 along the time axis and move vertically upward until the line is reached. From this point on the line, we move horizontally to the left until the pulse rate axis is reached. Reading the number of beats per minute, we find that the pulse rate is 80 beats per minute 15 minutes after a cigarette is lit.

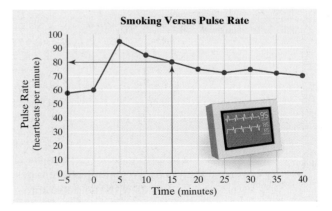

**b.** We find the lowest point of the line graph, which represents the lowest pulse rate. From this point, we move vertically downward to the time axis. We find that the pulse rate is the lowest at −5 minutes, which means 5 minutes *before* lighting a cigarette.

**c.** The pulse rate shows the greatest change during the 5 minutes between 0 and 5. Notice that the line graph is *steepest* between 0 and 5 minutes.    ☐

**PRACTICE**

**2**    Use the graph from Example 2 to answer the following.

**a.** What is the pulse rate 40 minutes after lighting a cigarette?

**b.** What is the pulse rate when the cigarette is being lit?

**c.** When is the pulse rate the highest?

**OBJECTIVE 2 ▶ Defining the rectangular coordinate system and plotting ordered pairs of numbers.** Notice in the previous graph that there are two numbers associated with each point of the graph. For example, we discussed earlier that 15 minutes after lighting a cigarette, the pulse rate is 80 beats per minute. If we agree to write the time first and the pulse rate second, we can say there is a point on the graph corresponding to the **ordered pair** of numbers (15, 80). A few more ordered pairs are listed alongside their corresponding points.

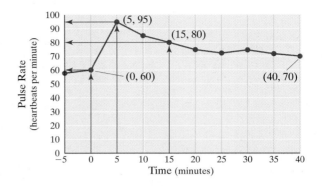

In general, we use this same ordered pair idea to describe the location of a point in a plane (such as a piece of paper). We start with a horizontal and a vertical axis. Each axis is a number line, and for the sake of consistency we construct our axes to intersect at the 0 coordinate of both. This point of intersection is called the **origin.** Notice that these two number lines or axes divide the plane into four regions called **quadrants.** The quadrants are usually numbered with Roman numerals as shown. The axes are not considered to be in any quadrant.

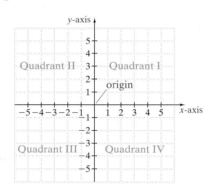

It is helpful to label axes, so we label the horizontal axis the **x-axis** and the vertical axis the **y-axis.** We call the system described above the **rectangular coordinate system.**

Just as with the pulse rate graph, we can then describe the locations of points by ordered pairs of numbers. We list the horizontal **x-axis** measurement first and the vertical **y-axis** measurement second.

To plot or graph the point corresponding to the ordered pair

$$(a, b)$$

we start at the origin. We then move $a$ units left or right (right if $a$ is positive, left if $a$ is negative). From there, we move $b$ units up or down (up if $b$ is positive, down if $b$ is negative). For example, to plot the point corresponding to the ordered pair (3, 2), we start at the origin, move 3 units right, and from there move 2 units up. (See the figure to the left.) The x-value, 3, is called the **x-coordinate** and the y-value, 2, is called the **y-coordinate.** From now on, we will call the point with coordinates (3, 2) simply the point (3, 2). The point (−2, 5) is graphed to the left also.

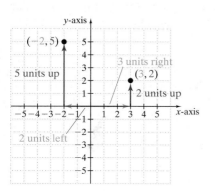

Does the order in which the coordinates are listed matter? Yes! Notice that the point corresponding to the ordered pair (2, 3) is in a different location than the point corresponding to (3, 2). These two ordered pairs of numbers describe two different points of the plane.

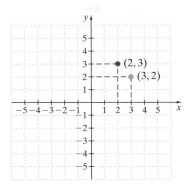

## Concept Check ☑

Is the graph of the point $(-5, 1)$ in the same location as the graph of the point $(1, -5)$? Explain.

> ▶ **Helpful Hint**
>
> Don't forget that **each ordered pair corresponds to exactly one point in the plane and that each point in the plane corresponds to exactly one ordered pair.**

### EXAMPLE 3    On a single coordinate system, plot each ordered pair. State in which quadrant, if any, each point lies.

**a.** $(5, 3)$      **b.** $(-5, 3)$      **c.** $(-2, -4)$      **d.** $(1, -2)$

**e.** $(0, 0)$      **f.** $(0, 2)$      **g.** $(-5, 0)$      **h.** $\left(0, -5\frac{1}{2}\right)$

*Solution*

Point $(5, 3)$ lies in quadrant I.
Point $(-5, 3)$ lies in quadrant II.
Point $(-2, -4)$ lies in quadrant III.
Point $(1, -2)$ lies in quadrant IV.

Points $(0, 0), (0, 2), (-5, 0)$, and $\left(0, -5\frac{1}{2}\right)$ lie on axes, so they are not in any quadrant.

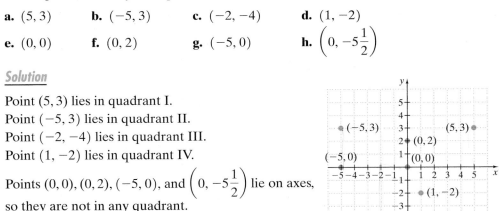

From Example 3, notice that the $y$-coordinate of any point on the $x$-axis is 0. For example, the point $(-5, 0)$ lies on the $x$-axis. Also, the $x$-coordinate of any point on the $y$-axis is 0. For example, the point $(0, 2)$ lies on the $y$-axis.    □

**PRACTICE**

**3**    On a single coordinate system, plot each ordered pair. State in which quadrant, if any, each point lies.

**a.** $(4, -3)$      **b.** $(-3, 5)$      **c.** $(0, 4)$      **d.** $(-6, 1)$

**e.** $(-2, 0)$      **f.** $(5, 5)$      **g.** $\left(3\frac{1}{2}, 1\frac{1}{2}\right)$      **h.** $(-4, -5)$

**Answer to Concept Check:**

The graph of point $(-5, 1)$ lies in quadrant II and the graph of point $(1, -5)$ lies in quadrant IV. They are *not* in the same location.

**Concept Check** ☑

For each description of a point in the rectangular coordinate system, write an ordered pair that represents it.

**a.** Point A is located three units to the left of the *y*-axis and five units above the *x*-axis.

**b.** Point B is located six units below the origin.

**OBJECTIVE 3 ▶ Graphing paired data.** Data that can be represented as an ordered pair is called **paired data.** Many types of data collected from the real world are paired data. For instance, the annual measurement of a child's height can be written as an ordered pair of the form (year, height in inches) and is paired data. The graph of paired data as points in the rectangular coordinate system is called a **scatter diagram.** Scatter diagrams can be used to look for patterns and trends in paired data.

**EXAMPLE 4** The table gives the annual net sales for Wal-Mart Stores for the years shown. (*Source:* Wal-Mart Stores, Inc.)

| Year | Wal-Mart Net Sales (in billions of dollars) |
|------|---------------------------------------------|
| 2000 | 181 |
| 2001 | 204 |
| 2002 | 230 |
| 2003 | 256 |
| 2004 | 285 |
| 2005 | 312 |
| 2006 | 345 |

**a.** Write this paired data as a set of ordered pairs of the form (year, sales in billions of dollars).

**b.** Create a scatter diagram of the paired data.

**c.** What trend in the paired data does the scatter diagram show?

*Solution*

**a.** The ordered pairs are (2000, 181), (2001, 204), (2002, 230), (2003, 256), (2004, 285), (2005, 312), and (2006, 345).

**b.** We begin by plotting the ordered pairs. Because the *x*-coordinate in each ordered pair is a year, we label the *x*-axis "Year" and mark the horizontal axis with the years given. Then we label the *y*-axis or vertical axis "Net Sales (in billions of dollars)." In this case it is convenient to mark the vertical axis in multiples of 20. Since no net sale is less than 180, we use the notation ⌇ to skip to 180, then proceed by multiples of 20.

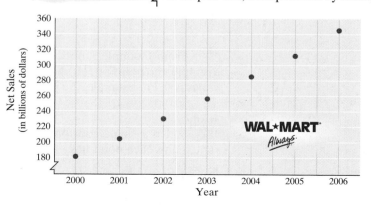

**c.** The scatter diagram shows that Wal-Mart net sales steadily increased over the years 2000–2006.                                                                                ☐

PRACTICE

**4**   The table gives the approximate annual number of wildfires (in the thousands) that have occurred in the United States for the years shown. (*Source:* National Interagency Fire Center)

| Year | Wildfires (in thousands) |
|------|--------------------------|
| 2000 | 92 |
| 2001 | 84 |
| 2002 | 73 |
| 2003 | 64 |
| 2004 | 65 |
| 2005 | 67 |
| 2006 | 96 |

**a.** Write this paired data as a set of ordered pairs of the form (year, number of wildfires in thousands).

**b.** Create a scatter diagram of the paired data.

---

**OBJECTIVE 4 ▶ Determining whether an ordered pair is a solution.** Let's see how we can use ordered pairs to record solutions of equations containing two variables. An equation in one variable such as $x + 1 = 5$ has one solution, which is 4: the number 4 is the value of the variable $x$ that makes the equation true.

An equation in two variables, such as $2x + y = 8$, has solutions consisting of two values, one for $x$ and one for $y$. For example, $x = 3$ and $y = 2$ is a solution of $2x + y = 8$ because, if $x$ is replaced with 3 and $y$ with 2, we get a true statement.

$$2x + y = 8$$
$$2(3) + 2 = 8$$
$$8 = 8 \quad \text{True}$$

The solution $x = 3$ and $y = 2$ can be written as $(3, 2)$, an **ordered pair** of numbers. The first number, 3, is the $x$-value and the second number, 2, is the $y$-value.

In general, an ordered pair is a **solution** of an equation in two variables if replacing the variables by the values of the ordered pair results in a true statement.

**EXAMPLE 5**   Determine whether each ordered pair is a solution of the equation $x - 2y = 6$.

**a.** $(6, 0)$         **b.** $(0, 3)$         **c.** $\left(1, -\dfrac{5}{2}\right)$

*Solution*

**a.** Let $x = 6$ and $y = 0$ in the equation $x - 2y = 6$.

$$x - 2y = 6$$
$$6 - 2(0) = 6 \quad \text{Replace } x \text{ with 6 and } y \text{ with 0.}$$
$$6 - 0 = 6 \quad \text{Simplify.}$$
$$6 = 6 \quad \text{True}$$

$(6, 0)$ is a solution, since $6 = 6$ is a true statement.

**b.** Let $x = 0$ and $y = 3$.

$$x - 2y = 6$$
$$0 - 2(3) = 6 \quad \text{Replace } x \text{ with 0 and } y \text{ with 3.}$$
$$0 - 6 = 6$$
$$-6 = 6 \quad \text{False}$$

$(0, 3)$ is *not* a solution, since $-6 = 6$ is a false statement.

**c.** Let $x = 1$ and $y = -\dfrac{5}{2}$ in the equation.

$$x - 2y = 6$$
$$1 - 2\left(-\dfrac{5}{2}\right) = 6 \quad \text{Replace } x \text{ with 1 and } y \text{ with } -\dfrac{5}{2}.$$
$$1 + 5 = 6$$
$$6 = 6 \quad \text{True}$$

$\left(1, -\dfrac{5}{2}\right)$ is a solution, since $6 = 6$ is a true statement.  □

**PRACTICE**

**5**  Determine whether each ordered pair is a solution of the equation $x + 3y = 6$.

**a.** $(3, 1)$   **b.** $(6, 0)$   **c.** $\left(-2, \dfrac{2}{3}\right)$

---

**OBJECTIVE 5 ▶ Completing ordered pair solutions.** If one value of an ordered pair solution of an equation is known, the other value can be determined. To find the unknown value, replace one variable in the equation by its known value. Doing so results in an equation with just one variable that can be solved for the variable using the methods of Chapter 2.

**EXAMPLE 6**  Complete the following ordered pair solutions for the equation $3x + y = 12$.

**a.** $(0, \ \ )$   **b.** $( \ \ , 6)$   **c.** $(-1, \ \ )$

*Solution*

**a.** In the ordered pair $(0, \ \ )$, the $x$-value is 0. Let $x = 0$ in the equation and solve for $y$.

$$3x + y = 12$$
$$3(0) + y = 12 \quad \text{Replace } x \text{ with 0.}$$
$$0 + y = 12$$
$$y = 12$$

The completed ordered pair is $(0, 12)$.

**b.** In the ordered pair $( \ \ , 6)$, the $y$-value is 6. Let $y = 6$ in the equation and solve for $x$.

$$3x + y = 12$$
$$3x + 6 = 12 \quad \text{Replace } y \text{ with 6.}$$
$$3x = 6 \quad \text{Subtract 6 from both sides.}$$
$$x = 2 \quad \text{Divide both sides by 3.}$$

The ordered pair is $(2, 6)$.

**c.** In the ordered pair $(-1, \quad)$, the $x$-value is $-1$. Let $x = -1$ in the equation and solve for $y$.

$$3x + y = 12$$
$$3(-1) + y = 12 \quad \text{Replace } x \text{ with } -1.$$
$$-3 + y = 12$$
$$y = 15 \quad \text{Add 3 to both sides.}$$

The ordered pair is $(-1, 15)$.    □

**PRACTICE**
**6**   Complete the following ordered pair solutions for the equation $2x - y = 8$.

**a.** $(0, \quad)$            **b.** $(\quad, 4)$            **c.** $(-3, \quad)$

Solutions of equations in two variables can also be recorded in a **table of values,** as shown in the next example.

**EXAMPLE 7**   Complete the table for the equation $y = 3x$.

|     | $x$ | $y$ |
|-----|-----|-----|
| **a.** | $-1$ | $-3$ |
| **b.** | $0$ | $0$ |
| **c.** | $-3$ | $-9$ |

**Solution**

**a.** Replace $x$ with $-1$ in the equation and solve for $y$.

$$y = 3x$$
$$y = 3(-1) \quad \text{Let } x = -1.$$
$$y = -3$$

The ordered pair is $(-1, -3)$.

**b.** Replace $y$ with 0 in the equation and solve for $x$.

$$y = 3x$$
$$0 = 3x \quad \text{Let } y = 0.$$
$$0 = x \quad \text{Divide both sides by 3.}$$

The ordered pair is $(0, 0)$.

**c.** Replace $y$ with $-9$ in the equation and solve for $x$.

| $x$ | $y$ |
|-----|-----|
| $-1$ | $-3$ |
| $0$ | $0$ |
| $-3$ | $-9$ |

$$y = 3x$$
$$-9 = 3x \quad \text{Let } y = -9.$$
$$-3 = x \quad \text{Divide both sides by 3.}$$

The ordered pair is $(-3, -9)$. The completed table is shown to the left.    □

**PRACTICE**
**7**   Complete the table for the equation $y = -4x$.

|     | $x$ | $y$ |
|-----|-----|-----|
| **a.** | $-2$ | $8$ |
| **b.** | $3$ | $-12$ |
| **c.** | $0$ | $0$ |

**EXAMPLE 8**  Complete the table for the equation

$$y = \frac{1}{2}x - 5.$$

| | x | y |
|---|---|---|
| **a.** | −2 | 6 |
| **b.** | 0 | 6 |
| **c.** | 2 | 0 |

*Solution*

**a.** Let $x = -2$.

$$y = \frac{1}{2}x - 5$$
$$y = \frac{1}{2}(-2) - 5$$
$$y = -1 - 5$$
$$y = -6$$

Ordered Pairs: $(-2, -6)$

**b.** Let $x = 0$.

$$y = \frac{1}{2}x - 5$$
$$y = \frac{1}{2}(0) - 5$$
$$y = 0 - 5$$
$$y = -5$$

$(0, -5)$

**c.** Let $y = 0$.

$$y = \frac{1}{2}x - 5$$
$$0 = \frac{1}{2}x - 5 \quad \text{Now, solve for } x.$$
$$5 = \frac{1}{2}x \quad \text{Add 5.}$$
$$10 = x \quad \text{Multiply by 2.}$$

$(10, 0)$

The completed table is

| x | y |
|---|---|
| −2 | −6 |
| 0 | −5 |
| 10 | 0 |

**PRACTICE**
**8**  Compute the table for the equation $y = \frac{1}{5}x - 2$.

| | x | y |
|---|---|---|
| **a.** | −10 | 4 |
| **b.** | 0 | −2 |
| **c.** | 1 | 0 |

**EXAMPLE 9**  Finding the Value of a Computer

A computer was recently purchased for a small business for $2000. The business manager predicts that the computer will be used for 5 years and the value in dollars $y$ of the computer in $x$ years is $y = -300x + 2000$. Complete the table.

| x | 0 | 1 | 2 | 3 | 4 | 5 |
|---|---|---|---|---|---|---|
| y | | | | | | |

*Solution*  To find the value of $y$ when $x$ is 0, replace $x$ with 0 in the equation. We use this same procedure to find $y$ when $x$ is 1 and when $x$ is 2.

**When x = 0,**

$$y = -300x + 2000$$
$$y = -300 \cdot 0 + 2000$$
$$y = 0 + 2000$$
$$y = 2000$$

**When x = 1,**

$$y = -300x + 2000$$
$$y = -300 \cdot 1 + 2000$$
$$y = -300 + 2000$$
$$y = 1700$$

**When x = 2,**

$$y = -300x + 2000$$
$$y = -300 \cdot 2 + 2000$$
$$y = -600 + 2000$$
$$y = 1400$$

We have the ordered pairs $(0, 2000)$, $(1, 1700)$, and $(2, 1400)$. This means that in 0 years the value of the computer is $2000, in 1 year the value of the computer is $1700, and in

Computers for the Home

2 years the value is $1400. To complete the table of values, we continue the procedure for $x = 3$, $x = 4$, and $x = 5$.

**When $x = 3$,**

$y = -300x + 2000$
$y = -300 \cdot 3 + 2000$
$y = -900 + 2000$
$y = 1100$

**When $x = 4$,**

$y = -300x + 2000$
$y = -300 \cdot 4 + 2000$
$y = -1200 + 2000$
$y = 800$

**When $x = 5$,**

$y = -300x + 2000$
$y = -300 \cdot 5 + 2000$
$y = -1500 + 2000$
$y = 500$

The completed table is

| $x$ | 0 | 1 | 2 | 3 | 4 | 5 |
|-----|------|------|------|------|-----|-----|
| $y$ | 2000 | 1700 | 1400 | 1100 | 800 | 500 |

**PRACTICE**

**9**   A college student purchased a used car for $12,000. The student predicted that she would need to use the car for four years and the value in dollars $y$ of the car in $x$ years is $y = -1800x + 12{,}000$. Complete this table.

| $x$ | 0 | 1 | 2 | 3 | 4 |
|-----|---|---|---|---|---|
| $y$ |   |   |   |   |   |

The ordered pair solutions recorded in the completed table for the example above are graphed below. Notice that the graph gives a visual picture of the decrease in value of the computer.

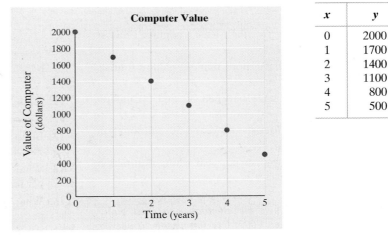

| $x$ | $y$ |
|-----|------|
| 0 | 2000 |
| 1 | 1700 |
| 2 | 1400 |
| 3 | 1100 |
| 4 | 800 |
| 5 | 500 |

# VOCABULARY & READINESS CHECK

*Use the choices below to fill in each blank. The exercises below all have to do with the rectangular coordinate system.*

origin            $x$-coordinate        $x$-axis          one                 four
quadrants     $y$-coordinate        $y$-axis          solution

**1.** The horizontal axis is called the _____.

**2.** The vertical axis is called the _____.

**3.** The intersection of the horizontal axis and the vertical axis is a point called the _____.

**4.** The axes divide the plane into regions, called _____. There are _____ of these regions.

**5.** In the ordered pair of numbers $(-2, 5)$, the number $-2$ is called the _____ and the number 5 is called the _____.

**6.** Each ordered pair of numbers corresponds to _____ point in the plane.

**7.** An ordered pair is a _____ of an equation in two variables if replacing the variables by the coordinates of the ordered pair results in a true statement.

## 3.1 EXERCISE SET

PRACTICE WATCH DOWNLOAD READ REVIEW

*The following bar graph shows the top 10 tourist destinations and the number of tourists that visit each country per year. Use this graph to answer Exercises 1 through 6. See Example 1.*

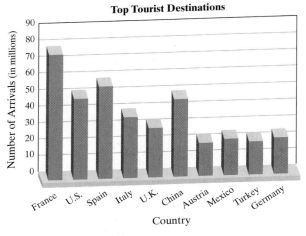

*Source*: World Tourism Organization

**1.** Which country shown is the most popular tourist destination?

**2.** Which country shown is the least popular tourist destination?

**3.** Which countries shown have more than 40 million tourists per year?

**4.** Which countries shown have between 40 and 50 million tourists per year?

**5.** Estimate the number of tourists per year whose destination is the United Kingdom.

**6.** Estimate the number of tourists per year whose destination is Turkey.

*The following line graph shows the attendance at each Super Bowl game from 2000 through 2007. Use this graph to answer Exercises 7 through 10. See Example 2.*

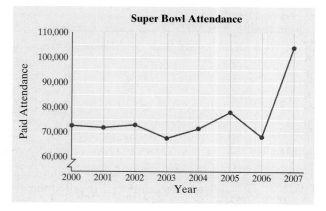

**7.** Estimate the Super Bowl attendance in 2000.

**8.** Estimate the Super Bowl attendance in 2004.

**9.** Find the year on the graph with the greatest Super Bowl attendance and approximate that attendance.

**10.** Find the year on the graph with the least Super Bowl attendance and approximate that attendance.

*The line graph below shows the number of students per teacher in U.S. public elementary and secondary schools. Use this graph for Exercises 11 through 16. See Example 2.*

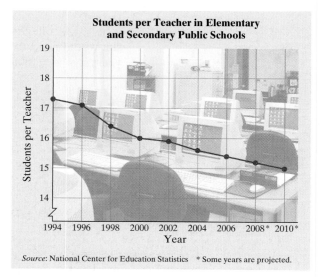

*Source*: National Center for Education Statistics   * Some years are projected.

**11.** Approximate the number of students per teacher in 2002.

**12.** Approximate the number of students per teacher in 2010.

**13.** Between what years shown did the greatest decrease in number of students per teacher occur?

**14.** What was the first year shown that the number of students per teacher fell below 17?

**15.** What was the first year shown that the number of students per teacher fell below 16?

**16.** Discuss any trends shown by this line graph.

*Plot each ordered pair. State in which quadrant or on which axis each point lies. See Example 3.*

**17. a.** $(1, 5)$      **b.** $(-5, -2)$
   **c.** $(-3, 0)$      **d.** $(0, -1)$
   **e.** $(2, -4)$      **f.** $\left(-1, 4\frac{1}{2}\right)$
   **g.** $(3.7, 2.2)$      **h.** $\left(\frac{1}{2}, -3\right)$

**18. a.** $(2, 4)$      **b.** $(0, 2)$
   **c.** $(-2, 1)$      **d.** $(-3, -3)$
   **e.** $\left(3\frac{3}{4}, 0\right)$      **f.** $(5, -4)$
   **g.** $(-3.4, 4.8)$      **h.** $\left(\frac{1}{3}, -5\right)$

*Find the x- and y-coordinates of each labeled point. See Example 3.*

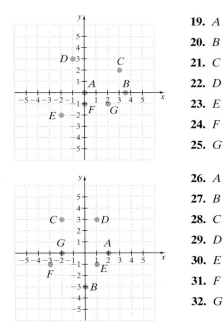

**19.** *A*

**20.** *B*

**21.** *C*

**22.** *D*

**23.** *E*

**24.** *F*

**25.** *G*

**26.** *A*

**27.** *B*

**28.** *C*

**29.** *D*

**30.** *E*

**31.** *F*

**32.** *G*

*Solve. See Example 4.*

**33.** The table shows the number of regular-season NFL football games won by the winner of the Super Bowl for the years shown. (*Source:* National Football League)

| Year | Regular-Season Games Won by Super Bowl Winner |
|------|------|
| 2002 | 12 |
| 2003 | 14 |
| 2004 | 14 |
| 2005 | 11 |
| 2006 | 12 |

  **a.** Write each paired data as an ordered pair of the form (year, games won).  $(x, y)$

  **b.** Draw a grid such as the one in Example 4 and create a scatter diagram of the paired data.

**34.** The table shows the average price of a gallon of regular unleaded gasoline (in dollars) for the years shown. (*Source:* Energy Information Administration)

| Year | Price per Gallon of Unleaded Gasoline (in dollars) |
|------|------|
| 2001 | 1.38 |
| 2002 | 1.31 |
| 2003 | 1.52 |
| 2004 | 1.81 |
| 2005 | 2.24 |
| 2006 | 2.53 |

  **a.** Write each paired data as an ordered pair of the form (year, gasoline price).

  **b.** Draw a grid such as the one in Example 4 and create a scatter diagram of the paired data.

**35.** The table shows the ethanol fuel production in the United States. (*Source:* Renewable Fuels Association; *some years projected)

| Year | Ethanol Fuel Production (in millions of gallons) |
|------|------|
| 2001 | 1770 |
| 2003 | 2800 |
| 2005 | 3904 |
| 2007* | 7500 |
| 2009* | 10,800 |

  **a.** Write each paired data as an ordered pair of the form (year, millions of gallons produced)

  **b.** Draw a grid such as the one in Example 4 and create a scatter diagram of the paired data.

  **c.** What trend in the paired data does the scatter diagram show?

**36.** The table shows the enrollment in college in the United States for the years shown. (*Source:* U.S. Department of Education)

| Year | Enrollment in College (in millions) |
|------|------|
| 1970 | 8.6 |
| 1980 | 12.1 |
| 1990 | 13.8 |
| 2000 | 15.3 |
| 2010* | 18.7 |

*projected

  **a.** Write each paired data as an ordered pair of the form (year, college enrollment in millions).

  **b.** Draw a grid such as the one in Example 4 and create a scatter diagram of the paired data.

  **c.** What trend in the paired data does the scatter diagram show?

**37.** The table shows the distance from the equator (in miles) and the average annual snowfall (in inches) for each of eight selected U.S. cities. (*Sources:* National Climatic Data Center, Wake Forest University Albatross Project)

| City | Distance from Equator (in miles) | Average Annual Snowfall (in inches) |
|---|---|---|
| 1. Atlanta, GA | 2313 | 2 |
| 2. Austin, TX | 2085 | 1 |
| 3. Baltimore, MD | 2711 | 21 |
| 4. Chicago, IL | 2869 | 39 |
| 5. Detroit, MI | 2920 | 42 |
| 6. Juneau, AK | 4038 | 99 |
| 7. Miami, FL | 1783 | 0 |
| 8. Winston-Salem, NC | 2493 | 9 |

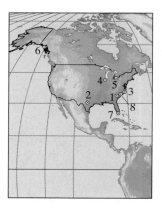

a. Write this paired data as a set of ordered pairs of the form (distance from equator, average annual snowfall).

b. Create a scatter diagram of the paired data. Be sure to label the axes appropriately.

c. What trend in the paired data does the scatter diagram show?

38. The table shows the average farm size (in acres) in the United States during the years shown. (*Source:* National Agricultural Statistics Service)

| Year | Average Farm Size (in acres) |
|---|---|
| 2001 | 438 |
| 2002 | 440 |
| 2003 | 441 |
| 2004 | 443 |
| 2005 | 445 |
| 2006 | 446 |

a. Write this paired data as a set of ordered pairs of the form (year, average farm size).

b. Create a scatter diagram of the paired data. Be sure to label the axes appropriately.

*Determine whether each ordered pair is a solution of the given linear equation. See Example 5.*

39. $2x + y = 7$; $(3, 1), (7, 0), (0, 7)$

40. $3x + y = 8$; $(2, 3), (0, 8), (8, 0)$

41. $x = -\frac{1}{3}y$; $(0, 0), (3, -9)$

42. $y = -\frac{1}{2}x$; $(0, 0), (4, 2)$

43. $x = 5$; $(4, 5), (5, 4), (5, 0)$

44. $y = -2$; $(-2, 2), (2, -2), (0, -2)$

*Complete each ordered pair so that it is a solution of the given linear equation. See Examples 6 through 8.*

45. $x - 4y = 4$; $(\quad, -2), (4, \quad)$

46. $x - 5y = -1$; $(\quad, -2), (4, \quad)$

47. $y = \frac{1}{4}x - 3$; $(-8, \quad), (\quad, 1)$

48. $y = \frac{1}{5}x - 2$; $(-10, \quad), (\quad, 1)$

*Complete the table of ordered pairs for each linear equation. See Examples 6 through 8.*

49. $y = -7x$

| x | y |
|---|---|
| 0 | |
| -1 | |
| | 2 |

50. $y = -9x$

| x | y |
|---|---|
| 0 | 0 |
| -3 | |
| | 2 |

51. $y = -x + 2$

| x | y |
|---|---|
| 0 | |
| | 0 |
| -3 | |

52. $x = -y + 4$

| x | y |
|---|---|
| | 0 |
| 0 | |
| | -3 |

53. $y = \frac{1}{2}x$

| x | y |
|---|---|
| 0 | |
| -6 | |
| | 1 |

54. $y = \frac{1}{3}x$

| x | y |
|---|---|
| 0 | |
| -6 | |
| | 1 |

55. $x + 3y = 6$

| x | y |
|---|---|
| 0 | |
| | 0 |
| | 1 |

56. $2x + y = 4$

| x | y |
|---|---|
| | 4 |
| 2 | |
| | 2 |

**57.** $y = 2x - 12$

| $x$ | $y$ |
|---|---|
| 0 | |
| | $-2$ |
| 3 | |

**58.** $y = 5x + 10$

| $x$ | $y$ |
|---|---|
| 0 | |
| | 5 |
| | 0 |

**59.** $2x + 7y = 5$

| $x$ | $y$ |
|---|---|
| 0 | |
| | 0 |
| | 1 |

**60.** $x - 6y = 3$

| $x$ | $y$ |
|---|---|
| 0 | |
| 1 | |
| | $-1$ |

## MIXED PRACTICE

*Complete the table of ordered pairs for each equation. Then plot the ordered pair solutions. See Examples 1 through 8.*

**61.** $x = -5y$

| $x$ | $y$ |
|---|---|
| | 0 |
| | 1 |
| 10 | |

**62.** $y = -3x$

| $x$ | $y$ |
|---|---|
| 0 | |
| $-2$ | |
| | 9 |

**63.** $y = \dfrac{1}{3}x + 2$

| $x$ | $y$ |
|---|---|
| 0 | |
| $-3$ | |
| | 0 |

**64.** $y = \dfrac{1}{2}x + 3$

| $x$ | $y$ |
|---|---|
| 0 | |
| $-4$ | |
| | 0 |

*Solve. See Example 9.*

**65.** The cost in dollars $y$ of producing $x$ computer desks is given by $y = 80x + 5000$.

**a.** Complete the table.

| $x$ | 100 | 200 | 300 |
|---|---|---|---|
| $y$ | | | |

**b.** Find the number of computer desks that can be produced for $8600. (*Hint:* Find $x$ when $y = 8600$.)

**66.** The hourly wage $y$ of an employee at a certain production company is given by $y = 0.25x + 9$ where $x$ is the number of units produced by the employee in an hour.

**a.** Complete the table.

| $x$ | 0 | 1 | 5 | 10 |
|---|---|---|---|---|
| $y$ | | | | |

**b.** Find the number of units that an employee must produce each hour to earn an hourly wage of $12.25. (*Hint:* Find $x$ when $y = 12.25$.)

**67.** The average amount of money $y$ spent per person on recorded music from 2001 to 2005 is given by $y = -2.35x + 55.92$. In this equation, $x$ represents the number of years after 2001. (*Source:* Veronis Suhler Stevenson)

**a.** Complete the table.

| $x$ | 1 | 3 | 5 |
|---|---|---|---|
| $y$ | | | |

**b.** Find the year in which the yearly average amount of money per person spent on recorded music was approximately $46. (*Hint:* Find $x$ when $y = 46$ and round to the nearest whole number.)

**68.** The amount $y$ of land operated by farms in the United States (in million acres) from 2000 through 2006 is given by $y = -2.18x + 944.68$. In the equation, $x$ represents the number of years after 2000. (*Source:* National Agricultural Statistics Service)

**a.** Complete the table.

| $x$ | 2 | 4 | 6 |
|---|---|---|---|
| $y$ | | | |

**b.** Find the year in which there were approximately 933 million acres of land operated by farms. (*Hint:* Find $x$ when $y = 933$ and round to the nearest whole number.)

*The graph below shows the number of Target stores for each year. Use this graph to answer Exercises 69 through 72.*

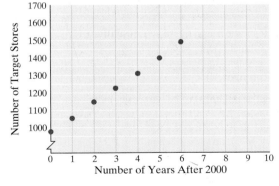

*Source*: Target

**69.** The ordered pair (4, 1308) is a point of the graph. Write a sentence describing the meaning of this ordered pair.

**70.** The ordered pair (6, 1488) is a point of the graph. Write a sentence describing the meaning of this ordered pair.

**71.** Estimate the increase in Target stores for years 1, 2, and 3.

**72.** Use a straightedge or ruler and this graph to predict the number of Target stores in the year 2009.

**73.** When is the graph of the ordered pair $(a, b)$ the same as the graph of the ordered pair $(b, a)$?

**74.** In your own words, describe how to plot an ordered pair.

## REVIEW AND PREVIEW

*Solve each equation for y. See Section 2.5.*

**75.** $x + y = 5$

**76.** $x - y = 3$

**77.** $2x + 4y = 5$

**78.** $5x + 2y = 7$

**79.** $10x = -5y$

**80.** $4y = -8x$

**81.** $x - 3y = 6$

**82.** $2x - 9y = -20$

## CONCEPT EXTENSIONS

*Answer each exercise with true or false.*

**83.** Point $(-1, 5)$ lies in quadrant IV.

**84.** Point $(3, 0)$ lies on the $y$-axis.

**85.** For the point $\left(-\frac{1}{2}, 1.5\right)$, the first value, $-\frac{1}{2}$, is the $x$-coordinate and the second value, 1.5, is the $y$-coordinate.

**86.** The ordered pair $\left(2, \frac{2}{3}\right)$ is a solution of $2x - 3y = 6$.

*For Exercises 87 through 91, fill in each blank with "0," "positive," or "negative." For Exercises 92 and 93, fill in each blank with "x" or "y."*

| | Point | Location |
|---|---|---|
| **87.** | (_____ , _____) | quadrant III |
| **88.** | (_____ , _____) | quadrant I |
| **89.** | (_____ , _____) | quadrant IV |
| **90.** | (_____ , _____) | quadrant II |
| **91.** | (_____ , _____) | origin |
| **92.** | (number, 0) | ___-axis |
| **93.** | (0, number) | ___-axis |

**94.** Give an example of an ordered pair whose location is in (or on)

    **a.** quadrant I          **b.** quadrant II

    **c.** quadrant III     **d.** quadrant IV

    **e.** $x$-axis           **f.** $y$-axis

*Solve. See the Concept Check in this section.*

**95.** Is the graph of $(3, 0)$ in the same location as the graph of $(0, 3)$? Explain why or why not.

**96.** Give the coordinates of a point such that if the coordinates are reversed, their location is the same.

**97.** In general, what points can have coordinates reversed and still have the same location?

**98.** In your own words, describe how to plot or graph an ordered pair of numbers.

*Write an ordered pair for each point described.*

**99.** Point $C$ is four units to the right of the $y$-axis and seven units below the $x$-axis.

**100.** Point $D$ is three units to the left of the origin.

**101.** Three vertices of a rectangle are $(-2, -3)$, $(-7, -3)$, and $(-7, 6)$.

    **a.** Find the coordinates of the fourth vertex of a rectangle.

    **b.** Find the perimeter of the rectangle.

    **c.** Find the area of the rectangle.

**102.** Three vertices of a square are $(-4, -1)$, $(-4, 8)$, and $(5, 8)$.

    **a.** Find the coordinates of the fourth vertex of the square.

    **b.** Find the perimeter of the square.

    **c.** Find the area of the square.

---

### 📖 STUDY SKILLS BUILDER

**Are You Satisfied with Your Performance in This Course Thus Far?**

To see if there is room for improvement, answer these questions:

**1.** Am I attending all classes and arriving on time?

**2.** Am I working and checking my homework assignments on time?

**3.** Am I getting help (from my instructor or a campus learning resource lab) when I need it?

**4.** In addition to my instructor, am I using the text supplements that might help me?

**5.** Am I satisfied with my performance on quizzes and exams?

If you answered no to any of these questions, read or reread Section 1.1 for suggestions in these areas. Also, you might want to contact your instructor for additional feedback.

# 3.2 GRAPHING LINEAR EQUATIONS

**OBJECTIVES**

1 Identify linear equations.

2 Graph a linear equation by finding and plotting ordered pair solutions.

**OBJECTIVE 1 ▶ Identifying linear equations.** In the previous section, we found that equations in two variables may have more than one solution. For example, both $(6, 0)$ and $(2, -2)$ are solutions of the equation $x - 2y = 6$. In fact, this equation has an infinite number of solutions. Other solutions include $(0, -3)$, $(4, -1)$, and $(-2, -4)$. If we graph these solutions, notice that a pattern appears.

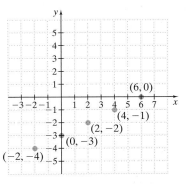

These solutions all appear to lie on the same line, which has been filled in below. It can be shown that every ordered pair solution of the equation corresponds to a point on this line, and every point on this line corresponds to an ordered pair solution. Thus, we say that this line is the **graph of the equation** $x - 2y = 6$.

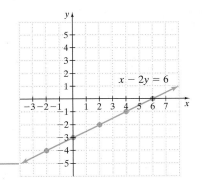

▶ **Helpful Hint**

Notice that we can only show a part of a line on a graph. The arrowheads on each end of the line remind us that the line actually extends indefinitely in both directions.

The equation $x - 2y = 6$ is called a **linear equation in two variables** and **the graph of every linear equation in two variables is a line.**

---

**Linear Equation in Two Variables**

A linear equation in two variables is an equation that can be written in the form

$$Ax + By = C$$

where $A$, $B$, and $C$ are real numbers and $A$ and $B$ are not both 0. **The graph of a linear equation in two variables is a straight line.**

---

The form $Ax + By = C$ is called **standard form.**

---

▶ **Helpful Hint**

Notice in the form $Ax + By = C$, the understood exponent on both $x$ and $y$ is 1.

***Examples of Linear Equations in Two Variables***

$$2x + y = 8 \qquad -2x = 7y \qquad y = \frac{1}{3}x + 2 \qquad y = 7$$

(Standard Form)

Before we graph linear equations in two variables, let's practice identifying these equations.

**EXAMPLE 1** Determine whether each equation is a linear equation in two variables.

**a.** $x - 1.5y = -1.6$     **b.** $y = -2x$     **c.** $x + y^2 = 9$     **d.** $x = 5$

*Solution*

**a.** This is a linear equation in two variables because it is written in the form $Ax + By = C$ with $A = 1$, $B = -1.5$, and $C = -1.6$.

**b.** This is a linear equation in two variables because it can be written in the form $Ax + By = C$.

$$y = -2x$$
$$2x + y = 0 \qquad \text{Add } 2x \text{ to both sides.}$$

**c.** This is *not* a linear equation in two variables because $y$ is squared.

**d.** This is a linear equation in two variables because it can be written in the form $Ax + By = C$.

$$x = 5$$
$$x + 0y = 5 \qquad \text{Add } 0 \cdot y.$$

**PRACTICE**

**1** Determine whether each equation is a linear equation in two variables.

**a.** $3x + 2.7y = -5.3$     **b.** $x^2 + y = 8$     **c.** $y = 12$     **d.** $5x = -3y$

**OBJECTIVE 2 ▶ Graphing linear equations by plotting ordered pair solutions.** From geometry, we know that a straight line is determined by just two points. Graphing a linear equation in two variables, then, requires that we find just two of its infinitely many solutions. Once we do so, we plot the solution points and draw the line connecting the points. Usually, we find a third solution as well, as a check.

**EXAMPLE 2** Graph the linear equation $2x + y = 5$.

*Solution* Find three ordered pair solutions of $2x + y = 5$. To do this, choose a value for one variable, $x$ or $y$, and solve for the other variable. For example, let $x = 1$. Then $2x + y = 5$ becomes

$$2x + y = 5$$
$$2(1) + y = 5 \qquad \text{Replace } x \text{ with 1.}$$
$$2 + y = 5 \qquad \text{Multiply.}$$
$$y = 3 \qquad \text{Subtract 2 from both sides.}$$

Since $y = 3$ when $x = 1$, the ordered pair $(1, 3)$ is a solution of $2x + y = 5$. Next, let $x = 0$.

$$2x + y = 5$$
$$2(0) + y = 5 \qquad \text{Replace } x \text{ with 0.}$$
$$0 + y = 5$$
$$y = 5$$

The ordered pair $(0, 5)$ is a second solution.

The two solutions found so far allow us to draw the straight line that is the graph of all solutions of $2x + y = 5$. However, we find a third ordered pair as a check. Let $y = -1$.

$$2x + y = 5$$
$$2x + (-1) = 5 \quad \text{Replace } y \text{ with } -1.$$
$$2x - 1 = 5$$
$$2x = 6 \quad \text{Add 1 to both sides.}$$
$$x = 3 \quad \text{Divide both sides by 2.}$$

The third solution is $(3, -1)$. These three ordered pair solutions are listed in table form as shown. The graph of $2x + y = 5$ is the line through the three points.

| $x$ | $y$ |
|-----|-----|
| 1 | 3 |
| 0 | 5 |
| 3 | −1 |

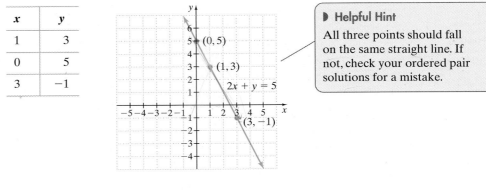

> ▶ **Helpful Hint**
> All three points should fall on the same straight line. If not, check your ordered pair solutions for a mistake.

**PRACTICE**
**2**   Graph the linear equation $x + 3y = 9$.

**EXAMPLE 3**   Graph the linear equation $-5x + 3y = 15$.

*Solution*   Find three ordered pair solutions of $-5x + 3y = 15$.

**Let $x = 0$.**

$$-5x + 3y = 15$$
$$-5 \cdot 0 + 3y = 15$$
$$0 + 3y = 15$$
$$3y = 15$$
$$y = 5$$

**Let $y = 0$.**

$$-5x + 3y = 15$$
$$-5x + 3 \cdot 0 = 15$$
$$-5x + 0 = 15$$
$$-5x = 15$$
$$x = -3$$

**Let $x = -2$.**

$$-5x + 3y = 15$$
$$-5(-2) + 3y = 15$$
$$10 + 3y = 15$$
$$3y = 5$$
$$y = \frac{5}{3}$$

The ordered pairs are $(0, 5)$, $(-3, 0)$, and $\left(-2, \dfrac{5}{3}\right)$. The graph of $-5x + 3y = 15$ is the line through the three points.

| $x$ | $y$ |
|-----|-----|
| 0 | 5 |
| −3 | 0 |
| −2 | $\dfrac{5}{3} = 1\dfrac{2}{3}$ |

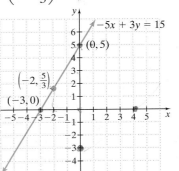

**PRACTICE**
**3**   Graph the linear equation $3x - 4y = 12$.

**EXAMPLE 4**   Graph the linear equation $y = 3x$.

_Solution_   To graph this linear equation, we find three ordered pair solutions. Since this equation is solved for $y$, choose three $x$ values.

| $x$ | $y$ |
|-----|-----|
| 2   | 6   |
| 0   | 0   |
| $-1$ | $-3$ |

If $x = 2$, $y = 3 \cdot 2 = 6$.

If $x = 0$, $y = 3 \cdot 0 = 0$.

If $x = -1$, $y = 3 \cdot -1 = -3$.

Next, graph the ordered pair solutions listed in the table above and draw a line through the plotted points as shown on the next page. The line is the graph of $y = 3x$. Every point on the graph represents an ordered pair solution of the equation and every ordered pair solution is a point on this line.

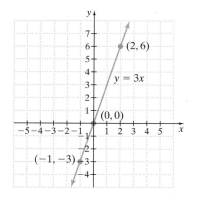

**PRACTICE**
**4**   Graph the linear equation $y = -2x$.

**EXAMPLE 5**   Graph the linear equation $y = -\frac{1}{3}x + 2$.

_Solution_   Find three ordered pair solutions, graph the solutions, and draw a line through the plotted solutions. To avoid fractions, choose $x$ values that are multiples of 3 to substitute in the equation. When a multiple of 3 is multiplied by $-\frac{1}{3}$, the result is an integer. See the calculations shown above the table.

If $x = 6$, then $y = -\frac{1}{3} \cdot \overset{2}{\cancel{6}} + 2 = -2 + 2 = 0$

If $x = 0$, then $y = -\frac{1}{3} \cdot 0 + 2 = 0 + 2 = 2$

If $x = -3$, then $y = -\frac{1}{3} \cdot -3 + 2 = 1 + 2 = 3$

| $x$ | $y$ |
|-----|-----|
| 6   | 0   |
| 0   | 2   |
| $-3$ | 3   |

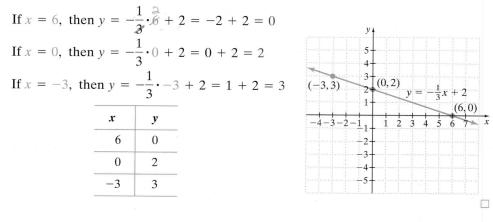

**PRACTICE**
**5**   Graph the linear equation $y = \frac{1}{2}x + 3$.

Let's compare the graphs in Examples 4 and 5. The graph of $y = 3x$ tilts upward (as we follow the line from left to right) and the graph of $y = -\dfrac{1}{3}x + 2$ tilts downward (as we follow the line from left to right). We will learn more about the tilt, or slope, of a line in Section 3.4.

*help*

**EXAMPLE 6**   Graph the linear equation $y = 3x + 6$ and compare this graph with the graph of $y = 3x$ in Example 4.

*Solution*   Find ordered pair solutions, graph the solutions, and draw a line through the plotted solutions. We choose $x$ values and substitute in the equation $y = 3x + 6$.

If $x = -3$, then $y = 3(-3) + 6 = -3$.

If $x = 0$, then $y = 3(0) + 6 = 6$.

If $x = 1$, then $y = 3(1) + 6 = 9$.

| $x$ | $y$ |
|-----|-----|
| $-3$ | $-3$ |
| $0$ | $6$ |
| $1$ | $9$ |

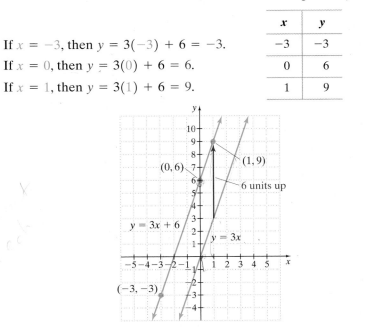

The most startling similarity is that both graphs appear to have the same upward tilt as we move from left to right. Also, the graph of $y = 3x$ crosses the $y$-axis at the origin, while the graph of $y = 3x + 6$ crosses the $y$-axis at 6. In fact, the graph of $y = 3x + 6$ is the same as the graph of $y = 3x$ moved vertically upward 6 units.  ☐

**PRACTICE**
**6**   Graph the linear equation $y = -2x + 3$ and compare this graph with the graph of $y = -2x$ in Practice 4.

*Ques*

Notice that the graph of $y = 3x + 6$ crosses the $y$-axis at 6. This happens because when $x = 0$, $y = 3x + 6$ becomes $y = 3 \cdot 0 + 6 = 6$. The graph contains the point $(0, 6)$, which is on the $y$-axis.

In general, if a linear equation in two variables is solved for $y$, we say that it is written in the form $y = mx + b$. The graph of this equation contains the point $(0, b)$ because when $x = 0$, $y = mx + b$ is $y = m \cdot 0 + b = b$.

> The graph of $y = mx + b$ crosses the $y$-axis at $(0, b)$.

We will review this again in Section 3.5.

Linear equations are often used to model real data as seen in the next example.

**EXAMPLE 7** **Estimating the Number of Medical Assistants**

One of the occupations expected to have the most growth in the next few years is medical assistant. The number of people $y$ (in thousands) employed as medical assistants in the United States can be estimated by the linear equation $y = 31.8x + 180$, where $x$ is the number of years after the year 1995. (*Source:* Based on data from the Bureau of Labor Statistics)

**a.** Graph the equation.

**b.** Use the graph to predict the number of medical assistants in the year 2010.

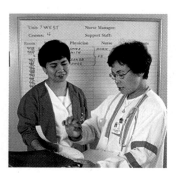

*Que)*

*Solution*

**a.** To graph $y = 31.8x + 180$, choose $x$-values and substitute in the equation.

If $x = 0$, then $y = 31.8(0) + 180 = 180$.

If $x = 2$, then $y = 31.8(2) + 180 = 243.6$.

If $x = 7$, then $y = 31.8(7) + 180 = 402.6$.

| $x$ | $y$ |
|-----|-----|
| 0 | 180 |
| 2 | 243.6 |
| 7 | 402.6 |

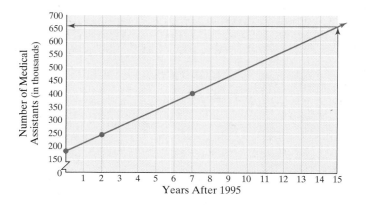

**b.** To use the graph to *predict* the number of medical assistants in the year 2010, we need to find the $y$-coordinate that corresponds to $x = 15$. (15 years after 1995 is the year 2010.) To do so, find 15 on the $x$-axis. Move vertically upward to the graphed line and then horizontally to the left. We approximate the number on the $y$-axis to be 655. Thus in the year 2010, we predict that there will be 655 thousand medical assistants. (The actual value, using 15 for $x$, is 657.)

**PRACTICE**
**7** One of the occupations expected to have the most growth in the next few years is computer software application engineers. The number of people $y$ (in thousands) employed as computer software application engineers in the United States can be estimated by the linear equation $y = 22.2x + 371$, where $x$ is the number of years after 2000. (*Source:* Based on data from the Bureau of Labor Statistics)

**a.** Graph the equation.

**b.** Use the graph to predict the number of computer software application engineers in the year 2015.

▶ **Helpful Hint**

Make sure you understand that models are mathematical approximations of the data for the known years. (For example, see the model in Example 7.) Any number of unknown factors can affect future years, so be cautious when using models to predict.

**Graphing Calculator Explorations**

In this section, we begin an optional study of graphing calculators and graphing software packages for computers. These graphers use the same point plotting technique that was introduced in this section. The advantage of this graphing technology is, of course, that graphing calculators and computers can find and plot ordered pair solutions much faster than we can. Note, however, that the features described in these boxes may not be available on all graphing calculators.

The rectangular screen where a portion of the rectangular coordinate system is displayed is called a **window.** We call it a **standard window** for graphing when both the $x$- and $y$-axes show coordinates between $-10$ and $10$. This information is often displayed in the window menu on a graphing calculator as

$$\text{Xmin} = -10$$
$$\text{Xmax} = 10$$
$$\text{Xscl} = 1 \qquad \text{The scale on the } x\text{-axis is one unit per tick mark.}$$
$$\text{Ymin} = -10$$
$$\text{Ymax} = 10$$
$$\text{Yscl} = 1 \qquad \text{The scale on the } y\text{-axis is one unit per tick mark.}$$

To use a graphing calculator to graph the equation $y = 2x + 3$, press the $\boxed{\text{Y=}}$ key and enter the keystrokes $\boxed{2}$ $\boxed{x}$ $\boxed{+}$ $\boxed{3}$. The top row should now read $Y_1 = 2x + 3$. Next press the $\boxed{\text{GRAPH}}$ key, and the display should look like this:

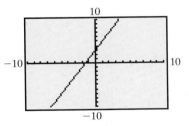

*Use a standard window and graph the following linear equations. (Unless otherwise stated, use a standard window when graphing.)*

**1.** $y = -3x + 7$

**2.** $y = -x + 5$

**3.** $y = 2.5x - 7.9$

**4.** $y = -1.3x + 5.2$

**5.** $y = -\dfrac{3}{10}x + \dfrac{32}{5}$

**6.** $y = \dfrac{2}{9}x - \dfrac{22}{3}$

---

**3.2 | EXERCISE SET**

*Determine whether each equation is a linear equation in two variables. See Example 1.*

**1.** $-x = 3y + 10$

**2.** $y = x - 15$

**3.** $x = y$

**4.** $x = y^3$

**5.** $x^2 + 2y = 0$

**6.** $0.01x - 0.2y = 8.8$

**7.** $y = -1$

**8.** $x = 25$

*For each equation, find three ordered pair solutions by completing the table. Then use the ordered pairs to graph the equation. See Examples 2 through 6.*

**9.** $x - y = 6$

| x | y |
|---|---|
| 6 | 0 |
| 4 | −2 |
| 5 | −1 |

**10.** $x - y = 4$

| x | y |
|---|---|
| 0 | |
| | 2 |
| −1 | |

**11.** $y = -4x$

| x | y |
|---|---|
| 1 | -4 |
| 0 | 0 |
| -1 | 4 |

**12.** $y = -5x$

| x | y |
|---|---|
| 1 | -5 |
| 0 | 0 |
| -1 | 5 |

**13.** $y = \frac{1}{3}x$

| x | y |
|---|---|
| 0 | |
| 6 | |
| -3 | |

**14.** $y = \frac{1}{2}x$

| x | y |
|---|---|
| 0 | |
| -4 | |
| 2 | |

**15.** $y = -4x + 3$

| x | y |
|---|---|
| 0 | 3 |
| 1 | -1 |
| 2 | -5 |

**16.** $y = -5x + 2$

| x | y |
|---|---|
| 0 | |
| 1 | |
| 2 | |

## MIXED PRACTICE

*Graph each linear equation. See Examples 2 through 6.*

**17.** $x + y = 1$

**18.** $x + y = 7$

**19.** $x - y = -2$

**20.** $-x + y = 6$

**21.** $x - 2y = 6$

**22.** $-x + 5y = 5$

**23.** $y = 6x + 3$

**24.** $y = -2x + 7$

**25.** $x = -4$

**26.** $y = 5$

**27.** $y = 3$

**28.** $x = -1$

**29.** $y = x$

**30.** $y = -x$

**31.** $x = -3y$

**32.** $x = -5y$

**33.** $x + 3y = 9$

**34.** $2x + y = 2$

**35.** $y = \frac{1}{2}x + 2$

**36.** $y = \frac{1}{4}x + 3$

**37.** $3x - 2y = 12$

**38.** $2x - 7y = 14$

**39.** $y = -3.5x + 4$

**40.** $y = -1.5x - 3$

*Graph each pair of linear equations on the same set of axes. Discuss how the graphs are similar and how they are different. See Example 6.*

**41.** $y = 5x; y = 5x + 4$

**42.** $y = 2x; y = 2x + 5$

**43.** $y = -2x; y = -2x - 3$

**44.** $y = x; y = x - 7$

**45.** $y = \frac{1}{2}x; y = \frac{1}{2}x + 2$

**46.** $y = -\frac{1}{4}x; y = -\frac{1}{4}x + 3$

*The graph of $y = 5x$ is given below as well as Figures a–d. For Exercises 47 through 50, match each equation with its graph. Hint: Recall that if an equation is written in the form $y = mx + b$, its graph crosses the y-axis at $(0, b)$.*

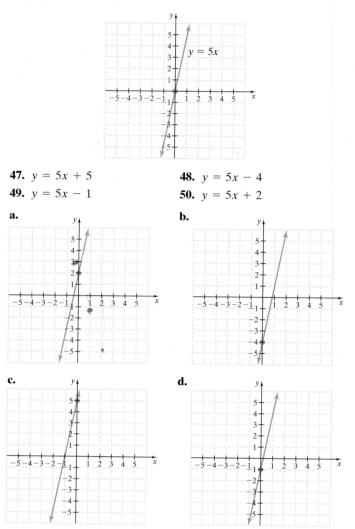

**47.** $y = 5x + 5$

**48.** $y = 5x - 4$

**49.** $y = 5x - 1$

**50.** $y = 5x + 2$

**a.**

**b.**

**c.**

**d.**

*Solve. See Example 7.*

**51.** Snowboarding is the fastest growing snow sport, and the number of participants has been increasing at a steady rate. The number of people involved in snowboarding (in millions) from the years 1997 to 2005 is given by the equation $y = 0.5x + 3$, where $x$ is the number of years after 1997. (*Source:* Based on data from the National Sporting Goods Association)

**a.** Use this equation or a graph of it to complete the ordered pair (8,   ).

**b.** Write a sentence explaining the meaning of the answer to part (a).

**c.** If this trend continues, how many snowboarders will there be in 2012?

**52.** The revenue $y$ (in billions of dollars) for Home Depot stores during the years 2000 through 2005 is given by the equation $y = 7x + 45$, where $x$ is the number of years after 2000. (*Source:* Based on data from Home Depot stores)

**a.** Use this equation or a graph of it to complete the ordered pair (5,   ).

**b.** Write a sentence explaining the meaning of the answer to part (a).

**c.** If this trend continues, predict the revenue for Home Depot stores for the year 2015.

**53.** The minimum salary for a player in the NFL is determined by the league and depends on the years of experience. In a recent year, the salary of a player earning only the minimum $y$ (in thousands of dollars) can be approximated by the equation $y = 54x + 275$, where $x$ is the number of years experience. If this trend continues, use this equation to predict the minimum salary for an NFL player after 5 years' experience. Write your answer as a sentence. (*Source:* Based on data from the National Football League)

**54.** The U.S. silver production (in metric tons) from 2000 to 2004 has been steadily dropping, and can be approximated by the equation $y = -196x + 1904$ where $x$ is the number of years after 2000. If this current trend continues, use the equation to estimate the U.S. silver production in 2009. Write your answer as a sentence. (*Source*: U.S. Geological Survey)

### REVIEW AND PREVIEW

**55.** The coordinates of three vertices of a rectangle are $(-2, 5)$, $(4, 5)$, and $(-2, -1)$. Find the coordinates of the fourth vertex. See Section 3.1.

**56.** The coordinates of two vertices of a square are $(-3, -1)$ and $(2, -1)$. Find the coordinates of two pairs of points possible for the third and fourth vertices. See Section 3.1.

*Solve the following equations. See Section 2.3.*

**57.** $3(x - 2) + 5x = 6x - 16$

**58.** $5 + 7(x + 1) = 12 + 10x$

**59.** $3x + \dfrac{2}{5} = \dfrac{1}{10}$

**60.** $\dfrac{1}{6} + 2x = \dfrac{2}{3}$

### CONCEPT EXTENSIONS

*Write each statement as an equation in two variables. Then graph the equation.*

**61.** The $y$-value is 5 more than the $x$-value.

**62.** The $y$-value is twice the $x$-value.

**63.** Two times the $x$-value, added to three times the $y$-value is 6.

**64.** Five times the $x$-value, added to twice the $y$-value is $-10$.

**65.** The perimeter of the trapezoid below is 22 centimeters. Write a linear equation in two variables for the perimeter. Find $y$ if $x$ is 3 cm.

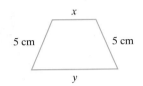

**66.** The perimeter of the rectangle below is 50 miles. Write a linear equation in two variables for this perimeter. Use this equation to find $x$ when $y$ is 20.

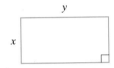

**67.** Explain how to find ordered pair solutions of linear equations in two variables.

**68.** If $(a, b)$ is an ordered pair solution of $x + y = 5$, is $(b, a)$ also a solution? Explain why or why not.

**69.** Graph the nonlinear equation $y = x^2$ by completing the table shown. Plot the ordered pairs and connect them with a smooth curve.

| $x$ | $y$ |
| --- | --- |
| 0 | |
| 1 | |
| $-1$ | |
| 2 | |
| $-2$ | |

**70.** Graph the nonlinear equation $y = |x|$ by completing the table shown. Plot the ordered pairs and connect them. This curve is "V" shaped.

$y = |x|$

| $x$ | $y$ |
| --- | --- |
| 0 | |
| 1 | |
| $-1$ | |
| 2 | |
| $-2$ | |

# 3.3 INTERCEPTS

**OBJECTIVES**

1 Identify intercepts of a graph.

2 Graph a linear equation by finding and plotting intercepts.

3 Identify and graph vertical and horizontal lines.

**OBJECTIVE 1 ▶ Identifying intercepts.** In this section, we graph linear equations in two variables by identifying intercepts. For example, the graph of $y = 4x - 8$ is shown on right. Notice that this graph crosses the $y$-axis at the point $(0, -8)$. This point is called the **$y$-intercept.** Likewise, the graph crosses the $x$-axis at $(2, 0)$, and this point is called the **$x$-intercept.**

The intercepts are $(2, 0)$ and $(0, -8)$.

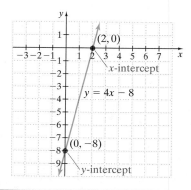

▶ **Helpful Hint**

If a graph crosses the $x$-axis at $(-3, 0)$ and the $y$-axis at $(0, 7)$, then

$$\underbrace{(-3, 0)}_{x\text{-intercept}} \qquad \underbrace{(0, 7)}_{y\text{-intercept}}$$

Notice that for the $y$-intercept, the $x$-value is 0 and for the $x$-intercept, the $y$-value is 0.

**Note:** Sometimes in mathematics, you may see just the number 7 stated as the $y$-intercept, and $-3$ stated as the $x$-intercept.

**EXAMPLES**    Identify the $x$- and $y$-intercepts.

**1.**

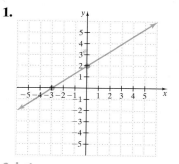

_Solution_

$x$-intercept: $(-3, 0)$

$y$-intercept: $(0, 2)$

**2.**

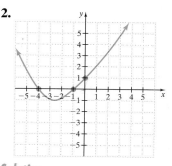

_Solution_

$x$-intercepts: $(-4, 0)$, $(-1, 0)$

$y$-intercept: $(0, 1)$

▶ **Helpful Hint**

Notice that any time $(0, 0)$ is a point of a graph, then it is an $x$-intercept and a $y$-intercept.

**3.**

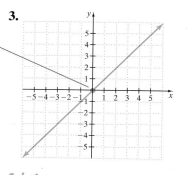

_Solution_

$x$-intercept: $(0, 0)$

$y$-intercept: $(0, 0)$

**4.**

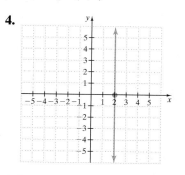

_Solution_

$x$-intercept: $(2, 0)$

$y$-intercept: none

**5.**

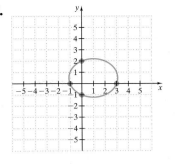

<u>**Solution**</u>

$x$-intercepts: $(-1, 0)$, $(3, 0)$

$y$-intercepts: $(0, 2)$, $(0, -1)$

◻

**PRACTICES**

**1–5**   Identify the $x$- and $y$-intercepts.

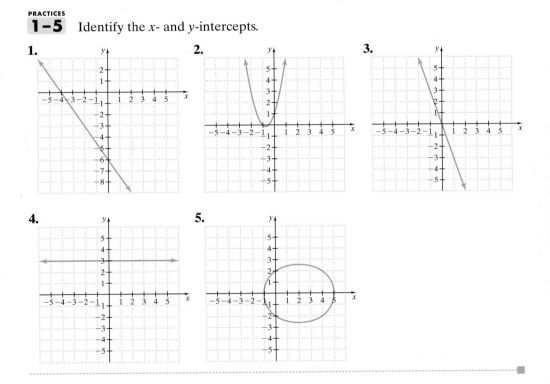

**OBJECTIVE 2 ▶ Using intercepts to graph a linear equation.** Given the equation of a line, intercepts are usually easy to find since one coordinate is 0.

One way to find the $y$-intercept of a line, given its equation, is to let $x = 0$, since a point on the $y$-axis has an $x$-coordinate of 0. To find the $x$-intercept of a line, let $y = 0$, since a point on the $x$-axis has a $y$-coordinate of 0.

---

**Finding x- and y-intercepts**

To find the $x$-intercept, let $y = 0$ and solve for $x$.

To find the $y$-intercept, let $x = 0$ and solve for $y$.

**EXAMPLE 6**    Graph $x - 3y = 6$ by finding and plotting intercepts.

*Solution*    Let $y = 0$ to find the $x$-intercept and let $x = 0$ to find the $y$-intercept.

| Let $y = 0$ | Let $x = 0$ |
|---|---|
| $x - 3y = 6$ | $x - 3y = 6$ |
| $x - 3(0) = 6$ | $0 - 3y = 6$ |
| $x - 0 = 6$ | $-3y = 6$ |
| $x = 6$ | $y = -2$ |

The $x$-intercept is $(6, 0)$ and the $y$-intercept is $(0, -2)$. We find a third ordered pair solution to check our work. If we let $y = -1$, then $x = 3$. Plot the points $(6, 0)$, $(0, -2)$, and $(3, -1)$. The graph of $x - 3y = 6$ is the line drawn through these points, as shown.

| $x$ | $y$ |
|---|---|
| 6 | 0 |
| 0 | -2 |
| 3 | -1 |

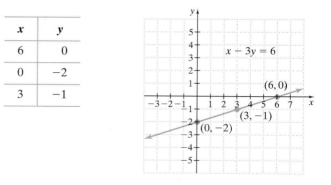

**PRACTICE**
**6**    Graph $x + 2y = -4$ by finding and plotting intercepts.

**EXAMPLE 7**    Graph $x = -2y$ by plotting intercepts.

*Solution*    Let $y = 0$ to find the $x$-intercept and $x = 0$ to find the $y$-intercept.

| Let $y = 0$ | Let $x = 0$ |
|---|---|
| $x = -2y$ | $x = -2y$ |
| $x = -2(0)$ | $0 = -2y$ |
| $x = 0$ | $0 = y$ |

Both the $x$-intercept and $y$-intercept are $(0, 0)$. In other words, when $x = 0$, then $y = 0$, which gives the ordered pair $(0, 0)$. Also, when $y = 0$, then $x = 0$, which gives the same ordered pair $(0, 0)$. This happens when the graph passes through the origin. Since two points are needed to determine a line, we must find at least one more ordered pair that satisfies $x = -2y$. Let $y = -1$ to find a second ordered pair solution and let $y = 1$ as a checkpoint.

| Let $y = -1$ | Let $y = 1$ |
|---|---|
| $x = -2(-1)$ | $x = -2(1)$ |
| $x = 2$ | $x = -2$ |

The ordered pairs are $(0, 0)$, $(2, -1)$, and $(-2, 1)$. Plot these points to graph $x = -2y$.

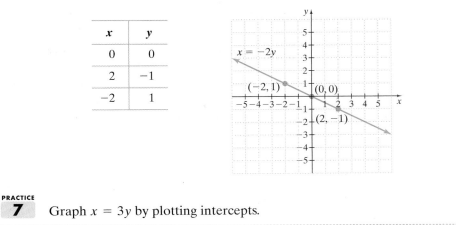

| x | y |
|---|---|
| 0 | 0 |
| 2 | −1 |
| −2 | 1 |

**PRACTICE**
**7**   Graph $x = 3y$ by plotting intercepts.

**EXAMPLE 8**   Graph $4x = 3y - 9$.

*Solution*   Find the $x$- and $y$-intercepts, and then choose $x = 2$ to find a third checkpoint.

Let $y = 0$
$$4x = 3(0) - 9$$
$$4x = -9$$
Solve for $x$.
$$x = -\frac{9}{4} \text{ or } -2\frac{1}{4}$$

Let $x = 0$
$$4 \cdot 0 = 3y - 9$$
$$9 = 3y$$
Solve for $y$.
$$3 = y$$

Let $x = 2$
$$4(2) = 3y - 9$$
$$8 = 3y - 9$$
Solve for $y$.
$$17 = 3y$$
$$\frac{17}{3} = y \text{ or } y = 5\frac{2}{3}$$

The ordered pairs are $\left(-2\frac{1}{4}, 0\right)$, $(0, 3)$, and $\left(2, 5\frac{2}{3}\right)$. The equation $4x = 3y - 9$ is graphed as follows.

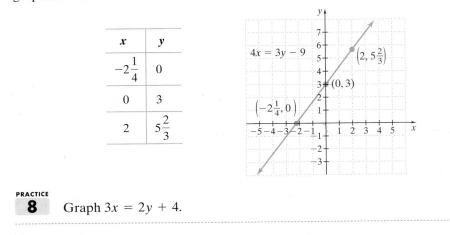

| x | y |
|---|---|
| $-2\frac{1}{4}$ | 0 |
| 0 | 3 |
| 2 | $5\frac{2}{3}$ |

**PRACTICE**
**8**   Graph $3x = 2y + 4$.

**OBJECTIVE 3 ▶ Graphing vertical and horizontal lines.** The equation $x = c$, where $c$ is a real number constant, is a linear equation in two variables because it can be written in the form $x + 0y = c$. The graph of this equation is a vertical line as shown in the next example.

**EXAMPLE 9**   Graph $x = 2$.

**Solution**   The equation $x = 2$ can be written as $x + 0y = 2$. For any $y$-value chosen, notice that $x$ is 2. No other value for $x$ satisfies $x + 0y = 2$. Any ordered pair whose $x$-coordinate is 2 is a solution of $x + 0y = 2$. We will use the ordered pair solutions $(2, 3)$, $(2, 0)$, and $(2, -3)$ to graph $x = 2$.

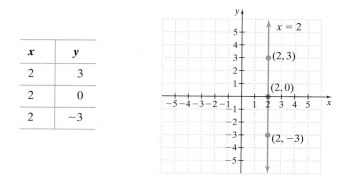

| $x$ | $y$ |
|-----|-----|
| 2   | 3   |
| 2   | 0   |
| 2   | -3  |

The graph is a vertical line with $x$-intercept $(2, 0)$. Note that this graph has no $y$-intercept because $x$ is never 0.

**PRACTICE**
**9**   Graph $y = 2$.

---

**Vertical Lines**
The graph of $x = c$, where $c$ is a real number, is a vertical line with $x$-intercept $(c, 0)$.

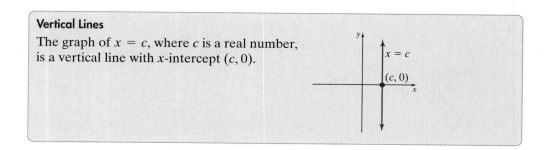

**EXAMPLE 10**   Graph $y = -3$.

**Solution**   The equation $y = -3$ can be written as $0x + y = -3$. For any $x$-value chosen, $y$ is $-3$. If we choose 4, 1, and $-2$ as $x$-values, the ordered pair solutions are $(4, -3)$, $(1, -3)$, and $(-2, -3)$. Use these ordered pairs to graph $y = -3$. The graph is a horizontal line with $y$-intercept $(0, -3)$ and no $x$-intercept.

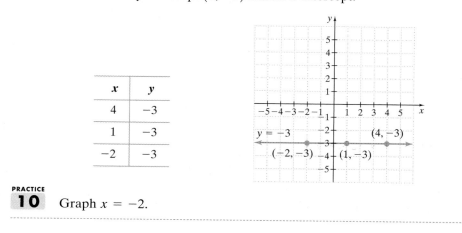

| $x$ | $y$ |
|-----|-----|
| 4   | -3  |
| 1   | -3  |
| -2  | -3  |

**PRACTICE**
**10**   Graph $x = -2$.

**Horizontal Lines**

The graph of $y = c$, where $c$ is a real number, is a horizontal line with $y$-intercept $(0, c)$.

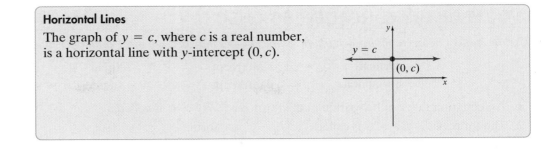

**Graphing Calculator Explorations**

You may have noticed that to use the $\boxed{Y=}$ key on a grapher to graph an equation, the equation must be solved for $y$. For example, to graph $2x + 3y = 7$, we solve this equation for $y$.

$$2x + 3y = 7$$
$$3y = -2x + 7 \quad \text{Subtract } 2x \text{ from both sides.}$$
$$\frac{3y}{3} = -\frac{2x}{3} + \frac{7}{3} \quad \text{Divide both sides by 3.}$$
$$y = -\frac{2}{3}x + \frac{7}{3} \quad \text{Simplify.}$$

To graph $2x + 3y = 7$ or $y = -\dfrac{2}{3}x + \dfrac{7}{3}$, press the $\boxed{Y=}$ key and enter

$$Y_1 = -\frac{2}{3}x + \frac{7}{3}$$

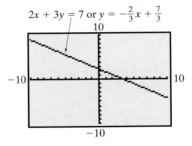

*Graph each linear equation.*

**1.** $x = 3.78y$

**2.** $-2.61y = x$

**3.** $3x + 7y = 21$

**4.** $-4x + 6y = 21$

**5.** $-2.2x + 6.8y = 15.5$

**6.** $5.9x - 0.8y = -10.4$

## VOCABULARY & READINESS CHECK

*Use the choices below to fill in each blank. Some choices may be used more than once. Exercises 1 and 2 come from Section 3.2.*

| | | | |
|---|---|---|---|
| $x$ | vertical | $x$-intercept | linear |
| $y$ | horizontal | $y$-intercept | standard |

1. An equation that can be written in the form $Ax + By = C$ is called a _____ equation in two variables.
2. The form $Ax + By = C$ is called _____ form.
3. The graph of the equation $y = -1$ is a _____ line.
4. The graph of the equation $x = 5$ is a _____ line.
5. A point where a graph crosses the $y$-axis is called a(n) _____.
6. A point where a graph crosses the $x$-axis is called a(n) _____.
7. Given an equation of a line, to find the $x$-intercept (if there is one), let _____ = 0 and solve for _____.
8. Given an equation of a line, to find the $y$-intercept (if there is one), let _____ = 0 and solve for _____.

*Answer the following true or false.*

9. All lines have an $x$-intercept *and* a $y$-intercept.
10. The graph of $y = 4x$ contains the point $(0, 0)$.
11. The graph of $x + y = 5$ has an $x$-intercept of $(5, 0)$ and a $y$-intercept of $(0, 5)$.
12. The graph of $y = 5x$ contains the point $(5, 1)$.

## 3.3 | EXERCISE SET

MyMathLab     Powered by CourseCompass™ and MathXL®

MathXL PRACTICE   WATCH   DOWNLOAD   READ   REVIEW

*Identify the intercepts. See Examples 1 through 5.*

 **1.**
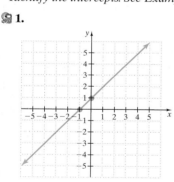

**2.**

**5.**
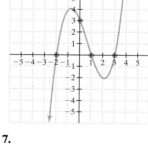

**6.**
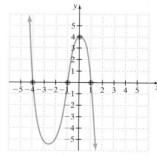

**3.**
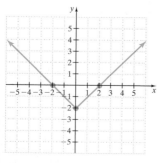

**4.**
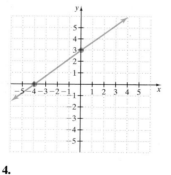

**7.**
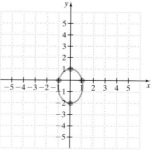

**8.**
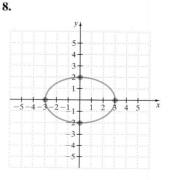

*Solve. See Example 1.*

**9.** What is the greatest number of intercepts for a line?

**10.** What is the least number of intercepts for a line?

**11.** What is the least number of intercepts for a circle?

**12.** What is the greatest number of intercepts for a circle?

*Graph each linear equation by finding and plotting its intercepts. See Examples 6 through 8.*

**13.** $x - y = 3$     **14.** $x - y = -4$     **15.** $x = 5y$

**16.** $x = 2y$     **17.** $-x + 2y = 6$     **18.** $x - 2y = -8$

**19.** $2x - 4y = 8$     **20.** $2x + 3y = 6$     **21.** $y = 2x$

**22.** $y = -2x$     **23.** $y = 3x + 6$     **24.** $y = 2x + 10$

*Graph each linear equation. See Examples 9 and 10.*

**25.** $x = -1$     **26.** $y = 5$     **27.** $y = 0$

**28.** $x = 0$     **29.** $y + 7 = 0$     **30.** $x - 2 = 0$

**31.** $x + 3 = 0$     **32.** $y - 6 = 0$

## MIXED PRACTICE

*Graph each linear equation. See Examples 6 through 10.*

**33.** $x = y$                 **34.** $x = -y$

**35.** $x + 8y = 8$          **36.** $x + 3y = 9$

**37.** $5 = 6x - y$          **38.** $4 = x - 3y$

**39.** $-x + 10y = 11$     **40.** $-x + 9y = 10$

**41.** $x = -4\frac{1}{2}$           **42.** $x = -1\frac{3}{4}$

**43.** $y = 3\frac{1}{4}$            **44.** $y = 2\frac{1}{2}$

**45.** $y = -\frac{2}{3}x + 1$      **46.** $y = -\frac{3}{5}x + 3$

**47.** $4x - 6y + 2 = 0$    **48.** $9x - 6y + 3 = 0$

*For Exercises 49 through 54, match each equation with its graph.*

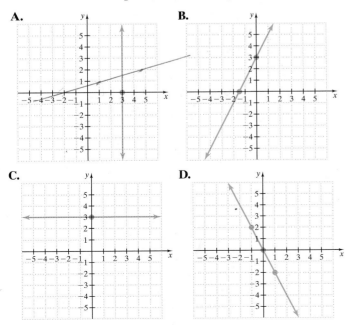

**A.**

**B.**

**C.**

**D.**

**E.**

**F.**

**49.** $y = 3$              **50.** $y = 2x + 2$

**51.** $x = -1$          **52.** $x = 3$

**53.** $y = 2x + 3$      **54.** $y = -2x$

## REVIEW AND PREVIEW

*Simplify. See Sections 1.5, 1.6, and 1.7.*

**55.** $\dfrac{-6 - 3}{2 - 8}$           **56.** $\dfrac{4 - 5}{-1 - 0}$

**57.** $\dfrac{-8 - (-2)}{-3 - (-2)}$     **58.** $\dfrac{12 - 3}{10 - 9}$

**59.** $\dfrac{0 - 6}{5 - 0}$           **60.** $\dfrac{2 - 2}{3 - 5}$

## CONCEPT EXTENSIONS

**61.** The revenue for the Disney Parks and Resorts $y$ (in millions) for the years 2003–2006 can be approximated by the equation $y = 1181x + 6505$, where $x$ represents the number of years after 2003. (*Source:* Based on data from The Walt Disney Company)

   **a.** Find the $y$-intercept of this equation.

   **b.** What does the $y$-intercept mean?

**62.** The average price of a digital camera $y$ (in dollars) can be modeled by the linear equation $y = -78.1x + 491.8$ where $x$ represents the number of years after 2000. (*Source*: NPDTechworld)

   **a.** Find the $y$-intercept of this equation.

   **b.** What does this $y$-intercept mean?

**63.** Since 2002, admissions at movie theaters have been in a decline. The number of people $y$ (in billions) who go to movie theaters each year can be estimated by the equation $y = -0.075x + 1.65$, where $x$ represents the number of years since 2002. (*Source:* Based on data from Motion Picture Association of America)

a. Find the *x*-intercept of this equation.

b. What does this *x*-intercept mean?

c. Use part (b) to comment on the limitations of using equations to model real data.

64. The price of admission to a movie theater has been steadily increasing. The price of regular admission *y* (in dollars) to a movie theater may be represented by the equation $y = 0.2x + 5.42$, where *x* is the number of years after 2000. (*Source:* Based on data from Motion Picture Association of America)

a. Find the *x*-intercept of this equation.

b. What does this *x*-intercept mean?

c. Use part (b) to comment on the limitation of using equations to model real data.

65. The production supervisor at Alexandra's Office Products finds that it takes 3 hours to manufacture a particular office chair and 6 hours to manufacture a computer desk. A total of 1200 hours is available to produce office chairs and desks of this style. The linear equation that models this situation is $3x + 6y = 1200$, where *x* represents the number of chairs produced and *y* the number of desks manufactured.

a. Complete the ordered pair solution (0, ) of this equation. Describe the manufacturing situation that corresponds to this solution.

b. Complete the ordered pair solution ( , 0) of this equation. Describe the manufacturing situation that corresponds to this solution.

c. Use the ordered pairs found above and graph the equation $3x + 6y = 1200$.

d. If 50 computer desks are manufactured, find the greatest number of chairs that they can make.

*Two lines in the same plane that do not intersect are called* **parallel lines.**

66. Draw a line parallel to the line $x = 5$ that intersects the *x*-axis at $(1, 0)$. What is the equation of this line?

67. Draw a line parallel to the line $y = -1$ that intersects the *y*-axis at $(0, -4)$. What is the equation of this line?

68. Discuss whether a vertical line ever has a *y*-intercept.

69. Explain why it is a good idea to use three points to graph a linear equation.

70. Discuss whether a horizontal line ever has an *x*-intercept.

71. Explain how to find intercepts.

---

### 📖 STUDY SKILLS BUILDER

**Are You Familiar with Your Textbook Supplements?**

Below is a review of some of the student supplements available for additional study. Check to see if you are using the ones most helpful to you.

• Chapter Test Prep Videos on CD. This CD is found with your textbook and contains video clip solutions to the Chapter Test exercises in this text. You will find this extremely useful when studying for chapter tests.

• Lecture Videos on CD-ROM. These are keyed to each section of the text. The material is presented by me, Elayn Martin-Gay, and I have placed a 🌐 by the exercises in the text that I have worked on the video.

• The *Student Solutions Manual*. This contains worked out solutions to odd-numbered exercises as well as every exercise in the Integrated Reviews, Chapter Reviews, Chapter Tests, and Cumulative Reviews.

• Pearson Tutor Center. Mathematics questions may be phoned, faxed, or emailed to this center.

• MyMathLab is a text-specific online course. MathXL is an online homework, tutorial, and assessment system. Take a moment and determine whether these are available to you.

As usual, your instructor is your best source of information.

*Let's see how you are doing with textbook supplements.*

1. Name one way the Lecture Videos can be helpful to you.

2. Name one way the Chapter Test Prep Video can help you prepare for a chapter test.

3. List any textbook supplements that you have found useful.

4. Have you located and visited a learning resource lab located on your campus?

5. List the textbook supplements that are currently housed in your campus' learning resource lab.

# 3.4 SLOPE AND RATE OF CHANGE

## OBJECTIVES

1 Find the slope of a line given two points of the line.

2 Find the slope of a line given its equation.

3 Find the slopes of horizontal and vertical lines.

4 Compare the slopes of parallel and perpendicular lines.

5 Slope as a rate of change.

**OBJECTIVE 1** ▶ **Finding the slope of a line given two points of the line.** Thus far, much of this chapter has been devoted to graphing lines. You have probably noticed by now that a key feature of a line is its slant or steepness. In mathematics, the slant or steepness of a line is formally known as its **slope**. We measure the slope of a line by the ratio of vertical change to the corresponding horizontal change as we move along the line.

On the line below, for example, suppose that we begin at the point $(1, 2)$ and move to the point $(4, 6)$. The vertical change is the change in $y$-coordinates: $6 - 2$ or 4 units. The corresponding horizontal change is the change in $x$-coordinates: $4 - 1 = 3$ units. The ratio of these changes is

$$\text{slope} = \frac{\text{change in } y \text{ (vertical change)}}{\text{change in } x \text{ (horizontal change)}} = \frac{4}{3}$$

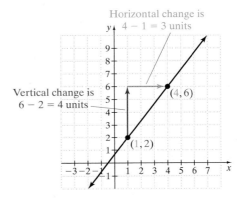

The slope of this line, then, is $\dfrac{4}{3}$. This means that for every 4 units of change in $y$-coordinates, there is a corresponding change of 3 units in $x$-coordinates.

▶ **Helpful Hint**

It makes no difference what two points of a line are chosen to find its slope. The slope of a line is the same everywhere on the line.

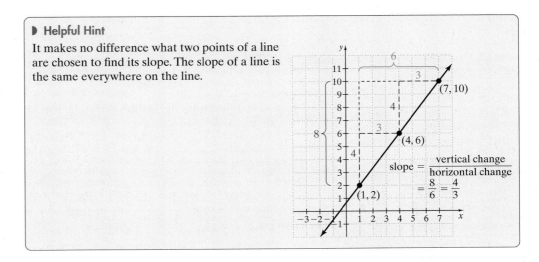

To find the slope of a line, then, choose two points of the line. Label the two $x$-coordinates of two points, $x_1$ and $x_2$ (read "$x$ sub one" and "$x$ sub two"), and label the corresponding $y$-coordinates $y_1$ and $y_2$.

The vertical change or **rise** between these points is the difference in the $y$-coordinates: $y_2 - y_1$. The horizontal change or **run** between the points is the difference of the $x$-coordinates: $x_2 - x_1$. The slope of the line is the ratio of $y_2 - y_1$ to $x_2 - x_1$, and we traditionally use the letter $m$ to denote slope $m = \dfrac{y_2 - y_1}{x_2 - x_1}$.

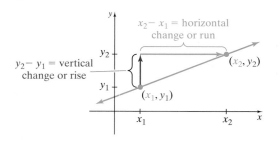

### Slope of a Line

The slope $m$ of the line containing the points $(x_1, y_1)$ and $(x_2, y_2)$ is given by

$$m = \frac{\text{rise}}{\text{run}} = \frac{\text{change in } y}{\text{change in } x} = \frac{y_2 - y_1}{x_2 - x_1}, \qquad \text{as long as } x_2 \neq x_1$$

**EXAMPLE 1** Find the slope of the line through $(-1, 5)$ and $(2, -3)$. Graph the line.

***Solution*** If we let $(x_1, y_1)$ be $(-1, 5)$, then $x_1 = -1$ and $y_1 = 5$. Also, let $(x_2, y_2)$ be $(2, -3)$ so that $x_2 = 2$ and $y_2 = -3$. Then, by the definition of slope,

$$m = \frac{y_2 - y_1}{x_2 - x_1}$$

$$= \frac{-3 - 5}{2 - (-1)}$$

$$= \frac{-8}{3} = -\frac{8}{3}$$

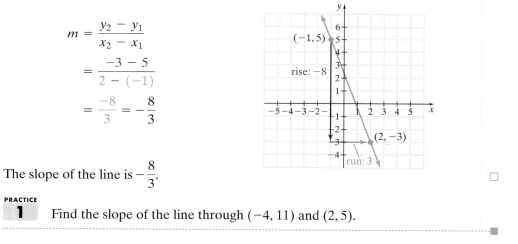

The slope of the line is $-\dfrac{8}{3}$.

**PRACTICE**

**1** Find the slope of the line through $(-4, 11)$ and $(2, 5)$.

▶ **Helpful Hint**

When finding slope, it makes no difference which point is identified as $(x_1, y_1)$ and which is identified as $(x_2, y_2)$. Just remember that whatever $y$-value is first in the numerator, its corresponding $x$-value is first in the denominator. Another way to calculate the slope in Example 1 is:

$$m = \frac{y_2 - y_1}{x_2 - x_1} = \frac{5 - (-3)}{-1 - 2} = \frac{8}{-3} \quad \text{or} \quad -\frac{8}{3} \quad \leftarrow \text{Same slope as found in Example 1.}$$

**Answer to Concept Check:**

$$m = \frac{3}{2}$$

**Concept Check** ☑

The points $(-2, -5)$, $(0, -2)$, $(4, 4)$, and $(10, 13)$ all lie on the same line. Work with a partner and verify that the slope is the same no matter which points are used to find slope.

**EXAMPLE 2**   Find the slope of the line through $(-1, -2)$ and $(2, 4)$. Graph the line.

*Solution*   Let $(x_1, y_1)$ be $(2, 4)$ and $(x_2, y_2)$ be $(-1, -2)$.

$$m = \frac{y_2 - y_1}{x_2 - x_1}$$

$$= \frac{-2 - 4}{-1 - 2} \quad \begin{array}{l} y\text{-value} \\ \text{corresponding } x\text{-value} \end{array}$$

$$= \frac{-6}{-3} = 2$$

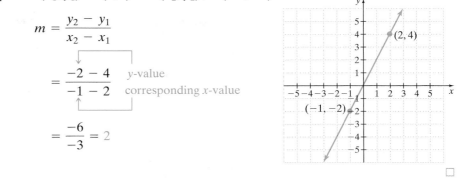

> **▶ Helpful Hint**
>
> The slope for Example 2 is the same if we let $(x_1, y_1)$ be $(-1, -2)$ and $(x_2, y_2)$ be $(2, 4)$.
>
> $$m = \frac{\overset{y\text{-value}}{\overset{\searrow}{4 - (-2)}}}{\underset{\uparrow}{2 - (-1)}} = \frac{6}{3} = 2$$
>
> corresponding $x$-value

**PRACTICE**

**2**   Find the slope of the line through $(-3, -1)$ and $(3, 1)$.

**Concept Check** ☑️

What is wrong with the following slope calculation for the points $(3, 5)$ and $(-2, 6)$?

$$m = \frac{5 - 6}{-2 - 3} = \frac{-1}{-5} = \frac{1}{5}$$

Notice that the slope of the line in Example 1 is negative, whereas the slope of the line in Example 2 is positive. Let your eye follow the line with negative slope from left to right and notice that the line "goes down." Following the line with positive slope from left to right, notice that the line "goes up." This is true in general.

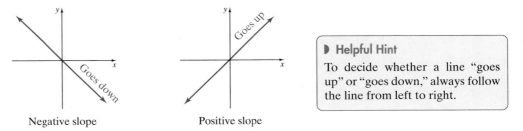

Negative slope                    Positive slope

> **▶ Helpful Hint**
>
> To decide whether a line "goes up" or "goes down," always follow the line from left to right.

**OBJECTIVE 2 ▶ Finding the slope of a line given its equation.** As we have seen, the slope of a line is defined by two points on the line. Thus, if we know the equation of a line, we can find its slope by finding two of its points. For example, let's find the slope of the line

$$y = 3x + 2$$

To find two points, we can choose two values for $x$ and substitute to find corresponding $y$-values. If $x = 0$, for example, $y = 3 \cdot 0 + 2$ or $y = 2$. If $x = 1$, $y = 3 \cdot 1 + 2$ or $y = 5$. This gives the ordered pairs $(0, 2)$ and $(1, 5)$. Using the definition for slope, we have

$$m = \frac{5 - 2}{1 - 0} = \frac{3}{1} = 3 \quad \text{The slope is 3.}$$

Notice that the slope, 3, is the same as the coefficient of $x$ in the equation $y = 3x + 2$.

Also, recall from Section 3.2 that the graph of an equation of the form $y = mx + b$ has $y$-intercept $(0, b)$.

This means that the $y$-intercept of the graph of $y = 3x + 2$ is $(0, 2)$. This is true in general.

**Answer to Concept Check:**

The order in which the $x$- and $y$-values are used must be the same.

$$m = \frac{5 - 6}{3 - (-2)} = \frac{-1}{5} = -\frac{1}{5}$$

When a linear equation is written in the form $y = mx + b$, not only is $(0, b)$ the $y$-intercept of the line, but $m$ is its slope. The form $y = mx + b$ is appropriately called the **slope-intercept form.**

$$\underset{\underset{(0,\, b)}{\underset{\uparrow}{\text{slope}}\quad\underset{\uparrow}{y\text{-intercept}}}}{}$$

---

**Slope-Intercept Form**

When a linear equation in two variables is written in slope-intercept form,

$$y = mx + b$$

$m$ is the slope of the line and $(0, b)$ is the $y$-intercept of the line.

---

**EXAMPLE 3**   Find the slope and $y$-intercept of the line whose equation is $y = \dfrac{3}{4}x + 6$.

**Solution**   The equation is in slope-intercept form, $y = mx + b$.

$$y = \dfrac{3}{4}x + 6$$

The coefficient of $x$, $\dfrac{3}{4}$, is the slope and the constant term, 6 is the $y$-value of the $y$-intercept, $(0, 6)$.   □

**PRACTICE**
**3**   Find the slope and $y$-intercept of the line whose equation is $y = \dfrac{2}{3}x - 2$.

---

**EXAMPLE 4**   Find the slope and the $y$-intercept of the line whose equation is $5x + y = 2$.

**Solution**   Write the equation in slope-intercept form by solving the equation for $y$.

$$5x + y = 2$$
$$y = -5x + 2 \quad \text{Subtract } 5x \text{ from both sides.}$$

The coefficient of $x$, $-5$, is the slope and the constant term, 2, is the $y$-value of the $y$-intercept, $(0, 2)$.   □

**PRACTICE**
**4**   Find the slope and $y$-intercept of the line whose equation is $6x - y = 5$.

---

**EXAMPLE 5**   Find the slope and the $y$-intercept of the line whose equation is $3x - 4y = 4$.

**Solution**   Write the equation in slope-intercept form by solving for $y$.

$$3x - 4y = 4$$
$$-4y = -3x + 4 \quad \text{Subtract } 3x \text{ from both sides.}$$
$$\dfrac{-4y}{-4} = \dfrac{-3x}{-4} + \dfrac{4}{-4} \quad \text{Divide both sides by } -4.$$
$$y = \dfrac{3}{4}x - 1 \quad \text{Simplify.}$$

The coefficient of $x$, $\dfrac{3}{4}$, is the slope, and the $y$-intercept is $(0, -1)$.   □

**PRACTICE**
**5**   Find the slope and the $y$-intercept of the line whose equation is $5x + 2y = 8$.

**OBJECTIVE 3 ▶ Finding slopes of horizontal and vertical li**
upward from left to right, its slope is positive. If a line tilts
its slope is negative. Let's now find the slopes of two
vertical lines.

**EXAMPLE 6**    Find the slope of the line $y =$

**Solution**    Recall that $y = -1$ is a horizontal line with
$y$-intercept $(0, -1)$. To find the slope, find two ordered
pair solutions of $y = -1$. Solutions of $y = -1$ must
have a $y$-value of $-1$. Let's use points $(2, -1)$ and
$(-3, -1)$, which are on the line.

$$m = \frac{y_2 - y_1}{x_2 - x_1} = \frac{-1 - (-1)}{-3 - 2} = \frac{0}{-5} = 0$$

The slope of the line $y = -1$ is 0 and its graph is shown.

**PRACTICE**
**6**    Find the slope of the line $y = 3$.

Any two points of a horizontal line will have the same $y$-values. This means that
the $y$-values will always have a difference of 0 for all horizontal lines. Thus, **all horizontal lines have a slope 0.**

**EXAMPLE 7**    Find the slope of the line $x = 5$.

**Solution**    Recall that the graph of $x = 5$ is a vertical line with $x$-intercept $(5, 0)$.

To find the slope, find two ordered pair solutions of $x = 5$. Solutions of $x = 5$
must have an $x$-value of 5. Let's use points $(5, 0)$ and $(5, 4)$, which are on the line.

$$m = \frac{y_2 - y_1}{x_2 - x_1} = \frac{4 - 0}{5 - 5} = \frac{4}{0}$$

Since $\dfrac{4}{0}$ is undefined, we say the slope of the vertical
line $x = 5$ is undefined, and its graph is shown.

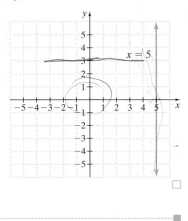

**PRACTICE**
**7**    Find the slope of the line $x = -4$.

Any two points of a vertical line will have the same $x$-values. This means that the
$x$-values will always have a difference of 0 for all vertical lines. Thus **all vertical lines
have undefined slope.**

▶ **Helpful Hint**
Slope of 0 and undefined slope are not the same. Vertical lines have undefined slope or no
slope, while horizontal lines have a slope of 0.

Here is a general review of slope.

## Summary of Slope

Slope $m$ of the line through $(x_1, y_1)$ and $(x_2, y_2)$ is given by the equation $m = \dfrac{y_2 - y_1}{x_2 - x_1}$.

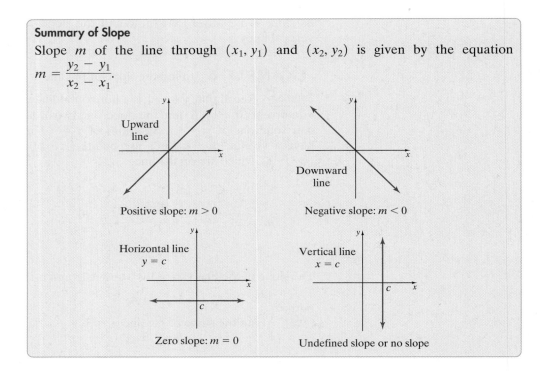

Upward line

Positive slope: $m > 0$

Downward line

Negative slope: $m < 0$

Horizontal line $y = c$

Zero slope: $m = 0$

Vertical line $x = c$

Undefined slope or no slope

**OBJECTIVE 4 ▶ Slopes of parallel and perpendicular lines.** Two lines in the same plane are **parallel** if they do not intersect. Slopes of lines can help us determine whether lines are parallel. Parallel lines have the same steepness, so it follows that they have the same slope.

For example, the graphs of

$$y = -2x + 4$$

and

$$y = -2x - 3$$

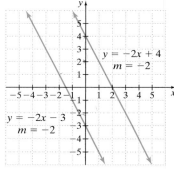

$y = -2x + 4$
$m = -2$

$y = -2x - 3$
$m = -2$

are shown. These lines have the same slope, $-2$. They also have different $y$-intercepts, so the lines are parallel. (If the $y$-intercepts were the same also, the lines would be the same.)

## Parallel Lines

Nonvertical parallel lines have the same slope and different $y$-intercepts.

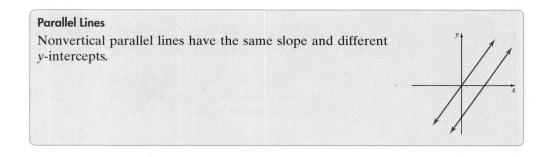

Two lines are **perpendicular** if they lie in the same plane and meet at a 90° (right) angle. How do the slopes of perpendicular lines compare? The product of the slopes of two perpendicular lines is $-1$.

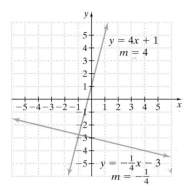

For example, the graphs of

$$y = 4x + 1$$

and

$$y = -\frac{1}{4}x - 3$$

are shown. The slopes of the lines are $4$ and $-\frac{1}{4}$. Their product is $4\left(-\frac{1}{4}\right) = -1$, so the lines are perpendicular.

---

**Perpendicular Lines**

If the product of the slopes of two lines is $-1$, then the lines are perpendicular.

    (Two nonvertical lines are perpendicular if the slopes of one is the negative reciprocal of the slope of the other.)

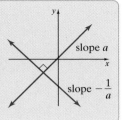

---

▶ **Helpful Hint**

Here are examples of numbers that are negative (opposite) reciprocals.

| Number | Negative Reciprocal | Their Product Is $-1$. |
|:---:|:---:|:---:|
| $\frac{2}{3}$ | $-\frac{3}{2}$ | $\frac{2}{3} \cdot -\frac{3}{2} = -\frac{6}{6} = -1$ |
| $-5$ or $-\frac{5}{1}$ | $\frac{1}{5}$ | $-5 \cdot \frac{1}{5} = -\frac{5}{5} = -1$ |

---

▶ **Helpful Hint**

Here are a few important facts about vertical and horizontal lines.

- Two distinct vertical lines are parallel.
- Two distinct horizontal lines are parallel.
- A horizontal line and a vertical line are always perpendicular.

---

△ **EXAMPLE 8**    Determine whether each pair of lines is parallel, perpendicular, or neither.

**a.** $y = -\frac{1}{5}x + 1$    **b.** $x + y = 3$    **c.** $3x + y = 5$

    $2x + 10y = 3$        $-x + y = 4$        $2x + 3y = 6$

*Solution*

**a.** The slope of the line $y = -\frac{1}{5}x + 1$ is $-\frac{1}{5}$. We find the slope of the second line by solving its equation for $y$.

$$2x + 10y = 3$$
$$10y = -2x + 3 \quad \text{Subtract } 2x \text{ from both sides.}$$
$$y = \frac{-2}{10}x + \frac{3}{10} \quad \text{Divide both sides by 10.}$$
$$y = -\frac{1}{5}x + \frac{3}{10} \quad \text{Simplify.}$$

The slope of this line is $-\frac{1}{5}$ also. Since the lines have the same slope and different $y$-intercepts, they are parallel, as shown in the figure on the next page.

**b.** To find each slope, we solve each equation for $y$.

$$x + y = 3 \qquad\qquad -x + y = 4$$
$$y = -x + 3 \qquad\qquad y = x + 4$$

The slope is $-1$.        The slope is $1$.

The slopes are not the same, so the lines are not parallel. Next we check the product of the slopes: $(-1)(1) = -1$. Since the product is $-1$, the lines are perpendicular, as shown in the figure.

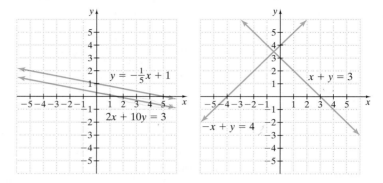

$y = -\frac{1}{5}x + 1$

$2x + 10y = 3$

$x + y = 3$

$-x + y = 4$

**c.** We solve each equation for $y$ to find each slope. The slopes are $-3$ and $-\frac{2}{3}$. The slopes are not the same and their product is not $-1$. Thus, the lines are neither parallel nor perpendicular.

**PRACTICE**
**8**    Determine whether each pair of lines is parallel, perpendicular, or neither.

**a.** $y = -5x + 1$

     $x - 5y = 10$

**b.** $x + y = 11$

     $2x + y = 11$

**c.** $2x + 3y = 21$

     $6y = -4x - 2$

**Concept Check** ☑

Consider the line $-6x + 2y = 1$.

**a.** Write the equations of two lines parallel to this line.

**b.** Write the equations of two lines perpendicular to this line.

**OBJECTIVE 5 ▶ Slope as a rate of change.** Slope can also be interpreted as a rate of change. In other words, slope tells us how fast $y$ is changing with respect to $x$. To see this, let's look at a few of the many real-world applications of slope. For example, the pitch of a roof, used by builders and architects, is its slope. The pitch of the roof on the left is $\frac{7}{10}\left(\frac{\text{rise}}{\text{run}}\right)$. This means that the roof rises vertically 7 feet for every horizontal 10 feet. The rate of change for the roof is 7 vertical feet ($y$) per 10 horizontal feet ($x$).

$\frac{7}{10}$ pitch

7 feet

10 feet

     The grade of a road is its slope written as a percent. A 7% grade, as shown below, means that the road rises (or falls) 7 feet for every horizontal 100 feet. $\Big($Recall that $7\% = \frac{7}{100}.\Big)$ Here, the slope of $\frac{7}{100}$ gives us the rate of change. The road rises (in our diagram) 7 vertical feet ($y$) for every 100 horizontal feet ($x$).

**Answers to Concept Check:**

**a.** any two lines with $m = 3$ and
y-intercept not $\left(0, \frac{1}{2}\right)$

**b.** any two lines with $m = -\frac{1}{3}$

$\frac{7}{100} = 7\%$ grade

7 feet

100 feet

**EXAMPLE 9**    **Finding the Grade of a Road**

At one part of the road to the summit of Pikes Peak, the road rises at a rate of 15 vertical feet for a horizontal distance of 250 feet. Find the grade of the road.

*Solution*    Recall that the grade of a road is its slope written as a percent.

$$\text{grade} = \frac{\text{rise}}{\text{run}} = \frac{15}{250} = 0.06 = 6\%$$

15 feet

250 feet

The grade is 6%.                                                                                          □

**PRACTICE**

**9**    One part of the Mt. Washington (New Hampshire) cog railway rises about 1794 feet over a horizontal distance of 7176 feet. Find the grade of this part of the railway.

- - - - - - - - - - - - - - - - - - - - - - - - - - - - - - - - - - - - - - - - - - - - - - ■

**EXAMPLE 10**    **Finding the Slope of a Line**

The following graph shows the cost $y$ (in cents) of a nationwide long-distance telephone call from Texas with a certain telephone-calling plan, where $x$ is the length of the call in minutes. Find the slope of the line and attach the proper units for the rate of change. Then write a sentence explaining the meaning of slope in this application.

*Solution*    Use $(2, 34)$ and $(6, 62)$ to calculate slope.

$$m = \frac{62 - 34}{6 - 2} = \frac{28}{4} = \frac{7 \text{ cents}}{1 \text{ minute}}$$

This means that the rate of change of a phone call is 7 cents per 1 minute or the cost of the phone call is 7 cents per minute.

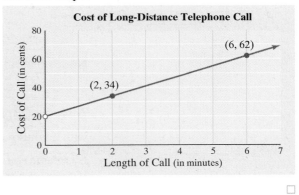

**Cost of Long-Distance Telephone Call**

$(6, 62)$

$(2, 34)$

Cost of Call (in cents)

Length of Call (in minutes)

□

**PRACTICE**

**10**    The following graph shows the cost $y$ (in dollars) of having laundry done at the Wash-n-Fold, where $x$ is the number of pounds of laundry. Find the slope of the line, and attach the proper units for the rate of change.

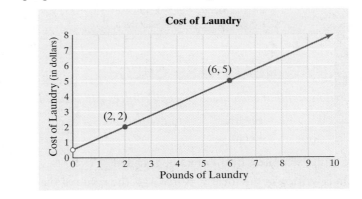

**Cost of Laundry**

$(6, 5)$

$(2, 2)$

Cost of Laundry (in dollars)

Pounds of Laundry

- - - - - - - - - - - - - - - - - - - - - - - - - - - - - - - - - - - - - - - - - - - - - - ■

### Graphing Calculator Explorations

It is possible to use a grapher to sketch the graph of more than one equation on the same set of axes. This feature can be used to confirm our findings from Section 3.2 when we learned that the graph of an equation written in the form $y = mx + b$ has a $y$-intercept of $b$. For example, graph the equations $y = \frac{2}{5}x$, $y = \frac{2}{5}x + 7$, and $y = \frac{2}{5}x - 4$ on the same set of axes. To do so, press the $\boxed{Y=}$ key and enter the equations on the first three lines.

$$Y_1 = \left(\frac{2}{5}\right)x$$

$$Y_2 = \left(\frac{2}{5}\right)x + 7$$

$$Y_3 = \left(\frac{2}{5}\right)x - 4$$

The screen should look like:

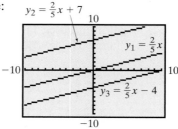

Notice that all three graphs appear to have the same positive slope. The graph of $y = \frac{2}{5}x + 7$ is the graph of $y = \frac{2}{5}x$ moved 7 units upward with a $y$-intercept of 7. Also, the graph of $y = \frac{2}{5}x - 4$ is the graph of $y = \frac{2}{5}x$ moved 4 units downward with a $y$-intercept of $-4$.

*Graph the equations on the same set of axes. Describe the similarities and differences in their graphs.*

1. $y = 3.8x$, $y = 3.8x - 3$, $y = 3.8x + 6$
2. $y = -4.9x$, $y = -4.9x + 1$, $y = -4.9x + 8$
3. $y = \frac{1}{4}x$; $y = \frac{1}{4}x + 5$, $y = \frac{1}{4}x - 8$
4. $y = -\frac{3}{4}x$, $y = -\frac{3}{4}x - 5$, $y = -\frac{3}{4}x + 6$

## VOCABULARY & READINESS CHECK

*Use the choices below to fill in each blank. Not all choices will be used.*

| | | | | |
|---|---|---|---|---|
| $m$ | $x$ | 0 | positive | undefined |
| $b$ | $y$ | slope | negative | |

1. The measure of the steepness or tilt of a line is called _____.
2. If an equation is written in the form $y = mx + b$, the value of the letter _____ is the value of the slope of the graph.
3. The slope of a horizontal line is _____.
4. The slope of a vertical line is _____.
5. If the graph of a line moves upward from left to right, the line has _____ slope.

**6.** If the graph of a line moves downward from left to right, the line has _____ slope.

**7.** Given two points of a line, slope $= \dfrac{\text{change in } \_\_}{\text{change in } \_\_}$.

*State whether the slope of the line is positive, negative, 0, or is undefined.*

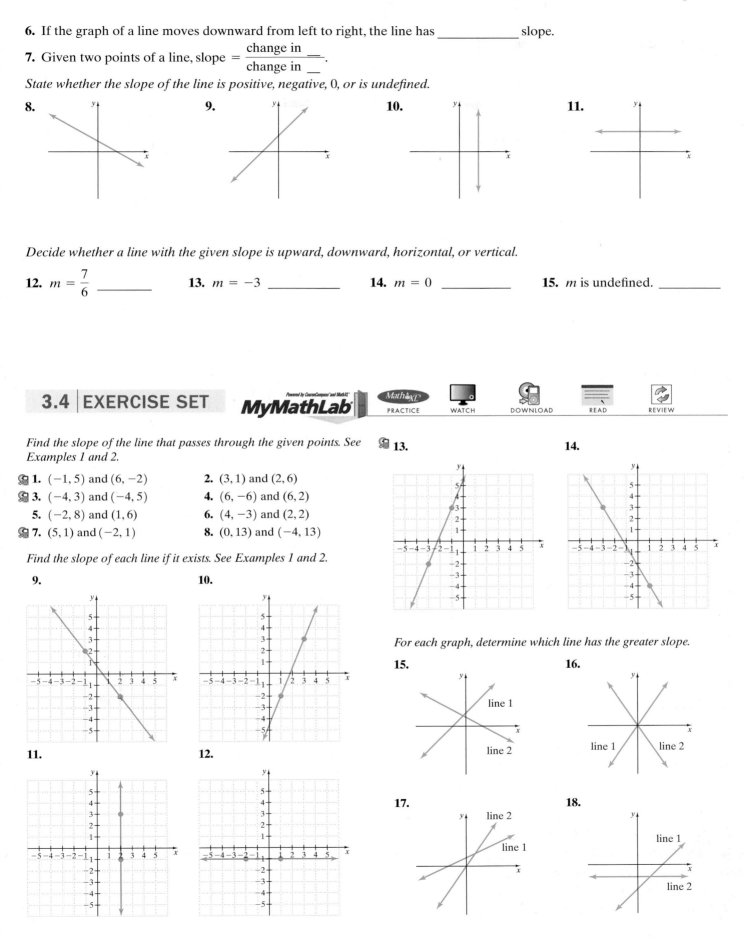

**8.**  **9.**  **10.**  **11.**

*Decide whether a line with the given slope is upward, downward, horizontal, or vertical.*

**12.** $m = \dfrac{7}{6}$ _____

**13.** $m = -3$ _____

**14.** $m = 0$ _____

**15.** $m$ is undefined. _____

## 3.4 EXERCISE SET

**MyMathLab** *Powered by CourseCompass™ and MathXL*

Math XL  PRACTICE      WATCH      DOWNLOAD      READ      REVIEW

*Find the slope of the line that passes through the given points. See Examples 1 and 2.*

**1.** $(-1, 5)$ and $(6, -2)$

**2.** $(3, 1)$ and $(2, 6)$

**3.** $(-4, 3)$ and $(-4, 5)$

**4.** $(6, -6)$ and $(6, 2)$

**5.** $(-2, 8)$ and $(1, 6)$

**6.** $(4, -3)$ and $(2, 2)$

**7.** $(5, 1)$ and $(-2, 1)$

**8.** $(0, 13)$ and $(-4, 13)$

*Find the slope of each line if it exists. See Examples 1 and 2.*

**9.**

**10.**

**11.**

**12.**

**13.**

**14.**

*For each graph, determine which line has the greater slope.*

**15.**

**16.**

**17.**

**18.**

*In Exercises 19 through 24, match each line with its slope.*

**A.** $m = 0$ **B.** undefined slope **C.** $m = 3$

**D.** $m = 1$ **E.** $m = -\dfrac{1}{2}$ **F.** $m = -\dfrac{3}{4}$

**19.**

**20.**

**21.**

**22.**

**23.**

**24.**

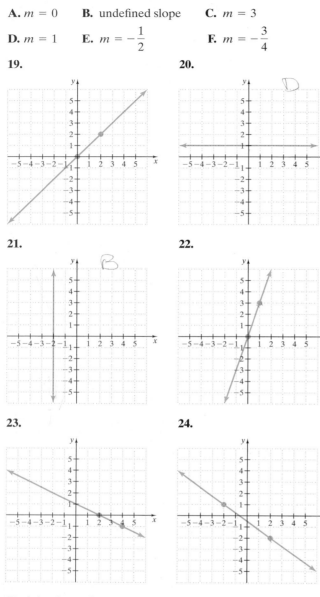

*Find the slope of each line. See Examples 6 and 7.*

**25.** $x = 6$ **26.** $y = 4$

**27.** $y = -4$ **28.** $x = 2$

**29.** $x = -3$ **30.** $y = -11$

**31.** $y = 0$ **32.** $x = 0$

**MIXED PRACTICE**

*Find the slope of each line. See Examples 3 through 7.*

**33.** $y = 5x - 2$ **34.** $y = -2x + 6$

**35.** $y = -0.3x + 2.5$ **36.** $y = -7.6x - 0.1$

**37.** $2x + y = 7$ **38.** $-5x + y = 10$

**39.** $2x - 3y = 10$ **40.** $3x - 5y = 1$

**41.** $x = 1$ **42.** $y = -2$

**43.** $x = 2y$ **44.** $x = -4y$

**45.** $y = -3$ **46.** $x = 5$

**47.** $-3x - 4y = 6$ **48.** $-4x - 7y = 9$

**49.** $20x - 5y = 1.2$ **50.** $24x - 3y = 5.7$

△ *Determine whether each pair of lines is parallel, perpendicular, or neither. See Example 8.*

**51.** $y = \dfrac{2}{9}x + 3$ **52.** $y = \dfrac{1}{5}x + 20$

$\quad y = -\dfrac{2}{9}x$ $\quad y = -\dfrac{1}{5}x$

**53.** $x - 3y = -6$ **54.** $y = 4x - 2$

$\quad y = 3x - 9$ $\quad 4x + y = 5$

**55.** $6x = 5y + 1$ **56.** $-x + 2y = -2$

$\quad -12x + 10y = 1$ $\quad 2x = 4y + 3$

**57.** $6 + 4x = 3y$ **58.** $10 + 3x = 5y$

$\quad 3x + 4y = 8$ $\quad 5x + 3y = 1$

*The pitch of a roof is its slope. Find the pitch of each roof shown. See Example 9.*

**59.**

**60.**

*The grade of a road is its slope written as a percent. Find the grade of each road shown. See Example 9.*

**61.**

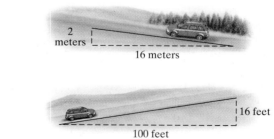

**62.**

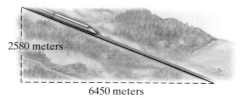

**63.** One of Japan's superconducting "bullet" trains is researched and tested at the Yamanashi Maglev Test Line near Otsuki City. The steepest section of the track has a rise of 2580 meters for a horizontal distance of 6450 meters. What is the grade of this section of track? (*Source:* Japan Railways Central Co.)

**64.** Professional plumbers suggest that a sewer pipe should rise 0.25 inch for every horizontal foot. Find the recommended slope for a sewer pipe. Round to the nearest hundredth.

**65.** The steepest street is Baldwin Street in Dunedin, New Zealand. It has a maximum rise of 10 meters for a horizontal distance of 12.66 meters. Find the grade of this section of road. Round to the nearest whole percent. (*Source: The Guinness Book of Records*)

**66.** According to federal regulations, a wheelchair ramp should rise no more than 1 foot for a horizontal distance of 12 feet. Write the slope as a grade. Round to the nearest tenth of a percent.

*Find the slope of each line and write the slope as a rate of change. Don't forget to attach the proper units. See Example 10.*

**67.** This graph approximates the number of U.S. households that have personal computers *y* (in millions) for year *x*.

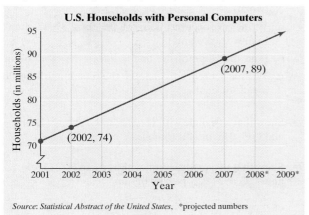

*Source: Statistical Abstract of the United States,* *projected numbers

**68.** This graph approximates the number *y* (per hundred population) of Attention Deficit Hyperactivity Disorder (ADHD) prescriptions for children under 18 for the year *x*. (*Source: Centers for Disease Control and Prevention (CDC) National Center for Health Statistics*)

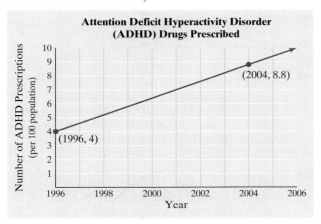

**69.** The graph below shows the total cost *y* (in dollars) of owning and operating a compact car where *x* is the number of miles driven.

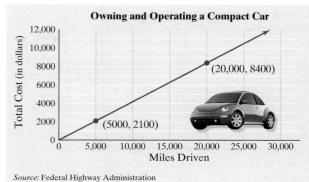

*Source*: Federal Highway Administration

**70.** The graph below shows the total cost *y* (in dollars) of owning and operating a standard pickup truck, where *x* is the number of miles driven.

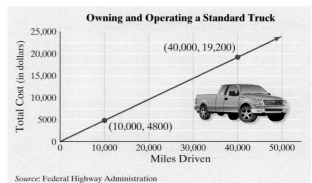

*Source*: Federal Highway Administration

**REVIEW AND PREVIEW**

*Solve each equation for y. See Section 2.5.*

**71.** $y - (-6) = 2(x - 4)$

**72.** $y - 7 = -9(x - 6)$

**73.** $y - 1 = -6(x - (-2))$

**74.** $y - (-3) = 4(x - (-5))$

**CONCEPT EXTENSIONS**

△ *Find the slope of the line that is (**a**) parallel and (**b**) perpendicular to the line through each pair of points.*

**75.** $(-3, -3)$ and $(0, 0)$

**76.** $(6, -2)$ and $(1, 4)$

**77.** $(-8, -4)$ and $(3, 5)$

**78.** $(6, -1)$ and $(-4, -10)$

*Solve. See a Concept Check in this section.*

**79.** Verify that the points $(2, 1), (0, 0), (-2, -1)$ and $(-4, -2)$ are all on the same line by computing the slope between each pair of points. (See the first Concept Check.)

**80.** Given the points $(2, 3)$ and $(-5, 1)$, can the slope of the line through these points be calculated by $\dfrac{1 - 3}{2 - (-5)}$? Why or why not? (See the second Concept Check.)

**81.** Write the equations of three lines parallel to $10x - 5y = -7$. (See the third Concept Check.)

**82.** Write the equations of two lines perpendicular to $10x - 5y = -7$. (See the third Concept Check.)

*The following line graph shows the average fuel economy (in miles per gallon) by mid-size passenger automobiles produced during each of the model years shown. Use this graph to answer Exercises 83 through 88.*

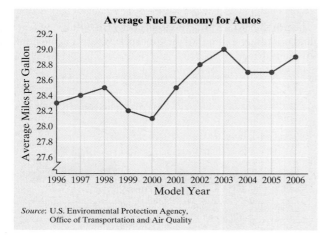

*Source*: U.S. Environmental Protection Agency, Office of Transportation and Air Quality

**83.** What was the average fuel economy (in miles per gallon) for automobiles produced during 2001?

**84.** Find the decrease in average fuel economy for automobiles between the years 2003 to 2004.

**85.** During which of the model years shown was average fuel economy the lowest?

What was the average fuel economy for that year?

**86.** During which of the model years shown was average fuel economy the highest?

What was the average fuel economy for that year?

**87.** What line segment has the greatest slope?

**88.** What line segment has the least positive slope?

*Solve.*

**89.** Find $x$ so that the pitch of the roof is $\dfrac{1}{3}$.

**90.** Find $x$ so that the pitch of the roof is $\dfrac{2}{5}$.

**91.** The average price of an acre of U.S. farmland was $1132 in 2001. In 2006, the price of an acre rose to approximately $1657. (*Source:* National Agricultural Statistics Service)

    **a.** Write two ordered pairs of the form (year, price of acre)

    **b.** Find the slope of the line through the two points.

    **c.** Write a sentence explaining the meaning of the slope as a rate of change.

**92.** There were approximately 14,774 kidney transplants performed in the United States in 2002. In 2006, the number of kidney transplants performed in the United States rose to 15,722. (*Source:* Organ Procurement and Transplantation Network)

    **a.** Write two ordered pairs of the form (year, number of kidney transplants).

    **b.** Find the slope of the line between the two points.

    **c.** Write a sentence explaining the meaning of the slope as a rate of change.

**93.** Show that a triangle with vertices at the points $(1, 1)$, $(-4, 4)$, and $(-3, 0)$ is a right triangle.

**94.** Show that the quadrilateral with vertices $(1, 3), (2, 1), (-4, 0)$, and $(-3, -2)$ is a parallelogram.

*Find the slope of the line through the given points.*

**95.** $(2.1, 6.7)$ and $(-8.3, 9.3)$

**96.** $(-3.8, 1.2)$ and $(-2.2, 4.5)$

**97.** $(2.3, 0.2)$ and $(7.9, 5.1)$

**98.** $(14.3, -10.1)$ and $(9.8, -2.9)$

**99.** The graph of $y = -\dfrac{1}{3}x + 2$ has a slope of $-\dfrac{1}{3}$. The graph of $y = -2x + 2$ has a slope of $-2$. The graph of $y = -4x + 2$ has a slope of $-4$. Graph all three equations on a single coordinate system. As the absolute value of the slope becomes larger, how does the steepness of the line change?

**100.** The graph of $y = \dfrac{1}{2}x$ has a slope of $\dfrac{1}{2}$. The graph of $y = 3x$ has a slope of 3. The graph of $y = 5x$ has a slope of 5. Graph all three equations on a single coordinate system. As slope becomes larger, how does the steepness of the line change?

## INTEGRATED REVIEW SUMMARY ON SLOPE & GRAPHING LINEAR EQUATIONS

Sections 3.1–3.4

*Find the slope of each line.*

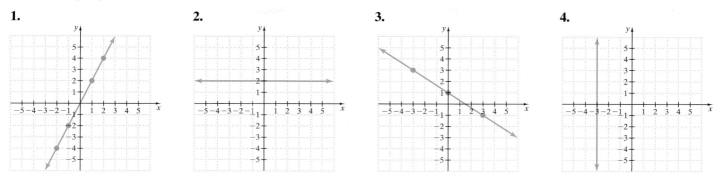

**1.**　　　　　　　　　**2.**　　　　　　　　　**3.**　　　　　　　　　**4.**

*Graph each linear equation.*

**5.** $y = -2x$　　　　　　**6.** $x + y = 3$　　　　　　**7.** $x = -1$　　　　　　**8.** $y = 4$

**9.** $x - 2y = 6$　　　　　**10.** $y = 3x + 2$　　　　**11.** $5x + 3y = 15$　　　**12.** $2x - 4y = 8$

*Determine whether the lines through the points are parallel, perpendicular, or neither.*

**13.** $y = -\dfrac{1}{5}x + \dfrac{1}{3}$

$3x = -15y$

**14.** $x - y = \dfrac{1}{2}$

$3x - y = \dfrac{1}{2}$

**15.** In the years 2002 through 2005 the number of admissions to movie theaters in the United States can be modeled by the linear equation $y = -75x + 1650$ where $x$ is years after 2002 and $y$ is admissions in millions. (*Source:* Motion Picture Assn. of America)

  **a.** Find the $y$-intercept of this line.

  **b.** Write a sentence explaining the meaning of this intercept.

  **c.** Find the slope of this line.

  **d.** Write a sentence explaining the meaning of the slope as a rate of change.

**16.** Online advertising is a means of promoting products and services using the Internet. The revenue (in billions of dollars) for online advertising for the years 2003 through a projected 2010 is given by $y = 3.3x - 3.1$, where $x$ is the number of years after 2000.

  **a.** Use this equation to complete the ordered pair (9,　　).

  **b.** Write a sentence explaining the meaning of the answer to part (a).

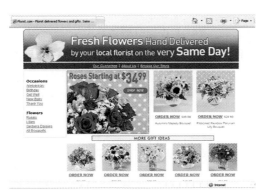

# 3.5 EQUATIONS OF LINES

Recall that the form $y = mx + b$ is appropriately called the *slope-intercept form* of a linear equation.

$y$-intercept is $(0, b)$

slope

---

**Slope-Intercept Form**

When a linear equation in two variables is written in **slope-intercept form,**

$$y = mx + b$$

slope    $(0, b)$, $y$-intercept

then $m$ is the slope of the line and $(0, b)$ is the $y$-intercept of the line.

---

**OBJECTIVE 1 ▶ Using the slope-intercept form to write an equation.** As we know from the previous section, writing an equation in slope-intercept form is a way to find the slope and $y$-intercept of its graph. The slope-intercept form can be used to write the equation of a line when we know its slope and $y$-intercept.

**EXAMPLE 1**  Find an equation of the line with $y$-intercept $(0, -3)$ and slope of $\frac{1}{4}$.

_Solution_  We are given the slope and the $y$-intercept. We let $m = \frac{1}{4}$ and $b = -3$ and write the equation in slope-intercept form, $y = mx + b$.

$$y = mx + b$$

$$y = \frac{1}{4}x + (-3) \quad \text{Let } m = \frac{1}{4} \text{ and } b = -3.$$

$$y = \frac{1}{4}x - 3 \qquad \text{Simplify.}$$

**PRACTICE**
**1**  Find an equation of the line with $y$-intercept $(0, 7)$ and slope of $\frac{1}{2}$.

**OBJECTIVE 2 ▶ Using the slope-intercept form to graph an equation.** We also can use the slope-intercept form of the equation of a line to graph a linear equation.

**EXAMPLE 2**  Use the slope-intercept form to graph the equation

$$y = \frac{3}{5}x - 2$$

_Solution_  Since the equation $y = \frac{3}{5}x - 2$ is written in slope-intercept form $y = mx + b$, the slope of its graph is $\frac{3}{5}$ and the $y$-intercept is $(0, -2)$. To graph this equation, we begin by plotting the point $(0, -2)$. From this point, we can find another point of the graph by using the slope $\frac{3}{5}$ and recalling that slope is $\frac{\text{rise}}{\text{run}}$. We start at the $y$-intercept and move 3 units up since the numerator of the slope is 3; then we move 5 units to the right since the denominator of the slope is 5. We stop at the point $(5, 1)$. The line through $(0, -2)$ and $(5, 1)$ is the graph of $y = \frac{3}{5}x - 2$.

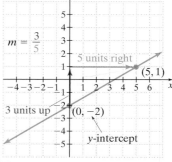

**PRACTICE**
**2**  Graph $y = \frac{2}{3}x - 5$.

**EXAMPLE 3**   Use the slope-intercept form to graph the equation $4x + y = 1$.

_Solution_   First we write the given equation in slope-intercept form.

$$4x + y = 1$$
$$y = -4x + 1$$

The graph of this equation will have slope $-4$ and $y$-intercept $(0, 1)$. To graph this line, we first plot the point $(0, 1)$. To find another point of the graph, we use the slope $-4$, which can be written as $\dfrac{-4}{1}\left(\dfrac{4}{-1}\text{ could also be used}\right)$. We start at the point $(0, 1)$ and move 4 units down (since the numerator of the slope is $-4$), and then 1 unit to the right (since the denominator of the slope is 1).

We arrive at the point $(1, -3)$. The line through $(0, 1)$ and $(1, -3)$ is the graph of $4x + y = 1$.

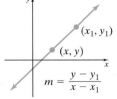

$m = \dfrac{-4}{1}$

$(0, 1)$

4 units down

$(1, -3)$

1 unit right

**PRACTICE**
**3**   Use the slope-intercept form to graph the equation $3x - y = 2$.

▶ **Helpful Hint**
In Example 3, if we interpret the slope of $-4$ as $\dfrac{4}{-1}$, we arrive at $(-1, 5)$ for a second point. Notice that this point is also on the line.

**OBJECTIVE 3** ▶ **Writing an equation given slope and a point.** Thus far, we have seen that we can write an equation of a line if we know its slope and $y$-intercept. We can also write an equation of a line if we know its slope and any point on the line. To see how we do this, let $m$ represent slope and $(x_1, y_1)$ represent the point on the line. Then if $(x, y)$ is any other point of the line, we have that

$$\frac{y - y_1}{x - x_1} = m$$
$$y - y_1 = m(x - x_1) \quad \text{Multiply both sides by } (x - x_1).$$

slope

This is the _point-slope form_ of the equation of a line.

$(x_1, y_1)$

$(x, y)$

$m = \dfrac{y - y_1}{x - x_1}$

---

**Point-Slope Form of the Equation of a Line**
The **point-slope form** of the equation of a line is          Variables

$$y - y_1 = m(x - x_1)$$

slope

$(x_1, y_1)$ point of the line

where $m$ is the slope of the line and $(x_1, y_1)$ is a point on the line.

---

**EXAMPLE 4**   Find an equation of the line with slope $-2$ that passes through $(-1, 5)$. Write the equation in slope-intercept form, $y = mx + b$, and in standard form, $Ax + By = C$.

_Solution_   Since the slope and a point on the line are given, we use point-slope form $y - y_1 = m(x - x_1)$ to write the equation. Let $m = -2$ and $(-1, 5) = (x_1, y_1)$.

$$y - y_1 = m(x - x_1)$$
$$y - 5 = -2[x - (-1)] \quad \text{Let } m = -2 \text{ and } (x_1, y_1) = (-1, 5).$$
$$y - 5 = -2(x + 1) \quad \text{Simplify.}$$
$$y - 5 = -2x - 2 \quad \text{Use the distributive property.}$$

To write the equation in slope-intercept form, $y = mx + b$, we simply solve the equation for $y$. To do this, we add 5 to both sides.

$$y - 5 = -2x - 2$$
$$y = -2x + 3 \quad \text{Slope-intercept form.}$$
$$2x + y = 3 \quad \text{Add } 2x \text{ to both sides and we have standard form.} \quad \square$$

**PRACTICE**
**4** Find an equation of the line passing through $(2, 3)$ with slope 4. Write the equation in standard form: $Ax + By = C$.

**OBJECTIVE 4** ▶ **Writing an equation given two points.** We can also find the equation of a line when we are given any two points of the line.

**EXAMPLE 5** Find an equation of the line through $(2, 5)$ and $(-3, 4)$. Write the equation in standard form.

**Solution** First, use the two given points to find the slope of the line.

$$m = \frac{4 - 5}{-3 - 2} = \frac{-1}{-5} = \frac{1}{5}$$

Next we use the slope $\frac{1}{5}$ and either one of the given points to write the equation in point-slope form. We use $(2, 5)$. Let $x_1 = 2$, $y_1 = 5$, and $m = \frac{1}{5}$.

$$y - y_1 = m(x - x_1) \quad \text{Use point-slope form.}$$
$$y - 5 = \frac{1}{5}(x - 2) \quad \text{Let } x_1 = 2, y_1 = 5, \text{ and } m = \frac{1}{5}.$$
$$5(y - 5) = 5 \cdot \frac{1}{5}(x - 2) \quad \text{Multiply both sides by 5 to clear fractions.}$$
$$5y - 25 = x - 2 \quad \text{Use the distributive property and simplify.}$$
$$-x + 5y - 25 = -2 \quad \text{Subtract } x \text{ from both sides.}$$
$$-x + 5y = 23 \quad \text{Add 25 to both sides.} \quad \square$$

**PRACTICE**
**5** Find an equation of the line through $(-1, 6)$ and $(3, 1)$. Write the equation in standard form.

> ▶ **Helpful Hint**
> Multiply both sides of the equation $-x + 5y = 23$ by $-1$, and it becomes $x - 5y = -23$.
> Both $-x + 5y = 23$ and $x - 5y = -23$ are in standard form, and they are equations of the same line.

**OBJECTIVE 5** ▶ **Finding equations of vertical and horizontal lines.** Recall from Section 3.3 that:

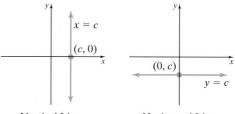

Vertical Line          Horizontal Line

**EXAMPLE 6**   Find an equation of the vertical line through $(-1, 5)$.

*Solution*   The equation of a vertical line can be written in the form $x = c$, so an equation for a vertical line passing through $(-1, 5)$ is $x = -1$.

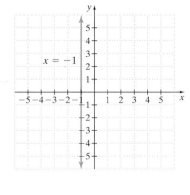

PRACTICE
**6**   Find an equation of the vertical line through $(3, -2)$.

**EXAMPLE 7**   Find an equation of the line parallel to the line $y = 5$ and passing through $(-2, -3)$.

*Solution*   Since the graph of $y = 5$ is a horizontal line, any line parallel to it is also horizontal. The equation of a horizontal line can be written in the form $y = c$. An equation for the horizontal line passing through

$$(-2, -3) \text{ is } y = -3.$$

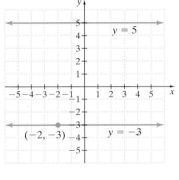

PRACTICE
**7**   Find an equation of the line parallel to the line $y = -2$ and passing through $(4, 3)$.

Note: Further discussion of parallel and perpendicular lines is in Section 8.1.

**OBJECTIVE 6 ▶ Using the point-slope form to solve problems.** Problems occurring in many fields can be modeled by linear equations in two variables. The next example is from the field of marketing and shows how consumer demand of a product depends on the price of the product.

**EXAMPLE 8**   **Predicting the Sales of T-Shirts**

A web-based T-shirt company has learned that by pricing a clearance-sale T-shirt at $6, sales will reach 2000 T-shirts per day. Raising the price to $8 will cause the sales to fall to 1500 T-shirts per day.

**a.** Assume that the relationship between sales price and number of T-shirts sold is linear and write an equation describing this relationship. Write the equation in slope-intercept form.

**b.** Predict the daily sales of T-shirts if the price is $7.50.

*Solution*

**a.** First, use the given information and write two ordered pairs. Ordered pairs will be in the form (sales price, number sold) so that our ordered pairs are (6, 2000) and

$(8, 1500)$. Use the point-slope form to write an equation. To do so, we find the slope of the line that contains these points.

$$m = \frac{2000 - 1500}{6 - 8} = \frac{500}{-2} = -250$$

Next, use the slope and either one of the points to write the equation in point-slope form. We use $(6, 2000)$.

$$y - y_1 = m(x - x_1) \qquad \text{Use point-slope form.}$$
$$y - 2000 = -250(x - 6) \qquad \text{Let } x_1 = 6, y_1 = 2000, \text{ and } m = -250.$$
$$y - 2000 = -250x + 1500 \qquad \text{Use the distributive property.}$$
$$y = -250x + 3500 \qquad \text{Write in slope-intercept form.}$$

**b.** To predict the sales if the price is \$7.50, we find $y$ when $x = 7.50$.

$$y = -250x + 3500$$
$$y = -250(7.50) + 3500 \qquad \text{Let } x = 7.50.$$
$$y = -1875 + 3500$$
$$y = 1625$$

If the price is \$7.50, sales will reach 1625 T-shirts per day. ☐

**PRACTICE**

**8**    The new *Camelot* condos were selling at a rate of 30 per month when they were priced at \$150,000 each. Lowering the price to \$120,000 caused the sales to rise to 50 condos per month.

**a.** Assume that the relationship between number of condos sold and price is linear, and write an equation describing this relationship. Write the equation in slope-intercept form.

**b.** What should the condos be priced at if the developer wishes to sell 60 condos per month?

The preceding example may also be solved by using ordered pairs of the form (number sold, sales price).

---

### Forms of Linear Equations

| | |
|---|---|
| $Ax + By = C$ | **Standard form** of a linear equation. |
| | $A$ and $B$ are not both 0. |
| $y = mx + b$ | **Slope-intercept form** of a linear equation. |
| | The slope is $m$ and the $y$-intercept is $(0, b)$. |
| $y - y_1 = m(x - x_1)$ | **Point-slope form** of a linear equation. |
| | The slope is $m$ and $(x_1, y_1)$ is a point on the line. |
| $y = c$ | **Horizontal line** |
| | The slope is 0 and the $y$-intercept is $(0, c)$. |
| $x = c$ | **Vertical line** |
| | The slope is undefined and the $x$-intercept is $(c, 0)$. |

### Parallel and Perpendicular Lines

Nonvertical parallel lines have the same slope.

The product of the slopes of two nonvertical perpendicular lines is $-1$.

**Graphing Calculator Explorations**

A grapher is a very useful tool for discovering patterns. To discover the change in the graph of a linear equation caused by a change in slope, try the following. Use a standard window and graph a linear equation in the form $y = mx + b$. Recall that the graph of such an equation will have slope $m$ and $y$-intercept $b$.

First graph $y = x + 3$. To do so, press the $\boxed{Y=}$ key and enter $Y_1 = x + 3$. Notice that this graph has slope 1 and that the $y$-intercept is 3. Next, on the same set of axes, graph $y = 2x + 3$ and $y = 3x + 3$ by pressing $\boxed{Y=}$ and entering $Y_2 = 2x + 3$ and $Y_3 = 3x + 3$.

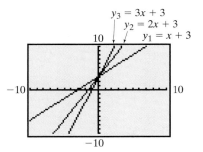

Notice the difference in the graph of each equation as the slope changes from 1 to 2 to 3. How would the graph of $y = 5x + 3$ appear? To see the change in the graph caused by a change in negative slope, try graphing $y = -x + 3$, $y = -2x + 3$, and $y = -3x + 3$ on the same set of axes.

*Use a grapher to graph the following equations. For each exercise, graph the first equation and use its graph to predict the appearance of the other equations. Then graph the other equations on the same set of axes and check your prediction.*

**1.** $y = x$; $y = 6x$, $y = -6x$

**2.** $y = -x$; $y = -5x$, $y = -10x$

**3.** $y = \dfrac{1}{2}x + 2$; $y = \dfrac{3}{4}x + 2$, $y = x + 2$

**4.** $y = x + 1$; $y = \dfrac{5}{4}x + 1$, $y = \dfrac{5}{2}x + 1$

**5.** $y = -7x + 5$; $y = 7x + 5$

**6.** $y = 3x - 1$; $y = -3x - 1$

# VOCABULARY & READINESS CHECK

*Use the choices below to fill in each blank. Some choices may be used more than once and some not at all.*

$b$          $(y_1, x_1)$      point-slope          vertical          standard

$m$          $(x_1, y_1)$      slope-intercept      horizontal

**1.** The form $y = mx + b$ is called _____ form. When a linear equation in two variables is written in this form, _____ is the slope of its graph and $(0,$ _____$)$ is its $y$-intercept.

**2.** The form $y - y_1 = m(x - x_1)$ is called _____ form. When a linear equation in two variables is written in this form, _____ is the slope of its graph and _____ is a point on the graph.

*For Exercises 3, 4, and 7, identify the form in which the linear equation in two variables is written. For Exercises 5 and 6, identify the appearance of the graph of the equation.*

**3.** $y - 7 = 4(x + 3)$; _____ form

**4.** $5x - 9y = 11$; _____ form

**5.** $y = \dfrac{1}{2}$; _____ line

**6.** $x = -17$; _____ line

**7.** $y = \dfrac{3}{4}x - \dfrac{1}{3}$; _____ form

*Write an equation of the line with each given slope, m, and y-intercept, (0, b). See Example 1.*

**1.** $m = 5, b = 3$

**2.** $m = -3, b = -3$

**3.** $m = -4, b = -\dfrac{1}{6}$

**4.** $m = 2, b = \dfrac{3}{4}$

**5.** $m = \dfrac{2}{3}, b = 0$

**6.** $m = -\dfrac{4}{5}, b = 0$

**7.** $m = 0, b = -8$

**8.** $m = 0, b = -2$

**9.** $m = -\dfrac{1}{5}, b = \dfrac{1}{9}$

**10.** $m = \dfrac{1}{2}, b = -\dfrac{1}{3}$     $Ax + Bg = C$

*Use the slope-intercept form to graph each equation. See Examples 2 and 3.*

**11.** $y = 2x + 1$

**12.** $y = -4x - 1$

**13.** $y = \dfrac{2}{3}x + 5$

**14.** $y = \dfrac{1}{4}x - 3$

**15.** $y = -5x$

**16.** $y = -6x$

**17.** $4x + y = 6$

**18.** $-3x + y = 2$

**19.** $4x - 7y = -14$

**20.** $3x - 4y = 4$

**21.** $x = \dfrac{5}{4}y$

**22.** $x = \dfrac{3}{2}y$

*Find an equation of each line with the given slope that passes through the given point. Write the equation in the form $Ax + By = C$. See Example 4.*

**23.** $m = 6; \quad (2, 2)$

**24.** $m = 4; \quad (1, 3)$

**25.** $m = -8; \quad (-1, -5)$

**26.** $m = -2; \quad (-11, -12)$

**27.** $m = \dfrac{3}{2}; \quad (5, -6)$

**28.** $m = \dfrac{2}{3}; \quad (-8, 9)$

**29.** $m = -\dfrac{1}{2}; \quad (-3, 0)$

**30.** $m = -\dfrac{1}{5}; \quad (4, 0)$

*Find an equation of the line passing through each pair of points. Write the equation in the form $Ax + By = C$. See Example 5.*

**31.** $(3, 2)$ and $(5, 6)$

**32.** $(6, 2)$ and $(8, 8)$

**33.** $(-1, 3)$ and $(-2, -5)$

**34.** $(-4, 0)$ and $(6, -1)$

**35.** $(2, 3)$ and $(-1, -1)$

**36.** $(7, 10)$ and $(-1, -1)$

**37.** $(0, 0)$ and $\left(-\dfrac{1}{8}, \dfrac{1}{13}\right)$

**38.** $(0, 0)$ and $\left(-\dfrac{1}{2}, \dfrac{1}{3}\right)$

*Find an equation of each line. See Example 6.*

**39.** Vertical line through $(0, 2)$

**40.** Horizontal line through $(1, 4)$

**41.** Horizontal line through $(-1, 3)$

**42.** Vertical line through $(-1, 3)$

**43.** Vertical line through $\left(-\dfrac{7}{3}, -\dfrac{2}{5}\right)$

**44.** Horizontal line through $\left(\dfrac{2}{7}, 0\right)$

*Find an equation of each line. See Example 7.*

**45.** Parallel to $y = 5$, through $(1, 2)$

**46.** Perpendicular to $y = 5$, through $(1, 2)$

**47.** Perpendicular to $x = -3$, through $(-2, 5)$

**48.** Parallel to $y = -4$, through $(0, -3)$

**49.** Parallel to $x = 0$, through $(6, -8)$

**50.** Perpendicular to $x = 7$, through $(-5, 0)$

**MIXED PRACTICE**

*See Examples 1 through 7. Find an equation of each line described. Write each equation in slope-intercept form (solved for y), when possible.*

**51.** With slope $-\dfrac{1}{2}$, through $\left(0, \dfrac{5}{3}\right)$

**52.** With slope $\dfrac{5}{7}$, through $(0, -3)$

**53.** Through $(10, 7)$ and $(7, 10)$

**54.** Through $(5, -6)$ and $(-6, 5)$

**55.** With undefined slope, through $\left(-\dfrac{3}{4}, 1\right)$

**56.** With slope 0, through $(6.7, 12.1)$

**57.** Slope 1, through $(-7, 9)$

**58.** Slope 5, through $(6, -8)$

**59.** Slope $-5$, $y$-intercept $(0, 7)$

**60.** Slope $-2$; $y$-intercept $(0, -4)$

**61.** Through $(6, 7)$, parallel to the $x$-axis

**62.** Through $(1, -5)$, parallel to the $y$-axis

**63.** Through $(2, 3)$ and $(0, 0)$

**64.** Through $(4, 7)$ and $(0, 0)$

**65.** Through $(-2, -3)$, perpendicular to the $y$-axis

**66.** Through $(0, 12)$, perpendicular to the $x$-axis

**67.** Slope $-\dfrac{4}{7}$, through $(-1, -2)$

**68.** Slope $-\dfrac{3}{5}$, through $(4, 4)$

*Solve. Assume each exercise describes a linear relationship. Write the equations in slope-intercept form. See Example 8.*

**69.** A rock is dropped from the top of a 400-foot cliff. After 1 second, the rock is traveling 32 feet per second. After 3 seconds, the rock is traveling 96 feet per second.

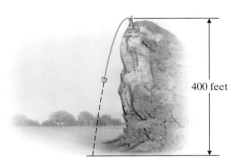

400 feet

a. Assume that the relationship between time and speed is linear and write an equation describing this relationship. Use ordered pairs of the form (time, speed).

b. Use this equation to determine the speed of the rock 4 seconds after it was dropped.

**70.** A Hawaiian fruit company is studying the sales of a pineapple sauce to see if this product is to be continued. At the end of its first year, profits on this product amounted to $30,000. At the end of the fourth year, profits were $66,000.

a. Assume that the relationship between years on the market and profit is linear and write an equation describing this relationship. Use ordered pairs of the form (years on the market, profit).

b. Use this equation to predict the profit at the end of 7 years.

**71.** In January 2007, there were 71,000 registered gasoline-electric hybrid cars in the United States. In 2004, there were only 29,000 registered gasoline-electric hybrids. (*Source:* U.S. Energy Information Administration)

a. Write an equation describing the relationship between time and number of registered gasoline-hybrid cars. Use ordered pairs of the form (years past 2004, number of cars).

b. Use this equation to predict the number of gasoline-electric hybrids in the year 2010.

**72.** In 2006, there were 935 thousand eating establishments in the United States. In 1996, there were 457 thousand eating establishments. (*Source:* National Restaurant Association)

a. Write an equation describing the relationship between time and number of eating establishments. Use ordered pairs of the form (years past 1996, number of eating establishments in thousands).

b. Use this equation to predict the number of eating establishments in 2010.

**73.** In 2006, the U.S. population per square mile of land area was 85. In 2000, the person per square mile population was 79.6.

a. Write an equation describing the relationship between year and persons per square mile. Use ordered pairs of the form (years past 2000, persons per square mile).

b. Use this equation to predict the person per square mile population in 2010.

**74.** In 2001, there were a total of 152 thousand apparel and accessory stores. In 2005, there were a total of 150 thousand apparel and accessory stores. (*Source:* U.S. Bureau of the Census. *County Business Patterns, annual*)

a. Write an equation describing this relationship. Use ordered pairs of the form (years past 2001, numbers of stores in thousand).

b. Use this equation to predict the number of apparel and accessory stores in 2011.

**75.** The birth rate in the United States in 1996 was 14.7 births per thousand population. In 2006, the birth rate was 14.14 births per thousand. (*Source:* Department of Health and Human Services, National Center for Health Statistics)

a. Write two ordered pairs of the form (years after 1996, birth rate per thousand population).

b. Assume that the relationship between years after 1996 and birth rate per thousand is linear over this period. Use the ordered pairs from part (a) to write an equation of the line relating years to birth rate.

c. Use the linear equation from part (b) to estimate the birth rate in the United States in the year 2016.

**76.** In 2002, crude oil production by OPEC countries was about 28.7 million barrels per day. In 2007, crude oil production had risen to about 34.5 million barrels per day. (*Source:* OPEC)

 **a.** Write two ordered pairs of the form (years after 2002, crude oil production) for this situation.

 **b.** Assume that crude oil production is linear between the years 2002 and 2007. Use the ordered pairs from part (a) to write an equation of the line relating year and crude oil production.

 **c.** Use the linear equation from part (b) to estimate the crude oil production by OPEC countries in 2004.

**77.** Better World Club is a relatively new automobile association which prides itself on its "green" philosophy. In 2003, the membership totaled 5 thousand. By 2006, there were 20 thousand members of this ecologically minded club. (*Source:* Better World Club)

 **a.** Write two ordered pairs of the form (years after 2003, membership in thousands)

 **b.** Assume that the membership is linear between the years 2003 and 2006. Use the ordered pairs from part (a) to write an equation of the line relating year and Better World membership.

 **c.** Use the linear equation from part (b) to predict the Better World Club membership in 2012.

**78.** In 2002, 9.9 million electronic bill statements were delivered and payment occurred. In 2005, that number rose to 26.9 million. (*Source:* Forrester Research)

 **a.** Write two ordered pairs of the form (years after 2002, millions of electronic bills).

 **b.** Assume that this method of delivery and payment between the years 2002 and 2005 is linear. Use the ordered pairs from part (a) to write an equation of the line relating year and number of electronic bills.

 **c.** Use the linear equation from part (b) to predict the number of electronic bills to be delivered and paid in 2011.

**REVIEW AND PREVIEW**

*Find the value of $x^2 - 3x + 1$ for each given value of x. See Section 1.7.*

**79.** 2          **80.** 5          **81.** −1          **82.** −3

*For each graph, determine whether any x-values correspond to two or more y-values. See Section 3.1.*

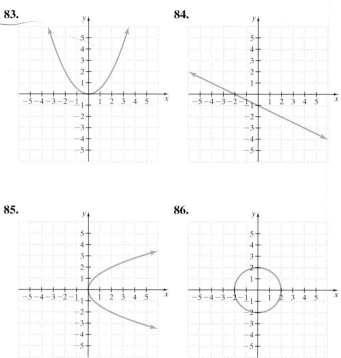

**83.**          **84.**

**85.**          **86.**

**CONCEPT EXTENSIONS**

**87.** Given the equation of a nonvertical line, explain how to find the slope without finding two points on the line.

**88.** Given two points on a nonvertical line, explain how to use the point-slope form to find the equation of the line.

**89.** Write an equation in standard form of the line that contains the point $(-1, 2)$ and is

 **a.** parallel to the line $y = 3x - 1$.

 **b.** perpendicular to the line $y = 3x - 1$.

**90.** Write an equation in standard form of the line that contains the point $(4, 0)$ and is

 **a.** parallel to the line $y = -2x + 3$.

 **b.** perpendicular to the line $y = -2x + 3$.

**91.** Write an equation in standard form of the line that contains the point $(3, -5)$ and is

 **a.** parallel to the line $3x + 2y = 7$.

 **b.** perpendicular to the line $3x + 2y = 7$.

**92.** Write an equation in standard form of the line that contains the point $(-2, 4)$ and is

 **a.** parallel to the line $x + 3y = 6$.

 **b.** perpendicular to the line $x + 3y = 6$.

# 3.6 FUNCTIONS

### OBJECTIVES

1  Identify relations, domains, and ranges.

2  Identify functions.

3  Use the vertical line test.

4  Use function notation.

**OBJECTIVE 1 ▶ Identifying relations, domains, and ranges.** In previous sections, we have discussed the relationships between two quantities. For example, the relationship between the length of the side of a square $x$ and its area $y$ is described by the equation $y = x^2$. Ordered pairs can be used to write down solutions of this equation. For example, $(2, 4)$ is a solution of $y = x^2$, and this notation tells us that the $x$-value 2 is related to the $y$-value 4 for this equation. In other words, when the length of the side of a square is 2 units, its area is 4 square units.

### Examples of Relationships Between Two Quantities

| Area of Square: $y = x^2$ | Equation of Line: $y = x + 2$ | Online Advertising Revenue |
|---|---|---|

**Some Ordered Pairs**

| $x$ | $y$ |
|---|---|
| 2 | 4 |
| 5 | 25 |
| 7 | 49 |
| 12 | 144 |

**Some Ordered Pairs**

| $x$ | $y$ |
|---|---|
| −3 | −1 |
| 0 | 2 |
| 2 | 4 |
| 9 | 11 |

**Ordered Pairs**

| Year | Billions of Dollars |
|---|---|
| 2006 | 16.7 |
| 2007 | 20.3 |
| 2008 | 23.5 |
| 2009 | 26.6 |
| 2010 | 29.4 |

A set of ordered pairs is called a **relation.** The set of all $x$-coordinates is called the **domain** of a relation, and the set of all $y$-coordinates is called the **range** of a relation. Equations such as $y = x^2$ are also called relations since equations in two variables define a set of ordered pair solutions.

**EXAMPLE 1**  Find the domain and the range of the relation $\{(0, 2), (3, 3), (-1, 0), (3, -2)\}$.

*Solution*  The domain is the set of all $x$-values or $\{-1, 0, 3\}$, and the range is the set of all $y$-values, or $\{-2, 0, 2, 3\}$.  □

**PRACTICE**
**1**  Find the domain and the range of the relation $\{(1, 3)(5, 0)(0, -2)(5, 4)\}$.

**OBJECTIVE 2 ▶ Identifying functions.** Some relations are also functions.

### Function
A function is a set of ordered pairs that assigns to each $x$-value exactly one $y$-value.

**EXAMPLE 2**   Which of the following relations are also functions?

**a.** $\{(-1, 1), (2, 3), (7, 3), (8, 6)\}$       **b.** $\{(0, -2), (1, 5), (0, 3), (7, 7)\}$

*Solution*

**a.** Although the ordered pairs $(2, 3)$ and $(7, 3)$ have the same $y$-value, each $x$-value is assigned to only one $y$-value so this set of ordered pairs is a function.

**b.** The $x$-value 0 is assigned to two $y$-values, $-2$ and 3, so this set of ordered pairs is not a function.  □

**PRACTICE**
**2**   Which of the following relations are also functions?

**a.** $\{(4, 1)(3, -2)(8, 5)(-5, 3)\}$       **b.** $\{(1, 2)(-4, 3)(0, 8)(1, 4)\}$

Relations and functions can be described by a graph of their ordered pairs.

**EXAMPLE 3**   Which graph is the graph of a function?

**a.**   **b.**

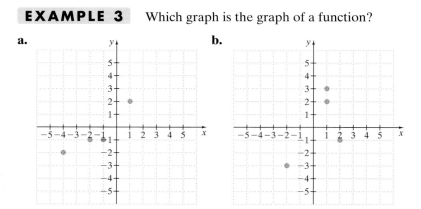

*Solution*

**a.** This is the graph of the relation $\{(-4, -2), (-2, -1)(-1, -1), (1, 2)\}$. Each $x$-coordinate has exactly one $y$-coordinate, so this is the graph of a function.

**b.** This is the graph of the relation $\{(-2, -3), (1, 2), (1, 3), (2, -1)\}$. The $x$-coordinate 1 is paired with two $y$-coordinates, 2 and 3, so this is not the graph of a function.  □

**PRACTICE**
**3**   Which graph is the graph of a function?

**a.**   **b.**

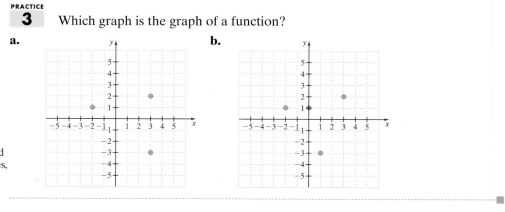

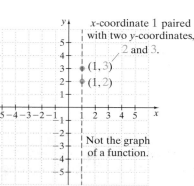

x-coordinate 1 paired with two y-coordinates, 2 and 3.

(1, 3)

(1, 2)

Not the graph of a function.

**OBJECTIVE 3 ▶ Using the vertical line test.** The graph in Example 3(b) was not the graph of a function because the $x$-coordinate 1 was paired with two $y$-coordinates, 2 and 3. Notice that when an $x$-coordinate is paired with more than one $y$-coordinate, a vertical line can be drawn that will intersect the graph at more than one point. We can use this fact to determine whether a relation is also a function. We call this the **vertical line test.**

> **Vertical Line Test**
> If a vertical line can be drawn so that it intersects a graph more than once, the graph is not the graph of a function.

This vertical line test works for all types of graphs on the rectangular coordinate system.

**EXAMPLE 4**   Use the vertical line test to determine whether each graph is the graph of a function.

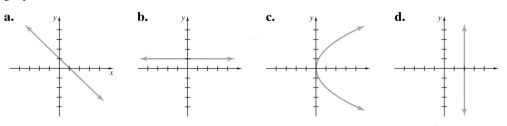

a.          b.          c.          d.

**Solution**

a. This graph is the graph of a function since no vertical line will intersect this graph more than once.

b. This graph is also the graph of a function; no vertical line will intersect it more than once.

c. This graph is not the graph of a function. Vertical lines can be drawn that intersect the graph in two points. An example of one is shown.

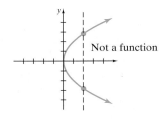

Not a function

d. This graph is not the graph of a function. A vertical line can be drawn that intersects this line at every point.

**PRACTICE**
**4**   Use the vertical line test to determine whether each graph is the graph of a function.

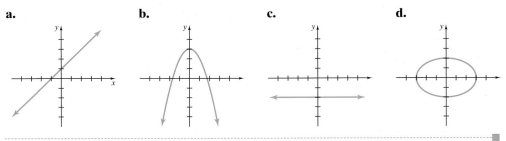

a.          b.          c.          d.

Recall that the graph of a linear equation is a line, and a line that is not vertical will pass the vertical line test. **Thus, all linear equations are functions except those of the form $x = c$, which are vertical lines.**

**EXAMPLE 5** Which of the following linear equations are functions?

**a.** $y = x$     **b.** $y = 2x + 1$     **c.** $y = 5$     **d.** $x = -1$

*Solution* **a**, **b**, and **c** are functions because their graphs are nonvertical lines. **d** is not a function because its graph is a vertical line. ☐

**PRACTICE**
**5** Which of the following linear equations are functions?

**a.** $y = 2x$     **b.** $y = -3x - 1$     **c.** $y = 8$     **d.** $x = 2$

Examples of functions can often be found in magazines, newspapers, books, and other printed material in the form of tables or graphs such as that in Example 6.

**EXAMPLE 6** The graph shows the sunrise time for Indianapolis, Indiana, for the year. Use this graph to answer the questions.

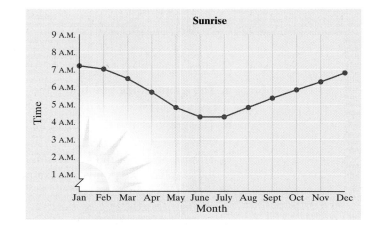

**a.** Approximate the time of sunrise on February 1.
**b.** Approximately when does the sun rise at 5 a.m.?
**c.** Is this the graph of a function?

*Solution*

**a.** To approximate the time of sunrise on February 1, we find the mark on the horizontal axis that corresponds to February 1. From this mark, we move vertically upward until the graph is reached. From that point on the graph, we move horizontally to the left until the vertical axis is reached. The vertical axis there reads 7 a.m.

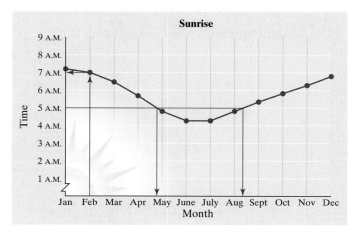

**b.** To approximate when the sun rises at 5 a.m., we find 5 a.m. on the time axis and move horizontally to the right. Notice that we will reach the graph twice, corresponding to two dates for which the sun rises at 5 a.m. We follow both points on the graph vertically downward until the horizontal axis is reached. The sun rises at 5 a.m. at approximately the end of the month of April and the middle of the month of August.

**c.** The graph is the graph of a function since it passes the vertical line test. In other words, for every day of the year in Indianapolis, there is exactly one sunrise time. ☐

**PRACTICE**

**6** The graph shows the average monthly temperature for Chicago, Illinois, for the year. Use this graph to answer the questions.

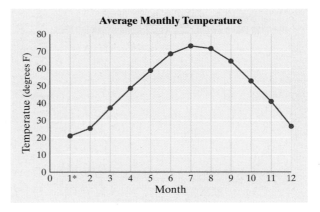

*(1 is Jan., 12 is Dec.)

**a.** Approximate the average monthly temperature for June.

**b.** Approximately when is the average monthly temperature 40°?

**c.** Is this the graph of a function?

**OBJECTIVE 4 ▶ Using function notation.** The graph of the linear equation $y = 2x + 1$ passes the vertical line test, so we say that $y = 2x + 1$ is a function. In other words, $y = 2x + 1$ gives us a rule for writing ordered pairs where every $x$-coordinate is paired with one $y$-coordinate.

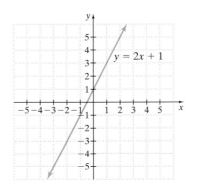

We often use letters such as $f$, $g$, and $h$ to name functions. For example, the symbol $f(x)$ means *function of x* and is read "$f$ of $x$." This notation is called **function notation.** The equation $y = 2x + 1$ can be written as $f(x) = 2x + 1$ using function notation, and these equations mean the same thing. In other words, $y = f(x)$.

The notation $f(1)$ means to replace $x$ with 1 and find the resulting $y$ or function value. Since

$$f(x) = 2x + 1$$

then

$$f(1) = 2(1) + 1 = 3$$

This means that, when $x = 1$, $y$ or $f(x) = 3$, and we have the ordered pair $(1, 3)$. Now let's find $f(2), f(0)$, and $f(-1)$.

<table>
<tr><td>$f(x) = 2x + 1$</td><td>$f(x) = 2x + 1$</td><td>$f(x) = 2x + 1$</td></tr>
<tr><td>$f(2) = 2(2) + 1$</td><td>$f(0) = 2(0) + 1$</td><td>$f(-1) = 2(-1) + 1$</td></tr>
<tr><td>$= 4 + 1$</td><td>$= 0 + 1$</td><td>$= -2 + 1$</td></tr>
<tr><td>$= 5$</td><td>$= 1$</td><td>$= -1$</td></tr>
</table>

> **Helpful Hint**
> Note that, for example, if $f(2) = 5$, the corresponding ordered pair is $(2, 5)$.

Ordered
Pair:     $(2, 5)$          $(0, 1)$          $(-1, -1)$

> **Helpful Hint**
> Note that $f(x)$ is a special symbol in mathematics used to denote a function. The symbol $f(x)$ is read "$f$ of $x$." It does **not** mean $f \cdot x$ ($f$ times $x$).

**EXAMPLE 7**  Given $g(x) = x^2 - 3$, find the following. Then write down the corresponding ordered pairs generated.

**a.** $g(2)$      **b.** $g(-2)$      **c.** $g(0)$

*Solution*

<table>
<tr><td>**a.** $g(x) = x^2 - 3$</td><td>**b.** $g(x) = x^2 - 3$</td><td>**c.** $g(x) = x^2 - 3$</td></tr>
<tr><td>$g(2) = 2^2 - 3$</td><td>$g(-2) = (-2)^2 - 3$</td><td>$g(0) = 0^2 - 3$</td></tr>
<tr><td>$= 4 - 3$</td><td>$= 4 - 3$</td><td>$= 0 - 3$</td></tr>
<tr><td>$= 1$</td><td>$= 1$</td><td>$= -3$</td></tr>
</table>

| Ordered Pairs: | $g(2) = 1$ gives $(2, 1)$ | $g(-2) = 1$ gives $(-2, 1)$ | $g(0) = -3$ gives $(0, -3)$ |
|---|---|---|---|

**PRACTICE**

**7**  Given $h(x) = x^2 + 5$, find the following. Then write the corresponding ordered pairs generated.

**a.** $h(2)$               **b.** $h(-5)$                    **c.** $h(0)$

We now practice finding the domain and the range of a function. The domain of our functions will be the set of all possible real numbers that $x$ can be replaced by. The range is the set of corresponding $y$-values.

**EXAMPLE 8**   Find the domain of each function.

**a.** $g(x) = \dfrac{1}{x}$     **b.** $f(x) = 2x + 1$

*Solution*

**a.** Recall that we cannot divide by 0 so that the domain of $g(x)$ is the set of all real numbers except 0. In interval notation, we can write $(-\infty, 0) \cup (0, \infty)$.

**b.** In this function, $x$ can be any real number. The domain of $f(x)$ is the set of all real numbers, or $(-\infty, \infty)$ in interval notation.  ☐

**PRACTICE**
**8**   Find the domain of each function.

**a.** $h(x) = 6x + 3$                    **b.** $f(x) = \dfrac{1}{x^2}$

**Concept Check** ☑

Suppose that the value of $f$ is $-7$ when the function is evaluated at 2. Write this situation in function notation.

**EXAMPLE 9**   Find the domain and the range of each function graphed. Use interval notation.

**a.**

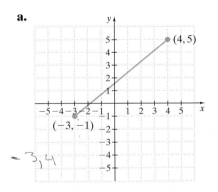

**b.**

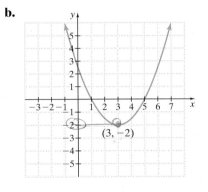

*Solution*

**a.**

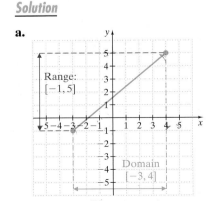

Range: $[-1, 5]$

Domain $[-3, 4]$

**b.**

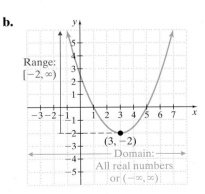

Range: $[-2, \infty)$

$(3, -2)$

Domain: All real numbers or $(-\infty, \infty)$

☐

**Answer to Concept Check:**
$f(2) = -7$

**PRACTICE**
**9** Find the domain and the range of each function graphed. Use interval notation.

a.

b.

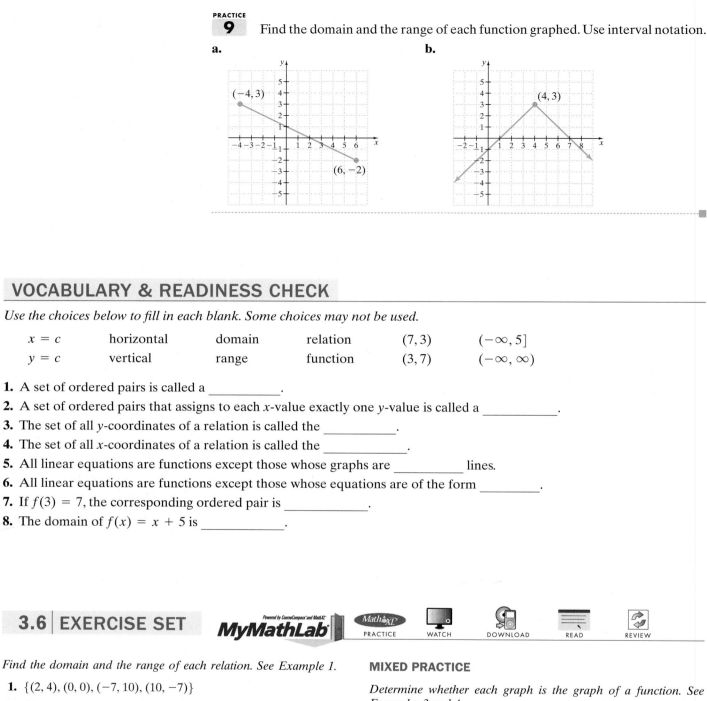

## VOCABULARY & READINESS CHECK

*Use the choices below to fill in each blank. Some choices may not be used.*

| | | | | | |
|---|---|---|---|---|---|
| $x = c$ | horizontal | domain | relation | $(7, 3)$ | $(-\infty, 5]$ |
| $y = c$ | vertical | range | function | $(3, 7)$ | $(-\infty, \infty)$ |

**1.** A set of ordered pairs is called a _____ .

**2.** A set of ordered pairs that assigns to each $x$-value exactly one $y$-value is called a _____ .

**3.** The set of all $y$-coordinates of a relation is called the _____ .

**4.** The set of all $x$-coordinates of a relation is called the _____ .

**5.** All linear equations are functions except those whose graphs are _____ lines.

**6.** All linear equations are functions except those whose equations are of the form _____ .

**7.** If $f(3) = 7$, the corresponding ordered pair is _____ .

**8.** The domain of $f(x) = x + 5$ is _____ .

## 3.6 EXERCISE SET

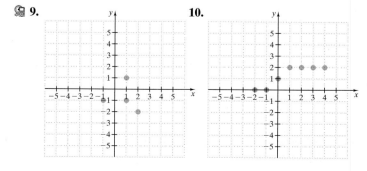

*Find the domain and the range of each relation. See Example 1.*

**1.** $\{(2, 4), (0, 0), (-7, 10), (10, -7)\}$

**2.** $\{(3, -6), (1, 4), (-2, -2)\}$

**3.** $\{(0, -2), (1, -2), (5, -2)\}$

**4.** $\{(5, 0), (5, -3), (5, 4), (5, 3)\}$

*Determine whether each relation is also a function. See Example 2.*

**5.** $\{(1, 1), (2, 2), (-3, -3), (0, 0)\}$

**6.** $\{(11, 6), (-1, -2), (0, 0), (3, -2)\}$

**7.** $\{(-1, 0), (-1, 6), (-1, 8)\}$

**8.** $\{(1, 2), (3, 2), (1, 4)\}$

**MIXED PRACTICE**

*Determine whether each graph is the graph of a function. See Examples 3 and 4.*

**9.**

**10.**

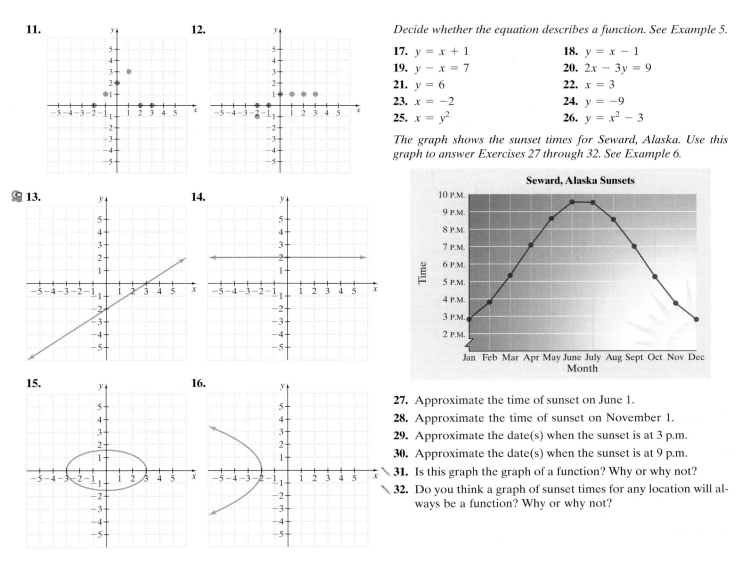

**11.**

**12.**

**13.**

**14.**

**15.**

**16.**

*Decide whether the equation describes a function. See Example 5.*

**17.** $y = x + 1$        **18.** $y = x - 1$

**19.** $y - x = 7$       **20.** $2x - 3y = 9$

**21.** $y = 6$          **22.** $x = 3$

**23.** $x = -2$          **24.** $y = -9$

**25.** $x = y^2$        **26.** $y = x^2 - 3$

*The graph shows the sunset times for Seward, Alaska. Use this graph to answer Exercises 27 through 32. See Example 6.*

**27.** Approximate the time of sunset on June 1.

**28.** Approximate the time of sunset on November 1.

**29.** Approximate the date(s) when the sunset is at 3 p.m.

**30.** Approximate the date(s) when the sunset is at 9 p.m.

**31.** Is this graph the graph of a function? Why or why not?

**32.** Do you think a graph of sunset times for any location will always be a function? Why or why not?

*This graph shows the U.S. hourly minimum wage for each year shown. Use this graph to answer Exercises 33 through 38. See Example 6.*

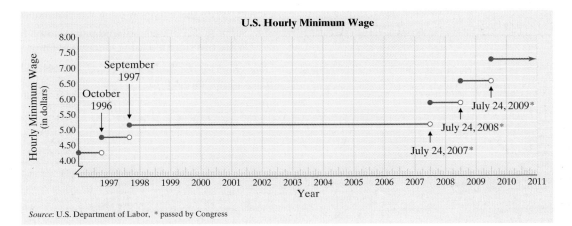

*Source*: U.S. Department of Labor,  * passed by Congress

**33.** Approximate the minimum wage before October, 1996.

**34.** Approximate the minimum wage in 2006.

**35.** Approximate the year when the minimum wage will increase to over $7.00 per hour.

**36.** According to the graph, what hourly wage was in effect for the greatest number of years?

**37.** Is this graph the graph of a function? Why or why not?

**38.** Do you think that a similar graph of your hourly wage on January 1 of every year (whether you are working or not) will be the graph of a function? Why or why not?

*Find $f(-2)$, $f(0)$, and $f(3)$ for each function. See Example 7.*

**39.** $f(x) = 2x - 5$

**40.** $f(x) = 3 - 7x$

**41.** $f(x) = x^2 + 2$

**42.** $f(x) = x^2 - 4$

**43.** $f(x) = 3x$

**44.** $f(x) = -3x$

**45.** $f(x) = |x|$

**46.** $f(x) = |2 - x|$

*Find $h(-1)$, $h(0)$, and $h(4)$ for each function. See Example 7.*

**47.** $h(x) = -5x$

**48.** $h(x) = -3x$

**49.** $h(x) = 2x^2 + 3$

**50.** $h(x) = 3x^2$

*For each given function value, write a corresponding ordered pair.*

**51.** $f(3) = 6$,

**52.** $f(7) = -2$,

**53.** $g(0) = -\dfrac{1}{2}$

**54.** $g(0) = -\dfrac{7}{8}$

**55.** $h(-2) = 9$

**56.** $h(-10) = 1$

*Find the domain of each function. See Example 8.*

**57.** $f(x) = 3x - 7$

**58.** $g(x) = 5 - 2x$

**59.** $h(x) = \dfrac{1}{x + 5}$

**60.** $f(x) = \dfrac{1}{x - 6}$

**61.** $g(x) = |x + 1|$

**62.** $h(x) = |2x|$

*Find the domain and the range of each relation graphed. See Example 9.*

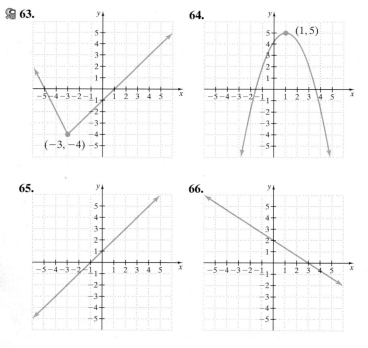

**63.**

**64.**

**65.**

**66.**

**67.**

**68.**

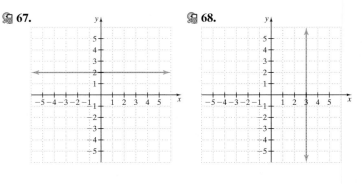

## REVIEW AND PREVIEW

*Find the coordinates of the point of intersection. See Section 3.1.*

**69.**

**70.**

**71.**

**72.**

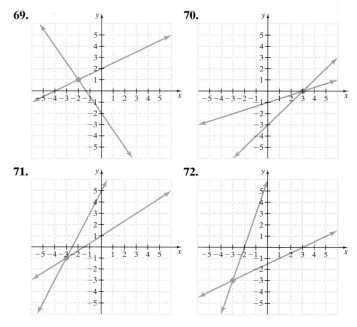

## CONCEPT EXTENSIONS

*Solve. See the Concept Check in this section.*

**73.** If a function $f$ is evaluated at $-5$, the value of the function is 12. Write this situation using function notation.

**74.** Suppose $(9, 20)$ is an ordered-pair solution for the function $g$. Write this situation using function notation.

*The graph of the function, f, is below. Use this graph to answer Exercises 75 through 78.*

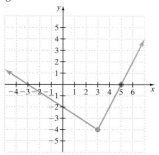

**75.** Write the coordinates of the lowest point of the graph.

**76.** Write the answer to Exercise 75 in function notation.

**77.** An $x$-intercept of this graph is $(5, 0)$. Write this using function notation.

**78.** Write the other $x$-intercept of this graph (see Exercise 77) using function notation.

 **79.** Forensic scientists use the function

$$H(x) = 2.59x + 47.24$$

to estimate the height of a woman in centimeters given the length $x$ of her femur bone.

    **a.** Estimate the height of a woman whose femur measures 46 centimeters.

    **b.** Estimate the height of a woman whose femur measures 39 centimeters.

**80.** The dosage in milligrams $D$ of Ivermectin, a heartworm preventive for a dog who weighs $x$ pounds, is given by the function

$$D(x) = \frac{136}{25}x$$

    **a.** Find the proper dosage for a dog that weighs 35 pounds.

    **b.** Find the proper dosage for a dog that weighs 70 pounds.

**81.** In your own words define **(a)** function; **(b)** domain; **(c)** range.

**82.** Explain the vertical line test and how it is used.

**83.** Since $y = x + 7$ is a function, rewrite the equation using function notation.

*See the example below for Exercises 84 through 87.*

Example

If $f(x) = x^2 + 2x + 1$, find $f(\pi)$.

*Solution:*

$$f(x) = x^2 + 2x + 1$$
$$f(\pi) = \pi^2 + 2\pi + 1$$

*Given the following functions, find the indicated values.*

**84.** $f(x) = 2x + 7$

    **a.** $f(2)$            **b.** $f(a)$

**85.** $g(x) = -3x + 12$

    **a.** $g(s)$            **b.** $g(r)$

**86.** $h(x) = x^2 + 7$

    **a.** $h(3)$            **b.** $h(a)$

**87.** $f(x) = x^2 - 12$

    **a.** $f(12)$           **b.** $f(a)$

# CHAPTER 3 GROUP ACTIVITY

## Financial Analysis

Investment analysts investigate a company's sales, net profit, debt, and assets to decide whether investing in it is a wise choice. One way to analyze this data is to graph it and look for trends over time. Another way is to find algebraically the rate at which the data changes over time.

    The following table gives the net incomes in millions of dollars for some of the leading U.S. businesses in the pharmaceutical industry for the years 2004 and 2005. In this project, you will analyze the performances of these companies and, based on this information alone, make an investment recommendation. This project may be completed by working in groups or individually.

**Pharmaceutical Industry Net Income (In Millions of Dollars)**

| Company | 2004 | 2005 |
|---|---|---|
| Merck | $5813.4 | $4631.3 |
| Pfizer | $11,361 | $8085 |
| Johnson & Johnson | $8509 | $10,411 |
| Bristol-Myers Squibb | $2378 | $2992 |
| Abbot Laboratories | $3175.8 | $3372.1 |
| Eli Lilly | $1819.1 | $1979.6 |
| Schering-Plough | $269 | -$947 |
| Wyeth | $1234 | $3656.3 |

*Source:* The 2006 annual report for each of the companies listed.

**1.** Scan the table. Did any of the companies have a loss during the years shown? If so, which company and when? What does this mean?

**2.** Write the data for each company as two ordered pairs of the form (year, net income). Assuming that the trends in net income are linear, use graph paper to graph the line represented by the ordered pairs for each company. Describe the trends shown by each graph.

**3.** Find the slope of the line for each company.

**4.** Which of the lines, if any, have positive slopes? What does that mean in this context? Which of the lines have negative slopes? What does that mean in this context?

**5.** Of these pharmaceutical companies, which one(s) would you recommend as an investment choice? Why?

**6.** Do you think it is wise to make a decision after looking at only two years of net profits? What other factors do you think should be taken into consideration when making an investment choice?

**(Optional)** *Use financial magazines, company annual reports, or online investing information to find net income information for two different years for two to four companies in the same industry. Analyze the net income and make an investment recommendation.*

## CHAPTER 3 VOCABULARY CHECK

*Fill in each blank with one of the words listed below.*

| relation | function | domain | range | standard | slope-intercept |
|---|---|---|---|---|---|
| y-axis | x-axis | solution | linear | slope | point-slope |
| x-intercept | y-intercept | y | x | | |

**1.** An ordered pair is a _____ of an equation in two variables if replacing the variables by the coordinates of the ordered pair results in a true statement.

**2.** The vertical number line in the rectangular coordinate system is called the _____ .

**3.** A _____ equation can be written in the form $Ax + By = C$.

**4.** A(n) _____ is a point of the graph where the graph crosses the x-axis.

**5.** The form $Ax + By = C$ is called _____ form.

**6.** A(n) _____ is a point of the graph where the graph crosses the y-axis.

**7.** The equation $y = 7x - 5$ is written in _____ form.

**8.** The equation $y + 1 = 7(x - 2)$ is written in _____ form.

**9.** To find an x-intercept of a graph, let _____ = 0.

**10.** The horizontal number line in the rectangular coordinate system is called the _____ .

**11.** To find a y-intercept of a graph, let _____ = 0.

**12.** The _____ of a line measures the steepness or tilt of a line.

**13.** A set of ordered pairs that assigns to each x-value exactly one y-value is called a _____ .

**14.** The set of all x-coordinates of a relation is called the _____ of the relation.

**15.** The set of all y-coordinates of a relation is called the _____ of the relation.

**16.** A set of ordered pairs is called a _____ .

> ▶ **Helpful Hint**
>
> Are you preparing for your test? Don't forget to take the Chapter 3 Test on page 242. Then check your answers at the back of the text and use the Chapter Test Prep Video CD to see the fully worked-out solutions to any of the exercises you want to review.

## CHAPTER 3 HIGHLIGHTS

| DEFINITIONS AND CONCEPTS | EXAMPLES |
|---|---|

### SECTION 3.1   READING GRAPHS AND THE RECTANGULAR COORDINATE SYSTEM

The **rectangular coordinate system** consists of a plane and a vertical and a horizontal number line intersecting at their 0 coordinates. The vertical number line is called the **y-axis** and the horizontal number line is called the **x-axis.** The point of intersection of the axes is called the **origin.**

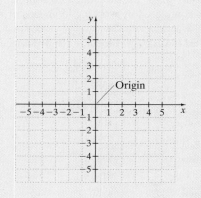

| DEFINITIONS AND CONCEPTS | EXAMPLES |
|---|---|

To **plot** or **graph** an ordered pair means to find its corresponding point on a rectangular coordinate system.

To plot or graph an ordered pair such as $(3, -2)$, start at the origin. Move 3 units to the right and from there, 2 units down.

To plot or graph $(-3, 4)$ start at the origin. Move 3 units to the left and from there, 4 units up.

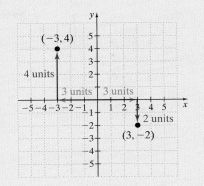

An ordered pair is a **solution** of an equation in two variables if replacing the variables by the coordinates of the ordered pair results in a true statement.

Determine whether $(-1, 5)$ is a solution of $2x + 3y = 13$.

$$2x + 3y = 13$$
$$2(-1) + 3 \cdot 5 = 13 \quad \text{Let } x = -1, y = 5$$
$$-2 + 15 = 13$$
$$13 = 13 \quad \text{True}$$

If one coordinate of an ordered pair solution is known, the other value can be determined by substitution.

Complete the ordered pair solution $(0, \quad)$ for the equation $x - 6y = 12$.

$$x - 6y = 12$$
$$0 - 6y = 12 \quad \text{Let } x = 0.$$
$$\frac{-6y}{-6} = \frac{12}{-6} \quad \text{Divide by } -6.$$
$$y = -2$$

The ordered pair solution is $(0, -2)$.

A **linear equation in two variables** is an equation that can be written in the form $Ax + By = C$ where $A$ and $B$ are not both 0. The form $Ax + By = C$ is called **standard form.**

*Linear Equations*

$$3x + 2y = -6 \qquad x = -5$$
$$y = 3 \qquad y = -x + 10$$

$x + y = 10$ is in standard form.

To graph a linear equation in two variables, find three ordered pair solutions. Plot the solution points and draw the line connecting the points.

Graph $x - 2y = 5$.

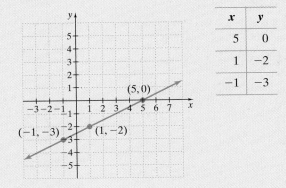

| x | y |
|---|---|
| 5 | 0 |
| 1 | -2 |
| -1 | -3 |

| DEFINITIONS AND CONCEPTS | EXAMPLES |
|---|---|

An **intercept** of a graph is a point where the graph intersects an axis. If a graph intersects the $x$-axis at $a$, then $(a, 0)$ is the **$x$-intercept.** If a graph intersects the $y$-axis at $b$, then $(0, b)$ is the **$y$-intercept.**

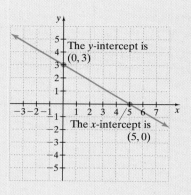

To find the **$x$-intercept,** let $y = 0$ and solve for $x$.

To find the **$y$-intercept,** let $x = 0$ and solve for $y$.

Graph $2x - 5y = -10$ by finding intercepts.

$$\text{If } y = 0, \text{ then} \qquad\qquad \text{If } x = 0, \text{ then}$$
$$2x - 5 \cdot 0 = -10 \qquad\qquad 2 \cdot 0 - 5y = -10$$
$$2x = -10 \qquad\qquad -5y = -10$$
$$\frac{2x}{2} = \frac{-10}{2} \qquad\qquad \frac{-5y}{-5} = \frac{-10}{-5}$$
$$x = -5 \qquad\qquad y = 2$$

The $x$-intercept is $(-5, 0)$. The $y$-intercept is $(0, 2)$.

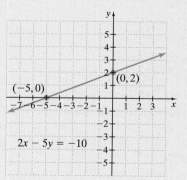

The graph of $x = c$ is a vertical line with $x$-intercept $(c, 0)$.

The graph of $y = c$ is a horizontal line with $y$-intercept $(0, c)$.

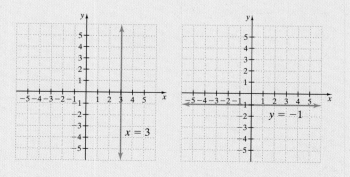

| **DEFINITIONS AND CONCEPTS** | **EXAMPLES** |

The **slope $m$** of the line through points $(x_1, y_1)$ and $(x_2, y_2)$ is given by

$$m = \frac{y_2 - y_1}{x_2 - x_1} \quad \text{as long as } x_2 \neq x_1$$

The slope of the line through points $(-1, 6)$ and $(-5, 8)$ is

$$m = \frac{y_2 - y_1}{x_2 - x_1} = \frac{8 - 6}{-5 - (-1)} = \frac{2}{-4} = -\frac{1}{2}$$

The slope of the line $y = -5$ is 0.

The line $x = 3$ has undefined slope.

A horizontal line has slope 0.

The slope of a vertical line is undefined.

Nonvertical parallel lines have the same slope.

Two nonvertical lines are perpendicular if the slope of one is the negative reciprocal of the slope of the other.

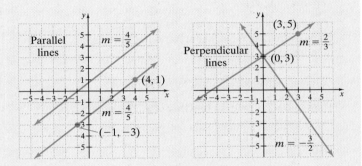

***Slope-Intercept Form***

$$y = mx + b$$

$m$ is the slope of the line.
$(0, b)$ is the $y$-intercept.

Find the slope and the $y$-intercept of the line whose equation is $2x + 3y = 6$.

Solve for $y$:

$$2x + 3y = 6$$
$$3y = -2x + 6 \quad \text{Subtract } 2x.$$
$$y = -\frac{2}{3}x + 2 \quad \text{Divide by 3.}$$

The slope of the line is $-\frac{2}{3}$ and the $y$-intercept is $(0, 2)$.

Find an equation of the line with slope 3 and $y$-intercept $(0, -1)$.

The equation is $y = 3x - 1$.

Find an equation of the line with slope $\frac{3}{4}$ that contains the point $(-1, 5)$.

***Point-Slope Form***

$$y - y_1 = m(x - x_1)$$

$m$ is the slope.
$(x_1, y_1)$ is a point on the line.

$$y - 5 = \frac{3}{4}[x - (-1)]$$

$$4(y - 5) = 3(x + 1) \quad \text{Multiply by 4.}$$
$$4y - 20 = 3x + 3 \quad \text{Distribute.}$$
$$-3x + 4y = 23 \quad \text{Subtract } 3x \text{ and add 20.}$$

| DEFINITIONS AND CONCEPTS | EXAMPLES |
|---|---|

A set of ordered pairs is a **relation.** The set of all *x*-coordinates is called the **domain** of the relation and the set of all *y*-coordinates is called the **range** of the relation.

The domain of the relation $\{(0, 5), (2, 5), (4, 5), (5, -2)\}$ is $\{0, 2, 4, 5\}$. The range is $\{-2, 5\}$.

A **function** is a set of ordered pairs that assigns to each *x*-value exactly one *y*-value.

Which are graphs of functions?

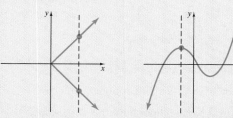

*Vertical Line Test*

If a vertical line can be drawn so that it intersects a graph more than once, the graph is not the graph of a function.

This graph is not the graph of a function.

This graph is the graph of a function.

The symbol $f(x)$ means **function of x.** This notation is called **function** notation.

If $f(x) = 2x^2 + 6x - 1$, find $f(3)$.

$$f(3) = 2(3)^2 + 6 \cdot 3 - 1$$
$$= 2 \cdot 9 + 18 - 1$$
$$= 18 + 18 - 1$$
$$= 35$$

# CHAPTER 3  REVIEW

*(3.1) Plot the following ordered pairs on a Cartesian coordinate system.*

1. $(-7, 0)$

2. $\left(0, 4\frac{4}{5}\right)$

3. $(-2, -5)$

4. $(1, -3)$

5. $(0.7, 0.7)$

6. $(-6, 4)$

7. A local lumberyard uses quantity pricing. The table shows the price per board for different amounts of lumber purchased.

| Price per Board (in dollars) | Number of Boards Purchased |
|---|---|
| 8.00 | 1 |
| 7.50 | 10 |
| 6.50 | 25 |
| 5.00 | 50 |
| 2.00 | 100 |

a. Write each paired data as an ordered pair of the form (price per board, number of boards purchased).

b. Create a scatter diagram of the paired data. Be sure to label the axes appropriately.

8. The table shows the annual overnight stays in national parks (*Source*: National Park Service)

| Year | Overnight Stays in National Parks (in millions) |
|---|---|
| 2001 | 9.8 |
| 2002 | 15.1 |
| 2003 | 14.6 |
| 2004 | 14.0 |
| 2005 | 13.8 |
| 2006 | 13.6 |

a. Write each paired data as an ordered pair of the form (year, number of overnight stays).

b. Create a scatter diagram of the paired data. Be sure to label the axes properly.

*Determine whether each ordered pair is a solution of the given equation.*

9. $7x - 8y = 56; (0, 56), (8, 0)$

10. $-2x + 5y = 10; (-5, 0), (1, 1)$

11. $x = 13; (13, 5), (13, 13)$

12. $y = 2; (7, 2), (2, 7)$

*Complete the ordered pairs so that each is a solution of the given equation.*

**13.** $-2 + y = 6x;\ (7,\ \ )$    **14.** $y = 3x + 5;\ \left(\ \ ,\, -8\right)$

*Complete the table of values for each given equation; then plot the ordered pairs. Use a single coordinate system for each exercise.*

**15.** $9 = -3x + 4y$

| x | y |
|---|---|
|  | 0 |
|  | 3 |
| 9 |  |

**16.** $y = 5$

| x | y |
|---|---|
| 7 |  |
| -7 |  |
| 0 |  |

**17.** $x = 2y$

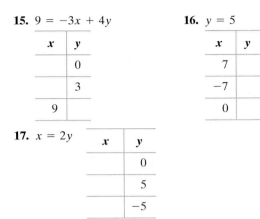

| x | y |
|---|---|
|  | 0 |
|  | 5 |
|  | -5 |

**18.** The cost in dollars of producing $x$ compact disk holders is given by $y = 5x + 2000$.

   **a.** Complete the following table.

| x | y |
|---|---|
| 1 |  |
| 100 |  |
| 1000 |  |

   **b.** Find the number of compact disk holders that can be produced for $6430.

*(3.2) Graph each linear equation.*

**19.** $x - y = 1$    **20.** $x + y = 6$

**21.** $x - 3y = 12$    **22.** $5x - y = -8$

**23.** $x = 3y$    **24.** $y = -2x$

**25.** $2x - 3y = 6$    **26.** $4x - 3y = 12$

**27.** The projected U.S. long-distance revenue (in billions of dollars) from 1999 to 2004 is given by the equation, $y = 3x + 111$ where $x$ is the number of years after 1999. Graph this equation and use it to estimate the amount of long-distance revenue in 2007. (*Source*: Giga Information Group)

*(3.3) Identify the intercepts.*

**28.**

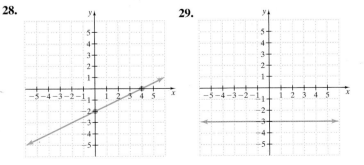

**29.**

**30.**

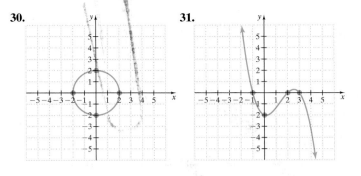

**31.**

*Graph each linear equation by finding its intercepts.*

**32.** $x - 3y = 12$    **33.** $-4x + y = 8$

**34.** $y = -3$    **35.** $x = 5$

**36.** $y = -3x$    **37.** $x = 5y$

**38.** $x - 2 = 0$    **39.** $y + 6 = 0$

*(3.4) Find the slope of each line.*

**40.**

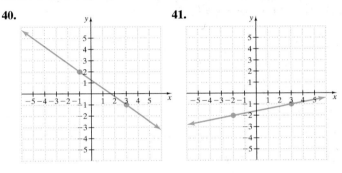

**41.**

*In Exercises 42 through 45, match each line with its slope.*

**a.**    **b.**

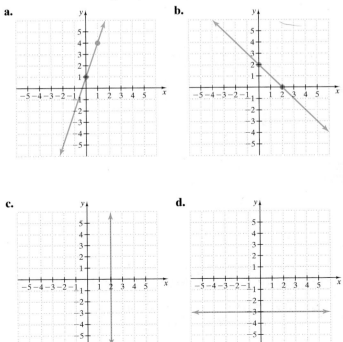

**c.**    **d.**

**e.**

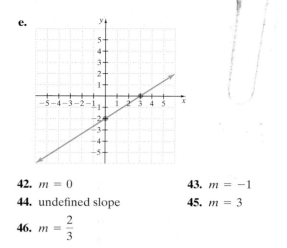

**42.** $m = 0$

**43.** $m = -1$

**44.** undefined slope

**45.** $m = 3$

**46.** $m = \dfrac{2}{3}$

*Find the slope of the line that goes through the given points.*

**47.** $(2, 5)$ and $(6, 8)$

**48.** $(4, 7)$ and $(1, 2)$

**49.** $(1, 3)$ and $(-2, -9)$

**50.** $(-4, 1)$ and $(3, -6)$

*Find the slope of each line.*

**51.** $y = 3x + 7$

**52.** $x - 2y = 4$

**53.** $y = -2$

**54.** $x = 0$

△ *Determine whether each pair of lines is parallel, perpendicular, or neither.*

**55.** $x - y = -6$
 $x + y = 3$

**56.** $3x + y = 7$
 $-3x - y = 10$

**57.** $y = 4x + \dfrac{1}{2}$
 $4x + 2y = 1$

**58.** $x = 4$
 $y = -2$

*Find the slope of each line and write the slope as a rate of change. Don't forget to attach the proper units.*

**59.** The graph below shows the average monthly day care cost for a 3-year-old attending 8 hours a day, 5 days a week.

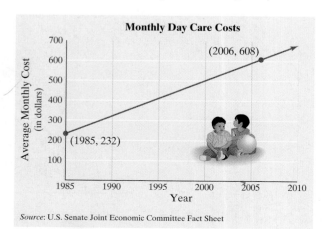

**Monthly Day Care Costs**

(2006, 608)

(1985, 232)

*Source*: U.S. Senate Joint Economic Committee Fact Sheet

**60.** The graph below shows the U.S. government's projected spending (in billions of dollars) on technology. (Some years projected.)

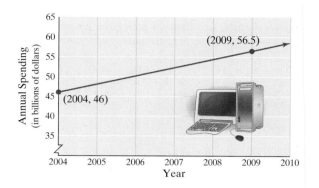

(2009, 56.5)

(2004, 46)

**(3.5)** *Determine the slope and the y-intercept of the graph of each equation.*

**61.** $3x + y = 7$

**62.** $x - 6y = -1$

**63.** $y = 2$

**64.** $x = -5$

*Write an equation of each line in slope-intercept form.*

**65.** slope $-5$; y-intercept $\dfrac{1}{2}$

**66.** slope $\dfrac{2}{3}$; y-intercept $6$

*Use the slope-intercept form to graph each equation.*

**67.** $y = 3x - 1$

**68.** $y = -3x$

**69.** $5x - 3y = 15$

**70.** $-x + 2y = 8$

*Match each equation with its graph.*

**71.** $y = -4x$

**72.** $y = -2x + 1$

**73.** $y = 2x - 1$

**74.** $y = 2x$

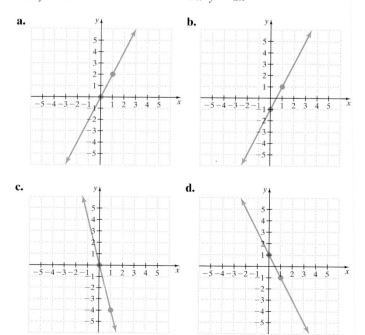

**a.**

**b.**

**c.**

**d.**

Write an equation of each line in standard form.

**75.** With slope $-3$, through $(0, -5)$

**76.** With slope $\dfrac{1}{2}$, through $\left(0, -\dfrac{7}{2}\right)$

**77.** With slope $0$, through $(-2, -3)$

**78.** With $0$ slope, through the origin

**79.** With slope $-6$, through $(2, -1)$

**80.** With slope $12$, through $\left(\dfrac{1}{2}, 5\right)$

**81.** Through $(0, 6)$ and $(6, 0)$

**82.** Through $(0, -4)$ and $(-8, 0)$

**83.** Vertical line, through $(5, 7)$

**84.** Horizontal line, through $(-6, 8)$

**85.** Through $(6, 0)$, perpendicular to $y = 8$

**86.** Through $(10, 12)$, perpendicular to $x = -2$

**(3.6)** *Determine which of the following are functions*

**87.** $\{(7, 1), (7, 5), (2, 6)\}$

**88.** $\{(0, -1), (5, -1), (2, 2)\}$

**89.** $7x - 6y = 1$

**90.** $y = 7$

**91.** $x = 2$

**92.** $y = x^3$

**93.**

**94.**

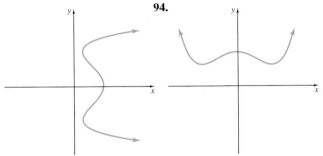

Given the following functions, find the indicated function values.

**95.** Given $f(x) = -2x + 6$, find

   **a.** $f(0)$      **b.** $f(-2)$      **c.** $f\left(\dfrac{1}{2}\right)$

**96.** Given $h(x) = -5 - 3x$, find

   **a.** $h(2)$      **b.** $h(-3)$      **c.** $h(0)$

**97.** Given $g(x) = x^2 + 12x$, find

   **a.** $g(3)$      **b.** $g(-5)$      **c.** $g(0)$

**98.** Given $h(x) = 6 - |x|$, find

   **a.** $h(-1)$      **b.** $h(1)$      **c.** $h(-4)$

Find the domain of each function.

**99.** $f(x) = 2x + 7$      **100.** $g(x) = \dfrac{7}{x - 2}$

Find the domain and the range of each function graphed.

**101.**

**102.**

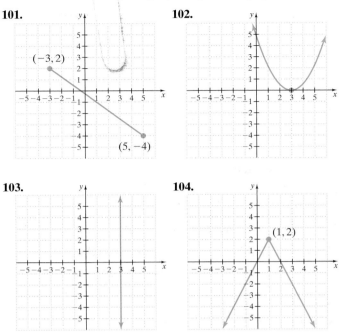

**103.**

**104.**

## MIXED REVIEW

Complete the table of values for each given equation.

**105.** $2x - 5y = 9$

**106.** $x = -3y$

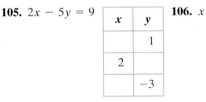

| $x$ | $y$ |
|-----|-----|
|     | 1   |
| 2   |     |
|     | $-3$ |

| $x$ | $y$ |
|-----|-----|
| 0   |     |
|     | 1   |
| 6   |     |

Find the intercepts for each equation.

**107.** $2x - 3y = 6$      **108.** $-5x + y = 10$

Graph each linear equation.

**109.** $x - 5y = 10$      **110.** $x + y = 4$

**111.** $y = -4x$      **112.** $2x + 3y = -6$

**113.** $x = 3$      **114.** $y = -2$

Find the slope of the line that passes through each pair of points.

**115.** $(3, -5)$ and $(-4, 2)$      **116.** $(1, 3)$ and $(-6, -8)$

Find the slope of each line.

**117.**

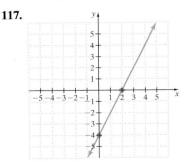

**118.**

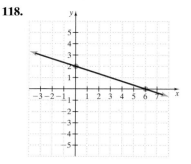

Determine the slope and y-intercept of the graph of each equation.

**119.** $-2x + 3y = -15$

**120.** $6x + y - 2 = 0$

Write an equation of the line with the given slope that passes through the given point. Write the equation in the form $Ax + By = C$.

**121.** $m = -5; (3, -7)$

**122.** $m = 3; (0, 6)$

Write an equation of the line passing through each pair of points. Write the equation in the form $Ax + By = C$.

**123.** $(-3, 9)$ and $(-2, 5)$

**124.** $(3, 1)$ and $(5, -9)$

*Use the line graph to answer Exercises 125 through 128.*

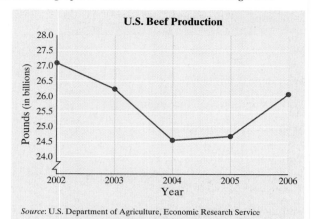

*Source*: U.S. Department of Agriculture, Economic Research Service

**125.** Which year shows the greatest production of beef? Estimate production for that year.

**126.** Which year shows the least production of beef? Estimate production for that year.

**127.** Which years had beef production greater then 25 billion pounds?

**128.** Which year shows the greatest increase in beef production?

# CHAPTER 3 TEST TEST PREP VIDEO

Remember to use the Chapter Test Prep Video CD to see the fully worked-out solutions to any of the exercises you want to review.

*Graph the following.*

**1.** $y = \frac{1}{2}x$

**2.** $2x + y = 8$

**3.** $5x - 7y = 10$

**4.** $y = -1$

**5.** $x - 3 = 0$

*Find the slopes of the following lines.*

**6.** **7.**

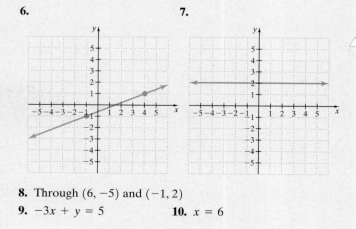

**8.** Through $(6, -5)$ and $(-1, 2)$

**9.** $-3x + y = 5$ **10.** $x = 6$

**11.** Determine the slope and the y-intercept of the graph of $7x - 3y = 2$.

△ **12.** Determine whether the graphs of $y = 2x - 6$ and $-4x = 2y$ are parallel lines, perpendicular lines, or neither.

*Find equations of the following lines. Write the equation in standard form.*

**13.** With slope of $-\frac{1}{4}$, through $(2, 2)$

**14.** Through the origin and $(6, -7)$

**15.** Through $(2, -5)$ and $(1, 3)$

△ **16.** Through $(-5, -1)$ and parallel to $x = 7$

**17.** With slope $\frac{1}{8}$ and y-intercept $(0, 12)$

*Which of the following are functions?*

**18.** **19.**

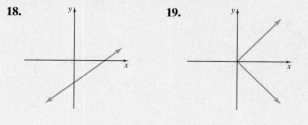

*Given the following functions, find the indicated function values.*

**20.** $h(x) = x^3 - x$

    **a.** $h(-1)$      **b.** $h(0)$      **c.** $h(4)$

**21.** Find the domain of $y = \dfrac{1}{x+1}$.

*Find the domain and the range of each function graphed.*

**22.**                            **23.**

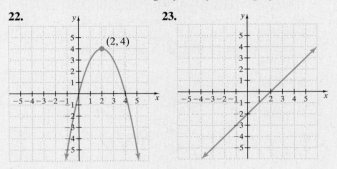

**24.** If $f(7) = 20$, write the corresponding ordered pair.

*Use the bar graph below to answer Exercises 25 and 26.*

**Average Water Use Per Person Per Day for Selected Countries**

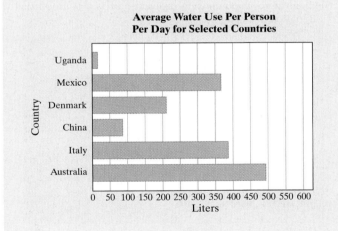

**25.** Estimate the average water use per person per day in Denmark.

**26.** Estimate the average water use per person per day in Australia.

*Use this graph to answer Exercises 27 through 29.*

**Average Monthly High Temperature: Portland, Oregon**

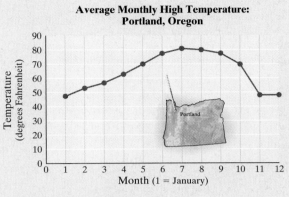

Source: The Weather Channel Enterprises, Inc.

**27.** During what month is the average high temperature the greatest?

**28.** Approximate the average high temperature for the month of April.

**29.** During what month(s) is the average high temperature below 60°F?

# CHAPTER 3 CUMULATIVE REVIEW

**1.** Insert $<$, $>$, or $=$ in the space between each pair of numbers to make each statement true.

    **a.** 2    3      **b.** 7    4      **c.** 72    27

**2.** Write the fraction $\dfrac{56}{64}$ in lowest terms.

**3.** Multiply $\dfrac{2}{15}$ and $\dfrac{5}{13}$. Write the product in lowest terms

**4.** Add: $\dfrac{10}{3} + \dfrac{5}{21}$

**5.** Simplify: $\dfrac{3 + |4 - 3| + 2^2}{6 - 3}$

**6.** Simplify: $16 - 3 \cdot 3 + 2^4$

**7.** Add.

    **a.** $-8 + (-11)$          **b.** $-5 + 35$

    **c.** $0.6 + (-1.1)$         **d.** $-\dfrac{7}{10} + \left(-\dfrac{1}{10}\right)$

    **e.** $11.4 + (-4.7)$        **f.** $-\dfrac{3}{8} + \dfrac{2}{5}$

**8.** Simplify: $|9 + (-20)| + |-10|$

**9.** Simplify each expression.

    **a.** $-14 - 8 + 10 - (-6)$

    **b.** $1.6 - (-10.3) + (-5.6)$

**10.** Simplify: $-9 - (3 - 8)$

**11.** If $x = -2$ and $y = -4$, evaluate each expression.

    **a.** $5x - y$             **b.** $x^4 - y^2$

    **c.** $\dfrac{3x}{2y}$

**12.** Is $-20$ a solution of $\dfrac{x}{-10} = 2$?

**13.** Simplify each expression.

    **a.** $10 + (x + 12)$       **b.** $-3(7x)$

**14.** Simplify: $(12 + x) - (4x - 7)$

**15.** Identify the numerical coefficient in each term.

    **a.** $-3y$             **b.** $22z^4$

    **c.** $y$              **d.** $-x$

    **e.** $\dfrac{x}{7}$

**16.** Multiply: $-5(x - 7)$

**17.** Solve $x - 7 = 10$ for $x$.

**18.** Solve: $5(3 + z) - (8z + 9) = -4$

**19.** Solve: $12a - 8a = 10 + 2a - 13 - 7$

**20.** Solve: $\dfrac{x}{4} - 1 = -7$

**21.** If $x$ is the first of three consecutive integers, express the sum of the three integers in terms of $x$. Simplify if possible.

**22.** Solve: $\dfrac{x}{3} - 2 = \dfrac{x}{3}$

**23.** Solve: $\dfrac{2(a + 3)}{3} = 6a + 2$

**24.** Solve: $x + 2y = 6$ for $y$.

**25.** In a recent year, the U.S. House of Representatives had a total of 435 Democrats and Republicans. There were 31 more Democratic representatives than Republican representatives. Find the number of representatives from each party. (*Source:* Office of the Clerk of the U.S. House of Representatives)

**26.** Solve $5(x + 4) \geq 4(2x + 3)$. Write the solution set in interval notation.

**27.** Charles Pecot can afford enough fencing to enclose a rectangular garden with a perimeter of 140 feet. If the width of his garden is to be 30 feet, find the length.

**28.** Solve $-3 < 4x - 1 \leq 2$. Write the solution set in interval notation.

**29.** Solve $y = mx + b$ for $x$.

**30.** Complete the table for $y = -5x$.

| $x$ | $y$ |
|-----|-----|
| 0 | |
| $-1$ | |
| | $-10$ |

**31.** A chemist working on his doctoral degree at Massachusetts Institute of Technology needs 12 liters of a 50% acid solution for a lab experiment. The stockroom has only 40% and 70% solutions. How much of each solution should be mixed together to form 12 liters of a 50% solution?

**32.** Graph: $y = -3x + 5$

**33.** Graph $x \geq -1$.

**34.** Find the $x$- and $y$-intercepts of $2x + 4y = -8$.

**35.** Solve $-1 \leq 2x - 3 < 5$. Graph the solution set and write it in interval notation.

**36.** Graph $x = 2$ on a rectangular coordinate system.

**37.** Determine whether each ordered pair is a solution of the equation $x - 2y = 6$.

    **a.** $(6, 0)$           **b.** $(0, 3)$

    **c.** $\left(1, -\dfrac{5}{2}\right)$

**38.** Find the slope of the line through $(0, 5)$ and $(-5, 4)$.

**39.** Determine whether each equation is a linear equation in two variables.

    **a.** $x - 1.5y = -1.6$       **b.** $y = -2x$

    **c.** $x + y^2 = 9$           **d.** $x = 5$

**40.** Find the slope of $x = -10$.

**41.** Find the slope of the line $y = -1$.

**42.** Find the slope and $y$-intercept of the line whose equation is $2x - 5y = 10$.

**43.** Find an equation of the line with $y$-intercept $(0, -3)$ and slope of $\dfrac{1}{4}$.

**44.** Write an equation of the line through $(2, 3)$ and $(0, 0)$. Write the equation in standard form.

# 4

# Solving Systems of Linear Equations

Many of the occupations predicted to have the largest percent increase in number of jobs are in the fields of medicine and computer science. For example, from 2004 to 2014, the job growth predicted for pharmacy technicians is 28.6%, and for network systems and data communication analysts it is 54.5%. Although the demand for both jobs is growing, these jobs are growing at different rates. In Section 4.3, Exercise 73, we will predict when these occupations might have the same number of jobs.

In Chapter 3, we graphed equations containing two variables. Equations like these are often needed to represent relationships between two different values. There are also many real-life opportunities to compare and contrast two such equations, called a system of equations. This chapter presents linear systems and ways we solve these systems and apply them to real-life situations.

| Job Title | Job Description | Employment (in thousands) | |
| --- | --- | --- | --- |
| | | 2004 | 2014 |
| Pharmacy technician | Prepare medications under the direction of a pharmacist | 258 | 332 |
| Network systems and data communications analyst | Analyze, design, test, and evaluate network systems, Internet, intranet, and other data communication systems | 231 | 357 |

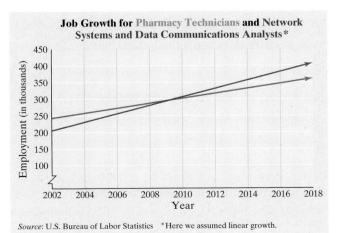

**Job Growth for Pharmacy Technicians and Network Systems and Data Communications Analysts\***

*Source*: U.S. Bureau of Labor Statistics   \*Here we assumed linear growth.

# 4.1 SOLVING SYSTEMS OF LINEAR EQUATIONS BY GRAPHING

**OBJECTIVES**

1 Determine if an ordered pair is a solution of a system of equations in two variables.

2 Solve a system of linear equations by graphing.

3 Without graphing, determine the number of solutions of a system.

**OBJECTIVE 1 ▶ Deciding whether an ordered pair is a solution.** A **system of linear equations** consists of two or more linear equations. In this section, we focus on solving systems of linear equations containing two equations in two variables. Examples of such linear systems are

$$\begin{cases} 3x - 3y = 0 \\ x = 2y \end{cases} \quad \begin{cases} x - y = 0 \\ 2x + y = 10 \end{cases} \quad \begin{cases} y = 7x - 1 \\ y = 4 \end{cases}$$

A **solution** of a system of two equations in two variables is an ordered pair of numbers that is a solution of both equations in the system.

**EXAMPLE 1** Determine whether $(12, 6)$ is a solution of the system

$$\begin{cases} 2x - 3y = 6 \\ x = 2y \end{cases}$$

*Solution* To determine whether $(12, 6)$ is a solution of the system, we replace $x$ with 12 and $y$ with 6 in both equations.

$$2x - 3y = 6 \quad \text{First equation} \qquad\qquad x = 2y \quad \text{Second equation}$$
$$2(12) - 3(6) \overset{?}{=} 6 \quad \text{Let } x = 12 \text{ and } y = 6. \qquad 12 \overset{?}{=} 2(6) \quad \text{Let } x = 12 \text{ and } y = 6.$$
$$24 - 18 \overset{?}{=} 6 \quad \text{Simplify.} \qquad\qquad 12 = 12 \quad \text{True}$$
$$6 = 6 \quad \text{True}$$

Since $(12, 6)$ is a solution of both equations, it is a solution of the system. □

**PRACTICE**
**1** Determine whether $(4, 12)$ is a solution of the system.

$$\begin{cases} 4x - y = 2 \\ y = 3x \end{cases}$$

**EXAMPLE 2** Determine whether $(-1, 2)$ is a solution of the system

$$\begin{cases} x + 2y = 3 \\ 4x - y = 6 \end{cases}$$

*Solution* We replace $x$ with $-1$ and $y$ with 2 in both equations.

$$x + 2y = 3 \quad \text{First equation} \qquad\qquad 4x - y = 6 \quad \text{Second equation}$$
$$-1 + 2(2) \overset{?}{=} 3 \quad \text{Let } x = -1 \text{ and } y = 2. \qquad 4(-1) - 2 \overset{?}{=} 6 \quad \text{Let } x = -1 \text{ and } y = 2.$$
$$-1 + 4 \overset{?}{=} 3 \quad \text{Simplify.} \qquad\qquad -4 - 2 \overset{?}{=} 6 \quad \text{Simplify.}$$
$$3 = 3 \quad \text{True} \qquad\qquad\qquad -6 = 6 \quad \text{False}$$

$(-1, 2)$ is not a solution of the second equation, $4x - y = 6$, so it is not a solution of the system. □

**PRACTICE**
**2** Determine whether $(-4, 1)$ is a solution of the system.

$$\begin{cases} x - 3y = -7 \\ 2x + 9y = 1 \end{cases}$$

**OBJECTIVE 2 ▶ Solving systems of equations by graphing.** Since a solution of a system of two equations in two variables is a solution common to both equations, it is also a point common to the graphs of both equations. Let's practice finding solutions of both equations in a system—that is, solutions of a system—by graphing and identifying points of intersection.

**EXAMPLE 3**   Solve the system of equations by graphing.

$$\begin{cases} -x + 3y = 10 \\ \phantom{-}x + \phantom{3}y = 2 \end{cases}$$

_Solution_   On a single set of axes, graph each linear equation.

$-x + 3y = 10$

| x | y |
|---|---|
| 0 | $\frac{10}{3}$ |
| −4 | 2 |
| 2 | 4 |

$x + y = 2$

| x | y |
|---|---|
| 0 | 2 |
| 2 | 0 |
| 1 | 1 |

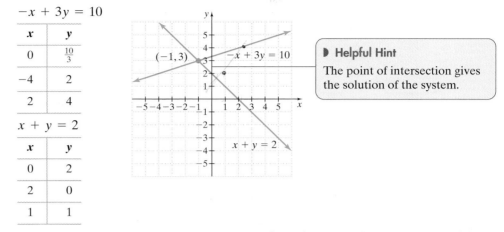

▶ **Helpful Hint**

The point of intersection gives the solution of the system.

The two lines appear to intersect at the point $(-1, 3)$. To check, we replace $x$ with $-1$ and $y$ with 3 in both equations.

| | |
|---|---|
| $-x + 3y = 10$   First equation | $x + y = 2$   Second equation |
| $-(-1) + 3(3) \stackrel{?}{=} 10$   Let $x = -1$ and $y = 3$. | $-1 + 3 \stackrel{?}{=} 2$   Let $x = -1$ and $y = 3$. |
| $1 + 9 \stackrel{?}{=} 10$   Simplify. | $2 = 2$   True |
| $10 = 10$   True | |

$(-1, 3)$ checks, so it is the solution of the system.   □

**PRACTICE**
**3**   Solve the system of equations by graphing:

$$\begin{cases} x - \phantom{2}y = 3 \\ x + 2y = 18 \end{cases}$$

---

▶ **Helpful Hint**

Neatly drawn graphs can help when you are estimating the solution of a system of linear equations by graphing.

In the example above, notice that the two lines intersected in a point. This means that the system has 1 solution.

**EXAMPLE 4**   Solve the system of equations by graphing.

$$\begin{cases} 2x + 3y = -2 \\ x = 2 \end{cases}$$

_Solution_   We graph each linear equation on a single set of axes.

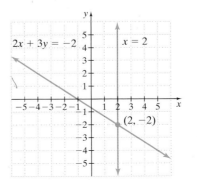

The two lines appear to intersect at the point $(2, -2)$. To determine whether $(2, -2)$ is the solution, we replace $x$ with 2 and $y$ with $-2$ in both equations.

$$2x + 3y = -2 \quad \text{First equation} \qquad\qquad x = 2 \quad \text{Second equation}$$
$$2(2) + 3(-2) \stackrel{?}{=} -2 \quad \text{Let } x = 2 \text{ and } y = -2. \qquad 2 \stackrel{?}{=} 2 \quad \text{Let } x = 2.$$
$$4 + (-6) \stackrel{?}{=} -2 \quad \text{Simplify.} \qquad\qquad\qquad 2 = 2 \quad \text{True}$$
$$-2 = -2 \quad \text{True}$$

Since a true statement results in both equations, $(2, -2)$ is the solution of the system. □

**PRACTICE**
**4** Solve the system of equations by graphing.

$$\begin{cases} -4x + 3y = -3 \\ y = -5 \end{cases}$$

A system of equations that has at least one solution as in Examples 3 and 4 is said to be a **consistent system.** A system that has no solution is said to be an **inconsistent system.**

**EXAMPLE 5** Solve the following system of equations by graphing.

$$\begin{cases} 2x + y = 7 \\ 2y = -4x \end{cases}$$

*Solution* Graph the two lines in the system.

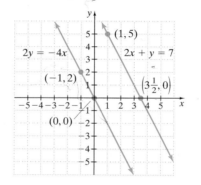

The lines **appear** to be parallel. To confirm this, write both equations in slope-intercept form by solving each equation for $y$.

$2x + y = 7$ First equation $\qquad\qquad\qquad 2y = -4x$ Second equation

$\quad y = -2x + 7$ Subtract $2x$ from both sides. $\qquad \dfrac{2y}{2} = \dfrac{-4x}{2}$ Divide both sides by 2.

$$y = -2x$$

Recall that when an equation is written in slope-intercept form, the coefficient of $x$ is the slope. Since both equations have the same slope, $-2$, but different $y$-intercepts, the lines are parallel and have no points in common. Thus, there is no solution of the system and the system is inconsistent. □

**PRACTICE**
**5** Solve the system of equations by graphing.

$$\begin{cases} 3y = 9x \\ 6x - 2y = 12 \end{cases}$$

In Examples 3, 4, and 5, the graphs of the two linear equations of each system are different. When this happens, we call these equations **independent equations.** If the graphs of the two equations in a system are identical, we call the equations **dependent equations.**

**EXAMPLE 6**   Solve the system of equations by graphing.

$$\begin{cases} x - y = 3 \\ -x + y = -3 \end{cases}$$

*Solution*   Graph each line.

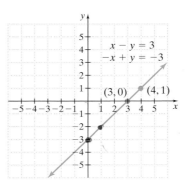

These graphs **appear** to be identical. To confirm this, write each equation in slope-intercept form.

| | |
|---|---|
| $x - y = 3$    First equation | $-x + y = -3$    Second equation |
| $-y = -x + 3$    Subtract $x$ from both sides. | $y = x - 3$    Add $x$ to both sides. |
| $\dfrac{-y}{-1} = \dfrac{-x}{-1} + \dfrac{3}{-1}$    Divide both sides by $-1$. | |
| $y = x - 3$ | |

The equations are identical and so must be their graphs. The lines have an infinite number of points in common. Thus, there is an infinite number of solutions of the system and this is a consistent system. The equations are dependent equations.   □

**PRACTICE**
**6**   Solve the system of equations by graphing.

$$\begin{cases} x - y = 4 \\ -2x + 2y = -8 \end{cases}$$

As we have seen, three different situations can occur when graphing the two lines associated with the equations in a linear system:

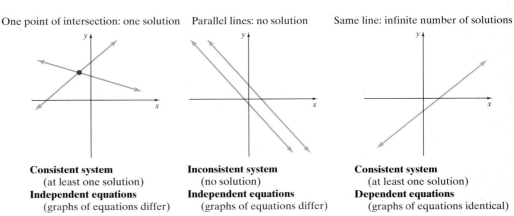

| One point of intersection: one solution | Parallel lines: no solution | Same line: infinite number of solutions |
|---|---|---|
| **Consistent system** (at least one solution) **Independent equations** (graphs of equations differ) | **Inconsistent system** (no solution) **Independent equations** (graphs of equations differ) | **Consistent system** (at least one solution) **Dependent equations** (graphs of equations identical) |

**OBJECTIVE 3 ▶ Finding the number of solutions of a system without graphing.** You may have suspected by now that graphing alone is not an accurate way to solve a system of linear equations. For example, a solution of $\left(\frac{1}{2}, \frac{2}{9}\right)$ is unlikely to be read correctly from a graph. The next two sections present two accurate methods of solving these systems. In the meantime, we can decide how many solutions a system has by writing each equation in the slope-intercept form.

**EXAMPLE 7** Without graphing, determine the number of solutions of the system.

$$\begin{cases} \frac{1}{2}x - y = 2 \\ x = 2y + 5 \end{cases}$$

**Solution** First write each equation in slope-intercept form.

$\frac{1}{2}x - y = 2$    First equation      $x = 2y + 5$    Second equation

$\frac{1}{2}x = y + 2$    Add $y$ to both sides.    $x - 5 = 2y$    Subtract 5 from both sides.

$\frac{1}{2}x - 2 = y$    Subtract 2 from both sides.    $\frac{x}{2} - \frac{5}{2} = \frac{2y}{2}$    Divide both sides by 2.

         $\frac{1}{2}x - \frac{5}{2} = y$    Simplify.

The slope of each line is $\frac{1}{2}$, but they have different $y$-intercepts. This tells us that the lines representing these equations are parallel. Since the lines are parallel, the system has no solution and is inconsistent. ☐

**PRACTICE 7** Without graphing, determine the number of solutions of the system.

$$\begin{cases} 5x + 4y = 6 \\ x - y = 3 \end{cases}$$

**EXAMPLE 8** Without graphing, determine the number of solutions of the system.

$$\begin{cases} 3x - y = 4 \\ x + 2y = 8 \end{cases}$$

**Solution** Once again, the slope-intercept form helps determine how many solutions this system has.

$3x - y = 4$    First equation    $x + 2y = 8$    Second equation

$3x = y + 4$    Add $y$ to both sides.    $x = -2y + 8$    Subtract $2y$ from both sides.

$3x - 4 = y$    Subtract 4 from both sides.    $x - 8 = -2y$    Subtract 8 from both sides.

     $\frac{x}{-2} - \frac{8}{-2} = \frac{-2y}{-2}$    Divide both sides by $-2$.

     $-\frac{1}{2}x + 4 = y$    Simplify.

The slope of the second line is $-\frac{1}{2}$, whereas the slope of the first line is 3. Since the slopes are not equal, the two lines are neither parallel nor identical and must intersect. Therefore, this system has one solution and is consistent. ☐

**PRACTICE 8** Without graphing, determine the number of solutions of the system.

$$\begin{cases} -\frac{2}{3}x + y = 6 \\ 3y = 2x + 5 \end{cases}$$

## Graphing Calculator Explorations

A graphing calculator may be used to approximate solutions of systems of equations. For example, to approximate the solution of the system

$$\begin{cases} y = -3.14x - 1.35 \\ y = 4.88x + 5.25, \end{cases}$$

first graph each equation on the same set of axes. Then use the intersect feature of your calculator to approximate the point of intersection.

The approximate point of intersection is $(-0.82, 1.23)$.

*Solve each system of equations. Approximate the solutions to two decimal places.*

1. $\begin{cases} y = -2.68x + 1.21 \\ y = 5.22x - 1.68 \end{cases}$

2. $\begin{cases} y = 4.25x + 3.89 \\ y = -1.88x + 3.21 \end{cases}$

3. $\begin{cases} 4.3x - 2.9y = 5.6 \\ 8.1x + 7.6y = -14.1 \end{cases}$

4. $\begin{cases} -3.6x - 8.6y = 10 \\ -4.5x + 9.6y = -7.7 \end{cases}$

## VOCABULARY & READINESS CHECK

*Fill in each blank with one of the words or phrases listed below.*

| | | |
|---|---|---|
| system of linear equations | solution | consistent |
| dependent | inconsistent | independent |

1. In a system of linear equations in two variables, if the graphs of the equations are the same, the equations are _____ equations.

2. Two or more linear equations are called a _____ .

3. A system of equations that has at least one solution is called a(n) _____ system.

4. A _____ of a system of two equations in two variables is an ordered pair of numbers that is a solution of both equations in the system.

5. A system of equations that has no solution is called a(n) _____ system.

6. In a system of linear equations in two variables, if the graphs of the equations are different, the equations are _____ equations.

*Each rectangular coordinate system shows the graph of the equations in a system of equations. Use each graph to determine the number of solutions for each associated system. If the system has only one solution, give its coordinates.*

**7.**   **8.**   **9.**   **10.**

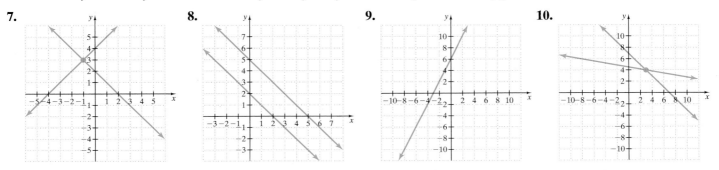

## 4.1 | EXERCISE SET

**MyMathLab** PRACTICE WATCH DOWNLOAD READ REVIEW

*Determine whether each ordered pair is a solution of the system of linear equations. See Examples 1 and 2.*

**1.** $\begin{cases} x + y = 8 \\ 3x + 2y = 21 \end{cases}$
 **a.** $(2, 4)$
 **b.** $(5, 3)$

**2.** $\begin{cases} 2x + y = 5 \\ x + 3y = 5 \end{cases}$
 **a.** $(5, 0)$
 **b.** $(2, 1)$

**3.** $\begin{cases} 3x - y = 5 \\ x + 2y = 11 \end{cases}$
 **a.** $(3, 4)$
 **b.** $(0, -5)$

**4.** $\begin{cases} 2x - 3y = 8 \\ x - 2y = 6 \end{cases}$
 **a.** $(-2, -4)$
 **b.** $(7, 2)$

**5.** $\begin{cases} 2y = 4x + 6 \\ 2x - y = -3 \end{cases}$
 **a.** $(-3, -3)$
 **b.** $(0, 3)$

**6.** $\begin{cases} x + 5y = -4 \\ -2x = 10y + 8 \end{cases}$
 **a.** $(-4, 0)$
 **b.** $(6, -2)$

**7.** $\begin{cases} -2 = x - 7y \\ 6x - y = 13 \end{cases}$
 **a.** $(-2, 0)$
 **b.** $\left(\frac{1}{2}, \frac{5}{14}\right)$

**8.** $\begin{cases} 4x = 1 - y \\ x - 3y = -8 \end{cases}$
 **a.** $(0, 1)$
 **b.** $\left(\frac{1}{6}, \frac{1}{3}\right)$

### MIXED PRACTICE

*Solve each system of linear equations by graphing. See Examples 3 through 6.*

**9.** $\begin{cases} x + y = 4 \\ x - y = 2 \end{cases}$

**10.** $\begin{cases} x + y = 3 \\ x - y = 5 \end{cases}$

**11.** $\begin{cases} x + y = 6 \\ -x + y = -6 \end{cases}$

**12.** $\begin{cases} x + y = 1 \\ -x + y = -3 \end{cases}$

**13.** $\begin{cases} y = 2x \\ 3x - y = -2 \end{cases}$

**14.** $\begin{cases} y = -3x \\ 2x - y = -5 \end{cases}$

**15.** $\begin{cases} y = x + 1 \\ y = 2x - 1 \end{cases}$

**16.** $\begin{cases} y = 3x - 4 \\ y = x + 2 \end{cases}$

**17.** $\begin{cases} 2x + y = 0 \\ 3x + y = 1 \end{cases}$

**18.** $\begin{cases} 2x + y = 1 \\ 3x + y = 0 \end{cases}$

**19.** $\begin{cases} y = -x - 1 \\ y = 2x + 5 \end{cases}$

**20.** $\begin{cases} y = x - 1 \\ y = -3x - 5 \end{cases}$

**21.** $\begin{cases} x + y = 5 \\ x + y = 6 \end{cases}$

**22.** $\begin{cases} x - y = 4 \\ x - y = 1 \end{cases}$

**23.** $\begin{cases} 2x - y = 6 \\ y = 2 \end{cases}$

**24.** $\begin{cases} x + y = 5 \\ x = 4 \end{cases}$

**25.** $\begin{cases} x - 2y = 2 \\ 3x + 2y = -2 \end{cases}$

**26.** $\begin{cases} x + 3y = 7 \\ 2x - 3y = -4 \end{cases}$

**27.** $\begin{cases} 2x + y = 4 \\ 6x = -3y + 6 \end{cases}$

**28.** $\begin{cases} y + 2x = 3 \\ 4x = 2 - 2y \end{cases}$

**29.** $\begin{cases} y - 3x = -2 \\ 6x - 2y = 4 \end{cases}$

**30.** $\begin{cases} x - 2y = -6 \\ -2x + 4y = 12 \end{cases}$

**31.** $\begin{cases} x = 3 \\ y = -1 \end{cases}$

**32.** $\begin{cases} x = -5 \\ y = 3 \end{cases}$

**33.** $\begin{cases} y = x - 2 \\ y = 2x + 3 \end{cases}$

**34.** $\begin{cases} y = x + 5 \\ y = -2x - 4 \end{cases}$

**35.** $\begin{cases} 2x - 3y = -2 \\ -3x + 5y = 5 \end{cases}$

**36.** $\begin{cases} 4x - y = 7 \\ 2x - 3y = -9 \end{cases}$

**37.** $\begin{cases} 6x - y = 4 \\ \frac{1}{2}y = -2 + 3x \end{cases}$

**38.** $\begin{cases} 3x - y = 6 \\ \frac{1}{3}y = -2 + x \end{cases}$

*Without graphing, decide. See Examples 7 and 8.*

**a.** Are the graphs of the equations identical lines, parallel lines, or lines intersecting at a single point?

**b.** How many solutions does the system have?

**39.** $\begin{cases} 4x + y = 24 \\ x + 2y = 2 \end{cases}$

**40.** $\begin{cases} 3x + y = 1 \\ 3x + 2y = 6 \end{cases}$

**41.** $\begin{cases} 2x + y = 0 \\ 2y = 6 - 4x \end{cases}$

**42.** $\begin{cases} 3x + y = 0 \\ 2y = -6x \end{cases}$

**43.** $\begin{cases} 6x - y = 4 \\ \frac{1}{2}y = -2 + 3x \end{cases}$

**44.** $\begin{cases} 3x - y = 2 \\ \frac{1}{3}y = -2 + 3x \end{cases}$

**45.** $\begin{cases} x = 5 \\ y = -2 \end{cases}$

**46.** $\begin{cases} y = 3 \\ x = -4 \end{cases}$

**47.** $\begin{cases} 3y - 2x = 3 \\ x + 2y = 9 \end{cases}$

**48.** $\begin{cases} 2y = x + 2 \\ y + 2x = 3 \end{cases}$

**49.** $\begin{cases} 6y + 4x = 6 \\ 3y - 3 = -2x \end{cases}$

**50.** $\begin{cases} 8y + 6x = 4 \\ 4y - 2 = 3x \end{cases}$

**51.** $\begin{cases} x + y = 4 \\ x + y = 3 \end{cases}$

**52.** $\begin{cases} 2x + y = 0 \\ y = -2x + 1 \end{cases}$

### REVIEW AND PREVIEW

*Solve each equation. See Section 2.3.*

**53.** $5(x - 3) + 3x = 1$

**54.** $-2x + 3(x + 6) = 17$

**55.** $4\left(\dfrac{y+1}{2}\right) + 3y = 0$

**56.** $-y + 12\left(\dfrac{y-1}{4}\right) = 3$

**57.** $8a - 2(3a - 1) = 6$

**58.** $3z - (4z - 2) = 9$

## CONCEPT EXTENSIONS

**59.** Draw a graph of two linear equations whose associated system has the solution $(-1, 4)$.

**60.** Draw a graph of two linear equations whose associated system has the solution $(3, -2)$.

**61.** Draw a graph of two linear equations whose associated system has no solution.

**62.** Draw a graph of two linear equations whose associated system has an infinite number of solutions.

**63.** Explain how to use a graph to determine the number of solutions of a system.

**64.** The ordered pair $(-2, 3)$ is a solution of all three independent equations:

$$x + y = 1$$
$$2x - y = -7$$
$$x + 3y = 7$$

Describe the graph of all three equations on the same axes.

*The double line graph below shows the number of pounds of fish and shellfish consumed per person in the United States for the years shown. Use the graph for Exercises 65 and 66.* (*Source:* Economic Research Service, U.S. Department of Agriculture)

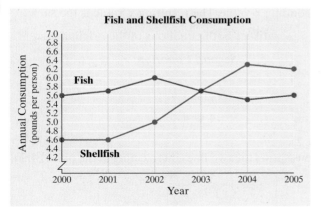

**65.** In what year(s) was the pounds per person of fish greater than the pounds per person of shellfish?

**66.** In what year(s) was the pounds per person of shellfish greater than or equal to the pounds per person of fish?

*The double line graph below shows the annual number of Toyota cars and General Motors cars sold in the United States for the years shown. Use this graph to answer Exercises 67–70.* (*Sources:* Toyota Corporation, General Motors Corporation)

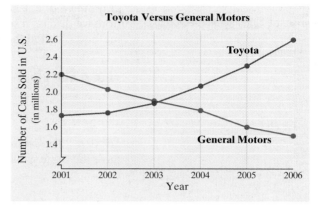

**67.** In what year(s) was the number of Toyota cars sold in the United States less than the number of GM cars sold in the United States?

**68.** In what year(s) was the number of GM cars sold in the United States less than the number of Toyota cars sold in the United States?

**69.** Describe any trends you see in this graph.

**70.** Approximate how many more cars Toyota sold in the United States than GM in 2005.

**71.** Construct a system of two linear equations that has $(1, 3)$ as a solution.

**72.** Construct a system of two linear equations that has $(0, 7)$ as a solution.

**73.** Below are two tables of values for two linear equations. Using the tables,

    **a.** find a solution of the corresponding system.

    **b.** graph several ordered pairs from each table and sketch the two lines.

    **c.** Does your graph confirm the solution from part a?

| $x$ | $y$ | $x$ | $y$ |
|---|---|---|---|
| 1 | 3 | 1 | 6 |
| 2 | 5 | 2 | 7 |
| 3 | 7 | 3 | 8 |
| 4 | 9 | 4 | 9 |
| 5 | 11 | 5 | 10 |

**74.** Explain how writing each equation in a linear system in slope-intercept form helps determine the number of solutions of a system.

**75.** Is it possible for a system of two linear equations in two variables to be inconsistent, but with dependent equations? Why or why not?

# 4.2 SOLVING SYSTEMS OF LINEAR EQUATIONS BY SUBSTITUTION

**OBJECTIVE**

1 Use the substitution method to solve a system of linear equations.

**OBJECTIVE 1 ▶ Using the substitution method.** As we stated in the preceding section, graphing alone is not an accurate way to solve a system of linear equations. In this section, we discuss a second, more accurate method for solving systems of equations. This method is called the **substitution method** and is introduced in the next example.

**EXAMPLE 1**    Solve the system:

$$\begin{cases} 2x + y = 10 & \text{First equation} \\ x = y + 2 & \text{Second equation} \end{cases}$$

*Solution*    The second equation in this system is $x = y + 2$. This tells us that $x$ and $y + 2$ have the same value. This means that we may substitute $y + 2$ for $x$ in the first equation.

$$2x + y = 10 \quad \text{First equation}$$

$$2\,\overbrace{(y + 2)} + y = 10 \quad \text{Substitute } y + 2 \text{ for } x \text{ since } x = y + 2.$$

Notice that this equation now has one variable, $y$. Let's now solve this equation for $y$.

**▶ Helpful Hint**

Don't forget the distributive property.

$$2(y + 2) + y = 10$$
$$2y + 4 + y = 10 \quad \text{Use the distributive property.}$$
$$3y + 4 = 10 \quad \text{Combine like terms.}$$
$$3y = 6 \quad \text{Subtract 4 from both sides.}$$
$$y = 2 \quad \text{Divide both sides by 3.}$$

Now we know that the $y$-value of the ordered pair solution of the system is 2. To find the corresponding $x$-value, we replace $y$ with 2 in the equation $x = y + 2$ and solve for $x$.

$$x = y + 2$$
$$x = 2 + 2 \quad \text{Let } y = 2.$$
$$x = 4$$

The solution of the system is the ordered pair $(4, 2)$. Since an ordered pair solution must satisfy both linear equations in the system, we could have chosen the equation $2x + y = 10$ to find the corresponding $x$-value. The resulting $x$-value is the same.

**Check:**    We check to see that $(4, 2)$ satisfies both equations of the original system.

| *First Equation* | *Second Equation* |
|---|---|
| $2x + y = 10$ | $x = y + 2$ |
| $2(4) + 2 \stackrel{?}{=} 10$ | $4 \stackrel{?}{=} 2 + 2$   Let $x = 4$ and $y = 2$. |
| $10 = 10$    True | $4 = 4$    True |

The solution of the system is $(4, 2)$.

A graph of the two equations shows the two lines intersecting at the point $(4, 2)$.

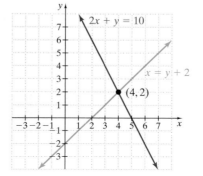

**PRACTICE**
**1**   Solve the system:

$$\begin{cases} 2x - y = 9 \\ x = y + 1 \end{cases}$$

**EXAMPLE 2**   Solve the system:

$$\begin{cases} 5x - y = -2 \\ y = 3x \end{cases}$$

_Solution_   The second equation is solved for $y$ in terms of $x$. We substitute $3x$ for $y$ in the first equation.

$$5x - y = -2 \quad \text{First equation}$$

$$5x - (3x) = -2 \quad \text{Substitute } 3x \text{ for } y.$$

Now we solve for $x$.

$$5x - 3x = -2$$

$$2x = -2 \quad \text{Combine like terms.}$$

$$x = -1 \quad \text{Divide both sides by 2.}$$

The $x$-value of the ordered pair solution is $-1$. To find the corresponding $y$-value, we replace $x$ with $-1$ in the second equation $y = 3x$.

$$y = 3x \quad \text{Second equation}$$

$$y = 3(-1) \quad \text{Let } x = -1.$$

$$y = -3$$

Check to see that the solution of the system is $(-1, -3)$.   ☐

**PRACTICE**
**2**   Solve the system:

$$\begin{cases} 7x - y = -15 \\ y = 2x \end{cases}$$

To solve a system of equations by substitution, we first need an equation solved for one of its variables, as in Examples 1 and 2. If neither equation in a system is solved for $x$ or $y$, this will be our first step.

**EXAMPLE 3**   Solve the system:

$$\begin{cases} x + 2y = 7 \\ 2x + 2y = 13 \end{cases}$$

_Solution_   We choose one of the equations and solve for $x$ or $y$. We will solve the first equation for $x$ by subtracting $2y$ from both sides.

$$x + 2y = 7 \quad \text{First equation}$$

$$x = 7 - 2y \quad \text{Subtract } 2y \text{ from both sides.}$$

Since $x = 7 - 2y$, we now substitute $7 - 2y$ for $x$ in the second equation and solve for $y$.

$$2x + 2y = 13 \quad \text{Second equation}$$

$$2(7 - 2y) + 2y = 13 \quad \text{Let } x = 7 - 2y.$$

$$14 - 4y + 2y = 13 \quad \text{Use the distributive property.}$$

$$14 - 2y = 13 \quad \text{Simplify.}$$

$$-2y = -1 \quad \text{Subtract 14 from both sides.}$$

$$y = \frac{1}{2} \quad \text{Divide both sides by } -2.$$

▶ **Helpful Hint**
Don't forget to insert parentheses when substituting $7 - 2y$ for $x$.

To find $x$, we let $y = \dfrac{1}{2}$ in the equation $x = 7 - 2y$.

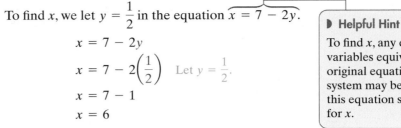

$$x = 7 - 2y$$

$$x = 7 - 2\left(\frac{1}{2}\right) \quad \text{Let } y = \frac{1}{2}.$$

$$x = 7 - 1$$

$$x = 6$$

> ▶ **Helpful Hint**
>
> To find $x$, any equation in two variables equivalent to the original equations of the system may be used. We used this equation since it is solved for $x$.

The solution is $\left(6, \dfrac{1}{2}\right)$. Check the solution in both equations of the original system. □

**PRACTICE**

**3** Solve the system:

$$\begin{cases} x + 3y = 6 \\ 2x + 3y = 10 \end{cases}$$

The following steps may be used to solve a system of equations by the substitution method.

> **Solving a System of Two Linear Equations by the Substitution Method**
>
> **STEP 1.** Solve one of the equations for one of its variables.
>
> **STEP 2.** Substitute the expression for the variable found in Step 1 into the other equation.
>
> **STEP 3.** Solve the equation from Step 2 to find the value of one variable.
>
> **STEP 4.** Substitute the value found in Step 3 in any equation containing both variables to find the value of the other variable.
>
> **STEP 5.** Check the proposed solution in the original system.

**Concept Check** ☑

As you solve the system $\begin{cases} 2x + y = -5 \\ x - y = 5 \end{cases}$ you find that $y = -5$. Is this the solution of the system?

**EXAMPLE 4** Solve the system:

$$\begin{cases} 7x - 3y = -14 \\ -3x + y = 6 \end{cases}$$

*Solution* To avoid introducing fractions, we will solve the second equation for $y$.

$$-3x + y = 6 \qquad \text{Second equation}$$

$$y = 3x + 6$$

Next, substitute $3x + 6$ for $y$ in the first equation.

$$7x - 3y = -14 \quad \text{First equation}$$

$$7x - 3\,(3x + 6) = -14 \quad \text{Let } y = 3x + 6.$$

$$7x - 9x - 18 = -14 \quad \text{Use the distributive property.}$$

$$-2x - 18 = -14 \quad \text{Simplify.}$$

$$-2x = 4 \quad \text{Add 18 to both sides.}$$

$$x = -2 \quad \text{Divide both sides by } -2.$$

**Answer to Concept Check:**

No, the solution will be an ordered pair.

To find the corresponding $y$-value, substitute $-2$ for $x$ in the equation $y = 3x + 6$. Then $y = 3(-2) + 6$ or $y = 0$. The solution of the system is $(-2, 0)$. Check this solution in both equations of the system. □

**PRACTICE**
**4** Solve the system:

$$\begin{cases} 5x + 3y = -9 \\ -2x + y = 8 \end{cases}$$

---

▶ **Helpful Hint**

When solving a system of equations by the substitution method, begin by solving an equation for one of its variables. If possible, solve for a variable that has a coefficient of 1 or $-1$. This way, we avoid working with time-consuming fractions.

**EXAMPLE 5** Solve the system:

$$\begin{cases} \dfrac{1}{2}x - y = 3 \\ x = 6 + 2y \end{cases}$$

**Solution** The second equation is already solved for $x$ in terms of $y$. Thus we substitute $6 + 2y$ for $x$ in the first equation and solve for $y$.

$$\frac{1}{2}x - y = 3 \quad \text{First equation}$$

$$\frac{1}{2}\overbrace{(6 + 2y)} - y = 3 \quad \text{Let } x = 6 + 2y.$$

$$3 + y - y = 3 \quad \text{Use the distributive property.}$$

$$3 = 3 \quad \text{Simplify.}$$

Arriving at a true statement such as $3 = 3$ indicates that the two linear equations in the original system are equivalent. This means that their graphs are identical and there are an infinite number of solutions of the system. Any solution of one equation is also a solution of the other.

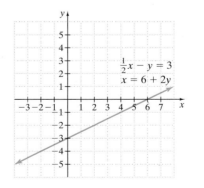

□

**PRACTICE**
**5** Solve the system:

$$\begin{cases} \dfrac{1}{4}x - y = 2 \\ x = 4y + 8 \end{cases}$$

---

**EXAMPLE 6**    Use substitution to solve the system.

$$\begin{cases} 6x + 12y = 5 \\ -4x - 8y = 0 \end{cases}$$

_Solution_    Choose the second equation and solve for $y$.

$$-4x - 8y = 0 \qquad \text{Second equation}$$

$$-8y = 4x \qquad \text{Add } 4x \text{ to both sides.}$$

$$\frac{-8y}{-8} = \frac{4x}{-8} \qquad \text{Divide both sides by } -8.$$

$$y = -\frac{1}{2}x \qquad \text{Simplify.}$$

Now replace $y$ with $-\dfrac{1}{2}x$ in the first equation.

$$6x + 12y = 5 \qquad \text{First equation}$$

$$6x + 12\left(-\frac{1}{2}x\right) = 5 \qquad \text{Let } y = -\frac{1}{2}x.$$

$$6x + (-6x) = 5 \qquad \text{Simplify.}$$

$$0 = 5 \qquad \text{Combine like terms.}$$

The false statement $0 = 5$ indicates that this system has no solution and is inconsistent. The graph of the linear equations in the system is a pair of parallel lines.

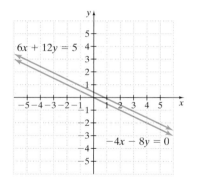

**PRACTICE**
**6**    Use substitution to solve the system.

$$\begin{cases} 4x - 3y = 12 \\ -8x + 6y = -30 \end{cases}$$

**Concept Check** ✓

Describe how the graphs of the equations in a system appear if the system has

**a.** no solution

**b.** one solution

**c.** an infinite number of solutions

**Answers to Concept Check:**

**a.** parallel lines
**b.** intersect at one point
**c.** identical graphs

# VOCABULARY & READINESS CHECK

Give the solution of each system. If the system has no solution or an infinite number of solutions, say so. If the system has one solution, find it.

**1.** $\begin{cases} y = 4x \\ -3x + y = 1 \end{cases}$

When solving, you obtain $x = 1$

**2.** $\begin{cases} 4x - y = 17 \\ -8x + 2y = 0 \end{cases}$

When solving, you obtain $0 = 34$

**3.** $\begin{cases} 4x - y = 17 \\ -8x + 2y = -34 \end{cases}$

When solving, you obtain $0 = 0$

**4.** $\begin{cases} 5x + 2y = 25 \\ x = y + 5 \end{cases}$

When solving, you obtain $y = 0$

**5.** $\begin{cases} x + y = 0 \\ 7x - 7y = 0 \end{cases}$

When solving, you obtain $x = 0$

**6.** $\begin{cases} y = -2x + 5 \\ 4x + 2y = 10 \end{cases}$

When solving, you obtain $0 = 0$

# 4.2 | EXERCISE SET

**MyMathLab** Powered by CourseCompass™ and MathXL®

Math XP PRACTICE    WATCH    DOWNLOAD    READ    REVIEW

*Solve each system of equations by the substitution method. See Examples 1 and 2.*

**1.** $\begin{cases} x + y = 3 \\ x = 2y \end{cases}$

**2.** $\begin{cases} x + y = 20 \\ x = 3y \end{cases}$

**3.** $\begin{cases} x + y = 6 \\ y = -3x \end{cases}$

**4.** $\begin{cases} x + y = 6 \\ y = -4x \end{cases}$

**5.** $\begin{cases} y = 3x + 1 \\ 4y - 8x = 12 \end{cases}$

**6.** $\begin{cases} y = 2x + 3 \\ 5y - 7x = 18 \end{cases}$

**7.** $\begin{cases} y = 2x + 9 \\ y = 7x + 10 \end{cases}$

**8.** $\begin{cases} y = 5x - 3 \\ y = 8x + 4 \end{cases}$

## MIXED PRACTICE

*Solve each system of equations by the substitution method. See Examples 1 through 6.*

**9.** $\begin{cases} 3x - 4y = 10 \\ y = x - 3 \end{cases}$

**10.** $\begin{cases} 4x - 3y = 10 \\ y = x - 5 \end{cases}$

**11.** $\begin{cases} x + 2y = 6 \\ 2x + 3y = 8 \end{cases}$

**12.** $\begin{cases} x + 3y = -5 \\ 2x + 2y = 6 \end{cases}$

**13.** $\begin{cases} 3x + 2y = 16 \\ x = 3y - 2 \end{cases}$

**14.** $\begin{cases} 2x + 3y = 18 \\ x = 2y - 5 \end{cases}$

**15.** $\begin{cases} 2x - 5y = 1 \\ 3x + y = -7 \end{cases}$

**16.** $\begin{cases} 3y - x = 6 \\ 4x + 12y = 0 \end{cases}$

**17.** $\begin{cases} 4x + 2y = 5 \\ -2x = y + 4 \end{cases}$

**18.** $\begin{cases} 2y = x + 2 \\ 6x - 12y = 0 \end{cases}$

**19.** $\begin{cases} 4x + y = 11 \\ 2x + 5y = 1 \end{cases}$

**20.** $\begin{cases} 3x + y = -14 \\ 4x + 3y = -22 \end{cases}$

**21.** $\begin{cases} x + 2y + 5 = -4 + 5y - x \\ \quad 2x + x = y + 4 \end{cases}$

(*Hint:* First simplify each equation.)

**22.** $\begin{cases} 5x + 4y - 2 = -6 + 7y - 3x \\ \quad 3x + 4x = y + 3 \end{cases}$

(*Hint:* See Exercise 21.)

**23.** $\begin{cases} 6x - 3y = 5 \\ x + 2y = 0 \end{cases}$

**24.** $\begin{cases} 10x - 5y = -21 \\ x + 3y = 0 \end{cases}$

**25.** $\begin{cases} 3x - y = 1 \\ 2x - 3y = 10 \end{cases}$

**26.** $\begin{cases} 2x - y = -7 \\ 4x - 3y = -11 \end{cases}$

**27.** $\begin{cases} -x + 2y = 10 \\ -2x + 3y = 18 \end{cases}$

**28.** $\begin{cases} -x + 3y = 18 \\ -3x + 2y = 19 \end{cases}$

**29.** $\begin{cases} 5x + 10y = 20 \\ 2x + 6y = 10 \end{cases}$

**30.** $\begin{cases} 6x + 3y = 12 \\ 9x + 6y = 15 \end{cases}$

**31.** $\begin{cases} 3x + 6y = 9 \\ 4x + 8y = 16 \end{cases}$

**32.** $\begin{cases} 2x + 4y = 6 \\ 5x + 10y = 16 \end{cases}$

**33.** $\begin{cases} \dfrac{1}{3}x - y = 2 \\ x - 3y = 6 \end{cases}$

**34.** $\begin{cases} \dfrac{1}{4}x - 2y = 1 \\ x - 8y = 4 \end{cases}$

**35.** $\begin{cases} x = \dfrac{3}{4}y - 1 \\ 8x - 5y = -6 \end{cases}$

**36.** $\begin{cases} x = \dfrac{5}{6}y - 2 \\ 12x - 5y = -9 \end{cases}$

*Solve each system by the substitution method. First simplify each equation by combining like terms.*

**37.** $\begin{cases} -5y + 6y = 3x + 2(x - 5) - 3x + 5 \\ 4(x + y) - x + y = -12 \end{cases}$

**38.** $\begin{cases} 5x + 2y - 4x - 2y = 2(2y + 6) - 7 \\ 3(2x - y) - 4x = 1 + 9 \end{cases}$

## REVIEW AND PREVIEW

*Write equivalent equations by multiplying both sides of the given equation by the given nonzero number. See Section 2.2.*

**39.** $3x + 2y = 6$ by $-2$

**40.** $-x + y = 10$ by $5$

**41.** $-4x + y = 3$ by $3$

**42.** $5a - 7b = -4$ by $-4$

*Add the binomials. See Section 2.1.*

**43.**  $\begin{aligned} 3n + 6m \\ 2n - 6m \end{aligned}$   **44.**  $\begin{aligned} -2x + 5y \\ 2x + 11y \end{aligned}$   **45.**  $\begin{aligned} -5a - 7b \\ 5a - 8b \end{aligned}$   **46.**  $\begin{aligned} 9q + p \\ -9q - p \end{aligned}$

**CONCEPT EXTENSIONS**

**47.** Explain how to identify a system with no solution when using the substitution method.

**48.** Occasionally, when using the substitution method, we obtain the equation $0 = 0$. Explain how this result indicates that the graphs of the equations in the system are identical.

*Solve. See a Concept Check in this section.*

**49.** As you solve the system $\begin{cases} 3x - y = -6 \\ -3x + 2y = 7 \end{cases}$, you find that $y = 1$. Is this the solution to the system?

**50.** As you solve the system $\begin{cases} x = 5y \\ y = 2x \end{cases}$, you find that $x = 0$ and $y = 0$. What is the solution to this system?

**51.** To avoid fractions, which of the equations below would you use if solving for $y$? Explain why.
  **a.** $\frac{1}{2}x - 4y = \frac{3}{4}$   **b.** $8x - 5y = 13$
  **c.** $7x - y = 19$

**52.** Give the number of solutions for a system if the graphs of the equations in the system are
  **a.** lines intersecting in one point
  **b.** parallel lines
  **c.** same line

**53.** The number of men and women receiving bachelor's degrees each year has been steadily increasing. For the years 1970 through the projection of 2014, the number of men receiving degrees (in thousands) is given by the equation $y = 3.9x + 443$, and for women, the equation is $y = 14.2x + 314$ where $x$ is the number of years after 1970. (*Source:* National Center for Education Statistics)
  **a.** Use the substitution method to solve this system of equations. (Round your final results to the nearest whole numbers.)

**b.** Explain the meaning of your answer to part (a).

**c.** Sketch a graph of the system of equations. Write a sentence describing the trends for men and women receiving bachelor degrees.

**54.** The number of Adult Contemporary Music radio stations in the United States from 2000 to 2006 is given by the equation $y = -6.17x + 719$, where $x$ is the number of years after 2000. The number of Spanish radio stations is given by $y = 33.9x + 534$ for the same time period. (*Source:* M Street Corporation)
  **a.** Use the substitution method to solve this system of equations. (Round your numbers to the nearest tenth.)
  **b.** Explain the meaning of your answer to part (a).
  **c.** Sketch a graph of the system of equations. Write a sentence describing the trends in the popularity of these two types of music format.

*Solve each system by substitution. When necessary, round answers to the nearest hundredth.*

**55.** $\begin{cases} y = 5.1x + 14.56 \\ y = -2x - 3.9 \end{cases}$

**56.** $\begin{cases} y = 3.1x - 16.35 \\ y = -9.7x + 28.45 \end{cases}$

**57.** $\begin{cases} 3x + 2y = 14.05 \\ 5x + y = 18.5 \end{cases}$

**58.** $\begin{cases} x + y = -15.2 \\ -2x + 5y = -19.3 \end{cases}$

---

## 📖 STUDY SKILLS BUILDER

**How Are Your Homework Assignments Going?**

Remember that it is important to keep up with homework. Why? Many concepts in mathematics build on each other. Often, your understanding of a day's lecture depends on an understanding of the previous day's material.

   To complete a homework assignment, remember these 4 things:

- Attempt all of it.
- Check it.
- Correct it.
- If needed, ask questions about it.

*Take a moment and review your completed homework assignments. Answer the exercises below based on this review.*

**1.** Approximate the fraction of your homework you have attempted.

**2.** Approximate the fraction of your homework you have checked (if possible).

**3.** If you are able to check your homework, have you corrected it when errors have been found?

**4.** When working homework, if you do not understand a concept, what do you personally do?

# 4.3 SOLVING SYSTEMS OF LINEAR EQUATIONS BY ADDITION

**OBJECTIVE**

1 Use the addition method to solve a system of linear equations.

**OBJECTIVE 1 ▶ Using the addition method.** We have seen that substitution is an accurate way to solve a linear system. Another method for solving a system of equations accurately is the **addition** or **elimination method.** The addition method is based on the addition property of equality: adding equal quantities to both sides of an equation does not change the solution of the equation. In symbols,

$$\text{if } A = B \text{ and } C = D, \text{ then } A + C = B + D.$$

**EXAMPLE 1**  Solve the system:

$$\begin{cases} x + y = 7 \\ x - y = 5 \end{cases}$$

**Solution**  Since the left side of each equation is equal to the right side, we add equal quantities by adding the left sides of the equations together and the right sides of the equations together. This adding eliminates the variable $y$ and gives us an equation in one variable, $x$. We can then solve for $x$.

▶ **Helpful Hint**

Our goal when solving a system of equations by the addition method is to eliminate a variable when adding the equations.

$$\begin{array}{ll} x + y = 7 & \text{First equation} \\ \underline{x - y = 5} & \text{Second equation} \\ 2x \quad\;\; = 12 & \text{Add the equations.} \\ \quad\;\; x = 6 & \text{Divide both sides by 2.} \end{array}$$

The $x$-value of the solution is $6$. To find the corresponding $y$-value, let $x = 6$ in either equation of the system. We will use the first equation.

$$\begin{array}{ll} x + y = 7 & \text{First equation} \\ 6 + y = 7 & \text{Let } x = 6. \\ \quad\;\; y = 7 - 6 & \text{Solve for } y. \\ \quad\;\; y = 1 & \text{Simplify.} \end{array}$$

**Check:**  The solution is $(6, 1)$. Check this in both equations.

| *First Equation* | *Second Equation* |
|---|---|
| $x + y = 7$ | $x - y = 5$ |
| $6 + 1 \overset{?}{=} 7$ | $6 - 1 \overset{?}{=} 5$  Let $x = 6$ and $y = 1$. |
| $7 = 7$  True | $5 = 5$  True |

Thus, the solution of the system is $(6, 1)$ and the graphs of the two equations intersect at the point $(6, 1)$ as shown next.

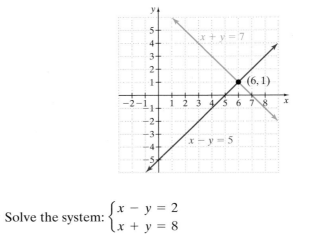

**PRACTICE**
**1**  Solve the system: $\begin{cases} x - y = 2 \\ x + y = 8 \end{cases}$

**EXAMPLE 2** Solve the system: $\begin{cases} -2x + y = 2 \\ -x + 3y = -4 \end{cases}$

*Solution* If we simply add the two equations, the result is still an equation in two variables. However, our goal is to eliminate one of the variables. Notice what happens if we multiply *both sides* of the first equation by $-3$, which we are allowed to do by the multiplication property of equality. The system

$\begin{cases} -3(-2x + y) = -3(2) \\ -x + 3y = -4 \end{cases}$ simplifies to $\begin{cases} 6x - 3y = -6 \\ -x + 3y = -4 \end{cases}$

Now add the resulting equations and the $y$-variable is eliminated.

$$\begin{array}{rcl} 6x - 3y &=& -6 \\ -x + 3y &=& -4 \\ \hline 5x &=& -10 \quad \text{Add.} \\ x &=& -2 \quad \text{Divide both sides by 5.} \end{array}$$

To find the corresponding $y$-value, let $x = -2$ in any of the preceding equations containing both variables. We use the first equation of the original system.

$$\begin{array}{rcl} -2x + y &=& 2 \quad \text{First equation} \\ -2(-2) + y &=& 2 \quad \text{Let } x = -2. \\ 4 + y &=& 2 \\ y &=& -2 \quad \text{Subtract 4 from both sides.} \end{array}$$

The solution is $(-2, -2)$. Check this ordered pair in both equations of the original system. □

**PRACTICE**
**2** Solve the system: $\begin{cases} x - 2y = 11 \\ 3x - y = 13 \end{cases}$

In Example 2, the decision to multiply the first equation by $-3$ was no accident. **To eliminate a variable** when adding two equations, **the coefficient of the variable in one equation must be the opposite of its coefficient in the other equation.**

> ▶ **Helpful Hint**
> Be sure to multiply *both sides* of an equation by a chosen number when solving by the addition method. A common mistake is to multiply only the side containing the variables.

**EXAMPLE 3** Solve the system: $\begin{cases} 2x - y = 7 \\ 8x - 4y = 1 \end{cases}$

*Solution* Multiply both sides of the first equation by $-4$ and the resulting coefficient of $x$ is $-8$, the opposite of 8, the coefficient of $x$ in the second equation. The system becomes

> ▶ **Helpful Hint**
> Don't forget to multiply both sides by $-4$.

$\begin{cases} -4(2x - y) = -4(7) \\ 8x - 4y = 1 \end{cases}$ simplifies to $\begin{cases} -8x + 4y = -28 \\ 8x - 4y = 1 \end{cases}$

Now add the resulting equations.

$$\begin{array}{rcl} -8x + 4y &=& -28 \\ 8x - 4y &=& 1 \\ \hline 0 &=& -27 \quad \text{Add the equations.} \\ && \qquad\quad \text{False} \end{array}$$

When we add the equations, both variables are eliminated and we have $0 = -27$, a false statement. This means that the system has no solution. The graphs of these equations are parallel lines. □

**PRACTICE**
**3** Solve the system: $\begin{cases} x - 3y = 5 \\ 2x - 6y = -3 \end{cases}$

**EXAMPLE 4**  Solve the system: $\begin{cases} 3x - 2y = 2 \\ -9x + 6y = -6 \end{cases}$

*Solution*  First we multiply both sides of the first equation by 3, then we add the resulting equations.

$\begin{cases} 3(3x - 2y) = 3(2) \\ -9x + 6y = -6 \end{cases}$  simplifies to  $\begin{cases} 9x - 6y = 6 \\ -9x + 6y = -6 \\ \overline{\phantom{-9x + 6y =} 0 = 0} \end{cases}$  Add the equations.
True

Both variables are eliminated and we have $0 = 0$, a true statement. Whenever you eliminate a variable and get the equation $0 = 0$, the system has an infinite number of solutions. The graphs of these equations are identical.  □

**PRACTICE**
**4**  Solve the system: $\begin{cases} 4x - 3y = 5 \\ -8x + 6y = -10 \end{cases}$

**Concept Check** ☑️

Suppose you are solving the system

$$\begin{cases} 3x + 8y = -5 \\ 2x - 4y = 3 \end{cases}$$

You decide to use the addition method and begin by multiplying both sides of the first equation by $-2$. In which of the following was the multiplication performed correctly? Explain.

**a.** $-6x - 16y = -5$      **b.** $-6x - 16y = 10$

**EXAMPLE 5**  Solve the system: $\begin{cases} 3x + 4y = 13 \\ 5x - 9y = 6 \end{cases}$

*Solution*  We can eliminate the variable $y$ by multiplying the first equation by 9 and the second equation by 4.

$\begin{cases} 9(3x + 4y) = 9(13) \\ 4(5x - 9y) = 4(6) \end{cases}$  simplifies to  $\begin{cases} 27x + 36y = 117 \\ 20x - 36y = \phantom{0}24 \\ \overline{47x \phantom{+ 36y} = 141} \end{cases}$  Add the equations.
$x = 3$      Divide both sides by 47.

To find the corresponding $y$-value, we let $x = 3$ in any equation in this example containing two variables. Doing so in any of these equations will give $y = 1$. The solution to this system is $(3, 1)$. Check to see that $(3, 1)$ satisfies each equation in the original system.  □

**PRACTICE**
**5**  Solve the system: $\begin{cases} 4x + 3y = 14 \\ 3x - 2y = 2 \end{cases}$

If we had decided to eliminate $x$ instead of $y$ in Example 5, the first equation could have been multiplied by 5 and the second by $-3$. Try solving the original system this way to check that the solution is $(3, 1)$.

The following steps summarize how to solve a system of linear equations by the addition method.

**Solving a System of Two Linear Equations by the Addition Method**

**STEP 1.** Rewrite each equation in standard form $Ax + By = C$.

**STEP 2.** If necessary, multiply one or both equations by a nonzero number so that the coefficients of a chosen variable in the system are opposites.

**STEP 3.** Add the equations.

**STEP 4.** Find the value of one variable by solving the resulting equation from Step 3.

**STEP 5.** Find the value of the second variable by substituting the value found in Step 4 into either of the original equations.

**STEP 6.** Check the proposed solution in the original system.

**Concept Check** ☑

Suppose you are solving the system

$$\begin{cases} -4x + 7y = 6 \\ x + 2y = 5 \end{cases}$$

by the addition method.

**a.** What step(s) should you take if you wish to eliminate $x$ when adding the equations?

**b.** What step(s) should you take if you wish to eliminate $y$ when adding the equations?

**EXAMPLE 6** Solve the system:

$$\begin{cases} -x - \dfrac{y}{2} = \dfrac{5}{2} \\ -\dfrac{x}{2} + \dfrac{y}{4} = 0 \end{cases}$$

**Solution** We begin by clearing each equation of fractions. To do so, we multiply both sides of the first equation by the LCD 2 and both sides of the second equation by the LCD 4. Then the system

$$\begin{cases} 2\left(-x - \dfrac{y}{2}\right) = 2\left(\dfrac{5}{2}\right) \\ 4\left(-\dfrac{x}{2} + \dfrac{y}{4}\right) = 4(0) \end{cases} \quad \text{simplifies to} \quad \begin{cases} -2x - y = 5 \\ -2x + y = 0 \end{cases}$$

Now we add the resulting equations in the simplified system.

$$\begin{array}{r} -2x - y = 5 \\ \underline{-2x + y = 0} \\ -4x \qquad\; = 5 \end{array} \quad \text{Add the equations.}$$

$$x = -\frac{5}{4}$$

To find $y$, we could replace $x$ with $-\dfrac{5}{4}$ in one of the equations with two variables.

Instead, let's go back to the simplified system and multiply by appropriate factors to eliminate the variable $x$ and solve for $y$. To do this, we multiply the first equation in the simplified system by $-1$. Then the system

$$\begin{cases} -1(-2x - y) = -1(5) \\ -2x + y = 0 \end{cases} \quad \text{simplifies to} \quad \begin{cases} 2x + y = -5 \\ \underline{-2x + y = \phantom{-}0} \\ \phantom{-2x +} 2y = -5 \quad \text{Add.} \\ \phantom{-2x + 2} y = -\dfrac{5}{2} \quad \text{Solve for } y. \end{cases}$$

Check the ordered pair $\left(-\dfrac{5}{4}, -\dfrac{5}{2}\right)$ in both equations of the original system. The solution is $\left(-\dfrac{5}{4}, -\dfrac{5}{2}\right)$.

**PRACTICE**
**6** Solve the system: $\begin{cases} -2x + \dfrac{3y}{2} = 5 \\ -\dfrac{x}{2} - \dfrac{y}{4} = \dfrac{1}{2} \end{cases}$

## 4.3 EXERCISE SET

*MyMathLab* — Powered by CourseCompass™ and MathXL®

| MathXP PRACTICE | WATCH | DOWNLOAD | READ | REVIEW |

Solve each system of equations by the addition method. See Example 1.

**1.** $\begin{cases} 3x + y = 5 \\ 6x - y = 4 \end{cases}$

**2.** $\begin{cases} 4x + y = 13 \\ 2x - y = 5 \end{cases}$

**3.** $\begin{cases} x - 2y = 8 \\ -x + 5y = -17 \end{cases}$

**4.** $\begin{cases} x - 2y = -11 \\ -x + 5y = 23 \end{cases}$

**MIXED PRACTICE**

Solve each system of equations by the addition method. If a system contains fractions or decimals, you may want to first clear each equation of fractions or decimals. See Examples 2 through 6.

**5.** $\begin{cases} 3x + y = -11 \\ 6x - 2y = -2 \end{cases}$

**6.** $\begin{cases} 4x + y = -13 \\ 6x - 3y = -15 \end{cases}$

**7.** $\begin{cases} 3x + 2y = 11 \\ 5x - 2y = 29 \end{cases}$

**8.** $\begin{cases} 4x + 2y = 2 \\ 3x - 2y = 12 \end{cases}$

**9.** $\begin{cases} x + 5y = 18 \\ 3x + 2y = -11 \end{cases}$

**10.** $\begin{cases} x + 4y = 14 \\ 5x + 3y = 2 \end{cases}$

**11.** $\begin{cases} x + y = 6 \\ x - y = 6 \end{cases}$

**12.** $\begin{cases} x - y = 1 \\ -x + 2y = 0 \end{cases}$

**13.** $\begin{cases} 2x + 3y = 0 \\ 4x + 6y = 3 \end{cases}$

**14.** $\begin{cases} 3x + y = 4 \\ 9x + 3y = 6 \end{cases}$

**15.** $\begin{cases} -x + 5y = -1 \\ 3x - 15y = 3 \end{cases}$

**16.** $\begin{cases} 2x + y = 6 \\ 4x + 2y = 12 \end{cases}$

**17.** $\begin{cases} 3x - 2y = 7 \\ 5x + 4y = 8 \end{cases}$

**18.** $\begin{cases} 6x - 5y = 25 \\ 4x + 15y = 13 \end{cases}$

**19.** $\begin{cases} 8x = -11y - 16 \\ 2x + 3y = -4 \end{cases}$

**20.** $\begin{cases} 10x + 3y = -12 \\ 5x = -4y - 16 \end{cases}$

**21.** $\begin{cases} 4x - 3y = 7 \\ 7x + 5y = 2 \end{cases}$

**22.** $\begin{cases} -2x + 3y = 10 \\ 3x + 4y = 2 \end{cases}$

**23.** $\begin{cases} 4x - 6y = 8 \\ 6x - 9y = 12 \end{cases}$

**24.** $\begin{cases} 9x - 3y = 12 \\ 12x - 4y = 18 \end{cases}$

**25.** $\begin{cases} 2x - 5y = 4 \\ 3x - 2y = 4 \end{cases}$

**26.** $\begin{cases} 6x - 5y = 7 \\ 4x - 6y = 7 \end{cases}$

**27.** $\begin{cases} \dfrac{x}{3} + \dfrac{y}{6} = 1 \\ \dfrac{x}{2} - \dfrac{y}{4} = 0 \end{cases}$

**28.** $\begin{cases} \dfrac{x}{2} + \dfrac{y}{8} = 3 \\ x - \dfrac{y}{4} = 0 \end{cases}$

**29.** $\begin{cases} \dfrac{10}{3}x + 4y = -4 \\ 5x + 6y = -6 \end{cases}$

**30.** $\begin{cases} \dfrac{3}{2}x + 4y = 1 \\ 9x + 24y = 5 \end{cases}$

**31.** $\begin{cases} x - \dfrac{y}{3} = -1 \\ -\dfrac{x}{2} + \dfrac{y}{8} = \dfrac{1}{4} \end{cases}$

**32.** $\begin{cases} 2x - \dfrac{3y}{4} = -3 \\ x + \dfrac{y}{9} = \dfrac{13}{3} \end{cases}$

**33.** $-4(x + 2) = 3y$
$2x - 2y = 3$

**34.** $-9(x + 3) = 8y$
$3x - 3y = 8$

**35.** $\begin{cases} \dfrac{x}{3} - y = 2 \\ -\dfrac{x}{2} + \dfrac{3y}{2} = -3 \end{cases}$

**36.** $\begin{cases} \dfrac{x}{2} + \dfrac{y}{4} = 1 \\ -\dfrac{x}{4} - \dfrac{y}{8} = 1 \end{cases}$

**37.** $\begin{cases} \dfrac{3}{5}x - y = -\dfrac{4}{5} \\ 3x + \dfrac{y}{2} = -\dfrac{9}{5} \end{cases}$

**38.** $\begin{cases} 3x + \dfrac{7}{2}y = \dfrac{3}{4} \\ -\dfrac{x}{2} + \dfrac{5}{3}y = -\dfrac{5}{4} \end{cases}$

**39.** $\begin{cases} 3.5x + 2.5y = 17 \\ -1.5x - 7.5y = -33 \end{cases}$

**40.** $\begin{cases} -2.5x - 6.5y = 47 \\ 0.5x - 4.5y = 37 \end{cases}$

**41.** $\begin{cases} 0.02x + 0.04y = 0.09 \\ -0.1x + 0.3y = 0.8 \end{cases}$

**42.** $\begin{cases} 0.04x - 0.05y = 0.105 \\ 0.2x - 0.6y = 1.05 \end{cases}$

## MIXED PRACTICE

*Solve each system by either the addition method or the substitution method.*

**43.** $\begin{cases} 2x - 3y = -11 \\ y = 4x - 3 \end{cases}$

**44.** $\begin{cases} 4x - 5y = 6 \\ y = 3x - 10 \end{cases}$

**45.** $\begin{cases} x + 2y = 1 \\ 3x + 4y = -1 \end{cases}$

**46.** $\begin{cases} x + 3y = 5 \\ 5x + 6y = -2 \end{cases}$

**47.** $\begin{cases} 2y = x + 6 \\ 3x - 2y = -6 \end{cases}$

**48.** $\begin{cases} 3y = x + 14 \\ 2x - 3y = -16 \end{cases}$

**49.** $\begin{cases} y = 2x - 3 \\ y = 5x - 18 \end{cases}$

**50.** $\begin{cases} y = 6x - 5 \\ y = 4x - 11 \end{cases}$

**51.** $\begin{cases} x + \dfrac{1}{6}y = \dfrac{1}{2} \\ 3x + 2y = 3 \end{cases}$

**52.** $\begin{cases} x + \dfrac{1}{3}y = \dfrac{5}{12} \\ 8x + 3y = 4 \end{cases}$

**53.** $\begin{cases} \dfrac{x + 2}{2} = \dfrac{y + 11}{3} \\ \dfrac{x}{2} = \dfrac{2y + 16}{6} \end{cases}$

**54.** $\begin{cases} \dfrac{x + 5}{2} = \dfrac{y + 14}{4} \\ \dfrac{x}{3} = \dfrac{2y + 2}{6} \end{cases}$

**55.** $\begin{cases} 2x + 3y = 14 \\ 3x - 4y = -69.1 \end{cases}$

**56.** $\begin{cases} 5x - 2y = -19.8 \\ -3x + 5y = -3.7 \end{cases}$

## REVIEW AND PREVIEW

*Rewrite the following sentences using mathematical symbols. Do not solve the equations. See Section 2.4.*

**57.** Twice a number, added to 6, is 3 less than the number.

**58.** The sum of three consecutive integers is 66.

**59.** Three times a number, subtracted from 20, is 2.

**60.** Twice the sum of 8 and a number is the difference of the number and 20.

**61.** The product of 4 and the sum of a number and 6 is twice the number.

**62.** The quotient of twice a number and 7 is subtracted from the reciprocal of the number.

## CONCEPT EXTENSIONS

*Solve. See a Concept Check in this section.*

**63.** To solve this system by the addition method and eliminate the variable $y$,

$$\begin{cases} 4x + 2y = -7 \\ 3x - y = -12 \end{cases}$$

by what value would you multiply the second equation? What do you get when you complete the multiplication?

*Given the system of linear equations* $\begin{cases} 3x - y = -8 \\ 5x + 3y = 2 \end{cases}$

**64.** Use the addition method and
  **a.** Solve the system by eliminating $x$.
  **b.** Solve the system by eliminating $y$.

**65.** Suppose you are solving the system

$$\begin{cases} 3x + 8y = -5 \\ 2x - 4y = 3. \end{cases}$$

You decide to use the addition method by multiplying both sides of the second equation by 2. In which of the following was the multiplication performed correctly? Explain.
  **a.** $4x - 8y = 3$
  **b.** $4x - 8y = 6$

**66.** Suppose you are solving the system

$$\begin{cases} -2x - y = 0 \\ -2x + 3y = 6. \end{cases}$$

You decide to use the addition method by multiplying both sides of the first equation by 3, then adding the resulting equation to the second equation. Which of the following is the correct sum? Explain.
  **a.** $-8x = 6$
  **b.** $-8x = 9$

**67.** When solving a system of equations by the addition method, how do we know when the system has no solution?

**68.** Explain why the addition method might be preferred over the substitution method for solving the system $\begin{cases} 2x - 3y = 5 \\ 5x + 2y = 6. \end{cases}$

**69.** Use the system of linear equations below to answer the questions.

$$\begin{cases} x + y = 5 \\ 3x + 3y = b \end{cases}$$

  **a.** Find the value of $b$ so that the system has an infinite number of solutions.
  **b.** Find a value of $b$ so that there are no solutions to the system.

**70.** Use the system of linear equations below to answer the questions.

$$\begin{cases} x + y = 4 \\ 2x + by = 8 \end{cases}$$

    **a.** Find the value of $b$ so that the system has an infinite number of solutions.

    **b.** Find a value of $b$ so that the system has a single solution.

*Solve each system by the addition method.*

**71.** $\begin{cases} 1.2x + 3.4y = 27.6 \\ 7.2x - 1.7y = -46.56 \end{cases}$      **72.** $\begin{cases} 5.1x - 2.4y = 3.15 \\ -15.3x + 1.2y = 27.75 \end{cases}$

**73.** Two occupations predicted to greatly increase in number of jobs are pharmacy technicians and network systems and data communication analysts. The number of pharmacy technician jobs predicted for 2004 through 2014 can be approximated by $7.4x - y = -258$. The number of network and data analyst jobs for the same years can be approximated by $12.6x - y = -231$. For both equations, $x$ is the number of years since 2004 and $y$ is the number of jobs in thousands.

    **a.** Use the addition method to solve this system of equations:

$$\begin{cases} 7.4x - y = -258 \\ 12.6x - y = -231 \end{cases}$$

    (Round answer to the nearest whole number.)

    **b.** Use your result from part (a) and estimate the year in which the number of both jobs is equal.

    **c.** Use your result from part (a) and estimate the number of pharmacy technician jobs (or the number of analyst jobs since they should be equal).

**74.** In recent years, the number of Americans (in millions) who have skateboarded at least once in a year has been increasing, while the number of Americans who have in-line roller skated has been decreasing. The number of skateboarders (people who have skateboarded at least once a year) from 1996 to 2006 is given by $-0.7x + y = 4.6$ and the number of in-line roller skaters can be given by the equation $1.3x + y = 26.9$. For both equations, $x$ is the number of years after 1996. (*Source:* National Sporting Goods Association)

    **a.** Use the addition method to solve this system of equations:

$$\begin{cases} -0.7x + y = 4.6 \\ 1.3x + y = 26.9 \end{cases}$$

    (Round to the nearest whole number. Because of rounding, the $y$-value of your ordered pair solution may vary.)

    **b.** Use your result from part (a) and estimate the year in which the number of skate boarders equals the number of in-line skaters.

    **c.** Use your result from part (a) and estimate the number of skateboarders (or the number of in-line skaters since they should be equal).

# INTEGRATED REVIEW   SOLVING SYSTEMS OF EQUATIONS

Sections 4.1–4.3

*Solve each system by either the addition method or the substitution method.*

**1.** $\begin{cases} 2x - 3y = -11 \\ y = 4x - 3 \end{cases}$     **2.** $\begin{cases} 4x - 5y = 6 \\ y = 3x - 10 \end{cases}$     **3.** $\begin{cases} x + y = 3 \\ x - y = 7 \end{cases}$     **4.** $\begin{cases} x - y = 20 \\ x + y = -8 \end{cases}$

**5.** $\begin{cases} x + 2y = 1 \\ 3x + 4y = -1 \end{cases}$     **6.** $\begin{cases} x + 3y = 5 \\ 5x + 6y = -2 \end{cases}$     **7.** $\begin{cases} y = x + 3 \\ 3x - 2y = -6 \end{cases}$     **8.** $\begin{cases} y = -2x \\ 2x - 3y = -16 \end{cases}$

**9.** $\begin{cases} y = 2x - 3 \\ y = 5x - 18 \end{cases}$     **10.** $\begin{cases} y = 6x - 5 \\ y = 4x - 11 \end{cases}$     **11.** $\begin{cases} x + \dfrac{1}{6}y = \dfrac{1}{2} \\ 3x + 2y = 3 \end{cases}$     **12.** $\begin{cases} x + \dfrac{1}{3}y = \dfrac{5}{12} \\ 8x + 3y = 4 \end{cases}$

**13.** $\begin{cases} x - 5y = 1 \\ -2x + 10y = 3 \end{cases}$  **14.** $\begin{cases} -x + 2y = 3 \\ 3x - 6y = -9 \end{cases}$  **15.** $\begin{cases} 0.2x - 0.3y = -0.95 \\ 0.4x + 0.1y = 0.55 \end{cases}$  **16.** $\begin{cases} 0.08x - 0.04y = -0.11 \\ 0.02x - 0.06y = -0.09 \end{cases}$

**17.** $\begin{cases} x = 3y - 7 \\ 2x - 6y = -14 \end{cases}$  **18.** $\begin{cases} y = \dfrac{x}{2} - 3 \\ 2x - 4y = 0 \end{cases}$  **19.** $\begin{cases} 2x + 5y = -1 \\ 3x - 4y = 33 \end{cases}$  **20.** $\begin{cases} 7x - 3y = 2 \\ 6x + 5y = -21 \end{cases}$

**21.** Which method, substitution or addition, would you prefer to use to solve the system below? Explain your reasoning.

$$\begin{cases} 3x + 2y = -2 \\ y = -2x \end{cases}$$

**22.** Which method, substitution or addition, would you prefer to use to solve the system below? Explain your reasoning.

$$\begin{cases} 3x - 2y = -3 \\ 6x + 2y = 12 \end{cases}$$

---

# 4.4 SOLVING SYSTEMS OF LINEAR EQUATIONS IN THREE VARIABLES

**OBJECTIVE**

**1** Solve a system of three linear equations in three variables.

In this section, the algebraic methods of solving systems of two linear equations in two variables are extended to systems of three linear equations in three variables. We call the equation $3x - y + z = -15$, for example, a **linear equation in three variables** since there are three variables and each variable is raised only to the power 1. A solution of this equation is an **ordered triple $(x, y, z)$** that makes the equation a true statement. For example, the ordered triple $(2, 0, -21)$ is a solution of $3x - y + z = -15$ since replacing $x$ with $2$, $y$ with $0$, and $z$ with $-21$ yields the true statement $3(2) - 0 + (-21) = -15$. The graph of this equation is a plane in three-dimensional space, just as the graph of a linear equation in two variables is a line in two-dimensional space.

Although we will not discuss the techniques for graphing equations in three variables, visualizing the possible patterns of intersecting planes gives us insight into the possible patterns of solutions of a system of three three-variable linear equations. There are four possible patterns.

**1.** Three planes have a single point in common. This point represents the single solution of the system. This system is **consistent.**

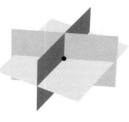

**2.** Three planes intersect at no point common to all three. This system has no solution. A few ways that this can occur are shown. This system is **inconsistent.**

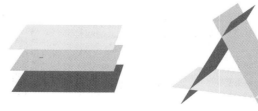

**3.** Three planes intersect at all the points of a single line. The system has infinitely many solutions. This system is **consistent.**

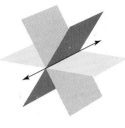

**4.** Three planes coincide at all points on the plane. The system is consistent, and the equations are **dependent.**

**OBJECTIVE 1 ▶ Solving a system of three linear equations in three variables.** Just as with systems of two equations in two variables, we can use the elimination or substitution method to solve a system of three equations in three variables. To use the elimination method, we eliminate a variable and obtain a system of two equations in two variables. Then we use the methods we learned in the previous section to solve the system of two equations.

**EXAMPLE 1**   Solve the system.

$$\begin{cases} 3x - y + z = -15 & \text{Equation (1)} \\ x + 2y - z = 1 & \text{Equation (2)} \\ 2x + 3y - 2z = 0 & \text{Equation (3)} \end{cases}$$

*Solution*   Add equations (1) and (2) to eliminate $z$.

$$\begin{array}{r} 3x - y + z = -15 \\ \underline{x + 2y - z = 1} \\ 4x + y \quad\ = -14 \quad \text{Equation (4)} \end{array}$$

Next, add two *other* equations and *eliminate z again*. To do so, multiply both sides of equation (1) by 2 and add this resulting equation to equation (3). Then

> **▶ Helpful Hint**
>
> Don't forget to add two other equations besides equations (1) and (2) *and* to **eliminate the same variable.**

$$\begin{cases} 2(3x - y + z) = 2(-15) \\ 2x + 3y - 2z = 0 \end{cases} \quad \text{simplifies to} \quad \begin{cases} 6x - 2y + 2z = -30 \\ \underline{2x + 3y - 2z = 0} \\ 8x + y \quad\ = -30 \quad \text{Equation (5)} \end{cases}$$

Now solve equations (4) and (5) for $x$ and $y$. To solve by elimination, multiply both sides of equation (4) by $-1$ and add this resulting equation to equation (5). Then

$$\begin{cases} -1(4x + y) = -1(-14) \\ 8x + y = -30 \end{cases} \quad \text{simplifies to} \quad \begin{cases} -4x - y = 14 \\ \underline{8x + y = -30} \\ 4x \quad\quad = -16 \quad \text{Add the equations.} \\ x = -4 \quad \text{Solve for } x. \end{cases}$$

Replace $x$ with $-4$ in equation (4) or (5).

$$4x + y = -14 \quad \text{Equation (4)}$$
$$4(-4) + y = -14 \quad \text{Let } x = -4.$$
$$y = 2 \quad \text{Solve for } y.$$

Finally, replace $x$ with $-4$ and $y$ with $2$ in equation (1), (2), or (3).

$$x + 2y - z = 1 \qquad \text{Equation (2)}$$
$$-4 + 2(2) - z = 1 \qquad \text{Let } x = -4 \text{ and } y = 2.$$
$$-4 + 4 - z = 1$$
$$-z = 1$$
$$z = -1$$

The solution is $(-4, 2, -1)$. To check, let $x = -4$, $y = 2$, and $z = -1$ in all three original equations of the system.

| **Equation (1)** | **Equation (2)** | **Equation (3)** |
|---|---|---|
| $3x - y + z = -15$ | $x + 2y - z = 1$ | $2x + 3y - 2z = 0$ |
| $3(-4) - 2 + (-1) \overset{?}{=} -15$ | $-4 + 2(2) - (-1) \overset{?}{=} 1$ | $2(-4) + 3(2) - 2(-1) \overset{?}{=} 0$ |
| $-12 - 2 - 1 \overset{?}{=} -15$ | $-4 + 4 + 1 \overset{?}{=} 1$ | $-8 + 6 + 2 \overset{?}{=} 0$ |
| $-15 = -15$ | $1 = 1$ | $0 = 0$ |
| True | True | True |

All three statements are true, so the solution is $(-4, 2, -1)$. $\square$

**PRACTICE**
**1** Solve the system. $\begin{cases} 3x + 2y - z = 0 \\ x - y + 5z = 2 \\ 2x + 3y + 3z = 7 \end{cases}$

**EXAMPLE 2** Solve the system.

$$\begin{cases} 2x - 4y + 8z = 2 & (1) \\ -x - 3y + z = 11 & (2) \\ x - 2y + 4z = 0 & (3) \end{cases}$$

**Solution** Add equations (2) and (3) to eliminate $x$, and the new equation is

$$-5y + 5z = 11 \quad (4)$$

To eliminate $x$ again, multiply both sides of equation (2) by 2, and add the resulting equation to equation (1). Then

$$\begin{cases} 2x - 4y + 8z = 2 \\ 2(-x - 3y + z) = 2(11) \end{cases} \begin{matrix} \text{simplifies} \\ \text{to} \end{matrix} \begin{cases} 2x - 4y + 8z = 2 \\ \underline{-2x - 6y + 2z = 22} \\ -10y + 10z = 24 \quad (5) \end{cases}$$

Next, solve for $y$ and $z$ using equations (4) and (5). Multiply both sides of equation (4) by $-2$, and add the resulting equation to equation (5).

$$\begin{cases} -2(-5y + 5z) = -2(11) \\ -10y + 10z = 24 \end{cases} \begin{matrix} \text{simplifies} \\ \text{to} \end{matrix} \begin{cases} 10y - 10z = -22 \\ \underline{-10y + 10z = 24} \\ 0 = 2 \quad \text{False} \end{cases}$$

Since the statement is false, this system is inconsistent and has no solution. The solution set is the empty set $\{\ \}$ or $\varnothing$. $\square$

**PRACTICE**
**2** Solve the system. $\begin{cases} 6x - 3y + 12z = 4 \\ -6x + 4y - 2z = 7 \\ -2x + y - 4z = 3 \end{cases}$

The elimination method is summarized next.

> **Solving a System of Three Linear Equations by the Elimination Method**
> **STEP 1.** Write each equation in standard form $Ax + By + Cz = D$.
>
> **STEP 2.** Choose a pair of equations and use the equations to eliminate a variable.
>
> **STEP 3.** Choose any **other** pair of equations and eliminate the **same variable** as in Step 2.
>
> **STEP 4.** Two equations in two variables should be obtained from Step 2 and Step 3. Use methods from Section 4.1 to solve this system for both variables.
>
> **STEP 5.** To solve for the third variable, substitute the values of the variables found in Step 4 into any of the original equations containing the third variable.
>
> **STEP 6.** Check the ordered triple solution in *all three* original equations.

▶ **Helpful Hint**
Make sure you read closely and follow Step 3.

**Concept Check** ☑

In the system

$$\begin{cases} x + y + z = 6 & \text{Equation (1)} \\ 2x - y + z = 3 & \text{Equation (2)} \\ x + 2y + 3z = 14 & \text{Equation (3)} \end{cases}$$

equations (1) and (2) are used to eliminate $y$. Which action could be used to best finish solving? Why?

**a.** Use (1) and (2) to eliminate $z$.      **b.** Use (2) and (3) to eliminate $y$.

**c.** Use (1) and (3) to eliminate $x$.

**EXAMPLE 3**   Solve the system.

$$\begin{cases} 2x + 4y \quad\quad = 1 & (1) \\ 4x \quad\quad - 4z = -1 & (2) \\ \quad\quad y - 4z = -3 & (3) \end{cases}$$

**Solution**   Notice that equation (2) has no term containing the variable $y$. Let us eliminate $y$ using equations (1) and (3). Multiply both sides of equation (3) by $-4$, and add the resulting equation to equation (1). Then

$$\begin{cases} 2x + 4y \quad\quad = 1 \\ -4(y - 4z) = -4(-3) \end{cases} \text{ simplifies to } \begin{cases} 2x + 4y \quad\quad = 1 \\ \underline{\quad\quad - 4y + 16z = 12} \\ 2x \quad\quad + 16z = 13 \quad (4) \end{cases}$$

Next, solve for $z$ using equations (4) and (2). Multiply both sides of equation (4) by $-2$ and add the resulting equation to equation (2).

$$\begin{cases} -2(2x + 16z) = -2(13) \\ 4x - 4z = -1 \end{cases} \text{ simplifies to } \begin{cases} -4x - 32z = -26 \\ \underline{\quad 4x - 4z = -1} \\ -36z = -27 \\ z = \dfrac{3}{4} \end{cases}$$

Replace $z$ with $\frac{3}{4}$ in equation (3) and solve for $y$.

$$y - 4\left(\frac{3}{4}\right) = -3 \quad \text{Let } z = \frac{3}{4} \text{ in equation (3).}$$
$$y - 3 = -3$$
$$y = 0$$

Replace $y$ with 0 in equation (1) and solve for $x$.

$$2x + 4(0) = 1$$
$$2x = 1$$
$$x = \frac{1}{2}$$

The solution is $\left(\frac{1}{2}, 0, \frac{3}{4}\right)$. Check to see that this solution satisfies all three equations of the system. □

**PRACTICE**
**3** Solve the system. $\begin{cases} 3x + 4y = 0 \\ 9x \quad - 4z = 6 \\ -2y + 7z = 1 \end{cases}$

**EXAMPLE 4** Solve the system.

$$\begin{cases} x - 5y - 2z = 6 & (1) \\ -2x + 10y + 4z = -12 & (2) \\ \frac{1}{2}x - \frac{5}{2}y - z = 3 & (3) \end{cases}$$

_Solution_ Multiply both sides of equation (3) by 2 to eliminate fractions, and multiply both sides of equation (2) by $-\frac{1}{2}$ so that the coefficient of $x$ is 1. The resulting system is then

$$\begin{cases} x - 5y - 2z = 6 & (1) \\ x - 5y - 2z = 6 & \text{Multiply (2) by } -\frac{1}{2}. \\ x - 5y - 2z = 6 & \text{Multiply (3) by 2.} \end{cases}$$

All three equations are identical, and therefore equations (1), (2), and (3) are all equivalent. There are infinitely many solutions of this system. The equations are dependent. The solution set can be written as $\{(x, y, z) \mid x - 5y - 2z = 6\}$. □

**PRACTICE**
**4** Solve the system. 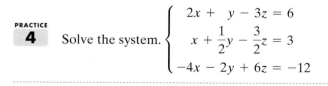 $\begin{cases} 2x + y - 3z = 6 \\ x + \frac{1}{2}y - \frac{3}{2}z = 3 \\ -4x - 2y + 6z = -12 \end{cases}$

As mentioned earlier, we can also use the substitution method to solve a system of linear equations in three variables.

**EXAMPLE 5** Solve the system:

$$\begin{cases} x - 4y - 5z = 35 & (1) \\ x - 3y \phantom{{}-5z} = 0 & (2) \\ \phantom{x}-y + z = -55 & (3) \end{cases}$$

**Solution** Notice in equations (2) and (3) that a variable is missing. Also notice that both equations contain the variable $y$. Let's use the substitution method by solving equation (2) for $x$ and equation (3) for $z$ and substituting the results in equation (1).

$$x - 3y = 0 \qquad (2)$$
$$x = 3y \qquad \text{Solve equation (2) for } x.$$
$$-y + z = -55 \qquad (3)$$
$$z = y - 55 \qquad \text{Solve equation (3) for } z.$$

Now substitute $3y$ for $x$ and $y - 55$ for $z$ in equation (1).

$$x - 4y - 5z = 35 \qquad (1)$$

▶ **Helpful Hint**

Do not forget to distribute.

$$3y - 4y - 5(y - 55) = 35 \qquad \text{Let } x = 3y \text{ and } z = y - 55.$$
$$3y - 4y - 5y + 275 = 35 \qquad \text{Use the distributive law and multiply.}$$
$$-6y + 275 = 35 \qquad \text{Combine like terms.}$$
$$-6y = -240 \qquad \text{Subtract 275 from both sides.}$$
$$y = 40 \qquad \text{Solve.}$$

To find $x$, recall that $x = 3y$ and substitute 40 for $y$. Then $x = 3y$ becomes $x = 3 \cdot 40 = 120$. To find $z$, recall that $z = y - 55$ and substitute 40 for $y$, also. Then $z = y - 55$ becomes $z = 40 - 55 = -15$. The solution is $(120, 40, -15)$. ☐

**PRACTICE**
**5** Solve the system. $\begin{cases} x + 2y + 4z = 16 \\ x \phantom{{}+2y} + 2z = -4 \\ \phantom{x}y - 3z = 30 \end{cases}$

## 4.4 EXERCISE SET

*Solve.*

**1.** Choose the equation(s) that has $(-1, 3, 1)$ as a solution.
   **a.** $x + y + z = 3$
   **b.** $-x + y + z = 5$
   **c.** $-x + y + 2z = 0$
   **d.** $x + 2y - 3z = 2$

**2.** Choose the equation(s) that has $(2, 1, -4)$ as a solution.
   **a.** $x + y + z = -1$
   **b.** $x - y - z = -3$
   **c.** $2x - y + z = -1$
   **d.** $-x - 3y - z = -1$

**3.** Use the result of Exercise 1 to determine whether $(-1, 3, 1)$ is a solution of the system below. Explain your answer.

$$\begin{cases} x + y + z = 3 \\ -x + y + z = 5 \\ x + 2y - 3z = 2 \end{cases}$$

**4.** Use the result of Exercise 2 to determine whether $(2, 1, -4)$ is a solution of the system below. Explain your answer.

$$\begin{cases} x + y + z = -1 \\ x - y - z = -3 \\ 2x - y + z = -1 \end{cases}$$

**MIXED PRACTICE**

*Solve each system. See Examples 1 through 5.*

**5.** $\begin{cases} x - y + z = -4 \\ 3x + 2y - z = 5 \\ -2x + 3y - z = 15 \end{cases}$

**6.** $\begin{cases} x + y - z = -1 \\ -4x - y + 2z = -7 \\ 2x - 2y - 5z = 7 \end{cases}$

**7.** $\begin{cases} x + y \phantom{{}- 3z} = 3 \\ \phantom{x+}2y \phantom{{}- 3z} = 10 \\ 3x + 2y - 3z = 1 \end{cases}$

**8.** $\begin{cases} 5x \phantom{{}+ y - 4z} = 5 \\ 2x + y \phantom{{}- 4z} = 4 \\ 3x + y - 4z = -15 \end{cases}$

**9.** $\begin{cases} 2x + 2y + z = 1 \\ -x + y + 2z = 3 \\ x + 2y + 4z = 0 \end{cases}$

**10.** $\begin{cases} 2x - 3y + z = 5 \\ x + y + z = 0 \\ 4x + 2y + 4z = 4 \end{cases}$

**11.** $\begin{cases} x - 2y + z = -5 \\ -3x + 6y - 3z = 15 \\ 2x - 4y + 2z = -10 \end{cases}$

**12.** $\begin{cases} 3x + y - 2z = 2 \\ -6x - 2y + 4z = -2 \\ 9x + 3y - 6z = 6 \end{cases}$

**13.** $\begin{cases} 4x - y + 2z = 5 \\ 2y + z = 4 \\ 4x + y + 3z = 10 \end{cases}$

**14.** $\begin{cases} 5y - 7z = 14 \\ 2x + y + 4z = 10 \\ 2x + 6y - 3z = 30 \end{cases}$

**15.** $\begin{cases} x + 5z = 0 \\ 5x + y = 0 \\ y - 3z = 0 \end{cases}$

**16.** $\begin{cases} x - 5y = 0 \\ x - z = 0 \\ -x + 5z = 0 \end{cases}$

**17.** $\begin{cases} 6x - 5z = 17 \\ 5x - y + 3z = -1 \\ 2x + y = -41 \end{cases}$

**18.** $\begin{cases} x + 2y = 6 \\ 7x + 3y + z = -33 \\ x - z = 16 \end{cases}$

**19.** $\begin{cases} x + y + z = 8 \\ 2x - y - z = 10 \\ x - 2y - 3z = 22 \end{cases}$

**20.** $\begin{cases} 5x + y + 3z = 1 \\ x - y + 3z = -7 \\ -x + y = 1 \end{cases}$

**21.** $\begin{cases} x + 2y - z = 5 \\ 6x + y + z = 7 \\ 2x + 4y - 2z = 5 \end{cases}$

**22.** $\begin{cases} 4x - y + 3z = 10 \\ x + y - z = 5 \\ 8x - 2y + 6z = 10 \end{cases}$

**23.** $\begin{cases} 2x - 3y + z = 2 \\ x - 5y + 5z = 3 \\ 3x + y - 3z = 5 \end{cases}$

**24.** $\begin{cases} 4x + y - z = 8 \\ x - y + 2z = 3 \\ 3x - y + z = 6 \end{cases}$

**25.** $\begin{cases} -2x - 4y + 6z = -8 \\ x + 2y - 3z = 4 \\ 4x + 8y - 12z = 16 \end{cases}$

**26.** $\begin{cases} -6x + 12y + 3z = -6 \\ 2x - 4y - z = 2 \\ -x + 2y + \dfrac{z}{2} = -1 \end{cases}$

**27.** $\begin{cases} 2x + 2y - 3z = 1 \\ y + 2z = -14 \\ 3x - 2y = -1 \end{cases}$

**28.** $\begin{cases} 7x + 4y = 10 \\ x - 4y + 2z = 6 \\ y - 2z = -1 \end{cases}$

**29.** $\begin{cases} x + 2y - z = 5 \\ -3x - 2y - 3z = 11 \\ 4x + 4y + 5z = -18 \end{cases}$

**30.** $\begin{cases} 3x - 3y + z = -1 \\ 3x - y - z = 3 \\ -6x + y + 2z = -6 \end{cases}$

**31.** $\begin{cases} \dfrac{3}{4}x - \dfrac{1}{3}y + \dfrac{1}{2}z = 9 \\ \dfrac{1}{6}x + \dfrac{1}{3}y - \dfrac{1}{2}z = 2 \\ \dfrac{1}{2}x - y + \dfrac{1}{2}z = 2 \end{cases}$

**32.** $\begin{cases} \dfrac{1}{3}x - \dfrac{1}{4}y + z = -9 \\ \dfrac{1}{2}x - \dfrac{1}{3}y - \dfrac{1}{4}z = -6 \\ x - \dfrac{1}{2}y - z = -8 \end{cases}$

## REVIEW AND PREVIEW

*Solve. See Section 2.4.*

**33.** The sum of two numbers is 45 and one number is twice the other. Find the numbers.

**34.** The difference between two numbers is 5. Twice the smaller number added to five times the larger number is 53. Find the numbers.

*Solve. See Section 2.3.*

**35.** $2(x - 1) - 3x = x - 12$

**36.** $7(2x - 1) + 4 = 11(3x - 2)$

**37.** $-y - 5(y + 5) = 3y - 10$

**38.** $z - 3(z + 7) = 6(2z + 1)$

## CONCEPT EXTENSIONS

**39.** Write a single linear equation in three variables that has $(-1, 2, -4)$ as a solution. (There are many possibilities.) Explain the process you used to write an equation.

**40.** Write a system of three linear equations in three variables that has $(2, 1, 5)$ as a solution. (There are many possibilities.) Explain the process you used to write an equation.

**41.** Write a system of linear equations in three variables that has the solution $(-1, 2, -4)$. Explain the process you used to write your system.

**42.** When solving a system of three equation in three unknowns, explain how to determine that a system has no solution.

**43.** The fraction $\dfrac{1}{24}$ can be written as the following sum:

$$\frac{1}{24} = \frac{x}{8} + \frac{y}{4} + \frac{z}{3}$$

where the numbers $x$, $y$, and $z$ are solutions of

$$\begin{cases} x + y + z = 1 \\ 2x - y + z = 0 \\ -x + 2y + 2z = -1 \end{cases}$$

Solve the system and see that the sum of the fractions is $\dfrac{1}{24}$.

**44.** The fraction $\dfrac{1}{18}$ can be written as the following sum:

$$\frac{1}{18} = \frac{x}{2} + \frac{y}{3} + \frac{z}{9}$$

where the numbers $x$, $y$, and $z$ are solutions of

$$\begin{cases} x + 3y + z = -3 \\ -x + y + 2z = -14 \\ 3x + 2y - z = 12 \end{cases}$$

Solve the system and see that the sum of the fractions is $\dfrac{1}{18}$.

Solving systems involving more than three variables can be accomplished with methods similar to those encountered in this section. Apply what you already know to solve each system of equations in four variables.

**45.** $\begin{cases} x + y \quad\;\; - w = 0 \\ \quad\;\; y + 2z + w = 3 \\ x \quad\;\; - z \quad\;\;\;\; = 1 \\ 2x - y \quad\;\; - w = -1 \end{cases}$

**46.** $\begin{cases} 5x + 4y \quad\quad\;\;\; = 29 \\ \quad\;\; y + z - w = -2 \\ 5x \quad\;\; + z \quad\;\;\;\; = 23 \\ \quad\;\; y - z + w = 4 \end{cases}$

**47.** $\begin{cases} x + y + z + w = 5 \\ 2x + y + z + w = 6 \\ x + y + z \quad\;\; = 2 \\ x + y \quad\quad\;\; = 0 \end{cases}$

**48.** $\begin{cases} 2x \quad\;\; - z \quad\quad\; = -1 \\ \quad\;\; y + z + w = 9 \\ \quad\;\; y \quad\;\; - 2w = -6 \\ x + y \quad\quad\;\; = 3 \end{cases}$

**49.** Write a system of three linear equations in three variables that are dependent equations.

**50.** What is the solution to the system in Exercise 49?

---

## 4.5   SYSTEMS OF LINEAR EQUATIONS AND PROBLEM SOLVING

**OBJECTIVES**

**1** Solve problems that can be modeled by a system of two linear equations.

**2** Solve problems with cost and revenue functions.

**3** Solve problems that can be modeled by a system of three linear equations.

**OBJECTIVE 1 ▶ Solving problem modeled by systems of two equations.** Thus far, we have solved problems by writing one-variable equations and solving for the variable. Some of these problems can be solved, perhaps more easily, by writing a system of equations, as illustrated in this section.

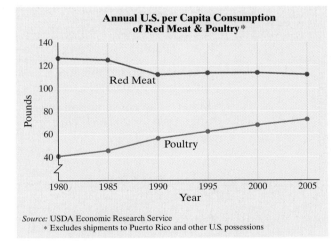

**Annual U.S. per Capita Consumption of Red Meat & Poultry***

*Source:* USDA Economic Research Service
* Excludes shipments to Puerto Rico and other U.S. possessions

**EXAMPLE 1**    **Predicting Equal Consumption of Red Meat and Poultry**

America's consumption of red meat has decreased most years since 1980 while consumption of poultry has increased. The function $y = -0.59x + 124.6$ approximates the annual pounds of red meat consumed per capita, where $x$ is the number of years since 1980. The function $y = 1.34x + 40.9$ approximates the annual pounds of poultry consumed per capita, where $x$ is also the number of years since 1980. If this trend continues, determine the year in which the annual consumption of red meat and poultry is equal. (*Source:* Based on data from Economic Research Service, U.S. Dept. of Agriculture)

*Solution*

**1. UNDERSTAND.** Read and reread the problem and guess a year. Let's guess the year 2020. This year is 40 years since 1980, so $x = 40$. Now let $x = 40$ in each given function.

Red meat :    $y = -0.59x + 124.6 = -0.59(40) + 124.6 = 101$ pounds
Poultry :     $y = 1.34x + 40.9 \quad = 1.34(40) + 40.9 \quad = 94.5$ pounds

Since the projected pounds in 2020 for red meat and poultry are not the same, we guessed incorrectly, but we do have a better understanding of the problem. We also know that the year will be later than 2020 since projected consumption of red meat is still greater than poultry that year.

**2. TRANSLATE.** We are already given the system of equations.

**3. SOLVE.** We want to know the year $x$ in which pounds $y$ are the same, so we solve the system:

$$\begin{cases} y = -0.59x + 124.6 \\ y = 1.34x + 40.9 \end{cases}.$$

Since both equations are solved for $y$, one way to solve is to use the substitution method.

$$y = -0.59x + 124.6 \quad \text{First equation}$$

$$1.34x + 40.9 = -0.59x + 124.6 \quad \text{Let } y = 1.34x + 40.9.$$

$$1.93x = 83.7$$

$$x = \frac{83.7}{1.93} \approx 43.37$$

**4. INTERPRET.** Since we are only asked to give the year, we need only solve for $x$.

**Check:** To check, see whether $x \approx 43.37$ gives approximately the same number of pounds of red meat and poultry.

Red meat: $-0.59x + 124.6 = -0.59(43.37) + 124.6 \approx 99.01$ pounds

Poultry: $1.34x + 40.9 = 1.34(43.37) + 40.9 \approx 99.02$ pounds

Since we rounded the number of years, the number of pounds do differ slightly. They differ only by 0.0041, so we can assume that we solved correctly.

**State:** The consumption of red meat and poultry will be the same about 43.37 years after 1980, or 2023.37. Thus, in the year 2023, we predict the consumption will be the same. □

**PRACTICE**

**1** Read Example 1. If we use the years 1995, 2000, and 2005 only to write functions approximating the consumption of red meat and poultry, we have the following:

Red meat: $y = -0.16x + 113.9$

Poultry: $y = 1.06x + 62.3$

where $x$ is the years since 1995 and $y$ is pounds per year consumed.

**a.** Assuming this trend continues, predict the year in which the consumption of red meat and poultry will be the same.

**b.** Does your answer differ from the example? Why or why not?

**EXAMPLE 2** **Finding Unknown Numbers**

A first number is 4 less than a second number. Four times the first number is 6 more than twice the second. Find the numbers.

*Solution*

**1. UNDERSTAND.** Read and reread the problem and guess a solution. If a first number is 10 and this is 4 less than a second number, the second number is 14. Four times the first number is 4(10), or 40. This is not equal to 6 more than twice the second number, which is 2(14) + 6 or 34. Although we guessed incorrectly, we now have a better understanding of the problem.

Since we are looking for two numbers, we will let

$$x = \text{first number}$$
$$y = \text{second number}$$

2. TRANSLATE. Since we have assigned two variables to this problem, we will translate the given facts into two equations. For the first statement we have

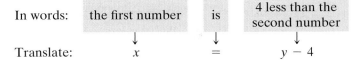

| In words: | the first number | is | 4 less than the second number |
|---|---|---|---|
| | ↓ | ↓ | ↓ |
| Translate: | $x$ | $=$ | $y - 4$ |

Next we translate the second statement into an equation.

| In words: | four times the first number | is | 6 more than twice the second number |
|---|---|---|---|
| | ↓ | ↓ | ↓ |
| Translate: | $4x$ | $=$ | $2y + 6$ |

3. SOLVE. Here we solve the system

$$\begin{cases} x = y - 4 \\ 4x = 2y + 6 \end{cases}$$

Since the first equation expresses $x$ in terms of $y$, we will use substitution. We substitute $y - 4$ for $x$ in the second equation and solve for $y$.

$$4x = 2y + 6 \quad \text{Second equation}$$

$$4\overbrace{(y - 4)} = 2y + 6 \quad \text{Let } x = y - 4.$$
$$4y - 16 = 2y + 6$$
$$2y = 22$$
$$y = 11$$

Now we replace $y$ with 11 in the equation $x = y - 4$ and solve for $x$. Then $x = y - 4$ becomes $x = 11 - 4 = 7$. The ordered pair solution of the system is $(7, 11)$.

4. INTERPRET. Since the solution of the system is $(7, 11)$, then the first number we are looking for is 7 and the second number is 11.

**Check:** Notice that 7 *is* 4 less than 11, and 4 times 7 *is* 6 more than twice 11. The proposed numbers, 7 and 11, are correct.

**State:** The numbers are 7 and 11.  □

**PRACTICE**

**2** A first number is 5 more than a second number. Twice the first number is 2 less than 3 times the second number. Find the numbers.

---

**EXAMPLE 3**  **Solving a Problem about Prices**

The Cirque du Soleil show Corteo is performing locally. Matinee admission for 4 adults and 2 children is $374, while admission for 2 adults and 3 children is $285.

**a.** What is the price of an adult's ticket?

**b.** What is the price of a child's ticket?

**c.** Suppose that a special rate of $1000 is offered for groups of 20 persons. Should a group of 4 adults and 16 children use the group rate? Why or why not?

*Solution*

1. UNDERSTAND. Read and reread the problem and guess a solution. Let's suppose that the price of an adult's ticket is $50 and the price of a child's ticket is $40. To check

our proposed solution, let's see if admission for 4 adults and 2 children is $374. Admission for 4 adults is 4($50) or $200 and admission for 2 children is 2($40) or $80. This gives a total admission of $200 + $80 = $280, not the required $374. Again though, we have accomplished the purpose of this process. We have a better understanding of the problem. To continue, we let

$A$ = the price of an adult's ticket and

$C$ = the price of a child's ticket

**2.** TRANSLATE. We translate the problem into two equations using both variables.

| In words: | admission for 4 adults | and | admission for 2 children | is | $374 |
|---|---|---|---|---|---|
| | ↓ | ↓ | ↓ | ↓ | ↓ |
| Translate: | $4A$ | $+$ | $2C$ | $=$ | $374$ |

| In words: | admission for 2 adults | and | admission for 3 children | is | $285 |
|---|---|---|---|---|---|
| | ↓ | ↓ | ↓ | ↓ | ↓ |
| Translate: | $2A$ | $+$ | $3C$ | $=$ | $285$ |

**3.** SOLVE. We solve the system.

$$\begin{cases} 4A + 2C = 374 \\ 2A + 3C = 285 \end{cases}$$

Since both equations are written in standard form, we solve by the addition method. First we multiply the second equation by $-2$ so that when we add the equations we eliminate the variable $A$. Then the system

$$\begin{cases} 4A + 2C = 374 \\ -2(2A + 3C) = -2(285) \end{cases} \quad \text{simplifies to} \quad \begin{cases} 4A + 2C = 374 \\ -4A - 6C = -570 \end{cases}$$

$$\begin{aligned} -4C &= -196 \quad \text{Add the equations.} \\ C &= 49 \quad \text{Divide by } -4. \end{aligned}$$

or $49, the children's ticket price.

To find $A$, we replace $C$ with 49 in the first equation.

$$\begin{aligned} 4A + 2C &= 374 \quad &&\text{First equation} \\ 4A + 2(49) &= 374 \quad &&\text{Let } C = 49. \\ 4A + 98 &= 374 \\ 4A &= 276 \\ A &= 69 \end{aligned}$$

or $69, the adult's ticket price.

**4.** INTERPRET.

**Check:** Notice that 4 adults and 2 children will pay 4($69) + 2($49) = $276 + $98 = $374, the required amount. Also, the price for 2 adults and 3 children is 2($69) + 3($49) = $138 + $147 = $285, the required amount.

**State:** Answer the three original questions.

**a.** Since $A = 69$, the price of an adult's ticket is $69.

**b.** Since $C = 49$, the price of a child's ticket is $49.

**c.** The regular admission price for 4 adults and 16 children is

$$\begin{aligned} 4(\$69) + 16(\$49) &= \$276 + \$784 \\ &= \$1060 \end{aligned}$$

This is $60 more than the special group rate of $1000, so they should request the group rate. □

PRACTICE
**3**    It is considered a premium game when the Red Sox or the Yankees come to Texas to play the Rangers. Admission for one of these games for three adults and three children under 14 is $75, while admission for two adults and 4 children is $62. (*Source:* MLB.com, Texas Rangers)

**a.** What is the price of an adult admission at Ameriquest Park?

**b.** What is the price of a child's admission?

**c.** Suppose that a special rate of $200 is offered for groups of 20 persons. Should a group of 5 adults and 15 children use the group rate? Why or why not?

---

### EXAMPLE 4    Finding the Rate of Speed

Two cars leave Indianapolis, one traveling east and the other west. After 3 hours they are 297 miles apart. If one car is traveling 5 mph faster than the other, what is the speed of each?

*Solution*

1. UNDERSTAND. Read and reread the problem. Let's guess a solution and use the formula $d = rt$ (distance = rate · time) to check. Suppose that one car is traveling at a rate of 55 miles per hour. This means that the other car is traveling at a rate of 50 miles per hour since we are told that one car is traveling 5 mph faster than the other. To find the distance apart after 3 hours, we will first find the distance traveled by each car. One car's distance is rate · time = 55(3) = 165 miles. The other car's distance is rate · time = 50(3) = 150 miles. Since one car is traveling east and the other west, their distance apart is the sum of their distances, or 165 miles + 150 miles = 315 miles. Although this distance apart is not the required distance of 297 miles, we now have a better understanding of the problem.

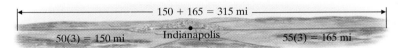

Let's model the problem with a system of equations. We will let

$$x = \text{speed of one car}$$
$$y = \text{speed of the other car}$$

We summarize the information on the following chart. Both cars have traveled 3 hours. Since distance = rate · time, their distances are $3x$ and $3y$ miles, respectively.

|  | *Rate* | • | *Time* | = | *Distance* |
|---|---|---|---|---|---|
| **One Car** | $x$ | | 3 | | $3x$ |
| **Other Car** | $y$ | | 3 | | $3y$ |

2. TRANSLATE. We can now translate the stated conditions into two equations.

| In words: | one car's distance | added to | the other car's distance | is | 297 |
|---|---|---|---|---|---|
| Translate: | $3x$ | $+$ | $3y$ | $=$ | 297 |

| In words: | one car's speed | is | 5 mph faster than the other |
|---|---|---|---|
| Translate: | $x$ | $=$ | $y + 5$ |

**3.** SOLVE. Here we solve the system

$$\begin{cases} 3x + 3y = 297 \\ x = y + 5 \end{cases}$$

Again, the substitution method is appropriate. We replace $x$ with $y + 5$ in the first equation and solve for $y$.

$$3x + 3y = 297 \quad \text{First equation}$$
$$3(y + 5) + 3y = 297 \quad \text{Let } x = y + 5.$$
$$3y + 15 + 3y = 297$$
$$6y = 282$$
$$y = 47$$

To find $x$, we replace $y$ with 47 in the equation $x = y + 5$. Then $x = 47 + 5 = 52$. The ordered pair solution of the system is (52, 47).

**4.** INTERPRET. The solution (52, 47) means that the cars are traveling at 52 mph and 47 mph, respectively.

**Check:** Notice that one car is traveling 5 mph faster than the other. Also, if one car travels 52 mph for 3 hours, the distance is 3(52) = 156 miles. The other car traveling for 3 hours at 47 mph travels a distance of 3(47) = 141 miles. The sum of the distances 156 + 141 is 297 miles, the required distance.

**State:** The cars are traveling at 52 mph and 47 mph.

> ▶ **Helpful Hint**
> Don't forget to attach units, if appropriate.

**PRACTICE**
**4** In 2007, the French train TGV V150 became the fastest conventional rail train in the world. It broke the 1990 record of the next fastest conventional rail train, the French TGV Atlantique. Assume the V150 and the Atlantique left the same station in Paris, with one heading west and one heading east. After 2 hours, they were 2150 kilometers apart. If the V150 is 75 kph faster than the Atlantique, what is the speed of each?

**EXAMPLE 5** **Mixing Solutions**

Lynn Pike, a pharmacist, needs 70 liters of a 50% alcohol solution. She has available a 30% alcohol solution and an 80% alcohol solution. How many liters of each solution should she mix to obtain 70 liters of a 50% alcohol solution?

*Solution*

**1.** UNDERSTAND. Read and reread the problem. Next, guess the solution. Suppose that we need 20 liters of the 30% solution. Then we need 70 − 20 = 50 liters of the 80% solution. To see if this gives us 70 liters of a 50% alcohol solution, let's find the amount of pure alcohol in each solution.

| number of liters | × | alcohol strength | = | amount of pure alcohol |
|---|---|---|---|---|
| 20 liters | × | 0.30 | = | 6 liters |
| 50 liters | × | 0.80 | = | 40 liters |
| 70 liters | × | 0.50 | = | 35 liters |

Since 6 liters + 40 liters = 46 liters and not 35 liters, our guess is incorrect, but we have gained some insight as to how to model and check this problem.

We will let

$$x = \text{amount of 30\% solution, in liters}$$
$$y = \text{amount of 80\% solution, in liters}$$

and use a table to organize the given data.

|  | Number of Liters | Alcohol Strength | Amount of Pure Alcohol |
|---|---|---|---|
| **30% Solution** | $x$ | 30% | $0.30x$ |
| **80% Solution** | $y$ | 80% | $0.80y$ |
| **50% Solution Needed** | 70 | 50% | $(0.50)(70)$ |

**2.** TRANSLATE. We translate the stated conditions into two equations.

In words:
$$\boxed{\text{amount of 30\% solution}} \quad + \quad \boxed{\text{amount of 80\% solution}} \quad = \quad \boxed{70}$$

Translate:
$$x \qquad\qquad + \qquad\qquad y \qquad\qquad = \qquad 70$$

In words:
$$\boxed{\begin{array}{c}\text{amount of pure}\\\text{alcohol in 30\%}\\\text{solution}\end{array}} + \boxed{\begin{array}{c}\text{amount of pure}\\\text{alcohol in 80\%}\\\text{solution}\end{array}} = \boxed{\begin{array}{c}\text{amount of pure}\\\text{alcohol in 50\%}\\\text{solution}\end{array}}$$

Translate:
$$0.30x \qquad + \qquad 0.80y \qquad = \qquad (0.50)(70)$$

**3.** SOLVE. Here we solve the system

$$\begin{cases} x + y = 70 \\ 0.30x + 0.80y = (0.50)(70) \end{cases}$$

To solve this system, we use the elimination method. We multiply both sides of the first equation by $-3$ and both sides of the second equation by 10. Then

$$\begin{cases} -3(x + y) = -3(70) \\ 10(0.30x + 0.80y) = 10(0.50)(70) \end{cases} \quad\begin{array}{c}\text{simplifies}\\\text{to}\end{array}\quad \begin{cases} -3x - 3y = -210 \\ \underline{3x + 8y = 350} \\ \phantom{3x + }5y = 140 \\ \phantom{3x + }y = 28 \end{cases}$$

Now we replace $y$ with 28 in the equation $x + y = 70$ and find that $x + 28 = 70$, or $x = 42$.
The ordered pair solution of the system is $(42, 28)$.

**4.** INTERPRET.

**Check:**   Check the solution in the same way that we checked our guess.

**State:**   The pharmacist needs to mix 42 liters of 30% solution and 28 liters of 80% solution to obtain 70 liters of 50% solution. ☐

**PRACTICE**
**5**   Keith Robinson is a chemistry teacher who needs 1 liter of a solution of 5% hydrochloric acid to carry out an experiment. If he only has a stock solution of 99% hydrochloric acid, how much water (0% acid) and how much stock solution (99%) of HCL must he mix to get 1 liter of 5% solution? Round answers to the nearest hundredth of a liter.

**Concept Check** ☑

Suppose you mix an amount of **25%** acid solution with an amount of **60%** acid solution. You then calculate the acid strength of the resulting acid mixture. For which of the following results should you suspect an error in your calculation? Why?

**a.** 14%          **b.** 32%          **c.** 55%

**OBJECTIVE 2 ▶ Solving problems with cost and revenue functions.** Recall that businesses are often computing cost and revenue functions or equations to predict sales, to determine whether prices need to be adjusted, and to see whether the company is making or losing money. Recall also that the value at which revenue equals cost is called the break-even point. When revenue is less than cost, the company is losing money; when revenue is greater than cost, the company is making money.

<u>**EXAMPLE 6**</u>    **Finding a Break-Even Point**

A manufacturing company recently purchased $3000 worth of new equipment to offer new personalized stationery to its customers. The cost of producing a package of personalized stationery is $3.00, and it is sold for $5.50. Find the number of packages that must be sold for the company to break even.

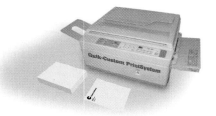

*Solution*

1. UNDERSTAND. Read and reread the problem. Notice that the cost to the company will include a one-time cost of $3000 for the equipment and then $3.00 per package produced. The revenue will be $5.50 per package sold.

    To model this problem, we will let

$$x = \text{number of packages of personalized stationery}$$
$$C(x) = \text{total cost for producing } x \text{ packages of stationery}$$
$$R(x) = \text{total revenue for selling } x \text{ packages of stationery}$$

2. TRANSLATE. The revenue equation is

| In words: | revenue for selling x packages of stationery | = | price per package | · | number of packages |
|---|---|---|---|---|---|
| Translate: | $R(x)$ | = | 5.5 | · | $x$ |

The cost equation is

| In words: | cost for producing x packages of stationery | = | cost per package | · | number of packages | + | cost for equipment |
|---|---|---|---|---|---|---|---|
| Translate: | $C(x)$ | = | 3 | · | $x$ | + | 3000 |

Since the break-even point is when $R(x) = C(x)$, we solve the equation

$$5.5x = 3x + 3000$$

**3.** SOLVE.

$$5.5x = 3x + 3000$$
$$2.5x = 3000 \qquad \text{Subtract } 3x \text{ from both sides.}$$
$$x = 1200 \qquad \text{Divide both sides by 2.5.}$$

**4.** INTERPRET.

**Check:**   To see whether the break-even point occurs when 1200 packages are produced and sold, see if revenue equals cost when $x = 1200$. When $x = 1200$, $R(x) = 5.5x = 5.5(1200) = 6600$ and $C(x) = 3x + 3000 = 3(1200) + 3000 = 6600$. Since $R(1200) = C(1200) = 6600$, the break-even point is 1200.

**State:**   The company must sell 1200 packages of stationery to break even. The graph of this system is shown.

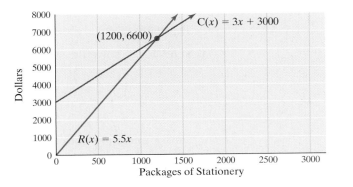

PRACTICE
**6**   An online-only electronics firm recently purchased $3000 worth of new equipment to create shock-proof packaging for its products. The cost of producing one shock-proof package is $2.50, and the firm charges the customer $4.50 for the packaging. Find the number of packages that must be sold for the company to break even.

**OBJECTIVE 3 ▶ Solving problems modeled by systems of three equations.** To introduce problem solving by writing a system of three linear equations in three variables, we solve a problem about triangles.

**EXAMPLE 7**   **Finding Angle Measures**

The measure of the largest angle of a triangle is 80° more than the measure of the smallest angle, and the measure of the remaining angle is 10° more than the measure of the smallest angle. Find the measure of each angle.

*Solution*

**1.** UNDERSTAND. Read and reread the problem. Recall that the sum of the measures of the angles of a triangle is 180°. Then guess a solution. If the smallest angle measures 20°, the measure of the largest angle is 80° more, or $20° + 80° = 100°$. The measure of the remaining angle is 10° more than the measure of the smallest angle, or $20° + 10° = 30°$. The sum of these three angles is $20° + 100° + 30° = 150°$, not the required 180°. We now know that the measure of the smallest angle is greater than 20°.

To model this problem we will let

$x$ = degree measure of the smallest angle
$y$ = degree measure of the largest angle
$z$ = degree measure of the remaining angle

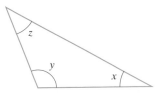

**2.** TRANSLATE. We translate the given information into three equations.

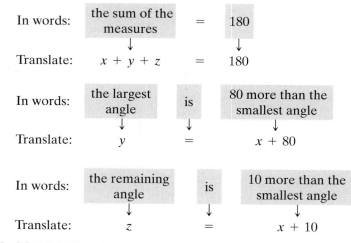

**3.** SOLVE. We solve the system

$$\begin{cases} x + y + z = 180 \\ y = x + 80 \\ z = x + 10 \end{cases}$$

Since $y$ and $z$ are both expressed in terms of $x$, we will solve using the substitution method. We substitute $y = x + 80$ and $z = x + 10$ in the first equation. Then

$$x + y + z = 180 \quad \text{First equation}$$

$$x + \overbrace{(x + 80)} + \overbrace{(x + 10)} = 180 \quad \text{Let } y = x + 80 \text{ and } z = x + 10.$$

$$3x + 90 = 180$$

$$3x = 90$$

$$x = 30$$

Then $y = x + 80 = 30 + 80 = 110$, and $z = x + 10 = 30 + 10 = 40$. The ordered triple solution is $(30, 110, 40)$.

**4.** INTERPRET.

**Check:** Notice that $30° + 40° + 110° = 180°$. Also, the measure of the largest angle, $110°$, is $80°$ more than the measure of the smallest angle, $30°$. The measure of the remaining angle, $40°$, is $10°$ more than the measure of the smallest angle, $30°$. ☐

**PRACTICE**

**7** The measure of the largest angle of a triangle is $40°$ more than the measure of the smallest angle, and the measure of the remaining angle is $20°$ more than the measure of the smallest angle. Find the measure of each angle.

## 4.5 | EXERCISE SET

*Without actually solving each problem, choose each correct solution by deciding which choice satisfies the given conditions.*

△ **1.** The length of a rectangle is 3 feet longer than the width. The perimeter is 30 feet. Find the dimensions of the rectangle.

    **a.** length = 8 feet; width = 5 feet

    **b.** length = 8 feet; width = 7 feet

    **c.** length = 9 feet; width = 6 feet

△ **2.** An isosceles triangle, a triangle with two sides of equal length, has a perimeter of 20 inches. Each of the equal sides is one inch longer than the third side. Find the lengths of the three sides.

    **a.** 6 inches, 6 inches, and 7 inches

    **b.** 7 inches, 7 inches, and 6 inches

    **c.** 6 inches, 7 inches, and 8 inches

**3.** Two computer disks and three notebooks cost $17. However, five computer disks and four notebooks cost $32. Find the price of each.

    **a.** notebook = $4; computer disk = $3

    **b.** notebook = $3; computer disk = $4

    **c.** notebook = $5; computer disk = $2

**4.** Two music CDs and four music cassette tapes cost a total of $40. However, three music CDs and five cassette tapes cost $55. Find the price of each.

    **a.** CD = $12; cassette = $4  **b.** CD = $15; cassette = $2

    **c.** CD = $10; cassette = $5

**5.** Kesha has a total of 100 coins, all of which are either dimes or quarters. The total value of the coins is $13.00. Find the number of each type of coin.

    **a.** 80 dimes; 20 quarters    **b.** 20 dimes; 44 quarters

    **c.** 60 dimes; 40 quarters

**6.** Samuel has 28 gallons of saline solution available in two large containers at his pharmacy. One container holds three times as much as the other container. Find the capacity of each container.

    **a.** 15 gallons; 5 gallons    **b.** 20 gallons; 8 gallons

    **c.** 21 gallons; 7 gallons

*Write a system of equations in x and y describing each situation. Do not solve the system. See Example 2.*

**7.** A smaller number and a larger number add up to 15 and have a difference of 7. (Let $x$ be the larger number.)

**8.** The total of two numbers is 16. The first number plus 2 more than 3 times the second equals 18. (Let $x$ be the first number.)

**9.** Keiko has a total of $6500, which she has invested in two accounts. The larger account is $800 greater than the smaller account. (Let $x$ be the amount of money in the larger account.)

**10.** Dominique has four times as much money in his savings account as in his checking account. The total amount is $2300. (Let $x$ be the amount of money in his checking account.)

## MIXED PRACTICE

*Solve. See Examples 1 through 5.*

**11.** Two numbers total 83 and have a difference of 17. Find the two numbers.

**12.** The sum of two numbers is 76 and their difference is 52. Find the two numbers.

**13.** A first number plus twice a second number is 8. Twice the first number, plus the second totals 25. Find the numbers.

**14.** One number is 4 more than twice the second number. Their total is 25. Find the numbers.

**15.** The highest scorer during the WNBA 2006 regular season was Diana Taurasi of the Phoenix Mercury. Over the season, Taurasi scored 116 more points than Seimone Augustus of the Minnesota Lynx. Together, Taurasi and Augustus scored 1604 points during the 2006 regular season. How many points did each player score over the course of the season? (*Source:* Women's National Basketball Association)

**16.** During the 2006 regular MLB season, Ryan Howard of the Philadelphia Phillies hit the most home runs of any player in the major leagues. Over the course of the season, he hit 4 more home runs than David Ortiz of the Boston Red Sox. Together, these batting giants hit 112 home runs. How many home runs did each player hit? (*Source:* Major League Baseball)

**17.** Ann Marie Jones has been pricing Amtrak train fares for a group trip to New York. Three adults and four children must pay $159. Two adults and three children must pay $112. Find the price of an adult's ticket, and find the price of a child's ticket.

**18.** Last month, Jerry Papa purchased five DVDs and two CDs at Wall-to-Wall Sound for $65. This month he bought three DVDs and four CDs for $81. Find the price of each DVD, and find the price of each CD.

**19.** Johnston and Betsy Waring have a jar containing 80 coins, all of which are either quarters or nickels. The total value of the coins is $14.60. How many of each type of coin do they have?

**20.** Sarah and Keith Robinson purchased 40 stamps, a mixture of 39¢ and 24¢ stamps. Find the number of each type of stamp if they spent $14.85.

**21.** Davie and Judi Mihaly own 50 shares of Apple stock and 60 shares of Microsoft stock. At the close of the markets on March 9, 2007, their stock portfolio was worth $6035.90. The closing price of the Microsoft stock was $60.68 less than the closing price of Apple stock on that day. What was the price of each stock on March 9, 2007? (*Source:* New York Stock Exchange)

**22.** Pho Lin has investments in eBay and Amazon stock. On March 9, 2007, eBay stock closed at $30.82 per share and Amazon stock closed at $38.84 per share. Pho's stock portfolio was worth $2866.60 at the end of the day. If Pho owns 20 more shares of Amazon stock than eBay stock, how many of each type of stock does she own?

**23.** Twice last month, Judy Carter rented a car from Enterprise in Fresno, California, and traveled around the Southwest on business. Enterprise rents its cars for a daily fee, plus an additional charge per mile driven. Judy recalls that her first trip lasted 4 days, she drove 450 miles, and the rental cost her $240.50. On her second business trip she drove 200 miles in 3 days, and paid $146.00 for the rental. Find the daily fee and the mileage charge.

**24.** Joan Gundersen rented a car from Hertz, which rents its cars for a daily fee plus an additional charge per mile driven. Joan recalls that a car rented for 5 days and driven for 300 miles cost her $178, while a car rented for 4 days and driven for 500 miles cost $197. Find the daily fee, and find the mileage charge.

**25.** Pratap Puri rowed 18 miles down the Delaware River in 2 hours, but the return trip took him $4\frac{1}{2}$ hours. Find the rate Pratap can row in still water, and find the rate of the current.
Let $x$ = rate Pratap can row in still water and
$\quad y$ = rate of the current

| $d =$ | $r$ | $\cdot$ $t$ |
|---|---|---|
| **Downstream** | $x + y$ | |
| **Upstream** | $x - y$ | |

**26.** The Jonathan Schultz family took a canoe 10 miles down the Allegheny River in $1\frac{1}{4}$ hours. After lunch it took them 4 hours to return. Find the rate of the current.
Let $x$ = rate the family can row in still water and
$\quad y$ = rate of the current

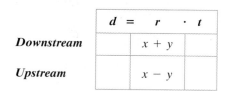

| $d =$ | $r$ | $\cdot$ $t$ |
|---|---|---|
| **Downstream** | $x + y$ | |
| **Upstream** | $x - y$ | |

**27.** Dave and Sandy Hartranft are frequent flyers with Delta Airlines. They often fly from Philadelphia to Chicago, a distance of 780 miles. On one particular trip they fly into the wind, and the flight takes 2 hours. The return trip, with the wind behind them, only takes $1\frac{1}{2}$ hours. If the wind speed is the same on each trip, find the speed of the wind and find the speed of the plane in still air.

**28.** With a strong wind behind it, a United Airlines jet flies 2400 miles from Los Angeles to Orlando in $4\frac{3}{4}$ hours. The return trip takes 6 hours, as the plane flies into the wind. If the wind speed is the same on each trip, find the speed of the plane in still air, and find the wind speed to the nearest tenth of a mile per hour.

**29.** Jim Williamson began a 96-mile bicycle trip to build up stamina for a triathlete competition. Unfortunately, his bicycle chain broke, so he finished the trip walking. The whole trip took 6 hours. If Jim walks at a rate of 4 miles per hour and rides at 20 miles per hour, find the amount of time he spent on the bicycle.

**30.** In Canada, eastbound and westbound trains travel along the same track, with sidings to pull onto to avoid accidents. Two trains are now 150 miles apart, with the westbound train traveling twice as fast as the eastbound train. A warning must be issued to pull one train onto a siding or else the trains will crash in $1\frac{1}{4}$ hours. Find the speed of the eastbound train and the speed of the westbound train.

**31.** Doreen Schmidt is a chemist with Gemco Pharmaceutical. She needs to prepare 12 ounces of a 9% hydrochloric acid solution. Find the amount of a 4% solution and the amount of a 12% solution she should mix to get this solution.

| Concentration Rate | Ounces of Solution | Ounces of Pure Acid |
|---|---|---|
| 0.04 | $x$ | $0.04x$ |
| 0.12 | $y$ | ? |
| 0.09 | 12 | ? |

**32.** Elise Everly is preparing 15 liters of a 25% saline solution. Elise has two other saline solutions with strengths of 40% and

10%. Find the amount of 40% solution and the amount of 10% solution she should mix to get 15 liters of a 25% solution.

| Concentration Rate | Liters of Solution | Liters of Pure Salt |
|---|---|---|
| 0.40 | $x$ | $0.40x$ |
| 0.10 | $y$ | ? |
| 0.25 | 15 | ? |

**33.** Wayne Osby blends coffee for a local coffee café. He needs to prepare 200 pounds of blended coffee beans selling for $3.95 per pound. He intends to do this by blending together a high-quality bean costing $4.95 per pound and a cheaper bean costing $2.65 per pound. To the nearest pound, find how much high-quality coffee bean and how much cheaper coffee bean he should blend.

**34.** Macadamia nuts cost an astounding $16.50 per pound, but research by an independent firm says that mixed nuts sell better if macadamias are included. The standard mix costs $9.25 per pound. Find how many pounds of macadamias and how many pounds of the standard mix should be combined to produce 40 pounds that will cost $10 per pound. Find the amounts to the nearest tenth of a pound.

△ **35.** Recall that two angles are complementary if the sum of their measures is 90°. Find the measures of two complementary angles if one angle is twice the other.

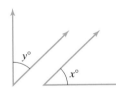

△ **36.** Recall that two angles are supplementary if the sum of their measures is 180°. Find the measures of two supplementary angles if one angle is 20° more than four times the other.

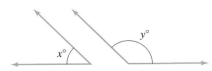

△ **37.** Find the measures of two complementary angles if one angle is 10° more than three times the other.

△ **38.** Find the measures of two supplementary angles if one angle is 18° more than twice the other.

**39.** Kathi and Robert Hawn had a pottery stand at the annual Skippack Craft Fair. They sold some of their pottery at the original price of $9.50 each, but later decreased the price of each by $2. If they sold all 90 pieces and took in $721, find how many they sold at the original price and how many they sold at the reduced price.

**40.** A charity fund-raiser consisted of a spaghetti supper where a total of 387 people were fed. They charged $6.80 for adults and half-price for children. If they took in $2444.60, find how many adults and how many children attended the supper.

**41.** The Santa Fe National Historic Trail is approximately 1200 miles between Old Franklin, Missouri, and Santa Fe, New Mexico. Suppose that a group of hikers start from each town and walk the trail toward each other. They meet after a total hiking time of 240 hours. If one group travels $\frac{1}{2}$ mile per hour slower than the other group, find the rate of each group. (*Source:* National Park Service)

**42.** California 1 South is a historic highway that stretches 123 miles along the coast from Monterey to Morro Bay. Suppose that two antique cars start driving this highway, one from each town. They meet after 3 hours. Find the rate of each car if one car travels 1 mile per hour faster than the other car. (*Source:* National Geographic)

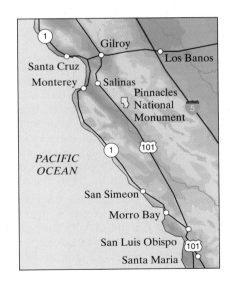

**43.** A 30% solution of fertilizer is to be mixed with a 60% solution of fertilizer in order to get 150 gallons of a 50% solution. How many gallons of the 30% solution and 60% solution should be mixed?

**44.** A 10% acid solution is to be mixed with a 50% acid solution in order to get 120 ounces of a 20% acid solution. How many ounces of the 10% solution and 50% solution should be mixed?

**45.** Traffic signs are regulated by the *Manual on Uniform Traffic Control Devices* (MUTCD). According to this manual, if the sign below is placed on a freeway, its perimeter must be 144 inches. Also, its length is 12 inches longer than its width. Find the dimensions of this sign.

**46.** According to the MUTCD (see Exercise 45), this sign must have a perimeter of 60 inches. Also, its length must be 6 inches longer than its width. Find the dimensions of this sign.

**47.** In the United States, the percent of women using the Internet is increasing faster than the percent of men. For the years 2000–2005, the function $y = 5.3x + 39.5$ can be used to estimate the percent of females using the Internet, while the function $y = 4.5x + 45.5$ can be used to estimate the percent of males. For both functions, $x$ is the number of years since 2000. If this trend continues, predict the year in which the percent of females using the Internet equals the percent of males. (*Source: Pew Internet & American Life Project*)

**48.** The percent of car vehicle sales has been decreasing over a ten-year period while the percent of light truck (pickups, sport-utility vans, and minivans) vehicles has been increasing. For the years 2000–2006, the function $y = -x + 54.2$ can be used to estimate the percent of new car vehicle sales in the U.S., while the function $y = x + 45.8$ can be used to estimate the percent of light truck vehicle sales. For both functions, $x$ is the number of years since 2000. (*Source: USA Today, Environmental Protection Agency, "Light-Duty Automotive Technology and fuel Economy Trends: 1975–2006"*)

   **a.** Calculate the year in which the percent of new car sales equaled the percent of light truck sales.

   **b.** Before the actual 2001 vehicle sales data was published, *USA Today* predicted that light truck sales would likely be greater than car sales in the year 2001. Does your finding in part (a) agree with this statement?

**49.** While it is said that trains opened up the American West to settlement, U.S. railroad miles have been on the decline for decades. On the other hand, the miles of roads in the U.S. highway system have been increasing. The function $y = -1379.4x + 150,604$ represents the U.S. railroad miles, while the function $y = 478.4x + 157,838$ models the number of U.S. highway miles, where $x$ is the number of years after 1995. For each function, $x$ is the number of years after 1995. (*Source:* Association of American Railroads, Federal Highway Administration)

   **a.** Explain how the decrease in railroad miles can be verified by their given function while the increase in highway miles can be verified by their given function.

   **b.** Find the year in which it is estimated that the number of U.S. railroad miles and the number of U.S. highway miles were the same.

**50.** The annual U.S. per capita consumption of whole milk has decreased since 1980, while the per capita consumption of lower fat milk has increased. For the years 1980–2005, the function $y = -0.40x + 15.9$ approximates the annual U.S. per capita consumption of whole milk in gallons, and the function $y = 0.14x + 11.9$ approximates the annual U.S. per capita consumption of lower fat milk in gallons. Determine the year in which the per capita consumption of whole milk equaled the per capita consumption of lower fat milk. (*Source:* Economic Research Service: U.S.D.A.)

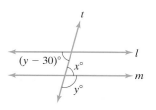

△ **51.** In the figure, line $l$ and line $m$ are parallel lines cut by transversal $t$. Find the values of $x$ and $y$.

△ **52.** Find the values of $x$ and $y$ in the following isosceles triangle.

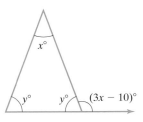

*Given the cost function $C(x)$ and the revenue function $R(x)$, find the number of units $x$ that must be sold to break even. See Example 6.*

**53.** $C(x) = 30x + 10,000 \ R(x) = 46x$

**54.** $C(x) = 12x + 15,000 \ R(x) = 32x$

**55.** $C(x) = 1.2x + 1500 \ R(x) = 1.7x$

**56.** $C(x) = 0.8x + 900 \ R(x) = 2x$

**57.** $C(x) = 75x + 160,000 \ R(x) = 200x$

**58.** $C(x) = 105x + 70,000 \ R(x) = 245x$

**59.** The planning department of Abstract Office Supplies has been asked to determine whether the company should introduce a new computer desk next year. The department estimates that $6000 of new manufacturing equipment will need to be purchased and that the cost of constructing each desk will be $200. The department also estimates that the revenue from each desk will be $450.

   **a.** Determine the revenue function $R(x)$ from the sale of $x$ desks.

   **b.** Determine the cost function $C(x)$ for manufacturing $x$ desks.

   **c.** Find the break-even point.

**60.** Baskets, Inc., is planning to introduce a new woven basket. The company estimates that $500 worth of new equipment will be needed to manufacture this new type of basket and that it will cost $15 per basket to manufacture. The company also estimates that the revenue from each basket will be $31.

   **a.** Determine the revenue function $R(x)$ from the sale of $x$ baskets.

   **b.** Determine the cost function $C(x)$ for manufacturing $x$ baskets.

   **c.** Find the break-even point.

*Solve. See Example 7.*

**61.** Rabbits in a lab are to be kept on a strict daily diet that includes 30 grams of protein, 16 grams of fat, and 24 grams of carbohydrates. The scientist has only three food mixes available with the following grams of nutrients per unit.

|        | Protein | Fat | Carbohydrate |
|--------|---------|-----|--------------|
| Mix A  | 4       | 6   | 3            |
| Mix B  | 6       | 1   | 2            |
| Mix C  | 4       | 1   | 12           |

Find how many units of each mix are needed daily to meet each rabbit's dietary need.

**62.** Gerry Gundersen mixes different solutions with concentrations of 25%, 40%, and 50% to get 200 liters of a 32% solution. If he uses twice as much of the 25% solution as of the 40% solution, find how many liters of each kind he uses.

△ **63.** The perimeter of a quadrilateral (four-sided polygon) is 29 inches. The longest side is twice as long as the shortest side. The other two sides are equally long and are 2 inches longer than the shortest side. Find the length of all four sides.

△ **64.** The measure of the largest angle of a triangle is 90° more than the measure of the smallest angle, and the measure of the remaining angle is 30° more than the measure of the smallest angle. Find the measure of each angle.

**65.** The sum of three numbers is 40. One number is five more than a second number. It is also twice the third. Find the numbers.

**66.** The sum of the digits of a three-digit number is 15. The tens-place digit is twice the hundreds-place digit, and the ones-place digit is 1 less than the hundreds-place digit. Find the three-digit number.

**67.** Diana Taurasi, of the Phoenix Mercury, was the WNBA's top scorer for the 2006 regular season, with a total of 860 points. The number of two-point field goals that Taurasi made was 65 less than double the number of three-point field goals she made. The number of free throws (each worth one point) she made was 34 less than the number of two-point field goals she made. Find how many free throws, two-point field goals, and three-point field goals Diana Taurasi made during the 2006 regular season. (*Source:* Women's National Basketball Association)

**68.** During the 2006 NBA playoffs, the top scoring player was Dwayne Wade of the Miami Heat. Wade scored a total of 654 points during the playoffs. The number of free throws (each worth one point) he made was three less than the number of two-point field goals he made. He also made 27 fewer three-point field goals than one-fifth the number of two-point field goals. How many free throws, two-point field goals, and three-point field goals did Dwayne Wade make during the 2006 playoffs? (*Source:* National Basketball Association)

△ **69.** Find the values of $x$, $y$, and $z$ in the following triangle.

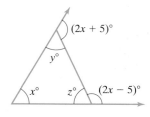

△ **70.** The sum of the measures of the angles of a quadrilateral is 360°. Find the value of $x$, $y$, and $z$ in the following quadrilateral.

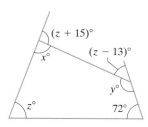

**REVIEW AND PREVIEW**

*Solve each linear inequality. See Section 2.8.*

**71.** $-3x < -9$

**72.** $2x - 7 \le 5x + 11$

**73.** $4(2x - 1) \ge 0$

**74.** $\dfrac{2}{3}x < \dfrac{1}{3}$

**CONCEPT EXTENSIONS**

*Solve. See the Concept Check in the section.*

**75.** Suppose you mix an amount of candy costing $0.49 a pound with candy costing $0.65 a pound. Which of the following costs per pound could result?

   **a.** $0.58    **b.** $0.72    **c.** $0.29

**76.** Suppose you mix a 50% acid solution with pure acid (100%). Which of the following acid strengths are possible for the resulting acid mixture?

   **a.** 25%    **b.** 150%    **c.** 62%    **d.** 90%

△ **77.** Dale and Sharon Mahnke have decided to fence off a garden plot behind their house, using their house as the "fence" along one side of the garden. The length (which runs parallel to the house) is 3 feet less than twice the width. Find the dimensions if 33 feet of fencing is used along the three sides requiring it.

△ **78.** Judy McElroy plans to erect 152 feet of fencing around her rectangular horse pasture. A river bank serves as one side length of the rectangle. If each width is 4 feet longer than half the length, find the dimensions.

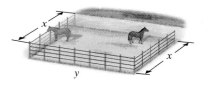

**79.** The percent of viewers who watch nightly network news can be approximated by the equation $y = 0.82x + 17.2$, where $x$ is the years of age over 18 of the viewer. The percent of viewers who watch cable TV news is approximated by the equation $y = 0.33x + 30.5$ where $x$ is also the years of age over 18 of the viewer. (*Source:* The Pew Research Center for The People & The Press)

   **a.** Solve the system of equations: $\begin{cases} y = 0.82x + 17.2 \\ y = 0.33x + 30.5 \end{cases}$

   Round $x$ and $y$ to the nearest tenth.

   **b.** Explain what the point of intersection means in terms of the context of the exercise.

   **c.** Look at the slopes of both equations of the system. What type of news attracts older viewers more? What type of news attracts younger viewers more?

**80.** In the triangle below, the measure of angle $x$ is 6 times the measure of angle $y$. Find the measure of $x$ and $y$ by writing a system of two equations in two unknowns and solving the system.

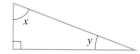

**81.** MySpace and Facebook are both popular social networking Web sites. The function $f(x) = 0.85x + 41.75$ represents the MySpace minutes (in thousands)/month while the function $f(x) = 1.13x + 10.49$ represents the Facebook minutes (in thousands)/month. In both of these functions $x = 1$ represents Feb. 2007, $x = 2$ represents March 2007 and so on.

   **a.** Solve the system formed by these functions. Round each coordinate to the nearest whole number.

   **b.** Use your answer to part **a** to predict the month and year in which the total pages viewed for MySpace and Facebook are the same.

**82.** Find the values of $a$, $b$, and $c$ such that the equation $y = ax^2 + bx + c$ has ordered pair solutions $(1, 6)$, $(-1, -2)$, and $(0, -1)$. To do so, substitute each ordered pair solution into the equation. Each time, the result is an equation in three unknowns: $a$, $b$, and $c$. Then solve the resulting system of three linear equations in three unknowns, $a$, $b$, and $c$.

# CHAPTER 4 GROUP ACTIVITY

## Break-Even Point
### Sections 4.1, 4.2, 4.3, 4.5

When a business sells a new product, it generally does not start making a profit right away. There are usually many expenses associated with creating a new product. These expenses might include an advertising blitz to introduce the product to the public. These start-up expenses might also include the cost of market research and product development or any brand-new equipment needed to manufacture the product. Start-up costs like these are generally called *fixed costs* because they don't depend on the number of items manufactured. Expenses that depend on the number of items manufactured, such as the cost of materials and shipping, are called *variable costs*. The total cost of manufacturing the new product is given by the cost equation: Total cost = Fixed costs + Variable costs.

For instance, suppose a greeting card company is launching a new line of greeting cards. The company spent $7000 doing product research and development for the new line and spent $15,000 on advertising the new line. The company does not need to buy any new equipment to manufacture the cards, but the paper and ink needed to make each card will cost $0.20 per card. The total cost $y$ in dollars for manufacturing $x$ cards is $y = 22,000 + 0.20x$.

Once a business sets a price for the new product, the company can find the product's expected *revenue*. Revenue is the amount of money the company takes in from the sales of its product. The revenue from selling a product is given by the revenue equation: Revenue = Price per item × Number of items sold.

For instance, suppose that the card company plans to sell its new cards for $1.50 each. The revenue $y$, in dollars, that the company can expect to receive from the sales of $x$ cards is $y = 1.50x$.

If the total cost and revenue equations are graphed on the same coordinate system, the graphs should intersect. The point of intersection is where total cost equals revenue and is called the *break-even point*. The break-even point gives the number of items $x$ that must be manufactured and sold for the company to recover its expenses. If fewer than this number of items are produced and sold, the company loses money. If more than this number of items are produced and sold, the company makes a profit. In the case of the greeting card company, approximately 16,923 cards must be manufactured and sold for the company to break even on this new card line. The total cost and revenue of producing and selling 16,923 cards is the same. It is approximately $25,385.

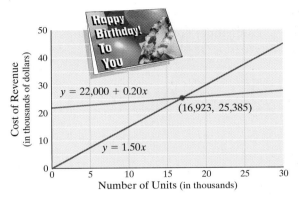

## Group Activity

Suppose your group is starting a small business near your campus.

**a.** Choose a business and decide what campus-related product or service you will provide.

**b.** Research the fixed costs of starting up such a business.

**c.** Research the variable costs of producing such a product or providing such a service.

**d.** Decide how much you would charge per unit of your product or service.

**e.** Find a system of equations for the total cost and revenue of your product or service.

**f.** How many units of your product or service must be sold before your business will break even?

## CHAPTER 4 VOCABULARY CHECK

*Fill in each blank with one of the words or phrases listed below.*

system of linear equations    solution    consistent    independent
dependent    inconsistent    substitution    addition

**1.** In a system of linear equations in two variables, if the graphs of the equations are the same, the equations are _____ equations.

**2.** Two or more linear equations are called a _____ .

**3.** A system of equations that has at least one solution is called a(n) _____ system.

**4.** A _____ of a system of two equations in two variables is an ordered pair of numbers that is a solution of both equations in the system.

**5.** Two algebraic methods for solving systems of equations are _____ and _____ .

**6.** A system of equations that has no solution is called a(n) _____ system.

**7.** In a system of linear equations in two variables, if the graphs of the equations are different, the equations are _____ equations.

---

### 📖 STUDY SKILLS BUILDER

**Are You Preparing for a Test on Chapter 4?**

Below I have listed some common trouble areas for topics covered in Chapter 4. After studying for your test—but before taking your test—read these.

- If you are having trouble drawing a neat graph, remember to ask your instructor if you can use graph paper on your test. This will save your time and keep your graphs neat.

- Do you remember how to check solutions of systems of equations? If $(-1, 5)$ is a solution of the system

$$\begin{cases} 3x - y = -8 \\ -x + y = 6, \end{cases}$$

then the ordered pair will make *both* equations a true statement.

$$3x - y = -8 \qquad -x + y = 6$$
$$3(-1) - 5 = -8 \qquad -(-1) + 5 = 6 \quad \text{Let } x = -1$$
$$\text{and } y = 5.$$
$$-8 = -8 \quad \text{True} \qquad 6 = 6 \quad \text{True}$$

Remember: This is simply a list of a few common trouble areas. For a review of Chapter 4, see the Highlights and Chapter Review at the end of this chapter.

> **▶ Helpful Hint**
>
> Are you preparing for your test? Don't forget to take the Chapter 4 Test on page 297. Then check your answers at the back of the text and use the Chapter Test Prep Video CD to see the fully worked-out solutions to any of the exercises you want to review.

## CHAPTER 4 HIGHLIGHTS

| DEFINITIONS AND CONCEPTS | EXAMPLES |
|---|---|

### SECTION 4.1 SOLVING SYSTEMS OF LINEAR EQUATIONS BY GRAPHING

A **solution** of a system of two equations in two variables is an ordered pair of numbers that is a solution of both equations in the system.

Determine whether $(-1, 3)$ is a solution of the system:

$$\begin{cases} 2x - y = -5 \\ x = 3y - 10 \end{cases}$$

Replace $x$ with $-1$ and $y$ with 3 in both equations.

$$2x - y = -5 \qquad\qquad x = 3y - 10$$
$$2(-1) - 3 \stackrel{?}{=} -5 \qquad -1 \stackrel{?}{=} 3 \cdot 3 - 10$$
$$-5 = -5 \quad \text{True} \quad -1 = -1 \quad \text{True}$$

$(-1, 3)$ is a solution of the system.

Graphically, a solution of a system is a point common to the graphs of both equations.

Solve by graphing. $\begin{cases} 3x - 2y = -3 \\ x + y = 4 \end{cases}$

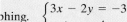

A system of equations with at least one solution is a **consistent system.** A system that has no solution is an **inconsistent system.**

If the graphs of two linear equations are identical, the equations are **dependent.** If their graphs are different, the equations are **independent.**

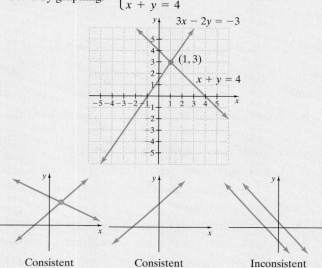

Consistent and independent     Consistent and dependent     Inconsistent and independent

| DEFINITIONS AND CONCEPTS | EXAMPLES |
|---|---|

**SECTION 4.2    SOLVING SYSTEMS OF LINEAR EQUATIONS BY SUBSTITUTION**

To solve a system of linear equations by the substitution method.

**Step  1.** Solve one equation for a variable.

**Step  2.** Substitute the expression for the variable into the other equation.

**Step  3.** Solve the equation from Step 2 to find the value of one variable.

**Step  4.** Substitute the value from Step 3 in either original equation to find the value of the other variable.

**Step  5.** Check the solution in both equations.

Solve by substitution.

$$\begin{cases} 3x + 2y = 1 \\ x = y - 3 \end{cases}$$

Substitute $y - 3$ for $x$ in the first equation.

$$3x + 2y = 1$$
$$3(y - 3) + 2y = 1$$
$$3y - 9 + 2y = 1$$
$$5y = 10$$
$$y = 2 \quad \text{Divide by 5.}$$

To find $x$, substitute $2$ for $y$ in $x = y - 3$ so that $x = 2 - 3$ or $-1$. The solution $(-1, 2)$ checks.

**SECTION 4.3    SOLVING SYSTEMS OF LINEAR EQUATIONS BY ADDITION**

To solve a system of linear equations by the addition method

**Step  1.** Rewrite each equation in standard form $Ax + By = C$.

**Step  2.** Multiply one or both equations by a nonzero number so that the coefficients of a variable are opposites.

**Step  3.** Add the equations.

**Step  4.** Find the value of one variable by solving the resulting equation.

**Step  5.** Substitute the value from Step 4 into either original equation to find the value of the other variable.

**Step  6.** Check the solution in both equations.

If solving a system of linear equations by substitution or addition yields a true statement such as $-2 = -2$, then the graphs of the equations in the system are identical and there is an infinite number of solutions of the system.

Solve by addition.

$$\begin{cases} x - 2y = 8 \\ 3x + y = -4 \end{cases}$$

Multiply both sides of the first equation by $-3$.

$$\begin{cases} -3x + 6y = -24 \\ \underline{\phantom{-}3x + \phantom{6}y = -4} \end{cases}$$
$$7y = -28 \quad \text{Add.}$$
$$y = -4 \quad \text{Divide by 7.}$$

To find $x$, let $y = -4$ in an original equation.

$$x - 2(-4) = 8 \quad \text{First equation}$$
$$x + 8 = 8$$
$$x = 0$$

The solution $(0, -4)$ checks.

Solve: $\begin{cases} 2x - 6y = -2 \\ x = 3y - 1 \end{cases}$

Substitute $3y - 1$ for $x$ in the first equation.

$$2(3y - 1) - 6y = -2$$
$$6y - 2 - 6y = -2$$
$$-2 = -2 \quad \text{True}$$

The system has an infinite number of solutions.

| **DEFINITIONS AND CONCEPTS** | **EXAMPLES** |
|---|---|

A **solution** of an equation in three variables $x$, $y$, and $z$ is an **ordered triple** $(x, y, z)$ that makes the equation a true statement.

Verify that $(-2, 1, 3)$ is a solution of $2x + 3y - 2z = -7$.
Replace $x$ with $-2$, $y$ with 1, and $z$ with 3.

$$2(-2) + 3(1) - 2(3) \stackrel{?}{=} -7$$
$$-4 + 3 - 6 \stackrel{?}{=} -7$$
$$-7 = -7 \quad \text{True}$$

$(-2, 1, 3)$ is a solution.

**Solving a System of Three Linear Equations by the Elimination Method**

**Step 1.** Write each equation in standard form, $Ax + By + Cz = D$.

**Step 2.** Choose a pair of equations and use them to eliminate a variable.

**Step 3.** Choose any other pair of equations and eliminate the same variable.

**Step 4.** Solve the system of two equations in two variables from Steps 2 and 3.

**Step 5.** Solve for the third variable by substituting the values of the variables from Step 4 into any of the original equations.

**Step 6.** Check the solution in all three original equations.

Solve:

$$\begin{cases} 2x + y - z = 0 & (1) \\ x - y - 2z = -6 & (2) \\ -3x - 2y + 3z = -22 & (3) \end{cases}$$

**1.** Each equation is written in standard form.

**2.**
$$\begin{aligned} 2x + y - z &= 0 \quad (1) \\ \underline{x - y - 2z} &= \underline{-6} \quad (2) \\ 3x \qquad - 3z &= -6 \quad (4) \quad \text{Add.} \end{aligned}$$

**3.** Eliminate $y$ from equations (1) and (3) also.

$$\begin{aligned} 4x + 2y - 2z &= 0 \qquad \text{Multiply equation} \\ \underline{-3x - 2y + 3z} &= \underline{-22} \quad (3) \quad \text{(1) by 2.} \\ x \qquad + z &= -22 \quad (5) \quad \text{Add.} \end{aligned}$$

**4.** Solve.

$$\begin{cases} 3x - 3z = -6 & (4) \\ x + z = -22 & (5) \end{cases}$$

$$\begin{aligned} x - z &= -2 \qquad \text{Divide equation (4) by 3.} \\ \underline{x + z} &= \underline{-22} \quad (5) \\ 2x \qquad &= -24 \\ x \qquad &= -12 \end{aligned}$$

To find $z$, use equation (5).

$$\begin{aligned} x + z &= -22 \\ -12 + z &= -22 \\ z &= -10 \end{aligned}$$

**5.** To find $y$, use equation (1).

$$\begin{aligned} 2x + y - z &= 0 \\ 2(-12) + y - (-10) &= 0 \\ -24 + y + 10 &= 0 \\ y &= 14 \end{aligned}$$

**6.** The solution $(-12, 14, -10)$ checks.

| DEFINITIONS AND CONCEPTS | EXAMPLES |
|---|---|

SECTION 4.5  SYSTEMS OF LINEAR EQUATIONS AND PROBLEM SOLVING

Problem-solving steps

**1.** UNDERSTAND. Read and reread the problem.

Two angles are supplementary if their sum is 180°.

The larger of two supplementary angles is three times the smaller, decreased by twelve. Find the measure of each angle. Let

$$x = \text{measure of smaller angle}$$
$$y = \text{measure of larger angle}$$

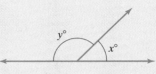

**2.** TRANSLATE.

In words:

| the sum of supplementary angles | is | 180° |
|---|---|---|
| ↓ | ↓ | ↓ |

Translate:  $x + y \quad = \quad 180$

In words:

| larger angle | is | 3 times smaller | decreased by | 12 |
|---|---|---|---|---|
| ↓ | ↓ | ↓ | ↓ | ↓ |

Translate:  $y \quad = \quad 3x \quad - \quad 12$

**3.** SOLVE.

Solve the system:

$$\begin{cases} x + y = 180 \\ y = 3x - 12 \end{cases}$$

Use the substitution method and replace $y$ with $3x - 12$ in the first equation.

$$x + y = 180$$
$$x + (3x - 12) = 180$$
$$4x = 192$$
$$x = 48$$

Since $y = 3x - 12$, then $y = 3 \cdot 48 - 12$ or 132.

**4.** INTERPRET.

The solution checks. The smaller angle measures 48° and the larger angle measures 132°.

# CHAPTER 4 REVIEW

*(4.1) Determine whether any of the following ordered pairs satisfy the system of linear equations.*

**1.** $\begin{cases} 2x - 3y = 12 \\ 3x + 4y = 1 \end{cases}$

    **a.** $(12, 4)$      **b.** $(3, -2)$      **c.** $(-3, 6)$

**2.** $\begin{cases} 4x + y = 0 \\ -8x - 5y = 9 \end{cases}$

    **a.** $\left(\dfrac{3}{4}, -3\right)$      **b.** $(-2, 8)$      **c.** $\left(\dfrac{1}{2}, -2\right)$

**3.** $\begin{cases} 5x - 6y = 18 \\ 2y - x = -4 \end{cases}$

    **a.** $(-6, -8)$      **b.** $\left(3, \dfrac{5}{2}\right)$      **c.** $\left(3, -\dfrac{1}{2}\right)$

**4.** $\begin{cases} 2x + 3y = 1 \\ 3y - x = 4 \end{cases}$

    **a.** $(2, 2)$      **b.** $(-1, 1)$      **c.** $(2, -1)$

*Solve each system of equations by graphing.*

**5.** $\begin{cases} x + y = 5 \\ x - y = 1 \end{cases}$ **6.** $\begin{cases} x + y = 3 \\ x - y = -1 \end{cases}$

**7.** $\begin{cases} x = 5 \\ y = -1 \end{cases}$ **8.** $\begin{cases} x = -3 \\ y = 2 \end{cases}$

**9.** $\begin{cases} 2x + y = 5 \\ x = -3y \end{cases}$ **10.** $\begin{cases} 3x + y = -2 \\ y = -5x \end{cases}$

**11.** $\begin{cases} y = 3x \\ -6x + 2y = 6 \end{cases}$ **12.** $\begin{cases} x - 2y = 2 \\ -2x + 4y = -4 \end{cases}$

**(4.2)** *Solve each system of equations by the substitution method.*

**13.** $\begin{cases} y = 2x + 6 \\ 3x - 2y = -11 \end{cases}$ **14.** $\begin{cases} y = 3x - 7 \\ 2x - 3y = 7 \end{cases}$

**15.** $\begin{cases} x + 3y = -3 \\ 2x + y = 4 \end{cases}$ **16.** $\begin{cases} 3x + y = 11 \\ x + 2y = 12 \end{cases}$

**17.** $\begin{cases} 4y = 2x + 6 \\ x - 2y = -3 \end{cases}$ **18.** $\begin{cases} 9x = 6y + 3 \\ 6x - 4y = 2 \end{cases}$

**19.** $\begin{cases} x + y = 6 \\ y = -x - 4 \end{cases}$ **20.** $\begin{cases} -3x + y = 6 \\ y = 3x + 2 \end{cases}$

**(4.3)** *Solve each system of equations by the addition method.*

**21.** $\begin{cases} 2x + 3y = -6 \\ x - 3y = -12 \end{cases}$ **22.** $\begin{cases} 4x + y = 15 \\ -4x + 3y = -19 \end{cases}$

**23.** $\begin{cases} 2x - 3y = -15 \\ x + 4y = 31 \end{cases}$ **24.** $\begin{cases} x - 5y = -22 \\ 4x + 3y = 4 \end{cases}$

**25.** $\begin{cases} 2x - 6y = -1 \\ -x + 3y = \dfrac{1}{2} \end{cases}$ **26.** $\begin{cases} 0.6x - 0.3y = -1.5 \\ 0.04x - 0.02y = -0.1 \end{cases}$

**27.** $\begin{cases} \dfrac{3}{4}x + \dfrac{2}{3}y = 2 \\ x + \dfrac{y}{3} = 6 \end{cases}$

**28.** $\begin{cases} 10x + 2y = 0 \\ 3x + 5y = 33 \end{cases}$

**(4.4)** *Solve each system of equations in three variables.*

**29.** $\begin{cases} x \quad + z = 4 \\ 2x - y \quad = 4 \\ x + y - z = 0 \end{cases}$

**30.** $\begin{cases} 2x + 5y \quad = 4 \\ x - 5y + z = -1 \\ 4x \quad - z = 11 \end{cases}$

**31.** $\begin{cases} 4y + 2z = 5 \\ 2x + 8y \quad = 5 \\ 6x + \quad 4z = 1 \end{cases}$

**32.** $\begin{cases} 5x + 7y \quad = 9 \\ 14y - z = 28 \\ 4x \quad + 2z = -4 \end{cases}$

**33.** $\begin{cases} 3x - 2y + 2z = 5 \\ -x + 6y + z = 4 \\ 3x + 14y + 7z = 20 \end{cases}$

**34.** $\begin{cases} x + 2y + 3z = 11 \\ y + 2z = 3 \\ 2x \quad + 2z = 10 \end{cases}$

**35.** $\begin{cases} 7x - 3y + 2z = 0 \\ 4x - 4y - z = 2 \\ 5x + 2y + 3z = 1 \end{cases}$

**36.** $\begin{cases} x - 3y - 5z = -5 \\ 4x - 2y + 3z = 13 \\ 5x + 3y + 4z = 22 \end{cases}$

**(4.5)** *Solve each problem by writing and solving a system of linear equations.*

**37.** The sum of two numbers is 16. Three times the larger number decreased by the smaller number is 72. Find the two numbers.

**38.** The Forrest Theater can seat a total of 360 people. They take in $15,150 when every seat is sold. If orchestra section tickets cost $45 and balcony tickets cost $35, find the number of seats in the orchestra section and the number of seats in the balcony.

**39.** A riverboat can head 340 miles upriver in 19 hours, but the return trip takes only 14 hours. Find the current of the river and find the speed of the riverboat in still water to the nearest tenth of a mile.

| | $d$ = | $r$ | $\cdot$ | $t$ |
|---|---|---|---|---|
| **Upriver** | 340 | $x - y$ | | 19 |
| **Downriver** | 340 | $x + y$ | | 14 |

**40.** Find the amount of a 6% acid solution and the amount of a 14% acid solution Pat Mayfield should combine to prepare 50 cc (cubic centimeters) of a 12% solution.

**41.** A deli charges $3.80 for a breakfast of three eggs and four strips of bacon. The charge is $2.75 for two eggs and three strips of bacon. Find the cost of each egg and the cost of each strip of bacon.

**42.** An exercise enthusiast alternates between jogging and walking. He traveled 15 miles during the past 3 hours. He jogs at a rate of 7.5 miles per hour and walks at a rate of 4 miles per hour. Find how much time, to the nearest hundredth of an hour, he actually spent jogging and how much time he spent walking.

**43.** Chris Kringler has $2.77 in her coin jar—all in pennies, nickels, and dimes. If she has 53 coins in all and four more nickels than dimes, find how many of each type of coin she has.

**44.** An employee at See's Candy Store needs a special mixture of candy. She has creme-filled chocolates that sell for $3.00 per pound, chocolate-covered nuts that sell for $2.70 per pound, and chocolate-covered raisins that sell for $2.25 per pound. She wants to have twice as many raisins as nuts in the mixture. Find how many pounds of each she should use to make 45 pounds worth $2.80 per pound.

**45.** The perimeter of an isosceles (two sides equal) triangle is 73 centimeters. If the unequal side is 7 centimeters longer than the two equal sides, find the lengths of the three sides.

**46.** The sum of three numbers is 295. One number is five more than a second and twice the third. Find the numbers.

## MIXED REVIEW

*Solve each system of equations by graphing.*

**47.** $\begin{cases} x - 2y = 1 \\ 2x + 3y = -12 \end{cases}$

**48.** $\begin{cases} 3x - y = -4 \\ 6x - 2y = -8 \end{cases}$

*Solve each system of equations.*

**49.** $\begin{cases} x + 4y = 11 \\ 5x - 9y = -3 \end{cases}$

**50.** $\begin{cases} x + 9y = 16 \\ 3x - 8y = 13 \end{cases}$

**51.** $\begin{cases} y = -2x \\ 4x + 7y = -15 \end{cases}$

**52.** $\begin{cases} 3y = 2x + 15 \\ -2x + 3y = 21 \end{cases}$

**53.** $\begin{cases} 3x - y = 4 \\ 4y = 12x - 16 \end{cases}$

**54.** $\begin{cases} x + y = 19 \\ x - y = -3 \end{cases}$

**55.** $\begin{cases} x - 3y = -11 \\ 4x + 5y = -10 \end{cases}$

**56.** $\begin{cases} -x - 15y = 44 \\ 2x + 3y = 20 \end{cases}$

**57.** $\begin{cases} x - 3y + 2z = 0 \\ 9y - z = 22 \\ 5x + 3z = 10 \end{cases}$

**58.** $\begin{cases} x - 4y = 4 \\ \frac{1}{8}x - \frac{1}{2}y = 3 \end{cases}$

*Solve each problem by writing and solving a system of linear equations.*

**59.** The sum of two numbers is 12. Three times the smaller number increased by the larger number is 20. Find the numbers.

**60.** The difference of two numbers is -18. Twice the smaller decreased by the larger is -23. Find the two numbers.

**61.** Emma Hodges has a jar containing 65 coins, all of which are either nickels or dimes. The total value of the coins is $5.30. How many of each type does she have?

**62.** Sarah and Owen Hebert purchased 26 stamps, a mixture of 13¢ and 22¢ stamps. Find the number of each type of stamp if they spent $4.19.

**63.** The perimeter of a triangle is 126 units. The length of one side is twice the length of the shortest side. The length of the third side is fourteen more than the length of the shortest side. Find the length of the sides of the triangles.

# CHAPTER 4 TEST TEST PREP VIDEO

Remember to use the Chapter Test Prep Video CD to see the fully worked-out solutions to any of the exercises you want to review.

*Answer each question true or false.*

**1.** A system of two linear equations in two variables can have exactly two solutions.

**2.** Although (1, 4) is not a solution of $x + 2y = 6$, it can still be a solution of the system $\begin{cases} x + 2y = 6 \\ x + y = 5 \end{cases}$.

**3.** If the two equations in a system of linear equations are added and the result is $3 = 0$, the system has no solution.

**4.** If the two equations in a system of linear equations are added and the result is $3x = 0$, the system has no solution.

*Is the ordered pair a solution of the given linear system?*

**5.** $\begin{cases} 2x - 3y = 5 \\ 6x + y = 1 \end{cases}$; (1, -1)

**6.** $\begin{cases} 4x - 3y = 24 \\ 4x + 5y = -8 \end{cases}$; (3, -4)

**7.** Use graphing to find the solutions of the system $\begin{cases} y - x = 6 \\ y + 2x = -6 \end{cases}$

**8.** Use the substitution method to solve the system $\begin{cases} 3x - 2y = -14 \\ x + 3y = -1 \end{cases}$

**9.** Use the substitution method to solve the system $\begin{cases} \frac{1}{2}x + 2y = -\frac{15}{4} \\ 4x = -y \end{cases}$

**10.** Use the addition method to solve the system $\begin{cases} 3x + 5y = 2 \\ 2x - 3y = 14 \end{cases}$

**11.** Use the addition method to solve the system $\begin{cases} 4x - 6y = 7 \\ -2x + 3y = 0 \end{cases}$

*Solve each system using the substitution method or the addition method.*

**12.** $\begin{cases} 3x + y = 7 \\ 4x + 3y = 1 \end{cases}$

**13.** $\begin{cases} 3(2x + y) = 4x + 20 \\ x - 2y = 3 \end{cases}$

**14.** $\begin{cases} \dfrac{x - 3}{2} = \dfrac{2 - y}{4} \\ \dfrac{7 - 2x}{3} = \dfrac{y}{2} \end{cases}$

*Solve.*

**15.** Two numbers have a sum of 124 and a difference of 32. Find the numbers.

**16.** Find the amount of a 12% saline solution a lab assistant should add to 80 cc (cubic centimeters) of a 22% saline solution in order to have a 16% solution.

**17.** Although the number of farms in the U.S. is still decreasing, small farms are making a comeback. Texas and Missouri are the states with the most number of farms. Texas has 116 thousand more farms than Missouri and the total number of farms for these two states is 336 thousand. Find the number of farms for each state.

**18.** $\begin{cases} 2x - 3y \quad\;\; = 4 \\ \quad\;\; 3y + 2z = 2 \\ x \qquad\;\; - z = -5 \end{cases}$

**19.** $\begin{cases} 3x - 2y - z = -1 \\ 2x - 2y \quad\;\; = 4 \\ 2x \qquad\; - 2z = -12 \end{cases}$

**20.** The measure of the largest angle of a triangle is three less than 5 times the measure of the smallest angle. The measure of the remaining angle is 1 less than twice the measure of the smallest angle. Find the measure of each angle.

# CHAPTER 4 CUMULATIVE REVIEW

**1.** Insert $<$, $>$, or $=$ in the space between the paired numbers to make each statement true.
   **a.** $-1 \quad 0$    **b.** $7 \quad \dfrac{14}{2}$    **c.** $-5 \quad -6$

**2.** Evaluate.
   **a.** $5^2$     **b.** $2^5$

**3.** Name the property or properties illustrated by each true statement.
   **a.** $3 \cdot y = y \cdot 3$
   **b.** $(x + 7) + 9 = x + (7 + 9)$
   **c.** $(b + 0) + 3 = b + 3$
   **d.** $0.2 \cdot (z \cdot 5) = 0.2 \cdot (5 \cdot z)$
   **e.** $-2 \cdot \left(-\dfrac{1}{2}\right) = 1$
   **f.** $-2 + 2 = 0$
   **g.** $-6 \cdot (y \cdot 2) = (-6 \cdot 2) \cdot y$

**4.** Evaluate $y^2 - 3x$ for $x = 8$ and $y = 5$.

**5.** Subtract $4x - 2$ from $2x - 3$.

**6.** Simplify: $7 - 12 + (-5) - 2 + (-2)$

**7.** Solve: $7 = -5(2a - 1) - (-11a + 6)$.

**8.** Evaluate $2y^2 - x^2$ for $x = -7$ and $y = -3$.

**9.** Solve: $\dfrac{5}{2}x = 15$

**10.** Simplify: $0.4y - 6.7 + y - 0.3 - 2.6y$

**11.** Solve: $\dfrac{x}{2} - 1 = \dfrac{2}{3}x - 3$

**12.** Solve: $7(x - 2) - 6(x + 1) = 20$

**13.** Twice the sum of a number and 4 is the same as four times the number, decreased by 12. Find the number.

**14.** Solve: $5(y - 5) = 5y + 10$

**15.** Solve $y = mx + b$ for $x$.

**16.** Five times the sum of a number and $-1$ is the same as 6 times the number. Find the number.

**17.** Solve $-2x \le -4$. Write the solution set in interval notation.

**18.** Solve $P = a + b + c$ for $b$.

**19.** Graph $x = -2y$ by plotting intercepts.

**20.** Solve $3x + 7 \geq x - 9$. Write the solution set in interval notation.

**21.** Find the slope of the line through $(-1, 5)$ and $(2, -3)$.

**22.** Complete the table of values for $x - 3y = 3$

| x | y |
|---|---|
|   | −1 |
| 3 |   |
|   | 2 |

**23.** Find the slope of the line whose equation is $y = \dfrac{3}{4}x + 6$.

**24.** Find the slope of the line parallel to the line passing through $(-1, 3)$ and $(2, -8)$.

**25.** Find the slope and the $y$-intercept of the line whose equation is $3x - 4y = 4$.

**26.** Find the slope and $y$-intercept of the line whose equation is $y = 7x$.

**27.** Find an equation of the line passing through $(-1, 5)$ with slope $-2$. Write the equation in standard form: $Ax + By = C$.

**28.** Determine whether the lines are parallel, perpendicular or neither.

$$y = 4x - 5$$
$$-4x + y = 7$$

**29.** Find an equation of the vertical line through $(-1, 5)$.

**30.** Write an equation of the line with slope $-5$, through $(-2, 3)$.

**31.** Find the domain and the range of the relation $\{(0, 2), (3, 3), (-1, 0), (3, -2)\}$.

**32.** If $f(x) = 5x^2 - 6$, find $f(0)$ and $f(-2)$.

**33.** Which of the following relations are also functions?
   **a.** $\{(-1, 1), (2, 3), (7, 3), (8, 6)\}$
   **b.** $\{(0, -2), (1, 5), (0, 3), (7, 7)\}$

**34.** Determine which graph(s) are graphs of functions.
   **a.**        **b.**        **c.**

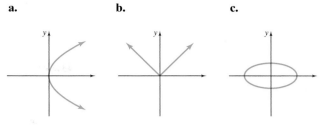

**35.** Determine the number of solutions of the system.

$$\begin{cases} 3x - y = 4 \\ x + 2y = 8 \end{cases}$$

**36.** Determine whether any ordered pairs satisfy the given system.

$$\begin{cases} 2x - y = 6 \\ 3x + 2y = -5 \end{cases}$$

   **a.** $(1, -4)$      **b.** $(0, 6)$      **c.** $(3, 0)$

*Solve each system.*

**37.** $\begin{cases} x + 2y = 7 \\ 2x + 2y = 13 \end{cases}$

**38.** $\begin{cases} 3x - 4y = 10 \\ \qquad y = 2x \end{cases}$

**39.** $\begin{cases} x + y = 7 \\ x - y = 5 \end{cases}$

**40.** $\begin{cases} x = 5y - 3 \\ x = 8y + 4 \end{cases}$

**41.** Solve the system.

$$\begin{cases} 3x - y + z = -15 \\ x + 2y - z = 1 \\ 2x + 3y - 2z = 0 \end{cases}$$

**42.** Solve the system.

$$\begin{cases} x - 2y + z = 0 \\ 3x - y - 2z = -15 \\ 2x - 3y + 3z = 7 \end{cases}$$

**43.** A first number is 4 less than a second number. Four times the first number is 6 more than twice the second. Find the numbers.

**44.** Find two numbers whose sum is 37 and whose difference is 21.

# 5

# Exponents and Polynomials

Recall from Chapter 1 that an exponent is a shorthand notation for repeated factors. This chapter explores additional concepts about exponents and exponential expressions. An especially useful type of exponential expression is a polynomial. Polynomials model many real-world phenomena. This chapter will focus on operations on polynomials.

A popular use of the Internet is the World Wide Web. The World Wide Web was invented in 1989–1990 as an environment originally by which scientists could share information. It has grown into a medium containing text, graphics, audio, animation, and video. In Section 5.5, Exercises 91 and 92, you will have the opportunity to estimate the number of visitors to the most popular Web sites.

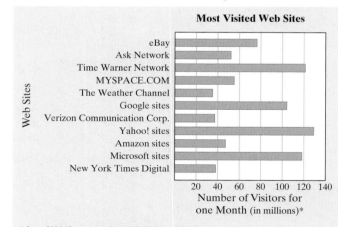

**Most Visited Web Sites**

Web Sites: eBay, Ask Network, Time Warner Network, MYSPACE.COM, The Weather Channel, Google sites, Verizon Communication Corp., Yahoo! sites, Amazon sites, Microsoft sites, New York Times Digital

Number of Visitors for one Month (in millions)*

20  40  60  80  100  120  140

* June, 2006 (*Source*: comScore Media Metrix, Inc.)

# 5.1 EXPONENTS

## OBJECTIVES

**1** Evaluate exponential expressions.

**2** Use the product rule for exponents.

**3** Use the power rule for exponents.

**4** Use the power rules for products and quotients.

**5** Use the quotient rule for exponents, and define a number raised to the 0 power.

**6** Decide which rule(s) to use to simplify an expression.

**OBJECTIVE 1 ▶ Evaluating exponential expressions.** As we reviewed in Section 1.4, an exponent is a shorthand notation for repeated factors. For example, $2 \cdot 2 \cdot 2 \cdot 2 \cdot 2$ can be written as $2^5$. The expression $2^5$ is called an **exponential expression.** It is also called the fifth **power** of 2, or we say that 2 is **raised** to the fifth power.

$$5^6 = \underbrace{5 \cdot 5 \cdot 5 \cdot 5 \cdot 5 \cdot 5}_{6 \text{ factors; each factor is } 5} \qquad \text{and} \qquad (-3)^4 = \underbrace{(-3) \cdot (-3) \cdot (-3) \cdot (-3)}_{4 \text{ factors; each factor is } -3}$$

The **base** of an exponential expression is the repeated factor. The **exponent** is the number of times that the base is used as a factor.

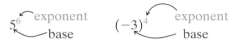

**EXAMPLE 1**   Evaluate each expression.

**a.** $2^3$   **b.** $3^1$   **c.** $(-4)^2$   **d.** $-4^2$   **e.** $\left(\dfrac{1}{2}\right)^4$   **f.** $(0.5)^3$   **g.** $4 \cdot 3^2$

*Solution*

**a.** $2^3 = 2 \cdot 2 \cdot 2 = 8$

**b.** To raise 3 to the first power means to use 3 as a factor only once. Therefore, $3^1 = 3$. Also, when no exponent is shown, the exponent is assumed to be 1.

**c.** $(-4)^2 = (-4)(-4) = 16$          **d.** $-4^2 = -(4 \cdot 4) = -16$

**e.** $\left(\dfrac{1}{2}\right)^4 = \dfrac{1}{2} \cdot \dfrac{1}{2} \cdot \dfrac{1}{2} \cdot \dfrac{1}{2} = \dfrac{1}{16}$      **f.** $(0.5)^3 = (0.5)(0.5)(0.5) = 0.125$

**g.** $4 \cdot 3^2 = 4 \cdot 9 = 36$

**PRACTICE**
**1**   Evaluate each expression.

**a.** $3^3$        **b.** $4^1$        **c.** $(-8)^2$        **d.** $-8^2$

**e.** $\left(\dfrac{3}{4}\right)^3$        **f.** $(0.3)^4$        **g.** $3 \cdot 5^2$

Notice how similar $-4^2$ is to $(-4)^2$ in the example above. The difference between the two is the parentheses. In $(-4)^2$, the parentheses tell us that the base, or repeated factor, is $-4$. In $-4^2$, only 4 is the base.

> ▶ **Helpful Hint**
> Be careful when identifying the base of an exponential expression. Pay close attention to the use of parentheses.
>
> $(-3)^2$            $-3^2$            $2 \cdot 3^2$
> The base is $-3$.      The base is 3.      The base is 3.
> $(-3)^2 = (-3)(-3) = 9$    $-3^2 = -(3 \cdot 3) = -9$    $2 \cdot 3^2 = 2 \cdot 3 \cdot 3 = 18$

An exponent has the same meaning whether the base is a number or a variable. If $x$ is a real number and $n$ is a positive integer, then $x^n$ is the product of $n$ factors, each of which is $x$.

$$x^n = \underbrace{x \cdot x \cdot x \cdot x \cdot x \cdot \ldots \cdot x}_{n \text{ factors of } x}$$

**EXAMPLE 2** Evaluate each expression for the given value of $x$.

**a.** $2x^3$; $x$ is 5

**b.** $\dfrac{9}{x^2}$; $x$ is $-3$

*Solution*   **a.** If $x$ is 5, $2x^3 = 2 \cdot (5)^3$
$$= 2 \cdot (5 \cdot 5 \cdot 5)$$
$$= 2 \cdot 125$$
$$= 250$$

**b.** If $x$ is $-3$, $\dfrac{9}{x^2} = \dfrac{9}{(-3)^2}$
$$= \dfrac{9}{(-3)(-3)}$$
$$= \dfrac{9}{9}$$
$$= 1$$

**PRACTICE**
**2** Evaluate each expression for the given value of $x$.

**a.** $3x^4$; $x$ is 3

**b.** $\dfrac{6}{x^2}$; $x$ is $-4$

**OBJECTIVE 2 ▶ Using the product rule.** Exponential expressions can be multiplied, divided, added, subtracted, and themselves raised to powers. By our definition of an exponent,

$$5^4 \cdot 5^3 = \underbrace{(5 \cdot 5 \cdot 5 \cdot 5)}_{\text{4 factors of 5}} \cdot \underbrace{(5 \cdot 5 \cdot 5)}_{\text{3 factors of 5}}$$
$$= \underbrace{5 \cdot 5 \cdot 5 \cdot 5 \cdot 5 \cdot 5 \cdot 5}_{\text{7 factors of 5}}$$
$$= 5^7$$

Also,

$$x^2 \cdot x^3 = (x \cdot x) \cdot (x \cdot x \cdot x)$$
$$= x \cdot x \cdot x \cdot x \cdot x$$
$$= x^5$$

In both cases, notice that the result is exactly the same if the exponents are added.

$$5^4 \cdot 5^3 = 5^{4+3} = 5^7 \qquad \text{and} \qquad x^2 \cdot x^3 = x^{2+3} = x^5$$

This suggests the following rule.

> **Product Rule for Exponents**
>
> If $m$ and $n$ are positive integers and $a$ is a real number, then
>
> $$a^m \cdot a^n = a^{m+n} \leftarrow \text{Add exponents.}$$
> $$\text{Keep common base.}$$

For example, $3^5 \cdot 3^7 = 3^{5+7} = 3^{12} \leftarrow$ Add exponents.
      Keep common base.

In other words, to multiply two exponential expressions with a **common base**, keep the base and add the exponents. We call this simplifying the exponential expression.

**EXAMPLE 3** Use the product rule to simplify.

**a.** $4^2 \cdot 4^5$    **b.** $x^4 \cdot x^6$    **c.** $y^3 \cdot y$    🔊 **d.** $y^3 \cdot y^2 \cdot y^7$    🔊 **e.** $(-5)^7 \cdot (-5)^8$    **f.** $a^2 \cdot b^2$

*Solution*

**a.** $4^2 \cdot 4^5 = 4^{2+5} = 4^7 \leftarrow$ Add exponents.
      Keep common base.

**b.** $x^4 \cdot x^6 = x^{4+6} = x^{10}$

**c.** $y^3 \cdot y = y^3 \cdot y^1$

$\qquad = y^{3+1}$

$\qquad = y^4$

**d.** $y^3 \cdot y^2 \cdot y^7 = y^{3+2+7} = y^{12}$

**e.** $(-5)^7 \cdot (-5)^8 = (-5)^{7+8} = (-5)^{15}$

**f.** $a^2 \cdot b^2$ Cannot be simplified because $a$ and $b$ are different bases.

> ▶ **Helpful Hint**
>
> Don't forget that if no exponent is written, it is assumed to be 1.

**PRACTICE**
**3** Use the product rule to simplify.

**a.** $3^4 \cdot 3^6$            **b.** $y^3 \cdot y^2$

**c.** $z \cdot z^4$             **d.** $x^3 \cdot x^2 \cdot x^6$

**e.** $(-2)^5 \cdot (-2)^3$     **f.** $b^3 \cdot t^5$

**Concept Check** ☑

Where possible, use the product rule to simplify the expression.

**a.** $z^2 \cdot z^{14}$     **b.** $x^2 \cdot y^{14}$     **c.** $9^8 \cdot 9^3$     **d.** $9^8 \cdot 2^7$

**EXAMPLE 4** Use the product rule to simplify $(2x^2)(-3x^5)$.

*Solution* Recall that $2x^2$ means $2 \cdot x^2$ and $-3x^5$ means $-3 \cdot x^5$.

$\qquad (2x^2)(-3x^5) = 2 \cdot x^2 \cdot -3 \cdot x^5$    Remove parentheses.

$\qquad\qquad\qquad\quad = 2 \cdot -3 \cdot x^2 \cdot x^5$    Group factors with common bases.

$\qquad\qquad\qquad\quad = -6x^7$           Simplify.

**PRACTICE**
**4** Use the product rule to simplify $(-5y^3)(-3y^4)$.

**EXAMPLE 5** Simplify.

**a.** $(x^2 y)(x^3 y^2)$            **b.** $(-a^7 b^4)(3ab^9)$

*Solution*

**a.** $(x^2 y)(x^3 y^2) = (x^2 \cdot x^3) \cdot (y^1 \cdot y^2)$    Group like bases and write $y$ as $y^1$.

$\qquad\qquad\qquad\quad = x^5 \cdot y^3$   or   $x^5 y^3$    Multiply.

**b.** $(-a^7 b^4)(3ab^9) = (-1 \cdot 3) \cdot (a^7 \cdot a^1) \cdot (b^4 \cdot b^9)$

$\qquad\qquad\qquad\quad = -3a^8 b^{13}$

**PRACTICE**
**5** Simplify.

**a.** $(y^7 z^3)(y^5 z)$        **b.** $(-m^4 n^4)(7mn^{10})$

> ▶ **Helpful Hint**
>
> These examples will remind you of the difference between adding and multiplying terms.
>
> **Addition**
>
> $\qquad 5x^3 + 3x^3 = (5 + 3)x^3 = 8x^3$    By the distributive property.
>
> $\qquad 7x + 4x^2 = 7x + 4x^2$          Cannot be combined.
>
> **Multiplication**
>
> $\qquad (5x^3)(3x^3) = 5 \cdot 3 \cdot x^3 \cdot x^3 = 15x^{3+3} = 15x^6$   By the product rule.
>
> $\qquad (7x)(4x^2) = 7 \cdot 4 \cdot x \cdot x^2 = 28x^{1+2} = 28x^3$   By the product rule.

**Answers to Concept Check:**

**a.** $z^{16}$   **b.** cannot be simplified

**c.** $9^{11}$   **d.** cannot be simplified

**OBJECTIVE 3 ▶ Using the power rule.** Exponential expressions can themselves be raised to powers. Let's try to discover a rule that simplifies an expression like $(x^2)^3$. By definition,

$$(x^2)^3 = \underbrace{(x^2)(x^2)(x^2)}_{3 \text{ factors of } x^2}$$

which can be simplified by the product rule for exponents.

$$(x^2)^3 = (x^2)(x^2)(x^2) = x^{2+2+2} = x^6$$

Notice that the result is exactly the same if we multiply the exponents.

$$(x^2)^3 = x^{2 \cdot 3} = x^6$$

The following property states this result.

---

**Power Rule for Exponents**

If $m$ and $n$ are positive integers and $a$ is a real number, then

$$(a^m)^n = a^{mn} \leftarrow \text{Multiply exponents.}$$
$$\uparrow \text{——— Keep common base.}$$

---

For example, $(7^2)^5 = 7^{2 \cdot 5} = 7^{10} \leftarrow$ Multiply exponents.
$\quad \uparrow$ ——— Keep common base.

To raise a power to a power, keep the base and multiply the exponents.

**EXAMPLE 6** Use the power rule to simplify.

**a.** $(x^2)^5$          **b.** $(y^8)^2$          **c.** $[(-5)^3]^7$

*Solution*    **a.** $(x^2)^5 = x^{2 \cdot 5} = x^{10}$      **b.** $(y^8)^2 = y^{8 \cdot 2} = y^{16}$      **c.** $[(-5)^3]^7 = (-5)^{21}$

**PRACTICE**
**6** Use the power rule to simplify.

**a.** $(x^4)^3$          **b.** $(z^3)^7$          **c.** $[(-2)^3]^5$

---

**▶ Helpful Hint**

Take a moment to make sure that you understand when to apply the product rule and when to apply the power rule.

| *Product Rule → Add Exponents* | *Power Rule → Multiply Exponents* |
|---|---|
| $x^5 \cdot x^7 = x^{5+7} = x^{12}$ | $(x^5)^7 = x^{5 \cdot 7} = x^{35}$ |
| $y^6 \cdot y^2 = y^{6+2} = y^8$ | $(y^6)^2 = y^{6 \cdot 2} = y^{12}$ |

---

**OBJECTIVE 4 ▶ Using the power rules for products and quotients.** When the base of an exponential expression is a product, the definition of $x^n$ still applies. To simplify $(xy)^3$, for example,

$$(xy)^3 = (xy)(xy)(xy) \qquad (xy)^3 \text{ means 3 factors of } (xy).$$
$$= x \cdot x \cdot x \cdot y \cdot y \cdot y \qquad \text{Group factors with common bases.}$$
$$= x^3 y^3 \qquad\qquad\qquad \text{Simplify.}$$

Notice that to simplify the expression $(xy)^3$, we raise each factor within the parentheses to a power of 3.

$$(xy)^3 = x^3y^3$$

In general, we have the following rule.

---

**Power of a Product Rule**

If $n$ is a positive integer and $a$ and $b$ are real numbers, then

$$(ab)^n = a^n b^n$$

---

For example, $(3x)^5 = 3^5 x^5$.

In other words, to raise a product to a power, we raise each factor to the power.

**EXAMPLE 7** Simplify each expression.

**a.** $(st)^4$ **b.** $(2a)^3$ **c.** $\left(\dfrac{1}{3}mn^3\right)^2$ **d.** $(-5x^2y^3z)^2$

*Solution*

**a.** $(st)^4 = s^4 \cdot t^4 = s^4t^4$      Use the power of a product rule.

**b.** $(2a)^3 = 2^3 \cdot a^3 = 8a^3$      Use the power of a product rule.

**c.** $\left(\dfrac{1}{3}mn^3\right)^2 = \left(\dfrac{1}{3}\right)^2 \cdot (m)^2 \cdot (n^3)^2 = \dfrac{1}{9}m^2n^6$      Use the power of a product rule.

**d.** $(-5x^2y^3z)^2 = (-5)^2 \cdot (x^2)^2 \cdot (y^3)^2 \cdot (z^1)^2$      Use the power rule for exponents.
$$= 25x^4y^6z^2$$

**PRACTICE**
**7** Simplify each expression.

**a.** $(pr)^5$ **b.** $(6b)^2$ **c.** $\left(\dfrac{1}{4}x^2y\right)^3$ **d.** $(-3a^3b^4c)^4$

---

Let's see what happens when we raise a quotient to a power. To simplify $\left(\dfrac{x}{y}\right)^3$, for example,

$$\left(\dfrac{x}{y}\right)^3 = \left(\dfrac{x}{y}\right)\left(\dfrac{x}{y}\right)\left(\dfrac{x}{y}\right) \quad \left(\dfrac{x}{y}\right)^3 \text{ means 3 factors of } \left(\dfrac{x}{y}\right)$$

$$= \dfrac{x \cdot x \cdot x}{y \cdot y \cdot y} \quad \text{Multiply fractions.}$$

$$= \dfrac{x^3}{y^3} \quad \text{Simplify.}$$

Notice that to simplify the expression $\left(\dfrac{x}{y}\right)^3$, we raise both the numerator and the denominator to a power of 3.

$$\left(\dfrac{x}{y}\right)^3 = \dfrac{x^3}{y^3}$$

In general, we have the following.

---

**Power of a Quotient Rule**

If $n$ is a positive integer and $a$ and $c$ are real numbers, then

$$\left(\dfrac{a}{c}\right)^n = \dfrac{a^n}{c^n}, \quad c \neq 0$$

---

For example, $\left(\dfrac{y}{7}\right)^4 = \dfrac{y^4}{7^4}$.

In other words, to raise a quotient to a power, we raise both the numerator and the denominator to the power.

**EXAMPLE 8** Simplify each expression.

**a.** $\left(\dfrac{m}{n}\right)^7$　　　　　　　**b.** $\left(\dfrac{x^3}{3y^5}\right)^4$

*Solution*

**a.** $\left(\dfrac{m}{n}\right)^7 = \dfrac{m^7}{n^7}, n \neq 0$　　　　　Use the power of a quotient rule.

**b.** $\left(\dfrac{x^3}{3y^5}\right)^4 = \dfrac{(x^3)^4}{3^4 \cdot (y^5)^4}, y \neq 0$　　　　Use the power of a product or quotient rule.

　　　$= \dfrac{x^{12}}{81y^{20}}$　　　　　　Use the power rule for exponents.□

**PRACTICE**
**8**　Simplify each expression.

**a.** $\left(\dfrac{x}{y^2}\right)^5$　　　　　　　**b.** $\left(\dfrac{2a^4}{b^3}\right)^5$

**OBJECTIVE 5** ▶ **Using the quotient rule and defining the zero exponent.** Another pattern for simplifying exponential expressions involves quotients.

To simplify an expression like $\dfrac{x^5}{x^3}$, in which the numerator and the denominator have a common base, we can apply the fundamental principle of fractions and divide the numerator and the denominator by the common base factors. Assume for the remainder of this section that denominators are not 0.

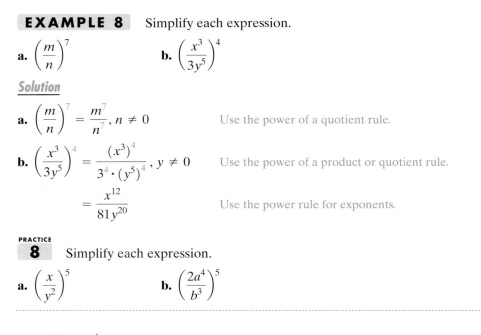

$$\frac{x^5}{x^3} = \frac{x \cdot x \cdot x \cdot x \cdot x}{x \cdot x \cdot x}$$
$$= \frac{x \cdot x \cdot x \cdot x \cdot x}{x \cdot x \cdot x}$$
$$= x \cdot x$$
$$= x^2$$

Notice that the result is exactly the same if we subtract exponents of the common bases.

$$\frac{x^5}{x^3} = x^{5-3} = x^2$$

The quotient rule for exponents states this result in a general way.

---

**Quotient Rule for Exponents**
If $m$ and $n$ are positive integers and $a$ is a real number, then

$$\frac{a^m}{a^n} = a^{m-n}$$

as long as $a$ is not 0.

---

For example, $\dfrac{x^6}{x^2} = x^{6-2} = x^4$.

In other words, to divide one exponential expression by another with a common base, keep the base and subtract exponents.

**EXAMPLE 9** Simplify each quotient.

**a.** $\dfrac{x^5}{x^2}$ **b.** $\dfrac{4^7}{4^3}$ **c.** $\dfrac{(-3)^5}{(-3)^2}$ **d.** $\dfrac{s^2}{t^3}$ **e.** $\dfrac{2x^5y^2}{xy}$

*Solution*

**a.** $\dfrac{x^5}{x^2} = x^{5-2} = x^3$     Use the quotient rule.

**b.** $\dfrac{4^7}{4^3} = 4^{7-3} = 4^4 = 256$     Use the quotient rule.

**c.** $\dfrac{(-3)^5}{(-3)^2} = (-3)^3 = -27$

**d.** $\dfrac{s^2}{t^3}$     Cannot be simplified because $s$ and $t$ are different bases.

**e.** Begin by grouping common bases.

$$\frac{2x^5y^2}{xy} = 2 \cdot \frac{x^5}{x^1} \cdot \frac{y^2}{y^1}$$

$$= 2 \cdot (x^{5-1}) \cdot (y^{2-1}) \quad \text{Use the quotient rule.}$$

$$= 2x^4y^1 \quad \text{or} \quad 2x^4y$$

**PRACTICE**
**9** Simplify each quotient.

**a.** $\dfrac{z^8}{z^4}$ **b.** $\dfrac{(-5)^5}{(-5)^3}$ **c.** $\dfrac{8^8}{8^6}$ **d.** $\dfrac{q^5}{t^2}$ **e.** $\dfrac{6x^3y^7}{xy^5}$

**Concept Check** ✓

Suppose you are simplifying each expression. Tell whether you would *add* the exponents, *subtract* the exponents, *multiply* the exponents, *divide* the exponents, or *none of these*.

**a.** $(x^{63})^{21}$ **b.** $\dfrac{y^{15}}{y^3}$ **c.** $z^{16} + z^8$ **d.** $w^{45} \cdot w^9$

Let's now give meaning to an expression such as $x^0$. To do so, we will simplify $\dfrac{x^3}{x^3}$ in two ways and compare the results.

$$\frac{x^3}{x^3} = x^{3-3} = x^0 \qquad \text{Apply the quotient rule.}$$

$$\frac{x^3}{x^3} = \frac{x \cdot x \cdot x}{x \cdot x \cdot x} = 1 \quad \text{Apply the fundamental principle for fractions.}$$

Since $\dfrac{x^3}{x^3} = x^0$ and $\dfrac{x^3}{x^3} = 1$, we define that $x^0 = 1$ as long as $x$ is not 0.

---

**Zero Exponent**
$a^0 = 1$, as long as $a$ is not 0.

---

In other words, any base raised to the 0 power is 1, as long as the base is not 0.

**Answers to Concept Check:**
**a.** multiply     **b.** subtract
**c.** none of these     **d.** add

**EXAMPLE 10** Simplify each expression.

**a.** $3^0$    **b.** $(ab)^0$    **c.** $(-5)^0$    **d.** $-5^0$    **e.** $\left(\dfrac{3}{100}\right)^0$

_Solution_

**a.** $3^0 = 1$

**b.** Assume that neither $a$ nor $b$ is zero.

$$(ab)^0 = a^0 \cdot b^0 = 1 \cdot 1 = 1$$

**c.** $(-5)^0 = 1$

**d.** $-5^0 = -1 \cdot 5^0 = -1 \cdot 1 = -1$

**e.** $\left(\dfrac{3}{100}\right)^0 = 1$

**PRACTICE**
**10**  Simplify the following expressions.

**a.** $-3^0$    **b.** $(-3)^0$    **c.** $8^0$    **d.** $(0.2)^0$    **e.** $(xz)^0$

**OBJECTIVE 6 ▶ Deciding which rule to use.** Let's practice deciding which rule(s) to use to simplify. We will continue this discussion with more examples in Section 5.5.

**EXAMPLE 11** Simplify each expression.

**a.** $\left(\dfrac{st}{2}\right)^4$    **b.** $(9y^5z^7)^2$    **c.** $\left(\dfrac{-5x^2}{y^3}\right)^2$

_Solution_

**a.** This is a quotient raised to a power, so we use the power of a quotient rule.

$$\left(\frac{st}{2}\right)^4 = \frac{s^4t^4}{2^4} = \frac{s^4t^4}{16}$$

**b.** This is a product raised to a power, so we use the power of a product rule.

$$(9y^5z^7)^2 = 9^2(y^5)^2(z^7)^2 = 81y^{10}z^{14}$$

**c.** Use the power of a product or quotient rule; then use the power rule for exponents.

$$\left(\frac{-5x^2}{y^3}\right)^2 = \frac{(-5)^2(x^2)^2}{(y^3)^2} = \frac{25x^4}{y^6}$$

**PRACTICE**
**11**  Simplify each expression.

**a.** $\left(\dfrac{5}{xz}\right)^3$    **b.** $(2z^8x^5)^4$    **c.** $\left(\dfrac{-3x^3}{y^4}\right)^3$

# VOCABULARY & READINESS CHECK

*Use the choices below to fill in each blank. Some choices may be used more than once.*

| | | |
|---|---|---|
| 0 | base | add |
| 1 | exponent | multiply |

1. Repeated multiplication of the same factor can be written using a(n) _____.
2. In $5^2$, the 2 is called the _____ and the 5 is called the _____.
3. To simplify $x^2 \cdot x^7$, keep the base and _____ the exponents.
4. To simplify $(x^3)^6$, keep the base and _____ the exponents.
5. The understood exponent on the term $y$ is _____.
6. If $x^{\square} = 1$, the exponent is _____.

*State the bases and the exponents for each of the following expressions.*

7. $3^2$

8. $(-3)^6$

9. $-4^2$

10. $5 \cdot 3^4$

11. $5x^2$

12. $(5x)^2$

## 5.1 | EXERCISE SET

**MyMathLab** Powered by CourseCompass™ and MathXL®

PRACTICE · WATCH · DOWNLOAD · READ · REVIEW

*Evaluate each expression. See Example 1.*

1. $7^2$

2. $-3^2$

3. $(-5)^1$

4. $(-3)^2$

5. $-2^4$

6. $-4^3$

7. $(-2)^4$

8. $(-4)^3$

9. $(0.1)^5$

10. $(0.2)^5$

11. $\left(\dfrac{1}{3}\right)^4$

12. $\left(-\dfrac{1}{9}\right)^2$

13. $7 \cdot 2^5$

14. $9 \cdot 1^7$

15. $-2 \cdot 5^3$

16. $-4 \cdot 3^3$

17. Explain why $(-5)^4 = 625$, while $-5^4 = -625$.

18. Explain why $5 \cdot 4^2 = 80$, while $(5 \cdot 4)^2 = 400$.

*Evaluate each expression given the replacement values for x. See Example 2.*

19. $x^2$; $x = -2$

20. $x^3$; $x = -2$

21. $5x^3$; $x = 3$

22. $4x^2$; $x = -1$

23. $2xy^2$; $x = 3$ and $y = 5$

24. $-4x^2y^3$; $x = 2$ and $y = -1$

25. $\dfrac{2z^4}{5}$; $z = -2$

26. $\dfrac{10}{3y^3}$; $y = 5$

*Use the product rule to simplify each expression. Write the results using exponents. See Examples 3 through 5.*

27. $x^2 \cdot x^5$

28. $y^2 \cdot y$

29. $(-3)^3 \cdot (-3)^9$

30. $(-5)^7 \cdot (-5)^6$

31. $(5y^4)(3y)$

32. $(-2z^3)(-2z^2)$

33. $(x^9y)(x^{10}y^5)$

34. $(a^2b)(a^{13}b^{17})$

35. $(-8mn^6)(9m^2n^2)$

36. $(-7a^3b^3)(7a^{19}b)$

37. $(4z^{10})(-6z^7)(z^3)$

38. $(12x^5)(-x^6)(x^4)$

39. The rectangle below has width $4x^2$ feet and length $5x^3$ feet. Find its area as an expression in $x$. $(A = l \cdot w)$

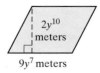

4x² feet

5x³ feet

40. The parallelogram below has base length $9y^7$ meters and height $2y^{10}$ meters. Find its area as an expression in $y$. $(A = b \cdot h)$

2y¹⁰ meters

9y⁷ meters

**MIXED PRACTICE**

*Use the power rule and the power of a product or quotient rule to simplify each expression. See Examples 6 through 8.*

41. $(x^9)^4$

42. $(y^7)^5$

**43.** $(pq)^8$

**44.** $(ab)^6$

**45.** $(2a^5)^3$

**46.** $(4x^6)^2$

**47.** $(x^2y^3)^5$

**48.** $(a^4b)^7$

**49.** $(-7a^2b^5c)^2$

**50.** $(-3x^7yz^2)^3$

**51.** $\left(\dfrac{r}{s}\right)^9$

**52.** $\left(\dfrac{q}{t}\right)^{11}$

**53.** $\left(\dfrac{mp}{n}\right)^5$

**54.** $\left(\dfrac{xy}{7}\right)^2$

**55.** $\left(\dfrac{-2xz}{y^5}\right)^2$

**56.** $\left(\dfrac{xy^4}{-3z^3}\right)^3$

**△ 57.** The square shown has sides of length $8z^5$ decimeters. Find its area. $(A = s^2)$

$8z^5$
decimeters

**△ 58.** Given the circle below with radius $5y$ centimeters, find its area. Do not approximate $\pi$. $(A = \pi r^2)$

$5y$ cm

**△ 59.** The vault below is in the shape of a cube. If each side is $3y^4$ feet, find its volume. $(V = s^3)$

$3y^4$ feet

$3y^4$ feet

$3y^4$ feet

**△ 60.** The silo shown is in the shape of a cylinder. If its radius is $4x$ meters and its height is $5x^3$ meters, find its volume. Do not approximate $\pi$. $(V = \pi r^2 h)$

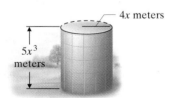

$4x$ meters

$5x^3$
meters

*Use the quotient rule and simplify each expression. See Example 9.*

**61.** $\dfrac{x^3}{x}$

**62.** $\dfrac{y^{10}}{y^9}$

**63.** $\dfrac{(-4)^6}{(-4)^3}$

**64.** $\dfrac{(-6)^{13}}{(-6)^{11}}$

**65.** $\dfrac{p^7q^{20}}{pq^{15}}$

**66.** $\dfrac{x^8y^6}{xy^5}$

**67.** $\dfrac{7x^2y^6}{14x^2y^3}$

**68.** $\dfrac{9a^4b^7}{27ab^2}$

*Simplify each expression. See Example 10.*

**69.** $7^0$

**70.** $23^0$

**71.** $(2x)^0$

**72.** $(4y)^0$

**73.** $-7x^0$

**74.** $-2x^0$

**75.** $5^0 + y^0$

**76.** $-3^0 + 4^0$

## MIXED PRACTICE

*Simplify each expression. See Examples 1 through 11.*

**77.** $-9^2$

**78.** $(-9)^2$

**79.** $\left(\dfrac{1}{4}\right)^3$

**80.** $\left(\dfrac{2}{3}\right)^3$

**81.** $\left(\dfrac{9}{qr}\right)^2$

**82.** $\left(\dfrac{pt}{3}\right)^3$

**83.** $a^2a^3a$

**84.** $x^2x^{15}x$

**85.** $(2x^3)(-8x^4)$

**86.** $(3y^4)(-5y)$

**87.** $(a^7b^{12})(a^4b^8)$

**88.** $(y^2z^2)(y^{15}z^{13})$

**89.** $(-2mn^6)(-13m^8n)$

**90.** $(-3s^5t)(-7st^{10})$

**91.** $(z^4)^{10}$

**92.** $(t^5)^{11}$

**93.** $(-6xyz^3)^2$

**94.** $(-3xy^2a^3)^3$

**95.** $\dfrac{3x^5}{x^4}$

**96.** $\dfrac{5x^9}{x^3}$

**97.** $(9xy)^2$

**98.** $(2ab)^5$

**99.** $2^0 + 2^5$

**100.** $7^2 - 7^0$

**101.** $\left(\dfrac{3y^5}{6x^4}\right)^3$

**102.** $\left(\dfrac{2ab}{6yz}\right)^4$

**103.** $\dfrac{2x^3y^2z}{xyz}$

**104.** $\dfrac{x^{12}y^{13}}{x^5y^7}$

## REVIEW AND PREVIEW

*Simplify each expression by combining any like terms. Use the distributive property to remove any parentheses. See Section 2.1.*

**105.** $y - 10 + y$

**106.** $-6z + 20 - 3z$

**107.** $7x + 2 - 8x - 6$

**108.** $10y - 14 - y - 14$

**109.** $2(x - 5) + 3(5 - x)$

**110.** $-3(w + 7) + 5(w + 1)$

## CONCEPT EXTENSIONS

*Solve. See the Concept Checks in this section. For Exercises 111 through 114, match the expression with the operation needed to simplify each. A letter may be used more than once and a letter may not be used at all.*

**111.** $(x^{14})^{23}$

**112.** $x^{14} \cdot x^{23}$

**113.** $x^{14} + x^{23}$

**114.** $\dfrac{x^{35}}{x^{17}}$

**a.** Add the exponents

**b.** Subtract the exponents

**c.** Multiply the exponents

**d.** Divide the exponents

**e.** None of these

*Fill in the boxes so that each statement is true. (More than one answer is possible for each exercise.)*

**115.** $x^\square \cdot x^\square = x^{12}$

**116.** $(x^\square)^\square = x^{20}$

**117.** $\dfrac{y^\square}{y^\square} = y^7$

**118.** $(y^\square)^\square \cdot (y^\square)^\square = y^{30}$

△ **119.** The formula $V = x^3$ can be used to find the volume $V$ of a cube with side length $x$. Find the volume of a cube with side length 7 meters. (Volume is measured in cubic units.)

△ **120.** The formula $S = 6x^2$ can be used to find the surface area $S$ of a cube with side length $x$. Find the surface area of a cube with side length 5 meters. (Surface area is measured in square units.)

△ **121.** To find the amount of water that a swimming pool in the shape of a cube can hold, do we use the formula for volume of the cube or surface area of the cube? (See Exercises 119 and 120.)

△ **122.** To find the amount of material needed to cover an ottoman in the shape of a cube, do we use the formula for volume of the cube or surface area of the cube? (See Exercises 119 and 120.)

**123.** In your own words, explain why $5^0 = 1$.

**124.** In your own words, explain when $(-3)^n$ is positive and when it is negative.

*Simplify each expression. Assume that variables represent positive integers.*

**125.** $x^{5a}x^{4a}$

**126.** $b^{9a}b^{4a}$

**127.** $(a^b)^5$

**128.** $(2a^{4b})^4$

**129.** $\dfrac{x^{9a}}{x^{4a}}$

**130.** $\dfrac{y^{15b}}{y^{6b}}$

**131.** Suppose you borrow money for 6 months. If the interest rate is compounded monthly, the formula $A = P\left(1 + \dfrac{r}{12}\right)^6$ gives the total amount $A$ to be repaid at the end of 6 months. For a loan of $P = \$1000$ and interest rate of 9% ($r = 0.09$), how much money will you need to pay off the loan?

**132.** On January 1, 2007, the Federal Reserve discount rate was set at $5\frac{1}{4}$%. (*Source:* Federal Reserve Board) The discount rate is the interest rate at which banks can borrow money from the Federal Reserve System. Suppose a bank needs to borrow money from the Federal Reserve System for 3 months. If the interest is compounded monthly, the formula $A = P\left(1 + \dfrac{r}{12}\right)^3$ gives the total amount $A$ to be repaid at the end of 3 months. For a loan of $P = \$500,000$ and interest rate of $r = 0.0525$, how much money will the bank repay to the Federal Reserve at the end of 3 months? Round to the nearest dollar.

---

 **STUDY SKILLS BUILDER**

### How Well Do You Know Your Textbook?

The questions below will help determine whether you are familiar with your textbook. For additional information, see Section 1.1 in this text.

**1.** What does the 🖥 icon mean?

**2.** What does the ＼ icon mean?

**3.** What does the △ icon mean?

**4.** Where can you find a review for each chapter? What answers to this review can be found in the back of your text?

**5.** Each chapter contains an overview of the chapter along with examples. What is this feature called?

**6.** Each chapter contains a review of vocabulary. What is this feature called?

**7.** There is a CD in your text. What content is contained on this CD?

**8.** What is the location of the section that is entirely devoted to study skills?

**9.** There are practice exercises that are contained in this text. What are they and how can they be used?

# 5.2 POLYNOMIAL FUNCTIONS AND ADDING AND SUBTRACTING POLYNOMIALS

**OBJECTIVES**

1 Define polynomial, monomial, binomial, trinomial, and degree.

2 Define polynomial functions.

3 Simplify a polynomial by combining like terms.

4 Add and subtract polynomials.

**OBJECTIVE 1 ▶ Defining polynomial, monomial, binomial, trinomial, and degree.** In this section, we introduce a special algebraic expression called a polynomial. Let's first review some definitions presented in Section 2.1.

Recall that a term is a number or the product of a number and variables raised to powers. The terms of the expression $4x^2 + 3x$ are $4x^2$ and $3x$. The terms of the expression $9x^4 - 7x - 1$ are $9x^4$, $-7x$, and $-1$.

| Expression | Terms |
|:---:|:---:|
| $4x^2 + 3x$ | $4x^2, 3x$ |
| $9x^4 - 7x - 1$ | $9x^4, -7x, -1$ |
| $7y^3$ | $7y^3$ |
| $5$ | $5$ |

The **numerical coefficient** of a term, or simply the **coefficient,** is the numerical factor of each term. If no numerical factor appears in the term, then the coefficient is understood to be 1. If the term is a number only, it is called a **constant** term or simply a constant.

| Term | Coefficient |
|:---:|:---:|
| $x^5$ | $1$ |
| $3x^2$ | $3$ |
| $-4x$ | $-4$ |
| $-x^2y$ | $-1$ |
| 3 (constant) | $3$ |

**Polynomial**

A **polynomial in $x$** is a finite sum of terms of the form $ax^n$, where $a$ is a real number and $n$ is a whole number.

For example,

$$x^5 - 3x^3 + 2x^2 - 5x + 1$$

is a polynomial. Notice that this polynomial is written in **descending powers** of $x$ because the powers of $x$ decrease from left to right. (Recall that the term 1 can be thought of as $1x^0$.)

On the other hand,

$$x^{-5} + 2x - 3$$

is **not** a polynomial because it contains an exponent, $-5$, that is not a whole number. (We study negative exponents in Section 5.5 of this chapter.)

Some polynomials are given special names.

**Types of Polynomials**

A **monomial** is a polynomial with exactly one term.

A **binomial** is a polynomial with exactly two terms.

A **trinomial** is a polynomial with exactly three terms.

The following are examples of monomials, binomials, and trinomials. Each of these examples is also a polynomial.

**POLYNOMIALS**

| Monomials | Binomials | Trinomials | None of These |
|-----------|-----------|------------|---------------|
| $ax^2$ | $x + y$ | $x^2 + 4xy + y^2$ | $5x^3 - 6x^2 + 3x - 6$ |
| $-3z$ | $3p + 2$ | $x^5 + 7x^2 - x$ | $-y^5 + y^4 - 3y^3 - y^2 + y$ |
| $4$ | $4x^2 - 7$ | $-q^4 + q^3 - 2q$ | $x^6 + x^4 - x^3 + 1$ |

Each term of a polynomial has a **degree.**

> **Degree of a Term**
> The degree of a term is the sum of the exponents on the variables contained in the term.

**EXAMPLE 1**    Find the degree of each term.

**a.** $3x^2$    **b.** $-2^3x^5$    **c.** $y$    **d.** $12x^2yz^3$    **e.** 5

_Solution_

**a.** The exponent on $x$ is 2, so the degree of the term is 2.

**b.** The exponent on $x$ is 5, so the degree of the term is 5. (Recall that the degree is the sum of the exponents on only the *variables*.)

**c.** The degree of $y$, or $y^1$, is 1.

**d.** The degree is the sum of the exponents on the variables, or $2 + 1 + 3 = 6$.

**e.** The degree of 5, which can be written as $5x^0$, is 0.    ☐

**PRACTICE**
**1**    Find the degree of each term.

**a.** $5y^3$    **b.** $10\,xy$    **c.** $z$    **d.** $-3a^2b^5c$    **e.** 8

From the preceding, we can say that **the degree of a constant is 0.**

Each polynomial also has a degree.

> **Degree of a Polynomial**
> The degree of a polynomial is the greatest degree of any term of the polynomial.

**EXAMPLE 2**    Find the degree of each polynomial and tell whether the polynomial is a monomial, binomial, trinomial, or none of these.

**a.** $-2t^2 + 3t + 6$    **b.** $15x - 10$    **c.** $7x + 3x^3 + 2x^2 - 1$

_Solution_

**a.** The degree of the trinomial $-2t^2 + 3t + 6$ is 2, the greatest degree of any of its terms.

**b.** The degree of the binomial $15x - 10$ or $15x^1 - 10$ is 1.

**c.** The degree of the polynomial $7x + 3x^3 + 2x^2 - 1$ is 3.    ☐

PRACTICE
**2** Find the degree of each polynomial and tell whether the polynomial is a monomial, binomial, trinomial, or none of these.

**a.** $5b^2 - 3b + 7$     **b.** $7t + 3$
**c.** $5x^2 + 3x - 6x^3 + 4$

**EXAMPLE 3** Complete the table for the polynomial
$$7x^2y - 6xy + x^2 - 3y + 7$$

Use the table to give the degree of the polynomial.

*Solution*

| Term | Numerical Coefficient | Degree of Term |
|------|------|------|
| $7x^2y$ | 7 | 3 |
| $-6xy$ | $-6$ | 2 |
| $x^2$ | 1 | 2 |
| $-3y$ | $-3$ | 1 |
| 7 | 7 | 0 |

The degree of the polynomial is 3.

PRACTICE
**3** Complete the table for the polynomial $-3x^3y^2 + 4xy^2 - y^2 + 3x - 2$.

| Term | Numerical Coefficient | Degree of Term |
|------|------|------|
| $-3x^3y^2$ | | |
| $4xy^2$ | | |
| $-y^2$ | | |
| $3x$ | | |
| $-2$ | | |

**OBJECTIVE 2** ▶ **Defining polynomial functions.** At times, it is convenient to use function notation to represent polynomials. For example, we may write $P(x)$ to represent the polynomial $3x^2 - 2x - 5$. In symbols, this is

$$P(x) = 3x^2 - 2x - 5$$

This function is called a **polynomial function** because the expression $3x^2 - 2x - 5$ is a polynomial.

> ▶ **Helpful Hint**
> Recall that the symbol $P(x)$ **does not mean** $P$ times $x$. It is a special symbol used to denote a function.

**EXAMPLE 4** If $P(x) = 3x^2 - 2x - 5$, find the following.

**a.** $P(1)$     **b.** $P(-2)$

*Solution*

**a.** Substitute 1 for $x$ in $P(x) = 3x^2 - 2x - 5$ and simplify.

$$P(x) = 3x^2 - 2x - 5$$
$$P(1) = 3(1)^2 - 2(1) - 5 = -4$$

**b.** Substitute $-2$ for $x$ in $P(x) = 3x^2 - 2x - 5$ and simplify.

$$P(x) = 3x^2 - 2x - 5$$
$$P(-2) = 3(-2)^2 - 2(-2) - 5 = 11$$

**PRACTICE**
**4**   If $P(x) = -2x^2 - x + 7$, find

**a.** $P(1)$   **b.** $P(-4)$

---

Many real-world phenomena are modeled by polynomial functions. If the polynomial function model is given, we can often find the solution of a problem by evaluating the function at a certain value.

**EXAMPLE 5**   **Finding the Height of a Dropped Object**

The Swiss Re Building, in London, is a unique building. Londoners often refer to it as the "pickle building." The building is 592.1 feet tall. An object is dropped from the highest point of this building. Neglecting air resistance, the height in feet of the object above ground at time $t$ seconds is given by the polynomial function $P(t) = -16t^2 + 592.1$. Find the height of the object when $t = 1$ second, and when $t = 6$ seconds.

*Solution*   To find each height, we find $P(1)$ and $P(6)$.

$$P(t) = -16t^2 + 592.1$$
$$P(1) = -16(1)^2 + 592.1 \quad \text{Replace } t \text{ with 1.}$$
$$P(1) = -16 + 592.1$$
$$P(1) = 576.1$$

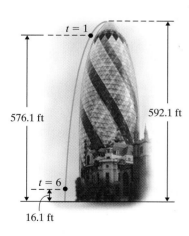

$t = 1$

576.1 ft                    592.1 ft

$t = 6$

16.1 ft

The height of the object at 1 second is 576.1 feet.

$$P(t) = -16t^2 + 592.1$$
$$P(6) = -16(6)^2 + 592.1 \quad \text{Replace } t \text{ with 6.}$$
$$P(6) = -576 + 592.1$$
$$P(6) = 16.1$$

The height of the object at 6 seconds is 16.1 feet.

**PRACTICE**
**5**   The cliff divers of Acapulco dive 130 feet into La Quebrada several times a day for the entertainment of the tourists. If a tourist is standing near the diving platform and drops his camera off the cliff, the height of the camera above the water at time $t$ seconds is given by the polynomial function $P(t) = -16t^2 + 130$. Find the height of the camera when $t = 1$ second and when $t = 2$ seconds.

---

**OBJECTIVE 3 ▶ Simplifying polynomials by combining like terms.** Polynomials with like terms can be simplified by combining the like terms. Recall that like terms are terms that contain exactly the same variables raised to exactly the same powers.

| *Like Terms* | *Unlike Terms* |
|---|---|
| $5x^2, -7x^2$ | $3x, 3y$ |
| $y, 2y$ | $-2x^2, -5x$ |
| $\frac{1}{2}a^2b, -a^2b$ | $6st^2, 4s^2t$ |

**Only like terms can be combined.** We combine like terms by applying the distributive property.

**EXAMPLE 6**  Simplify each polynomial by combining any like terms.

**a.** $-3x + 7x$

**b.** $x + 3x^2$

**c.** $11x^2 + 5 + 2x^2 - 7$

**d.** $\frac{2}{5}x^4 + \frac{2}{3}x^3 - x^2 + \frac{1}{10}x^4 - \frac{1}{6}x^3$

*Solution*

**a.** $-3x + 7x = (-3 + 7)x = 4x$

**b.** $x + 3x^2$   These terms cannot be combined because $x$ and $3x^2$ are not like terms.

**c.** $11x^2 + 5 + 2x^2 - 7 = 11x^2 + 2x^2 + 5 - 7$
$$= 13x^2 - 2 \qquad \text{Combine like terms.}$$

**d.** $\frac{2}{5}x^4 + \frac{2}{3}x^3 - x^2 + \frac{1}{10}x^4 - \frac{1}{6}x^3$

$$= \left(\frac{2}{5} + \frac{1}{10}\right)x^4 + \left(\frac{2}{3} - \frac{1}{6}\right)x^3 - x^2$$

$$= \left(\frac{4}{10} + \frac{1}{10}\right)x^4 + \left(\frac{4}{6} - \frac{1}{6}\right)x^3 - x^2$$

$$= \frac{5}{10}x^4 + \frac{3}{6}x^3 - x^2$$

$$= \frac{1}{2}x^4 + \frac{1}{2}x^3 - x^2$$

**PRACTICE**
**6**  Simplify each polynomial by combining any like terms.

**a.** $-4y + 2y$

**b.** $z + 5z^3$

**c.** $7a^2 - 5 - 3a^2 - 7$

**d.** $\frac{3}{8}x^3 - x^2 + \frac{5}{6}x^4 + \frac{1}{12}x^3 - \frac{1}{2}x^4$

**Concept Check** ☑

When combining like terms in the expression, $5x - 8x^2 - 8x$, which of the following is the proper result?

**a.** $-11x^2$  **b.** $-8x^2 - 3x$  **c.** $-11x$  **d.** $-11x^4$

**EXAMPLE 7**  Combine like terms to simplify.

$$-9x^2 + 3xy - 5y^2 + 7yx$$

> **Helpful Hint**
> This term can be written as $7yx$ or $7xy$.

*Solution*   $-9x^2 + 3xy - 5y^2 + 7yx = -9x^2 + (3 + 7)xy - 5y^2$
$$= -9x^2 + 10xy - 5y^2$$

**PRACTICE**
**7**  Combine like terms to simplify: $9xy - 3x^2 - 4yx + 5y^2$.

**Answer to Concept Check:**  b

⚠ **EXAMPLE 8** Write a polynomial that describes the total area of the squares and rectangles shown below. Then simplify the polynomial.

*Solution*

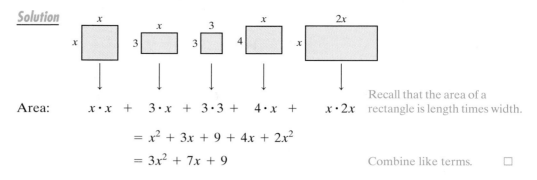

Area:     $x \cdot x$  +   $3 \cdot x$  +  $3 \cdot 3$ +   $4 \cdot x$  +      $x \cdot 2x$   Recall that the area of a rectangle is length times width.

$$= x^2 + 3x + 9 + 4x + 2x^2$$

$$= 3x^2 + 7x + 9$$   Combine like terms.   □

**PRACTICE**
**8**   Write a polynomial that describes the total area of the squares and rectangles shown below. Then simplify the polynomial.

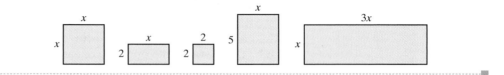

**OBJECTIVE 4 ▶ Adding and subtracting polynomials.** We now practice adding and subtracting polynomials.

**Adding Polynomials**
To add polynomials, combine all like terms.

**EXAMPLE 9**   Add.

**a.** $(7x^3y - xy^3 + 11) + (6x^3y - 4)$     **b.** $(3a^3 - b + 2a - 5) + (a + b + 5)$

*Solution*

**a.** To add, remove the parentheses and group like terms.

$$(7x^3y - xy^3 + 11) + (6x^3y - 4)$$
$$= 7x^3y - xy^3 + 11 + 6x^3y - 4$$
$$= 7x^3y + 6x^3y - xy^3 + 11 - 4 \quad \text{Group like terms.}$$
$$= 13x^3y - xy^3 + 7 \quad\quad\quad \text{Combine like terms.}$$

**b.**      $(3a^3 - b + 2a - 5) + (a + b + 5)$
$$= 3a^3 - b + 2a - 5 + a + b + 5$$
$$= 3a^3 - b + b + 2a + a - 5 + 5 \quad \text{Group like terms.}$$
$$= 3a^3 + 3a \quad\quad\quad\quad\quad\quad \text{Combine like terms.}\quad □$$

**PRACTICE**
**9**   Add.

**a.** $(4y^2 + x - 3y - 7) + (x + y^2 - 2)$
**b.** $(-8a^2b - ab^2 + 10) + (-2ab^2 - 10)$

**EXAMPLE 10** Add $(11x^3 - 12x^2 + x - 3)$ and $(x^3 - 10x + 5)$.

**Solution**

$(11x^3 - 12x^2 + x - 3) + (x^3 - 10x + 5)$
$$= 11x^3 + x^3 - 12x^2 + x - 10x - 3 + 5 \quad \text{Group like terms.}$$
$$= 12x^3 - 12x^2 - 9x + 2 \quad\quad\quad\quad \text{Combine like terms.} \quad \square$$

**PRACTICE**
**10** Add $(3x^2 - 9x + 11)$ and $(-3x^2 + 7x^3 + 3x - 4)$

Sometimes it is more convenient to add polynomials vertically. To do this, line up like terms beneath one another and add like terms.

The definition of subtraction of real numbers can be extended to apply to polynomials. To subtract a number, we add its opposite.

$$a - b = a + (-b)$$

Likewise, to subtract a polynomial, we add its opposite. In other words, if $P$ and $Q$ are polynomials, then
$$P - Q = P + (-Q)$$

The polynomial $-Q$ is the **opposite,** or **additive inverse,** of the polynomial $Q$. We can find $-Q$ by writing the opposite of each term of $Q$.

> **Subtracting Polynomials**
> To subtract two polynomials, change the signs of the terms of the polynomial being subtracted and then add.

**Concept Check** ☑

Which polynomial is the opposite of $16x^3 - 5x + 7$?

**a.** $-16x^3 - 5x + 7$            **b.** $-16x^3 + 5x - 7$

**c.** $16x^3 + 5x + 7$            **d.** $-16x^3 + 5x + 7$

**EXAMPLE 11** Subtract: $(2x^3 + 8x^2 - 6x) - (2x^3 - x^2 + 1)$.

**Solution** First, change the sign of each term of the second polynomial and then add.

> **Helpful Hint**
> Notice the sign of each term is changed.

$(2x^3 + 8x^2 - 6x) - (2x^3 - x^2 + 1) = (2x^3 + 8x^2 - 6x) + (-2x^3 + x^2 - 1)$
$$= 2x^3 - 2x^3 + 8x^2 + x^2 - 6x - 1$$
$$= 9x^2 - 6x - 1 \quad \text{Combine like terms.} \quad \square$$

**PRACTICE**
**11** Subtract: $(3x^3 - 5x^2 + 4x) - (x^3 - x^2 + 6)$.

**EXAMPLE 12** Subtract $(5z - 7)$ from the sum of $(8z + 11)$ and $(9z - 2)$.

**Solution** Notice that $(5z - 7)$ is to be subtracted **from** a sum. The translation is
$$[(8z + 11) + (9z - 2)] - (5z - 7)$$
$$= 8z + 11 + 9z - 2 - 5z + 7 \quad \text{Remove grouping symbols.}$$
$$= 8z + 9z - 5z + 11 - 2 + 7 \quad \text{Group like terms.}$$
$$= 12z + 16 \quad\quad\quad\quad\quad\quad \text{Combine like terms.} \quad \square$$

**PRACTICE**
**12** Subtract $(3x + 5)$ from the sum of $(8x - 11)$ and $(2x + 5)$.

**Answer to Concept Check:**
b

**EXAMPLE 13**   Add or subtract as indicated.

**a.** $(3x^2 - 6xy + 5y^2) + (-2x^2 + 8xy - y^2)$

**b.** $(9a^2b^2 + 6ab - 3ab^2) - (5b^2a + 2ab - 3 - 9b^2)$

*Solution*

**a.** $(3x^2 - 6xy + 5y^2) + (-2x^2 + 8xy - y^2)$

$= 3x^2 - 6xy + 5y^2 - 2x^2 + 8xy - y^2$

$= x^2 + 2xy + 4y^2$   Combine like terms.

**b.** $(9a^2b^2 + 6ab - 3ab^2) - (5b^2a + 2ab - 3 - 9b^2)$   Change the sign of each term

$= 9a^2b^2 + 6ab - 3ab^2 - 5b^2a - 2ab + 3 + 9b^2$   of the polynomial being subtracted.

$= 9a^2b^2 + 4ab - 8ab^2 + 3 + 9b^2$   Combine like terms.   □

**PRACTICE**
**13**   Add or subtract as indicated.

**a.** $(3a^2 - 4ab + 7b^2) + (-8a^2 + 3ab - b^2)$

**b.** $(5x^2y^2 - 6xy - 4xy^2) - (2x^2y^2 + 4xy - 5 + 6y^2)$

**Concept Check** ☑

If possible, simplify each expression by performing the indicated operation.

**a.** $2y + y$    **b.** $2y \cdot y$    **c.** $-2y - y$    **d.** $(-2y)(-y)$    **e.** $2x + y$

To add or subtract polynomials vertically, just remember to line up like terms. For example, perform the subtraction $(10x^3y^2 - 7x^2y^2) - (4x^3y^2 - 3x^2y^2 + 2y^2)$ vertically. Add the opposite of the second polynomial.

$$
\begin{array}{r}
10x^3y^2 - 7x^2y^2 \\
-(4x^3y^2 - 3x^2y^2 + 2y^2) \\
\end{array}
\quad \text{is equivalent to} \quad
\begin{array}{r}
10x^3y^2 - 7x^2y^2 \\
-4x^3y^2 + 3x^2y^2 - 2y^2 \\
\hline
6x^3y^2 - 4x^2y^2 - 2y^2 \\
\end{array}
$$

**Concept Check** ☑

Why is the following subtraction incorrect?

$$(7z - 5) - (3z - 4)$$
$$= 7z - 5 - 3z - 4$$
$$= 4z - 9$$

Polynomial functions, like polynomials, can be added, subtracted, multiplied, and divided. For example, if

$$P(x) = x^2 + x + 1$$

then

$$2P(x) = 2(x^2 + x + 1) = 2x^2 + 2x + 2$$

Also, if $Q(x) = 5x^2 - 1$, then $P(x) + Q(x) = (x^2 + x + 1) + (5x^2 - 1)$
$$= 6x^2 + x.$$

A useful business and economics application of subtracting polynomial functions is finding the profit function $P(x)$ when given a revenue function $R(x)$ and a cost function $C(x)$. In business, it is true that

$$\text{profit} = \text{revenue} - \text{cost, or}$$
$$P(x) = R(x) - C(x)$$

**Answers to Concept Check:**

**a.** $3y$  **b.** $2y^2$  **c.** $-3y$  **d.** $2y^2$
**e.** cannot be simplified; With parentheses removed, the expression should be
$$7z - 5 - 3z + 4 = 4z - 1$$

For example, if the revenue function is $R(x) = 7x$ and the cost function is $C(x) = 2x + 5000$, then the profit function is

$$P(x) = R(x) - C(x)$$

or

$$P(x) = 7x - (2x + 5000) \quad \text{Substitute } R(x) = 7x$$
$$P(x) = 5x - 5000 \qquad\qquad \text{and } C(x) = 2x + 5000.$$

Problem-solving exercises involving profit are in the exercise set.

---

## Graphing Calculator Explorations

A graphing calculator may be used to visualize addition and subtraction of polynomials in one variable. For example, to visualize the following polynomial subtraction statement

$$(3x^2 - 6x + 9) - (x^2 - 5x + 6) = 2x^2 - x + 3$$

graph both

$$Y_1 = (3x^2 - 6x + 9) - (x^2 - 5x + 6) \quad \text{Left side of equation}$$

and

$$Y_2 = 2x^2 - x + 3 \quad \text{Right side of equation}$$

on the same screen and see that their graphs coincide. (*Note:* If the graphs do not coincide, we can be sure that a mistake has been made in combining polynomials or in calculator keystrokes. If the graphs appear to coincide, we cannot be sure that our work is correct. This is because it is possible for the graphs to differ so slightly that we do not notice it.)

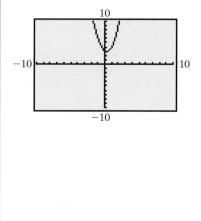

The graphs of $Y_1$ and $Y_2$ are shown. The graphs appear to coincide, so the subtraction statement

$$(3x^2 - 6x + 9) - (x^2 - 5x + 6) = 2x^2 - x + 3$$

appears to be correct.

*Perform the indicated operations. Then visualize by using the procedure described above.*

**1.** $(2x^2 + 7x + 6) + (x^3 - 6x^2 - 14)$

**2.** $(-14x^3 - x + 2) + (-x^3 + 3x^2 + 4x)$

**3.** $(1.8x^2 - 6.8x - 1.7) - (3.9x^2 - 3.6x)$

**4.** $(-4.8x^2 + 12.5x - 7.8) - (3.1x^2 - 7.8x)$

**5.** $(1.29x - 5.68) + (7.69x^2 - 2.55x + 10.98)$

**6.** $(-0.98x^2 - 1.56x + 5.57) + (4.36x - 3.71)$

---

## VOCABULARY & READINESS CHECK

*Use the choices below to fill in each blank. Not all choices will be used.*

| | | | |
|---|---|---|---|
| least | monomial | trinomial | coefficient |
| greatest | binomial | constant | |

**1.** A(n) _____ is a polynomial with exactly 2 terms.

**2.** A(n) _____ is a polynomial with exactly one term.

**3.** A(n) _____ is a polynomial with exactly three terms.

**4.** The numerical factor of a term is called the _____.

**5.** A number term is also called a _____.

**6.** The degree of a polynomial is the _____ degree of any term of the polynomial.

*Simplify by combining like terms if possible.*

**7.** $-9y - 5y$

**8.** $6m^5 + 7m^5$

**9.** $x + 6x$

**10.** $7z - z$

**11.** $5m^2 + 2m$

**12.** $8p^3 + 3p^2$

## 5.2 EXERCISE SET

*Find the degree of each of the following polynomials and determine whether it is a monomial, binomial, trinomial, or none of these. See Examples 1 through 3.*

**1.** $x + 2$

**2.** $-6y + y^2 + 4$

**3.** $9m^3 - 5m^2 + 4m - 8$

**4.** $5a^2 + 3a^3 - 4a^4$

**5.** $12x^4y - x^2y^2 - 12x^2y^4$

**6.** $7r^2s^2 + 2r - 3s^5$

**7.** $3zx - 5x^2$

**8.** $5y + 2$

*In the second column, write the degree of the polynomial in the first column. See Examples 1 through 3.*

| Polynomial | Degree |
|---|---|
| **9.** $3xy^2 - 4$ | |
| **10.** $8x^2y^2 - 7x^3 + 6xy - 1$ | |
| **11.** $5a^2 - 2a + 1$ | |
| **12.** $4z^6 + 3z^2$ | |

*If $P(x) = x^2 + x + 1$ and $Q(x) = 5x^2 - 1$, find the following. See Examples 4 and 5.*

**13.** $P(7)$

**14.** $Q(4)$

**15.** $Q(-10)$

**16.** $P(-4)$

**17.** $P(0)$

**18.** $Q(0)$

**19.** $Q\left(\dfrac{1}{4}\right)$

**20.** $P\left(\dfrac{1}{2}\right)$

*The CN Tower in Toronto, Ontario, is 1821 feet tall and is the world's tallest self-supporting structure. An object is dropped from the Skypod of the Tower which is at 1150 feet. Neglecting air resistance, the height of the object at time t seconds is given by the polynomial function $P(t) = -16t^2 + 1150$. Find the height of the object at the given times. See Example 5.*

1150 ft

**21.** $t = 1$ second

**22.** $t = 7$ seconds

**23.** $t = 3$ seconds

**24.** $t = 6$ seconds

*Simplify each of the following by combining like terms. See Examples 6 and 7.*

**25.** $14x^2 + 9x^2$

**26.** $18x^3 - 4x^3$

**27.** $15x^2 - 3x^2 - y$

**28.** $12k^3 - 9k^3 + 11$

**29.** $8s - 5s + 4s$

**30.** $5y + 7y - 6y$

**31.** $0.1y^2 - 1.2y^2 + 6.7 - 1.9$

**32.** $7.6y + 3.2y^2 - 8y - 2.5y^2$

**33.** $\dfrac{2}{5}x^2 - \dfrac{1}{3}x^3 + x^2 - \dfrac{1}{4}x^3 + 6$

**34.** $\dfrac{1}{6}x^4 - \dfrac{1}{7}x^2 + 5 - \dfrac{3}{2}x^4 - \dfrac{3}{7}x^2 + \dfrac{1}{3}$

**35.** $6a^2 - 4ab + 7b^2 - a^2 - 5ab + 9b^2$

**36.** $x^2y + xy - y + 10x^2y - 2y + xy$

*Perform the indicated operations. See Examples 9 through 13.*

**37.** $(-7x + 5) + (-3x^2 + 7x + 5)$

**38.** $(3x - 8) + (4x^2 - 3x + 3)$

**39.** $(2x^2 + 5) - (3x^2 - 9)$

**40.** $(5x^2 + 4) - (-2y^2 + 4)$

**41.** $3x - (5x - 9)$

**42.** $4 - (-y - 4)$

**43.** $(2x^2 + 3x - 9) - (-4x + 7)$

**44.** $(-7x^2 + 4x + 7) - (-8x + 2)$

**45.** $\begin{array}{r} 3t^2 + 4 \\ +5t^2 - 8 \\ \hline \end{array}$

**46.** $\begin{array}{r} 7x^3 + 3 \\ +2x^3 + 1 \\ \hline \end{array}$

**47.** $\begin{array}{r} 4z^2 - 8z + 3 \\ -(6z^2 + 8z - 3) \\ \hline \end{array}$

**48.** $\begin{array}{r} 5u^5 - 4u^2 + 3u - 7 \\ -(3u^5 + 6u^2 - 8u + 2) \\ \hline \end{array}$

**49.** $\begin{array}{r} 5x^3 - 4x^2 + 6x - 2 \\ -(3x^3 - 2x^2 - x - 4) \\ \hline \end{array}$

**50.** $\begin{array}{r} 7a^2 - 9a + 6 \\ -(11a^2 - 4a + 2) \\ \hline \end{array}$

**51.** Subtract $(19x^2 + 5)$ from $(81x^2 + 10)$.

**52.** Subtract $(2x + xy)$ from $(3x - 9xy)$.

**53.** Subtract $(2x + 2)$ from the sum of $(8x + 1)$ and $(6x + 3)$.

**54.** Subtract $(-12x - 3)$ from the sum of $(-5x - 7)$ and $(12x + 3)$.

## MIXED PRACTICE

*Perform the indicated operations.*

**55.** $(-3y^2 - 4y) + (2y^2 + y - 1)$

**56.** $(7x^2 + 2x - 9) + (-3x^2 + 5)$

**57.** $(5x + 8) - (-2x^2 - 6x + 8)$

**58.** $(-6y^2 + 3y - 4) - (9y^2 - 3y)$

**59.** $(-8x^4 + 7x) + (-8x^4 + x + 9)$

**60.** $(6y^5 - 6y^3 + 4) + (-2y^5 - 8y^3 - 7)$

**61.** $(3x^2 + 5x - 8) + (5x^2 + 9x + 12) - (x^2 - 14)$

**62.** $(-a^2 + 1) - (a^2 - 3) + (5a^2 - 6a + 7)$

**63.** Subtract $4x$ from $7x - 3$.

**64.** Subtract $y$ from $y^2 - 4y + 1$.

**65.** Subtract $(5x + 7)$ from $(7x^2 + 3x + 9)$.

**66.** Subtract $(5y^2 + 8y + 2)$ from $(7y^2 + 9y - 8)$.

**67.** Subtract $(4y^2 - 6y - 3)$ from the sum of $(8y^2 + 7)$ and $(6y + 9)$.

**68.** Subtract $(5y + 7x^2)$ from the sum of $(8y - x)$ and $(3 + 8x^2)$.

**69.** Subtract $(-2x^2 + 4x - 12)$ from the sum of $(-x^2 - 2x)$ and $(5x^2 + x + 9)$.

**70.** Subtract $(4x^2 - 2x + 2)$ from the sum of $(x^2 + 7x + 1)$ and $(7x + 5)$.

*Find the area of each figure. Write a polynomial that describes the total area of the rectangles and squares shown in Exercises 71–72. Then simplify the polynomial. See Example 8.*

△ **71.**

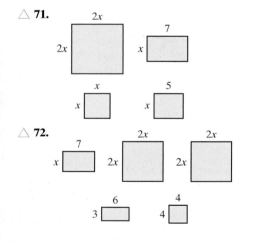

△ **72.**

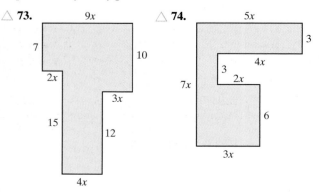

*Recall that the perimeter of a figure such as the ones shown in Exercises 73 through 76 is the sum of the lengths of its sides. Find the perimeter of each figure.*

△ **73.**   △ **74.**

△ **75.**

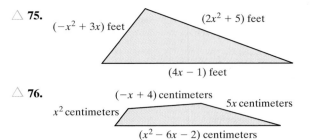

△ **76.**

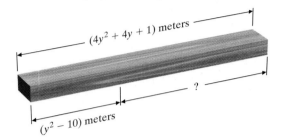

△ **77.** A wooden beam is $(4y^2 + 4y + 1)$ meters long. If a piece $(y^2 - 10)$ meters is cut, express the length of the remaining piece of beam as a polynomial in $y$.

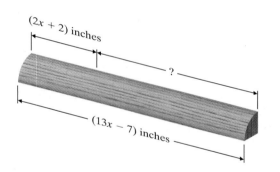

△ **78.** A piece of quarter-round molding is $(13x - 7)$ inches long. If a piece $(2x + 2)$ inches is removed, express the length of the remaining piece of molding as a polynomial in $x$.

*Add or subtract as indicated. See Examples 9 through 13.*

**79.** $(9a + 6b - 5) + (-11a - 7b + 6)$

**80.** $(3x - 2 + 6y) + (7x - 2 - y)$

**81.** $(4x^2 + y^2 + 3) - (x^2 + y^2 - 2)$

**82.** $(7a^2 - 3b^2 + 10) - (-2a^2 + b^2 - 12)$

**83.** $(x^2 + 2xy - y^2) + (5x^2 - 4xy + 20y^2)$

**84.** $(a^2 - ab + 4b^2) + (6a^2 + 8ab - b^2)$

**85.** $(11r^2s + 16rs - 3 - 2r^2s^2) - (3sr^2 + 5 - 9r^2s^2)$

**86.** $(3x^2y - 6xy + x^2y^2 - 5) - (11x^2y^2 - 1 + 5yx^2)$

*Simplify each polynomial by combining like terms.*

**87.** $7.75x + 9.16x^2 - 1.27 - 14.58x^2 - 18.34$

**88.** $1.85x^2 - 3.76x + 9.25x^2 + 10.76 - 4.21x$

*Perform each indicated operation.*

**89.** $[(7.9y^4 - 6.8y^3 + 3.3y) + (6.1y^3 - 5)] - (4.2y^4 + 1.1y - 1)$

**90.** $[(1.2x^2 - 3x + 9.1) - (7.8x^2 - 3.1 + 8)] + (1.2x - 6)$

## REVIEW AND PREVIEW

*Multiply. See Section 5.1.*

**91.** $3x(2x)$

**92.** $-7x(x)$

**93.** $(12x^3)(-x^5)$

**94.** $6r^3(7r^{10})$

**95.** $10x^2(20xy^2)$

**96.** $-z^2y(11zy)$

## CONCEPT EXTENSIONS

**97.** Describe how to find the degree of a term.

**98.** Describe how to find the degree of a polynomial.

**99.** Explain why $xyz$ is a monomial while $x + y + z$ is a trinomial.

**100.** Explain why the degree of the term $5y^3$ is 3 and the degree of the polynomial $2y + y + 2y$ is 1.

*Match each expression on the left with its simplification on the right. Not all letters on the right must be used and a letter may be used more than once.*

**101.** $10y - 6y^2 - y$

**102.** $5x + 5x$

**103.** $(5x - 3) + (5x - 3)$

**104.** $(15x - 3) - (5x - 3)$

    **a.** $3y$
    **b.** $9y - 6y^2$
    **c.** $10x$
    **d.** $25x^2$
    **e.** $10x - 6$
    **f.** none of these

*Simplify each expression by performing the indicated operation. Explain how you arrived at each answer. See the Concept Check in this section.*

**105.** **a.** $z + 3z$
    **b.** $z \cdot 3z$
    **c.** $-z - 3z$
    **d.** $(-z)(-3z)$

**106.** **a.** $x + x$
    **b.** $x \cdot x$
    **c.** $-x - x$
    **d.** $(-x)(-x)$

*Perform indicated operations.*

**107.** $(4x^{2a} - 3x^a + 0.5) - (x^{2a} - 5x^a - 0.2)$

**108.** $(9y^{5a} - 4y^{3a} + 1.5y) - (6y^{5a} - y^{3a} + 4.7y)$

**109.** $(8x^{2y} - 7x^y + 3) + (-4x^{2y} + 9x^y - 14)$

**110.** $(14z^{5x} + 3z^{2x} + z) - (2z^{5x} - 10z^{2x} + 3z)$

*If $P(x) = 3x + 3, Q(x) = 4x^2 - 6x + 3$, and $R(x) = 5x^2 - 7$, find the following.*

**111.** $P(x) + Q(x)$

**112.** $R(x) + P(x)$

**113.** $Q(x) - R(x)$

**114.** $P(x) - Q(x)$

**115.** $2[Q(x)] - R(x)$

**116.** $-5[P(x)] - Q(x)$

*If $P(x)$ is the polynomial given, find a. $P(a)$, b. $P(-x)$, and c. $P(x + h)$.*

**117.** $P(x) = 2x - 3$

**118.** $P(x) = 8x + 3$

**119.** $P(x) = 4x$

**120.** $P(x) = -4x$

**121.** The average tuition, fees, and room and board rates charged per year for full-time students in degree-granting two year public colleges is approximated by the polynomial function $f(x) = 6.4x^2 + 37.9x + 2856.8$ for the years 1984 through 2006. (*Source:* National Center for Education Statistics & The College Board) Use this model to predict what the costs will be for a student at a public two-year institution in 2010. ($x = 26$). Round to the nearest dollar.

**122.** The number of wireless telephone subscribers (in millions) $x$ years after 1990 is given by the polynomial function $f(x) = 0.74x^2 + 2.6x + 3.2$ for 1990 to 2005. Use this model to predict the number of wireless telephone subscribers in 2010 ($x = 20$). (*Source:* CTIA—The Wireless Association)

**123.** The polynomial $2.13x^2 + 21.89x + 1190$ represents the sale of electricity (in billion kilowatt-hours) in the U.S. residential sector during 2000–2005. The polynomial $8.71x^2 - 1.46x + 2095$ represents the sale of electricity (in billion kilowatt-hours) in all other U.S. sectors during 2000–2005. In both polynomials, $x$ represents the number of years after 2000. Find a polynomial for the total sales of electricity (in billion kilowatt hours) to all sectors in the United States during this period. (*Source:* Based on data from the Energy Information Administration)

**124.** The polynomial $-3.5x^2 + 33.3x + 392$ represents the number of prescriptions (in millions) purchased from a supermarket for the years 2000–2005. The polynomial $19x + 141$ represents the number of prescriptions (in millions) purchased through the mail for the same years. In both polynomials, $x$ represents the number of years since 2000. Find a polynomial for the total number of prescriptions purchased from a supermarket or mail order. (*Source:* National Association of Chain Drug Stores)

# 5.3 MULTIPLYING POLYNOMIALS

**OBJECTIVES**

1 Use the distributive property to multiply polynomials.

2 Multiply polynomials vertically.

**OBJECTIVE 1 ▶ Using the distributive property to multiply polynomials.** To multiply polynomials, we apply our knowledge of the rules and definitions of exponents.

Recall from Section 5.1 that to multiply two monomials such as $(-5x^3)$ and $(-2x^4)$, we use the associative and commutative properties and regroup. Remember, also, that to multiply exponential expressions with a common base we use the product rule for exponents and add exponents.

$$(-5x^3)(-2x^4) = (-5)(-2)(x^3)(x^4) = 10x^7$$

**EXAMPLES** Multiply.

**1.** $6x \cdot 4x = (6 \cdot 4)(x \cdot x)$   Use the commutative and associative properties.

$= 24x^2$   Multiply.

**2.** $-7x^2 \cdot 0.2x^5 = (-7 \cdot 0.2)(x^2 \cdot x^5)$

$= -1.4x^7$

**3.** $\left(-\dfrac{1}{3}x^5\right)\left(-\dfrac{2}{9}x\right) = \left(-\dfrac{1}{3} \cdot -\dfrac{2}{9}\right) \cdot \left(x^5 \cdot x\right)$

$= \dfrac{2}{27}x^6$

**PRACTICES**

**1–3** Multiply.

**1.** $5y \cdot 2y$          **2.** $(5z^3) \cdot (-0.4z^5)$          **3.** $\left(-\dfrac{1}{9}b^6\right)\left(-\dfrac{7}{8}b^3\right)$

**Concept Check** ☑

Simplify.

**a.** $3x \cdot 2x$          **b.** $3x + 2x$

To multiply polynomials that are not monomials, use the distributive property.

**EXAMPLE 4** Use the distributive property to find each product.

**a.** $5x(2x^3 + 6)$          **b.** $-3x^2(5x^2 + 6x - 1)$

*Solution*

**a.** $5x(2x^3 + 6) = 5x(2x^3) + 5x(6)$   Use the distributive property.

$= 10x^4 + 30x$   Multiply.

**b.** $-3x^2(5x^2 + 6x - 1)$

$= (-3x^2)(5x^2) + (-3x^2)(6x) + (-3x^2)(-1)$   Use the distributive property.

$= -15x^4 - 18x^3 + 3x^2$   Multiply.

**PRACTICE**

**4** Use the distributive property to find each product.

**a.** $3x(5x^5 + 5)$          **b.** $-5x^3(2x^2 - 9x + 2)$

**Answers to Concept Check:**

**a.** $6x^2$          **b.** $5x$

We also use the distributive property to multiply two binomials. To multiply $(x + 3)$ by $(x + 1)$, distribute the factor $(x + 3)$ first.

$$(x + 3)(x + 1) = x(x + 1) + 3(x + 1) \qquad \text{Distribute } (x + 3).$$
$$= x(x) + x(1) + 3(x) + 3(1) \qquad \text{Apply distributive property a second time.}$$
$$= x^2 + x + 3x + 3 \qquad \text{Multiply.}$$
$$= x^2 + 4x + 3 \qquad \text{Combine like terms.}$$

This idea can be expanded so that we can multiply any two polynomials.

> **To Multiply Two Polynomials**
> Multiply each term of the first polynomial by each term of the second polynomial, and then combine like terms.

**EXAMPLE 5**  Multiply $(3x + 2)(2x - 5)$.

*Solution*  Multiply each term of the first binomial by each term of the second.

$$(3x + 2)(2x - 5) = 3x(2x) + 3x(-5) + 2(2x) + 2(-5)$$
$$= 6x^2 - 15x + 4x - 10 \qquad \text{Multiply.}$$
$$= 6x^2 - 11x - 10 \qquad \text{Combine like terms.} \quad \square$$

**PRACTICE**
**5**  Multiply $(5x - 2)(2x + 3)$

**EXAMPLE 6**  Multiply $(2x - y)^2$.

*Solution*  Recall that $a^2 = a \cdot a$, so $(2x - y)^2 = (2x - y)(2x - y)$. Multiply each term of the first polynomial by each term of the second.

$$(2x - y)(2x - y) = 2x(2x) + 2x(-y) + (-y)(2x) + (-y)(-y)$$
$$= 4x^2 - 2xy - 2xy + y^2 \qquad \text{Multiply.}$$
$$= 4x^2 - 4xy + y^2 \qquad \text{Combine like terms.} \quad \square$$

**PRACTICE**
**6**  Multiply $(5x - 3y)^2$

**Concept Check** ☑
Square where indicated. Simplify if possible.
**a.** $(4a)^2 + (3b)^2$  **b.** $(4a + 3b)^2$

**EXAMPLE 7**  Multiply $(t + 2)$ by $(3t^2 - 4t + 2)$.

*Solution*  Multiply each term of the first polynomial by each term of the second.

$$(t + 2)(3t^2 - 4t + 2) = t(3t^2) + t(-4t) + t(2) + 2(3t^2) + 2(-4t) + 2(2)$$
$$= 3t^3 - 4t^2 + 2t + 6t^2 - 8t + 4$$
$$= 3t^3 + 2t^2 - 6t + 4 \qquad \text{Combine like terms.} \quad \square$$

**Answers to Concept Check:**
**a.** $16a^2 + 9b^2$
**b.** $16a^2 + 24ab + 9b^2$

**PRACTICE**
**7**  Multiply $(y + 4)$ by $(2y^2 - 3y + 5)$

**EXAMPLE 8**   Multiply $(3a + b)^3$.

*Solution*   Write $(3a + b)^3$ as $(3a + b)(3a + b)(3a + b)$.

$$(3a + b)(3a + b)(3a + b) = (9a^2 + 3ab + 3ab + b^2)(3a + b)$$
$$= (9a^2 + 6ab + b^2)(3a + b)$$
$$= (9a^2 + 6ab + b^2)3a + (9a^2 + 6ab + b^2)b$$
$$= 27a^3 + 18a^2b + 3ab^2 + 9a^2b + 6ab^2 + b^3$$
$$= 27a^3 + 27a^2b + 9ab^2 + b^3$$

**PRACTICE
8**   Multiply $(s + 2t)^3$

**OBJECTIVE 2 ▶ Multiplying polynomials vertically.** Another convenient method for multiplying polynomials is to use a vertical format similar to the format used to multiply real numbers. We demonstrate this method by multiplying $(3y^2 - 4y + 1)$ by $(y + 2)$.

**EXAMPLE 9**   Multiply $(3y^2 - 4y + 1)(y + 2)$. Use a vertical format.

*Solution*

$$
\begin{array}{r}
3y^2 - 4y + 1 \\
\times \qquad y + 2 \\
\hline
6y^2 - 8y + 2 \\
3y^3 - 4y^2 + \ y \\
\hline
3y^3 + 2y^2 - 7y + 2
\end{array}
$$

1st, Multiply $3y^2 - 4y + 1$ by 2.
2nd, Multiply $3y^2 - 4y + 1$ by $y$. Line up like terms.
3rd, Combine like terms.

▶ **Helpful Hint**
Make sure like terms are lined up.

Thus, $(y + 2)(3y^2 - 4y + 1) = 3y^3 + 2y^2 - 7y + 2$.

**PRACTICE
9**   Multiply $(5x^2 - 3x + 5)(x - 4)$

When multiplying vertically, be careful if a power is missing, you may want to leave space in the partial products and take care that like terms are lined up.

**EXAMPLE 10**   Multiply $(2x^3 - 3x + 4)(x^2 + 1)$. Use a vertical format.

*Solution*

$$
\begin{array}{r}
2x^3 - 3x + 4 \\
\times \qquad x^2 + 1 \\
\hline
2x^3 \qquad - 3x + 4 \\
2x^5 - 3x^3 + 4x^2 \\
\hline
2x^5 - x^3 + 4x^2 - 3x + 4
\end{array}
$$

Leave space for missing powers of $x$.
← Line up like terms.
Combine like terms.

**PRACTICE
10**   Multiply $(x^3 - 2x^2 + 1)(x^2 + 2)$.

**EXAMPLE 11**   Find the product of $(2x^2 - 3x + 4)$ and $(x^2 + 5x - 2)$ using a vertical format.

*Solution*   First, we arrange the polynomials in a vertical format. Then we multiply each term of the second polynomial by each term of the first polynomial.

$$
\begin{array}{r}
2x^2 - 3x + 4 \\
x^2 + 5x - 2 \\
\hline
-4x^2 + 6x - 8 \\
10x^3 - 15x^2 + 20x \\
2x^4 - 3x^3 + 4x^2 \\
\hline
2x^4 + 7x^3 - 15x^2 + 26x - 8
\end{array}
$$

Multiply $2x^2 - 3x + 4$ by $-2$.
Multiply $2x^2 - 3x + 4$ by $5x$.
Multiply $2x^2 - 3x + 4$ by $x^2$.
Combine like terms.

**PRACTICE
11**   Find the product of $(5x^2 + 2x - 2)$ and $(x^2 - x + 3)$ using a vertical format.

## VOCABULARY & READINESS CHECK

*Fill in each blank with the correct choice.*

**1.** The expression $5x(3x + 2)$ equals $5x \cdot 3x + 5x \cdot 2$ by the _____ property.
   **a.** commutative   **b.** associative   **c.** distributive

**2.** The expression $(x + 4)(7x - 1)$ equals $x(7x - 1) + 4(7x - 1)$ by the _____ property.
   **a.** commutative   **b.** associative   **c.** distributive

**3.** The expression $(5y - 1)^2$ equals _____.
   **a.** $2(5y - 1)$   **b.** $(5y - 1)(5y + 1)$   **c.** $(5y - 1)(5y - 1)$

**4.** The expression $9x \cdot 3x$ equals _____.
   **a.** $27x$   **b.** $27x^2$   **c.** $12x$   **d.** $12x^2$

*Perform the indicated operation, if possible.*

**5.** $x^3 \cdot x^5$

**6.** $x^2 \cdot x^6$

**7.** $x^3 + x^5$

**8.** $x^2 + x^6$

**9.** $x^7 \cdot x^7$

**10.** $x^{11} \cdot x^{11}$

**11.** $x^7 + x^7$

**12.** $x^{11} + x^{11}$

## 5.3 | EXERCISE SET

MyMathLab   Powered by CourseCompass™ and MathXL®
Math XL   PRACTICE   WATCH   DOWNLOAD   READ   REVIEW

*Multiply. See Examples 1 through 3.*

**1.** $-4n^3 \cdot 7n^7$

**2.** $9t^6(-3t^5)$

**3.** $(-3.1x^3)(4x^9)$

**4.** $(-5.2x^4)(3x^4)$

**5.** $\left(-\dfrac{1}{3}y^2\right)\left(\dfrac{2}{5}y\right)$

**6.** $\left(-\dfrac{3}{4}y^7\right)\left(\dfrac{1}{7}y^4\right)$

**7.** $(2x)(-3x^2)(4x^5)$

**8.** $(x)(5x^4)(-6x^7)$

*Multiply. See Example 4.*

**9.** $3x(2x + 5)$

**10.** $2x(6x + 3)$

**11.** $-2a(a + 4)$

**12.** $-3a(2a + 7)$

**13.** $3x(2x^2 - 3x + 4)$

**14.** $4x(5x^2 - 6x - 10)$

**15.** $-2a^2(3a^2 - 2a + 3)$

**16.** $-4b^2(3b^3 - 12b^2 - 6)$

**17.** $-y(4x^3 - 7x^2y + xy^2 + 3y^3)$

**18.** $-x(6y^3 - 5xy^2 + x^2y - 5x^3)$

**19.** $\dfrac{1}{2}x^2(8x^2 - 6x + 1)$

**20.** $\dfrac{1}{3}y^2(9y^2 - 6y + 1)$

*Multiply. See Examples 5 and 6.*

**21.** $(x + 4)(x + 3)$

**22.** $(x + 2)(x + 9)$

**23.** $(a + 7)(a - 2)$

**24.** $(y - 10)(y + 11)$

**25.** $\left(x + \dfrac{2}{3}\right)\left(x - \dfrac{1}{3}\right)$

**26.** $\left(x + \dfrac{3}{5}\right)\left(x - \dfrac{2}{5}\right)$

**27.** $(3x^2 + 1)(4x^2 + 7)$

**28.** $(5x^2 + 2)(6x^2 + 2)$

**29.** $(2y - 4)^2$

**30.** $(6x - 7)^2$

**31.** $(4x - 3)(3x - 5)$

**32.** $(8x - 3)(2x - 4)$

**33.** $(3x^2 + 1)^2$

**34.** $(x^2 + 4)^2$

**35.** Perform the indicated operations.
   **a.** $(3x + 5) + (3x + 7)$
   **b.** $(3x + 5)(3x + 7)$
   **c.** Explain the difference between the two expressions.

**36.** Perform the indicated operations.
   **a.** $9x^2(-10x^2)$
   **b.** $9x^2 - 10x^2$
   **c.** Explain the difference between the two expressions.

*Multiply. See Example 7.*

**37.** $(x - 2)(x^2 - 3x + 7)$

**38.** $(x + 3)(x^2 + 5x - 8)$

**39.** $(x + 5)(x^3 - 3x + 4)$

**40.** $(a + 2)(a^3 - 3a^2 + 7)$

**41.** $(2a - 3)(5a^2 - 6a + 4)$

**42.** $(3 + b)(2 - 5b - 3b^2)$

*Multiply. See Example 8.*

**43.** $(x + 2)^3$

**44.** $(y - 1)^3$

**45.** $(2y - 3)^3$

**46.** $(3x + 4)^3$

*Multiply vertically. See Examples 9 through 11.*

**47.** $(2x - 11)(6x + 1)$

**48.** $(4x - 7)(5x + 1)$

**49.** $(5x + 1)(2x^2 + 4x - 1)$

**50.** $(4x - 5)(8x^2 + 2x - 4)$

**51.** $(x^2 + 5x - 7)(2x^2 - 7x - 9)$

**52.** $(3x^2 - x + 2)(x^2 + 2x + 1)$

## MIXED PRACTICE

*Multiply. See Examples 1 through 11.*

**53.** $-1.2y(-7y^6)$

**54.** $-4.2x(-2x^5)$

**55.** $-3x(x^2 + 2x - 8)$

**56.** $-5x(x^2 - 3x + 10)$

**57.** $(x + 19)(2x + 1)$

**58.** $(3y + 4)(y + 11)$

**59.** $\left(x + \dfrac{1}{7}\right)\left(x - \dfrac{3}{7}\right)$

**60.** $\left(m + \dfrac{2}{9}\right)\left(m - \dfrac{1}{9}\right)$

**61.** $(3y + 5)^2$

**62.** $(7y + 2)^2$

**63.** $(a + 4)(a^2 - 6a + 6)$

**64.** $(t + 3)(t^2 - 5t + 5)$

**65.** $(2x - 5)^3$

**66.** $(3y - 1)^3$

**67.** $(4x + 5)(8x^2 + 2x - 4)$

**68.** $(5x + 4)(x^2 - x + 4)$

**69.** $(3x^2 + 2x - 4)(2x^2 - 4x + 3)$

**70.** $(a^2 + 3a - 2)(2a^2 - 5a - 1)$

*Express as the product of polynomials. Then multiply.*

**71.** Find the area of the rectangle.

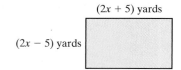

(2x + 5) yards

(2x − 5) yards

**72.** Find the area of the square field.

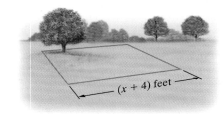

(x + 4) feet

**73.** Find the area of the triangle.

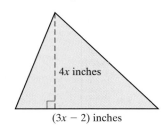

4x inches

(3x − 2) inches

**74.** Find the volume of the cube.

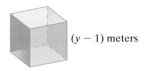

(y − 1) meters

## REVIEW AND PREVIEW

*Perform the indicated operation. See Section 5.1.*

**75.** $(5x)^2$

**76.** $(4p)^2$

**77.** $(-3y^3)^2$

**78.** $(-7m^2)^2$

## CONCEPT EXTENSIONS

**79.** The area of the larger rectangle below is $x(x + 3)$. Find another expression for this area by finding the sum of the areas of the two smaller rectangles.

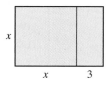

x

x    3

**80.** Write an expression for the area of the larger rectangle below in two different ways.

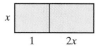

x

1    2x

**81.** The area of the figure below is $(x + 2)(x + 3)$. Find another expression for this area by finding the sum of the areas of the four smaller rectangles.

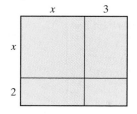

x    3

x

2

△ **82.** Write an expression for the area of the figure in two different ways.

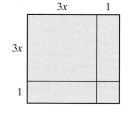

*Simplify.*

*See the Concept Checks in this section.*

**83.** $5a + 6a$

**84.** $5a \cdot 6a$

*Square where indicated. Simplify if possible.*

**85.** $(5x)^2 + (2y)^2$

**86.** $(5x + 2y)^2$

**MIXED PRACTICE**

*See Sections 5.2, 5.3. Perform the indicated operations.*

**87.** $(3x - 1) + (10x - 6)$

**88.** $(2x - 1) + (10x - 7)$

**89.** $(3x - 1)(10x - 6)$

**90.** $(2x - 1)(10x - 7)$

**91.** $(3x - 1) - (10x - 6)$

**92.** $(2x - 1) - (10x - 7)$

**93.** Multiply each of the following polynomials.
   **a.** $(a + b)(a - b)$
   **b.** $(2x + 3y)(2x - 3y)$
   **c.** $(4x + 7)(4x - 7)$
   **d.** Can you make a general statement about all products of the form $(x + y)(x - y)$?

**94.** Evaluate each of the following.
   **a.** $(2 + 3)^2; 2^2 + 3^2$
   **b.** $(8 + 10)^2; 8^2 + 10^2$
      Does $(a + b)^2 = a^2 + b^2$ no matter what the values of $a$ and $b$ are? Why or why not?

△ **95.** Write a polynomial that describes the area of the shaded region. (Find the area of the larger square minus the area of the smaller square.)

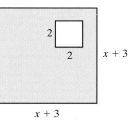

△ **96.** Write a polynomial that describes the area of the shaded region. (See Exercise 95.)

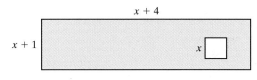

---

📖 **STUDY SKILLS BUILDER**

**Are You Organized?**

Have you ever had trouble finding a completed assignment? When it's time to study for a test, are your notes neat and organized? Have you ever had trouble reading your own mathematics handwriting? (Be honest—I have.)

   When any of these things happen, it's time to get organized. Here are a few suggestions:

• Write your notes and complete your homework assignment in a notebook with pockets (spiral or ring binder.)

• Take class notes in this notebook, and then follow the notes with your completed homework assignment.

• When you receive graded papers or handouts, place them in the notebook pocket so that you will not lose them.

• Mark (possibly with an exclamation point) any note(s) that seem extra important to you.

• Mark (possibly with a question mark) any notes or homework that you are having trouble with.

• See your instructor or a math tutor to help you with the concepts or exercises that you are having trouble understanding.

• If you are having trouble reading your own handwriting, *slow down* and write your mathematics work clearly!

*Exercises*

**1.** Have you been completing your assignments on time?

**2.** Have you been correcting any exercises you may be having difficulty with?

**3.** If you are having trouble with a mathematical concept or correcting any homework exercises, have you visited your instructor, a tutor, or your campus math lab?

**4.** Are you taking lecture notes in your mathematics course? (By the way, these notes should include worked-out examples solved by your instructor.)

**5.** Is your mathematics course material (handouts, graded papers, lecture notes) organized?

**6.** If your answer to Exercise 5 is no, take a moment and review your course material. List at least two ways that you might better organize it. Then read the Study Skills Builder on organizing a notebook in Chapter 2.

# 5.4 SPECIAL PRODUCTS

**OBJECTIVES**

1 Multiply two binomials using the FOIL method.

2 Square a binomial.

3 Multiply the sum and difference of two terms.

**OBJECTIVE 1** ▶ **Using the FOIL method.** In this section, we multiply binomials using special products. First, a special order for multiplying binomials called the FOIL order or method is introduced. This method is demonstrated by multiplying $(3x + 1)$ by $(2x + 5)$.

**The FOIL Method**

**F** stands for the product of the **First** terms.     $(3x + 1)(2x + 5)$
$$(3x)(2x) = 6x^2 \quad \textbf{F}$$

**O** stands for the product of the **Outer** terms.     $(3x + 1)(2x + 5)$
$$(3x)(5) = 15x \quad \textbf{O}$$

**I** stands for the product of the **Inner** terms.     $(3x + 1)(2x + 5)$
$$(1)(2x) = 2x \quad \textbf{I}$$

**L** stands for the product of the **Last** terms.     $(3x + 1)(3x + 5)$
$$(1)(5) = 5 \quad \textbf{L}$$

$$\overset{\text{F} \quad\quad \text{O} \quad\quad \text{I} \quad \text{L}}{(3x + 1)(2x + 5) = 6x^2 + 15x + 2x + 5}$$
$$= 6x^2 + 17x + 5 \quad \text{Combine like terms.}$$

**Concept Check** ✓

Multiply $(3x + 1)(2x + 5)$ using methods from the last section. Show that the product is still $6x^2 + 17x + 5$.

**EXAMPLE 1**    Multiply $(x - 3)(x + 4)$ by the FOIL method.

*Solution*

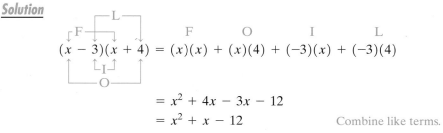

$$(x - 3)(x + 4) = (x)(x) + (x)(4) + (-3)(x) + (-3)(4)$$

$$= x^2 + 4x - 3x - 12$$
$$= x^2 + x - 12 \quad \text{Combine like terms.} \quad \square$$

**PRACTICE**
**1**    Multiply $(x + 2)(x - 5)$ by the FOIL method.

**EXAMPLE 2**    Multiply $(5x - 7)(x - 2)$ by the FOIL method.

*Solution*

$$\overset{\text{F} \quad\quad\quad \text{O} \quad\quad\quad \text{I} \quad\quad\quad \text{L}}{(5x - 7)(x - 2) = 5x(x) + 5x(-2) + (-7)(x) + (-7)(-2)}$$

$$= 5x^2 - 10x - 7x + 14$$
$$= 5x^2 - 17x + 14 \quad \text{Combine like terms.} \quad \square$$

**Answer to Concept Check:**
Multiply and simplify:
$3x(2x + 5) + 1(2x + 5)$

**PRACTICE**
**2**    Multiply $(4x - 9)(x - 1)$ by the FOIL method.

**EXAMPLE 3**   Multiply $2(y + 6)(2y - 1)$.

<u>*Solution*</u>   $2(y + 6)(2y - 1) = 2(\overset{\text{F}}{2y^2} - \overset{\text{O}}{1y} + \overset{\text{I}}{12y} - \overset{\text{L}}{6})$

$$= 2(2y^2 + 11y - 6) \qquad \text{Simplify inside parentheses.}$$

$$= 4y^2 + 22y - 12 \qquad \text{Now use the distributive property.} \ \square$$

**PRACTICE**
**3**   Multiply $3(x + 5)(3x - 1)$.

**OBJECTIVE 2 ▶ Squaring binomials.** Now, try squaring a binomial using the FOIL method.

**EXAMPLE 4**   Multiply $(3y + 1)^2$.

<u>*Solution*</u>   $(3y + 1)^2 = (3y + 1)(3y + 1)$

$$= \overset{\text{F}}{(3y)(3y)} + \overset{\text{O}}{(3y)(1)} + \overset{\text{I}}{1(3y)} + \overset{\text{L}}{1(1)}$$

$$= 9y^2 + 3y + 3y + 1$$

$$= 9y^2 + 6y + 1 \qquad \square$$

**PRACTICE**
**4**   Multiply $(4x - 1)^2$.

Notice the pattern that appears in Example 4.

$(3y + 1)^2 = 9y^2 + 6y + 1$   $9y^2$ is the first term of the binomial squared. $(3y)^2 = 9y^2$.

6y is 2 times the product of both terms of the binomial. $(2)(3y)(1) = 6y$.

1 is the second term of the binomial squared. $(1)^2 = 1$.

This pattern leads to the following, which can be used when squaring a binomial. We call these **special products.**

> **Squaring a Binomial**
> A binomial squared is equal to the square of the first term plus or minus twice the product of both terms plus the square of the second term.
>
> $$(a + b)^2 = a^2 + 2ab + b^2$$
> $$(a - b)^2 = a^2 - 2ab + b^2$$

This product can be visualized geometrically.

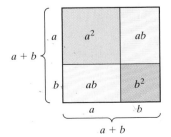

The area of the large square is side · side.

$$\text{Area} = (a + b)(a + b) = (a + b)^2$$

The area of the large square is also the sum of the areas of the smaller rectangles.

$$\text{Area} = a^2 + ab + ab + b^2 = a^2 + 2ab + b^2$$

Thus, $(a + b)^2 = a^2 + 2ab + b^2$.

**EXAMPLE 5** Use a special product to square each binomial.

**a.** $(t + 2)^2$ **b.** $(p - q)^2$ **c.** $(2x + 5)^2$ **d.** $(x^2 - 7y)^2$

*Solution*

| | first term squared | plus or minus | twice the product of the terms | plus | second term squared |
|---|---|---|---|---|---|

**a.** $(t + 2)^2 = t^2 + 2(t)(2) + 2^2 = t^2 + 4t + 4$

**b.** $(p - q)^2 = p^2 - 2(p)(q) + q^2 = p^2 - 2pq + q^2$

**c.** $(2x + 5)^2 = (2x)^2 + 2(2x)(5) + 5^2 = 4x^2 + 20x + 25$

**d.** $(x^2 - 7y)^2 = (x^2)^2 - 2(x^2)(7y) + (7y^2) = x^4 - 14x^2y + 49y^2$ □

**PRACTICE**

**5** Use a special product to square each binomial.

**a.** $(b + 3)^2$ **b.** $(x - y)^2$

**c.** $(3y + 2)^2$ **d.** $(a^2 - 5b)^2$

---

▶ **Helpful Hint**

Notice that

$$(a + b)^2 \neq a^2 + b^2 \quad \text{The middle term } 2ab \text{ is missing.}$$
$$(a + b)^2 = (a + b)(a + b) = a^2 + 2ab + b^2$$

Likewise,

$$(a - b)^2 \neq a^2 - b^2$$
$$(a - b)^2 = (a - b)(a - b) = a^2 - 2ab + b^2$$

---

**OBJECTIVE 3** ▶ **Multiplying the sum and difference of two terms.** Another special product is the product of the sum and difference of the same two terms, such as $(x + y)(x - y)$. Finding this product by the FOIL method, we see a pattern emerge.

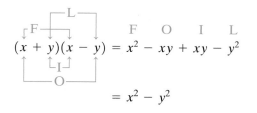

$$(x + y)(x - y) = x^2 - xy + xy - y^2$$

$$= x^2 - y^2$$

Notice that the middle two terms subtract out. This is because the **O**uter product is the opposite of the **I**nner product. Only the **difference of squares** remains.

---

**Multiplying the Sum and Difference of Two Terms**

The product of the sum and difference of two terms is the square of the first term minus the square of the second term.

$$(a + b)(a - b) = a^2 - b^2$$

**EXAMPLE 6** Use a special product to multiply.

**a.** $4(x + 4)(x - 4)$     **b.** $(6t + 7)(6t - 7)$     **c.** $\left(x - \dfrac{1}{4}\right)\left(x + \dfrac{1}{4}\right)$
**d.** $(2p - q)(2p + q)$     **e.** $(3x^2 - 5y)(3x^2 + 5y)$

*Solution*

first term squared   minus   second term squared
↓   ↓   ↓

**a.** $4(x + 4)(x - 4) = 4(x^2 - 4^2) = 4(x^2 - 16) = 4x^2 - 64$
**b.** $(6t + 7)(6t - 7) = (6t)^2 - 7^2 = 36t^2 - 49$

**c.** $\left(x - \dfrac{1}{4}\right)\left(x + \dfrac{1}{4}\right) = x^2 - \left(\dfrac{1}{4}\right)^2 = x^2 - \dfrac{1}{16}$
**d.** $(2p - q)(2p + q) = (2p)^2 - q^2 = 4p^2 - q^2$
**e.** $(3x^2 - 5y)(3x^2 + 5y) = (3x^2)^2 - (5y)^2 = 9x^4 - 25y^2$                □

**PRACTICE**
**6** Use a special product to multiply.

**a.** $3(x + 5)(x - 5)$                     **b.** $(4b - 3)(4b + 3)$

**c.** $\left(x + \dfrac{2}{3}\right)\left(x - \dfrac{2}{3}\right)$                     **d.** $(5s + t)(5s - t)$

**e.** $(2y - 3z^2)(2y + 3z^2)$

**Concept Check** ☑

Match each expression on the left to the equivalent expression or expressions in the list below.

**1.** $(a + b)^2$                     **2.** $(a + b)(a - b)$
  **a.** $(a + b)(a + b)$   **b.** $a^2 - b^2$   **c.** $a^2 + b^2$   **d.** $a^2 - 2ab + b^2$   **e.** $a^2 + 2ab + b^2$

Let's now practice multiplying polynomials in general. If possible, use a special product.

**EXAMPLE 7** Use a special product to multiply, if possible.

**a.** $(x - 5)(3x + 4)$     **b.** $(7x + 4)^2$     **c.** $(y - 0.6)(y + 0.6)$
**d.** $(y^4 + 2)(3y^2 - 1)$     **e.** $(a - 3)(a^2 + 2a - 1)$

*Solution*

**a.** $(x - 5)(3x + 4) = 3x^2 + 4x - 15x - 20$       FOIL.
$= 3x^2 - 11x - 20$
**b.** $(7x + 4)^2 = (7x)^2 + 2(7x)(4) + 4^2$       Squaring a binomial.
$= 49x^2 + 56x + 16$
**c.** $(y - 0.6)(y + 0.6) = y^2 - (0.6)^2 = y^2 - 0.36$   Multiplying the sum and difference of 2 terms.
**d.** $(y^4 + 2)(3y^2 - 1) = 3y^6 - y^4 + 6y^2 - 2$       FOIL.
**e.** I've inserted this product as a reminder that since it is not a binomial times a binomial, the FOIL order may not be used.

$(a - 3)(a^2 + 2a - 1) = a(a^2 + 2a - 1) - 3(a^2 + 2a - 1)$   Multiplying each term of the binomial by each term
$= a^3 + 2a^2 - a - 3a^2 - 6a + 3$   of the trinomial.
$= a^3 - a^2 - 7a + 3$                □

**PRACTICE**
**7** Use a special product to multiply, if possible.

a. $(4x + 3)(x - 6)$                    b. $(7b - 2)^2$

c. $(x + 0.4)(x - 0.4)$                  d. $(x^2 - 3)(3x^4 + 2)$

e. $(x + 1)(x^2 + 5x - 2)$

> **▶ Helpful Hint**
>
> - When multiplying two binomials, you may always use the FOIL order or method.
> - When multiplying any two polynomials, you may always use the distributive property to find the product.

## VOCABULARY & READINESS CHECK

*Answer each exercise true or false.*

**1.** $(x + 4)^2 = x^2 + 16$
**3.** $(x + 4)(x - 4) = x^2 + 16$

**2.** For $(x + 6)(2x - 1)$ the product of the first terms is $2x^2$.
**4.** The product $(x - 1)(x^3 + 3x - 1)$ is a polynomial of degree 5.

## 5.4 EXERCISE SET

*Multiply using the FOIL method. See Examples 1 through 3.*

**1.** $(x + 3)(x + 4)$
**2.** $(x + 5)(x - 1)$
**3.** $(x - 5)(x + 10)$
**4.** $(y - 12)(y + 4)$
**5.** $(5x - 6)(x + 2)$
**6.** $(3y - 5)(2y - 7)$
**7.** $(y - 6)(4y - 1)$
**8.** $(2x - 9)(x - 11)$
**9.** $(2x + 5)(3x - 1)$
**10.** $(6x + 2)(x - 2)$

*Multiply. See Examples 4 and 5.*

**11.** $(x - 2)^2$                    **12.** $(x + 7)^2$
**13.** $(2x - 1)^2$                   **14.** $(7x - 3)^2$
**15.** $(3a - 5)^2$                   **16.** $(5a + 2)^2$
**17.** $(5x + 9)^2$                   **18.** $(6s - 2)^2$

**19.** Using your own words, explain how to square a binomial such as $(a + b)^2$.

**20.** Explain how to find the product of two binomials using the FOIL method.

*Multiply. See Example 6.*

**21.** $(a - 7)(a + 7)$              **22.** $(b + 3)(b - 3)$
**23.** $(3x - 1)(3x + 1)$            **24.** $(4x - 5)(4x + 5)$

**25.** $\left(3x - \dfrac{1}{2}\right)\left(3x + \dfrac{1}{2}\right)$

**26.** $\left(10x + \dfrac{2}{7}\right)\left(10x - \dfrac{2}{7}\right)$

**27.** $(9x + y)(9x - y)$
**28.** $(2x - y)(2x + y)$
**29.** $(2x + 0.1)(2x - 0.1)$
**30.** $(5x - 1.3)(5x + 1.3)$

**MIXED PRACTICE**

*Multiply. See Example 7.*

**31.** $(a + 5)(a + 4)$
**32.** $(a - 5)(a - 7)$
**33.** $(a + 7)^2$
**34.** $(b - 2)^2$
**35.** $(4a + 1)(3a - 1)$
**36.** $(6a + 7)(6a + 5)$
**37.** $(x + 2)(x - 2)$
**38.** $(x - 10)(x + 10)$
**39.** $(3a + 1)^2$
**40.** $(4a - 2)^2$
**41.** $(x^2 + y)(4x - y^4)$
**42.** $(x^3 - 2)(5x + y)$
**43.** $(x + 3)(x^2 - 6x + 1)$
**44.** $(x - 2)(x^2 - 4x + 2)$

45. $(2a - 3)^2$

46. $(5b - 4x)^2$

47. $(5x - 6z)(5x + 6z)$

48. $(11x - 7y)(11x + 7y)$

49. $(x^5 - 3)(x^5 - 5)$

50. $(a^4 + 5)(a^4 + 6)$

51. $\left(x - \dfrac{1}{3}\right)\left(x + \dfrac{1}{3}\right)$

52. $\left(3x + \dfrac{1}{5}\right)\left(3x - \dfrac{1}{5}\right)$

53. $(a^3 + 11)(a^4 - 3)$

54. $(x^5 + 5)(x^2 - 8)$

55. $3(x - 2)^2$

56. $2(3b + 7)^2$

🖥 57. $(3b + 7)(2b - 5)$

58. $(3y - 13)(y - 3)$

59. $(7p - 8)(7p + 8)$

60. $(3s - 4)(3s + 4)$

🖥 61. $\left(\dfrac{1}{3}a^2 - 7\right)\left(\dfrac{1}{3}a^2 + 7\right)$

62. $\left(\dfrac{2}{3}a - b^2\right)\left(\dfrac{2}{3}a - b^2\right)$

63. $5x^2(3x^2 - x + 2)$

64. $4x^3(2x^2 + 5x - 1)$

65. $(2r - 3s)(2r + 3s)$

66. $(6r - 2x)(6r + 2x)$

67. $(3x - 7y)^2$

68. $(4s - 2y)^2$

🖥 69. $(4x + 5)(4x - 5)$

70. $(3x + 5)(3x - 5)$

71. $(8x + 4)^2$

72. $(3x + 2)^2$

73. $\left(a - \dfrac{1}{2}y\right)\left(a + \dfrac{1}{2}y\right)$

74. $\left(\dfrac{a}{2} + 4y\right)\left(\dfrac{a}{2} - 4y\right)$

75. $\left(\dfrac{1}{5}x - y\right)\left(\dfrac{1}{5}x + y\right)$

76. $\left(\dfrac{y}{6} - 8\right)\left(\dfrac{y}{6} + 8\right)$

77. $(a + 1)(3a^2 - a + 1)$

78. $(b + 3)(2b^2 + b - 3)$

*Express each as a product of polynomials in x. Then multiply and simplify.*

△ 79. Find the area of the square rug shown if its side is $(2x + 1)$ feet.

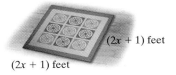

$(2x + 1)$ feet

$(2x + 1)$ feet

△ 80. Find the area of the rectangular canvas if its length is $(3x - 2)$ inches and its width is $(x - 4)$ inches.

$(x - 4)$ inches

$(3x - 2)$ inches

**REVIEW AND PREVIEW**

*Simplify each expression. See Section 5.1.*

81. $\dfrac{50b^{10}}{70b^5}$

82. $\dfrac{x^3 y^6}{xy^2}$

83. $\dfrac{8a^{17}b^{15}}{-4a^7 b^{10}}$

84. $\dfrac{-6a^8 y}{3a^4 y}$

85. $\dfrac{2x^4 y^{12}}{3x^4 y^4}$

86. $\dfrac{-48ab^6}{32ab^3}$

*Find the slope of each line. See Section 3.4.*

87.
88.

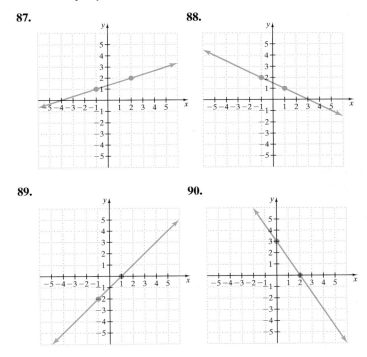

89.
90.

**CONCEPT EXTENSIONS**

*Match each expression on the left to the equivalent expression on the right. See the Concept Check in this section.*

91. $(a - b)^2$

92. $(a - b)(a + b)$

93. $(a + b)^2$

94. $(a + b)^2(a - b)^2$

 a. $a^2 - b^2$

 b. $a^2 + b^2$

 c. $a^2 - 2ab + b^2$

 d. $a^2 + 2ab + b^2$

 e. none of these

*Fill in the squares so that a true statement forms.*

95. $(x^{\square} + 7)(x^{\square} + 3) = x^4 + 10x^2 + 21$

96. $(5x^{\square} - 2)^2 = 25x^6 - 20x^3 + 4$

*Find the area of each shaded region.*

△ **97.**

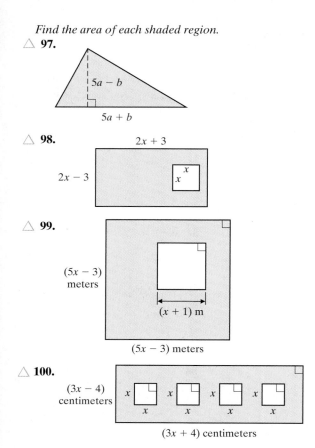

5a − b

5a + b

△ **98.**

2x + 3

2x − 3

x
x

△ **99.**

(5x − 3) meters

(x + 1) m

(5x − 3) meters

△ **100.**

(3x − 4) centimeters

x ▢ x ▢ x ▢ x ▢
x    x    x    x

(3x + 4) centimeters

△ **101.**

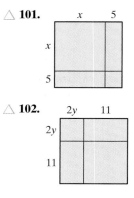

x    5

x

5

△ **102.**

2y    11

2y

11

✎ **103.** In your own words, describe the different methods that can be used to find the product: $(2x - 5)(3x + 1)$.

✎ **104.** In your own words, describe the different methods that can be used to find the product: $(5x + 1)^2$.

*Find each product. For example,*

$$[(a + b) - 2][(a + b) + 2] = (a + b)^2 - 2^2$$
$$= a^2 + 2ab + b^2 - 4$$

**105.** $[(x + y) - 3][(x + y) + 3]$
**106.** $[(a + c) - 5][(a + c) + 5]$
**107.** $[(a - 3) + b][(a - 3) - b]$
**108.** $[(x - 2) + y][(x - 2) - y]$

## INTEGRATED REVIEW   EXPONENTS AND OPERATIONS ON POLYNOMIALS

Sections 5.1–5.4

*Perform the indicated operations and simplify.*

**1.** $(5x^2)(7x^3)$

**2.** $(4y^2)(8y^7)$

**3.** $-4^2$

**4.** $(-4)^2$

**5.** $(x - 5)(2x + 1)$

**6.** $(3x - 2)(x + 5)$

**7.** $(x - 5) + (2x + 1)$

**8.** $(3x - 2) + (x + 5)$

**9.** $\dfrac{7x^9y^{12}}{x^3y^{10}}$

**10.** $\dfrac{20a^2b^8}{14a^2b^2}$

**11.** $(12m^7n^6)^2$

**12.** $(4y^9z^{10})^3$

**13.** $3(4y - 3)(4y + 3)$

**14.** $2(7x - 1)(7x + 1)$

**15.** $(x^7y^5)^9$

**16.** $(3^1x^9)^3$

**17.** $(7x^2 - 2x + 3) - (5x^2 + 9)$

**18.** $(10x^2 + 7x - 9) - (4x^2 - 6x + 2)$

**19.** $0.7y^2 - 1.2 + 1.8y^2 - 6y + 1$

**20.** $7.8x^2 - 6.8x + 3.3 + 0.6x^2 - 9$

**21.** $(x + 4y)^2$

**22.** $(y - 9z)^2$

**23.** $(x + 4y) + (x + 4y)$

**24.** $(y - 9z) + (y - 9z)$

**25.** $7x^2 - 6xy + 4(y^2 - xy)$

**26.** $5a^2 - 3ab + 6(b^2 - a^2)$

**27.** $(x - 3)(x^2 + 5x - 1)$

**28.** $(x + 1)(x^2 - 3x - 2)$

**29.** $(2x^3 - 7)(3x^2 + 10)$

**30.** $(5x^3 - 1)(4x^4 + 5)$

**31.** $(2x - 7)(x^2 - 6x + 1)$

**32.** $(5x - 1)(x^2 + 2x - 3)$

*Perform exercises and simplify, if possible.*

**33.** $5x^3 + 5y^3$

**34.** $(5x^3)(5y^3)$

**35.** $(5x^3)^3$

**36.** $\dfrac{5x^3}{5y^3}$

**37.** $x + x$

**38.** $x \cdot x$

# 5.5 NEGATIVE EXPONENTS AND SCIENTIFIC NOTATION

## OBJECTIVES

1 Simplify expressions containing negative exponents.

2 Use all the rules and definitions for exponents to simplify exponential expressions.

3 Write numbers in scientific notation.

4 Convert numbers from scientific notation to standard form.

**OBJECTIVE 1 ▶ Simplifying expressions containing negative exponents.** Our work with exponential expressions so far has been limited to exponents that are positive integers or 0. Here we expand to give meaning to an expression like $x^{-3}$.

Suppose that we wish to simplify the expression $\dfrac{x^2}{x^5}$. If we use the quotient rule for exponents, we subtract exponents:

$$\frac{x^2}{x^5} = x^{2-5} = x^{-3}, \quad x \neq 0$$

But what does $x^{-3}$ mean? Let's simplify $\dfrac{x^2}{x^5}$ using the definition of $x^n$.

$$\frac{x^2}{x^5} = \frac{x \cdot x}{x \cdot x \cdot x \cdot x \cdot x}$$

$$= \frac{x \cdot x}{x \cdot x \cdot x \cdot x \cdot x} \qquad \text{Divide numerator and denominator by common factors by applying the fundamental principle for fractions.}$$

$$= \frac{1}{x^3}$$

If the quotient rule is to hold true for negative exponents, then $x^{-3}$ must equal $\dfrac{1}{x^3}$. From this example, we state the definition for negative exponents.

---

**Negative Exponents**

If $a$ is a real number other than 0 and $n$ is an integer, then

$$a^{-n} = \frac{1}{a^n}$$

---

For example, $x^{-3} = \dfrac{1}{x^3}$.

In other words, another way to write $a^{-n}$ is to take its reciprocal and change the sign of its exponent.

**EXAMPLE 1**   Simplify by writing each expression with positive exponents only.

**a.** $3^{-2}$      **b.** $2x^{-3}$      **c.** $2^{-1} + 4^{-1}$      **d.** $(-2)^{-4}$      **e.** $\dfrac{1}{y^{-4}}$      **f.** $\dfrac{1}{7^{-2}}$

*Solution*

**a.** $3^{-2} = \dfrac{1}{3^2} = \dfrac{1}{9}$      Use the definition of negative exponents.

**b.** $2x^{-3} = 2 \cdot \dfrac{1}{x^3} = \dfrac{2}{x^3}$      Use the definition of negative exponents.

**c.** $2^{-1} + 4^{-1} = \dfrac{1}{2} + \dfrac{1}{4} = \dfrac{2}{4} + \dfrac{1}{4} = \dfrac{3}{4}$

**d.** $(-2)^{-4} = \dfrac{1}{(-2)^4} = \dfrac{1}{(-2)(-2)(-2)(-2)} = \dfrac{1}{16}$

**e.** $\dfrac{1}{y^{-4}} = \dfrac{1}{\dfrac{1}{y^4}} = y^4$      **f.** $\dfrac{1}{7^{-2}} = \dfrac{1}{\dfrac{1}{7^2}} = \dfrac{7^2}{1}$ or 49

> **▶ Helpful Hint**
>
> Don't forget that since there are no parentheses, only $x$ is the base for the exponent $-3$.

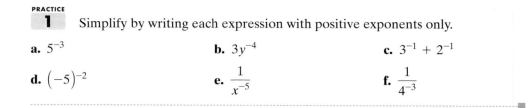

**PRACTICE**
**1** Simplify by writing each expression with positive exponents only.

**a.** $5^{-3}$        **b.** $3y^{-4}$        **c.** $3^{-1} + 2^{-1}$

**d.** $(-5)^{-2}$        **e.** $\dfrac{1}{x^{-5}}$        **f.** $\dfrac{1}{4^{-3}}$

---

> ▶ **Helpful Hint**
>
> A negative exponent *does not affect* the sign of its base.
> Remember: Another way to write $a^{-n}$ is to take its reciprocal and change the sign of its exponent: $a^{-n} = \dfrac{1}{a^n}$. For example,
>
> $$x^{-2} = \frac{1}{x^2}, \qquad 2^{-3} = \frac{1}{2^3} \ \text{ or } \ \frac{1}{8}$$
>
> $$\frac{1}{y^{-4}} = \frac{1}{\frac{1}{y^4}} = y^4, \qquad \frac{1}{5^{-2}} = 5^2 \ \text{ or } \ 25$$

From the preceding Helpful Hint, we know that $x^{-2} = \dfrac{1}{x^2}$ and $\dfrac{1}{y^{-4}} = y^4$. We can use this to include another statement in our definition of negative exponents.

**Negative Exponents**
If $a$ is a real number other than 0 and $n$ is an integer, then

$$a^{-n} = \frac{1}{a^n} \quad \text{and} \quad \frac{1}{a^{-n}} = a^n$$

**EXAMPLE 2** Simplify each expression. Write results using positive exponents only.

**a.** $\dfrac{1}{x^{-3}}$      **b.** $\dfrac{1}{3^{-4}}$      **c.** $\dfrac{p^{-4}}{q^{-9}}$      **d.** $\dfrac{5^{-3}}{2^{-5}}$

*Solution*

**a.** $\dfrac{1}{x^{-3}} = \dfrac{x^3}{1} = x^3$    **b.** $\dfrac{1}{3^{-4}} = \dfrac{3^4}{1} = 81$    **c.** $\dfrac{p^{-4}}{q^{-9}} = \dfrac{q^9}{p^4}$    **d.** $\dfrac{5^{-3}}{2^{-5}} = \dfrac{2^5}{5^3} = \dfrac{32}{125}$

**PRACTICE**
**2** Simplify each expression. Write results using positive exponents only.

**a.** $\dfrac{1}{s^{-5}}$      **b.** $\dfrac{1}{2^{-3}}$      **c.** $\dfrac{x^{-7}}{y^{-5}}$      **d.** $\dfrac{4^{-3}}{3^{-2}}$

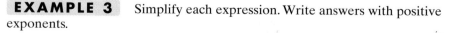

**EXAMPLE 3** Simplify each expression. Write answers with positive exponents.

**a.** $\dfrac{y}{y^{-2}}$      **b.** $\dfrac{3}{x^{-4}}$      **c.** $\dfrac{x^{-5}}{x^7}$

*Solution*

**a.** $\dfrac{y}{y^{-2}} = \dfrac{y^1}{y^{-2}} = y^{1-(-2)} = y^3$   Remember that $\dfrac{a^m}{a^n} = a^{m-n}$.

**b.** $\dfrac{3}{x^{-4}} = 3 \cdot \dfrac{1}{x^{-4}} = 3 \cdot x^4$  or  $3x^4$

**c.** $\dfrac{x^{-5}}{x^7} = x^{-5-7} = x^{-12} = \dfrac{1}{x^{12}}$

**PRACTICE**

**3**   Simplify each expression. Write answers with positive exponents.

**a.** $\dfrac{x^{-3}}{x^2}$ 　　　　　　　　**b.** $\dfrac{5}{y^{-7}}$ 　　　　　　　　**c.** $\dfrac{z}{z^{-4}}$

---

**OBJECTIVE 2 ▶ Simplifying exponential expressions.** All the previously stated rules for exponents apply for negative exponents also. Here is a summary of the rules and definitions for exponents.

> **Summary of Exponent Rules**
> If $m$ and $n$ are integers and $a$, $b$, and $c$ are real numbers, then:
> Product rule for exponents: $a^m \cdot a^n = a^{m+n}$
> Power rule for exponents: $(a^m)^n = a^{m \cdot n}$
> Power of a product: $(ab)^n = a^n b^n$
> Power of a quotient: $\left(\dfrac{a}{c}\right)^n = \dfrac{a^n}{c^n}$,   $c \neq 0$
> Quotient rule for exponents: $\dfrac{a^m}{a^n} = a^{m-n}$,   $a \neq 0$
> Zero exponent: $a^0 = 1$,   $a \neq 0$
> Negative exponent: $a^{-n} = \dfrac{1}{a^n}$,   $a \neq 0$

**EXAMPLE 4**   Simplify the following expressions. Write each result using positive exponents only.

**a.** $\left(\dfrac{2}{3}\right)^{-3}$  **b.** $\dfrac{(x^3)^4 x}{x^7}$  **c.** $\left(\dfrac{3a^2}{b}\right)^{-3}$  **d.** $\dfrac{4^{-1}x^{-3}y}{4^{-3}x^2 y^{-6}}$  **e.** $(y^{-3}z^6)^{-6}$  **f.** $\left(\dfrac{-2x^3 y}{xy^{-1}}\right)^3$

*Solution*

**a.** $\left(\dfrac{2}{3}\right)^{-3} = \dfrac{2^{-3}}{3^{-3}} = \dfrac{3^3}{2^3} = \dfrac{27}{8}$

**b.** $\dfrac{(x^3)^4 x}{x^7} = \dfrac{x^{12} \cdot x}{x^7} = \dfrac{x^{12+1}}{x^7} = \dfrac{x^{13}}{x^7} = x^{13-7} = x^6$   Use the power rule.

**c.** $\left(\dfrac{3a^2}{b}\right)^{-3} = \dfrac{3^{-3}(a^2)^{-3}}{b^{-3}}$   Raise each factor in the numerator and the denominator to the $-3$ power.

$= \dfrac{3^{-3}a^{-6}}{b^{-3}}$   Use the power rule.

$= \dfrac{b^3}{3^3 a^6}$   Use the negative exponent rule.

$= \dfrac{b^3}{27a^6}$   Write $3^3$ as 27.

**d.** $\dfrac{4^{-1}x^{-3}y}{4^{-3}x^{2}y^{-6}} = 4^{-1-(-3)}x^{-3-2}y^{1-(-6)} = 4^{2}x^{-5}y^{7} = \dfrac{4^{2}y^{7}}{x^{5}} = \dfrac{16y^{7}}{x^{5}}$

**e.** $(y^{-3}z^{6})^{-6} = y^{18}\cdot z^{-36} = \dfrac{y^{18}}{z^{36}}$

**f.** $\left(\dfrac{-2x^{3}y}{xy^{-1}}\right)^{3} = \dfrac{(-2)^{3}x^{9}y^{3}}{x^{3}y^{-3}} = \dfrac{-8x^{9}y^{3}}{x^{3}y^{-3}} = -8x^{9-3}y^{3-(-3)} = -8x^{6}y^{6}$

**PRACTICE**

**4** Simplify the following expression. Write each result using positive exponents only.

**a.** $\left(\dfrac{3}{4}\right)^{-2}$  

**b.** $\dfrac{x^{2}(x^{5})^{3}}{x^{7}}$  

**c.** $\left(\dfrac{5p^{8}}{q}\right)^{-2}$

**d.** $\dfrac{6^{-2}x^{-4}y^{-7}}{6^{-3}x^{3}y^{-9}}$  

**e.** $(a^{4}b^{-3})^{-5}$  

**f.** $\left(\dfrac{-3x^{4}y}{x^{2}y^{-2}}\right)^{3}$

**OBJECTIVE 3** ▶ **Writing numbers in scientific notation.** Both very large and very small numbers frequently occur in many fields of science. For example, the distance between the sun and the dwarf planet Pluto is approximately 5,906,000,000 kilometers, and the mass of a proton is approximately 0.000000000000000000000000165 gram. It can be tedious to write these numbers in this standard decimal notation, so **scientific notation** is used as a convenient shorthand for expressing very large and very small numbers.

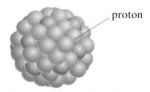

Mass of proton is approximately
0.000 000 000 000 000 000 000 001 65 gram

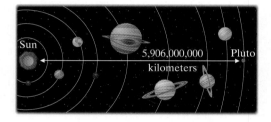

**Scientific Notation**

A positive number is written in scientific notation if it is written as the product of a number $a$, where $1 \le a < 10$, and an integer power $r$ of 10:

$$a \times 10^{r}$$

The numbers below are written in scientific notation. The $\times$ sign for multiplication is used as part of the notation.

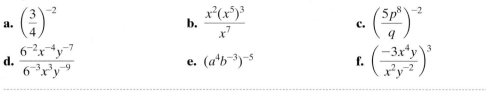

$2.03 \times 10^{2}$   $7.362 \times 10^{7}$   $5.906 \times 10^{9}$   (Distance between the sun and Pluto)

$1 \times 10^{-3}$   $8.1 \times 10^{-5}$   $1.65 \times 10^{-24}$   (Mass of a proton)

The following steps are useful when writing numbers in scientific notation.

**To Write a Number in Scientific Notation**

**STEP 1.** Move the decimal point in the original number to the left or right so that the new number has a value between 1 and 10.

**STEP 2.** Count the number of decimal places the decimal point is moved in Step 1. If the original number is 10 or greater, the count is positive. If the original number is less than 1, the count is negative.

**STEP 3.** Multiply the new number in Step 1 by 10 raised to an exponent equal to the count found in Step 2.

**EXAMPLE 5**   Write each number in scientific notation.

**a.** 367,000,000     **b.** 0.000003     **c.** 20,520,000,000     **d.** 0.00085

_Solution_

**a. STEP 1.** Move the decimal point until the number is between 1 and 10.

367,000,000.
8 places

**STEP 2.** The decimal point is moved 8 places, and the original number is 10 or greater, so the count is positive 8.

**STEP 3.** $367,000,000 = 3.67 \times 10^8$.

**b. STEP 1.** Move the decimal point until the number is between 1 and 10.

0.000003
6 places

**STEP 2.** The decimal point is moved 6 places, and the original number is less than 1, so the count is $-6$.

**STEP 3.** $0.000003 = 3.0 \times 10^{-6}$

**c.** $20,520,000,000 = 2.052 \times 10^{10}$

**d.** $0.00085 = 8.5 \times 10^{-4}$

**PRACTICE**
**5**   Write each number in scientific notation.

**a.** 0.000007     **b.** 20,700,000     **c.** 0.0043     **d.** 812,000,000

**OBJECTIVE 4 ▶ Converting numbers to standard form.** A number written in scientific notation can be rewritten in standard form. For example, to write $8.63 \times 10^3$ in standard form, recall that $10^3 = 1000$.

$$8.63 \times 10^3 = 8.63(1000) = 8630$$

Notice that the exponent on the 10 is positive 3, and we moved the decimal point 3 places to the right. To write $7.29 \times 10^{-3}$ in standard form, recall that $10^{-3} = \dfrac{1}{10^3} = \dfrac{1}{1000}$.

$$7.29 \times 10^{-3} = 7.29\left(\frac{1}{1000}\right) = \frac{7.29}{1000} = 0.00729$$

The exponent on the 10 is negative 3, and we moved the decimal to the left 3 places.

In general, **to write a scientific notation number in standard form,** move the decimal point the same number of places as the exponent on 10. If the exponent is positive, move the decimal point to the right; if the exponent is negative, move the decimal point to the left.

**EXAMPLE 6**   Write each number in standard notation, without exponents.

**a.** $1.02 \times 10^5$     **b.** $7.358 \times 10^{-3}$     **c.** $8.4 \times 10^7$     **d.** $3.007 \times 10^{-5}$

_Solution_

**a.** Move the decimal point 5 places to the right.

$$1.02 \times 10^5 = 102,000.$$

**b.** Move the decimal point 3 places to the left.

$$7.358 \times 10^{-3} = 0.007358$$

**c.** $8.4 \times 10^7 = 84,000,000.$   7 places to the right

**d.** $3.007 \times 10^{-5} = 0.00003007$   5 places to the left

**PRACTICE**
**6**   Write each number in standard notation, without exponents.

**a.** $3.67 \times 10^{-4}$   **b.** $8.954 \times 10^6$   **c.** $2.009 \times 10^{-5}$   **d.** $4.054 \times 10^3$

**Concept Check** ☑

Which number in each pair is larger?

**a.** $7.8 \times 10^3$ or $2.1 \times 10^5$   **b.** $9.2 \times 10^{-2}$ or $2.7 \times 10^4$   **c.** $5.6 \times 10^{-4}$ or $6.3 \times 10^{-5}$

Performing operations on numbers written in scientific notation makes use of the rules and definitions for exponents.

**EXAMPLE 7**   Perform each indicated operation. Write each result in standard decimal notation.

**a.** $(8 \times 10^{-6})(7 \times 10^3)$

**b.** $\dfrac{12 \times 10^2}{6 \times 10^{-3}}$

*Solution*

**a.** $(8 \times 10^{-6})(7 \times 10^3) = (8 \cdot 7) \times (10^{-6} \cdot 10^3)$

$$= 56 \times 10^{-3}$$

$$= 0.056$$

**b.** $\dfrac{12 \times 10^2}{6 \times 10^{-3}} = \dfrac{12}{6} \times 10^{2-(-3)} = 2 \times 10^5 = 200,000$

**PRACTICE**
**7**   Perform each indicated operation. Write each result in standard decimal notation.

**a.** $(5 \times 10^{-4})(8 \times 10^6)$

**b.** $\dfrac{64 \times 10^3}{32 \times 10^{-7}}$

---

### Calculator Explorations

**Scientific Notation**

To enter a number written in scientific notation on a scientific calculator, locate the scientific notation key, which may be marked $\boxed{\text{EE}}$ or $\boxed{\text{EXP}}$. To enter $3.1 \times 10^7$, press $\boxed{3.1}$ $\boxed{\text{EE}}$ $\boxed{7}$. The display should read $\boxed{3.1 \quad 07}$.

*Enter each number written in scientific notation on your calculator.*

**1.** $5.31 \times 10^3$   **2.** $-4.8 \times 10^{14}$
**3.** $6.6 \times 10^{-9}$   **4.** $-9.9811 \times 10^{-2}$

*Multiply each of the following on your calculator. Notice the form of the result.*

**5.** $3,000,000 \times 5,000,000$   **6.** $230,000 \times 1000$

*Multiply each of the following on your calculator. Write the product in scientific notation.*

**7.** $(3.26 \times 10^6)(2.5 \times 10^{13})$   **8.** $(8.76 \times 10^{-4})(1.237 \times 10^9)$

# VOCABULARY & READINESS CHECK

*Fill in each blank with the correct choice.*

**1.** The expression $x^{-3}$ equals _____.

  **a.** $-x^3$  **b.** $\dfrac{1}{x^3}$  **c.** $\dfrac{-1}{x^3}$  **d.** $\dfrac{1}{x^{-3}}$

**2.** The expression $5^{-4}$ equals _____.

  **a.** $-20$  **b.** $-625$  **c.** $\dfrac{1}{20}$  **d.** $\dfrac{1}{625}$

**3.** The number $3.021 \times 10^{-3}$ is written in _____.

  **a.** standard form    **b.** expanded form
  **c.** scientific notation

**4.** The number 0.0261 is written in _____.

  **a.** standard form    **b.** expanded form
  **c.** scientific notation

*Write each expression using positive exponents only.*

**5.** $5x^{-2}$    **6.** $3x^{-3}$    **7.** $\dfrac{1}{y^{-6}}$    **8.** $\dfrac{1}{x^{-3}}$    **9.** $\dfrac{4}{y^{-3}}$    **10.** $\dfrac{16}{y^{-7}}$

# 5.5 | EXERCISE SET

MyMathLab  PRACTICE  WATCH  DOWNLOAD  READ  REVIEW

*Simplify each expression. Write each result using positive exponents only. See Examples 1 through 3.*

**1.** $4^{-3}$    **2.** $6^{-2}$    **3.** $(-2)^{-4}$

**4.** $(-3)^{-5}$    **5.** $7x^{-3}$    **6.** $(7x)^{-3}$

**7.** $\left(\dfrac{1}{2}\right)^{-5}$    **8.** $\left(\dfrac{1}{8}\right)^{-2}$    **9.** $\left(-\dfrac{1}{4}\right)^{-3}$

**10.** $\left(-\dfrac{1}{8}\right)^{-2}$    **11.** $3^{-1} + 2^{-1}$    **12.** $4^{-1} + 4^{-2}$

**13.** $\dfrac{1}{p^{-3}}$    **14.** $\dfrac{1}{q^{-5}}$    **15.** $\dfrac{p^{-5}}{q^{-4}}$

**16.** $\dfrac{r^{-5}}{s^{-2}}$    **17.** $\dfrac{x^{-2}}{x}$    **18.** $\dfrac{y}{y^{-3}}$

**19.** $\dfrac{z^{-4}}{z^{-7}}$    **20.** $\dfrac{x^{-4}}{x^{-1}}$    **21.** $3^{-2} + 3^{-1}$

**22.** $4^{-2} - 4^{-3}$    **23.** $\dfrac{-1}{p^{-4}}$    **24.** $\dfrac{-1}{y^{-6}}$

**25.** $-2^0 - 3^0$    **26.** $5^0 + (-5)^0$

## MIXED PRACTICE

*Simplify each expression. Write each result using positive exponents only. See Examples 1 through 4.*

**27.** $\dfrac{x^2 x^5}{x^3}$    **28.** $\dfrac{y^4 y^5}{y^6}$    **29.** $\dfrac{p^2 p}{p^{-1}}$

**30.** $\dfrac{y^3 y}{y^{-2}}$    **31.** $\dfrac{(m^5)^4 m}{m^{10}}$    **32.** $\dfrac{(x^2)^8 x}{x^9}$

**33.** $\dfrac{r}{r^{-3} r^{-2}}$    **34.** $\dfrac{p}{p^{-3} q^{-5}}$    **35.** $(x^5 y^3)^{-3}$

**36.** $(z^5 x^5)^{-3}$    **37.** $\dfrac{(x^2)^3}{x^{10}}$    **38.** $\dfrac{(y^4)^2}{y^{12}}$

**39.** $\dfrac{(a^5)^2}{(a^3)^4}$    **40.** $\dfrac{(x^2)^5}{(x^4)^3}$    **41.** $\dfrac{8k^4}{2k}$

**42.** $\dfrac{27r^4}{3r^6}$    **43.** $\dfrac{-6m^4}{-2m^3}$    **44.** $\dfrac{15a^4}{-15a^5}$

**45.** $\dfrac{-24a^6 b}{6ab^2}$    **46.** $\dfrac{-5x^4 y^5}{15x^4 y^2}$

**47.** $(-2x^3 y^{-4})(3x^{-1} y)$    **48.** $(-5a^4 b^{-7})(-a^{-4} b^3)$

**49.** $(a^{-5} b^2)^{-6}$    **50.** $(4^{-1} x^5)^{-2}$

**51.** $\left(\dfrac{x^{-2} y^4}{x^3 y^7}\right)^2$    **52.** $\left(\dfrac{a^5 b}{a^7 b^{-2}}\right)^{-3}$

**53.** $\dfrac{4^2 z^{-3}}{4^3 z^{-5}}$    **54.** $\dfrac{3^{-1} x^4}{3^3 x^{-7}}$

**55.** $\dfrac{2^{-3} x^{-4}}{2^2 x}$    **56.** $\dfrac{5^{-1} z^7}{5^{-2} z^9}$

**57.** $\dfrac{7ab^{-4}}{7^{-1} a^{-3} b^2}$    **58.** $\dfrac{6^{-5} x^{-1} y^2}{6^{-2} x^{-4} y^4}$

**59.** $\left(\dfrac{a^{-5} b}{ab^3}\right)^{-4}$    **60.** $\left(\dfrac{r^{-2} s^{-3}}{r^{-4} s^{-3}}\right)^{-3}$

**61.** $\dfrac{(xy^3)^5}{(xy)^{-4}}$    **62.** $\dfrac{(rs)^{-3}}{(r^2 s^3)^2}$

**63.** $\dfrac{(-2xy^{-3})^{-3}}{(xy^{-1})^{-1}}$    **64.** $\dfrac{(-3x^2 y^2)^{-2}}{(xyz)^{-2}}$

**65.** $\dfrac{6x^2 y^3}{-7xy^5}$    **66.** $\dfrac{-8xa^2 b}{-5xa^5 b}$

**67.** $\dfrac{(a^4 b^{-7})^{-5}}{(5a^2 b^{-1})^{-2}}$    **68.** $\dfrac{(a^6 b^{-2})^4}{(4a^{-3} b^{-3})^3}$

*Write each number in scientific notation. See Example 5.*

**69.** 78,000    **70.** 9,300,000,000

**71.** 0.00000167    **72.** 0.00000017

**73.** 0.00635    **74.** 0.00194

**75.** 1,160,000    **76.** 700,000

**77.** More than 2,000,000,000 pencils are manufactured in the United States annually. Write this number in scientific notation (*Source*: AbsoluteTrivia.com)

**78.** The temperature at the interior of the Earth is 20,000,000 degrees Celsius. Write 20,000,000 in scientific notation.

**79.** The Cassini-Huygens Space Mission to Saturn was launched October 15, 1997, with a goal of reaching and orbiting Saturn and its moons. When the Cassini spacecraft disconnected the Huygens probe, which landed on the surface of the moon Titan on January 14, 2005, it was approximately 1,212,000,000 km from earth. Write 1,212,000,000 in scientific notation. (*Source:* Jet Propulsion Laboratory, California Institute of Technology)

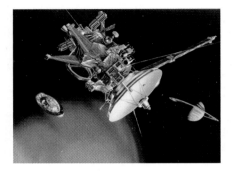

**80.** At this writing, the world's largest optical telescopes are the twin Keck Telescopes located near the summit of Mauna Kea in Hawaii. The elevation of the Keck Telescopes is about 13,600 feet above sea level. Write 13,600 in scientific notation. (*Source:* W.M. Keck Observatory)

*Write each number in standard notation. See Example 6.*

**81.** $8.673 \times 10^{-10}$
**82.** $9.056 \times 10^{-4}$
**83.** $3.3 \times 10^{-2}$
**84.** $4.8 \times 10^{-6}$
**85.** $2.032 \times 10^{4}$
**86.** $9.07 \times 10^{10}$
**87.** Each second, the Sun converts $7.0 \times 10^{8}$ tons of hydrogen into helium and energy in the form of gamma rays. Write this number in standard notation. (*Source:* Students for the Exploration and Development of Space)
**88.** In chemistry, Avogadro's number is the number of atoms in one mole of an element. Avogadro's number is $6.02214199 \times 10^{23}$. Write this number in standard notation. (*Source:* National Institute of Standards and Technology)
**89.** The distance light travels in 1 year is $9.460 \times 10^{12}$ kilometers. Write this number in standard notation.

**90.** The population of the world is $6.067 \times 10^{9}$. Write this number in standard notation. (*Source:* U.S. Bureau of the Census)

**MIXED PRACTICE**

*See Examples 5 and 6. Below are some interesting facts about the Internet. If a number is written in standard form, write it in scientific notation. If a number is written in scientific notation, write it in standard form.*

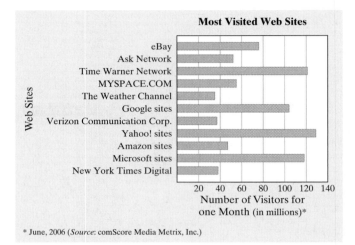

**Most Visited Web Sites**

* June, 2006 (*Source*: comScore Media Metrix, Inc.)

*The bar graph above shows the most visited Web sites on the computer.*

**91.** Estimate the length of the longest bar. Then write the number in scientific notation.
**92.** Estimate the length of the shortest bar. Then write the number in scientific notation.
**93.** The total number of Internet users exceeds 1,000,000,000. (*Source: Computer Industry, Almanac*)
**94.** In a recent year, the retail sales generated by the Internet was $1.08 \times 10^{11}$ dollars. (*Source*: U.S. Census Bureau)
**95.** An estimated $5.7 \times 10^{7}$ American adults read online weblogs (blogs). (*Source*: PEW Internet & American Life Project)
**96.** Junk e-mail (SPAM) costs consumers and businesses an estimated $23,000,000,000.

*Evaluate each expression using exponential rules. Write each result in standard notation. See Example 7.*

**97.** $(1.2 \times 10^{-3})(3 \times 10^{-2})$
**98.** $(2.5 \times 10^{6})(2 \times 10^{-6})$
**99.** $(4 \times 10^{-10})(7 \times 10^{-9})$
**100.** $(5 \times 10^{6})(4 \times 10^{-8})$
**101.** $\dfrac{8 \times 10^{-1}}{16 \times 10^{5}}$
**102.** $\dfrac{25 \times 10^{-4}}{5 \times 10^{-9}}$
**103.** $\dfrac{1.4 \times 10^{-2}}{7 \times 10^{-8}}$
**104.** $\dfrac{0.4 \times 10^{5}}{0.2 \times 10^{11}}$

**REVIEW AND PREVIEW**

*Simplify the following. See Section 5.1.*

**105.** $\dfrac{5x^7}{3x^4}$

**106.** $\dfrac{27y^{14}}{3y^7}$

**107.** $\dfrac{15z^4y^3}{21zy}$

**108.** $\dfrac{18a^7b^{17}}{30a^7b}$

*Use the distributive property and multiply. See Sections 5.3 and 5.5.*

**109.** $\dfrac{1}{y}(5y^2 - 6y + 5)$

**110.** $\dfrac{2}{x}(3x^5 + x^4 - 2)$

**CONCEPT EXTENSIONS**

△ **111.** Find the volume of the cube.

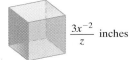

$\dfrac{3x^{-2}}{z}$ inches

△ **112.** Find the area of the triangle.

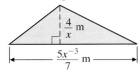

$\dfrac{4}{x}$ m

$\dfrac{5x^{-3}}{7}$ m

*Simplify.*

**113.** $(2a^3)^3a^4 + a^5a^8$

**114.** $(2a^3)^3a^{-3} + a^{11}a^{-5}$

*Fill in the boxes so that each statement is true. (More than one answer is possible for these exercises.)*

**115.** $x^{\square} = \dfrac{1}{x^5}$

**116.** $7^{\square} = \dfrac{1}{49}$

**117.** $z^{\square} \cdot z^{\square} = z^{-10}$

**118.** $(x^{\square})^{\square} = x^{-15}$

**119.** Which is larger? See the Concept Check in this section.
  **a.** $9.7 \times 10^{-2}$ or $1.3 \times 10^1$
  **b.** $8.6 \times 10^5$ or $4.4 \times 10^7$
  **c.** $6.1 \times 10^{-2}$ or $5.6 \times 10^{-4}$

**120.** It was stated earlier that for an integer $n$,
$$x^{-n} = \dfrac{1}{x^n}, \quad x \neq 0$$
Explain why $x$ may not equal 0.

**121.** Determine whether each statement is true or false.
  **a.** $5^{-1} < 5^{-2}$
  **b.** $\left(\dfrac{1}{5}\right)^{-1} < \left(\dfrac{1}{5}\right)^{-2}$
  **c.** $a^{-1} < a^{-2}$ for all nonzero numbers.

*Simplify each expression. Assume that variables represent positive integers.*

**122.** $a^{-4m} \cdot a^{5m}$

**123.** $(x^{-3s})^3$

**124.** $(3y^{2z})^3$

**125.** $a^{4m+1} \cdot a^4$

*Simplify each expression. Write each result in standard notation.*

**126.** $(2.63 \times 10^{12})(-1.5 \times 10^{-10})$

**127.** $(6.785 \times 10^{-4})(4.68 \times 10^{10})$

*Light travels at a rate of $1.86 \times 10^5$ miles per second. Use this information and the distance formula $d = r \cdot t$ to answer Exercises 128 and 129.*

**128.** If the distance from the moon to the Earth is 238,857 miles, find how long it takes the reflected light of the moon to reach the Earth. (Round to the nearest tenth of a second.)

**129.** If the distance from the sun to the Earth is 93,000,000 miles, find how long it takes the light of the sun to reach the Earth. (Round to the nearest tenth of a second.)

---

# 5.6 DIVIDING POLYNOMIALS

**OBJECTIVES**

1 Divide a polynomial by a monomial.

2 Use long division to divide a polynomial by another polynomial.

**OBJECTIVE 1 ▶ Dividing by a monomial.** Now that we know how to add, subtract, and multiply polynomials, we practice dividing polynomials.

To divide a polynomial by a monomial, recall addition of fractions. Fractions that have a common denominator are added by adding the numerators:

$$\frac{a}{c} + \frac{b}{c} = \frac{a + b}{c}$$

If we read this equation from right to left and let $a$, $b$, and $c$ be monomials, $c \neq 0$, we have the following:

> **Dividing a Polynomial By a Monomial**
> Divide each term of the polynomial by the monomial.
> $$\frac{a + b}{c} = \frac{a}{c} + \frac{b}{c}, \quad c \neq 0$$

Throughout this section, we assume that denominators are not 0.

**EXAMPLE 1** Divide $6m^2 + 2m$ by $2m$.

_Solution_ We begin by writing the quotient in fraction form. Then we divide each term of the polynomial $6m^2 + 2m$ by the monomial $2m$.

$$\frac{6m^2 + 2m}{2m} = \frac{6m^2}{2m} + \frac{2m}{2m}$$
$$= 3m + 1 \qquad \text{Simplify.}$$

**Check:** We know that if $\dfrac{6m^2 + 2m}{2m} = 3m + 1$, then $2m \cdot (3m + 1)$ must equal $6m^2 + 2m$. Thus, to check, we multiply.

$$2m(3m + 1) = 2m(3m) + 2m(1) = 6m^2 + 2m$$

The quotient $3m + 1$ checks.

**PRACTICE**
**1** Divide $8t^3 + 4t^2$ by $4t^2$

**EXAMPLE 2** Divide $\dfrac{9x^5 - 12x^2 + 3x}{3x^2}$.

_Solution_ 
$$\frac{9x^5 - 12x^2 + 3x}{3x^2} = \frac{9x^5}{3x^2} - \frac{12x^2}{3x^2} + \frac{3x}{3x^2} \qquad \text{Divide each term by } 3x^2.$$
$$= 3x^3 - 4 + \frac{1}{x} \qquad \text{Simplify.}$$

Notice that the quotient is not a polynomial because of the term $\dfrac{1}{x}$. This expression is called a rational expression—we will study rational expressions further in Chapter 7. Although the quotient of two polynomials is not always a polynomial, we may still check by multiplying.

**Check:** 
$$3x^2\left(3x^3 - 4 + \frac{1}{x}\right) = 3x^2(3x^3) - 3x^2(4) + 3x^2\left(\frac{1}{x}\right)$$
$$= 9x^5 - 12x^2 + 3x$$

**PRACTICE**
**2** Divide $\dfrac{16x^6 + 20x^3 - 12x}{4x^2}$.

**EXAMPLE 3** Divide $\dfrac{8x^2y^2 - 16xy + 2x}{4xy}$.

_Solution_ 
$$\frac{8x^2y^2 - 16xy + 2x}{4xy} = \frac{8x^2y^2}{4xy} - \frac{16xy}{4xy} + \frac{2x}{4xy} \qquad \text{Divide each term by } 4xy.$$
$$= 2xy - 4 + \frac{1}{2y} \qquad \text{Simplify.}$$

**Check:**     $4xy\left(2xy - 4 + \dfrac{1}{2y}\right) = 4xy(2xy) - 4xy(4) + 4xy\left(\dfrac{1}{2y}\right)$

$$= 8x^2y^2 - 16xy + 2x \qquad \square$$

**PRACTICE**
**3**    Divide $\dfrac{15x^4y^4 - 10xy + y}{5xy}$.

---

**Concept Check** ☑

In which of the following is $\dfrac{x + 5}{5}$ simplified correctly?

**a.** $\dfrac{x}{5} + 1$      **b.** $x$      **c.** $x + 1$

**OBJECTIVE 2 ▶ Using long division to divide by a polynomial.**  To divide a polynomial by a polynomial other than a monomial, we use a process known as long division. Polynomial long division is similar to number long division, so we review long division by dividing 13 into 3660.

> **▶ Helpful Hint**
> Recall that 3660 is called the dividend.

$$
\begin{array}{r}
281 \\
13\overline{)3660} \\
\underline{26}\downarrow\downarrow \\
106 \\
\underline{104}\downarrow \\
20 \\
\underline{13} \\
7
\end{array}
$$

$2 \cdot 13 = 26$
Subtract and bring down the next digit in the dividend.
$8 \cdot 13 = 104$
Subtract and bring down the next digit in the dividend.
$1 \cdot 13 = 13$
Subtract. There are no more digits to bring down, so the remainder is 7.

The quotient is 281  R  7, which can be written as $281\dfrac{7}{13}$ $\begin{array}{l}\leftarrow \text{remainder}\\ \leftarrow \text{divisor}\end{array}$

Recall that division can be checked by multiplication. To check a division problem such as this one, we see that

$$13 \cdot 281 + 7 = 3660$$

Now we demonstrate long division of polynomials.

**EXAMPLE 4**    Divide $x^2 + 7x + 12$ by $x + 3$ using long division.

*Solution*

To subtract, change the signs of these terms and add.

$$
\begin{array}{r}
x \phantom{+\ 12} \\
x + 3\overline{)x^2 + 7x + 12} \\
\underline{x^2 \mp 3x}\phantom{+} \downarrow \\
4x + 12
\end{array}
$$

How many times does $x$ divide $x^2$? $\dfrac{x^2}{x} = x$.
Multiply: $x(x + 3)$.
Subtract and bring down the next term.

Now we repeat this process.

$$
\begin{array}{r}
x + 4 \\
x + 3\overline{)x^2 + 7x + 12} \\
\underline{x^2 \mp 3x}\phantom{+ 12} \\
4x + 12 \\
\underline{4x \mp 12} \\
0
\end{array}
$$

How many times does $x$ divide $4x$? $\dfrac{4x}{x} = 4$.

To subtract, change the signs of these terms and add.

Multiply: $4(x + 3)$.
Subtract. The remainder is 0.

The quotient is $x + 4$.

**Check:** We check by multiplying.

| divisor | · | quotient | + | remainder | = | dividend |

or
↓             ↓                    ↓              ↓

$(x + 3)$  ·  $(x + 4)$  +  0  =  $x^2 + 7x + 12$

The quotient checks.  □

**PRACTICE**
**4**  Divide $x^2 + 5x + 6$ by $x + 2$ using long division.

---

**EXAMPLE 5**  Divide $6x^2 + 10x - 5$ by $3x - 1$ using long division.

*Solution*

$$
\begin{array}{r}
2x + 4 \\
3x - 1 \overline{)6x^2 + 10x - 5} \\
\underline{6x^2 - 2x} \phantom{xx} \downarrow \\
12x - 5 \\
\underline{12x - 4} \\
-1
\end{array}
$$

$\dfrac{6x^2}{3x} = 2x$, so $2x$ is a term of the quotient.
Multiply $2x(3x - 1)$.
Subtract and bring down the next term.
$\dfrac{12x}{3x} = 4$, multiply $4(3x - 1)$
Subtract. The remainder is $-1$.

Thus $(6x^2 + 10x - 5)$ divided by $(3x - 1)$ is $(2x + 4)$ with a remainder of $-1$. This can be written as

$$\frac{6x^2 + 10x - 5}{3x - 1} = 2x + 4 + \frac{-1}{3x - 1} \quad \begin{array}{l} \leftarrow \text{remainder} \\ \leftarrow \text{divisor} \end{array}$$

**Check:** To check, we multiply $(3x - 1)(2x + 4)$. Then we add the remainder, $-1$, to this product.

$$(3x - 1)(2x + 4) + (-1) = (6x^2 + 12x - 2x - 4) - 1$$
$$= 6x^2 + 10x - 5$$

The quotient checks.  □

**PRACTICE**
**5**  Divide $4x^2 + 8x - 7$ by $2x + 1$ using long division.

---

In Example 5, the degree of the divisor, $3x - 1$, is 1 and the degree of the remainder, $-1$, is 0. The division process is continued until the degree of the remainder polynomial is less than the degree of the divisor polynomial.

**EXAMPLE 6**  Divide $\dfrac{4x^2 + 7 + 8x^3}{2x + 3}$.

*Solution*  Before we begin the division process, we rewrite

$$4x^2 + 7 + 8x^3 \quad \text{as} \quad 8x^3 + 4x^2 + 0x + 7$$

Notice that we have written the polynomial in descending order and have represented the missing $x$ term by $0x$.

$$
\begin{array}{r}
4x^2 - 4x + 6 \\
2x + 3 \overline{)\,8x^3 + 4x^2 + 0x + 7} \\
\underline{8x^3 \mp 12x^2} \\
-8x^2 + 0x \\
\underline{\mp 8x^2 \mp 12x} \\
12x + 7 \\
\underline{12x \mp 18} \\
-11 \quad \text{Remainder}
\end{array}
$$

Thus, $\dfrac{4x^2 + 7 + 8x^3}{2x + 3} = 4x^2 - 4x + 6 + \dfrac{-11}{2x + 3}.$

$\square$

**PRACTICE**
**6**   Divide $\dfrac{11x - 3 + 9x^3}{3x + 2}.$

---

**EXAMPLE 7**   Divide $\dfrac{2x^4 - x^3 + 3x^2 + x - 1}{x^2 + 1}.$

**Solution**   Before dividing, rewrite the divisor polynomial

$$x^2 + 1 \quad \text{as} \quad x^2 + 0x + 1$$

The $0x$ term represents the missing $x^1$ term in the divisor.

$$
\begin{array}{r}
2x^2 - x + 1 \\
x^2 + 0x + 1 \overline{)\,2x^4 - x^3 + 3x^2 + x - 1} \\
\underline{2x^4 \mp 0x^3 \mp 2x^2} \\
-x^3 + x^2 + x \\
\underline{\mp x^3 \mp 0x^2 \mp x} \\
x^2 + 2x - 1 \\
\underline{x^2 \mp 0x \mp 1} \\
2x - 1 \quad \text{Remainder}
\end{array}
$$

Thus, $\dfrac{2x^4 - x^3 + 3x^2 + x - 1}{x^2 + 1} = 2x^2 - x + 1 + \dfrac{2x - 2}{x^2 + 1}.$

$\square$

**PRACTICE**
**7**   Divide $\dfrac{3x^4 - 2x^3 - 3x^2 + x + 4}{x^2 + 2}.$

---

## VOCABULARY & READINESS CHECK

*Use the choices below to fill in each blank. Choices may be used more than once.*

dividend          divisor          quotient

**1.** In $\dfrac{3}{6\,)\,18}$, the 18 is the _____ , the 3 is the _____ and the 6 is the _____ .

**2.** In $\dfrac{x + 2}{x + 1\,)\,x^2 + 3x + 2}$, the $x + 1$ is the _____ , the $x^2 + 3x + 2$ is the _____ and the $x + 2$ is the _____ .

*Simplify each expression mentally.*

**3.** $\dfrac{a^6}{a^4}$          **4.** $\dfrac{p^8}{p^3}$          **5.** $\dfrac{y^2}{y}$          **6.** $\dfrac{a^3}{a}$

## 5.6 | EXERCISE SET

**MyMathLab**
Powered by CourseCompass™ and MathXL™

Math XL PRACTICE • WATCH • DOWNLOAD • READ • REVIEW

*Perform each division. See Examples 1 through 3.*

1. $\dfrac{12x^4 + 3x^2}{x}$

2. $\dfrac{15x^2 - 9x^5}{x}$

3. $\dfrac{20x^3 - 30x^2 + 5x + 5}{5}$

4. $\dfrac{8x^3 - 4x^2 + 6x + 2}{2}$

5. $\dfrac{15p^3 + 18p^2}{3p}$

6. $\dfrac{14m^2 - 27m^3}{7m}$

7. $\dfrac{-9x^4 + 18x^5}{6x^5}$

8. $\dfrac{6x^5 + 3x^4}{3x^4}$

9. $\dfrac{-9x^5 + 3x^4 - 12}{3x^3}$

10. $\dfrac{6a^2 - 4a + 12}{-2a^2}$

11. $\dfrac{4x^4 - 6x^3 + 7}{-4x^4}$

12. $\dfrac{-12a^3 + 36a - 15}{3a}$

*Find each quotient using long division. See Examples 4 and 5.*

13. $\dfrac{x^2 + 4x + 3}{x + 3}$

14. $\dfrac{x^2 + 7x + 10}{x + 5}$

15. $\dfrac{2x^2 + 13x + 15}{x + 5}$

16. $\dfrac{3x^2 + 8x + 4}{x + 2}$

17. $\dfrac{2x^2 - 7x + 3}{x - 4}$

18. $\dfrac{3x^2 - x - 4}{x - 1}$

19. $\dfrac{9a^3 - 3a^2 - 3a + 4}{3a + 2}$

20. $\dfrac{4x^3 + 12x^2 + x - 14}{2x + 3}$

21. $\dfrac{8x^2 + 10x + 1}{2x + 1}$

22. $\dfrac{3x^2 + 17x + 7}{3x + 2}$

23. $\dfrac{2x^3 + 2x^2 - 17x + 8}{x - 2}$

24. $\dfrac{4x^3 + 11x^2 - 8x - 10}{x + 3}$

*Find each quotient using long division. Don't forget to write the polynomials in descending order and fill in any missing terms. See Examples 6 and 7.*

25. $\dfrac{x^2 - 36}{x - 6}$

26. $\dfrac{a^2 - 49}{a - 7}$

27. $\dfrac{x^3 - 27}{x - 3}$

28. $\dfrac{x^3 + 64}{x + 4}$

29. $\dfrac{1 - 3x^2}{x + 2}$

30. $\dfrac{7 - 5x^2}{x + 3}$

31. $\dfrac{-4b + 4b^2 - 5}{2b - 1}$

32. $\dfrac{-3y + 2y^2 - 15}{2y + 5}$

## MIXED PRACTICE

*Divide. If the divisor contains 2 or more terms, use long division. See Examples 1 through 7.*

33. $\dfrac{a^2b^2 - ab^3}{ab}$

34. $\dfrac{m^3n^2 - mn^4}{mn}$

35. $\dfrac{8x^2 + 6x - 27}{2x - 3}$

36. $\dfrac{18w^2 + 18w - 8}{3w + 4}$

37. $\dfrac{2x^2y + 8x^2y^2 - xy^2}{2xy}$

38. $\dfrac{11x^3y^3 - 33xy + x^2y^2}{11xy}$

39. $\dfrac{2b^3 + 9b^2 + 6b - 4}{b + 4}$

40. $\dfrac{2x^3 + 3x^2 - 3x + 4}{x + 2}$

41. $\dfrac{5x^2 + 28x - 10}{x + 6}$

42. $\dfrac{2x^2 + x - 15}{x + 3}$

43. $\dfrac{10x^3 - 24x^2 - 10x}{10x}$

44. $\dfrac{2x^3 + 12x^2 + 16}{4x^2}$

45. $\dfrac{6x^2 + 17x - 4}{x + 3}$

46. $\dfrac{2x^2 - 9x + 15}{x - 6}$

47. $\dfrac{30x^2 - 17x + 2}{5x - 2}$

48. $\dfrac{4x^2 - 13x - 12}{4x + 3}$

49. $\dfrac{3x^4 - 9x^3 + 12}{-3x}$

50. $\dfrac{8y^6 - 3y^2 - 4y}{4y}$

**51.** $\dfrac{x^3 + 6x^2 + 18x + 27}{x + 3}$

**52.** $\dfrac{x^3 - 8x^2 + 32x - 64}{x - 4}$

**53.** $\dfrac{y^3 + 3y^2 + 4}{y - 2}$

**54.** $\dfrac{3x^3 + 11x + 12}{x + 4}$

**55.** $\dfrac{5 - 6x^2}{x - 2}$

**56.** $\dfrac{3 - 7x^2}{x - 3}$

*Divide.*

**57.** $\dfrac{x^5 + x^2}{x^2 + x}$

**58.** $\dfrac{x^6 - x^4}{x^3 + 1}$

## REVIEW AND PREVIEW

*Multiply each expression. See Section 5.3.*

**59.** $2a(a^2 + 1)$

**60.** $-4a(3a^2 - 4)$

**61.** $2x(x^2 + 7x - 5)$

**62.** $4y(y^2 - 8y - 4)$

**63.** $-3xy(xy^2 + 7x^2y + 8)$

**64.** $-9xy(4xyz + 7xy^2z + 2)$

**65.** $9ab(ab^2c + 4bc - 8)$

**66.** $-7sr(6s^2r + 9sr^2 + 9rs + 8)$

*Use the bar graph below to answer Exercises 67 through 70. See Section 3.1.*

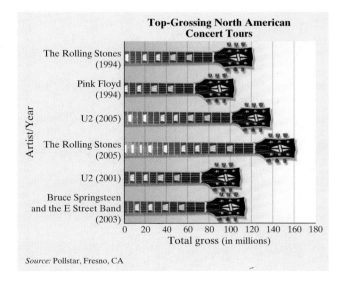

**Top-Grossing North American Concert Tours**

*Source:* Pollstar, Fresno, CA

**67.** Which artist has grossed the most money on one tour?

**68.** Estimate the amount of money made by the 2005 concert tour of The Rolling Stones.

**69.** Estimate the amount of money made by the 2005 concert tour of U2.

**70.** Which artist shown has grossed the least amount of money on a tour?

## CONCEPT EXTENSIONS

△ **71.** The perimeter of a square is $(12x^3 + 4x - 16)$ feet. Find the length of its side.

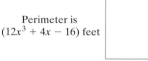

Perimeter is $(12x^3 + 4x - 16)$ feet

△ **72.** The volume of the swimming pool shown is $(36x^5 - 12x^3 + 6x^2)$ cubic feet. If its height is $2x$ feet and its width is $3x$ feet, find its length.

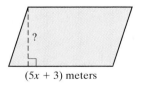

3x feet

2x feet

**73.** In which of the following is $\dfrac{a + 7}{7}$ simplified correctly? See the Concept Check in this section.

   **a.** $a + 1$      **b.** $a$      **c.** $\dfrac{a}{7} + 1$

✎ **74.** Explain how to check a polynomial long division result when the remainder is 0.

✎ **75.** Explain how to check a polynomial long division result when the remainder is not 0.

△ **76.** The area of the following parallelogram is $(10x^2 + 31x + 15)$ square meters. If its base is $(5x + 3)$ meters, find its height.

?

$(5x + 3)$ meters

△ **77.** The area of the top of the Ping-Pong table is $(49x^2 + 70x - 200)$ square inches. If its length is $(7x + 20)$ inches, find its width.

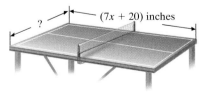

?   $(7x + 20)$ inches

**78.** $(18x^{10a} - 12x^{8a} + 14x^{5a} - 2x^{3a}) \div 2x^{3a}$

**79.** $(25y^{11b} + 5y^{6b} - 20y^{3b} + 100y^b) \div 5y^b$

## THE BIGGER PICTURE    SIMPLIFYING EXPRESSIONS AND SOLVING EQUATIONS AND INEQUALITIES

Now we continue our outline from Sections 1.7, 2.3, and 2.8. Although suggestions are given, this outline should be in your own words. Once you complete this new portion, try the exercises below.

**I.** Simplifying Expressions

   **A.** Real Numbers

     **1.** Add (Section 1.5)

     **2.** Subtract (Section 1.6)

     **3.** Multiply or Divide (Section 1.7)

   **B.** Exponents— $x^7 \cdot x^5 = x^{12}; (x^7)^5 = x^{35}; \dfrac{x^7}{x^5} = x^2;$

   $x^0 = 1; 8^{-2} = \dfrac{1}{8^2} = \dfrac{1}{64}$

   **C.** Polynomials

     **1.** Add:    Combine like terms.

       $(3y^2 + 6y + 7) + (9y^2 - 11y - 15)$

       $= 3y^2 + 6y + 7 + 9y^2 - 11y - 15$

       $= 12y^2 - 5y - 8$

     **2.** Subtract:    Change the sign of the terms of the polynomial being subtracted, then add.

       $(3y^2 + 6y + 7) - (9y^2 - 11y - 15)$

       $= 3y^2 + 6y + 7 - 9y^2 + 11y + 15$

       $= -6y^2 + 17y + 22$

     **3.** Multiply:    Multiply each term of one polynomial by each term of the other polynomial.

       $(x + 5)(2x^2 - 3x + 4)$

       $= x(2x^2 - 3x + 4) + 5(2x^2 - 3x + 4)$

       $= 2x^3 - 3x^2 + 4x + 10x^2 - 15x + 20$

       $= 2x^3 + 7x^2 - 11x + 20$

     **4.** Divide:

       **a.** To divide by a monomial, divide each term of the polynomial by the monomial.

   $$\frac{8x^2 + 2x - 6}{2x} = \frac{8x^2}{2x} + \frac{2x}{2x} - \frac{6}{2x}$$

   $$= 4x + 1 - \frac{3}{x}$$

**b.** To divide by a polynomial other than a monomial, use long division.

$$
\begin{array}{r}
x - 6 + \dfrac{40}{2x + 5} \\[4pt]
2x + 5 \overline{\smash{\big)}\ 2x^2 - 7x + 10} \\
\underline{2x^2 + 5x}\phantom{xxxxxxx} \\
-12x + 10 \\
\underline{-12x - 30} \\
40
\end{array}
$$

**II.** Solving Equations

   **A.** Linear Equations (Section 2.3)

**III.** Solving Inequalities

   **A.** Linear Inequalities (Section 2.8)

*Simplify the expressions.*

**1.** $-5.7 + (-0.23)$

**2.** $\dfrac{1}{2} - \dfrac{9}{10}$

**3.** $(-5x^2y^3)(-x^7y)$

**4.** $2^{-3}a^{-7}a^3$

**5.** $(7y^3 - 6y + 2) - (y^3 + 2y^2 + 2)$

**6.** Subtract $(y^2 + 7)$ from $(9y^2 - 3y)$

**7.** Multiply:   $(x - 3)(4x^2 - x + 7)$

**8.** Multiply:   $(6m - 5)^2$

**9.** Divide:   $\dfrac{20n^2 - 5n + 10}{5n}$

**10.** Divide:   $\dfrac{6x^2 - 20x + 20}{3x - 1}$

*Solve the equations or inequalities.*

**11.** $-6x = 3.6$

**12.** $-6x < 3.6$

**13.** $6x + 6 \geq 8x + 2$

**14.** $7y + 3(y - 1) = 4(y + 1) - 3$

# 5.7 SYNTHETIC DIVISION AND THE REMAINDER THEOREM

**OBJECTIVES**

**1** Use synthetic division to divide a polynomial by a binomial.

**2** Use the remainder theorem to evaluate polynomials.

**OBJECTIVE 1 ▶ Using synthetic division.** When a polynomial is to be divided by a binomial of the form $x - c$, a shortcut process called **synthetic division** may be used. On the left is an example of long division, and on the right, the same example showing the coefficients of the variables only.

$$
\begin{array}{r}
2x^2 + 5x + 2 \\
x - 3{\overline{\smash{\big)}\,2x^3 - x^2 - 13x + 1}} \\
\underline{2x^3 - 6x^2} \\
5x^2 - 13x \\
\underline{5x^2 - 15x} \\
2x + 1 \\
\underline{2x - 6} \\
7
\end{array}
\qquad
\begin{array}{r}
2 \quad 5 \quad 2 \\
1 - 3{\overline{\smash{\big)}\,2 - 1 - 13 + 1}} \\
\underline{2 - 6} \\
5 - 13 \\
\underline{5 - 15} \\
2 + 1 \\
\underline{2 - 6} \\
7
\end{array}
$$

Notice that as long as we keep coefficients of powers of $x$ in the same column, we can perform division of polynomials by performing algebraic operations on the coefficients only. This shortcut process of dividing with coefficients only in a special format is called synthetic division. To find $(2x^3 - x^2 - 13x + 1) \div (x - 3)$ by synthetic division, follow the next example.

**EXAMPLE 1**   Use synthetic division to divide $2x^3 - x^2 - 13x + 1$ by $x - 3$.

*Solution*   To use synthetic division, the divisor must be in the form $x - c$. Since we are dividing by $x - 3$, $c$ is 3. Write down 3 and the coefficients of the dividend.

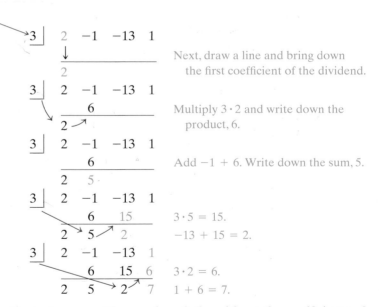

The quotient is found in the bottom row. The numbers 2, 5, and 2 are the coefficients of the quotient polynomial, and the number 7 is the remainder. The degree of the quotient polynomial is one less than the degree of the dividend. In our example, the degree of the dividend is 3, so the degree of the quotient polynomial is 2. As we found when we performed the long division, the quotient is

$$2x^2 + 5x + 2, \quad \text{remainder } 7$$

or

$$2x^2 + 5x + 2 + \frac{7}{x - 3}.$$

**PRACTICE**

**1**   Use synthetic division to divide $4x^3 - 3x^2 + 6x + 5$ by $x - 1$

**EXAMPLE 2** Use synthetic division to divide $x^4 - 2x^3 - 11x^2 + 5x + 34$ by $x + 2$.

_Solution_ The divisor is $x + 2$, which we write in the form $x - c$ as $x - (-2)$. Thus, $c$ is $-2$. The dividend coefficients are $1, -2, -11, 5$, and $34$.

$$
c \longrightarrow -2 \;\bigg|\; \begin{array}{rrrrr} 1 & -2 & -11 & 5 & 34 \\ & -2 & 8 & 6 & -22 \\ \hline 1 & -4 & -3 & 11 & 12 \end{array}
$$

The dividend is a fourth-degree polynomial, so the quotient polynomial is a third-degree polynomial. The quotient is $x^3 - 4x^2 - 3x + 11$ with a remainder of 12. Thus,

$$
\frac{x^4 - 2x^3 - 11x^2 + 5x + 34}{x + 2} = x^3 - 4x^2 - 3x + 11 + \frac{12}{x + 2}.
$$

**PRACTICE**

**2** Use synthetic division to divide $x^4 + 3x^3 - 5x^2 + 6x + 12$ by $x + 3$.

**Concept Check** ☑

Which division problems are candidates for the synthetic division process?

**a.** $(3x^2 + 5) \div (x + 4)$
**b.** $(x^3 - x^2 + 2) \div (3x^3 - 2)$
**c.** $(y^4 + y - 3) \div (x^2 + 1)$
**d.** $x^5 \div (x - 5)$

> **▶ Helpful Hint**
>
> Before dividing by synthetic division, write the dividend in descending order of variable exponents. Any "missing powers" of the variable should be represented by 0 times the variable raised to the missing power.

**EXAMPLE 3** If $P(x) = 2x^3 - 4x^2 + 5$

**a.** Find $P(2)$ by substitution.
**b.** Use synthetic division to find the remainder when $P(x)$ is divided by $x - 2$.

_Solution_

**a.** $P(x) = 2x^3 - 4x^2 + 5$
$P(2) = 2(2)^3 - 4(2)^2 + 5$
$\qquad = 2(8) - 4(4) + 5 = 16 - 16 + 5 = 5$
Thus, $P(2) = 5$.

**b.** The coefficients of $P(x)$ are $2, -4, 0$, and $5$. The number 0 is a coefficient of the missing power of $x^1$. The divisor is $x - 2$, so $c$ is 2.

$$
c \longrightarrow 2 \;\bigg|\; \begin{array}{rrrr} 2 & -4 & 0 & 5 \\ & 4 & 0 & 0 \\ \hline 2 & 0 & 0 & 5 \end{array} \quad \text{remainder}
$$

The remainder when $P(x)$ is divided by $x - 2$ is 5.

**PRACTICE**

**3** If $P(x) = x^3 - 5x - 2$,

**a.** Find $P(2)$ by substitution.
**b.** Use synthetic division to find the remainder when $P(x)$ is divided by $x - 2$.

**OBJECTIVE 2** ▶ **Using the remainder theorem.** Notice in the preceding example that $P(2) = 5$ and that the remainder when $P(x)$ is divided by $x - 2$ is 5. This is no accident. This illustrates the **remainder theorem.**

> **Remainder Theorem**
>
> If a polynomial $P(x)$ is divided by $x - c$, then the remainder is $P(c)$.

**EXAMPLE 4**   Use the remainder theorem and synthetic division to find $P(4)$ if

$$P(x) = 4x^6 - 25x^5 + 35x^4 + 17x^2.$$

*Solution*   To find $P(4)$ by the remainder theorem, we divide $P(x)$ by $x - 4$. The coefficients of $P(x)$ are $4, -25, 35, 0, 17, 0,$ and $0$. Also, $c$ is 4.

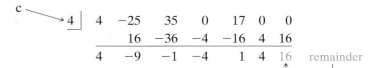

$$
\begin{array}{r|rrrrrrr}
4 & 4 & -25 & 35 & 0 & 17 & 0 & 0 \\
  &   & 16 & -36 & -4 & -16 & 4 & 16 \\
\hline
  & 4 & -9 & -1 & -4 & 1 & 4 & 16 \quad \text{remainder}
\end{array}
$$

Thus, $P(4) = 16$, the remainder.    □

**PRACTICE**

**4**   Use the remainder theorem and synthetic division to find $P(3)$ if $P(x) = 2x^5 - 18x^4 + 90x^2 + 59x$.

---

## 5.7 EXERCISE SET

*Use synthetic division to divide. See Examples 1 and 2.*

1. $(x^2 + 3x - 40) \div (x - 5)$
2. $(x^2 - 14x + 24) \div (x - 2)$
3. $(x^2 + 5x - 6) \div (x + 6)$
4. $(x^2 + 12x + 32) \div (x + 4)$
5. $(x^3 - 7x^2 - 13x + 5) \div (x - 2)$
6. $(x^3 + 6x^2 + 4x - 7) \div (x + 5)$
7. $(4x^2 - 9) \div (x - 2)$
8. $(3x^2 - 4) \div (x - 1)$

*For the given polynomial $P(x)$ and the given $c$, find $P(c)$ by* **(a)** *direct substitution and* **(b)** *the remainder theorem. See Examples 3 and 4.*

9. $P(x) = 3x^2 - 4x - 1; P(2)$
10. $P(x) = x^2 - x + 3; P(5)$
11. $P(x) = 4x^4 + 7x^2 + 9x - 1; P(-2)$
12. $P(x) = 8x^5 + 7x + 4; P(-3)$
13. $P(x) = x^5 + 3x^4 + 3x - 7; P(-1)$
14. $P(x) = 5x^4 - 4x^3 + 2x - 1; P(-1)$

**MIXED PRACTICE**

*Use synthetic division to divide.*

15. $(x^3 - 3x^2 + 2) \div (x - 3)$
16. $(x^2 + 12) \div (x + 2)$
17. $(6x^2 + 13x + 8) \div (x + 1)$
18. $(x^3 - 5x^2 + 7x - 4) \div (x - 3)$
19. $(2x^4 - 13x^3 + 16x^2 - 9x + 20) \div (x - 5)$
20. $(3x^4 + 5x^3 - x^2 + x - 2) \div (x + 2)$
21. $(3x^2 - 15) \div (x + 3)$
22. $(3x^2 + 7x - 6) \div (x + 4)$
23. $(3x^3 - 6x^2 + 4x + 5) \div \left(x - \dfrac{1}{2}\right)$
24. $(8x^3 - 6x^2 - 5x + 3) \div \left(x + \dfrac{3}{4}\right)$
25. $(3x^3 + 2x^2 - 4x + 1) \div \left(x - \dfrac{1}{3}\right)$
26. $(9y^3 + 9y^2 - y + 2) \div \left(y + \dfrac{2}{3}\right)$

**27.** $(7x^2 - 4x + 12 + 3x^3) \div (x + 1)$

**28.** $(x^4 + 4x^3 - x^2 - 16x - 4) \div (x - 2)$

**29.** $(x^3 - 1) \div (x - 1)$

**30.** $(y^3 - 8) \div (y - 2)$

**31.** $(x^2 - 36) \div (x + 6)$

**32.** $(4x^3 + 12x^2 + x - 12) \div (x + 3)$

*For the given polynomial $P(x)$ and the given c, use the remainder theorem to find $P(c)$.*

**33.** $P(x) = x^3 + 3x^2 - 7x + 4; 1$

**34.** $P(x) = x^3 + 5x^2 - 4x - 6; 2$

**35.** $P(x) = 3x^3 - 7x^2 - 2x + 5; -3$

**36.** $P(x) = 4x^3 + 5x^2 - 6x - 4; -2$

**37.** $P(x) = 4x^4 + x^2 - 2; -1$

**38.** $P(x) = x^4 - 3x^2 - 2x + 5; -2$

**39.** $P(x) = 2x^4 - 3x^2 - 2; \dfrac{1}{3}$

**40.** $P(x) = 4x^4 - 2x^3 + x^2 - x - 4; \dfrac{1}{2}$

**41.** $P(x) = x^5 + x^4 - x^3 + 3; \dfrac{1}{2}$

**42.** $P(x) = x^5 - 2x^3 + 4x^2 - 5x + 6; \dfrac{2}{3}$

**43.** Explain an advantage of using the remainder theorem instead of direct substitution.

**44.** Explain an advantage of using synthetic division instead of long division.

### REVIEW AND PREVIEW

*Solve each equation for x. See Section 2.3.*

**45.** $7x + 2 = x - 3$

**46.** $4 - 2x = 17 - 5x$

**47.** $\dfrac{x}{3} - 5 = 13$

**48.** $\dfrac{2x}{9} + 1 = \dfrac{7}{9}$

*Evaluate. See Section 5.1.*

**49.** $2^3$

**50.** $3^4$

**51.** $(-2)^5$

**52.** $-2^5$

**53.** $3 \cdot 4^2$

**54.** $4 \cdot 3^3$

*Evaluate each expression for the given replacement value. See Section 5.1.*

**55.** $x^2$ if $x$ is $-5$

**56.** $x^3$ if $x$ is $-5$

**57.** $2x^3$ if $x$ is $-1$

**58.** $3x^2$ if $x$ is $-1$

### CONCEPT EXTENSIONS

*Which division problems are candidates for the synthetic division process? See the Concept Checks in this section.*

**59.** $(5x^2 - 3x + 2) \div (x + 2)$

**60.** $(x^4 - 6) \div (x^3 + 3x - 1)$

**61.** $(x^7 - 2) \div (x^5 + 1)$

**62.** $(3x^2 + 7x - 1) \div \left( x - \dfrac{1}{3} \right)$

**63.** If the area of a parallelogram is $(x^4 - 23x^2 + 9x - 5)$ square centimeters and its base is $(x + 5)$ centimeters, find its height.

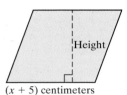

$(x + 5)$ centimeters

**64.** If the volume of a box is $(x^4 + 6x^3 - 7x^2)$ cubic meters, its height is $x^2$ meters, and its length is $(x + 7)$ meters, find its width.

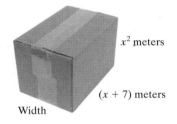

$x^2$ meters

$(x + 7)$ meters

Width

*Divide.*

**65.** $\left( x^4 + \dfrac{2}{3}x^3 + x \right) \div (x - 1)$

**66.** $\left( 2x^3 + \dfrac{9}{2}x^2 - 4x - 10 \right) \div (x + 2)$

*We say that 2 is a factor of 8 because 2 divides 8 evenly, or with a remainder of 0. In the same manner, the polynomial $x - 2$ is a factor of the polynomial $x^3 - 14x^2 + 24x$ because the remainder is 0 when $x^3 - 14x^2 + 24x$ is divided by $x - 2$. Use this information for Exercises 67 through 69.*

**67.** Use synthetic division to show that $x + 3$ is a factor of $x^3 + 3x^2 + 4x + 12$.

**68.** Use synthetic division to show that $x - 2$ is a factor of $x^3 - 2x^2 - 3x + 6$.

**69.** From the remainder theorem, the polynomial $x - c$ is a factor of a polynomial function $P(x)$ if $P(c)$ is what value?

**70.** If a polynomial is divided by $x - 5$, the quotient is $2x^2 + 5x - 6$ and the remainder is 3. Find the original polynomial.

**71.** If a polynomial is divided by $x + 3$, the quotient is $x^2 - x + 10$ and the remainder is $-2$. Find the original polynomial.

**72.** Explain an advantage of using the remainder theorem instead of direct substitution.

**73.** Explain an advantage of using synthetic division instead of long division.

# CHAPTER 5 GROUP ACTIVITY

## Modeling with Polynomials

The polynomial $-2.02x^2 + 51.60x + 674.60$ dollars represents consumer spending per person per year on all U.S. media from 2000 to 2006. This includes spending on subscription TV services, recorded music, newspapers, magazines, books, home video, theater movies, video games, and educational software. The polynomial model $-1.39x^2 + 22.61x + 206.53$ dollars represents consumer spending per person per year on subscription TV services alone during this same period. In both models, $x$ is the number of years after 2000. (*Source:* Based on data from *Statistical Abstract of the United States, 2007*).

   In this project, you will have the opportunity to investigate these polynomial models numerically, algebraically, and graphically. This project may be completed by working in groups or individually.

1. Use the polynomials to complete the following table showing the annual consumer spending per person over the period 2000–2006 by evaluating each polynomial at the given values of $x$. Then subtract each value in the fourth column from the corresponding value in the third column. Record the result in the last column, "Difference." What do you think these values represent? What trends do you notice in the data?

| Year | $x$ | Consumer Spending per Person per Year on All U.S. Media | Consumer Spending per Person per Year on Subscription TV | Difference |
|---|---|---|---|---|
| 2000 | 0 | | | |
| 2002 | 2 | | | |
| 2004 | 4 | | | |
| 2006 | 6 | | | |

2. Use the polynomial models to find a new polynomial model representing the amount of consumer spending per person on U.S. media other than subscription TV services (such as recorded music, newspapers, magazines, books, home video, theater movies, video games, and educational software). Then use this new polynomial to complete the following table.

| Year | $x$ | Consumer Spending per Person per Year on Media Other Than Subscription TV |
|---|---|---|
| 2000 | 0 | |
| 2002 | 2 | |
| 2004 | 4 | |
| 2006 | 6 | |

3. Compare the values in the last column of the table in Question 1 to the values in the last column of the table in Question 2. What do you notice? What can you conclude?

4. Use the polynomial models to estimate consumer spending on
   a. all U.S. media,
   b. subscription TV,
   c. media other than subscription TV for the year 2008.

5. Use the polynomial models to estimate consumer spending on
   a. all U.S. media,
   b. subscription TV,
   c. media other than subscription TV for the year 2010.

6. Create a bar graph that represents the data for consumer spending on all U.S. media in the years 2000, 2002, 2004, and 2006 along with your estimates for 2008 and 2010. Study your bar graph. Discuss what the graph implies about the future.

# CHAPTER 5 VOCABULARY CHECK

*Fill in each blank with one of the words or phrases listed below.*

term          coefficient          monomial          binomial          trinomial
polynomials          degree of a term          degree of a polynomial          FOIL

1. A _____ is a number or the product of numbers and variables raised to powers.
2. The _____ method may be used when multiplying two binomials.
3. A polynomial with exactly 3 terms is called a _____.
4. The _____ is the greatest degree of any term of the polynomial.
5. A polynomial with exactly 2 terms is called a _____.
6. The _____ of a term is its numerical factor.
7. The _____ is the sum of the exponents on the variables in the term.
8. A polynomial with exactly 1 term is called a _____.
9. Monomials, binomials, and trinomials are all examples of _____.

---

📖 **STUDY SKILLS BUILDER**

---

### Are You Preparing for a Test on Chapter 5?

Below is a list of some *common trouble areas* for topics covered in Chapter 5. After studying for your test—but before taking your test—read these.

- Do you know that a negative exponent does not make the base a negative number? For example,

$$3^{-2} = \frac{1}{3^2} = \frac{1}{9}$$

- Make sure you remember that $x$ has an understood coefficient of 1 and an understood exponent of 1. For example,

$$2x + x = 2x + 1x = 3x; \quad x^5 \cdot x = x^5 \cdot x^1 = x^6$$

- Do you know the difference between $5x^2$ and $(5x)^2$?

$$5x^2 \text{ is } 5 \cdot x^2; \quad (5x)^2 = 5^2 \cdot x^2 \text{ or } 25 \cdot x^2$$

- Can you evaluate $x^2 - x$ when $x = -2$?

$$x^2 - x = (-2)^2 - (-2) = 4 - (-2) = 4 + 2 = 6$$

- Can you subtract $5x^2 + 1$ from $3x^2 - 6$?

$$(3x^2 - 6) - (5x^2 + 1) = 3x^2 - 6 - 5x^2 - 1 = -2x^2 - 7$$

- Make sure you are familiar with squaring a binomial and other special products.

$$(3x - 4)^2 = (3x)^2 - 2(3x)(4) + 4^2 = 9x^2 - 24x + 16$$

or

$$(3x - 4)^2 = (3x - 4)(3x - 4) = 9x^2 - 24x + 16$$
$$(2x^2 + 1)(2x^2 - 1) = (2x^2)^2 - 1^2 = 4x^4 - 1$$

Remember: This is simply a checklist of common trouble areas. For a review of Chapter 5, see the Highlights and Chapter Review.

---

▶ **Helpful Hint**

Are you preparing for your test? Don't forget to take the Chapter 5 Test on page 363. Then check your answers at the back of the text and use the Chapter Test Prep Video CD to see the fully worked-out solutions to any of the exercises you want to review.

---

# CHAPTER 5 HIGHLIGHTS

| DEFINITIONS AND CONCEPTS | EXAMPLES |
|---|---|
| **SECTION 5.1 EXPONENTS** | |
| $a^n$ means the product of $n$ factors, each of which is $a$. | $3^2 = 3 \cdot 3 = 9$ <br> $(-5)^3 = (-5)(-5)(-5) = -125$ <br> $\left(\frac{1}{2}\right)^4 = \frac{1}{2} \cdot \frac{1}{2} \cdot \frac{1}{2} \cdot \frac{1}{2} = \frac{1}{16}$ |
| If $m$ and $n$ are integers and no denominators are 0, | |
| Product Rule: $a^m \cdot a^n = a^{m+n}$ | $x^2 \cdot x^7 = x^{2+7} = x^9$ |
| Power Rule: $(a^m)^n = a^{mn}$ | $(5^3)^8 = 5^{3 \cdot 8} = 5^{24}$ |
| Power of a Product Rule: $(ab)^n = a^n b^n$ | $(7y)^4 = 7^4 y^4$ |
| Power of a Quotient Rule: $\left(\frac{a}{b}\right)^n = \frac{a^n}{b^n}$ | $\left(\frac{x}{8}\right)^3 = \frac{x^3}{8^3}$ |
| Quotient Rule: $\dfrac{a^m}{a^n} = a^{m-n}$ | $\dfrac{x^9}{x^4} = x^{9-4} = x^5$ |
| Zero Exponent: $a^0 = 1, a \neq 0$. | $5^0 = 1, x^0 = 1, x \neq 0$ |

| **DEFINITIONS AND CONCEPTS** | **EXAMPLES** |
|---|---|

*Terms*

A **term** is a number or the product of numbers and variables raised to powers.

$-5x, 7a^2b, \dfrac{1}{4}y^4, 0.2$

The **numerical coefficient** or **coefficient** of a term is its numerical factor.

| *Term* | *Coefficient* |
|---|---|
| $7x^2$ | 7 |
| $y$ | 1 |
| $-a^2b$ | $-1$ |

*Polynomials*

A **polynomial** is a finite sum of terms in which all variables have exponents raised to nonnegative integer powers and no variables appear in the denominator.

$$1.3x^2 \quad \text{(monomial)}$$
$$-\frac{1}{3}y + 5 \quad \text{(binomial)}$$
$$6z^2 - 5z + 7 \quad \text{(trinomial)}$$

A function $P$ is a **polynomial function** if $P(x)$ is a polynomial.

For the polynomial function
$$P(x) = -x^2 + 6x - 12, \text{ find } P(-2)$$
$$P(-2) = -(-2)^2 + 6(-2) - 12 = -28.$$

The **degree of a term** is the sum of the exponents on the variables in the term.

| *Term* | *Degree* |
|---|---|
| $-5x^3$ | 3 |
| 3 (or $3x^0$) | 0 |
| $2a^2b^2c$ | 5 |

The **degree of a polynomial** is the greatest degree of any term of the polynomial.

| *Polynomial* | *Degree* |
|---|---|
| $5x^2 - 3x + 2$ | 2 |
| $7y + 8y^2z^3 - 12$ | $2 + 3 = 5$ |

Add:

To add polynomials, add or combine like terms.

$$(7x^2 - 3x + 2) + (-5x - 6) = 7x^2 - 3x + 2 - 5x - 6$$
$$= 7x^2 - 8x - 4$$

Subtract:

**To subtract two polynomials,** change the signs of the terms of the second polynomial, then add.

$$(17y^2 - 2y + 1) - (-3y^3 + 5y - 6)$$
$$= (17y^2 - 2y + 1) + (3y^3 - 5y + 6)$$
$$= 17y^2 - 2y + 1 + 3y^3 - 5y + 6$$
$$= 3y^3 + 17y^2 - 7y + 7$$

**To multiply two polynomials,** multiply each term of one polynomial by each term of the other polynomial, and then combine like terms.

Multiply:

$$(2x + 1)(5x^2 - 6x + 2)$$

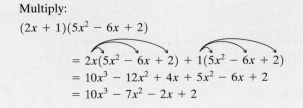

$$= 2x(5x^2 - 6x + 2) + 1(5x^2 - 6x + 2)$$
$$= 10x^3 - 12x^2 + 4x + 5x^2 - 6x + 2$$
$$= 10x^3 - 7x^2 - 2x + 2$$

| **DEFINITIONS AND CONCEPTS** | **EXAMPLES** |
|---|---|

The **FOIL method** may be used when multiplying two binomials.

Multiply: $(5x - 3)(2x + 3)$

First / Last / Outer / Inner

$$(5x - 3)(2x + 3) = \overset{F}{(5x)(2x)} + \overset{O}{(5x)(3)} + \overset{I}{(-3)(2x)} + \overset{L}{(-3)(3)}$$
$$= 10x^2 + 15x - 6x - 9$$
$$= 10x^2 + 9x - 9$$

**Squaring a Binomial**

$(a + b)^2 = a^2 + 2ab + b^2$

$(a - b)^2 = a^2 - 2ab + b^2$

Square each binomial.

$$(x + 5)^2 = x^2 + 2(x)(5) + 5^2$$
$$= x^2 + 10x + 25$$
$$(3x - 2y)^2 = (3x)^2 - 2(3x)(2y) + (2y)^2$$
$$= 9x^2 - 12xy + 4y^2$$

**Multiplying the Sum and Difference of Two Terms**

$(a + b)(a - b) = a^2 - b^2$

Multiply:

$$(6y + 5)(6y - 5) = (6y)^2 - 5^2$$
$$= 36y^2 - 25$$

If $a \neq 0$ and $n$ is an integer,

$$a^{-n} = \frac{1}{a^n}$$

Rules for exponents are true for positive and negative integers.

$$3^{-2} = \frac{1}{3^2} = \frac{1}{9}; 5x^{-2} = \frac{5}{x^2}$$

Simplify:
$$\left(\frac{x^{-2}y}{x^5}\right)^{-2} = \frac{x^4 y^{-2}}{x^{-10}}$$
$$= x^{4-(-10)}y^{-2}$$
$$= \frac{x^{14}}{y^2}$$

A positive number is written in scientific notation if it is as the product of a number $a$, $1 \leq a < 10$, and an integer power $r$ of 10.

$$a \times 10^r$$

Write each number in scientific notation.

$$12{,}000 = 1.2 \times 10^4$$

$$0.00000568 = 5.68 \times 10^{-6}$$

To divide a polynomial by a monomial:

$$\frac{a + b}{c} = \frac{a}{c} + \frac{b}{c}$$

To divide a polynomial by a polynomial other than a monomial, use long division.

Divide:

$$\frac{15x^5 - 10x^3 + 5x^2 - 2x}{5x^2} = \frac{15x^5}{5x^2} - \frac{10x^3}{5x^2} + \frac{5x^2}{5x^2} - \frac{2x}{5x^2}$$
$$= 3x^3 - 2x + 1 - \frac{2}{5x}$$

$$
\begin{array}{r}
5x - 1 + \dfrac{-4}{2x + 3} \\
2x + 3 \overline{)\,10x^2 + 13x - 7} \\
\underline{10x^2 + 15x} \\
-2x - 7 \\
\underline{-2x - 3} \\
-4
\end{array}
$$

A shortcut method called **synthetic division** may be used to divide a polynomial by a binomial of the form $x - c$.

Use synthetic division to divide $2x^3 - x^2 - 8x - 1$ by $x - 2$.

$$
\begin{array}{r|rrrr}
2 & 2 & -1 & -8 & -1 \\
  &   & 4 & 6 & -4 \\
\hline
  & 2 & 3 & -2 & -5
\end{array}
$$

The quotient is $2x^2 + 3x - 2 - \dfrac{5}{x - 2}$.

# CHAPTER 5 REVIEW

*(5.1) State the base and the exponent for each expression.*

**1.** $7^9$                           **2.** $(-5)^4$

**3.** $-5^4$                          **4.** $x^6$

*Evaluate each expression.*

**5.** $8^3$                           **6.** $(-6)^2$

**7.** $-6^2$                          **8.** $-4^3 - 4^0$

**9.** $(3b)^0$                        **10.** $\dfrac{8b}{8b}$

*Simplify each expression.*

**11.** $y^2 \cdot y^7$                **12.** $x^9 \cdot x^5$

**13.** $(2x^5)(-3x^6)$                **14.** $(-5y^3)(4y^4)$

**15.** $(x^4)^2$                      **16.** $(y^3)^5$

**17.** $(3y^6)^4$                     **18.** $(2x^3)^3$

**19.** $\dfrac{x^9}{x^4}$             **20.** $\dfrac{z^{12}}{z^5}$

**21.** $\dfrac{a^5 b^4}{ab}$          **22.** $\dfrac{x^4 y^6}{xy}$

**23.** $\dfrac{12xy^6}{3x^4 y^{10}}$   **24.** $\dfrac{2x^7 y^8}{8xy^2}$

**25.** $5a^7(2a^4)^3$                 **26.** $(2x)^2(9x)$

**27.** $(-5a)^0 + 7^0 + 8^0$          **28.** $8x^0 + 9^0$

*Simplify the given expression and choose the correct result.*

**29.** $\left(\dfrac{3x^4}{4y}\right)^3$

   **a.** $\dfrac{27x^{64}}{64y^3}$   **b.** $\dfrac{27x^{12}}{64y^3}$   **c.** $\dfrac{9x^{12}}{12y^3}$   **d.** $\dfrac{3x^{12}}{4y^3}$

**30.** $\left(\dfrac{5a^6}{b^3}\right)^2$

   **a.** $\dfrac{10a^{12}}{b^6}$   **b.** $\dfrac{25a^{36}}{b^9}$   **c.** $\dfrac{25a^{12}}{b^6}$   **d.** $25a^{12}b^6$

*(5.2) Find the degree of each term.*

**31.** $-5x^4 y^3$                    **32.** $10x^3 y^2 z$

**33.** $35a^5 bc^2$                   **34.** $95xyz$

*Find the degree of each polynomial.*

**35.** $y^5 + 7x - 8x^4$

**36.** $9y^2 + 30y + 25$

**37.** $-14x^2 y - 28x^2 y^3 - 42x^2 y^2$

**38.** $6x^2 y^2 z^2 + 5x^2 y^3 - 12xyz$

**39. a.** Complete the table for the polynomial
   $x^2 y^2 + 5x^2 - 7y^2 + 11xy - 1$.

| Term | Numerical Coefficient | Degree of Term |
|------|----------------------|----------------|
| $x^2 y^2$ | | |
| $5x^2$ | | |
| $-7y^2$ | | |
| $11xy$ | | |
| $-1$ | | |

   **b.** What is the degree of the polynomial?

△ **40.** The surface area of a box with a square base and a height of 5 units is given by the polynomial $2x^2 + 20x$. Fill in the table below by evaluating $2x^2 + 20x$ for the given values of $x$.

| $x$ | 1 | 3 | 5.1 | 10 |
|-----|---|---|-----|-----|
| $2x^2 + 20x$ | | | | |

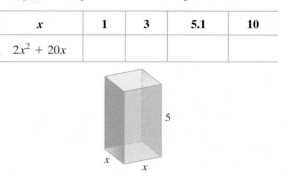

*Combine like terms in each expression.*

**41.** $6a^2 + 4a + 9a^2$

**42.** $21x^2 + 3x + x^2 + 6$

**43.** $4a^2 b - 3b^2 - 8q^2 - 10a^2 b + 7q^2$

**44.** $2s^{14} + 3s^{13} + 12s^{12} - s^{10}$

*Add or subtract as indicated.*

**45.** $(3x^2 + 2x + 6) + (5x^2 + x)$

**46.** $(2x^5 + 3x^4 + 4x^3 + 5x^2) + (4x^2 + 7x + 6)$

**47.** $(-5y^2 + 3) - (2y^2 + 4)$

**48.** $(3x^2 - 7xy + 7y^2) - (4x^2 - xy + 9y^2)$

**49.** Subtract $(3x - y)$ from $(7x - 14y)$.

**50.** Subtract $(4x^2 + 8x - 7)$ from the sum of $(x^2 + 7x + 9)$ and $(x^2 + 4)$.

*If $P(x) = 9x^2 - 7x + 8$, find the following.*

**51.** $P(6)$                         **52.** $P(-2)$

**53.** Find the perimeter of the rectangle.

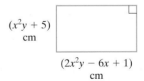

$(x^2 y + 5)$ cm

$(2x^2 y - 6x + 1)$ cm

**54.** With the ownership of computers growing rapidly, the market for new software is also increasing. The revenue for software publishers (in millions of dollars) in the United States from 2001 to 2006 can be represented by the polynomial function $f(x) = 754x^2 - 228x + 80{,}134$ where $x$ is the number of years since 2001. Use this model to predict the revenues from software sales in 2009. (*Source:* Software & Information Industry Association)

*(5.3) Multiply each expression.*

**55.** $4(2a + 7)$

**56.** $9(6a - 3)$

**57.** $-7x(x^2 + 5)$

**58.** $-8y(4y^2 - 6)$

**59.** $(3a^3 - 4a + 1)(-2a)$

**60.** $(6b^3 - 4b + 2)(7b)$

**61.** $(2x + 2)(x - 7)$

**62.** $(2x - 5)(3x + 2)$

**63.** $(x - 9)^2$

**64.** $(x - 12)^2$

**65.** $(4a - 1)(a + 7)$

**66.** $(6a - 1)(7a + 3)$

**67.** $(5x + 2)^2$

**68.** $(3x + 5)^2$

**69.** $(x + 7)(x^3 + 4x - 5)$

**70.** $(x + 2)(x^5 + x + 1)$

**71.** $(x^2 + 2x + 4)(x^2 + 2x - 4)$

**72.** $(x^3 + 4x + 4)(x^3 + 4x - 4)$

**73.** $(x + 7)^3$

**74.** $(2x - 5)^3$

*(5.4)* *Use special products to multiply each of the following.*

**75.** $(x + 7)^2$

**76.** $(x - 5)^2$

**77.** $(3x - 7)^2$

**78.** $(4x + 2)^2$

**79.** $(5x - 9)^2$

**80.** $(5x + 1)(5x - 1)$

**81.** $(7x + 4)(7x - 4)$

**82.** $(a + 2b)(a - 2b)$

**83.** $(2x - 6)(2x + 6)$

**84.** $(4a^2 - 2b)(4a^2 + 2b)$

*Express each as a product of polynomials in x. Then multiply and simplify.*

△ **85.** Find the area of the square if its side is $(3x - 1)$ meters.

$(3x - 1)$ meters

△ **86.** Find the area of the rectangle.

$(x - 1)$ miles
$(5x + 2)$ miles

*(5.5) Simplify each expression.*

**87.** $7^{-2}$   **88.** $-7^{-2}$   **89.** $2x^{-4}$   **90.** $(2x)^{-4}$

**91.** $\left(\dfrac{1}{5}\right)^{-3}$   **92.** $\left(\dfrac{-2}{3}\right)^{-2}$

**93.** $2^0 + 2^{-4}$   **94.** $6^{-1} - 7^{-1}$

*Simplify each expression. Write each answer using positive exponents only.*

**95.** $\dfrac{x^5}{x^{-3}}$   **96.** $\dfrac{z^4}{z^{-4}}$

**97.** $\dfrac{r^{-3}}{r^{-4}}$   **98.** $\dfrac{y^{-2}}{y^{-5}}$

**99.** $\left(\dfrac{bc^{-2}}{bc^{-3}}\right)^4$   **100.** $\left(\dfrac{x^{-3}y^{-4}}{x^{-2}y^{-5}}\right)^{-3}$

**101.** $\dfrac{x^{-4}y^{-6}}{x^2y^7}$   **102.** $\dfrac{a^5b^{-5}}{a^{-5}b^5}$

**103.** $a^{6m}a^{5m}$   **104.** $\dfrac{(x^{5+h})^3}{x^5}$

**105.** $(3xy^{2z})^3$   **106.** $a^{m+2}a^{m+3}$

*Write each number in scientific notation.*

**107.** 0.00027   **108.** 0.8868

**109.** 80,800,000   **110.** 868,000

**111.** Google.com is an Internet search engine that handles 91,000,000 searches every day. Write 91,000,000 in scientific notation. (*Source:* Google, Inc.)

**112.** The approximate diameter of the Milky Way galaxy is 150,000 light years. Write this number in scientific notation. (*Source:* NASA IMAGE/POETRY Education and Public Outreach Program)

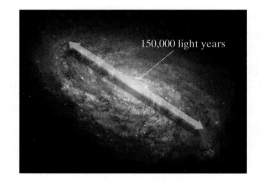

150,000 light years

*Write each number in standard form.*

**113.** $8.67 \times 10^5$   **114.** $3.86 \times 10^{-3}$

**115.** $8.6 \times 10^{-4}$   **116.** $8.936 \times 10^5$

**117.** The volume of the planet Jupiter is $1.43128 \times 10^{15}$ cubic kilometers. Write this number in standard notation. (*Source:* National Space Science Data Center)

**118.** An angstrom is a unit of measure, equal to $1 \times 10^{-10}$ meter, used for measuring wavelengths or the diameters of atoms. Write this number in standard notation. (*Source:* National Institute of Standards and Technology)

*Simplify. Express each result in standard form.*

**119.** $(8 \times 10^4)(2 \times 10^{-7})$

**120.** $\dfrac{8 \times 10^4}{2 \times 10^{-7}}$

**(5.6)** *Divide.*

**121.** $\dfrac{x^2 + 21x + 49}{7x^2}$

**122.** $\dfrac{5a^3b - 15ab^2 + 20ab}{-5ab}$

**123.** $(a^2 - a + 4) \div (a - 2)$

**124.** $(4x^2 + 20x + 7) \div (x + 5)$

**125.** $\dfrac{a^3 + a^2 + 2a + 6}{a - 2}$

**126.** $\dfrac{9b^3 - 18b^2 + 8b - 1}{3b - 2}$

**127.** $\dfrac{4x^4 - 4x^3 + x^2 + 4x - 3}{2x - 1}$

**128.** $\dfrac{-10x^2 - x^3 - 21x + 18}{x - 6}$

△ **129.** The area of the rectangle below is $(15x^3 - 3x^2 + 60)$ square feet. If its length is $3x^2$ feet, find its width.

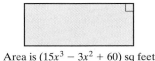

Area is $(15x^3 - 3x^2 + 60)$ sq feet

△ **130.** The perimeter of the equilateral triangle below is $(21a^3b^6 + 3a - 3)$ units. Find the length of a side.

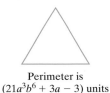

Perimeter is
$(21a^3b^6 + 3a - 3)$ units

**(5.7)** *Use synthetic division to find each quotient.*

**131.** $(3x^3 + 12x - 4) \div (x - 2)$

**132.** $(3x^3 + 2x^2 - 4x - 1) \div \left(x + \dfrac{3}{2}\right)$

**133.** $(x^5 - 1) \div (x + 1)$

**134.** $(x^3 - 81) \div (x - 3)$

**135.** $(x^3 - x^2 + 3x^4 - 2) \div (x - 4)$

**136.** $(3x^4 - 2x^2 + 10) \div (x + 2)$

*If $P(x) = 3x^5 - 9x + 7$, use the remainder theorem to find the following.*

**137.** $P(4)$

**138.** $P(-5)$

## MIXED REVIEW

*Evaluate.*

**139.** $\left(-\dfrac{1}{2}\right)^3$

*Simplify each expression. Write each answer using positive exponents only.*

**140.** $(4xy^2)(x^3y^5)$

**141.** $\dfrac{18x^9}{27x^3}$

**142.** $\left(\dfrac{3a^4}{b^2}\right)^3$

**143.** $(2x^{-4}y^3)^{-4}$

**144.** $\dfrac{a^{-3}b^6}{9^{-1}a^{-5}b^{-2}}$

*Perform the indicated operations and simplify.*

**145.** $(6x + 2) + (5x - 7)$

**146.** $(-y^2 - 4) + (3y^2 - 6)$

**147.** $(8y^2 - 3y + 1) - (3y^2 + 2)$

**148.** $(5x^2 + 2x - 6) - (-x - 4)$

**149.** $4x(7x^2 + 3)$

**150.** $(2x + 5)(3x - 2)$

**151.** $(x - 3)(x^2 + 4x - 6)$

**152.** $(7x - 2)(4x - 9)$

*Use special products to multiply.*

**153.** $(5x + 4)^2$

**154.** $(6x + 3)(6x - 3)$

*Divide.*

**155.** $\dfrac{8a^4 - 2a^3 + 4a - 5}{2a^3}$

**156.** $\dfrac{x^2 + 2x + 10}{x + 5}$

**157.** $\dfrac{4x^3 + 8x^2 - 11x + 4}{2x - 3}$

**CHAPTER 5 TEST** TEST PREP [VIDEO]  Remember to use your Chapter Test Prep Video CD to see the fully worked-out solutions to any of the exercises you want to review.

*Evaluate each expression.*

**1.** $2^5$     **2.** $(-3)^4$     **3.** $-3^4$     **4.** $4^{-3}$

**7.** $\dfrac{r^{-8}}{r^{-3}}$

**8.** $\left(\dfrac{x^2y^3}{x^3y^{-4}}\right)^2$

*Simplify each exponential expression. Write the result using only positive exponents.*

**5.** $(3x^2)(-5x^9)$     **6.** $\dfrac{y^7}{y^2}$

**9.** $\dfrac{6^2x^{-4}y^{-1}}{6^3x^{-3}y^7}$

*Express each number in scientific notation.*

**10.** 563,000

**11.** 0.0000863

*Write each number in standard form.*

**12.** $1.5 \times 10^{-3}$

**13.** $6.23 \times 10^4$

**14.** Simplify. Write the answer in standard form.

$$(1.2 \times 10^5)(3 \times 10^{-7})$$

**15. a.** Complete the table for the polynomial $4xy^2 + 7xyz + x^3y - 2$.

| Term | Numerical Coefficient | Degree of Term |
|------|-----------------------|----------------|
| $4xy^2$ | | |
| $7xyz$ | | |
| $x^3y$ | | |
| $-2$ | | |

**b.** What is the degree of the polynomial?

**16.** Simplify by combining like terms.

$$5x^2 + 4xy - 7x^2 + 11 + 8xy$$

*Perform each indicated operation.*

**17.** $(8x^3 + 7x^2 + 4x - 7) + (8x^3 - 7x - 6)$

**18.** $5x^3 + x^2 + 5x - 2 - (8x^3 - 4x^2 + x - 7)$

**19.** Subtract $(4x + 2)$ from the sum of $(8x^2 + 7x + 5)$ and $(x^3 - 8)$.

*Multiply.*

**20.** $(3x + 7)(x^2 + 5x + 2)$

**21.** $3x^2(2x^2 - 3x + 7)$

**22.** $(x + 7)(3x - 5)$

**23.** $\left(3x - \dfrac{1}{5}\right)\left(3x + \dfrac{1}{5}\right)$

**24.** $(4x - 2)^2$

**25.** $(x^2 - 9b)(x^2 + 9b)$

**26.** The height of the Bank of China in Hong Kong is 1001 feet. Neglecting air resistance, the height of an object dropped from this building at time $t$ seconds is given by the polynomial $-16t^2 + 1001$. Find the height of the object at the given times below.

| $t$ | 0 seconds | 1 second | 3 seconds | 5 seconds |
|-----|-----------|----------|-----------|-----------|
| $-16t^2 + 1001$ | | | | |

△ **27.** Find the area of the top of the table. Express the area as a product, then multiply and simplify.

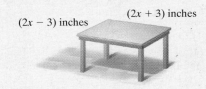

$(2x - 3)$ inches    $(2x + 3)$ inches

*Divide.*

**28.** $\dfrac{4x^2 + 24xy - 7x}{8xy}$

**29.** $(x^2 + 7x + 10) \div (x + 5)$

**30.** $\dfrac{27x^3 - 8}{3x + 2}$

**31.** A pebble is hurled upward from the top of the Canada Trust Tower, which is 880 feet tall, with an initial velocity of 96 feet per second. Neglecting air resistance, the height $h(t)$ of the pebble after $t$ seconds is given by the polynomial function

$$h(t) = -16t^2 + 96t + 880$$

**a.** Find the height of the pebble when $t = 1$.

**b.** Find the height of the pebble when $t = 5.1$.

**c.** When will the pebble hit the ground?

**32.** Use synthetic division to divide $(4x^4 - 3x^3 - x - 1)$ by $(x + 3)$.

**33.** If $P(x) = 4x^4 + 7x^2 - 2x - 5$, use the remainder theorem to find $P(-2)$.

# CHAPTER 5 CUMULATIVE REVIEW

**1.** Tell whether each statement is true or false.

**a.** $8 \geq 8$

**b.** $8 \leq 8$

**c.** $23 \leq 0$

**d.** $23 \geq 0$

**2.** Find the absolute value of each number.

**a.** $|-7.2|$

**b.** $|0|$

**c.** $\left|-\dfrac{1}{2}\right|$

**3.** Divide. Write all quotients in lowest terms.

**a.** $\dfrac{4}{5} \div \dfrac{5}{16}$

**b.** $\dfrac{7}{10} \div 14$

**c.** $\dfrac{3}{8} \div \dfrac{3}{10}$

**4.** Multiply. Write products in lowest terms.

**a.** $\dfrac{3}{4} \cdot \dfrac{7}{21}$

**b.** $\dfrac{1}{2} \cdot 4\dfrac{5}{6}$

**5.** Evaluate the following:

**a.** $3^2$

**b.** $5^3$

**c.** $2^4$

**d.** $7^1$

**e.** $\left(\dfrac{3}{7}\right)^2$

**6.** Evaluate $\dfrac{2x - 7y}{x^2}$ for $x = 5$ and $y = 1$.

**7.** Add.

 **a.** $-3 + (-7)$    **b.** $-1 + (-20)$

 **c.** $-2 + (-10)$

**8.** Simplify: $8 + 3(2 \cdot 6 - 1)$

**9.** Subtract 8 from $-4$.

**10.** Is $x = 1$ a solution of $5x^2 + 2 = x - 8$.

**11.** Find the reciprocal of each number.

 **a.** 22    **b.** $\dfrac{3}{16}$

 **c.** $-10$    **d.** $-\dfrac{9}{13}$

**12.** Subtract:

 **a.** $7 - 40$    **b.** $-5 - (-10)$

**13.** Use an associative property to complete each statement.

 **a.** $5 + (4 + 6) = $ _____

 **b.** $(-1 \cdot 2) \cdot 5 = $ _____

**14.** Simplify: $\dfrac{4(-3) + (-8)}{5 + (-5)}$

**15.** Simplify each expression.

 **a.** $10 + (x + 12)$

 **b.** $-3(7x)$

**16.** Use the distributive property to write $-2(x + 3y - z)$ without parentheses.

**17.** Find each product by using the distributive property to remove parentheses.

 **a.** $5(x + 2)$

 **b.** $-2(y + 0.3z - 1)$

 **c.** $-(x + y - 2z + 6)$

**18.** Simplify: $2(6x - 1) - (x - 7)$

**19.** Solve $x - 7 = 10$ for $x$.

**20.** Write the phrase as an algebraic expression: double a number, subtracted from the sum of a number and seven

**21.** Solve: $\dfrac{5}{2}x = 15$

**22.** Solve: $2x + \dfrac{1}{8} = x - \dfrac{3}{8}$

**23.** Twice a number, added to seven, is the same as three subtracted from the number. Find the number.

**24.** Solve: $10 = 5j - 2$

**25.** Twice the sum of a number and 4 is the same as four times the number, decreased by 12. Find the number.

**26.** Solve: $\dfrac{7x + 5}{3} = x + 3$

**27.** The length of a rectangular road sign is 2 feet less than three times its width. Find the dimensions if the perimeter is 28 feet.

**28.** Graph $x < 5$ and write in interval notation.

**29.** Solve $F = \dfrac{9}{5}C + 32$ for $C$.

**30.** Find the slope of each line.

 **a.** $x = -1$

 **b.** $y = 7$

**31.** Graph $2 < x \le 4$.

**32.** Recall that the grade of a road is its slope written as a percent. Find the grade of the road shown.

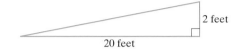

**33.** Complete the following ordered-pair solutions for the equation $3x + y = 12$.

 **a.** $(0, \ )$

 **b.** $( \ , 6)$

 **c.** $(-1, \ )$

**34.** Solve the system: $\begin{cases} 3x + 2y = -8 \\ 2x - 6y = -9 \end{cases}$

**35.** Graph the linear equation $2x + y = 5$.

**36.** Solve the system: $\begin{cases} x = -3y + 3 \\ 2x + 9y = 5 \end{cases}$

**37.** Graph $x = 2$.

**38.** Evaluate.

 **a.** $(-5)^2$

 **b.** $-5^2$

 **c.** $2 \cdot 5^2$

**39.** Find the slope of the line $x = 5$.

**40.** Simplify: $\dfrac{(z^2)^3 \cdot z^7}{z^9}$

**41.** Subtract: $(2x^3 + 8x^2 - 6x) - (2x^3 - x^2 + 1)$.

**42.** Subtract $(5y^2 - 6) - (y^2 + 2)$.

**43.** Use the product rule to simplify $(2x^2)(-3x^5)$.

**44.** Find the value of $-x^2$ when

 **a.** $x = 2$    **b.** $x = -2$

**45.** Add $(11x^3 - 12x^2 + x - 3)$ and $(x^3 - 10x + 5)$.

**46.** Multiply $(10x^2 - 3)(10x^2 + 3)$.

**47.** Multiply $(2x - y)^2$.

**48.** Multiply $(10x^2 + 3)^2$.

**49.** Divide $6m^2 + 2m$ by $2m$.

**50.** Evaluate.

 **a.** $5^{-1}$

 **b.** $7^{-2}$

# Factoring Polynomials

In Chapter 5, you learned how to multiply polynomials. This chapter deals with an operation that is the reverse process of multiplying, called *factoring*. Factoring is an important algebraic skill because this process allows us to write a sum as a product.

At the end of this chapter, we use factoring to help us solve equations other than linear equations, and in Chapter 7 we use factoring to simplify and perform arithmetic operations on rational expressions.

The majority of college students hold credit cards. According to the Nellie May Corporation, 56% of final year undergraduate students carry four or more cards with an average balance of $2864. The circle graph below shows when students with credit cards obtained a card.

American Consumer Credit Counseling (ACCC) released the following guidelines to help protect college students from the lasting effects of credit card debt.

- Pay at least the minimum payment on all bills before the due date.
- Keep a budget. Record all income and record all outgoings for at least two months so you know where your money is going.
- Avoid borrowing money to pay off other creditors.
- Only carry as much cash as your weekly budget allows.

In Section 6.5, Exercises 77 through 80, you will have the opportunity to calculate percents of students with credit cards during selected years.

**76% of Undergraduate College Students Have Credit Cards—
When Do These Students Obtain a Card?**

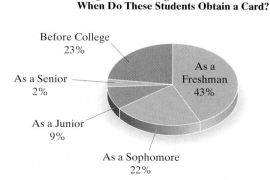

Before College 23%

As a Senior 2%

As a Junior 9%

As a Sophomore 22%

As a Freshman 43%

# 6.1 THE GREATEST COMMON FACTOR AND FACTORING BY GROUPING

**OBJECTIVES**

1  Find the greatest common factor of a list of integers.

2  Find the greatest common factor of a list of terms.

3  Factor out the greatest common factor from a polynomial.

4  Factor a polynomial by grouping.

In the product $2 \cdot 3 = 6$, the numbers 2 and 3 are called **factors** of 6 and $2 \cdot 3$ is a **factored form** of 6. This is true of polynomials also. Since $(x + 2)(x + 3) = x^2 + 5x + 6$, then $(x + 2)$ and $(x + 3)$ are factors of $x^2 + 5x + 6$, and $(x + 2)(x + 3)$ is a factored form of the polynomial.

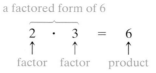

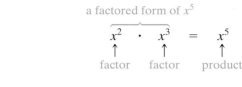

a factored form of $x^2 + 5x + 6$

$$\underset{\substack{\uparrow \\ \text{factor}}}{(x + 2)}\underset{\substack{\uparrow \\ \text{factor}}}{(x + 3)} = \underset{\substack{\uparrow \\ \text{product}}}{x^2 + 5x + 6}$$

Do you see that factoring is the reverse process of multiplying?

$$x^2 + 5x + 6 \underset{\text{multiplying}}{\overset{\text{factoring}}{=}} (x + 2)(x + 3)$$

**Concept Check** ☑

Multiply: $2(x - 4)$

What do you think the result of factoring $2x - 8$ would be? Why?

The first step in factoring a polynomial is to see whether the terms of the polynomial have a common factor. If there is one, we can write the polynomial as a product by **factoring out** the common factor. We will usually factor out the **greatest common factor (GCF).**

**OBJECTIVE 1 ▶ Finding the greatest common factor of a list of integers.** The GCF of a list of integers is the largest integer that is a factor of all the integers in the list. For example, the GCF of 12 and 20 is 4 because 4 is the largest integer that is a factor of both 12 and 20. With large integers, the GCF may not be easily found by inspection. When this happens, use the following steps.

> **Finding the GCF of a List of Integers**
>
> **STEP 1.** Write each number as a product of prime numbers.
>
> **STEP 2.** Identify the common prime factors.
>
> **STEP 3.** The product of all common prime factors found in Step 2 is the greatest common factor. If there are no common prime factors, the greatest common factor is 1.

Recall from Section 1.3 that a prime number is a whole number other than 1, whose only factors are 1 and itself.

**Answers to Concept Check:**

$2x - 8$; The result would be $2(x - 4)$ because factoring is the reverse process of multiplying.

**EXAMPLE 1** Find the GCF of each list of numbers.

**a.** 28 and 40 **b.** 55 and 21 **c.** 15, 18, and 66

*Solution*

**a.** Write each number as a product of primes.

$$28 = 2 \cdot 2 \cdot 7 = 2^2 \cdot 7$$
$$40 = 2 \cdot 2 \cdot 2 \cdot 5 = 2^3 \cdot 5$$

There are two common factors, each of which is 2, so the GCF is

$$\text{GCF} = 2 \cdot 2 = 4$$

**b.** $55 = 5 \cdot 11$
$21 = 3 \cdot 7$

There are no common prime factors; thus, the GCF is 1.

**c.** $15 = 3 \cdot 5$
$18 = 2 \cdot 3 \cdot 3 = 2 \cdot 3^2$
$66 = 2 \cdot 3 \cdot 11$

The only prime factor common to all three numbers is 3, so the GCF is

$$\text{GCF} = 3$$

**PRACTICE**
**1** Find the GCF of each list of numbers.

**a.** 36 and 42 **b.** 35 and 44 **c.** 12, 16, and 40

**OBJECTIVE 2 ▶ Finding the greatest common factor of a list of terms.** The greatest common factor of a list of variables raised to powers is found in a similar way. For example, the GCF of $x^2$, $x^3$, and $x^5$ is $x^2$ because each term contains a factor of $x^2$ and no higher power of $x$ is a factor of each term.

$$x^2 = x \cdot x$$
$$x^3 = x \cdot x \cdot x$$
$$x^5 = x \cdot x \cdot x \cdot x \cdot x$$

There are two common factors, each of which is $x$, so the GCF $= x \cdot x$ or $x^2$.

From this example, we see that **the GCF of a list of common variables raised to powers is the variable raised to the smallest exponent in the list.**

**EXAMPLE 2** Find the GCF of each list of terms.

**a.** $x^3$, $x^7$, and $x^5$ **b.** $y$, $y^4$, and $y^7$

*Solution*

**a.** The GCF is $x^3$, since 3 is the smallest exponent to which $x$ is raised.
**b.** The GCF is $y^1$ or $y$, since 1 is the smallest exponent on $y$.

**PRACTICE**
**2** Find the GCF of each list of terms.

**a.** $y^7$, $y^4$, and $y^6$ **b.** $x$, $x^4$, and $x^2$

In general, the **greatest common factor (GCF) of a list of terms** is the product of the GCF of the numerical coefficients and the GCF of the variable factors.

**EXAMPLE 3**    Find the GCF of each list of terms.

**a.** $6x^2$, $10x^3$, and $-8x$         **b.** $-18y^2$, $-63y^3$, and $27y^4$         **c.** $a^3b^2$, $a^5b$, and $a^6b^2$

*Solution*

**a.**  $6x^2 = 2 \cdot 3 \cdot x^2$
   $10x^3 = 2 \cdot 5 \cdot x^3$
   $-8x = -1 \cdot 2 \cdot 2 \cdot 2 \cdot x^1$          $\rightarrow$ The GCF of $x^2$, $x^3$, and $x^1$ is $x^1$ or $x$.
   $\text{GCF} = 2 \cdot x^1$   or   $2x$

**b.**  $-18y^2 = -1 \cdot 2 \cdot 3 \cdot 3 \cdot y^2$
   $-63y^3 = -1 \cdot 3 \cdot 3 \cdot 7 \cdot y^3$          $\rightarrow$ The GCF of $y^2$, $y^3$, and $y^4$ is $y^2$.
   $27y^4 = 3 \cdot 3 \cdot 3 \cdot y^4$
   $\text{GCF} = 3 \cdot 3 \cdot y^2$   or   $9y^2$

**c.** The GCF of $a^3$, $a^5$, and $a^6$ is $a^3$.
   The GCF of $b^2$, $b$, and $b^2$ is $b$. Thus,
   the GCF of $a^3b^2$, $a^5b$, and $a^6b^2$ is $a^3b$.

> ▶ **Helpful Hint**
>
> Remember that the GCF of a list of terms contains the smallest exponent on each common variable.
>
>                   Smallest exponent on $x$.
>                           |
> The GCF of $x^5y^6$, $x^2y^7$ and $x^3y^4$ is $x^2y^4$.
>                           |
>           Smallest exponent on $y$.

**PRACTICE**
**3**    Find the GCF of each list of terms.

**a.** $5y^4$, $15y^2$, and $-20y^3$         **b.** $4x^2$, $x^3$, and $3x^8$         **c.** $a^4b^2$, $a^3b^5$, and $a^2b^3$

---

**OBJECTIVE 3** ▶ **Factoring out the greatest common factor.**  The first step in factoring a polynomial is to find the GCF of its terms. Once we do so, we can write the polynomial as a product by **factoring out** the GCF.

   The polynomial $8x + 14$, for example, contains two terms: $8x$ and $14$. The GCF of these terms is 2. We factor out 2 from each term by writing each term as a product of 2 and the term's remaining factors.

$$8x + 14 = 2 \cdot 4x + 2 \cdot 7$$

Using the distributive property, we can write

$$8x + 14 = 2 \cdot 4x + 2 \cdot 7$$
$$= 2(4x + 7)$$

Thus, a factored form of $8x + 14$ is $2(4x + 7)$. We can check by multiplying:

$$2(4x + 7) = 2 \cdot 4x + 2 \cdot 7 = 8x + 14.$$

> ▶ **Helpful Hint**
>
> A factored form of $8x + 14$ is *not*
>
> $$2 \cdot 4x + 2 \cdot 7$$
>
> Although the *terms* have been factored (written as a product), the *polynomial* $8x + 14$ has not been factored (written as a product). A factored form of $8x + 14$ is the *product* $2(4x + 7)$.

**Concept Check** ☑

Which of the following is/are factored form(s) of $7t + 21$?

**a.** $7$         **b.** $7 \cdot t + 7 \cdot 3$         **c.** $7(t + 3)$         **d.** $7(t + 21)$

**EXAMPLE 4**   Factor each polynomial by factoring out the GCF.

**a.** $6t + 18$      **b.** $y^5 - y^7$

*Solution*

**a.** The GCF of terms $6t$ and $18$ is $6$.

$$6t + 18 = 6 \cdot t + 6 \cdot 3$$
$$= 6(t + 3) \qquad \text{Apply the distributive property.}$$

Our work can be checked by multiplying $6$ and $(t + 3)$.

$$6(t + 3) = 6 \cdot t + 6 \cdot 3 = 6t + 18, \text{ the original polynomial.}$$

**b.** The GCF of $y^5$ and $y^7$ is $y^5$. Thus,

$$y^5 - y^7 = y^5(1) - y^5(y^2)$$
$$= y^5(1 - y^2)$$

> ▶ **Helpful Hint**
> Don't forget the 1.

**PRACTICE**
**4**   Factor each polynomial by factoring out the GCF.

**a.** $4t + 12$         **b.** $y^8 + y^4$

---

**EXAMPLE 5**   Factor: $-9a^5 + 18a^2 - 3a$

*Solution*

$$-9a^5 + 18a^2 - 3a = (3a)(-3a^4) + (3a)(6a) + (3a)(-1)$$
$$= 3a(-3a^4 + 6a - 1)$$

> ▶ **Helpful Hint**
> Don't forget the $-1$.

**PRACTICE**
**5**   Factor $-8b^6 + 16b^4 - 8b^2$.

---

In Example 5 we could have chosen to factor out a $-3a$ instead of $3a$. If we factor out a $-3a$, we have

$$-9a^5 + 18a^2 - 3a = (-3a)(3a^4) + (-3a)(-6a) + (-3a)(1)$$
$$= -3a(3a^4 - 6a + 1)$$

> ▶ **Helpful Hint**
> Notice the changes in signs when factoring out $-3a$.

**EXAMPLES**   Factor.

**6.** $6a^4 - 12a = 6a(a^3 - 2)$

**7.** $\dfrac{3}{7}x^4 + \dfrac{1}{7}x^3 - \dfrac{5}{7}x^2 = \dfrac{1}{7}x^2(3x^2 + x - 5)$

**8.** $15p^2q^4 + 20p^3q^5 + 5p^3q^3 = 5p^2q^3(3q + 4pq^2 + p)$

**PRACTICES**
**6-8**   Factor.

**6.** $5x^4 - 20x$        **7.** $\dfrac{5}{9}z^5 + \dfrac{1}{9}z^4 - \dfrac{2}{9}z^3$        **8.** $8a^2b^4 - 20a^3b^3 + 12ab^3$

**EXAMPLE 9**   Factor: $5(x + 3) + y(x + 3)$

_Solution_   The binomial $(x + 3)$ is the greatest common factor. Use the distributive property to factor out $(x + 3)$.

$$5(x + 3) + y(x + 3) = (x + 3)(5 + y) \qquad \square$$

PRACTICE
**9**   Factor $8(y - 2) + x(y - 2)$.

_Be careful_

**EXAMPLE 10**   Factor: $3m^2n(a + b) - (a + b)$

_Solution_   The greatest common factor is $(a + b)$.

$$3m^2n(a + b) - 1(a + b) = (a + b)(3m^2n - 1) \qquad \square$$

PRACTICE
**10**   Factor $7xy^3(p + q) - (p + q)$

**OBJECTIVE 4 ▶ Factoring by grouping.**   Once the GCF is factored out, we can often continue to factor the polynomial, using a variety of techniques. We discuss here a technique for factoring polynomials called **grouping.**

**EXAMPLE 11**   Factor $xy + 2x + 3y + 6$ by grouping. Check by multiplying.

_Solution_   The GCF of the first two terms is $x$, and the GCF of the last two terms is 3.

$$xy + 2x + 3y + 6 = (xy + 2x) + (3y + 6) \qquad \text{Group terms.}$$
$$= x(y + 2) + 3(y + 2) \qquad \text{Factor out GCF from each grouping.}$$

> ▶ Helpful Hint
>
> Notice that this form, $x(y + 2) + 3(y + 2)$, is _not_ a factored form of the original polynomial. It is a sum, not a product.

Next we factor out the common binomial factor, $(y + 2)$.

$$x(y + 2) + 3(y + 2) = (y + 2)(x + 3)$$

Now the result is a factored form because it is a product. We were able to write the polynomial as a product because of the common binomial factor, $(y + 2)$, that appeared. If this does not happen, try rearranging the terms of the original polynomial.

**Check:**   Multiply $(y + 2)$ by $(x + 3)$.

$$(y + 2)(x + 3) = xy + 2x + 3y + 6,$$

the original polynomial.
Thus, the factored form of $xy + 2x + 3y + 6$ is the product $(y + 2)(x + 3)$. $\qquad \square$

PRACTICE
**11**   Factor $xy + 3y + 4x + 12$ by grouping. Check by multiplying.

You may want to try these steps when factoring by grouping.

> **To Factor a Four-Term Polynomial by Grouping**
>
> **STEP 1.** Group the terms in two groups of two terms so that each group has a common factor.
>
> **STEP 2.** Factor out the GCF from each group.
>
> **STEP 3.** If there is now a common binomial factor in the groups, factor it out.
>
> **STEP 4.** If not, rearrange the terms and try these steps again.

**EXAMPLES** Factor by grouping.

**12.** $3x^2 + 4xy - 3x - 4y$

$= (3x^2 + 4xy) + (-3x - 4y)$

$= x(3x + 4y) - 1(3x + 4y)$     Factor each group. A $-1$ is factored from the second pair of terms so that there is a common factor, $(3x + 4y)$.

$= (3x + 4y)(x - 1)$     Factor out the common factor, $(3x + 4y)$.

**13.** $2a^2 + 5ab + 2a + 5b$

$= (2a^2 + 5ab) + (2a + 5b)$     Factor each group. An understood 1 is written before $(2a + 5b)$ to help remember that $(2a + 5b)$ is $1(2a + 5b)$.

$= a(2a + 5b) + 1(2a + 5b)$

$= (2a + 5b)(a + 1)$     Factor out the common factor, $(2a + 5b)$.     □

> ▶ **Helpful Hint**
> Notice the factor of 1 is written when $(2a + 5b)$ is factored out.

**PRACTICES**

**12–13**

**12.** Factor $2xy + 3y^2 - 2x - 3y$ by grouping.

**13.** Factor $7a^3 + 5a^2 + 7a + 5$ by grouping.

- - - - - - - - - - - - - - - - - - - - - - - - - - - - - - - - - - - - - - - - - - ▪

**EXAMPLES** Factor by grouping.

**14.** $3xy + 2 - 3x - 2y$

Notice that the first two terms have no common factor other than 1. However, if we rearrange these terms, a grouping emerges that does lead to a common factor.

$3xy + 2 - 3x - 2y$

$= (3xy - 3x) + (-2y + 2)$

$= 3x(y - 1) - 2(y - 1)$     Factor $-2$ from the second group so that there is a common factor $(y - 1)$.

$= (y - 1)(3x - 2)$     Factor out the common factor, $(y - 1)$.

**15.** $5x - 10 + x^3 - x^2 = 5(x - 2) + x^2(x - 1)$

There is no common binomial factor that can now be factored out. No matter how we rearrange the terms, no grouping will lead to a common factor. Thus, this polynomial is not factorable by grouping.     □

**PRACTICES**

**14–15**

**14.** Factor $4xy + 15 - 12x - 5y$ by grouping.

**15.** Factor $9y - 18 + y^3 - 4y^2$.

- - - - - - - - - - - - - - - - - - - - - - - - - - - - - - - - - - - - - - - - - - ▪

> ▶ **Helpful Hint**
> One more reminder: When **factoring** a polynomial, make sure the polynomial is written as a **product.** For example, it is true that
>
> $$3x^2 + 4xy - 3x - 4y = \underbrace{x(3x + 4y) - 1(3x + 4y)}_{\text{but is not a } \textbf{factored form}},$$
>
> since it is a **sum (difference)**, not a **product.** A factored form of $3x^2 + 4xy - 3x - 4y$ is the product $(3x + 4y)(x - 1)$.

Factoring out a greatest common factor first makes factoring by any method easier, as we see in the next example.

**EXAMPLE 16**   Factor: $4ax - 4ab - 2bx + 2b^2$

**Solution**   First, factor out the common factor 2 from all four terms.

$$4ax - 4ab - 2bx + 2b^2$$
$$= 2(2ax - 2ab - bx + b^2) \quad \text{Factor out 2 from all four terms.}$$
$$= 2[2a(x - b) - b(x - b)] \quad \text{Factor each pair of terms. A "}-b\text{" is factored from the second pair so that there is a common factor, } x - b.$$
$$= 2(x - b)(2a - b) \quad \text{Factor out the common binomial.} \qquad \square$$

**PRACTICE**
**16**   Factor $3xy - 3ay - 6ax + 6a^2$

> **▶ Helpful Hint**
>
> Throughout this chapter, we will be factoring polynomials. Even when the instructions do not so state, it is always a good idea to check your answers by multiplying.

## VOCABULARY & READINESS CHECK

*Use the choices below to fill in each blank. Some choices may be used more than once and some may not be used at all.*

greatest common factor      factors      factoring      true      false      least      greatest

1. Since $5 \cdot 4 = 20$, the numbers 5 and 4 are called _____ of 20.
2. The _____ of a list of integers is the largest integer that is a factor of all the integers in the list.
3. The greatest common factor of a list of common variables raised to powers is the variable raised to the _____ exponent in the list.
4. The process of writing a polynomial as a product is called _____.
5. True or false: A factored form of $7x + 21 + xy + 3y$ is $7(x + 3) + y(x + 3)$. _____
6. True or false: A factored form of $3x^3 + 6x + x^2 + 2$ is $3x(x^2 + 2)$. _____

*Write the prime factorization of the following integers.*

7. 14                8. 15

*Write the GCF of the following pairs of integers.*

9. 18, 3          10. 7, 35          11. 20, 15          12. 6, 15

## 6.1 EXERCISE SET

MyMathLab®   |   Math XP PRACTICE   |   WATCH   |   DOWNLOAD   |   READ   |   REVIEW

*Find the GCF for each list. See Examples 1 through 3.*

1. 32, 36
2. 36, 90
3. 18, 42, 84
4. 30, 75, 135
5. 24, 14, 21
6. 15, 25, 27
7. $y^2, y^4, y^7$
8. $x^3, x^2, x^5$
9. $z^7, z^9, z^{11}$
10. $y^8, y^{10}, y^{12}$
11. $x^{10}y^2, xy^2, x^3y^3$
12. $p^7q, p^8q^2, p^9q^3$
13. $14x, 21$
14. $20y, 15$
15. $12y^4, 20y^3$
16. $32x^5, 18x^2$
17. $-10x^2, 15x^3$
18. $-21x^3, 14x$
19. $12x^3, -6x^4, 3x^5$
20. $15y^2, 5y^7, -20y^3$
21. $-18x^2y, 9x^3y^3, 36x^3y$
22. $7x^3y^3, -21x^2y^2, 14xy^4$
23. $20a^6b^2c^8, 50a^7b$
24. $40x^7y^2z, 64x^9y$

*Factor out the GCF from each polynomial. See Examples 4 through 10.*

25. $3a + 6$
26. $18a + 12$
27. $30x - 15$
28. $42x - 7$
29. $x^3 + 5x^2$
30. $y^5 + 6y^4$
31. $6y^4 + 2y^3$
32. $5x^2 + 10x^6$
33. $4x - 8y + 4$
34. $7x + 21y - 7$
35. $6x^3 - 9x^2 + 12x$
36. $12x^3 + 16x^2 - 8x$
37. $a^7b^6 - a^3b^2 + a^2b^5 - a^2b^2$
38. $x^9y^6 + x^3y^5 - x^4y^3 + x^3y^3$
39. $8x^5 + 16x^4 - 20x^3 + 12$
40. $9y^6 - 27y^4 + 18y^2 + 6$

**41.** $\frac{1}{3}x^4 + \frac{2}{3}x^3 - \frac{4}{3}x^5 + \frac{1}{3}x$

**42.** $\frac{2}{5}y^7 - \frac{4}{5}y^5 + \frac{3}{5}y^2 - \frac{2}{5}y$

**43.** $y(x^2 + 2) + 3(x^2 + 2)$

**44.** $x(y^2 + 1) - 3(y^2 + 1)$

**45.** $z(y + 4) - 3(y + 4)$

**46.** $8(x + 2) - y(x + 2)$

**47.** $r(z^2 - 6) + (z^2 - 6)$

**48.** $q(b^3 - 5) + (b^3 - 5)$

*Factor a negative number or a GCF with a negative coefficient from each polynomial. See Example 5.*

**49.** $-2x - 14$

**50.** $-7y - 21$

**51.** $-2x^5 + x^7$

**52.** $-5y^3 + y^6$

**53.** $-6a^4 + 9a^3 - 3a^2$

**54.** $-5m^6 + 10m^5 - 5m^3$

*Factor each four-term polynomial by grouping. See Examples 11 through 16.*

**55.** $x^3 + 2x^2 + 5x + 10$

**56.** $x^3 + 4x^2 + 3x + 12$

**57.** $5x + 15 + xy + 3y$

**58.** $xy + y + 2x + 2$

**59.** $6x^3 - 4x^2 + 15x - 10$

**60.** $16x^3 - 28x^2 + 12x - 21$

**61.** $5m^3 + 6mn + 5m^2 + 6n$

**62.** $8w^2 + 7wv + 8w + 7v$

**63.** $2y - 8 + xy - 4x$

**64.** $6x - 42 + xy - 7y$

**65.** $2x^3 - x^2 + 8x - 4$

**66.** $2x^3 - x^2 - 10x + 5$

**67.** $4x^2 - 8xy - 3x + 6y$

**68.** $5xy - 15x - 6y + 18$

**69.** $5q^2 - 4pq - 5q + 4p$

**70.** $6m^2 - 5mn - 6m + 5n$

**71.** $2x^4 + 5x^3 + 2x^2 + 5x$

**72.** $4y^4 + y^2 + 20y^3 + 5y$

**73.** $12x^2y - 42x^2 - 4y + 14$

**74.** $90 + 15y^2 - 18x - 3xy^2$

**MIXED PRACTICE**

*Factor. See Examples 4 through 16.*

**75.** $32xy - 18x^2$

**76.** $10xy - 15x^2$

**77.** $y(x + 2) - 3(x + 2)$

**78.** $z(y - 4) + 3(y - 4)$

**79.** $14x^3y + 7x^2y - 7xy$

**80.** $5x^3y - 15x^2y + 10xy$

**81.** $28x^3 - 7x^2 + 12x - 3$

**82.** $15x^3 + 5x^2 - 6x - 2$

**83.** $-40x^8y^6 - 16x^9y^5$

**84.** $-21x^3y - 49x^2y^2$

**85.** $6a^2 + 9ab^2 + 6ab + 9b^3$

**86.** $16x^2 + 4xy^2 + 8xy + 2y^3$

**REVIEW AND PREVIEW**

*Multiply. See Section 5.3.*

**87.** $(x + 2)(x + 5)$

**88.** $(y + 3)(y + 6)$

**89.** $(b + 1)(b - 4)$

**90.** $(x - 5)(x + 10)$

*Fill in the chart by finding two numbers that have the given product and sum. The first column is filled in for you.*

|  | | **91.** | **92.** | **93.** | **94.** | **95.** | **96.** |
|---|---|---|---|---|---|---|---|
| *Two Numbers* | 4, 7 | | | | | | |
| *Their Product* | 28 | 12 | 20 | 8 | 16 | −10 | −24 |
| *Their Sum* | 11 | 8 | 9 | −9 | −10 | 3 | −5 |

**CONCEPT EXTENSIONS**

*See the Concept Checks in this section.*

**97.** Which of the following is/are factored form(s) of $8a - 24$?

   **a.** $8 \cdot a - 24$       **b.** $8(a - 3)$

   **c.** $4(2a - 12)$      **d.** $8 \cdot a - 2 \cdot 12$

*Which of the following expressions are factored?*

**98.** $(a + 6)(a + 2)$     **99.** $(x + 5)(x + y)$

**100.** $5(2y + z) - b(2y + z)$   **101.** $3x(a + 2b) + 2(a + 2b)$

**102.** Construct a binomial whose greatest common factor is $5a^3$. (*Hint:* Multiply $5a^3$ by a binomial whose terms contain no common factor other than 1. $5a^3(\square + \square)$.)

**103.** Construct a trinomial whose greatest common factor is $2x^2$. See the hint for Exercise 102.

**104.** Explain how you can tell whether a polynomial is written in factored form.

**105.** Construct a four-term polynomial that can be factored by grouping.

**106.** The number (in millions) of single digital downloads annually in the United States each year during 2003–2005 can be modeled by the polynomial $45x^2 + 95x$, where $x$ is the number of years since 2003. (*Source:* Recording Industry Association of America)

   **a.** Find the number of single digital downloads in 2005. To do so, let $x = 2$ and evaluate $45x^2 + 95x$.

   **b.** Use this expression to predict the number of single digital downloads in 2009.

   **c.** Factor the polynomial $45x^2 + 95x$.

**107.** The number (in thousands) of students who graduated from U.S. high schools each year during 2003–2005 can be modeled by $-8x^2 + 50x + 3020$, where $x$ is the number of years since 2003. (*Source:* National Center for Education Statistics)

   **a.** Find the number of students who graduated from U.S. high schools in 2005. To do so, let $x = 2$ and evaluate $-8x^2 + 50x + 3020$.

   **b.** Use this expression to predict the number of students who graduated from U.S. high schools in 2007.

   **c.** Factor the polynomial $-8x^2 + 50x + 3020$.

*Write an expression for the length of each rectangle. (**Hint:** Factor the area binomial and recall that Area = width·length.)*

△ **110.**  △ **111.**

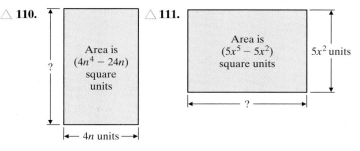

*Write an expression for the area of each shaded region. Then write the expression as a factored polynomial.*

△ **108.**  △ **109.**

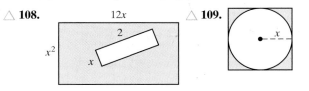

*Factor each polynomial by grouping.*

**112.** $x^{2n} + 2x^n + 3x^n + 6$
   (**Hint:** Don't forget that $x^{2n} = x^n \cdot x^n$.)

**113.** $x^{2n} + 6x^n + 10x^n + 60$

**114.** $3x^{2n} + 21x^n - 5x^n - 35$

**115.** $12x^{2n} - 10x^n - 30x^n + 25$

---

# 6.2 FACTORING TRINOMIALS OF THE FORM $x^2 + bx + c$

**OBJECTIVES**

**1** Factor trinomials of the form $x^2 + bx + c$.

**2** Factor out the greatest common factor and then factor a trinomial of the form $x^2 + bx + c$.

**OBJECTIVE 1 ▶ Factoring trinomials of the form $x^2 + bx + c$.** In this section, we factor trinomials of the form $x^2 + bx + c$, such as

$$x^2 + 4x + 3, \qquad x^2 - 8x + 15, \qquad x^2 + 4x - 12, \qquad r^2 - r - 42$$

Notice that for these trinomials, the coefficient of the squared variable is 1.

Recall that factoring means to write as a product and that factoring and multiplying are reverse processes. Using the FOIL method of multiplying binomials, we have that

$$\begin{aligned} (x + 3)(x + 1) &= x^2 + 1x + 3x + 3 \\ &= x^2 + 4x + 3 \end{aligned}$$

Thus, a factored form of $x^2 + 4x + 3$ is $(x + 3)(x + 1)$.

Notice that the product of the first terms of the binomials is $x \cdot x = x^2$, the first term of the trinomial. Also, the product of the last two terms of the binomials is $3 \cdot 1 = 3$, the third term of the trinomial. The sum of these same terms is $3 + 1 = 4$, the coefficient of the middle term, $x$, of the trinomial.

The product of these numbers is 3.

$$x^2 + 4x + 3 = (x + 3)(x + 1)$$

The sum of these numbers is 4.

Many trinomials, such as the one above, factor into two binomials. To factor $x^2 + 7x + 10$, let's assume that it factors into two binomials and begin by writing two pairs of parentheses. The first term of the trinomial is $x^2$, so we use $x$ and $x$ as the first terms of the binomial factors.

$$x^2 + 7x + 10 = (x + \square)(x + \square)$$

To determine the last term of each binomial factor, we look for two integers whose product is 10 and whose sum is 7. Since our numbers must have a positive product and a positive sum, we list pairs of positive integer factors of 10 only.

| *Positive Factors of 10* | *Sum of Factors* |
|---|---|
| 1, 10 | 1 + 10 = 11 |
| 2, 5 | 2 + 5 = 7 |

The correct pair of numbers is 2 and 5 because their product is 10 and their sum is 7. Now we can fill in the last terms of the binomial factors.

$$x^2 + 7x + 10 = (x + 2)(x + 5)$$

**Check:**   To see if we have factored correctly, multiply.

$$(x + 2)(x + 5) = x^2 + 5x + 2x + 10$$
$$= x^2 + 7x + 10 \qquad \text{Combine like terms.}$$

> ❱ **Helpful Hint**
>
> Since multiplication is commutative, the factored form of $x^2 + 7x + 10$ can be written as either $(x + 2)(x + 5)$ or $(x + 5)(x + 2)$.

---

**Factoring a Trinomial of the Form $x^2 + bx + c$**

The factored form of $x^2 + bx + c$ is

The product of these numbers is $c$.

$$x^2 + bx + c = (x + \square)(x + \square)$$

The sum of these numbers is $b$.

---

**EXAMPLE 1**   Factor: $x^2 + 7x + 12$

*Solution*   We begin by writing the first terms of the binomial factors.

$$(x + \square)(x + \square)$$

Next we look for two numbers whose product is 12 and whose sum is 7. Since our numbers must have a positive product and a positive sum, we look at pairs of positive factors of 12 only.

| *Positive Factors of 12* | *Sum of Factors* | |
|---|---|---|
| 1, 12 | 13 | |
| 2, 6 | 8 | |
| 3, 4 | 7 | Correct sum, so the numbers are 3 and 4. |

Thus, $x^2 + 7x + 12 = (x + 3)(x + 4)$

**Check:**   $(x + 3)(x + 4) = x^2 + 4x + 3x + 12 = x^2 + 7x + 12.$   □

**PRACTICE**

**1**   Factor $x^2 + 5x + 6$.

**EXAMPLE 2**   Factor: $x^2 - 12x + 35$

*Solution*   Again, we begin by writing the first terms of the binomials.

$$(x + \square)(x + \square)$$

Now we look for two numbers whose product is 35 and whose sum is $-12$. Since our numbers must have a positive product and a negative sum, we look at pairs of negative factors of 35 only.

| Negative Factors of 35 | Sum of Factors | |
|---|---|---|
| $-1, -35$ | $-36$ | |
| $-5, -7$ | $-12$ | Correct sum, so the numbers are $-5$ and $-7$. |

Thus, $x^2 - 12x + 35 = (x - 5)(x - 7)$

**Check:**   To check, multiply $(x - 5)(x - 7)$.   □

**PRACTICE**
**2**   Factor $x^2 - 17x + 70$.

**EXAMPLE 3**   Factor: $x^2 + 4x - 12$

*Solution*   $x^2 + 4x - 12 = (x + \square)(x + \square)$

We look for two numbers whose product is $-12$ and whose sum is 4. Since our numbers must have a negative product, we look at pairs of factors with opposite signs.

| Factors of $-12$ | Sum of Factors | |
|---|---|---|
| $-1, 12$ | $11$ | |
| $1, -12$ | $-11$ | |
| $-2, 6$ | $4$ | Correct sum, so the numbers are $-2$ and 6. |
| $2, -6$ | $-4$ | |
| $-3, 4$ | $1$ | |
| $3, -4$ | $-1$ | |

Thus, $x^2 + 4x - 12 = (x - 2)(x + 6)$   □

**PRACTICE**
**3**   Factor $x^2 + 5x - 14$.

**EXAMPLE 4**   Factor: $r^2 - r - 42$

*Solution*   Because the variable in this trinomial is $r$, the first term of each binomial factor is $r$.

$$r^2 - r - 42 = (r + \square)(r + \square)$$

Now we look for two numbers whose product is $-42$ and whose sum is $-1$, the numerical coefficient of $r$. The numbers are 6 and $-7$. Therefore,

$$r^2 - r - 42 = (r + 6)(r - 7)$$   □

**PRACTICE**
**4**   Factor $p^2 - 2p - 63$.

**EXAMPLE 5**  Factor: $a^2 + 2a + 10$

_Solution_  Look for two numbers whose product is 10 and whose sum is 2. Neither 1 and 10 nor 2 and 5 give the required sum, 2. We conclude that $a^2 + 2a + 10$ is not factorable with integers. A polynomial such as $a^2 + 2a + 10$ is called a **prime polynomial.** ☐

**PRACTICE**
**5**  Factor $b^2 + 5b + 1$.

---

**EXAMPLE 6**  Factor: $x^2 + 7xy + 6y^2$

_Solution_                  $x^2 + 7xy + 6y^2 = (x + \square)(x + \square)$

Recall that the middle term $7xy$ is the same as $7yx$. Thus, we can see that $7y$ is the "coefficient" of $x$. We then look for two terms whose product is $6y^2$ and whose sum is $7y$. The terms are $6y$ and $1y$ or $6y$ and $y$ because $6y \cdot y = 6y^2$ and $6y + y = 7y$. Therefore,

$$x^2 + 7xy + 6y^2 = (x + 6y)(x + y)$$  ☐

**PRACTICE**
**6**  Factor $x^2 + 7xy + 12y^2$.

---

**EXAMPLE 7**  Factor: $x^4 + 5x^2 + 6$

_Solution_  As usual, we begin by writing the first terms of the binomials. Since the greatest power of $x$ in this polynomial is $x^4$, we write

$$(x^2 + \square)(x^2 + \square) \quad \text{since } x^2 \cdot x^2 = x^4$$

Now we look for two factors of 6 whose sum is 5. The numbers are 2 and 3. Thus,

$$x^4 + 5x^2 + 6 = (x^2 + 2)(x^2 + 3)$$  ☐

**PRACTICE**
**7**  Factor $x^4 + 13x^2 + 12$.

---

If the terms of a polynomial are not written in descending powers of the variable, you may want to do so before factoring.

**EXAMPLE 8**  Factor: $40 - 13t + t^2$

_Solution_  First, we rearrange terms so that the trinomial is written in descending powers of $t$.

$$40 - 13t + t^2 = t^2 - 13t + 40$$

Next, try to factor.

$$t^2 - 13t + 40 = (t + \square)(t + \square)$$

Now we look for two factors of 40 whose sum is $-13$. The numbers are $-8$ and $-5$. Thus,

$$t^2 - 13t + 40 = (t - 8)(t - 5)$$  ☐

**PRACTICE**
**8**  Factor $48 - 14x + x^2$.

The following sign patterns may be useful when factoring trinomials.

> **▶ Helpful Hint**
>
> A positive constant in a trinomial tells us to look for two numbers with the same sign. The sign of the coefficient of the middle term tells us whether the signs are both positive or both negative.
>
> both   same                     both   same
> positive  sign                 negative  sign
>
> $x^2 + 10x + 16 = (x + 2)(x + 8)$      $x^2 - 10x + 16 = (x - 2)(x - 8)$
>
> A negative constant in a trinomial tells us to look for two numbers with opposite signs.
>
> opposite                         opposite
> signs                            signs
>
> $x^2 + 6x - 16 = (x + 8)(x - 2)$      $x^2 - 6x - 16 = (x - 8)(x + 2)$

**OBJECTIVE 2 ▶ Factoring out the greatest common factor.** Remember that the first step in factoring any polynomial is to factor out the greatest common factor (if there is one other than 1 or $-1$).

**EXAMPLE 9**   Factor: $3m^2 - 24m - 60$

_Solution_   First we factor out the greatest common factor, 3, from each term.

$$3m^2 - 24m - 60 = 3(m^2 - 8m - 20)$$

Now we factor $m^2 - 8m - 20$ by looking for two factors of $-20$ whose sum is $-8$. The factors are $-10$ and 2. Therefore, the complete factored form is

$$3m^2 - 24m - 60 = 3(m + 2)(m - 10)$$

> **▶ Helpful Hint**
>
> Remember to write the common factor 3 as part of the factored form.

**PRACTICE**
**9**   Factor $4x^2 - 24x + 36$.

**EXAMPLE 10**   Factor: $2x^4 - 26x^3 + 84x^2$

_Solution_

$$\begin{aligned} 2x^4 - 26x^3 + 84x^2 &= 2x^2(x^2 - 13x + 42) &&\text{Factor out common factor, } 2x^2. \\ &= 2x^2(x - 6)(x - 7) &&\text{Factor } x^2 - 13x + 42. \end{aligned}$$

**PRACTICE**
**10**   Factor $3y^4 - 18y^3 - 21y^2$.

## VOCABULARY & READINESS CHECK

_Fill in each blank with "true or false."_

**1.** To factor $x^2 + 7x + 6$, we look for two numbers whose product is 6 and whose sum is 7. _____

**2.** We can write the factorization $(y + 2)(y + 4)$ also as $(y + 4)(y + 2)$. _____

**3.** The factorization $(4x - 12)(x - 5)$ is completely factored. _____

**4.** The factorization $(x + 2y)(x + y)$ may also be written as $(x + 2y)^2$. _____

*Complete each factored form.*

**5.** $x^2 + 9x + 20 = (x + 4)(x \quad )$

**6.** $x^2 + 12x + 35 = (x + 5)(x \quad )$

**7.** $x^2 - 7x + 12 = (x - 4)(x \quad )$

**8.** $x^2 - 13x + 22 = (x - 2)(x \quad )$

**9.** $x^2 + 4x + 4 = (x + 2)(x \quad )$

**10.** $x^2 + 10x + 24 = (x + 6)(x \quad )$

## 6.2 EXERCISE SET

MyMathLab   MathXL PRACTICE   WATCH   DOWNLOAD   READ   REVIEW

*Factor each trinomial completely. If a polynomial can't be factored, write "prime." See Examples 1 through 8.*

**1.** $x^2 + 7x + 6$

**2.** $x^2 + 6x + 8$

**3.** $y^2 - 10y + 9$

**4.** $y^2 - 12y + 11$

**5.** $x^2 - 6x + 9$

**6.** $x^2 - 10x + 25$

**7.** $x^2 - 3x - 18$

**8.** $x^2 - x - 30$

**9.** $x^2 + 3x - 70$

**10.** $x^2 + 4x - 32$

**11.** $x^2 + 5x + 2$

**12.** $x^2 - 7x + 5$

**13.** $x^2 + 8xy + 15y^2$

**14.** $x^2 + 6xy + 8y^2$

**15.** $a^4 - 2a^2 - 15$

**16.** $y^4 - 3y^2 - 70$

**17.** $13 + 14m + m^2$

**18.** $17 + 18n + n^2$

**19.** $10t - 24 + t^2$

**20.** $6q - 27 + q^2$

**21.** $a^2 - 10ab + 16b^2$

**22.** $a^2 - 9ab + 18b^2$

### MIXED PRACTICE

*Factor each trinomial completely. Some of these trinomials contain a greatest common factor (other than 1). Don't forget to factor out the GCF first. See Examples 1 through 10.*

**23.** $2z^2 + 20z + 32$

**24.** $3x^2 + 30x + 63$

**25.** $2x^3 - 18x^2 + 40x$

**26.** $3x^3 - 12x^2 - 36x$

**27.** $x^2 - 3xy - 4y^2$

**28.** $x^2 - 4xy - 77y^2$

**29.** $x^2 + 15x + 36$

**30.** $x^2 + 19x + 60$

**31.** $x^2 - x - 2$

**32.** $x^2 - 5x - 14$

**33.** $r^2 - 16r + 48$

**34.** $r^2 - 10r + 21$

**35.** $x^2 + xy - 2y^2$

**36.** $x^2 - xy - 6y^2$

**37.** $3x^2 + 9x - 30$

**38.** $4x^2 - 4x - 48$

**39.** $3x^2 - 60x + 108$

**40.** $2x^2 - 24x + 70$

**41.** $x^2 - 18x - 144$

**42.** $x^2 + x - 42$

**43.** $r^2 - 3r + 6$

**44.** $x^2 + 4x - 10$

**45.** $x^2 - 8x + 15$

**46.** $x^2 - 9x + 14$

**47.** $6x^3 + 54x^2 + 120x$

**48.** $3x^3 + 3x^2 - 126x$

**49.** $4x^2y + 4xy - 12y$

**50.** $3x^2y - 9xy + 45y$

**51.** $x^2 - 4x - 21$

**52.** $x^2 - 4x - 32$

**53.** $x^2 + 7xy + 10y^2$

**54.** $x^2 - 3xy - 4y^2$

**55.** $64 + 24t + 2t^2$

**56.** $50 + 20t + 2t^2$

**57.** $x^3 - 2x^2 - 24x$

**58.** $x^3 - 3x^2 - 28x$

**59.** $2t^5 - 14t^4 + 24t^3$

**60.** $3x^6 + 30x^5 + 72x^4$

**61.** $5x^3y - 25x^2y^2 - 120xy^3$

**62.** $7a^3b - 35a^2b^2 + 42ab^3$

**63.** $162 - 45m + 3m^2$

**64.** $48 - 20n + 2n^2$

**65.** $-x^2 + 12x - 11$ (Factor out $-1$ first.)

**66.** $-x^2 + 8x - 7$ (Factor out $-1$ first.)

**67.** $\frac{1}{2}y^2 - \frac{9}{2}y - 11$ (Factor out $\frac{1}{2}$ first.)

**68.** $\frac{1}{3}y^2 - \frac{5}{3}y - 8$ (Factor out $\frac{1}{3}$ first.)

**69.** $x^3y^2 + x^2y - 20x$

**70.** $a^2b^3 + ab^2 - 30b$

## REVIEW AND PREVIEW

*Multiply. See Section 5.4.*

**71.** $(2x + 1)(x + 5)$

**72.** $(3x + 2)(x + 4)$

**73.** $(5y - 4)(3y - 1)$

**74.** $(4z - 7)(7z - 1)$

**75.** $(a + 3b)(9a - 4b)$

**76.** $(y - 5x)(6y + 5x)$

## CONCEPT EXTENSIONS

**77.** Write a polynomial that factors as $(x - 3)(x + 8)$.

**78.** To factor $x^2 + 13x + 42$, think of two numbers whose _____ is 42 and whose _____ is 13.

*Complete each sentence in your own words.*

**79.** If $x^2 + bx + c$ is factorable and $c$ is negative, then the signs of the last-term factors of the binomials are opposite because ....

**80.** If $x^2 + bx + c$ is factorable and $c$ is positive, then the signs of the last-term factors of the binomials are the same because ....

*Remember that perimeter means distance around. Write the perimeter of each rectangle as a simplified polynomial. Then factor the polynomial.*

△ **81.**

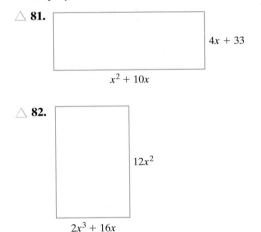

$4x + 33$

$x^2 + 10x$

△ **82.**

$12x^2$

$2x^3 + 16x$

**83.** An object is thrown upward from the top of an 80-foot building with an initial velocity of 64 feet per second. The height of the object after $t$ seconds is given by $-16t^2 + 64t + 80$. Factor this polynomial.

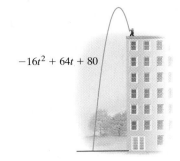

$-16t^2 + 64t + 80$

*Factor each trinomial completely.*

**84.** $x^2 + x + \dfrac{1}{4}$

**85.** $x^2 + \dfrac{1}{2}x + \dfrac{1}{16}$

**86.** $y^2(x + 1) - 2y(x + 1) - 15(x + 1)$

**87.** $z^2(x + 1) - 3z(x + 1) - 70(x + 1)$

*Factor each trinomial.* (**Hint:** *Notice that $x^{2n} + 4x^n + 3$ factors as* $(x^n + 1)(x^n + 3)$. **Remember:** $x^n \cdot x^n = x^{n+n}$ *or* $x^{2n}$.)

**88.** $x^{2n} + 5x^n + 6$

**89.** $x^{2n} + 8x^n - 20$

*Find a positive value of $c$ so that each trinomial is factorable.*

**90.** $x^2 + 6x + c$

**91.** $t^2 + 8t + c$

**92.** $y^2 - 4y + c$

**93.** $n^2 - 16n + c$

*Find a positive value of $b$ so that each trinomial is factorable.*

**94.** $x^2 + bx + 15$

**95.** $y^2 + by + 20$

**96.** $m^2 + bm - 27$

**97.** $x^2 + bx - 14$

---

## 6.3 FACTORING TRINOMIALS OF THE FORM $ax^2 + bx + c$ AND PERFECT SQUARE TRINOMIALS

### OBJECTIVES

**1** Factor trinomials of the form $ax^2 + bx + c$, where $a \neq 1$.

**2** Factor out a GCF before factoring a trinomial of the form $ax^2 + bx + c$.

**3** Factor perfect square trinomials.

**OBJECTIVE 1 ▶ Factoring trinomials of the form $ax^2 + bx + c$.** In this section, we factor trinomials of the form $ax^2 + bx + c$, such as

$$3x^2 + 11x + 6, \quad 8x^2 - 22x + 5, \quad \text{and} \quad 2x^2 + 13x - 7$$

Notice that the coefficient of the squared variable in these trinomials is a number other than 1. We will factor these trinomials using a trial-and-check method based on our work in the last section.

To begin, let's review the relationship between the numerical coefficients of the trinomial and the numerical coefficients of its factored form. For example, since

$$(2x + 1)(x + 6) = 2x^2 + 13x + 6,$$

a factored form of $2x^2 + 13x + 6$ is $(2x + 1)(x + 6)$

Notice that $2x$ and $x$ are factors of $2x^2$, the first term of the trinomial. Also, 6 and 1 are factors of 6, the last term of the trinomial, as shown:

$$2x^2 + 13x + 6 = (2x + 1)(x + 6)$$

with $2x \cdot x$ and $1 \cdot 6$ indicated.

Also notice that $13x$, the middle term, is the sum of the following products:

$$2x^2 + 13x + 6 = (2x + 1)(x + 6)$$

$$
\begin{array}{r}
1x \\
+12x \\
\hline
13x
\end{array}
\quad \text{Middle term}
$$

Let's use this pattern to factor $5x^2 + 7x + 2$. First, we find factors of $5x^2$. Since all numerical coefficients in this trinomial are positive, we will use factors with positive numerical coefficients only. Thus, the factors of $5x^2$ are $5x$ and $x$. Let's try these factors as first terms of the binomials. Thus far, we have

$$5x^2 + 7x + 2 = (5x + \square)(x + \square)$$

Next, we need to find positive factors of 2. Positive factors of 2 are 1 and 2. Now we try possible combinations of these factors as second terms of the binomials until we obtain a middle term of $7x$.

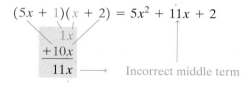

$$(5x + 1)(x + 2) = 5x^2 + 11x + 2$$

$$
\begin{array}{r}
1x \\
+10x \\
\hline
11x
\end{array}
\longrightarrow \quad \text{Incorrect middle term}
$$

Let's try switching factors 2 and 1.

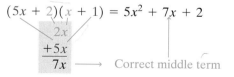

$$(5x + 2)(x + 1) = 5x^2 + 7x + 2$$

$$
\begin{array}{r}
2x \\
+5x \\
\hline
7x
\end{array}
\longrightarrow \quad \text{Correct middle term}
$$

Thus the factored form of $5x^2 + 7x + 2$ is $(5x + 2)(x + 1)$. To check, we multiply $(5x + 2)$ and $(x + 1)$. The product is $5x^2 + 7x + 2$.

**EXAMPLE 1**    Factor: $3x^2 + 11x + 6$

*Solution*    Since all numerical coefficients are positive, we use factors with positive numerical coefficients. We first find factors of $3x^2$.

$$\text{Factors of } 3x^2: \quad 3x^2 = 3x \cdot x$$

If factorable, the trinomial will be of the form

$$3x^2 + 11x + 6 = (3x + \square)(x + \square)$$

Next we factor 6.

$$\text{Factors of } 6: \quad 6 = 1 \cdot 6, \quad 6 = 2 \cdot 3$$

Now we try combinations of factors of 6 until a middle term of $11x$ is obtained. Let's try 1 and 6 first.

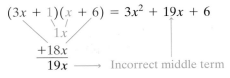

$$(3x + 1)(x + 6) = 3x^2 + 19x + 6$$

$$\begin{array}{r} 1x \\ +18x \\ \hline 19x \end{array} \longrightarrow \text{Incorrect middle term}$$

Now let's next try 6 and 1.

$$(3x + 6)(x + 1)$$

Before multiplying, notice that the terms of the factor $3x + 6$ have a common factor of 3. The terms of the original trinomial $3x^2 + 11x + 6$ have no common factor other than 1, so the terms of its factors will also contain no common factor other than 1. This means that $(3x + 6)(x + 1)$ is not a factored form.

Next let's try 2 and 3 as last terms.

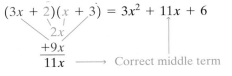

$$(3x + 2)(x + 3) = 3x^2 + 11x + 6$$

$$\begin{array}{r} 2x \\ +9x \\ \hline 11x \end{array} \longrightarrow \text{Correct middle term}$$

Thus a factored form of $3x^2 + 11x + 6$ is $(3x + 2)(x + 3)$. □

**PRACTICE**

**1** Factor: $2x^2 + 11x + 15$.

---

▶ **Helpful Hint**

If the terms of a trinomial have no common factor (other than 1), then the terms of neither of its binomial factors will contain a common factor (other than 1).

---

**Concept Check** ☑

Do the terms of $3x^2 + 29x + 18$ have a common factor? Without multiplying, decide which of the following factored forms could not be a factored form of $3x^2 + 29x + 18$.

**a.** $(3x + 18)(x + 1)$          **b.** $(3x + 2)(x + 9)$

**c.** $(3x + 6)(x + 3)$          **d.** $(3x + 9)(x + 2)$

**EXAMPLE 2** Factor: $8x^2 - 22x + 5$

**Solution** Factors of $8x^2$:  $8x^2 = 8x \cdot x$,  $8x^2 = 4x \cdot 2x$

We'll try $8x$ and $x$.

$$8x^2 - 22x + 5 = (8x + \square)(x + \square)$$

Since the middle term, $-22x$, has a negative numerical coefficient, we factor 5 into negative factors.

$$\text{Factors of 5:}\quad 5 = -1 \cdot -5$$

Let's try $-1$ and $-5$.

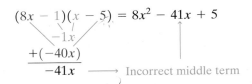

$$(8x - 1)(x - 5) = 8x^2 - 41x + 5$$

$$\begin{array}{r} -1x \\ +(-40x) \\ \hline -41x \end{array} \longrightarrow \text{Incorrect middle term}$$

Now let's try $-5$ and $-1$.

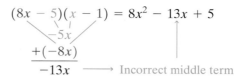

$$(8x - 5)(x - 1) = 8x^2 - 13x + 5$$

$$-5x$$
$$+(-8x)$$
$$\overline{-13x} \longrightarrow \text{Incorrect middle term}$$

Don't give up yet! We can still try other factors of $8x^2$. Let's try $4x$ and $2x$ with $-1$ and $-5$.

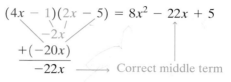

$$(4x - 1)(2x - 5) = 8x^2 - 22x + 5$$

$$-2x$$
$$+(-20x)$$
$$\overline{-22x} \longrightarrow \text{Correct middle term}$$

A factored form of $8x^2 - 22x + 5$ is $(4x - 1)(2x - 5)$.

**PRACTICE**
**2** Factor: $15x^2 - 22x + 8$.

---

**EXAMPLE 3** Factor: $2x^2 + 13x - 7$

*Solution* Factors of $2x^2$: $2x^2 = 2x \cdot x$

Factors of $-7$: $-7 = -1 \cdot 7$, $-7 = 1 \cdot -7$

We try possible combinations of these factors:

$(2x + 1)(x - 7) = 2x^2 - 13x - 7$    Incorrect middle term
$(2x - 1)(x + 7) = 2x^2 + 13x - 7$    Correct middle term

A factored form of $2x^2 + 13x - 7$ is $(2x - 1)(x + 7)$.

**PRACTICE**
**3** Factor: $4x^2 + 11x - 3$.

---

**EXAMPLE 4** Factor: $10x^2 - 13xy - 3y^2$

*Solution* Factors of $10x^2$: $10x^2 = 10x \cdot x$, $10x^2 = 2x \cdot 5x$

Factors of $-3y^2$: $-3y^2 = -3y \cdot y$, $-3y^2 = 3y \cdot -y$

We try some combinations of these factors:

$$\overset{\text{Correct}}{\downarrow} \qquad \overset{\text{Correct}}{\downarrow}$$

$(10x - 3y)(x + y) = 10x^2 + 7xy - 3y^2$
$(x + 3y)(10x - y) = 10x^2 + 29xy - 3y^2$
$(5x + 3y)(2x - y) = 10x^2 + xy - 3y^2$
$(2x - 3y)(5x + y) = 10x^2 - 13xy - 3y^2$    Correct middle term

A factored form of $10x^2 - 13xy - 3y^2$ is $(2x - 3y)(5x + y)$.

**PRACTICE**
**4** Factor: $21x^2 + 11xy - 2y^2$.

**EXAMPLE 5**   Factor: $3x^4 - 5x^2 - 8$

*Solution*   Factors of $3x^4$:   $3x^4 = 3x^2 \cdot x^2$
          Factors of $-8$:   $-8 = -2 \cdot 4, 2 \cdot -4, -1 \cdot 8, 1 \cdot -8$

Try combinations of these factors:

$$\overset{\text{Correct}}{\downarrow} \qquad \overset{\text{Correct}}{\downarrow}$$

$$(3x^2 - 2)(x^2 + 4) = 3x^4 + 10x^2 - 8$$
$$(3x^2 + 4)(x^2 - 2) = 3x^4 - 2x^2 - 8$$
$$(3x^2 + 8)(x^2 - 1) = 3x^4 + 5x^2 - 8 \qquad \text{Incorrect sign on middle term,}$$
$$(3x^2 - 8)(x^2 + 1) = 3x^4 - 5x^2 - 8 \qquad \text{so switch signs in binomial factors.}$$
$$\text{Correct middle term.}$$

A factored form of $3x^4 - 5x^2 - 8$ is $(3x^2 - 8)(x^2 + 1)$.   □

**PRACTICE**
**5**   Factor: $2x^4 - 5x^2 - 7$.

▶ **Helpful Hint**

Study the last two lines of Example 5. If a factoring attempt gives you a middle term whose numerical coefficient is the opposite of the desired numerical coefficient, try switching the signs of the last terms in the binomials.

Switched signs $\begin{cases} (3x^2 + 8)(x^2 - 1) = 3x^4 + 5x^2 - 8 & \text{Middle term: } +5x \\ (3x^2 - 8)(x^2 + 1) = 3x^4 - 5x^2 - 8 & \text{Middle term: } -5x \end{cases}$

**OBJECTIVE 2** ▶ **Factoring out the greatest common factor.**  Don't forget that the first step in factoring any polynomial is to look for a common factor to factor out.

**EXAMPLE 6**   Factor: $24x^4 + 40x^3 + 6x^2$

*Solution*   Notice that all three terms have a common factor of $2x^2$. Thus we factor out $2x^2$ first.

$$24x^4 + 40x^3 + 6x^2 = 2x^2(12x^2 + 20x + 3)$$

Next we factor $12x^2 + 20x + 3$.

Factors of $12x^2$:   $12x^2 = 4x \cdot 3x$,   $12x^2 = 12x \cdot x$,   $12x^2 = 6x \cdot 2x$

Since all terms in the trinomial have positive numerical coefficients, we factor 3 using positive factors only.

Factors of 3:   $3 = 1 \cdot 3$

We try some combinations of the factors.

$$2x^2(4x + 3)(3x + 1) = 2x^2(12x^2 + 13x + 3)$$
$$2x^2(12x + 1)(x + 3) = 2x^2(12x^2 + 37x + 3)$$
$$2x^2(2x + 3)(6x + 1) = 2x^2(12x^2 + 20x + 3) \qquad \text{Correct middle term}$$

▶ **Helpful Hint**

Don't forget to include the common factor in the factored form.

A factored form of $24x^4 + 40x^3 + 6x^2$ is $2x^2(2x + 3)(6x + 1)$.   □

**PRACTICE**
**6**   Factor: $3x^3 + 17x^2 + 10x$

When the term containing the squared variable has a negative coefficient, you may want to first factor out a common factor of $-1$.

**EXAMPLE 7** Factor: $-6x^2 - 13x + 5$

**Solution** We begin by factoring out a common factor of $-1$.

$$-6x^2 - 13x + 5 = -1(6x^2 + 13x - 5) \quad \text{Factor out } -1.$$
$$= -1(3x - 1)(2x + 5) \quad \text{Factor } 6x^2 + 13x - 5. \qquad \square$$

**PRACTICE**
**7** Factor: $-8x^2 + 2x + 3$

**OBJECTIVE 3 ▶ Factoring perfect square trinomials.** A trinomial that is the square of a binomial is called a **perfect square trinomial.** For example,

$$(x + 3)^2 = (x + 3)(x + 3)$$
$$= x^2 + 6x + 9$$

Thus $x^2 + 6x + 9$ is a perfect square trinomial.

In Chapter 5, we discovered special product formulas for squaring binomials.

$$(a + b)^2 = a^2 + 2ab + b^2 \quad \text{and} \quad (a - b)^2 = a^2 - 2ab + b^2$$

Because multiplication and factoring are reverse processes, we can now use these special products to help us factor perfect square trinomials. If we reverse these equations, we have the following.

**Factoring Perfect Square Trinomials**

$$a^2 + 2ab + b^2 = (a + b)^2$$
$$a^2 - 2ab + b^2 = (a - b)^2$$

▶ **Helpful Hint**

Notice that for both given forms of a perfect square trinomial, the last term is positive. This is because the last term is a square.

To use these equations to help us factor, we must first be able to recognize a perfect square trinomial. A trinomial is a perfect square when

1. two terms, $a^2$ and $b^2$, are squares and
2. another term is $2 \cdot a \cdot b$ or $-2 \cdot a \cdot b$. That is, this term is twice the product of $a$ and $b$, or its opposite.

When a trinomial fits this description, its factored form is $(a + b)^2$.

**EXAMPLE 8** Factor: $x^2 + 12x + 36$

**Solution** First, is this a perfect square trinomial?

$$x^2 + 12x + 36$$

1. $x^2 = (x)^2$ and $36 = 6^2$.
2. Is $2 \cdot x \cdot 6$ the middle term? Yes, $2 \cdot x \cdot 6 = 12x$.

Thus, $x^2 + 12x + 36$ factors as $(x + 6)^2$. $\qquad \square$

**PRACTICE**
**8** Factor: $x^2 + 14x + 49$

**EXAMPLE 9** Factor: $25x^2 + 25xy + 4y^2$

*Solution* Is this a perfect square trinomial?

$$25x^2 + 25xy + 4y^2$$
$$\swarrow$$

1. $25x^2 = (5x)^2$ and $4y^2 = (2y)^2$.
2. Is $2 \cdot 5x \cdot 2y$ the middle term? **No**, $2 \cdot 5x \cdot 2y = 20xy$, **not** $25xy$.

Therefore, $25x^2 + 25xy + 4y^2$ is not a perfect square trinomial. It is factorable, though. Using earlier techniques, we find that $25x^2 + 25xy + 4y^2$ factors as $(5x + 4y)(5x + y)$. □

> ▶ **Helpful Hint**
> A perfect square trinomial can also be factored by other methods.

**PRACTICE**
**9** Factor $4x^2 + 20xy + 9y^2$.

**EXAMPLE 10** Factor: $4m^4 - 4m^2 + 1$

*Solution* Is this a perfect square trinomial?

$$4m^4 - 4m^2 + 1$$
$$\swarrow$$

1. $4m^4 = (2m^2)^2$ and $1 = 1^2$.
2. Is $2 \cdot 2m^2 \cdot 1$ the middle term? Yes, $2 \cdot 2m^2 \cdot 1 = 4m^2$, the opposite of the middle term.

Thus, $4m^4 - 4m^2 + 1$ factors as $(2m^2 - 1)^2$. □

**PRACTICE**
**10** Factor $36n^4 - 12n^2 + 1$.

**EXAMPLE 11** Factor: $162x^3 - 144x^2 + 32x$

*Solution* Don't forget to first look for a common factor. There is a greatest common factor of $2x$ in this trinomial.

$$\begin{aligned}
162x^3 - 144x^2 + 32x &= 2x(81x^2 - 72x + 16) \\
&= 2x[(9x)^2 - 2 \cdot 9x \cdot 4 + 4^2] \\
&= 2x(9x - 4)^2
\end{aligned}$$ □

**PRACTICE**
**11** Factor $12x^3 - 84x^2 + 147x$.

## VOCABULARY & READINESS CHECK

*Use the choices below to fill in each blank. Some choices will be used more than once and some not used at all.*

| | | | |
|---|---|---|---|
| $5y^2$ | $(x + 5y)^2$ | yes | perfect square trinomial |
| $(5y)^2$ | $(x - 5y)^2$ | no | perfect square binomial |

1. A(n) _____ is a trinomial that is the square of a binomial.
2. The term $25y^2$ written as a square is _____.
3. The expression $x^2 + 10xy + 25y^2$ is called a(n)_____.
4. The factorization $(x + 5y)(x + 5y)$ may also be written as _____.

*Answer 5 and 6 with yes or no.*

5. The factorization $(x - 5y)(x + 5y)$ may also be written as $(x - 5y)^2$. _____
6. The greatest common factor of $10x^3 - 45x^2 + 20x$ is $5x$. _____

*Write each number or term as a square. For example, 16 written as a square is $4^2$.*

**7.** 64      **8.** 9      **9.** $121a^2$      **10.** $81b^2$      **11.** $36p^4$      **12.** $4q^4$

## 6.3 | EXERCISE SET

PRACTICE    WATCH    DOWNLOAD    READ    REVIEW

*Complete each factored form. See Examples 1 through 5, and 8 through 10.*

1. $5x^2 + 22x + 8 = (5x + 2)(\quad)$

2. $2y^2 + 27y + 25 = (2y + 25)(\quad)$

3. $50x^2 + 15x - 2 = (5x + 2)(\quad)$

4. $6y^2 + 11y - 10 = (2y + 5)(\quad)$

5. $25x^2 - 20x + 4 = (5x - 2)(\quad)$

6. $4y^2 - 20y + 25 = (2y - 5)(\quad)$

*Factor completely. See Examples 1 through 5.*

7. $2x^2 + 13x + 15$

8. $3x^2 + 8x + 4$

9. $8y^2 - 17y + 9$

10. $21x^2 - 31x + 10$

11. $2x^2 - 9x - 5$

12. $36r^2 - 5r - 24$

13. $20r^2 + 27r - 8$

14. $3x^2 + 20x - 63$

15. $10x^2 + 31x + 3$

16. $12x^2 + 17x + 5$

17. $2m^2 + 17m + 10$

18. $3n^2 + 20n + 5$

19. $6x^2 - 13xy + 5y^2$

20. $8x^2 - 14xy + 3y^2$

21. $15m^2 - 16m - 15$

22. $25n^2 - 5n - 6$

*Factor completely. See Examples 1 through 7.*

23. $12x^3 + 11x^2 + 2x$

24. $8a^3 + 14a^2 + 3a$

25. $21b^2 - 48b - 45$

26. $12x^2 - 14x - 10$

27. $7z + 12z^2 - 12$

28. $16t + 15t^2 - 15$

29. $6x^2y^2 - 2xy^2 - 60y^2$

30. $8x^2y + 34xy - 84y$

31. $4x^2 - 8x - 21$

32. $6x^2 - 11x - 10$

33. $-x^2 + 2x + 24$

34. $-x^2 + 4x + 21$

35. $4x^3 - 9x^2 - 9x$

36. $6x^3 - 31x^2 + 5x$

37. $24x^2 - 58x + 9$

38. $36x^2 + 55x - 14$

*Factor each perfect square trinomial completely. See Examples 8 through 11.*

39. $x^2 + 22x + 121$

40. $x^2 + 18x + 81$

41. $x^2 - 16x + 64$

42. $x^2 - 12x + 36$

43. $16a^2 - 24a + 9$

44. $25x^2 - 20x + 4$

45. $x^4 + 4x^2 + 4$

46. $m^4 + 10m^2 + 25$

47. $2n^2 - 28n + 98$

48. $3y^2 - 6y + 3$

49. $16y^2 + 40y + 25$

50. $9y^2 + 48y + 64$

### MIXED PRACTICE

*Factor each trinomial completely. See Examples 1 through 11 and Section 6.2.*

51. $2x^2 - 7x - 99$

52. $2x^2 + 7x - 72$

53. $24x^2 + 41x + 12$

54. $24x^2 - 49x + 15$

55. $3a^2 + 10ab + 3b^2$

56. $2a^2 + 11ab + 5b^2$

57. $-9x + 20 + x^2$

58. $-7x + 12 + x^2$

59. $p^2 + 12pq + 36q^2$

60. $m^2 + 20mn + 100n^2$

61. $x^2y^2 - 10xy + 25$

62. $x^2y^2 - 14xy + 49$

63. $40a^2b + 9ab - 9b$

64. $24y^2x + 7yx - 5x$

65. $30x^3 + 38x^2 + 12x$

66. $6x^3 - 28x^2 + 16x$

67. $6y^3 - 8y^2 - 30y$

68. $12x^3 - 34x^2 + 24x$

69. $10x^4 + 25x^3y - 15x^2y^2$

70. $42x^4 - 99x^3y - 15x^2y^2$

71. $-14x^2 + 39x - 10$

72. $-15x^2 + 26x - 8$

73. $16p^4 - 40p^3 + 25p^2$

74. $9q^4 - 42q^3 + 49q^2$

75. $x + 3x^2 - 2$

76. $y + 8y^2 - 9$

77. $8x^2 + 6xy - 27y^2$

78. $54a^2 + 39ab - 8b^2$

79. $1 + 6x^2 + x^4$

80. $1 + 16x^2 + x^4$

81. $9x^2 - 24xy + 16y^2$

82. $25x^2 - 60xy + 36y^2$

83. $18x^2 - 9x - 14$

84. $42a^2 - 43a + 6$

**85.** $-27t + 7t^2 - 4$

**86.** $-3t + 4t^2 - 7$

**87.** $49p^2 - 7p - 2$

**88.** $3r^2 + 10r - 8$

**89.** $m^3 + 18m^2 + 81m$

**90.** $y^3 + 12y^2 + 36y$

**91.** $5x^2y^2 + 20xy + 1$

**92.** $3a^2b^2 + 12ab + 1$

**93.** $6a^5 + 37a^3b^2 + 6ab^4$

**94.** $5m^5 + 26m^3h^2 + 5mh^4$

## REVIEW AND PREVIEW

*Multiply the following. See Section 5.4.*

**95.** $(x - 2)(x + 2)$

**96.** $(y^2 + 3)(y^2 - 3)$

**97.** $(a + 3)(a^2 - 3a + 9)$

**98.** $(z - 2)(z^2 + 2z + 4)$

*As of 2006, approximately 80% of U.S. households have access to the Internet. The following graph shows the percent of households having Internet access grouped according to household income. See Section 3.1.*

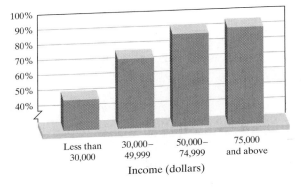

*Source*: Pew Internet Research

**99.** Which range of household income corresponds to the highest percent of households having access to the Internet?

**100.** Which range of household income corresponds to the greatest increase in percent of households having access to the Internet?

**101.** Describe any trend you see.

**102.** Why don't the percents shown in the graph add to 100%?

## CONCEPT EXTENSIONS

*See the Concept Check in this section.*

**103.** Do the terms of $4x^2 + 19x + 12$ have a common factor (other than 1)?

**104.** Without multiplying, decide which of the following factored forms is not a factored form of $4x^2 + 19x + 12$.

   **a.** $(2x + 4)(2x + 3)$     **b.** $(4x + 4)(x + 3)$

   **c.** $(4x + 3)(x + 4)$     **d.** $(2x + 2)(2x + 6)$

**105.** Describe a perfect square trinomial.

**106.** Write the perfect square trinomial that factors as $(x + 3y)^2$.

*Write the perimeter of each figure as a simplified polynomial. Then factor the polynomial.*

**107.**

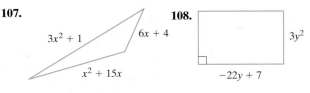

**108.**

*Factor each trinomial completely.*

**109.** $4x^2 + 2x + \dfrac{1}{4}$

**110.** $27x^2 + 2x - \dfrac{1}{9}$

**111.** $4x^2(y - 1)^2 + 10x(y - 1)^2 + 25(y - 1)^2$

**112.** $3x^2(a + 3)^3 - 10x(a + 3)^3 + 25(a + 3)^3$

**113.** Fill in the blank so that $x^2 + \underline{\quad} x + 16$ is a perfect square trinomial.

**114.** Fill in the blank so that $9x^2 + \underline{\quad} x + 25$ is a perfect square trinomial.

*The area of the largest square in the figure is $(a + b)^2$. Use this figure to answer Exercises 115 and 116.*

**115.** Write the area of the largest square as the sum of the areas of the smaller squares and rectangles.

**116.** What factoring formula from this section is visually represented by this square?

*Find a positive value of b so that each trinomial is factorable.*

**117.** $3x^2 + bx - 5$     **118.** $2y^2 + by + 3$

*Find a positive value of c so that each trinomial is factorable.*

**119.** $5x^2 + 7x + c$     **120.** $11y^2 - 40y + c$

*Factor completely. Don't forget to first factor out the greatest common factor.*

**121.** $-12x^3y^2 + 3x^2y^2 + 15xy^2$

**122.** $-12r^3x^2 + 38r^2x^2 + 14rx^2$

**123.** $4x^2(y - 1)^2 + 20x(y - 1)^2 + 25(y - 1)^2$

**124.** $3x^2(a + 3)^3 - 28x(a + 3)^3 + 25(a + 3)^3$

*Factor.*

**125.** $3x^{2n} + 17x^n + 10$

**126.** $2x^{2n} + 5x^n - 12$

**127.** In your own words, describe the steps you will use to factor a trinomial.

**Are You Satisfied with Your Performance on a Particular Quiz or Exam?**

If not, don't forget to analyze your quiz or exam and look for common errors. Were most of your errors a result of:

- *Carelessness?* Did you turn in your quiz or exam before the allotted time expired? If so, resolve to use any extra time to check your work.

- *Running out of time?* Try completing any questions that you are unsure of last and delay checking your work until all questions have been answered.

- *Not understanding a concept?* If so, review that concept and correct your work so that you make sure it doesn't happen before the next quiz or the final exam.

- *Test conditions?* When studying for a quiz or exam, make sure you place yourself in conditions similar to test conditions. For example, before your next quiz or exam, take a sample test without the aid of your notes or text.

(For a sample test, see your instructor or use the Chapter Test at the end of each chapter.)

*Exercises*

1. Have you corrected all your previous quizzes and exams?

2. List any errors you have found common to two or more of your graded papers.

3. Is one of your common errors not understanding a concept? If so, are you making sure you understand all the concepts for the next quiz or exam?

4. Is one of your common errors making careless mistakes? If so, are you now taking all the time allotted to check over your work so that you can minimize the number of careless mistakes?

5. Are you satisfied with your grades thus far on quizzes and tests?

6. If your answer to Exercise 5 is no, are there any more suggestions you can make to your instructor or yourself to help? If so, list them here and share these with your instructor.

## 6.4 FACTORING TRINOMIALS OF THE FORM $ax^2 + bx + c$ BY GROUPING

**OBJECTIVE**

1 Use the grouping method to factor trinomials of the form $ax^2 + bx + c$.

**OBJECTIVE 1 ▶ Using the grouping method.** There is an alternative method that can be used to factor trinomials of the form $ax^2 + bx + c$, $a \neq 1$. This method is called the **grouping method** because it uses factoring by grouping as we learned in Section 6.1.

To see how this method works, recall from Section 6.1 that to factor a trinomial such as $x^2 + 11x + 30$, we find two numbers such that

$$\text{Product is 30}$$
$$\downarrow$$
$$x^2 + 11x + 30$$
$$\downarrow$$
$$\text{Sum is 11.}$$

To factor a trinomial such as $2x^2 + 11x + 12$ by grouping, we use an extension of the method in Section 6.1. Here we look for two numbers such that

$$\text{Product is } 2 \cdot 12 = 24$$
$$\downarrow \qquad \qquad$$
$$2x^2 + 11x + 12$$
$$\downarrow$$
$$\text{Sum is 11.}$$

This time, we use the two numbers to write

$$2x^2 + 11x + 12 \text{ as}$$
$$= 2x^2 + \square x + \square x + 12$$

Then we factor by grouping. Since we want a positive product, 24, and a positive sum, 11, we consider pairs of positive factors of 24 only.

| Factors of 24 | Sum of Factors | |
|---|---|---|
| 1, 24 | 25 | |
| 2, 12 | 14 | |
| 3, 8 | 11 | Correct sum |

The factors are 3 and 8. Now we use these factors to write the middle term $11x$ as $3x + 8x$ (or $8x + 3x$). We replace $11x$ with $3x + 8x$ in the original trinomial and then we can factor by grouping.

$$\begin{aligned} 2x^2 + 11x + 12 &= 2x^2 + 3x + 8x + 12 \\ &= (2x^2 + 3x) + (8x + 12) && \text{Group the terms.} \\ &= x(2x + 3) + 4(2x + 3) && \text{Factor each group.} \\ &= (2x + 3)(x + 4) && \text{Factor out } (2x + 3). \end{aligned}$$

In general, we have the following procedure.

> **To Factor Trinomials by Grouping**
>
> **STEP 1.** Factor out a greatest common factor, if there is one other than 1.
>
> **STEP 2.** For the resulting trinomial $ax^2 + bx + c$, find two numbers whose product is $a \cdot c$ and whose sum is $b$.
>
> **STEP 3.** Write the middle term, $bx$, using the factors found in Step 2.
>
> **STEP 4.** Factor by grouping.

**EXAMPLE 1**   Factor $3x^2 + 31x + 10$ by grouping.

*Solution*

**STEP 1.** The terms of this trinomial contain no greatest common factor other than 1 (or $-1$).

**STEP 2.** In $3x^2 + 31x + 10$, $a = 3$, $b = 31$, and $c = 10$.

Let's find two numbers whose product is $a \cdot c$ or $3(10) = 30$ and whose sum is $b$ or 31. The numbers are 1 and 30.

| Factors of 30 | Sum of factors | |
|---|---|---|
| 5, 6 | 11 | |
| 3, 10 | 13 | |
| 2, 15 | 17 | |
| 1, 30 | 31 | Correct sum |

**STEP 3.** Write $31x$ as $1x + 30x$ so that $3x^2 + 31x + 10 = 3x^2 + 1x + 30x + 10$.

**STEP 4.** Factor by grouping.

$$\begin{aligned} 3x^2 + 1x + 30x + 10 &= x(3x + 1) + 10(3x + 1) \\ &= (3x + 1)(x + 10) \end{aligned}$$

**PRACTICE**

**1**   Factor $5x^2 + 61x + 12$ by grouping.

**EXAMPLE 2** Factor $8x^2 - 14x + 5$ by grouping.

_Solution_

**STEP 1.** The terms of this trinomial contain no greatest common factor other than 1.

**STEP 2.** This trinomial is of the form $ax^2 + bx + c$ with $a = 8$, $b = -14$, and $c = 5$. Find two numbers whose product is $a \cdot c$ or $8 \cdot 5 = 40$, and whose sum is $b$ or $-14$.

The numbers are $-4$ and $-10$.

**STEP 3.** Write $-14x$ as $-4x - 10x$ so that

$$8x^2 - 14x + 5 = 8x^2 - 4x - 10x + 5$$

| Factors of 40 | Sum of Factors |
|---|---|
| $-40, -1$ | $-41$ |
| $-20, -2$ | $-22$ |
| $-10, -4$ | $-14$ |

Correct sum

**STEP 4.** Factor by grouping.

$$8x^2 - 4x - 10x + 5 = 4x(2x - 1) - 5(2x - 1)$$
$$= (2x - 1)(4x - 5)$$

□

**PRACTICE**
**2** Factor $12x^2 - 19x + 5$ by grouping.

**EXAMPLE 3** Factor $6x^2 - 2x - 20$ by grouping.

_Solution_

**STEP 1.** First factor out the greatest common factor, 2.
$$6x^2 - 2x - 20 = 2(3x^2 - x - 10)$$

**STEP 2.** Next notice that $a = 3$, $b = -1$, and $c = -10$ in the resulting trinomial. Find two numbers whose product is $a \cdot c$ or $3(-10) = -30$ and whose sum is $b$, $-1$. The numbers are $-6$ and 5.

**STEP 3.** $3x^2 - x - 10 = 3x^2 - 6x + 5x - 10$

**STEP 4.** $3x^2 - 6x + 5x - 10 = 3x(x - 2) + 5(x - 2)$
$$= (x - 2)(3x + 5)$$

The factored form of $6x^2 - 2x - 20 = 2(x - 2)(3x + 5)$.

└ Don't forget to include the common factor of 2.

□

**PRACTICE**
**3** Factor $30x^2 - 14x - 4$ by grouping.

**EXAMPLE 4** Factor $18y^4 + 21y^3 - 60y^2$ by grouping.

_Solution_

**STEP 1.** First factor out the greatest common factor, $3y^2$.
$$18y^4 + 21y^3 - 60y^2 = 3y^2(6y^2 + 7y - 20)$$

**STEP 2.** Notice that $a = 6$, $b = 7$, and $c = -20$ in the resulting trinomial. Find two numbers whose product is $a \cdot c$ or $6(-20) = -120$ and whose sum is 7. It may help to factor $-120$ as a product of primes and $-1$.

$$-120 = 2 \cdot 2 \cdot 2 \cdot 3 \cdot 5 \cdot (-1)$$

Then choose pairings of factors until you have two pairings whose sum is 7.

$2 \cdot 2 \cdot 2 \cdot 3 \cdot 5 \cdot (-1)$   The numbers are $-8$ and 15.

**STEP 3.** $6y^2 + 7y - 20 = 6y^2 - 8y + 15y - 20$

**STEP 4.** $6y^2 - 8y + 15y - 20 = 2y(3y - 4) + 5(3y - 4)$
$$= (3y - 4)(2y + 5)$$

The factored form of $18y^4 + 21y^3 - 60y^2$ is $3y^2(3y - 4)(2y + 5)$

⌐ Don't forget to include the common
factor of $3y^2$.

**PRACTICE**
**4**    Factor $40m^4 + 5m^3 - 35m^2$ by grouping.

---

**EXAMPLE 5**    Factor $4x^2 + 20x + 25$ by grouping.

*Solution*

**STEP 1.** The terms of this trinomial contain no greatest common factor other than 1 (or $-1$).

**STEP 2.** In $4x^2 + 20x + 25$, $a = 4$, $b = 20$, and $c = 25$. Find two numbers whose product is $a \cdot c$ or $4 \cdot 25 = 100$ and whose sum is 20. The numbers are 10 and 10.

**STEP 3.** Write $20x$ as $10x + 10x$ so that
$$4x^2 + 20x + 25 = 4x^2 + 10x + 10x + 25$$

**STEP 4.** Factor by grouping.
$$4x^2 + 10x + 10x + 25 = 2x(2x + 5) + 5(2x + 5)$$
$$= (2x + 5)(2x + 5)$$

The factored form of $4x^2 + 20x + 25$ is $(2x + 5)(2x + 5)$ or $(2x + 5)^2$

**PRACTICE**
**5**    Factor $16x^2 + 24x + 9$ by grouping.

---

A trinomial that is the square of a binomial, such as the trinomial in Example 5, is called a **perfect square trinomial.** From Chapter 5, there are special product formulas we can use to help us recognize and factor these trinomials. To study these formulas further, see Section 6.3, Objective 3. **Remember:** A perfect square trinomial, such as the one in Example 5, may be factored by special product formulas or by other methods of factoring trinomials, such as by grouping.

## 6.4 | EXERCISE SET

*Factor each polynomial by grouping. Notice that Step 3 has already been done in these exercises. See Examples 1 through 5.*

**1.** $x^2 + 3x + 2x + 6$

**2.** $x^2 + 5x + 3x + 15$

**3.** $y^2 + 8y - 2y - 16$

**4.** $z^2 + 10z - 7z - 70$

**5.** $8x^2 - 5x - 24x + 15$

**6.** $4x^2 - 9x - 32x + 72$

**7.** $5x^4 - 3x^2 + 25x^2 - 15$

**8.** $2y^4 - 10y^2 + 7y^2 - 35$

**MIXED PRACTICE**

*Factor each trinomial by grouping. Exercises 9–12 are broken into parts to help you get started. See Examples 1 through 5.*

**9.** $6x^2 + 11x + 3$

   **a.** Find two numbers whose product is $6 \cdot 3 = 18$ and whose sum is 11.

   **b.** Write $11x$ using the factors from part (a).

   **c.** Factor by grouping.

**10.** $8x^2 + 14x + 3$

   **a.** Find two numbers whose product is $8 \cdot 3 = 24$ and whose sum is 14.

   **b.** Write $14x$ using the factors from part (a).

   **c.** Factor by grouping.

11. $15x^2 - 23x + 4$

   **a.** Find two numbers whose product is $15 \cdot 4 = 60$ and whose sum is $-23$.

   **b.** Write $-23x$ using the factors from part (a).

   **c.** Factor by grouping.

12. $6x^2 - 13x + 5$

   **a.** Find two numbers whose product is $6 \cdot 5 = 30$ and whose sum is $-13$.

   **b.** Write $-13x$ using the factors from part (a).

   **c.** Factor by grouping.

🔳 13. $21y^2 + 17y + 2$

14. $15x^2 + 11x + 2$

15. $7x^2 - 4x - 11$

16. $8x^2 - x - 9$

🔳 17. $10x^2 - 9x + 2$

18. $30x^2 - 23x + 3$

19. $2x^2 - 7x + 5$

20. $2x^2 - 7x + 3$

21. $12x + 4x^2 + 9$

22. $20x + 25x^2 + 4$

23. $4x^2 - 8x - 21$

24. $6x^2 - 11x - 10$

25. $10x^2 - 23x + 12$

26. $21x^2 - 13x + 2$

27. $2x^3 + 13x^2 + 15x$

28. $3x^3 + 8x^2 + 4x$

29. $16y^2 - 34y + 18$

30. $4y^2 - 2y - 12$

31. $-13x + 6 + 6x^2$

32. $-25x + 12 + 12x^2$

33. $54a^2 - 9a - 30$

34. $30a^2 + 38a - 20$

35. $20a^3 + 37a^2 + 8a$

36. $10a^3 + 17a^2 + 3a$

🔳 37. $12x^3 - 27x^2 - 27x$

38. $30x^3 - 155x^2 + 25x$

39. $3x^2y + 4xy^2 + y^3$

40. $6r^2t + 7rt^2 + t^3$

41. $20z^2 + 7z + 1$

42. $36z^2 + 6z + 1$

43. $5x^2 + 50xy + 125y^2$

44. $3x^2 + 42xy + 147y^2$

45. $24a^2 - 6ab - 30b^2$

46. $30a^2 + 5ab - 25b^2$

47. $15p^4 + 31p^3q + 2p^2q^2$

48. $20s^4 + 61s^3t + 3s^2t^2$

49. $162a^4 - 72a^2 + 8$

50. $32n^4 - 112n^2 + 98$

51. $35 + 12x + x^2$

52. $33 + 14x + x^2$

53. $6 - 11x + 5x^2$

54. $5 - 12x + 7x^2$

## REVIEW AND PREVIEW

*Multiply. See Section 5.4.*

55. $(x - 2)(x + 2)$

56. $(y - 5)(y + 5)$

57. $(y + 4)(y + 4)$

58. $(x + 7)(x + 7)$

59. $(9z + 5)(9z - 5)$

60. $(8y + 9)(8y - 9)$

61. $(x - 3)(x^2 + 3x + 9)$

62. $(2z - 1)(4z^2 + 2z + 1)$

## CONCEPT EXTENSIONS

*Write the perimeter of each figure as a simplified polynomial. Then factor the polynomial.*

63.

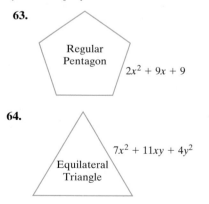

64.

*Factor each polynomial by grouping.*

65. $x^{2n} + 2x^n + 3x^n + 6$

   (***Hint:*** Don't forget that $x^{2n} = x^n \cdot x^n$.)

66. $x^{2n} + 6x^n + 10x^n + 60$

67. $3x^{2n} + 16x^n - 35$

68. $12x^{2n} - 40x^n + 25$

69. In your own words, explain how to factor a trinomial by grouping.

# 6.5 FACTORING BINOMIALS

**OBJECTIVES**

1 Factor the difference of two squares.

2 Factor the sum or difference of two cubes.

**OBJECTIVE 1 ▶ Factoring the difference of two squares.** When learning to multiply binomials in Chapter 5, we studied a special product, the product of the sum and difference of two terms, $a$ and $b$:

$$(a + b)(a - b) = a^2 - b^2$$

For example, the product of $x + 3$ and $x - 3$ is

$$(x + 3)(x - 3) = x^2 - 9$$

The binomial $x^2 - 9$ is called a **difference** **of squares.** In this section, we use the pattern for the product of a sum and difference to factor the binomial difference of squares.

> **Factoring the Difference of Two Squares**
> $$a^2 - b^2 = (a + b)(a - b)$$

> ▶ **Helpful Hint**
> Since multiplication is commutative, remember that the order of factors does not matter. In other words,
> $$a^2 - b^2 = (a + b)(a - b) \text{ or } (a - b)(a + b)$$

**EXAMPLE 1**   Factor: $x^2 - 25$

**Solution**   $x^2 - 25$ is the difference of two squares since $x^2 - 25 = x^2 - 5^2$. Therefore,

$$x^2 - 25 = x^2 - 5^2 = (x + 5)(x - 5)$$

Multiply to check.   □

**PRACTICE**
**1**   Factor $x^2 - 81$.

**EXAMPLE 2**   Factor each difference of squares.

**a.** $4x^2 - 1$        **b.** $25a^2 - 9b^2$        **c.** $y^2 - \dfrac{4}{9}$

**Solution**

**a.** $4x^2 - 1 = (2x)^2 - 1^2 = (2x + 1)(2x - 1)$

**b.** $25a^2 - 9b^2 = (5a)^2 - (3b)^2 = (5a + 3b)(5a - 3b)$

**c.** $y^2 - \dfrac{4}{9} = y^2 - \left(\dfrac{2}{3}\right)^2 = \left(y + \dfrac{2}{3}\right)\left(y - \dfrac{2}{3}\right)$   □

**PRACTICE**
**2**   Factor each difference of squares.

**a.** $9x^2 - 1$        **b.** $36a^2 - 49b^2$        **c.** $p^2 - \dfrac{25}{36}$

**EXAMPLE 3** Factor: $x^4 - y^6$

*Solution* This is a difference of squares since $x^4 = (x^2)^2$ and $y^6 = (y^3)^2$. Thus,

$$x^4 - y^6 = (x^2)^2 - (y^3)^2 = (x^2 + y^3)(x^2 - y^3) \qquad \square$$

**PRACTICE**
**3** Factor $p^4 - q^{10}$.

**EXAMPLE 4** Factor each binomial.

**a.** $y^4 - 16$ **b.** $x^2 + 4$

*Solution*

**a.** $y^4 - 16 = (y^2)^2 - 4^2$
$= (y^2 + 4)\underbrace{(y^2 - 4)}.$ Factor the difference of two squares.
This binomial can be factored further since it is the difference of two squares.
$= (y^2 + 4)(y + 2)(y - 2)$ Factor the difference of two squares.

**b.** $x^2 + 4$

Note that the binomial $x^2 + 4$ is the *sum* of two squares since we can write $x^2 + 4$ as $x^2 + 2^2$. We might try to factor using $(x + 2)(x + 2)$ or $(x - 2)(x - 2)$. But when we multiply to check, we find that neither factoring is correct.

$$(x + 2)(x + 2) = x^2 + 4x + 4$$
$$(x - 2)(x - 2) = x^2 - 4x + 4$$

In both cases, the product is a trinomial, not the required binomial. In fact, $x^2 + 4$ is a prime polynomial. $\qquad \square$

**PRACTICE**
**4** Factor each binomial.

**a.** $z^4 - 81$ **b.** $m^2 + 49$

---

> ▶ **Helpful Hint**
> When factoring, don't forget:
>
> • See whether the terms have a greatest common factor (GCF) (other than 1) that can be factored out.
> • Other than a GCF, the **sum** of two squares cannot be factored using real numbers.
> • Factor completely. Always check to see whether any factors can be factored further.

---

**EXAMPLES** Factor each difference of two squares.

**5.** $4x^3 - 49x = x(4x^2 - 49)$ Factor out the common factor, $x$.
$= x[(2x)^2 - 7^2]$
$= x(2x + 7)(2x - 7)$ Factor the difference of two squares.

**6.** $162x^4 - 2 = 2(81x^4 - 1)$ Factor out the common factor, 2.
$= 2(9x^2 + 1)(9x^2 - 1)$ Factor the difference of two squares.
$= 2(9x^2 + 1)(3x + 1)(3x - 1)$ Factor the difference of two squares. $\qquad \square$

**PRACTICES**
**5–6** Factor each difference of two squares.

**5.** $36y^3 - 25y$ **6.** $80y^4 - 5$

**EXAMPLE 7**   Factor: $-49x^2 + 16$

*Solution*   Factor as is, or, if you like, rearrange terms.

Factor as is: $-49x^2 + 16 = -1(49x^2 - 16)$     Factor out $-1$.

$= -1(7x + 4)(7x - 4)$   Factor the difference of two squares.

Rewrite binomial: $-49x^2 + 16 = 16 - 49x^2 = 4^2 - (7x)^2$

$= (4 + 7x)(4 - 7x)$

Both factorizations are correct and are equal. To see this, factor $-1$ from $(4 - 7x)$ in the second factorization.     ☐

**PRACTICE**
**7**   Factor: $-9x^2 + 100$

---

**OBJECTIVE 2** ▶ **Factoring the sum or difference of two cubes.** Although the sum of two squares usually does not factor, the sum or difference of two cubes can be factored and reveals factoring patterns. The pattern for the sum of cubes is illustrated by multiplying the binomial $x + y$ and the trinomial $x^2 - xy + y^2$.

$$
\begin{array}{r}
x^2 - xy + y^2 \\
\underline{x + y} \\
x^2y - xy^2 + y^3 \\
\underline{x^3 - x^2y + xy^2 \phantom{+ y^3}} \\
x^3 \phantom{- x^2y + xy^2} + y^3
\end{array}
$$

Thus, $(x + y)(x^2 - xy + y^2) = x^3 + y^3$   Sum of cubes

The pattern for the difference of two cubes is illustrated by multiplying the binomial $x - y$ by the trinomial $x^2 + xy + y^2$. The result is

$(x - y)(x^2 + xy + y^2) = x^3 - y^3$   Difference of cubes

---

**Factoring the Sum or Difference of Two Cubes**

$$a^3 + b^3 = (a + b)(a^2 - ab + b^2)$$
$$a^3 - b^3 = (a - b)(a^2 + ab + b^2)$$

---

Recall that "factor" means "to write as a product." Above are patterns for writing sums and differences as products.

**EXAMPLE 8**   Factor: $x^3 + 8$

*Solution*   First, write the binomial in the form $a^3 + b^3$.

$x^3 + 8 = x^3 + 2^3$   Write in the form $a^3 + b^3$.

If we replace $a$ with $x$ and $b$ with 2 in the formula above, we have

$x^3 + 2^3 = (x + 2)[x^2 - (x)(2) + 2^2]$

$= (x + 2)(x^2 - 2x + 4)$     ☐

**PRACTICE**
**8**   Factor $x^3 + 64$.

▶ **Helpful Hint**

When factoring sums or differences of cubes, notice the sign patterns.

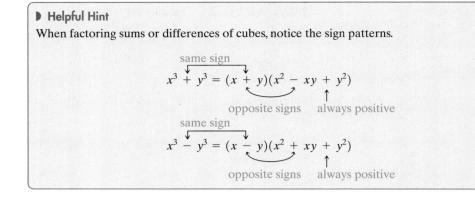

**EXAMPLE 9** Factor: $y^3 - 27$

*Solution*

$$y^3 - 27 = y^3 - 3^3 \qquad \text{Write in the form } a^3 - b^3.$$
$$= (y - 3)[y^2 + (y)(3) + 3^2]$$
$$= (y - 3)(y^2 + 3y + 9)$$

□

**PRACTICE**
**9** Factor $x^3 - 125$.

**EXAMPLE 10** Factor: $64x^3 + 1$

*Solution*

$$64x^3 + 1 = (4x)^3 + 1^3$$
$$= (4x + 1)[(4x)^2 - (4x)(1) + 1^2]$$
$$= (4x + 1)(16x^2 - 4x + 1)$$

□

**PRACTICE**
**10** Factor $27y^3 + 1$.

**EXAMPLE 11** Factor: $54a^3 - 16b^3$

*Solution* Remember to factor out common factors first before using other factoring methods.

$$54a^3 - 16b^3 = 2(27a^3 - 8b^3) \qquad \text{Factor out the GCF 2.}$$
$$= 2[(3a)^3 - (2b)^3] \qquad \text{Difference of two cubes}$$
$$= 2(3a - 2b)[(3a)^2 + (3a)(2b) + (2b)^2]$$
$$= 2(3a - 2b)(9a^2 + 6ab + 4b^2)$$

□

**PRACTICE**
**11** Factor $32x^3 - 500y^3$.

### Calculator Explorations

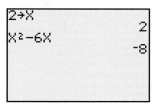

### Graphing

A graphing calculator is a convenient tool for evaluating an expression at a given replacement value. For example, let's evaluate $x^2 - 6x$ when $x = 2$. To do so, store the value 2 in the variable $x$ and then enter and evaluate the algebraic expression.

```
2→X
               2
X²-6X
              -8
```

The value of $x^2 - 6x$ when $x = 2$ is $-8$. You may want to use this method for evaluating expressions as you explore the following.

We can use a graphing calculator to explore factoring patterns numerically. Use your calculator to evaluate $x^2 - 2x + 1$, $x^2 - 2x - 1$, and $(x - 1)^2$ for each value of $x$ given in the table. What do you observe?

|           | $x^2 - 2x + 1$ | $x^2 - 2x - 1$ | $(x - 1)^2$ |
|-----------|----------------|----------------|-------------|
| $x = 5$   |                |                |             |
| $x = -3$  |                |                |             |
| $x = 2.7$ |                |                |             |
| $x = -12.1$ |              |                |             |
| $x = 0$   |                |                |             |

Notice in each case that $x^2 - 2x - 1 \neq (x - 1)^2$. Because for each $x$ in the table the value of $x^2 - 2x + 1$ and the value of $(x - 1)^2$ are the same, we might guess that $x^2 - 2x + 1 = (x - 1)^2$. We can verify our guess algebraically with multiplication:

$$(x - 1)(x - 1) = x^2 - x - x + 1 = x^2 - 2x + 1$$

## VOCABULARY & READINESS CHECK

*Use the choices below to fill in each blank. Some choices may be used more than once and some choices may not be used at all.*

    true           difference of two squares       sum of two cubes

    false         difference of two cubes

**1.** The expression $x^3 - 27$ is called a(n) _____.

**2.** The expression $x^2 - 49$ is called a(n) _____.

**3.** The expression $z^3 + 1$ is called a(n) _____.

**4.** True or false: The binomial $y^2 + 9$ factors as $(y + 3)^2$. _____

*Write each number or term as a square.*

**5.** 64            **6.** 100                 **7.** $49x^2$             **8.** $25y^4$

*Write each number or term as a cube.*

**9.** 64            **10.** 1                  **11.** $8y^3$             **12.** $x^6$

## 6.5 | EXERCISE SET

PRACTICE   WATCH   DOWNLOAD   READ   REVIEW

*Factor each binomial completely. See Examples 1 through 7.*

1. $x^2 - 4$

2. $x^2 - 36$

3. $81p^2 - 1$

4. $49m^2 - 1$

5. $25y^2 - 9$

6. $49a^2 - 16$

7. $121m^2 - 100n^2$

8. $169a^2 - 49b^2$

9. $x^2y^2 - 1$

10. $a^2b^2 - 16$

11. $x^2 - \dfrac{1}{4}$

12. $y^2 - \dfrac{1}{16}$

13. $-4r^2 + 1$

14. $-9t^2 + 1$

15. $16r^2 + 1$

16. $49y^2 + 1$

17. $-36 + x^2$

18. $-1 + y^2$

19. $m^4 - 1$

20. $n^4 - 16$

21. $m^4 - n^{18}$

22. $n^4 - r^6$

*Factor the sum or difference of two cubes. See Examples 8 through 11.*

23. $x^3 + 125$

24. $p^3 + 1$

25. $8a^3 - 1$

26. $27y^3 - 1$

27. $m^3 + 27n^3$

28. $y^3 + 64z^3$

29. $5k^3 + 40$

30. $6r^3 + 162$

31. $x^3y^3 - 64$

32. $a^3b^3 - 8$

33. $250r^3 - 128t^3$

34. $24x^3 - 81y^3$

### MIXED PRACTICE

*Factor each binomial completely. See Examples 1 through 11.*

35. $r^2 - 64$

36. $q^2 - 121$

37. $x^2 - 169y^2$

38. $x^2 - 225y^2$

39. $27 - t^3$

40. $125 - r^3$

41. $18r^2 - 8$

42. $32t^2 - 50$

43. $9xy^2 - 4x$

44. $36x^2y - 25y$

45. $8m^3 + 64$

46. $2x^3 + 54$

47. $xy^3 - 9xyz^2$

48. $x^3y - 4xy^3$

49. $36x^2 - 64y^2$

50. $225a^2 - 81b^2$

51. $144 - 81x^2$

52. $12x^2 - 27$

53. $x^3y^3 - z^6$

54. $a^3b^3 - c^9$

55. $49 - \dfrac{9}{25}m^2$

56. $100 - \dfrac{4}{81}n^2$

57. $t^3 + 343$

58. $s^3 + 216$

59. $n^3 + 49n$

60. $y^3 + 64y$

61. $x^6 - 81x^2$

62. $n^9 - n^5$

63. $64p^3q - 81pq^3$

64. $100x^3y - 49xy^3$

65. $27x^2y^3 + xy^2$

66. $8x^3y^3 + x^3y$

67. $125a^4 - 64ab^3$

68. $64m^4 - 27mn^3$

69. $16x^4 - 64x^2$

70. $25y^4 - 100y^2$

### REVIEW AND PREVIEW

*Solve each equation. See Section 2.3.*

71. $x - 6 = 0$

72. $y + 5 = 0$

73. $2m + 4 = 0$

**74.** $3x - 9 = 0$

**75.** $5z - 1 = 0$

**76.** $4a + 2 = 0$

*Solve. See Section 6.1. The percent of undergraduate college students who have credit cards each year from 2000 through 2006 can be approximately modeled by the polynomial $-1.2x^2 + 4x + 80$, where x is the number of years since 2000.*

**77.** Find the percent of college students who had credit cards in 2003.

**78.** Find the percent of college students who had credit cards in 2006.

**79.** Write a factored form of $-1.2x^2 + 4x + 80$ by factoring $-4$ from the terms of this polynomial.

**80.** Use your answers to Exercises 77 and 78 to write down any trends.

**CONCEPT EXTENSIONS**

*Factor each expression completely.*

**81.** $(x + 2)^2 - y^2$

**82.** $(y - 6)^2 - z^2$

**83.** $a^2(b - 4) - 16(b - 4)$

**84.** $m^2(n + 8) - 9(n + 8)$

**85.** $(x^2 + 6x + 9) - 4y^2$ (***Hint:*** Factor the trinomial in parentheses first.)

**86.** $(x^2 + 2x + 1) - 36y^2$

**87.** $x^{2n} - 100$

**88.** $x^{2n} - 81$

**89.** What binomial multiplied by $(x - 6)$ gives the difference of two squares?

**90.** What binomial multiplied by $(5 + y)$ gives the difference of two squares?

**91.** In your own words, explain how to tell whether a binomial is a difference of squares. Then explain how to factor a difference of squares.

**92.** In your own words, explain how to tell whether a binomial is a sum of cubes. Then explain how to factor a sum of cubes.

**93.** An object is dropped from the top of Pittsburgh's USX Tower, which is 841 feet tall. (*Source: World Almanac*

research) The height of the object after $t$ seconds is given by the expression $841 - 16t^2$.

**a.** Find the height of the object after 2 seconds.

**b.** Find the height of the object after 5 seconds.

**c.** To the nearest whole second, estimate when the object hits the ground.

**d.** Factor $841 - 16t^2$.

841 feet

**94.** A worker on the top of the Aetna Life Building in San Francisco accidentally drops a bolt. The Aetna Life Building is 529 feet tall. (*Source: World Almanac* research) The height of the bolt after $t$ seconds is given by the expression $529 - 16t^2$.

**a.** Find the height of the bolt after 1 second.

**b.** Find the height of the bolt after 4 seconds.

**c.** To the nearest whole second, estimate when the bolt hits the ground.

**d.** Factor $529 - 16t^2$.

**95.** At this writing, the world's tallest building is the Taipei 101 in Taipei, Taiwan, at a height of 1671 feet. (*Source:* Council on Tall Buildings and Urban Habitat) Suppose a worker is suspended 71 feet below the top of the pinnacle atop the building, at a height of 1600 feet above the ground. If the worker accidentally drops a bolt, the height of the bolt after $t$ seconds is given by the expression $1600 - 16t^2$.

**a.** Find the height of the bolt after 3 seconds.

**b.** Find the height of the bolt after 7 seconds.

**c.** To the nearest whole second, estimate when the bolt hits the ground.

**d.** Factor $1600 - 16t^2$.

**96.** A performer with the Moscow Circus is planning a stunt involving a free fall from the top of the Moscow State University building, which is 784 feet tall. (*Source:* Council on Tall

Buildings and Urban Habitat) Neglecting air resistance, the performer's height above gigantic cushions positioned at ground level after $t$ seconds is given by the expression $784 - 16t^2$.

**a.** Find the performer's height after 2 seconds.

**b.** Find the performer's height after 5 seconds.

**c.** To the nearest whole second, estimate when the performer reaches the cushions positioned at ground level.

**d.** Factor $784 - 16t^2$.

---

### STUDY SKILLS BUILDER

**Are You Getting All the Mathematics Help That You Need?**

Remember that, in addition to your instructor, there are many places to get help with your mathematics course. For example:

- This text has an accompanying video lesson for every section and the CD in this text contains worked out solutions to every Chapter Test exercise.

- The back of the book contains answers to odd-numbered exercises and selected solutions.

- A student *Solutions Manual* is available that contains worked-out solutions to odd-numbered exercises as well as solutions to every exercise in the Integrated Reviews, Chapter Reviews, Chapter Tests, and Cumulative Reviews.

- Don't forget to check with your instructor for other local resources available to you, such as a tutor center.

*Exercises*

**1.** List items you find helpful in the text and all student supplements to this text.

**2.** List all the campus help that is available to you for this course.

**3.** List any help (besides the textbook) from Exercises 1 and 2 above that you are using.

**4.** List any help (besides the textbook) that you feel you should try.

**5.** Write a goal for yourself that includes trying anything you listed in Exercise 4 during the next week.

---

## INTEGRATED REVIEW CHOOSING A FACTORING STRATEGY

Sections 6.1–6.5

*The following steps may be helpful when factoring polynomials.*

**Factoring a Polynomial**

**STEP 1.** Are there any common factors? If so, factor out the GCF.

**STEP 2.** How many terms are in the polynomial?

**a.** If there are **two** terms, decide if one of the following can be applied.

   **i.** Difference of two squares: $a^2 - b^2 = (a + b)(a - b)$.

   **ii.** Difference of two cubes: $a^3 - b^3 = (a - b)(a^2 + ab + b^2)$.

   **iii.** Sum of two cubes: $a^3 + b^3 = (a + b)(a^2 - ab + b^2)$.

**b.** If there are **three** terms, try one of the following.

   **i.** Perfect square trinomial: $a^2 + 2ab + b^2 = (a + b)^2$
$$a^2 - 2ab + b^2 = (a - b)^2.$$

   **ii.** If not a perfect square trinomial, factor using the methods presented in Sections 6.2 through 6.4.

**c.** If there are **four** or more terms, try factoring by grouping.

**STEP 3.** See if any factors in the factored polynomial can be factored further.

**STEP 4.** Check by multiplying.

Study the next five examples to help you use the steps on the previous page.

**EXAMPLE 1**  Factor $10t^2 - 17t + 3$.

*Solution*

**STEP 1.**  The terms of this polynomial have no common factor (other than 1).

**STEP 2.**  There are three terms, so this polynomial is a trinomial. This trinomial is not a perfect square trinomial, so factor using methods from earlier sections.

$$\text{Factors of } 10t^2: \quad 10t^2 = 2t \cdot 5t, \qquad 10t^2 = t \cdot 10t$$

Since the middle term, $-17t$, has a negative numerical coefficient, find negative factors of 3.

$$\text{Factors of 3:} \quad 3 = -1 \cdot -3$$

Try different combinations of these factors. The correct combination is

$$(2t - 3)(5t - 1) = 10t^2 - 17t + 3$$
$$\underbrace{\phantom{xxxxx}}_{-15t}$$
$$-2t$$
$$\overline{-17t} \quad \text{Correct middle term}$$

**STEP 3.**  No factor can be factored further, so we have factored completely.

**STEP 4.**  To check, multiply $2t - 3$ and $5t - 1$.

$$(2t - 3)(5t - 1) = 10t^2 - 2t - 15t + 3 = 10t^2 - 17t + 3$$

The factored form of $10t^2 - 17t + 3$ is $(2t - 3)(5t - 1)$.  $\square$

**PRACTICE**
**1**  Factor $6x^2 - 11x + 3$.

---

**EXAMPLE 2**  Factor $2x^3 + 3x^2 - 2x - 3$.

*Solution*

**STEP 1.**  There are no factors common to all terms.

**STEP 2.**  Try factoring by grouping since this polynomial has four terms.

$$2x^3 + 3x^2 - 2x - 3 = x^2(2x + 3) - 1(2x + 3) \quad \text{Factor out the greatest common factor for each pair of terms.}$$
$$= (2x + 3)(x^2 - 1) \quad \text{Factor out } 2x + 3.$$

**STEP 3.**  The binomial $x^2 - 1$ can be factored further. It is the difference of two squares.

$$= (2x + 3)(x + 1)(x - 1) \quad \text{Factor } x^2 - 1 \text{ as a difference of squares.}$$

**STEP 4.**  Check by finding the product of the three binomials. The polynomial factored completely is $(2x + 3)(x + 1)(x - 1)$.  $\square$

**PRACTICE**
**2**  Factor $3x^3 + x^2 - 12x - 4$.

---

**EXAMPLE 3**  Factor $12m^2 - 3n^2$.

*Solution*

**STEP 1.**  The terms of this binomial contain a greatest common factor of 3.

$$12m^2 - 3n^2 = 3(4m^2 - n^2) \quad \text{Factor out the greatest common factor.}$$

**STEP 2.** The binomial $4m^2 - n^2$ is a difference of squares.

$$= 3(2m + n)(2m - n) \quad \text{Factor the difference of squares.}$$

**STEP 3.** No factor can be factored further.

**STEP 4.** We check by multiplying.

$$3(2m + n)(2m - n) = 3(4m^2 - n^2) = 12m^2 - 3n^2$$

The factored form of $12m^2 - 3n^2$ is $3(2m + n)(2m - n)$.

**PRACTICE**
**3** Factor $27x^2 - 3y^2$.

**EXAMPLE 4** Factor $x^3 + 27y^3$.

*Solution*

**STEP 1.** The terms of this binomial contain no common factor (other than 1).

**STEP 2.** This binomial is the sum of two cubes.

$$x^3 + 27y^3 = (x)^3 + (3y)^3$$
$$= (x + 3y)[x^2 - x(3y) + (3y)^2]$$
$$= (x + 3y)(x^2 - 3xy + 9y^2)$$

**STEP 3.** No factor can be factored further.

**STEP 4.** We check by multiplying.

$$(x + 3y)(x^2 - 3xy + 9y^2) = x(x^2 - 3xy + 9y^2) + 3y(x^2 - 3xy + 9y^2)$$
$$= x^3 - 3x^2y + 9xy^2 + 3x^2y - 9xy^2 + 27y^3$$
$$= x^3 + 27y^3$$

Thus, $x^3 + 27y^3$ factored completely is $(x + 3y)(x^2 - 3xy + 9y^2)$.

**PRACTICE**
**4** Factor $8a^3 + b^3$.

**EXAMPLE 5** Factor $30a^2b^3 + 55a^2b^2 - 35a^2b$.

*Solution*

**STEP 1.** $30a^2b^3 + 55a^2b^2 - 35a^2b = 5a^2b(6b^2 + 11b - 7)$ \quad Factor out the GCF.

**STEP 2.** $= 5a^2b(2b - 1)(3b + 7)$ \quad Factor the resulting trinomial.

**STEP 3.** No factor can be factored further.

**STEP 4.** Check by multiplying.
The trinomial factored completely is $5a^2b(2b - 1)(3b + 7)$.

**PRACTICE**
**5** Factor $60x^3y^2 - 66x^2y^2 - 36xy^2$.

*Factor the following completely.*

**1.** $x^2 + 2xy + y^2$
**2.** $x^2 - 2xy + y^2$
**3.** $a^2 + 11a - 12$
**4.** $a^2 - 11a + 10$
**5.** $a^2 - a - 6$
**6.** $a^2 - 2a + 1$
**7.** $x^2 + 2x + 1$
**8.** $x^2 + x - 2$
**9.** $x^2 + 4x + 3$
**10.** $x^2 + x - 6$
**11.** $x^2 + 7x + 12$
**12.** $x^2 + x - 12$
**13.** $x^2 + 3x - 4$
**14.** $x^2 - 7x + 10$
**15.** $x^2 + 2x - 15$
**16.** $x^2 + 11x + 30$
**17.** $x^2 - x - 30$
**18.** $x^2 + 11x + 24$
**19.** $2x^2 - 98$
**20.** $3x^2 - 75$
**21.** $x^2 + 3x + xy + 3y$

**22.** $3y - 21 + xy - 7x$

**23.** $x^2 + 6x - 16$

**24.** $x^2 - 3x - 28$

**25.** $4x^3 + 20x^2 - 56x$

**26.** $6x^3 - 6x^2 - 120x$

**27.** $12x^2 + 34x + 24$

**28.** $8a^2 + 6ab - 5b^2$

**29.** $4a^2 - b^2$

**30.** $28 - 13x - 6x^2$

**31.** $20 - 3x - 2x^2$

**32.** $x^2 - 2x + 4$

**33.** $a^2 + a - 3$

**34.** $6y^2 + y - 15$

**35.** $4x^2 - x - 5$

**36.** $x^2y - y^3$

**37.** $4t^2 + 36$

**38.** $x^2 + x + xy + y$

**39.** $ax + 2x + a + 2$

**40.** $18x^3 - 63x^2 + 9x$

**41.** $12a^3 - 24a^2 + 4a$

**42.** $x^2 + 14x - 32$

**43.** $x^2 - 14x - 48$

**44.** $16a^2 - 56ab + 49b^2$

**45.** $25p^2 - 70pq + 49q^2$

**46.** $7x^2 + 24xy + 9y^2$

**47.** $125 - 8y^3$

**48.** $64x^3 + 27$

**49.** $-x^2 - x + 30$

**50.** $-x^2 + 6x - 8$

**51.** $14 + 5x - x^2$

**52.** $3 - 2x - x^2$

**53.** $3x^4y + 6x^3y - 72x^2y$

**54.** $2x^3y + 8x^2y^2 - 10xy^3$

**55.** $5x^3y^2 - 40x^2y^3 + 35xy^4$

**56.** $4x^4y - 8x^3y - 60x^2y$

**57.** $12x^3y + 243xy$

**58.** $6x^3y^2 + 8xy^2$

**59.** $4 - x^2$

**60.** $9 - y^2$

**61.** $3rs - s + 12r - 4$

**62.** $x^3 - 2x^2 + 3x - 6$

**63.** $4x^2 - 8xy - 3x + 6y$

**64.** $4x^2 - 2xy - 7yz + 14xz$

**65.** $6x^2 + 18xy + 12y^2$

**66.** $12x^2 + 46xy - 8y^2$

**67.** $xy^2 - 4x + 3y^2 - 12$

**68.** $x^2y^2 - 9x^2 + 3y^2 - 27$

**69.** $5(x + y) + x(x + y)$

**70.** $7(x - y) + y(x - y)$

**71.** $14t^2 - 9t + 1$

**72.** $3t^2 - 5t + 1$

**73.** $3x^2 + 2x - 5$

**74.** $7x^2 + 19x - 6$

**75.** $x^2 + 9xy - 36y^2$

**76.** $3x^2 + 10xy - 8y^2$

**77.** $1 - 8ab - 20a^2b^2$

**78.** $1 - 7ab - 60a^2b^2$

**79.** $9 - 10x^2 + x^4$

**80.** $36 - 13x^2 + x^4$

**81.** $x^4 - 14x^2 - 32$

**82.** $x^4 - 22x^2 - 75$

**83.** $x^2 - 23x + 120$

**84.** $y^2 + 22y + 96$

**85.** $6x^3 - 28x^2 + 16x$

**86.** $6y^3 - 8y^2 - 30y$

**87.** $27x^3 - 125y^3$

**88.** $216y^3 - z^3$

**89.** $x^3y^3 + 8z^3$

**90.** $27a^3b^3 + 8$

**91.** $2xy - 72x^3y$

**92.** $2x^3 - 18x$

**93.** $x^3 + 6x^2 - 4x - 24$

**94.** $x^3 - 2x^2 - 36x + 72$

**95.** $6a^3 + 10a^2$

**96.** $4n^2 - 6n$

**97.** $a^2(a + 2) + 2(a + 2)$

**98.** $a - b + x(a - b)$

**99.** $x^3 - 28 + 7x^2 - 4x$

**100.** $a^3 - 45 - 9a + 5a^2$

## CONCEPT EXTENSIONS

*Factor.*

**101.** $(x - y)^2 - z^2$

**102.** $(x + 2y)^2 - 9$

**103.** $81 - (5x + 1)^2$

**104.** $b^2 - (4a + c)^2$

**105.** Explain why it makes good sense to factor out the GCF first, before using other methods of factoring.

**106.** The sum of two squares usually does not factor. Is the sum of two squares $9x^2 + 81y^2$ factorable?

**107.** Which of the following are equivalent to $(x + 10)(x - 7)$?

    **a.** $(x - 7)(x + 10)$     **b.** $-1(x + 10)(x - 7)$

    **c.** $-1(x + 10)(7 - x)$     **d.** $-1(-x - 10)(7 - x)$

# 6.6 SOLVING QUADRATIC EQUATIONS BY FACTORING

**OBJECTIVES**

1 Solve quadratic equations by factoring.

2 Solve equations with degree greater than 2 by factoring.

3 Find the *x*-intercepts of the graph of a quadratic equation in two variables.

In this section, we introduce a new type of equation—the **quadratic equation.**

> **Quadratic Equation**
> A quadratic equation is one that can be written in the form
> $$ax^2 + bx + c = 0$$
> where $a$, $b$, and $c$ are real numbers and $a \neq 0$.

Some examples of quadratic equations are shown below.

$$x^2 - 9x - 22 = 0 \qquad 4x^2 - 28 = -49 \qquad x(2x - 7) = 4$$

The form $ax^2 + bx + c = 0$ is called the **standard form** of a quadratic equation. The quadratic equation $x^2 - 9x - 22 = 0$ is the only equation above that is in standard form.

Quadratic equations model many real-life situations. For example, let's suppose we want to know how long before a person diving from a 144-foot cliff reaches the ocean. The answer to this question is found by solving the quadratic equation $-16t^2 + 144 = 0$. (See Example 1 in Section 6.7.)

144 feet

**OBJECTIVE 1 ▶ Solving quadratic equations by factoring.** Some quadratic equations can be solved by making use of factoring and the **zero factor property.**

> **Zero Factor Theorem**
> If $a$ and $b$ are real numbers and if $ab = 0$, then $a = 0$ or $b = 0$.

This theorem states that if the product of two numbers is 0 then at least one of the numbers must be 0.

**EXAMPLE 1**　Solve: $(x - 3)(x + 1) = 0$.

*Solution*　If this equation is to be a true statement, then either the factor $x - 3$ must be 0 or the factor $x + 1$ must be 0. In other words, either

$$x - 3 = 0 \qquad \text{or} \qquad x + 1 = 0$$

If we solve these two linear equations, we have

$$x = 3 \qquad \text{or} \qquad x = -1$$

Thus, 3 and $-1$ are both solutions of the equation $(x - 3)(x + 1) = 0$. To check, we replace $x$ with 3 in the original equation. Then we replace $x$ with $-1$ in the original equation.

**Check:**　Let $x = 3$.

$(x - 3)(x + 1) = 0$

$(3 - 3)(3 + 1) \overset{?}{=} 0$　Replace $x$ with 3.

$0(4) = 0$　True

Let $x = -1$.

$(x - 3)(x + 1) = 0$

$(-1 - 3)(-1 + 1) \overset{?}{=} 0$　Replace $x$ with $-1$.

$(-4)(0) = 0$　True

The solutions are 3 and $-1$, or we say that the solution set is $\{-1, 3\}$.　□

**PRACTICE**

**1**　Solve: $(x + 4)(x - 5) = 0$.

> ▶ **Helpful Hint**
>
> The zero factor property says that *if a product is 0, then a factor is 0.*
>
> If $a \cdot b = 0$, then $a = 0$ or $b = 0$.
> If $x(x + 5) = 0$, then $x = 0$ or $x + 5 = 0$.
> If $(x + 7)(2x - 3) = 0$, then $x + 7 = 0$ or $2x - 3 = 0$.
>
> Use this property only when the product is 0. For example, if $a \cdot b = 8$, we do not know the value of $a$ or $b$. The values may be $a = 2, b = 4$ or $a = 8, b = 1$, or any other two numbers whose product is 8.

**EXAMPLE 2**  Solve: $x(5x - 2) = 0$

*Solution*
$$x(5x - 2) = 0$$
$$x = 0 \quad \text{or} \quad 5x - 2 = 0 \quad \text{Use the zero factor property.}$$
$$5x = 2$$
$$x = \frac{2}{5}$$

**Check:**  Let $x = 0$.

$x(5x - 2) = 0$

$0(5 \cdot 0 - 2) \stackrel{?}{=} 0$   Replace $x$ with 0.

$0(-2) \stackrel{?}{=} 0$

$0 = 0$  True

Let $x = \frac{2}{5}$.

$x(5x - 2) = 0$

$\frac{2}{5}\left(5 \cdot \frac{2}{5} - 2\right) \stackrel{?}{=} 0$   Replace $x$ with $\frac{2}{5}$.

$\frac{2}{5}(2 - 2) \stackrel{?}{=} 0$

$\frac{2}{5}(0) \stackrel{?}{=} 0$

$0 = 0$  True

The solutions are 0 and $\frac{2}{5}$.

**PRACTICE**
**2**  Solve: $x(7x - 6) = 0$.

**EXAMPLE 3**  Solve: $x^2 - 9x - 22 = 0$

*Solution*  One side of the equation is 0. However, to use the zero factor property, one side of the equation must be 0 *and* the other side must be written as a product (must be factored). Thus, we must first factor this polynomial.

$$x^2 - 9x - 22 = 0$$
$$(x - 11)(x + 2) = 0 \quad \text{Factor.}$$

Now we can apply the zero factor property.

$$x - 11 = 0 \quad \text{or} \quad x + 2 = 0$$
$$x = 11 \qquad\qquad x = -2$$

**Check:**  Let $x = 11$.

$x^2 - 9x - 22 = 0$

$11^2 - 9 \cdot 11 - 22 \stackrel{?}{=} 0$

$121 - 99 - 22 \stackrel{?}{=} 0$

$22 - 22 \stackrel{?}{=} 0$

$0 = 0$  True

Let $x = -2$.

$x^2 - 9x - 22 = 0$

$(-2)^2 - 9(-2) - 22 \stackrel{?}{=} 0$

$4 + 18 - 22 \stackrel{?}{=} 0$

$22 - 22 \stackrel{?}{=} 0$

$0 = 0$  True

The solutions are 11 and $-2$.

**PRACTICE**
**3**  Solve: $x^2 - 8x - 48 = 0$.

**EXAMPLE 4** Solve: $4x^2 - 28x = -49$

**Solution** First we rewrite the equation in standard form so that one side is 0. Then we factor the polynomial.

$$4x^2 - 28x = -49$$
$$4x^2 - 28x + 49 = 0 \qquad \text{Write in standard form by adding 49 to both sides.}$$
$$(2x - 7)(2x - 7) = 0 \qquad \text{Factor.}$$

Next we use the zero factor property and set each factor equal to 0. Since the factors are the same, the related equations will give the same solution.

$$2x - 7 = 0 \quad \text{or} \quad 2x - 7 = 0 \qquad \text{Set each factor equal to 0.}$$
$$2x = 7 \qquad\qquad 2x = 7 \qquad \text{Solve.}$$
$$x = \frac{7}{2} \qquad\qquad x = \frac{7}{2}$$

**Check:** Although $\dfrac{7}{2}$ occurs twice, there is a single solution. Check this solution in the original equation. The solution is $\dfrac{7}{2}$.  □

**PRACTICE**
**4** Solve: $9x^2 - 24x = -16$.

The following steps may be used to solve a quadratic equation by factoring.

---

**To Solve Quadratic Equations by Factoring**

**STEP 1.** Write the equation in standard form so that one side of the equation is 0.

**STEP 2.** Factor the quadratic expression completely.

**STEP 3.** Set each factor containing a variable equal to 0.

**STEP 4.** Solve the resulting equations.

**STEP 5.** Check each solution in the original equation.

---

Since it is not always possible to factor a quadratic polynomial, not all quadratic equations can be solved by factoring. Other methods of solving quadratic equations are presented in Chapter 11.

**EXAMPLE 5** Solve: $x(2x - 7) = 4$

**Solution** First we write the equation in standard form; then we factor.

$$x(2x - 7) = 4$$
$$2x^2 - 7x = 4 \qquad\qquad \text{Multiply.}$$
$$2x^2 - 7x - 4 = 0 \qquad\qquad \text{Write in standard form.}$$
$$(2x + 1)(x - 4) = 0 \qquad\qquad \text{Factor.}$$
$$2x + 1 = 0 \quad \text{or} \quad x - 4 = 0 \qquad \text{Set each factor equal to zero.}$$
$$2x = -1 \qquad\qquad x = 4 \qquad \text{Solve.}$$
$$x = -\frac{1}{2}$$

Check the solutions in the original equation. The solutions are $-\dfrac{1}{2}$ and 4.  □

**PRACTICE**
**5** Solve: $x(3x + 7) = 6$.

▶ **Helpful Hint**

To solve the equation $x(2x - 7) = 4$, do **not** set each factor equal to 4. Remember that to apply the zero factor property, one side of the equation must be 0 and the other side of the equation must be in factored form.

**Concept Check** ☑

Explain the error and solve the equation correctly.

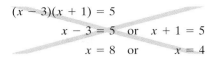

$$(x - 3)(x + 1) = 5$$
$$x - 3 = 5 \quad \text{or} \quad x + 1 = 5$$
$$x = 8 \quad \text{or} \quad x = 4$$

**EXAMPLE 6**   Solve: $-2x^2 - 4x + 30 = 0$.

*Solution*   The equation is in standard form so we begin by factoring out a common factor of $-2$.

$$-2x^2 - 4x + 30 = 0$$
$$-2(x^2 + 2x - 15) = 0 \quad \text{Factor out } -2.$$
$$-2(x + 5)(x - 3) = 0 \quad \text{Factor the quadratic.}$$

Next, set each factor **containing a variable** equal to 0.

$$x + 5 = 0 \qquad \text{or} \qquad x - 3 = 0 \quad \text{Set each factor containing a variable equal to 0.}$$
$$x = -5 \qquad \text{or} \qquad x = 3 \quad \text{Solve.}$$

**Note:** The factor $-2$ is a constant term containing no variables and can never equal 0. The solutions are $-5$ and 3.   ☐

**PRACTICE**
**6**   Solve: $-3x^2 - 6x + 72 = 0$.

**OBJECTIVE 2** ▶ **Solving equations with degree greater than two by factoring.** Some equations involving polynomials of degree higher than 2 may also be solved by factoring and then applying the zero factor theorem.

**EXAMPLE 7**   Solve: $3x^3 - 12x = 0$.

*Solution*   Factor the left side of the equation. Begin by factoring out the common factor of $3x$.

$$3x^3 - 12x = 0$$
$$3x(x^2 - 4) = 0 \quad \text{Factor out the GCF } 3x.$$
$$3x(x + 2)(x - 2) = 0 \quad \text{Factor } x^2 - 4, \text{ a difference of squares.}$$

$$3x = 0 \qquad \text{or} \qquad x + 2 = 0 \qquad \text{or} \qquad x - 2 = 0 \quad \text{Set each factor equal to 0.}$$
$$x = 0 \qquad \text{or} \qquad x = -2 \qquad \text{or} \qquad x = 2 \quad \text{Solve.}$$

Thus, the equation $3x^3 - 12x = 0$ has three solutions: $0, -2,$ and 2. To check, replace $x$ with each solution in the original equation.

| ***Let x = 0.*** | ***Let x = −2.*** | ***Let x = 2.*** |
|---|---|---|
| $3(0)^3 - 12(0) \overset{?}{=} 0$ | $3(-2)^3 - 12(-2) \overset{?}{=} 0$ | $3(2)^3 - 12(2) \overset{?}{=} 0$ |
| $0 = 0$ | $3(-8) + 24 \overset{?}{=} 0$ | $3(8) - 24 \overset{?}{=} 0$ |
| | $0 = 0$ | $0 = 0$ |

Substituting $0, -2,$ or 2 into the original equation results each time in a true equation. The solutions are $0, -2,$ and 2.   ☐

**PRACTICE**
**7**   Solve: $7x^3 - 63x = 0$.

**Answer to Concept Check:**

To use the zero factor property, one side of the equation must be 0, not 5. Correctly, $(x - 3)(x + 1) = 5$, $x^2 - 2x - 3 = 5, x^2 - 2x - 8 = 0$, $(x - 4)(x + 2) = 0, x - 4 = 0$ or $x + 2 = 0, x = 4$ or $x = -2$.

**EXAMPLE 8**  Solve: $(5x - 1)(2x^2 + 15x + 18) = 0$.

*Solution*
$$(5x - 1)(2x^2 + 15x + 18) = 0$$
$$(5x - 1)(2x + 3)(x + 6) = 0 \quad \text{Factor the trinomial.}$$
$$5x - 1 = 0 \quad \text{or} \quad 2x + 3 = 0 \quad \text{or} \quad x + 6 = 0 \quad \text{Set each factor equal to 0.}$$
$$5x = 1 \quad \text{or} \quad 2x = -3 \quad \text{or} \quad x = -6 \quad \text{Solve.}$$
$$x = \frac{1}{5} \quad \text{or} \quad x = -\frac{3}{2}$$

The solutions are $\frac{1}{5}, -\frac{3}{2}$, and $-6$. Check by replacing $x$ with each solution in the original equation. The solutions are $-6, -\frac{3}{2}$, and $\frac{1}{5}$. □

**PRACTICE**
**8**  Solve: $(3x - 2)(2x^2 - 13x + 15) = 0$.

**EXAMPLE 9**  Solve: $2x^3 - 4x^2 - 30x = 0$.

*Solution*  Begin by factoring out the GCF $2x$.
$$2x^3 - 4x^2 - 30x = 0$$
$$2x(x^2 - 2x - 15) = 0 \quad \text{Factor out the GCF } 2x.$$
$$2x(x - 5)(x + 3) = 0 \quad \text{Factor the quadratic.}$$
$$2x = 0 \quad \text{or} \quad x - 5 = 0 \quad \text{or} \quad x + 3 = 0 \quad \text{\begin{tabular}{l} Set each factor containing a \\ variable equal to 0. \end{tabular}}$$
$$x = 0 \quad \text{or} \quad x = 5 \quad \text{or} \quad x = -3 \quad \text{Solve.}$$

Check by replacing $x$ with each solution in the cubic equation. The solutions are $-3, 0$, and 5. □

**PRACTICE**
**9**  Solve: $5x^3 + 5x^2 - 30x = 0$.

**OBJECTIVE 3 ▶ Finding *x*-intercepts of the graph of a quadratic equation.** In Chapter 3, we graphed linear equations in two variables, such as $y = 5x - 6$. Recall that to find the $x$-intercept of the graph of a linear equation, let $y = 0$ and solve for $x$. This is also how to find the $x$-intercepts of the graph of a **quadratic equation in two variables,** such as $y = x^2 - 5x + 4$.

**EXAMPLE 10**  Find the $x$-intercepts of the graph of $y = x^2 - 5x + 4$.

*Solution*  Let $y = 0$ and solve for $x$.
$$y = x^2 - 5x + 4$$
$$0 = x^2 - 5x + 4 \quad \text{Let } y = 0.$$
$$0 = (x - 1)(x - 4) \quad \text{Factor.}$$
$$x - 1 = 0 \quad \text{or} \quad x - 4 = 0 \quad \text{Set each factor equal to 0.}$$
$$x = 1 \quad \text{or} \quad x = 4 \quad \text{Solve.}$$

The $x$-intercepts of the graph of $y = x^2 - 5x + 4$ are $(1, 0)$ and $(4, 0)$.
  The graph of $y = x^2 - 5x + 4$ is shown in the margin. □

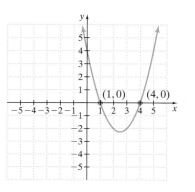

**PRACTICE**
**10**  Find the $x$-intercepts of the graph of $y = x^2 - 6x + 8$.

  In general, a quadratic equation in two variables is one that can be written in the form $y = ax^2 + bx + c$ where $a \neq 0$. The graph of such an equation is called a **parabola** and will open up or down depending on the sign of $a$.

Notice that the $x$-intercepts of the graph of $y = ax^2 + bx + c$ are the real number solutions of $0 = ax^2 + bx + c$. Also, the real number solutions of $0 = ax^2 + bx + c$ are the $x$-intercepts of the graph of $y = ax^2 + bx + c$. We study more about graphs of quadratic equations in two variables in Chapters 8 and 11.

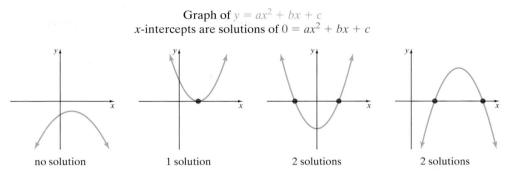

Graph of $y = ax^2 + bx + c$
$x$-intercepts are solutions of $0 = ax^2 + bx + c$

no solution        1 solution        2 solutions        2 solutions

### Graphing Calculator Explorations

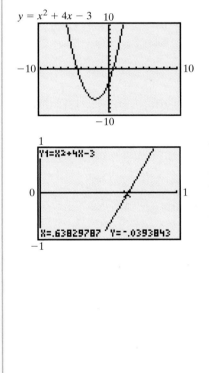

$y = x^2 + 4x - 3$

A grapher may be used to find solutions of a quadratic equation whether the related quadratic polynomial is factorable or not. For example, let's use a grapher to approximate the solutions of $0 = x^2 + 4x - 3$. To do so, graph $y_1 = x^2 + 4x - 3$. Recall that the $x$-intercepts of this graph are the solutions of $0 = x^2 + 4x - 3$.

Notice that the graph appears to have an $x$-intercept between $-5$ and $-4$ and one between $0$ and $1$. Many graphers contain a TRACE feature. This feature activates a graph cursor that can be used to *trace* along a graph while the corresponding $x$- and $y$-coordinates are shown on the screen. Use the TRACE feature to confirm that $x$-intercepts lie between $-5$ and $-4$ and also $0$ and $1$. To approximate the $x$-intercepts to the nearest tenth, use a ROOT or a ZOOM feature on your grapher or redefine the viewing window. (A ROOT feature calculates the $x$-intercept. A ZOOM feature magnifies the viewing window around a specific location such as the graph cursor.) If we redefine the window to $[0, 1]$ on the $x$-axis and $[-1, 1]$ on the $y$-axis, the following graph is generated.

By using the TRACE feature, we can conclude that one $x$-intercept is approximately $0.6$ to the nearest tenth. By repeating these steps for the other $x$-intercept, we find that it is approximately $-4.6$.

*Use a grapher to approximate the real number solutions to the nearest tenth. If an equation has no real number solution, state so.*

**1.** $3x^2 - 4x - 6 = 0$

**2.** $x^2 - x - 9 = 0$

**3.** $2x^2 + x + 2 = 0$

**4.** $-4x^2 - 5x - 4 = 0$

**5.** $-x^2 + x + 5 = 0$

**6.** $10x^2 + 6x - 3 = 0$

## VOCABULARY & READINESS CHECK

*Use the choices below to fill in each blank. Not all choices will be used.*

$-3, 5$        $a = 0$  or  $b = 0$        $0$        linear

$3, -5$        quadratic                $1$

**1.** An equation that can be written in the form $ax^2 + bx + c = 0$, (with $a \neq 0$), is called a(n) _____ equation.

**2.** If the product of two numbers is $0$, then at least one of the numbers must be _____.

**3.** The solutions to $(x - 3)(x + 5) = 0$ are _____.

**4.** If $a \cdot b = 0$, then _____.

*Solve each equation by inspection.*

**5.** $(a - 3)(a - 7) = 0$

**6.** $(a - 5)(a - 2) = 0$

**7.** $(x + 8)(x + 6) = 0$

**8.** $(x + 2)(x + 3) = 0$

**9.** $(x + 1)(x - 3) = 0$

**10.** $(x - 1)(x + 2) = 0$

## 6.6 EXERCISE SET

*Solve each equation. See Examples 1 and 2.*

**1.** $(x - 2)(x + 1) = 0$

**2.** $(x + 4)(x - 10) = 0$

**3.** $(x + 9)(x + 17) = 0$

**4.** $(x + 11)(x + 1) = 0$

**5.** $x(x + 6) = 0$

**6.** $x(x - 7) = 0$

**7.** $3x(x - 8) = 0$

**8.** $2x(x + 12) = 0$

**9.** $(2x + 3)(4x - 5) = 0$

**10.** $(3x - 2)(5x + 1) = 0$

**11.** $(2x - 7)(7x + 2) = 0$

**12.** $(9x + 1)(4x - 3) = 0$

**13.** $\left(x - \dfrac{1}{2}\right)\left(x + \dfrac{1}{3}\right) = 0$

**14.** $\left(x + \dfrac{2}{9}\right)\left(x - \dfrac{1}{4}\right) = 0$

**15.** $(x + 0.2)(x + 1.5) = 0$

**16.** $(x + 1.7)(x + 2.3) = 0$

**17.** Write a quadratic equation that has two solutions, 6 and $-1$. Leave the polynomial in the equation in factored form.

**18.** Write a quadratic equation that has two solutions, 0 and $-2$. Leave the polynomial in the equation in factored form.

*Solve. See Examples 3 through 6.*

**19.** $x^2 - 13x + 36 = 0$

**20.** $x^2 + 2x - 63 = 0$

**21.** $x^2 + 2x - 8 = 0$

**22.** $x^2 - 5x + 6 = 0$

**23.** $x^2 - 7x = 0$

**24.** $x^2 - 3x = 0$

**25.** $x^2 - 4x = 32$

**26.** $x^2 - 5x = 24$

**27.** $x^2 = 16$

**28.** $x^2 = 9$

**29.** $(x + 4)(x - 9) = 4x$

**30.** $(x + 3)(x + 8) = x$

**31.** $x(3x - 1) = 14$

**32.** $x(4x - 11) = 3$

**33.** $-3x^2 + 75 = 0$

**34.** $-2y^2 + 72 = 0$

**35.** $24x^2 + 44x = 8$

**36.** $6x^2 + 57x = 30$

*Solve each equation. See Examples 7 through 9.*

**37.** $x^3 - 12x^2 + 32x = 0$

**38.** $x^3 - 14x^2 + 49x = 0$

**39.** $(4x - 3)(16x^2 - 24x + 9) = 0$

**40.** $(2x + 5)(4x^2 + 20x + 25) = 0$

**41.** $4x^3 - x = 0$

**42.** $4y^3 - 36y = 0$

**43.** $32x^3 - 4x^2 - 6x = 0$

**44.** $15x^3 + 24x^2 - 63x = 0$

## MIXED PRACTICE

*Solve each equation. See Examples 1 through 9. (A few exercises are linear equations.)*

**45.** $(x + 3)(x - 2) = 0$

**46.** $(x - 6)(x + 7) = 0$

**47.** $x^2 + 20x = 0$

**48.** $x^2 + 15x = 0$

**49.** $4(x - 7) = 6$

**50.** $5(3 - 4x) = 9$

**51.** $4y^2 - 1 = 0$

**52.** $4y^2 - 81 = 0$

**53.** $(2x + 3)(2x^2 - 5x - 3) = 0$

**54.** $(2x - 9)(x^2 + 5x - 36) = 0$

**55.** $x^2 - 15 = -2x$

**56.** $x^2 - 26 = -11x$

**57.** $30x^2 - 11x - 30 = 0$

**58.** $12x^2 + 7x - 12 = 0$

**59.** $5x^2 - 6x - 8 = 0$

**60.** $9x^2 + 7x = 2$

**61.** $6y^2 - 22y - 40 = 0$

**62.** $3x^2 - 6x - 9 = 0$

**63.** $(y - 2)(y + 3) = 6$

**64.** $(y - 5)(y - 2) = 28$

**65.** $3x^3 + 19x^2 - 72x = 0$

**66.** $36x^3 + x^2 - 21x = 0$

**67.** $x^2 + 14x + 49 = 0$

**68.** $x^2 + 22x + 121 = 0$

**69.** $12y = 8y^2$

**70.** $9y = 6y^2$

**71.** $7x^3 - 7x = 0$

**72.** $3x^3 - 27x = 0$

**73.** $3x^2 + 8x - 11 = 13 - 6x$

**74.** $2x^2 + 12x - 1 = 4 + 3x$

**75.** $3x^2 - 20x = -4x^2 - 7x - 6$

**76.** $4x^2 - 20x = -5x^2 - 6x - 5$

*Find the x-intercepts of the graph of each equation. See Example 10.*

**77.** $y = (3x + 4)(x - 1)$

**78.** $y = (5x - 3)(x - 4)$

**79.** $y = x^2 - 3x - 10$

**80.** $y = x^2 + 7x + 6$

**81.** $y = 2x^2 + 11x - 6$

**82.** $y = 4x^2 + 11x + 6$

*For Exercises 83 through 88, match each equation with its graph. See Example 10.*

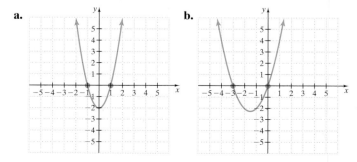

**c.**

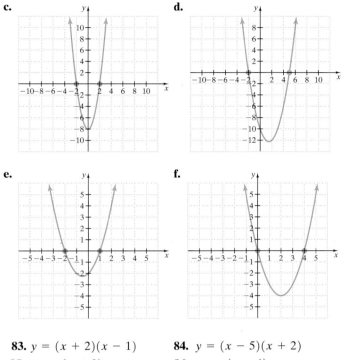

**d.**

**e.**

**f.**

**83.** $y = (x + 2)(x - 1)$     **84.** $y = (x - 5)(x + 2)$

**85.** $y = x(x + 3)$     **86.** $y = x(x - 4)$

**87.** $y = 2x^2 - 8$     **88.** $y = 2x^2 - 2$

### REVIEW AND PREVIEW

*Perform the following operations. Write all results in lowest terms. See Section 1.3.*

**89.** $\dfrac{3}{5} + \dfrac{4}{9}$     **90.** $\dfrac{2}{3} + \dfrac{3}{7}$

**91.** $\dfrac{7}{10} - \dfrac{5}{12}$     **92.** $\dfrac{5}{9} - \dfrac{5}{12}$

**93.** $\dfrac{7}{8} \div \dfrac{7}{15}$     **94.** $\dfrac{5}{12} - \dfrac{3}{10}$

**95.** $\dfrac{4}{5} \cdot \dfrac{7}{8}$     **96.** $\dfrac{3}{7} \cdot \dfrac{12}{17}$

### CONCEPT EXTENSIONS

*For Exercises 97 and 98, see the Concept Check in this section.*

**97.** Explain the error and solve correctly:

$$x(x - 2) = 8$$
$$x = 8 \quad \text{or} \quad x - 2 = 8$$
$$x = 10$$

**98.** Explain the error and solve correctly:

$$(x - 4)(x + 2) = 0$$
$$x = -4 \quad \text{or} \quad x = 2$$

**99.** Write a quadratic equation in standard form that has two solutions, 5 and 7.

**100.** Write an equation that has three solutions, 0, 1, and 2.

**101.** A compass is accidentally thrown upward and out of an air balloon at a height of 300 feet. The height, $y$, of the compass at time $x$ in seconds is given by the equation

$$y = -16x^2 + 20x + 300$$

300 ft

**a.** Find the height of the compass at the given times by filling in the table below.

| time, x | 0 | 1 | 2 | 3 | 4 | 5 | 6 |
|---------|---|---|---|---|---|---|---|
| height, y | | | | | | | |

**b.** Use the table to determine when the compass strikes the ground.

**c.** Use the table to approximate the maximum height of the compass.

**d.** Plot the points $(x, y)$ on a rectangular coordinate system and connect them with a smooth curve. Explain your results.

**102.** A rocket is fired upward from the ground with an initial velocity of 100 feet per second. The height, $y$, of the rocket at any time $x$ is given by the equation

$$y = -16x^2 + 100x$$

y

**a.** Find the height of the rocket at the given times by filling in the table below.

| time, x | 0 | 1 | 2 | 3 | 4 | 5 | 6 | 7 |
|---------|---|---|---|---|---|---|---|---|
| height, y | | | | | | | | |

  **b.** Use the table to approximate when the rocket strikes the ground to the nearest second.

  **c.** Use the table to approximate the maximum height of the rocket.

  **d.** Plot the points $(x, y)$ on a rectangular coordinate system and connect them with a smooth curve. Explain your results.

*Solve each equation. First, multiply the binomial(s).*

To solve $(x - 6)(2x - 3) = (x + 2)(x + 9)$, see below.

$$(x - 6)(2x - 3) = (x + 2)(x + 9)$$
$$2x^2 - 15x + 18 = x^2 + 11x + 18$$
$$x^2 - 26x = 0$$
$$x(x - 26) = 0$$
$$x = 0 \quad \text{or} \quad x - 26 = 0$$
$$x = 26$$

**103.** $(x - 3)(3x + 4) = (x + 2)(x - 6)$

**104.** $(2x - 3)(x + 6) = (x - 9)(x + 2)$

**105.** $(2x - 3)(x + 8) = (x - 6)(x + 4)$

**106.** $(x + 6)(x - 6) = (2x - 9)(x + 4)$

---

## THE BIGGER PICTURE   SIMPLIFYING EXPRESSIONS AND SOLVING EQUATIONS AND INEQUALITIES

Now we continue our outline from Sections 1.7, 2.3, 2.8, and 5.6. Although suggestions are given, this outline should be in your own words. Once you complete this new portion, try the exercises below.

  **I.** Simplifying Expressions

    **A.** Real Numbers

      **1.** Add (Section 1.5)

      **2.** Subtract (Section 1.6)

      **3.** Multiply or Divide (Section 1.7)

    **B.** Exponents (Section 5.1 and 5.5)

    **C.** Polynomials

      **1.** Add (Section 5.2)

      **2.** Subtract (Section 5.2)

      **3.** Multiply (Section 5.3 and 5.4)

      **4.** Divide (Section 5.6)

    **D.** Factoring Polynomials—see the Chapter 6 Integrated Review for steps.

$$3x^4 - 78x^2 + 75$$
$$= 3(x^4 - 26x^2 + 25) \qquad \text{Factor out GCF—always first step.}$$
$$= 3(x^2 - 25)(x^2 - 1) \qquad \text{Factor trinomial.}$$
$$= 3(x + 5)(x - 5)(x + 1)(x - 1) \qquad \text{Factor further—each difference of squares.}$$

  **II.** Solving Equations

    **A.** Linear Equations (Section 2.3)

    **B.** Quadratic & Higher Degree Equations (Solving by Factoring)—highest power on variable is at least 2 when equation is written in standard form (set equal to 0).

$$x^2 + x = 6$$
$$x^2 + x - 6 = 0 \qquad \text{Write the equation in standard form (set it equal to 0).}$$
$$(x - 2)(x + 3) = 0 \qquad \text{Factor.}$$
$$x = 2 \quad \text{or} \quad x = -3 \qquad \text{Set each factor equal to 0 and solve.}$$

  **III.** Solving Inequalities

    **A.** Linear Inequalities (Section 2.8)

*Simplify each expression.*

**1.** $-7 + (-27)$

**2.** $\dfrac{(x^3)^4}{(x^{-2})^5}$

**3.** $(x^3 - 6x^2 + 2) - (5x^3 - 6)$

**4.** $\dfrac{3y^3 - 3y^2 + 9}{3y^2}$

*Factor each expression.*

**5.** $10x^3 - 250x$

**6.** $x^2 - 36x + 35$

**7.** $6xy + 15x - 6y - 15$

**8.** $5xy^2 - 2xy - 7x$

*Solve each equation. Remember to use your outline to determine whether the equation is linear or quadratic and how to proceed with solving.*

**9.** $(x - 5)(2x + 1) = 0$

**10.** $5x - 5 = 0$

**11.** $x(x - 12) = 28$

**12.** $7(x - 3) + 2(5x + 1) = 14$

# 6.7 QUADRATIC EQUATIONS AND PROBLEM SOLVING

**OBJECTIVE 1 ▶ Solving problems modeled by quadratic equations.** Some problems may be modeled by quadratic equations. To solve these problems, we use the same problem-solving steps that were introduced in Section 2.4. When solving these problems, keep in mind that a solution of an equation that models a problem may not be a solution to the problem. For example, a person's age or the length of a rectangle is always a positive number. Discard solutions that do not make sense as solutions of the problem.

**EXAMPLE 1**   **Finding Free-Fall Time**

Since the 1940s, one of the top tourist attractions in Acapulco, Mexico, is watching the cliff divers off the La Quebrada. The divers' platform is about 144 feet above the sea. These divers must time their descent just right, since they land in the crashing Pacific, in an inlet that is at most $9\frac{1}{2}$ feet deep. Neglecting air resistance, the height $h$ in feet of a cliff diver above the ocean after $t$ seconds is given by the quadratic equation $h = -16t^2 + 144$.

Find out how long it takes the diver to reach the ocean.

*Solution*

1. UNDERSTAND. Read and reread the problem. Then draw a picture of the problem.

   The equation $h = -16t^2 + 144$ models the height of the falling diver at time $t$. Familiarize yourself with this equation by find the height of the diver at time $t = 1$ second and $t = 2$ seconds.

   When $t = 1$ second, the height of the diver is $h = -16(1)^2 + 144 = 128$ feet.
   When $t = 2$ seconds, the height of the diver is $h = -16(2)^2 + 144 = 80$ feet.

2. TRANSLATE. To find out how long it takes the diver to reach the ocean, we want to know the value of $t$ for which $h = 0$.

$$0 = -16t^2 + 144$$
$$0 = -16(t^2 - 9) \qquad \text{Factor out } -16.$$
$$0 = -16(t - 3)(t + 3) \qquad \text{Factor completely.}$$
$$t - 3 = 0 \quad \text{or} \quad t + 3 = 0 \qquad \text{Set each factor containing a variable equal to 0.}$$
$$t = 3 \quad \text{or} \qquad t = -3 \quad \text{Solve.}$$

3. INTERPRET. Since the time $t$ cannot be negative, the proposed solution is 3 seconds.

**Check:**   Verify that the height of the diver when $t$ is 3 seconds is 0.

When $t = 3$ seconds, $h = -16(3)^2 + 144 = -144 + 144 = 0$.

**State:**   It takes the diver 3 seconds to reach the ocean.   □

**PRACTICE**

**1**    Cliff divers also frequent the falls at Waimea Falls Park in Oahu, Hawaii. One of the popular diving spots is 64 feet high. Neglecting air resistance, the height of a diver above the pool after $t$ seconds is $h = -16t^2 + 64$. Find how long it takes a diver to reach the pool.

### EXAMPLE 2  Finding an Unknown Number

The square of a number plus three times the number is 70. Find the number.

*Solution*

1. UNDERSTAND. Read and reread the problem. Suppose that the number is 5. The square of 5 is $5^2$ or 25. Three times 5 is 15. Then $25 + 15 = 40$, not 70, so the number must be greater than 5. Remember, the purpose of proposing a number, such as 5, is to better understand the problem. Now that we do, we will let $x =$ the number.

2. TRANSLATE.

| the square of a number | plus | three times the number | is | 70 |
|:---:|:---:|:---:|:---:|:---:|
| ↓ | ↓ | ↓ | ↓ | ↓ |
| $x^2$ | $+$ | $3x$ | $=$ | 70 |

3. SOLVE.

$$x^2 + 3x = 70$$
$$x^2 + 3x - 70 = 0 \qquad \text{Subtract 70 from both sides.}$$
$$(x + 10)(x - 7) = 0 \qquad \text{Factor.}$$
$$x + 10 = 0 \quad \text{or} \quad x - 7 = 0 \quad \text{Set each factor equal to 0.}$$
$$x = -10 \qquad\qquad x = 7 \quad \text{Solve.}$$

4. INTERPRET.

**Check:**  The square of $-10$ is $(-10)^2$, or 100. Three times $-10$ is $3(-10)$ or $-30$. Then $100 + (-30) = 70$, the correct sum, so $-10$ checks.

The square of 7 is $7^2$ or 49. Three times 7 is $3(7)$, or 21. Then $49 + 21 = 70$, the correct sum, so 7 checks.

**State:**  There are two numbers. They are $-10$ and 7. □

**PRACTICE**

**2**  The square of a number minus eight times the number is equal to forty-eight. Find the number.

---

### ⚠ EXAMPLE 3  Finding the Dimensions of a Sail

The height of a triangular sail is 2 meters less than twice the length of the base. If the sail has an area of 30 square meters, find the length of its base and the height.

*Solution*

1. UNDERSTAND. Read and reread the problem. Since we are finding the length of the base and the height, we let

$$x = \text{the length of the base}$$

and since the height is 2 meters less than twice the base,

$$2x - 2 = \text{the height}$$

An illustration is shown to the right.

Height = $2x - 2$

Base = $x$

**2.** TRANSLATE. We are given that the area of the triangle is 30 square meters, so we use the formula for area of a triangle.

$$\begin{array}{ccccccc}
\text{area of triangle} & = & \dfrac{1}{2} & \cdot & \text{base} & \cdot & \text{height} \\
\downarrow & & \downarrow & & \downarrow & & \downarrow \\
30 & = & \dfrac{1}{2} & \cdot & x & \cdot & (2x - 2)
\end{array}$$

**3.** SOLVE. Now we solve the quadratic equation.

$$30 = \frac{1}{2}x(2x - 2)$$

$$30 = x^2 - x \qquad \text{Multiply.}$$

$$x^2 - x - 30 = 0 \qquad \text{Write in standard form.}$$

$$(x - 6)(x + 5) = 0 \qquad \text{Factor.}$$

$$x - 6 = 0 \quad \text{or} \quad x + 5 = 0 \qquad \text{Set each factor equal to 0.}$$

$$x = 6 \qquad\qquad x = -5$$

**4.** INTERPRET. Since $x$ represents the length of the base, we discard the solution $-5$. The base of a triangle cannot be negative. The base is then 6 meters and the height is $2(6) - 2 = 10$ meters.

**Check:**   To check this problem, we recall that $\dfrac{1}{2}$ base $\cdot$ height $=$ area, or

$$\frac{1}{2}(6)(10) = 30 \quad \text{The required area}$$

**State:**   The base of the triangular sail is 6 meters and the height is 10 meters.   □

**PRACTICE**
**3**   An engineering team from Georgia Tech earned second place in a recent flight competition, with their triangular shaped paper hang glider. The base of their prize-winning entry was 1 foot less than three times the height. If the area of the triangular glider wing was 210 square feet, find the dimensions of the wing. (*Source: The Technique* [Georgia Tech's newspaper], April 18, 2003)

The next examples make use of the **Pythagorean theorem** and consecutive integers. Before we review this theorem, recall that a **right triangle** is a triangle that contains a 90° or right angle. The **hypotenuse** of a right triangle is the side opposite the right angle and is the longest side of the triangle. The **legs** of a right triangle are the other sides of the triangle.

▶ **Helpful Hint**
If you use this formula, don't forget that $c$ represents the length of the hypotenuse.

**Pythagorean Theorem**
In a right triangle, the sum of the squares of the lengths of the two legs is equal to the square of the length of the hypotenuse.

$$(\text{leg})^2 + (\text{leg})^2 = (\text{hypotenuse})^2 \qquad \text{or} \qquad a^2 + b^2 = c^2$$

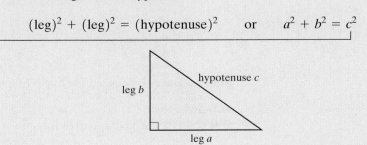

Study the following diagrams for a review of consecutive integers.

**Examples**

If $x$ is the first integer, then consecutive integers are $x, x + 1, x + 2, \ldots$

If $x$ is the first even integer, then consecutive even integers are $x, x + 2, x + 4, \ldots$

If $x$ is the first odd integer, then consecutive odd integers are $x, x + 2, x + 4, \ldots$

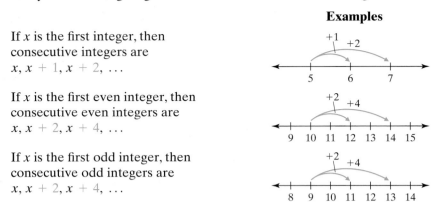

### EXAMPLE 4  Finding Consecutive Even Integers

Find two consecutive even integers whose product is 34 more than their sum.

*Solution*

1. UNDERSTAND. Read and reread the problem. Let's just choose two consecutive even integers to help us better understand the problem. Let's choose 10 and 12. Their product is $10(12) = 120$ and their sum is $10 + 12 = 22$. The product is $120 - 22$, or 98 greater than the sum. Thus our guess is incorrect, but we have a better understanding of this example.

   Let's let $x$ and $x + 2$ be the consecutive even integers.

2. TRANSLATE.

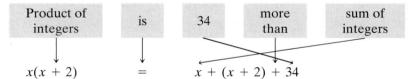

3. SOLVE. Now we solve the equation.

$$
\begin{aligned}
x(x + 2) &= x + (x + 2) + 34 & \\
x^2 + 2x &= x + x + 2 + 34 & \text{Multiply.} \\
x^2 + 2x &= 2x + 36 & \text{Combine like terms.} \\
x^2 - 36 &= 0 & \text{Write in standard form.} \\
(x + 6)(x - 6) &= 0 & \text{Factor.} \\
x + 6 = 0 \quad \text{or} \quad x - 6 &= 0 & \text{Set each factor equal to 0.} \\
x = -6 \qquad\qquad x &= 6 & \text{Solve.}
\end{aligned}
$$

4. INTERPRET. If $x = -6$, then $x + 2 = -6 + 2$, or $-4$.
   If $x = 6$, then $x + 2 = 6 + 2$, or 8.

**Check:** $-6, -4$ $\qquad\qquad\qquad\qquad$ $6, 8$

$$
\begin{aligned}
-6(-4) &\stackrel{?}{=} -6 + (-4) + 34 & \qquad 6(8) &\stackrel{?}{=} 6 + 8 + 34 \\
24 &\stackrel{?}{=} -10 + 34 & 48 &\stackrel{?}{=} 14 + 34 \\
24 &= 24 \qquad \text{True} & 48 &= 48 \qquad \text{True}
\end{aligned}
$$

**State:** The two consecutive even integers are $-6$ and $-4$ or 6 and 8. □

**PRACTICE**

**4** Find two consecutive integers whose product is 41 more than their sum.

⚠️ **EXAMPLE 5**   **Finding the Dimensions of a Triangle**

Find the lengths of the sides of a right triangle if the lengths can be expressed as three consecutive even integers.

*Solution*

1. UNDERSTAND. Read and reread the problem. Let's suppose that the length of one leg of the right triangle is 4 units. Then the other leg is the next even integer, or 6 units, and the hypotenuse of the triangle is the next even integer, or 8 units. Remember that the hypotenuse is the longest side. Let's see if a triangle with sides of these lengths forms a right triangle. To do this, we check to see whether the Pythagorean theorem holds true.

$$4^2 + 6^2 \overset{?}{=} 8^2$$
$$16 + 36 \overset{?}{=} 64$$
$$52 = 64 \quad \text{False}$$

Our proposed numbers do not check, but we now have a better understanding of the problem.

We let $x$, $x + 2$, and $x + 4$ be three consecutive even integers. Since these integers represent lengths of the sides of a right triangle, we have the following.

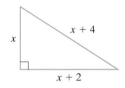

$$x = \text{one leg}$$
$$x + 2 = \text{other leg}$$
$$x + 4 = \text{hypotenuse (longest side)}$$

2. TRANSLATE. By the Pythagorean theorem, we have that

$$(\text{leg})^2 + (\text{leg})^2 = (\text{hypotenuse})^2$$
$$(x)^2 + (x + 2)^2 = (x + 4)^2$$

3. SOLVE. Now we solve the equation.

$$x^2 + (x + 2)^2 = (x + 4)^2$$

| | |
|---|---|
| $x^2 + x^2 + 4x + 4 = x^2 + 8x + 16$ | Multiply. |
| $2x^2 + 4x + 4 = x^2 + 8x + 16$ | Combine like terms. |
| $x^2 - 4x - 12 = 0$ | Write in standard form. |
| $(x - 6)(x + 2) = 0$ | Factor. |
| $x - 6 = 0 \quad \text{or} \quad x + 2 = 0$ | Set each factor equal to 0. |
| $x = 6 \qquad\qquad x = -2$ | |

4. INTERPRET. We discard $x = -2$ since length cannot be negative. If $x = 6$, then $x + 2 = 8$ and $x + 4 = 10$.

**Check:**   Verify that

$$(\text{leg})^2 + (\text{leg})^2 = (\text{hypotenuse})^2$$
$$6^2 + 8^2 \overset{?}{=} 10^2$$
$$36 + 64 \overset{?}{=} 100$$
$$100 = 100 \quad \text{True}$$

**State:**   The sides of the right triangle have lengths 6 units, 8 units, and 10 units.

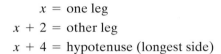

**PRACTICE**

**5**   Find the dimensions of a right triangle where the second leg is 1 unit less than double the first leg, and the hypotenuse is 1 unit more than double the length of the first leg.

## 6.7 | EXERCISE SET

### MIXED PRACTICE

*See Examples 1 through 5 for all exercises. For Exercises 1 through 6, represent each given condition using a single variable, x.*

△ **1.** The length and width of a rectangle whose length is 4 centimeters more than its width

△ **2.** The length and width of a rectangle whose length is twice its width

**3.** Two consecutive odd integers

**4.** Two consecutive even integers

△ **5.** The base and height of a triangle whose height is one more than four times its base

△ **6.** The base and height of a trapezoid whose base is three less than five times its height

*Use the information given to find the dimensions of each figure.*

△ **7.** The *area* of the square is 121 square units. Find the length of its sides.

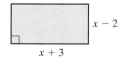

△ **8.** The *area* of the rectangle is 84 square inches. Find its length and width.

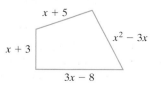

△ **9.** The *perimeter* of the quadrilateral is 120 centimeters. Find the lengths of the sides.

△ **10.** The *perimeter* of the triangle is 85 feet. Find the lengths of its sides.

△ **11.** The *area* of the parallelogram is 96 square miles. Find its base and height.

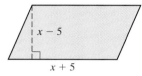

△ **12.** The *area* of the circle is $25\pi$ square kilometers. Find its radius.

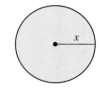

*Solve.*

**13.** An object is thrown upward from the top of an 80-foot building with an initial velocity of 64 feet per second. The height $h$ of the object after $t$ seconds is given by the quadratic equation $h = -16t^2 + 64t + 80$. When will the object hit the ground?

**14.** A hang glider pilot accidentally drops her compass from the top of a 400-foot cliff. The height $h$ of the compass after $t$ seconds is given by the quadratic equation $h = -16t^2 + 400$. When will the compass hit the ground?

△ **15.** The length of a rectangle is 7 centimeters less than twice its width. Its area is 30 square centimeters. Find the dimensions of the rectangle.

△ **16.** The length of a rectangle is 9 inches more than its width. Its area is 112 square inches. Find the dimensions of the rectangle.

*The equation $D = \frac{1}{2}n(n-3)$ gives the number of diagonals D for a polygon with n sides. For example, a polygon with 6 sides has $D = \frac{1}{2} \cdot 6(6-3)$ or $D = 9$ diagonals. (See if you can count all 9 diagonals. Some are shown in the figure.) Use this equation, $D = \frac{1}{2}n(n-3)$, for Exercises 17 through 20.*

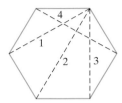

△ **17.** Find the number of diagonals for a polygon that has 12 sides.

△ **18.** Find the number of diagonals for a polygon that has 15 sides.

△ **19.** Find the number of sides $n$ for a polygon that has 35 diagonals.

△ **20.** Find the number of sides $n$ for a polygon that has 14 diagonals.

*Solve.*

**21.** The sum of a number and its square is 132. Find the number(s).

**22.** The sum of a number and its square is 182. Find the number(s).

**23.** The product of two consecutive room numbers is 210. Find the room numbers.

**24.** The product of two consecutive page numbers is 420. Find the page numbers.

**25.** A ladder is leaning against a building so that the distance from the ground to the top of the ladder is one foot less than the length of the ladder. Find the length of the ladder if the distance from the bottom of the ladder to the building is 5 feet.

**26.** Use the given figure to find the length of the guy wire.

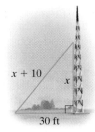

△ **27.** If the sides of a square are increased by 3 inches, the area becomes 64 square inches. Find the length of the sides of the original square.

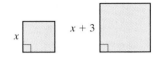

△ **28.** If the sides of a square are increased by 5 meters, the area becomes 100 square meters. Find the length of the sides of the original square.

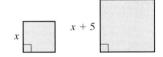

△ **29.** One leg of a right triangle is 4 millimeters longer than the smaller leg and the hypotenuse is 8 millimeters longer than the smaller leg. Find the lengths of the sides of the triangle.

△ **30.** One leg of a right triangle is 9 centimeters longer than the other leg and the hypotenuse is 45 centimeters. Find the lengths of the legs of the triangle.

△ **31.** The length of the base of a triangle is twice its height. If the area of the triangle is 100 square kilometers, find the height.

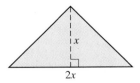

△ **32.** The height of a triangle is 2 millimeters less than the base. If the area is 60 square millimeters, find the base.

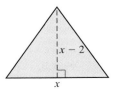

△ **33.** Find the length of the shorter leg of a right triangle if the longer leg is 12 feet more than the shorter leg and the hypotenuse is 12 feet less than twice the shorter leg.

△ **34.** Find the length of the shorter leg of a right triangle if the longer leg is 10 miles more than the shorter leg and the hypotenuse is 10 miles less than twice the shorter leg.

**35.** An object is dropped from 39 feet below the tip of the pinnacle atop one of the 1483-foot-tall Petronas Twin Towers in Kuala Lumpur, Malaysia. (*Source:* Council on Tall Buildings and Urban Habitat) The height $h$ of the object after $t$ seconds is given by the equation $h = -16t^2 + 1444$. Find how many seconds pass before the object reaches the ground.

**36.** An object is dropped from the top of 311 South Wacker Drive, a 961-foot-tall office building in Chicago. (*Source:* Council on Tall Buildings and Urban Habitat) The height $h$ of the object after $t$ seconds is given by the equation $h = -16t^2 + 961$. Find how many seconds pass before the object reaches the ground.

**37.** At the end of 2 years, $P$ dollars invested at an interest rate $r$ compounded annually increases to an amount, $A$ dollars, given by

$$A = P(1 + r)^2$$

Find the interest rate if \$100 increased to \$144 in 2 years. Write your answer as a percent.

**38.** At the end of 2 years, $P$ dollars invested at an interest rate $r$ compounded annually increases to an amount, $A$ dollars, given by

$$A = P(1 + r)^2$$

Find the interest rate if \$2000 increased to \$2420 in 2 years. Write your answer as a percent.

△ **39.** Find the dimensions of a rectangle whose width is 7 miles less than its length and whose area is 120 square miles.

△ **40.** Find the dimensions of a rectangle whose width is 2 inches less than half its length and whose area is 160 square inches.

**41.** If the cost, $C$, for manufacturing $x$ units of a certain product is given by $C = x^2 - 15x + 50$, find the number of units manufactured at a cost of \$9500.

**42.** If a switchboard handles $n$ telephones, the number $C$ of telephone connections it can make simultaneously is given by the equation $C = \dfrac{n(n - 1)}{2}$. Find how many telephones are handled by a switchboard making 120 telephone connections simultaneously.

## REVIEW AND PREVIEW

*The following double line graph shows a comparison of the number of farms in the United States and the size of the average farm. Use this graph to answer Exercises 43–49. See Section 3.1.*

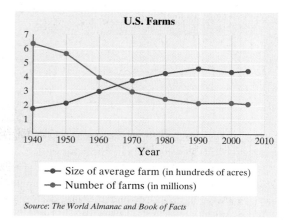

**U.S. Farms**

Year

→ Size of average farm (in hundreds of acres)
→ Number of farms (in millions)

*Source: The World Almanac and Book of Facts*

△ **43.** Approximate the size of the average farm in 1940.

△ **44.** Approximate the size of the average farm in 2005.

**45.** Approximate the number of farms in 1940.

**46.** Approximate the number of farms in 2005.

**47.** Approximate the year that the colored lines in this graph intersect.

✎ **48.** In your own words, explain the meaning of the point of intersection in the graph.

✎ **49.** Describe the trends shown in this graph and speculate as to why these trends have occurred.

*Write each fraction in simplest form. See Section 1.3.*

**50.** $\dfrac{20}{35}$  **51.** $\dfrac{24}{32}$

**52.** $\dfrac{27}{18}$  **53.** $\dfrac{15}{27}$

**54.** $\dfrac{14}{42}$  **55.** $\dfrac{45}{50}$

## CONCEPT EXTENSIONS

△ **56.** Two boats travel at right angles to each other after leaving the same dock at the same time. One hour later the boats are 17 miles apart. If one boat travels 7 miles per hour faster than the other boat, find the rate of each boat.

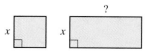

17 miles

△ **57.** The side of a square equals the width of a rectangle. The length of the rectangle is 6 meters longer than its width. The sum of the areas of the square and the rectangle is 176 square meters. Find the side of the square.

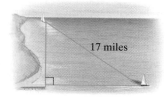

**58.** The sum of two numbers is 20, and the sum of their squares is 218. Find the numbers.

**59.** The sum of two numbers is 25, and the sum of their squares is 325. Find the numbers.

△ **60.** According to the International America's Cup Class (IACC) rule, a sailboat competing in the America's Cup match must have a 110-foot-tall mast and a combined mainsail and jib sail area of 3000 square feet. (*Source:* America's Cup Organizing Committee) A design for an IACC-class sailboat calls for the mainsail to be 60% of the combined sail area. If the height of the triangular mainsail is 28 feet more than twice the length of the boom, find the length of the boom and the height of the mainsail.

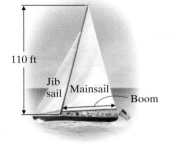

110 ft

Jib sail / Mainsail — Boom

△ **61.** A rectangular pool is surrounded by a walk 4 meters wide. The pool is 6 meters longer than its width. If the total area of the pool and walk is 576 square meters more than the area of the pool, find the dimensions of the pool.

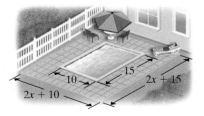

$x + 6$   $x$

4

4

△ **62.** A rectangular garden is surrounded by a walk of uniform width. The area of the garden is 180 square yards. If the dimensions of the garden plus the walk are 16 yards by 24 yards, find the width of the walk.

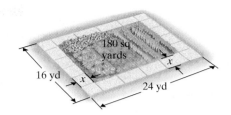

180 sq yards

$x$

16 yd   $x$

24 yd

╲ **63.** Write down two numbers whose sum is 10. Square each number and find the sum of the squares. Use this work to write a word problem like Exercise 59. Then give the word problem to a classmate to solve.

# CHAPTER 6 GROUP ACTIVITY

## Choosing Among Building Options

Whether putting in a new floor, hanging new wallpaper, or retiling a bathroom, it may be necessary to choose among several different materials with different pricing schemes. If a fixed amount of money is available for projects like these, it can be helpful to compare the choices by calculating how much area can be covered by a fixed dollar-value of material.

In this project, you will have the opportunity to choose among three different choices of materials for building a patio around a swimming pool. This project may be completed by working in groups or individually.

[Figure of pool and patio with labeled dimensions: 10, 15, $2x + 15$, $2x + 10$]

**Situation:** Suppose you have just had a 10-foot-by-15-foot in-ground swimming pool installed in your backyard. You have $3000 left from the building project that you would like to spend on surrounding the pool with a patio, equally wide on all sides (see figure). You have talked to several local suppliers about options for building this patio and must choose among the following.

| Option | Material | Price |
|--------|----------|-------|
| A | Poured cement | $5 per square foot |
| B | Brick | $7.50 per square foot plus a $30 flat fee for delivering the bricks |
| C | Outdoor carpeting | $4.50 per square foot plus $10.86 per foot of the pool's perimeter to install edging |

△ **1.** Find the area of the swimming pool.

△ **2.** Write an algebraic expression for the total area of the region containing both the pool and the patio.

△ **3.** Use subtraction to find an algebraic expression for the area of just the patio (not including the area of the pool).

△ **4.** Find the perimeter of the swimming pool alone.

**5.** For each patio material option, write an algebraic expression for the total cost of installing the patio based on its area and the given price information.

**6.** If you plan to spend the entire $3000 on the patio, how wide would the patio in option A be?

**7.** If you plan to spend the entire $3000 on the patio, how wide would the patio in option B be?

**8.** If you plan to spend the entire $3000 on the patio, how wide would the patio in option C be?

**9.** Which option would you choose? Why? Discuss the pros and cons of each option.

# CHAPTER 6 VOCABULARY CHECK

*Fill in each blank with one of the words or phrases listed below. Not all choices will be used.*

| | | |
|---|---|---|
| factoring | quadratic equation | perfect square trinomial |
| greatest common factor | 0 | sum of two cubes |
| difference of two cubes | difference of two squares | 1 |

1. An equation that can be written in the form $ax^2 + bx + c = 0$ (with $a$ not 0) is called a _____ .
2. _____ is the process of writing an expression as a product.
3. The _____ of a list of common variables raised to powers is the variable raised to the smallest exponent in the list.
4. A trinomial that is the square of some binomial is called a _____ .
5. The expression $a^2 - b^2$ is called a(n) _____ .
6. The expression $a^3 - b^3$ is called a(n) _____ .
7. The expression $a^3 + b^3$ is called a(n) _____ .
8. By the zero factor property, if the product of two numbers is 0, then at least one of the numbers must be _____ .

> ▶ **Helpful Hint**
> Are you preparing for your test? Don't forget to take the Chapter 6 Test on page 430. Then check your answers at the back of the text and use the Chapter Test Prep Video CD to see the fully worked-out solutions to any of the exercises you want to review.

# CHAPTER 6 HIGHLIGHTS

| **DEFINITIONS AND CONCEPTS** | **EXAMPLES** |
|---|---|

### SECTION 6.1   THE GREATEST COMMON FACTOR AND FACTORING BY GROUPING

| | |
|---|---|
| **Factoring** is the process of writing an expression as a product. | Factor: $6 = 2 \cdot 3$ $$x^2 + 5x + 6 = (x + 2)(x + 3)$$ |
| *To Find the GCF of a List of Integers* | Find the GCF of 12, 36, and 48. |
| **Step 1.** Write each number as a product of primes. | $12 = 2 \cdot 2 \cdot 3$ |
| **Step 2.** Identify the common prime factors. | $36 = 2 \cdot 2 \cdot 3 \cdot 3$ |
| **Step 3.** The product of all common factors is the greatest common factor. If there are no common prime factors, the GCF is 1. | $48 = 2 \cdot 2 \cdot 2 \cdot 2 \cdot 3$ $\text{GCF} = 2 \cdot 2 \cdot 3 = 12$ |
| **The GCF of a list of common variables raised to powers** is the variable raised to the smallest exponent in the list. | The GCF of $z^5$, $z^3$, and $z^{10}$ is $z^3$. |
| **The GCF of a list of terms** is the product of all common factors. | Find the GCF of $8x^2y$, $10x^3y^2$, and $26x^2y^3$. The GCF of 8, 10, and 26 is 2. The GCF of $x^2$, $x^3$, and $x^2$ is $x^2$. The GCF of $y$, $y^2$, and $y^3$ is $y$. The GCF of the terms is $2x^2y$. |

| **DEFINITIONS AND CONCEPTS** | **EXAMPLES** |
|---|---|

SECTION 6.1    THE GREATEST COMMON FACTOR AND FACTORING BY GROUPING (continued)

| | |
|---|---|
| ***To Factor by Grouping*** | Factor $10ax + 15a - 6xy - 9y$. |
| **Step** **1.** Arrange the terms so that the first two terms have a common factor and the last two have a common factor. | **Step** **1.** $10ax + 15a - 6xy - 9y$ |
| **Step** **2.** For each pair of terms, factor out the pair's GCF. | **Step** **2.** $5a(2x + 3) - 3y(2x + 3)$ |
| **Step** **3.** If there is now a common binomial factor, factor it out. | **Step** **3.** $(2x + 3)(5a - 3y)$ |
| **Step** **4.** If there is no common binomial factor, begin again, rearranging the terms differently. If no rearrangement leads to a common binomial factor, the polynomial cannot be factored. | |

SECTION 6.2    FACTORING TRINOMIALS OF THE FORM $x^2 + bx + c$

| | |
|---|---|
| **To factor a trinomial of the form $x^2 + bx + c$,** look for two numbers whose product is $c$ and whose sum is $b$. The factored form is <br><br> $(x + \text{one number})(x + \text{other number})$ | Factor: $x^2 + 7x + 12$ <br><br> $3 + 4 = 7 \qquad 3 \cdot 4 = 12$ <br> $(x + 3)(x + 4)$ |

SECTION 6.3    FACTORING TRINOMIALS OF THE FORM $ax^2 + bx + c$

| | |
|---|---|
| To factor $ax^2 + bx + c$, try various combinations of factors of $ax^2$ and $c$ until a middle term of $bx$ is obtained when checking. | Factor: $3x^2 + 14x - 5$ <br><br> Factors of $3x^2$: $3x, x$ <br> Factors of $-5$: $-1, 5$ and $1, -5$. <br><br> $(3x - \underline{1})(x + 5)$ <br> $\underbrace{\qquad -1x \qquad}$ <br> $15x$ <br> $14x$     Correct middle term |
| A **perfect square trinomial** is a trinomial that is the square of some binomial. | Perfect square trinomial = square of binomial <br><br> $x^2 + 4x + 4 = (x + 2)^2$ <br> $25x^2 - 10x + 1 = (5x - 1)^2$ |
| ***Factoring Perfect Square Trinomials:*** <br><br> $a^2 + 2ab + b^2 = (a + b)^2$ <br> $a^2 - 2ab + b^2 = (a - b)^2$ | Factor: <br><br> $x^2 + 6x + 9 = x^2 + 2 \cdot x \cdot 3 + 3^2 = (x + 3)^2$ <br> $4x^2 - 12x + 9 = (2x)^2 - 2 \cdot 2x \cdot 3 + 3^2 = (2x - 3)^2$ |

SECTION 6.4    FACTORING TRINOMIALS OF THE FORM $ax^2 + bx + c$ BY GROUPING

| | |
|---|---|
| ***To Factor $ax^2 + bx + c$ by Grouping*** | Factor: $3x^2 + 14x - 5$ |
| **Step** **1.** Find two numbers whose product is $a \cdot c$ and whose sum is $b$. | **Step** **1.** Find two numbers whose product is $3 \cdot (-5)$ or $-15$ and whose sum is 14. They are 15 and $-1$. |
| **Step** **2.** Rewrite $bx$, using the factors found in Step 1. | **Step** **2.** $3x^2 + 14x - 5$ <br> $= 3x^2 + 15x - 1x - 5$ |
| **Step** **3.** Factor by grouping. | **Step** **3.** $= 3x(x + 5) - 1(x + 5)$ <br> $= (x + 5)(3x - 1)$ |

| DEFINITIONS AND CONCEPTS | EXAMPLES |
|---|---|

### SECTION 6.5 FACTORING BINOMIALS

**Difference of Squares**

$$a^2 - b^2 = (a + b)(a - b)$$

**Sum or Difference of Cubes**

$$a^3 + b^3 = (a + b)(a^2 - ab + b^2)$$
$$a^3 - b^3 = (a - b)(a^2 + ab + b^2)$$

Factor:

$$x^2 - 9 = x^2 - 3^2 = (x + 3)(x - 3)$$

$$y^3 + 8 = y^3 + 2^3 = (y + 2)(y^2 - 2y + 4)$$
$$125z^3 - 1 = (5z)^3 - 1^3 = (5z - 1)(25z^2 + 5z + 1)$$

### INTEGRATED REVIEW—CHOOSING A FACTORING STRATEGY

**To Factor a Polynomial,**

**Step 1.** Factor out the GCF.

**Step 2. a.** If two terms,

   **i.** $a^2 - b^2 = (a + b)(a - b)$

   **ii.** $a^3 - b^3 = (a - b)(a^2 + ab + b^2)$

   **iii.** $a^3 + b^3 = (a + b)(a^2 - ab + b^2)$

   **b.** If three terms,

   **i.** $a^2 + 2ab + b^2 = (a + b)^2$

   **ii.** Methods in Sections 6.2 and 6.3

   **c.** If four or more terms, try factoring by grouping.

**Step 3.** See if any factors can be factored further.

**Step 4.** Check by multiplying.

Factor: $2x^4 - 6x^2 - 8$

**Step 1.** $2x^4 - 6x^2 - 8 = 2(x^4 - 3x^2 - 4)$

**Step 2. b. ii.** $\qquad = 2(x^2 + 1)(x^2 - 4)$

**Step 3.** $= 2(x^2 + 1)(x + 2)(x - 2)$

**Step 4.** Check by multiplying.

$$2(x^2 + 1)(x + 2)(x - 2) = 2(x^2 + 1)(x^2 - 4)$$
$$= 2(x^4 - 3x^2 - 4)$$
$$= 2x^4 - 6x^2 - 8$$

### SECTION 6.6 SOLVING QUADRATIC EQUATIONS BY FACTORING

A **quadratic equation** is an equation that can be written in the form $ax^2 + bx + c = 0$ with $a$ not 0.

The form $ax^2 + bx + c = 0$ is called the **standard form** of a quadratic equation.

**Zero Factor Theorem**

If $a$ and $b$ are real numbers and if $ab = 0$, then $a = 0$ or $b = 0$.

**To solve quadratic equations by factoring,**

**Step 1.** Write the equation in standard form: $ax^2 + bx + c = 0$.

**Step 2.** Factor the quadratic.

**Step 3.** Set each factor containing a variable equal to 0.

**Step 4.** Solve the equations.

**Step 5.** Check in the original equation.

| **Quadratic Equation** | **Standard Form** |
|---|---|
| $x^2 = 16$ | $x^2 - 16 = 0$ |
| $y = -2y^2 + 5$ | $2y^2 + y - 5 = 0$ |

If $(x + 3)(x - 1) = 0$, then $x + 3 = 0$ or $x - 1 = 0$

Solve: $3x^2 = 13x - 4$

**Step 1.** $3x^2 - 13x + 4 = 0$

**Step 2.** $(3x - 1)(x - 4) = 0$

**Step 3.** $3x - 1 = 0$ or $x - 4 = 0$

**Step 4.** $3x = 1$ or $x = 4$

$$x = \frac{1}{3}$$

**Step 5.** Check both $\frac{1}{3}$ and 4 in the original equation.

| **DEFINITIONS AND CONCEPTS** | **EXAMPLES** |
|---|---|

SECTION 6.7   QUADRATIC EQUATIONS AND PROBLEM SOLVING

| | |
|---|---|
| ***Problem-Solving Steps*** | A garden is in the shape of a rectangle whose length is two feet more than its width. If the area of the garden is 35 square feet, find its dimensions. |
| **1.** UNDERSTAND the problem. | **1.** Read and reread the problem. Guess a solution and check your guess.<br>Let $x$ be the width of the rectangular garden. Then $x + 2$ is the length.<br><br>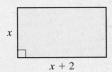 |
| **2.** TRANSLATE. | **2.** In words:  length · width = area<br>                          ↓         ↓        ↓<br>Translate:  $(x + 2)$ · $x$ = $35$ |
| **3.** SOLVE. | **3.** $(x + 2)x = 35$<br>$x^2 + 2x - 35 = 0$<br>$(x - 5)(x + 7) = 0$<br>$x - 5 = 0$  or  $x + 7 = 0$<br>$x = 5$  or  $x = -7$ |
| **4.** INTERPRET. | **4.** Discard the solution of $-7$ since $x$ represents width.<br>***Check:*** If $x$ is 5 feet then $x + 2 = 5 + 2 = 7$ feet. The area of a rectangle whose width is 5 feet and whose length is 7 feet is (5 feet)(7 feet) or 35 square feet.<br>***State:*** The garden is 5 feet by 7 feet. |

---

## 📖 STUDY SKILLS BUILDER

### Are You Prepared for a Test on Chapter 6?

Below is a list of some *common trouble areas* for students in Chapter 6. After studying for your test—but before taking your test—read these.

- The difference of two squares such as $x^2 - 25$ factors as $x^2 - 25 = (x + 5)(x - 5)$.
- The sum of two squares, for example, $x^2 + 25$, cannot be factored using real numbers.
- Don't forget that the first step to factor any polynomial is to first factor out any common factors.

$$9x^2 - 36 = 9(x^2 - 4) = 9(x + 2)(x - 2)$$

- Can you completely factor $x^4 - 24x^2 - 25$?

$$x^4 - 24x^2 - 25 = (x^2 - 25)(x^2 + 1)$$
$$= (x + 5)(x - 5)(x^2 + 1)$$

- Remember that to use the zero factor property to solve a quadratic equation, one side of the equation must be 0 and the other side must be a factored polynomial.

$$x(x - 2) = 3 \quad \text{Cannot use zero factor property.}$$
$$x^2 - 2x - 3 = 0$$
$$(x - 3)(x + 1) = 0 \quad \text{Now we can use zero factor property.}$$
$$x - 3 = 0 \quad \text{or} \quad x + 1 = 0$$
$$x = 3 \quad \text{or} \quad x = -1$$

**Remember:** This is simply a sampling of selected topics given to check your understanding. For a review of Chapter 6 in your text, see the material at the end of this chapter.

# CHAPTER 6 REVIEW

**(6.1)** *Complete the factoring.*

**1.** $6x^2 - 15x = 3x(\quad)$

**2.** $2x^3y - 6x^2y^2 - 8xy^3 = 2xy(\quad)$

*Factor the GCF from each polynomial.*

**3.** $20x^2 + 12x$

**4.** $6x^2y^2 - 3xy^3$

**5.** $-8x^3y + 6x^2y^2$

**6.** $3x(2x + 3) - 5(2x + 3)$

**7.** $5x(x + 1) - (x + 1)$

*Factor.*

**8.** $3x^2 - 3x + 2x - 2$

**9.** $6x^2 + 10x - 3x - 5$

**10.** $3a^2 + 9ab + 3b^2 + ab$

**(6.2)** *Factor each trinomial.*

**11.** $x^2 + 6x + 8$

**12.** $x^2 - 11x + 24$

**13.** $x^2 + x + 2$

**14.** $x^2 - 5x - 6$

**15.** $x^2 + 2x - 8$

**16.** $x^2 + 4xy - 12y^2$

**17.** $x^2 + 8xy + 15y^2$

**18.** $3x^2y + 6xy^2 + 3y^3$

**19.** $72 - 18x - 2x^2$

**20.** $32 + 12x - 4x^2$

**(6.3)** *or* **(6.4)** *Factor each trinomial.*

**21.** $2x^2 + 11x - 6$

**22.** $4x^2 - 7x + 4$

**23.** $4x^2 + 4x - 3$

**24.** $6x^2 + 5xy - 4y^2$

**25.** $6x^2 - 25xy + 4y^2$

**26.** $18x^2 - 60x + 50$

**27.** $2x^2 - 23xy - 39y^2$

**28.** $4x^2 - 28xy + 49y^2$

**29.** $18x^2 - 9xy - 20y^2$

**30.** $36x^3y + 24x^2y^2 - 45xy^3$

**(6.5)** *Factor each binomial.*

**31.** $4x^2 - 9$

**32.** $9t^2 - 25s^2$

**33.** $16x^2 + y^2$

**34.** $x^3 - 8y^3$

**35.** $8x^3 + 27$

**36.** $2x^3 + 8x$

**37.** $54 - 2x^3y^3$

**38.** $9x^2 - 4y^2$

**39.** $16x^4 - 1$

**40.** $x^4 + 16$

**(6.6)** *Solve the following equations.*

**41.** $(x + 6)(x - 2) = 0$

**42.** $3x(x + 1)(7x - 2) = 0$

**43.** $4(5x + 1)(x + 3) = 0$

**44.** $x^2 + 8x + 7 = 0$

**45.** $x^2 - 2x - 24 = 0$

**46.** $x^2 + 10x = -25$

**47.** $x(x - 10) = -16$

**48.** $(3x - 1)(9x^2 + 3x + 1) = 0$

**49.** $56x^2 - 5x - 6 = 0$

**50.** $20x^2 - 7x - 6 = 0$

**51.** $5(3x + 2) = 4$

**52.** $6x^2 - 3x + 8 = 0$

**53.** $12 - 5t = -3$

**54.** $5x^3 + 20x^2 + 20x = 0$

**55.** $4t^3 - 5t^2 - 21t = 0$

**56.** Write a quadratic equation that has the two solutions 4 and 5.

**(6.7)** *Use the given information to choose the correct dimensions.*

△ **57.** The perimeter of a rectangle is 24 inches. The length is twice the width. Find the dimensions of the rectangle.

    **a.** 5 inches by 7 inches

    **b.** 5 inches by 10 inches

    **c.** 4 inches by 8 inches

    **d.** 2 inches by 10 inches

△ **58.** The area of a rectangle is 80 meters. The length is one more than three times the width. Find the dimensions of the rectangle.

    **a.** 8 meters by 10 meters

    **b.** 4 meters by 13 meters

    **c.** 4 meters by 20 meters

    **d.** 5 meters by 16 meters

*Use the given information to find the dimensions of each figure.*

△ **59.** The *area* of the square is 81 square units. Find the length of a side.

△ **60.** The *perimeter* of the quadrilateral is 47 units. Find the lengths of the sides.

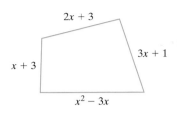

△ **61.** A flag for a local organization is in the shape of a rectangle whose length is 15 inches less than twice its width. If the area of the flag is 500 square inches, find its dimensions.

△ **62.** The base of a triangular sail is four times its height. If the area of the triangle is 162 square yards, find the base.

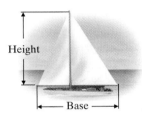

**63.** Find two consecutive positive integers whose product is 380.

**64.** A rocket is fired from the ground with an initial velocity of 440 feet per second. Its height $h$ after $t$ seconds is given by the equation

$$h = -16t^2 + 440t$$

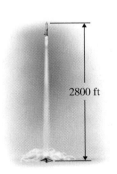

2800 ft

✎ **a.** Find how many seconds pass before the rocket reaches a height of 2800 feet. Explain why two answers are obtained.

**b.** Find how many seconds pass before the rocket reaches the ground again.

**65.** An object is dropped from the top of the 625-foot-tall Waldorf-Astoria Hotel on Park Avenue in New York City. (*Source: World Almanac* research) The height $h$ of the object after $t$ seconds is given by the equation $h = -16t^2 + 625$. Find how many seconds pass before the object reaches the ground.

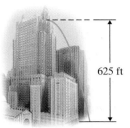

625 ft

△ **66.** An architect's squaring instrument is in the shape of a right triangle. Find the length of the long leg of the right triangle if the hypotenuse is 8 centimeters longer than the long leg and the short leg is 8 centimeters shorter than the long leg.

## MIXED REVIEW

*Factor completely.*

**67.** $7x - 63$

**68.** $11x(4x - 3) - 6(4x - 3)$

**69.** $m^2 - \dfrac{4}{25}$

**70.** $3x^3 - 4x^2 + 6x - 8$

**71.** $xy + 2x - y - 2$

**72.** $2x^2 + 2x - 24$

**73.** $3x^3 - 30x^2 + 27x$

**74.** $4x^2 - 81$

**75.** $2x^2 - 18$

**76.** $16x^2 - 24x + 9$

**77.** $5x^2 + 20x + 20$

**78.** $2x^2 + 5x - 12$

**79.** $4x^2y - 6xy^2$

**80.** $8x^2 - 15x - x^3$

**81.** $125x^3 + 27$

**82.** $24x^2 - 3x - 18$

**83.** $(x + 7)^2 - y^2$

**84.** $x^2(x + 3) - 4(x + 3)$

**85.** $54a^3b - 2b$

**86.** To factor $x^2 + 2x - 48$, think of two numbers whose product is _____ and whose sum is _____.

**87.** What is the first step to factoring $3x^2 + 15x + 30$?

*Write the perimeter of each figure as a simplified polynomial. Then factor each polynomial.*

△ **88.**

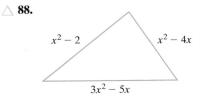

△ **89.**

$2x^2 + 3$

$6x^2 - 14x$

*Solve.*

**90.** $2x^2 - x - 28 = 0$

**91.** $x^2 - 2x = 15$

**92.** $2x(x + 7)(x + 4) = 0$

**93.** $x(x - 5) = -6$

**94.** $x^2 = 16x$

*Solve.*

**95.** The perimeter of the following triangle is 48 inches. Find the lengths of its sides.

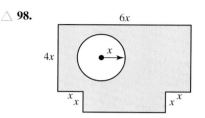

$x^2 + 3$     $4x + 5$

$2x$

**96.** The width of a rectangle is 4 inches less than its length. Its area is 12 square inches. Find the dimensions of the rectangle.

**97.** A 6-foot-tall person drops an object from the top of the Westin Peachtree Plaza in Atlanta, Georgia. The Westin building is 723 feet tall. (*Source:World Almanac* research) The height $h$ of the object after $t$ seconds is given by the equation $h = -16t^2 + 729$. Find how many seconds pass before the object reaches the ground.

723 ft

*Write an expression for the area of the shaded region. Then write the expression as a factored polynomial.*

△ **98.**

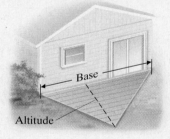

$6x$

$4x$

$x$

$x$ $x$      $x$ $x$

# CHAPTER 6 TEST ![TEST PREP VIDEO]

Remember to use the Chapter Test Prep Video CD to see the fully worked-out solutions to any of the exercises you want to review.

*Factor each polynomial completely. If a polynomial cannot be factored, write "prime."*

**1.** $x^2 + 11x + 28$

**2.** $49 - m^2$

**3.** $y^2 + 22y + 121$

**4.** $4(a + 3) - y(a + 3)$

**5.** $x^2 + 4$

**6.** $y^2 - 8y - 48$

**7.** $x^2 + x - 10$

**8.** $9x^3 + 39x^2 + 12x$

**9.** $3a^2 + 3ab - 7a - 7b$

**10.** $3x^2 - 5x + 2$

**11.** $x^2 + 14xy + 24y^2$

**12.** $180 - 5x^2$

**13.** $6t^2 - t - 5$

**14.** $xy^2 - 7y^2 - 4x + 28$

**15.** $x - x^5$

**16.** $-xy^3 - x^3y$

**17.** $64x^3 - 1$

**18.** $8y^3 - 64$

*Solve each equation.*

**19.** $(x - 3)(x + 9) = 0$

**20.** $x^2 + 5x = 14$

**21.** $x(x + 6) = 7$

**22.** $3x(2x - 3)(3x + 4) = 0$

**23.** $5t^3 - 45t = 0$

**24.** $t^2 - 2t - 15 = 0$

**25.** $6x^2 = 15x$

*Solve each problem.*

△ **26.** A deck for a home is in the shape of a triangle. The length of the base of the triangle is 9 feet longer than its altitude. If the area of the triangle is 68 square feet, find the length of the base.

Base

Altitude

**27.** The sum of two numbers is 17 and the sum of their squares is 145. Find the numbers.

**28.** An object is dropped from the top of the Woolworth Building on Broadway in New York City. The height $h$ of the object after $t$ seconds is given by the equation

$$h = -16t^2 + 784$$

Find how many seconds pass before the object reaches the ground.

△ **29.** Find the lengths of the sides of a right triangle if the hypotenuse is 10 centimeters longer than the shorter leg and 5 centimeters longer than the longer leg.

# CHAPTER 6 CUMULATIVE REVIEW

1. Translate each sentence into a mathematical statement.
   a. Nine is less than or equal to eleven.
   b. Eight is greater than one.
   c. Three is not equal to four.

2. Insert $<$ or $>$ in the space to make each statement true.
   a. $|-5|\ \_\ |-3|$
   b. $|0|\ \_\ |-2|$

3. Write each fraction in lowest terms.
   a. $\dfrac{42}{49}$   b. $\dfrac{11}{27}$   c. $\dfrac{88}{20}$

4. Evaluate $\dfrac{x}{y} + 5x$ if $x = 20$ and $y = 10$.

5. Simplify: $\dfrac{8 + 2 \cdot 3}{2^2 - 1}$

6. Evaluate $\dfrac{x}{y} + 5x$ if $x = -20$ and $y = 10$.

7. Add.
   a. $3 + (-7) + (-8)$
   b. $[7 + (-10)] + [-2 + |-4|]$

8. Evaluate $\dfrac{x}{y} + 5x$ if $x = -20$ and $y = -10$.

9. Multiply.
   a. $(-8)(4)$   b. $14(-1)$
   c. $-9(-10)$

10. Simplify: $5 - 2(3x - 7)$

11. Simplify each expression by combining like terms.
    a. $7x - 3x$   b. $10y^2 + y^2$
    c. $8x^2 + 2x - 3x$

12. Solve: $0.8y + 0.2(y - 1) = 1.8$

*Solve.*

13. $\dfrac{y}{7} = 20$

14. $\dfrac{x}{-7} = -4$

15. $-3x = 33$

16. $-\dfrac{2}{3}x = -22$

17. $8(2 - t) = -5t$

18. $-z = \dfrac{7z + 3}{5}$

19. Balsa wood sticks are commonly used to build models (for example, bridge models). A 48-inch Balsa wood stick is to be cut into two pieces so that the longer piece is 3 times the shorter. Find the length of each piece.

20. Solve $3x + 9 \le 5(x - 1)$. Write the solution set using interval notation.

21. Graph the linear equation $y = -\dfrac{1}{3}x + 2$.

22. Is the ordered pair $(-1, 2)$ a solution of $-7x - 8y = -9$?

23. Find the slope and y-intercept of the line whose equation is $3x - 4y = 4$.

24. Find the slope of the line through $(5, -6)$ and $(5, 2)$.

25. Evaluate each expression for the given value of x.
    a. $2x^3$; $x$ is 5   b. $\dfrac{9}{x^2}$; $x$ is $-3$

26. Find the slope and y-intercept of the line whose equation is $7x - 3y = 2$.

27. Find the degree of each term.
    a. $3x^2$   b. $-2^3x^5$   c. $y$
    d. $12x^2yz^3$   e. 5

28. Find an equation of the vertical line through $(0, 7)$.

29. Subtract: $(2x^3 + 8x^2 - 6x) - (2x^3 - x^2 + 1)$

30. Find an equation of the line with slope 4 and y-intercept $\left(0, \dfrac{1}{2}\right)$. Write the equation in standard form.

31. Multiply $(3x + 2)(2x - 5)$.

32. Write an equation of the line through $(-4, 0)$ and $(6, -1)$. Write the equation in standard form.

33. Multiply $(3y + 1)^2$.

34. Solve the system: $\begin{cases} -x + 3y = 18 \\ -3x + 2y = 19 \end{cases}$

35. Simplify by writing each expression with positive exponents only.
    a. $3^{-2}$   b. $2x^{-3}$
    c. $2^{-1} + 4^{-1}$   d. $(-2)^{-4}$
    e. $\dfrac{1}{y^{-4}}$   f. $\dfrac{1}{7^{-2}}$

36. Simplify: $\dfrac{(5a^7)^2}{a^5}$

37. Write each number in scientific notation.
    a. 367,000,000   b. 0.000003
    c. 20,520,000,000
    d. 0.00085

38. Multiply: $(3x - 7y)^2$

39. Divide $x^2 + 7x + 12$ by $x + 3$ using long division.

40. Simplify: $\dfrac{(xy)^{-3}}{(x^5y^6)^3}$

41. Find the GCF of each list of terms.
    a. $x^3$, $x^7$, and $x^5$   b. $y$, $y^4$, and $y^7$

*Factor.*

42. $z^3 + 7z + z^2 + 7$
43. $x^2 + 7x + 12$
44. $2x^3 + 2x^2 - 84x$
45. $8x^2 - 22x + 5$
46. $-4x^2 - 23x + 6$
47. $25a^2 - 9b^2$
48. $9xy^2 - 16x$
49. Solve $(x - 3)(x + 1) = 0$.
50. Solve $x^2 - 13x = -36$.

# 7 Rational Expressions

In this chapter, we expand our knowledge of algebraic expressions to include another category called rational expressions, such as $\dfrac{x+1}{x}$. We explore the operations of addition, subtraction, multiplication, and division for these algebraic fractions, using principles similar to the principles for number fractions.

**C**ephalic index is the ratio of the maximum width of the head to its maximum length, sometimes multiplied by 100. It is used by anthropologists and forensic scientists on human skulls, but it is used especially on animal skulls to categorize animals such as dogs and cats. In Section 7.1, Exercise 59, you will have the opportunity to calculate this index for a human skull.

Cephalic Index Formula: $C = \dfrac{100\,W}{L}$

where $W$ is width of the skull and $L$ is length of the skull.

| Cephalic Index for Dogs | | |
|---|---|---|
| Value | Scientific Term | Meaning |
| <80 or <75 | *dolichocephalic* | "long-headed" |
| | *mesocephalic* | "medium-headed" |
| >80 | *brachycephalic* | "short-headed" |

A brachycephalic skull is relatively broad and short, as in the Pug.
A mesocephalic skull is of intermediate length and width, as in the Cocker Spaniel.
A dolichocephalic skull is relatively long, as in the Afghan Hound.

# 7.1 RATIONAL FUNCTIONS AND SIMPLIFYING RATIONAL EXPRESSIONS

**OBJECTIVES**

1 Find the domain of a rational expression.

2 Simplify rational expressions.

3 Write equivalent forms of rational expressions.

4 Use rational functions in applications.

Recall that a *rational number*, or *fraction*, is a number that can be written as the quotient $\frac{p}{q}$ of two integers $p$ and $q$ as long as $q$ is not 0. A **rational expression** is an expression that can be written as the quotient $\frac{P}{Q}$ of two polynomials $P$ and $Q$ as long as $Q$ is not 0.

### Examples of Rational Expressions

$$\frac{3x + 7}{2} \qquad \frac{5x^2 - 3}{x - 1} \qquad \frac{7x - 2}{2x^2 + 7x + 6}$$

Rational expressions are sometimes used to describe functions. For example, we call the function $f(x) = \dfrac{x^2 + 2}{x - 3}$ a **rational function** since $\dfrac{x^2 + 2}{x - 3}$ is a rational expression.

**OBJECTIVE 1 ▶ Finding the domain of a rational expression.** As with fractions, a rational expression is **undefined** if the denominator is 0. If a variable in a rational expression is replaced with a number that makes the denominator 0, we say that the rational expression is **undefined** for this value of the variable. For example, the rational expression $\dfrac{x^2 + 2}{x - 3}$ is undefined when $x$ is 3, because replacing $x$ with 3 results in a denominator of 0. For this reason, we must exclude 3 from the domain of the function $f(x) = \dfrac{x^2 + 2}{x - 3}$.

The domain of $f$ is then

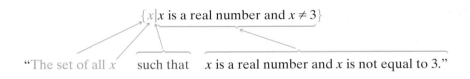

$$\{x \mid x \text{ is a real number and } x \neq 3\}$$

"The set of all $x$     such that     $x$ is a real number and $x$ is not equal to 3."

In this section, we will use this set builder notations to write domains. Unless told otherwise, we assume that the domain of a function described by an equation is the set of all real numbers for which the equation is defined.

**EXAMPLE 1**    Find the domain of each rational function.

**a.** $f(x) = \dfrac{8x^3 + 7x^2 + 20}{2}$   **b.** $g(x) = \dfrac{5x^2 - 3}{x - 1}$   **c.** $f(x) = \dfrac{7x - 2}{x^2 - 2x - 15}$

<u>Solution</u>    The domain of each function will contain all real numbers except those values that make the denominator 0.

**a.** No matter what the value of $x$, the denominator of $f(x) = \dfrac{8x^3 + 7x^2 + 20}{2}$ is never 0, so the domain of $f$ is $\{x \mid x \text{ is a real number}\}$.

**b.** To find the values of $x$ that make the denominator of $g(x)$ equal to 0, we solve the equation "denominator = 0":

$$x - 1 = 0, \quad \text{or} \quad x = 1$$

The domain must exclude 1 since the rational expression is undefined when $x$ is 1. The domain of $g$ is $\{x \mid x \text{ is a real number and } x \neq 1\}$.

**c.** We find the domain by setting the denominator equal to 0.

$$x^2 - 2x - 15 = 0 \quad \text{Set the denominator equal to 0 and solve.}$$
$$(x - 5)(x + 3) = 0$$
$$x - 5 = 0 \quad \text{or} \quad x + 3 = 0$$
$$x = 5 \quad \text{or} \quad x = -3$$

If $x$ is replaced with $5$ or with $-3$, the rational expression is undefined.

The domain of $f$ is $\{x \mid x \text{ is a real number and } x \neq 5, x \neq -3\}$. □

**PRACTICE**

**1** Find the domain of each rational function.

**a.** $f(x) = \dfrac{4x^5 - 3x^2 + 2}{-6}$     **b.** $g(x) = \dfrac{6x^2 + 1}{x + 3}$     **c.** $h(x) = \dfrac{8x - 3}{x^2 - 5x + 6}$

**Concept Check** ☑

For which of these values (if any) is the rational expression $\dfrac{x - 3}{x^2 + 2}$ undefined?

**a.** 2     **b.** 3     **c.** −2     **d.** 0     **e.** None of these

**OBJECTIVE 2 ▶ Simplifying rational expressions.** Recall that a fraction is in lowest terms or simplest form if the numerator and denominator have no common factors other than 1 (or −1). For example, $\dfrac{3}{13}$ is in lowest terms since 3 and 13 have no common factors other than 1 (or −1).

To **simplify** a rational expression, or to write it in lowest terms, we use a method similar to simplifying a fraction.

Recall that to simplify a fraction, we essentially "remove factors of 1." Our ability to do this comes from these facts:

• If $c \neq 0$, then $\dfrac{c}{c} = 1$. For example, $\dfrac{7}{7} = 1$ and $\dfrac{-8.65}{-8.65} = 1$.

• $n \cdot 1 = n$. For example, $-5 \cdot 1 = -5$, $\quad 126.8 \cdot 1 = 126.8$, $\quad$ and $\dfrac{a}{b} \cdot 1 = \dfrac{a}{b}, b \neq 0$.

In other words, we have the following:

$$\frac{a \cdot c}{b \cdot c} = \frac{a}{b} \cdot \frac{c}{c} = \frac{a}{b}$$

$$\text{Since } \tfrac{a}{b} \cdot 1 = \tfrac{a}{b}$$

Let's practice simplifying a fraction by simplifying $\dfrac{15}{65}$.

$$\frac{15}{65} = \frac{3 \cdot 5}{13 \cdot 5} = \frac{3}{13} \cdot \frac{5}{5} = \frac{3}{13} \cdot 1 = \frac{3}{13}$$

Let's use the same technique and simplify the rational expression $\dfrac{(x + 2)^2}{x^2 - 4}$.

$$\frac{(x + 2)^2}{x^2 - 4} = \frac{(x + 2)(x + 2)}{(x - 2)(x + 2)}$$

$$= \frac{(x + 2)}{(x - 2)} \cdot \frac{x + 2}{x + 2}$$

$$= \frac{x + 2}{x - 2} \cdot 1$$

$$= \frac{x + 2}{x - 2}$$

This means that the rational expression $\dfrac{(x + 2)^2}{x^2 - 4}$ has the same value as the rational expression $\dfrac{x + 2}{x - 2}$ for all values of $x$ except 2 and $-2$. (Remember that when $x$ is 2, the denominators of both rational expressions are 0 and that when $x$ is $-2$, the original rational expression has a denominator of 0.)

As we simplify rational expressions, we will assume that the simplified rational expression is equivalent to the original rational expression for all real numbers except those for which either denominator is 0.

Just as for numerical fractions, we can use a shortcut notation. Remember that as long as exact factors in both the numerator and denominator are divided out, we are "removing a factor of 1." We can use the following notation:

$$\frac{(x + 2)^2}{x^2 - 4} = \frac{(x + 2)\,(x + 2)}{(x - 2)\,(x + 2)} \qquad \text{A factor of 1 is identified by the shading.}$$

$$= \frac{x + 2}{x - 2} \qquad \text{"Remove" the factor of 1.}$$

This "removing a factor of 1" is stated in the principle below:

---

**Fundamental Principle of Rational Expressions**

For any rational expression $\dfrac{P}{Q}$ and any polynomial $R$, where $R \neq 0$,

$$\frac{PR}{QR} = \frac{P}{Q} \cdot \frac{R}{R} = \frac{P}{Q} \cdot 1 = \frac{P}{Q}$$

or, simply,

$$\frac{PR}{QR} = \frac{P}{Q}$$

---

In general, the following steps may be used to simplify rational expressions or to write a rational expression in lowest terms.

---

**Simplifying or Writing a Rational Expression in Lowest Terms**

**STEP 1.** Completely factor the numerator and denominator of the rational expression.

**STEP 2.** Divide out factors common to the numerator and denominator. (This is the same as "removing a factor of 1.")

---

For now, we assume that variables in a rational expression do not represent values that make the denominator 0.

**EXAMPLE 2**  Simplify each rational expression.

**a.** $\dfrac{2x^2}{10x^3 - 2x^2}$   **b.** $\dfrac{9x^2 + 13x + 4}{8x^2 + x - 7}$

**Solution**

**a.** $\dfrac{2x^2}{10x^3 - 2x^2} = \dfrac{2x^2 \cdot 1}{2x^2\,(5x - 1)} = 1 \cdot \dfrac{1}{5x - 1} = \dfrac{1}{5x - 1}$

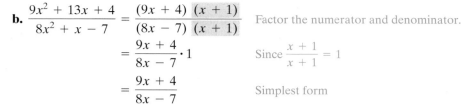

**b.** $\dfrac{9x^2 + 13x + 4}{8x^2 + x - 7} = \dfrac{(9x + 4)\ (x + 1)}{(8x - 7)\ (x + 1)}$   Factor the numerator and denominator.

$= \dfrac{9x + 4}{8x - 7} \cdot 1$   Since $\dfrac{x + 1}{x + 1} = 1$

$= \dfrac{9x + 4}{8x - 7}$   Simplest form

**PRACTICE**
**2**   Simplify each rational expressions.

**a.** $\dfrac{5z^4}{10z^5 - 5z^4}$

**b.** $\dfrac{5x^2 + 13x + 6}{6x^2 + 7x - 10}$

**EXAMPLE 3**   Simplify each rational expression.

**a.** $\dfrac{2 + x}{x + 2}$

**b.** $\dfrac{2 - x}{x - 2}$

*Solution*

**a.** $\dfrac{2 + x}{x + 2} = \dfrac{x + 2}{x + 2} = 1$   By the commutative property of addition, $2 + x = x + 2$.

**b.** $\dfrac{2 - x}{x - 2}$

The terms in the numerator of $\dfrac{2 - x}{x - 2}$ differ by sign from the terms of the denominator, so the polynomials are opposites of each other and the expression simplifies to $-1$. To see this, we factor out $-1$ from the numerator or the denominator. If $-1$ is factored from the numerator, then

$$\dfrac{2 - x}{x - 2} = \dfrac{-1(-2 + x)}{x - 2} = \dfrac{-1\ (x - 2)}{x - 2} = \dfrac{-1}{1} = -1$$

If $-1$ is factored from the denominator, the result is the same.

$$\dfrac{2 - x}{x - 2} = \dfrac{2 - x}{-1(-x + 2)} = \dfrac{2 - x}{-1\ (2 - x)} = \dfrac{1}{-1} = -1$$

> ▶ **Helpful Hint**
>
> When the numerator and the denominator of a rational expression are opposites of each other, the expression simplifies to $-1$.

**PRACTICE**
**3**   Simplify each rational expression.

**a.** $\dfrac{x + 3}{3 + x}$

**b.** $\dfrac{3 - x}{x - 3}$

**EXAMPLE 4**   Simplify $\dfrac{18 - 2x^2}{x^2 - 2x - 3}$.

*Solution*   $\dfrac{18 - 2x^2}{x^2 - 2x - 3} = \dfrac{2(9 - x^2)}{(x + 1)(x - 3)}$   Factor.

$= \dfrac{2(3 + x)(3 - x)}{(x + 1)(x - 3)}$   Factor completely.

$= \dfrac{2(3 + x) \cdot -1\ (x - 3)}{(x + 1)\ (x - 3)}$   Notice the opposites $3 - x$ and $x - 3$. Write $3 - x$ as $-1(x - 3)$ and simplify.

$= -\dfrac{2(3 + x)}{x + 1}$

**PRACTICE**
**4**   Simplify $\dfrac{20 - 5x^2}{x^2 + x - 6}$.

> ▶ **Helpful Hint**
> When simplifying a rational expression, we look for **common *factors*, not common *terms*.**

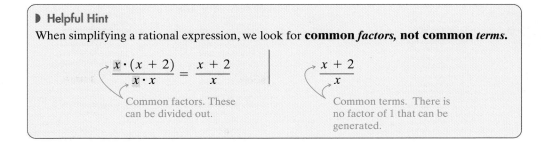

| | |
|---|---|
| $\dfrac{x \cdot (x + 2)}{x \cdot x} = \dfrac{x + 2}{x}$ | $\dfrac{x + 2}{x}$ |
| Common factors. These can be divided out. | Common terms. There is no factor of 1 that can be generated. |

**Concept Check** ☑

Recall that we can only remove *factors* of 1. Which of the following are *not* true? Explain why.

**a.** $\dfrac{3 - 1}{3 + 5}$ simplifies to $-\dfrac{1}{5}$?

**b.** $\dfrac{2x + 10}{2}$ simplifies to $x + 5$?

**c.** $\dfrac{37}{72}$ simplifies to $\dfrac{3}{2}$?

**d.** $\dfrac{2x + 3}{2}$ simplifies to $x + 3$?

**EXAMPLE 5**    Simplify each rational expression.

**a.** $\dfrac{x^3 + 8}{2 + x}$

**b.** $\dfrac{2y^2 + 2}{y^3 - 5y^2 + y - 5}$

*Solution*

**a.** $\dfrac{x^3 + 8}{2 + x} = \dfrac{(x + 2)(x^2 - 2x + 4)}{x + 2}$      Factor the sum of the two cubes.

$= x^2 - 2x + 4$      Divide out common factors.

**b.** $\dfrac{2y^2 + 2}{y^3 - 5y^2 + y - 5} = \dfrac{2(y^2 + 1)}{(y^3 - 5y^2) + (y - 5)}$      Factor the numerator.

$= \dfrac{2(y^2 + 1)}{y^2(y - 5) + 1(y - 5)}$      Factor the denominator by grouping.

$= \dfrac{2(y^2 + 1)}{(y - 5)(y^2 + 1)}$

$= \dfrac{2}{y - 5}$      Divide out common factors.      □

**PRACTICE**
**5**    Simplify each rational expression.

**a.** $\dfrac{x^3 + 64}{4 + x}$

**b.** $\dfrac{5z^2 + 10}{z^3 - 3z^2 + 2z - 6}$

**Concept Check** ☑

Does $\dfrac{n}{n + 2}$ simplify to $\dfrac{1}{2}$? Why or why not?

For a negative fraction such as $\dfrac{2}{-7}$, recall from Section 1.7 that

$$\dfrac{2}{-7} = \dfrac{-2}{7} = -\dfrac{2}{7}$$

**Answers to Concept Check:**
a, c, d
no; answers may vary.

In general, for any fraction,

$$\frac{-a}{b} = \frac{a}{-b} = -\frac{a}{b}, \qquad b \neq 0$$

This is also true for rational expressions. For example,

$$\underbrace{\frac{-(x + 2)}{x}}_{} = \frac{x + 2}{-x} = -\frac{x + 2}{x}$$

↑
Notice the parentheses.

**OBJECTIVE 3 ▶ Writing equivalent forms of rational expressions.** From Example 3, we have

$$\frac{2 + x}{x + 2} = \boxed{\frac{x + 2}{x + 2}} = 1 \qquad \text{and} \qquad \frac{2 - x}{x - 2} = \boxed{\frac{2 - x}{-1(2 - x)}} = \frac{1}{-1} = -1.$$

When performing operations on rational expressions, equivalent forms of answers often result. For this reason, it is very important to be able to recognize equivalent answers.

**EXAMPLE 6**   List some equivalent forms of $-\dfrac{5x - 1}{x + 9}$.

*Solution*   To do so, recall that $-\dfrac{a}{b} = \dfrac{-a}{b} = \dfrac{a}{-b}$. Thus

$$-\frac{5x - 1}{x + 9} = \frac{-(5x - 1)}{x + 9} = \frac{-5x + 1}{x + 9} \quad \text{or} \quad \frac{1 - 5x}{x + 9}$$

Also,

$$-\frac{5x - 1}{x + 9} = \frac{5x - 1}{-(x + 9)} = \frac{5x - 1}{-x - 9} \quad \text{or} \quad \frac{5x - 1}{-9 - x}$$

Thus $-\dfrac{5x - 1}{x + 9} = \dfrac{-(5x - 1)}{x + 9} = \dfrac{-5x + 1}{x + 9} = \dfrac{5x - 1}{-(x + 9)} = \dfrac{5x - 1}{-x - 9}$   ☐

**PRACTICE**
**6**   List some equivalent forms of $-\dfrac{x + 3}{6x - 11}$.

▶ **Helpful Hint**
Remember, a negative sign in front of a fraction or rational expression may be moved to the numerator or the denominator, but *not* both.

Keep in mind that many rational expressions may look different, but in fact be equivalent.

**OBJECTIVE 4 ▶ Using rational functions in applications.** Rational functions occur often in real-life situations.

**EXAMPLE 7**   **Cost for Pressing Compact Discs**

For the ICL Production Company, the rational function $C(x) = \dfrac{2.6x + 10,000}{x}$ describes the company's cost per disc of pressing $x$ compact discs. Find the cost per disc for pressing:

**a.** 100 compact discs     **b.** 1000 compact discs

*Solution*

**a.** $C(100) = \dfrac{2.6(100) + 10,000}{100} = \dfrac{10,260}{100} = 102.6$

The cost per disc for pressing 100 compact discs is $102.60.

**b.** $C(1000) = \dfrac{2.6(1000) + 10,000}{1000} = \dfrac{12,600}{1000} = 12.6$

The cost per disc for pressing 1000 compact discs is $12.60. Notice that as more compact discs are produced, the cost per disc decreases.   □

**PRACTICE**

**7**   A company's cost per tee shirt for silk screening $x$ tee shirts is given by the rational function $C(x) = \dfrac{3.2x + 400}{x}$. Find the cost per tee shirt for printing:

**a.** 100 tee shirts            **b.** 1000 tee shirts

---

## Graphing Calculator Explorations

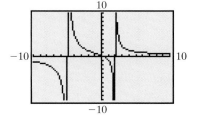

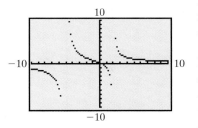

Recall that since the rational expression $\dfrac{7x - 2}{(x - 2)(x + 5)}$ is not defined when $x = 2$ or when $x = -5$, we say that the domain of the rational function $f(x) = \dfrac{7x - 2}{(x - 2)(x + 5)}$ is all real numbers except 2 and $-5$. This domain can be written as $\{x \mid x$ is a real number and $x \neq 2, x \neq -5\}$. This means that the graph of $f(x)$ should not cross the vertical lines $x = 2$ and $x = -5$. The graph of $f(x)$ in *connected* mode is to the left. In connected mode the graphing calculator tries to connect all dots of the graph so that the result is a smooth curve. This is what has happened in the graph. Notice that the graph appears to contain vertical lines at $x = 2$ and at $x = -5$. We know that this cannot happen because the function is not defined at $x = 2$ and at $x = -5$. We also know that this cannot happen because the graph of this function would not pass the vertical line test.

The graph of $f(x)$ in *dot* mode, is to the left. In dot mode the graphing calculator will not connect dots with a smooth curve. Notice that the vertical lines have disappeared, and we have a better picture of the graph. The graph, however, actually appears more like the hand-drawn graph below. By using a Table feature, a Calculate Value feature, or by tracing, we can see that the function is not defined at $x = 2$ and at $x = -5$.

*Find the domain of each rational function. Then graph each rational function and use the graph to confirm the domain.*

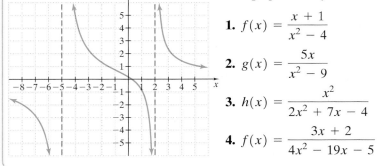

**1.** $f(x) = \dfrac{x + 1}{x^2 - 4}$

**2.** $g(x) = \dfrac{5x}{x^2 - 9}$

**3.** $h(x) = \dfrac{x^2}{2x^2 + 7x - 4}$

**4.** $f(x) = \dfrac{3x + 2}{4x^2 - 19x - 5}$

# VOCABULARY & READINESS CHECK

*Use the choices below to fill in each blank. Some choices may not be used.*

| | | | | | | |
|---|---|---|---|---|---|---|
| 1 | true | rational | simplified | $\dfrac{-a}{-b}$ | $\dfrac{-a}{b}$ | $\dfrac{a}{-b}$ |
| $-1$ | false | domain | 0 | | | |

1. A _____ expression is an expression that can be written as the quotient $\dfrac{P}{Q}$ of two polynomials $P$ and $Q$ as long as $Q \neq 0$.

2. A rational expression is undefined if the denominator is _____.

3. The _____ of the rational function $f(x) = \dfrac{2}{x}$ is $\{x | x$ is a real number and $x \neq 0\}$.

4. A rational expression is _____ if the numerator and denominator have no common factors other than 1 or $-1$.

5. The expression $\dfrac{x^2 + 2}{2 + x^2}$ simplifies to _____.

6. The expression $\dfrac{y - z}{z - y}$ simplifies to _____.

7. For a rational expression, $-\dfrac{a}{b} = $ _____ $ = $ _____.

8. True or false: $\dfrac{a - 6}{a + 2} = \dfrac{-(a - 6)}{-(a + 2)} = \dfrac{-a + 6}{-a - 2}$. _____

*Decide which rational expression can be simplified. (Do not actually simplify.)*

9. $\dfrac{x}{x + 7}$

10. $\dfrac{3 + x}{x + 3}$

11. $\dfrac{5 - x}{x - 5}$

12. $\dfrac{x + 2}{x + 8}$

## 7.1 EXERCISE SET

*Find the domain of each rational expression. See Example 1.*

1. $f(x) = \dfrac{5x - 7}{4}$

2. $g(x) = \dfrac{4 - 3x}{2}$

3. $s(t) = \dfrac{t^2 + 1}{2t}$

4. $v(t) = -\dfrac{5t + t^2}{3t}$

5. $f(x) = \dfrac{3x}{7 - x}$

6. $f(x) = \dfrac{-4x}{-2 + x}$

7. $f(x) = \dfrac{x}{3x - 1}$

8. $g(x) = \dfrac{-2}{2x + 5}$

9. $R(x) = \dfrac{3 + 2x}{x^3 + x^2 - 2x}$

10. $h(x) = \dfrac{5 - 3x}{2x^2 - 14x + 20}$

11. $C(x) = \dfrac{x + 3}{x^2 - 4}$

12. $R(x) = \dfrac{5}{x^2 - 7x}$

*Study Example 6. Then list four equivalent forms for each rational expression.*

13. $-\dfrac{x - 10}{x + 8}$

14. $-\dfrac{x + 11}{x - 4}$

15. $-\dfrac{5y - 3}{y - 12}$

16. $-\dfrac{8y - 1}{y - 15}$

**MIXED PRACTICE**

*Simplify each expression. See Examples 2 through 6.*

17. $\dfrac{x + 7}{7 + x}$

18. $\dfrac{y + 9}{9 + y}$

19. $\dfrac{x - 7}{7 - x}$

20. $\dfrac{y - 9}{9 - y}$

21. $\dfrac{2}{8x + 16}$

22. $\dfrac{3}{9x + 6}$

23. $\dfrac{-5a - 5b}{a + b}$

24. $\dfrac{-4x - 4y}{x + y}$

25. $\dfrac{7x + 35}{x^2 + 5x}$

26. $\dfrac{9x + 99}{x^2 + 11x}$

**27.** $\dfrac{x+5}{x^2-4x-45}$

**28.** $\dfrac{x-3}{x^2-6x+9}$

**29.** $\dfrac{5x^2+11x+2}{x+2}$

**30.** $\dfrac{12x^2+4x-1}{2x+1}$

**31.** $\dfrac{x^3+7x^2}{x^2+5x-14}$

**32.** $\dfrac{x^4-10x^3}{x^2-17x+70}$

**33.** $\dfrac{2x^2-8}{4x-8}$

**34.** $\dfrac{5x^2-500}{35x+350}$

**35.** $\dfrac{4-x^2}{x-2}$

**36.** $\dfrac{49-y^2}{y-7}$

**37.** $\dfrac{11x^2-22x^3}{6x-12x^2}$

**38.** $\dfrac{24y^2-8y^3}{15y-5y^2}$

**39.** $\dfrac{x^2+xy+2x+2y}{x+2}$

**40.** $\dfrac{ab+ac+b^2+bc}{b+c}$

**41.** $\dfrac{x^3+8}{x+2}$

**42.** $\dfrac{x^3+64}{x+4}$

**43.** $\dfrac{x^3-1}{1-x}$

**44.** $\dfrac{3-x}{x^3-27}$

**45.** $\dfrac{2xy+5x-2y-5}{3xy+4x-3y-4}$

**46.** $\dfrac{2xy+2x-3y-3}{2xy+4x-3y-6}$

**47.** $\dfrac{3x^2-5x-2}{6x^3+2x^2+3x+1}$

**48.** $\dfrac{2x^2-x-3}{2x^3-3x^2+2x-3}$

**49.** $\dfrac{9x^2-15x+25}{27x^3+125}$

**50.** $\dfrac{8x^3-27}{4x^2+6x+9}$

*Find each function value. See Example 7.*

**51.** If $f(x)=\dfrac{x+8}{2x-1}$, find $f(2)$, $f(0)$, and $f(-1)$.

**52.** If $f(x)=\dfrac{x-2}{-5+x}$, find $f(-5)$, $f(0)$, and $f(10)$.

**53.** If $g(x)=\dfrac{x^2+8}{x^3-25x}$, find $g(3)$, $g(-2)$, and $g(1)$.

**54.** If $s(t)=\dfrac{t^3+1}{t^2+1}$, find $s(-1)$, $s(1)$, and $s(2)$.

*Solve. See Example 7.*

**55.** The total revenue from the sale of a popular book is approximated by the rational function $R(x)=\dfrac{1000x^2}{x^2+4}$, where $x$ is the number of years since publication and $R(x)$ is the total revenue in millions of dollars.

   **a.** Find the total revenue at the end of the first year.

   **b.** Find the total revenue at the end of the second year.

   **c.** Find the revenue during the second year only.

   **d.** Find the domain of function $R$.

**56.** The function $f(x)=\dfrac{100,000x}{100-x}$ models the cost in dollars for removing $x$ percent of the pollutants from a bayou in which a nearby company dumped creosol.

   **a.** Find the cost of removing 20% of the pollutants from the bayou. [*Hint:* Find $f(20)$.]

   **b.** Find the cost of removing 60% of the pollutants and then 80% of the pollutants.

   **c.** Find $f(90)$, then $f(95)$, and then $f(99)$. What happens to the cost as $x$ approaches 100%?

   **d.** Find the domain of function $f$.

**57.** The dose of medicine prescribed for a child depends on the child's age $A$ in years and the adult dose $D$ for the medication. Young's Rule is a formula used by pediatricians that gives a child's dose $C$ as

$$C=\dfrac{DA}{A+12}$$

Suppose that an 8-year-old child needs medication, and the normal adult dose is 1000 mg. What size dose should the child receive?

**58.** Calculating body-mass index is a way to gauge whether a person should lose weight. Doctors recommend that body-mass index values fall between 18.5 and 25. The formula for body-mass index $B$ is

$$B=\dfrac{703w}{h^2}$$

where $w$ is weight in pounds and $h$ is height in inches. Should a 148-pound person who is 5 feet 6 inches tall lose weight?

**59.** Anthropologists and forensic scientists use a measure called the cephalic index to help classify skulls. The cephalic index of a skull with width $W$ and length $L$ from front to back is given by the formula

$$C=\dfrac{100W}{L}$$

A long skull has an index value less than 75, a medium skull has an index value between 75 and 85, and a broad skull has an index value over 85. Find the cephalic index of a skull that is 5 inches wide and 6.4 inches long. Classify the skull.

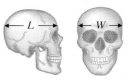

**60.** During a storm, water treatment engineers monitor how quickly rain is falling. If too much rain comes too fast, there is a danger of sewers backing up. A formula that gives the rainfall intensity $i$ in millimeters per hour for a certain strength storm in eastern Virginia is

$$i = \frac{5840}{t + 29}$$

where $t$ is the duration of the storm in minutes. What rainfall intensity should engineers expect for a storm of this strength in eastern Virginia that lasts for 80 minutes? Round your answer to one decimal place.

**61.** For a certain model fax machine, the manufacturing cost $C(x)$ per machine is given by the function

$$C(x) = \frac{250x + 10,000}{x}$$

where $x$ is the number of fax machines manufactured and cost $C(x)$ is in dollars per machine.

**a.** Find the cost per fax machine when manufacturing 100 fax machines.

**b.** Find the cost per fax machine when manufacturing 1000 fax machines.

**c.** Does the cost per machine decrease or increase when more machines are manufactured? Explain why this is so.

**62.** The total revenue $R(x)$ from the sale of a popular music compact disc is approximately given by the function

$$R(x) = \frac{150x^2}{x^2 + 3}$$

where $x$ is the number of years since the CD has been released and revenue $R$ is in millions of dollars.

**a.** Find the total revenue generated by the end of the first year.

**b.** Find the total revenue generated by the end of the second year.

**c.** Find the total revenue generated in the second year only.

**REVIEW AND PREVIEW**

*Perform each indicated operation. See Section 1.3.*

**63.** $\frac{1}{3} \cdot \frac{9}{11}$

**64.** $\frac{5}{27} \cdot \frac{2}{5}$

**65.** $\frac{1}{3} \div \frac{1}{4}$

**66.** $\frac{7}{8} \div \frac{1}{2}$

**67.** $\frac{13}{20} \div \frac{2}{9}$

**68.** $\frac{8}{15} \div \frac{5}{8}$

**CONCEPT EXTENSIONS**

*Which of the following are incorrect and why? See the Concept Check in this section.*

**69.** $\frac{5a - 15}{5}$ simplifies to $a - 3$?

**70.** $\frac{7m - 9}{7}$ simplifies to $m - 9$?

**71.** $\frac{1 + 2}{1 + 3}$ simplifies to $\frac{2}{3}$?

**72.** $\frac{46}{54}$ simplifies to $\frac{6}{5}$?

**73.** Does $\frac{x}{x + 5}$ simplify to $\frac{1}{5}$? Why or why not?

**74.** Does $\frac{x + 7}{x}$ simplify to 7? Why or why not?

**75.** In your own words explain how to simplify a rational expression.

**76.** In your own words, explain how to find the domain of a rational function.

**77.** Decide whether each rational expression equals 1, −1, or neither.

**a.** $\frac{x + 5}{5 + x}$

**b.** $\frac{x - 5}{5 - x}$

**c.** $\frac{x + 5}{x - 5}$

**d.** $\frac{-x - 5}{x + 5}$

**e.** $\frac{x - 5}{-x + 5}$

**f.** $\frac{-5 + x}{x - 5}$

**78.** The domain of the function $f(x) = \frac{1}{x}$ is all real numbers except 0. This means that the graph of this function will be in two pieces: one piece corresponding to $x$ values less than 0 and one piece corresponding to $x$ values greater than 0. Graph the function by completing the following tables, separately plotting the points, and connecting each set of plotted points with a smooth curve.

| $x$ | $\frac{1}{4}$ | $\frac{1}{2}$ | 1 | 2 | 4 |
|---|---|---|---|---|---|
| $y$ or $f(x)$ | | | | | |

| $x$ | −4 | −2 | −1 | $-\frac{1}{2}$ | $-\frac{1}{4}$ |
|---|---|---|---|---|---|
| $y$ or $f(x)$ | | | | | |

**79.** Graph a portion of the function $f(x) = \frac{20x}{100 - x}$. To do so, complete the given table, plot the points, and then connect the plotted points with a smooth curve.

| $x$ | 0 | 10 | 30 | 50 | 70 | 90 | 95 | 99 |
|---|---|---|---|---|---|---|---|---|
| $y$ or $f(x)$ | | | | | | | | |

How does the graph of $y = \dfrac{x^2 - 9}{x - 3}$ compare to the graph of $y = x + 3$? Recall that $\dfrac{x^2 - 9}{x - 3} = \dfrac{(x + 3)(x - 3)}{x - 3} = x + 3$ as long as x is not 3. This means that the graph of $y = \dfrac{x^2 - 9}{x - 3}$ is the same as the graph of $y = x + 3$ with $x \neq 3$. To graph $y = \dfrac{x^2 - 9}{x - 3}$, then, graph the linear equation $y = x + 3$ and place an open dot on the graph at 3. This open dot or interruption of the line at 3 means $x \neq 3$.

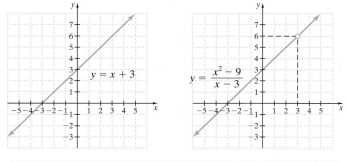

**80.** Graph $y = \dfrac{x^2 - 25}{x + 5}$.

**81.** Graph $y = \dfrac{x^2 - 16}{x - 4}$.

**82.** Graph $y = \dfrac{x^2 + x - 12}{x + 4}$.

**83.** Graph $y = \dfrac{x^2 - 6x + 8}{x - 2}$.

---

### 📖 STUDY SKILLS BUILDER

**Is Your Notebook Still Organized?**

It's never too late to organize your material in a course. Let's see how you are doing.

**1.** Are all your graded papers in one place in your math notebook or binder?

**2.** Flip through the pages of your notebook. Are your notes neat and readable?

**3.** Are your notes complete with no sections missing?

**4.** Are important notes marked in some way (like an exclamation point) so that you will know to review them before a quiz or test?

**5.** Are your assignments complete?

**6.** Do exercises that have given you trouble have a mark (like a question mark) so that you will remember to talk to your instructor or a tutor about them?

**7.** Describe your attitude toward this course.

**8.** List ways your attitude can improve and make a commitment to work on at least one of those during the next week.

---

## 7.2 MULTIPLYING AND DIVIDING RATIONAL EXPRESSIONS

**OBJECTIVES**

**1** Multiply rational expressions.

**2** Divide rational expressions.

**3** Multiply or divide rational expressions.

**OBJECTIVE 1 ▶ Multiplying rational expressions.** Just as simplifying rational expressions is similar to simplifying number fractions, multiplying and dividing rational expressions is similar to multiplying and dividing number fractions.

| *Fractions* | *Rational Expressions* |
|---|---|
| Multiply: $\dfrac{3}{5} \cdot \dfrac{10}{11}$ | Multiply: $\dfrac{x - 3}{x + 5} \cdot \dfrac{2x + 10}{x^2 - 9}$ |

Multiply numerators and then multiply denominators.

$$\dfrac{3}{5} \cdot \dfrac{10}{11} = \dfrac{3 \cdot 10}{5 \cdot 11} \qquad \dfrac{x - 3}{x + 5} \cdot \dfrac{2x + 10}{x^2 - 9} = \dfrac{(x - 3) \cdot (2x + 10)}{(x + 5) \cdot (x^2 - 9)}$$

Simplify by factoring numerators and denominators.

$$= \dfrac{3 \cdot 2 \cdot 5}{5 \cdot 11} \qquad\qquad = \dfrac{(x - 3) \cdot 2 \, (x + 5)}{(x + 5)(x + 3)(x - 3)}$$

Apply the fundamental principle.

$$= \dfrac{3 \cdot 2}{11} \quad \text{or} \quad \dfrac{6}{11} \qquad\qquad = \dfrac{2}{x + 3}$$

---

**Multiplying Rational Expressions**

If $\dfrac{P}{Q}$ and $\dfrac{R}{S}$ are rational expressions, then

$$\frac{P}{Q} \cdot \frac{R}{S} = \frac{PR}{QS}$$

To multiply rational expressions, multiply the numerators and then multiply the denominators.

---

*Note:* Recall that for Sections 7.1 through 7.4, we assume variables in rational expressions have only those replacement values for which the expressions are defined.

**EXAMPLE 1** Multiply.

**a.** $\dfrac{25x}{2} \cdot \dfrac{1}{y^3}$

**b.** $\dfrac{-7x^2}{5y} \cdot \dfrac{3y^5}{14x^2}$

**Solution** To multiply rational expressions, multiply the numerators and then multiply the denominators of both expressions. Then simplify if possible.

**a.** $\dfrac{25x}{2} \cdot \dfrac{1}{y^3} = \dfrac{25x \cdot 1}{2 \cdot y^3} = \dfrac{25x}{2y^3}$

The expression $\dfrac{25x}{2y^3}$ is in simplest form.

**b.** $\dfrac{-7x^2}{5y} \cdot \dfrac{3y^5}{14x^2} = \dfrac{-7x^2 \cdot 3y^5}{5y \cdot 14x^2}$   Multiply.

The expression $\dfrac{-7x^2 \cdot 3y^5}{5y \cdot 14x^2}$ is not in simplest form, so we factor the numerator and the denominator and divide out common factors.

$$= \frac{-1 \cdot \boxed{7} \cdot 3 \cdot \boxed{x^2} \cdot \boxed{y} \cdot y^4}{5 \cdot 2 \cdot \boxed{7} \cdot \boxed{x^2} \cdot \boxed{y}}$$

$$= -\frac{3y^4}{10}$$

**PRACTICE**

**1** Multiply.

**a.** $\dfrac{4a}{5} \cdot \dfrac{3}{b^2}$

**b.** $\dfrac{-3p^4}{q^2} \cdot \dfrac{2q^3}{9p^4}$

---

When multiplying rational expressions, it is usually best to factor each numerator and denominator. This will help us when we divide out common factors to write the product in lowest terms.

**EXAMPLE 2** Multiply: $\dfrac{x^2 + x}{3x} \cdot \dfrac{6}{5x + 5}$

**Solution**

$$\frac{x^2 + x}{3x} \cdot \frac{6}{5x + 5} = \frac{x(x + 1)}{3x} \cdot \frac{2 \cdot 3}{5(x + 1)} \quad \text{Factor numerators and denominators.}$$

$$= \frac{x(x + 1) \cdot 2 \cdot 3}{3x \cdot 5(x + 1)} \quad \text{Multiply.}$$

$$= \frac{2}{5} \quad \begin{array}{l}\text{Simplify by dividing}\\\text{out common factors.}\end{array}$$

**PRACTICE**

**2** Multiply: $\dfrac{x^2 - x}{5x} \cdot \dfrac{15}{x^2 - 1}$.

The following steps may be used to multiply rational expressions.

> **Multiplying Rational Expressions**
> **STEP 1.** Completely factor numerators and denominators.
> **STEP 2.** Multiply numerators and multiply denominators.
> **STEP 3.** Simplify or write the product in lowest terms by dividing out common factors.

**Concept Check** ☑

Which of the following is a true statement?

**a.** $\dfrac{1}{3} \cdot \dfrac{1}{2} = \dfrac{1}{5}$    **b.** $\dfrac{2}{x} \cdot \dfrac{5}{x} = \dfrac{10}{x}$    **c.** $\dfrac{3}{x} \cdot \dfrac{1}{2} = \dfrac{3}{2x}$    **d.** $\dfrac{x}{7} \cdot \dfrac{x+5}{4} = \dfrac{2x+5}{28}$

**EXAMPLE 3**    Multiply: $\dfrac{3x+3}{5x-5x^2} \cdot \dfrac{2x^2+x-3}{4x^2-9}$

**_Solution_**

$$\dfrac{3x+3}{5x-5x^2} \cdot \dfrac{2x^2+x-3}{4x^2-9} = \dfrac{3(x+1)}{5x(1-x)} \cdot \dfrac{(2x+3)(x-1)}{(2x-3)(2x+3)} \quad \text{Factor.}$$

$$= \dfrac{3(x+1)(2x+3)(x-1)}{5x(1-x)(2x-3)(2x+3)} \quad \text{Multiply.}$$

$$= \dfrac{3(x+1)(x-1)}{5x(1-x)(2x-3)} \quad \text{Divide out common factors.}$$

Next, recall that $x-1$ and $1-x$ are opposites so that $x-1 = -1(1-x)$.

$$= \dfrac{3(x+1)(-1)(1-x)}{5x(1-x)(2x-3)} \quad \text{Write } x-1 \text{ as } -1(1-x).$$

$$= \dfrac{-3(x+1)}{5x(2x-3)} \quad \text{or} \quad -\dfrac{3(x+1)}{5x(2x-3)} \quad \text{Divide out common factors.} \qquad \square$$

**PRACTICE**
**3**    Multiply: $\dfrac{6-3x}{6x+6x^2} \cdot \dfrac{3x^2-2x-5}{x^2-4}$.

**OBJECTIVE 2 ▶ Dividing rational expressions.** We can divide by a rational expression in the same way we divide by a fraction. To divide by a fraction, multiply by its reciprocal.

> ▶ **Helpful Hint**
> Don't forget how to find reciprocals. The reciprocal of $\dfrac{a}{b}$ is $\dfrac{b}{a}$, $a \neq 0, b \neq 0$.

For example, to divide $\dfrac{3}{2}$ by $\dfrac{7}{8}$, multiply $\dfrac{3}{2}$ by $\dfrac{8}{7}$.

$$\dfrac{3}{2} \div \dfrac{7}{8} = \dfrac{3}{2} \cdot \dfrac{8}{7} = \dfrac{3 \cdot 4 \cdot 2}{2 \cdot 7} = \dfrac{12}{7}$$

**Dividing Rational Expressions**

If $\dfrac{P}{Q}$ and $\dfrac{R}{S}$ are rational expressions and $\dfrac{R}{S}$ is not 0, then

$$\frac{P}{Q} \div \frac{R}{S} = \frac{P}{Q} \cdot \frac{S}{R} = \frac{PS}{QR}$$

To divide two rational expressions, multiply the first rational expression by the reciprocal of the second rational expression.

**EXAMPLE 4**   Divide: $\dfrac{3x^3y^7}{40} \div \dfrac{4x^3}{y^2}$

**Solution**

$$\frac{3x^3y^7}{40} \div \frac{4x^3}{y^2} = \frac{3x^3y^7}{40} \cdot \frac{y^2}{4x^3} \qquad \text{Multiply by the reciprocal of } \frac{4x^3}{y^2}.$$

$$= \frac{3x^3y^9}{160x^3}$$

$$= \frac{3y^9}{160} \qquad \text{Simplify.} \qquad \square$$

**PRACTICE 4**   Divide: $\dfrac{5a^3b^2}{24} \div \dfrac{10a^5}{6}$.

**EXAMPLE 5**   Divide: $\dfrac{(x-1)(x+2)}{10}$ by $\dfrac{2x+4}{5}$.

**Solution**

$$\frac{(x-1)(x+2)}{10} \div \frac{2x+4}{5} = \frac{(x-1)(x+2)}{10} \cdot \frac{5}{2x+4} \qquad \begin{array}{l}\text{Multiply by the reciprocal} \\ \text{of } \dfrac{2x+4}{5}.\end{array}$$

$$= \frac{(x-1)(x+2) \cdot 5}{5 \cdot 2 \cdot 2 \cdot (x+2)} \qquad \text{Factor and multiply.}$$

$$= \frac{x-1}{4} \qquad \text{Simplify.} \qquad \square$$

**PRACTICE 5**   Divide $\dfrac{(3x+1)(x-5)}{3}$ by $\dfrac{4x-20}{9}$.

The following may be used to divide by a rational expression.

**Dividing by a Rational Expression**
Multiply by its reciprocal.

**EXAMPLE 6**   Divide: $\dfrac{6x+2}{x^2-1} \div \dfrac{3x^2+x}{x-1}$

**Solution**

$$\frac{6x + 2}{x^2 - 1} \div \frac{3x^2 + x}{x - 1} = \frac{6x + 2}{x^2 - 1} \cdot \frac{x - 1}{3x^2 + x} \qquad \text{Multiply by the reciprocal.}$$

$$= \frac{2(3x + 1)(x - 1)}{(x + 1)(x - 1) \cdot x(3x + 1)} \qquad \text{Factor and multiply.}$$

$$= \frac{2}{x(x + 1)} \qquad \text{Simplify.} \qquad \square$$

**PRACTICE**
**6**   Divide $\dfrac{10x - 2}{x^2 - 9} \div \dfrac{5x^2 - x}{x + 3}$.

---

**EXAMPLE 7**   Divide: $\dfrac{2x^2 - 11x + 5}{5x - 25} \div \dfrac{4x - 2}{10}$

**Solution**

$$\frac{2x^2 - 11x + 5}{5x - 25} \div \frac{4x - 2}{10} = \frac{2x^2 - 11x + 5}{5x - 25} \cdot \frac{10}{4x - 2} \qquad \text{Multiply by the reciprocal.}$$

$$= \frac{(2x - 1)(x - 5) \cdot 2 \cdot 5}{5(x - 5) \cdot 2(2x - 1)} \qquad \text{Factor and multiply.}$$

$$= \frac{1}{1} \quad \text{or} \quad 1 \qquad \text{Simplify.} \qquad \square$$

**PRACTICE**
**7**   Divide $\dfrac{3x^2 - 11x - 4}{2x - 8} \div \dfrac{9x + 3}{6}$.

---

**OBJECTIVE 3 ▶ Multiplying or dividing rational expressions.** Let's make sure that we understand the difference between multiplying and dividing rational expressions.

| *Rational Expressions* | |
|---|---|
| Multiplication | Multiply the numerators and multiply the denominators. |
| Division | Multiply by the reciprocal of the divisor. |

**EXAMPLE 8**   Multiply or divide as indicated.

a. $\dfrac{x - 4}{5} \cdot \dfrac{x}{x - 4}$     b. $\dfrac{x - 4}{5} \div \dfrac{x}{x - 4}$     c. $\dfrac{x^2 - 4}{2x + 6} \cdot \dfrac{x^2 + 4x + 3}{2 - x}$

**Solution**

a. $\dfrac{x - 4}{5} \cdot \dfrac{x}{x - 4} = \dfrac{(x - 4) \cdot x}{5 \cdot (x - 4)} = \dfrac{x}{5}$

b. $\dfrac{x - 4}{5} \div \dfrac{x}{x - 4} = \dfrac{x - 4}{5} \cdot \dfrac{x - 4}{x} = \dfrac{(x - 4)^2}{5x}$

**c.** $\dfrac{x^2 - 4}{2x + 6} \cdot \dfrac{x^2 + 4x + 3}{2 - x} = \dfrac{(x - 2)(x + 2) \cdot (x + 1)(x + 3)}{2(x + 3) \cdot (2 - x)}$  Factor and multiply.

$$= \dfrac{(x - 2)(x + 2) \cdot (x + 1)(x + 3)}{2(x + 3) \cdot (2 - x)}$$

$$= \dfrac{-1(x + 2)(x + 1)}{2}$$  Divide out common factors. Recall that $\dfrac{x - 2}{2 - x} = -1$.

$$= -\dfrac{(x + 2)(x + 1)}{2}$$

**PRACTICE**
**8**  Multiply or divide as indicated.

**a.** $\dfrac{y + 9}{8x} \cdot \dfrac{y + 9}{2x}$    **b.** $\dfrac{y + 9}{8x} \div \dfrac{y + 9}{2}$    **c.** $\dfrac{35x - 7x^2}{x^2 - 25} \cdot \dfrac{x^2 + 3x - 10}{x^2 + 4x}$

## VOCABULARY & READINESS CHECK

*Use one of the choices below to fill in the blank.*

opposites        reciprocals

**1.** The expressions $\dfrac{x}{2y}$ and $\dfrac{2y}{x}$ are called _____.

*Multiply or divide as indicated.*

**2.** $\dfrac{a}{b} \cdot \dfrac{c}{d} =$ _____

**3.** $\dfrac{a}{b} \div \dfrac{c}{d} =$ _____

**4.** $\dfrac{x}{7} \cdot \dfrac{x}{6} =$ _____

**5.** $\dfrac{x}{7} \div \dfrac{x}{6} =$ _____

## 7.2 EXERCISE SET

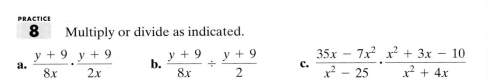

*Find each product and simplify if possible. See Examples 1 through 3.*

**1.** $\dfrac{3x}{y^2} \cdot \dfrac{7y}{4x}$

**2.** $\dfrac{9x^2}{y} \cdot \dfrac{4y}{3x^3}$

**3.** $\dfrac{8x}{2} \cdot \dfrac{x^5}{4x^2}$

**4.** $\dfrac{6x^2}{10x^3} \cdot \dfrac{5x}{12}$

**5.** $-\dfrac{5a^2b}{30a^2b^2} \cdot b^3$

**6.** $-\dfrac{9x^3y^2}{18xy^5} \cdot y^3$

**7.** $\dfrac{x}{2x - 14} \cdot \dfrac{x^2 - 7x}{5}$

**8.** $\dfrac{4x - 24}{20x} \cdot \dfrac{5}{x - 6}$

**9.** $\dfrac{6x + 6}{5} \cdot \dfrac{10}{36x + 36}$

**10.** $\dfrac{x^2 + x}{8} \cdot \dfrac{16}{x + 1}$

**11.** $\dfrac{(m + n)^2}{m - n} \cdot \dfrac{m}{m^2 + mn}$

**12.** $\dfrac{(m - n)^2}{m + n} \cdot \dfrac{m}{m^2 - mn}$

**13.** $\dfrac{x^2 - 25}{x^2 - 3x - 10} \cdot \dfrac{x + 2}{x}$

**14.** $\dfrac{a^2 - 4a + 4}{a^2 - 4} \cdot \dfrac{a + 3}{a - 2}$

**15.** $\dfrac{x^2 + 6x + 8}{x^2 + x - 20} \cdot \dfrac{x^2 + 2x - 15}{x^2 + 8x + 16}$

**16.** $\dfrac{x^2 + 9x + 20}{x^2 - 15x + 44} \cdot \dfrac{x^2 - 11x + 28}{x^2 + 12x + 35}$

*Find each quotient and simplify. See Examples 4 through 7.*

**17.** $\dfrac{5x^7}{2x^5} \div \dfrac{15x}{4x^3}$

**18.** $\dfrac{9y^4}{6y} \div \dfrac{y^2}{3}$

**19.** $\dfrac{8x^2}{y^3} \div \dfrac{4x^2y^3}{6}$

**20.** $\dfrac{7a^2b}{3ab^2} \div \dfrac{21a^2b^2}{14ab}$

**21.** $\dfrac{(x - 6)(x + 4)}{4x} \div \dfrac{2x - 12}{8x^2}$

**22.** $\dfrac{(x + 3)^2}{5} \div \dfrac{5x + 15}{25}$

**23.** $\dfrac{3x^2}{x^2 - 1} \div \dfrac{x^5}{(x + 1)^2}$

24. $\dfrac{9x^5}{a^2 - b^2} \div \dfrac{27x^2}{3b - 3a}$

25. $\dfrac{m^2 - n^2}{m + n} \div \dfrac{m}{m^2 + nm}$

26. $\dfrac{(m - n)^2}{m + n} \div \dfrac{m^2 - mn}{m}$

27. $\dfrac{x + 2}{7 - x} \div \dfrac{x^2 - 5x + 6}{x^2 - 9x + 14}$

28. $\dfrac{x - 3}{2 - x} \div \dfrac{x^2 + 3x - 18}{x^2 + 2x - 8}$

29. $\dfrac{x^2 + 7x + 10}{x - 1} \div \dfrac{x^2 + 2x - 15}{x - 1}$

30. $\dfrac{x + 1}{(x + 1)(2x + 3)} \div \dfrac{20x + 100}{2x + 3}$

## MIXED PRACTICE

*Multiply or divide as indicated. See Examples 1 through 8.*

31. $\dfrac{5x - 10}{12} \div \dfrac{4x - 8}{8}$

32. $\dfrac{6x + 6}{5} \div \dfrac{9x + 9}{10}$

33. $\dfrac{x^2 + 5x}{8} \cdot \dfrac{9}{3x + 15}$

34. $\dfrac{3x^2 + 12x}{6} \cdot \dfrac{9}{2x + 8}$

35. $\dfrac{7}{6p^2 + q} \div \dfrac{14}{18p^2 + 3q}$

36. $\dfrac{3x + 6}{20} \div \dfrac{4x + 8}{8}$

37. $\dfrac{3x + 4y}{x^2 + 4xy + 4y^2} \cdot \dfrac{x + 2y}{2}$

38. $\dfrac{x^2 - y^2}{3x^2 + 3xy} \cdot \dfrac{3x^2 + 6x}{3x^2 - 2xy - y^2}$

39. $\dfrac{(x + 2)^2}{x - 2} \div \dfrac{x^2 - 4}{2x - 4}$

40. $\dfrac{x + 3}{x^2 - 9} \div \dfrac{5x + 15}{(x - 3)^2}$

41. $\dfrac{x^2 - 4}{24x} \div \dfrac{2 - x}{6xy}$

42. $\dfrac{3y}{3 - x} \div \dfrac{12xy}{x^2 - 9}$

43. $\dfrac{a^2 + 7a + 12}{a^2 + 5a + 6} \cdot \dfrac{a^2 + 8a + 15}{a^2 + 5a + 4}$

44. $\dfrac{b^2 + 2b - 3}{b^2 + b - 2} \cdot \dfrac{b^2 - 4}{b^2 + 6b + 8}$

45. $\dfrac{5x - 20}{3x^2 + x} \cdot \dfrac{3x^2 + 13x + 4}{x^2 - 16}$

46. $\dfrac{9x + 18}{4x^2 - 3x} \cdot \dfrac{4x^2 - 11x + 6}{x^2 - 4}$

47. $\dfrac{8n^2 - 18}{2n^2 - 5n + 3} \div \dfrac{6n^2 + 7n - 3}{n^2 - 9n + 8}$

48. $\dfrac{36n^2 - 64}{3n^2 + 10n + 8} \div \dfrac{3n^2 - 13n + 12}{n^2 - 5n - 14}$

49. Find the quotient of $\dfrac{x^2 - 9}{2x}$ and $\dfrac{x + 3}{8x^4}$.

50. Find the quotient of $\dfrac{4x^2 + 4x + 1}{4x + 2}$ and $\dfrac{4x + 2}{16}$.

*Multiply or divide as indicated. Some of these expressions contain 4-term polynomials and sums and differences of cubes. See Examples 1 through 8.*

51. $\dfrac{a^2 + ac + ba + bc}{a - b} \div \dfrac{a + c}{a + b}$

52. $\dfrac{x^2 + 2x - xy - 2y}{x^2 - y^2} \div \dfrac{2x + 4}{x + y}$

53. $\dfrac{3x^2 + 8x + 5}{x^2 + 8x + 7} \cdot \dfrac{x + 7}{x^2 + 4}$

54. $\dfrac{16x^2 + 2x}{16x^2 + 10x + 1} \cdot \dfrac{1}{4x^2 + 2x}$

55. $\dfrac{x^3 + 8}{x^2 - 2x + 4} \cdot \dfrac{4}{x^2 - 4}$

56. $\dfrac{9y}{3y - 3} \cdot \dfrac{y^3 - 1}{y^3 + y^2 + y}$

57. $\dfrac{a^2 - ab}{6a^2 + 6ab} \div \dfrac{a^3 - b^3}{a^2 - b^2}$

58. $\dfrac{x^3 + 27y^3}{6x} \div \dfrac{x^2 - 9y^2}{x^2 - 3xy}$

## REVIEW AND PREVIEW

*Perform each indicated operation. See Section 1.3.*

59. $\dfrac{1}{5} + \dfrac{4}{5}$

60. $\dfrac{3}{15} + \dfrac{6}{15}$

61. $\dfrac{9}{9} - \dfrac{19}{9}$

62. $\dfrac{4}{3} - \dfrac{8}{3}$

63. $\dfrac{6}{5} + \left(\dfrac{1}{5} - \dfrac{8}{5}\right)$

64. $-\dfrac{3}{2} + \left(\dfrac{1}{2} - \dfrac{3}{2}\right)$

*Graph each linear equation. See Section 3.2.*

65. $x - 2y = 6$

66. $5x - y = 10$

## CONCEPT EXTENSIONS

*Identify each statement as true or false. If false, correct the multiplication. See the Concept Check in this section.*

67. $\dfrac{4}{a} \cdot \dfrac{1}{b} = \dfrac{4}{ab}$

68. $\dfrac{2}{3} \cdot \dfrac{2}{4} = \dfrac{2}{7}$

69. $\dfrac{x}{5} \cdot \dfrac{x + 3}{4} = \dfrac{2x + 3}{20}$

70. $\dfrac{7}{a} \cdot \dfrac{3}{a} = \dfrac{21}{a}$

△ **71.** Find the area of the rectangle.

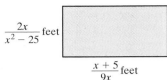

$\frac{2x}{x^2 - 25}$ feet

$\frac{x + 5}{9x}$ feet

△ **72.** Find the area of the square.

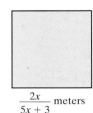

$\frac{2x}{5x + 3}$ meters

*Multiply or divide as indicated.*

**73.** $\left( \dfrac{x^2 - y^2}{x^2 + y^2} \div \dfrac{x^2 - y^2}{3x} \right) \cdot \dfrac{x^2 + y^2}{6}$

**74.** $\left( \dfrac{x^2 - 9}{x^2 - 1} \cdot \dfrac{x^2 + 2x + 1}{2x^2 + 9x + 9} \right) \div \dfrac{2x + 3}{1 - x}$

**75.** $\left( \dfrac{2a + b}{b^2} \cdot \dfrac{3a^2 - 2ab}{ab + 2b^2} \right) \div \dfrac{a^2 - 3ab + 2b^2}{5ab - 10b^2}$

**76.** $\left( \dfrac{x^2y^2 - xy}{4x - 4y} \div \dfrac{3y - 3x}{8x - 8y} \right) \cdot \dfrac{y - x}{8}$

**77.** In your own words, explain how you multiply rational expressions.

**78.** Explain how dividing rational expressions is similar to dividing rational numbers.

## 7.3 ADDING AND SUBTRACTING RATIONAL EXPRESSIONS WITH COMMON DENOMINATORS AND LEAST COMMON DENOMINATOR

**OBJECTIVES**

1 Add and subtract rational expressions with the same denominator.

2 Find the least common denominator of a list of rational expressions.

3 Write a rational expression as an equivalent expression whose denominator is given.

**OBJECTIVE 1 ▶ Adding and subtracting rational expressions with the same denominator.** Like multiplication and division, addition and subtraction of rational expressions is similar to addition and subtraction of rational numbers. In this section, we add and subtract rational expressions with a common (or the same) denominator.

Add: $\dfrac{6}{5} + \dfrac{2}{5}$ | Add: $\dfrac{9}{x + 2} + \dfrac{3}{x + 2}$

Add the numerators and place the sum over the common denominator.

$\dfrac{6}{5} + \dfrac{2}{5} = \dfrac{6 + 2}{5}$ | $\dfrac{9}{x + 2} + \dfrac{3}{x + 2} = \dfrac{9 + 3}{x + 2}$

$= \dfrac{8}{5}$    Simplify. | $= \dfrac{12}{x + 2}$    Simplify.

---

**Adding and Subtracting Rational Expressions with Common Denominators**

If $\dfrac{P}{R}$ and $\dfrac{Q}{R}$ are rational expressions, then

$$\frac{P}{R} + \frac{Q}{R} = \frac{P + Q}{R} \quad \text{and} \quad \frac{P}{R} - \frac{Q}{R} = \frac{P - Q}{R}$$

To add or subtract rational expressions, add or subtract numerators and place the sum or difference over the common denominator.

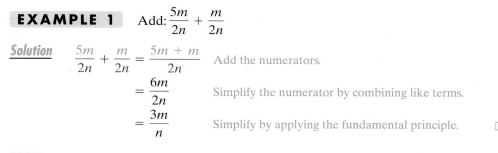

**EXAMPLE 1**   Add: $\dfrac{5m}{2n} + \dfrac{m}{2n}$

*Solution*

$$\dfrac{5m}{2n} + \dfrac{m}{2n} = \dfrac{5m + m}{2n} \qquad \text{Add the numerators.}$$

$$= \dfrac{6m}{2n} \qquad \text{Simplify the numerator by combining like terms.}$$

$$= \dfrac{3m}{n} \qquad \text{Simplify by applying the fundamental principle.} \qquad \square$$

**PRACTICE**
**1**   Add: $\dfrac{7a}{4b} + \dfrac{a}{4b}$.

---

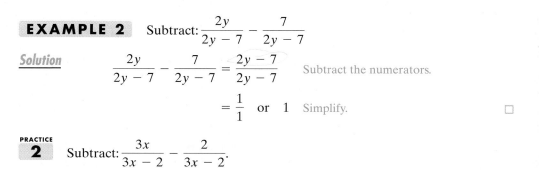

**EXAMPLE 2**   Subtract: $\dfrac{2y}{2y - 7} - \dfrac{7}{2y - 7}$

*Solution*

$$\dfrac{2y}{2y - 7} - \dfrac{7}{2y - 7} = \dfrac{2y - 7}{2y - 7} \qquad \text{Subtract the numerators.}$$

$$= \dfrac{1}{1} \quad \text{or} \quad 1 \quad \text{Simplify.} \qquad \square$$

**PRACTICE**
**2**   Subtract: $\dfrac{3x}{3x - 2} - \dfrac{2}{3x - 2}$.

---

**EXAMPLE 3**   Subtract: $\dfrac{3x^2 + 2x}{x - 1} - \dfrac{10x - 5}{x - 1}$.

*Solution*   $\dfrac{3x^2 + 2x}{x - 1} - \dfrac{10x - 5}{x - 1} = \dfrac{(3x^2 + 2x) - (10x - 5)}{x - 1}$   Subtract the numerators. Notice the parentheses.

> **▶ Helpful Hint**
> Parentheses are inserted so that the entire numerator, $10x - 5$, is subtracted.

$$= \dfrac{3x^2 + 2x - 10x + 5}{x - 1} \qquad \text{Use the distributive property.}$$

$$= \dfrac{3x^2 - 8x + 5}{x - 1} \qquad \text{Combine like terms.}$$

$$= \dfrac{(x - 1)(3x - 5)}{x - 1} \qquad \text{Factor.}$$

$$= 3x - 5 \qquad \text{Simplify.} \qquad \square$$

**PRACTICE**
**3**   Subtract: $\dfrac{4x^2 + 15x}{x + 3} - \dfrac{8x + 15}{x + 3}$

---

> **▶ Helpful Hint**
> Notice how the numerator $10x - 5$ has been subtracted in Example 3.
>
> This $-$ sign applies to the entire numerator of $10x - 5$.      So parentheses are inserted here to indicate this.
>
> $$\dfrac{3x^2 + 2x}{x - 1} - \dfrac{10x - 5}{x - 1} = \dfrac{3x^2 + 2x - (10x - 5)}{x - 1}$$

**OBJECTIVE 2 ▶ Finding the least common denominator.** To add and subtract fractions with **unlike** denominators, first find a least common denominator (LCD), and then write all fractions as equivalent fractions with the LCD.

For example, suppose we add $\frac{8}{3}$ and $\frac{2}{5}$. The LCD of denominators 3 and 5 is 15, since 15 is the least common multiple (LCM) of 3 and 5. That is, 15 is the smallest number that both 3 and 5 divide into evenly.

Next, rewrite each fraction so that its denominator is 15.

$$\frac{8}{3} + \frac{2}{5} = \frac{8(5)}{3(5)} + \frac{2(3)}{5(3)} = \frac{40}{15} + \frac{6}{15} = \frac{40 + 6}{15} = \frac{46}{15}$$

We are multiplying by 1.

To add or subtract rational expressions with unlike denominators, we also first find an LCD and then write all rational expressions as equivalent expressions with the LCD. The **least common denominator (LCD) of a list of rational expressions** is a polynomial of least degree whose factors include all the factors of the denominators in the list.

> **Finding the Least Common Denominator (LCD)**
>
> **STEP 1.** Factor each denominator completely.
>
> **STEP 2.** The least common denominator (LCD) is the product of all unique factors found in Step 1, each raised to a power equal to the greatest number of times that the factor appears in any one factored denominator.

**EXAMPLE 4**  Find the LCD for each pair.

**a.** $\frac{1}{8}, \frac{3}{22}$        **b.** $\frac{7}{5x}, \frac{6}{15x^2}$

**Solution**

**a.** Start by finding the prime factorization of each denominator.

$$8 = 2 \cdot 2 \cdot 2 = 2^3 \quad \text{and}$$
$$22 = 2 \cdot 11$$

Next, write the product of all the unique factors, each raised to a power equal to the greatest number of times that the factor appears in any denominator.

The greatest number of times that the factor 2 appears is 3.

The greatest number of times that the factor 11 appears is 1.

$$LCD = 2^3 \cdot 11^1 = 8 \cdot 11 = 88$$

**b.** Factor each denominator.

$$5x = 5 \cdot x \quad \text{and}$$
$$15x^2 = 3 \cdot 5 \cdot x^2$$

The greatest number of times that the factor 5 appears is 1.

The greatest number of times that the factor 3 appears is 1.

The greatest number of times that the factor $x$ appears is 2.

$$LCD = 3^1 \cdot 5^1 \cdot x^2 = 15x^2$$

**PRACTICE**

**4**  Find the LCD for each pair.

**a.** $\frac{3}{14}, \frac{5}{21}$        **b.** $\frac{4}{9y}, \frac{11}{15y^3}$

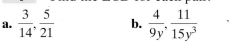

**EXAMPLE 5**  Find the LCD of

**a.** $\dfrac{7x}{x+2}$  and  $\dfrac{5x^2}{x-2}$  **b.** $\dfrac{3}{x}$  and  $\dfrac{6}{x+4}$

*Solution*

**a.** The denominators $x + 2$ and $x - 2$ are completely factored already. The factor $x + 2$ appears once and the factor $x - 2$ appears once.

$$\text{LCD} = (x+2)(x-2)$$

**b.** The denominators $x$ and $x + 4$ cannot be factored further. The factor $x$ appears once and the factor $x + 4$ appears once.

$$\text{LCD} = x(x+4)$$

**PRACTICE**
**5**  Find the LCD of

**a.** $\dfrac{16}{y-5}$  and  $\dfrac{3y^3}{y-4}$  **b.** $\dfrac{8}{a}$  and  $\dfrac{5}{a+2}$

**EXAMPLE 6**  Find the LCD of $\dfrac{6m^2}{3m+15}$  and  $\dfrac{2}{(m+5)^2}$.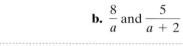

*Solution*  We factor each denominator.

$$3m + 15 = 3(m+5)$$

$$(m+5)^2 = (m+5)^2 \quad \text{This denominator is already factored.}$$

The greatest number of times that the factor 3 appears is 1.

The greatest number of times that the factor $m + 5$ appears *in any one denominator* is 2.

$$\text{LCD} = 3(m+5)^2$$

**PRACTICE**
**6**  Find the LCD of $\dfrac{2x^3}{(2x-1)^2}$  and  $\dfrac{5x}{6x-3}$.

**Concept Check** ☑

Choose the correct LCD of $\dfrac{x}{(x+1)^2}$ and $\dfrac{5}{x+1}$.

**a.** $x + 1$  **b.** $(x+1)^2$  **c.** $(x+1)^3$  **d.** $5x(x+1)^2$

**EXAMPLE 7**  Find the LCD of $\dfrac{t-10}{t^2-t-6}$  and  $\dfrac{t+5}{t^2+3t+2}$.

*Solution*  Start by factoring each denominator.

$$t^2 - t - 6 = (t-3)(t+2)$$

$$t^2 + 3t + 2 = (t+1)(t+2)$$

$$\text{LCD} = (t-3)(t+2)(t+1)$$

**PRACTICE**
**7**  Find the LCD of $\dfrac{x-5}{x^2+5x+4}$  and  $\dfrac{x+8}{x^2-16}$.

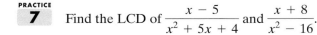

**Answer to Concept Check:**  b

**EXAMPLE 8** Find the LCD of $\dfrac{2}{x-2}$ and $\dfrac{10}{2-x}$.

**Solution** The denominators $x-2$ and $2-x$ are opposites. That is, $2-x = -1(x-2)$. Use $x-2$ or $2-x$ as the LCD.

$$\text{LCD} = x-2 \qquad \text{or} \qquad \text{LCD} = 2-x$$ ☐

**PRACTICE**
**8** Find the LCD of $\dfrac{5}{3-x}$ and $\dfrac{4}{x-3}$.

**OBJECTIVE 3 ▶ Writing equivalent rational expressions.** Next we practice writing a rational expression as an equivalent rational expression with a given denominator. To do this, we multiply by a form of 1. Recall that multiplying an expression by 1 produces an equivalent expression. In other words,

$$\frac{P}{Q} = \frac{P}{Q} \cdot 1 = \frac{P}{Q} \cdot \frac{R}{R} = \frac{PR}{QR}.$$

**EXAMPLE 9** Write each rational expression as an equivalent rational expression with the given denominator.

**a.** $\dfrac{4b}{9a} = \dfrac{}{27a^2b}$  **b.** $\dfrac{7x}{2x+5} = \dfrac{}{6x+15}$

**Solution**

**a.** We can ask ourselves: "What do we multiply $9a$ by to get $27a^2b$?" The answer is $3ab$, since $9a(3ab) = 27a^2b$. So we multiply by 1 in the form of $\dfrac{3ab}{3ab}$.

$$\frac{4b}{9a} = \frac{4b}{9a} \cdot 1 = \frac{4b}{9a} \cdot \frac{3ab}{3ab}$$

$$= \frac{4b(3ab)}{9a(3ab)} = \frac{12ab^2}{27a^2b}$$

**b.** First, factor the denominator on the right.

$$\frac{7x}{2x+5} = \frac{}{3(2x+5)}$$

To obtain the denominator on the right from the denominator on the left, we multiply by 1 in the form of $\dfrac{3}{3}$.

$$\frac{7x}{2x+5} = \frac{7x}{2x+5} \cdot \frac{3}{3} = \frac{7x \cdot 3}{(2x+5) \cdot 3} = \frac{21x}{3(2x+5)} \text{ or } \frac{21x}{6x+15}$$ ☐

**PRACTICE**
**9** Write each rational expression as an equivalent fraction with the given denominator.

**a.** $\dfrac{3x}{5y} = \dfrac{}{35xy^2}$  **b.** $\dfrac{9x}{4x+7} = \dfrac{}{8x+14}$

**EXAMPLE 10** Write the rational expression as an equivalent rational expression with the given denominator.

$$\frac{5}{x^2-4} = \frac{}{(x-2)(x+2)(x-4)}$$

 *where did this come from? unless this diff of sq*

<u>**Solution**</u>   First, factor the denominator $x^2 - 4$ as $(x - 2)(x + 2)$.

If we multiply the original denominator $(x - 2)(x + 2)$ by $x - 4$, the result is the new denominator $(x + 2)(x - 2)(x - 4)$. Thus, we multiply by 1 in the form of $\dfrac{x - 4}{x - 4}$.

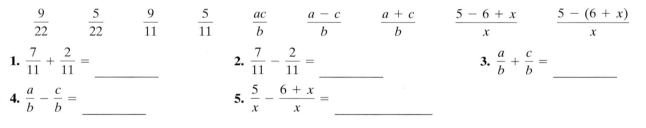

$$\frac{5}{x^2 - 4} = \frac{5}{(x - 2)(x + 2)} = \frac{5}{(x - 2)(x + 2)} \cdot \frac{x - 4}{x - 4}$$

$$= \frac{5(x - 4)}{(x - 2)(x + 2)(x - 4)}$$

$$= \frac{5x - 20}{(x - 2)(x + 2)(x - 4)}$$

**PRACTICE**
**10**   Write the rational expression as an equivalent rational expression with the given denominator.

$$\frac{3}{x^2 - 2x - 15} = \frac{}{(x - 2)(x + 3)(x - 5)}$$

## VOCABULARY & READINESS CHECK

*Use the choices below to fill in each blank. Not all choices will be used.*

$$\frac{9}{22} \qquad \frac{5}{22} \qquad \frac{9}{11} \qquad \frac{5}{11} \qquad \frac{ac}{b} \qquad \frac{a - c}{b} \qquad \frac{a + c}{b} \qquad \frac{5 - 6 + x}{x} \qquad \frac{5 - (6 + x)}{x}$$

**1.** $\dfrac{7}{11} + \dfrac{2}{11} = $ _____

**2.** $\dfrac{7}{11} - \dfrac{2}{11} = $ _____

**3.** $\dfrac{a}{b} + \dfrac{c}{b} = $ _____

**4.** $\dfrac{a}{b} - \dfrac{c}{b} = $ _____

**5.** $\dfrac{5}{x} - \dfrac{6 + x}{x} = $ _____

## 7.3 | EXERCISE SET

*Add or subtract as indicated. Simplify the result if possible. See Examples 1 through 3.*

**1.** $\dfrac{a + 1}{13} + \dfrac{8}{13}$

**2.** $\dfrac{x + 1}{7} + \dfrac{6}{7}$

**3.** $\dfrac{4m}{3n} + \dfrac{5m}{3n}$

**4.** $\dfrac{3p}{2q} + \dfrac{11p}{2q}$

**5.** $\dfrac{4m}{m - 6} - \dfrac{24}{m - 6}$

**6.** $\dfrac{8y}{y - 2} - \dfrac{16}{y - 2}$

**7.** $\dfrac{9}{3 + y} + \dfrac{y + 1}{3 + y}$

**8.** $\dfrac{9}{y + 9} + \dfrac{y - 5}{y + 9}$

**9.** $\dfrac{5x^2 + 4x}{x - 1} - \dfrac{6x + 3}{x - 1}$

**10.** $\dfrac{x^2 + 9x}{x + 7} - \dfrac{4x + 14}{x + 7}$

**11.** $\dfrac{4a}{a^2 + 2a - 15} - \dfrac{12}{a^2 + 2a - 15}$

**12.** $\dfrac{3y}{y^2 + 3y - 10} - \dfrac{6}{y^2 + 3y - 10}$

**13.** $\dfrac{2x + 3}{x^2 - x - 30} - \dfrac{x - 2}{x^2 - x - 30}$

**14.** $\dfrac{3x - 1}{x^2 + 5x - 6} - \dfrac{2x - 7}{x^2 + 5x - 6}$

**15.** $\dfrac{2x + 1}{x - 3} + \dfrac{3x + 6}{x - 3}$

**16.** $\dfrac{4p - 3}{2p + 7} + \dfrac{3p + 8}{2p + 7}$

**17.** $\dfrac{2x^2}{x - 5} - \dfrac{25 + x^2}{x - 5}$

**18.** $\dfrac{6x^2}{2x - 5} - \dfrac{25 + 2x^2}{2x - 5}$

**19.** $\dfrac{5x + 4}{x - 1} - \dfrac{2x + 7}{x - 1}$

**20.** $\dfrac{7x + 1}{x - 4} - \dfrac{2x + 21}{x - 4}$

*Find the LCD for each list of rational expressions. See Examples 4 through 8.*

**21.** $\dfrac{19}{2x}$, $\dfrac{5}{4x^3}$

**22.** $\dfrac{17x}{4y^5}$, $\dfrac{2}{8y}$

**23.** $\dfrac{9}{8x}$, $\dfrac{3}{2x + 4}$

**24.** $\dfrac{1}{6y}$, $\dfrac{3x}{4y + 12}$

**25.** $\dfrac{2}{x + 3}$, $\dfrac{5}{x - 2}$

**26.** $\dfrac{-6}{x - 1}$, $\dfrac{4}{x + 5}$

**27.** $\dfrac{x}{x + 6}$, $\dfrac{10}{3x + 18}$

**28.** $\dfrac{12}{x + 5}$, $\dfrac{x}{4x + 20}$

**29.** $\dfrac{8x^2}{(x - 6)^2}$, $\dfrac{13x}{5x - 30}$

**30.** $\dfrac{9x^2}{7x - 14}$, $\dfrac{6x}{(x - 2)^2}$

**31.** $\dfrac{1}{3x + 3}$, $\dfrac{8}{2x^2 + 4x + 2}$

**32.** $\dfrac{19x + 5}{4x - 12}$, $\dfrac{3}{2x^2 - 12x + 18}$

**33.** $\dfrac{5}{x - 8}$, $\dfrac{3}{8 - x}$

**34.** $\dfrac{2x + 5}{3x - 7}$, $\dfrac{5}{7 - 3x}$

**35.** $\dfrac{5x + 1}{x^2 + 3x - 4}$, $\dfrac{3x}{x^2 + 2x - 3}$

**36.** $\dfrac{4}{x^2 + 4x + 3}$, $\dfrac{4x - 2}{x^2 + 10x + 21}$

**37.** $\dfrac{2x}{3x^2 + 4x + 1}$, $\dfrac{7}{2x^2 - x - 1}$

**38.** $\dfrac{3x}{4x^2 + 5x + 1}$, $\dfrac{5}{3x^2 - 2x - 1}$

**39.** $\dfrac{1}{x^2 - 16}$, $\dfrac{x + 6}{2x^3 - 8x^2}$

**40.** $\dfrac{5}{x^2 - 25}$, $\dfrac{x + 9}{3x^3 - 15x^2}$

*Rewrite each rational expression as an equivalent rational expression with the given denominator. See Examples 9 and 10.*

**41.** $\dfrac{3}{2x} = \dfrac{}{4x^2}$

**42.** $\dfrac{3}{9y^5} = \dfrac{}{72y^9}$

**43.** $\dfrac{6}{3a} = \dfrac{}{12ab^2}$

**44.** $\dfrac{5}{4y^2x} = \dfrac{}{32y^3x^2}$

**45.** $\dfrac{9}{2x + 6} = \dfrac{}{2y(x + 3)}$

**46.** $\dfrac{4x + 1}{3x + 6} = \dfrac{}{3y(x + 2)}$

**47.** $\dfrac{9a + 2}{5a + 10} = \dfrac{}{5b(a + 2)}$

**48.** $\dfrac{5 + y}{2x^2 + 10} = \dfrac{}{4(x^2 + 5)}$

**49.** $\dfrac{x}{x^3 + 6x^2 + 8x} = \dfrac{}{x(x + 4)(x + 2)(x + 1)}$

**50.** $\dfrac{5x}{x^3 + 2x^2 - 3x} = \dfrac{}{x(x - 1)(x - 5)(x + 3)}$

**51.** $\dfrac{9y - 1}{15x^2 - 30} = \dfrac{}{30x^2 - 60}$

**52.** $\dfrac{6m - 5}{3x^2 - 9} = \dfrac{}{12x^2 - 36}$

**MIXED PRACTICE**

*Perform the indicated operations.*

**53.** $\dfrac{5x}{7} + \dfrac{9x}{7}$

**54.** $\dfrac{5x}{7} \cdot \dfrac{9x}{7}$

**55.** $\dfrac{x + 3}{4} \div \dfrac{2x - 1}{4}$

**56.** $\dfrac{x + 3}{4} - \dfrac{2x - 1}{4}$

**57.** $\dfrac{x^2}{x - 6} - \dfrac{5x + 6}{x - 6}$

**58.** $\dfrac{x^2 + 5x}{x^2 - 25} \cdot \dfrac{3x - 15}{x^2}$

**59.** $\dfrac{-2x}{x^3 - 8x} + \dfrac{3x}{x^3 - 8x}$

**60.** $\dfrac{-2x}{x^3 - 8x} \div \dfrac{3x}{x^3 - 8x}$

**61.** $\dfrac{12x - 6}{x^2 + 3x} \cdot \dfrac{4x^2 + 13x + 3}{4x^2 - 1}$

**62.** $\dfrac{x^3 + 7x^2}{3x^3 - x^2} \div \dfrac{5x^2 + 36x + 7}{9x^2 - 1}$

**REVIEW AND PREVIEW**

*Perform each indicated operation. See Section 1.3.*

**63.** $\dfrac{2}{3} + \dfrac{5}{7}$

**64.** $\dfrac{9}{10} - \dfrac{3}{5}$

**65.** $\dfrac{2}{6} - \dfrac{3}{4}$

**66.** $\dfrac{11}{15} + \dfrac{5}{9}$

**67.** $\dfrac{1}{12} + \dfrac{3}{20}$

**68.** $\dfrac{7}{30} + \dfrac{3}{18}$

**CONCEPT EXTENSIONS**

**69.** Choose the correct LCD of $\dfrac{11a^3}{4a - 20}$ and $\dfrac{15a^3}{(a - 5)^2}$. See the Concept Check in this section.

    **a.** $4a(a - 5)(a + 5)$    **b.** $a - 5$

    **c.** $(a - 5)^2$    **d.** $4(a - 5)^2$

    **e.** $(4a - 20)(a - 5)^2$

**70.** An algebra student approaches you with a problem. He's tried to subtract two rational expressions, but his result does not match the book's. Check to see if the student has made an error. If so, correct his work shown below.

$$\dfrac{2x - 6}{x - 5} - \dfrac{x + 4}{x - 5}$$

$$= \dfrac{2x - 6 - x + 4}{x - 5}$$

$$= \dfrac{x - 2}{x - 5}$$

*Multiple choice. Select the correct result.*

**71.** $\dfrac{3}{x} + \dfrac{y}{x} =$

    **a.** $\dfrac{3+y}{x^2}$    **b.** $\dfrac{3+y}{2x}$    **c.** $\dfrac{3+y}{x}$

**72.** $\dfrac{3}{x} - \dfrac{y}{x} =$

    **a.** $\dfrac{3-y}{x^2}$    **b.** $\dfrac{3-y}{2x}$    **c.** $\dfrac{3-y}{x}$

**73.** $\dfrac{3}{x} \cdot \dfrac{y}{x} =$

    **a.** $\dfrac{3y}{x}$    **b.** $\dfrac{3y}{x^2}$    **c.** $3y$

**74.** $\dfrac{3}{x} \div \dfrac{y}{x} =$

    **a.** $\dfrac{3}{y}$    **b.** $\dfrac{y}{3}$    **c.** $\dfrac{3}{x^2 y}$

*Write each rational expression as an equivalent expression with a denominator of* $x - 2$.

**75.** $\dfrac{5}{2-x}$          **76.** $\dfrac{8y}{2-x}$

**77.** $-\dfrac{7+x}{2-x}$      **78.** $\dfrac{x-3}{-(x-2)}$

△ **79.** A square has a side of length $\dfrac{5}{x-2}$ meters. Express its perimeter as a rational expression.

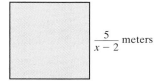

$\dfrac{5}{x-2}$ meters

△ **80.** A trapezoid has sides of the indicated lengths. Find its perimeter.

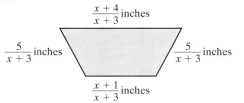

$\dfrac{x+4}{x+3}$ inches

$\dfrac{5}{x+3}$ inches                     $\dfrac{5}{x+3}$ inches

$\dfrac{x+1}{x+3}$ inches

**81.** Write two rational expressions with the same denominator whose sum is $\dfrac{5}{3x-1}$.

**82.** Write two rational expressions with the same denominator whose difference is $\dfrac{x-7}{x^2+1}$.

**83.** The planet Mercury revolves around the sun in 88 Earth days. It takes Jupiter 4332 Earth days to make one revolution around the sun. (*Source:* National Space Science Data Center) If the two planets are aligned as shown in the figure, how long will it take for them to align again?

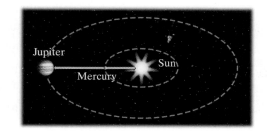

**84.** You are throwing a barbecue and you want to make sure that you purchase the same number of hot dogs as hot dog buns. Hot dogs come 8 to a package and hot dog buns come 12 to a package. What is the least number of each type of package you should buy?

**85.** Write some instructions to help a friend who is having difficulty finding the LCD of two rational expressions.

**86.** Explain why the LCD of the rational expressions $\dfrac{7}{x+1}$ and $\dfrac{9x}{(x+1)^2}$ is $(x+1)^2$ and not $(x+1)^3$.

**87.** In your own words, describe how to add or subtract two rational expressions with the same denominators.

**88.** Explain the similarities between subtracting $\dfrac{3}{8}$ from $\dfrac{7}{8}$ and subtracting $\dfrac{6}{x+3}$ from $\dfrac{9}{x+3}$.

---

### 📖 STUDY SKILLS BUILDER

**How Are You Doing?**

If you haven't done so yet, take a few moments and think about how you are doing in this course. Are you working toward your goal of successfully completing this course? Is your performance on homework, quizzes, and tests satisfactory? If not, you might want to see your instructor to see if he/she has any suggestions on how you can improve your performance. Reread Section 1.1 for ideas on places to get help with your mathematics course.

*Answer the following.*

**1.** List any textbook supplements you are using to help you through this course.

**2.** List any campus resources you are using to help you through this course.

**3.** Write a short paragraph describing how you are doing in your mathematics course.

**4.** If improvement is needed, list ways that you can work toward improving your situation as described in Exercise 3.

# 7.4 ADDING AND SUBTRACTING RATIONAL EXPRESSIONS WITH UNLIKE DENOMINATORS

**OBJECTIVE**

1 Add and subtract rational expressions with unlike denominators.

**OBJECTIVE 1 ▶ Adding and subtracting rational expressions with unlike denominators.** In the previous section, we practiced all the skills we need to add and subtract rational expressions with unlike or different denominators. We add or subtract rational expressions the same way as we add or subtract fractions. You may want to use the steps below.

---

**Adding or Subtracting Rational Expressions with Unlike Denominators**

**STEP 1.** Find the LCD of the rational expressions.

**STEP 2.** Rewrite each rational expression as an equivalent expression whose denominator is the LCD found in Step 1.

**STEP 3.** Add or subtract numerators and write the sum or difference over the common denominator.

**STEP 4.** Simplify or write the rational expression in simplest form.

---

**EXAMPLE 1**  Perform each indicated operation.

**a.** $\dfrac{a}{4} - \dfrac{2a}{8}$   **b.** $\dfrac{3}{10x^2} + \dfrac{7}{25x}$

**Solution**

**a.** First, we must find the LCD. Since $4 = 2^2$ and $8 = 2^3$, the LCD $= 2^3 = 8$. Next we write each fraction as an equivalent fraction with the denominator 8, then we subtract.

$$\frac{a}{4} - \frac{2a}{8} = \frac{a(2)}{4(2)} - \frac{2a}{8} = \frac{2a}{8} - \frac{2a}{8} = \frac{2a - 2a}{8} = \frac{0}{8} = 0$$

Multiplying the numerator and denominator by 2 is the same as multiplying by $\dfrac{2}{2}$ or 1.

**b.** Since $10x^2 = 2 \cdot 5 \cdot x \cdot x$ and $25x = 5 \cdot 5 \cdot x$, the LCD $= 2 \cdot 5^2 \cdot x^2 = 50x^2$. We write each fraction as an equivalent fraction with a denominator of $50x^2$.

$$\frac{3}{10x^2} + \frac{7}{25x} = \frac{3(5)}{10x^2(5)} + \frac{7(2x)}{25x(2x)}$$

$$= \frac{15}{50x^2} + \frac{14x}{50x^2}$$

$$= \frac{15 + 14x}{50x^2} \qquad \text{Add numerators. Write the sum over the common denominator.}$$

**PRACTICE**

**1**  Perform each indicated operation.

**a.** $\dfrac{2x}{5} - \dfrac{6x}{15}$   **b.** $\dfrac{7}{8a} + \dfrac{5}{12a^2}$

**EXAMPLE 2**   Subtract: $\dfrac{6x}{x^2 - 4} - \dfrac{3}{x + 2}$

**_Solution_**   Since $x^2 - 4 = (x + 2)(x - 2)$, the LCD = $(x - 2)(x + 2)$. We write equivalent expressions with the LCD as denominators.

$$\dfrac{6x}{x^2 - 4} - \dfrac{3}{x + 2} = \dfrac{6x}{(x - 2)(x + 2)} - \dfrac{3(x - 2)}{(x + 2)(x - 2)}$$

$$= \dfrac{6x - 3(x - 2)}{(x + 2)(x - 2)} \qquad \text{Subtract numerators. Write the difference over the common denominator.}$$

$$= \dfrac{6x - 3x + 6}{(x + 2)(x - 2)} \qquad \text{Apply the distributive property in the numerator.}$$

$$= \dfrac{3x + 6}{(x + 2)(x - 2)} \qquad \text{Combine like terms in the numerator.}$$

Next we factor the numerator to see if this rational expression can be simplified.

$$= \dfrac{3(x + 2)}{(x + 2)(x - 2)} \qquad \text{Factor.}$$

$$= \dfrac{3}{x - 2} \qquad \text{Divide out common factors to simplify.} \qquad \square$$

**PRACTICE**
**2**   Subtract: $\dfrac{12x}{x^2 - 25} - \dfrac{6}{x + 5}$

---

**EXAMPLE 3**   Add: $\dfrac{2}{3t} + \dfrac{5}{t + 1}$

**_Solution_**   The LCD is $3t(t + 1)$. We write each rational expression as an equivalent rational expression with a denominator of $3t(t + 1)$.

$$\dfrac{2}{3t} + \dfrac{5}{t + 1} = \dfrac{2(t + 1)}{3t(t + 1)} + \dfrac{5(3t)}{(t + 1)(3t)}$$

$$= \dfrac{2(t + 1) + 5(3t)}{3t(t + 1)} \qquad \text{Add numerators. Write the sum over the common denominator.}$$

$$= \dfrac{2t + 2 + 15t}{3t(t + 1)} \qquad \text{Apply the distributive property in the numerator.}$$

$$= \dfrac{17t + 2}{3t(t + 1)} \qquad \text{Combine like terms in the numerator.} \quad \square$$

**PRACTICE**
**3**   Add: $\dfrac{3}{5y} + \dfrac{2}{y + 1}$

---

**EXAMPLE 4**   Subtract: $\dfrac{7}{x - 3} - \dfrac{9}{3 - x}$

**_Solution_**   To find a common denominator, we notice that $x - 3$ and $3 - x$ are opposites. That is, $3 - x = -(x - 3)$. We write the denominator $3 - x$ as $-(x - 3)$ and simplify.

$$\frac{7}{x-3} - \frac{9}{3-x} = \frac{7}{x-3} - \frac{9}{-(x-3)}$$

$$= \frac{7}{x-3} - \frac{-9}{x-3} \qquad \text{Apply } \frac{a}{-b} = \frac{-a}{b}.$$

$$= \frac{7-(-9)}{x-3} \qquad \text{Subtract numerators. Write the difference over the common denominator.}$$

$$= \frac{16}{x-3}$$

**PRACTICE**
**4**   Subtract: $\dfrac{6}{x-5} - \dfrac{7}{5-x}$

**EXAMPLE 5**   Add: $1 + \dfrac{m}{m+1}$

**Solution**   Recall that 1 is the same as $\dfrac{1}{1}$. The LCD of $\dfrac{1}{1}$ and $\dfrac{m}{m+1}$ is $m+1$.

$$1 + \frac{m}{m+1} = \frac{1}{1} + \frac{m}{m+1} \qquad \text{Write 1 as } \frac{1}{1}.$$

$$= \frac{1(m+1)}{1(m+1)} + \frac{m}{m+1} \qquad \text{Multiply both the numerator and the denominator of } \frac{1}{1} \text{ by } m+1.$$

$$= \frac{m+1+m}{m+1} \qquad \text{Add numerators. Write the sum over the common denominator.}$$

$$= \frac{2m+1}{m+1} \qquad \text{Combine like terms in the numerator.}$$

**PRACTICE**
**5**   Add: $2 + \dfrac{b}{b+3}$

**EXAMPLE 6**   Subtract: $\dfrac{3}{2x^2+x} - \dfrac{2x}{6x+3}$

**Solution**   First, we factor the denominators.

$$\frac{3}{2x^2+x} - \frac{2x}{6x+3} = \frac{3}{x(2x+1)} - \frac{2x}{3(2x+1)}$$

The LCD is $3x(2x+1)$. We write equivalent expressions with denominators of $3x(2x+1)$.

$$= \frac{3(3)}{x(2x+1)(3)} - \frac{2x(x)}{3(2x+1)(x)}$$

$$= \frac{9-2x^2}{3x(2x+1)} \qquad \text{Subtract numerators. Write the difference over the common denominator.}$$

**PRACTICE**
**6**   Subtract: $\dfrac{5}{2x^2+3x} - \dfrac{3x}{4x+6}$

**EXAMPLE 7**    Add: $\dfrac{2x}{x^2 + 2x + 1} + \dfrac{x}{x^2 - 1}$

*Solution*    First we factor the denominators.

$$\frac{2x}{x^2 + 2x + 1} + \frac{x}{x^2 - 1} = \frac{2x}{(x + 1)(x + 1)} + \frac{x}{(x + 1)(x - 1)}$$

Now we write the rational expressions as equivalent expressions with denominators of $(x + 1)(x + 1)(x - 1)$, the LCD.

$$= \frac{2x(x - 1)}{(x + 1)(x + 1)(x - 1)} + \frac{x(x + 1)}{(x + 1)(x - 1)(x + 1)}$$

$$= \frac{2x(x - 1) + x(x + 1)}{(x + 1)^2(x - 1)}$$    Add numerators. Write the sum over the common denominator.

$$= \frac{2x^2 - 2x + x^2 + x}{(x + 1)^2(x - 1)}$$    Apply the distributive property in the numerator.

$$= \frac{3x^2 - x}{(x + 1)^2(x - 1)}$$    or    $\dfrac{x(3x - 1)}{(x + 1)^2(x - 1)}$

☐

**PRACTICE**
**7**    Add: $\dfrac{2x}{x^2 + 7x + 12} + \dfrac{3x}{x^2 - 9}$

The numerator was factored as a last step to see if the rational expression could be simplified further. Since there are no factors common to the numerator and the denominator, we can't simplify further.

## VOCABULARY & READINESS CHECK

*Match each exercise with the first step needed to perform the operation. Do not actually perform the operation.*

**1.** $\dfrac{3}{4} - \dfrac{y}{4}$    **2.** $\dfrac{2}{a} \cdot \dfrac{3}{(a + 6)}$    **3.** $\dfrac{x + 1}{x} \div \dfrac{x - 1}{x}$    **4.** $\dfrac{9}{x - 2} - \dfrac{x}{x + 2}$

**a.** Multiply the first rational expression by the reciprocal of the second rational expression.
**b.** Find the LCD. Write each expression as an equivalent expression with the LCD as denominator.
**c.** Multiply numerators, then multiply denominators.
**d.** Subtract numerators. Place the difference over a common denominator.

## 7.4 | EXERCISE SET    MyMathLab®    *Powered by CourseCompass® and MathXL®*    Math XP PRACTICE    WATCH    DOWNLOAD    READ    REVIEW

### MIXED PRACTICE

*Perform each indicated operation. Simplify if possible. See Examples 1 through 7.*

**1.** $\dfrac{4}{2x} + \dfrac{9}{3x}$

**2.** $\dfrac{15}{7a} + \dfrac{8}{6a}$

**3.** $\dfrac{15a}{b} - \dfrac{6b}{5}$

**4.** $\dfrac{4c}{d} - \dfrac{8d}{5}$

**5.** $\dfrac{3}{x} + \dfrac{5}{2x^2}$

**6.** $\dfrac{14}{3x^2} + \dfrac{6}{x}$

**7.** $\dfrac{6}{x + 1} + \dfrac{10}{2x + 2}$

**8.** $\dfrac{8}{x + 4} - \dfrac{3}{3x + 12}$

**9.** $\dfrac{3}{x + 2} - \dfrac{2x}{x^2 - 4}$

**10.** $\dfrac{5}{x - 4} + \dfrac{4x}{x^2 - 16}$

**11.** $\dfrac{3}{4x} + \dfrac{8}{x - 2}$

**12.** $\dfrac{5}{y^2} - \dfrac{y}{2y + 1}$

**13.** $\dfrac{6}{x - 3} + \dfrac{8}{3 - x}$

**14.** $\dfrac{15}{y - 4} + \dfrac{20}{4 - y}$

**15.** $\dfrac{9}{x-3} + \dfrac{9}{3-x}$

*(handwritten: $x - 3$)*

*(handwritten: $-1(x-3)$)*

**16.** $\dfrac{5}{a-7} + \dfrac{5}{7-a}$

**17.** $\dfrac{-8}{x^2-1} - \dfrac{7}{1-x^2}$

**18.** $\dfrac{-9}{25x^2-1} + \dfrac{7}{1-25x^2}$

**19.** $\dfrac{5}{x} + 2$

**20.** $\dfrac{7}{x^2} - 5x$

**21.** $\dfrac{5}{x-2} + 6$

**22.** $\dfrac{6y}{y+5} + 1$

**23.** $\dfrac{y+2}{y+3} - 2$

**24.** $\dfrac{7}{2x-3} - 3$

**25.** $\dfrac{-x+2}{x} - \dfrac{x-6}{4x}$

**26.** $\dfrac{-y+1}{y} - \dfrac{2y-5}{3y}$

**27.** $\dfrac{5x}{x+2} - \dfrac{3x-4}{x+2}$

**28.** $\dfrac{7x}{x-3} - \dfrac{4x+9}{x-3}$

**29.** $\dfrac{3x^4}{7} - \dfrac{4x^2}{21}$

**30.** $\dfrac{5x}{6} + \dfrac{11x^2}{2}$

**31.** $\dfrac{1}{x+3} - \dfrac{1}{(x+3)^2}$

**32.** $\dfrac{5x}{(x-2)^2} - \dfrac{3}{x-2}$

**33.** $\dfrac{4}{5b} + \dfrac{1}{b-1}$

**34.** $\dfrac{1}{y+5} + \dfrac{2}{3y}$

**35.** $\dfrac{2}{m} + 1$

**36.** $\dfrac{6}{x} - 1$

**37.** $\dfrac{2x}{x-7} - \dfrac{x}{x-2}$

**38.** $\dfrac{9x}{x-10} - \dfrac{x}{x-3}$

**39.** $\dfrac{6}{1-2x} - \dfrac{4}{2x-1}$

**40.** $\dfrac{10}{3n-4} - \dfrac{5}{4-3n}$

**41.** $\dfrac{7}{(x+1)(x-1)} + \dfrac{8}{(x+1)^2}$

**42.** $\dfrac{5}{(x+1)(x+5)} - \dfrac{2}{(x+5)^2}$

**43.** $\dfrac{x}{x^2-1} - \dfrac{2}{x^2-2x+1}$

**44.** $\dfrac{x}{x^2-4} - \dfrac{5}{x^2-4x+4}$

**45.** $\dfrac{3a}{2a+6} - \dfrac{a-1}{a+3}$

**46.** $\dfrac{1}{x+y} - \dfrac{y}{x^2-y^2}$

**47.** $\dfrac{y-1}{2y+3} + \dfrac{3}{(2y+3)^2}$

**48.** $\dfrac{x-6}{5x+1} + \dfrac{6}{(5x+1)^2}$

**49.** $\dfrac{5}{2-x} + \dfrac{x}{2x-4}$

**50.** $\dfrac{-1}{a-2} + \dfrac{4}{4-2a}$

**51.** $\dfrac{15}{x^2+6x+9} + \dfrac{2}{x+3}$

**52.** $\dfrac{2}{x^2+4x+4} + \dfrac{1}{x+2}$

**53.** $\dfrac{13}{x^2-5x+6} - \dfrac{5}{x-3}$

**54.** $\dfrac{-7}{y^2-3y+2} - \dfrac{2}{y-1}$

**55.** $\dfrac{70}{m^2-100} + \dfrac{7}{2(m+10)}$

**56.** $\dfrac{27}{y^2-81} + \dfrac{3}{2(y+9)}$

**57.** $\dfrac{x+8}{x^2-5x-6} + \dfrac{x+1}{x^2-4x-5}$

**58.** $\dfrac{x+4}{x^2+12x+20} + \dfrac{x+1}{x^2+8x-20}$

**59.** $\dfrac{5}{4n^2-12n+8} - \dfrac{3}{3n^2-6n}$

**60.** $\dfrac{6}{5y^2-25y+30} - \dfrac{2}{4y^2-8y}$

## MIXED PRACTICE

*Perform the indicated operations. Addition, subtraction, multiplication, and division of rational expressions are included here.*

**61.** $\dfrac{15x}{x+8} \cdot \dfrac{2x+16}{3x}$

**62.** $\dfrac{9z+5}{15} \cdot \dfrac{5z}{81z^2-25}$

**63.** $\dfrac{8x+7}{3x+5} - \dfrac{2x-3}{3x+5}$

**64.** $\dfrac{2z^2}{4z-1} - \dfrac{z-2z^2}{4z-1}$

**65.** $\dfrac{5a+10}{18} \div \dfrac{a^2-4}{10a}$

**66.** $\dfrac{9}{x^2-1} \div \dfrac{12}{3x+3}$

**67.** $\dfrac{5}{x^2-3x+2} + \dfrac{1}{x-2}$

**68.** $\dfrac{4}{2x^2+5x-3} + \dfrac{2}{x+3}$

## REVIEW AND PREVIEW

*Solve the following linear and quadratic equations. See Sections 2.3 and 6.6.*

**69.** $3x+5=7$

**70.** $5x-1=8$

**71.** $2x^2-x-1=0$

**72.** $4x^2-9=0$

**73.** $4(x+6)+3=-3$

**74.** $2(3x+1)+15=-7$

## CONCEPT EXTENSIONS

*Perform each indicated operation.*

**75.** $\dfrac{3}{x} - \dfrac{2x}{x^2-1} + \dfrac{5}{x+1}$

**76.** $\dfrac{5}{x-2} + \dfrac{7x}{x^2-4} - \dfrac{11}{x}$

**77.** $\dfrac{5}{x^2-4} + \dfrac{2}{x^2-4x+4} - \dfrac{3}{x^2-x-6}$

**78.** $\dfrac{8}{x^2+6x+5} - \dfrac{3x}{x^2+4x-5} + \dfrac{2}{x^2-1}$

**79.** $\dfrac{9}{x^2+9x+14} - \dfrac{3x}{x^2+10x+21} + \dfrac{x+4}{x^2+5x+6}$

**80.** $\dfrac{x+10}{x^2-3x-4} - \dfrac{8}{x^2+6x+5} - \dfrac{9}{x^2+x-20}$

**81.** A board of length $\dfrac{3}{x+4}$ inches was cut into two pieces. If one piece is $\dfrac{1}{x-4}$ inches, express the length of the other piece as a rational expression.

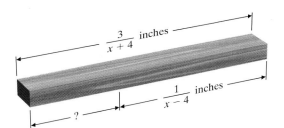

**82.** The length of a rectangle is $\dfrac{3}{y-5}$ feet, while its width is $\dfrac{2}{y}$ feet. Find its perimeter and then find its area.

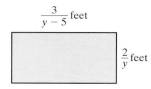

**83.** In ice hockey, penalty killing percentage is a statistic calculated as $1 - \dfrac{G}{P}$, where $G$ = opponent's power play goals and $P$ = opponent's power play opportunities. Simplify this expression.

**84.** The dose of medicine prescribed for a child depends on the child's age $A$ in years and the adult dose $D$ for the medication. Two expressions that give a child's dose are Young's Rule, $\dfrac{DA}{A+12}$, and Cowling's Rule, $\dfrac{D(A+1)}{24}$. Find an expression for the difference in the doses given by these expressions.

**85.** Explain when the LCD of the rational expressions in a sum is the product of the denominators.

**86.** Explain when the LCD is the same as one of the denominators of a rational expression to be added or subtracted.

**87.** Two angles are said to be complementary if the sum of their measures is 90°. If one angle measures $\dfrac{40}{x}$ degrees, find the measure of its complement.

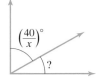

**88.** Two angles are said to be supplementary if the sum of their measures is 180°. If one angle measures $\dfrac{x+2}{x}$ degrees, find the measure of its supplement.

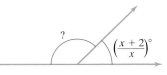

**89.** In your own words, explain how to add two rational expressions with different denominators.

**90.** In your own words, explain how to subtract two rational expressions with different denominators.

---

## THE BIGGER PICTURE   SIMPLIFYING EXPRESSIONS AND SOLVING EQUATIONS AND INEQUALITIES

---

Now we continue our outline from Sections 1.7, 2.3, 2.8, 5.6, and 6.6. Although suggestions are given, this outline should be in your own words. Once you complete this new portion, try the exercises below.

**I.** Simplifying Expressions

   **A.** Real Numbers

      **1.** Add (Section 1.5)

      **2.** Subtract (Section 1.6)

      **3.** Multiply or Divide (Section 1.7)

   **B.** Exponents (Section 5.1 and 5.5)

   **C.** Polynomials

      **1.** Add (Section 5.2)

      **2.** Subtract (Section 5.2)

      **3.** Multiply (Section 5.3 and 5.4)

      **4.** Divide (Section 5.6)

   **D.** Factoring Polynomials (Chapter 6 Integrated Review)

   **E.** Rational Expressions

      **1.** Simplify: Factor the numerator and denominator. Then divide out factors of 1 by dividing out common factors in the numerator and denominator.

$$\frac{x^2 - 9}{7x^2 - 21x} = \frac{(x + 3)(x - 3)}{7x(x - 3)} = \frac{x + 3}{7x}$$

      **2.** Multiply: Multiply numerators, then multiply denominators.

$$\frac{5z}{2z^2 - 9z - 18} \cdot \frac{22z + 33}{10z}$$

$$= \frac{5 \cdot z}{(2z + 3)(z - 6)} \cdot \frac{11(2z + 3)}{2 \cdot 5 \cdot z} = \frac{11}{2(z - 6)}$$

      **3.** Divide: First fraction times the reciprocal of the second fraction.

$$\frac{14}{x + 5} \div \frac{x + 1}{2} = \frac{14}{x + 5} \cdot \frac{2}{x + 1}$$

$$= \frac{28}{(x + 5)(x + 1)}$$

      **4.** Add or Subtract: Must have same denominator. If not find the LCD and write each fraction as an equivalent fraction with the LCD as denominator.

$$\frac{9}{10} - \frac{x + 1}{x + 5} = \frac{9(x + 5)}{10(x + 5)} - \frac{10(x + 1)}{10(x + 5)}$$

$$= \frac{9x + 45 - 10x - 10}{10(x + 5)}$$

$$= \frac{-x + 35}{10(x + 5)}$$

**II.** Solving Equations

   **A.** Linear Equations (Section 2.3)

   **B.** Quadratic & Higher Degree Equations (Section 6.6)

**III.** Solving Inequalities

   **A.** Linear Inequalities (Section 2.8)

*Perform indicated operations and simplify.*

**1.** $-8.6 + (-9.1)$

**2.** $(-8.6)(-9.1)$

**3.** $14 - (-14)$

**4.** $3x^4 - 7 + x^4 - x^2 - 10$

**5.** $\dfrac{5x^2 - 5}{25x + 25}$

**6.** $\dfrac{7x}{x^2 + 4x + 3} \div \dfrac{x}{2x + 6}$

**7.** $\dfrac{2}{9} - \dfrac{5}{6}$

**8.** $\dfrac{x}{9} - \dfrac{x + 3}{5}$

*Factor.*

**9.** $9x^3 - 2x^2 - 11x$

**10.** $12xy - 21x + 4y - 7$

*Solve.*

**11.** $7x - 14 = 5x + 10$

**12.** $\dfrac{-x + 2}{5} < \dfrac{3}{10}$

**13.** $1 + 4(x + 4) = 3^2 + x$

**14.** $x(x - 2) = 24$

## 7.5 SOLVING EQUATIONS CONTAINING RATIONAL EXPRESSIONS

**OBJECTIVES**

1 Solve equations containing rational expressions.

2 Solve equations containing rational expressions for a specified variable.

**OBJECTIVE 1** ▶ **Solving equations containing rational expressions.** In Chapter 2, we solved equations containing fractions. In this section, we continue the work we began in Chapter 2 by solving equations containing rational expressions.

*Examples of Equations Containing Rational Expressions*

$$\frac{x}{2} + \frac{8}{3} = \frac{1}{6} \quad \text{and} \quad \frac{4x}{x^2 + x - 30} + \frac{2}{x - 5} = \frac{1}{x + 6}$$

To solve equations such as these, use the multiplication property of equality to clear the equation of fractions by multiplying both sides of the equation by the LCD.

**EXAMPLE 1**   Solve: $\dfrac{x}{2} + \dfrac{8}{3} = \dfrac{1}{6}$

**Solution**   The LCD of denominators 2, 3, and 6 is 6, so we multiply both sides of the equation by 6.

$$6\left(\frac{x}{2} + \frac{8}{3}\right) = 6\left(\frac{1}{6}\right)$$

▶ **Helpful Hint**

Make sure that *each* term is multiplied by the LCD, 6.

$$6\left(\frac{x}{2}\right) + 6\left(\frac{8}{3}\right) = 6\left(\frac{1}{6}\right) \quad \text{Use the distributive property.}$$

$$3 \cdot x + 16 = 1 \quad \text{Multiply and simplify.}$$

$$3x = -15 \quad \text{Subtract 16 from both sides.}$$

$$x = -5 \quad \text{Divide both sides by 3.}$$

**Check:**   To check, we replace $x$ with $-5$ in the original equation.

$$\frac{x}{2} + \frac{8}{3} = \frac{1}{6}$$

$$\frac{-5}{2} + \frac{8}{3} \stackrel{?}{=} \frac{1}{6} \quad \text{Replace } x \text{ with } -5.$$

$$\frac{1}{6} = \frac{1}{6} \quad \text{True}$$

This number checks, so the solution is $-5$.   □

**PRACTICE**
**1**   Solve: $\dfrac{x}{3} + \dfrac{4}{5} = \dfrac{2}{15}$

**EXAMPLE 2**   Solve: $\dfrac{t - 4}{2} - \dfrac{t - 3}{9} = \dfrac{5}{18}$

**Solution**   The LCD of denominators 2, 9, and 18 is 18, so we multiply both sides of the equation by 18.

$$18\left(\frac{t - 4}{2} - \frac{t - 3}{9}\right) = 18\left(\frac{5}{18}\right)$$

▶ **Helpful Hint**

Multiply *each* term by 18.

$$18\left(\frac{t - 4}{2}\right) - 18\left(\frac{t - 3}{9}\right) = 18\left(\frac{5}{18}\right) \quad \text{Use the distributive property.}$$

$$9(t - 4) - 2(t - 3) = 5 \quad \text{Simplify.}$$

$$9t - 36 - 2t + 6 = 5 \quad \text{Use the distributive property.}$$

$$7t - 30 = 5 \quad \text{Combine like terms.}$$

$$7t = 35$$

$$t = 5 \quad \text{Solve for } t.$$

**Check:**

$$\frac{t-4}{2} - \frac{t-3}{9} = \frac{5}{18}$$

$$\frac{5-4}{2} - \frac{5-3}{9} \overset{?}{=} \frac{5}{18} \qquad \text{Replace } t \text{ with 5.}$$

$$\frac{1}{2} - \frac{2}{9} \overset{?}{=} \frac{5}{18} \qquad \text{Simplify.}$$

$$\frac{5}{18} = \frac{5}{18} \qquad \text{True}$$

The solution is 5.

**PRACTICE**
**2** Solve: $\dfrac{x+4}{4} - \dfrac{x-3}{3} = \dfrac{11}{12}$

Recall from Section 7.1 that a rational expression is defined for all real numbers except those that make the denominator of the expression 0. This means that if an equation contains *rational expressions with variables in the denominator,* we must be certain that the proposed solution does not make the denominator 0. If replacing the variable with the proposed solution makes the denominator 0, the rational expression is undefined and this proposed solution must be rejected.

**EXAMPLE 3**  Solve: $3 - \dfrac{6}{x} = x + 8$

**Solution**  In this equation, 0 cannot be a solution because if $x$ is 0, the rational expression $\dfrac{6}{x}$ is undefined. The LCD is $x$, so we multiply both sides of the equation by $x$.

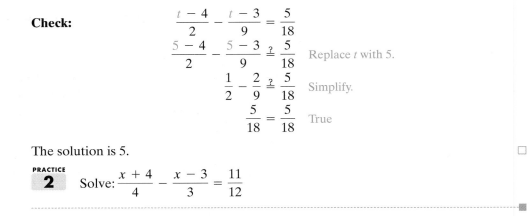

$$x\left(3 - \frac{6}{x}\right) = x(x+8)$$

> **Helpful Hint**
> Multiply *each* term by $x$.

$$x(3) - x\left(\frac{6}{x}\right) = x \cdot x + x \cdot 8 \qquad \text{Use the distributive property.}$$

$$3x - 6 = x^2 + 8x \qquad \text{Simplify.}$$

Now we write the quadratic equation in standard form and solve for $x$.

$$0 = x^2 + 5x + 6$$
$$0 = (x+3)(x+2) \qquad \text{Factor.}$$
$$x + 3 = 0 \quad \text{or} \quad x + 2 = 0 \qquad \text{Set each factor equal to 0 and solve.}$$
$$x = -3 \qquad\qquad x = -2$$

Notice that neither $-3$ nor $-2$ makes the denominator in the original equation equal to 0.

**Check:**  To check these solutions, we replace $x$ in the original equation by $-3$, and then by $-2$.

If $x = -3$:

$$3 - \frac{6}{x} = x + 8$$

$$3 - \frac{6}{-3} \overset{?}{=} -3 + 8$$

$$3 - (-2) \overset{?}{=} 5$$

$$5 = 5 \qquad \text{True}$$

If $x = -2$:

$$3 - \frac{6}{x} = x + 8$$

$$3 - \frac{6}{-2} \overset{?}{=} -2 + 8$$

$$3 - (-3) \overset{?}{=} 6$$

$$6 = 6 \qquad \text{True}$$

Both $-3$ and $-2$ are solutions.

**PRACTICE**
**3** Solve: $8 + \dfrac{7}{x} = x + 2$

The following steps may be used to solve an equation containing rational expressions.

---

**Solving an Equation Containing Rational Expressions**

**STEP 1.** Multiply both sides of the equation by the LCD of all rational expressions in the equation.

**STEP 2.** Remove any grouping symbols and solve the resulting equation.

**STEP 3.** Check the solution in the original equation.

---

**EXAMPLE 4**   Solve: $\dfrac{4x}{x^2 + x - 30} + \dfrac{2}{x - 5} = \dfrac{1}{x + 6}$

*Solution*

The denominator $x^2 + x - 30$ factors as $(x + 6)(x - 5)$. The LCD is then $(x + 6)(x - 5)$, so we multiply both sides of the equation by this LCD.

$$(x + 6)(x - 5)\left(\dfrac{4x}{x^2 + x - 30} + \dfrac{2}{x - 5}\right) = (x + 6)(x - 5)\left(\dfrac{1}{x + 6}\right) \quad \text{Multiply by the LCD.}$$

$$(x + 6)(x - 5) \cdot \dfrac{4x}{x^2 + x - 30} + (x + 6)(x - 5) \cdot \dfrac{2}{x - 5} \quad \text{Apply the distributive property.}$$

$$= (x + 6)(x - 5) \cdot \dfrac{1}{x + 6}$$

$$4x + 2(x + 6) = x - 5 \quad \text{Simplify.}$$

$$4x + 2x + 12 = x - 5 \quad \text{Apply the distributive property.}$$

$$6x + 12 = x - 5 \quad \text{Combine like terms.}$$

$$5x = -17$$

$$x = -\dfrac{17}{5} \quad \text{Divide both sides by 5.}$$

**Check:**   Check by replacing $x$ with $-\dfrac{17}{5}$ in the original equation. The solution is $-\dfrac{17}{5}$.

□

**PRACTICE**
**4**   Solve: $\dfrac{6x}{x^2 - 5x - 14} - \dfrac{3}{x + 2} = \dfrac{1}{x - 7}$

---

**EXAMPLE 5**   Solve: $\dfrac{2x}{x - 4} = \dfrac{8}{x - 4} + 1$

*Solution*   Multiply both sides by the LCD, $x - 4$.

$$(x - 4)\left(\dfrac{2x}{x - 4}\right) = (x - 4)\left(\dfrac{8}{x - 4} + 1\right) \quad \begin{array}{l}\text{Multiply by the LCD.}\\ \text{Notice that 4 cannot be a}\\ \text{solution.}\end{array}$$

$$(x - 4) \cdot \dfrac{2x}{x - 4} = (x - 4) \cdot \dfrac{8}{x - 4} + (x - 4) \cdot 1 \quad \text{Use the distributive property.}$$

$$2x = 8 + (x - 4) \quad \text{Simplify.}$$

$$2x = 4 + x$$

$$x = 4$$

Notice that 4 makes the denominator 0 in the original equation. Therefore, 4 is *not* a solution.

This equation has *no solution*.

□

**PRACTICE**
**5**   Solve: $\dfrac{7}{x - 2} = \dfrac{3}{x - 2} + 4$

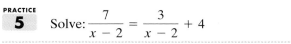

> ▶ **Helpful Hint**
> As we can see from Example 5, it is important to check the proposed solution(s) in the *original* equation.

**Concept Check** ☑

When can we clear fractions by multiplying through by the LCD?

**a.** When adding or subtracting rational expressions

**b.** When solving an equation containing rational expressions

**c.** Both of these

**d.** Neither of these

**EXAMPLE 6** Solve: $x + \dfrac{14}{x-2} = \dfrac{7x}{x-2} + 1$

**Solution** Notice the denominators in this equation. We can see that 2 can't be a solution. The LCD is $x - 2$, so we multiply both sides of the equation by $x - 2$.

$$(x-2)\left(x + \frac{14}{x-2}\right) = (x-2)\left(\frac{7x}{x-2} + 1\right)$$

$$(x-2)(x) + (x-2)\left(\frac{14}{x-2}\right) = (x-2)\left(\frac{7x}{x-2}\right) + (x-2)(1)$$

$$x^2 - 2x + 14 = 7x + x - 2 \quad \text{Simplify.}$$

$$x^2 - 2x + 14 = 8x - 2 \quad \text{Combine like terms.}$$

$$x^2 - 10x + 16 = 0 \quad \text{Write the quadratic equation in standard form.}$$

$$(x-8)(x-2) = 0 \quad \text{Factor.}$$

$$x - 8 = 0 \quad \text{or} \quad x - 2 = 0 \quad \text{Set each factor equal to 0.}$$

$$x = 8 \qquad\qquad x = 2 \quad \text{Solve.}$$

As we have already noted, 2 can't be a solution of the original equation. So we need only replace $x$ with 8 in the original equation. We find that 8 is a solution; the only solution is 8. ☐

**PRACTICE**
**6** Solve: $x + \dfrac{x}{x-5} = \dfrac{5}{x-5} - 7$

**OBJECTIVE 2** ▶ **Solving equations for a specified variable.** The last example in this section is an equation containing several variables, and we are directed to solve for one of the variables. The steps used in the preceding examples can be applied to solve equations for a specified variable as well.

**EXAMPLE 7** Solve: $\dfrac{1}{a} + \dfrac{1}{b} = \dfrac{1}{x}$ for $x$.

**Solution** (This type of equation often models a work problem, as we shall see in Section 7.6.) The LCD is $abx$, so we multiply both sides by $abx$.

$$abx\left(\frac{1}{a} + \frac{1}{b}\right) = abx\left(\frac{1}{x}\right)$$

$$abx\left(\frac{1}{a}\right) + abx\left(\frac{1}{b}\right) = abx \cdot \frac{1}{x}$$

$$bx + ax = ab \quad \text{Simplify.}$$

$$x(b + a) = ab \quad \text{Factor out } x \text{ from each term on the left side.}$$

**Answer to Concept Check:** b

$$\frac{x(b + a)}{b + a} = \frac{ab}{b + a} \qquad \text{Divide both sides by } b + a.$$

$$x = \frac{ab}{b + a} \qquad \text{Simplify.}$$

This equation is now solved for $x$.

**PRACTICE**
**7**  Solve: $\frac{1}{a} + \frac{1}{b} = \frac{1}{x}$ for $b$

---

## Graphing Calculator Explorations

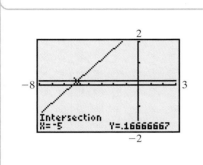

Intersection
X=-5          Y=.16666667

A graphing calculator may be used to check solutions of equations containing rational expressions. For example, to check the solution of Example 1, $\frac{x}{2} + \frac{8}{3} = \frac{1}{6}$, graph $y_1 = \frac{x}{2} + \frac{8}{3}$ and $y_2 = \frac{1}{6}$.

Use TRACE and ZOOM, or use INTERSECT, to find the point of intersection. The point of intersection has an $x$-value of $-5$, so the solution of the equation is $-5$.

*Use a graphing calculator to check the examples of this section.*

**1.** Example 2

**2.** Example 3

**3.** Example 5

**4.** Example 6

---

## 7.5 EXERCISE SET

MyMathLab  *Powered by CourseCompass™ and MathXL®*

Math▲XP  PRACTICE    WATCH    DOWNLOAD    READ    REVIEW

*Solve each equation and check each solution. See Examples 1 through 3.*

**1.** $\frac{x}{5} + 3 = 9$

**2.** $\frac{x}{5} - 2 = 9$

**3.** $\frac{x}{2} + \frac{5x}{4} = \frac{x}{12}$

**4.** $\frac{x}{6} + \frac{4x}{3} = \frac{x}{18}$

**5.** $2 - \frac{8}{x} = 6$

**6.** $5 + \frac{4}{x} = 1$

**7.** $2 + \frac{10}{x} = x + 5$

**8.** $6 + \frac{5}{y} = y - \frac{2}{y}$

**9.** $\frac{a}{5} = \frac{a - 3}{2}$

**10.** $\frac{b}{5} = \frac{b + 2}{6}$

**11.** $\frac{x - 3}{5} + \frac{x - 2}{2} = \frac{1}{2}$

**12.** $\frac{a + 5}{4} + \frac{a + 5}{2} = \frac{a}{8}$

*Solve each equation and check each proposed solution. See Examples 4 through 6.*

**13.** $\frac{3}{2a - 5} = -1$

**14.** $\frac{6}{4 - 3x} = -3$

**15.** $\frac{4y}{y - 4} + 5 = \frac{5y}{y - 4}$

**16.** $\frac{2a}{a + 2} - 5 = \frac{7a}{a + 2}$

**17.** $2 + \frac{3}{a - 3} = \frac{a}{a - 3}$

**18.** $\frac{2y}{y - 2} - \frac{4}{y - 2} = 4$

**19.** $\frac{1}{x + 3} + \frac{6}{x^2 - 9} = 1$

**20.** $\frac{1}{x + 2} + \frac{4}{x^2 - 4} = 1$

**21.** $\frac{2y}{y + 4} + \frac{4}{y + 4} = 3$

**22.** $\frac{5y}{y + 1} - \frac{3}{y + 1} = 4$

**23.** $\frac{2x}{x + 2} - 2 = \frac{x - 8}{x - 2}$

**24.** $\frac{4y}{y - 3} - 3 = \frac{3y - 1}{y + 3}$

### MIXED PRACTICE

*Solve each equation. See Examples 1 through 6.*

**25.** $\frac{2}{y} + \frac{1}{2} = \frac{5}{2y}$

**26.** $\frac{6}{3y} + \frac{3}{y} = 1$

**27.** $\frac{a}{a - 6} = \frac{-2}{a - 1}$

**28.** $\frac{5}{x - 6} = \frac{x}{x - 2}$

**29.** $\frac{11}{2x} + \frac{2}{3} = \frac{7}{2x}$

**30.** $\frac{5}{3} - \frac{3}{2x} = \frac{3}{2}$

**31.** $\frac{2}{x - 2} + 1 = \frac{x}{x + 2}$

**32.** $1 + \frac{3}{x + 1} = \frac{x}{x - 1}$

**33.** $\frac{x + 1}{3} - \frac{x - 1}{6} = \frac{1}{6}$

**34.** $\frac{3x}{5} - \frac{x - 6}{3} = -\frac{2}{5}$

**35.** $\dfrac{t}{t-4} = \dfrac{t+4}{6}$

**36.** $\dfrac{15}{x+4} = \dfrac{x-4}{x}$

**37.** $\dfrac{y}{2y+2} + \dfrac{2y-16}{4y+4} = \dfrac{2y-3}{y+1}$

**38.** $\dfrac{1}{x+2} = \dfrac{4}{x^2-4} - \dfrac{1}{x-2}$

**39.** $\dfrac{4r-4}{r^2+5r-14} + \dfrac{2}{r+7} = \dfrac{1}{r-2}$

**40.** $\dfrac{3}{x+3} = \dfrac{12x+19}{x^2+7x+12} - \dfrac{5}{x+4}$

**41.** $\dfrac{x+1}{x+3} = \dfrac{x^2-11x}{x^2+x-6} - \dfrac{x-3}{x-2}$

**42.** $\dfrac{2t+3}{t-1} - \dfrac{2}{t+3} = \dfrac{5-6t}{t^2+2t-3}$

*Solve each equation for the indicated variable. See Example 7.*

**43.** $R = \dfrac{E}{I}$ for $I$ (Electronics: resistance of a circuit)

**44.** $T = \dfrac{V}{Q}$ for $Q$ (Water purification: settling time)

**45.** $T = \dfrac{2U}{B+E}$ for $B$ (Merchandising: stock turnover rate)

**46.** $i = \dfrac{A}{t+B}$ for $t$ (Hydrology: rainfall intensity)

**47.** $B = \dfrac{705w}{h^2}$ for $w$ (Health: body-mass index)

**48.** $\dfrac{A}{W} = L$ for $W$ (Geometry: area of a rectangle)

**49.** $N = R + \dfrac{V}{G}$ for $G$ (Urban forestry: tree plantings per year)

**50.** $C = \dfrac{D(A+1)}{24}$ for $A$ (Medicine: Cowling's Rule for child's dose)

**51.** $\dfrac{C}{\pi r} = 2$ for $r$ (Geometry: circumference of a circle)

**52.** $W = \dfrac{CE^2}{2}$ for $C$ (Electronics: energy stored in a capacitor)

**53.** $\dfrac{1}{y} + \dfrac{1}{3} = \dfrac{1}{x}$ for $x$

**54.** $\dfrac{1}{5} + \dfrac{2}{y} = \dfrac{1}{x}$ for $x$

**REVIEW AND PREVIEW**

*Write each phrase as an expression. See Sections 1.3 and 7.2.*

**55.** The reciprocal of $x$

**56.** The reciprocal of $x+1$

**57.** The reciprocal of $x$, added to the reciprocal of 2

**58.** The reciprocal of $x$, subtracted from the reciprocal of 5

*Answer each question.*

**59.** If a tank is filled in 3 hours, what fractional part of the tank is filled in 1 hour?

**60.** If a strip of beach is cleaned in 4 hours, what fractional part of the beach is cleaned in 1 hour?

*Identify the x- and y-intercepts. See Section 3.3.*

**61.** **62.** **63.** **64.**

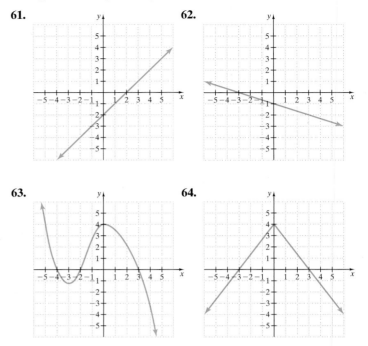

**CONCEPT EXTENSIONS**

**65.** Explain the difference between solving an equation such as $\dfrac{x}{2} + \dfrac{3}{4} = \dfrac{x}{4}$ for $x$ and performing an operation such as adding $\dfrac{x}{2} + \dfrac{3}{4}$.

**66.** When solving an equation such as $\dfrac{y}{4} = \dfrac{y}{2} - \dfrac{1}{4}$, we may multiply all terms by 4. When subtracting two rational expressions such as $\dfrac{y}{2} - \dfrac{1}{4}$, we may not. Explain why.

*Determine whether each of the following is an equation or an expression. If it is an equation, then solve it for its variable. If it is an expression, perform the indicated operation.*

**67.** $\dfrac{1}{x} + \dfrac{5}{9}$

**68.** $\dfrac{1}{x} + \dfrac{5}{9} = \dfrac{2}{3}$

**69.** $\dfrac{5}{x-1} - \dfrac{2}{x} = \dfrac{5}{x(x-1)}$

**70.** $\dfrac{5}{x-1} - \dfrac{2}{x}$

*Recall that two angles are supplementary if the sum of their measures is 180°. Find the measures of the following supplementary angles.*

**71.** **72.**

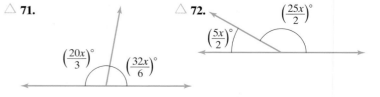

Recall that two angles are complementary if the sum of their measures is 90°. Find the measures of the following complementary angles.

△ **73.**

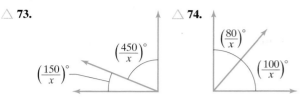

△ **74.**

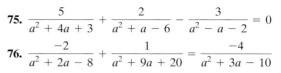

Solve each equation.

**75.** $\dfrac{5}{a^2 + 4a + 3} + \dfrac{2}{a^2 + a - 6} - \dfrac{3}{a^2 - a - 2} = 0$

**76.** $\dfrac{-2}{a^2 + 2a - 8} + \dfrac{1}{a^2 + 9a + 20} = \dfrac{-4}{a^2 + 3a - 10}$

## INTEGRATED REVIEW    SUMMARY ON RATIONAL EXPRESSIONS

Sections 7.1–7.5

It is important to know the difference between performing operations with rational expressions and solving an equation containing rational expressions. Study the examples below.

**PERFORMING OPERATIONS WITH RATIONAL EXPRESSIONS**

Adding: $\dfrac{1}{x} + \dfrac{1}{x + 5} = \dfrac{1 \cdot (x + 5)}{x(x + 5)} + \dfrac{1 \cdot x}{x(x + 5)} = \dfrac{x + 5 + x}{x(x + 5)} = \dfrac{2x + 5}{x(x + 5)}$

Subtracting: $\dfrac{3}{x} - \dfrac{5}{x^2 y} = \dfrac{3 \cdot xy}{x \cdot xy} - \dfrac{5}{x^2 y} = \dfrac{3xy - 5}{x^2 y}$

Multiplying: $\dfrac{2}{x} \cdot \dfrac{5}{x - 1} = \dfrac{2 \cdot 5}{x(x - 1)} = \dfrac{10}{x(x - 1)}$

Dividing: $\dfrac{4}{2x + 1} \div \dfrac{x - 3}{x} = \dfrac{4}{2x + 1} \cdot \dfrac{x}{x - 3} = \dfrac{4x}{(2x + 1)(x - 3)}$

**SOLVING AN EQUATION CONTAINING RATIONAL EXPRESSIONS**

To solve an equation containing rational expressions, we clear the equation of fractions by multiplying both sides by the LCD.

$$\dfrac{3}{x} - \dfrac{5}{x - 1} = \dfrac{1}{x(x - 1)} \qquad \text{Note that } x \text{ can't be 0 or 1.}$$

$$x(x - 1)\left(\dfrac{3}{x}\right) - x(x - 1)\left(\dfrac{5}{x - 1}\right) = x(x - 1) \cdot \dfrac{1}{x(x - 1)} \qquad \text{Multiply both sides by the LCD.}$$

$$3(x - 1) - 5x = 1 \qquad \text{Simplify.}$$

$$3x - 3 - 5x = 1 \qquad \text{Use the distributive property.}$$

$$-2x - 3 = 1 \qquad \text{Combine like terms.}$$

$$-2x = 4 \qquad \text{Add 3 to both sides.}$$

$$x = -2 \qquad \text{Divide both sides by } -2.$$

Determine whether each of the following is an equation or an expression. If it is an equation, solve it for its variable. If it is an expression, perform the indicated operation.

**1.** $\dfrac{1}{x} + \dfrac{2}{3}$

**2.** $\dfrac{3}{a} + \dfrac{5}{6}$

**3.** $\dfrac{1}{x} + \dfrac{2}{3} = \dfrac{3}{x}$

**4.** $\dfrac{3}{a} + \dfrac{5}{6} = 1$

**5.** $\dfrac{2}{x - 1} - \dfrac{1}{x}$

**6.** $\dfrac{4}{x - 3} - \dfrac{1}{x}$

**7.** $\dfrac{2}{x + 1} - \dfrac{1}{x} = 1$

**8.** $\dfrac{4}{x - 3} - \dfrac{1}{x} = \dfrac{6}{x(x - 3)}$

**9.** $\dfrac{15x}{x+8} \cdot \dfrac{2x+16}{3x}$

**10.** $\dfrac{9z+5}{15} \cdot \dfrac{5z}{81z^2-25}$

**11.** $\dfrac{2x+1}{x-3} + \dfrac{3x+6}{x-3}$

**12.** $\dfrac{4p-3}{2p+7} + \dfrac{3p+8}{2p+7}$

**13.** $\dfrac{x+5}{7} = \dfrac{8}{2}$

**14.** $\dfrac{1}{2} = \dfrac{x-1}{8}$

**15.** $\dfrac{5a+10}{18} \div \dfrac{a^2-4}{10a}$

**16.** $\dfrac{9}{x^2-1} + \dfrac{12}{3x+3}$

**17.** $\dfrac{x+2}{3x-1} + \dfrac{5}{(3x-1)^2}$

**18.** $\dfrac{4}{(2x-5)^2} + \dfrac{x+1}{2x-5}$

**19.** $\dfrac{x-7}{x} - \dfrac{x+2}{5x}$

**20.** $\dfrac{9}{x^2-4} + \dfrac{2}{x+2} = \dfrac{-1}{x-2}$

**21.** $\dfrac{3}{x+3} = \dfrac{5}{x^2-9} - \dfrac{2}{x-3}$

**22.** $\dfrac{10x-9}{x} - \dfrac{x-4}{3x}$

---

# 7.6 PROPORTION AND PROBLEM SOLVING WITH RATIONAL EQUATIONS

**OBJECTIVES**

1 Solve proportions.

2 Use proportions to solve problems.

3 Solve problems about numbers.

4 Solve problems about work.

5 Solve problems about distance.

**OBJECTIVE 1 ▶ Solving proportions.** A **ratio** is the quotient of two numbers or two quantities. For example, the ratio of 2 to 5 can be written as $\dfrac{2}{5}$, the quotient of 2 and 5.

If two ratios are equal, we say the ratios are **in proportion** to each other. A **proportion** is a mathematical statement that two ratios are equal.

For example, the equation $\dfrac{1}{2} = \dfrac{4}{8}$ is a proportion, as is $\dfrac{x}{5} = \dfrac{8}{10}$, because both sides of the equations are ratios. When we want to emphasize the equation as a proportion, we

> **read the proportion $\dfrac{1}{2} = \dfrac{4}{8}$ as "one is to two as four is to eight"**

In a proportion, cross products are equal. To understand cross products, let's start with the proportion

$$\frac{a}{b} = \frac{c}{d}$$

and multiply both sides by the LCD, $bd$.

$$bd\left(\frac{a}{b}\right) = bd\left(\frac{c}{d}\right) \quad \text{Multiply both sides by the LCD, } bd.$$

$$\underbrace{ad}_{\text{Cross product}} = \underbrace{bc}_{\text{Cross product}} \quad \text{Simplify.}$$

Notice why $ad$ and $bc$ are called cross products.

Cross Products

If $\dfrac{a}{b} = \dfrac{c}{d}$, then $ad = bc$.

For example, if

$$\frac{1}{2} = \frac{4}{8}, \quad \text{then} \quad 1 \cdot 8 = 2 \cdot 4 \quad \text{or}$$
$$8 = 8$$

Notice that a proportion contains four numbers (or expressions). If any three numbers are known, we can solve and find the fourth number.

**EXAMPLE 1**   Solve for $x$: $\dfrac{45}{x} = \dfrac{5}{7}$

**Solution**   This is an equation with rational expressions, and also a proportion. Below are two ways to solve.

Since this is a rational equation, we can use the methods of the previous section.

$$\frac{45}{x} = \frac{5}{7}$$

$$7x \cdot \frac{45}{x} = 7x \cdot \frac{5}{7} \quad \text{Multiply both sides by LCD } 7x.$$

$$7 \cdot 45 = x \cdot 5 \quad \text{Divide out common factors.}$$

$$315 = 5x \quad \text{Multiply.}$$

$$\frac{315}{5} = \frac{5x}{5} \quad \text{Divide both sides by 5.}$$

$$63 = x \quad \text{Simplify.}$$

Since this is also a proportion, we may set cross products equal.

$$\frac{45}{x} = \frac{5}{7}$$

$$45 \cdot 7 = x \cdot 5 \quad \text{Set cross products equal.}$$

$$315 = 5x \quad \text{Multiply.}$$

$$\frac{315}{5} = \frac{5x}{5} \quad \text{Divide both sides by 5.}$$

$$63 = x \quad \text{Simplify.}$$

**Check:**   Both methods give us a solution of 63. To check, substitute 63 for $x$ in the original proportion. The solution is 63.   □

**PRACTICE**
**1**   Solve for $x$: $\dfrac{36}{x} = \dfrac{4}{11}$

In this section, if the rational equation is a proportion, we will use cross products to solve.

**EXAMPLE 2**   Solve for $x$: $\dfrac{x - 5}{3} = \dfrac{x + 2}{5}$

**Solution**

$$\frac{x - 5}{3} = \frac{x + 2}{5}$$

$$5(x - 5) = 3(x + 2) \quad \text{Set cross products equal.}$$

$$5x - 25 = 3x + 6 \quad \text{Multiply.}$$

$$5x = 3x + 31 \quad \text{Add 25 to both sides.}$$

$$2x = 31 \quad \text{Subtract } 3x \text{ from both sides.}$$

$$\frac{2x}{2} = \frac{31}{2} \quad \text{Divide both sides by 2.}$$

$$x = \frac{31}{2}$$

**Check:**   Verify that $\dfrac{31}{2}$ is the solution.   □

**PRACTICE**
**2**   Solve for $x$: $\dfrac{3x + 2}{9} = \dfrac{x - 1}{2}$

**OBJECTIVE 2 ▶ Using proportions to solve problems.** Proportions can be used to model and solve many real-life problems. When using proportions in this way, it is important to judge whether the solution is reasonable. Doing so helps us to decide if the proportion has been formed correctly. We use the same problem-solving steps that were introduced in Section 2.4.

**EXAMPLE 3**    **Calculating the Cost of Recordable Compact Discs**

Three boxes of CD-Rs (recordable compact discs) cost $37.47. How much should 5 boxes cost?

*Solution*

1. UNDERSTAND. Read and reread the problem. We know that the cost of 5 boxes is more than the cost of 3 boxes, or $37.47, and less than the cost of 6 boxes, which is double the cost of 3 boxes, or 2($37.47) = $74.94. Let's suppose that 5 boxes cost $60.00. To check, we see if 3 boxes is to 5 boxes as the *price* of 3 boxes is to the *price* of 5 boxes. In other words, we see if

$$\frac{3 \text{ boxes}}{5 \text{ boxes}} = \frac{\text{price of 3 boxes}}{\text{price of 5 boxes}}$$

or

$$\frac{3}{5} = \frac{37.47}{60.00}$$

$3(60.00) = 5(37.47)$    Set cross products equal.

or

$180.00 = 187.35$    Not a true statement.

Thus, $60 is not correct, but we now have a better understanding of the problem.

Let $x$ = price of 5 boxes of CD-Rs.

2. TRANSLATE.

$$\frac{3 \text{ boxes}}{5 \text{ boxes}} = \frac{\text{price of 3 boxes}}{\text{price of 5 boxes}}$$

$$\frac{3}{5} = \frac{37.47}{x}$$

3. SOLVE.

$$\frac{3}{5} = \frac{37.47}{x}$$

$3x = 5(37.47)$    Set cross products equal.

$3x = 187.35$

$x = 62.45$    Divide both sides by 3.

4. INTERPRET.

**Check:**   Verify that 3 boxes is to 5 boxes as $37.47 is to $62.45. Also, notice that our solution is a reasonable one as discussed in Step 1.

**State:**   Five boxes of CD-Rs cost $62.45.    ☐

**PRACTICE**
**3**    Four 2-liter bottles of Diet Pepsi cost $5.16. How much will seven 2-liter bottles cost?

> ▶ **Helpful Hint**
>
> The proportion $\dfrac{5 \text{ boxes}}{3 \text{ boxes}} = \dfrac{\text{price of 5 boxes}}{\text{price of 3 boxes}}$ could also have been used to solve Example 3.
>
> Notice that the cross products are the same.

**Similar triangles** have the same shape but not necessarily the same size. In similar triangles, the measures of corresponding angles are equal, and corresponding sides are in proportion.

If triangle $ABC$ and triangle $XYZ$ shown are similar, then we know that the measure of angle $A$ = the measure of angle $X$, the measure of angle $B$ = the measure of angle $Y$, and the measure of angle $C$ = the measure of angle $Z$. We also know that corresponding sides are in proportion: $\dfrac{a}{x} = \dfrac{b}{y} = \dfrac{c}{z}$.

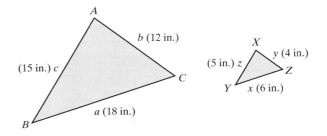

In this section, we will position similar triangles so that they have the same orientation.

To show that corresponding sides are in proportion for the triangles above, we write the ratios of the corresponding sides.

$$\frac{a}{x} = \frac{18}{6} = 3 \qquad \frac{b}{y} = \frac{12}{4} = 3 \qquad \frac{c}{z} = \frac{15}{5} = 3$$

△ **EXAMPLE 4**   **Finding the Length of a Side of a Triangle**

If the following two triangles are similar, find the missing length $x$.

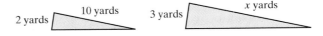

**Solution**

1. UNDERSTAND. Read the problem and study the figure.

2. TRANSLATE. Since the triangles are similar, their corresponding sides are in proportion and we have

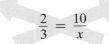

3. SOLVE. To solve, we multiply both sides by the LCD, $3x$, or cross multiply.

$$2x = 30$$

$$x = 15 \quad \text{Divide both sides by 2.}$$

4. INTERPRET.

**Check:**   To check, replace $x$ with 15 in the original proportion and see that a true statement results.

**State:**   The missing length is 15 yards.   □

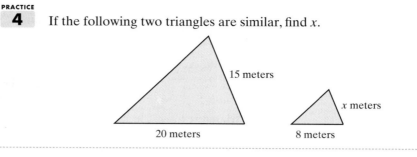

**PRACTICE**
**4** If the following two triangles are similar, find $x$.

15 meters

$x$ meters

20 meters

8 meters

**OBJECTIVE 3 ▶ Solving problems about numbers.** Let's continue to solve problems. The remaining problems are all modeled by rational equations.

### EXAMPLE 5 Finding an Unknown Number

The quotient of a number and 6, minus $\frac{5}{3}$, is the quotient of the number and 2. Find the number.

*Solution*

1. **UNDERSTAND.** Read and reread the problem. Suppose that the unknown number is 2, then we see if the quotient of 2 and 6, or $\frac{2}{6}$, minus $\frac{5}{3}$ is equal to the quotient of 2 and 2, or $\frac{2}{2}$.

$$\frac{2}{6} - \frac{5}{3} = \frac{1}{3} - \frac{5}{3} = -\frac{4}{3}, \text{ not } \frac{2}{2}$$

Don't forget that the purpose of a proposed solution is to better understand the problem.

Let $x$ = the unknown number.

2. **TRANSLATE.**

| In words: | the quotient of $x$ and 6 | minus | $\frac{5}{3}$ | is | the quotient of $x$ and 2 |
|---|---|---|---|---|---|
| | ↓ | ↓ | ↓ | ↓ | ↓ |
| Translate: | $\frac{x}{6}$ | $-$ | $\frac{5}{3}$ | $=$ | $\frac{x}{2}$ |

3. **SOLVE.** Here, we solve the equation $\frac{x}{6} - \frac{5}{3} = \frac{x}{2}$. We begin by multiplying both sides of the equation by the LCD, 6.

$$6\left(\frac{x}{6} - \frac{5}{3}\right) = 6\left(\frac{x}{2}\right)$$

$$6\left(\frac{x}{6}\right) - 6\left(\frac{5}{3}\right) = 6\left(\frac{x}{2}\right) \quad \text{Apply the distributive property.}$$

$$x - 10 = 3x \quad \text{Simplify.}$$

$$-10 = 2x \quad \text{Subtract } x \text{ from both sides.}$$

$$\frac{-10}{2} = \frac{2x}{2} \quad \text{Divide both sides by 2.}$$

$$-5 = x \quad \text{Simplify.}$$

4. **INTERPRET.**

**Check:** To check, we verify that "the quotient of $-5$ and 6 minus $\frac{5}{3}$ is the quotient of $-5$ and 2," or $-\frac{5}{6} - \frac{5}{3} = -\frac{5}{2}$.

**State:** The unknown number is $-5$.

PRACTICE
**5**  The quotient of a number and 5, minus $\frac{3}{2}$, is the quotient of the number and 10.

---

**OBJECTIVE 4 ▶ Solving problems about work.** The next example is often called a work problem. Work problems usually involve people or machines doing a certain task.

**EXAMPLE 6**  **Finding Work Rates**

Sam Waterton and Frank Schaffer work in a plant that manufactures automobiles. Sam can complete a quality control tour of the plant in 3 hours while his assistant, Frank, needs 7 hours to complete the same job. The regional manager is coming to inspect the plant facilities, so both Sam and Frank are directed to complete a quality control tour together. How long will this take?

*Solution*

1. UNDERSTAND. Read and reread the problem. The key idea here is the relationship between the **time** (hours) it takes to complete the job and the **part of the job** completed in 1 unit of time (hour). For example, if the **time** it takes Sam to complete the job is 3 hours, the **part of the job** he can complete in 1 hour is $\frac{1}{3}$. Similarly, Frank can complete $\frac{1}{7}$ of the job in 1 hour.

   Let $x$ = the **time** in hours it takes Sam and Frank to complete the job together. Then $\frac{1}{x}$ = the **part of the job** they complete in 1 hour.

| | *Hours to Complete Total Job* | *Part of Job Completed in 1 Hour* |
|---|---|---|
| Sam | 3 | $\frac{1}{3}$ |
| Frank | 7 | $\frac{1}{7}$ |
| Together | $x$ | $\frac{1}{x}$ |

2. TRANSLATE.

| In words: | part of job Sam completed in 1 hour | added to | part of job Frank completed in 1 hour | is equal to | part of job they completed together in 1 hour |
|---|---|---|---|---|---|
| | ↓ | ↓ | ↓ | ↓ | ↓ |
| Translate: | $\frac{1}{3}$ | $+$ | $\frac{1}{7}$ | $=$ | $\frac{1}{x}$ |

3. SOLVE. Here, we solve the equation $\frac{1}{3} + \frac{1}{7} = \frac{1}{x}$. We begin by multiplying both sides of the equation by the LCD, $21x$.

$$21x\left(\frac{1}{3}\right) + 21x\left(\frac{1}{7}\right) = 21x\left(\frac{1}{x}\right)$$

$$7x + 3x = 21 \qquad \text{Simplify.}$$

$$10x = 21$$

$$x = \frac{21}{10} \quad \text{or} \quad 2\frac{1}{10} \text{ hours}$$

**4. INTERPRET.**

**Check:** Our proposed solution is $2\frac{1}{10}$ hours. This proposed solution is reasonable since $2\frac{1}{10}$ hours is more than half of Sam's time and less than half of Frank's time. Check this solution in the originally *stated* problem.

**State:** Sam and Frank can complete the quality control tour in $2\frac{1}{10}$ hours. □

**PRACTICE**
**6** Cindy Liu and Mary Beckwith own a landscaping company. Cindy can complete a certain garden planting in 3 hours, while Mary takes 4 hours to complete the same job. If both of them work together, how long will it take to plant the garden?

**Concept Check** ☑

Solve $E = mc^2$

**a.** for $m$. **b.** for $c^2$.

**OBJECTIVE 5** ▶ **Solving problems about distance.** Next we look at a problem solved by the distance formula,

$$d = r \cdot t$$

**EXAMPLE 7** **Finding Speeds of Vehicles**

A car travels 180 miles in the same time that a truck travels 120 miles. If the car's speed is 20 miles per hour faster than the truck's, find the car's speed and the truck's speed.

**Solution**

**1. UNDERSTAND.** Read and reread the problem. Suppose that the truck's speed is 45 miles per hour. Then the car's speed is 20 miles per hour more, or 65 miles per hour.

We are given that the car travels 180 miles in the same time that the truck travels 120 miles. To find the time it takes the car to travel 180 miles, remember that since $d = rt$, we know that $\frac{d}{r} = t$.

| *Car's Time* | *Truck's Time* |
|---|---|
| $t = \dfrac{d}{r} = \dfrac{180}{65} = 2\dfrac{50}{65} = 2\dfrac{10}{13}$ hours | $t = \dfrac{d}{r} = \dfrac{120}{45} = 2\dfrac{30}{45} = 2\dfrac{2}{3}$ hours |

Since the times are not the same, our proposed solution is not correct. But we have a better understanding of the problem.

Let $x$ = the speed of the truck.

Since the car's speed is 20 miles per hour faster than the truck's, then

$$x + 20 = \text{the speed of the car}$$

Use the formula $d = r \cdot t$ or **d**istance = **r**ate $\cdot$ **t**ime. Prepare a chart to organize the information in the problem.

▶ **Helpful Hint**

If $d = r \cdot t$,

then $t = \dfrac{d}{r}$ ⎫
          ⎬
or *time* $= \dfrac{distance}{rate}$. ⎭

|  | **Distance** | **=** | **Rate** | **·** | **Time** |
|---|---|---|---|---|---|
| **Truck** | 120 | | $x$ | | $\begin{cases} \dfrac{120}{x} \leftarrow \text{distance} \\ \phantom{\dfrac{120}{x}} \leftarrow \text{rate} \end{cases}$ |
| **Car** | 180 | | $x + 20$ | | $\begin{cases} \dfrac{180}{x+20} \leftarrow \text{distance} \\ \phantom{\dfrac{180}{x+20}} \leftarrow \text{rate} \end{cases}$ |

**2. TRANSLATE.** Since the car and the truck traveled the same amount of time, we have that

In words:  | car's time | = | truck's time |

$$\downarrow \qquad\qquad \downarrow$$

Translate:  $\dfrac{180}{x + 20} = \dfrac{120}{x}$

**3. SOLVE.** We begin by multiplying both sides of the equation by the LCD, $x(x + 20)$, or cross multiplying.

$$\frac{180}{x + 20} = \frac{120}{x}$$

$$180x = 120(x + 20)$$

$$180x = 120x + 2400 \qquad \text{Use the distributive property.}$$

$$60x = 2400 \qquad\qquad \text{Subtract } 120x \text{ from both sides.}$$

$$x = 40 \qquad\qquad \text{Divide both sides by 60.}$$

**4. INTERPRET.** The speed of the truck is 40 miles per hour. The speed of the car must then be $x + 20$ or 60 miles per hour.

**Check:**  Find the time it takes the car to travel 180 miles and the time it takes the truck to travel 120 miles.

*Car's Time*                *Truck's Time*

$$t = \frac{d}{r} = \frac{180}{60} = 3 \text{ hours} \qquad t = \frac{d}{r} = \frac{120}{40} = 3 \text{ hours}$$

Since both travel the same amount of time, the proposed solution is correct.

**State:**  The car's speed is 60 miles per hour and the truck's speed is 40 miles per hour.

□

**PRACTICE**
**7**    A bus travels 180 miles in the same time that a car travels 240 miles. If the car's speed is 15 mph faster than the speed of the bus, find the speed of the car and the speed of the bus.

---

# VOCABULARY & READINESS CHECK

*Without solving algebraically, select the best choice for each exercise.*

**1.** One person can complete a job in 7 hours. A second person can complete the same job in 5 hours. How long will it take them to complete the job if they work together?

**a.** more than 7 hours
**b.** between 5 and 7 hours
**c.** less than 5 hours

**2.** One inlet pipe can fill a pond in 30 hours. A second inlet pipe can fill the same pond in 25 hours. How long before the pond is filled if both inlet pipes are on?

**a.** less than 25 hours
**b.** between 25 and 30 hours
**c.** more than 30 hours

---

## 7.6 | EXERCISE SET

*Solve each proportion. See Examples 1 and 2.*

**1.** $\dfrac{2}{3} = \dfrac{x}{6}$

**2.** $\dfrac{x}{2} = \dfrac{16}{6}$

**3.** $\dfrac{x}{10} = \dfrac{5}{9}$

**4.** $\dfrac{9}{4x} = \dfrac{6}{2}$

**5.** $\dfrac{x + 1}{2x + 3} = \dfrac{2}{3}$

**6.** $\dfrac{x + 1}{x + 2} = \dfrac{5}{3}$

**7.** $\dfrac{9}{5} = \dfrac{12}{3x + 2}$

**8.** $\dfrac{6}{11} = \dfrac{27}{3x - 2}$

*Solve. See Example 3.*

**9.** The ratio of the weight of an object on Earth to the weight of the same object on Pluto is 100 to 3. If an elephant weighs 4100 pounds on Earth, find the elephant's weight on Pluto.

**10.** If a 170-pound person weighs approximately 65 pounds on Mars, about how much does a 9000-pound satellite weigh? Round your answer to the nearest pound.

**11.** There are 110 calories per 28.8 grams of Frosted Flakes cereal. Find how many calories are in 43.2 grams of this cereal.

**12.** On an architect's blueprint, 1 inch corresponds to 4 feet. Find the length of a wall represented by a line that is $3\frac{7}{8}$ inches long on the blueprint.

*Find the unknown length x or y in the following pairs of similar triangles. See Example 4.*

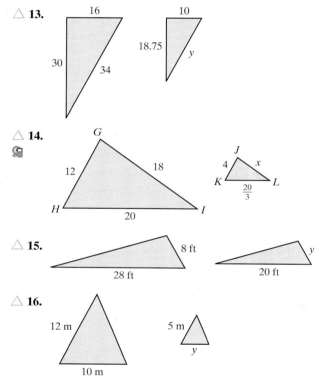

**13.**

**14.**

**15.**

**16.**

*Solve the following. See Example 5.*

**17.** Three times the reciprocal of a number equals 9 times the reciprocal of 6. Find the number.

**18.** Twelve divided by the sum of $x$ and 2 equals the quotient of 4 and the difference of $x$ and 2. Find $x$.

**19.** If twice a number added to 3 is divided by the number plus 1, the result is three halves. Find the number.

**20.** A number added to the product of 6 and the reciprocal of the number equals −5. Find the number.

*See Example 6.*

**21.** Smith Engineering found that an experienced surveyor surveys a roadbed in 4 hours. An apprentice surveyor needs 5 hours to survey the same stretch of road. If the two work together, find how long it takes them to complete the job.

**22.** An experienced bricklayer constructs a small wall in 3 hours. The apprentice completes the job in 6 hours. Find how long it takes if they work together.

**23.** In 2 minutes, a conveyor belt moves 300 pounds of recyclable aluminum from the delivery truck to a storage area. A smaller belt moves the same quantity of cans the same distance in 6 minutes. If both belts are used, find how long it takes to move the cans to the storage area.

**24.** Find how long it takes the conveyor belts described in Exercise 23 to move 1200 pounds of cans. (*Hint:* Think of 1200 pounds as four 300-pound jobs.)

*See Example 7.*

**25.** A jogger begins her workout by jogging to the park, a distance of 12 miles. She then jogs home at the same speed but along a different route. This return trip is 18 miles and her time is one hour longer. Find her jogging speed. Complete the accompanying chart and use it to find her jogging speed.

|  | Distance | = | Rate | · | Time |
|---|---|---|---|---|---|
| **Trip to Park** | 12 |  |  |  |  |
| **Return Trip** | 18 |  |  |  |  |

**26.** A boat can travel 9 miles upstream in the same amount of time it takes to travel 11 miles downstream. If the current of the river is 3 miles per hour, complete the chart below and use it to find the speed of the boat in still water.

|  | Distance | = | Rate | · | Time |
|---|---|---|---|---|---|
| **Upstream** | 9 |  | $r - 3$ |  |  |
| **Downstream** | 11 |  | $r + 3$ |  |  |

**27.** A cyclist rode the first 20-mile portion of his workout at a constant speed. For the 16-mile cooldown portion of his workout, he reduced his speed by 2 miles per hour. Each portion of the workout took the same time. Find the cyclist's speed during the first portion and find his speed during the cooldown portion.

**28.** A semi-truck travels 300 miles through the flatland in the same amount of time that it travels 180 miles through mountains. The rate of the truck is 20 miles per hour slower in the mountains than in the flatland. Find both the flatland rate and mountain rate.

**MIXED PRACTICE**

*Solve the following. See Examples 1 through 7.* (Note: *Some exercises can be modeled by equations without rational expressions.*)

**29.** A human factors expert recommends that there be at least 9 square feet of floor space in a college classroom for every student in the class. Find the minimum floor space that 40 students need.

**30.** Due to space problems at a local university, a 20-foot by 12-foot conference room is converted into a classroom. Find the maximum number of students the room can accommodate. (See Exercise 29.)

**31.** One-fourth equals the quotient of a number and 8. Find the number.

**32.** Four times a number added to 5 is divided by 6. The result is $\frac{7}{2}$. Find the number.

**33.** Marcus and Tony work for Lombardo's Pipe and Concrete. Mr. Lombardo is preparing an estimate for a customer. He knows that Marcus lays a slab of concrete in 6 hours. Tony lays the same size slab in 4 hours. If both work on the job and the cost of labor is $45.00 per hour, decide what the labor estimate should be.

**34.** Mr. Dodson can paint his house by himself in 4 days. His son needs an additional day to complete the job if he works by himself. If they work together, find how long it takes to paint the house.

**35.** A pilot can travel 400 miles with the wind in the same amount of time as 336 miles against the wind. Find the speed of the wind if the pilot's speed in still air is 230 miles per hour.

**36.** A fisherman on Pearl River rows 9 miles downstream in the same amount of time he rows 3 miles upstream. If the current is 6 miles per hour, find how long it takes him to cover the 12 miles.

**37.** Find the unknown length $y$.

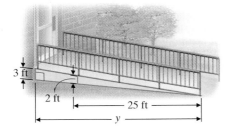

**△ 38.** Find the unknown length $y$.

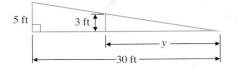

**39.** Ken Hall, a tailback, holds the high school sports record for total yards rushed in a season. In 1953, he rushed for 4045 total yards in 12 games. Find his average rushing yards per game. Round your answer to the nearest whole yard.

**40.** To estimate the number of people in Jackson, population 50,000, who have no health insurance, 250 people were polled. Of those polled, 39 had no insurance. How many people in the city might we expect to be uninsured?

**41.** Two divided by the difference of a number and 3 minus 4 divided by a number plus 3, equals 8 times the reciprocal of the difference of the number squared and 9. What is the number?

**42.** If 15 times the reciprocal of a number is added to the ratio of 9 times a number minus 7 and the number plus 2, the result is 9. What is the number?

**43.** A pilot flies 630 miles with a tail wind of 35 miles per hour. Against the wind, he flies only 455 miles in the same amount of time. Find the rate of the plane in still air.

**44.** A marketing manager travels 1080 miles in a corporate jet and then an additional 240 miles by car. If the car ride takes one hour longer than the jet ride takes, and if the rate of the jet is 6 times the rate of the car, find the time the manager travels by jet and find the time the manager travels by car.

**45.** To mix weed killer with water correctly, it is necessary to mix 8 teaspoons of weed killer with 2 gallons of water. Find how many gallons of water are needed to mix with the entire box if it contains 36 teaspoons of weed killer.

**46.** The directions for a certain bug spray concentrate is to mix 3 ounces of concentrate with 2 gallons of water. How many ounces of concentrate are needed to mix with 5 gallons of water?

**47.** A boater travels 16 miles per hour on the water on a still day. During one particular windy day, he finds that he travels 48 miles with the wind behind him in the same amount of time that he travels 16 miles into the wind. Find the rate of the wind.

Let $x$ be the rate of the wind.

|  | $r$ | × | $t$ | = | $d$ |
|---|---|---|---|---|---|
| ***with wind*** | $16 + x$ |  |  |  | 48 |
| ***into wind*** | $16 - x$ |  |  |  | 16 |

**48.** The current on a portion of the Mississippi River is 3 miles per hour. A barge can go 6 miles upstream in the same amount of time it takes to go 10 miles downstream. Find the speed of the boat in still water.

Let $x$ be the speed of the boat in still water.

|  | $r$ | × | $t$ | = | $d$ |
|---|---|---|---|---|---|
| ***upstream*** | $x - 3$ |  |  |  | 6 |
| ***downstream*** | $x + 3$ |  |  |  | 10 |

**49.** The best selling two-seater sports car is the Mazda Miata. A driver of this car took a day-trip around the California coastline driving at two different speeds. He drove 70 miles at a slower speed and 300 miles at a speed 40 miles per hour faster. If the time spent during the faster speed was twice that spent at a slower speed, find the two speeds during the trip. (*Source: Guinness World Records*)

**50.** Currently, the Toyota Corolla is the most produced car in the world. Suppose that during a drive test of two Corollas, one car travels 224 miles in the same time that the second car travels 175 miles. If the speed of one car is 14 miles per hour faster than the speed of the second car, find the speed of both cars. (*Source: Guinness World Records*)

**51.** One custodian cleans a suite of offices in 3 hours. When a second worker is asked to join the regular custodian, the job takes only $1\frac{1}{2}$ hours. How long does it take the second worker to do the same job alone?

**52.** One person proofreads a copy for a small newspaper in 4 hours. If a second proofreader is also employed, the job can be done in $2\frac{1}{2}$ hours. How long does it take for the second proofreader to do the same job alone?

△ **53.** An architect is completing the plans for a triangular deck. Use the diagram below to find the missing dimension.

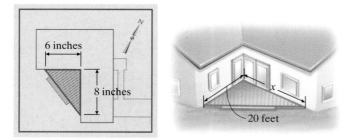

△ **54.** A student wishes to make a small model of a triangular mainsail in order to study the effects of wind on the sail. The smaller model will be the same shape as a regular-size sailboat's mainsail. Use the following diagram to find the missing dimensions.

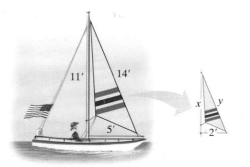

**55.** The manufacturers of cans of salted mixed nuts state that the ratio of peanuts to other nuts is 3 to 2. If 324 peanuts are in a can, find how many other nuts should also be in the can.

**56.** There are 1280 calories in a 14-ounce portion of Eagle Brand Milk. Find how many calories are in 2 ounces of Eagle Brand Milk.

**57.** A pilot can fly an MD-11 2160 miles with the wind in the same time as she can fly 1920 miles against the wind. If the speed of the wind is 30 mph, find the speed of the plane in still air. (*Source*: Air Transport Association of America)

**58.** A pilot can fly a DC-10 1365 miles against the wind in the same time as he can fly 1575 miles with the wind. If the speed of the plane in still air is 490 miles per hour, find the speed of the wind. (*Source*: Air Transport Association of America)

**59.** One pipe fills a storage pool in 20 hours. A second pipe fills the same pool in 15 hours. When a third pipe is added and all three are used to fill the pool, it takes only 6 hours. Find how long it takes the third pipe to do the job.

**60.** One pump fills a tank 2 times as fast as another pump. If the pumps work together, they fill the tank in 18 minutes. How long does it take for each pump to fill the tank?

**61.** A car travels 280 miles in the same time that a motorcycle travels 240 miles. If the car's speed is 10 miles per hour more than the motorcycle's, find the speed of the car and the speed of the motorcycle.

**62.** A walker travels 3.6 miles in the same time that a jogger travels 6 miles. If the walker's speed is 2 miles per hour less than the jogger's, find the speed of the walker and the speed of the jogger.

**63.** In 6 hours, an experienced cook prepares enough pies to supply a local restaurant's daily order. Another cook prepares the same number of pies in 7 hours. Together with a third cook, they prepare the pies in 2 hours. Find how long it takes the third cook to prepare the pies alone.

**64.** It takes 9 hours for pump A to fill a tank alone. Pump B takes 15 hours to fill the same tank alone. If pumps A, B, and C are used, the tank fills in 5 hours. How long does it take pump C to fill the tank alone?

**65.** One pump fills a tank 3 times as fast as another pump. If the pumps work together, they fill the tank in 21 minutes. How long does it take for each pump to fill the tank?

**66.** Mrs. Smith balances the company books in 8 hours. It takes her assistant 12 hours to do the same job. If they work together, find how long it takes them to balance the books.

*Given that the following pairs of triangles are similar, find each missing length.*

△ **67.**

△ **68.**

△ **69.**

△ **70.**

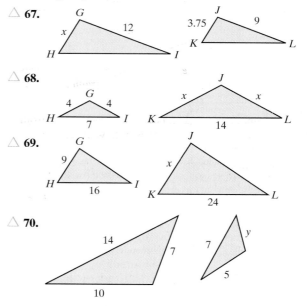

**REVIEW AND PREVIEW**

*Find the slope of the line through each pair of points. Use the slope to determine whether the line is vertical, horizontal, or moves upward or downward from left to right. See Section 3.4.*

**71.** $(-2, 5), (4, -3)$

**72.** $(0, 4), (2, 10)$

*Simplify. Follow the circled steps in the order shown. See Section 1.3.*

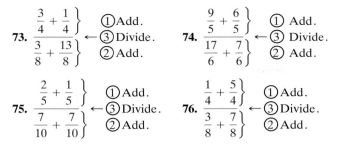

**73.** $\dfrac{\dfrac{3}{4}+\dfrac{1}{4}}{\dfrac{3}{8}+\dfrac{13}{8}}$  ←③Add. ①Add. ②Divide. ②Add.

**74.** $\dfrac{\dfrac{9}{5}+\dfrac{6}{5}}{\dfrac{17}{6}+\dfrac{7}{6}}$  ←③Add. ① Add. ③ Divide. ② Add.

**75.** $\dfrac{\dfrac{2}{5}+\dfrac{1}{5}}{\dfrac{7}{10}+\dfrac{7}{10}}$  ←③Add. ①Add. ③Divide. ②Add.

**76.** $\dfrac{\dfrac{1}{4}+\dfrac{5}{4}}{\dfrac{3}{8}+\dfrac{7}{8}}$  ←③Add. ①Add. ③Divide. ②Add.

## CONCEPT EXTENSIONS

*The following bar graph shows the capacity of the United States to generate electricity from the wind in the years shown. Use this graph for Exercises 77 and 78.*

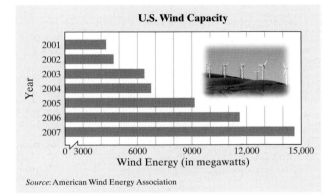

**U.S. Wind Capacity**

*Source*: American Wind Energy Association

**77.** Find the approximate increase in megawatt capacity during the 2-year period from 2001 to 2003.

**78.** Find the approximate increase in megawatt capacity during the 2-year period from 2004 to 2006.

*In general, 1000 megawatts will serve the average electricity needs of 560,000 people. Use this fact and the preceding graph to answer Exercises 79 and 80.*

**79.** In 2007, the number of megawatts that were generated from wind would serve the electricity needs of how many people? (Round to the nearest ten-thousand.)

**80.** How many megawatts of electricity are needed to serve the city or town in which you live?

**81.** Person A can complete a job in 5 hours, and person B can complete the same job in 3 hours. Without solving algebraically, discuss reasonable and unreasonable answers for how long it would take them to complete the job together.

**82.** For which of the following equations can we immediately use cross products to solve for $x$?

**a.** $\dfrac{2-x}{5}=\dfrac{1+x}{3}$   **b.** $\dfrac{2}{5}-x=\dfrac{1+x}{3}$

**83.** For what value of $x$ is $\dfrac{x}{x-1}$ in proportion to $\dfrac{x+1}{x}$? Explain your result.

**84.** If $x$ is 10, is $\dfrac{2}{x}$ in proportion to $\dfrac{x}{50}$? Explain why or why not.

One of the great algebraists of ancient times was a man named *Diophantus. Little is known of his life other than that he lived and worked in Alexandria. Some historians believe he lived during the first century of the Christian era, about the time of Nero. The only clue to his personal life is the following epigram found in a collection called the Palatine Anthology.*

God granted him youth for a sixth of his life and added a twelfth part to this. He clothed his cheeks in down. He lit him the light of wedlock after a seventh part and five years after his marriage, He granted him a son. Alas, lateborn wretched child. After attaining the measure of half his father's life, cruel fate overtook him, thus leaving Diophantus during the last four years of his life only such consolation as the science of numbers. How old was Diophantus at his death?*

We are looking for Diophantus' age when he died, so let $x$ represent that age. If we sum the parts of his life, we should get the total age.

Parts of his life $\begin{cases} \dfrac{1}{6}x \ + \ \dfrac{1}{12}x \text{ is the time of his youth.} \\[4pt] \dfrac{1}{7}x \text{ is the time between his youth and when he married.} \\[4pt] 5 \text{ years is the time between his marriage and the birth of his son.} \\[4pt] \dfrac{1}{2}x \text{ is the time Diophantus had with his son.} \\[4pt] 4 \text{ years is the time between his son's death and his own.} \end{cases}$

The sum of these parts should equal Diophantus' age when he died.

$$\dfrac{1}{6}\cdot x + \dfrac{1}{12}\cdot x + \dfrac{1}{7}\cdot x + 5 + \dfrac{1}{2}\cdot x + 4 = x$$

**85.** Solve the epigram.

**86.** How old was Diophantus when his son was born? How old was the son when he died?

**87.** Solve the following epigram:

I was four when my mother packed my lunch and sent me off to school. Half my life was spent in school and another sixth was spent on a farm. Alas, hard times befell me. My crops and cattle fared poorly and my land was sold. I returned to school for 3 years and have spent one tenth of my life teaching. How old am I?

**88.** Write an epigram describing your life. Be sure that none of the time periods in your epigram overlap.

**89.** A hyena spots a giraffe 0.5 mile away and begins running toward it. The giraffe starts running away from the hyena just as the hyena begins running toward it. A hyena can run at a speed of 40 mph and a giraffe can run at 32 mph. How long will it take for the hyena to overtake the giraffe? (*Source: World Almanac* and *Book of Facts*)

H                                    G

|← —————— 0.5 mile —————— →|

*From *The Nature and Growth of Modern Mathematics*, Edna Kramer, 1970, Fawcett Premier Books, Vol. 1, pages 107–108.

## THE BIGGER PICTURE   SIMPLIFYING EXPRESSIONS AND SOLVING EQUATIONS

Now we continue our outline from Sections 1.7, 2.3, 2.8, 5.6, 6.6, and 7.4. Although suggestions are given, this outline should be in your own words. Once you complete this new portion, try the exercises below.

**I.** Simplifying Expressions
   **A.** Real Numbers
      **1.** Add (Section 1.5)
      **2.** Subtract (Section 1.6)
      **3.** Multiply or Divide (Section 1.7)
   **B.** Exponents (Section 5.1 and 5.5)
   **C.** Polynomials
      **1.** Add (Section 5.2)
      **2.** Subtract (Section 5.2)
      **3.** Multiply (Section 5.3 and 5.4)
      **4.** Divide (Section 5.6)
   **D.** Factoring Polynomials (Chapter 6 Integrated Review)
   **E.** Rational Expressions
      **1.** Simplify (Section 7.1)
      **2.** Multiply (Section 7.2)
      **3.** Divide (Section 7.2)
      **4.** Add or Subtract (Section 7.4)

**II.** Solving Equations
   **A.** Linear Equations (Section 2.3)
   **B.** Quadratic and Higher Degree Equations (Section 6.6)
   **C.** Equations with Rational Expressions—solving equations with rational expressions

$$\frac{3}{x} - \frac{1}{x-1} = \frac{4}{x-1} \quad \text{Equation with rational expressions.}$$

$$x(x-1)\cdot\frac{3}{x} - x(x-1)\frac{1}{x-1} \quad \text{Multiply through by } x(x-1).$$

$$= x(x-1)\frac{4}{x-1}$$

$$3(x-1) - x\cdot 1 = x\cdot 4 \quad \text{Simplify.}$$

$$3x - 3 - x = 4x \quad \text{Use the distributive property.}$$

$$-3 = 2x \quad \text{Simplify and move variable terms to right side.}$$

$$-\frac{3}{2} = x \quad \text{Divide both sides by 2.}$$

   **D.** Proportions—an equation with two ratios equal. Set cross products equal, then solve.

$$\frac{5}{x} = \frac{9}{2x-3}, \text{ or } 5(2x-3) = 9\cdot x$$
$$\text{or } 10x - 15 = 9x \text{ or } x = 15$$

**III.** Solving Inequalities
   **A.** Linear Inequalities (Section 2.8)

*Multiply.*
**1.** $(3x - 2)(4x^2 - x - 5)$
**2.** $(2x - y)^2$

*Factor.*
**3.** $8y^3 - 20y^5$
**4.** $9m^2 - 11mn + 2n^2$

*Simplify or solve.*

*If an expression, perform indicated operations and simplify. If an equation or inequality, solve it.*

**5.** $\dfrac{7}{x} = \dfrac{9}{x-10}$

**6.** $\dfrac{7}{x} + \dfrac{9}{x-10}$

**7.** $(-3x^5)\left(\dfrac{1}{2}x^7\right)(8x)$

**8.** $5x - 1 = |-4| + |-5|$

**9.** $\dfrac{8 - 12}{12 \div 3 \cdot 2}$

**10.** $-2(3y - 4) \le 5y - 7 - 7y - 1$

**11.** $\dfrac{7}{x} + \dfrac{5}{2x+3} = \dfrac{-2}{x}$

**12.** $\dfrac{(a^{-3}b^2)^{-5}}{ab^4}$

## 7.7 SIMPLIFYING COMPLEX FRACTIONS

**OBJECTIVES**

1 Simplify complex fractions by simplifying the numerator and denominator and then dividing.

2 Simplify complex fractions by multiplying by a common denominator.

3 Simplify expressions with negative exponents.

**OBJECTIVE 1 ▶ Simplifying complex fractions: Method 1.** A rational expression whose numerator, denominator, or both contain one or more rational expressions is called a **complex rational expression** or a **complex fraction.**

***Complex Fractions***

$$\frac{\dfrac{1}{a}}{\dfrac{b}{2}} \qquad \frac{\dfrac{x}{2y^2}}{\dfrac{6x-2}{9y}} \qquad \frac{x+\dfrac{1}{y}}{y+1}$$

The parts of a complex fraction are

$$\left.\begin{array}{c} \dfrac{x}{y+2} \end{array}\right\} \leftarrow \text{Numerator of complex fraction}$$
$$\phantom{xxxxxx} \leftarrow \text{Main fraction bar}$$
$$\left.7 + \dfrac{1}{y}\right\} \leftarrow \text{Denominator of complex fraction}$$

Our goal in this section is to simplify complex fractions. A complex fraction is simplified when it is in the form $\dfrac{P}{Q}$, where $P$ and $Q$ are polynomials that have no common factors. Two methods of simplifying complex fractions are introduced. The first method evolves from the definition of a fraction as a quotient.

---

**Simplifying a Complex Fraction: Method I**

**STEP 1.** Simplify the numerator and the denominator of the complex fraction so that each is a single fraction.

**STEP 2.** Perform the indicated division by multiplying the numerator of the complex fraction by the reciprocal of the denominator of the complex fraction.

**STEP 3.** Simplify if possible.

---

**EXAMPLE 1** Simplify each complex fraction.

**a.** $\dfrac{\dfrac{2x}{27y^2}}{\dfrac{6x^2}{9}}$
  **b.** $\dfrac{\dfrac{5x}{x+2}}{\dfrac{10}{x-2}}$
  **c.** $\dfrac{\dfrac{x}{y^2}+\dfrac{1}{y}}{\dfrac{y}{x^2}+\dfrac{1}{x}}$

*Solution*

**a.** The numerator of the complex fraction is already a single fraction, and so is the denominator. Perform the indicated division by multiplying the numerator, $\dfrac{2x}{27y^2}$, by the reciprocal of the denominator, $\dfrac{6x^2}{9}$. Then simplify.

$$\dfrac{\dfrac{2x}{27y^2}}{\dfrac{6x^2}{9}} = \dfrac{2x}{27y^2} \div \dfrac{6x^2}{9}$$

$$= \dfrac{2x}{27y^2} \cdot \dfrac{9}{6x^2} \qquad \text{Multiply by the reciprocal of } \dfrac{6x^2}{9}.$$

$$= \dfrac{2x \cdot 9}{27y^2 \cdot 6x^2}$$

$$= \dfrac{1}{9xy^2}$$

> **▶ Helpful Hint**
> Both the numerator and denominator are single fractions, so we perform the indicated division.

**b.** $\dfrac{\dfrac{5x}{x+2}}{\dfrac{10}{x-2}} = \dfrac{5x}{x+2} \div \dfrac{10}{x-2} = \dfrac{5x}{x+2} \cdot \dfrac{x-2}{10}$  Multiply by the reciprocal of $\dfrac{10}{x-2}$.

$$= \dfrac{5x(x-2)}{2 \cdot 5(x+2)}$$

$$= \dfrac{x(x-2)}{2(x+2)} \qquad\qquad \text{Simplify.}$$

**c.** First simplify the numerator and the denominator of the complex fraction separately so that each is a single fraction. Then perform the indicated division.

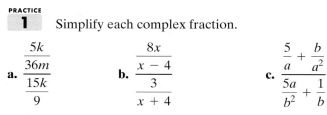

$$\frac{\dfrac{x}{y^2} + \dfrac{1}{y}}{\dfrac{y}{x^2} + \dfrac{1}{x}} = \frac{\dfrac{x}{y^2} + \dfrac{1 \cdot y}{y \cdot y}}{\dfrac{y}{x^2} + \dfrac{1 \cdot x}{x \cdot x}}$$

Simplify the numerator. The LCD is $y^2$.

Simplify the denominator. The LCD is $x^2$.

$$= \frac{\dfrac{x + y}{y^2}}{\dfrac{y + x}{x^2}}$$   Add.

$$= \frac{x + y}{y^2} \cdot \frac{x^2}{y + x}$$   Multiply by the reciprocal of $\dfrac{y + x}{x^2}$.

$$= \frac{x^2 (x + y)}{y^2 (y + x)}$$

$$= \frac{x^2}{y^2}$$   Simplify.   □

**PRACTICE**

**1**  Simplify each complex fraction.

**a.** $\dfrac{\dfrac{5k}{36m}}{\dfrac{15k}{9}}$

**b.** $\dfrac{\dfrac{8x}{x - 4}}{\dfrac{3}{x + 4}}$

**c.** $\dfrac{\dfrac{5}{a} + \dfrac{b}{a^2}}{\dfrac{5a}{b^2} + \dfrac{1}{b}}$

**Concept Check** ☑

Which of the following are equivalent to $\dfrac{\dfrac{1}{x}}{\dfrac{3}{y}}$?

**a.** $\dfrac{1}{x} \div \dfrac{3}{y}$

**b.** $\dfrac{1}{x} \cdot \dfrac{y}{3}$

**c.** $\dfrac{1}{x} \div \dfrac{y}{3}$

**OBJECTIVE 2 ▶ Simplifying complex fractions: Method 2.** Next we look at another method of simplifying complex fractions. With this method we multiply the numerator and the denominator of the complex fraction by the LCD of all fractions in the complex fraction.

> **Simplifying a Complex Fraction: Method II**
> **STEP 1.** Multiply the numerator and the denominator of the complex fraction by the LCD of the fractions in both the numerator and the denominator.
> **STEP 2.** Simplify.

**Answer to Concept Check:**
a and b

**EXAMPLE 2**   Simplify each complex fraction.

a. $\dfrac{\dfrac{5x}{x+2}}{\dfrac{10}{x-2}}$

b. $\dfrac{\dfrac{x}{y^2}+\dfrac{1}{y}}{\dfrac{y}{x^2}+\dfrac{1}{x}}$

_Solution_

**a.** The least common denominator of $\dfrac{5x}{x+2}$ and $\dfrac{10}{x-2}$ is $(x+2)(x-2)$. Multiply both the numerator, $\dfrac{5x}{x+2}$, and the denominator, $\dfrac{10}{x-2}$, by the LCD.

$$\dfrac{\dfrac{5x}{x+2}}{\dfrac{10}{x-2}}=\dfrac{\left(\dfrac{5x}{x+2}\right)\cdot(x+2)(x-2)}{\left(\dfrac{10}{x-2}\right)\cdot(x+2)(x-2)}$$   Multiply numerator and denominator by the LCD.

$$=\dfrac{5x\cdot(x-2)}{2\cdot5\cdot(x+2)}$$   Simplify.

$$=\dfrac{x(x-2)}{2(x+2)}$$   Simplify.

**b.** The least common denominator of $\dfrac{x}{y^2},\dfrac{1}{y},\dfrac{y}{x^2}$, and $\dfrac{1}{x}$ is $x^2y^2$.

$$\dfrac{\dfrac{x}{y^2}+\dfrac{1}{y}}{\dfrac{y}{x^2}+\dfrac{1}{x}}=\dfrac{\left(\dfrac{x}{y^2}+\dfrac{1}{y}\right)\cdot x^2y^2}{\left(\dfrac{y}{x^2}+\dfrac{1}{x}\right)\cdot x^2y^2}$$   Multiply the numerator and denominator by the LCD.

$$=\dfrac{\dfrac{x}{y^2}\cdot x^2y^2+\dfrac{1}{y}\cdot x^2y^2}{\dfrac{y}{x^2}\cdot x^2y^2+\dfrac{1}{x}\cdot x^2y^2}$$   Use the distributive property.

$$=\dfrac{x^3+x^2y}{y^3+xy^2}$$   Simplify.

$$=\dfrac{x^2(x+y)}{y^2(y+x)}$$   Factor.

$$=\dfrac{x^2}{y^2}$$   Simplify.   □

**PRACTICE**
**2**   Use Method 2 to simplify:

a. $\dfrac{\dfrac{8x}{x-4}}{\dfrac{3}{x+4}}$

b. $\dfrac{\dfrac{b}{a^2}+\dfrac{1}{a}}{\dfrac{a}{b^2}+\dfrac{1}{b}}$

**OBJECTIVE 3 ▶ Simplifying expressions with negative exponents.** If an expression contains negative exponents, write the expression as an equivalent expression with positive exponents.

**EXAMPLE 3** Simplify.

$$\frac{x^{-1} + 2xy^{-1}}{x^{-2} - x^{-2}y^{-1}}$$

**Solution** This fraction does not appear to be a complex fraction. If we write it by using only positive exponents, however, we see that it is a complex fraction.

$$\frac{x^{-1} + 2xy^{-1}}{x^{-2} - x^{-2}y^{-1}} = \frac{\dfrac{1}{x} + \dfrac{2x}{y}}{\dfrac{1}{x^2} - \dfrac{1}{x^2y}}$$

The LCD of $\dfrac{1}{x}$, $\dfrac{2x}{y}$, $\dfrac{1}{x^2}$, and $\dfrac{1}{x^2y}$ is $x^2y$. Multiply both the numerator and denominator by $x^2y$.

$$= \frac{\left(\dfrac{1}{x} + \dfrac{2x}{y}\right) \cdot x^2y}{\left(\dfrac{1}{x^2} - \dfrac{1}{x^2y}\right) \cdot x^2y}$$

$$= \frac{\dfrac{1}{x} \cdot x^2y + \dfrac{2x}{y} \cdot x^2y}{\dfrac{1}{x^2} \cdot x^2y - \dfrac{1}{x^2y} \cdot x^2y} \qquad \text{Apply the distributive property.}$$

$$= \frac{xy + 2x^3}{y - 1} \quad \text{or} \quad \frac{x(y + 2x^2)}{y - 1} \qquad \text{Simplify.} \qquad \square$$

**PRACTICE**
**3** Simplify: $\dfrac{3x^{-1} + x^{-2}y^{-1}}{y^{-2} + xy^{-1}}$.

**EXAMPLE 4** Simplify: $\dfrac{(2x)^{-1} + 1}{2x^{-1} - 1}$

**Solution** 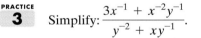 $\dfrac{(2x)^{-1} + 1}{2x^{-1} - 1} = \dfrac{\dfrac{1}{2x} + 1}{\dfrac{2}{x} - 1}$     Write using positive exponents.

> **Helpful Hint**
>
> Don't forget that $(2x)^{-1} = \dfrac{1}{2x}$, but $2x^{-1} = 2 \cdot \dfrac{1}{x} = \dfrac{2}{x}$.

$$= \frac{\left(\dfrac{1}{2x} + 1\right) \cdot 2x}{\left(\dfrac{2}{x} - 1\right) \cdot 2x} \qquad \text{The LDC of } \dfrac{1}{2x} \text{ and } \dfrac{2}{x} \text{ is } 2x.$$

$$= \frac{\dfrac{1}{2x} \cdot 2x + 1 \cdot 2x}{\dfrac{2}{x} \cdot 2x - 1 \cdot 2x} \qquad \text{Use distributive property.}$$

$$= \frac{1 + 2x}{4 - 2x} \quad \text{or} \quad \frac{1 + 2x}{2(2 - x)} \qquad \text{Simplify.} \qquad \square$$

**PRACTICE**
**4** Simplify: $\dfrac{(3x)^{-1} - 2}{5x^{-1} + 2}$.

## VOCABULARY & READINESS CHECK

*Complete the steps by writing the simplified complex fraction.*

**1.** $\dfrac{\dfrac{7}{x}}{\dfrac{1}{x} + \dfrac{z}{x}} = \dfrac{x\left(\dfrac{7}{x}\right)}{x\left(\dfrac{1}{x}\right) + x\left(\dfrac{z}{x}\right)} = $ _____

**2.** $\dfrac{\dfrac{x}{4}}{\dfrac{x^2}{2} + \dfrac{1}{4}} = \dfrac{4\left(\dfrac{x}{4}\right)}{4\left(\dfrac{x^2}{2}\right) + 4\left(\dfrac{1}{4}\right)} = $ _____

*Write each with positive exponents.*

**3.** $x^{-2} =$ _____

**4.** $y^{-3} =$ _____

**5.** $2x^{-1} =$ _____

**6.** $(2x)^{-1} =$ _____

**7.** $(9y)^{-1} =$ _____

**8.** $9y^{-2} =$ _____

## 7.7 EXERCISE SET

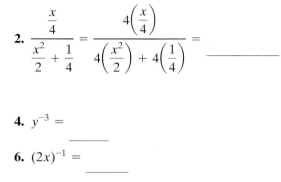

*Simplify each complex fraction. See Examples 1 and 2.*

**1.** $\dfrac{\dfrac{10}{3x}}{\dfrac{5}{6x}}$

**2.** $\dfrac{\dfrac{15}{2x}}{\dfrac{5}{6x}}$

**3.** $\dfrac{1 + \dfrac{2}{5}}{2 + \dfrac{3}{5}}$

**4.** $\dfrac{2 + \dfrac{1}{7}}{3 - \dfrac{4}{7}}$

**5.** $\dfrac{\dfrac{4}{x - 1}}{\dfrac{x}{x - 1}}$

**6.** $\dfrac{\dfrac{x}{x + 2}}{\dfrac{2}{x + 2}}$

**7.** $\dfrac{1 - \dfrac{2}{x}}{x + \dfrac{4}{9x}}$

**8.** $\dfrac{5 - \dfrac{3}{x}}{x + \dfrac{2}{3x}}$

**9.** $\dfrac{\dfrac{4x^2 - y^2}{xy}}{\dfrac{2}{y} - \dfrac{1}{x}}$

**10.** $\dfrac{\dfrac{x^2 - 9y^2}{xy}}{\dfrac{1}{y} - \dfrac{3}{x}}$

**11.** $\dfrac{\dfrac{x + 1}{3}}{\dfrac{2x - 1}{6}}$

**12.** $\dfrac{\dfrac{x + 3}{12}}{\dfrac{4x - 5}{15}}$

**13.** $\dfrac{\dfrac{2}{x} + \dfrac{3}{x^2}}{\dfrac{4}{x^2} - \dfrac{9}{x}}$

**14.** $\dfrac{\dfrac{2}{x^2} + \dfrac{1}{x}}{\dfrac{4}{x^2} - \dfrac{1}{x}}$

**15.** $\dfrac{\dfrac{1}{x} + \dfrac{2}{x^2}}{x + \dfrac{8}{x^2}}$

**16.** $\dfrac{\dfrac{1}{y} + \dfrac{3}{y^2}}{y + \dfrac{27}{y^2}}$

**17.** $\dfrac{\dfrac{4}{5 - x} + \dfrac{5}{x - 5}}{\dfrac{2}{x} + \dfrac{3}{x - 5}}$

**18.** $\dfrac{\dfrac{3}{x - 4} - \dfrac{2}{4 - x}}{\dfrac{2}{x - 4} - \dfrac{2}{x}}$

**19.** $\dfrac{\dfrac{x + 2}{x} - \dfrac{2}{x - 1}}{\dfrac{x + 1}{x} + \dfrac{x + 1}{x - 1}}$

**20.** $\dfrac{\dfrac{5}{a + 2} - \dfrac{1}{a - 2}}{\dfrac{3}{2 + a} + \dfrac{6}{2 - a}}$

**21.** $\dfrac{\dfrac{2}{x} + 3}{\dfrac{4}{x^2} - 9}$

**22.** $\dfrac{2 + \dfrac{1}{x}}{4x - \dfrac{1}{x}}$

**23.** $\dfrac{1 - \dfrac{x}{y}}{\dfrac{x^2}{y^2} - 1}$

**24.** $\dfrac{1 - \dfrac{2}{x}}{x - \dfrac{4}{x}}$

**25.** $\dfrac{\dfrac{-2x}{x - y}}{\dfrac{y}{x^2}}$

**26.** $\dfrac{\dfrac{7y}{x^2 + xy}}{\dfrac{y^2}{x^2}}$

**27.** $\dfrac{\dfrac{2}{x} + \dfrac{1}{x^2}}{\dfrac{y}{x^2}}$

**28.** $\dfrac{\dfrac{5}{x^2} - \dfrac{2}{x}}{\dfrac{1}{x} + 2}$

**29.** $\dfrac{\dfrac{x}{9} - \dfrac{1}{x}}{1 + \dfrac{3}{x}}$

**30.** $\dfrac{\dfrac{x}{4} - \dfrac{4}{x}}{1 - \dfrac{4}{x}}$

**31.** $\dfrac{\dfrac{x - 1}{x^2 - 4}}{1 + \dfrac{1}{x - 2}}$

**32.** $\dfrac{\dfrac{x + 3}{x^2 - 9}}{1 + \dfrac{1}{x - 3}}$

**33.** $\dfrac{\dfrac{2}{x+5} + \dfrac{4}{x+3}}{\dfrac{3x+13}{x^2+8x+15}}$

**34.** $\dfrac{\dfrac{2}{x+2} + \dfrac{6}{x+7}}{\dfrac{4x+13}{x^2+9x+14}}$

*Simplify. See Examples 3 and 4.*

**35.** $\dfrac{x^{-1}}{x^{-2}+y^{-2}}$

**36.** $\dfrac{a^{-3}+b^{-1}}{a^{-2}}$

**37.** $\dfrac{2a^{-1}+3b^{-2}}{a^{-1}-b^{-1}}$

**38.** $\dfrac{x^{-1}+y^{-1}}{3x^{-2}+5y^{-2}}$

**39.** $\dfrac{1}{x-x^{-1}}$

**40.** $\dfrac{x^{-2}}{x+3x^{-1}}$

**41.** $\dfrac{a^{-1}+1}{a^{-1}-1}$

**42.** $\dfrac{a^{-1}-4}{4+a^{-1}}$

**43.** $\dfrac{3x^{-1}+(2y)^{-1}}{x^{-2}}$

**44.** $\dfrac{5x^{-2}-3y^{-1}}{x^{-1}+y^{-1}}$

**45.** $\dfrac{2a^{-1}+(2a)^{-1}}{a^{-1}+2a^{-2}}$

**46.** $\dfrac{a^{-1}+2a^{-2}}{2a^{-1}+(2a)^{-1}}$

**47.** $\dfrac{5x^{-1}+2y^{-1}}{x^{-2}y^{-2}}$

**48.** $\dfrac{x^{-2}y^{-2}}{5x^{-1}+2y^{-1}}$

**49.** $\dfrac{5x^{-1}-2y^{-1}}{25x^{-2}-4y^{-2}}$

**50.** $\dfrac{3x^{-1}+3y^{-1}}{4x^{-2}-9y^{-2}}$

## REVIEW AND PREVIEW

*Simplify. See Sections 5.1 and 5.5.*

**51.** $\dfrac{3x^3y^2}{12x}$

**52.** $\dfrac{-36xb^3}{9xb^2}$

**53.** $\dfrac{144x^5y^5}{-16x^2y}$

**54.** $\dfrac{48x^3y^2}{-4xy}$

*Solve the following. See Section 3.6.*

**55.** If $P(x) = -x^2$, find $P(-3)$.

**56.** If $f(x) = x^2 - 6$, find $f(-1)$.

## CONCEPT EXTENSIONS

*Solve. See the Concept Check in the Section.*

**57.** Which of the following are equivalent to $\dfrac{\dfrac{x+1}{9}}{\dfrac{y-2}{5}}$?

**a.** $\dfrac{x+1}{9} \div \dfrac{y-2}{5}$  **b.** $\dfrac{x+1}{9} \cdot \dfrac{y-2}{5}$  **c.** $\dfrac{x+1}{9} \cdot \dfrac{5}{y-2}$

**58.** Which of the following are equivalent to $\dfrac{\dfrac{a}{7}}{\dfrac{b}{13}}$?

**a.** $\dfrac{a}{7} \cdot \dfrac{b}{13}$  **b.** $\dfrac{a}{7} \div \dfrac{b}{13}$  **c.** $\dfrac{a}{7} \div \dfrac{13}{b}$  **d.** $\dfrac{a}{7} \cdot \dfrac{13}{b}$

**59.** When the source of a sound is traveling toward a listener, the pitch that the listener hears due to the Doppler effect is given by the complex rational compression $\dfrac{a}{1 - \dfrac{s}{770}}$, where $a$ is the

actual pitch of the sound and $s$ is the speed of the sound source. Simplify this expression.

**60.** In baseball, the earned run average (ERA) statistic gives the average number of earned runs scored on a pitcher per game. It is computed with the following expression: $\dfrac{E}{\dfrac{I}{9}}$, where $E$ is the number of earned runs scored on a pitcher and $I$ is the total number of innings pitched by the pitcher. Simplify this expression.

**61.** Which of the following are equivalent to $\dfrac{\dfrac{1}{x}}{\dfrac{3}{y}}$?

**a.** $\dfrac{1}{x} \div \dfrac{3}{y}$  **b.** $\dfrac{1}{x} \cdot \dfrac{y}{3}$  **c.** $\dfrac{1}{x} \div \dfrac{y}{3}$

**62.** In your own words, explain one method for simplifying a complex fraction.

*Simplify.*

**63.** $\dfrac{1}{1 + (1+x)^{-1}}$

**64.** $\dfrac{(x+2)^{-1} + (x-2)^{-1}}{(x^2-4)^{-1}}$

**65.** $\dfrac{x}{1 - \dfrac{1}{1 + \dfrac{1}{x}}}$

**66.** $\dfrac{x}{1 - \dfrac{1}{1 - \dfrac{1}{x}}}$

**67.** $\dfrac{\dfrac{2}{y^2} - \dfrac{5}{xy} - \dfrac{3}{x^2}}{\dfrac{2}{y^2} + \dfrac{7}{xy} + \dfrac{3}{x^2}}$

**68.** $\dfrac{\dfrac{2}{x^2} - \dfrac{1}{xy} - \dfrac{1}{y^2}}{\dfrac{1}{x^2} - \dfrac{3}{xy} + \dfrac{2}{y^2}}$

**69.** $\dfrac{3(a + 1)^{-1} + 4a^{-2}}{(a^3 + a^2)^{-1}}$

**70.** $\dfrac{9x^{-1} - 5(x - y)^{-1}}{4(x - y)^{-1}}$

In the study of calculus, the difference quotient $\dfrac{f(a + h) - f(a)}{h}$ is often found and simplified. Find and simplify this quotient for each function f(x) by following steps **a** through **d**.

**a.** Find $(a + h)$.

**b.** Find $f(a)$.

**c.** Use steps **a** and **b** to find $\dfrac{f(a + h) - f(a)}{h}$

**d.** Simplify the result of step **c**.

**71.** $f(x) = \dfrac{1}{x}$

**72.** $f(x) = \dfrac{5}{x}$

**73.** $\dfrac{3}{x + 1}$

**74.** $\dfrac{2}{x^2}$

---

### 📖 STUDY SKILLS BUILDER

**Learning New Terms**

Many of the terms used in this text may be new to you. It will be helpful to make a list of new mathematical terms and symbols as you encounter them and to review them frequently. Placing these new terms (including page references) on 3 × 5 index cards might help you later when you're preparing for a quiz.

*Answer the following.*

**1.** Name one way you might place a word and its definition on a 3 × 5 card.

**2.** How do new terms stand out in this text so that they can be found?

---

## CHAPTER 7 GROUP ACTIVITY

### Comparing Dosage Formulas

In this project, you will have the opportunity to investigate two well-known formulas for predicting the correct doses of medication for children. This project may be completed by working in groups or individually.

Young's Rule and Cowling's Rule are dose formulas for prescribing medicines to children. Unlike formulas for, say area or distance, these dose formulas describe only an approximate relationship. The formulas relate a child's age $A$ in years and an adult dose $D$ of medication to the proper child's dose $C$. The formulas are most accurate when applied to children between the ages of 2 and 13.

$$\text{Young's Rule:} \quad C = \frac{DA}{A + 12}$$

$$\text{Cowling's Rule:} \quad C = \frac{D(A + 1)}{24}$$

**1.** Let the adult dose $D = 1000 \text{ mg}$. Complete the Young's Rule and Cowling's Rule columns of the following table comparing the doses predicted by both formulas for ages 2 through 13.

**2.** Use the data from the table in Question 1 to form sets of ordered pairs of the form (age, child's dose) for each formula. Graph the ordered pairs for each formula on the same graph. Describe the shapes of the graphed data.

**3.** Use your table, graph, or both, to decide whether either formula will consistently predict a larger dose than the other. If so, which one? If not, is there an age at which the doses predicted by one becomes greater than the doses predicted by the other? If so, estimate that age.

**4.** Use your graph to estimate for what age the difference in the two predicted doses is greatest.

**5.** Return to the table in Question 1 and complete the last column, titled "Difference," by finding the absolute value of the difference between the Young's dose and the Cowling's dose for each age. Use this column in the table to verify your graphical estimate found in Question 4.

**6.** Does Cowling's Rule ever predict exactly the adult dose? If so, at what age? Explain. Does Young's Rule ever predict exactly the adult dose? If so, at what age? Explain.

**7.** Many doctors prefer to use formulas that relate doses to factors other than a child's age. Why is age not necessarily the most important factor when predicting a child's dose? What other factors might be used?

| Age A | Young's Rule | Cowling's Rule | Difference | Age A | Young's Rule | Cowling's Rule | Difference |
|-------|--------------|----------------|------------|-------|--------------|----------------|------------|
| 2 |  |  |  | 8 |  |  |  |
| 3 |  |  |  | 9 |  |  |  |
| 4 |  |  |  | 10 |  |  |  |
| 5 |  |  |  | 11 |  |  |  |
| 6 |  |  |  | 12 |  |  |  |
| 7 |  |  |  | 13 |  |  |  |

## CHAPTER 7 VOCABULARY CHECK

*Fill in each blank with one of the words or phrases listed below.*

rational expression    complex fraction    ratio    proportion

cross products    domain    reciprocal

**1.** A _____ is the quotient of two numbers.

**2.** $\dfrac{x}{2} = \dfrac{7}{16}$ is an example of a _____.

**3.** If $\dfrac{a}{b} = \dfrac{c}{d}$, then $ad$ and $bc$ are called _____.

**4.** A _____ is an expression that can be written in the form $\dfrac{P}{Q}$, where $P$ and $Q$ are polynomials and $Q$ is not 0.

**5.** In a _____, the numerator or denominator or both may contain fractions.

**6.** The _____ of the rational function $f(x) = \dfrac{1}{x-3}$ is $\{x \mid x$ is a real number, $x \neq 3\}$.

**7.** The _____ of $\dfrac{9}{7}$ is $\dfrac{7}{9}$.

▶ **Helpful Hint**
Are you preparing for your test? Don't forget to take the Chapter 7 Test on page 499. Then check your answers at the back of the text and use the Chapter Test Prep Video CD to see the fully worked-out solutions to any of the exercises you want to review.

# CHAPTER 7 HIGHLIGHTS

| DEFINITIONS AND CONCEPTS | EXAMPLES |
|---|---|

**SECTION 7.1   RATIONAL FUNCTIONS AND SIMPLIFYING RATIONAL EXPRESSIONS**

A **rational expression** is an expression that can be written in the form $\dfrac{P}{Q}$, where $P$ and $Q$ are polynomials and $Q$ does not equal 0.

To find values for which a rational expression is undefined, find values for which the denominator is 0.

$$\frac{7y^3}{4}, \quad \frac{x^2 + 6x + 1}{x - 3}, \quad \frac{-5}{s^3 + 8}$$

Find any values for which the expression $\dfrac{5y}{y^2 - 4y + 3}$ is undefined.

$$y^2 - 4y + 3 = 0 \quad \text{Set the denominator equal to 0.}$$
$$(y - 3)(y - 1) = 0 \quad \text{Factor.}$$
$$y - 3 = 0 \quad \text{or} \quad y - 1 = 0 \quad \text{Set each factor equal to 0.}$$
$$y = 3 \qquad\qquad y = 1 \quad \text{Solve.}$$

The expression is undefined when $y$ is 3 and when $y$ is 1.

***To Simplify a Rational Expression***

**Step 1.** Factor the numerator and denominator.

**Step 2.** Divide out factors common to the numerator and denominator. (This is the same as removing a factor of 1.)

A **rational function** is a function described by a rational expression.

Simplify: $\dfrac{4x + 20}{x^2 - 25}$

$$\frac{4x + 20}{x^2 - 25} = \frac{4(x + 5)}{(x + 5)(x - 5)} = \frac{4}{x - 5}$$

$$f(x) = \frac{2x - 6}{7}, \quad h(t) = \frac{t^2 - 3t + 5}{t - 1}$$

**SECTION 7.2   MULTIPLYING AND DIVIDING RATIONAL EXPRESSIONS**

***To multiply rational expressions,***

**Step 1.** Factor numerators and denominators.

**Step 2.** Multiply numerators and multiply denominators.

**Step 3.** Write the product in simplest form.

$$\frac{P}{Q} \cdot \frac{R}{S} = \frac{PR}{QS}$$

Multiply: $\dfrac{4x + 4}{2x - 3} \cdot \dfrac{2x^2 + x - 6}{x^2 - 1}$

$$\frac{4x + 4}{2x - 3} \cdot \frac{2x^2 + x - 6}{x^2 - 1} = \frac{4(x + 1)}{2x - 3} \cdot \frac{(2x - 3)(x + 2)}{(x + 1)(x - 1)}$$
$$= \frac{4(x + 1)(2x - 3)(x + 2)}{(2x - 3)(x + 1)(x - 1)}$$
$$= \frac{4(x + 2)}{x - 1}$$

**To divide by a rational expression,** multiply by the reciprocal.

$$\frac{P}{Q} \div \frac{R}{S} = \frac{P}{Q} \cdot \frac{S}{R} = \frac{PS}{QR}$$

Divide: $\dfrac{15x + 5}{3x^2 - 14x - 5} \div \dfrac{15}{3x - 12}$

$$\frac{15x + 5}{3x^2 - 14x - 5} \div \frac{15}{3x - 12} = \frac{5(3x + 1)}{(3x + 1)(x - 5)} \cdot \frac{3(x - 4)}{3 \cdot 5}$$
$$= \frac{x - 4}{x - 5}$$

| **DEFINITIONS AND CONCEPTS** | **EXAMPLES** |
|---|---|

**To add or subtract rational expressions with the same denominator,** add or subtract numerators, and place the sum or difference over a common denominator.

$$\frac{P}{R} + \frac{Q}{R} = \frac{P + Q}{R}$$

$$\frac{P}{R} - \frac{Q}{R} = \frac{P - Q}{R}$$

Perform indicated operations.

$$\frac{5}{x + 1} + \frac{x}{x + 1} = \frac{5 + x}{x + 1}$$

$$\frac{2y + 7}{y^2 - 9} - \frac{y + 4}{y^2 - 9} = \frac{(2y + 7) - (y + 4)}{y^2 - 9}$$

$$= \frac{2y + 7 - y - 4}{y^2 - 9}$$

$$= \frac{y + 3}{(y + 3)(y - 3)}$$

$$= \frac{1}{y - 3}$$

**To find the least common denominator (LCD),**

**Step 1.** Factor the denominators.

**Step 2.** The LCD is the product of all unique factors, each raised to a power equal to the greatest number of times that it appears in any one factored denominator.

Find the LCD for

$$\frac{7x}{x^2 + 10x + 25} \quad \text{and} \quad \frac{11}{3x^2 + 15x}$$

$$x^2 + 10x + 25 = (x + 5)(x + 5)$$

$$3x^2 + 15x = 3x(x + 5)$$

LCD is $3x(x + 5)(x + 5)$ or $3x(x + 5)^2$

**To add or subtract rational expressions with unlike denominators,**

**Step 1.** Find the LCD.

**Step 2.** Rewrite each rational expression as an equivalent expression whose denominator is the LCD.

**Step 3.** Add or subtract numerators and place the sum or difference over the common denominator.

**Step 4.** Write the result in simplest form.

Perform the indicated operation.

$$\frac{9x + 3}{x^2 - 9} - \frac{5}{x - 3}$$

$$= \frac{9x + 3}{(x + 3)(x - 3)} - \frac{5}{x - 3}$$

LCD is $(x + 3)(x - 3)$.

$$= \frac{9x + 3}{(x + 3)(x - 3)} - \frac{5(x + 3)}{(x - 3)(x + 3)}$$

$$= \frac{9x + 3 - 5(x + 3)}{(x + 3)(x - 3)}$$

$$= \frac{9x + 3 - 5x - 15}{(x + 3)(x - 3)}$$

$$= \frac{4x - 12}{(x + 3)(x - 3)}$$

$$= \frac{4(x - 3)}{(x + 3)(x - 3)} = \frac{4}{x + 3}$$

**To solve an equation containing rational expressions,**

**Step 1.** Multiply both sides of the equation by the LCD of all rational expressions in the equation.

**Step 2.** Remove any grouping symbols and solve the resulting equation.

**Step 3.** Check the solution in the original equation.

Solve: $\dfrac{5x}{x + 2} + 3 = \dfrac{4x - 6}{x + 2}$

$$(x + 2)\left(\frac{5x}{x + 2} + 3\right) = (x + 2)\left(\frac{4x - 6}{x + 2}\right)$$

$$(x + 2)\left(\frac{5x}{x + 2}\right) + (x + 2)(3) = (x + 2)\left(\frac{4x - 6}{x + 2}\right)$$

$$5x + 3x + 6 = 4x - 6$$

$$4x = -12$$

$$x = -3$$

The solution checks and the solution is $-3$.

| DEFINITIONS AND CONCEPTS | EXAMPLES |
|---|---|

A **ratio** is the quotient of two numbers or two quantities. A **proportion** is a mathematical statement that two ratios are equal.

*Proportions*

$$\frac{2}{3} = \frac{8}{12} \qquad \frac{x}{7} = \frac{15}{35}$$

**Cross products:**

If $\frac{a}{b} = \frac{c}{d}$, then $ad = bc$.

*Cross Products*

$2 \cdot 12$ or $24$ $\qquad\qquad\qquad$ $3 \cdot 8$ or $24$

$$\frac{2}{3} = \frac{8}{12}$$

Solve: $\dfrac{3}{4} = \dfrac{x}{x - 1}$

$$\frac{3}{4} = \frac{x}{x - 1}$$

$3(x - 1) = 4x$    Set cross products equal.

$3x - 3 = 4x$

$-3 = x$

**Problem-Solving Steps**

**1.** UNDERSTAND. Read and reread the problem.

A small plane and a car leave Kansas City, Missouri, and head for Minneapolis, Minnesota, a distance of 450 miles. The speed of the plane is 3 times the speed of the car, and the plane arrives 6 hours ahead of the car. Find the speed of the car.

Let $x$ = the speed of the car.
Then $3x$ = the speed of the plane.

|  | Distance | = | Rate · Time |
|---|---|---|---|
| **Car** | 450 | $x$ | $\dfrac{450}{x} \left(\dfrac{\text{distance}}{\text{rate}}\right)$ |
| **Plane** | 450 | $3x$ | $\dfrac{450}{3x} \left(\dfrac{\text{distance}}{\text{rate}}\right)$ |

**2.** TRANSLATE.

In words: $\quad\boxed{\begin{matrix}\text{plane's}\\\text{time}\end{matrix}} + \boxed{6 \text{ hours}} = \boxed{\begin{matrix}\text{car's}\\\text{time}\end{matrix}}$

$\qquad\qquad\qquad\downarrow\qquad\qquad\downarrow\qquad\quad\downarrow$

Translate: $\quad\dfrac{450}{3x} \quad + \quad 6 \quad = \quad \dfrac{450}{x}$

**3.** SOLVE.

$$\frac{450}{3x} + 6 = \frac{450}{x}$$

$$3x\left(\frac{450}{3x}\right) + 3x(6) = 3x\left(\frac{450}{x}\right)$$

$$450 + 18x = 1350$$

$$18x = 900$$

$$x = 50$$

**4.** INTERPRET.

**Check** this solution in the originally stated problem. **State** the conclusion: The speed of the car is 50 miles per hour.

| DEFINITIONS AND CONCEPTS | EXAMPLES |
|---|---|

SECTION 7.7 SIMPLIFYING COMPLEX FRACTIONS

**Method 1: To Simplify a Complex Fraction**

**Step 1.** Add or subtract fractions in the numerator and the denominator of the complex fraction.

**Step 2.** Perform the indicated division.

**Step 3.** Write the result in lowest terms.

Simplify:

$$\frac{\dfrac{1}{x}+2}{\dfrac{1}{x}-\dfrac{1}{y}} = \frac{\dfrac{1}{x}+\dfrac{2x}{x}}{\dfrac{y}{xy}-\dfrac{x}{xy}}$$

$$= \frac{\dfrac{1+2x}{x}}{\dfrac{y-x}{xy}}$$

$$= \frac{1+2x}{x}\cdot\frac{x\,y}{y-x}$$

$$= \frac{y(1+2x)}{y-x}$$

**Method 2: To Simplify a Complex Fraction**

**Step 1.** Find the LCD of all fractions in the complex fraction.

**Step 2.** Multiply the numerator and the denominator of the complex fraction by the LCD.

**Step 3.** Perform the indicated operations and write the result in lowest terms.

$$\frac{\dfrac{1}{x}+2}{\dfrac{1}{x}-\dfrac{1}{y}} = \frac{xy\left(\dfrac{1}{x}+2\right)}{xy\left(\dfrac{1}{x}-\dfrac{1}{y}\right)}$$

$$= \frac{xy\left(\dfrac{1}{x}\right)+xy(2)}{xy\left(\dfrac{1}{x}\right)-xy\left(\dfrac{1}{y}\right)}$$

$$= \frac{y+2xy}{y-x} \quad \text{or} \quad \frac{y(1+2x)}{y-x}$$

---

### 📖 STUDY SKILLS BUILDER

**Are You Preparing for a Test on Chapter 7?**

Below I have listed some common trouble areas for students in Chapter 7. After studying for your test—but before taking your test—read these.

- Make sure you know the difference in the following:

Simplify: $\dfrac{\dfrac{3}{x}}{\dfrac{1}{x}-\dfrac{5}{y}}$

Multiply numerator and denominator by the LCD.

$$\frac{\dfrac{3}{x}\cdot xy}{\dfrac{1}{x}\cdot xy-\dfrac{5}{y}\cdot xy}$$

$$= \frac{3y}{y-5x}$$

Solve: $\dfrac{5x}{6}-\dfrac{1}{2}=\dfrac{5x}{12}$

Multiply both sides by the LCD.

$$12\cdot\frac{5x}{6}-12\cdot\frac{1}{2}=12\cdot\frac{5x}{12}$$

$$2\cdot 5x-6=5x$$

$$10x-6=5x$$

$$5x=6$$

$$x=\frac{6}{5}$$

Subtract: $\dfrac{1}{2x}-\dfrac{7}{x-3}$

Write each expression as an equivalent expression with the LCD.

$$\frac{1\cdot(x-3)}{2x\cdot(x-3)}-\frac{7\cdot 2x}{(x-3)\cdot 2x}$$

$$= \frac{x-3}{2x(x-3)}-\frac{14x}{2x(x-3)}$$

$$= \frac{-13x-3}{2x(x-3)}$$

Remember: This is simply a checklist of common trouble areas. For a review of Chapter 7, see the Highlights and Chapter Review at the end of this chapter.

# CHAPTER 7 REVIEW

*(7.1) Find the domain for each rational function.*

**1.** $f(x) = \dfrac{3 - 5x}{7}$

**2.** $g(x) = \dfrac{2x + 4}{11}$

**3.** $F(x) = \dfrac{-3x^2}{x - 5}$

**4.** $h(x) = \dfrac{4x}{3x - 12}$

**5.** $f(x) = \dfrac{x^3 + 2}{x^2 + 8x}$

**6.** $G(x) = \dfrac{20}{3x^2 - 48}$

*Write each rational expression in lowest terms.*

**7.** $\dfrac{x - 12}{12 - x}$

**8.** $\dfrac{5x - 15}{25x - 75}$

**9.** $\dfrac{2x}{2x^2 - 2x}$

**10.** $\dfrac{x + 7}{x^2 - 49}$

**11.** $\dfrac{2x^2 + 4x - 30}{x^2 + x - 20}$

**12.** The average cost (per bookcase) of manufacturing $x$ bookcases is given by the rational function.

$$C(x) = \dfrac{35x + 4200}{x}$$

  **a.** Find the average cost per bookcase of manufacturing 50 bookcases.

  **b.** Find the average cost per bookcase of manufacturing 100 bookcases.

  **c.** As the number of bookcases increases, does the average cost per bookcase increase or decrease? (See parts (a) and (b).)

*Simplify each expression. This section contains four-term polynomials and sums and differences of two cubes.*

**13.** $\dfrac{x^2 + xa + xb + ab}{x^2 - xc + bx - bc}$

**14.** $\dfrac{x^2 + 5x - 2x - 10}{x^2 - 3x - 2x + 6}$

**15.** $\dfrac{4 - x}{x^3 - 64}$

**16.** $\dfrac{x^2 - 4}{x^3 + 8}$

*(7.2) Perform each indicated operation and simplify.*

**17.** $\dfrac{15x^3 y^2}{z} \cdot \dfrac{z}{5xy^3}$

**18.** $\dfrac{-y^3}{8} \cdot \dfrac{9x^2}{y^3}$

**19.** $\dfrac{x^2 - 9}{x^2 - 4} \cdot \dfrac{x - 2}{x + 3}$

**20.** $\dfrac{2x + 5}{x - 6} \cdot \dfrac{2x}{-x + 6}$

**21.** $\dfrac{x^2 - 5x - 24}{x^2 - x - 12} \div \dfrac{x^2 - 10x + 16}{x^2 + x - 6}$

**22.** $\dfrac{4x + 4y}{xy^2} \div \dfrac{3x + 3y}{x^2 y}$

**23.** $\dfrac{x^2 + x - 42}{x - 3} \cdot \dfrac{(x - 3)^2}{x + 7}$

**24.** $\dfrac{2a + 2b}{3} \cdot \dfrac{a - b}{a^2 - b^2}$

**25.** $\dfrac{2x^2 - 9x + 9}{8x - 12} \div \dfrac{x^2 - 3x}{2x}$

**26.** $\dfrac{x^2 - y^2}{x^2 + xy} \div \dfrac{3x^2 - 2xy - y^2}{3x^2 + 6x}$

**27.** $\dfrac{x - y}{4} \div \dfrac{y^2 - 2y - xy + 2x}{16x + 24}$

**28.** $\dfrac{5 + x}{7} \div \dfrac{xy + 5y - 3x - 15}{7y - 35}$

*(7.3) Perform each indicated operation and simplify.*

**29.** $\dfrac{x}{x^2 + 9x + 14} + \dfrac{7}{x^2 + 9x + 14}$

**30.** $\dfrac{x}{x^2 + 2x - 15} + \dfrac{5}{x^2 + 2x - 15}$

**31.** $\dfrac{4x - 5}{3x^2} - \dfrac{2x + 5}{3x^2}$

**32.** $\dfrac{9x + 7}{6x^2} - \dfrac{3x + 4}{6x^2}$

*Find the LCD of each pair of rational expressions.*

**33.** $\dfrac{x + 4}{2x}, \dfrac{3}{7x}$

**34.** $\dfrac{x - 2}{x^2 - 5x - 24}, \dfrac{3}{x^2 + 11x + 24}$

*Rewrite each rational expression as an equivalent expression whose denominator is the given polynomial.*

**35.** $\dfrac{5}{7x} = \dfrac{}{14x^3 y}$

**36.** $\dfrac{9}{4y} = \dfrac{}{16y^3 x}$

**37.** $\dfrac{x + 2}{x^2 + 11x + 18} = \dfrac{}{(x + 2)(x - 5)(x + 9)}$

**38.** $\dfrac{3x - 5}{x^2 + 4x + 4} = \dfrac{}{(x + 2)^2(x + 3)}$

*(7.4) Perform each indicated operation and simplify.*

**39.** $\dfrac{4}{5x^2} - \dfrac{6}{y}$

**40.** $\dfrac{2}{x - 3} - \dfrac{4}{x - 1}$

**41.** $\dfrac{4}{x + 3} - 2$

**42.** $\dfrac{3}{x^2 + 2x - 8} + \dfrac{2}{x^2 - 3x + 2}$

**43.** $\dfrac{2x - 5}{6x + 9} - \dfrac{4}{2x^2 + 3x}$

**44.** $\dfrac{x - 1}{x^2 - 2x + 1} - \dfrac{x + 1}{x - 1}$

*Find the perimeter and the area of each figure.*

△ **45.**

rectangle with sides $\dfrac{x + 2}{4x}$ and $\dfrac{x}{8}$

△ **46.**

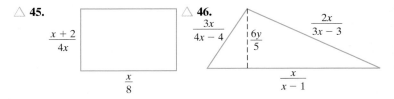

triangle with sides $\dfrac{3x}{4x - 4}$, $\dfrac{6y}{5}$, $\dfrac{2x}{3x - 3}$, and base $\dfrac{x}{x - 1}$

*(7.5) Solve each equation.*

**47.** $\dfrac{n}{10} = 9 - \dfrac{n}{5}$

**48.** $\dfrac{2}{x + 1} - \dfrac{1}{x - 2} = -\dfrac{1}{2}$

**49.** $\dfrac{y}{2y + 2} + \dfrac{2y - 16}{4y + 4} = \dfrac{y - 3}{y + 1}$

**50.** $\dfrac{2}{x-3} - \dfrac{4}{x+3} = \dfrac{8}{x^2-9}$    **51.** $\dfrac{x-3}{x+1} - \dfrac{x-6}{x+5} = 0$

**52.** $x + 5 = \dfrac{6}{x}$

*Solve the equation for the indicated variable.*

**53.** $\dfrac{4A}{5b} = x^2$, for $b$    **54.** $\dfrac{x}{7} + \dfrac{y}{8} = 10$, for $y$

**(7.6)** *Solve each proportion.*

**55.** $\dfrac{x}{2} = \dfrac{12}{4}$    **56.** $\dfrac{20}{1} = \dfrac{x}{25}$

**57.** $\dfrac{2}{x-1} = \dfrac{3}{x+3}$    **58.** $\dfrac{4}{y-3} = \dfrac{2}{y-3}$

*Solve.*

**59.** A machine can process 300 parts in 20 minutes. Find how many parts can be processed in 45 minutes.

**60.** As his consulting fee, Mr. Visconti charges \$90.00 per day. Find how much he charges for 3 hours of consulting. Assume an 8-hour work day.

**61.** Five times the reciprocal of a number equals the sum of $\dfrac{3}{2}$ the reciprocal of the number and $\dfrac{7}{6}$. What is the number?

**62.** The reciprocal of a number equals the reciprocal of the difference of 4 and the number. Find the number.

**63.** A car travels 90 miles in the same time that a car traveling 10 miles per hour slower travels 60 miles. Find the speed of each car.

**64.** The current in a bayou near Lafayette, Louisiana, is 4 miles per hour. A paddle boat travels 48 miles upstream in the same amount of time it takes to travel 72 miles downstream. Find the speed of the boat in still water.

**65.** When Mark and Maria manicure Mr. Stergeon's lawn, it takes them 5 hours. If Mark works alone, it takes 7 hours. Find how long it takes Maria alone.

**66.** It takes pipe A 20 days to fill a fish pond. Pipe B takes 15 days. Find how long it takes both pipes together to fill the pond.

*Given that the pairs of triangles are similar, find each missing length x.*

△ **67.**

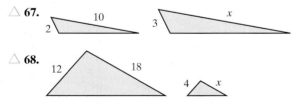

△ **68.**

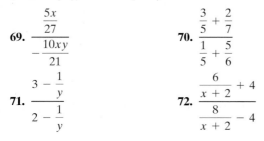

**(7.7)** *Simplify each complex fraction.*

**69.** $\dfrac{\dfrac{5x}{27}}{-\dfrac{10xy}{21}}$    **70.** $\dfrac{\dfrac{3}{5} + \dfrac{2}{7}}{\dfrac{1}{5} + \dfrac{5}{6}}$

**71.** $\dfrac{3 - \dfrac{1}{y}}{2 - \dfrac{1}{y}}$    **72.** $\dfrac{\dfrac{6}{x+2} + 4}{\dfrac{8}{x+2} - 4}$

**73.** $\dfrac{\dfrac{x-3}{x+3} + \dfrac{x+3}{x-3}}{\dfrac{x-3}{x+3} - \dfrac{x+3}{x-3}}$    **74.** $\dfrac{\dfrac{3}{x-1} - \dfrac{2}{1-x}}{\dfrac{2}{x-1} - \dfrac{2}{x}}$

**75.** $\dfrac{x + y^{-1}}{\dfrac{x}{y}}$    **76.** $\dfrac{x - xy^{-1}}{\dfrac{1+x}{y}}$

## MIXED REVIEW

*Simplify each rational expression.*

**77.** $\dfrac{4x+12}{8x^2+24x}$    **78.** $\dfrac{x^3-6x^2+9x}{x^2+4x-21}$

*Perform the indicated operations and simplify.*

**79.** $\dfrac{x^2+9x+20}{x^2-25} \cdot \dfrac{x^2-9x+20}{x^2+8x+16}$

**80.** $\dfrac{x^2-x-72}{x^2-x-30} \div \dfrac{x^2+6x-27}{x^2-9x+18}$

**81.** $\dfrac{x}{x^2-36} + \dfrac{6}{x^2-36}$

**82.** $\dfrac{5x-1}{4x} - \dfrac{3x-2}{4x}$

**83.** $\dfrac{4}{3x^2+8x-3} + \dfrac{2}{3x^2-7x+2}$

**84.** $\dfrac{3x}{x^2+9x+14} - \dfrac{6x}{x^2+4x-21}$

*Solve.*

**85.** $\dfrac{4}{a-1} + 2 = \dfrac{3}{a-1}$    **86.** $\dfrac{x}{x+3} + 4 = \dfrac{x}{x+3}$

*Solve.*

**87.** The quotient of twice a number and three, minus one-sixth is the quotient of the number and two. Find the number.

**88.** Mr. Crocker can paint his house by himself in three days. His son will need an additional day to complete the job if he works alone. If they work together, find how long it takes to paint the house.

*Given that the following pairs of triangles are similar, find each missing length.*

**89.**

**90.**

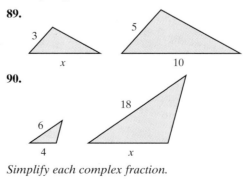

*Simplify each complex fraction.*

**91.** $\dfrac{\dfrac{1}{4}}{\dfrac{1}{3} + \dfrac{1}{2}}$    **92.** $\dfrac{4 + \dfrac{2}{x}}{6 + \dfrac{3}{x}}$

**93.** $\dfrac{y^{-2}}{1 - y^{-2}}$    **94.** $\dfrac{4 + x^{-1}}{3 + x^{-1}}$

# CHAPTER 7 TEST TEST PREP VIDEO

Remember to use the Chapter Test Prep Video CD to see the fully worked-out solutions to any of the exercises you want to review.

1. Find the domain of the rational function

$$g(x) = \frac{9x^2 - 9}{x^2 + 4x + 3}$$

2. For a certain computer desk, the average cost $C$ (in dollars) per desk manufactured is

$$C = \frac{100x + 3000}{x}$$

where $x$ is the number of desks manufactured.

   a. Find the average cost per desk when manufacturing 200 computer desks.
   b. Find the average cost per desk when manufacturing 1000 computer desks.

*Simplify each rational expression.*

3. $\dfrac{3x - 6}{5x - 10}$

4. $\dfrac{x + 6}{x^2 + 12x + 36}$

5. $\dfrac{x + 3}{x^3 + 27}$

6. $\dfrac{2m^3 - 2m^2 - 12m}{m^2 - 5m + 6}$

7. $\dfrac{ay + 3a + 2y + 6}{ay + 3a + 5y + 15}$

8. $\dfrac{y - x}{x^2 - y^2}$

*Perform the indicated operation and simplify if possible.*

9. $\dfrac{3}{x - 1} \cdot (5x - 5)$

10. $\dfrac{y^2 - 5y + 6}{2y + 4} \cdot \dfrac{y + 2}{2y - 6}$

11. $\dfrac{15x}{2x + 5} - \dfrac{6 - 4x}{2x + 5}$

12. $\dfrac{5a}{a^2 - a - 6} - \dfrac{2}{a - 3}$

13. $\dfrac{6}{x^2 - 1} + \dfrac{3}{x + 1}$

14. $\dfrac{x^2 - 9}{x^2 - 3x} \div \dfrac{xy + 5x + 3y + 15}{2x + 10}$

15. $\dfrac{x + 2}{x^2 + 11x + 18} + \dfrac{5}{x^2 - 3x - 10}$

*Solve each equation.*

16. $\dfrac{4}{y} - \dfrac{5}{3} = \dfrac{-1}{5}$

17. $\dfrac{5}{y + 1} = \dfrac{4}{y + 2}$

18. $\dfrac{a}{a - 3} = \dfrac{3}{a - 3} - \dfrac{3}{2}$

19. $x - \dfrac{14}{x - 1} = 4 - \dfrac{2x}{x - 1}$

20. $\dfrac{10}{x^2 - 25} = \dfrac{3}{x + 5} + \dfrac{1}{x - 5}$

*Simplify each complex fraction.*

21. $\dfrac{\dfrac{5x^2}{yz^2}}{\dfrac{10x}{z^3}}$

22. $\dfrac{5 - \dfrac{1}{y^2}}{\dfrac{1}{y} + \dfrac{2}{y^2}}$

23. In a sample of 85 fluorescent bulbs, 3 were found to be defective. At this rate, how many defective bulbs should be found in 510 bulbs?

24. One number plus five times its reciprocal is equal to six. Find the number.

25. A pleasure boat traveling down the Red River takes the same time to go 14 miles upstream as it takes to go 16 miles downstream. If the current of the river is 2 miles per hour, find the speed of the boat in still water.

26. An inlet pipe can fill a tank in 12 hours. A second pipe can fill the tank in 15 hours. If both pipes are used, find how long it takes to fill the tank.

△ 27. Given that the two triangles are similar, find $x$.

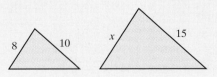

# CHAPTER 7 CUMULATIVE REVIEW

1. Write each sentence as an equation. Let $x$ represent the unknown number.
   a. The quotient of 15 and a number is 4.
   b. Three subtracted from 12 is a number.
   c. Four times a number, added to 17, is not equal to 21.
   d. Triple a number is less than 48.

2. Write each sentence as an equation. Let $x$ represent the unknown number.
   a. The difference of 12 and a number is $-45$.
   b. The product of 12 and a number is $-45$.
   c. A number less 10 is twice the number.

**3.** Rajiv Puri invested part of his $20,000 inheritance in a mutual funds account that pays 7% simple interest yearly and the rest in a certificate of deposit that pays 9% simple interest yearly. At the end of one year, Rajiv's investments earned $1550. Find the amount he invested at each rate.

**4.** The number of non-business bankruptcies has increased over the years. In 2002, the number of non-business bankruptcies was 80,000 less than twice the number in 1994. If the total of non-business bankruptcies for these two years is 2,290,000 find the number of non-business bankruptcies for each year. (*Source:* American Bankruptcy Institute)

**5.** Graph $x - 3y = 6$ by finding and plotting intercepts.

**6.** Find the slope of the line whose equation is $7x + 2y = 9$.

**7.** Use the product rule to simplify each expression.
   **a.** $4^2 \cdot 4^5$         **b.** $x^4 \cdot x^6$
   **c.** $y^3 \cdot y$          **d.** $y^3 \cdot y^2 \cdot y^7$
   **e.** $(-5)^7 \cdot (-5)^8$     **f.** $a^2 \cdot b^2$

**8.** Simplify.
   **a.** $\dfrac{x^9}{x^7}$          **b.** $\dfrac{x^{19}y^5}{xy}$
   **c.** $(x^5y^2)^3$          **d.** $(-3a^2b)(5a^3b)$

**9.** Subtract $(5z - 7)$ from the sum of $(8z + 11)$ and $(9z - 2)$.

**10.** Subtract $(9x^2 - 6x + 2)$ from $(x + 1)$.

**11.** Multiply: $(3a + b)^3$

**12.** Multiply: $(2x + 1)(5x^2 - x + 2)$

**13.** Use a special product to square each binomial.
   **a.** $(t + 2)^2$
   **b.** $(p - q)^2$
   **c.** $(2x + 5)^2$
   **d.** $(x^2 - 7y)^2$

**14.** Multiply.
   **a.** $(x + 9)^2$
   **b.** $(2x + 1)(2x - 1)$
   **c.** $8x(x^2 + 1)(x^2 - 1)$

**15.** Simplify each expression. Write results using positive exponents only.
   **a.** $\dfrac{1}{x^{-3}}$          **b.** $\dfrac{1}{3^{-4}}$
   **c.** $\dfrac{p^{-4}}{q^{-9}}$          **d.** $\dfrac{5^{-3}}{2^{-5}}$

**16.** Simplify. Write results with positive exponents.
   **a.** $5^{-3}$      **b.** $\dfrac{9}{x^{-7}}$      **c.** $\dfrac{11^{-1}}{7^{-2}}$

**17.** Divide: $\dfrac{4x^2 + 7 + 8x^3}{2x + 3}$

**18.** Divide $(4x^3 - 9x + 2)$ by $(x - 4)$.

**19.** Find the GCF of each list of numbers.
   **a.** 28 and 40
   **b.** 55 and 21
   **c.** 15, 18, and 66

**20.** Find the GCF of $9x^2$, $6x^3$, and $21x^5$.

*Factor.*

**21.** $-9a^5 + 18a^2 - 3a$

**22.** $7x^6 - 7x^5 + 7x^4$

**23.** $3m^2 - 24m - 60$

**24.** $-2a^2 + 10a + 12$

**25.** $3x^2 + 11x + 6$

**26.** $10m^2 - 7m + 1$

**27.** $x^2 + 12x + 36$

**28.** $4x^2 + 12x + 9$

**29.** $x^2 + 4$

**30.** $x^2 - 4$

**31.** $x^3 + 8$

**32.** $27y^3 - 1$

**33.** $2x^3 + 3x^2 - 2x - 3$

**34.** $3x^3 + 5x^2 - 12x - 20$

**35.** $12m^2 - 3n^2$

**36.** $x^5 - x$

**37.** Solve: $x(2x - 7) = 4$

**38.** Solve: $3x^2 + 5x = 2$

**39.** Find the $x$-intercepts of the graph of $y = x^2 - 5x + 4$.

**40.** Find the $x$-intercepts of the graph of $y = x^2 - x - 6$.

**41.** The height of a triangular sail is 2 meters less than twice the length of the base. If the sail has an area of 30 square meters, find the length of its base and the height.

**42.** The height of a parallelogram is 5 feet more than three times its base. If the area of the parallelogram is 182 square feet, find the length of its base and height.

**43.** Simplify: $\dfrac{18 - 2x^2}{x^2 - 2x - 3}$

**44.** Simplify: $\dfrac{2x^2 - 50}{4x^4 - 20x^3}$

**45.** Divide: $\dfrac{6x + 2}{x^2 - 1} \div \dfrac{3x^2 + x}{x - 1}$

**46.** Multiply: $\dfrac{6x^2 - 18x}{3x^2 - 2x} \cdot \dfrac{15x - 10}{x^2 - 9}$

**47.** Simplify: $\dfrac{(2x)^{-1} + 1}{2x^{-1} - 1}$

**48.** Simplify: $\dfrac{\dfrac{m}{3} + \dfrac{n}{6}}{\dfrac{m + n}{12}}$

# 8 More on Functions and Graphs

Over the past few years, diamonds have gained much higher visibility by increased media advertising. Strong consumer demand has caused the industry to increase production and is the basis for the bar graph below. By a method called least squares (Section 8.2), the function $f(x) = 0.42x + 10.5$ approximates the data below where $f(x)$ is world diamond production value (in billions of dollars) and where $x$ is the number of years past 2000. In Section 8.2, Exercises 45 and 46, page 516, we will use this linear equation to predict diamond production.

In Section 3.6, we introduced the notion of relation and to the notion of function, perhaps the single most important and useful concept in all of mathematics. In this chapter, we explore this concept further.

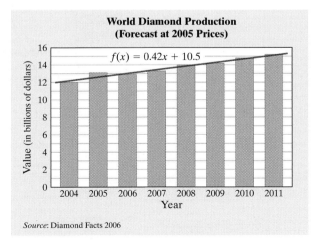

**World Diamond Production (Forecast at 2005 Prices)**

$f(x) = 0.42x + 10.5$

Value (in billions of dollars)

Year

*Source*: Diamond Facts 2006

# 8.1 GRAPHING AND WRITING LINEAR FUNCTIONS

**OBJECTIVE 1 ▶ Graphing linear functions.** In this section, we identify and graph linear functions. By the vertical line test, Section 3.6, we know that all linear equations except those whose graphs are vertical lines are functions. Thus, all linear equations except those of the form $x = c$ (vertical lines) are linear functions. For example, we know from Section 3.2 that $y = 2x$ is a linear equation in two variables. Its graph is shown.

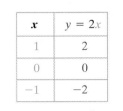

| $x$ | $y = 2x$ |
|-----|----------|
| 1 | 2 |
| 0 | 0 |
| -1 | -2 |

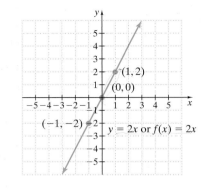

Because this graph passes the vertical line test, we know that $y = 2x$ is a function. If we want to emphasize that this equation describes a function, we may write $y = 2x$ as $f(x) = 2x$.

**EXAMPLE 1** Graph $g(x) = 2x + 1$. Compare this graph with the graph of $f(x) = 2x$.

**Solution** To graph $g(x) = 2x + 1$, find three ordered pair solutions.

| $x$ | $f(x) = 2x$ | $g(x) = 2x + 1$ |
|-----|-------------|------------------|
| 0 | 0 | 1 |
| -1 | -2 | -1 |
| 1 | 2 | 3 |

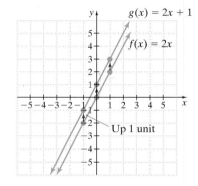

Notice that $y$-values for the graph of $g(x) = 2x + 1$ are obtained by adding 1 to each $y$-value of each corresponding point of the graph of $f(x) = 2x$. The graph of $g(x) = 2x + 1$ is the same as the graph of $f(x) = 2x$ shifted upward 1 unit. ☐

**PRACTICE**

**1** Graph $g(x) = 4x - 3$ and $f(x) = 4x$ on the same axes.

If a linear function is solved for $y$, we can easily use function notation to describe it by replacing $y$ with $f(x)$. Recall the slope-intercept form of a linear equation, $y = mx + b$, where $m$ is the slope of the line and $(0, b)$ is the $y$-intercept. Since this form is solved for $y$, we use it to define a linear function.

In general, a **linear function** is a function that can be written in the form $f(x) = mx + b$. For example, $g(x) = 2x + 1$ is in this form, with $m = 2$ and $b = 1$. Thus, the slope of the linear function $g(x)$ is 2 and the $y$-intercept is $(0, 1)$.

**EXAMPLE 2**   Graph the linear functions $f(x) = -3x$ and $g(x) = -3x - 6$ on the same set of axes.

*Solution*   To graph $f(x)$ and $g(x)$, find ordered pair solutions.

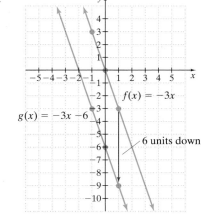

|     |              | ⎴ subtract 6 ⎴   |
| :-: | :----------: | :--------------: |
| $x$ | $f(x) = -3x$ | $g(x) = -3x - 6$ |
|  0  |      0       |       $-6$       |
|  1  |     $-3$     |       $-9$       |
| $-1$ |     3       |       $-3$       |
| $-2$ |     6       |        0         |

⎵ subtract 6 ⎵

Each $y$-value for the graph of $g(x) = -3x - 6$ is obtained by subtracting 6 from the $y$-value of the corresponding point of the graph of $f(x) = -3x$. The graph of $g(x) = -3x - 6$ is the same as the graph of $f(x) = -3x$ shifted down 6 units.   □

**PRACTICE**
**2**   Graph the linear functions $f(x) = -2x$ and $g(x) = -2x + 5$ on the same set of axes.

- - - - - - - - - - - - - - - - - - - - - - - - - - - - - - - - - - - - - - - - - - - - - - ■

**OBJECTIVE 2 ▶ Writing equations of lines using function notation.**

We now practice writing linear functions.

    This means the graph is a line that passes the vertical line test.

Below is a review of some tools we can use.

| | |
| :-- | :-- |
| $y = mx + b$ | **Slope-intercept form** of a linear equation. The slope is $m$, and the $y$-intercept is $(0, b)$. |
| $y - y_1 = m(x - x_1)$ | **Point-slope form** of a linear equation. The slope is $m$, and $(x_1, y_1)$ is a point on the line. |
| $y = c$ | **Horizontal line** The slope is 0, and the $y$-intercept is $(0, c)$. |

Note: $x = c$, whose graph is a vertical line is not included above as these equations do **not** define functions.

*Could use*

*Point - slope form*

**EXAMPLE 3**   Find an equation of the line with slope $-3$ and $y$-intercept $(0, -5)$. Write the equation using function notation.

*Solution*   Because we know the slope and the $y$-intercept, we use the slope-intercept form with $m = -3$ and $b = -5$.

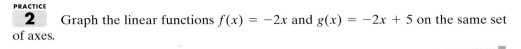

$$y = mx + b \qquad \text{Slope-intercept form}$$

$$y = -3 \cdot x + (-5) \quad \text{Let } m = -3 \text{ and } b = -5.$$

$$y = -3x - 5 \qquad \text{Simplify.}$$

This equation is solved for $y$. To write using function notation, we replace $y$ with $f(x)$.

$$f(x) = -3x - 5 \qquad\qquad\qquad □$$

**3** Find an equation of the line with slope $-4$ and $y$-intercept $(0, -3)$. Write the equation using function notation.

---

**EXAMPLE 4** Find an equation of the line through points $(4, 0)$ and $(-4, -5)$. Write the equation using function notation.

*Solution* First, find the slope of the line.

$$m = \frac{-5 - 0}{-4 - 4} = \frac{-5}{-8} = \frac{5}{8}$$

Next, make use of the point–slope form. Replace $(x_1, y_1)$ by either $(4, 0)$ or $(-4, -5)$ in the point–slope equation. We will choose the point $(4, 0)$. The line through $(4, 0)$ with slope $\frac{5}{8}$ is

$$y - y_1 = m(x - x_1) \quad \text{Point–slope form.}$$

$$y - 0 = \frac{5}{8}(x - 4) \quad \text{Let } m = \frac{5}{8} \text{ and } (x_1, y_1) = (4, 0).$$

$$8y = 5(x - 4) \quad \text{Multiply both sides by 8.}$$

$$8y = 5x - 20 \quad \text{Apply the distributive property.}$$

To write the equation using function notation, we solve for $y$, then replace $y$ with $f(x)$.

$$8y = 5x - 20$$

$$y = \frac{5}{8}x - \frac{20}{8} \quad \text{Divide both sides by 8.}$$

$$f(x) = \frac{5}{8}x - \frac{5}{2} \quad \text{Write using function notation.} \qquad \square$$

**4** Find an equation of the line through points $(-1, 2)$ and $(2, 0)$. Write the equation using function notation.

---

> ▶ **Helpful Hint**
> If two points of a line are given, either one may be used with the point-slope form to write an equation of the line.

**EXAMPLE 5** Find an equation of the horizontal line containing the point $(2, 3)$. Write the equation using function notation.

*Solution* A horizontal line has an equation of the form $y = c$. Since the line contains the point $(2, 3)$, the equation is $y = 3$, as shown to the right.

Using function notation, the equation is

$$f(x) = 3.$$

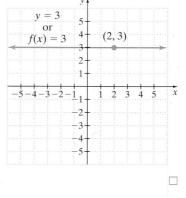

$\square$

**5** Find the equation of the horizontal line containing the point $(6, -2)$. Use function notation.

**OBJECTIVE 3 ▶ Finding equations of parallel and perpendicular lines.** Next, we find equations of parallel and perpendicular lines.

△  **EXAMPLE 6**   Find an equation of the line containing the point (4, 4) and parallel to the line $2x + 3y = -6$. Write the equation in standard form.

**Solution**   Because the line we want to find is *parallel* to the line $2x + 3y = -6$, the two lines must have equal slopes. Find the slope of $2x + 3y = -6$ by writing it in the form $y = mx + b$. In other words, solve the equation for $y$.

$$2x + 3y = -6$$
$$3y = -2x - 6 \quad \text{Subtract } 2x \text{ from both sides.}$$
$$y = \frac{-2x}{3} - \frac{6}{3} \quad \text{Divide by 3.}$$
$$y = -\frac{2}{3}x - 2 \quad \text{Write in slope-intercept form.}$$

The slope of this line is $-\frac{2}{3}$. Thus, a line parallel to this line will also have a slope of $-\frac{2}{3}$. The equation we are asked to find describes a line containing the point (4, 4) with a slope of $-\frac{2}{3}$. We use the point-slope form.

$$y - y_1 = m(x - x_1)$$
$$y - 4 = -\frac{2}{3}(x - 4) \quad \text{Let } m = -\frac{2}{3}, x_1 = 4, \text{ and } y_1 = 4.$$
$$3(y - 4) = -2(x - 4) \quad \text{Multiply both sides by 3.}$$
$$3y - 12 = -2x + 8 \quad \text{Apply the distributive property.}$$
$$2x + 3y = 20 \quad \text{Write in standard form.} \qquad \square$$

**▶ Helpful Hint**
Multiply both sides of the equation $2x + 3y = 20$ by $-1$ and it becomes $-2x - 3y = -20$. Both equations are in standard form, and their graphs are the same line.

**PRACTICE**
**6**   Find an equation of the line containing the point (8, −3) and parallel to the line $3x + 4y = 1$. Write the equation in standard form.

**EXAMPLE 7**   Write a function that describes the line containing the point (4, 4) and is perpendicular to the line $2x + 3y = -6$.

**Solution**   In the previous example, we found that the slope of the line $2x + 3y = -6$ is $-\frac{2}{3}$. A line perpendicular to this line will have a slope that is the negative reciprocal of $-\frac{2}{3}$, or $\frac{3}{2}$. From the point-slope equation, we have

$$y - y_1 = m(x - x_1)$$
$$y - 4 = \frac{3}{2}(x - 4) \quad \text{Let } x_1 = 4, y_1 = 4 \text{ and } m = \frac{3}{2}.$$
$$2(y - 4) = 3(x - 4) \quad \text{Multiply both sides by 2.}$$
$$2y - 8 = 3x - 12 \quad \text{Apply the distributive property.}$$
$$2y = 3x - 4 \quad \text{Add 8 to both sides.}$$
$$y = \frac{3}{2}x - 2 \quad \text{Divide both sides by 2.}$$
$$f(x) = \frac{3}{2}x - 2 \quad \text{Write using function notation.} \qquad \square$$

**PRACTICE**
**7**   Write a function that describes the line containing the point (8, −3) and is perpendicular to the line $3x + 4y = 1$.

## Graphing Calculator Explorations 🖩📱

You may have noticed by now that to use the [ Y = ] key on a graphing calculator to graph an equation, the equation must be solved for $y$.

*Graph each function by first solving the function for y.*

1. $x = 3.5y$

2. $-2.7y = x$

3. $5.78x + 2.31y = 10.98$

4. $-7.22x + 3.89y = 12.57$

5. $y - |x| = 3.78$

6. $3y - 5x^2 = 6x - 4$

7. $y - 5.6x^2 = 7.7x + 1.5$

8. $y + 2.6|x| = -3.2$

## VOCABULARY & READINESS CHECK

*Use the choices given to fill in each blank. Some choices may not be used.*

| | | |
|---|---|---|
| linear | $(0, b)$ | $m$ |
| quadratic | $(b, 0)$ | $mx$ |

1. A _____ function can be written in the form $f(x) = mx + b$.

2. In the form $f(x) = mx + b$, the $y$-intercept is _____ and the slope is _____.

*State the slope and the y-intercept of the graph of each function.*

3. $f(x) = -4x + 12$

4. $g(x) = \dfrac{2}{3}x - \dfrac{7}{2}$

5. $g(x) = 5x$

6. $f(x) = -x$

*Decide whether the lines are parallel, perpendicular, or neither.*

7. $y = 12x + 6$
   $y = 12x - 2$

8. $y = -5x + 8$
   $y = -5x - 8$

9. $y = -9x + 3$
   $y = \dfrac{3}{2}x - 7$

10. $y = 2x - 12$
    $y = \dfrac{1}{2}x - 6$

## 8.1 | EXERCISE SET

**MyMathLab**

PRACTICE  WATCH  DOWNLOAD  READ  REVIEW

*Graph each linear function. See Examples 1 and 2.*

**1.** $f(x) = -2x$

**2.** $f(x) = 2x$

**3.** $f(x) = -2x + 3$

**4.** $f(x) = 2x + 6$

**5.** $f(x) = \dfrac{1}{2}x$

**6.** $f(x) = \dfrac{1}{3}x$

**7.** $f(x) = \dfrac{1}{2}x - 4$

**8.** $f(x) = \dfrac{1}{3}x - 2$

*The graph of $f(x) = 5x$ follows. Use this graph to match each linear function with its graph. See Examples 1 and 2.*

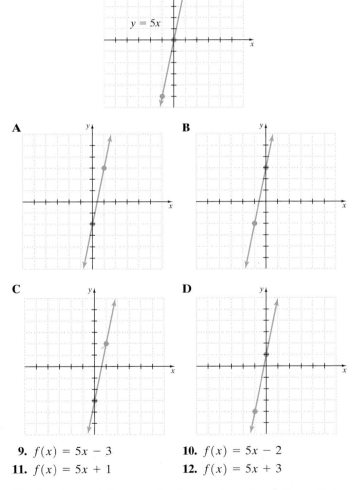

$y = 5x$

A

B

C

D

**9.** $f(x) = 5x - 3$

**10.** $f(x) = 5x - 2$

**11.** $f(x) = 5x + 1$

**12.** $f(x) = 5x + 3$

*Use function notation to write the equation of each line with the given slope and y-intercept. See Example 3.*

**13.** Slope $-1$; $y$-intercept $(0, 1)$

**14.** Slope $\dfrac{1}{2}$; $y$-intercept $(0, -6)$

**15.** Slope $2$; $y$-intercept $\left(0, \dfrac{3}{4}\right)$

**16.** Slope $-3$; $y$-intercept $\left(0, -\dfrac{1}{5}\right)$

**17.** Slope $\dfrac{2}{7}$; $y$-intercept $(0, 0)$

**18.** Slope $-\dfrac{4}{5}$; $y$-intercept $(0, 0)$

*Find an equation of the line with the given slope and containing the given point. Write the equation using function notation. See Example 3.*

**19.** Slope $3$; through $(1, 2)$

**20.** Slope $4$; through $(5, 1)$

**21.** Slope $-2$; through $(1, -3)$

**22.** Slope $-4$; through $(2, -4)$

**23.** Slope $\dfrac{1}{2}$; through $(-6, 2)$

**24.** Slope $\dfrac{2}{3}$; through $(-9, 4)$

**25.** Slope $-\dfrac{9}{10}$; through $(-3, 0)$

**26.** Slope $-\dfrac{1}{5}$; through $(4, -6)$

*Find an equation of the line passing through the given points. Use function notation to write the equation. See Example 4.*

**27.** $(2, 0), (4, 6)$

**28.** $(3, 0), (7, 8)$

**29.** $(-2, 5), (-6, 13)$

**30.** $(7, -4), (2, 6)$

**31.** $(-2, -4), (-4, -3)$

**32.** $(-9, -2), (-3, 10)$

**33.** $(-3, -8), (-6, -9)$

**34.** $(8, -3), (4, -8)$

**35.** $\left(\dfrac{3}{5}, \dfrac{4}{10}\right)$ and $\left(-\dfrac{1}{5}, \dfrac{7}{10}\right)$

**36.** $\left(\dfrac{1}{2}, -\dfrac{1}{4}\right)$ and $\left(\dfrac{3}{2}, \dfrac{3}{4}\right)$

*Write an equation of each line using function notation. See Example 5.*

**37.** Slope $0$; through $(-2, -4)$

**38.** Horizontal; through $(-3, 1)$

**39.** Horizontal; through $(0, 5)$

**40.** Slope $0$; through $(-10, 23)$

*Find an equation of each line. Write the equation using function notation. See Examples 6 and 7.*

**41.** Through $(3, 8)$; parallel to $f(x) = 4x - 2$

**42.** Through $(1, 5)$; parallel to $f(x) = 3x - 4$

**43.** Through $(2, -5)$; perpendicular to $3y = x - 6$

**44.** Through $(-4, 8)$; perpendicular to $2x - 3y = 1$

**45.** Through $(-2, -3)$; parallel to $3x + 2y = 5$

**46.** Through $(-2, -3)$; perpendicular to $3x + 2y = 5$

## MIXED PRACTICE

*Find the equation of each line. Write the equation using standard notation unless indicated otherwise. See Examples 3 through 7.*

**47.** Slope 2; through $(-2, 3)$

**48.** Slope 3; through $(-4, 2)$

**49.** Through $(1, 6)$ and $(5, 2)$; use function notation.

**50.** Through $(2, 9)$ and $(8, 6)$; use function notation

**51.** With slope $-\dfrac{1}{2}$; $y$-intercept 11; use function notation

**52.** With slope $-4$; $y$-intercept $\dfrac{2}{9}$; use function notation.

**53.** Through $(-7, -4)$ and $(0, -6)$

**54.** Through $(2, -8)$ and $(-4, -3)$

**55.** Slope $-\dfrac{4}{3}$; through $(-5, 0)$

**56.** Slope $-\dfrac{3}{5}$; through $(4, -1)$

**57.** Horizontal line; through $(-2, -10)$; use function notation

**58.** Horizontal line; through $(1, 0)$; use function notation

△ **59.** Through $(6, -2)$; parallel to the line $2x + 4y = 9$

△ **60.** Through $(8, -3)$; parallel to the line $6x + 2y = 5$

**61.** Slope 0; through $(-9, 12)$; use function notation

**62.** Slope 0; through $(10, -8)$; use function notation

△ **63.** Through $(6, 1)$; parallel to the line $8x - y = 9$

△ **64.** Through $(3, 5)$; perpendicular to the line $2x - y = 8$

△ **65.** Through $(5, -6)$; perpendicular to $y = 9$

△ **66.** Through $(-3, -5)$; parallel to $y = 9$

**67.** Through $(2, -8)$ and $(-6, -5)$; use function notation.

**68.** Through $(-4, -2)$ and $(-6, 5)$; use function notation.

## REVIEW AND PREVIEW

*Solve. Write the solution in interval notation. See Section 2.8.*

**69.** $2x - 7 \le 21$

**70.** $-3x + 1 > 0$

**71.** $5(x - 2) \ge 3(x - 1)$

**72.** $-2(x + 1) \le -x + 10$

**73.** $\dfrac{x}{2} + \dfrac{1}{4} < \dfrac{1}{8}$

**74.** $\dfrac{x}{5} - \dfrac{3}{10} \ge \dfrac{x}{2} - 1$

## CONCEPT EXTENSIONS

*Find an equation of each line graphed. Write the equation using function notation. (Hint: Use each graph to write 2 ordered-pair solutions. Find the slope of each line, then Examples 3 or 4 to complete.)*

**75.**

**76.**

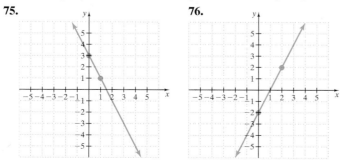

**77.**

**78.**

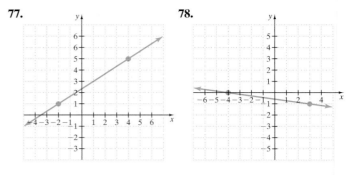

*Solve.*

**79.** Del Monte Fruit Company recently released a new applesauce. By the end of its first year, profits on this product amounted to $30,000. The anticipated profit for the end of the fourth year is $66,000. The ratio of change in time to change in profit is constant. Let $x$ be years and $P$ be profit.

　**a.** Write a linear function $P(x)$ that expresses profit as a function of time.

　**b.** Use this function to predict the company's profit at the end of the seventh year.

　**c.** Predict when the profit should reach $126,000.

**80.** The value of a computer bought in 2003 depreciates, or decreases, as time passes. Two years after the computer was bought, it was worth $2000; 4 years after it was bought, it was worth $800.

　**a.** If this relationship between number of years past 2003 and value of computer is linear, write an equation describing this relationship. [Use ordered pairs of the form (years past 2003, value of computer).]

　**b.** Use this equation to estimate the value of the computer in the year 2008.

**81.** The Pool Fun Company has learned that, by pricing a newly released Fun Noodle at $3, sales will reach 10,000 Fun Noodles per day during the summer. Raising the price to $5 will cause the sales to fall to 8000 Fun Noodles per day.

　**a.** Assume that the relationship between sales price and number of Fun Noodles sold is linear and write an equation describing this relationship.

　**b.** Predict the daily sales of Fun Noodles if the price is $3.50.

**82.** The value of a building bought in 1990 appreciates, or increases, as time passes. Seven years after the building was bought, it was worth $165,000; 12 years after it was bought, it was worth $180,000.

　**a.** If this relationship between number of years past 1990 and value of building is linear, write an equation describing this relationship. [Use ordered pairs of the form (years past 1990, value of building).]

　**b.** Use this equation to estimate the value of the building in the year 2010.

**83.** In 2006, the median price of an existing home in the United States was approximately $222,000. In 2001, the median price of an existing home was $150,900. Let $y$ be the median price of an existing home in the year $x$, where $x = 0$ represents 2001. (*Source:* National Association of REALTORS®)

　**a.** Write a linear equation that models the median existing home price in terms of the year $x$. [*Hint:* The line must pass through the points $(0, 150{,}900)$ and $(5, 222{,}000)$.]

**b.** Use this equation to predict the median existing home price for the year 2010.

**c.** Interpret the slope of the equation found in part **a.**

**84.** The number of births (in thousands) in the United States in 2000 was 4060. The number of births (in thousands) in the United States in 2004 was 4116. Let $y$ be the number of births (in thousands) in the year $x$, where $x = 0$ represents 2000. (*Source:* National Center for Health Statistics)

  **a.** Write a linear equation that models the number of births (in thousands) in terms of the year $x$. (See hint for Exercise 83a.)

  **b.** Use this equation to predict the number of births in the United States for the year 2013.

  **c.** Interpret the slope of the equation in part a.

**85.** The number of people employed in the United States as medical assistants was 387 thousand in 2004. By the year 2014, this number is expected to rise to 589 thousand. Let $y$ be the number of medical assistants (in thousands) employed in the United States in the year $x$, where $x = 0$ represents 2004. (*Source:* Bureau of Labor Statistics)

Ophthalmic medical assistant

  **a.** Write a linear equation that models the number of people (in thousands) employed as medical assistants in the year $x$. (See hint for Exercise 83a.)

  **b.** Use this equation to estimate the number of people who will be employed as medical assistants in the year 2013.

**86.** The number of people employed in the United States as systems analysts was 487 thousand in 2004. By the year 2014, this number is expected to rise to 640 thousand. Let $y$ be the number of systems analysts (in thousands) employed in the United States in the year $x$, where $x = 0$ represents 2004. (*Source:* Bureau of Labor Statistics)

  **a.** Write a linear equation that models the number of people (in thousands) employed as systems analysts in the year $x$. (See hint for Exercise 83a.)

  **b.** Use this equation to estimate the number of people who will be employed as systems analysts in the year 2012.

*Answer true or false.*

**87.** A vertical line is always perpendicular to a horizontal line.

**88.** A vertical line is always parallel to a vertical line.

*Example:*

*Find an equation of the perpendicular bisector of the line segment whose endpoints are (2, 6) and (0, −2).*

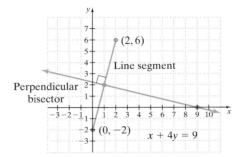

*Solution:*

A perpendicular bisector is a line that contains the midpoint of the given segment and is perpendicular to the segment.

**Step 1:** The midpoint of the segment with endpoints $(2, 6)$ and $(0, −2)$ is $(1, 2)$.

**Step 2:** The slope of the segment containing points $(2, 6)$ and $(0, −2)$ is 4.

**Step 3:** A line perpendicular to this line segment will have slope of $-\dfrac{1}{4}$.

**Step 4:** The equation of the line through the midpoint $(1, 2)$ with a slope of $-\dfrac{1}{4}$ will be the equation of the perpendicular bisector. This equation in standard form is $x + 4y = 9$.

*Find an equation of the perpendicular bisector of the line segment whose endpoints are given. See the previous example.*

△ **89.** $(3, −1); (−5, 1)$

△ **90.** $(−6, −3); (−8, −1)$

△ **91.** $(−2, 6); (−22, −4)$

△ **92** $(5, 8); (7, 2)$

△ **93.** $(2, 3); (−4, 7)$

△ **94.** $(−6, 8); (−4, −2)$

↘ **95.** Describe how to check to see if the graph of $2x − 4y = 7$ passes through the points $(1.4, −1.05)$ and $(0, −1.75)$. Then follow your directions and check these points.

## 8.2 REVIEWING FUNCTION NOTATION AND GRAPHING NONLINEAR FUNCTIONS

**OBJECTIVES**

1  Review function notation.

2  Find square roots of numbers.

3  Graph nonlinear functions.

In the previous section, we studied linear equations that described functions. Not all equations in two variables are linear equations, and not all graphs of equations in two variables are lines. In Chapter 3, we saw graphs of nonlinear equations, some of which were functions since they passed the vertical line test. In this section, we study the functions whose graphs may not be lines. First, let's review function notation.

**OBJECTIVE 1 ▶ Reviewing function notation.** Suppose we have a function $f$ such that $f(2) = -1$. Recall this means that when $x = 2$, $f(x)$, or $y = -1$. Thus, the graph of $f$ passes through $(2, -1)$.

▶ **Helpful Hint**

Remember that $f(x)$ is a special symbol in mathematics used to denote a function. The symbol $f(x)$ is read "$f$ of $x$." It does *not* mean $f \cdot x$ ($f$ times $x$).

**Concept Check** ☑

Suppose $y = f(x)$ and we are told that $f(3) = 9$. Which is not true?

**a.** When $x = 3$, $y = 9$.
**b.** A possible function is $f(x) = x^2$.
**c.** A point on the graph of the function is $(3, 9)$.
**d.** A possible function is $f(x) = 2x + 4$.

If it helps, think of a function, $f$, as a machine that has been programmed with a certain correspondence or rule. An input value (a member of the domain) is then fed into the machine, the machine does the correspondence or rule and the result is the output (a member of the range).

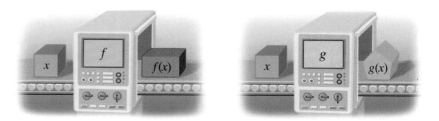

**EXAMPLE 1**    Given the graphs of the functions $f$ and $g$, find each function value by inspecting the graphs.

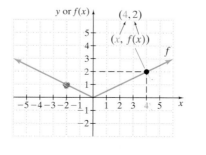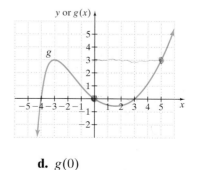

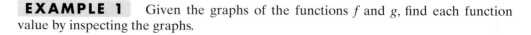

**a.** $f(4)$          **b.** $f(-2)$          **c.** $g(5)$          **d.** $g(0)$
**e.** Find all $x$-values such that $f(x) = 1$.
**f.** Find all $x$-values such that $g(x) = 0$.

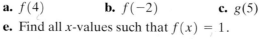

**Answer to Concept Check:**  d

## Solution

**a.** To find $f(4)$, find the $y$-value when $x = 4$. We see from the graph that when $x = 4$, $y$ or $f(x) = 2$. Thus, $f(4) = 2$.

**b.** $f(-2) = 1$ from the ordered pair $(-2, 1)$.

**c.** $g(5) = 3$ from the ordered pair $(5, 3)$.

**d.** $g(0) = 0$ from the ordered pair $(0, 0)$.

**e.** To find $x$-values such that $f(x) = 1$, we are looking for any ordered pairs on the graph of $f$ whose $f(x)$ or $y$-value is 1. They are $(2, 1)$ and $(-2, 1)$. Thus $f(2) = 1$ and $f(-2) = 1$. The $x$-values are 2 and $-2$.

**f.** Find ordered pairs on the graph of $g$ whose $g(x)$ or $y$-value is 0. They are $(3, 0)$ $(0, 0)$ and $(-4, 0)$. Thus $g(3) = 0$, $g(0) = 0$, and $g(-4) = 0$. The $x$-values are 3, 0, and $-4$.   ☐

**PRACTICE**

**1**    Given the graphs of the functions $f$ and $g$, find each function value by inspecting the graphs.

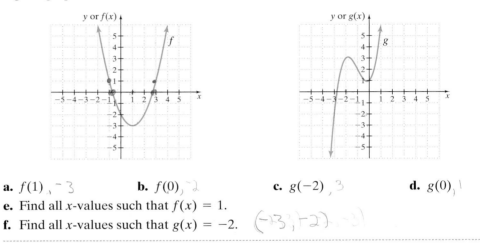

**a.** $f(1)$  $^-3$           **b.** $f(0)$  $2$           **c.** $g(-2)$  $3$           **d.** $g(0)$  $1$

**e.** Find all $x$-values such that $f(x) = 1$.

**f.** Find all $x$-values such that $g(x) = -2$.   $(-3, -2)$   $^-3$

Many types of real-world paired data form functions. The broken-line graph below shows the research and development spending by the Pharmaceutical Manufacturers Association.

**EXAMPLE 2**    The following graph shows the research and development expenditures by the Pharmaceutical Manufacturers Association as a function of time.

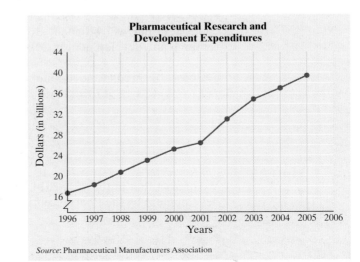

**Pharmaceutical Research and Development Expenditures**

*Source*: Pharmaceutical Manufacturers Association

**a.** Approximate the money spent on research and development in 2002.

**b.** In 1958, research and development expenditures were $200 million. Find the increase in expenditures from 1958 to 2004.

*Solution*

**a.** Find the year 2002 and move upward until you reach the graph. From the point on the graph move horizontally, to the left, until the other axis is reached. In 2002, approximately $31 billion was spent.

**b.** In 2004, approximately $37 billion, or $37,000 million was spent. The increase in spending from 1958 to 2004 is $37,000 − $200 = $36,800 million or $36.8 billion. □

**PRACTICE**
**2** Use the graph in Example 2 and approximate the money spent in 2003.

Notice that the graph in Example 2 is the graph of a function since for each year there is only one total amount of money spent by the Pharmaceutical Manufacturers Association on research and development. Also notice that the graph resembles the graph of a line. Often, businesses depend on equations that "closely fit" data-defined functions like this one in order to model the data and predict future trends. For example, by a method called **least squares,** the function $f(x) = 2.602x - 5178$ approximates the data shown. For this function, $x$ is the year and $f(x)$ is total money spent. Its graph and the actual data function are shown next.

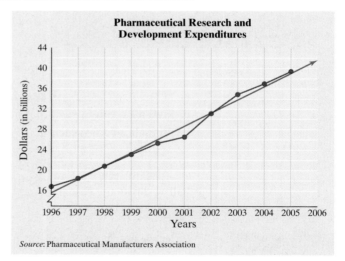

**Pharmaceutical Research and Development Expenditures**

*Source*: Pharmaceutical Manufacturers Association

**EXAMPLE 3** Use the function $f(x) = 2.602x - 5178$ to predict the amount of money that will be spent by the Pharmaceutical Manufacturers Association on research and development in 2014.

*Solution* To predict the amount of money that will be spent in the year 2014 we use $f(x) = 2.602x - 5178$ and find $f(2014)$.

$$f(x) = 2.602x - 5178$$
$$f(2014) = 2.602(2014) - 5178$$
$$= 62.428$$

We predict that in the year 2014, $62.428 billion dollars will be spent on research and development by the Pharmaceutical Manufacturers Association. □

**PRACTICE**
**3** Use $f(x) = 2.602x - 5178$ to approximate the money spent in 2012.

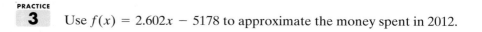

**OBJECTIVE 2** ▶ ~~Finding square roots of numbers.~~ Later in this section, we graph the square root function, $f(x) = \sqrt{x}$. To prepare for this graph, let's review finding square roots of numbers.

The opposite of squaring a number is taking the **square root** of a number. For example, since the square of 4, or $4^2$, is 16, we say that a square root of 16 is 4. The notation $\sqrt{a}$ is used to denote the **positive,** or **principal, square root** of a nonnegative number $a$. We then have in symbols that $\sqrt{16} = 4$. The negative square root of 16 is written $-\sqrt{16} = -4$. The square root of a negative number, such as $\sqrt{-16}$ is not a real number. Why? There is no real number, that when squared gives a negative number.

**EXAMPLE 4**   Find the square roots.

**a.** $\sqrt{9}$      **b.** $\sqrt{25}$      **c.** $\sqrt{\dfrac{1}{4}}$      **d.** $-\sqrt{36}$      **e.** $\sqrt{-36}$      **f.** $\sqrt{0}$

*Solution*

**a.** $\sqrt{9} = 3$ since 3 is positive and $3^2 = 9$.      **b.** $\sqrt{25} = 5$ since $5^2 = 25$.

**c.** $\sqrt{\dfrac{1}{4}} = \dfrac{1}{2}$ since $\left(\dfrac{1}{2}\right)^2 = \dfrac{1}{4}$.      **d.** $-\sqrt{36} = -6$

**e.** $\sqrt{-36}$ is not a real number.      **f.** $\sqrt{0} = 0$ since $0^2 = 0$.      □

**PRACTICE**
**4**   Find the square roots.

**a.** $\sqrt{121}$      **b.** $\sqrt{\dfrac{1}{16}}$      **c.** $-\sqrt{64}$      **d.** $\sqrt{-64}$      **e.** $\sqrt{100}$

---

We can find roots other than square roots. Also, not all roots simplify to rational numbers. For example, $\sqrt{3} \approx 1.7$ using a calculator. We study radicals further in Chapter 10.

**OBJECTIVE 3** ▶ **Graphing nonlinear functions.**
Let's practice graphing nonlinear functions.

**EXAMPLE 5**   Graph $f(x) = x^2$.

*Solution*   This equation is not linear because the $x^2$ term does not allow us to write it in the form $Ax + By = C$. Its graph is not a line. We begin by finding ordered pair solutions. Because $f(x) = y$, feel free to think of this equation as $f(x) = x^2$ or $y = x^2$. This graph is solved for $y$, so we choose $x$-values and find corresponding $y$-values.

If $x = -3$, then $f(-3) = (-3)^2$, or 9.

If $x = -2$, then $f(-2) = (-2)^2$, or 4.

If $x = -1$, then $f(-1) = (-1)^2$, or 1.

If $x = 0$, then $f(0) = 0^2$, or 0.

If $x = 1$, then $f(1) = 1^2$, or 1.

If $x = 2$, then $f(2) = 2^2$, or 4.

If $x = 3$, then $f(3) = 3^2$, or 9.

| $x$ | $y$ or $f(x)$ |
|-----|-----|
| $-3$ | 9 |
| $-2$ | 4 |
| $-1$ | 1 |
| 0 | 0 |
| 1 | 1 |
| 2 | 4 |
| 3 | 9 |

Study the table a moment and look for patterns. Notice that the ordered pair solution $(0, 0)$ contains the smallest $y$-value because any other $x$-value squared will give a positive result. This means that the point $(0, 0)$ will be the lowest point on the graph. Also notice that all other $y$-values correspond to two different $x$-values. For example, $3^2 = 9$ and also $(-3)^2 = 9$. This means that the graph will be a mirror image of itself across the $y$-axis. Connect the plotted points with a smooth curve to sketch the graph.

This curve is given a special name, a **parabola.** We will study more about parabolas in later chapters.

**PRACTICE**

**5**    Graph $f(x) = 2x^2$.

---

**EXAMPLE 6**    Graph the nonlinear function $f(x) = |x|$.

**Solution**   This is not a linear equation since it cannot be written in the form $Ax + By = C$. Its graph is not a line. Because we do not know the shape of this graph, we find many ordered pair solutions. We will choose $x$-values and substitute to find corresponding $y$-values.

If $x = -3$, then $f(-3) = |-3|$, or 3.

If $x = -2$, then $f(-2) = |-2|$, or 2.

If $x = -1$, then $f(-1) = |-1|$, or 1.

If $x = 0$, then $f(\ ) = |0|$, or 0.

If $x = 1$, then $f(1) = |1|$, or 1.

If $x = 2$, then $f(2) = |2|$, or 2.

If $x = 3$, then $f(3) = |3|$, or 3.

| $x$ | $y$ or $f(x)$ |
|-----|---------------|
| $-3$ | 3 |
| $-2$ | 2 |
| $-1$ | 1 |
| 0 | 0 |
| 1 | 1 |
| 2 | 2 |
| 3 | 3 |

Again, study the table of values for a moment and notice any patterns.

From the plotted ordered pairs, we see that the graph of this absolute value equation is V-shaped.

**PRACTICE**

**6**    Graph $f(x) = -|x|$.

---

**EXAMPLE 7**    Graph the nonlinear function $f(x) = \sqrt{x}$.

**Solution**   To graph, this square root function, we identify the domain, evaluate the function for several values of $x$, plot the resulting points, and connect the points with a smooth curve. Since $\sqrt{x}$ represents the nonnegative square root of $x$, the domain of this function is the set of all nonnegative numbers, $\{x | x \geq 0\}$, or $[0, \infty)$. We have approximated $\sqrt{3}$ below to help us locate the point corresponding to $(3, \sqrt{3})$.

If $x = 0$, then $f(0) = \sqrt{0}$, or 0.

If $x = 1$, then $f(1) = \sqrt{1}$, or 1.

If $x = 3$, then $f(3) = \sqrt{3}$, or 1.7.

If $x = 4$, then $f(4) = \sqrt{4}$, or 2.

If $x = 9$, then $f(9) = \sqrt{9}$, or 3.

| $x$ | $y$ or $f(x)$ |
|-----|---------------|
| 0 | 0 |
| 1 | 1 |
| 3 | $\sqrt{3} \approx 1.7$ |
| 4 | 2 |
| 9 | 3 |

**PRACTICE**

**7**    Graph $f(x) = \sqrt{x} + 1$.

**Graphing Calculator Explorations**

It is possible to use a graphing calculator to sketch the graph of more than one equation on the same set of axes. For example, graph the functions $f(x) = x^2$ and $g(x) = x^2 + 4$ on the same set of axes.

To graph on the same set of axes, press the $\boxed{Y =}$ key and enter the equations on the first two lines.

$$Y_1 = x^2$$
$$Y_2 = x^2 + 4$$

Then press the $\boxed{\text{GRAPH}}$ key as usual. The screen should look like this.

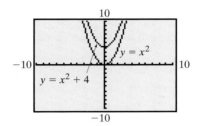

Notice that the graph of $y$ or $g(x) = x^2 + 4$ is the graph of $y = x^2$ moved 4 units upward.

*Graph each pair of functions on the same set of axes. Describe the similarities and differences in their graphs.*

**1.** $f(x) = |x|$
   $g(x) = |x| + 1$

**2.** $f(x) = x^2$
   $h(x) = x^2 - 5$

**3.** $f(x) = x$
   $H(x) = x - 6$

**4.** $f(x) = |x|$
   $G(x) = |x| + 3$

**5.** $f(x) = -x^2$
   $F(x) = -x^2 + 7$

**6.** $f(x) = x$
   $F(x) = x + 2$

## VOCABULARY & READINESS CHECK

*Use the choices below to fill in each blank. Some choices may not be used.*

| $(1.7, -2)$ | line | parabola | $-6$ | $-9$ |
| $(-2, 1.7)$ | V-shaped | 6 | 9 |

**1.** The graph of $y = |x|$ looks _____ .

**2.** The graph of $y = x^2$ is a _____ .

**3.** If $f(-2) = 1.7$, the corresponding ordered pair is _____ .

**4.** If $f(x) = x^2$, then $f(-3) = $ __ .

## 8.2 EXERCISE SET

*Use the graph of the following function f(x) to find each value. See Examples 1 and 2.*

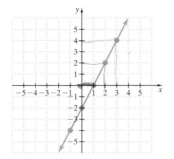

**1.** $f(1) \approx 0$

**2.** $f(0) = 0$

**3.** $f(-1) = -4$

**4.** $f(2) = 2$

**5.** Find $x$ such that $f(x) = 4.$ 3.5

**6.** Find $x$ such that $f(x) = -6.$ $\frac{-5}{4} = -2$

*Use the graph of the functions below to answer Exercises 7 through 18. See Examples 1 and 2.*

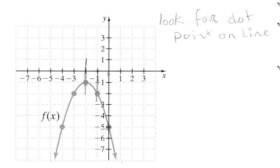

look for dot
point on Line

$f(x)$

**7.** If $f(1) = -10$, write the corresponding ordered pair. $(1, -10)$

**8.** If $f(-5) = -10$, write the corresponding ordered pair. $(-5, -10)$

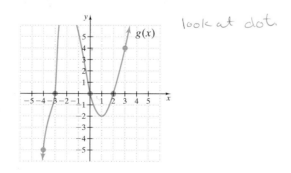

look at dot.

$g(x)$

**9.** If $g(4) = 56$, write the corresponding ordered pair. $(4, 56)$

**10.** If $g(-2) = 8$, write the corresponding ordered pair. $(-2, 8)$

**11.** Find $f(-1).$ $= -2$

**12.** Find $f(-2).$ $= -1$

**13.** Find $g(2).$ $= 0$

**14.** Find $g(-4).$ $= -5$

**15.** Find all values of $x$ such that $f(x) = -5.$ $(-4, -5) (0, -5)$

**16.** Find all values of $x$ such that $f(x) = -2.$ $-3$

**17.** Find all positive values of $x$ such that $g(x) = 4.$ $(3, 4)$

**18.** Find all values of $x$ such that $g(x) = 0.$ $0$

start at 4

*Find the following roots. See Example 4.*

**19.** $\sqrt{49}$

**20.** $\sqrt{81}$

**21.** $-\sqrt{\dfrac{4}{9}}$

**22.** $-\sqrt{\dfrac{4}{25}}$

**23.** $\sqrt{64}$

**24.** $\sqrt{4}$

**25.** $\sqrt{81}$

**26.** $\sqrt{1}$

**27.** $\sqrt{-100}$

**28.** $\sqrt{-25}$

### MIXED PRACTICE

*Graph each function by finding and plotting ordered pair solutions. See Examples 5 through 7.*

**29.** $f(x) = x^2 + 3$

**30.** $g(x) = (x + 2)^2$

**31.** $h(x) = |x| - 2$

**32.** $f(x) = |x - 2|$

**33.** $g(x) = 2x^2$

**34.** $h(x) = 5x^2$

**35.** $f(x) = 5x - 1$

**36.** $g(x) = -3x + 2$

**37.** $f(x) = \sqrt{x + 1}$

**38.** $f(x) = \sqrt{x} - 1$

**39.** $g(x) = -2|x|$

**40.** $g(x) = -3|x|$

**41.** $h(x) = \sqrt{x} + 2$

**42.** $h(x) = \sqrt{x + 2}$

*Use the graph in Example 2 to answer the following. Also see Example 3.*

**43. a.** Use the graph to approximate the money spent on research and development in 1996.

**b.** Recall that the function $f(x) = 2.602x - 5178$ approximates the graph in Example 2. Use this equation to approximate the money spent on research and development in 1996.

**44. a.** Use the graph to approximate the money spent on research and development in 1999.

**b.** Use the function $f(x) = 2.602x - 5178$ to approximate the money spent on research and development in 1999.

*The function $f(x) = 0.42x + 10.5$, can be used to predict diamond production. For this function, x is the number of years after 2000, and f(x) is the value (in billions of dollars) of diamond production. (See the Chapter 8 opener.)*

**45.** Use the function in the directions above to predict diamond production in 2012.

**46.** Use the function in the directions above to predict diamond production in 2015.

*The function $A(r) = \pi r^2$ may be used to find the area of a circle if we are given its radius.*

△ **47.** Find the area of a circle whose radius is 5 centimeters. (Do not approximate $\pi$.)

△ **48.** Find the area of a circular garden whose radius is 8 feet. (Do not approximate $\pi$.)

*The function $V(x) = x^3$ may be used to find the volume of a cube if we are given the length x of a side.*

**49.** Find the volume of a cube whose side is 14 inches.

**50.** Find the volume of a die whose side is 1.7 centimeters.

*Forensic scientists use the following functions to find the height of a woman if they are given the height of her femur bone f or her tibia bone t in centimeters.*

$$H(f) = 2.59f + 47.24$$
$$H(t) = 2.72t + 61.28$$

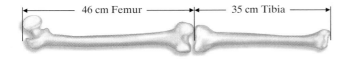

**51.** Find the height of a woman whose femur measures 46 centimeters.

**52.** Find the height of a woman whose tibia measures 35 centimeters.

*The dosage in milligrams D of Ivermectin, a heartworm preventive, for a dog who weighs x pounds is given by*

$$D(x) = \frac{136}{25}x$$

**53.** Find the proper dosage for a dog that weighs 30 pounds.

**54.** Find the proper dosage for a dog that weighs 50 pounds.

**55.** What is the greatest number of x-intercepts that a function may have? Explain your answer.

**56.** What is the greatest number of y-intercepts that a function may have? Explain your answer.

## REVIEW AND PREVIEW

*Solve the following equations. See Section 2.3.*

**57.** $3(x - 2) + 5x = 6x - 16$

**58.** $5 + 7(x + 1) = 12 + 10x$

**59.** $3x + \dfrac{2}{5} = \dfrac{1}{10}$   **60.** $\dfrac{1}{6} + 2x = \dfrac{2}{3}$

## CONCEPT EXTENSIONS

*For Exercises 61 through 64, match each description with the graph that best illustrates it.*

**61.** Moe worked 40 hours per week until the fall semester started. He quit and didn't work again until he worked 60 hours a week during the holiday season starting mid-December.

**62.** Kawana worked 40 hours a week for her father during the summer. She slowly cut back her hours to not working at all during the fall semester. During the holiday season in December, she started working again and increased her hours to 60 hours per week.

**63.** Wendy worked from July through February, never quitting. She worked between 10 and 30 hours per week.

**64.** Bartholomew worked from July through February. During the holiday season between mid-November and the beginning of January, he worked 40 hours per week. The rest of the time, he worked between 10 and 40 hours per week.

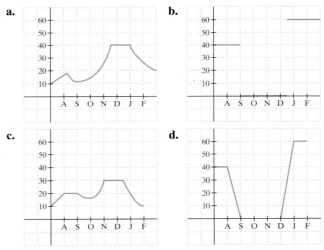

*The graph below shows first-class postal rates and the years it increased. Use this graph for Exercises 65 through 68.*

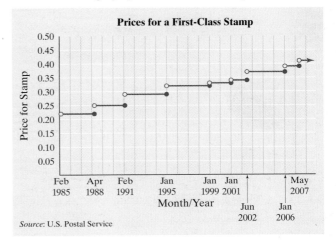

*Source*: U.S. Postal Service

**65.** What was the first year that the price for a first-class stamp rose above $0.25?

**66.** What was the first year that the price for a first-class stamp rose above $0.30?

**67.** Why do you think that this graph is shaped the way it is?

**68.** The U.S. Postal Service issued first-class stamps as far back as 1885. The cost for a first-class stamp then was $0.02. By how much had it increased by 2007?

**69.** Graph $y = x^2 - 4x + 7$. Let $x = 0, 1, 2, 3, 4$ to generate ordered pair solutions.

**70.** Graph $y = x^2 + 2x + 3$. Let $x = -3, -2, -1, 0, 1$ to generate ordered pair solutions.

**71.** The function $f(x) = [x]$ is called the greatest integer function and its graph is similar to the graph on the previous page. The value of $[x]$ is the greatest integer less than or equal to $x$. For example, $f(1.5) = [1.5] = 1$ since the greatest integer $\leq 1.5$ is 1. Find the graph of $f(x) = [x]$.

## INTEGRATED REVIEW  SUMMARY ON FUNCTIONS AND EQUATIONS OF LINES

Sections 8.1–8.2

*Find the slope and y-intercept of the graph of each function.*

**1.** $f(x) = 3x - 5$

**2.** $f(x) = \dfrac{5}{2}x - \dfrac{7}{2}$

*Determine whether each pair of lines is parallel, perpendicular, or neither.*

**3.** $f(x) = 8x - 6$
   $g(x) = 8x + 6$

**4.** $f(x) = \dfrac{2}{3}x + 1$
   $2y + 3x = 1$

*Find the equation of each line. Write the equation using function notation.*

**5.** Through $(1, 6)$ and $(5, 2)$

**6.** Through $(2, -8)$ and $(-6, -5)$

**7.** Through $(-1, -5)$; parallel to $3x - y = 5$

**8.** Through $(0, 4)$; perpendicular to $4x - 5y = 10$

**9.** Through $(2, -3)$; perpendicular to $4x + y = \dfrac{2}{3}$

**10.** Through $(-1, 0)$; parallel to $5x + 2y = 2$

*Determine whether each function is linear or not. Then graph the function.*

**11.** $f(x) = 4x - 2$

**12.** $f(x) = 6x - 5$

**13.** $g(x) = |x| + 3$

**14.** $h(x) = |x| + 2$

**15.** $f(x) = 2x^2$

**16.** $F(x) = 3x^2$

**17.** $h(x) = x^2 - 3$

**18.** $G(x) = x^2 + 3$

**19.** $F(x) = -2x$

**20.** $H(x) = -3x$

**21.** $G(x) = |x + 2|$

**22.** $g(x) = |x - 1|$

**23.** $f(x) = \dfrac{1}{3}x - 1$

**24.** $f(x) = \dfrac{1}{2}x - 3$

**25.** $g(x) = -\dfrac{3}{2}x + 1$

**26.** $G(x) = -\dfrac{2}{3}x + 1$

## 8.3 GRAPHING PIECEWISE-DEFINED FUNCTIONS AND SHIFTING AND REFLECTING GRAPHS OF FUNCTIONS

**OBJECTIVES**

1 Graph piecewise-defined functions.

2 Vertical and horizontal shifts.

3 Reflect graphs.

**OBJECTIVE 1 ▶ Graphing piecewise-defined functions.** Throughout Chapter 8, we have graphed functions. There are many special functions. In this objective, we study functions defined by two or more expressions. The expression used to complete the function varies with, and depends upon the value of $x$. Before we actually graph these piecewise-defined functions, let's practice finding function values.

**EXAMPLE 1** Evaluate $f(2)$, $f(-6)$, and $f(0)$ for the function

$$f(x) = \begin{cases} 2x + 3 & \text{if } x \le 0 \\ -x - 1 & \text{if } x > 0 \end{cases}$$

Then write your results in ordered-pair form.

*Solution* Take a moment and study this function. It is a single function defined by two expressions depending on the value of $x$. From above, if $x \le 0$, use $f(x) = 2x + 3$. If $x > 0$, use $f(x) = -x - 1$. Thus

$$f(2) = -(2) - 1$$
$$= -3 \quad \text{since } 2 > 0$$
$$f(2) = -3$$
Ordered pairs: $(2, -3)$

$$f(-6) = 2(-6) + 3$$
$$= -9 \quad \text{since } -6 \le 0$$
$$f(-6) = -9$$
$$(-6, -9)$$

$$f(0) = 2(0) + 3$$
$$= 3 \quad \text{since } 0 \le 0$$
$$f(0) = 3$$
$$(0, 3)$$

**PRACTICE**

**1** Evaluate $f(4)$, $f(-2)$, and $f(0)$ for the function

$$f(x) = \begin{cases} -4x - 2 & \text{if } x \le 0 \\ x + 1 & \text{if } x > 0. \end{cases}$$

Now, let's graph a piecewise-defined function.

**EXAMPLE 2** Graph $f(x) = \begin{cases} 2x + 3 & \text{if } x \le 0 \\ -x - 1 & \text{if } x > 0 \end{cases}$

*Solution* Let's graph each piece.

If $x \le 0$,          If $x > 0$,
$f(x) = 2x + 3$        $f(x) = -x - 1$

Values $\le 0$

| $x$ | $f(x) = 2x + 3$ |
|-----|------------------|
| 0 | 3 Closed circle |
| $-1$ | 1 |
| $-2$ | $-1$ |

Values $> 0$

| $x$ | $f(x) = -x - 1$ |
|-----|------------------|
| 1 | $-2$ |
| 2 | $-3$ |
| 3 | $-4$ |

The graph of the first part of $f(x)$ listed will look like a ray with a closed-circle endpoint at $(0, 3)$. The graph of the second part of $f(x)$ listed will look like a ray with an open-circle endpoint. To find the exact location of the open-circle endpoint, use

$f(x) = -x - 1$ and find $f(0)$. Since $f(0) = -0 - 1 = -1$, we graph the second table and place an open circle at $(0, -1)$.

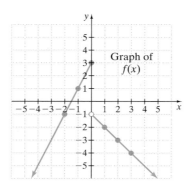

Notice that this graph is the graph of a function because it passes the vertical line test. The domain of this function is $(-\infty, \infty)$ and the range is $(-\infty, 3]$. □

**PRACTICE**
**2** Graph

$$f(x) = \begin{cases} -4x - 2 & \text{if } x \le 0 \\ x + 1 & \text{if } x > 0 \end{cases}$$

**OBJECTIVE 2 ▶ Vertical and horizontal shifting.** Your knowledge of the slope-intercept form, $f(x) = mx + b$, will help you understand simple shifting of transformations such as vertical shifts. For example, what is the difference between the graphs of $f(x) = x$ and $g(x) = x + 3$?

$f(x) = x$
slope, $m = 1$
$y$-intercept is $(0, 0)$

$g(x) = x + 3$
slope, $m = 1$
$y$-intercept is $(0, 3)$

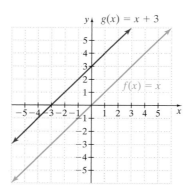

***Review of Common Graphs***

We now take common graphs and learn how more complicated graphs are actually formed by shifting and reflecting these common graphs. These shifts and reflections are called transformations, and it is possible to combine transformations. A knowledge of these transformations will help you simplify future graphs.

Let's begin with a review of the graphs of four common functions. Many of these functions we graphed in earlier sections.

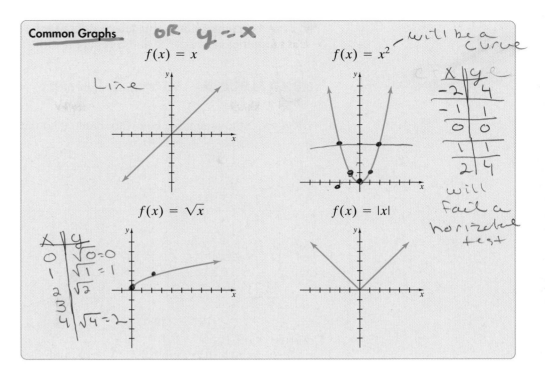

**Common Graphs**   OR $y = x$

$f(x) = x$        $f(x) = x^2$ — will be a curve

Line

| X | y |
|---|---|
| -2 | 4 |
| -1 | 1 |
| 0 | 0 |
| 1 | 1 |
| 2 | 4 |

will fail a horizontal test

$f(x) = \sqrt{x}$        $f(x) = |x|$

| X | y |
|---|---|
| 0 | $\sqrt{0} = 0$ |
| 1 | $\sqrt{1} = 1$ |
| 2 | $\sqrt{2}$ |
| 3 | |
| 4 | $\sqrt{4} = 2$ |

Notice that the graph of $g(x) = x + 3$ is the same as the graph of $f(x) = x$, but moved upward 3 units. This is an example of a **vertical shift** and is true for graphs in general.

**Vertical Shifts (Upward and Downward)**
**Let $k$ be a Positive Number**

| Graph of | Same As | Moved |
|---|---|---|
| $g(x) = f(x) + k$ | $f(x)$ | $k$ units upward |
| $g(x) = f(x) - k$ | $f(x)$ | $k$ units downward |

**EXAMPLES**   Without plotting points, sketch the graph of each pair of functions on the same set of axes.

**3.** $f(x) = x^2$ and $g(x) = x^2 + 2$        **4.** $f(x) = \sqrt{x}$ and $g(x) = \sqrt{x} - 3$

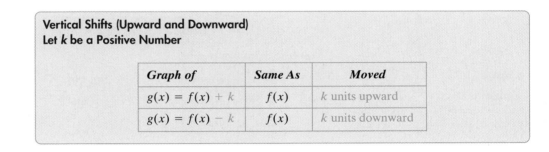

**PRACTICES**

**3–4**   Without plotting points, sketch the graphs of each pair of functions on the same set of axes.

**3.** $f(x) = x^2$ and $g(x) = x^2 - 3$        **4.** $f(x) = \sqrt{x}$ and $g(x) = \sqrt{x} + 1$

A horizontal shift to the left or right may be slightly more difficult to understand. Let's graph $g(x) = |x - 2|$ and compare it with $f(x) = |x|$.

**EXAMPLE 5** Sketch the graphs of $f(x) = |x|$ and $g(x) = |x - 2|$ on the same set of axes.

*Solution* Study the table below to understand the placement of both graphs.

V-function

| $x$ | $f(x) = |x|$ | $g(x) = |x - 2|$ |
|---|---|---|
| $-3$ | 3 | 5 |
| $-2$ | 2 | 4 |
| $-1$ | 1 | 3 |
| 0 | 0 | 2 |
| 1 | 1 | 1 |
| 2 | 2 | 0 |
| 3 | 3 | 1 |

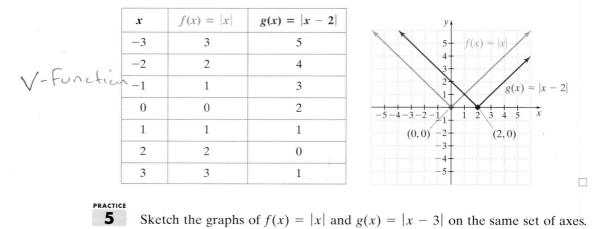

**PRACTICE**
**5** Sketch the graphs of $f(x) = |x|$ and $g(x) = |x - 3|$ on the same set of axes.

The graph of $g(x) = |x - 2|$ is the same as the graph of $f(x) = |x|$, but moved 2 units to the right. This is an example of a **horizontal shift** and is true for graphs in general.

**Horizontal Shift (To the Left or Right)**
**Let $h$ be a Positive Number**

| Graph of | Same as | Moved |
|---|---|---|
| $g(x) = f(x - h)$ | $f(x)$ | $h$ units to the right |
| $g(x) = f(x + h)$ | $f(x)$ | $h$ units to the left |

> ▶ **Helpful Hint**
> Notice that $f(x - h)$ corresponds to a shift to the right and $f(x + h)$ corresponds to a shift to the left.

Vertical and horizontal shifts can be combined.

**EXAMPLE 6** Sketch the graphs of $f(x) = x^2$ and $g(x) = (x - 2)^2 + 1$ on the same set of axes.

*Solution* The graph of $g(x)$ is the same as the graph of $f(x)$ shifted 2 units to the right and 1 unit up.

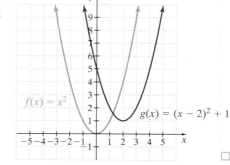

**PRACTICE**
**6** Sketch the graphs of $f(x) = |x|$ and $g(x) = |x - 2| + 3$ on the same set of axes.

**OBJECTIVE 3 ▶ Reflecting graphs.** Another type of transformation is called a **reflection.** In this section, we will study reflections (mirror images) about the $x$-axis only. For example, take a moment and study these two graphs. The graph of $g(x) = -x^2$ can be verified, as usual, by plotting points.

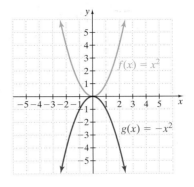

**Reflection about the x-axis**

The graph of $g(x) = -f(x)$ is the graph of $f(x)$ reflected about the $x$-axis.

**EXAMPLE 7**   Sketch the graph of $h(x) = -|x - 3| + 2$.

**Solution**   The graph of $h(x) = -|x - 3| + 2$ is the same as the graph of $f(x) = |x|$ reflected about the $x$-axis, then moved three units to the right and two units upward.

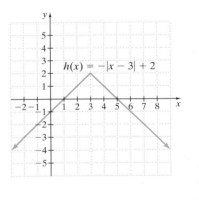

**PRACTICE**
**7**   Sketch the graph of $h(x) = -(x + 2)^2 - 1$.

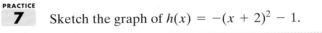

There are other transformations, such as stretching that won't be covered in this section. For a review of this transformation, see the Appendix.

# VOCABULARY & READINESS CHECK

*Match each equation with its graph.*

**1.** $y = \sqrt{x}$       **2.** $y = x^2$       **3.** $y = x$       **4.** $y = |x|$

A                    B                    C                    D

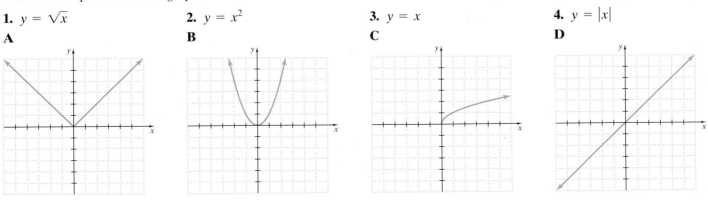

## 8.3 | EXERCISE SET

*Graph each piecewise-defined function. See Examples 1 and 2.*

**1.** $f(x) = \begin{cases} 2x & \text{if } x < 0 \\ x + 1 & \text{if } x \geq 0 \end{cases}$

**2.** $f(x) = \begin{cases} 3x & \text{if } x < 0 \\ x + 2 & \text{if } x \geq 0 \end{cases}$

**3.** $f(x) = \begin{cases} 4x + 5 & \text{if } x \leq 0 \\ \frac{1}{4}x + 2 & \text{if } x > 0 \end{cases}$

**4.** $f(x) = \begin{cases} 5x + 4 & \text{if } x \leq 0 \\ \frac{1}{3}x - 1 & \text{if } x > 0 \end{cases}$

**5.** $g(x) = \begin{cases} -x & \text{if } x \leq 1 \\ 2x + 1 & \text{if } x > 1 \end{cases}$

**6.** $g(x) = \begin{cases} 3x - 1 & \text{if } x \leq 2 \\ -x & \text{if } x > 2 \end{cases}$

**7.** $f(x) = \begin{cases} 5 & \text{if } x < -2 \\ 3 & \text{if } x \geq -2 \end{cases}$

**8.** $f(x) = \begin{cases} 4 & \text{if } x < -3 \\ -2 & \text{if } x \geq -3 \end{cases}$

### MIXED PRACTICE

*Graph each piecewise-defined function. Use the graph to determine the domain and range of the function. See Examples 1 and 2.*

**9.** $f(x) = \begin{cases} -2x & \text{if } x \leq 0 \\ 2x + 1 & \text{if } x > 0 \end{cases}$

**10.** $g(x) = \begin{cases} -3x & \text{if } x \leq 0 \\ 3x + 2 & \text{if } x > 0 \end{cases}$

**11.** $h(x) = \begin{cases} 5x - 5 & \text{if } x < 2 \\ -x + 3 & \text{if } x \geq 2 \end{cases}$

**12.** $f(x) = \begin{cases} 4x - 4 & \text{if } x < 2 \\ -x + 1 & \text{if } x \geq 2 \end{cases}$

**13.** $f(x) = \begin{cases} x + 3 & \text{if } x < -1 \\ -2x + 4 & \text{if } x \geq -1 \end{cases}$

**14.** $h(x) = \begin{cases} x + 2 & \text{if } x < 1 \\ 2x + 1 & \text{if } x \geq 1 \end{cases}$

**15.** $g(x) = \begin{cases} -2 & \text{if } x \leq 0 \\ -4 & \text{if } x \geq 1 \end{cases}$

**16.** $f(x) = \begin{cases} -1 & \text{if } x \leq 0 \\ -3 & \text{if } x \geq 2 \end{cases}$

### MIXED PRACTICE

*Sketch the graph of function. See Examples 3 through 6.*

**17.** $f(x) = |x| + 3$

**18.** $f(x) = |x| - 2$

**19.** $f(x) = \sqrt{x} - 2$

**20.** $f(x) = \sqrt{x} + 3$

**21.** $f(x) = |x - 4|$

**22.** $f(x) = |x + 3|$

**23.** $f(x) = \sqrt{x + 2}$

**24.** $f(x) = \sqrt{x - 2}$

**25.** $y = (x - 4)^2$

**26.** $y = (x + 4)^2$

**27.** $f(x) = x^2 + 4$

**28.** $f(x) = x^2 - 4$

**29.** $f(x) = \sqrt{x - 2} + 3$

**30.** $f(x) = \sqrt{x - 1} + 3$

**31.** $f(x) = |x - 1| + 5$

**32.** $f(x) = |x - 3| + 2$

**33.** $f(x) = \sqrt{x + 1} + 1$

**34.** $f(x) = \sqrt{x + 3} + 2$

**35.** $f(x) = |x + 3| - 1$

**36.** $f(x) = |x + 1| - 4$

**37.** $g(x) = (x - 1)^2 - 1$

**38.** $h(x) = (x + 2)^2 + 2$

**39.** $f(x) = (x + 3)^2 - 2$

**40.** $f(x) = (x + 2)^2 + 4$

*Sketch the graph of each function. See Examples 3 through 7.*

**41.** $f(x) = -(x - 1)^2$

**42.** $g(x) = -(x + 2)^2$

**43.** $h(x) = -\sqrt{x} + 3$

**44.** $f(x) = -\sqrt{x + 3}$

**45.** $h(x) = -|x + 2| + 3$

**46.** $g(x) = -|x + 1| + 1$

**47.** $f(x) = (x - 3) + 2$

**48.** $f(x) = (x - 1) + 4$

### REVIEW AND PREVIEW

*Match each equation with its graph. See Section 3.3.*

**49.** $y = -1$

**50.** $x = -1$

**51.** $x = 3$

**52.** $y = 3$

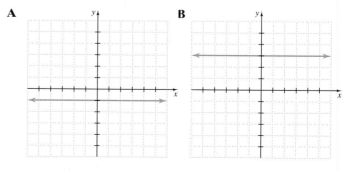

A       B

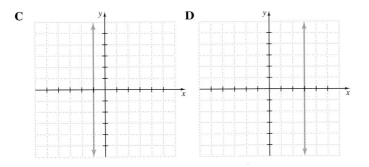

**C**                    **D**

## CONCEPT EXTENSIONS

**53.** Draw a graph whose domain is $(-\infty, 5]$ and whose range is $[2, \infty)$.

**54.** In your own words, describe how to graph a piecewise-defined function.

**55.** Graph: $f(x) = \begin{cases} -\dfrac{1}{2}x & \text{if } x \le 0 \\ x + 1 & \text{if } 0 < x \le 2 \\ 2x - 1 & \text{if } x > 2 \end{cases}$

**56.** Graph: $f(x) = \begin{cases} -\dfrac{1}{3}x & \text{if } x \le 0 \\ x + 2 & \text{if } 0 < x \le 4 \\ 3x - 4 & \text{if } x > 4 \end{cases}$

*Write the domain and range of the following exercises.*

**57.** Exercise 29

**58.** Exercise 30

**59.** Exercise 45

**60.** Exercise 46

*Without graphing, find the domain of each function.*

**61.** $f(x) = 5\sqrt{x - 20} + 1$

**62.** $g(x) = -3\sqrt{x + 5}$

**63.** $h(x) = 5|x - 20| + 1$

**64.** $f(x) = -3|x + 5.7|$

**65.** $g(x) = 9 - \sqrt{x + 103}$

**66.** $h(x) = \sqrt{x - 17} - 3$

*Sketch the graph of each piecewise-defined function. Write the domain and range of each function.*

**67.** $f(x) = \begin{cases} |x| & \text{if } x \le 0 \\ x^2 & \text{if } x > 0 \end{cases}$

**68.** $f(x) = \begin{cases} x^2 & \text{if } x < 0 \\ \sqrt{x} & \text{if } x \ge 0 \end{cases}$

**69.** $g(x) = \begin{cases} |x - 2| & \text{if } x < 0 \\ -x^2 & \text{if } x \ge 0 \end{cases}$

**70.** $g(x) = \begin{cases} -|x + 1| - 1 & \text{if } x < -2 \\ \sqrt{x + 2} - 4 & \text{if } x \ge -2 \end{cases}$

---

### 📖 STUDY SKILLS BUILDER

**Tips for Studying for an Exam**

To prepare for an exam, try the following study techniques:

- Start the study process days before your exam.
- Make sure that you are up-to-date on your assignments.
- If there is a topic that you are unsure of, use one of the many resources that are available to you. For example,

  See your instructor.

  Visit a learning resource center on campus.

  Read the textbook material and examples on the topic.

  View a video on the topic.

- Reread your notes and carefully review the Chapter Highlights at the end of any chapter.
- Work the review exercises at the end of the chapter. Check your answers and correct any mistakes. If you have trouble, use a resource listed above.
- Find a quiet place to take the Chapter Test found at the end of the chapter. Do not use any resources when taking this sample test. This way, you will have a clear indication of how prepared you are for your exam. Check your answers and make sure that you correct any missed exercises.
- Get lots of rest the night before the exam. It's hard to show how well you know the material if your brain is foggy from lack of sleep.

Good luck and keep a positive attitude.

*Let's see how you did on your last exam.*

**1.** How many days before your last exam did you start studying for that exam?

**2.** Were you up-to-date on your assignments at that time or did you need to catch up on assignments?

**3.** List the most helpful text supplement (if you used one).

**4.** List the most helpful campus supplement (if you used one).

**5.** List your process for preparing for a mathematics test.

**6.** Was this process helpful? In other words, were you satisfied with your performance on your exam?

**7.** If not, what changes can you make in your process that will make it more helpful to you?

# 8.4 VARIATION AND PROBLEM SOLVING

## OBJECTIVES

1 Solve problems involving direct variation.

2 Solve problems involving inverse variation.

3 Solve problems involving joint variation.

4 Solve problems involving combined variation.

**OBJECTIVE 1 ▶ Solving problems involving direct variation.** A very familiar example of direct variation is the relationship of the circumference $C$ of a circle to its radius $r$. The formula $C = 2\pi r$ expresses that the circumference is always $2\pi$ times the radius. In other words, $C$ is always a constant multiple $(2\pi)$ of $r$. Because it is, we say that $C$ **varies directly as $r$**, that **$C$ varies directly with $r$**, or that **$C$ is directly proportional to $r$.**

> **Direct Variation**
>
> **$y$ varies directly as $x$**, or **$y$ is directly proportional to $x$**, if there is a nonzero constant $k$ such that
>
> $$y = kx$$
>
> The number $k$ is called the **constant of variation** or the **constant of proportionality.**

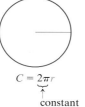

$C = 2\pi r$
↑
constant

In the above definition, the relationship described between $x$ and $y$ is a linear one. In other words, the graph of $y = kx$ is a line. The slope of the line is $k$, and the line passes through the origin.

For example, the graph of the direct variation equation $C = 2\pi r$ is shown. The horizontal axis represents the radius $r$, and the vertical axis is the circumference $C$. From the graph we can read that when the radius is 6 units, the circumference is approximately 38 units. Also, when the circumference is 45 units, the radius is between 7 and 8 units. Notice that as the radius increases, the circumference increases.

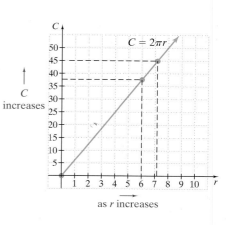

**EXAMPLE 1** Suppose that $y$ varies directly as $x$. If $y$ is 5 when $x$ is 30, find the constant of variation and the direct variation equation.

**Solution** Since $y$ varies directly as $x$, we write $y = kx$. If $y = 5$ when $x = 30$, we have that

$$y = kx$$
$$5 = k(30) \quad \text{Replace } y \text{ with 5 and } x \text{ with 30.}$$
$$\frac{1}{6} = k \quad \text{Solve for } k.$$

The constant of variation is $\frac{1}{6}$.

After finding the constant of variation $k$, the direct variation equation can be written as $y = \frac{1}{6}x$. ☐

**PRACTICE**

**1** Suppose that $y$ varies directly as $x$. If $y$ is 20 when $x$ is 15, find the constant of variation and the direct variation equation.

**EXAMPLE 2**   **Using Direct Variation and Hooke's Law**

Hooke's law states that the distance a spring stretches is directly proportional to the weight attached to the spring. If a 40-pound weight attached to the spring stretches the spring 5 inches, find the distance that a 65-pound weight attached to the spring stretches the spring.

**Solution**

1. UNDERSTAND. Read and reread the problem. Notice that we are given that the distance a spring stretches is **directly proportional** to the weight attached. We let

   $d$ = the distance stretched

   $w$ = the weight attached

   The constant of variation is represented by $k$.

2. TRANSLATE. Because $d$ is directly proportional to $w$, we write

$$d = kw$$

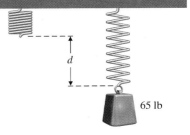

65 lb

3. SOLVE. When a weight of 40 pounds is attached, the spring stretches 5 inches. That is, when $w = 40$, $d = 5$.

$$d = kw$$
$$5 = k(40) \quad \text{Replace } d \text{ with 5 and } w \text{ with 40.}$$
$$\frac{1}{8} = k \quad \text{Solve for } k.$$

Now when we replace $k$ with $\frac{1}{8}$ in the equation

$$d = kw, \text{ we have}$$
$$d = \frac{1}{8}w$$

To find the stretch when a weight of 65 pounds is attached, we replace $w$ with 65 to find $d$.

$$d = \frac{1}{8}(65)$$
$$= \frac{65}{8} = 8\frac{1}{8} \quad \text{or} \quad 8.125$$

4. INTERPRET.

**Check:**   Check the proposed solution of 8.125 inches in the original problem.

**State:**   The spring stetches 8.125 inches when a 65-pound weight is attached.   □

**PRACTICE**
**2**   Use Hooke's law as stated in Example 2. If a 36-pound weight attached to a spring stretches the spring 9 inches, find the distance that a 75-pound weight attached to the spring stretches the spring.

**OBJECTIVE 2 ▶ Solving problems involving inverse variation.** When $y$ is proportional to the **reciprocal** of another variable $x$, we say that **$y$ varies inversely as $x$**, or that **$y$ is inversely proportional to $x$**. An example of the inverse variation relationship is the relationship between the pressure that a gas exerts and the volume of its container. As the volume of a container decreases, the pressure of the gas it contains increases.

**Inverse Variation**

**y varies inversely as *x*,** or **y is inversely proportional to *x*,** if there is a nonzero constant *k* such that

$$y = \frac{k}{x}$$

The number *k* is called the **constant of variation** or the **constant of proportionality.**

Notice that $y = \frac{k}{x}$ is a rational equation. Its graph for $k > 0$ and $x > 0$ is shown. From the graph, we can see that as *x* increases, *y* decreases.

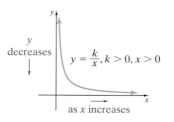

**EXAMPLE 3** Suppose that *u* varies inversely as *w*. If *u* is 3 when *w* is 5, find the constant of variation and the inverse variation equation.

**Solution** Since *u* varies inversely as *w*, we have $u = \frac{k}{w}$. We let $u = 3$ and $w = 5$, and we solve for *k*.

$$u = \frac{k}{w}$$

$$3 = \frac{k}{5} \qquad \text{Let } u = 3 \text{ and } w = 5.$$

$$15 = k \qquad \text{Multiply both sides by 5.}$$

The constant of variation *k* is 15. This gives the inverse variation equation

$$u = \frac{15}{w}$$

**PRACTICE**
**3** Suppose that *b* varies inversely as *a*. If *b* is 5 when *a* is 9, find the constant of variation and the inverse variation equation.

**EXAMPLE 4** Using Inverse Variation and Boyle's Law

Boyle's law says that if the temperature stays the same, the pressure *P* of a gas is inversely proportional to the volume *V*. If a cylinder in a steam engine has a pressure of 960 kilopascals when the volume is 1.4 cubic meters, find the pressure when the volume increases to 2.5 cubic meters.

**Solution**

1. UNDERSTAND. Read and reread the problem. Notice that we are given that the pressure of a gas is *inversely proportional* to the volume. We will let *P* = the pressure and *V* = the volume. The constant of variation is represented by *k*.
2. TRANSLATE. Because *P* is inversely proportional to *V*, we write

$$P = \frac{k}{V}$$

When $P = 960$ kilopascals, the volume $V = 1.4$ cubic meters. We use this information to find $k$.

$$960 = \frac{k}{1.4} \quad \text{Let } P = 960 \text{ and } V = 1.4.$$
$$1344 = k \quad \text{Multiply both sides by 1.4.}$$

Thus, the value of $k$ is 1344. Replacing $k$ with 1344 in the variation equation, we have

$$P = \frac{1344}{V}$$

Next we find $P$ when $V$ is 2.5 cubic meters.

**3.** SOLVE.

$$P = \frac{1344}{2.5} \quad \text{Let } V = 2.5.$$
$$= 537.6$$

**4.** INTERPRET.

**Check:**   Check the proposed solution in the original problem.

**State:**   When the volume is 2.5 cubic meters, the pressure is 537.6 kilopascals.   □

**PRACTICE**

**4**   Use Boyle's law as stated in Example 4. When $P = 350$ kilopascals and $V = 2.8$ cubic meters, find the pressure when the volume decreases to 1.5 cubic meters.

**OBJECTIVE 3 ▶ Solving problems involving joint variation.** Sometimes the ratio of a variable to the product of many other variables is constant. For example, the ratio of distance traveled to the product of speed and time traveled is always 1.

$$\frac{d}{rt} = 1 \quad \text{or} \quad d = rt$$

Such a relationship is called **joint variation.**

> **Joint Variation**
>
> If the ratio of a variable $y$ to the product of two or more variables is constant, then **$y$ varies jointly as,** or **is jointly proportional to,** the other variables. If
>
> $$y = kxz$$
>
> then the number $k$ is the **constant of variation** or the **constant of proportionality.**

**Concept Check ☑**

Which type of variation is represented by the equation $xy = 8$? Explain.

**a.** Direct variation   **b.** Inverse variation   **c.** Joint variation

△ **EXAMPLE 5**   **Expressing Surface Area**

The lateral surface area of a cylinder varies jointly as its radius and height. Express this surface area $S$ in terms of radius $r$ and height $h$.

*Solution* Because the surface area varies jointly as the radius $r$ and the height $h$, we equate $S$ to a constant multiple of $r$ and $h$.

$$S = krh$$

In the equation, $S = krh$, it can be determined that the constant $k$ is $2\pi$, and we then have the formula $S = 2\pi rh$. (The lateral surface area formula does not include the areas of the two circular bases.) ☐

**PRACTICE**

**5** The area of a regular polygon varies jointly as its apothem and its perimeter. Express the area in terms of the apothem $a$ and the perimeter $p$.

**OBJECTIVE 4 ▶ Solving problems involving combined variation.** Some examples of variation involve combinations of direct, inverse, and joint variation. We will call these variations **combined variation.**

**EXAMPLE 6** Suppose that $y$ varies directly as the square of $x$. If $y$ is 24 when $x$ is 2, find the constant of variation and the variation equation.

*Solution* Since $y$ varies directly as the square of $x$, we have

$$y = kx^2$$

Now let $y = 24$ and $x = 2$ and solve for $k$.

$$y = kx^2$$
$$24 = k \cdot 2^2$$
$$24 = 4k$$
$$6 = k$$

The constant of variation is 6, so the variation equation is

$$y = 6x^2$$
☐

**PRACTICE**

**6** Suppose that $y$ varies inversely as the cube of $x$. If $y$ is $\dfrac{1}{2}$ when $x$ is 2, find the constant of variation and the variation equation.

△ **EXAMPLE 7** **Finding Column Weight**

The maximum weight that a circular column can support is directly proportional to the fourth power of its diameter and is inversely proportional to the square of its height. A 2-meter-diameter column that is 8 meters in height can support 1 ton. Find the weight that a 1-meter-diameter column that is 4 meters in height can support.

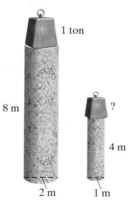

1 ton

8 m

?

4 m

2 m

1 m

*Solution*

1. UNDERSTAND. Read and reread the problem. Let $w$ = weight, $d$ = diameter, $h$ = height, and $k$ = the constant of variation.

2. TRANSLATE. Since $w$ is directly proportional to $d^4$ and inversely proportional to $h^2$, we have

$$w = \frac{kd^4}{h^2}$$

3. SOLVE. To find $k$, we are given that a 2-meter-diameter column that is 8 meters in height can support 1 ton. That is, $w = 1$ when $d = 2$ and $h = 8$, or

$$1 = \frac{k \cdot 2^4}{8^2} \quad \text{Let } w = 1, d = 2, \text{ and } h = 8.$$

$$1 = \frac{k \cdot 16}{64}$$

$$4 = k \qquad \text{Solve for } k.$$

Now replace $k$ with 4 in the equation $w = \dfrac{kd^4}{h^2}$ and we have

$$w = \frac{4d^4}{h^2}$$

To find weight $w$ for a 1-meter-diameter column that is 4 meters in height, let $d = 1$ and $h = 4$.

$$w = \frac{4 \cdot 1^4}{4^2}$$

$$w = \frac{4}{16} = \frac{1}{4}$$

4. INTERPRET.

**Check:**   Check the proposed solution in the original problem.

**State:**   The 1-meter-diameter column that is 4 meters in height can hold $\dfrac{1}{4}$ ton of weight.   ☐

**PRACTICE**

**7**   Suppose that $y$ varies directly as $z$ and inversely as the cube of $x$. If $y$ is 15 when $z = 5$ and $x = 3$, find the constant of variation and the variation equation.

---

## VOCABULARY & READINESS CHECK

*State whether each equation represents direct, inverse, or joint variation.*

**1.** $y = 5x$

**2.** $y = \dfrac{700}{x}$

**3.** $y = 5xz$

**4.** $y = \dfrac{1}{2}abc$

**5.** $y = \dfrac{9.1}{x}$

**6.** $y = 2.3x$

**7.** $y = \dfrac{2}{3}x$

**8.** $y = 3.1\, st$

---

## 8.4 | EXERCISE SET

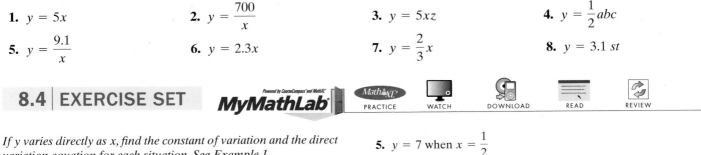

*If y varies directly as x, find the constant of variation and the direct variation equation for each situation. See Example 1.*

**1.** $y = 4$ when $x = 20$

**2.** $y = 5$ when $x = 30$

**3.** $y = 6$ when $x = 4$

**4.** $y = 12$ when $x = 8$

**5.** $y = 7$ when $x = \dfrac{1}{2}$

**6.** $y = 11$ when $x = \dfrac{1}{3}$

**7.** $y = 0.2$ when $x = 0.8$

**8.** $y = 0.4$ when $x = 2.5$

*Solve. See Example 2.*

**9.** The weight of a synthetic ball varies directly with the cube of its radius. A ball with a radius of 2 inches weighs 1.20 pounds. Find the weight of a ball of the same material with a 3-inch radius.

**10.** At sea, the distance to the horizon is directly proportional to the square root of the elevation of the observer. If a person who is 36 feet above the water can see 7.4 miles, find how far a person 64 feet above the water can see. Round to the nearest tenth of a mile.

**11.** The amount *P* of pollution varies directly with the population *N* of people. Kansas City has a population of 442,000 and produces 260,000 tons of pollutants. Find how many tons of pollution we should expect St. Louis to produce, if we know that its population is 348,000. Round to the nearest whole ton. (*Population Source: The World Almanac, 2005*)

**12.** Charles's law states that if the pressure *P* stays the same, the volume *V* of a gas is directly proportional to its temperature *T*. If a balloon is filled with 20 cubic meters of a gas at a temperature of 300 K, find the new volume if the temperature rises 360 K while the pressure stays the same.

*If y varies inversely as x, find the constant of variation and the inverse variation equation for each situation. See Example 3.*

**13.** $y = 6$ when $x = 5$

**14.** $y = 20$ when $x = 9$

**15.** $y = 100$ when $x = 7$

**16.** $y = 63$ when $x = 3$

**17.** $y = \frac{1}{8}$ when $x = 16$

**18.** $y = \frac{1}{10}$ when $x = 40$

**19.** $y = 0.2$ when $x = 0.7$

**20.** $y = 0.6$ when $x = 0.3$

*Solve. See Example 4.*

**21.** Pairs of markings a set distance apart are made on highways so that police can detect drivers exceeding the speed limit. Over a fixed distance, the speed *R* varies inversely with the time *T*. In one particular pair of markings, *R* is 45 mph when *T* is 6 seconds. Find the speed of a car that travels the given distance in 5 seconds.

**22.** The weight of an object on or above the surface of Earth varies inversely as the square of the distance between the object and Earth's center. If a person weighs 160 pounds on

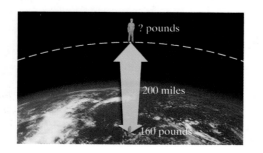

Earth's surface, find the individual's weight if he moves 200 miles above Earth. Round to the nearest whole pound. (Assume that Earth's radius is 4000 miles.)

**23.** If the voltage *V* in an electric circuit is held constant, the current *I* is inversely proportional to the resistance *R*. If the current is 40 amperes when the resistance is 270 ohms, find the current when the resistance is 150 ohms.

**24.** Because it is more efficient to produce larger numbers of items, the cost of producing Dysan computer disks is inversely proportional to the number produced. If 4000 can be produced at a cost of $1.20 each, find the cost per disk when 6000 are produced.

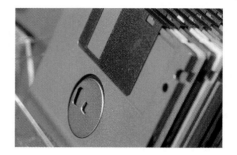

**25.** The intensity *I* of light varies inversely as the square of the distance *d* from the light source. If the distance from the light source is doubled (see the figure), determine what happens to the intensity of light at the new location.

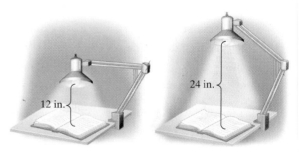

**26.** The maximum weight that a circular column can hold is inversely proportional to the square of its height. If an 8-foot column can hold 2 tons, find how much weight a 10-foot column can hold.

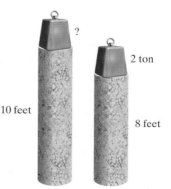

*Write each statement as an equation. Use k as the constant of variation. See Example 5.*

**27.** *x* varies jointly as *y* and *z*.

**28.** *P* varies jointly as *R* and the square of *S*.

**29.** $r$ varies jointly as $s$ and the cube of $t$.

**30.** $a$ varies jointly as $b$ and $c$.

*For each statement, find the constant of variation and the variation equation. See Examples 5 and 6.*

**31.** $y$ varies directly as the cube of $x$; $y = 9$ when $x = 3$

**32.** $y$ varies directly as the cube of $x$; $y = 32$ when $x = 4$

**33.** $y$ varies directly as the square root of $x$; $y = 0.4$ when $x = 4$

**34.** $y$ varies directly as the square root of $x$; $y = 2.1$ when $x = 9$

**35.** $y$ varies inversely as the square of $x$; $y = 0.052$ when $x = 5$

**36.** $y$ varies inversely as the square of $x$; $y = 0.011$ when $x = 10$

**37.** $y$ varies jointly as $x$ and the cube of $z$; $y = 120$ when $x = 5$ and $z = 2$

**38.** $y$ varies jointly as $x$ and the square of $z$; $y = 360$ when $x = 4$ and $z = 3$

*Solve. See Example 7.*

**39.** The maximum weight that a rectangular beam can support varies jointly as its width and the square of its height and inversely as its length. If a beam $\frac{1}{2}$ foot wide, $\frac{1}{3}$ foot high, and 10 feet long can support 12 tons find how much a similar beam can support if the beam is $\frac{2}{3}$ foot wide, $\frac{1}{2}$ foot high, and 16 feet long.

**40.** The number of cars manufactured on an assembly line at a General Motors plant varies jointly as the number of workers and the time they work. If 200 workers can produce 60 cars in 2 hours, find how many cars 240 workers should be able to make in 3 hours.

**41.** The volume of a cone varies jointly as its height and the square of its radius. If the volume of a cone is $32\pi$ cubic inches when the radius is 4 inches and the height is 6 inches, find the volume of a cone when the radius is 3 inches and the height is 5 inches.

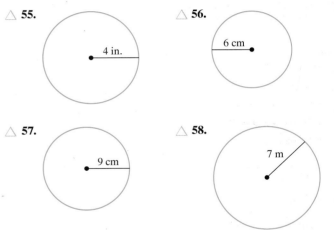

**42.** When a wind blows perpendicularly against a flat surface, its force is jointly proportional to the surface area and the speed of the wind. A sail whose surface area is 12 square feet experiences a 20-pound force when the wind speed is 10 miles per hour. Find the force on an 8-square-foot sail if the wind speed is 12 miles per hour.

**43.** The intensity of light (in foot-candles) varies inversely as the square of $x$, the distance in feet from the light source. The intensity of light 2 feet from the source is 80 foot-candles. How far away is the source if the intensity of light is 5 foot-candles?

**44.** The horsepower that can be safely transmitted to a shaft varies jointly as the shaft's angular speed of rotation (in revolutions per minute) and the cube of its diameter. A 2-inch shaft making 120 revolutions per minute safely transmits 40 horsepower. Find how much horsepower can be safely transmitted by a 3-inch shaft making 80 revolutions per minute.

**MIXED PRACTICE**

*Write an equation to describe each variation. Use k for the constant of proportionality. See Examples 1 through 7.*

**45.** $y$ varies directly as $x$

**46.** $p$ varies directly as $q$

**47.** $a$ varies inversely as $b$

**48.** $y$ varies inversely as $x$

**49.** $y$ varies jointly as $x$ and $z$

**50.** $y$ varies jointly as $q$, $r$, and $t$

**51.** $y$ varies inversely as $x^3$

**52.** $y$ varies inversely as $a^4$

**53.** $y$ varies directly as $x$ and inversely as $p^2$

**54.** $y$ varies directly as $a^5$ and inversely as $b$

**REVIEW AND PREVIEW**

*Find the exact circumference and area of each circle. See the inside cover for a list of geometric formulas.*

**55.**

**56.**

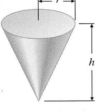

**57.**

**58.**

*Simplify. See Sections 1.5 through 1.7.*

**59.** $|-1.2|$

**60.** $|-3|$

**61.** $-|7|$

**62.** $|0|$

**63.** $-\left|-\frac{1}{2}\right|$

**64.** $-\left|\frac{1}{5}\right|$

**65.** $\left(\frac{2}{3}\right)^3$

**66.** $\left(\frac{5}{11}\right)^2$

**CONCEPT EXTENSIONS**

*Solve. See the Concept Check in this section. Choose the type of variation that each equation represents.* **a.** *Direct variation* **b.** *Inverse variation* **c.** *Joint variation*

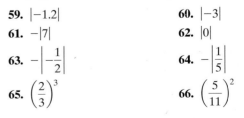

**67.** $y = \frac{2}{3}x$

**68.** $y = \frac{0.6}{x}$

**69.** $y = 9ab$

**70.** $xy = \frac{2}{11}$

**71.** The horsepower to drive a boat varies directly as the cube of the speed of the boat. If the speed of the boat is to double, determine the corresponding increase in horsepower required.

**72.** The volume of a cylinder varies jointly as the height and the square of the radius. If the height is halved and the radius is doubled, determine what happens to the volume.

**73.** Suppose that $y$ varies directly as $x$. If $x$ is doubled, what is the effect on $y$?

**74.** Suppose that $y$ varies directly as $x^2$. If $x$ is doubled, what is the effect on $y$?

*Complete the following table for the inverse variation $y = \dfrac{k}{x}$ over each given value of k. Plot the points on a rectangular coordinate system.*

| $x$ | $\dfrac{1}{4}$ | $\dfrac{1}{2}$ | 1 | 2 | 4 |
|---|---|---|---|---|---|
| $y = \dfrac{k}{x}$ | | | | | |

**75.** $k = 3$    **76.** $k = 1$    **77.** $k = \dfrac{1}{2}$    **78.** $k = 5$

# CHAPTER 8 GROUP ACTIVITY

## Modeling Real Data

The number of children who live with only one parent has been steadily increasing in the United States since the 1960s. According to the U.S. Bureau of the Census, the percent of children living with both parents is declining. The following table shows the percent of children (under age 18) living with *both* parents during selected years from 1980 to 2005. In this project, you will have the opportunity to use the data in the table to find a linear function $f(x)$ that represents the data, reflecting the change in living arrangements for children. This project may be completed by working in groups or individually.

**Percent of U.S. Children Who Live with Both Parents**

| Year | 1980 | 1985 | 1990 | 1995 | 2000 | 2005 |
|---|---|---|---|---|---|---|
| $x$ | 0 | 5 | 10 | 15 | 20 | 25 |
| Percent, $y$ | 77 | 74 | 73 | 69 | 67 | 68 |

*Source:* U.S. Bureau of the Census

**1.** Plot the data given in the table as ordered pairs.

**2.** Use a straight edge to draw on your graph what appears to be the line that "best fits" the data you plotted.

**3.** Estimate the coordinates of two points that fall on your best-fitting line. Use these points to find a linear function $f(x)$ for the line.

**4.** What is the slope of your line? Interpret its meaning. Does it make sense in the context of this situation?

**5.** Find the value of $f(50)$. Write a sentence interpreting its meaning in context.

**6.** Compare your linear function with that of another student or group. Are they different? If so, explain why.

(Optional) Enter the data from the table into a graphing calculator. Use the linear regression feature of the calculator to find a linear function for the data. Compare this function to the one you found in Question 3. How are they alike or different? Find the value of $f(50)$ using the model you found with the graphing calculator. Compare it to the value of $f(50)$ you found in Question 5.

# CHAPTER 8 VOCABULARY CHECK

*Fill in each blank with one of the words or phrases listed below.*

slope–intercept    directly    slope
jointly    parallel    perpendicular
function    inversely    linear function

**1.** _____ lines have the same slope and different $y$-intercepts.

**2.** _____ form of a linear equation in two variables is $y = mx + b$.

**3.** A _____ is a relation in which each first component in the ordered pairs corresponds to exactly one second component.

**4.** In the equation $y = 4x - 2$, the coefficient of $x$ is the _____ of its corresponding graph.

**5.** Two lines are _____ if the product of their slopes is $-1$.

**6.** A _____ is a function that can be written in the form $f(x) = mx + b$.

**7.** In the equation $y = kx$, $y$ varies _____ as $x$.

**8.** In the equation $y = \dfrac{k}{x}$, $y$ varies _____ as $x$.

**9.** In the equation $y = kxz$, $y$ varies _____ as $x$ and $z$.

> ▶ **Helpful Hint**
> Are you preparing for your test? Don't forget to take the Chapter 8 Test on page 538. Then check your answers at the back of the text and use the Chapter Test Prep Video CD to see the fully worked-out solutions to any of the exercises you want to review.

# CHAPTER 8  HIGHLIGHTS

| DEFINITIONS AND CONCEPTS | EXAMPLES |
|---|---|

A **linear function** is a function that can be written in the form $f(x) = mx + b$.

To graph a linear function, find three ordered pair solutions. Graph the solutions and draw a line through the plotted points.

**Linear Functions**

$$f(x) = -3, g(x) = 5x, h(x) = -\frac{1}{3}x - 7$$

Graph $f(x) = -2x$.

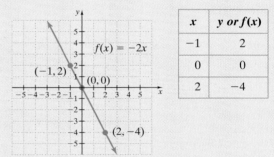

| $x$ | $y$ or $f(x)$ |
|---|---|
| $-1$ | $2$ |
| $0$ | $0$ |
| $2$ | $-4$ |

The point–slope form of the equation of a line is $y - y_1 = m(x - x_1)$, where $m$ is the slope of the line and $(x_1, y_1)$ is a point on the line.

Find an equation of the line parallel to $g(x) = 2x - 1$ and containing the point $(1, -4)$. Write the equation using function notation. Since we want a parallel line, use the same slope of $g(x)$, which is 2.

$$y - y_1 = m(x - x_1)$$
$$y - (-4) = 2(x - 1)$$
$$y + 4 = 2x - 2$$
$$y = 2x - 6 \qquad \text{Solve for } y.$$
$$f(x) = 2x - 6 \qquad \text{Let } y = f(x)$$

The graph of $y = mx + b$ is the same as the graph of $y = mx$, but shifted $|b|$ units up if $b$ is positive and $|b|$ units down if $b$ is negative.

Graph $g(x) = -2x + 3$.

This is the same as the graph of $f(x) = -2x$ shifted 3 units up.

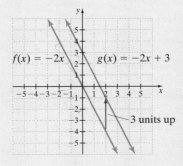

*(continued)*

| DEFINITIONS AND CONCEPTS | EXAMPLES |
|---|---|

To graph a function that is not linear, find a sufficient number of ordered pair solutions so that a pattern may be discovered.

Graph $f(x) = x^2 + 2$.

| $x$ | $y$ or $f(x)$ |
|---|---|
| $-2$ | 6 |
| $-1$ | 3 |
| 0 | 2 |
| 1 | 3 |
| 2 | 6 |

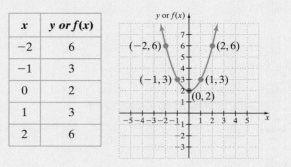

Vertical shifts (upward and downward) let $k$ be a positive number.

| Graph of | Same as | Moved |
|---|---|---|
| $g(x) = f(x) + k$ | $f(x)$ | $k$ units upward |
| $g(x) = f(x) + (-k)$ | $f(x)$ | $k$ units downward |

Horizontal shift (to the left or right) let $h$ be a positive number.

| Graph of | Same as | Moved |
|---|---|---|
| $g(x) = f(x - h)$ | $f(x)$ | $h$ units to the right |
| $g(x) = f(x + h)$ | $f(x)$ | $h$ units to the left |

Reflection about the $x$-axis

The graph of $g(x) = -f(x)$ is the graph of $f(x)$ reflected about the $x$-axis.

The graph of $h(x) = -|x - 3| + 1$ is the same as the graph of $f(x) = |x|$, reflected about the $x$-axis, shifted 3 units right, then 1 unit up.

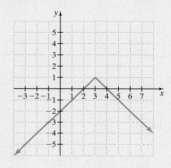

$y$ **varies directly as** $x$, or $y$ is **directly proportional to** $x$, if there is a nonzero constant $k$ such that

$$y = kx$$

$y$ **varies inversely as** $x$, or $y$ is **inversely proportional to** $x$, if there is a nonzero constant $k$ such that

$$y = \frac{k}{x}$$

$y$ **varies jointly as** $x$ and $z$ or $y$ is **jointly proportional to** $x$ and $z$ if there is a nonzero constant $k$ such that

$$y = kxz$$

The circumference of a circle $C$ varies directly as its radius $r$.

$$C = \underset{k}{2\pi} r$$

Pressure $P$ varies inversely with volume $V$.

$$P = \frac{k}{V}$$

The lateral surface area $S$ of a cylinder varies jointly as its radius $r$ and height $h$.

$$S = \underset{k}{2\pi} rh$$

# CHAPTER 8 REVIEW

*(8.1)* *Graph each linear function.*

**1.** $f(x) = x$

**2.** $f(x) = -\dfrac{1}{3}x$

**3.** $g(x) = 4x - 1$

**4.** $F(x) = -\dfrac{2}{3}x + 2$

*The graph of $f(x) = 3x$ is sketched below. Use this graph to match each linear function with its graph.*

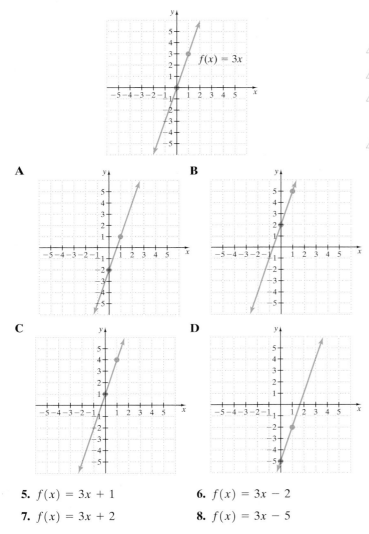

**5.** $f(x) = 3x + 1$

**6.** $f(x) = 3x - 2$

**7.** $f(x) = 3x + 2$

**8.** $f(x) = 3x - 5$

*Find the slope and y-intercept of each function.*

**9.** $f(x) = \dfrac{2}{5}x - \dfrac{4}{3}$

**10.** $f(x) = -\dfrac{2}{7}x + \dfrac{3}{2}$

*Find the standard form equation of each line satisfying the given conditions.*

**11.** Slope 2; through $(5, -2)$

**12.** Through $(-3, 5)$; slope 3

**13.** Through $(-5, 3)$ and $(-4, -8)$

**14.** Through $(-6, -1)$ and $(-4, -2)$

**15.** Through $(-2, -5)$; parallel to $y = 8$

**16.** Through $(-2, 3)$; perpendicular to $x = 4$

*Find the equation of each line satisfying the given conditions. Write each equation using function notation.*

**17.** Horizontal; through $(3, -1)$

**18.** Slope $-\dfrac{2}{3}$; y-intercept $(0, 4)$

**19.** Slope $-1$; y-intercept $(0, -2)$

△ **20.** Through $(2, -6)$; parallel to $6x + 3y = 5$

△ **21.** Through $(-4, -2)$; parallel to $3x + 2y = 8$

△ **22.** Through $(-6, -1)$; perpendicular to $4x + 3y = 5$

**23.** Through $(-4, 5)$; perpendicular to $2x - 3y = 6$

△ **24.** In 2005, the percent of U.S. drivers wearing seat belts was 82%. The number of drivers wearing seat belts in 2000 was 71%. Let $y$ be the number of drivers wearing seat belts in the year $x$, where $x = 0$ represents 2000. (*Source:* Strategis Group for Personal Communications Asso.)

**a.** Write a linear equation that models the percent of U.S. drivers wearing seat belts in terms of the year $x$. [*Hint:* Write 2 ordered pairs of the form (years past 2000, percent of drivers).]

**b.** Use this equation to predict the number of U.S. drivers wearing seat belts in the year 2009. (Round to the nearest percent.)

**25.** In 1998, the number of people (in millions) reporting arthritis was 43. The number of people (in millions) predicted to be reporting arthritis in 2020 is 60. Let $y$ be the number of people (in millions) reporting arthritis in the year $x$, where $x = 0$ represents 1998. (*Source:* Arthritis Foundation)

**a.** Write a linear equation that models the number of people (in millions) reporting arthritis in terms of the year $x$ (See the hint for Exercise 24.)

**b.** Use this equation to predict the number of people reporting arthritis in 2010. (Round to the nearest million.)

*Decide whether the lines are parallel, perpendicular, or neither.*

**26.** $\begin{aligned} -x + 3y &= 2 \\ 6x - 18y &= 3 \end{aligned}$

**(8.2)** *Use the graph of the function below to answer Exercises 27 through 30.*

**27.** Find $f(-1)$.

**28.** Find $f(1)$.

**29.** Find all values of $x$ such that $f(x) = 1$.

**30.** Find all values of $x$ such that $f(x) = -1$.

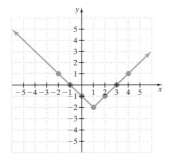

*Determine whether each function is linear or not. Then graph the function.*

**31.** $f(x) = 3x$

**32.** $f(x) = 5x$

**33.** $g(x) = |x| + 4$

**34.** $h(x) = x^2 + 4$

**35.** $F(x) = -\dfrac{1}{2}x + 2$

**36.** $G(x) = -x + 5$

**37.** $y = -1.36x$

**38.** $y = 2.1x + 5.9$

**39.** $H(x) = (x - 2)^2$

**40.** $f(x) = -|x - 3|$

**(8.3)** *Graph each function.*

**41.** $g(x) = \begin{cases} -\dfrac{1}{5}x & \text{if } x \le -1 \\ -4x + 2 & \text{if } x > -1 \end{cases}$

**42.** $f(x) = \begin{cases} -3x & \text{if } x < 0 \\ x - 3 & \text{if } x \ge 0 \end{cases}$

*Graph each function.*

**43.** $f(x) = \sqrt{x - 4}$

**44.** $y = \sqrt{x} - 4$

**45.** $h(x) = -(x + 3)^2 - 1$

**46.** $g(x) = |x - 2| - 2$

**(8.4)** *Solve each variation problem.*

**47.** $A$ is directly proportional to $B$. If $A = 6$ when $B = 14$, find $A$ when $B = 21$.

**48.** $C$ is inversely proportional to $D$. If $C = 12$ when $D = 8$, find $C$ when $D = 24$.

**49.** According to Boyle's law, the pressure exerted by a gas is inversely proportional to the volume, as long as the temperature stays the same. If a gas exerts a pressure of 1250 pounds per square inch when the volume is 2 cubic feet, find the volume when the pressure is 800 pounds per square inch.

△ **50.** The surface area of a sphere varies directly as the square of its radius. If the surface area is $36\pi$ square inches when the radius is 3 inches, find the surface area when the radius is 4 inches.

**MIXED REVIEW**

*Write an equation of the line satisfying each set of conditions. Write the equation in the form $f(x) = mx + b$.*

**51.** Slope 0; through $\left(-4, \dfrac{9}{2}\right)$

**52.** Slope $\dfrac{3}{4}$; through $(-8, -4)$

**53.** Through $(-3, 8)$ and $(-2, 3)$

**54.** Through $(-6, 1)$; parallel to $y = -\dfrac{3}{2}x + 11$

**55.** Through $(-5, 7)$; perpendicular to $5x - 4y = 10$

*Graph each piecewise-defined function.*

**56.** $g(x) = \begin{cases} 4x - 3 & \text{if } x \le 1 \\ 2x & \text{if } x > 1 \end{cases}$

**57.** $f(x) = \begin{cases} x - 2 & \text{if } x \le 0 \\ -\dfrac{x}{3} & \text{if } x \ge 3 \end{cases}$

*Graph each function.*

**58.** $f(x) = |x + 1| - 3$

**59.** $f(x) = \sqrt{x - 2}$

**60.** $y$ is inversely proportional to $x$. If $y = 14$ when $x = 6$, find $y$ when $x = 21$.

# CHAPTER 8 TEST TEST PREP VIDEO

Remember to use the Chapter Test Prep Video CD to see the fully worked-out solutions to any of the exercises you want to review.

*Use the graph of the function f to find each value.*

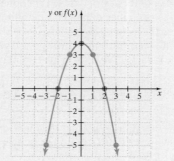

**1.** Find $f(1)$.

**2.** Find $f(-3)$.

**3.** Find all values of $x$ such that $f(x) = 0$.

**4.** Find all values of $x$ such that $f(x) = 4$.

*Graph each line.*

**5.** $2x - 3y = -6$

**6.** $f(x) = \dfrac{2}{3}x$

*Find an equation of each line satisfying the given conditions. Write Exercises 7–9 in standard form. Write Exercises 10–12 using function notation.*

**7.** Horizontal; through $(2, -8)$

**8.** Through $(4, -1)$; slope $-3$

**9.** Through $(0, -2)$; slope 5

**10.** Through $(4, -2)$ and $(6, -3)$

△ **11.** Through $(-1, 2)$; perpendicular to $3x - y = 4$

△ **12.** Parallel to $2y + x = 3$; through $(3, -2)$

**13.** Line $L_1$ has the equation $2x - 5y = 8$. Line $L_2$ passes through the points $(1, 4)$ and $(-1, -1)$. Determine whether these lines are parallel lines, perpendicular lines, or neither.

*Find the domain and range of each relation. Also determine whether the relation is a function.*

**14.**

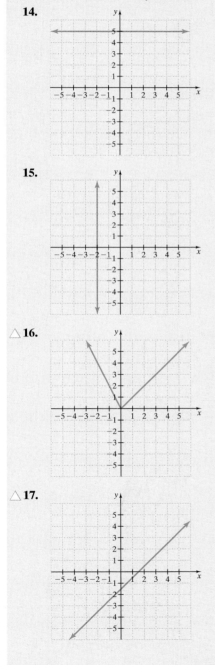

**15.**

**16.**

**17.**

△ **18.** The average yearly earnings for high school graduates age 18 and older is given by the linear function

$$f(x) = 1031x + 25{,}193$$

where $x$ is the number of years since 2000 that a person graduated. (*Source:* U.S. Census Bureau)

   **a.** Find the average earnings in 2000 for high school graduates.

   **b.** Find the average earnings for high school graduates in the year 2007.

   **c.** Predict the first whole year that the average earnings for high school graduates will be greater than $40,000.

   **d.** Find and interpret the slope of this equation.

   **e.** Find and interpret the $y$-intercept of this equation.

*Graph each function. For Exercises 19 and 21, state the domain and the range of the function.*

**19.** $f(x) = \begin{cases} -\dfrac{1}{2}x & \text{if } x \leq 0 \\ 2x - 3 & \text{if } x > 0 \end{cases}$

**20.** $f(x) = (x - 4)^2$

**21.** $g(x) = -|x + 2| - 1$

**22.** $h(x) = \sqrt{x} - 1$

**23.** Suppose that $W$ is inversely proportional to $V$. If $W = 20$ when $V = 12$, find $W$ when $V = 15$.

**24.** Suppose that $Q$ is jointly proportional to $R$ and the square of $S$. If $Q = 24$ when $R = 3$ and $S = 4$, find $Q$ when $R = 2$ and $S = 3$.

**25.** When an anvil is dropped into a gorge, the speed with which it strikes the ground is directly proportional to the square root of the distance it falls. An anvil that falls 400 feet hits the ground at a speed of 160 feet per second. Find the height of a cliff over the gorge if a dropped anvil hits the ground at a speed of 128 feet per second.

# CHAPTER 8 CUMULATIVE REVIEW

1. Simplify $3[4 + 2(10 - 1)]$.

2. Simplify $5[3 + 6(8 - 5)]$.

*Find the value of each expression when $x = 2$ and $y = -5$.*

3. **a.** $\dfrac{x - y}{12 + x}$        **b.** $x^2 - 3y$

4. **a.** $\dfrac{x + y}{3y}$        **b.** $y^2 - x$

*Solve.*

5. $-3x = 33$

6. $\dfrac{2}{3}y = 7$

7. $8(2 - t) = -5t$

8. $5x - 9 = 5x - 29$

9. Solve $y = mx + b$ for $x$.

10. Solve $y = 7x - 2$ for $x$.

11. $-4x + 7 \geq -9$ Write the solution using interval notation rotation.

12. $-5x - 6 < 3x + 1$ Write the solution using interval notation.

13. Find the slope of a line perpendicular to the line passing through the points $(-1, 7)$ and $(2, 2)$.

14. Find the slope of a line parallel to the line passing through the points $(0, 7)$ and $(-1, 0)$.

15. If $g(x) = x^2 - 3$, find
    **a.** $g(2)$        **b.** $g(-2)$
    **c.** $g(0)$.

16. If $f(x) = 3 - x^2$, find
    **a.** $f(2)$        **b.** $f(-2)$
    **c.** $f(0)$

17. Solve the system: $\begin{cases} 2x + y = 10 \\ x = y + 2 \end{cases}$

18. Solve the system: $\begin{cases} 3y = x + 10 \\ 2x + 5y = 24 \end{cases}$

19. Solve the system: $\begin{cases} -x - \dfrac{y}{2} = \dfrac{5}{2} \\ -\dfrac{x}{2} + \dfrac{y}{4} = 0 \end{cases}$

20. Solve the system: $\begin{cases} \dfrac{x}{2} + y = \dfrac{5}{6} \\ 2x - y = \dfrac{5}{6} \end{cases}$

21. Divide $x^2 + 7x + 12$ by $x + 3$ using long division.

22. Divide: $\dfrac{5x^2y - 6xy + 2}{6xy}$

*Factor completely.*

23. **a.** $6t + 18$
    **b.** $y^5 - y^7$

24. **a.** $5y - 20$
    **b.** $z^{10} - z^3$

25. $x^2 + 4x - 12$

26. $x^2 - 10x + 21$

27. $10x^2 - 13xy - 3y^2$

28. $12a^2 + 5ab - 2b^2$

29. $x^3 + 8$

30. $y^3 - 27$

*Solve.*

31. $x^2 - 9x - 22 = 0$

32. $y^2 - 5y = -6$

*Simplify.*

33. **a.** $\dfrac{2x^2}{10x^3 - 2x^2}$

    **b.** $\dfrac{9x^2 + 13x + 4}{8x^2 + x - 7}$

34. **a.** $\dfrac{33x^4y^2}{3xy}$

    **b.** $\dfrac{9y}{90y^2 + 9y}$

*Perform indicated operations.*

35. $\dfrac{3x + 3}{5x - 5x^2} \cdot \dfrac{2x^2 + x - 3}{4x^2 - 9}$

36. $\dfrac{2x}{x - 6} - \dfrac{x + 6}{x - 6}$

37. $\dfrac{3x^2 + 2x}{x - 1} - \dfrac{10x - 5}{x - 1}$

38. $\dfrac{9}{y^2} - 4y$

*Solve.*

39. $3 - \dfrac{6}{x} = x + 8$

40. $\dfrac{x}{2} + \dfrac{x}{5} = \dfrac{x - 7}{20}$

*Find an equation of the line through the given points. Write the equation using function notation.*

41. $(4, 0)$ and $(-4, -5)$

42. $(-1, 3)$ and $(-2, 7)$

# 9

# Inequalities and Absolute Value

Our ability to solve problems using mathematics depends in part on our ability to solve equations and inequalities. In this chapter, we solve absolute value equations and inequalities in one variable and graph their solutions on number lines. We conclude this chapter by graphing linear inequalities in two variables.

The federal Bureau of Labor Statistics (BLS) has issued its projections for job growth in the United States between 2000 and 2012. These projections are based on employer surveys distributed and collected by BLS. In Exercise Set 9.1, Exercises 83 and 84, you will answer questions about the number of some jobs.

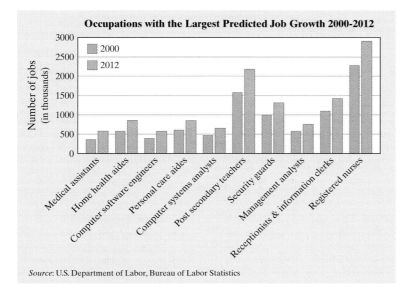

*Source*: U.S. Department of Labor, Bureau of Labor Statistics

# 9.1 COMPOUND INEQUALITIES

Two inequalities joined by the words **and** or **or** are called **compound inequalities.**

*Compound Inequalities*

$$x + 3 < 8 \quad \text{and} \quad x > 2$$

$$\frac{2x}{3} \geq 5 \quad \text{or} \quad -x + 10 < 7$$

**OBJECTIVE 1 ▶ Finding the intersection of two sets.** The solution set of a compound inequality formed by the word **and** is the **intersection** of the solution sets of the two inequalities. We use the symbol ∩ to represent "intersection."

> **Intersection of Two Sets**
> The intersection of two sets, $A$ and $B$, is the set of all elements common to both sets. $A$ intersect $B$ is denoted by
>
>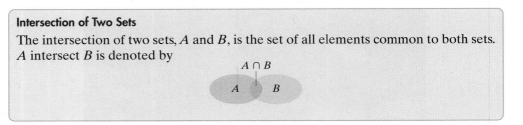

**EXAMPLE 1** If $A = \{x \mid x$ is an even number greater than 0 and less than 10$\}$ and $B = \{3, 4, 5, 6\}$, find $A \cap B$.

*Solution* Let's list the elements in set $A$.

$$A = \{2, 4, 6, 8\}$$

The numbers 4 and 6 are in sets $A$ and $B$. The intersection is $\{4, 6\}$. ☐

**PRACTICE**

**1** If $A = \{x \mid x$ is an odd number greater than 0 and less than 10$\}$ and $B = \{1, 2, 3, 4\}$, find $A \cap B$.

**OBJECTIVE 2 ▶ Solving compound inequalities containing "and."** A value is a solution of a compound inequality formed by the word **and** if it is a solution of *both* inequalities. For example, the solution set of the compound inequality $x \leq 5$ and $x \geq 3$ contains all values of $x$ that make the inequality $x \leq 5$ a true statement **and** the inequality $x \geq 3$ a true statement. The first graph shown below is the graph of $x \leq 5$, the second graph is the graph of $x \geq 3$, and the third graph shows the intersection of the two graphs. The third graph is the graph of $x \leq 5$ **and** $x \geq 3$.

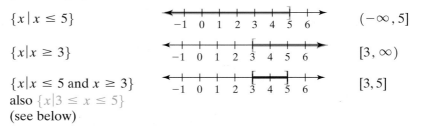

| | | |
|---|---|---|
| $\{x \mid x \leq 5\}$ | | $(-\infty, 5]$ |
| $\{x \mid x \geq 3\}$ | | $[3, \infty)$ |
| $\{x \mid x \leq 5$ and $x \geq 3\}$ also $\{x \mid 3 \leq x \leq 5\}$ (see below) | | $[3, 5]$ |

Since $x \geq 3$ is the same as $3 \leq x$, the compound inequality $3 \leq x$ and $x \leq 5$ can be written in a more compact form as $3 \leq x \leq 5$. The solution set $\{x \mid 3 \leq x \leq 5\}$ includes all numbers that are greater than or equal to 3 and at the same time less than or equal to 5.

In interval notation, the set $\{x \mid x \leq 5$ and $x \geq 3\}$ or $\{x \mid 3 \leq x \leq 5\}$ is written as $[3, 5]$.

> ▶ **Helpful Hint**
>
> Don't forget that some compound inequalities containing "and" can be written in a more compact form.
>
> | *Compound Inequality* | *Compact Form* | *Interval Notation* |
> |---|---|---|
> | $2 \leq x$ *and* $x \leq 6$ | $2 \leq x \leq 6$ | $[2, 6]$ |
> | Graph: | | |
>
> $$\xleftarrow{\hspace{1cm}}\begin{array}{cccccccc} & & [ & & & & ] & \\ 0 & 1 & 2 & 3 & 4 & 5 & 6 & 7 \end{array}\xrightarrow{\hspace{1cm}}$$

**EXAMPLE 2**   Solve: $x - 7 < 2$ *and* $2x + 1 < 9$

*Solution*   First we solve each inequality separately.

$$\begin{array}{ccc} x - 7 < 2 & and & 2x + 1 < 9 \\ x < 9 & and & 2x < 8 \\ x < 9 & and & x < 4 \end{array}$$

Now we can graph the two intervals on two number lines and find their intersection. Their intersection is shown on the third number line.

$\{x \mid x < 9\}$

$$\xleftarrow{\hspace{0.5cm}}\begin{array}{cccccccc} 3 & 4 & 5 & 6 & 7 & 8 & 9 & 10 \end{array}\xrightarrow{\hspace{0.5cm}}\qquad (-\infty, 9)$$

$\{x \mid x < 4\}$

$$\xleftarrow{\hspace{0.5cm}}\begin{array}{cccccccc} 3 & 4 & 5 & 6 & 7 & 8 & 9 & 10 \end{array}\xrightarrow{\hspace{0.5cm}}\qquad (-\infty, 4)$$

$\{x \mid x < 9 \text{ and } x < 4\} = \{x \mid x < 4\}$

$$\xleftarrow{\hspace{0.5cm}}\begin{array}{cccccccc} 3 & 4 & 5 & 6 & 7 & 8 & 9 & 10 \end{array}\xrightarrow{\hspace{0.5cm}}\qquad (-\infty, 4)$$

The solution set is $(-\infty, 4)$.  $\square$

**PRACTICE**

**2**   Solve: $x + 3 < 8$ *and* $2x - 1 < 3$. Write the solution set in interval notation.

**EXAMPLE 3**   Solve: $2x \geq 0$ *and* $4x - 1 \leq -9$.

*Solution*   First we solve each inequality separately.

$$\begin{array}{ccc} 2x \geq 0 & and & 4x - 1 \leq -9 \\ x \geq 0 & and & 4x \leq -8 \\ x \geq 0 & and & x \leq -2 \end{array}$$

Now we can graph the two intervals and find their intersection.

$\{x \mid x \geq 0\}$

$$\xleftarrow{\hspace{0.5cm}}\begin{array}{cccccccc} -3 & -2 & -1 & 0 & 1 & 2 & 3 & 4 \end{array}\xrightarrow{\hspace{0.5cm}}\qquad [0, \infty)$$

$\{x \mid x \leq -2\}$

$$\xleftarrow{\hspace{0.5cm}}\begin{array}{cccccccc} -3 & -2 & -1 & 0 & 1 & 2 & 3 & 4 \end{array}\xrightarrow{\hspace{0.5cm}}\qquad (-\infty, -2]$$

$\{x \mid x \geq 0 \text{ and } x \leq -2\} = \varnothing$

$$\xleftarrow{\hspace{1cm}}\xrightarrow{\hspace{1cm}}\qquad \varnothing$$

There is no number that is greater than or equal to 0 *and* less than or equal to $-2$. The solution set is $\varnothing$.  $\square$

**PRACTICE**

**3**   Solve: $4x \leq 0$ *and* $3x + 2 > 8$. Write the solution set in interval notation.

> ▶ **Helpful Hint**
>
> Example 3 shows that some compound inequalities have no solution. Also, some have all real numbers as solutions.

To solve a compound inequality written in a compact form, such as $2 < 4 - x < 7$, we get $x$ alone in the "middle part." Since a compound inequality is really two inequalities in one statement, we must perform the same operations on all three parts of the inequality.

**EXAMPLE 4**    Solve: $2 < 4 - x < 7$

*Solution*    To get $x$ alone, we first subtract 4 from all three parts.

$$2 < 4 - x < 7$$
$$2 - 4 < 4 - x - 4 < 7 - 4 \quad \text{Subtract 4 from all three parts.}$$
$$-2 < -x < 3 \quad \text{Simplify.}$$
$$\frac{-2}{-1} > \frac{-x}{-1} > \frac{3}{-1} \quad \begin{array}{l}\text{Divide all three parts by } -1 \text{ and}\\ \text{reverse the inequality symbols.}\end{array}$$
$$2 > x > -3$$

> ▶ **Helpful Hint**
> Don't forget to reverse both inequality symbols.

This is equivalent to $-3 < x < 2$.
     The solution set in interval notation is $(-3, 2)$, and its graph is shown.

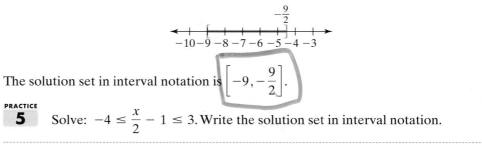

**PRACTICE**
**4**    Solve: $3 < 5 - x < 9$. Write the solution set in interval notation.

**EXAMPLE 5**    Solve: $-1 \leq \dfrac{2x}{3} + 5 \leq 2$.

*Solution*    First, clear the inequality of fractions by multiplying all three parts by the LCD of 3.

$$-1 \leq \frac{2x}{3} + 5 \leq 2$$
$$3(-1) \leq 3\left(\frac{2x}{3} + 5\right) \leq 3(2) \quad \text{Multiply all three parts by the LCD of 3.}$$
$$-3 \leq 2x + 15 \leq 6 \quad \text{Use the distributive property and multiply.}$$
$$-3 - 15 \leq 2x + 15 - 15 \leq 6 - 15 \quad \text{Subtract 15 from all three parts.}$$
$$-18 \leq 2x \leq -9 \quad \text{Simplify.}$$
$$\frac{-18}{2} \leq \frac{2x}{2} \leq \frac{-9}{2} \quad \text{Divide all three parts by 2.}$$
$$-9 \leq x \leq -\frac{9}{2} \quad \text{Simplify.}$$

The graph of the solution is shown.

The solution set in interval notation is $\left[-9, -\dfrac{9}{2}\right]$.

**PRACTICE**
**5**    Solve: $-4 \leq \dfrac{x}{2} - 1 \leq 3$. Write the solution set in interval notation.

**OBJECTIVE 3** ▶ **Finding the union of two sets.** The solution set of a compound inequality formed by the word **or** is the **union** of the solution sets of the two inequalities. We use the symbol $\cup$ to denote "union."

▶ **Helpful Hint**

The word "either" in this definition means "one or the other
or both."

**Union of Two Sets**

The **union** of two sets, $A$ and $B$, is the set of elements that belong to *either* of the sets.
$A$ union $B$ is denoted by

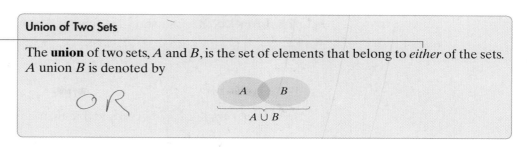

$$A \cup B$$

**EXAMPLE 6**   If $A = \{x \mid x \text{ is an even number greater than 0 and less than 10}\}$
and $B = \{3, 4, 5, 6\}$. Find $A \cup B$.

***Solution***   Recall from Example 1 that $A = \{2, 4, 6, 8\}$. The numbers that are in either
set or both sets are $\{2, 3, 4, 5, 6, 8\}$. This set is the union.   □

**PRACTICE**
**6**   If $A = \{x \mid x \text{ is an odd number greater than 0 and less than 10}\}$ and
$B = \{2, 3, 4, 5, 6\}$. Find $A \cup B$.

**OBJECTIVE 4** ▶ **Solving compound inequalities containing "or."**   A value is a solution
of a compound inequality formed by the word **or** if it is a solution of **either** inequality.
For example, the solution set of the compound inequality $x \leq 1$ **or** $x \geq 3$ contains all
numbers that make the inequality $x \leq 1$ a true statement **or** the inequality $x \geq 3$ a
true statement.

$\{x \mid x \leq 1\}$        $(-\infty, 1]$

$\{x \mid x \geq 3\}$        $[3, \infty)$

$\{x \mid x \leq 1 \text{ or } x \geq 3\}$        $(-\infty, 1] \cup [3, \infty)$

In interval notation, the set $\{x \mid x \leq 1 \text{ or } x \geq 3\}$ is written as $(-\infty, 1] \cup [3, \infty)$.

**EXAMPLE 7**   Solve: $5x - 3 \leq 10 \text{ or } x + 1 \geq 5$.

***Solution***   First we solve each inequality separately.

$$
\begin{array}{rclcrcl}
5x - 3 & \leq & 10 & \text{or} & x + 1 & \geq & 5 \\
5x & \leq & 13 & \text{or} & x & \geq & 4 \\
x & \leq & \dfrac{13}{5} & \text{or} & x & \geq & 4
\end{array}
$$

Now we can graph each interval and find their union.

$\left\{ x \mid x \leq \dfrac{13}{5} \right\}$        $\left( -\infty, \dfrac{13}{5} \right]$

$\{x \mid x \geq 4\}$        $[4, \infty)$

$\left\{ x \mid x \leq \dfrac{13}{5} \text{ or } x \geq 4 \right\}$        $\left( -\infty, \dfrac{13}{5} \right] \cup [4, \infty)$

The solution set is $\left( -\infty, \dfrac{13}{5} \right] \cup [4, \infty)$.   □

**PRACTICE**
**7**   Solve: $8x + 5 \leq 8 \text{ or } x - 1 \geq 2$. Write the solution set in interval notation.

**EXAMPLE 8** Solve: $-2x - 5 < -3 \text{ or } 6x < 0$.

*Solution* First we solve each inequality separately.

$$-2x - 5 < -3 \quad or \quad 6x < 0$$
$$-2x < 2 \quad or \quad x < 0$$
$$x > -1 \quad or \quad x < 0$$

Now we can graph each interval and find their union.

$\{x|x > -1\}$         $(-1, \infty)$

$\{x|x < 0\}$         $(-\infty, 0)$

$\{x|x > -1 \ or \ x < 0\}$         $(-\infty, \infty)$
= all real numbers

The solution set is $(-\infty, \infty)$.

**PRACTICE**
**8**   Solve: $-3x - 2 > -8 \text{ or } 5x > 0$. Write the solution set in interval notation.

**Concept Check** ☑

Which of the following is *not* a correct way to represent the set of all numbers between $-3$ and $5$?

**a.** $\{x | -3 < x < 5\}$        **b.** $-3 < x \text{ or } x < 5$
**c.** $(-3, 5)$                **d.** $x > -3 \text{ and } x < 5$

## VOCABULARY & READINESS CHECK

*Use the choices below to fill in each blank. Some choices may be used more than once.*

| or | ∪ | ∅ |
|---|---|---|
| and | ∩ | compound |

**1.** Two inequalities joined by the words "and" or "or" are called _____ inequalities.

**2.** The word _____ means intersection.

**3.** The word _____ means union.

**4.** The symbol _____ represents intersection.

**5.** The symbol _____ represents union.

**6.** The symbol _____ is the empty set.

**7.** The inequality $-2 \le x < 1$ means $-2 \le x$ _____ $x < 1$.

**8.** $\{x|x < 0 \text{ and } x > 0\} =$ _____

## 9.1 EXERCISE SET

**MyMathLab** *Powered by CourseCompass™ and MathXL®*   Math XL   PRACTICE   WATCH   DOWNLOAD   READ   REVIEW

**MIXED PRACTICE**

*If* $A = \{x|x \text{ is an even integer}\}$, $B = \{x|x \text{ is an odd integer}\}$, $C = \{2, 3, 4, 5\}$, *and* $D = \{4, 5, 6, 7\}$, *list the elements of each set. See Examples 1 and 6.*

**1.** $C \cup D$           **2.** $C \cap D$

**3.** $A \cap D$           **4.** $A \cup D$

**5.** $A \cup B$           **6.** $A \cap B$

**7.** $B \cap D$           **8.** $B \cup D$

**9.** $B \cup C$           **10.** $B \cap C$

**11.** $A \cap C$           **12.** $A \cup C$

*Solve each compound inequality. Graph the solution set and write it in interval notation. See Examples 2 and 3.*

**13.** $x < 1 \text{ and } x > -3$       **14.** $x \le 0 \text{ and } x \ge -2$

**15.** $x \le -3 \text{ and } x \ge -2$     **16.** $x < 2 \text{ and } x > 4$

**17.** $x < -1 \text{ and } x < 1$       **18.** $x \ge -4 \text{ and } x > 1$

Solve each compound inequality. Write solutions in interval notation. See Examples 2 and 3.

**19.** $x + 1 \geq 7$ and $3x - 1 \geq 5$

**20.** $x + 2 \geq 3$ and $5x - 1 \geq 9$

**21.** $4x + 2 \leq -10$ and $2x \leq 0$

**22.** $2x + 4 > 0$ and $4x > 0$

**23.** $-2x < -8$ and $x - 5 < 5$

**24.** $-7x \leq -21$ and $x - 20 \leq -15$

Solve each compound inequality. See Examples 4 and 5.

**25.** $5 < x - 6 < 11$

**26.** $-2 \leq x + 3 \leq 0$

**27.** $-2 \leq 3x - 5 \leq 7$

**28.** $1 < 4 + 2x < 7$

**29.** $1 \leq \frac{2}{3}x + 3 \leq 4$

**30.** $-2 < \frac{1}{2}x - 5 < 1$

**31.** $-5 \leq \frac{-3x + 1}{4} \leq 2$

**32.** $-4 \leq \frac{-2x + 5}{3} \leq 1$

Solve each compound inequality. Graph the solution set and write it in interval notation. See Examples 7 and 8.

**33.** $x < 4 \ or \ x < 5$

**34.** $x \geq -2 \ or \ x \leq 2$

**35.** $x \leq -4 \ or \ x \geq 1$

**36.** $x < 0 \ or \ x < 1$

**37.** $x > 0 \ or \ x < 3$

**38.** $x \geq -3 \ or \ x \leq -4$

Solve each compound inequality. Write solutions in interval notation. See Examples 7 and 8.

**39.** $-2x \leq -4$ or $5x - 20 \geq 5$

**40.** $-5x \leq 10$ or $3x - 5 \geq 1$

**41.** $x + 4 < 0$ or $6x > -12$

**42.** $x + 9 < 0$ or $4x > -12$

**43.** $3(x - 1) < 12$ or $x + 7 > 10$

**44.** $5(x - 1) \geq -5$ or $5 - x \leq 11$

**MIXED PRACTICE**

Solve each compound inequality. Write solutions in interval notation. See Examples 1 through 8.

**45.** $x < \frac{2}{3}$ and $x > -\frac{1}{2}$

**46.** $x < \frac{5}{7}$ and $x < 1$

**47.** $x < \frac{2}{3}$ or $x > -\frac{1}{2}$

**48.** $x < \frac{5}{7}$ or $x < 1$

**49.** $0 \leq 2x - 3 \leq 9$

**50.** $3 < 5x + 1 < 11$

**51.** $\frac{1}{2} < x - \frac{3}{4} < 2$

**52.** $\frac{2}{3} < x + \frac{1}{2} < 4$

**53.** $x + 3 \geq 3$ and $x + 3 \leq 2$

**54.** $2x - 1 \geq 3$ and $-x > 2$

**55.** $3x \geq 5$ or $-\frac{5}{8}x - 6 > 1$

**56.** $\frac{3}{8}x + 1 \leq 0$ or $-2x < -4$

**57.** $0 < \frac{5 - 2x}{3} < 5$

**58.** $-2 < \frac{-2x - 1}{3} < 2$

**59.** $-6 < 3(x - 2) \leq 8$

**60.** $-5 < 2(x + 4) < 8$

**61.** $-x + 5 > 6$ and $1 + 2x \leq -5$

**62.** $5x \leq 0$ and $-x + 5 < 8$

**63.** $3x + 2 \leq 5$ or $7x > 29$

**64.** $-x < 7$ or $3x + 1 < -20$

**65.** $5 - x > 7$ and $2x + 3 \geq 13$

**66.** $-2x < -6$ or $1 - x > -2$

**67.** $-\frac{1}{2} \leq \frac{4x - 1}{6} < \frac{5}{6}$

**68.** $-\frac{1}{2} \leq \frac{3x - 1}{10} < \frac{1}{2}$

**69.** $\frac{1}{15} < \frac{8 - 3x}{15} < \frac{4}{5}$

**70.** $-\frac{1}{4} < \frac{6 - x}{12} < \frac{1}{6}$

**71.** $0.3 < 0.2x - 0.9 < 1.5$

**72.** $-0.7 \leq 0.4x + 0.8 < 0.5$

**REVIEW AND PREVIEW**

Evaluate the following. See Sections 1.5 and 1.6.

**73.** $|-7| - |19|$

**74.** $|-7 - 19|$

**75.** $-(-6) - |-10|$

**76.** $|-4| - (-4) + |-20|$

Find by inspection all values for x that make each equation true. See Sections 1.2 and 1.4.

**77.** $|x| = 7$

**78.** $|x| = 5$

**79.** $|x| = 0$

**80.** $|x| = -2$

**CONCEPT EXTENSIONS**

Use the graph to answer Exercises 81 and 82.

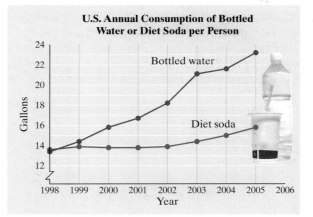

**81.** For what years was the consumption of bottled water greater than 20 gallons per person *and* the consumption of diet soda greater than 14 gallons per person?

**82.** For what years was the consumption of bottled water less than 15 gallons per person *or* the consumption of diet soda greater than 14 gallons per person?

*Use the Chapter 9 Opener graph, page 541, to answer Exercises 83 and 84.*

**83.** For which occupation(s) is the number of jobs in 2000 *and* the predicted number of jobs in 2012 greater than 1800 thousand?

**84.** For which occupation(s) is the number of jobs in 2000 *or* the predicted number of jobs in 2012 greater than 1800 thousand?

*The formula for converting Fahrenheit temperatures to Celsius temperatures is $C = \dfrac{5}{9}(F - 32)$. Use this formula for Exercises 85 and 86.*

**85.** During a recent year, the temperatures in Chicago ranged from $-29°$ to $35°C$. Use a compound inequality to convert these temperatures to Fahrenheit temperatures.

**86.** In Oslo, the average temperature ranges from $-10°$ to $18°$ Celsius. Use a compound inequality to convert these temperatures to the Fahrenheit scale.

*Solve.*

**87.** Christian D'Angelo has scores of 68, 65, 75, and 78 on his algebra tests. Use a compound inequality to find the scores he can make on his final exam to receive a C in the course. The final exam counts as two tests, and a C is received if the final course average is from 70 to 79.

**88.** Wendy Wood has scores of 80, 90, 82, and 75 on her chemistry tests. Use a compound inequality to find the range of scores she can make on her final exam to receive a B in the course. The final exam counts as two tests, and a B is received if the final course average is from 80 to 89.

*Solve each compound inequality for x. See the example below. To solve $x - 6 < 3x < 2x + 5$, notice that this inequality contains a variable not only in the middle, but also on the left and the right. When this occurs, we solve by rewriting the inequality using the word **and**.*

$$x - 6 < 3x \quad \text{and} \quad 3x < 2x + 5$$
$$-6 < 2x \quad \text{and} \quad x < 5$$
$$-3 < x$$
$$x > -3 \quad \text{and} \quad x < 5$$

$x > -3$

$x < 5$

$-3 < x < 5$, or $(-3, 5)$

**89.** $2x - 3 < 3x + 1 < 4x - 5$

**90.** $x + 3 < 2x + 1 < 4x + 6$

**91.** $-3(x - 2) \le 3 - 2x \le 10 - 3x$

**92.** $7x - 1 \le 7 + 5x \le 3(1 + 2x)$

**93.** $5x - 8 < 2(2 + x) < -2(1 + 2x)$

**94.** $1 + 2x < 3(2 + x) < 1 + 4x$

---

## THE BIGGER PICTURE   SIMPLIFYING EXPRESSIONS AND SOLVING EQUATIONS AND INEQUALITIES

We now continue the outline from Sections 1.7, 2.3, 2.8, 5.6, 6.6, 7.4, and 7.6. Although suggestions will be given, this outline should be in your own words. Once you complete this new portion, try the exercises. (*Note: From now on, the breakdown of Part I, Simplifying Expressions, will be shown when insertions are made. This breakdown can always be seen in Section 7.6, page 484 or Appendix A.*)

  **I.** Simplifying Expressions (See Section 7.6, page 484.)

  **II.** Solving Equations

    **A.** Linear Equations (Section 2.3)

    **B.** Quadratic and Higher Degree Equations (Section 6.6)

    **C.** Equations with Rational Expressions (Section 7.5)

    **D.** Proportions (Section 7.6)

  **III.** Solving Inequalities

    **A.** Linear Inequalities (Section 2.8)

**B.** Compound Inequalities: Two inequality signs or 2 inequalities separated by "and" or "or." *Or* means *union* and *and* means *intersection*.

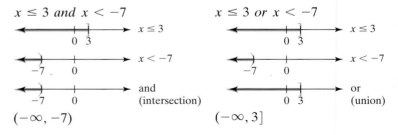

*Perform indicated operations and simplify.*

**1.** $-\dfrac{1}{2} - \left(-\dfrac{3}{8}\right)$

**2.** $(8xy - 7y^2) - (4xy - y^2)$

3. $\dfrac{x+2}{xy-z^2} - \dfrac{x+1}{xy-z^2}$

4. $\dfrac{x^3-8}{x-2} \cdot \dfrac{x^2-4}{x-2}$

*Solve. Write inequality solutions in interval notation.*

5. $x - 2 \le 1$ and $3x - 1 \ge -4$

6. $-2 < x - 1 < 5$

7. $-2x + 2.5 = -7.7$

8. $-5x > 20$

9. $x \le -3$ or $x \le -5$

10. $5x < -10$ or $3x - 4 > 2$

11. $\dfrac{5t}{2} - \dfrac{3t}{4} = 7$

12. $5(x - 3) + x + 2 \ge 3(x + 2) + 2x$

---

# 9.2 ABSOLUTE VALUE EQUATIONS

## OBJECTIVE

**1** Solve absolute value equations.

**OBJECTIVE 1 ▶ Solving absolute equations.** In Chapter 1, we defined the absolute value of a number as its distance from 0 on a number line.

$$|-2| = 2 \text{ and } |3| = 3$$

In this section, we concentrate on solving equations containing the absolute value of a variable or a variable expression. Examples of absolute value equations are

$$|x| = 3 \qquad -5 = |2y + 7| \qquad |z - 6.7| = |3z + 1.2|$$

Since distance and absolute value are so closely related, absolute value equations and inequalities (see Section 2.7) are extremely useful in solving distance-type problems, such as calculating the possible error in a measurement.

For the absolute value equation $|x| = 3$, its solution set will contain all numbers whose distance from 0 is 3 units. Two numbers are 3 units away from 0 on the number line: 3 and $-3$.

Thus, the solution set of the equation $|x| = 3$ is $\{3, -3\}$. This suggests the following:

*Rule 1*

### Solving Equations of the Form $|X| = a$

If $a$ is a positive number, then $|X| = a$ is equivalent to $X = a$ or $X = -a$.

**EXAMPLE 1** Solve: $|p| = 2$.

**Solution** Since 2 is positive, $|p| = 2$ is equivalent to $p = 2$ or $p = -2$.

To check, let $p = 2$ and then $p = -2$ in the original equation.

| | | | |
|---|---|---|---|
| $|p| = 2$ | Original equation | $|p| = 2$ | Original equation |
| $|2| = 2$ | Let $p = 2$. | $|-2| = 2$ | Let $p = -2$. |
| $2 = 2$ | True | $2 = 2$ | True |

The solutions are 2 and $-2$ or the solution set is $\{2, -2\}$. □

**PRACTICE**

**1** Solve: $|q| = 7$.

▶ **Helpful Hint**

For the equation $|X| = a$ in the box above, $X$ can be a single variable or a variable expression.

If the expression inside the absolute value bars is more complicated than a single variable, we can still apply the absolute value property.

**EXAMPLE 2** Solve: $|5w + 3| = 7$.

*Solution* Here the expression inside the absolute value bars is $5w + 3$. If we think of the expression $5w + 3$ as $X$ in the absolute value property, we see that $|X| = 7$ is equivalent to

$$X = 7 \quad \text{or} \quad X = -7$$

Then substitute $5w + 3$ for $X$, and we have

$$5w + 3 = 7 \quad \text{or} \quad 5w + 3 = -7$$

Solve these two equations for $w$.

$$5w + 3 = 7 \quad \text{or} \quad 5w + 3 = -7$$
$$5w = 4 \quad \text{or} \quad 5w = -10$$
$$w = \frac{4}{5} \quad \text{or} \quad w = -2$$

**Check:** To check, let $w = -2$ and then $w = \frac{4}{5}$ in the original equation.

| Let $w = -2$ | Let $w = \frac{4}{5}$ |
|---|---|
| $|5(-2) + 3| = 7$ | $\left|5\left(\frac{4}{5}\right) + 3\right| = 7$ |
| $|-10 + 3| = 7$ | $|4 + 3| = 7$ |
| $|-7| = 7$ | $|7| = 7$ |
| $7 = 7$   True | $7 = 7$   True |

Both solutions check, and the solutions are $-2$ and $\frac{4}{5}$ or the solution set is $\left\{-2, \frac{4}{5}\right\}$. □

**PRACTICE**
**2** Solve: $|2x - 3| = 5$.

**EXAMPLE 3** Solve: $\left|\frac{x}{2} - 1\right| = 11$.

*Solution* $\left|\frac{x}{2} - 1\right| = 11$ is equivalent to

$$\frac{x}{2} - 1 = 11 \quad \text{or} \quad \frac{x}{2} - 1 = -11$$
$$2\left(\frac{x}{2} - 1\right) = 2(11) \quad \text{or} \quad 2\left(\frac{x}{2} - 1\right) = 2(-11) \quad \text{Clear fractions.}$$
$$x - 2 = 22 \quad \text{or} \quad x - 2 = -22 \quad \text{Apply the distributive property.}$$
$$x = 24 \quad \text{or} \quad x = -20$$

The solutions are 24 and $-20$. □

**PRACTICE**
**3** Solve: $\left|\frac{x}{5} + 1\right| = 15$.

To apply the absolute value rule, first make sure that the absolute value expression is isolated.

> ▶ **Helpful Hint**
> If the equation has a single absolute value expression containing variables, isolate the absolute value expression first.

**EXAMPLE 4**    Solve: $|2x| + 5 = 7$.

*Solution*    We want the absolute value expression alone on one side of the equation, so begin by subtracting 5 from both sides. Then apply the absolute value property.

$$|2x| + 5 = 7$$
$$|2x| = 2 \qquad \text{Subtract 5 from both sides.}$$
$$2x = 2 \quad \text{or} \quad 2x = -2$$
$$x = 1 \quad \text{or} \quad x = -1$$

The solutions are $-1$ and $1$.    □

**PRACTICE**
**4**    Solve: $|3x| + 8 = 14$.

**EXAMPLE 5**    Solve: $|y| = 0$.

*Solution*    We are looking for all numbers whose distance from 0 is zero units. The only number is 0. The solution is 0.    □

**PRACTICE**
**5**    Solve: $|z| = 0$.

The next two examples illustrate a special case for absolute value equations. This special case occurs when an isolated absolute value is equal to a negative number.

**EXAMPLE 6**    Solve: $2|x| + 25 = 23$.

*Solution*    First, isolate the absolute value.

$$2|x| + 25 = 23$$
$$2|x| = -2 \quad \text{Subtract 25 from both sides.}$$
$$|x| = -1 \quad \text{Divide both sides by 2.}$$

The absolute value of a number is never negative, so this equation has no solution. The solution set is $\{\ \}$ or $\varnothing$.    □

**PRACTICE**
**6**    Solve: $3|z| + 9 = 7$.

**EXAMPLE 7**    Solve: $\left|\dfrac{3x + 1}{2}\right| = -2$.

*Solution*    Again, the absolute value of any expression is never negative, so no solution exists. The solution set is $\{\ \}$ or $\varnothing$.    □

**PRACTICE**
**7**    Solve: $\left|\dfrac{5x + 3}{4}\right| = -8$.

Given two absolute value expressions, we might ask, when are the absolute values of two expressions equal? To see the answer, notice that

$$|2| = |2|, \quad |-2| = |-2|, \quad |-2| = |2|, \quad \text{and} \quad |2| = |-2|$$

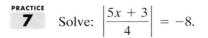
same            same            opposites            opposites

Two absolute value expressions are equal when the expressions inside the absolute value bars are equal to or are opposites of each other.

**EXAMPLE 8** Solve: $|3x + 2| = |5x - 8|$.

**Solution** This equation is true if the expressions inside the absolute value bars are equal to or are opposites of each other.

$$3x + 2 = 5x - 8 \quad \text{or} \quad 3x + 2 = -(5x - 8)$$

Next, solve each equation.

$$3x + 2 = 5x - 8 \quad \text{or} \quad 3x + 2 = -5x + 8$$
$$-2x + 2 = -8 \quad \text{or} \quad 8x + 2 = 8$$
$$-2x = -10 \quad \text{or} \quad 8x = 6$$
$$x = 5 \quad \text{or} \quad x = \frac{3}{4}$$

The solutions are $\frac{3}{4}$ and 5.

**PRACTICE**
**8** Solve: $|2x + 4| = |3x - 1|$.

**EXAMPLE 9** Solve: $|x - 3| = |5 - x|$.

**Solution**
$$x - 3 = 5 - x \quad \text{or} \quad x - 3 = -(5 - x)$$
$$2x - 3 = 5 \quad \text{or} \quad x - 3 = -5 + x$$
$$2x = 8 \quad \text{or} \quad x - 3 - x = -5 + x - x$$
$$x = 4 \quad \text{or} \quad -3 = -5 \qquad \text{False}$$

When an equation simplifies to a false statement, the equation has no solution. Thus, the only solution for the original absolute value equation is 4.

**PRACTICE**
**9** Solve: $|x - 2| = |8 - x|$.

**Concept Check** ☑

True or false? Absolute value equations always have two solutions. Explain your answer.

The following box summarizes the methods shown for solving absolute value equations.

---

**Absolute Value Equations**

$|X| = a$ 
- If $a$ is positive, then solve $X = a$ or $X = -a$.
- If $a$ is 0, solve $X = 0$.
- If $a$ is negative, the equation $|X| = a$ has no solution.

$|X| = |Y|$    Solve $X = Y$ or $X = -Y$.

---

**Answer to Concept Check:**
false; answers may vary

## VOCABULARY & READINESS CHECK

*Match each absolute value equation with an equivalent statement.*

**1.** $|x - 2| = 5$

**2.** $|x - 2| = 0$

**3.** $|x - 2| = |x + 3|$

**4.** $|x + 3| = 5$

**5.** $|x + 3| = -5$

**A.** $x - 2 = 0$

**B.** $x - 2 = x + 3$ or $x - 2 = -(x + 3)$

**C.** $x - 2 = 5$ or $x - 2 = -5$

**D.** $\varnothing$

**E.** $x + 3 = 5$ or $x + 3 = -5$

## 9.2 | EXERCISE SET

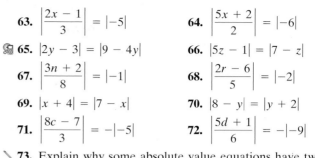

*Solve each absolute value equation. See Examples 1 through 7.*

**1.** $|x| = 7$

**2.** $|y| = 15$

**3.** $|3x| = 12.6$

**4.** $|6n| = 12.6$

**5.** $|2x - 5| = 9$

**6.** $|6 + 2n| = 4$

**7.** $\left|\dfrac{x}{2} - 3\right| = 1$

**8.** $\left|\dfrac{n}{3} + 2\right| = 4$

**9.** $|z| + 4 = 9$

**10.** $|x| + 1 = 3$

**11.** $|3x| + 5 = 14$

**12.** $|2x| - 6 = 4$

**13.** $|2x| = 0$

**14.** $|7z| = 0$

**15.** $|4n + 1| + 10 = 4$

**16.** $|3z - 2| + 8 = 1$

**17.** $|5x - 1| = 0$

**18.** $|3y + 2| = 0$

**19.** Write an absolute value equation representing all numbers $x$ whose distance from 0 is 5 units.

**20.** Write an absolute value equation representing all numbers $x$ whose distance from 0 is 2 units.

*Solve. See Examples 8 and 9.*

**21.** $|5x - 7| = |3x + 11|$

**22.** $|9y + 1| = |6y + 4|$

**23.** $|z + 8| = |z - 3|$

**24.** $|2x - 5| = |2x + 5|$

**25.** Describe how solving an absolute value equation such as $|2x - 1| = 3$ is similar to solving an absolute value equation such as $|2x - 1| = |x - 5|$.

**26.** Describe how solving an absolute value equation such as $|2x - 1| = 3$ is different from solving an absolute value equation such as $|2x - 1| = |x - 5|$.

### MIXED PRACTICE

*Solve each absolute value equation. See Examples 1 through 9.*

**27.** $|x| = 4$

**28.** $|x| = 1$

**29.** $|y| = 0$

**30.** $|y| = 8$

**31.** $|z| = -2$

**32.** $|y| = -9$

**33.** $|7 - 3x| = 7$

**34.** $|4m + 5| = 5$

**35.** $|6x| - 1 = 11$

**36.** $|7z| + 1 = 22$

**37.** $|4p| = -8$

**38.** $|5m| = -10$

**39.** $|x - 3| + 3 = 7$

**40.** $|x + 4| - 4 = 1$

**41.** $\left|\dfrac{z}{4} + 5\right| = -7$

**42.** $\left|\dfrac{c}{5} - 1\right| = -2$

**43.** $|9v - 3| = -8$

**44.** $|1 - 3b| = -7$

**45.** $|8n + 1| = 0$

**46.** $|5x - 2| = 0$

**47.** $|1 + 6c| - 7 = -3$

**48.** $|2 + 3m| - 9 = -7$

**49.** $|5x + 1| = 11$

**50.** $|8 - 6c| = 1$

**51.** $|4x - 2| = |-10|$

**52.** $|3x + 5| = |-4|$

**53.** $|5x + 1| = |4x - 7|$

**54.** $|3 + 6n| = |4n + 11|$

**55.** $|6 + 2x| = -|-7|$

**56.** $|4 - 5y| = -|-3|$

**57.** $|2x - 6| = |10 - 2x|$

**58.** $|4n + 5| = |4n + 3|$

**59.** $\left|\dfrac{2x - 5}{3}\right| = 7$

**60.** $\left|\dfrac{1 + 3n}{4}\right| = 4$

**61.** $2 + |5n| = 17$

**62.** $8 + |4m| = 24$

**63.** $\left|\dfrac{2x - 1}{3}\right| = |-5|$

**64.** $\left|\dfrac{5x + 2}{2}\right| = |-6|$

**65.** $|2y - 3| = |9 - 4y|$

**66.** $|5z - 1| = |7 - z|$

**67.** $\left|\dfrac{3n + 2}{8}\right| = |-1|$

**68.** $\left|\dfrac{2r - 6}{5}\right| = |-2|$

**69.** $|x + 4| = |7 - x|$

**70.** $|8 - y| = |y + 2|$

**71.** $\left|\dfrac{8c - 7}{3}\right| = -|-5|$

**72.** $\left|\dfrac{5d + 1}{6}\right| = -|-9|$

**73.** Explain why some absolute value equations have two solutions.

**74.** Explain why some absolute value equations have one solution.

### REVIEW AND PREVIEW

*The circle graph shows the U.S. Cheese consumption for 2005. Use this graph to answer Exercises 75–77. See Section 2.6. (Source: National Agriculture Statistics Service, USDA)*

**U.S. Cheese Consumption**

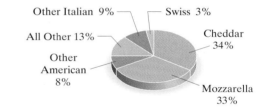

Other Italian 9% — Swiss 3%
All Other 13% — Cheddar 34%
Other American 8%
Mozzarella 33%

**75.** What percent of cheese consumption came from cheddar cheese?

**76.** A circle contains $360°$. Find the number of degrees in the 3% sector for swiss cheese.

**77.** If a family consumed 120 pounds of cheese in 2005, find the amount of mozzarella we might expect they consumed.

*List five integer solutions of each inequality. See Section 2.8.*

**78.** $|x| \le 3$

**79.** $|x| \ge -2$

**80.** $|y| > -10$

**81.** $|y| < 0$

### CONCEPT EXTENSIONS

**82.** Write an absolute value equation representing all numbers $x$ whose distance from 1 is 5 units.

**83.** Write an absolute value equation representing all numbers $x$ whose distance from 7 is 2 units.

*Write each as an equivalent absolute value.*

**84.** $x = 6$ or $x = -6$

**85.** $2x - 1 = 4$ or $2x - 1 = -4$

**86.** $x - 2 = 3x - 4$ or $x - 2 = -(3x - 4)$

**87.** For what value(s) of $c$ will an absolute value equation of the form $|ax + b| = c$ have

 **a.** one solution?

 **b.** no solution?

 **c.** two solutions?

📖 **STUDY SKILLS BUILDER**

**What to Do the Day of an Exam**

On the day of an exam, don't forget to try the following:

- Allow yourself plenty of time to arrive.
- Read the directions on the test carefully.
- Read each problem carefully as you take your test. Make sure that you answer the question asked.
- Watch your time and pace yourself so that you may attempt each problem on your test.
- Check your work and answers.
- ***Do not turn your test in early.*** If you have extra time, spend it double-checking your work.

Good luck!

*Answer the following questions based on your most recent mathematics exam, whenever that was.*

1. How soon before class did you arrive?
2. Did you read the directions on the test carefully?
3. Did you make sure you answered the question asked for each problem on the exam?
4. Were you able to attempt each problem on your exam?
5. If your answer to Question 4 is no, list reasons why.
6. Did you have extra time on your exam?
7. If your answer to Question 6 is yes, describe how you spent that extra time.

---

# 9.3 ABSOLUTE VALUE INEQUALITIES

**OBJECTIVES**

1. Solve absolute value inequalities of the form $|x| < a$.

2. Solve absolute value inequalities of the form $|x| > a$.

**OBJECTIVE 1 ▶ Solving absolute value inequalities of the form $|x| < a$.** The solution set of an absolute value inequality such as $|x| < 2$ contains all numbers whose distance from 0 is less than 2 units, as shown below.

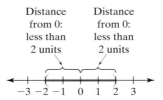

The solution set is $\{x \mid -2 < x < 2\}$, or $(-2, 2)$ in interval notation.

**EXAMPLE 1** Solve: $|x| \leq 3$.

*Solution* The solution set of this inequality contains all numbers whose distance from 0 is less than or equal to 3. Thus 3, −3, and all numbers between 3 and −3 are in the solution set.

The solution set is $[-3, 3]$.

**PRACTICE**

**1** Solve: $|x| < 2$ and graph the solution set.

In general, we have the following.

> **Solving Absolute Value Inequalities of the Form $|X| < a$**
> If $a$ is a positive number, then $|X| < a$ is equivalent to $-a < X < a$.

This property also holds true for the inequality symbol $\leq$.

**EXAMPLE 2** Solve for $m$: $|m - 6| < 2$.

*Solution* Replace $X$ with $m - 6$ and $a$ with 2 in the preceding property, and we see that

$$|m - 6| < 2 \quad \text{is equivalent to} \quad -2 < m - 6 < 2$$

Solve this compound inequality for $m$ by adding 6 to all three parts.

$$
\begin{aligned}
-2 &< m - 6 < 2 \\
-2 + 6 &< m - 6 + 6 < 2 + 6 \quad \text{Add 6 to all three parts.} \\
4 &< m < 8 \quad\quad\quad\quad\quad\;\; \text{Simplify.}
\end{aligned}
$$

The solution set is $(4, 8)$, and its graph is shown.

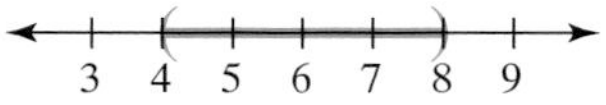

**PRACTICE**
**2** Solve for $b$: $|b + 1| < 3$. Graph the solution set.

---

> ▶ **Helpful Hint**
>
> Before using an absolute value inequality property, isolate the absolute value expression on one side of the inequality.

---

**EXAMPLE 3** Solve for $x$: $|5x + 1| + 1 \leq 10$.

*Solution* First, isolate the absolute value expression by subtracting 1 from both sides.

$$
\begin{aligned}
|5x + 1| + 1 &\leq 10 \\
|5x + 1| &\leq 10 - 1 \quad \text{Subtract 1 from both sides.} \\
|5x + 1| &\leq 9 \quad\quad\;\; \text{Simplify.}
\end{aligned}
$$

Since 9 is positive, we apply the absolute value property for $|X| \leq a$.

$$
\begin{aligned}
-9 &\leq 5x + 1 \leq 9 \\
-9 - 1 &\leq 5x + 1 - 1 \leq 9 - 1 \quad \text{Subtract 1 from all three parts.} \\
-10 &\leq 5x \leq 8 \quad\quad\quad\quad\quad\;\; \text{Simplify.} \\
-2 &\leq x \leq \frac{8}{5} \quad\quad\quad\quad\quad\;\;\; \text{Divide all three parts by 5.}
\end{aligned}
$$

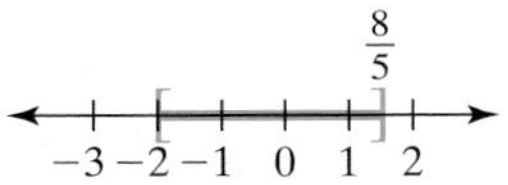

The solution set is $\left[ -2, \dfrac{8}{5} \right]$, and the graph is shown above.

**PRACTICE**
**3** Solve for $x$: $|3x - 2| + 5 \leq 9$. Graph the solution set.

**EXAMPLE 4** Solve for $x$: $\left|2x - \dfrac{1}{10}\right| < -13$.

*Solution* The absolute value of a number is always nonnegative and can never be less than $-13$. Thus this absolute value inequality has no solution. The solution set is $\{\ \}$ or $\varnothing$. ☐

**PRACTICE**
**4** Solve for $x$: $\left|3x + \dfrac{5}{8}\right| < -4$.

**OBJECTIVE 2 ▶ Solving absolute value inequalities of the form $|x| > a$.** Let us now solve an absolute value inequality of the form $|X| > a$, such as $|x| \geq 3$. The solution set contains all numbers whose distance from 0 is 3 or more units. Thus the graph of the solution set contains 3 and all points to the right of 3 on the number line or $-3$ and all points to the left of $-3$ on the number line.

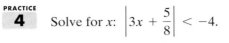

This solution set is written as $\{x \mid x \leq -3 \text{ or } x \geq 3\}$. In interval notation, the solution is $(-\infty, -3] \cup [3, \infty)$, since "or" means "union." In general, we have the following.

---

**Solving Absolute Value Inequalities of the Form $|X| > a$**
If $a$ is a positive number, then $|X| > a$ is equivalent to $X < -a$ or $X > a$.

---

This property also holds true for the inequality symbol $\geq$.

**EXAMPLE 5** Solve for $y$: $|y - 3| > 7$.

*Solution* Since 7 is positive, we apply the property for $|X| > a$.

$$|y - 3| > 7 \text{ is equivalent to } y - 3 < -7 \text{ or } y - 3 > 7$$

Next, solve the compound inequality.

$$
\begin{array}{ccl}
y - 3 < -7 & \text{or} & y - 3 > 7 \\
y - 3 + 3 < -7 + 3 & \text{or} & y - 3 + 3 > 7 + 3 \quad \text{Add 3 to both sides.} \\
y < -4 & \text{or} & y > 10 \quad \text{Simplify.}
\end{array}
$$

The solution set is $(-\infty, -4) \cup (10, \infty)$, and its graph is shown.

☐

**PRACTICE**
**5** Solve for $y$: $|y + 4| \geq 6$.

Examples 6 and 8 illustrate special cases of absolute value inequalities. These special cases occur when an isolated absolute value expression is less than, less than or equal to, greater than, or greater than or equal to a negative number or 0.

**EXAMPLE 6**   Solve: $|2x + 9| + 5 > 3$.

_Solution_   First isolate the absolute value expression by subtracting 5 from both sides.

$$|2x + 9| + 5 > 3$$

$$|2x + 9| + 5 - 5 > 3 - 5 \quad \text{Subtract 5 from both sides.}$$

$$|2x + 9| > -2 \quad \text{Simplify.}$$

The absolute value of any number is always nonnegative and thus is always greater than $-2$. This inequality and the original inequality are true for all values of $x$. The solution set is $\{x \mid x \text{ is a real number}\}$ or $(-\infty, \infty)$ and its graph is shown.

<!-- number line graph from -3 to 4 -->

**PRACTICE**
**6**   Solve: $|4x + 3| + 5 > 3$. Graph the solution set.

**Concept Check** ☑

Without taking any solution steps, how do you know that the absolute value inequality $|3x - 2| > -9$ has a solution? What is its solution?

**EXAMPLE 7**   Solve: $\left|\dfrac{x}{3} - 1\right| - 7 \geq -5$.

_Solution_   First, isolate the absolute value expression by adding 7 to both sides.

$$\left|\frac{x}{3} - 1\right| - 7 \geq -5$$

$$\left|\frac{x}{3} - 1\right| - 7 + 7 \geq -5 + 7 \quad \text{Add 7 to both sides.}$$

$$\left|\frac{x}{3} - 1\right| \geq 2 \quad \text{Simplify.}$$

Next, write the absolute value inequality as an equivalent compound inequality and solve.

$$\frac{x}{3} - 1 \leq -2 \qquad \text{or} \qquad \frac{x}{3} - 1 \geq 2$$

$$3\left(\frac{x}{3} - 1\right) \leq 3(-2) \qquad \text{or} \qquad 3\left(\frac{x}{3} - 1\right) \geq 3(2) \quad \text{Clear the inequalities of fractions.}$$

$$x - 3 \leq -6 \qquad \text{or} \qquad x - 3 \geq 6 \qquad \text{Apply the distributive property.}$$

$$x \leq -3 \qquad \text{or} \qquad x \geq 9 \qquad \text{Add 3 to both sides.}$$

The solution set is $(-\infty, -3] \cup [9, \infty)$, and its graph is shown.

<!-- number line graph from -6 to 12 showing -3 and 9 -->

**PRACTICE**
**7**   Solve: $\left|\dfrac{x}{2} - 3\right| - 5 > -2$. Graph the solution set.

**EXAMPLE 8** Solve for $x$: $\left|\dfrac{2(x+1)}{3}\right| \le 0$.

**Solution** Recall that "≤" means "less than or equal to." The absolute value of any expression will never be less than 0, but it may be equal to 0. Thus, to solve $\left|\dfrac{2(x+1)}{3}\right| \le 0$ we solve $\left|\dfrac{2(x+1)}{3}\right| = 0$

$$\frac{2(x+1)}{3} = 0$$

$$3\left[\frac{2(x+1)}{3}\right] = 3(0) \quad \text{Clear the equation of fractions.}$$

$$2x + 2 = 0 \quad \text{Apply the distributive property.}$$

$$2x = -2 \quad \text{Subtract 2 from both sides.}$$

$$x = -1 \quad \text{Divide both sides by 2.}$$

The solution set is $\{-1\}$.

**PRACTICE**
**8** Solve for $x$: $\left|\dfrac{3(x-2)}{5}\right| \le 0$.

---

The following box summarizes the types of absolute value equations and inequalities.

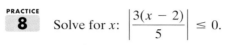

**Solving Absolute Value Equations and Inequalities with $a > 0$**

| **Algebraic Solution** | **Solution Graph** |
|---|---|
| $\|X\| = a$ is equivalent to $X = a$ or $X = -a$. | |
| $\|X\| < a$ is equivalent to $-a < X < a$. | |
| $\|X\| > a$ is equivalent to $X < -a$ or $X > a$. | |

# VOCABULARY & READINESS CHECK

*Match each absolute value statement with an equivalent statement.*

**1.** $|2x + 1| = 3$

**2.** $|2x + 1| \le 3$

**3.** $|2x + 1| < 3$

**4.** $|2x + 1| \ge 3$

**5.** $|2x + 1| > 3$

**A.** $2x + 1 > 3$ or $2x + 1 < -3$

**B.** $2x + 1 \ge 3$ or $2x + 1 \le -3$

**C.** $-3 < 2x + 1 < 3$

**D.** $2x + 1 = 3$ or $2x + 1 = -3$

**E.** $-3 \le 2x + 1 \le 3$

# 9.3 EXERCISE SET

*Solve each inequality. Then graph the solution set and write it in interval notation. See Examples 1 through 4.*

**1.** $|x| \le 4$

**2.** $|x| < 6$

**3.** $|x - 3| < 2$

**4.** $|y - 7| \le 5$

**5.** $|x + 3| < 2$

**6.** $|x + 4| < 6$

**7.** $|2x + 7| \le 13$

**8.** $|5x - 3| \le 18$

**9.** $|x| + 7 \le 12$

**10.** $|x| + 6 \le 7$

**11.** $|3x - 1| < -5$

**12.** $|8x - 3| < -2$

**13.** $|x - 6| - 7 \le -1$

**14.** $|z + 2| - 7 < -3$

*Solve each inequality. Graph the solution set and write it in interval notation. See Examples 5 through 7.*

**15.** $|x| > 3$ 　　　　　　　　　**16.** $|y| \geq 4$

**17.** $|x + 10| \geq 14$ 　　　　　**18.** $|x - 9| \geq 2$

**19.** $|x| + 2 > 6$ 　　　　　　　**20.** $|x| - 1 > 3$

**21.** $|5x| > -4$ 　　　　　　　　**22.** $|4x - 11| > -1$

**23.** $|6x - 8| + 3 > 7$ 　　　　**24.** $|10 + 3x| + 1 > 2$

*Solve each inequality. Graph the solution set and write it in interval notation. See Example 8.*

**25.** $|x| \leq 0$ 　　　　　　　　　**26.** $|x| \geq 0$

**27.** $|8x + 3| > 0$ 　　　　　　**28.** $|5x - 6| < 0$

## MIXED PRACTICE

*Solve each inequality. Graph the solution set and write it in interval notation. See Examples 1 through 8.*

**29.** $|x| \leq 2$ 　　　　　　　　　**30.** $|z| < 8$

**31.** $|y| > 1$ 　　　　　　　　　**32.** $|x| \geq 10$

**33.** $|x - 3| < 8$ 　　　　　　　**34.** $|-3 + x| \leq 10$

**35.** $|0.6x - 3| > 0.6$ 　　　　**36.** $|1 + 0.3x| \geq 0.1$

**37.** $5 + |x| \leq 2$ 　　　　　　**38.** $8 + |x| < 1$

**39.** $|x| > -4$ 　　　　　　　　**40.** $|x| \leq -7$

**41.** $|2x - 7| \leq 11$ 　　　　　**42.** $|5x + 2| < 8$

**43.** $|x + 5| + 2 \geq 8$ 　　　　**44.** $|-1 + x| - 6 > 2$

**45.** $|x| > 0$ 　　　　　　　　　**46.** $|x| < 0$

**47.** $9 + |x| > 7$ 　　　　　　　**48.** $5 + |x| \geq 4$

**49.** $6 + |4x - 1| \leq 9$ 　　　　**50.** $-3 + |5x - 2| \leq 4$

**51.** $\left|\dfrac{2}{3}x + 1\right| > 1$ 　　　　**52.** $\left|\dfrac{3}{4}x - 1\right| \geq 2$

**53.** $|5x + 3| < -6$ 　　　　　**54.** $|4 + 9x| \geq -6$

**55.** $\left|\dfrac{8x - 3}{4}\right| \leq 0$ 　　　　**56.** $\left|\dfrac{5x + 6}{2}\right| \leq 0$

**57.** $|1 + 3x| + 4 < 5$ 　　　　**58.** $|7x - 3| - 1 \leq 10$

**59.** $\left|\dfrac{x + 6}{3}\right| > 2$

**60.** $\left|\dfrac{7 + x}{2}\right| \geq 4$

**61.** $-15 + |2x - 7| \leq -6$

**62.** $-9 + |3 + 4x| < -4$

**63.** $\left|2x + \dfrac{3}{4}\right| - 7 \leq -2$

**64.** $\left|\dfrac{3}{5} + 4x\right| - 6 < -1$

## MIXED PRACTICE

*Solve each equation or inequality for x. (Sections 9.2, 9.3)*

**65.** $|2x - 3| < 7$

**66.** $|2x - 3| > 7$

**67.** $|2x - 3| = 7$

**68.** $|5 - 6x| = 29$

**69.** $|x - 5| \geq 12$

**70.** $|x + 4| \geq 20$

**71.** $|9 + 4x| = 0$

**72.** $|9 + 4x| \geq 0$

**73.** $|2x + 1| + 4 < 7$

**74.** $8 + |5x - 3| \geq 11$

**75.** $|3x - 5| + 4 = 5$

**76.** $|5x - 3| + 2 = 4$

**77.** $|x + 11| = -1$

**78.** $|4x - 4| = -3$

**79.** $\left|\dfrac{2x - 1}{3}\right| = 6$

**80.** $\left|\dfrac{6 - x}{4}\right| = 5$

**81.** $\left|\dfrac{3x - 5}{6}\right| > 5$

**82.** $\left|\dfrac{4x - 7}{5}\right| < 2$

## REVIEW AND PREVIEW

*Consider the equation $3x - 4y = 12$. For each value of x or y given, find the corresponding value of the other variable that makes the statement true. See Section 2.5.*

**83.** If $x = 2$, find $y$ 　　　　**84.** If $y = -1$, find $x$

**85.** If $y = -3$, find $x$ 　　　**86.** If $x = 4$, find $y$

## CONCEPT EXTENSIONS

**87.** Write an absolute value inequality representing all numbers $x$ whose distance from 0 is less than 7 units.

**88.** Write an absolute value inequality representing all numbers $x$ whose distance from 0 is greater than 4 units.

**89.** Write $-5 \leq x \leq 5$ as an equivalent inequality containing an absolute value.

**90.** Write $x > 1$ or $x < -1$ as an equivalent inequality containing an absolute value.

**91.** Describe how solving $|x - 3| = 5$ is different from solving $|x - 3| < 5$.

**92.** Describe how solving $|x + 4| = 0$ is similar to solving $|x + 4| \leq 0$.

*The expression $|x_T - x|$ is defined to be the absolute error in x, where $x_T$ is the true value of a quantity and x is the measured value or value as stored in a computer.*

**93.** If the true value of a quantity is 3.5 and the absolute error must be less than 0.05, find the acceptable measured values.

**94.** If the true value of a quantity is 0.2 and the approximate value stored in a computer is $\dfrac{51}{256}$, find the absolute error.

---

## THE BIGGER PICTURE  SIMPLIFYING EXPRESSIONS AND SOLVING EQUATIONS AND INEQUALITIES

We now continue the outline from Sections 1.7, 2.3, 2.8, 5.6, 6.6, 7.4, 7.6 and 9.1. Although suggestions will be given, this outline should be in your own words and you should include "how to recognize" and "how to begin to solve." Once you complete this new portion, try the exercises.

I. Simplifying Expressions (See Section 7.6, page 484.)

II. Solving Equations

   A. Linear Equations (Section 2.3)

   B. Quadratic and Higher Degree Equations (Section 6.6)

   C. Equations with Rational Expressions (Section 7.5)

   D. Proportions (Section 7.6)

   E. Absolute Value Equations: Equation contains the absolute value of a variable expression.

$$|3x - 1| - 12 = -4 \quad \text{Absolute value equation.}$$
$$|3x - 1| = 8 \quad \text{Isolate absolute value.}$$
$$3x - 1 = 8 \quad \text{or} \quad 3x - 1 = -8$$
$$3x = 9 \quad \text{or} \quad 3x = -7$$
$$x = 3 \quad \text{or} \quad x = -\frac{7}{3}$$

$$|x - 5| = |x + 1| \quad \text{Absolute value equation.}$$
$$x - 5 = x + 1 \quad \text{or} \quad x - 5 = -(x + 1)$$
$$\underbrace{-5 = 1}_{\text{No solution}} \quad \text{or} \quad x - 5 = -x - 1$$
$$\text{or} \quad 2x = 4$$
$$x = 2$$

III. Solving Inequalities

   A. Linear Inequalities (Section 2.8)

   B. Compound Inequalities (Section 9.1)

   C. Absolute Value Inequalities: Inequality with absolute value bars about variable expression.

| $|x - 5| - 8 < -2$ | $|2x + 1| \geq 17$ |
|---|---|
| $|x - 5| < 6$ | $2x + 1 \geq 17$ or $2x + 1 \leq -17$ |
| $-6 < x - 5 < 6$ | $2x \geq 16$ or $2x \leq -18$ |
| $-1 < x < 11$ | $x \geq 8$ or $x \leq -9$ |
| $(-1, 11)$ | $(-\infty, -9] \cup [8, \infty)$ |

*Solve. If an inequality, write your solutions in interval notation.*

**1.** $9x - 14 = 11x + 2$

**2.** $|x - 4| = 17$

**3.** $x - 1 \leq 5$ or $3x - 2 \leq 10$

**4.** $-x < 7$ and $4x \leq 20$

**5.** $|x - 2| = |x + 15|$

**6.** $9y - 6y + 1 = 4y + 10 - y + 3$

**7.** $1.5x - 3 = 1.2x - 18$

**8.** $\dfrac{7x + 1}{8} - 3 = x + \dfrac{2x + 1}{4}$

**9.** $|5x + 2| - 10 \leq -3$

**10.** $|x + 11| > 2$

**11.** $|9x + 2| - 1 = 24$

**12.** $\left|\dfrac{3x - 1}{2}\right| = |2x + 5|$

---

## INTEGRATED REVIEW  SOLVING COMPOUND INEQUALITIES AND ABSOLUTE VALUE EQUATIONS AND INEQUALITIES

*Solve each equation or inequality. Write inequality solution sets in interval rotation. For inequalities containing "and" or "or" also graph the solution set and write it in interval rotation.*

**1.** $x < 7$ and $x > -5$

**2.** $x < 7$ or $x > -5$

**3.** $|4x - 3| = 1$

**4.** $|2x + 1| < 5$

**5.** $|6x| - 9 \geq -3$

**6.** $|x - 7| = |2x + 11|$

**7.** $-5 \leq \dfrac{3x - 8}{2} \leq 2$

**8.** $|9x - 1| = -3$

**9.** $3x + 2 \leq 5$ or $-3x \geq 0$

**10.** $3x + 2 \leq 5$ and $-3x \geq 0$

**11.** $|3 - x| - 5 \leq -2$

**12.** $\left|\dfrac{4x + 1}{5}\right| = |-1|$

*Match each equation or inequality on the left with an equivalent statement on the right.*

**13.** $|2x + 1| = 5$

**14.** $|2x + 1| < 5$

**15.** $|2x + 1| > 5$

**16.** $x < 3$ or $x < 5$

**17.** $x < 3$ and $x < 5$

**A.** $2x + 1 > 5$ or $2x + 1 < -5$

**B.** $2x + 1 = 5$ or $2x + 1 = -5$

**C.** $x < 5$

**D.** $x < 3$

**E.** $-5 < 2x + 1 < 5$

# 9.4 GRAPHING LINEAR INEQUALITIES IN TWO VARIABLES AND SYSTEMS OF LINEAR INEQUALITIES

**OBJECTIVE**

1  Graph a linear inequality in two variables.

2  Solve a system of linear inequalities.

In this section, we first need to learn to graph a single linear inequality in two variables. We continue our work with systems by solving systems of linear inequalities.

Recall that a linear equation in two variables is an equation that can be written in the form $Ax + By = C$ where $A$, $B$, and $C$ are real numbers and $A$ and $B$ are not both 0. The definition of a linear inequality is the same except that the equal sign is replaced with an inequality sign.

A **linear inequality in two variables** is an inequality that can be written in one of the forms:

$$Ax + By < C \qquad Ax + By \leq C$$
$$Ax + By > C \qquad Ax + By \geq C$$

where $A$, $B$, and $C$ are real numbers and $A$ and $B$ are not both 0. Just as for linear equations in $x$ and $y$, an ordered pair is a **solution** of an inequality in $x$ and $y$ if replacing the variables by coordinates of the ordered pair results in a true statement.

**OBJECTIVE 1** ▶ **Graphing linear inequalities in two variables.** The linear equation $x - y = 1$ is graphed next. Recall that all points on the line correspond to ordered pairs that satisfy the equation $x - y = 1$.

Notice the line defined by $x - y = 1$ divides the rectangular coordinate system plane into 2 sides. All points on one side of the line satisfy the inequality $x - y < 1$ and all points on the other side satisfy the inequality $x - y > 1$. The graph below shows a few examples of this.

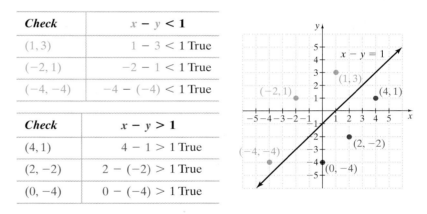

| Check | $x - y < 1$ |
|-------|-------------|
| $(1, 3)$ | $1 - 3 < 1$ True |
| $(-2, 1)$ | $-2 - 1 < 1$ True |
| $(-4, -4)$ | $-4 - (-4) < 1$ True |

| Check | $x - y > 1$ |
|-------|-------------|
| $(4, 1)$ | $4 - 1 > 1$ True |
| $(2, -2)$ | $2 - (-2) > 1$ True |
| $(0, -4)$ | $0 - (-4) > 1$ True |

The graph of $x - y < 1$ is the region shaded blue and the graph of $x - y > 1$ is the region shaded red below.

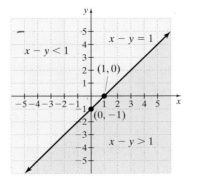

The region to the left of the line and the region to the right of the line are called **half-planes.** Every line divides the plane (similar to a sheet of paper extending indefinitely in all directions) into two half-planes; the line is called the **boundary.**

Recall that the inequality $x - y \leq 1$ means

$$x - y = 1 \quad \text{or} \quad x - y < 1$$

Thus, the graph of $x - y \leq 1$ is the half-plane $x - y < 1$ along with the boundary line $x - y = 1$.

---

**Graphing a Linear Inequality in Two Variables**

**STEP 1.** Graph the boundary line found by replacing the inequality sign with an equal sign. If the inequality sign is $>$ or $<$, graph a dashed boundary line (indicating that the points on the line are not solutions of the inequality). If the inequality sign is $\geq$ or $\leq$, graph a solid boundary line (indicating that the points on the line are solutions of the inequality).

**STEP 2.** Choose a point, *not* on the boundary line, as a test point. Substitute the coordinates of this test point into the *original* inequality.

**STEP 3.** If a true statement is obtained in Step 2, shade the half-plane that contains the test point. If a false statement is obtained, shade the half-plane that does not contain the test point.

---

**EXAMPLE 1** Graph: $x + y < 7$

*Solution*

**STEP 1.** First we graph the boundary line by graphing the equation $x + y = 7$. We graph this boundary as a dashed line because the inequality sign is $<$, and thus the points on the line are not solutions of the inequality $x + y < 7$.

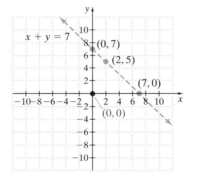

**STEP 2.** Next, choose a test point, being careful not to choose a point on the boundary line. We choose $(0, 0)$. Substitute the coordinates of $(0, 0)$ into $x + y < 7$.

$$x + y < 7 \quad \text{Original inequality}$$
$$0 + 0 \overset{?}{<} 7 \quad \text{Replace } x \text{ with 0 and } y \text{ with 0.}$$
$$0 < 7 \quad \text{True}$$

**STEP 3.** Since the result is a true statement, $(0, 0)$ is a solution of $x + y < 7$, and every point in the same half-plane as $(0, 0)$ is also a solution. To indicate this, shade the entire half-plane containing $(0, 0)$, as shown.

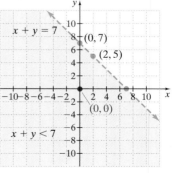

**PRACTICE**

**1** Graph: $x + y > 5$

**Concept Check** ☑

Determine whether $(0, 0)$ is included in the graph of

**a.** $y \geq 2x + 3$      **b.** $x < 7$      **c.** $2x - 3y < 6$

---

**EXAMPLE 2**   Graph: $2x - y \geq 3$

*Solution*

**STEP 1.** We graph the boundary line by graphing $2x - y = 3$. We draw this line as a solid line because the inequality sign is $\geq$, and thus the points on the line are solutions of $2x - y \geq 3$.

**STEP 2.** Once again, $(0, 0)$ is a convenient test point since it is not on the boundary line. We substitute 0 for $x$ and 0 for $y$ into the original inequality.

$$2x - y \geq 3$$
$$2(0) - 0 \geq 3 \quad \text{Let } x = 0 \text{ and } y = 0.$$
$$0 \geq 3 \quad \text{False}$$

**STEP 3.** Since the statement is false, no point in the half-plane containing $(0, 0)$ is a solution. Therefore, we shade the half-plane that does not contain $(0, 0)$. Every point in the shaded half-plane and every point on the boundary line is a solution of $2x - y \geq 3$.

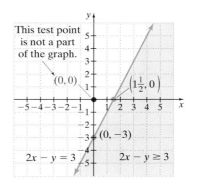

□

**PRACTICE**
**2**   Graph: $3x - y \geq 4$

---

▶ **Helpful Hint**

When graphing an inequality, make sure the test point is substituted into the **original inequality.** For Example 2, we substituted the test point $(0, 0)$ into the **original inequality** $2x - y \geq 3$, *not* $2x - y = 3$.

---

**EXAMPLE 3**   Graph: $x > 2y$

*Solution*

**STEP 1.** We find the boundary line by graphing $x = 2y$. The boundary line is a dashed line since the inequality symbol is $>$.

**STEP 2.** We cannot use $(0, 0)$ as a test point because it is a point on the boundary line. We choose instead $(0, 2)$.

$$x > 2y$$
$$0 > 2(2) \quad \text{Let } x = 0 \text{ and } y = 2.$$
$$0 > 4 \quad \text{False}$$

**Answers to Concept Check:**
**a.** no   **b.** yes   **c.** yes

**STEP 3.** Since the statement is false, we shade the half-plane that does not contain the test point $(0, 2)$, as shown.

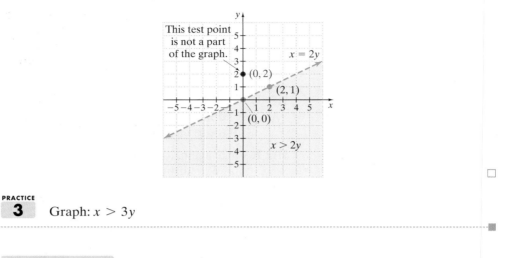

**PRACTICE**
**3**    Graph: $x > 3y$

---

**EXAMPLE 4**    Graph: $5x + 4y \le 20$

*Solution*    We graph the solid boundary line $5x + 4y = 20$ and choose $(0, 0)$ as the test point.

$$5x + 4y \le 20$$
$$5(0) + 4(0) \overset{?}{\le} 20 \quad \text{Let } x = 0 \text{ and } y = 0.$$
$$0 \le 20 \quad \text{True}$$

We shade the half-plane that contains $(0, 0)$, as shown.

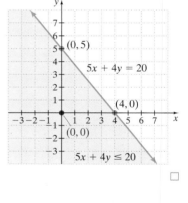

**PRACTICE**
**4**    Graph: $3x + 4y \ge 12$

---

**EXAMPLE 5**    Graph: $y > 3$

*Solution*    We graph the dashed boundary line $y = 3$ and choose $(0, 0)$ as the test point. (Recall that the graph of $y = 3$ is a horizontal line with $y$-intercept 3.)

$$y > 3$$
$$0 \overset{?}{>} 3 \quad \text{Let } y = 0.$$
$$0 > 3 \quad \text{False}$$

We shade the half-plane that does not contain $(0, 0)$, as shown.

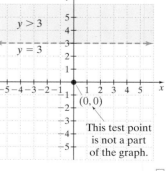

**PRACTICE**
**5**    Graph: $x > 3$

**OBJECTIVE 2** ▶ **Solving systems of linear inequalities.** Just as two linear equations make a system of linear equations, two linear inequalities make a **system of linear inequalities.** Systems of inequalities are very important in a process called linear programming. Many businesses use linear programming to find the most profitable way to use limited resources such as employees, machines, or buildings.

A **solution of a system of linear inequalities** is an ordered pair that satisfies each inequality in the system. The set of all such ordered pairs is the solution set of the system. Graphing this set gives us a picture of the solution set. We can graph a system of inequalities by graphing each inequality in the system and identifying the region of overlap.

**EXAMPLE 6** Graph the solution of the system: $\begin{cases} 3x \ge y \\ x + 2y \le 8 \end{cases}$

*Solution* We begin by graphing each inequality on the same set of axes. The graph of the solution of the system is the region contained in the graphs of both inequalities. It is their intersection.

First, graph $3x \ge y$. The boundary line is the graph of $3x = y$. Sketch a solid boundary line since the inequality $3x \ge y$ means $3x > y$ or $3x = y$. The test point $(1, 0)$ satisfies the inequality, so shade the half-plane that includes $(1, 0)$.

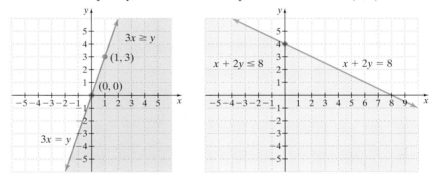

Next, sketch a solid boundary line $x + 2y = 8$ on the same set of axes. The test point $(0, 0)$ satisfies the inequality $x + 2y \le 8$, so shade the half-plane that includes $(0, 0)$. (For clarity, the graph of $x + 2y \le 8$ is shown on a separate set of axes.)

An ordered pair solution of the system must satisfy both inequalities. These solutions are points that lie in both shaded regions. The solution of the system is the purple shaded region as seen below. This solution includes parts of both boundary lines.

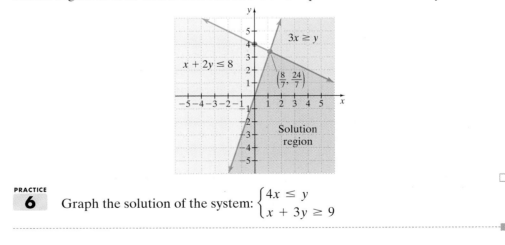

**PRACTICE**
**6** Graph the solution of the system: $\begin{cases} 4x \le y \\ x + 3y \ge 9 \end{cases}$

In linear programming, it is sometimes necessary to find the coordinates of the **corner point:** the point at which the two boundary lines intersect. To find the point of intersection, solve the related linear system

$$\begin{cases} 3x = y \\ x + 2y = 8 \end{cases}$$

by the substitution method or the addition method. The lines intersect at $\left(\dfrac{8}{7}, \dfrac{24}{7}\right)$, the corner point of the graph.

> **Graphing the Solution of a System of Linear Inequalities**
>
> **STEP 1.** Graph each inequality in the system on the same set of axes.
>
> **STEP 2.** The solutions of the system are the points common to the graphs of all the inequalities in the system.

**EXAMPLE 7**  Graph the solution of the system: $\begin{cases} x - y < 2 \\ x + 2y > -1 \end{cases}$

*Solution*  Graph both inequalities on the same set of axes. Both boundary lines are dashed lines since the inequality symbols are $<$ and $>$. The solution of the system is the region shown by the purple shading. In this example, the boundary lines are not a part of the solution.

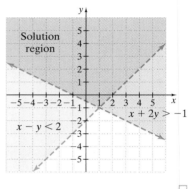

**PRACTICE**
**7**  Graph the solution of the system: $\begin{cases} x - y > 4 \\ x + 3y < -4 \end{cases}$

**EXAMPLE 8**  Graph the solution of the system: $\begin{cases} -3x + 4y < 12 \\ x \ge 2 \end{cases}$

*Solution*  Graph both inequalities on the same set of axes.

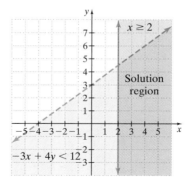

The solution of the system is the purple shaded region, including a portion of the line $x = 2$.

**PRACTICE**
**8**  Graph the solution of the system: $\begin{cases} y \le 6 \\ -2x + 5y > 10 \end{cases}$

## VOCABULARY & READINESS CHECK

*Use the choices below to fill in each blank. Some choices may be used more than once, and some not at all.*

| | | | | |
|---|---|---|---|---|
| true | $x < 3$ | $y < 3$ | half-planes | yes |
| false | $x \leq 3$ | $y \leq 3$ | linear inequality in two variables | no |

1. The statement $5x - 6y < 7$ is an example of a(n) _____.
2. A boundary line divides a plane into two regions called _____.
3. True or false: The graph of $5x - 6y < 7$ includes its corresponding boundary line. _____
4. True or false: When graphing a linear inequality, to determine which side of the boundary line to shade, choose a point *not* on the boundary line. _____
5. True or false: The boundary line for the inequality $5x - 6y < 7$ is the graph of $5x - 6y = 7$. _____
6. The graph of _____ is

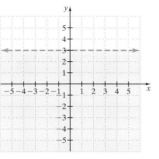

*State whether the graph of each inequality includes its corresponding boundary line. Answer yes or no.*

7. $y \geq x + 4$      8. $x - y > -7$      9. $y \geq x$      10. $x > 0$

*Decide whether (0, 0) is a solution of each given inequality.*

11. $x + y > -5$    12. $2x + 3y < 10$    13. $x - y \leq -1$    14. $\dfrac{2}{3}x + \dfrac{5}{6}y > 4$

## 9.4 EXERCISE SET

MyMathLab  *Powered by CourseCompass™ and MathXL®*

MathXL PRACTICE | WATCH | DOWNLOAD | READ | REVIEW

*Determine which ordered pairs given are solutions of the linear inequality in two variables. See Example 1.*

1. $x - y > 3; (2, -1), (5, 1)$
2. $y - x < -2; (2, 1), (5, -1)$
3. $3x - 5y \leq -4; (-1, -1), (4, 0)$
4. $2x + y \geq 10; (-1, -4), (5, 0)$
5. $x < -y; (0, 2), (-5, 1)$
6. $y > 3x; (0, 0), (-1, -4)$

**MIXED PRACTICE**

*Graph each inequality. See Examples 2 through 5.*

7. $x + y \leq 1$      8. $x + y \geq -2$
9. $2x + y > -4$      10. $x + 3y \leq 3$
11. $x + 6y \leq -6$      12. $7x + y > -14$
13. $2x + 5y > -10$      14. $5x + 2y \leq 10$
15. $x + 2y \leq 3$      16. $2x + 3y > -5$
17. $2x + 7y > 5$      18. $3x + 5y \leq -2$
19. $x - 2y \geq 3$      20. $4x + y \leq 2$
21. $5x + y < 3$      22. $x + 2y > -7$
23. $4x + y < 8$      24. $9x + 2y \geq -9$

25. $y \geq 2x$    26. $x < 5y$    27. $x \geq 0$
28. $y \leq 0$    29. $y \leq -3$    30. $x > -\dfrac{2}{3}$
31. $2x - 7y > 0$    32. $5x + 2y \leq 0$    33. $3x - 7y \geq 0$
34. $-2x - 9y > 0$    35. $x > y$    36. $x \leq -y$
37. $x - y \leq 6$    38. $x - y > 10$    39. $-\dfrac{1}{4}y + \dfrac{1}{3}x > 1$
40. $\dfrac{1}{2}x - \dfrac{1}{3}y \leq -1$    41. $-x < 0.4y$    42. $0.3x \geq 0.1y$

*In Exercises 43 through 48, match each inequality with its graph.*

a. $x > 2$      b. $y < 2$      c. $y < 2x$
d. $y \leq -3x$    e. $2x + 3y < 6$    f. $3x + 2y > 6$

43.              44.

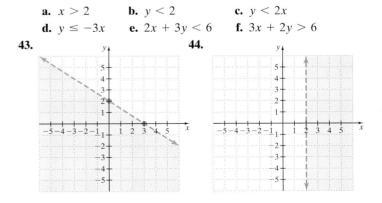

45. 46.

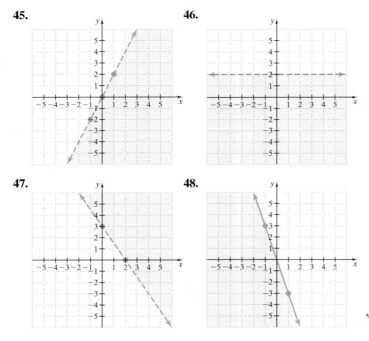

47. 48.

## MIXED PRACTICE

Graph the solution of each system of linear inequalities. See Examples 6 through 8.

49. $\begin{cases} y \geq x + 1 \\ y \geq 3 - x \end{cases}$

50. $\begin{cases} y \geq x - 3 \\ y \geq -1 - x \end{cases}$

51. $\begin{cases} y < 3x - 4 \\ y \leq x + 2 \end{cases}$

52. $\begin{cases} y \leq 2x + 1 \\ y > x + 2 \end{cases}$

53. $\begin{cases} y \leq -2x - 2 \\ y \geq x + 4 \end{cases}$

54. $\begin{cases} y \leq 2x + 4 \\ y \geq -x - 5 \end{cases}$

55. $\begin{cases} y \geq -x + 2 \\ y \leq 2x + 5 \end{cases}$

56. $\begin{cases} y \geq x - 5 \\ y \leq -3x + 3 \end{cases}$

57. $\begin{cases} x \geq 3y \\ x + 3y \leq 6 \end{cases}$

58. $\begin{cases} -2x < y \\ x + 2y < 3 \end{cases}$

59. $\begin{cases} y + 2x \geq 0 \\ 5x - 3y \leq 12 \end{cases}$

60. $\begin{cases} y + 2x \leq 0 \\ 5x + 3y \geq -2 \end{cases}$

61. $\begin{cases} 3x - 4y \geq -6 \\ 2x + y \leq 7 \end{cases}$

62. $\begin{cases} 4x - y \geq -2 \\ 2x + 3y \leq -8 \end{cases}$

63. $\begin{cases} x \leq 2 \\ y \geq -3 \end{cases}$

64. $\begin{cases} x \geq -3 \\ y \geq -2 \end{cases}$

65. $\begin{cases} y \geq 1 \\ x < -3 \end{cases}$

66. $\begin{cases} y > 2 \\ x \geq -1 \end{cases}$

67. $\begin{cases} 2x + 3y < -8 \\ x \geq -4 \end{cases}$

68. $\begin{cases} 3x + 2y \leq 6 \\ x < 2 \end{cases}$

69. $\begin{cases} 2x - 5y \leq 9 \\ y \leq -3 \end{cases}$

70. $\begin{cases} 2x + 5y \leq -10 \\ y \geq 1 \end{cases}$

71. $\begin{cases} y \geq \dfrac{1}{2}x + 2 \\ y \leq \dfrac{1}{2}x - 3 \end{cases}$

72. $\begin{cases} y \geq -\dfrac{3}{2}x + 3 \\ y < -\dfrac{3}{2}x + 6 \end{cases}$

## REVIEW AND PREVIEW

Evaluate each expression for the given replacement value. See Section 5.1.

73. $x^2$ if $x$ is $-5$

74. $x^3$ if $x$ is $-5$

75. $2x^3$ if $x$ is $-1$

76. $3x^2$ if $x$ is $-1$

## CONCEPT EXTENSIONS

Determine whether $(1, 1)$ is included in each graph. See the Concept Check in this section.

77. $3x + 4y < 8$

78. $y > 5x$

79. $y \geq -\dfrac{1}{2}x$

80. $x > 3$

81. Write an inequality whose solutions are all pairs of numbers $x$ and $y$ whose sum is at least 13. Graph the inequality.

82. Write an inequality whose solutions are all the pairs of numbers $x$ and $y$ whose sum is at most $-4$. Graph the inequality.

83. Explain why a point on the boundary line should not be chosen as the test point.

84. Describe the graph of a linear inequality.

85. The price for a taxi cab in a small city is $2.50 per mile, $x$, while traveling, and $.25 every minute, $y$, while waiting. If you have $20 to spend on a cab ride, the inequality

$$2.5x + 0.25y \leq 20$$

represents your situation. Graph this inequality in the first quadrant only.

86. A word processor charges $22 per hour, $x$, for typing a first draft, and $15 per hour, $y$, for making changes and typing a second draft. If you need a document typed and have $100, the inequality

$$22x + 15y \leq 100$$

represents your situation. Graph the inequality in the first quadrant only.

87. In Exercises 85 and 86, why were you instructed to graph each inequality in the first quadrant only?

88. Scott Sambracci and Sara Thygeson are planning their wedding. They have calculated that they want the cost of their wedding ceremony $x$ plus the cost of their reception $y$ to be no more than $5000.

   a. Write an inequality describing this relationship.

   b. Graph this inequality.

   c. Why should we be interested in only quadrant I of this graph?

89. It's the end of the budgeting period for Dennis Fernandes and he has $500 left in his budget for car rental expenses. He plans to spend this budget on a sales trip throughout southern Texas. He will rent a car that costs $30 per day and $0.15 per mile and he can spend no more than $500.

   a. Write an inequality describing this situation. Let $x =$ number of days and let $y =$ number of miles.

   b. Graph this inequality.

   c. Why should we be interested in only quadrant I of this graph?

**90.** Explain how to decide which region to shade to show the solution region of the following system.

$$\begin{cases} x \geq 3 \\ y \geq -2 \end{cases}$$

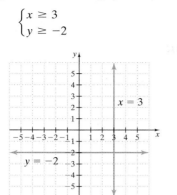

*For each system of inequalities, choose the corresponding graph.*

**91.** $\begin{cases} y < 5 \\ x > 3 \end{cases}$    **92.** $\begin{cases} y > 5 \\ x < 3 \end{cases}$    **93.** $\begin{cases} y \leq 5 \\ x < 3 \end{cases}$    **94.** $\begin{cases} y > 5 \\ x \geq 3 \end{cases}$

**A.**

**B.**

**C.**

**D.**

**95.** Graph the solution of $\begin{cases} 2x - y \leq 6 \\ x \geq 3 \\ y > 2 \end{cases}$

**96.** Graph the solution of $\begin{cases} x + y < 5 \\ y < 2x \\ x \geq 0 \\ y \geq 0 \end{cases}$

**97.** Describe the location of the solution region of the system

$$\begin{cases} x > 0 \\ y > 0. \end{cases}$$

# CHAPTER 9 GROUP ACTIVITY

## Analyzing Municipal Budgets

Nearly all cities, towns, and villages operate with an annual budget. Budget items might include expenses for fire and police protection as well as for street maintenance and parks. No matter how big or small the budget, city officials need to know if municipal spending is over or under budget. In this project, you will have the opportunity to analyze a municipal budget and make budgetary recommendations. This project may be completed by working in groups or individually.

Suppose that each year your town creates a municipal budget. The next year's annual municipal budget is submitted for approval by the town's citizens at the annual town meeting. This year's budget was printed in the town newspaper earlier in the year.

You have joined a group of citizens who are concerned about your town's budgeting and spending processes. Your group plans to analyze this year's budget along with what was actually spent by the town this year. You hope to present your findings at the annual town meeting and make some budgetary recommendations for next year's budget. The municipal budget contains many different areas of spending. To help focus your group's analysis, you have decided to research spending habits only for categories in which the actual expenses differ from the budgeted amount by more than 12% of the budgeted amount.

**1.** For each category in the budget, write a specific absolute value inequality that describes the condition that must be met

before your group will research spending habits for that category. In each case, let the variable $x$ represent the actual expense for a budget category.

**2.** For each category in the budget, write an equivalent compound inequality for the condition described in Question 1. Again, let the variable $x$ represent the actual expense for a budget category.

**3.** On the next page is a listing of the actual expenditures made this year for each budget category. Use the inequalities from either Question 1 or Question 2 to complete the Budget Worksheet given at the end of this project. (The first category has been filled in.) From the Budget Worksheet, decide which categories must be researched.

**4.** Can you think of possible reasons why spending in the categories that must be researched were over or under budget?

**5.** Based on this year's municipal budget and actual expenses, what recommendations would you make for next year's budget? Explain your reasoning.

**6.** (Optional) Research the annual budget used by your own town or your college or university. Conduct a similar analysis of the budget with respect to actual expenses. What can you conclude?

| | Department/Program | Actual Expenditure |
|---|---|---|
| I. | **Board of Health** | |
| | Immunization Programs | $14,800 |
| | Inspections | $41,900 |
| II. | **Fire Department** | |
| | Equipment | $375,000 |
| | Salaries | $268,500 |
| III. | **Libraries** | |
| | Book/Periodical Purchases | $107,300 |
| | Equipment | $29,000 |
| | Salaries | $118,400 |
| IV. | **Parks and Recreation** | |
| | Maintenance | $82,500 |
| | Playground Equipment | $45,000 |
| | Salaries | $118,000 |
| | Summer Programs | $96,200 |
| V. | **Police Department** | |
| | Equipment | $328,000 |
| | Salaries | $405,000 |
| VI. | **Public Works** | |
| | Recycling | $48,100 |
| | Sewage | $92,500 |
| | Snow Removal & Road Salt | $268,300 |
| | Street Maintenance | $284,000 |
| | Water Treatment | $94,100 |
| | **TOTAL** | $2,816,600 |

## THE TOWN CRIER
### Annual Budget Set at Town Meeting
ANYTOWN, USA (MG)—This year's annual budget is as follows:

| | Amount Budgeted |
|---|---|
| **BOARD OF HEALTH** | |
| Immunization Programs | $15,000 |
| Inspections | $50,000 |
| **FIRE DEPARTMENT** | |
| Equipment | $450,000 |
| Salaries | $275,000 |
| **LIBRARIES** | |
| Book/Periodical Purchases | $90,000 |
| Equipment | $30,000 |
| Salaries | $120,000 |
| **PARKS AND RECREATION** | |
| Maintenance | $70,000 |
| Playground Equipment | $50,000 |
| Salaries | $140,000 |
| Summer Programs | $80,000 |
| **POLICE DEPARTMENT** | |
| Equipment | $300,000 |
| Salaries | $400,000 |
| **PUBLIC WORKS** | |
| Recycling | $50,000 |
| Sewage | $100,000 |
| Snow Removal & Road Salt | $200,000 |
| Street Maintenance | $250,000 |
| Water Treatment | $100,000 |
| **TOTAL** | **$2,770,000** |

## BUDGET WORKSHEET

| Budget category | Budgeted amount | Minimum allowed | Actual expense | Maximum allowed | Within budget? | Amt over/ under budget |
|---|---|---|---|---|---|---|
| Immunization Programs | $15,000 | $13,200 | $14,800 | $16,800 | Yes | Under $200 |
| | | | | | | |
| | | | | | | |
| | | | | | | |

## CHAPTER 9 VOCABULARY CHECK

*Fill in each blank with one of the words or phrases listed below.*

| compound inequality | solution | system of linear inequalities |
|---|---|---|
| absolute value | union | intersection |

**1.** The statement "$x < 5$ or $x > 7$" is called a(n) _____ .

**2.** The _____ of two sets is the set of all elements common to both sets.

**3.** The _____ of two sets is the set of all elements that belong to either of the sets.

**4.** A number's distance from 0 is called its _____ .

**5.** When a variable in an equation is replaced by a number and the resulting equation is true, then that number is called a(n) _____ of the equation.

**6.** Two or more linear inequalities are called a(n) _____ .

▶ **Helpful Hint**

Are you preparing for your test? Don't forget to take the Chapter 9 Test on page 574. Then check your answers at the back of the text and use the Chapter Test Prep Video CD to see the fully worked-out solutions to any of the exercises you want to review.

# CHAPTER 9 HIGHLIGHTS

| DEFINITIONS AND CONCEPTS | EXAMPLES |
|---|---|

## SECTION 9.1   COMPOUND INEQUALITIES

Two inequalities joined by the words **and** or **or** are called **compound inequalities.**

The solution set of a compound inequality formed by the word **and** is the **intersection** ∩ of the solution sets of the two inequalities.

Compound inequalities:

$$x - 7 \le 4 \quad \text{and} \quad x \ge -21$$
$$2x + 7 > x - 3 \quad \text{or} \quad 5x + 2 > -3$$

Solve for $x$:

$$x < 5 \text{ and } x < 3$$

$\{x \mid x < 5\}$ $(-\infty, 5)$

$\{x \mid x < 3\}$ $(-\infty, 3)$

$\{x \mid x < 3$ and $x < 5\}$ $(-\infty, 3)$

The solution set of a compound inequality formed by the word **or** is the **union**, ∪, of the solution sets of the two inequalities.

Solve for $x$:

$$x - 2 \ge -3 \quad \text{or} \quad 2x \le -4$$
$$x \ge -1 \quad \text{or} \quad x \le -2$$

$\{x \mid x \ge -1\}$ $[-1, \infty)$

$\{x \mid x \le -2\}$ $(-\infty, -2]$

$\{x \mid x \le -2$ or $x \ge -1\}$ $(-\infty, -2] \cup [-1, \infty)$

## SECTION 9.2   ABSOLUTE VALUE EQUATIONS

If $a$ is a positive number, then $|x| = a$ is equivalent to $x = a$ or $x = -a$.

Solve for $y$:

$$|5y - 1| - 7 = 4$$

$|5y - 1| = 11$

$5y - 1 = 11 \quad \text{or} \quad 5y - 1 = -11$    Add 7.

$5y = 12 \quad \text{or} \quad\quad 5y = -10$    Add 1.

$y = \dfrac{12}{5} \quad\quad \text{or} \quad y = -2$    Divide by 5.

The solutions are $-2$ and $\dfrac{12}{5}$.

If $a$ is negative, then $|x| = a$ has no solution.

Solve for $x$:

$$\left| \frac{x}{2} - 7 \right| = -1$$

The solution set is $\{\ \}$ or $\varnothing$.

*(continued)*

| **DEFINITIONS AND CONCEPTS** | **EXAMPLES** |

If an absolute value equation is of the form $|x| = |y|$, solve $x = y$ or $x = -y$.

Solve for $x$:

$$|x - 7| = |2x + 1|$$

$$x - 7 = 2x + 1 \quad \text{or} \quad x - 7 = -(2x + 1)$$
$$x = 2x + 8 \qquad\qquad x - 7 = -2x - 1$$
$$-x = 8 \qquad\qquad\qquad x = -2x + 6$$
$$x = -8 \quad \text{or} \qquad 3x = 6$$
$$x = 2$$

The solutions are $-8$ and $2$.

If $a$ is a positive number, then $|x| < a$ is equivalent to $-a < x < a$.

Solve for $y$:

$$|y - 5| \leq 3$$
$$-3 \leq y - 5 \leq 3$$
$$-3 + 5 \leq y - 5 + 5 \leq 3 + 5 \quad \text{Add 5.}$$
$$2 \leq y \leq 8$$

The solution set is $[2, 8]$.

If $a$ is a positive number, then $|x| > a$ is equivalent to $x < -a$ or $x > a$.

Solve for $x$:

$$\left|\frac{x}{2} - 3\right| > 7$$

$$\frac{x}{2} - 3 < -7 \quad \text{or} \quad \frac{x}{2} - 3 > 7$$

$$x - 6 < -14 \quad \text{or} \quad x - 6 > 14 \quad \text{Multiply by 2.}$$

$$x < -8 \quad \text{or} \quad x > 20 \qquad \text{Add 6.}$$

The solution set is $(-\infty, -8) \cup (20, \infty)$.

A **linear inequality in two variables** is an inequality that can be written in one of the forms:

$$Ax + By < C \qquad Ax + By \leq C$$
$$Ax + By > C \qquad Ax + By \geq C$$

To graph a linear inequality

1. Graph the boundary line by graphing the related equation. Draw the line solid if the inequality symbol is $\leq$ or $\geq$. Draw the line dashed if the inequality symbol is $<$ or $>$.

2. Choose a test point not on the line. Substitute its coordinates into the original inequality.

3. If the resulting inequality is true, shade the half-plane that contains the test point. If the inequality is not true, shade the half-plane that does not contain the test point.

*Linear Inequalities*

$$2x - 5y < 6 \qquad x \geq -5$$
$$y > -8x \qquad y \leq 2$$

Graph $2x - y \leq 4$.

1. Graph $2x - y = 4$. Draw a solid line because the inequality symbol is $\leq$.

2. Check the test point $(0, 0)$ in the inequality $2x - y \leq 4$.

$$2 \cdot 0 - 0 \leq 4 \quad \text{Let } x = 0 \text{ and } y = 0.$$
$$0 \leq 4 \quad \text{True}$$

3. The inequality is true so we shade the half-plane containing $(0, 0)$.

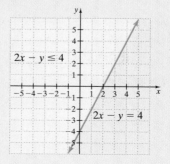

| **DEFINITIONS AND CONCEPTS** | **EXAMPLES** |
|---|---|

SECTION 9.4    GRAPHING LINEAR INEQUALITIES IN TWO VARIABLES AND SYSTEMS OF LINEAR INEQUALITIES (continued)

A system of linear inequalities consists of two or more linear inequalities.

To graph a system of inequalities, graph each inequality in the system. The overlapping region is the solution of the system.

System of Linear Inequalities

$$\begin{cases} x - y \geq 3 \\ y \leq -2x \end{cases}$$

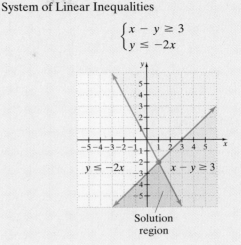

## STUDY SKILLS BUILDER

### Are You Prepared for a Test on Chapter 9?

Below I have listed some common trouble areas for students in Chapter 9. After studying for your test—but before taking your test—read these.

- Remember to reverse the direction of the inequality symbol when multiplying or dividing both sides of an inequality by a negative number.

$$-11x < 33$$
$$\frac{-11x}{-11} > \frac{33}{-11} \quad \text{Direction of arrow is reversed.}$$
$$x > -3$$

- Remember the differences when solving absolute value equations and inequalities.

| $\lvert x + 1 \rvert = 3$ | $\lvert x + 1 \rvert < 3$ |
|---|---|
| $x + 1 = 3$  or  $x + 1 = -3$ | $-3 < x + 1 < 3$ |
| $x = 2$  or  $x = -4$ | $-3 - 1 < x < 3 - 1$ |
| $\{2, -4\}$ | $-4 < x < 2$ |
|  | $(-4, 2)$ |

- Remember that an equation is not solved for a specified variable unless the variable is alone on one side of an equation *and* the other side contains *no* specified variables.

$$y = 10x + 6 - y \quad \text{Equation is \textit{not} solved for } y.$$
$$2y = 10x + 6 \qquad \text{Add } y \text{ to both sides.}$$
$$y = 5x + 3 \qquad\ \text{Divide both sides by 2.}$$

Remember: This is simply a checklist of common trouble areas. For a review of Chapter 9, see the Highlights and Chapter Review at the end of this chapter.

$\lvert x + 1 \rvert > 3$
$x + 1 < -3$  or  $x + 1 > 3$
$x < -4$  or  $x > 2$
$(-\infty, -4) \cup (2, \infty)$

# CHAPTER 9 REVIEW

*(9.1) Solve each inequality. Write your answers in interval notation.*

1. $-3 < 4(2x - 1) < 12$

2. $-2 \leq 8 + 5x < -1$

3. $\dfrac{1}{6} < \dfrac{4x - 3}{3} \leq \dfrac{4}{5}$

4. $-6 < x - (3 - 4x) < -3$

5. $3x - 5 > 6$ or $-x < -5$

6. $x \leq 2$ and $x > -5$

*(9.2)* Solve each absolute value equation.

**7.** $|8 - x| = 3$

**8.** $|x - 7| = 9$

**9.** $|-3x + 4| = 7$

**10.** $|2x + 9| = 9$

**11.** $5 + |6x + 1| = 5$

**12.** $|3x - 2| + 6 = 10$

**13.** $|5 - 6x| + 8 = 3$

**14.** $-5 = |4x - 3|$

**15.** $\left|\dfrac{3x - 7}{4}\right| = 2$

**16.** $-8 = |x - 3| - 10$

**17.** $|6x + 1| = |15 + 4x|$

**18.** $|x - 3| = |x + 5|$

*(9.3)* Solve each absolute value inequality. Graph the solution set and write it in interval notation.

**19.** $|5x - 1| < 9$

**20.** $|6 + 4x| \geq 10$

**21.** $|3x| - 8 > 1$

**22.** $9 + |5x| < 24$

**23.** $|6x - 5| \leq -1$

**24.** $|6x - 5| \leq 5$

**25.** $\left|3x + \dfrac{2}{5}\right| \geq 4$

**26.** $|5x - 3| > 2$

**27.** $\left|\dfrac{x}{3} + 6\right| - 8 > -5$

**28.** $\left|\dfrac{4(x - 1)}{7}\right| + 10 < 2$

*(9.4)* Graph the following inequalities.

**29.** $3x - 4y \leq 0$

**30.** $3x - 4y \geq 0$

**31.** $x + 6y < 6$

**32.** $y \leq -4$

**33.** $y \geq -7$

**34.** $x \geq -y$

Graph the solutions of the following systems of linear inequalities.

**35.** $\begin{cases} y \geq 2x - 3 \\ y \leq -2x + 1 \end{cases}$

**36.** $\begin{cases} y \leq -3x - 3 \\ y \leq 2x + 7 \end{cases}$

**37.** $\begin{cases} x + 2y > 0 \\ x - y \leq 6 \end{cases}$

**38.** $\begin{cases} 4x - y \leq 0 \\ 3x - 2y \geq -5 \end{cases}$

**39.** $\begin{cases} 3x - 2y \leq 4 \\ 2x + y \geq 5 \end{cases}$

**40.** $\begin{cases} -2x + 3y > -7 \\ x \geq -2 \end{cases}$

**MIXED REVIEW**

Solve. If an inequality, write your solutions in interval notation.

**41.** $0 \leq \dfrac{2(3x + 4)}{5} \leq 3$

**42.** $x \leq 2$ or $x > -5$

**43.** $-2x \leq 6$ and $-2x + 3 < -7$

**44.** $|7x| - 26 = -5$

**45.** $\left|\dfrac{9 - 2x}{5}\right| = -3$

**46.** $|x - 3| = |7 + 2x|$

**47.** $|6x - 5| \geq -1$

**48.** $\left|\dfrac{4x - 3}{5}\right| < 1$

Graph the solutions.

**49.** $-x \leq y$

**50.** $x + y > -2$

**51.** $\begin{cases} -3x + 2y > -1 \\ y < -2 \end{cases}$

**52.** $\begin{cases} x - 2y \geq 7 \\ x + y \leq -5 \end{cases}$

# CHAPTER 9 TEST ~ TEST PREP VIDEO

Remember to use the Chapter Test Prep Video CD to see the fully worked-out solutions to any of the exercises you want to review.

Solve each equation or inequality.

**1.** $|6x - 5| - 3 = -2$

**2.** $|8 - 2t| = -6$

**3.** $|x - 5| = |x + 2|$

**4.** $-3 < 2(x - 3) \leq 4$

**5.** $|3x + 1| > 5$

**6.** $|x - 5| - 4 < -2$

**7.** $x \leq -2$ and $x \leq -5$

**8.** $x \leq -2$ or $x \leq -5$

**9.** $-x > 1$ and $3x + 3 \geq x - 3$

**10.** $6x + 1 > 5x + 4$ or $1 - x > -4$

**11.** $\left|\dfrac{5x - 7}{2}\right| = 4$

**12.** $\left|17x - \dfrac{1}{5}\right| > -2$

**13.** $-1 \leq \dfrac{2x - 5}{3} < 2$

Graph each linear equality.

**14.** $y > -4x$

**15.** $2x - 3y > -6$

Graph the solutions of the following systems of linear inequalities.

**16.** $\begin{cases} y + 2x \leq 4 \\ y \geq 2 \end{cases}$

**17.** $\begin{cases} 2y - x \geq 1 \\ x + y \geq -4 \end{cases}$

# CHAPTER 9 CUMULATIVE REVIEW

**1.** Find the value of each expression when $x = 2$ and $y = -5$.

  **a.** $\dfrac{x - y}{12 + x}$      **b.** $x^2 - 3y$

**2.** Find the value of each expression when $x = -4$ and $y = 7$.

  **a.** $\dfrac{x - y}{7 - x}$      **b.** $x^2 + 2y$

**3.** Simplify each expression.

  **a.** $\dfrac{(-12)(-3) + 3}{-7 - (-2)}$      **b.** $\dfrac{2(-3)^2 - 20}{-5 + 4}$

**4.** Simplify each expression.

  **a.** $\dfrac{4(-3) - (-6)}{-8 + 4}$      **b.** $\dfrac{3 + (-3)(-2)^3}{-1 - (-4)}$

**5.** Simplify each expression by combining like terms.

  **a.** $2x + 3x + 5 + 2$
  **b.** $-5a - 3 + a + 2$
  **c.** $4y - 3y^2$
  **d.** $2.3x + 5x - 6$
  **e.** $-\dfrac{1}{2}b + b$

**6.** Simplify each expression by combining like terms.

  **a.** $4x - 3 + 7 - 5x$
  **b.** $-6y + 3y - 8 + 8y$
  **c.** $2 + 8.1a + a - 6$
  **d.** $2x^2 - 2x$

**7.** Solve $2x + 3x - 5 + 7 = 10x + 3 - 6x - 4$.

**8.** Solve $6y - 11 + 4 + 2y = 8 + 15y - 8y$.

**9.** Complete the table for the equation $y = 3x$.

| $x$ | $y$ |
|---|---|
| $-1$ | |
| | $0$ |
| | $-9$ |

**10.** Complete the table for the equation $2x + y = 6$.

| $x$ | $y$ |
|---|---|
| $0$ | |
| | $-2$ |
| $3$ | |

**11.** Identify the $x$- and $y$-intercepts.

**a.**      **b.**

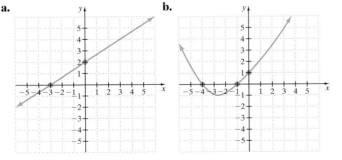

**c.**      **d.**

**e.**

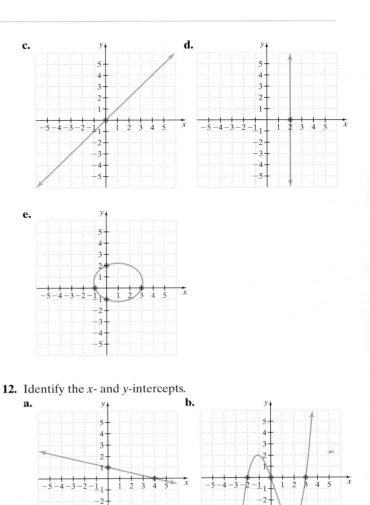

**12.** Identify the $x$- and $y$-intercepts.

  **a.**      **b.**

  **c.**      **d.**

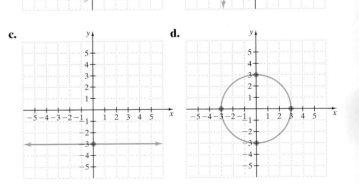

**13.** Determine whether the graphs of $y = -\dfrac{1}{5}x + 1$ and $2x + 10y = 3$ are parallel lines, perpendicular lines, or neither.

**14.** Determine whether the graphs of $y = 3x + 7$ and $x + 3y = -15$ are parallel lines, perpendicular lines, or neither.

**15.** Find an equation of the line with $y$-intercept $(0, -3)$ and slope of $\frac{1}{4}$.

**16.** Find an equation of a line with $y$-intercept $(0, 4)$ and slope of $-2$.

**17.** Find an equation of the line parallel to the line $y = 5$ and passing through $(-2, -3)$.

**18.** Find an equation of the line perpendicular to $y = 2x + 4$ and passing through $(1, 5)$.

**19.** Which of the following linear equations are functions?
  **a.** $y = x$
  **b.** $y = 2x + 1$
  **c.** $y = 5$
  **d.** $x = -1$

**20.** Which of the following linear equations are functions?
  **a.** $2x + 3 = y$
  **b.** $x + 4 = 0$
  **c.** $\frac{1}{2}y = 2x$
  **d.** $y = 0$

**21.** Determine whether $(12, 6)$ is a solution of the given system.
$$\begin{cases} 2x - 3y = 6 \\ x = 2y \end{cases}$$

**22.** Which of the following ordered pairs is a solution of the given system?
$$\begin{cases} 2x + y = 4 \\ x + y = 2 \end{cases}$$
  **a.** $(1, 1)$
  **b.** $(2, 0)$

**23.** Add $11x^3 - 12x^2 + x - 3$ and $x^3 - 10x + 5$.

**24.** Combine like terms to simplify.
$4a^2 + 3a - 2a^2 + 7a - 5$.

**25.** Factor $x^2 + 7yx + 6y^2$.

**26.** Factor $3x^2 + 15x + 18$.

**27.** Divide $\frac{3x^3 y^7}{40} \div \frac{4x^3}{y^2}$.

**28.** Divide $\frac{12x^2 y^3}{5} \div \frac{3y^2}{x}$.

**29.** Subtract $\frac{2y}{2y - 7} - \frac{7}{2y - 7}$.

**30.** Subtract $\frac{-4x^2}{x + 1} - \frac{4x}{x + 1}$.

**31.** Add $\frac{2x}{x^2 + 2x + 1} + \frac{x}{x^2 - 1}$.

**32.** Add $\frac{3x}{x^2 + 5x + 6} + \frac{1}{x^2 + 2x - 3}$.

**33.** Solve $\frac{x}{2} + \frac{8}{3} = \frac{1}{6}$.

**34.** Solve $\frac{1}{21} + \frac{x}{7} = \frac{5}{3}$.

**35.** Solve the following system of equations by graphing.
$$\begin{cases} 2x + y = 7 \\ 2y = -4x \end{cases}$$

**36.** Solve the following system by graphing.
$$\begin{cases} y = x + 2 \\ 2x + y = 5 \end{cases}$$

**37.** Solve the system.
$$\begin{cases} 7x - 3y = -14 \\ -3x + y = 6 \end{cases}$$

**38.** Solve the system.
$$\begin{cases} 5x + y = 3 \\ y = -5x \end{cases}$$

**39.** Solve the system.
$$\begin{cases} 3x - 2y = 2 \\ -9x + 6y = -6 \end{cases}$$

**40.** Solve the system.
$$\begin{cases} -2x + y = 7 \\ 6x - 3y = -21 \end{cases}$$

**41.** Graph the solution of the system
$$\begin{cases} -3x + 4y < 12 \\ x \geq 2 \end{cases}$$

**42.** Graph the solution of the system.
$$\begin{cases} 2x - y \leq 6 \\ y \geq 2 \end{cases}$$

**43.** Simplify the following.
  **a.** $\left(\frac{st}{2}\right)^4$
  **b.** $(9y^5 z^7)^2$
  **c.** $\left(\frac{-5x^2}{y^3}\right)^2$

**44.** Simplify.
  **a.** $\left(\frac{-6x}{y^3}\right)^3$
  **b.** $\frac{a^2 b^7}{(2b^2)^5}$
  **c.** $\frac{(3y)^2}{y^2}$
  **d.** $\frac{(x^2 y^4)^2}{xy^3}$

**45.** Solve $(5x - 1)(2x^2 + 15x + 18) = 0$.

**46.** Solve $(x + 1)(2x^2 - 3x - 5) = 0$.

**47.** Solve $\frac{45}{x} = \frac{5}{7}$.

**48.** Solve $\frac{2x + 7}{3} = \frac{x - 6}{2}$.

# 10 Rational Exponents, Radicals, and Complex Numbers

What is a zorb? Simply put, a zorb is a large inflated ball within a ball, and zorbing is a recreational activity which may involve rolling down a hill while strapped in a zorb. Zorbing started in New Zealand (as well as bungee jumping) and was invented by Andrew Akers and Dwane van der Sluis. The first site was set up in New Zealand's North Island. This downhill course has a length of about 490 feet, and you can reach speeds of up to 20 mph.

An example of a course is shown in the diagram below, and you can see the mathematics involved. In Section 10.3, Exercise 115, page 601, you will calculate the outer radius of a zorb, which would certainly be closely associated with the cost of production.

In this chapter, radical notation is reviewed, and then rational exponents are introduced. As the name implies, rational exponents are exponents that are rational numbers. We present an interpretation of rational exponents that is consistent with the meaning and rules already established for integer exponents, and we present two forms of notation for roots: radical and exponent. We conclude this chapter with complex numbers, a natural extension of the real number system.

7 meters

9 feet

# 10.1 RADICALS AND RADICAL FUNCTIONS

**OBJECTIVES**

**1** Find square roots.

**2** Approximate roots.

**3** Find cube roots.

**4** Find $n$th roots.

**5** Find $\sqrt[n]{a^n}$ where $a$ is a real number.

**6** Graph square and cube root functions.

**OBJECTIVE 1 ▶ Finding square roots.** Recall from Section 8.2 that to find a **square root** of a number $a$, we find a number that was squared to get $a$.

Thus, because

$$5^2 = 25 \quad \text{and} \quad (-5)^2 = 25, \text{then}$$

both $5$ and $-5$ are square roots of 25.

Recall that we denote the **nonnegative, or principal, square root** with the **radical sign.**

$$\sqrt{25} = 5$$

We denote the **negative square root** with the **negative radical sign.**

$$-\sqrt{25} = -5$$

An expression containing a radical sign is called a **radical expression.** An expression within, or "under," a radical sign is called a **radicand.**

$$\text{radical expression:} \quad \overset{\nearrow \text{radical sign}}{\underset{\nwarrow \text{radicand}}{\sqrt{a}}}$$

---

**Principal and Negative Square Roots**

If $a$ is a nonnegative number, then

$\sqrt{a}$ is the **principal, or nonnegative square root** of $a$

$-\sqrt{a}$ is the **negative square root** of $a$

---

**EXAMPLE 1** Simplify. Assume that all variables represent positive numbers.

**a.** $\sqrt{36}$     **b.** $\sqrt{0}$     **c.** $\sqrt{\dfrac{4}{49}}$     **d.** $\sqrt{0.25}$

**e.** $\sqrt{x^6}$     **f.** $\sqrt{9x^{12}}$     **g.** $-\sqrt{81}$     **h.** $\sqrt{-81}$

*Solution*

**a.** $\sqrt{36} = 6$ because $6^2 = 36$ and 6 is not negative.

**b.** $\sqrt{0} = 0$ because $0^2 = 0$ and 0 is not negative.

**c.** $\sqrt{\dfrac{4}{49}} = \dfrac{2}{7}$ because $\left(\dfrac{2}{7}\right)^2 = \dfrac{4}{49}$ and $\dfrac{2}{7}$ is not negative.

**d.** $\sqrt{0.25} = 0.5$ because $(0.5)^2 = 0.25$.

**e.** $\sqrt{x^6} = x^3$ because $(x^3)^2 = x^6$.

**f.** $\sqrt{9x^{12}} = 3x^6$ because $(3x^6)^2 = 9x^{12}$.

**g.** $-\sqrt{81} = -9$. The negative in front of the radical indicates the negative square root of 81.

**h.** $\sqrt{-81}$ is not a real number.                                                                        □

**PRACTICE**

**1** Simplify. Assume that all variables represent positive numbers.

**a.** $\sqrt{49}$     **b.** $\sqrt{\dfrac{0}{1}}$     **c.** $\sqrt{\dfrac{16}{81}}$     **d.** $\sqrt{0.64}$

**e.** $\sqrt{z^8}$     **f.** $\sqrt{16b^4}$     **g.** $-\sqrt{36}$     **h.** $\sqrt{-36}$

Recall from Section 8.2 our discussion of the square root of a negative number. For example, can we simplify $\sqrt{-4}$? That is, can we find a real number whose square is $-4$? No, there is no real number whose square is $-4$, and we say that $\sqrt{-4}$ is not a real number. In general:

*The square root of a negative number is not a real number.*

> ▶ **Helpful Hint**
> - Remember: $\sqrt{0} = 0$
> - Don't forget, the square root of a negative number, such as $\sqrt{-9}$, is not a real number. In Section 10.7, we will see what kind of a number $\sqrt{-9}$ is.

**OBJECTIVE 2 ▶ Approximating roots.** Recall that numbers such as 1, 4, 9, and 25 are called **perfect squares,** since $1 = 1^2, 4 = 2^2, 9 = 3^2$, and $25 = 5^2$. Square roots of perfect square radicands simplify to rational numbers. What happens when we try to simplify a root such as $\sqrt{3}$? Since there is no rational number whose square is 3, then $\sqrt{3}$ is not a rational number. It is called an **irrational number,** and we can find a decimal **approximation** of it. To find decimal approximations, use a calculator. For example, an approximation for $\sqrt{3}$ is

$$\sqrt{3} \approx 1.732$$
$$\uparrow$$
approximation symbol

To see if the approximation is reasonable, notice that since

$$1 < 3 < 4, \text{ then}$$
$$\sqrt{1} < \sqrt{3} < \sqrt{4}, \text{ or}$$
$$1 < \sqrt{3} < 2.$$

We found $\sqrt{3} \approx 1.732$, a number between 1 and 2, so our result is reasonable.

**EXAMPLE 2**   Use a calculator to approximate $\sqrt{20}$. Round the approximation to 3 decimal places and check to see that your approximation is reasonable.

$$\sqrt{20} \approx 4.472$$

*Solution*   Is this reasonable? Since $16 < 20 < 25$, then $\sqrt{16} < \sqrt{20} < \sqrt{25}$, or $4 < \sqrt{20} < 5$. The approximation is between 4 and 5 and thus is reasonable.   ☐

**PRACTICE**
**2**   Use a calculator to approximate $\sqrt{45}$. Round the approximation to three decimal places and check to see that your approximation is reasonable.

**OBJECTIVE 3 ▶ Finding cube roots.** Finding roots can be extended to other roots such as cube roots. For example, since $2^3 = 8$, we call 2 the **cube root** of 8. In symbols, we write

$$\sqrt[3]{8} = 2$$

> **Cube Root**
> The **cube root** of a real number $a$ is written as $\sqrt[3]{a}$, and
> $$\sqrt[3]{a} = b \text{ only if } b^3 = a$$

From this definition, we have

$$\sqrt[3]{64} = 4 \text{ since } 4^3 = 64$$
$$\sqrt[3]{-27} = -3 \text{ since } (-3)^3 = -27$$
$$\sqrt[3]{x^3} = x \text{ since } x^3 = x^3$$

Notice that, unlike with square roots, *it is possible to have a negative radicand when finding a cube root.* This is so because the *cube* of a negative number is a negative number. Therefore, the *cube root* of a negative number is a negative number.

**EXAMPLE 3** Find the cube roots.

**a.** $\sqrt[3]{1}$  **b.** $\sqrt[3]{-64}$  **c.** $\sqrt[3]{\dfrac{8}{125}}$  **d.** $\sqrt[3]{x^6}$  **e.** $\sqrt[3]{-27x^9}$

*Solution*

**a.** $\sqrt[3]{1} = 1$ because $1^3 = 1$.
**b.** $\sqrt[3]{-64} = -4$ because $(-4)^3 = -64$.
**c.** $\sqrt[3]{\dfrac{8}{125}} = \dfrac{2}{5}$ because $\left(\dfrac{2}{5}\right)^3 = \dfrac{8}{125}$.
**d.** $\sqrt[3]{x^6} = x^2$ because $(x^2)^3 = x^6$.
**e.** $\sqrt[3]{-27x^9} = -3x^3$ because $(-3x^3)^3 = -27x^9$.

**PRACTICE**
**3** Find the cube roots.

**a.** $\sqrt[3]{-1}$  **b.** $\sqrt[3]{27}$  **c.** $\sqrt[3]{\dfrac{27}{64}}$  **d.** $\sqrt[3]{x^{12}}$  **e.** $\sqrt[3]{-8x^3}$

**OBJECTIVE 4 ▶ Finding nth roots.** Just as we can raise a real number to powers other than 2 or 3, we can find roots other than square roots and cube roots. In fact, we can find the **nth root** of a number, where $n$ is any natural number. In symbols, the $n$th root of $a$ is written as $\sqrt[n]{a}$, where $n$ is called the **index.** The index 2 is usually omitted for square roots.

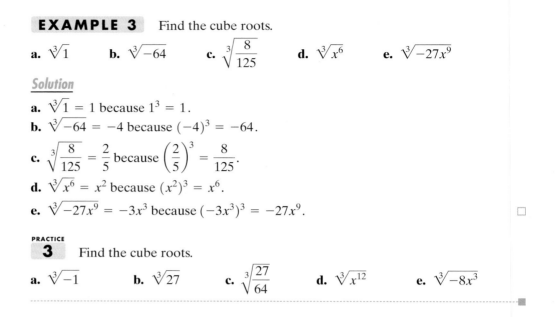

▶ **Helpful Hint**
If the index is even, such as $\sqrt{\phantom{x}}, \sqrt[4]{\phantom{x}}, \sqrt[6]{\phantom{x}}$, and so on, the radicand must be non-negative for the root to be a real number. For example,

$$\sqrt[4]{16} = 2, \text{ but } \sqrt[4]{-16} \text{ is not a real number.}$$
$$\sqrt[6]{64} = 2, \text{ but } \sqrt[6]{-64} \text{ is not a real number.}$$

If the index is odd, such as $\sqrt[3]{\phantom{x}}, \sqrt[5]{\phantom{x}}$, and so on, the radicand may be any real number. For example,

$$\sqrt[3]{64} = 4 \quad \text{and} \quad \sqrt[3]{-64} = -4$$
$$\sqrt[5]{32} = 2 \quad \text{and} \quad \sqrt[5]{-32} = -2$$

**Concept Check** ✓
Which one is not a real number?
**a.** $\sqrt[3]{-15}$  **b.** $\sqrt[4]{-15}$  **c.** $\sqrt[5]{-15}$  **d.** $\sqrt{(-15)^2}$

**EXAMPLE 4** Simplify the following expressions.

**a.** $\sqrt[4]{81}$  **b.** $\sqrt[5]{-243}$  **c.** $-\sqrt{25}$  **d.** $\sqrt[4]{-81}$  **e.** $\sqrt[3]{64x^3}$

*Solution*

**a.** $\sqrt[4]{81} = 3$ because $3^4 = 81$ and 3 is positive.
**b.** $\sqrt[5]{-243} = -3$ because $(-3)^5 = -243$.

**c.** $-\sqrt{25} = -5$ because $-5$ is the opposite of $\sqrt{25}$.

**d.** $\sqrt[4]{-81}$ is not a real number. There is no real number that, when raised to the fourth power, is $-81$.

**e.** $\sqrt[3]{64x^3} = 4x$ because $(4x)^3 = 64x^3$.

PRACTICE

**4**   Simplify the following expressions.

**a.** $\sqrt[4]{10,000}$   **b.** $\sqrt[5]{-1}$   **c.** $-\sqrt{81}$   **d.** $\sqrt[4]{-625}$   **e.** $\sqrt[3]{27x^9}$

**OBJECTIVE 5 ▶ Finding $\sqrt[n]{a^n}$ where $a$ is a real number.** Recall that the notation $\sqrt{a^2}$ indicates the positive square root of $a^2$ only. For example,

$$\sqrt{(-5)^2} = \sqrt{25} = 5$$

When variables are present in the radicand and it is unclear whether the variable represents a positive number or a negative number, absolute value bars are sometimes needed to ensure that the result is a positive number. For example,

$$\sqrt{x^2} = |x|$$

This ensures that the result is positive. This same situation may occur when the index is any *even* positive integer. When the index is any *odd* positive integer, absolute value bars are not necessary.

> **Finding $\sqrt[n]{a^n}$**
>
> If $n$ is an *even* positive integer, then $\sqrt[n]{a^n} = |a|$.
>
> If $n$ is an *odd* positive integer, then $\sqrt[n]{a^n} = a$.

**EXAMPLE 5**   Simplify.

**a.** $\sqrt{(-3)^2}$   **b.** $\sqrt{x^2}$   **c.** $\sqrt[4]{(x-2)^4}$   **d.** $\sqrt[3]{(-5)^3}$

**e.** $\sqrt[5]{(2x-7)^5}$   **f.** $\sqrt{25x^2}$   **g.** $\sqrt{x^2 + 2x + 1}$

*Solution*

**a.** $\sqrt{(-3)^2} = |-3| = 3$      When the index is even, the absolute value bars ensure us that our result is not negative.

**b.** $\sqrt{x^2} = |x|$

**c.** $\sqrt[4]{(x-2)^4} = |x-2|$

**d.** $\sqrt[3]{(-5)^3} = -5$

**e.** $\sqrt[5]{(2x-7)^5} = 2x - 7$   Absolute value bars are not needed when the index is odd.

**f.** $\sqrt{25x^2} = 5|x|$

**g.** $\sqrt{x^2 + 2x + 1} = \sqrt{(x+1)^2} = |x+1|$

PRACTICE

**5**   Simplify.

**a.** $\sqrt{(-4)^2}$   **b.** $\sqrt{x^{14}}$   **c.** $\sqrt[4]{(x+7)^4}$   **d.** $\sqrt[3]{(-7)^3}$

**e.** $\sqrt[5]{(3x-5)^5}$   **f.** $\sqrt{49x^2}$   **g.** $\sqrt{x^2 + 4x + 4}$

**OBJECTIVE 6 ▶ Graphing square and cube root functions.** Recall that an equation in $x$ and $y$ describes a function if each $x$-value is paired with exactly one $y$-value. With this in mind, does the equation

$$y = \sqrt{x}$$

describe a function? First, notice that replacement values for $x$ must be nonnegative real numbers, since $\sqrt{x}$ is not a real number if $x < 0$. The notation $\sqrt{x}$ denotes the principal square root of $x$, so for every nonnegative number $x$, there is exactly one number, $\sqrt{x}$. Therefore, $y = \sqrt{x}$ describes a function, and we may write it as

$$f(x) = \sqrt{x}$$

In general, radical functions are functions of the form

$$f(x) = \sqrt[n]{x}.$$

Recall that the domain of a function in $x$ is the set of all possible replacement values of $x$. This means that if $n$ is even, the domain is the set of all nonnegative numbers, or $\{x \mid x \geq 0\}$. If $n$ is odd, the domain is the set of all real numbers. Keep this in mind as we find function values.

**EXAMPLE 6**  If $f(x) = \sqrt{x - 4}$ and $g(x) = \sqrt[3]{x + 2}$, find each function value.

**a.** $f(8)$      **b.** $f(6)$      **c.** $g(-1)$      **d.** $g(1)$

*Solution*

**a.** $f(8) = \sqrt{8 - 4} = \sqrt{4} = 2$      **b.** $f(6) = \sqrt{6 - 4} = \sqrt{2}$

**c.** $g(-1) = \sqrt[3]{-1 + 2} = \sqrt[3]{1} = 1$      **d.** $g(1) = \sqrt[3]{1 + 2} = \sqrt[3]{3}$ ☐

**PRACTICE**
**6**  If $f(x) = \sqrt{x + 5}$ and $g(x) = \sqrt[3]{x - 3}$, find each function value.

**a.** $f(11)$      **b.** $f(-1)$      **c.** $g(11)$      **d.** $g(-5)$

▶ **Helpful Hint**

Notice that for the function $f(x) = \sqrt{x - 4}$, the domain includes all real numbers that make the radicand $\geq 0$. To see what numbers these are, solve $x - 4 \geq 0$ and find that $x \geq 4$. The domain is $\{x \mid x \geq 4\}$.

The domain of the cube root function $g(x) = \sqrt[3]{x + 2}$ is the set of real numbers.

**EXAMPLE 7**  Graph the square root function $f(x) = \sqrt{x}$.

*Solution*  To graph, we identify the domain, evaluate the function for several values of $x$, plot the resulting points, and connect the points with a smooth curve. Since $\sqrt{x}$ represents the nonnegative square root of $x$, the domain of this function is the set of all nonnegative numbers, $\{x \mid x \geq 0\}$, or $[0, \infty)$. We have approximated $\sqrt{3}$ below to help us locate the point corresponding to $\left(3, \sqrt{3}\right)$.

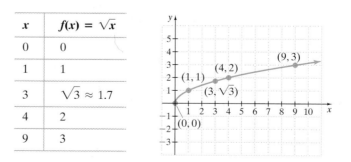

| $x$ | $f(x) = \sqrt{x}$ |
|---|---|
| 0 | 0 |
| 1 | 1 |
| 3 | $\sqrt{3} \approx 1.7$ |
| 4 | 2 |
| 9 | 3 |

Notice that the graph of this function passes the vertical line test, as expected. ☐

**PRACTICE**
**7**  Graph the square root function $h(x) = \sqrt{x + 2}$.

The equation $f(x) = \sqrt[3]{x}$ also describes a function. Here $x$ may be any real number, so the domain of this function is the set of all real numbers, or $(-\infty, \infty)$. A few function values are given next.

$$f(0) = \sqrt[3]{0} = 0$$
$$f(1) = \sqrt[3]{1} = 1$$
$$f(-1) = \sqrt[3]{-1} = -1$$
$$\left.\begin{array}{l} f(6) = \sqrt[3]{6} \\ f(-6) = \sqrt[3]{-6} \end{array}\right\}$$ Here, there is no rational number whose cube is 6. Thus, the radicals do not simplify to rational numbers.
$$f(8) = \sqrt[3]{8} = 2$$
$$f(-8) = \sqrt[3]{-8} = -2$$

**EXAMPLE 8** Graph the function $f(x) = \sqrt[3]{x}$.

*Solution* To graph, we identify the domain, plot points, and connect the points with a smooth curve. The domain of this function is the set of all real numbers. The table comes from the function values obtained earlier. We have approximated $\sqrt[3]{6}$ and $\sqrt[3]{-6}$ for graphing purposes.

| $x$ | $f(x) = \sqrt[3]{x}$ |
|---|---|
| 0 | 0 |
| 1 | 1 |
| $-1$ | $-1$ |
| 6 | $\sqrt[3]{6} \approx 1.8$ |
| $-6$ | $\sqrt[3]{-6} \approx -1.8$ |
| 8 | 2 |
| $-8$ | $-2$ |

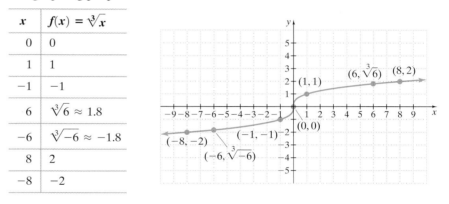

The graph of this function passes the vertical line test, as expected.

**PRACTICE**
**8** Graph the function $f(x) = \sqrt[3]{x} - 4$.

# VOCABULARY & READINESS CHECK

*Use the choices below to fill in each blank. Not all choices will be used.*

| is | cubes | $-\sqrt{a}$ | radical sign | index |
|---|---|---|---|---|
| is not | squares | $\sqrt{-a}$ | radicand | |

1. In the expression $\sqrt[n]{a}$, the $n$ is called the _index_, the $\sqrt{\phantom{x}}$ is called the _R-sign_, and $a$ is called the _rad_.
2. If $\sqrt{a}$ is the positive square root of $a$, $a \neq 0$, then _____ is the negative square root of $a$.
3. The square root of a negative number _____ a real number.
4. Numbers such as 1, 4, 9, and 25 are called perfect _root_ where numbers such as 1, 8, 27, and 125 are called perfect _Sq._

*Fill in the blank.*

5. The domain of the function $f(x) = \sqrt{x}$ is _____. $[0, \infty)$
6. The domain of the function $f(x) = \sqrt[3]{x}$ is _all real #_
7. If $f(16) = 4$, the corresponding ordered pair is _(16, 4)_
8. If $g(-8) = -2$, the corresponding ordered pair is _____.

*Choose the correct letter or letters. No pencil is needed, just think your way through these.*

**9.** Which radical is not a real number?

    **a.** $\sqrt{3}$     **b.** $-\sqrt{11}$     **c.** $\sqrt[3]{-10}$     **d.** $\sqrt{-10}$

**10.** Which radical(s) simplify to 3?

    **a.** $\sqrt{9}$     **b.** $\sqrt{-9}$     **c.** $\sqrt[3]{27}$     **d.** $\sqrt[3]{-27}$

**11.** Which radical(s) simplify to $-3$?

    **a.** $\sqrt{9}$     **b.** $\sqrt{-9}$     **c.** $\sqrt[3]{27}$     **d.** $\sqrt[3]{-27}$

**12.** Which radical does not simplify to a whole number?

    **a.** $\sqrt{64}$     **b.** $\sqrt[3]{64}$     **c.** $\sqrt{8}$     **d.** $\sqrt[3]{8}$

## 10.1 | EXERCISE SET

MyMathLab  PRACTICE  WATCH  DOWNLOAD  READ  REVIEW

*Simplify. Assume that variables represent positive real numbers. See Example 1.*

**1.** $\sqrt{100}$          **2.** $\sqrt{400}$

**3.** $\sqrt{\dfrac{1}{4}}$          **4.** $\sqrt{\dfrac{9}{25}}$

**5.** $\sqrt{0.0001}$          **6.** $\sqrt{0.04}$

**7.** $-\sqrt{36}$          **8.** $-\sqrt{9}$

**9.** $\sqrt{x^{10}}$          **10.** $\sqrt{x^{16}}$

**11.** $\sqrt{16y^6}$          **12.** $\sqrt{64y^{20}}$

*Use a calculator to approximate each square root to 3 decimal places. Check to see that each approximation is reasonable. See Example 2.*

**13.** $\sqrt{7}$          **14.** $\sqrt{11}$

**15.** $\sqrt{38}$          **16.** $\sqrt{56}$

**17.** $\sqrt{200}$          **18.** $\sqrt{300}$

*Find each cube root. See Example 3.*

**19.** $\sqrt[3]{64}$          **20.** $\sqrt[3]{27}$

**21.** $\sqrt[3]{\dfrac{1}{8}}$          **22.** $\sqrt[3]{\dfrac{27}{64}}$

**23.** $\sqrt[3]{-1}$          **24.** $\sqrt[3]{-125}$

**25.** $\sqrt[3]{x^{12}}$          **26.** $\sqrt[3]{x^{15}}$

**27.** $\sqrt[3]{-27x^9}$          **28.** $\sqrt[3]{-64x^6}$

*Find each root. Assume that all variables represent nonnegative real numbers. See Example 4.*

**29.** $-\sqrt[4]{16}$          **30.** $\sqrt[5]{-243}$

**31.** $\sqrt[4]{-16}$          **32.** $\sqrt{-16}$

**33.** $\sqrt[5]{-32}$          **34.** $\sqrt[5]{-1}$

**35.** $\sqrt[5]{x^{20}}$          **36.** $\sqrt[4]{x^{20}}$

**37.** $\sqrt[6]{64x^{12}}$          **38.** $\sqrt[5]{-32x^{15}}$

**39.** $\sqrt{81x^4}$          **40.** $\sqrt[4]{81x^4}$

**41.** $\sqrt[4]{256x^8}$          **42.** $\sqrt{256x^8}$

*Simplify. Assume that the variables represent any real number. See Example 5.*

**43.** $\sqrt{(-8)^2}$          **44.** $\sqrt{(-7)^2}$

**45.** $\sqrt[3]{(-8)^3}$          **46.** $\sqrt[5]{(-7)^5}$

**47.** $\sqrt{4x^2}$          **48.** $\sqrt[4]{16x^4}$

**49.** $\sqrt[3]{x^3}$          **50.** $\sqrt[5]{x^5}$

**51.** $\sqrt{(x-5)^2}$          **52.** $\sqrt{(y-6)^2}$

**53.** $\sqrt{x^2 + 4x + 4}$

    (*Hint:* Factor the polynomial first.)

**54.** $\sqrt{x^2 - 8x + 16}$

    (*Hint:* Factor the polynomial first.)

**MIXED PRACTICE**

*Simplify each radical. Assume that all variables represent positive real numbers.*

**55.** $-\sqrt{121}$          **56.** $-\sqrt[3]{125}$

**57.** $\sqrt[3]{8x^3}$          **58.** $\sqrt{16x^8}$

**59.** $\sqrt{y^{12}}$          **60.** $\sqrt[3]{y^{12}}$

**61.** $\sqrt{25a^2b^{20}}$          **62.** $\sqrt{9x^4y^6}$

**63.** $\sqrt[3]{-27x^{12}y^9}$          **64.** $\sqrt[3]{-8a^{21}b^6}$

**65.** $\sqrt[4]{a^{16}b^4}$          **66.** $\sqrt[4]{x^8y^{12}}$

**67.** $\sqrt[5]{-32x^{10}y^5}$          **68.** $\sqrt[5]{-243z^{15}}$

**69.** $\sqrt{\dfrac{25}{49}}$          **70.** $\sqrt{\dfrac{4}{81}}$

**71.** $\sqrt{\dfrac{x^2}{4y^2}}$          **72.** $\sqrt{\dfrac{y^{10}}{9x^6}}$

**73.** $-\sqrt[3]{\dfrac{z^{21}}{27x^3}}$          **74.** $-\sqrt[3]{\dfrac{64a^3}{b^9}}$

**75.** $\sqrt[4]{\dfrac{x^4}{16}}$          **76.** $\sqrt[4]{\dfrac{y^4}{81x^4}}$

*If $f(x) = \sqrt{2x + 3}$ and $g(x) = \sqrt[3]{x - 8}$, find the following function values. See Example 6.*

**77.** $f(0)$          **78.** $g(0)$

**79.** $g(7)$          **80.** $f(-1)$

**81.** $g(-19)$                **82.** $f(3)$

**83.** $f(2)$                  **84.** $g(1)$

*Identify the domain and then graph each function. See Example 7.*

**85.** $f(x) = \sqrt{x} + 2$

**86.** $f(x) = \sqrt{x} - 2$

**87.** $f(x) = \sqrt{x - 3}$; use the following table.

| $x$ | $f(x)$ |
|-----|--------|
| 3   |        |
| 4   |        |
| 7   |        |
| 12  |        |

**88.** $f(x) = \sqrt{x + 1}$; use the following table.

| $x$ | $f(x)$ |
|-----|--------|
| $-1$ |        |
| 0   |        |
| 3   |        |
| 8   |        |

*Identify the domain and then graph each function. See Example 8.*

**89.** $f(x) = \sqrt[3]{x} + 1$

**90.** $f(x) = \sqrt[3]{x} - 2$

**91.** $g(x) = \sqrt[3]{x - 1}$; use the following table.

| $x$ | $g(x)$ |
|-----|--------|
| 1   |        |
| 2   |        |
| 0   |        |
| 9   |        |
| $-7$ |        |

**92.** $g(x) = \sqrt[3]{x + 1}$; use the following table.

| $x$ | $g(x)$ |
|-----|--------|
| $-1$ |        |
| 0   |        |
| $-2$ |        |
| 7   |        |
| $-9$ |        |

**REVIEW AND PREVIEW**

*Simplify each exponential expression. See Sections 5.1 and 5.5.*

**93.** $(-2x^3y^2)^5$

**94.** $(4y^6z^7)^3$

**95.** $(-3x^2y^3z^5)(20x^5y^7)$

**96.** $(-14a^5bc^2)(2abc^4)$

**97.** $\dfrac{7x^{-1}y}{14(x^5y^2)^{-2}}$

**98.** $\dfrac{(2a^{-1}b^2)^3}{(8a^2b)^{-2}}$

**CONCEPT EXTENSIONS**

*Which of the following are not real numbers? See the Concept Check in this section.*

**99.** $\sqrt{-17}$                **100.** $\sqrt[3]{-17}$

**101.** $\sqrt[10]{-17}$           **102.** $\sqrt[15]{-17}$

**103.** Explain why $\sqrt{-64}$ is not a real number.

**104.** Explain why $\sqrt[3]{-64}$ is a real number.

*For Exercises 105 through 108, do not use a calculator.*

**105.** $\sqrt{160}$ is closest to

  **a.** 10    **b.** 13    **c.** 20    **d.** 40

**106.** $\sqrt{1000}$ is closest to

  **a.** 10    **b.** 30    **c.** 100    **d.** 500

**107.** The perimeter of the triangle is closest to

  **a.** 12    **b.** 18

  **c.** 66    **d.** 132

**108.** The length of the bent wire is closest to

  **a.** 5     **b.** $\sqrt{28}$

  **c.** 7     **d.** 14

*The Mosteller formula for calculating adult body surface area is $B = \sqrt{\dfrac{hw}{3131}}$, where B is an individual's body surface area in square meters, h is the individual's height in inches, and w is the individual's weight in pounds. Use this information to answer Exercises 109 and 110. Round answers to 2 decimal places.*

**109.** Find the body surface area of an individual who is 66 inches tall and who weighs 135 pounds.

**110.** Find the body surface area of an individual who is 74 inches tall and who weighs 225 pounds.

**111.** Suppose that a friend tells you that $\sqrt{13} \approx 5.7$. Without a calculator, how can you convince your friend that he or she must have made an error?

**112.** Escape velocity is the minimum speed that an object must reach to escape a planet's pull of gravity. Escape velocity $v$ is given by the equation $v = \sqrt{\dfrac{2Gm}{r}}$, where $m$ is the mass of the planet, $r$ is its radius, and $G$ is the universal gravitational constant, which has a value of $G = 6.67 \times 10^{-11}$ m³/ kg·sec². The mass of Earth is $5.97 \times 10^{24}$ kg and its radius is $6.37 \times 10^6$ m. Use this information to find the escape velocity for Earth. Round to the nearest whole number. (*Source: National Space Science Data Center*)

*Use a graphing calculator to verify the domain of each function and its graph.*

**113.** Exercise 85           **114.** Exercise 86

**115.** Exercise 89           **116.** Exercise 90

---

📖 **STUDY SKILLS BUILDER**

---

**How Are Your Homework Assignments Going?**

Remember that it is important to keep up with homework. Why? Many concepts in mathematics build on each other. Often, your understanding of a day's lecture depends on an understanding of the previous day's material.

To complete a homework assignment, remember these four things:

- Attempt all of it.
- Check it.
- Correct it.
- If needed, ask questions about it.

*Take a moment and review your completed homework assignments. Answer the exercises below based on this review.*

1. Approximate the fraction of your homework you have attempted.

2. Approximate the fraction of your homework you have checked (if possible).

3. If you are able to check your homework, have you corrected it when errors have been found?

4. What do you do, if you do not understand a concept while working on homework?

---

## 10.2 RATIONAL EXPONENTS

**OBJECTIVES**

1 Understand the meaning of $a^{1/n}$.

2 Understand the meaning of $a^{m/n}$.

3 Understand the meaning of $a^{-m/n}$.

4 Use rules for exponents to simplify expressions that contain rational exponents.

5 Use rational exponents to simplify radical expressions.

**OBJECTIVE 1 ▶ Understanding the meaning of $a^{1/n}$.** So far in this text, we have not defined expressions with rational exponents such as $3^{1/2}$, $x^{2/3}$, and $-9^{-1/4}$. We will define these expressions so that the rules for exponents will apply to these rational exponents as well.

Suppose that $x = 5^{1/3}$. Then

$$x^3 = (5^{1/3})^3 = 5^{1/3 \cdot 3} = 5^1 \text{ or } 5$$

$$\underset{\substack{\text{using rules} \\ \text{for exponents}}}{\uparrow}$$

Since $x^3 = 5$, then $x$ is the number whose cube is 5, or $x = \sqrt[3]{5}$. Notice that we also know that $x = 5^{1/3}$. This means

$$5^{1/3} = \sqrt[3]{5}$$

---

**Definition of $a^{1/n}$**

If $n$ is a positive integer greater than 1 and $\sqrt[n]{a}$ is a real number, then

$$a^{1/n} = \sqrt[n]{a}$$

---

Notice that the denominator of the rational exponent corresponds to the index of the radical.

**EXAMPLE 1** Use radical notation to write the following. Simplify if possible.

**a.** $4^{1/2}$    **b.** $64^{1/3}$    **c.** $x^{1/4}$    **d.** $0^{1/6}$    **e.** $-9^{1/2}$    **f.** $(81x^8)^{1/4}$    **g.** $(5y)^{1/3}$

*Solution*

**a.** $4^{1/2} = \sqrt{4} = 2$          **b.** $64^{1/3} = \sqrt[3]{64} = 4$

**c.** $x^{1/4} = \sqrt[4]{x}$          **d.** $0^{1/6} = \sqrt[6]{0} = 0$

**e.** $-9^{1/2} = -\sqrt{9} = -3$          **f.** $(81x^8)^{1/4} = \sqrt[4]{81x^8} = 3x^2$

**g.** $(5y)^{1/3} = \sqrt[3]{5y}$

**1**    Use radical notation to write the following. Simplify if possible.

**a.** $36^{1/2}$      **b.** $1000^{1/3}$      **c.** $x^{1/5}$      **d.** $1^{1/4}$      **e.** $-64^{1/2}$

**f.** $(125x^9)^{1/3}$      **g.** $(3x)^{1/4}$

---

**OBJECTIVE 2** ▶ **Understanding the meaning of $a^{m/n}$.** As we expand our use of exponents to include $\dfrac{m}{n}$, we define their meaning so that rules for exponents still hold true. For example, by properties of exponents,

$$8^{2/3} = (8^{1/3})^2 = \left(\sqrt[3]{8}\right)^2 \qquad \text{or}$$
$$8^{2/3} = (8^2)^{1/3} = \sqrt[3]{8^2}$$

> **Definition of $a^{m/n}$**
>
> If $m$ and $n$ are positive integers greater than 1 with $\dfrac{m}{n}$ in lowest terms, then
> $$a^{m/n} = \sqrt[n]{a^m} = \left(\sqrt[n]{a}\right)^m$$
> as long as $\sqrt[n]{a}$ is a real number.

Notice that the denominator $n$ of the rational exponent corresponds to the index of the radical. The numerator $m$ of the rational exponent indicates that the base is to be raised to the $m$th power. This means

$$8^{2/3} = \sqrt[3]{8^2} = \sqrt[3]{64} = 4 \qquad \text{or}$$
$$8^{2/3} = \left(\sqrt[3]{8}\right)^2 = 2^2 = 4$$

From simplifying $8^{2/3}$, can you see that it doesn't matter whether you raise to a power first and then take the $n$th root or you take the $n$th root first and then raise to a power?

> ▶ **Helpful Hint**
> Most of the time, $\left(\sqrt[n]{a}\right)^m$ will be easier to calculate than $\sqrt[n]{a^m}$.

**EXAMPLE 2**    Use radical notation to write the following. Then simplify if possible.

**a.** $4^{3/2}$      **b.** $-16^{3/4}$      **c.** $(-27)^{2/3}$

**d.** $\left(\dfrac{1}{9}\right)^{3/2}$      **e.** $(4x - 1)^{3/5}$

*Solution*

**a.** $4^{3/2} = \left(\sqrt{4}\right)^3 = 2^3 = 8$      **b.** $-16^{3/4} = -\left(\sqrt[4]{16}\right)^3 = -(2)^3 = -8$

**c.** $(-27)^{2/3} = \left(\sqrt[3]{-27}\right)^2 = (-3)^2 = 9$      **d.** $\left(\dfrac{1}{9}\right)^{3/2} = \left(\sqrt{\dfrac{1}{9}}\right)^3 = \left(\dfrac{1}{3}\right)^3 = \dfrac{1}{27}$

**e.** $(4x - 1)^{3/5} = \sqrt[5]{(4x - 1)^3}$      ☐

**2**    Use radical notation to write the following. Simplify if possible.

**a.** $16^{3/2}$      **b.** $-1^{3/5}$      **c.** $-(81)^{3/4}$

**d.** $\left(\dfrac{1}{25}\right)^{3/2}$      **e.** $(3x + 2)^{5/9}$

CHAPTER 10 Rational Exponents, Radicals, and Complex Numbers

> ▶ **Helpful Hint**
>
> The *denominator* of a rational exponent is the index of the corresponding radical. For example, $x^{1/5} = \sqrt[5]{x}$ and $z^{2/3} = \sqrt[3]{z^2}$, or $z^{2/3} = \left(\sqrt[3]{z}\right)^2$.

**OBJECTIVE 3 ▶ Understanding the meaning of $a^{-m/n}$.** The rational exponents we have given meaning to exclude negative rational numbers. To complete the set of definitions, we define $a^{-m/n}$.

> **Definition of $a^{-m/n}$**
>
> $$a^{-m/n} = \frac{1}{a^{m/n}}$$
>
> as long as $a^{m/n}$ is a nonzero real number.

**EXAMPLE 3**   Write each expression with a positive exponent, and then simplify.

**a.** $16^{-3/4}$    **b.** $(-27)^{-2/3}$

<u>Solution</u>

**a.** $16^{-3/4} = \dfrac{1}{16^{3/4}} = \dfrac{1}{\left(\sqrt[4]{16}\right)^3} = \dfrac{1}{2^3} = \dfrac{1}{8}$

**b.** $(-27)^{-2/3} = \dfrac{1}{(-27)^{2/3}} = \dfrac{1}{\left(\sqrt[3]{-27}\right)^2} = \dfrac{1}{(-3)^2} = \dfrac{1}{9}$

**PRACTICE**
**3**   Write each expression with a positive exponent; then simplify.

**a.** $9^{-3/2}$                **b.** $(-64)^{-2/3}$

> ▶ **Helpful Hint**
>
> If an expression contains a negative rational exponent, such as $9^{-3/2}$, you may want to first write the expression with a positive exponent and then interpret the rational exponent. Notice that the sign of the base is not affected by the sign of its exponent. For example,
>
> $$9^{-3/2} = \frac{1}{9^{3/2}} = \frac{1}{\left(\sqrt{9}\right)^3} = \frac{1}{27}$$
>
> Also, .
>
> $$(-27)^{-1/3} = \frac{1}{(-27)^{1/3}} = -\frac{1}{3}$$

**Concept Check** ✓

Which one is correct?

**a.** $-8^{2/3} = \dfrac{1}{4}$    **b.** $8^{-2/3} = -\dfrac{1}{4}$    **c.** $8^{-2/3} = -4$    **d.** $-8^{-2/3} = -\dfrac{1}{4}$

**OBJECTIVE 4 ▶ Using rules for exponents to simplify expressions.** It can be shown that the properties of integer exponents hold for rational exponents. By using these properties and definitions, we can now simplify expressions that contain rational exponents.

These rules are repeated here for review.

*Note:* For the remainder of this chapter, we will assume that variables represent positive real numbers. Since this is so, we need not insert absolute value bars when we simplify even roots.

**Answer to Concept Check:**  d

**Summary of Exponent Rules**

If $m$ and $n$ are rational numbers, and $a$, $b$, and $c$ are numbers for which the expressions below exist, then

Product rule for exponents: $\qquad\qquad a^m \cdot a^n = a^{m+n}$

Power rule for exponents: $\qquad\qquad\;\; (a^m)^n = a^{m \cdot n}$

Power rules for products and quotients: $\quad (ab)^n = a^n b^n \qquad\qquad$ and

$$\left(\frac{a}{c}\right)^n = \frac{a^n}{c^n}, c \neq 0$$

Quotient rule for exponents: $\qquad\qquad \dfrac{a^m}{a^n} = a^{m-n}, a \neq 0$

Zero exponent: $\qquad\qquad\qquad\qquad\quad a^0 = 1, a \neq 0$

Negative exponent: $\qquad\qquad\qquad\quad a^{-n} = \dfrac{1}{a^n}, a \neq 0$

**EXAMPLE 4**   Use properties of exponents to simplify. Write results with only positive exponents.

**a.** $b^{1/3} \cdot b^{5/3}$ $\qquad$ **b.** $x^{1/2} x^{1/3}$ $\qquad\qquad$ **c.** $\dfrac{7^{1/3}}{7^{4/3}}$

**d.** $y^{-4/7} \cdot y^{6/7}$ $\qquad$ **e.** $\dfrac{(2x^{2/5} y^{-1/3})^5}{x^2 y}$

*Solution*

**a.** $b^{1/3} \cdot b^{5/3} = b^{(1/3 + 5/3)} = b^{6/3} = b^2$

**b.** $x^{1/2} x^{1/3} = x^{(1/2 + 1/3)} = x^{3/6 + 2/6} = x^{5/6}$ $\quad$ Use the product rule.

*Be careful* **c.** $\dfrac{7^{1/3}}{7^{4/3}} = 7^{1/3 - 4/3} = 7^{-3/3} = 7^{-1} = \dfrac{1}{7}$ $\quad$ Use the quotient rule.

**d.** $y^{-4/7} \cdot y^{6/7} = y^{-4/7 + 6/7} = y^{2/7}$ $\qquad$ Use the product rule.

**e.** We begin by using the power rule $(ab)^m = a^m b^m$ to simplify the numerator.

$$\frac{(2x^{2/5} y^{-1/3})^5}{x^2 y} = \frac{2^5 (x^{2/5})^5 (y^{-1/3})^5}{x^2 y} = \frac{32 x^2 y^{-5/3}}{x^2 y} \quad \text{Use the power rule and simplify}$$

$$= 32 x^{2-2} y^{-5/3 - 3/3} \qquad\qquad \text{Apply the quotient rule.}$$

$$= 32 x^0 y^{-8/3}$$

*when subtracting exponent Fractions subtract exponents*

$$= \frac{32}{y^{8/3}}$$

**PRACTICE**
**4**   Use properties of exponents to simplify.

**a.** $y^{2/3} \cdot y^{8/3}$ $\qquad$ **b.** $x^{3/5} \cdot x^{1/4}$ $\qquad\qquad$ **c.** $\dfrac{9^{2/7}}{9^{9/7}}$

**d.** $b^{4/9} \cdot b^{-2/9}$ $\qquad$ **e.** $\dfrac{\left(3x^{1/4} y^{-2/3}\right)^4}{x^4 y}$

**EXAMPLE 5**   Multiply.

**a.** $z^{2/3}(z^{1/3} - z^5)$ $\qquad\qquad\qquad$ **b.** $(x^{1/3} - 5)(x^{1/3} + 2)$

*Solution*

**a.** $z^{2/3}(z^{1/3} - z^5) = z^{2/3}z^{1/3} - z^{2/3}z^5$    Apply the distributive property.

$= z^{(2/3+1/3)} - z^{(2/3+5)}$    Use the product rule.

$= z^{3/3} - z^{(2/3+15/3)}$

$= z - z^{17/3}$

**b.** $(x^{1/3} - 5)(x^{1/3} + 2) = x^{2/3} + 2x^{1/3} - 5x^{1/3} - 10$    Think of $(x^{1/3} - 5)$ and $(x^{1/3} + 2)$

$= x^{2/3} - 3x^{1/3} - 10$    as 2 binomials, and FOIL.

**PRACTICE**

**5**   Multiply.

**a.** $x^{3/5}(x^{1/3} - x^2)$          **b.** $(x^{1/2} + 6)(x^{1/2} - 2)$

---

**EXAMPLE 6**   Factor $x^{-1/2}$ from the expression $3x^{-1/2} - 7x^{5/2}$. Assume that all variables represent positive numbers.

*Solution*

$$3x^{-1/2} - 7x^{5/2} = (x^{-1/2})(3) - (x^{-1/2})(7x^{6/2})$$
$$= x^{-1/2}(3 - 7x^3)$$

To check, multiply $x^{-1/2}(3 - 7x^3)$ to see that the product is $3x^{-1/2} - 7x^{5/2}$.

**PRACTICE**

**6**   Factor $x^{-1/5}$ from the expression $2x^{-1/5} - 7x^{4/5}$.

---

**OBJECTIVE 5** ▶ **Using rational exponents to simplify radical expressions.** Some radical expressions are easier to simplify when we first write them with rational exponents. We can simplify some radical expressions by first writing the expression with rational exponents. Use properties of exponents to simplify, and then convert back to radical notation.

✔ **EXAMPLE 7**   Use rational exponents to simplify. Assume that variables represent positive numbers.

**a.** $\sqrt[8]{x^4}$    **b.** $\sqrt[6]{25}$    **c.** $\sqrt[4]{r^2 s^6}$

*Solution*

$\frac{4}{8} =$

**a.** $\sqrt[8]{x^4} = x^{4/8} = x^{1/2} = \sqrt{x}$

**b.** $\sqrt[6]{25} = 25^{1/6} = (5^2)^{1/6} = 5^{2/6} = 5^{1/3} = \sqrt[3]{5}$

**c.** $\sqrt[4]{r^2 s^6} = (r^2 s^6)^{1/4} = r^{2/4} s^{6/4} = r^{1/2} s^{3/2} = (rs^3)^{1/2} = \sqrt{rs^3}$

**PRACTICE**

**7**   Use rational exponents to simplify. Assume that the variables represent positive numbers.

**a.** $\sqrt[9]{x^3}$        **b.** $\sqrt[4]{36}$        **c.** $\sqrt[8]{a^4 b^2}$

---

**EXAMPLE 8**   Use rational exponents to write as a single radical.

**a.** $\sqrt{x} \cdot \sqrt[4]{x}$    **b.** $\dfrac{\sqrt{x}}{\sqrt[3]{x}}$    **c.** $\sqrt[3]{3} \cdot \sqrt{2}$

*Solution*

**a.** $\sqrt{x} \cdot \sqrt[4]{x} = x^{1/2} \cdot x^{1/4} = x^{1/2+1/4}$
$$= x^{3/4} = \sqrt[4]{x^3}$$

**b.** $\dfrac{\sqrt{x}}{\sqrt[3]{x}} = \dfrac{x^{1/2}}{x^{1/3}} = x^{1/2-1/3} = x^{3/6-2/6}$
$$= x^{1/6} = \sqrt[6]{x}$$

**c.** $\sqrt[3]{3} \cdot \sqrt{2} = 3^{1/3} \cdot 2^{1/2}$     Write with rational exponents.
$$= 3^{2/6} \cdot 2^{3/6} \quad \text{Write the exponents so that they have the same denominator.}$$
$$= (3^2 \cdot 2^3)^{1/6} \quad \text{Use } a^n b^n = (ab)^n$$
$$= \sqrt[6]{3^2 \cdot 2^3} \quad \text{Write with radical notation.}$$
$$= \sqrt[6]{72} \quad \text{Multiply } 3^2 \cdot 2^3.$$

**PRACTICE**
**8**    Use rational expressions to write each of the following as a single radical.

**a.** $\sqrt[3]{x} \cdot \sqrt[4]{x}$       **b.** $\dfrac{\sqrt[3]{y}}{\sqrt[5]{y}}$       **c.** $\sqrt[3]{5} \cdot \sqrt{3}$

## VOCABULARY & READINESS CHECK

*Answer each true or false.*

**1.** $9^{-1/2}$ is a positive number. _____

**2.** $9^{-1/2}$ is a whole number. _____

**3.** $\dfrac{1}{a^{-m/n}} = a^{m/n}$ (where $a^{m/n}$ is a nonzero real number). _____

*Fill in the blank with the correct choice.*

**4.** To simplify $x^{2/3} \cdot x^{1/5}$, _____ the exponents.
   **a.** add    **b.** subtract    **c.** multiply    **d.** divide

**5.** To simplify $(x^{2/3})^{1/5}$, _____ the exponents.
   **a.** add    **b.** subtract    **c.** multiply    **d.** divide

**6.** To simplify $\dfrac{x^{2/3}}{x^{1/5}}$, _____ the exponents.
   **a.** add    **b.** subtract    **c.** multiply    **d.** divide

*Choose the correct letter for each exercise. Letters will be used more than once. No pencil is needed. Just think about the meaning of each expression.*

$A = 2, B = -2, C = \text{not a real number}$

**7.** $4^{1/2}$ ____     **8.** $-4^{1/2}$ ____     **9.** $(-4)^{1/2}$ ____     **10.** $8^{1/3}$ ____     **11.** $-8^{1/3}$ ____     **12.** $(-8)^{1/3}$ ____

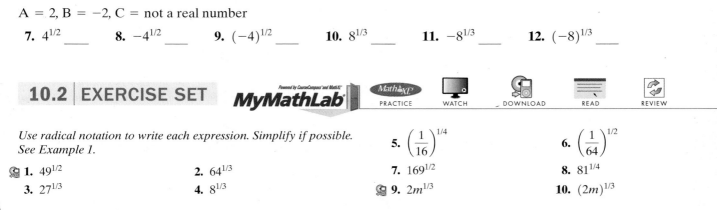

**10.2 EXERCISE SET**    **MyMathLab**    *Powered by CourseCompass™ and MathXL®*    MathXL PRACTICE    WATCH    DOWNLOAD    READ    REVIEW

*Use radical notation to write each expression. Simplify if possible. See Example 1.*

**1.** $49^{1/2}$       **2.** $64^{1/3}$

**3.** $27^{1/3}$       **4.** $8^{1/3}$

**5.** $\left(\dfrac{1}{16}\right)^{1/4}$       **6.** $\left(\dfrac{1}{64}\right)^{1/2}$

**7.** $169^{1/2}$       **8.** $81^{1/4}$

**9.** $2m^{1/3}$       **10.** $(2m)^{1/3}$

**11.** $(9x^4)^{1/2}$

**12.** $(16x^8)^{1/2}$

**13.** $(-27)^{1/3}$

**14.** $-64^{1/2}$

**15.** $-16^{1/4}$

**16.** $(-32)^{1/5}$

*Use radical notation to write each expression. Simplify if possible. See Example 2.*

**17.** $16^{3/4}$

**18.** $4^{5/2}$

**19.** $(-64)^{2/3}$

**20.** $(-8)^{4/3}$

**21.** $(-16)^{3/4}$

**22.** $(-9)^{3/2}$

**23.** $(2x)^{3/5}$

**24.** $2x^{3/5}$

**25.** $(7x + 2)^{2/3}$

**26.** $(x - 4)^{3/4}$

**27.** $\left(\dfrac{16}{9}\right)^{3/2}$

**28.** $\left(\dfrac{49}{25}\right)^{3/2}$

*Write with positive exponents. Simplify if possible. See Example 3.*

**29.** $8^{-4/3}$

**30.** $64^{-2/3}$

**31.** $(-64)^{-2/3}$

**32.** $(-8)^{-4/3}$

**33.** $(-4)^{-3/2}$

**34.** $(-16)^{-5/4}$

**35.** $x^{-1/4}$

**36.** $y^{-1/6}$

**37.** $\dfrac{1}{a^{-2/3}}$

**38.** $\dfrac{1}{n^{-8/9}}$

**39.** $\dfrac{5}{7x^{-3/4}}$

**40.** $\dfrac{2}{3y^{-5/7}}$

*Use the properties of exponents to simplify each expression. Write with positive exponents. See Example 4.*

**41.** $a^{2/3}a^{5/3}$

**42.** $b^{9/5}b^{8/5}$

**43.** $x^{-2/5} \cdot x^{7/5}$

**44.** $y^{4/3} \cdot y^{-1/3}$

**45.** $3^{1/4} \cdot 3^{3/8}$

**46.** $5^{1/2} \cdot 5^{1/6}$

**47.** $\dfrac{y^{1/3}}{y^{1/6}}$

**48.** $\dfrac{x^{3/4}}{x^{1/8}}$

**49.** $(4u^2)^{3/2}$

**50.** $(32^{1/5}x^{2/3})^3$

**51.** $\dfrac{b^{1/2}b^{3/4}}{-b^{1/4}}$

**52.** $\dfrac{a^{1/4}a^{-1/2}}{a^{2/3}}$

**53.** $\dfrac{(x^3)^{1/2}}{x^{7/2}}$

**54.** $\dfrac{y^{11/3}}{(y^5)^{1/3}}$

**55.** $\dfrac{(3x^{1/4})^3}{x^{1/12}}$

**56.** $\dfrac{(2x^{1/5})^4}{x^{3/10}}$

**57.** $\dfrac{(y^3z)^{1/6}}{y^{-1/2}z^{1/3}}$

**58.** $\dfrac{(m^2n)^{1/4}}{m^{-1/2}n^{5/8}}$

**59.** $\dfrac{(x^3y^2)^{1/4}}{(x^{-5}y^{-1})^{-1/2}}$

**60.** $\dfrac{(a^{-2}b^3)^{1/8}}{(a^{-3}b)^{-1/4}}$

*Multiply. See Example 5.*

**61.** $y^{1/2}(y^{1/2} - y^{2/3})$

**62.** $x^{1/2}(x^{1/2} + x^{3/2})$

**63.** $x^{2/3}(x - 2)$

**64.** $3x^{1/2}(x + y)$

**65.** $(2x^{1/3} + 3)(2x^{1/3} - 3)$

**66.** $(y^{1/2} + 5)(y^{1/2} + 5)$

*Factor the common factor from the given expression. See Example 6.*

**67.** $x^{8/3}$; $x^{8/3} + x^{10/3}$

**68.** $x^{3/2}$; $x^{5/2} - x^{3/2}$

**69.** $x^{1/5}$; $x^{2/5} - 3x^{1/5}$

**70.** $x^{2/7}$; $x^{3/7} - 2x^{2/7}$

**71.** $x^{-1/3}$; $5x^{-1/3} + x^{2/3}$

**72.** $x^{-3/4}$; $x^{-3/4} + 3x^{1/4}$

*Use rational exponents to simplify each radical. Assume that all variables represent positive numbers. See Example 7.*

**73.** $\sqrt[6]{x^3}$

**74.** $\sqrt[9]{a^3}$

**75.** $\sqrt[6]{4}$

**76.** $\sqrt[4]{36}$

**77.** $\sqrt[4]{16x^2}$

**78.** $\sqrt[8]{4y^2}$

**79.** $\sqrt[8]{x^4y^4}$

**80.** $\sqrt[9]{y^6z^3}$

**81.** $\sqrt[12]{a^8b^4}$

**82.** $\sqrt[10]{a^5b^5}$

**83.** $\sqrt[4]{(x + 3)^2}$

**84.** $\sqrt[8]{(y + 1)^4}$

*Use rational expressions to write as a single radical expression. See Example 8.*

**85.** $\sqrt[3]{y} \cdot \sqrt[5]{y^2}$

**86.** $\sqrt[3]{y^2} \cdot \sqrt[6]{y}$

**87.** $\dfrac{\sqrt[3]{b^2}}{\sqrt[4]{b}}$

**88.** $\dfrac{\sqrt[4]{a}}{\sqrt[5]{a}}$

**89.** $\sqrt[3]{x} \cdot \sqrt[4]{x} \cdot \sqrt[8]{x^3}$

**90.** $\sqrt[6]{y} \cdot \sqrt[3]{y} \cdot \sqrt[5]{y^2}$

**91.** $\dfrac{\sqrt[3]{a^2}}{\sqrt[6]{a}}$

**92.** $\dfrac{\sqrt[5]{b^2}}{\sqrt[10]{b^3}}$

**93.** $\sqrt{3} \cdot \sqrt[3]{4}$

**94.** $\sqrt[3]{5} \cdot \sqrt{2}$

**95.** $\sqrt[5]{7} \cdot \sqrt[3]{y}$

**96.** $\sqrt[4]{5} \cdot \sqrt[3]{x}$

**97.** $\sqrt{5r} \cdot \sqrt[3]{s}$

**98.** $\sqrt[3]{b} \cdot \sqrt[5]{4a}$

## REVIEW AND PREVIEW

*Write each integer as a product of two integers such that one of the factors is a perfect square. For example, write 18 as $9 \cdot 2$, because 9 is a perfect square. See Section 10.1.*

**99.** 75

**100.** 20

**101.** 48

**102.** 45

*Write each integer as a product of two integers such that one of the factors is a perfect cube. For example, write 24 as $8 \cdot 3$, because 8 is a perfect cube. See Section 10.1.*

**103.** 16

**104.** 56

**105.** 54

**106.** 80

## CONCEPT EXTENSIONS

*Basal metabolic rate (BMR) is the number of calories per day a person needs to maintain life. A person's basal metabolic rate $B(w)$ in calories per day can be estimated with the function $B(w) = 70w^{3/4}$, where $w$ is the person's weight in kilograms. Use this information to answer Exercises 107 and 108.*

**107.** Estimate the BMR for a person who weighs 60 kilograms. Round to the nearest calorie. (*Note:* 60 kilograms is approximately 132 pounds.)

**108.** Estimate the BMR for a person who weighs 90 kilograms. Round to the nearest calorie. (*Note:* 90 kilograms is approximately 198 pounds.)

*The number of cellular telephone subscriptions in the United States from 1996 through 2006 can be modeled by the function* $f(x) = 33.3x^{4/5}$, *where y is the number of cellular telephone subscriptions in millions, x years after 1996. (Source: Based on data from the Cellular Telecommunications & Internet Association, 1994–2000) Use this information to answer Exercises 109 and 110.*

**109.** Use this model to estimate the number of cellular telephone subscriptions in the United States in 2006. Round to the nearest tenth of a million.

**110.** Predict the number of cellular telephone subscriptions in the United States in 2010. Round to the nearest tenth of a million.

*Fill in each box with the correct expression.*

**111.** $\square \cdot a^{2/3} = a^{3/3}$, or $a$

**112.** $\square \cdot x^{1/8} = x^{4/8}$, or $x^{1/2}$

**113.** $\dfrac{\square}{x^{-2/5}} = x^{3/5}$

**114.** $\dfrac{\square}{y^{-3/4}} = y^{4/4}$, or $y$

*Use a calculator to write a four-decimal-place approximation of each number.*

**115.** $8^{1/4}$          **116.** $20^{1/5}$

**117.** $18^{3/5}$          **118.** $76^{5/7}$

**119.** In physics, the speed of a wave traveling over a stretched string with tension $t$ and density $u$ is given by the expression $\dfrac{\sqrt{t}}{\sqrt{u}}$. Write this expression with rational exponents.

**120.** In electronics, the angular frequency of oscillations in a certain type of circuit is given by the expression $(LC)^{-1/2}$. Use radical notation to write this expression.

---

## 10.3  SIMPLIFYING RADICAL EXPRESSIONS

### OBJECTIVES

1  Use the product rule for radicals.

2  Use the quotient rule for radicals.

3  Simplify radicals.

4  Use the distance and midpoint formula.

**OBJECTIVE 1 ▶ Using the product rule.**  It is possible to simplify some radicals that do not evaluate to rational numbers. To do so, we use a product rule and a quotient rule for radicals. To discover the product rule, notice the following pattern.

$$\sqrt{9} \cdot \sqrt{4} = 3 \cdot 2 = 6$$
$$\sqrt{9 \cdot 4} = \sqrt{36} = 6$$

Since both expressions simplify to 6, it is true that

$$\sqrt{9} \cdot \sqrt{4} = \sqrt{9 \cdot 4}$$

This pattern suggests the following product rule for r

**Product Rule for Radicals**

If $\sqrt[n]{a}$ and $\sqrt[n]{b}$ are real numbers, then

$$\sqrt[n]{a} \cdot \sqrt[n]{b} = \sqrt[n]{ab}$$

Notice that the product rule is the relationship $a^{1/n} \cdot b^{1/n} = (ab)^{1/n}$ stated in radical notation.

**EXAMPLE 1** Multiply.

**a.** $\sqrt{3} \cdot \sqrt{5}$ **b.** $\sqrt{21} \cdot \sqrt{x}$ **c.** $\sqrt[3]{4} \cdot \sqrt[3]{2}$

**d.** $\sqrt[4]{5y^2} \cdot \sqrt[4]{2x^3}$ **e.** $\sqrt{\dfrac{2}{a}} \cdot \sqrt{\dfrac{b}{3}}$

**Solution**

**a.** $\sqrt{3} \cdot \sqrt{5} = \sqrt{3 \cdot 5} = \sqrt{15}$

**b.** $\sqrt{21} \cdot \sqrt{x} = \sqrt{21x}$

**c.** $\sqrt[3]{4} \cdot \sqrt[3]{2} = \sqrt[3]{4 \cdot 2} = \sqrt[3]{8} = 2$

**d.** $\sqrt[4]{5y^2} \cdot \sqrt[4]{2x^3} = \sqrt[4]{5y^2 \cdot 2x^3} = \sqrt[4]{10y^2x^3}$

**e.** $\sqrt{\dfrac{2}{a}} \cdot \sqrt{\dfrac{b}{3}} = \sqrt{\dfrac{2}{a} \cdot \dfrac{b}{3}} = \sqrt{\dfrac{2b}{3a}}$ □

**PRACTICE**
**1** Multiply.

**a.** $\sqrt{5} \cdot \sqrt{7}$ **b.** $\sqrt{13} \cdot \sqrt{z}$ **c.** $\sqrt[4]{125} \cdot \sqrt[4]{5}$

**d.** $\sqrt[3]{5y} \cdot \sqrt[3]{3x^2}$ **e.** $\sqrt{\dfrac{5}{m}} \cdot \sqrt{\dfrac{t}{2}}$

**OBJECTIVE 2 ▶ Using the quotient rule.** To discover a quotient rule for radicals, notice the following pattern.

$$\sqrt{\frac{4}{9}} = \frac{2}{3}$$

$$\frac{\sqrt{4}}{\sqrt{9}} = \frac{2}{3}$$

Since both expressions simplify to $\dfrac{2}{3}$, it is true that

$$\sqrt{\frac{4}{9}} = \frac{\sqrt{4}}{\sqrt{9}}$$

This pattern suggests the following quotient rule for radicals.

---

**Quotient Rule for Radicals**

If $\sqrt[n]{a}$ and $\sqrt[n]{b}$ are real numbers and $\sqrt[n]{b}$ is not zero, then

$$\sqrt[n]{\frac{a}{b}} = \frac{\sqrt[n]{a}}{\sqrt[n]{b}}.$$

---

Notice that the quotient rule is the relationship $\left(\dfrac{a}{b}\right)^{1/n} = \dfrac{a^{1/n}}{b^{1/n}}$ stated in radical notation. We can use the quotient rule to simplify radical expressions by reading the rule from left to right, or to divide radicals by reading the rule from right to left.

For example,

$$\sqrt{\frac{x}{16}} = \frac{\sqrt{x}}{\sqrt{16}} = \frac{\sqrt{x}}{4} \qquad \text{Using } \sqrt[n]{\frac{a}{b}} = \frac{\sqrt[n]{a}}{\sqrt[n]{b}}$$

$$\frac{\sqrt{75}}{\sqrt{3}} = \sqrt{\frac{75}{3}} = \sqrt{25} = 5 \qquad \text{Using } \frac{\sqrt[n]{a}}{\sqrt[n]{b}} = \sqrt[n]{\frac{a}{b}}$$

Note: *Recall that from Section 10.2 on, we assume that variables represent positive real numbers. Since this is so, we need not insert absolute value bars when we simplify even roots.*

### EXAMPLE 2   Use the quotient rule to simplify.

**a.** $\sqrt{\frac{25}{49}}$  **b.** $\sqrt{\frac{x}{9}}$  **c.** $\sqrt[3]{\frac{8}{27}}$  **d.** $\sqrt[4]{\frac{3}{16y^4}}$

*Solution*

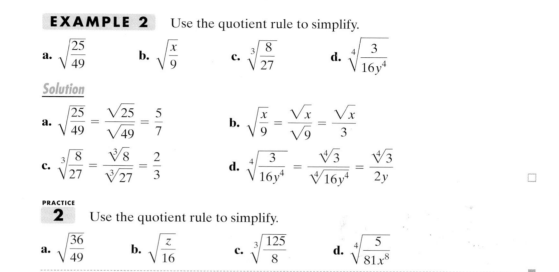

**a.** $\sqrt{\frac{25}{49}} = \frac{\sqrt{25}}{\sqrt{49}} = \frac{5}{7}$  **b.** $\sqrt{\frac{x}{9}} = \frac{\sqrt{x}}{\sqrt{9}} = \frac{\sqrt{x}}{3}$

**c.** $\sqrt[3]{\frac{8}{27}} = \frac{\sqrt[3]{8}}{\sqrt[3]{27}} = \frac{2}{3}$  **d.** $\sqrt[4]{\frac{3}{16y^4}} = \frac{\sqrt[4]{3}}{\sqrt[4]{16y^4}} = \frac{\sqrt[4]{3}}{2y}$

**PRACTICE**
**2**  Use the quotient rule to simplify.

**a.** $\sqrt{\frac{36}{49}}$  **b.** $\sqrt{\frac{z}{16}}$  **c.** $\sqrt[3]{\frac{125}{8}}$  **d.** $\sqrt[4]{\frac{5}{81x^8}}$

**OBJECTIVE 3 ▶ Simplifying radicals.**  Both the product and quotient rules can be used to simplify a radical. If the product rule is read from right to left, we have that

$$\sqrt[n]{ab} = \sqrt[n]{a} \cdot \sqrt[n]{b}.$$

This is used to simplify the following radicals.

### EXAMPLE 3   Simplify the following.

**a.** $\sqrt{50}$  **b.** $\sqrt[3]{24}$  **c.** $\sqrt{26}$  **d.** $\sqrt[4]{32}$

*Solution*

**a.** Factor 50 such that one factor is the largest perfect square that divides 50. The largest perfect square factor of 50 is 25, so we write 50 as $25 \cdot 2$ and use the product rule for radicals to simplify.

$$\sqrt{50} = \sqrt{25 \cdot 2} = \sqrt{25} \cdot \sqrt{2} = 5\sqrt{2}$$

↑ The largest perfect square factor of 50

▶ **Helpful Hint**
Don't forget that, for example, $5\sqrt{2}$ means $5 \cdot \sqrt{2}$.

**b.** $\sqrt[3]{24} = \sqrt[3]{8 \cdot 3} = \sqrt[3]{8} \cdot \sqrt[3]{3} = 2\sqrt[3]{3}$

↑ The largest perfect cube factor of 24

**c.** $\sqrt{26}$ The largest perfect square factor of 26 is 1, so $\sqrt{26}$ cannot be simplified further.

**d.** $\sqrt[4]{32} = \sqrt[4]{16 \cdot 2} = \sqrt[4]{16} \cdot \sqrt[4]{2} = 2\sqrt[4]{2}$

↑ The largest fourth power factor of 32

**PRACTICE**
**3**  Simplify the following.

**a.** $\sqrt{98}$  **b.** $\sqrt[3]{54}$  **c.** $\sqrt{35}$  **d.** $\sqrt[4]{243}$

After simplifying a radical such as a square root, always check the radicand to see that it contains no other perfect square factors. It may, if the largest perfect square factor of the radicand was not originally recognized. For example,

$$\sqrt{200} = \sqrt{4 \cdot 50} = \sqrt{4} \cdot \sqrt{50} = 2\sqrt{50}$$

Notice that the radicand 50 still contains the perfect square factor 25. This is because 4 is not the largest perfect square factor of 200. We continue as follows.

$$2\sqrt{50} = 2\sqrt{25 \cdot 2} = 2 \cdot \sqrt{25} \cdot \sqrt{2} = 2 \cdot 5 \cdot \sqrt{2} = 10\sqrt{2}$$

The radical is now simplified since 2 contains no perfect square factors (other than 1).

---

▶ **Helpful Hint**

To help you recognize largest perfect power factors of a radicand, it will help if you are familiar with some perfect powers. A few are listed below.

Perfect Squares     1,   4,   9,    16,   25,   36,   49,   64,   81,   100,   121,   144
$1^2$   $2^2$   $3^2$   $4^2$   $5^2$   $6^2$   $7^2$   $8^2$   $9^2$   $10^2$   $11^2$   $12^2$

Perfect Cubes     1,   8,   27,   64,   125
$1^3$   $2^3$   $3^3$   $4^3$   $5^3$

Perfect Fourth    1,   16,   81,   256
Powers          $1^4$   $2^4$   $3^4$   $4^4$

---

In general, we say that a radicand of the form $\sqrt[n]{a}$ is simplified when the radicand $a$ contains no factors that are perfect $n$th powers (other than 1 or $-1$).

---

**EXAMPLE 4**    Use the product rule to simplify.

**a.** $\sqrt{25x^3}$      **b.** $\sqrt[3]{54x^6y^8}$      **c.** $\sqrt[4]{81z^{11}}$

*Solution*

**a.** $\sqrt{25x^3} = \sqrt{25x^2 \cdot x}$           Find the largest perfect square factor.

$= \sqrt{25x^2} \cdot \sqrt{x}$         Apply the product rule.

$= 5x\sqrt{x}$             Simplify.

**b.** $\sqrt[3]{54x^6y^8} = \sqrt[3]{27 \cdot 2 \cdot x^6 \cdot y^6 \cdot y^2}$    Factor the radicand and identify perfect cube factors.

$= \sqrt[3]{27x^6y^6 \cdot 2y^2}$

$= \sqrt[3]{27x^6y^6} \cdot \sqrt[3]{2y^2}$     Apply the product rule.

$= 3x^2y^2\sqrt[3]{2y^2}$        Simplify.

**c.** $\sqrt[4]{81z^{11}} = \sqrt[4]{81 \cdot z^8 \cdot z^3}$      Factor the radicand and identify perfect fourth power factors.

$= \sqrt[4]{81z^8} \cdot \sqrt[4]{z^3}$      Apply the product rule.

$= 3z^2\sqrt[4]{z^3}$         Simplify.

**PRACTICE**
**4**    Use the product rule to simplify.

**a.** $\sqrt{36z^7}$       **b.** $\sqrt[3]{32p^4q^7}$       **c.** $\sqrt[4]{16x^{15}}$

**EXAMPLE 5**   Use the quotient rule to divide, and simplify if possible.

**a.** $\dfrac{\sqrt{20}}{\sqrt{5}}$     **b.** $\dfrac{\sqrt{50x}}{2\sqrt{2}}$     **c.** $\dfrac{7\sqrt[3]{48x^4y^8}}{\sqrt[3]{6y^2}}$     **d.** $\dfrac{2\sqrt[4]{32a^8b^6}}{\sqrt[4]{a^{-1}b^2}}$

*Solution*

**a.** $\dfrac{\sqrt{20}}{\sqrt{5}} = \sqrt{\dfrac{20}{5}}$       Apply the quotient rule.

$= \sqrt{4}$       Simplify.
$= 2$

**b.** $\dfrac{\sqrt{50x}}{2\sqrt{2}} = \dfrac{1}{2} \cdot \sqrt{\dfrac{50x}{2}}$       Apply the quotient rule.

$= \dfrac{1}{2} \cdot \sqrt{25x}$       Simplify.

$= \dfrac{1}{2} \cdot \sqrt{25} \cdot \sqrt{x}$       Factor $25x$.

$= \dfrac{1}{2} \cdot 5 \cdot \sqrt{x}$       Simplify.

$= \dfrac{5}{2}\sqrt{x}$

**c.** $\dfrac{7\sqrt[3]{48x^4y^8}}{\sqrt[3]{6y^2}} = 7 \cdot \sqrt[3]{\dfrac{48x^4y^8}{6y^2}}$       Apply the quotient rule.

$= 7 \cdot \sqrt[3]{8x^4y^6}$       Simplify.

$= 7\sqrt[3]{8x^3y^6 \cdot x}$       Factor.

$= 7 \cdot \sqrt[3]{8x^3y^6} \cdot \sqrt[3]{x}$       Apply the product rule.

$= 7 \cdot 2xy^2 \cdot \sqrt[3]{x}$       Simplify.

$= 14xy^2\sqrt[3]{x}$

**d.** $\dfrac{2\sqrt[4]{32a^8b^6}}{\sqrt[4]{a^{-1}b^2}} = 2\sqrt[4]{\dfrac{32a^8b^6}{a^{-1}b^2}} = 2\sqrt[4]{32a^9b^4} = 2\sqrt[4]{16 \cdot a^8 \cdot b^4 \cdot 2 \cdot a}$

$= 2\sqrt[4]{16a^8b^4} \cdot \sqrt[4]{2a} = 2 \cdot 2a^2b \cdot \sqrt[4]{2a} = 4a^2b\sqrt[4]{2a}$   □

**PRACTICE**
**5**   Use the quotient rule to divide and simplify.

**a.** $\dfrac{\sqrt{80}}{\sqrt{5}}$     **b.** $\dfrac{\sqrt{98z}}{3\sqrt{2}}$     **c.** $\dfrac{5\sqrt[3]{40x^5y^7}}{\sqrt[3]{5y}}$     **d.** $\dfrac{3\sqrt[5]{64x^9y^8}}{\sqrt[5]{x^{-1}y^2}}$

**Concept Check** ☑
Find and correct the error:

$$\dfrac{\sqrt[3]{27}}{\sqrt{9}} = \sqrt[3]{\dfrac{27}{9}} = \sqrt[3]{3}$$

Answer to Concept Check:
$\dfrac{\sqrt[3]{27}}{\sqrt{9}} = \dfrac{3}{3} = 1$

**OBJECTIVE 4** ▶ **Using the distance and midpoint formulas.**   Now that we know how to simplify radicals, we can derive and use the distance formula. The midpoint formula is often confused with the distance formula, so to clarify both, we will also review the midpoint formula.

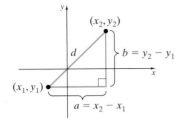

The Cartesian coordinate system helps us visualize a distance between points. To find the distance between two points, we use the distance formula, which is derived from the Pythagorean theorem.

To find the distance $d$ between two points $(x_1, y_1)$ and $(x_2, y_2)$ as shown to the left, notice that the length of leg $a$ is $x_2 - x_1$ and that the length of leg $b$ is $y_2 - y_1$.

Thus, the Pythagorean theorem tells us that

$$d^2 = a^2 + b^2$$

or

$$d^2 = (x_2 - x_1)^2 + (y_2 - y_1)^2$$

or

$$d = \sqrt{(x_2 - x_1)^2 + (y_2 - y_1)^2}$$

This formula gives us the distance between any two points on the real plane.

---

**Distance Formula**

The distance $d$ between two points $(x_1, y_1)$ and $(x_2, y_2)$ is given by

$$d = \sqrt{(x_2 - x_1)^2 + (y_2 - y_1)^2}$$

---

**EXAMPLE 6** Find the distance between $(2, -5)$ and $(1, -4)$. Give an exact distance and a three-decimal-place approximation.

**Solution** To use the distance formula, it makes no difference which point we call $(x_1, y_1)$ and which point we call $(x_2, y_2)$. We will let $(x_1, y_1) = (2, -5)$ and $(x_2, y_2) = (1, -4)$.

$$d = \sqrt{(x_2 - x_1)^2 + (y_2 - y_1)^2}$$
$$= \sqrt{(1 - 2)^2 + [-4 - (-5)]^2}$$
$$= \sqrt{(-1)^2 + (1)^2}$$
$$= \sqrt{1 + 1}$$
$$= \sqrt{2} \approx 1.414$$

The distance between the two points is exactly $\sqrt{2}$ units, or approximately 1.414 units.

□

PRACTICE

**6** Find the distance between $P(-3, 7)$ and $Q(-2, 3)$. Give an exact distance and a three-decimal-place approximation.

---

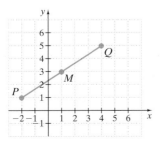

The **midpoint** of a line segment is the **point** located exactly halfway between the two endpoints of the line segment. On the graph to the left, the point $M$ is the midpoint of line segment $PQ$. Thus, the distance between $M$ and $P$ equals the distance between $M$ and $Q$.

*Note:* We usually need no knowledge of roots to calculate the midpoint of a line segment. We review midpoint here only because it is often confused with the distance between two points.

The $x$-coordinate of $M$ is at half the distance between the $x$-coordinates of $P$ and $Q$, and the $y$-coordinate of $M$ is at half the distance between the $y$-coordinates of $P$ and $Q$. That is, the $x$-coordinate of $M$ is the average of the $x$-coordinates of $P$ and $Q$; the $y$-coordinate of $M$ is the average of the $y$-coordinates of $P$ and $Q$.

> **Midpoint Formula**
> The midpoint of the line segment whose endpoints are $(x_1, y_1)$ and $(x_2, y_2)$ is the point with coordinates
> $$\left(\frac{x_1 + x_2}{2}, \frac{y_1 + y_2}{2}\right)$$

**EXAMPLE 7**   Find the midpoint of the line segment that joins points $P(-3, 3)$ and $Q(1, 0)$.

_Solution_   Use the midpoint formula. It makes no difference which point we call $(x_1, y_1)$ or which point we call $(x_2, y_2)$. Let $(x_1, y_1) = (-3, 3)$ and $(x_2, y_2) = (1, 0)$.

$$\text{midpoint} = \left(\frac{x_1 + x_2}{2}, \frac{y_1 + y_2}{2}\right)$$

$$= \left(\frac{-3 + 1}{2}, \frac{3 + 0}{2}\right)$$

$$= \left(\frac{-2}{2}, \frac{3}{2}\right)$$

$$= \left(-1, \frac{3}{2}\right)$$

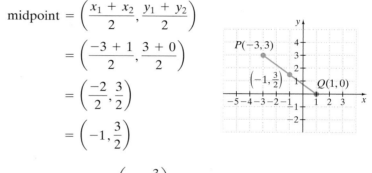

The midpoint of the segment is $\left(-1, \frac{3}{2}\right)$.

**PRACTICE**
**7**   Find the midpoint of the line segment that joins points $P(5, -2)$ and $Q(8, -6)$.

> ▶ **Helpful Hint**
> The distance between two points is a distance. The midpoint of a line segment is the point halfway between the endpoints of the segment.
>
>
> distance—measured in units
> midpoint—it is a point

# VOCABULARY & READINESS CHECK

_Use the choices below to fill in each blank. Some choices may be used more than once._

distance          midpoint          point

**1.** The _____ of a line segment is a _____ exactly halfway between the two endpoints of the line segment.

**2.** The _____ formula is $d = \sqrt{(x_2 - x_1)^2 + (y_2 - y_1)^2}$.

**3.** The _____ formula is $\left(\dfrac{x_1 + x_2}{2}, \dfrac{y_1 + y_2}{2}\right)$.

*Answer true or false. Assume all radicals represent nonzero real numbers.*

**4.** $\sqrt[n]{a} \cdot \sqrt[n]{b} = \sqrt[n]{ab}$ _____

**5.** $\sqrt[3]{7} \cdot \sqrt[3]{11} = \sqrt[3]{18}$ _____

**6.** $\sqrt[3]{7} \cdot \sqrt{11} = \sqrt{77}$ _____

**7.** $\sqrt{x^7 y^8} = \sqrt{x^7} \cdot \sqrt{y^8}$ _____

**8.** $\dfrac{\sqrt[n]{a}}{\sqrt[n]{b}} = \sqrt[n]{\dfrac{a}{b}}$ _____

**9.** $\dfrac{\sqrt[3]{12}}{\sqrt[3]{4}} = \sqrt[3]{8}$ _____

**10.** $\dfrac{\sqrt[n]{x^7}}{\sqrt[n]{x}} = \sqrt[n]{x^6}$ _____

## 10.3 | EXERCISE SET

MyMathLab | PRACTICE | WATCH | DOWNLOAD | READ | REVIEW

*Use the product rule to multiply. See Example 1.*

**1.** $\sqrt{7} \cdot \sqrt{2}$

**2.** $\sqrt{11} \cdot \sqrt{10}$

**3.** $\sqrt[4]{8} \cdot \sqrt[4]{2}$

**4.** $\sqrt[4]{27} \cdot \sqrt[4]{3}$

**5.** $\sqrt[3]{4} \cdot \sqrt[3]{9}$

**6.** $\sqrt[3]{10} \cdot \sqrt[3]{5}$

**7.** $\sqrt{2} \cdot \sqrt{3x}$

**8.** $\sqrt{3y} \cdot \sqrt{5x}$

**9.** $\sqrt{\dfrac{7}{x}} \cdot \sqrt{\dfrac{2}{y}}$

**10.** $\sqrt{\dfrac{6}{m}} \cdot \sqrt{\dfrac{n}{5}}$

**11.** $\sqrt[4]{4x^3} \cdot \sqrt[4]{5}$

**12.** $\sqrt[4]{ab^2} \cdot \sqrt[4]{27ab}$

*Use the quotient rule to simplify. See Example 2.*

**13.** $\sqrt{\dfrac{6}{49}}$

**14.** $\sqrt{\dfrac{8}{81}}$

**15.** $\sqrt{\dfrac{2}{49}}$

**16.** $\sqrt{\dfrac{5}{121}}$

**17.** $\sqrt[4]{\dfrac{x^3}{16}}$

**18.** $\sqrt[4]{\dfrac{y}{81x^4}}$

**19.** $\sqrt[3]{\dfrac{4}{27}}$

**20.** $\sqrt[3]{\dfrac{3}{64}}$

**21.** $\sqrt[4]{\dfrac{8}{x^8}}$

**22.** $\sqrt[4]{\dfrac{a^3}{81}}$

**23.** $\sqrt[3]{\dfrac{2x}{81y^{12}}}$

**24.** $\sqrt[3]{\dfrac{3}{8x^6}}$

**25.** $\sqrt{\dfrac{x^2 y}{100}}$

**26.** $\sqrt{\dfrac{y^2 z}{36}}$

**27.** $\sqrt{\dfrac{5x^2}{4y^2}}$

**28.** $\sqrt{\dfrac{y^{10}}{9x^6}}$

**29.** $-\sqrt[3]{\dfrac{z^7}{27x^3}}$

**30.** $-\sqrt[3]{\dfrac{64a}{b^9}}$

*Simplify. See Examples 3 and 4.*

**31.** $\sqrt{32}$

**32.** $\sqrt{27}$

**33.** $\sqrt[3]{192}$

**34.** $\sqrt[3]{108}$

**35.** $5\sqrt{75}$

**36.** $3\sqrt{8}$

**37.** $\sqrt{24}$

**38.** $\sqrt{20}$

**39.** $\sqrt{100x^5}$

**40.** $\sqrt{64y^9}$

**41.** $\sqrt[3]{16y^7}$

**42.** $\sqrt[3]{64y^9}$

**43.** $\sqrt[4]{a^8 b^7}$

**44.** $\sqrt[5]{32z^{12}}$

**45.** $\sqrt{y^5}$

**46.** $\sqrt[3]{y^5}$

**47.** $\sqrt{25a^2 b^3}$

**48.** $\sqrt{9x^5 y^7}$

**49.** $\sqrt[5]{-32x^{10}y}$

**50.** $\sqrt[5]{-243z^9}$

**51.** $\sqrt[3]{50x^{14}}$

**52.** $\sqrt[3]{40y^{10}}$

**53.** $-\sqrt{32a^8 b^7}$

**54.** $-\sqrt{20ab^6}$

**55.** $\sqrt{9x^7 y^9}$

**56.** $\sqrt{12r^9 s^{12}}$

**57.** $\sqrt[3]{125r^9 s^{12}}$

**58.** $\sqrt[3]{8a^6 b^9}$

*Use the quotient rule to divide. Then simplify if possible. See Example 5.*

**59.** $\dfrac{\sqrt{14}}{\sqrt{7}}$

**60.** $\dfrac{\sqrt{45}}{\sqrt{9}}$

**61.** $\dfrac{\sqrt[3]{24}}{\sqrt[3]{3}}$

**62.** $\dfrac{\sqrt[3]{10}}{\sqrt[3]{2}}$

**63.** $\dfrac{5\sqrt[4]{48}}{\sqrt[4]{3}}$

**64.** $\dfrac{7\sqrt[4]{162}}{\sqrt[4]{2}}$

**65.** $\dfrac{\sqrt{x^5 y^3}}{\sqrt{xy}}$

**66.** $\dfrac{\sqrt{a^7 b^6}}{\sqrt{a^3 b^2}}$

**67.** $\dfrac{8\sqrt[3]{54m^7}}{\sqrt[3]{2m}}$

**68.** $\dfrac{\sqrt[3]{128x^3}}{-3\sqrt[3]{2x}}$

**69.** $\dfrac{3\sqrt{100x^2}}{2\sqrt{2x^{-1}}}$

**70.** $\dfrac{\sqrt{270y^2}}{5\sqrt{3y^{-4}}}$

**71.** $\dfrac{\sqrt[4]{96a^{10}b^3}}{\sqrt[4]{3a^2 b^3}}$

**72.** $\dfrac{\sqrt[5]{64x^{10}y^3}}{\sqrt[5]{2x^3 y^{-7}}}$

*Find the distance between each pair of points. Give an exact distance and a three-decimal-place approximation. See Example 6.*

**73.** $(5, 1)$ and $(8, 5)$

**74.** $(2, 3)$ and $(14, 8)$

**75.** $(-3, 2)$ and $(1, -3)$

**76.** $(3, -2)$ and $(-4, 1)$

**77.** $(-9, 4)$ and $(-8, 1)$

**78.** $(-5, -2)$ and $(-6, -6)$

**79.** $\left(0, -\sqrt{2}\right)$ and $\left(\sqrt{3}, 0\right)$

**80.** $\left(-\sqrt{5}, 0\right)$ and $\left(0, \sqrt{7}\right)$

**81.** $(1.7, -3.6)$ and $(-8.6, 5.7)$

**82.** $(9.6, 2.5)$ and $(-1.9, -3.7)$

*Find the midpoint of the line segment whose endpoints are given. See Example 7.*

**83.** $(6, -8), (2, 4)$

**84.** $(3, 9), (7, 11)$

**85.** $(-2, -1), (-8, 6)$

**86.** $(-3, -4), (6, -8)$

**87.** $(7, 3), (-1, -3)$

**88.** $(-2, 5), (-1, 6)$

**89.** $\left(\dfrac{1}{2}, \dfrac{3}{8}\right), \left(-\dfrac{3}{2}, \dfrac{5}{8}\right)$

**90.** $\left(-\dfrac{2}{5}, \dfrac{7}{15}\right), \left(-\dfrac{2}{5}, -\dfrac{4}{15}\right)$

**91.** $\left(\sqrt{2}, 3\sqrt{5}\right), \left(\sqrt{2}, -2\sqrt{5}\right)$

**92.** $\left(\sqrt{8}, -\sqrt{12}\right), \left(3\sqrt{2}, 7\sqrt{3}\right)$

**93.** $(4.6, -3.5), (7.8, -9.8)$

**94.** $(-4.6, 2.1), (-6.7, 1.9)$

**REVIEW AND PREVIEW**

*Perform each indicated operation. See Sections 2.1 and 5.4.*

**95.** $6x + 8x$

**96.** $(6x)(8x)$

**97.** $(2x + 3)(x - 5)$

**98.** $(2x + 3) + (x - 5)$

**99.** $9y^2 - 8y^2$

**100.** $(9y^2)(-8y^2)$

**101.** $-3(x + 5)$

**102.** $-3 + x + 5$

**103.** $(x - 4)^2$

**104.** $(2x + 1)^2$

**CONCEPT EXTENSIONS**

*Find and correct the error. See a Concept Check in this section.*

**105.** 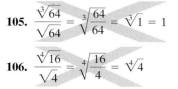 $\dfrac{\sqrt[3]{64}}{\sqrt{64}} = \sqrt[3]{\dfrac{64}{64}} = \sqrt[3]{1} = 1$

**106.** $\dfrac{\sqrt[4]{16}}{\sqrt{4}} = \sqrt[4]{\dfrac{16}{4}} = \sqrt[4]{4}$

*Simplify. See a Concept Check in this section. Assume variables represent positive numbers.*

**107.** $\sqrt[5]{x^{35}}$

**108.** $\sqrt[6]{y^{48}}$

**109.** $\sqrt[4]{a^{12}b^4c^{20}}$

**110.** $\sqrt[3]{a^9b^{21}c^3}$

**111.** $\sqrt[3]{z^{32}}$

**112.** $\sqrt[5]{x^{49}}$

**113.** $\sqrt[7]{q^{17}r^{40}s^7}$

**114.** $\sqrt[9]{p^{11}q^4r^{45}}$

**115.** The formula for the radius $r$ of a sphere with surface area $A$ is given by $r = \sqrt{\dfrac{A}{4\pi}}$. Calculate the radius of a standard zorb whose outside surface area is 32.17 sq m. Round to the nearest tenth. (See the chapter opener, page 577. *Source:* Zorb, Ltd.)

**116.** The formula for the surface area $A$ of a cone with height $h$ and radius $r$ is given by
$$A = \pi r \sqrt{r^2 + h^2}$$

**a.** Find the surface area of a cone whose height is 3 centimeters and whose radius is 4 centimeters.

**b.** Approximate to two decimal places the surface area of a cone whose height is 7.2 feet and whose radius is 6.8 feet.

**117.** The owner of Knightime Video has determined that the demand equation for renting older releases is given by the equation $F(x) = 0.6\sqrt{49 - x^2}$, where $x$ is the price in dollars per two-day rental and $F(x)$ is the number of times the video is demanded per week.

**a.** Approximate to one decimal place the demand per week of an older release if the rental price is \$3 per two-day rental.

**b.** Approximate to one decimal place the demand per week of an older release if the rental price is \$5 per two-day rental.

**c.** Explain how the owner of the video store can use this equation to predict the number of copies of each tape that should be in stock.

**118.** Before Mount Vesuvius, a volcano in Italy, erupted violently in 79 A.D., its height was 4190 feet. Vesuvius was roughly cone-shaped, and its base had a radius of approximately 25,200 feet. Use the formula for the surface area of a cone, given in Exercise 116, to approximate the surface area this volcano had before it erupted. (*Source:* Global Volcanism Network)

4190 ft

25,200 ft

# 10.4 ADDING, SUBTRACTING, AND MULTIPLYING RADICAL EXPRESSIONS

**OBJECTIVES**

1 Add or subtract radical expressions.

2 Multiply radical expressions.

**OBJECTIVE 1 ▶ Adding or subtracting radical expressions.** We have learned that sums or differences of like terms can be simplified. To simplify these sums or differences, we use the distributive property. For example,

$$2x + 3x = (2 + 3)x = 5x \quad \text{and} \quad 7x^2y - 4x^2y = (7 - 4)x^2y = 3x^2y$$

The distributive property can also be used to add **like radicals.**

**Like Radicals**

Radicals with the same index and the same radicand are like radicals.

For example, $2\sqrt{7} + 3\sqrt{7} = (2 + 3)\sqrt{7} = 5\sqrt{7}$. Also,

Like radicals

$$5\sqrt{3x} - 7\sqrt{3x} = (5 - 7)\sqrt{3x} = -2\sqrt{3x}$$

The expression $2\sqrt{7} + 2\sqrt[3]{7}$ cannot be simplified further since $2\sqrt{7}$ and $2\sqrt[3]{7}$ are not like radicals.

Unlike radicals

**EXAMPLE 1** Add or subtract as indicated. Assume all variables represent positive real numbers.

**a.** $4\sqrt{11} + 8\sqrt{11}$ **b.** $5\sqrt[3]{3x} - 7\sqrt[3]{3x}$ **c.** $2\sqrt{7} + 2\sqrt[3]{7}$

*Solution*

**a.** $4\sqrt{11} + 8\sqrt{11} = (4 + 8)\sqrt{11} = 12\sqrt{11}$

**b.** $5\sqrt[3]{3x} - 7\sqrt[3]{3x} = (5 - 7)\sqrt[3]{3x} = -2\sqrt[3]{3x}$

**c.** $2\sqrt{7} + 2\sqrt[3]{7}$

This expression cannot be simplified since $2\sqrt{7}$ and $2\sqrt[3]{7}$ do not contain like radicals. □

**PRACTICE**

**1** Add or subtract as indicated.

**a.** $3\sqrt{17} + 5\sqrt{17}$ **b.** $7\sqrt[3]{5z} - 12\sqrt[3]{5z}$ **c.** $3\sqrt{2} + 5\sqrt[3]{2}$

When adding or subtracting radicals, always check first to see whether any radicals can be simplified.

**Concept Check** ☑

True or false? Explain.

$$\sqrt{a} + \sqrt{b} = \sqrt{a + b}$$

**Answer to Concept Check:**
false; answers may vary

**EXAMPLE 2**   Add or subtract. Assume that variables represent positive real numbers.

**a.** $\sqrt{20} + 2\sqrt{45}$     **b.** $\sqrt[3]{54} - 5\sqrt[3]{16} + \sqrt[3]{2}$     **c.** $\sqrt{27x} - 2\sqrt{9x} + \sqrt{72x}$
**d.** $\sqrt[3]{98} + \sqrt{98}$     **e.** $\sqrt[3]{48y^4} + \sqrt[3]{6y^4}$

*Solution*   First, simplify each radical. Then add or subtract any like radicals.

**a.** $\sqrt{20} + 2\sqrt{45} = \sqrt{4 \cdot 5} + 2\sqrt{9 \cdot 5}$        Factor 20 and 45.
$\qquad\qquad\qquad = \sqrt{4} \cdot \sqrt{5} + 2 \cdot \sqrt{9} \cdot \sqrt{5}$        Use the product rule.
$\qquad\qquad\qquad = 2 \cdot \sqrt{5} + 2 \cdot 3 \cdot \sqrt{5}$        Simplify $\sqrt{4}$ and $\sqrt{9}$.
$\qquad\qquad\qquad = 2\sqrt{5} + 6\sqrt{5}$        Add like radicals.
$\qquad\qquad\qquad = 8\sqrt{5}$

**b.** $\sqrt[3]{54} - 5\sqrt[3]{16} + \sqrt[3]{2}$
$\qquad = \sqrt[3]{27} \cdot \sqrt[3]{2} - 5 \cdot \sqrt[3]{8} \cdot \sqrt[3]{2} + \sqrt[3]{2}$        Factor and use the product rule.
$\qquad = 3 \cdot \sqrt[3]{2} - 5 \cdot 2 \cdot \sqrt[3]{2} + \sqrt[3]{2}$        Simplify $\sqrt[3]{27}$ and $\sqrt[3]{8}$.
$\qquad = 3\sqrt[3]{2} - 10\sqrt[3]{2} + \sqrt[3]{2}$        Write $5 \cdot 2$ as 10.
$\qquad = -6\sqrt[3]{2}$        Combine like radicals.

**c.** $\sqrt{27x} - 2\sqrt{9x} + \sqrt{72x}$
$\qquad = \sqrt{9} \cdot \sqrt{3x} - 2 \cdot \sqrt{9} \cdot \sqrt{x} + \sqrt{36} \cdot \sqrt{2x}$        Factor and use the product rule.
$\qquad = 3 \cdot \sqrt{3x} - 2 \cdot 3 \cdot \sqrt{x} + 6 \cdot \sqrt{2x}$        Simplify $\sqrt{9}$ and $\sqrt{36}$.
$\qquad = 3\sqrt{3x} - 6\sqrt{x} + 6\sqrt{2x}$        Write $2 \cdot 3$ as 6.

**d.** $\sqrt[3]{98} + \sqrt{98} = \sqrt[3]{98} + \sqrt{49} \cdot \sqrt{2}$        Factor and use the product rule.
$\qquad\qquad\qquad = \sqrt[3]{98} + 7\sqrt{2}$        No further simplification is possible.

**e.** $\sqrt[3]{48y^4} + \sqrt[3]{6y^4} = \sqrt[3]{8y^3} \cdot \sqrt[3]{6y} + \sqrt[3]{y^3} \cdot \sqrt[3]{6y}$        Factor and use the product rule.
$\qquad\qquad\qquad = 2y\sqrt[3]{6y} + y\sqrt[3]{6y}$        Simplify $\sqrt[3]{8y^3}$ and $\sqrt[3]{y^3}$.
$\qquad\qquad\qquad = 3y\sqrt[3]{6y}$        Combine like radicals.   □

> **Helpful Hint**
> None of these terms contain like radicals. We can simplify no further.

*(handwritten note: Seems like you pick a factor that goes along with the 2nd sq root)*

**PRACTICE**
**2**   Add or subtract.

**a.** $\sqrt{24} + 3\sqrt{54}$     **b.** $\sqrt[3]{24} - 4\sqrt[3]{81} + \sqrt[3]{3}$     **c.** $\sqrt{75x} - 3\sqrt{27x} + \sqrt{12x}$
**d.** $\sqrt{40} + \sqrt[3]{40}$     **e.** $\sqrt[3]{81x^4} + \sqrt[3]{3x^4}$

---

Let's continue to assume that variables represent positive real numbers.

**EXAMPLE 3**   Add or subtract as indicated.

**a.** $\dfrac{\sqrt{45}}{4} - \dfrac{\sqrt{5}}{3}$     **b.** $\sqrt[3]{\dfrac{7x}{8}} + 2\sqrt[3]{7x}$

*Solution*

**a.** $\dfrac{\sqrt{45}}{4} - \dfrac{\sqrt{5}}{3} = \dfrac{3\sqrt{5}}{4} - \dfrac{\sqrt{5}}{3}$        To subtract, notice that the LCD is 12.

$\qquad\qquad = \dfrac{3\sqrt{5} \cdot 3}{4 \cdot 3} - \dfrac{\sqrt{5} \cdot 4}{3 \cdot 4}$        Write each expression as an equivalent expression with a denominator of 12.

$\qquad\qquad = \dfrac{9\sqrt{5}}{12} - \dfrac{4\sqrt{5}}{12}$        Multiply factors in the numerator and the denominator.

$\qquad\qquad = \dfrac{5\sqrt{5}}{12}$        Subtract.

**b.** $\sqrt[3]{\dfrac{7x}{8}} + 2\sqrt[3]{7x} = \dfrac{\sqrt[3]{7x}}{\sqrt[3]{8}} + 2\sqrt[3]{7x}$    Apply the quotient rule for radicals.

$\qquad\qquad\qquad = \dfrac{\sqrt[3]{7x}}{2} + 2\sqrt[3]{7x}$    Simplify.

$\qquad\qquad\qquad = \dfrac{\sqrt[3]{7x}}{2} + \dfrac{2\sqrt[3]{7x}\cdot 2}{2}$    Write each expression as an equivalent expression with a denominator of 2.

$\qquad\qquad\qquad = \dfrac{\sqrt[3]{7x}}{2} + \dfrac{4\sqrt[3]{7x}}{2}$

$\qquad\qquad\qquad = \dfrac{5\sqrt[3]{7x}}{2}$    Add.    $\square$

**PRACTICE**

**3**    Add or subtract as indicated.

**a.** $\dfrac{\sqrt{28}}{3} - \dfrac{\sqrt{7}}{4}$    **b.** $\sqrt[3]{\dfrac{6y}{64}} + 3\sqrt[3]{6y}$

**OBJECTIVE 2 ▶ Multiplying radical expressions.** We can multiply radical expressions by using many of the same properties used to multiply polynomial expressions. For instance, to multiply $\sqrt{2}(\sqrt{6} - 3\sqrt{2})$, we use the distributive property and multiply $\sqrt{2}$ by each term inside the parentheses.

$\sqrt{2}(\sqrt{6} - 3\sqrt{2}) = \sqrt{2}(\sqrt{6}) - \sqrt{2}(3\sqrt{2})$    Use the distributive property.

$\qquad\qquad\qquad = \sqrt{2\cdot 6} - 3\sqrt{2\cdot 2}$

$\qquad\qquad\qquad = \sqrt{2\cdot 2\cdot 3} - 3\cdot 2$    Use the product rule for radicals.

$\qquad\qquad\qquad = 2\sqrt{3} - 6$

**EXAMPLE 4**    Multiply.

**a.** $\sqrt{3}(5 + \sqrt{30})$    **b.** $(\sqrt{5} - \sqrt{6})(\sqrt{7} + 1)$    **c.** $(7\sqrt{x} + 5)(3\sqrt{x} - \sqrt{5})$

**d.** $(4\sqrt{3} - 1)^2$    **e.** $(\sqrt{2x} - 5)(\sqrt{2x} + 5)$    **f.** $(\sqrt{x - 3} + 5)^2$

**Solution**

**a.** $\sqrt{3}(5 + \sqrt{30}) = \sqrt{3}(5) + \sqrt{3}(\sqrt{30})$

$\qquad\qquad\qquad = 5\sqrt{3} + \sqrt{3\cdot 30}$

$\qquad\qquad\qquad = 5\sqrt{3} + \sqrt{3\cdot 3\cdot 10}$

$\qquad\qquad\qquad = 5\sqrt{3} + 3\sqrt{10}$

**b.** To multiply, we can use the FOIL method.

$\qquad\qquad\qquad\qquad\qquad\quad$ First $\qquad$ Outer $\qquad$ Inner $\qquad$ Last

$(\sqrt{5} - \sqrt{6})(\sqrt{7} + 1) = \sqrt{5}\cdot\sqrt{7} + \sqrt{5}\cdot 1 - \sqrt{6}\cdot\sqrt{7} - \sqrt{6}\cdot 1$

$\qquad\qquad\qquad\qquad = \sqrt{35} + \sqrt{5} - \sqrt{42} - \sqrt{6}$

**c.** $(7\sqrt{x} + 5)(3\sqrt{x} - \sqrt{5}) = 7\sqrt{x}(3\sqrt{x}) - 7\sqrt{x}(\sqrt{5}) + 5(3\sqrt{x}) - 5(\sqrt{5})$

$\qquad\qquad\qquad\qquad = 21x - 7\sqrt{5x} + 15\sqrt{x} - 5\sqrt{5}$

**d.** $(4\sqrt{3} - 1)^2 = (4\sqrt{3} - 1)(4\sqrt{3} - 1)$

$\qquad\qquad\qquad = 4\sqrt{3}(4\sqrt{3}) - 4\sqrt{3}(1) - 1(4\sqrt{3}) - 1(-1)$

$\qquad\qquad\qquad = 16\cdot 3 - 4\sqrt{3} - 4\sqrt{3} + 1$

$\qquad\qquad\qquad = 48 - 8\sqrt{3} + 1$

$\qquad\qquad\qquad = 49 - 8\sqrt{3}$

**e.** $\left(\sqrt{2x} - 5\right)\left(\sqrt{2x} + 5\right) = \sqrt{2x} \cdot \sqrt{2x} + 5\sqrt{2x} - 5\sqrt{2x} - 5 \cdot 5$

$$= 2x - 25$$

**f.** $\underbrace{\left(\underbrace{\sqrt{x-3}}_{a} + \underbrace{5}_{b}\right)^2}_{} = \underbrace{\left(\sqrt{x-3}\right)^2}_{a^2} \underbrace{+\ 2\cdot}_{+\ 2\cdot} \underbrace{\sqrt{x-3}}_{a} \underbrace{\cdot 5}_{\cdot b} + \underbrace{5^2}_{b^2}$

$$= x - 3 + 10\sqrt{x-3} + 25 \qquad \text{Simplify.}$$

$$= x + 22 + 10\sqrt{x-3} \qquad \text{Combine like terms.} \quad \square$$

**PRACTICE**
**4**   Multiply.

**a.** $\sqrt{5}\left(2 + \sqrt{15}\right)$   **b.** $\left(\sqrt{2} - \sqrt{5}\right)\left(\sqrt{6} + 2\right)$

**c.** $\left(3\sqrt{z} - 4\right)\left(2\sqrt{z} + 3\right)$   **d.** $\left(\sqrt{6} - 3\right)^2$

**e.** $\left(\sqrt{5x} + 3\right)\left(\sqrt{5x} - 3\right)$   **f.** $\left(\sqrt{x+2} + 3\right)^2$

## VOCABULARY & READINESS CHECK

*Complete the table with "Like" or "Unlike."*

| | Terms | Like or Unlike Radical Terms? |
|---|---|---|
| **1.** | $\sqrt{7}, \sqrt[3]{7}$ | |
| **2.** | $\sqrt[3]{x^2 y}, \sqrt[3]{yx^2}$ | |
| **3.** | $\sqrt[3]{abc}, \sqrt[3]{cba}$ | |
| **4.** | $2x\sqrt{5}, 2x\sqrt{10}$ | |

*Simplify. Assume that all variables represent positive real numbers.*

**5.** $2\sqrt{3} + 4\sqrt{3} = $ _____   **6.** $5\sqrt{7} + 3\sqrt{7} = $ _____   **7.** $8\sqrt{x} - \sqrt{x} = $ _____

**8.** $3\sqrt{y} - \sqrt{y} = $ _____   **9.** $7\sqrt[3]{x} + \sqrt[3]{x} = $ _____   **10.** $8\sqrt[3]{z} + \sqrt[3]{z} = $ _____

*Add or subtract if possible.*

**11.** $\sqrt{11} + \sqrt[3]{11} = $ _____   **12.** $9\sqrt{13} - \sqrt[4]{13} = $ _____

**13.** $8\sqrt[3]{2x} + 3\sqrt[3]{2x} - \sqrt[3]{2x} = $ _____   **14.** $8\sqrt[3]{2x} + 3\sqrt[3]{2x^2} - \sqrt[3]{2x} = $ _____

## 10.4 | EXERCISE SET

*MyMathLab*   Powered by CourseCompass™ and MathXL®

*Math XL*  PRACTICE   WATCH   DOWNLOAD   READ   REVIEW

*Add or subtract. See Examples 1 through 3.*

**1.** $\sqrt{8} - \sqrt{32}$

**2.** $\sqrt{27} - \sqrt{75}$

**3.** $2\sqrt{2x^3} + 4x\sqrt{8x}$

**4.** $3\sqrt{45x^3} + x\sqrt{5x}$

**5.** $2\sqrt{50} - 3\sqrt{125} + \sqrt{98}$

**6.** $4\sqrt{32} - \sqrt{18} + 2\sqrt{128}$

**7.** $\sqrt[3]{16x} - \sqrt[3]{54x}$

**8.** $2\sqrt[3]{3a^4} - 3a\sqrt[3]{81a}$

**9.** $\sqrt{9b^3} - \sqrt{25b^3} + \sqrt{49b^3}$

**10.** $\sqrt{4x^7} + 9x^2\sqrt{x^3} - 5x\sqrt{x^5}$

**11.** $\dfrac{5\sqrt{2}}{3} + \dfrac{2\sqrt{2}}{5}$

**12.** $\dfrac{\sqrt{3}}{2} + \dfrac{4\sqrt{3}}{3}$

**13.** $\sqrt[3]{\dfrac{11}{8}} - \dfrac{\sqrt[3]{11}}{6}$

**14.** $\dfrac{2\sqrt[3]{4}}{7} - \dfrac{\sqrt[3]{4}}{14}$

**15.** $\dfrac{\sqrt{20x}}{9} + \sqrt{\dfrac{5x}{9}}$

**16.** $\dfrac{3x\sqrt{7}}{5} + \sqrt{\dfrac{7x^2}{100}}$

**17.** $7\sqrt{9} - 7 + \sqrt{3}$

**18.** $\sqrt{16} - 5\sqrt{10} + 7$

**19.** $2 + 3\sqrt{y^2} - 6\sqrt{y^2} + 5$

**20.** $3\sqrt{7} - \sqrt[3]{x} + 4\sqrt{7} - 3\sqrt[3]{x}$

**21.** $3\sqrt{108} - 2\sqrt{18} - 3\sqrt{48}$

**22.** $-\sqrt{75} + \sqrt{12} - 3\sqrt{3}$

**23.** $-5\sqrt[3]{625} + \sqrt[3]{40}$

**24.** $-2\sqrt[3]{108} - \sqrt[3]{32}$

**25.** $\sqrt{9b^3} - \sqrt{25b^3} + \sqrt{16b^3}$

**26.** $\sqrt{4x^7y^5} + 9x^2\sqrt{x^3y^5} - 5xy\sqrt{x^5y^3}$

**27.** $5y\sqrt{8y} + 2\sqrt{50y^3}$

**28.** $3\sqrt{8x^2y^3} - 2x\sqrt{32y^3}$

**29.** $\sqrt[3]{54xy^3} - 5\sqrt[3]{2xy^3} + y\sqrt[3]{128x}$

**30.** $2\sqrt[3]{24x^3y^4} + 4x\sqrt[3]{81y^4}$

**31.** $6\sqrt[3]{11} + 8\sqrt{11} - 12\sqrt{11}$

**32.** $3\sqrt[3]{5} + 4\sqrt{5}$

**33.** $-2\sqrt[4]{x^7} + 3\sqrt[4]{16x^7}$

**34.** $6\sqrt[3]{24x^3} - 2\sqrt[3]{81x^3} - x\sqrt[3]{3}$

**35.** $\dfrac{4\sqrt{3}}{3} - \dfrac{\sqrt{12}}{3}$

**36.** $\dfrac{\sqrt{45}}{10} + \dfrac{7\sqrt{5}}{10}$

**37.** $\dfrac{\sqrt[3]{8x^4}}{7} + \dfrac{3x\sqrt[3]{x}}{7}$

**38.** $\dfrac{\sqrt[4]{48}}{5x} - \dfrac{2\sqrt[4]{3}}{10x}$

**39.** $\sqrt{\dfrac{28}{x^2}} + \sqrt{\dfrac{7}{4x^2}}$

**40.** $\dfrac{\sqrt{99}}{5x} - \sqrt{\dfrac{44}{x^2}}$

**41.** $\sqrt[3]{\dfrac{16}{27}} - \dfrac{\sqrt[3]{54}}{6}$

**42.** $\dfrac{\sqrt[3]{3}}{10} + \sqrt[3]{\dfrac{24}{125}}$

**43.** $-\dfrac{\sqrt[3]{2x^4}}{9} + \sqrt[3]{\dfrac{250x^4}{27}}$

**44.** $\dfrac{\sqrt[3]{y^5}}{8} + \dfrac{5y\sqrt[3]{y^2}}{4}$

△ **45.** Find the perimeter of the trapezoid.

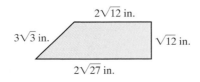

$2\sqrt{12}$ in.

$3\sqrt{3}$ in.     $\sqrt{12}$ in.

$2\sqrt{27}$ in.

△ **46.** Find the perimeter of the triangle.

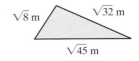

$\sqrt{8}$ m     $\sqrt{32}$ m

$\sqrt{45}$ m

*Multiply, and then simplify if possible. See Example 4.*

**47.** $\sqrt{7}\left(\sqrt{5} + \sqrt{3}\right)$

**48.** $\sqrt{5}\left(\sqrt{15} - \sqrt{35}\right)$

**49.** $\left(\sqrt{5} - \sqrt{2}\right)^2$

**50.** $\left(3x - \sqrt{2}\right)\left(3x - \sqrt{2}\right)$

**51.** $\sqrt{3x}\left(\sqrt{3} - \sqrt{x}\right)$ .

**52.** $\sqrt{5y}\left(\sqrt{y} + \sqrt{5}\right)$

**53.** $\left(2\sqrt{x} - 5\right)\left(3\sqrt{x} + 1\right)$

**54.** $\left(8\sqrt{y} + z\right)\left(4\sqrt{y} - 1\right)$

**55.** $\left(\sqrt[3]{a} - 4\right)\left(\sqrt[3]{a} + 5\right)$

**56.** $\left(\sqrt[3]{a} + 2\right)\left(\sqrt[3]{a} + 7\right)$

**57.** $6\left(\sqrt{2} - 2\right)$

**58.** $\sqrt{5}\left(6 - \sqrt{5}\right)$

**59.** $\sqrt{2}\left(\sqrt{2} + x\sqrt{6}\right)$

**60.** $\sqrt{3}\left(\sqrt{3} - 2\sqrt{5x}\right)$

**61.** $\left(2\sqrt{7} + 3\sqrt{5}\right)\left(\sqrt{7} - 2\sqrt{5}\right)$

**62.** $\left(\sqrt{6} - 4\sqrt{2}\right)\left(3\sqrt{6} + \sqrt{2}\right)$

**63.** $\left(\sqrt{x} - y\right)\left(\sqrt{x} + y\right)$

**64.** $\left(\sqrt{3x} + 2\right)\left(\sqrt{3x} - 2\right)$

**65.** $\left(\sqrt{3} + x\right)^2$

**66.** $\left(\sqrt{y} - 3x\right)^2$

**67.** $\left(\sqrt{5x} - 2\sqrt{3x}\right)\left(\sqrt{5x} - 3\sqrt{3x}\right)$

**68.** $\left(5\sqrt{7x} - \sqrt{2x}\right)\left(4\sqrt{7x} + 6\sqrt{2x}\right)$

**69.** $\left(\sqrt[3]{4} + 2\right)\left(\sqrt[3]{2} - 1\right)$

**70.** $\left(\sqrt[3]{3} + \sqrt[3]{2}\right)\left(\sqrt[3]{9} - \sqrt[3]{4}\right)$

**71.** $\left(\sqrt[3]{x} + 1\right)\left(\sqrt[3]{x^2} - \sqrt[3]{x} + 1\right)$

**72.** $\left(\sqrt[3]{3x} + 2\right)\left(\sqrt[3]{9x^2} - 2\sqrt[3]{3x} + 4\right)$

**73.** $\left(\sqrt{x - 1} + 5\right)^2$

**74.** $\left(\sqrt{3x + 1} + 2\right)^2$

**75.** $\left(\sqrt{2x + 5} - 1\right)^2$

**76.** $\left(\sqrt{x - 6} - 7\right)^2$

**REVIEW AND PREVIEW**

*Factor each numerator and denominator. Then simplify if possible. See Section 7.1.*

**77.** $\dfrac{2x - 14}{2}$

**78.** $\dfrac{8x - 24y}{4}$

**79.** $\dfrac{7x - 7y}{x^2 - y^2}$

**80.** $\dfrac{x^3 - 8}{4x - 8}$

**81.** $\dfrac{6a^2b - 9ab}{3ab}$

**82.** $\dfrac{14r - 28r^2s^2}{7rs}$

**83.** $\dfrac{-4 + 2\sqrt{3}}{6}$

**84.** $\dfrac{-5 + 10\sqrt{7}}{5}$

## CONCEPT EXTENSIONS

△ **85.** Find the perimeter and area of the rectangle.

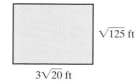

$\sqrt{125}$ ft

$3\sqrt{20}$ ft

△ **86.** Find the area and perimeter of the trapezoid. (*Hint:* The area of a trapezoid is the product of half the height $6\sqrt{3}$ meters and the sum of the bases $2\sqrt{63}$ and $7\sqrt{7}$ meters.)

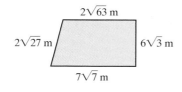

$2\sqrt{63}$ m

$2\sqrt{27}$ m        $6\sqrt{3}$ m

$7\sqrt{7}$ m

**87. a.** Add: $\sqrt{3} + \sqrt{3}$.

    **b.** Multiply: $\sqrt{3} \cdot \sqrt{3}$.

    **c.** Describe the differences in parts **a** and **b**.

**88.** Multiply: $\left(\sqrt{2} + \sqrt{3} - 1\right)^2$.

**89.** Explain how simplifying $2x + 3x$ is similar to simplifying $2\sqrt{x} + 3\sqrt{x}$.

**90.** Explain how multiplying $(x - 2)(x + 3)$ is similar to multiplying $\left(\sqrt{x} - \sqrt{2}\right)\left(\sqrt{x} + 3\right)$.

---

## 📖 STUDY SKILLS BUILDER

### Have You Decided to Successfully Complete This Course?

Hopefully by now, one of your current goals is to successfully complete this course.

    If it is not a goal of yours, ask yourself why? One common reason is fear of failure. Amazingly enough, fear of failure alone can be strong enough to keep many of us from doing our best in any endeavor. Another common reason is that you simply haven't taken the time to make successfully completing this course one of your goals.

    Anytime you are registered for a course, successfully completing this course should probably be a goal. How do you do this? Start by writing this goal in your mathematics notebook. Then list steps you will take to ensure success. A great first step is to read or reread Section 1.1 and make a commitment to try the suggestions in this section.

    Good luck, and don't forget that a positive attitude will make a big difference.

*Let's see how you are doing.*

**1.** Have you made the decision to make "successfully completing this course" a goal of yours? If not, please list reasons that this has not happened. Study your list and talk to your instructor about this.

**2.** If your answer to Exercise 1 is yes, take a moment and list, in your notebook, further specific goals that will help you achieve this major goal of successfully completing this course. (For example, my goal this semester is not to miss any of my mathematics classes.)

**3.** Rate your commitment to this course with a number between 1 and 5. Use the diagram below to help.

| High Commitment | | Average Commitment | | Not Committed at All |
|---|---|---|---|---|
| 5 | 4 | 3 | 2 | 1 |

**4.** If you have rated your personal commitment level (from the exercise above) as a 1, 2, or 3, list the reasons why this is so. Then determine whether it is possible to increase your commitment level to a 4 or 5.

# 10.5 RATIONALIZING DENOMINATORS AND NUMERATORS OF RADICAL EXPRESSIONS

**OBJECTIVES**

1 Rationalize denominators.

2 Rationalize denominators having two terms.

3 Rationalize numerators.

**OBJECTIVE 1 ▶ Rationalizing denominators of radical expressions.** Often in mathematics, it is helpful to write a radical expression such as $\dfrac{\sqrt{3}}{\sqrt{2}}$ either without a radical in the denominator or without a radical in the numerator. The process of writing this expression as an equivalent expression but without a radical in the denominator is called **rationalizing the denominator.** To rationalize the denominator of $\dfrac{\sqrt{3}}{\sqrt{2}}$, we use the fundamental principle of fractions and multiply the numerator and the denominator by $\sqrt{2}$. Recall that this is the same as multiplying by $\dfrac{\sqrt{2}}{\sqrt{2}}$, which simplifies to 1.

$$\frac{\sqrt{3}}{\sqrt{2}} = \frac{\sqrt{3}\cdot\sqrt{2}}{\sqrt{2}\cdot\sqrt{2}} = \frac{\sqrt{6}}{\sqrt{4}} = \frac{\sqrt{6}}{2}$$

In this section, we continue to assume that variables represent positive real numbers.

**EXAMPLE 1** Rationalize the denominator of each expression.

a. $\dfrac{2}{\sqrt{5}}$

b. $\dfrac{2\sqrt{16}}{\sqrt{9x}}$

c. $\sqrt[3]{\dfrac{1}{2}}$

*Solution*

a. To rationalize the denominator, we multiply the numerator and denominator by a factor that makes the radicand in the denominator a perfect square.

$$\frac{2}{\sqrt{5}} = \frac{2\cdot\sqrt{5}}{\sqrt{5}\cdot\sqrt{5}} = \frac{2\sqrt{5}}{5} \qquad \text{The denominator is now rationalized.}$$

b. First, we simplify the radicals and then rationalize the denominator.

$$\frac{2\sqrt{16}}{\sqrt{9x}} = \frac{2(4)}{3\sqrt{x}} = \frac{8}{3\sqrt{x}}$$

To rationalize the denominator, multiply the numerator and denominator by $\sqrt{x}$. Then

$$\frac{8}{3\sqrt{x}} = \frac{8\cdot\sqrt{x}}{3\sqrt{x}\cdot\sqrt{x}} = \frac{8\sqrt{x}}{3x}$$

c. $\sqrt[3]{\dfrac{1}{2}} = \dfrac{\sqrt[3]{1}}{\sqrt[3]{2}} = \dfrac{1}{\sqrt[3]{2}}$. Now we rationalize the denominator. Since $\sqrt[3]{2}$ is a cube root, we want to multiply by a value that will make the radicand 2 a perfect cube. If we multiply $\sqrt[3]{2}$ by $\sqrt[3]{2^2}$, we get $\sqrt[3]{2^3} = \sqrt[3]{8} = 2$.

$$\frac{1\cdot\sqrt[3]{2^2}}{\sqrt[3]{2}\cdot\sqrt[3]{2^2}} = \frac{\sqrt[3]{4}}{\sqrt[3]{2^3}} = \frac{\sqrt[3]{4}}{2} \qquad \text{Multiply the numerator and denominator by } \sqrt[3]{2^2} \text{ and then simplify.}$$

**PRACTICE**

**1** Rationalize the denominator of each expression.

a. $\dfrac{5}{\sqrt{3}}$

b. $\dfrac{3\sqrt{25}}{\sqrt{4x}}$

c. $\sqrt[3]{\dfrac{2}{9}}$

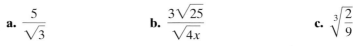

### Concept Check ✓

Determine by which number both the numerator and denominator can be multiplied to rationalize the denominator of the radical expression.

**a.** $\dfrac{1}{\sqrt[3]{7}}$ **b.** $\dfrac{1}{\sqrt[4]{8}}$

**EXAMPLE 2**   Rationalize the denominator of $\sqrt{\dfrac{7x}{3y}}$.

*Solution*

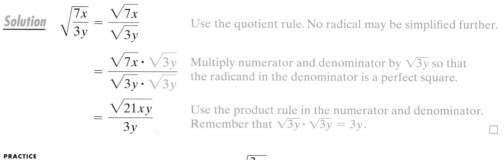

$$\sqrt{\dfrac{7x}{3y}} = \dfrac{\sqrt{7x}}{\sqrt{3y}} \qquad \text{Use the quotient rule. No radical may be simplified further.}$$

$$= \dfrac{\sqrt{7x} \cdot \sqrt{3y}}{\sqrt{3y} \cdot \sqrt{3y}} \qquad \text{Multiply numerator and denominator by } \sqrt{3y} \text{ so that the radicand in the denominator is a perfect square.}$$

$$= \dfrac{\sqrt{21xy}}{3y} \qquad \text{Use the product rule in the numerator and denominator. Remember that } \sqrt{3y} \cdot \sqrt{3y} = 3y.$$

**PRACTICE**
**2**   Rationalize the denominator of $\sqrt{\dfrac{3z}{5y}}$.

**EXAMPLE 3**   Rationalize the denominator of $\dfrac{\sqrt[4]{x}}{\sqrt[4]{81y^5}}$.

*Solution*   First, simplify each radical if possible.

$$\dfrac{\sqrt[4]{x}}{\sqrt[4]{81y^5}} = \dfrac{\sqrt[4]{x}}{\sqrt[4]{81y^4} \cdot \sqrt[4]{y}} \qquad \text{Use the product rule in the denominator.}$$

$$= \dfrac{\sqrt[4]{x}}{3y\sqrt[4]{y}} \qquad \text{Write } \sqrt[4]{81y^4} \text{ as } 3y.$$

$$= \dfrac{\sqrt[4]{x} \cdot \sqrt[4]{y^3}}{3y\sqrt[4]{y} \cdot \sqrt[4]{y^3}} \qquad \text{Multiply numerator and denominator by } \sqrt[4]{y^3} \text{ so that the radicand in the denominator is a perfect fourth power.}$$

$$= \dfrac{\sqrt[4]{xy^3}}{3y\sqrt[4]{y^4}} \qquad \text{Use the product rule in the numerator and denominator.}$$

$$= \dfrac{\sqrt[4]{xy^3}}{3y^2} \qquad \text{In the denominator, } \sqrt[4]{y^4} = y \text{ and } 3y \cdot y = 3y^2.$$

**PRACTICE**
**3**   Rationalize the denominator of $\dfrac{\sqrt[3]{z^2}}{\sqrt[3]{27x^4}}$.

**OBJECTIVE 2 ▶ Rationalizing denominators having two terms.**  Remember the product of the sum and difference of two terms?

$$(a + b)(a - b) = a^2 - b^2$$

These two expressions are called **conjugates** of each other.

To rationalize a numerator or denominator that is a sum or difference of two terms, we use conjugates. To see how and why this works, let's rationalize the denominator of the expression $\dfrac{5}{\sqrt{3}-2}$. To do so, we multiply both the numerator and the denominator by $\sqrt{3}+2$, the **conjugate** of the denominator $\sqrt{3}-2$, and see what happens.

$$\frac{5}{\sqrt{3}-2} = \frac{5(\sqrt{3}+2)}{(\sqrt{3}-2)(\sqrt{3}+2)}$$

$$= \frac{5(\sqrt{3}+2)}{(\sqrt{3})^2 - 2^2} \qquad \text{Multiply the sum and difference of two terms: } (a+b)(a-b) = a^2 - b^2.$$

$$= \frac{5(\sqrt{3}+2)}{3-4}$$

$$= \frac{5(\sqrt{3}+2)}{-1}$$

$$= -5(\sqrt{3}+2) \quad \text{or} \quad -5\sqrt{3} - 10$$

Notice in the denominator that the product of $(\sqrt{3}-2)$ and its conjugate, $(\sqrt{3}+2)$, is $-1$. In general, the product of an expression and its conjugate will contain no radical terms. This is why, when rationalizing a denominator or a numerator containing two terms, we multiply by its conjugate. Examples of conjugates are

$$\sqrt{a} - \sqrt{b} \quad \text{and} \quad \sqrt{a} + \sqrt{b}$$
$$x + \sqrt{y} \quad \text{and} \quad x - \sqrt{y}$$

**EXAMPLE 4**  Rationalize each denominator.

**a.** $\dfrac{2}{3\sqrt{2}+4}$  **b.** $\dfrac{\sqrt{6}+2}{\sqrt{5}-\sqrt{3}}$  **c.** $\dfrac{2\sqrt{m}}{3\sqrt{x}+\sqrt{m}}$

*Solution*

**a.** Multiply the numerator and denominator by the conjugate of the denominator, $3\sqrt{2}+4$.

$$\frac{2}{3\sqrt{2}+4} = \frac{2(3\sqrt{2}-4)}{(3\sqrt{2}+4)(3\sqrt{2}-4)}$$

$$= \frac{2(3\sqrt{2}-4)}{(3\sqrt{2})^2 - 4^2}$$

$$= \frac{2(3\sqrt{2}-4)}{18-16}$$

$$= \frac{2(3\sqrt{2}-4)}{2}, \quad \text{or} \quad 3\sqrt{2}-4$$

It is often useful to leave a numerator in factored form to help determine whether the expression can be simplified.

**b.** Multiply the numerator and denominator by the conjugate of $\sqrt{5} - \sqrt{3}$.

$$\frac{\sqrt{6} + 2}{\sqrt{5} - \sqrt{3}} = \frac{(\sqrt{6} + 2)(\sqrt{5} + \sqrt{3})}{(\sqrt{5} - \sqrt{3})(\sqrt{5} + \sqrt{3})}$$

$$= \frac{\sqrt{6}\sqrt{5} + \sqrt{6}\sqrt{3} + 2\sqrt{5} + 2\sqrt{3}}{(\sqrt{5})^2 - (\sqrt{3})^2}$$

$$= \frac{\sqrt{30} + \sqrt{18} + 2\sqrt{5} + 2\sqrt{3}}{5 - 3}$$

$$= \frac{\sqrt{30} + 3\sqrt{2} + 2\sqrt{5} + 2\sqrt{3}}{2}$$

**c.** Multiply by the conjugate of $3\sqrt{x} + \sqrt{m}$ to eliminate the radicals from the denominator.

$$\frac{2\sqrt{m}}{3\sqrt{x} + \sqrt{m}} = \frac{2\sqrt{m}(3\sqrt{x} - \sqrt{m})}{(3\sqrt{x} + \sqrt{m})(3\sqrt{x} - \sqrt{m})} = \frac{6\sqrt{mx} - 2m}{(3\sqrt{x})^2 - (\sqrt{m})^2}$$

$$= \frac{6\sqrt{mx} - 2m}{9x - m}$$

**PRACTICE**
**4**   Rationalize the denominator.

**a.** $\dfrac{5}{3\sqrt{5} + 2}$     **b.** $\dfrac{\sqrt{2} + 5}{\sqrt{3} - \sqrt{5}}$     **c.** $\dfrac{3\sqrt{x}}{2\sqrt{x} + \sqrt{y}}$

---

**OBJECTIVE 3 ▶ Rationalizing numerators.** As mentioned earlier, it is also often helpful to write an expression such as $\dfrac{\sqrt{3}}{\sqrt{2}}$ as an equivalent expression without a radical in the numerator. This process is called **rationalizing the numerator.** To rationalize the numerator of $\dfrac{\sqrt{3}}{\sqrt{2}}$, we multiply the numerator and the denominator by $\sqrt{3}$.

$$\frac{\sqrt{3}}{\sqrt{2}} = \frac{\sqrt{3} \cdot \sqrt{3}}{\sqrt{2} \cdot \sqrt{3}} = \frac{\sqrt{9}}{\sqrt{6}} = \frac{3}{\sqrt{6}}$$

**EXAMPLE 5**   Rationalize the numerator of $\dfrac{\sqrt{7}}{\sqrt{45}}$.

**Solution**   First we simplify $\sqrt{45}$.

$$\frac{\sqrt{7}}{\sqrt{45}} = \frac{\sqrt{7}}{\sqrt{9 \cdot 5}} = \frac{\sqrt{7}}{3\sqrt{5}}$$

Next we rationalize the numerator by multiplying the numerator and the denominator by $\sqrt{7}$.

$$\frac{\sqrt{7}}{3\sqrt{5}} = \frac{\sqrt{7} \cdot \sqrt{7}}{3\sqrt{5} \cdot \sqrt{7}} = \frac{7}{3\sqrt{5 \cdot 7}} = \frac{7}{3\sqrt{35}}$$

**PRACTICE**
**5**   Rationalize the numerator of $\dfrac{\sqrt{32}}{\sqrt{80}}$.

**EXAMPLE 6** Rationalize the numerator of $\dfrac{\sqrt[3]{2x^2}}{\sqrt[3]{5y}}$.

**Solution** The numerator and the denominator of this expression are already simplified. To rationalize the numerator, $\sqrt[3]{2x^2}$, we multiply the numerator and denominator by a factor that will make the radicand a perfect cube. If we multiply $\sqrt[3]{2x^2}$ by $\sqrt[3]{4x}$, we get $\sqrt[3]{8x^3} = 2x$.

$$\frac{\sqrt[3]{2x^2}}{\sqrt[3]{5y}} = \frac{\sqrt[3]{2x^2} \cdot \sqrt[3]{4x}}{\sqrt[3]{5y} \cdot \sqrt[3]{4x}} = \frac{\sqrt[3]{8x^3}}{\sqrt[3]{20xy}} = \frac{2x}{\sqrt[3]{20xy}}$$ □

**PRACTICE 6** Rationalize the numerator of $\dfrac{\sqrt[3]{5b}}{\sqrt[3]{2a}}$.

**EXAMPLE 7** Rationalize the numerator of $\dfrac{\sqrt{x} + 2}{5}$.

**Solution** We multiply the numerator and the denominator by the conjugate of the numerator, $\sqrt{x} + 2$.

$$\frac{\sqrt{x} + 2}{5} = \frac{\left(\sqrt{x} + 2\right)\left(\sqrt{x} - 2\right)}{5\left(\sqrt{x} - 2\right)} \qquad \text{Multiply by } \sqrt{x} - 2, \text{ the conjugate of } \sqrt{x} + 2.$$

$$= \frac{\left(\sqrt{x}\right)^2 - 2^2}{5\left(\sqrt{x} - 2\right)} \qquad (a + b)(a - b) = a^2 - b^2$$

$$= \frac{x - 4}{5\left(\sqrt{x} - 2\right)}$$ □

**PRACTICE 7** Rationalize the numerator of $\dfrac{\sqrt{x} - 3}{4}$.

# VOCABULARY & READINESS CHECK

*Use the choices below to fill in each blank. Not all choices will be used.*

rationalizing the numerator       conjugate      $\dfrac{\sqrt{3}}{\sqrt{3}}$

rationalizing the denominator      $\dfrac{5}{5}$

**1.** The _____ of $a + b$ is $a - b$.

**2.** The process of writing an equivalent expression, but without a radical in the denominator is called _____.

**3.** The process of writing an equivalent expression, but without a radical in the numerator is called _____.

**4.** To rationalize the denominator of $\dfrac{5}{\sqrt{3}}$, we multiply by _____.

*Find the conjugate of each expression.*

**5.** $\sqrt{2} + x$      **6.** $\sqrt{3} + y$      **7.** $5 - \sqrt{a}$      **8.** $6 - \sqrt{b}$

**9.** $-7\sqrt{5} + 8\sqrt{x}$      **10.** $-9\sqrt{2} - 6\sqrt{y}$

## 10.5 | EXERCISE SET

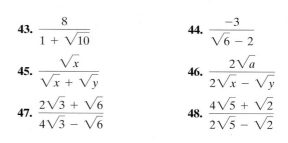

*Rationalize each denominator. See Examples 1 through 3.*

**1.** $\dfrac{\sqrt{2}}{\sqrt{7}}$

**2.** $\dfrac{\sqrt{3}}{\sqrt{2}}$

**3.** $\sqrt{\dfrac{1}{5}}$

**4.** $\sqrt{\dfrac{1}{2}}$

**5.** $\sqrt{\dfrac{4}{x}}$

**6.** $\sqrt{\dfrac{25}{y}}$

**7.** $\dfrac{4}{\sqrt[3]{3}}$

**8.** $\dfrac{6}{\sqrt[3]{9}}$

**9.** $\dfrac{3}{\sqrt{8x}}$

**10.** $\dfrac{5}{\sqrt{27a}}$

**11.** $\dfrac{3}{\sqrt[3]{4x^2}}$

**12.** $\dfrac{5}{\sqrt[3]{3y}}$

**13.** $\dfrac{9}{\sqrt{3a}}$

**14.** $\dfrac{x}{\sqrt{5}}$

**15.** $\dfrac{3}{\sqrt[3]{2}}$

**16.** $\dfrac{5}{\sqrt[3]{9}}$

**17.** $\dfrac{2\sqrt{3}}{\sqrt{7}}$

**18.** $\dfrac{-5\sqrt{2}}{\sqrt{11}}$

**19.** $\sqrt{\dfrac{2x}{5y}}$

**20.** $\sqrt{\dfrac{13a}{2b}}$

**21.** $\sqrt[3]{\dfrac{3}{5}}$

**22.** $\sqrt[3]{\dfrac{7}{10}}$

**23.** $\sqrt{\dfrac{3x}{50}}$

**24.** $\sqrt{\dfrac{11y}{45}}$

**25.** $\dfrac{1}{\sqrt{12z}}$

**26.** $\dfrac{1}{\sqrt{32x}}$

**27.** $\dfrac{\sqrt[3]{2y^2}}{\sqrt[3]{9x^2}}$

**28.** $\dfrac{\sqrt[3]{3x}}{\sqrt[3]{4y^4}}$

**29.** $\sqrt[4]{\dfrac{81}{8}}$

**30.** $\sqrt[4]{\dfrac{1}{9}}$

**31.** $\sqrt[4]{\dfrac{16}{9x^7}}$

**32.** $\sqrt[5]{\dfrac{32}{m^6 n^{13}}}$

**33.** $\dfrac{5a}{\sqrt[5]{8a^9 b^{11}}}$

**34.** $\dfrac{9y}{\sqrt[4]{4y^9}}$

*Rationalize each denominator. See Example 4.*

**35.** $\dfrac{6}{2 - \sqrt{7}}$

**36.** $\dfrac{3}{\sqrt{7} - 4}$

**37.** $\dfrac{-7}{\sqrt{x} - 3}$

**38.** $\dfrac{-8}{\sqrt{y} + 4}$

**39.** $\dfrac{\sqrt{2} - \sqrt{3}}{\sqrt{2} + \sqrt{3}}$

**40.** $\dfrac{\sqrt{3} + \sqrt{4}}{\sqrt{2} - \sqrt{3}}$

**41.** $\dfrac{\sqrt{a} + 1}{2\sqrt{a} - \sqrt{b}}$

**42.** $\dfrac{2\sqrt{a} - 3}{2\sqrt{a} + \sqrt{b}}$

**43.** $\dfrac{8}{1 + \sqrt{10}}$

**44.** $\dfrac{-3}{\sqrt{6} - 2}$

**45.** $\dfrac{\sqrt{x}}{\sqrt{x} + \sqrt{y}}$

**46.** $\dfrac{2\sqrt{a}}{2\sqrt{x} - \sqrt{y}}$

**47.** $\dfrac{2\sqrt{3} + \sqrt{6}}{4\sqrt{3} - \sqrt{6}}$

**48.** $\dfrac{4\sqrt{5} + \sqrt{2}}{2\sqrt{5} - \sqrt{2}}$

*Rationalize each numerator. See Examples 5 and 6.*

**49.** $\sqrt{\dfrac{5}{3}}$

**50.** $\sqrt{\dfrac{3}{2}}$

**51.** $\sqrt{\dfrac{18}{5}}$

**52.** $\sqrt{\dfrac{12}{7}}$

**53.** $\dfrac{\sqrt{4x}}{7}$

**54.** $\dfrac{\sqrt{3x^5}}{6}$

**55.** $\dfrac{\sqrt[3]{5y^2}}{\sqrt[3]{4x}}$

**56.** $\dfrac{\sqrt[3]{4x}}{\sqrt[3]{z^4}}$

**57.** $\sqrt{\dfrac{2}{5}}$

**58.** $\sqrt{\dfrac{3}{7}}$

**59.** $\dfrac{\sqrt{2x}}{11}$

**60.** $\dfrac{\sqrt{y}}{7}$

**61.** $\sqrt[3]{\dfrac{7}{8}}$

**62.** $\sqrt[3]{\dfrac{25}{2}}$

**63.** $\dfrac{\sqrt[3]{3x^5}}{10}$

**64.** $\sqrt[3]{\dfrac{9y}{7}}$

**65.** $\sqrt{\dfrac{18x^4 y^6}{3z}}$

**66.** $\sqrt{\dfrac{8x^5 y}{2z}}$

**67.** When rationalizing the denominator of $\dfrac{\sqrt{5}}{\sqrt{7}}$, explain why both the numerator and the denominator must be multiplied by $\sqrt{7}$.

**68.** When rationalizing the numerator of $\dfrac{\sqrt{5}}{\sqrt{7}}$, explain why both the numerator and the denominator must be multiplied by $\sqrt{5}$.

*Rationalize each numerator. See Example 7.*

**69.** $\dfrac{2 - \sqrt{11}}{6}$

**70.** $\dfrac{\sqrt{15} + 1}{2}$

**71.** $\dfrac{2 - \sqrt{7}}{-5}$

**72.** $\dfrac{\sqrt{5} + 2}{\sqrt{2}}$

**73.** $\dfrac{\sqrt{x} + 3}{\sqrt{x}}$

**74.** $\dfrac{5 + \sqrt{2}}{\sqrt{2x}}$

**75.** $\dfrac{\sqrt{2} - 1}{\sqrt{2} + 1}$

**76.** $\dfrac{\sqrt{8} - \sqrt{3}}{\sqrt{2} + \sqrt{3}}$

**77.** $\dfrac{\sqrt{x} + 1}{\sqrt{x} - 1}$

**78.** $\dfrac{\sqrt{x} + \sqrt{y}}{\sqrt{x} - \sqrt{y}}$

## REVIEW AND PREVIEW

*Solve each equation. See Sections 2.3 and 6.6.*

**79.** $2x - 7 = 3(x - 4)$

**80.** $9x - 4 = 7(x - 2)$

**81.** $(x - 6)(2x + 1) = 0$

**82.** $(y + 2)(5y + 4) = 0$

**83.** $x^2 - 8x = -12$

**84.** $x^3 = x$

## CONCEPTS EXTENSIONS

*Determine the smallest number both the numerator and denominator should be multiplied by to rationalize the denominator of the radical expression. See the Concept Check in this section.*

**85.** $\dfrac{9}{\sqrt[3]{5}}$

**86.** $\dfrac{5}{\sqrt{27}}$

△ **87.** The formula of the radius $r$ of a sphere with surface area $A$ is

$$r = \sqrt{\frac{A}{4\pi}}$$

Rationalize the denominator of the radical expression in this formula.

△ **88.** The formula for the radius $r$ of a cone with height 7 centimeters and volume $V$ is

$$r = \sqrt{\frac{3V}{7\pi}}$$

Rationalize the numerator of the radical expression in this formula.

7 cm

$r$

✎ **89.** Explain why rationalizing the denominator does not change the value of the original expression.

✎ **90.** Explain why rationalizing the numerator does not change the value of the original expression.

---

## THE BIGGER PICTURE   SIMPLIFYING EXPRESSIONS AND SOLVING EQUATIONS AND INEQUALITIES

Now we continue our outline from Sections 1.7, 2.3, 2.8, 5.6, 6.6, 7.4, 7.6, 9.1, and 9.3. Although suggestions are given, this outline should be in your own words. Once you complete this new portion, try the exercises below.

**I.** Simplifying Expressions
  **A.** Real Numbers
   **1.** Add (Section 1.5)
   **2.** Subtract (Section 1.6)
   **3.** Multiply or Divide (Section 1.7)
  **B.** Exponents (Section 5.1 and 5.5)
  **C.** Polynomials
   **1.** Add (Section 5.2)
   **2.** Subtract (Section 5.2)
   **3.** Multiply (Section 5.3 and 5.4)
   **4.** Divide (Section 5.6)
  **D.** Factoring Polynomials (Chapter 6 Integrated Review)
  **E.** Rational Expressions
   **1.** Simplify (Section 7.1)
   **2.** Multiply (Section 7.2)
   **3.** Divide (Section 7.2)
   **4.** Add or Subtract (Section 7.4)

**F.** Radicals
  **1.** Simplify square roots: If possible, factor the radicand so that one factor is a perfect square. Then use the product rule, and simplify.

$$\sqrt{75} = \sqrt{25 \cdot 3} = \sqrt{25} \cdot \sqrt{3} = 5\sqrt{3}$$

  **2.** Add or subtract: Only like radicals (same index and radicand) can be added or subtracted.

$$8\sqrt{10} - \sqrt{40} + \sqrt{5a}$$
$$= 8\sqrt{10} - 2\sqrt{10} + \sqrt{5a}$$
$$= 6\sqrt{10} + \sqrt{5a}$$

  **3.** Multiply or divide:

$$\sqrt{a} \cdot \sqrt{b} = \sqrt{ab}; \quad \frac{\sqrt{a}}{\sqrt{b}} = \sqrt{\frac{a}{b}}.$$

$$\sqrt{11} \cdot \sqrt{3x} = \sqrt{33x};$$

$$\frac{\sqrt{140y}}{\sqrt{7}} = \sqrt{\frac{140y}{7}} = \sqrt{20y} = \sqrt{4 \cdot 5y} = 2\sqrt{5y}$$

  **4.** Rationalizing the denominator:
   **a.** If the denominator is one term,

$$\frac{5z}{\sqrt{11}} = \frac{5z \cdot \sqrt{11}}{\sqrt{11} \cdot \sqrt{11}} = \frac{5z\sqrt{11}}{11}$$

**b.** If the denominator has two terms, multiply by 1 in the form of $\dfrac{\text{conjugate of denominator}}{\text{conjugate of denominator}}$.

$$\frac{13}{3 + \sqrt{2}} = \frac{13}{3 + \sqrt{2}} \cdot \frac{3 - \sqrt{2}}{3 - \sqrt{2}}$$

$$= \frac{13(3 - \sqrt{2})}{9 - 2} = \frac{13(3 - \sqrt{2})}{7}$$

**II.** Solving Equations

    **A.** Linear Equations (Section 2.3)

    **B.** Quadratic and Higher Degree Equations (Section 6.6)

    **C.** Equations with Rational Expressions (Section 7.5)

    **D.** Proportions (Section 7.6)

    **E.** Absolute Value Equations (Section 9.2)

**III.** Solving Inequalities

    **A.** Linear Inequalities (Section 2.8)

    **B.** Compound Inequalities (Section 9.1)

    **C.** Absolute Value Inequalities (Section 9.3)

*Perform indicated operations and simplify. If necessary, rationalize the denominator.*

**1.** $\sqrt{56}$

**2.** $\sqrt{\dfrac{20x^5}{49}}$

**3.** $(-5x^{12}y^{-3})(3x^{-7}y^{14})$

**4.** $\sqrt{\dfrac{10}{11}}$

**5.** $\dfrac{8}{\sqrt{5} - 1}$

**6.** $\dfrac{1}{2}(6x^2 - 4) + \dfrac{1}{3}(6x^2 - 9) - 14$

**7.** $\dfrac{\sqrt{13}}{\sqrt{2x^5}}$

**8.** $\dfrac{y}{y^2 + 1} - \dfrac{2y - 6}{y^2 + 1}$

**9.** $\dfrac{5x^3 + 20x}{10x - 10y} \div \dfrac{2x^2 + 8}{x^2 - y^2}$

**10.** $\sqrt[3]{16y^{20}}$

# INTEGRATED REVIEW RADICALS AND RATIONAL EXPONENTS

Sections 10.1–10.5

*Find each root. Throughout this review, assume that all variables represent positive real numbers.*

**1.** $\sqrt{81}$      **2.** $\sqrt[3]{-8}$      **3.** $\sqrt[4]{\dfrac{1}{16}}$      **4.** $\sqrt{x^6}$

**5.** $\sqrt[3]{y^9}$      **6.** $\sqrt{4y^{10}}$      **7.** $\sqrt[5]{-32y^5}$      **8.** $\sqrt[4]{81b^{12}}$

*Use radical notation to rewrite each expression. Simplify if possible.*

**9.** $36^{1/2}$      **10.** $(3y)^{1/4}$      **11.** $64^{-2/3}$      **12.** $(x + 1)^{3/5}$

*Use the properties of exponents to simplify each expression. Write with positive exponents.*

**13.** $y^{-1/6} \cdot y^{7/6}$      **14.** $\dfrac{(2x^{1/3})^4}{x^{5/6}}$      **15.** $\dfrac{x^{1/4}x^{3/4}}{x^{-1/4}}$      **16.** $4^{1/3} \cdot 4^{2/5}$

*Use rational exponents to simplify each radical.*

**17.** $\sqrt[3]{8x^6}$      **18.** $\sqrt[12]{a^9b^6}$

*Use rational exponents to write each as a single radical expression.*

**19.** $\sqrt[4]{x} \cdot \sqrt{x}$      **20.** $\sqrt{5} \cdot \sqrt[3]{2}$

*Simplify.*

**21.** $\sqrt{40}$      **22.** $\sqrt[4]{16x^7y^{10}}$      **23.** $\sqrt[3]{54x^4}$      **24.** $\sqrt[5]{-64b^{10}}$

*Multiply or divide. Then simplify if possible.*

**25.** $\sqrt{5} \cdot \sqrt{x}$        **26.** $\sqrt[3]{8x} \cdot \sqrt[3]{8x^2}$        **27.** $\dfrac{\sqrt{98y^6}}{\sqrt{2y}}$        **28.** $\dfrac{\sqrt[4]{48a^9b^3}}{\sqrt[4]{ab^3}}$

*Perform each indicated operation.*

**29.** $\sqrt{20} - \sqrt{75} + 5\sqrt{7}$        **30.** $\sqrt[3]{54y^4} - y\sqrt[3]{16y}$        **31.** $\sqrt{3}\left(\sqrt{5} - \sqrt{2}\right)$

**32.** $\left(\sqrt{7} + \sqrt{3}\right)^2$        **33.** $\left(2x - \sqrt{5}\right)\left(2x + \sqrt{5}\right)$        **34.** $\left(\sqrt{x+1} - 1\right)^2$

*Rationalize each denominator.*

**35.** $\sqrt{\dfrac{7}{3}}$        **36.** $\dfrac{5}{\sqrt[3]{2x^2}}$        **37.** $\dfrac{\sqrt{3} - \sqrt{7}}{2\sqrt{3} + \sqrt{7}}$

*Rationalize each numerator.*

**38.** $\sqrt{\dfrac{7}{3}}$        **39.** $\sqrt[3]{\dfrac{9y}{11}}$        **40.** $\dfrac{\sqrt{x} - 2}{\sqrt{x}}$

---

## 10.6 RADICAL EQUATIONS AND PROBLEM SOLVING

**OBJECTIVES**

1 Solve equations that contain radical expressions.

2 Use the Pythagorean theorem to model problems.

**OBJECTIVE 1 ▶ Solving equations that contain radical expressions.** In this section, we present techniques to solve equations containing radical expressions such as

$$\sqrt{2x - 3} = 9$$

We use the power rule to help us solve these radical equations.

> **Power Rule**
>
> If both sides of an equation are raised to the same power, **all** solutions of the original equation are **among** the solutions of the new equation.

This property *does not* say that raising both sides of an equation to a power yields an equivalent equation. A solution of the new equation *may or may not* be a solution of the original equation. For example, $(-2)^2 = 2^2$, but $-2 \neq 2$. Thus, *each solution of the new equation must be checked* to make sure it is a solution of the original equation. Recall that a proposed solution that is not a solution of the original equation is called an **extraneous solution.**

**EXAMPLE 1**    Solve: $\sqrt{2x - 3} = 9$.

*Solution*    We use the power rule to square both sides of the equation to eliminate the radical.

$$\sqrt{2x - 3} = 9$$
$$\left(\sqrt{2x - 3}\right)^2 = 9^2$$
$$2x - 3 = 81$$
$$2x = 84$$
$$x = 42$$

Now we, check the solution in the original equation.

**Check:**

$$\sqrt{2x - 3} = 9$$
$$\sqrt{2(42) - 3} \stackrel{?}{=} 9 \quad \text{Let } x = 42.$$
$$\sqrt{84 - 3} \stackrel{?}{=} 9$$
$$\sqrt{81} \stackrel{?}{=} 9$$
$$9 = 9 \quad \text{True}$$

The solution checks, so we conclude that the solution is 42 or the solution set is $\{42\}$.  □

**PRACTICE**
**1**   Solve: $\sqrt{3x - 5} = 7$.

To solve a radical equation, first isolate a radical on one side of the equation.

**EXAMPLE 2**   Solve: $\sqrt{-10x - 1} + 3x = 0$.

*Solution*   First, isolate the radical on one side of the equation. To do this, we subtract $3x$ from both sides.

$$\sqrt{-10x - 1} + 3x = 0$$
$$\sqrt{-10x - 1} + 3x - 3x = 0 - 3x$$
$$\sqrt{-10x - 1} = -3x$$

Next we use the power rule to eliminate the radical.

$$\left(\sqrt{-10x - 1}\right)^2 = (-3x)^2$$
$$-10x - 1 = 9x^2$$

Since this is a quadratic equation, we can set the equation equal to 0 and try to solve by factoring.

$$9x^2 + 10x + 1 = 0$$
$$(9x + 1)(x + 1) = 0 \quad \text{Factor.}$$
$$9x + 1 = 0 \quad \text{or} \quad x + 1 = 0 \quad \text{Set each factor equal to 0.}$$
$$x = -\frac{1}{9} \quad \text{or} \quad x = -1$$

**Check:**  Let $x = -\frac{1}{9}$.                                Let $x = -1$.

$$\sqrt{-10x - 1} + 3x = 0 \qquad\qquad \sqrt{-10x - 1} + 3x = 0$$

$$\sqrt{-10\left(-\frac{1}{9}\right) - 1} + 3\left(-\frac{1}{9}\right) \stackrel{?}{=} 0 \qquad \sqrt{-10(-1) - 1} + 3(-1) \stackrel{?}{=} 0$$

$$\sqrt{\frac{10}{9} - \frac{9}{9}} - \frac{3}{9} \stackrel{?}{=} 0 \qquad\qquad \sqrt{10 - 1} - 3 \stackrel{?}{=} 0$$

$$\sqrt{\frac{1}{9}} - \frac{1}{3} \stackrel{?}{=} 0 \qquad\qquad\qquad \sqrt{9} - 3 \stackrel{?}{=} 0$$

$$\frac{1}{3} - \frac{1}{3} = 0 \quad \text{True} \qquad\qquad\qquad 3 - 3 = 0 \quad \text{True}$$

Both solutions check. The solutions are $-\frac{1}{9}$ and $-1$ or the solution set is $\left\{-\frac{1}{9}, -1\right\}$.

□

**PRACTICE**
**2**   Solve: $\sqrt{3 - 2x} - 4x = 0$.

The following steps may be used to solve a radical equation.

> **Solving a Radical Equation**
> **STEP 1.** Isolate one radical on one side of the equation.
> **STEP 2.** Raise each side of the equation to a power equal to the index of the radical and simplify.
> **STEP 3.** If the equation still contains a radical term, repeat Steps 1 and 2. If not, solve the equation.
> **STEP 4.** Check all proposed solutions in the original equation.

**EXAMPLE 3**   Solve: $\sqrt[3]{x + 1} + 5 = 3$.

*Solution*   First we isolate the radical by subtracting 5 from both sides of the equation.

$$\sqrt[3]{x + 1} + 5 = 3$$
$$\sqrt[3]{x + 1} = -2$$

Next we raise both sides of the equation to the third power to eliminate the radical.

$$\left(\sqrt[3]{x + 1}\right)^3 = (-2)^3$$
$$x + 1 = -8$$
$$x = -9$$

The solution checks in the original equation, so the solution is $-9$.     □

**PRACTICE**
**3**   Solve: $\sqrt[3]{x - 2} + 1 = 3$.

**EXAMPLE 4**   Solve: $\sqrt{4 - x} = x - 2$.

*Solution*

$$\sqrt{4 - x} = x - 2$$
$$\left(\sqrt{4 - x}\right)^2 = (x - 2)^2$$
$$4 - x = x^2 - 4x + 4$$
$$x^2 - 3x = 0 \qquad \text{Write the quadratic equation in standard form.}$$
$$x(x - 3) = 0 \qquad \text{Factor.}$$
$$x = 0 \quad \text{or} \quad x - 3 = 0 \qquad \text{Set each factor equal to 0.}$$
$$x = 3$$

**Check:**

| $\sqrt{4 - x} = x - 2$ | $\sqrt{4 - x} = x - 2$ |
|---|---|
| $\sqrt{4 - 0} \overset{?}{=} 0 - 2$   Let $x = 0$. | $\sqrt{4 - 3} \overset{?}{=} 3 - 2$   Let $x = 3$. |
| $2 = -2$   False | $1 = 1$   True |

The proposed solution 3 checks, but 0 does not. Since 0 is an extraneous solution, the only solution is 3.     □

**PRACTICE**
**4**   Solve: $\sqrt{16 + x} = x - 4$.

> ▶ **Helpful Hint**
> In Example 4, notice that $(x - 2)^2 = x^2 - 4x + 4$. Make sure binomials are squared correctly.

**Concept Check** ✓

How can you immediately tell that the equation $\sqrt{2y + 3} = -4$ has no real solution?

**EXAMPLE 5**   Solve: $\sqrt{2x + 5} + \sqrt{2x} = 3$.

_Solution_   We get one radical alone by subtracting $\sqrt{2x}$ from both sides.

$$\sqrt{2x + 5} + \sqrt{2x} = 3$$
$$\sqrt{2x + 5} = 3 - \sqrt{2x}$$

Now we use the power rule to begin eliminating the radicals. First we square both sides.

$$\left(\sqrt{2x + 5}\right)^2 = \left(3 - \sqrt{2x}\right)^2$$
$$2x + 5 = 9 - 6\sqrt{2x} + 2x \quad \text{Multiply } \left(3 - \sqrt{2x}\right)\left(3 - \sqrt{2x}\right).$$

There is still a radical in the equation, so we get a radical alone again. Then we square both sides.

$$\begin{aligned} 2x + 5 &= 9 - 6\sqrt{2x} + 2x \quad && \text{Get the radical alone.}\\ 6\sqrt{2x} &= 4\\ 36(2x) &= 16 \quad && \text{Square both sides of the equation to eliminate the radical.}\\ 72x &= 16 \quad && \text{Multiply.}\\ x &= \frac{16}{72} \quad && \text{Solve.}\\ x &= \frac{2}{9} \quad && \text{Simplify.} \end{aligned}$$

The proposed solution, $\frac{2}{9}$, checks in the original equation. The solution is $\frac{2}{9}$.   □

**PRACTICE**
**5**   Solve: $\sqrt{8x + 1} + \sqrt{3x} = 2$.

▶ **Helpful Hint**
Make sure expressions are squared correctly. In Example 5, we squared $\left(3 - \sqrt{2x}\right)$ as

$$\left(3 - \sqrt{2x}\right)^2 = \left(3 - \sqrt{2x}\right)\left(3 - \sqrt{2x}\right)$$
$$= 3 \cdot 3 - 3\sqrt{2x} - 3\sqrt{2x} + \sqrt{2x} \cdot \sqrt{2x}$$
$$= 9 - 6\sqrt{2x} + 2x$$

**Concept Check** ✓

What is wrong with the following solution?

$$\begin{aligned} \sqrt{2x + 5} + \sqrt{4 - x} &= 8\\ \left(\sqrt{2x + 5} + \sqrt{4 - x}\right)^2 &= 8^2\\ (2x + 5) + (4 - x) &= 64\\ x + 9 &= 64\\ x &= 55 \end{aligned}$$

**Answers to Concept Checks:**
answers may vary
$\left(\sqrt{2x + 5} + \sqrt{4 - x}\right)^2$ is not $(2x + 5) + (4 - x)$.

**OBJECTIVE 2** ▶ **Using the pythagorean theorem.**   Recall that the Pythagorean theorem states that in a right triangle, the length of the hypotenuse squared equals the sum of the lengths of each of the legs squared.

**Pythagorean Theorem**

If $a$ and $b$ are the lengths of the legs of a right triangle and $c$ is the length of the hypotenuse, then $a^2 + b^2 = c^2$.

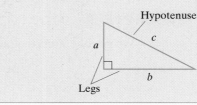

Hypotenuse

$c$

$a$

$b$

Legs

△ **EXAMPLE 6**  Find the length of the unknown leg of the right triangle.

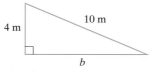

10 m

4 m

$b$

*Solution*  In the formula $a^2 + b^2 = c^2$, $c$ is the hypotenuse. Here, $c = 10$, the length of the hypotenuse, and $a = 4$. We solve for $b$. Then $a^2 + b^2 = c^2$ becomes

$$4^2 + b^2 = 10^2$$
$$16 + b^2 = 100$$
$$b^2 = 84 \quad \text{Subtract 16 from both sides.}$$
$$b = \pm\sqrt{84} = \pm\sqrt{4 \cdot 21} = \pm 2\sqrt{21}$$

Since $b$ is a length and thus is positive, we will use the positive value only. The unknown leg of the triangle is $2\sqrt{21}$ meters long.  □

**PRACTICE**
**6**  Find the length of the unknown leg of the right triangle.

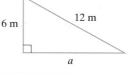

12 m

6 m

$a$

△ **EXAMPLE 7**  **Calculating Placement of a Wire**

A 50-foot supporting wire is to be attached to a 75-foot antenna. Because of surrounding buildings, sidewalks, and roadways, the wire must be anchored exactly 20 feet from the base of the antenna.

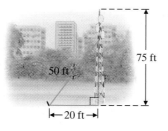

50 ft

75 ft

←20 ft→

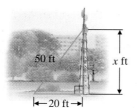

50 ft

$x$ ft

←20 ft→

**a.** How high from the base of the antenna is the wire attached?

**b.** Local regulations require that a supporting wire be attached at a height no less than $\frac{3}{5}$ of the total height of the antenna. From part **a,** have local regulations been met?

### *Solution*

1. UNDERSTAND. Read and reread the problem. From the diagram we notice that a right triangle is formed with hypotenuse 50 feet and one leg 20 feet. Let $x$ be the height from the base of the antenna to the attached wire.

2. TRANSLATE. Use the Pythagorean theorem.

$$a^2 + b^2 = c^2$$
$$20^2 + x^2 = 50^2 \quad a = 20, c = 50$$

3. SOLVE.

$$20^2 + x^2 = 50^2$$
$$400 + x^2 = 2500$$
$$x^2 = 2100 \qquad \text{Subtract 400 from both sides.}$$
$$x = \pm\sqrt{2100}$$
$$= \pm 10\sqrt{21}$$

4. INTERPRET. *Check* the work and *state* the solution.

**Check:**   We will use only the positive value, $x = 10\sqrt{21}$ because $x$ represents length. The wire is attached exactly $10\sqrt{21}$ feet from the base of the pole, or approximately 45.8 feet.

**State:**   The supporting wire must be attached at a height no less than $\frac{3}{5}$ of the total height of the antenna. This height is $\frac{3}{5}$ (75 feet), or 45 feet. Since we know from part **a** that the wire is to be attached at a height of approximately 45.8 feet, local regulations have been met.                    ☐

**PRACTICE**

**7**   Keith Robinson bought two Siamese fighting fish, but when he got home he found he only had one rectangular tank that was 12 in. long, 7 in. wide, and 5 in. deep. Since the fish must be kept separated, he needed to insert a plastic divider in the diagonal of the tank. He already has a piece that is 5 in. in one dimension, but how long must it be to fit corner to corner in the tank?

---

## Graphing Calculator Explorations

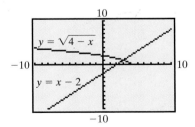

We can use a graphing calculator to solve radical equations. For example, to use a graphing calculator to approximate the solutions of the equation solved in Example 4, we graph the following.

$$Y_1 = \sqrt{4 - x} \qquad \text{and} \qquad Y_2 = x - 2$$

The $x$-value of the point of intersection is the solution. Use the Intersect feature or the Zoom and Trace features of your graphing calculator to see that the solution is 3.

*Use a graphing calculator to solve each radical equation. Round all solutions to the nearest hundredth.*

1. $\sqrt{x + 7} = x$

2. $\sqrt{3x + 5} = 2x$

3. $\sqrt{2x + 1} = \sqrt{2x + 2}$

4. $\sqrt{10x - 1} = \sqrt{-10x + 10} - 1$

5. $1.2x = \sqrt{3.1x + 5}$

6. $\sqrt{1.9x^2 - 2.2} = -0.8x + 3$

## VOCABULARY & READINESS CHECK

*Use the choices below to fill in each blank. Not all choices will be used.*

| | | | |
|---|---|---|---|
| hypotenuse | right | $x^2 + 25$ | $16 - 8\sqrt{7x} + 7x$ |
| extraneous solution | legs | $x^2 - 10x + 25$ | $16 + 7x$ |

**1.** A proposed solution that is not a solution of the original equation is called an _____ .

**2.** The Pythagorean Theorem states that $a^2 + b^2 = c^2$ where $a$ and $b$ are the lengths of the _____ of a _____ triangle and $c$ is the length of the _____ .

**3.** The square of $x - 5$, or $(x - 5)^2 = $ _____ .

**4.** The square of $4 - \sqrt{7x}$, or $(4 - \sqrt{7x})^2 = $ _____ .

## 10.6 | EXERCISE SET   MyMathLab   MathXL PRACTICE  WATCH  DOWNLOAD  READ  REVIEW

*Solve. See Examples 1 and 2.*

**1.** $\sqrt{2x} = 4$

**2.** $\sqrt{3x} = 3$

**3.** $\sqrt{x - 3} = 2$

**4.** $\sqrt{x + 1} = 5$

**5.** $\sqrt{2x} = -4$

**6.** $\sqrt{5x} = -5$

**7.** $\sqrt{4x - 3} - 5 = 0$

**8.** $\sqrt{x - 3} - 1 = 0$

**9.** $\sqrt{2x - 3} - 2 = 1$

**10.** $\sqrt{3x + 3} - 4 = 8$

*Solve. See Example 3.*

**11.** $\sqrt[3]{6x} = -3$

**12.** $\sqrt[3]{4x} = -2$

**13.** $\sqrt[3]{x - 2} - 3 = 0$

**14.** $\sqrt[3]{2x - 6} - 4 = 0$

*Solve. See Examples 4 and 5.*

**15.** $\sqrt{13 - x} = x - 1$

**16.** $\sqrt{2x - 3} = 3 - x$

**17.** $x - \sqrt{4 - 3x} = -8$

**18.** $2x + \sqrt{x + 1} = 8$

**19.** $\sqrt{y + 5} = 2 - \sqrt{y - 4}$

**20.** $\sqrt{x + 3} + \sqrt{x - 5} = 3$

**21.** $\sqrt{x - 3} + \sqrt{x + 2} = 5$

**22.** $\sqrt{2x - 4} - \sqrt{3x + 4} = -2$

### MIXED PRACTICE

*Solve. See Examples 1 through 5.*

**23.** $\sqrt{3x - 2} = 5$

**24.** $\sqrt{5x - 4} = 9$

**25.** $-\sqrt{2x} + 4 = -6$

**26.** $-\sqrt{3x + 9} = -12$

**27.** $\sqrt{3x + 1} + 2 = 0$

**28.** $\sqrt{3x + 1} - 2 = 0$

**29.** $\sqrt[4]{4x + 1} - 2 = 0$

**30.** $\sqrt[4]{2x - 9} - 3 = 0$

**31.** $\sqrt{4x - 3} = 7$

**32.** $\sqrt{3x + 9} = 6$

**33.** $\sqrt[3]{6x - 3} - 3 = 0$

**34.** $\sqrt[3]{3x} + 4 = 7$

**35.** $\sqrt[3]{2x - 3} - 2 = -5$

**36.** $\sqrt[3]{x - 4} - 5 = -7$

**37.** $\sqrt{x + 4} = \sqrt{2x - 5}$

**38.** $\sqrt{3y + 6} = \sqrt{7y - 6}$

**39.** $x - \sqrt{1 - x} = -5$

**40.** $x - \sqrt{x - 2} = 4$

**41.** $\sqrt[3]{-6x - 1} = \sqrt[3]{-2x - 5}$

**42.** $\sqrt[3]{-4x - 3} = \sqrt[3]{-x - 15}$

**43.** $\sqrt{5x - 1} - \sqrt{x + 2} = 3$

**44.** $\sqrt{2x - 1} - 4 = -\sqrt{x - 4}$

**45.** $\sqrt{2x - 1} = \sqrt{1 - 2x}$

**46.** $\sqrt{7x - 4} = \sqrt{4 - 7x}$

**47.** $\sqrt{3x + 4} - 1 = \sqrt{2x + 1}$

**48.** $\sqrt{x - 2} + 3 = \sqrt{4x + 1}$

**49.** $\sqrt{y + 3} - \sqrt{y - 3} = 1$

**50.** $\sqrt{x + 1} - \sqrt{x - 1} = 2$

*Find the length of the unknown side of each triangle. See Example 6.*

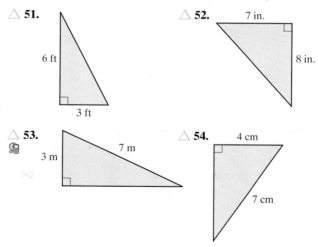

△ **51.**    6 ft    3 ft

△ **52.**    7 in.    8 in.

△ **53.**    3 m    7 m

△ **54.**    4 cm    7 cm

*Find the length of the unknown side of each triangle. Give the exact length and a one-decimal-place approximation. See Example 6.*

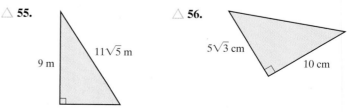

△ **55.**    9 m    $11\sqrt{5}$ m

△ **56.**    $5\sqrt{3}$ cm    10 cm

**57.**

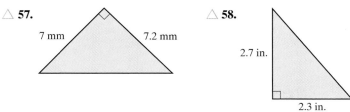

7 mm            7.2 mm

**58.**

2.7 in.

2.3 in.

*Solve. See Example 7. Give exact answers and two-decimal-place approximations where appropriate.*

△ **59.** A wire is needed to support a vertical pole 15 feet high. The cable will be anchored to a stake 8 feet from the base of the pole. How much cable is needed?

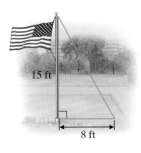

15 ft

8 ft

△ **60.** The tallest structure in the United States is a TV tower in Blanchard, North Dakota. Its height is 2063 feet. A 2382-foot length of wire is to be used as a guy wire attached to the top of the tower. Approximate to the nearest foot how far from the base of the tower the guy wire must be anchored. (*Source:* U.S. Geological Survey)

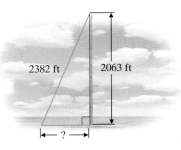

2382 ft        2063 ft

? 

△ **61.** A spotlight is mounted on the eaves of a house 12 feet above the ground. A flower bed runs between the house and the sidewalk, so the closest the ladder can be placed to the house is 5 feet. How long a ladder is needed so that an electrician can reach the place where the light is mounted?

12 ft

5 ft

△ **62.** A wire is to be attached to support a telephone pole. Because of surrounding buildings, sidewalks, and roadways, the wire must be anchored exactly 15 feet from the base of the pole. Telephone company workers have only 30 feet of cable, and 2 feet of that must be used to attach the cable to the pole and

to the stake on the ground. How high from the base of the pole can the wire be attached?

15 ft

△ **63.** The radius of the Moon is 1080 miles. Use the formula for the radius *r* of a sphere given its surface area *A*,

$$r = \sqrt{\frac{A}{4\pi}}$$

to find the surface area of the Moon. Round to the nearest square mile. (*Source:* National Space Science Data Center)

**64.** Police departments find it very useful to be able to approximate the speed of a car when they are given the distance that the car skidded before it came to a stop. If the road surface is wet concrete, the function $S(x) = \sqrt{10.5x}$ is used, where $S(x)$ is the speed of the car in miles per hour and *x* is the distance skidded in feet. Find how fast a car was moving if it skidded 280 feet on wet concrete.

**65.** The formula $v = \sqrt{2gh}$ gives the velocity *v*, in feet per second, of an object when it falls *h* feet accelerated by gravity *g*, in feet per second squared. If *g* is approximately 32 feet per second squared, find how far an object has fallen if its velocity is 80 feet per second.

**66.** Two tractors are pulling a tree stump from a field. If two forces *A* and *B* pull at right angles (90°) to each other, the size of the resulting force *R* is given by the formula $R = \sqrt{A^2 + B^2}$. If tractor *A* is exerting 600 pounds of force and the resulting force is 850 pounds, find how much force tractor *B* is exerting.

600 lb        ?

*In psychology, it has been suggested that the number S of nonsense syllables that a person can repeat consecutively depends on his or her IQ score I according to the equation $S = 2\sqrt{I} - 9$.*

**67.** Use this relationship to estimate the IQ of a person who can repeat 11 nonsense syllables consecutively.

**68.** Use this relationship to estimate the IQ of a person who can repeat 15 nonsense syllables consecutively.

*The **period** of a pendulum is the time it takes for the pendulum to make one full back-and-forth swing. The period of a pendulum depends on the length of the pendulum. The formula for the period P, in seconds, is $P = 2\pi\sqrt{\dfrac{l}{32}}$, where l is the length of the pendulum in feet. Use this formula for Exercises 69 through 74.*

**69.** Find the period of a pendulum whose length is 2 feet. Give an exact answer and a two-decimal-place approximation.

2 feet

**70.** Klockit sells a 43-inch lyre pendulum. Find the period of this pendulum. Round your answer to 2 decimal places. (*Hint:* First convert inches to feet.)

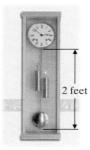

**71.** Find the length of a pendulum whose period is 4 seconds. Round your answer to 2 decimal places.

**72.** Find the length of a pendulum whose period is 3 seconds. Round your answer to 2 decimal places.

**73.** Study the relationship between period and pendulum length in Exercises 69 through 72 and make a conjecture about this relationship.

**74.** Galileo experimented with pendulums. He supposedly made conjectures about pendulums of equal length with different bob weights. Try this experiment. Make two pendulums 3 feet long. Attach a heavy weight (lead) to one and a light weight (a cork) to the other. Pull both pendulums back the same angle measure and release. Make a conjecture from your observations. (There is more about pendulums in the Chapter 7 Group Activity.)

*If the three lengths of the sides of a triangle are known, Heron's formula can be used to find its area. If a, b, and c are the three lengths of the sides, Heron's formula for area is*

$$A = \sqrt{s(s - a)(s - b)(s - c)}$$

*where s is half the perimeter of the triangle, or $s = \dfrac{1}{2}(a + b + c)$. Use this formula to find the area of each triangle. Give an exact answer and then a two-decimal place approximation.*

△ **75.**

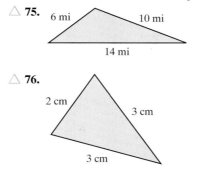

6 mi    10 mi

14 mi

△ **76.**

2 cm    3 cm

3 cm

**77.** Describe when Heron's formula might be useful.

**78.** In your own words, explain why you think s in Heron's formula is called the *semiperimeter*.

*The maximum distance D(h) in kilometers that a person can see from a height h kilometers above the ground is given by the function $D(h) = 111.7\sqrt{h}$. Use this function for Exercises 79 and 80. Round your answers to two decimal places.*

**79.** Find the height that would allow a person to see 80 kilometers.

**80.** Find the height that would allow a person to see 40 kilometers.

**REVIEW AND PREVIEW**

*Use the vertical line test to determine whether each graph represents the graph of a function. See Section 3.6.*

**81.**

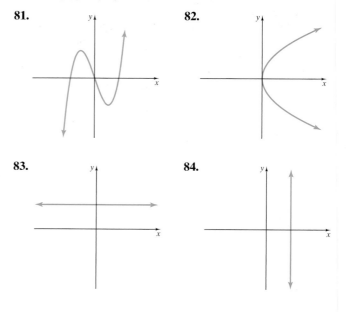

**82.**

**83.**

**84.**

**85.**

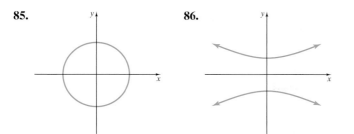

**86.**

*Simplify. See Section 7.7.*

**87.** $\dfrac{\dfrac{x}{6}}{\dfrac{2x}{3} + \dfrac{1}{2}}$

**88.** $\dfrac{\dfrac{1}{y} + \dfrac{4}{5}}{\dfrac{-3}{20}}$

**89.** $\dfrac{\dfrac{z}{5} + \dfrac{1}{10}}{\dfrac{z}{20} - \dfrac{z}{5}}$

**90.** $\dfrac{\dfrac{1}{y} + \dfrac{1}{x}}{\dfrac{1}{y} - \dfrac{1}{x}}$

### CONCEPT EXTENSIONS

**91.** Find the error in the following solution and correct. See the Concept Check in this section.

$$\sqrt{5x - 1} + 4 = 7$$
$$(\sqrt{5x - 1} + 4)^2 = 7^2$$
$$5x - 1 + 16 = 49$$
$$5x = 34$$
$$x = \frac{34}{5}$$

**92.** Explain why proposed solutions of radical equations must be checked.

**93.** Solve: $\sqrt{\sqrt{x + 3} + \sqrt{x}} = \sqrt{3}$

**94.** The cost $C(x)$ in dollars per day to operate a small delivery service is given by $C(x) = 80\sqrt[3]{x} + 500$, where $x$ is the number of deliveries per day. In July, the manager decides that it is necessary to keep delivery costs below \$1620.00. Find the greatest number of deliveries this company can make per day and still keep overhead below \$1620.00.

**95.** Consider the equations $\sqrt{2x} = 4$ and $\sqrt[3]{2x} = 4$.

  **a.** Explain the difference in solving these equations.

  **b.** Explain the similarity in solving these equations.

**Example**

For Exercises 96 through 99, see the example below.

Solve $(t^2 - 3t) - 2\sqrt{t^2 - 3t} = 0$.

*Solution*

Substitution can be used to make this problem somewhat simpler. Since $t^2 - 3t$ occurs more than once, let $x = t^2 - 3t$.

$$(t^2 - 3t) - 2\sqrt{t^2 - 3t} = 0$$
$$x - 2\sqrt{x} = 0$$
$$x = 2\sqrt{x}$$
$$x^2 = (2\sqrt{x})^2$$
$$x^2 = 4x$$
$$x^2 - 4x = 0$$
$$x(x - 4) = 0$$
$$x = 0 \quad \text{or} \quad x - 4 = 0$$
$$x = 4$$

Now we "undo" the substitution.

$x = 0$  Replace $x$ with $t^2 - 3t$.

$$t^2 - 3t = 0$$
$$t(t - 3) = 0$$
$$t = 0 \quad \text{or} \quad t - 3 = 0$$
$$t = 3$$

$x = 4$  Replace $x$ with $t^2 - 3t$.

$$t^2 - 3t = 4$$
$$t^2 - 3t - 4 = 0$$
$$(t - 4)(t + 1) = 0$$
$$t - 4 = 0 \quad \text{or} \quad t + 1 = 0$$
$$t = 4 \qquad\qquad t = -1$$

In this problem, we have four possible solutions: $0, 3, 4$, and $-1$. All four solutions check in the original equation, so the solutions are $-1, 0, 3, 4$.

*Solve. See the preceding example.*

**96.** $3\sqrt{x^2 - 8x} = x^2 - 8x$

**97.** $\sqrt{(x^2 - x) + 7} = 2(x^2 - x) - 1$

**98.** $7 - (x^2 - 3x) = \sqrt{(x^2 - 3x)} + 5$

**99.** $x^2 + 6x = 4\sqrt{x^2 + 6x}$

---

### THE BIGGER PICTURE    SIMPLIFYING EXPRESSIONS AND SOLVING EQUATIONS AND INEQUALITIES

We now continue the outline from Sections 1.7, 2.3, 2.8, 5.6, 6.6, 7.4, 7.6, 9.1, 9.3, and 10.5. Although suggestions are given, as usual, this outline should be in your own words. Try to include "how to recognize" and "how to begin to solve" in your outline. Once you complete this new portion, try the exercises.

I. Simplifying Expressions (See Section 10.5, page 614)

II. Solving Equations
   A. Linear Equations (Section 2.3)
   B. Quadratic and Higher Degree Equations (Section 6.6)
   C. Equations with Rational Expressions (Section 7.5)
   D. Proportions (Section 7.6)
   E. Absolute Value Equations (Section 9.2)
   **F. Equations with Radicals:** Equation contains at least one root of a variable expression.

$$\sqrt{5x + 10} - 2 = x \qquad \text{Radical equation}$$
$$\sqrt{5x + 10} = x + 2 \qquad \text{Isolate the radical.}$$
$$\left(\sqrt{5x + 10}\right)^2 = (x + 2)^2 \qquad \text{Square both sides.}$$
$$5x + 10 = x^2 + 4x + 4 \qquad \text{Simplify.}$$
$$0 = x^2 - x - 6 \qquad \text{Write in standard form.}$$
$$0 = (x - 3)(x + 2) \qquad \text{Factor.}$$
$$x - 3 = 0 \quad \text{or} \quad x + 2 = 0 \qquad \text{Set each factor equal to 0.}$$
$$x = 3 \quad \text{or} \quad x = -2 \qquad \text{Solve.}$$

Both solutions check.

III. Solving Inequalities
   A. Linear Inequalities (Sec. 2.8)
   B. Compound Inequalities (Sec. 9.1)
   C. Absolute Value Inequalities (Sec. 9.3)

*Solve. Write inequality solutions in interval notation.*

1. $\dfrac{x}{4} + \dfrac{x + 18}{20} = \dfrac{x - 5}{5}$

2. $|3x - 5| = 10$

3. $2x^2 - x = 45$

4. $-6 \le -5x - 1 \le 10$

5. $4(x - 1) + 3x > 1 + 2(x - 6)$

6. $\sqrt{x} + 14 = x - 6$

7. $x \ge 10 \quad \text{or} \quad -x < 5$

8. $\sqrt{3x - 1} + 4 = 1$

9. $|x - 2| > 15$

10. $5x - 4[x - 2(3x + 1)] = 25$

---

## 10.7 COMPLEX NUMBERS

### OBJECTIVES

1 Write square roots of negative numbers in the form *bi.*

2 Add or subtract complex numbers.

3 Multiply complex numbers.

4 Divide complex numbers.

5 Raise *i* to powers.

**OBJECTIVE 1 ▶ Writing numbers in the form *bi.*** Our work with radical expressions has excluded expressions such as $\sqrt{-16}$ because $\sqrt{-16}$ is not a real number; there is no real number whose square is $-16$. In this section, we discuss a number system that includes roots of negative numbers. This number system is the **complex number system,** and it includes the set of real numbers as a subset. The complex number system allows us to solve equations such as $x^2 + 1 = 0$ that have no real number solutions. The set of complex numbers includes the **imaginary unit.**

> **Imaginary Unit**
> The imaginary unit, written *i*, is the number whose square is $-1$. That is,
> $$i^2 = -1 \quad \text{and} \quad i = \sqrt{-1}$$

To write the square root of a negative number in terms of *i*, use the property that if *a* is a positive number, then

$$\sqrt{-a} = \sqrt{-1} \cdot \sqrt{a}$$
$$= i \cdot \sqrt{a}$$

Using *i*, we can write $\sqrt{-16}$ as

$$\sqrt{-16} = \sqrt{-1 \cdot 16} = \sqrt{-1} \cdot \sqrt{16} = i \cdot 4, \text{ or } 4i$$

**EXAMPLE 1**   Write with $i$ notation.

**a.** $\sqrt{-36}$          **b.** $\sqrt{-5}$          **c.** $-\sqrt{-20}$

*Solution*

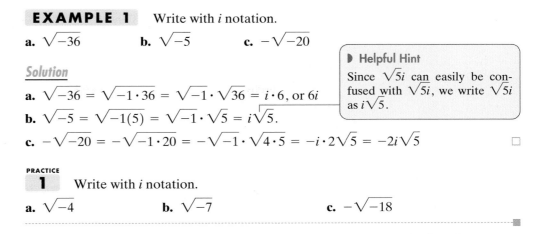

**a.** $\sqrt{-36} = \sqrt{-1 \cdot 36} = \sqrt{-1} \cdot \sqrt{36} = i \cdot 6$, or $6i$

**b.** $\sqrt{-5} = \sqrt{-1(5)} = \sqrt{-1} \cdot \sqrt{5} = i\sqrt{5}$.

**c.** $-\sqrt{-20} = -\sqrt{-1 \cdot 20} = -\sqrt{-1} \cdot \sqrt{4 \cdot 5} = -i \cdot 2\sqrt{5} = -2i\sqrt{5}$

> **▶ Helpful Hint**
> Since $\sqrt{5}i$ can easily be confused with $\sqrt{5i}$, we write $\sqrt{5}i$ as $i\sqrt{5}$.

**PRACTICE**
**1**   Write with $i$ notation.

**a.** $\sqrt{-4}$          **b.** $\sqrt{-7}$          **c.** $-\sqrt{-18}$

The product rule for radicals does not necessarily hold true for imaginary numbers. *To multiply square roots of negative numbers, first we write each number in terms of the imaginary unit i.* For example, to multiply $\sqrt{-4}$ and $\sqrt{-9}$, we first write each number in the form $bi$.

$$\sqrt{-4}\sqrt{-9} = 2i(3i) = 6i^2 = 6(-1) = -6 \quad \text{Correct}$$

We will also use this method to simplify quotients of square roots of negative numbers. Why? The product rule does not work for this example. In other words,

$$\sqrt{-4} \cdot \sqrt{-9} \neq \sqrt{(-4)(-9)} = \sqrt{36} = 6 \quad \text{Incorrect}$$

**EXAMPLE 2**   Multiply or divide as indicated.

**a.** $\sqrt{-3} \cdot \sqrt{-5}$      **b.** $\sqrt{-36} \cdot \sqrt{-1}$      **c.** $\sqrt{8} \cdot \sqrt{-2}$      **d.** $\dfrac{\sqrt{-125}}{\sqrt{5}}$

*Solution*

**a.** $\sqrt{-3} \cdot \sqrt{-5} = i\sqrt{3}(i\sqrt{5}) = i^2\sqrt{15} = -1\sqrt{15} = -\sqrt{15}$

**b.** $\sqrt{-36} \cdot \sqrt{-1} = 6i(i) = 6i^2 = 6(-1) = -6$

**c.** $\sqrt{8} \cdot \sqrt{-2} = 2\sqrt{2}(i\sqrt{2}) = 2i(\sqrt{2}\sqrt{2}) = 2i(2) = 4i$

**d.** $\dfrac{\sqrt{-125}}{\sqrt{5}} = \dfrac{i\sqrt{125}}{\sqrt{5}} = i\sqrt{25} = 5i$

**PRACTICE**
**2**   Multiply or divide as indicated.

**a.** $\sqrt{-5} \cdot \sqrt{-6}$      **b.** $\sqrt{-9} \cdot \sqrt{-1}$      **c.** $\sqrt{125} \cdot \sqrt{-5}$      **d.** $\dfrac{\sqrt{-27}}{\sqrt{3}}$

Now that we have practiced working with the imaginary unit, we define complex numbers.

> **Complex Numbers**
> A **complex number** is a number that can be written in the form $a + bi$, where $a$ and $b$ are real numbers.

Notice that the set of real numbers is a subset of the complex numbers since any real number can be written in the form of a complex number. For example,

$$16 = 16 + 0i$$

In general, a complex number $a + bi$ is a real number if $b = 0$. Also, a complex number is called a **pure imaginary number** if $a = 0$ and $b \neq 0$. For example,

$$3i = 0 + 3i \quad \text{and} \quad i\sqrt{7} = 0 + i\sqrt{7}$$

are pure imaginary numbers.

The following diagram shows the relationship between complex numbers and their subsets.

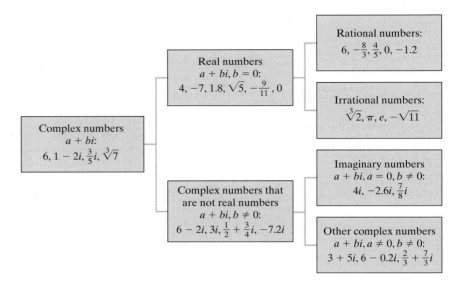

## Concept Check ✓

True or false? Every complex number is also a real number.

**OBJECTIVE 2 ▶ Adding or subtracting complex numbers.** Two complex numbers $a + bi$ and $c + di$ are equal if and only if $a = c$ and $b = d$. Complex numbers can be added or subtracted by adding or subtracting their real parts and then adding or subtracting their imaginary parts.

---

**Sum or Difference of Complex Numbers**

If $a + bi$ and $c + di$ are complex numbers, then their sum is

$$(a + bi) + (c + di) = (a + c) + (b + d)i$$

Their difference is

$$(a + bi) - (c + di) = a + bi - c - di = (a - c) + (b - d)i$$

---

**EXAMPLE 3** Add or subtract the complex numbers. Write the sum or difference in the form $a + bi$.

**a.** $(2 + 3i) + (-3 + 2i)$   **b.** $5i - (1 - i)$   **c.** $(-3 - 7i) - (-6)$

*Solution*

**a.** $(2 + 3i) + (-3 + 2i) = (2 - 3) + (3 + 2)i = -1 + 5i$

**b.** $5i - (1 - i) = 5i - 1 + i$

$\qquad\qquad = -1 + (5 + 1)i$

$\qquad\qquad = -1 + 6i$

**c.** $(-3 - 7i) - (-6) = -3 - 7i + 6$
$$= (-3 + 6) - 7i$$
$$= 3 - 7i \qquad \square$$

**PRACTICE**
**3**   Add or subtract the complex numbers. Write the sum or difference in the form $a + bi$.

**a.** $(3 - 5i) + (-4 + i)$ **b.** $4i - (3 - i)$ **c.** $(-5 - 2i) - (-8)$

**OBJECTIVE 3** ▶ **Multiplying complex numbers.** To multiply two complex numbers of the form $a + bi$, we multiply as though they are binomials. Then we use the relationship $i^2 = -1$ to simplify.

**EXAMPLE 4**   Multiply the complex numbers. Write the product in the form $a + bi$.

**a.** $-7i \cdot 3i$ **b.** $3i(2 - i)$ **c.** $(2 - 5i)(4 + i)$
**d.** $(2 - i)^2$ **e.** $(7 + 3i)(7 - 3i)$

*Solution*

**a.** $-7i \cdot 3i = -21i^2$
$$= -21(-1) \qquad \text{Replace } i^2 \text{ with } -1.$$
$$= 21$$

**b.** $3i(2 - i) = 3i \cdot 2 - 3i \cdot i$   Use the distributive property.
$$= 6i - 3i^2 \qquad \text{Multiply.}$$
$$= 6i - 3(-1) \quad \text{Replace } i^2 \text{ with } -1.$$
$$= 6i + 3$$
$$= 3 + 6i$$

Use the FOIL order below. (First, Outer, Inner, Last)
**c.** $(2 - 5i)(4 + i) = 2(4) + 2(i) - 5i(4) - 5i(i)$
$$\qquad\qquad\qquad\ \ \text{F} \quad\ \text{O} \quad\ \text{I} \quad\ \text{L}$$
$$= 8 + 2i - 20i - 5i^2$$
$$= 8 - 18i - 5(-1) \qquad\qquad i^2 = -1$$
$$= 8 - 18i + 5$$
$$= 13 - 18i$$

**d.** $(2 - i)^2 = (2 - i)(2 - i)$
$$= 2(2) - 2(i) - 2(i) + i^2$$
$$= 4 - 4i + (-1) \qquad\qquad i^2 = -1$$
$$= 3 - 4i$$

**e.** $(7 + 3i)(7 - 3i) = 7(7) - 7(3i) + 3i(7) - 3i(3i)$
$$= 49 - 21i + 21i - 9i^2$$
$$= 49 - 9(-1) \qquad\qquad\qquad i^2 = -1$$
$$= 49 + 9$$
$$= 58 \qquad \square$$

**PRACTICE**
**4**   Multiply the complex numbers. Write the product in the form $a + bi$.

**a.** $-4i \cdot 5i$ **b.** $5i(2 + i)$ **c.** $(2 + 3i)(6 - i)$
**d.** $(3 - i)^2$ **e.** $(9 + 2i)(9 - 2i)$

Notice that if you add, subtract, or multiply two complex numbers, just like real numbers, the result is a complex number.

**OBJECTIVE 4 ▶ Dividing complex numbers.** From Example 4e, notice that the product of $7 + 3i$ and $7 - 3i$ is a real number. These two complex numbers are called **complex conjugates** of one another. In general, we have the following definition.

> **Complex Conjugates**
> The complex numbers $(a + bi)$ and $(a - bi)$ are called **complex conjugates** of each other, and $(a + bi)(a - bi) = a^2 + b^2$.

To see that the product of a complex number $a + bi$ and its conjugate $a - bi$ is the real number $a^2 + b^2$, we multiply.

$$(a + bi)(a - bi) = a^2 - abi + abi - b^2i^2$$
$$= a^2 - b^2(-1)$$
$$= a^2 + b^2$$

We use complex conjugates to divide by a complex number.

**EXAMPLE 5** Divide. Write in the form $a + bi$.

a. $\dfrac{2 + i}{1 - i}$  b. $\dfrac{7}{3i}$

*Solution*

a. Multiply the numerator and denominator by the complex conjugate of $1 - i$ to eliminate the imaginary number in the denominator.

$$\frac{2 + i}{1 - i} = \frac{(2 + i)(1 + i)}{(1 - i)(1 + i)}$$
$$= \frac{2(1) + 2(i) + 1(i) + i^2}{1^2 - i^2}$$
$$= \frac{2 + 3i - 1}{1 + 1} \qquad \text{Here, } i^2 = -1.$$
$$= \frac{1 + 3i}{2} \quad \text{or} \quad \frac{1}{2} + \frac{3}{2}i$$

b. Multiply the numerator and denominator by the conjugate of $3i$. Note that $3i = 0 + 3i$, so its conjugate is $0 - 3i$ or $-3i$.

$$\frac{7}{3i} = \frac{7(-3i)}{(3i)(-3i)} = \frac{-21i}{-9i^2} = \frac{-21i}{-9(-1)} = \frac{-21i}{9} = \frac{-7i}{3} \quad \text{or} \quad 0 - \frac{7}{3}i \qquad \square$$

**PRACTICE**
**5** Divide. Write in the form $a + bi$.

a. $\dfrac{4 - i}{3 + i}$  b. $\dfrac{5}{2i}$

--------

> ▶ **Helpful Hint**
> Recall that division can be checked by multiplication.
> To check that $\dfrac{2 + i}{1 - i} = \dfrac{1}{2} + \dfrac{3}{2}i$, in Example 5a, multiply $\left(\dfrac{1}{2} + \dfrac{3}{2}i\right)(1 - i)$ to verify that the product is $2 + i$.

**OBJECTIVE 5 ▶ Finding powers of i.** We can use the fact that $i^2 = -1$ to find higher powers of $i$. To find $i^3$, we rewrite it as the product of $i^2$ and $i$.

$$i^3 = i^2 \cdot i = (-1)i = -i$$
$$i^4 = i^2 \cdot i^2 = (-1) \cdot (-1) = 1$$

We continue this process and use the fact that $i^4 = 1$ and $i^2 = -1$ to simplify $i^5$ and $i^6$.

$$i^5 = i^4 \cdot i = 1 \cdot i = i$$
$$i^6 = i^4 \cdot i^2 = 1 \cdot (-1) = -1$$

If we continue finding powers of $i$, we generate the following pattern. Notice that the values $i$, $-1$, $-i$, and $1$ repeat as $i$ is raised to higher and higher powers.

| | | |
|---|---|---|
| $i^1 = i$ | $i^5 = i$ | $i^9 = i$ |
| $i^2 = -1$ | $i^6 = -1$ | $i^{10} = -1$ |
| $i^3 = -i$ | $i^7 = -i$ | $i^{11} = -i$ |
| $i^4 = 1$ | $i^8 = 1$ | $i^{12} = 1$ |

This pattern allows us to find other powers of $i$. To do so, we will use the fact that $i^4 = 1$ and rewrite a power of $i$ in terms of $i^4$. For example,

$$i^{22} = i^{20} \cdot i^2 = (i^4)^5 \cdot i^2 = 1^5 \cdot (-1) = 1 \cdot (-1) = -1.$$

**EXAMPLE 6**   Find the following powers of $i$.

**a.** $i^7$     **b.** $i^{20}$     **c.** $i^{46}$     **d.** $i^{-12}$

*Solution*

**a.** $i^7 = i^4 \cdot i^3 = 1(-i) = -i$
**b.** $i^{20} = (i^4)^5 = 1^5 = 1$
**c.** $i^{46} = i^{44} \cdot i^2 = (i^4)^{11} \cdot i^2 = 1^{11}(-1) = -1$
**d.** $i^{-12} = \dfrac{1}{i^{12}} = \dfrac{1}{(i^4)^3} = \dfrac{1}{(1)^3} = \dfrac{1}{1} = 1$     □

**PRACTICE**
**6**   Find the following powers of $i$.

**a.** $i^9$     **b.** $i^{16}$     **c.** $i^{34}$     **d.** $i^{-24}$

# VOCABULARY & READINESS CHECK

*Use the choices below to fill in each blank. Not all choices will be used.*

| | | | |
|---|---|---|---|
| $-1$ | $\sqrt{-1}$ | real | imaginary unit |
| $1$ | $\sqrt{1}$ | complex | pure imaginary |

**1.** A _____ number is one that can be written in the form $a + bi$ where $a$ and $b$ are real numbers.
**2.** In the complex number system, $i$ denotes the _____.
**3.** $i^2 =$ _____
**4.** $i =$ _____
**5.** A complex number, $a + bi$, is a _____ number if $b = 0$.
**6.** A complex number, $a + bi$, is a _____ number if $a = 0$ and $b \neq 0$.

*Simplify. See Example 1.*

**7.** $\sqrt{-81}$          **8.** $\sqrt{-49}$          **9.** $\sqrt{-7}$          **10.** $\sqrt{-3}$
**11.** $-\sqrt{16}$         **12.** $-\sqrt{4}$          **13.** $\sqrt{-64}$         **14.** $\sqrt{-100}$

## 10.7 | EXERCISE SET

*Write in terms of i. See Example 1.*

**1.** $\sqrt{-24}$

**2.** $\sqrt{-32}$

**3.** $-\sqrt{-36}$

**4.** $-\sqrt{-121}$

**5.** $8\sqrt{-63}$

**6.** $4\sqrt{-20}$

**7.** $-\sqrt{54}$

**8.** $\sqrt{-63}$

*Multiply or divide. See Example 2.*

**9.** $\sqrt{-2} \cdot \sqrt{-7}$

**10.** $\sqrt{-11} \cdot \sqrt{-3}$

**11.** $\sqrt{-5} \cdot \sqrt{-10}$

**12.** $\sqrt{-2} \cdot \sqrt{-6}$

**13.** $\sqrt{16} \cdot \sqrt{-1}$

**14.** $\sqrt{3} \cdot \sqrt{-27}$

**15.** $\dfrac{\sqrt{-9}}{\sqrt{3}}$

**16.** $\dfrac{\sqrt{49}}{\sqrt{-10}}$

**17.** $\dfrac{\sqrt{-80}}{\sqrt{-10}}$

**18.** $\dfrac{\sqrt{-40}}{\sqrt{-8}}$

*Add or subtract. Write the sum or difference in the form a + bi. See Example 3.*

**19.** $(4 - 7i) + (2 + 3i)$

**20.** $(2 - 4i) - (2 - i)$

**21.** $(6 + 5i) - (8 - i)$

**22.** $(8 - 3i) + (-8 + 3i)$

**23.** $6 - (8 + 4i)$

**24.** $(9 - 4i) - 9$

*Multiply. Write the product in the form a + bi. See Example 4.*

**25.** $-10i \cdot -4i$

**26.** $-2i \cdot -11i$

**27.** $6i(2 - 3i)$

**28.** $5i(4 - 7i)$

**29.** $\left(\sqrt{3} + 2i\right)\left(\sqrt{3} - 2i\right)$

**30.** $\left(\sqrt{5} - 5i\right)\left(\sqrt{5} + 5i\right)$

**31.** $(4 - 2i)^2$

**32.** $(6 - 3i)^2$

*Write each quotient in the form a + bi. See Example 5.*

**33.** $\dfrac{4}{i}$

**34.** $\dfrac{5}{6i}$

**35.** $\dfrac{7}{4 + 3i}$

**36.** $\dfrac{9}{1 - 2i}$

**37.** $\dfrac{3 + 5i}{1 + i}$

**38.** $\dfrac{6 + 2i}{4 - 3i}$

**39.** $\dfrac{5 - i}{3 - 2i}$

**40.** $\dfrac{6 - i}{2 + i}$

### MIXED PRACTICE

*Perform each indicated operation. Write the result in the form a + bi.*

**41.** $(7i)(-9i)$

**42.** $(-6i)(-4i)$

**43.** $(6 - 3i) - (4 - 2i)$

**44.** $(-2 - 4i) - (6 - 8i)$

**45.** $-3i(-1 + 9i)$

**46.** $-5i(-2 + i)$

**47.** $\dfrac{4 - 5i}{2i}$

**48.** $\dfrac{6 + 8i}{3i}$

**49.** $(4 + i)(5 + 2i)$

**50.** $(3 + i)(2 + 4i)$

**51.** $(6 - 2i)(3 + i)$

**52.** $(2 - 4i)(2 - i)$

**53.** $(8 - 3i) + (2 + 3i)$

**54.** $(7 + 4i) + (4 - 4i)$

**55.** $(1 - i)(1 + i)$

**56.** $(6 + 2i)(6 - 2i)$

**57.** $\dfrac{16 + 15i}{-3i}$

**58.** $\dfrac{2 - 3i}{-7i}$

**59.** $(9 + 8i)^2$

**60.** $(4 - 7i)^2$

**61.** $\dfrac{2}{3 + i}$

**62.** $\dfrac{5}{3 - 2i}$

**63.** $(5 - 6i) - 4i$

**64.** $(6 - 2i) + 7i$

**65.** $\dfrac{2 - 3i}{2 + i}$

**66.** $\dfrac{6 + 5i}{6 - 5i}$

**67.** $(2 + 4i) + (6 - 5i)$

**68.** $(5 - 3i) + (7 - 8i)$

**69.** $(\sqrt{3} + 2i)(\sqrt{3} - 2i)$

**70.** $(\sqrt{5} - 5i)(\sqrt{5} + 5i)$

**71.** $(4 - 2i)^2$

**72.** $(6 - 3i)^2$

*Find each power of i. See Example 6.*

**73.** $i^8$

**74.** $i^{10}$

**75.** $i^{21}$

**76.** $i^{15}$

**77.** $i^{11}$

**78.** $i^{40}$

**79.** $i^{-6}$

**80.** $i^{-9}$

**81.** $(2i)^6$

**82.** $(5i)^4$

**83.** $(-3i)^5$

**84.** $(-2i)^7$

### REVIEW AND PREVIEW

*Recall that the sum of the measures of the angles of a triangle is 180°. Find the unknown angle in each triangle.*

**85.**

**86.**

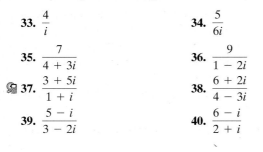

*Use synthetic division to divide the following. See Section 5.7.*

**87.** $(x^3 - 6x^2 + 3x - 4) \div (x - 1)$

**88.** $(5x^4 - 3x^2 + 2) \div (x + 2)$

*Thirty people were recently polled about their average monthly balance in their checking accounts. The results of this poll are shown in the following histogram. Use this graph to answer Exercises 89 through 94. See Section 3.1.*

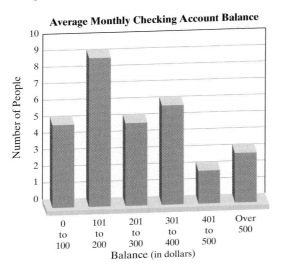

**Average Monthly Checking Account Balance**

Number of People

Balance (in dollars)

89. How many people polled reported an average checking balance of $201 to $300?

90. How many people polled reported an average checking balance of $0 to $100?

91. How many people polled reported an average checking balance of $200 or less?

92. How many people polled reported an average checking balance of $301 or more?

93. What percent of people polled reported an average checking balance of $201 to $300?

94. What percent of people polled reported an average checking balance of $0 to $100?

**CONCEPT EXTENSIONS**

*Write in the form $a + bi$.*

95. $i^3 - i^4$

96. $i^8 - i^7$

97. $i^6 + i^8$

98. $i^4 + i^{12}$

99. $2 + \sqrt{-9}$

100. $5 - \sqrt{-16}$

101. $\dfrac{6 + \sqrt{-18}}{3}$

102. $\dfrac{4 - \sqrt{-8}}{2}$

103. $\dfrac{5 - \sqrt{-75}}{10}$

104. Describe how to find the conjugate of a complex number.

105. Explain why the product of a complex number and its complex conjugate is a real number.

*Simplify.*

106. $\left(8 - \sqrt{-3}\right) - \left(2 + \sqrt{-12}\right)$

107. $\left(8 - \sqrt{-4}\right) - \left(2 + \sqrt{-16}\right)$

108. Determine whether $2i$ is a solution of $x^2 + 4 = 0$.

109. Determine whether $-1 + i$ is a solution of $x^2 + 2x = -2$.

# CHAPTER 10 GROUP ACTIVITY

## Heron of Alexandria

Heron (also Hero) was a Greek mathematician and engineer. He lived and worked in Alexandria, Egypt, around 75 A.D. During his prolific work life, Heron developed a rotary steam engine called an aeolipile, a surveying tool called a dioptra, as well as a wind organ and a fire engine. As an engineer, he must have had the need to approximate square roots because he described an iterative method for doing so in his work *Metrica*. Heron's method for approximating a square root can be summarized as follows:

Suppose that $x$ is not a perfect square and $a^2$ is the nearest perfect square to $x$. For a rough estimate of the value of $\sqrt{x}$, find the value of $y_1 = \dfrac{1}{2}\left(a + \dfrac{x}{a}\right)$. This estimate can be improved by calculating a second estimate using the first estimate $y_1$ in place of $a$: $y_2 = \dfrac{1}{2}\left(y_1 + \dfrac{x}{y_1}\right)$.

Repeating this process several times will give more and more accurate estimates of $\sqrt{x}$.

*Critical Thinking*

1. **a.** Which perfect square is closest to 80?

   **b.** Use Heron's method for approximating square roots to calculate the first estimate of the square root of 80. Give an exact decimal answer.

   **c.** Use the first estimate of the square root of 80 to find a more refined second estimate. Round this second estimate to 6 decimal places.

   **d.** Use a calculator to find the actual value of the square root of 80. List all digits shown on your calculator's display.

   **e.** Compare the actual value from part (d) to the values of the first and second estimates. What do you notice?

   **f.** How many iterations of this process are necessary to get an estimate that differs no more than one digit from the actual value recorded in part (d)?

2. Repeat Question 1 for finding an estimate of the square root of 30.

3. Repeat Question 1 for finding an estimate of the square root of 4572.

4. Why would this iterative method have been important to people of Heron's era? Would you say that this method is as important today? Why or why not?

---

### 📖 STUDY SKILLS BUILDER

**Are You Prepared for a Test on Chapter 10?**

Below I have listed some common trouble areas for students in Chapter 10. After studying for your test, but before taking your test, read these.

- Remember how to convert an expression with rational expressions to one with radicals and one with radicals to one with rational expressions.

$$7^{2/3} = \sqrt[3]{7^2} \text{ or } (\sqrt[3]{7})^2$$

$$\sqrt[5]{4^3} = 4^{3/5}$$

- Remember the difference between $\sqrt{x} + \sqrt{x}$ and $\sqrt{x} \cdot \sqrt{x}, x > 0$.

$$\sqrt{x} + \sqrt{x} = 2\sqrt{x}$$
$$\sqrt{x} \cdot \sqrt{x} = x$$

- Don't forget the difference between rationalizing the denominator of $\sqrt{\dfrac{2}{x}}$ and rationalizing the denominator of $\dfrac{\sqrt{2}}{\sqrt{x} + 1}, x > 0$.

$$\sqrt{\frac{2}{x}} = \frac{\sqrt{2}}{\sqrt{x}} = \frac{\sqrt{2} \cdot \sqrt{x}}{\sqrt{x} \cdot \sqrt{x}} = \frac{\sqrt{2x}}{x}$$

$$\frac{\sqrt{2}}{\sqrt{x} + 1} = \frac{\sqrt{2}(\sqrt{x} - 1)}{(\sqrt{x} + 1)(\sqrt{x} - 1)} = \frac{\sqrt{2}(\sqrt{x} - 1)}{x - 1}$$

- Remember that the midpoint of a segment is a *point*. The $x$-coordinate is the average of the $x$-coordinates of the endpoints of the segment and the $y$-coordinate is the average of the $y$-coordinates of the endpoints of the segment.

The midpoint of the segment joining $(-1, 5)$ and $(3, 4)$ is $\left( \dfrac{-1 + 3}{2}, \dfrac{5 + 4}{2} \right)$ or $\left( 1, \dfrac{9}{2} \right)$.

- Remember that the distance formula gives the *distance* between two points. The distance between $(-1, 5)$ and $(3, 4)$ is

$$\sqrt{(3 - (-1))^2 + (4 - 5)^2} = \sqrt{4^2 + (-1)^2}$$
$$= \sqrt{16 + 1} = \sqrt{17} \text{ units}$$

Remember: This is simply a checklist of common trouble areas. For a review of Chapter 10, see the Highlights and Chapter Review at the end of this chapter.

---

## CHAPTER 10 VOCABULARY CHECK

*Fill in each blank with one of the words or phrases listed below.*

| index | rationalizing | conjugate | principal square root | cube root | midpoint |
|---|---|---|---|---|---|
| complex number | like radicals | radicand | imaginary unit | distance | |

1. The _____ of $\sqrt{3} + 2$ is $\sqrt{3} - 2$.

2. The _____ of a nonnegative number $a$ is written as $\sqrt{a}$.

3. The process of writing a radical expression as an equivalent expression but without a radical in the denominator is called _____ the denominator.

4. The _____ written $i$, is the number whose square is $-1$.

5. The _____ of a number is written as $\sqrt[3]{a}$.

6. In the notation $\sqrt[n]{a}$, $n$ is called the _____ and $a$ is called the _____.

7. Radicals with the same index and the same radicand are called _____.

8. A _____ is a number that can be written in the form $a + bi$, where $a$ and $b$ are real numbers.

9. The _____ formula is $d = \sqrt{(x_2 - x_1)^2 + (y_2 - y_1)^2}$.

10. The _____ formula is $\left( \dfrac{x_1 + x_2}{2}, \dfrac{y_1 + y_2}{2} \right)$.

▶ **Helpful Hint**

Are you preparing for your test? Don't forget to take the Chapter 10 Test on page 641. Then check your answers at the back of the text and use the Chapter Test Prep Video CD to see the fully worked-out solutions to any of the exercises you want to review.

# CHAPTER 10 HIGHLIGHTS

| DEFINITIONS AND CONCEPTS | EXAMPLES |
|---|---|

## SECTION 10.1 RADICALS AND RADICAL FUNCTIONS

The **positive**, or **principal, square root** of a nonnegative number $a$ is written as $\sqrt{a}$.

$$\sqrt{a} = b \text{ only if } b^2 = a \text{ and } b \geq 0$$

The **negative square root of** $a$ is written as $-\sqrt{a}$.

The **cube root** of a real number $a$ is written as $\sqrt[3]{a}$.

$$\sqrt[3]{a} = b \text{ only if } b^3 = a$$

If $n$ is an even positive integer, then $\sqrt[n]{a^n} = |a|$.

If $n$ is an odd positive integer, then $\sqrt[n]{a^n} = a$.

A **radical function** in $x$ is a function defined by an expression containing a root of $x$.

$$\sqrt{36} = 6 \qquad \sqrt{\frac{9}{100}} = \frac{3}{10}$$

$$-\sqrt{36} = -6 \qquad \sqrt{0.04} = 0.2$$

$$\sqrt[3]{27} = 3 \qquad \sqrt[3]{-\frac{1}{8}} = -\frac{1}{2}$$

$$\sqrt[3]{y^6} = y^2 \qquad \sqrt[3]{64x^9} = 4x^3$$

$$\sqrt{(-3)^2} = |-3| = 3$$

$$\sqrt[3]{(-7)^3} = -7$$

If $f(x) = \sqrt{x} + 2$,

$$f(1) = \sqrt{(1)} + 2 = 1 + 2 = 3$$

$$f(3) = \sqrt{(3)} + 2 \approx 3.73$$

## SECTION 10.2 RATIONAL EXPONENTS

$a^{1/n} = \sqrt[n]{a}$ if $\sqrt[n]{a}$ is a real number.

If $m$ and $n$ are positive integers greater than 1 with $\dfrac{m}{n}$ in lowest terms and $\sqrt[n]{a}$ is a real number, then

$$a^{m/n} = (a^{1/n})^m = \left(\sqrt[n]{a}\right)^m$$

$a^{-m/n} = \dfrac{1}{a^{m/n}}$ as long as $a^{m/n}$ is a nonzero number.

**Exponent rules** are true for rational exponents.

$$81^{1/2} = \sqrt{81} = 9$$

$$(-8x^3)^{1/3} = \sqrt[3]{-8x^3} = -2x$$

$$4^{5/2} = \left(\sqrt{4}\right)^5 = 2^5 = 32$$

$$27^{2/3} = \left(\sqrt[3]{27}\right)^2 = 3^2 = 9$$

$$16^{-3/4} = \frac{1}{16^{3/4}} = \frac{1}{\left(\sqrt[4]{16}\right)^3} = \frac{1}{2^3} = \frac{1}{8}$$

$$x^{2/3} \cdot x^{-5/6} = x^{2/3 - 5/6} = x^{-1/6} = \frac{1}{x^{1/6}}$$

$$(8^4)^{1/2} = 8^2 = 64$$

$$\frac{a^{4/5}}{a^{-2/5}} = a^{4/5 - (-2/5)} = a^{6/5}$$

| **DEFINITIONS AND CONCEPTS** | **EXAMPLES** |
|---|---|

SECTION 10.3  SIMPLIFYING RADICAL EXPRESSIONS

**Product and Quotient Rules**

If $\sqrt[n]{a}$ and $\sqrt[n]{b}$ are real numbers,

$$\sqrt[n]{a} \cdot \sqrt[n]{b} = \sqrt[n]{a \cdot b}$$

$$\frac{\sqrt[n]{a}}{\sqrt[n]{b}} = \sqrt[n]{\frac{a}{b}}, \text{ provided } \sqrt[n]{b} \neq 0$$

A radical of the form $\sqrt[n]{a}$ is **simplified** when $a$ contains no factors that are perfect $n$th powers.

Multiply or divide as indicated:

$$\sqrt{11} \cdot \sqrt{3} = \sqrt{33}$$

$$\frac{\sqrt[3]{40x}}{\sqrt[3]{5x}} = \sqrt[3]{8} = 2$$

$$\sqrt{40} = \sqrt{4 \cdot 10} = 2\sqrt{10}$$

$$\sqrt{36x^5} = \sqrt{36x^4 \cdot x} = 6x^2\sqrt{x}$$

$$\sqrt[3]{24x^7y^3} = \sqrt[3]{8x^6y^3 \cdot 3x} = 2x^2y\sqrt[3]{3x}$$

$$\sqrt{36x^4 \cdot x} = 6x^2\sqrt{x}$$

**Distance Formula**

The distance $d$ between two points $(x_1, y_1)$ and $(x_2, y_2)$ is given by

$$d = \sqrt{(x_2 - x_1)^2 + (y_2 - y_1)^2}$$

Find the distance between points $(-1, 6)$ and $(-2, -4)$. Let $(x_1, y_1) = (-1, 6)$ and $(x_2, y_2) = (-2, -4)$.

$$d = \sqrt{(x_2 - x_1)^2 + (y_2 - y_1)^2}$$
$$= \sqrt{(-2 - (-1))^2 + (-4 - 6)^2}$$
$$= \sqrt{1 + 100} = \sqrt{101}$$

**Midpoint Formula**

The midpoint of the line segment whose endpoints are $(x_1, y_1)$ and $(x_2, y_2)$ is the point with coordinates

$$\left( \frac{x_1 + x_2}{2}, \frac{y_1 + y_2}{2} \right)$$

Find the midpoint of the line segment whose endpoints are $(-1, 6)$ and $(-2, -4)$.

$$\left( \frac{-1 + (-2)}{2}, \frac{6 + (-4)}{2} \right)$$

The midpoint is $\left( -\frac{3}{2}, 1 \right)$.

SECTION 10.4  ADDING, SUBTRACTING, AND MULTIPLYING RADICAL EXPRESSIONS

Radicals with the same index and the same radicand are **like radicals.**

The distributive property can be used to add like radicals.

$$5\sqrt{6} + 2\sqrt{6} = (5 + 2)\sqrt{6} = 7\sqrt{6}$$

$$= \sqrt[3]{3x} - 10\sqrt[3]{3x} + 3\sqrt[3]{10x}$$

$$= (-1 - 10)\sqrt[3]{3x} + 3\sqrt[3]{10x}$$

$$= -11\sqrt[3]{3x} + 3\sqrt[3]{10x}$$

Radical expressions are multiplied by using many of the same properties used to multiply polynomials.

Multiply:

$$(\sqrt{5} - \sqrt{2x})(\sqrt{2} + \sqrt{2x})$$

$$= \sqrt{10} + \sqrt{10x} - \sqrt{4x} - 2x$$

$$= \sqrt{10} + \sqrt{10x} - 2\sqrt{x} - 2x$$

$$(2\sqrt{3} - \sqrt{8x})(2\sqrt{3} + \sqrt{8x})$$

$$= 4(3) - 8x = 12 - 8x$$

| **DEFINITIONS AND CONCEPTS** | **EXAMPLES** |
|---|---|

The **conjugate** of $a + b$ is $a - b$.

The conjugate of $\sqrt{7} + \sqrt{3}$ is $\sqrt{7} - \sqrt{3}$.

The process of writing the denominator of a radical expression without a radical is called **rationalizing the denominator.**

Rationalize each denominator.

$$\frac{\sqrt{5}}{\sqrt{3}} = \frac{\sqrt{5} \cdot \sqrt{3}}{\sqrt{3} \cdot \sqrt{3}} = \frac{\sqrt{15}}{3}$$

$$\frac{6}{\sqrt{7} + \sqrt{3}} = \frac{6(\sqrt{7} - \sqrt{3})}{(\sqrt{7} + \sqrt{3})(\sqrt{7} - \sqrt{3})}$$

$$= \frac{6(\sqrt{7} - \sqrt{3})}{7 - 3}$$

$$= \frac{6(\sqrt{7} - \sqrt{3})}{4} = \frac{3(\sqrt{7} - \sqrt{3})}{2}$$

The process of writing the numerator of a radical expression without a radical is called **rationalizing the numerator.**

Rationalize each numerator:

$$\frac{\sqrt[3]{9}}{\sqrt[3]{5}} = \frac{\sqrt[3]{9} \cdot \sqrt[3]{3}}{\sqrt[3]{5} \cdot \sqrt[3]{3}} = \frac{\sqrt[3]{27}}{\sqrt[3]{15}} = \frac{3}{\sqrt[3]{15}}$$

$$\frac{\sqrt{9} + \sqrt{3x}}{12} = \frac{(\sqrt{9} + \sqrt{3x})(\sqrt{9} - \sqrt{3x})}{12(\sqrt{9} - \sqrt{3x})}$$

$$= \frac{9 - 3x}{12(\sqrt{9} - \sqrt{3x})}$$

$$= \frac{3(3 - x)}{3 \cdot 4(3 - \sqrt{3x})} = \frac{3 - x}{4(3 - \sqrt{3x})}$$

***To Solve a Radical Equation***

**Step 1.** Write the equation so that one radical is by itself on one side of the equation.

**Step 2.** Raise each side of the equation to a power equal to the index of the radical and simplify.

**Step 3.** If the equation still contains a radical, repeat Steps 1 and 2. If not, solve the equation.

**Step 4.** Check all proposed solutions in the original equation.

Solve: $x = \sqrt{4x + 9} + 3$.

**1.** $x - 3 = \sqrt{4x + 9}$

**2.** $(x - 3)^2 = (\sqrt{4x + 9})^2$
$x^2 - 6x + 9 = 4x + 9$

**3.** $x^2 - 10x = 0$
$x(x - 10) = 0$
$x = 0 \quad \text{or} \quad x = 10$

**4.** The proposed solution 10 checks, but 0 does not. The solution is 10.

| DEFINITIONS AND CONCEPTS | EXAMPLES |
|---|---|

$i^2 = -1$ and $i = \sqrt{-1}$

A **complex number** is a number that can be written in the form $a + bi$, where $a$ and $b$ are real numbers.

Simplify: $\sqrt{-9}$.

$\sqrt{-9} = \sqrt{-1 \cdot 9} = \sqrt{-1} \cdot \sqrt{9} = i \cdot 3$ or $3i$

| *Complex Numbers* | *Written in Form a + bi* |
|---|---|
| 12 | $12 + 0i$ |
| $-5i$ | $0 + (-5)i$ |
| $-2 - 3i$ | $-2 + (-3)i$ |

Multiply,

$$\sqrt{-3} \cdot \sqrt{-7} = i\sqrt{3} \cdot i\sqrt{7}$$
$$= i^2\sqrt{21}$$
$$= -\sqrt{21}$$

**To add or subtract complex numbers,** add or subtract their real parts and then add or subtract their imaginary parts.

**To multiply complex numbers,** multiply as though they are binomials.

Perform each indicated operation.

$$(-3 + 2i) - (7 - 4i) = -3 + 2i - 7 + 4i$$
$$= -10 + 6i$$

$$(-7 - 2i)(6 + i) = -42 - 7i - 12i - 2i^2$$
$$= -42 - 19i - 2(-1)$$
$$= -42 - 19i + 2$$
$$= -40 - 19i$$

The complex numbers $(a + bi)$ and $(a - bi)$ are called **complex conjugates.**

The complex conjugate of
$$(3 + 6i) \text{ is } (3 - 6i).$$
Their product is a real number:
$$(3 - 6i)(3 + 6i) = 9 - 36i^2$$
$$= 9 - 36(-1) = 9 + 36 = 45$$

**To divide complex numbers,** multiply the numerator and the denominator by the conjugate of the denominator.

Divide.
$$\frac{4}{2 - i} = \frac{4(2 + i)}{(2 - i)(2 + i)}$$
$$= \frac{4(2 + i)}{4 - i^2}$$
$$= \frac{4(2 + i)}{5}$$
$$= \frac{8 + 4i}{5} = \frac{8}{5} + \frac{4}{5}i$$

# CHAPTER 10 REVIEW

**(10.1)** *Find the root. Assume that all variables represent positive numbers.*

**1.** $\sqrt{81}$

**2.** $\sqrt[4]{81}$

**3.** $\sqrt[3]{-8}$

**4.** $\sqrt[4]{-16}$

**5.** $-\sqrt{\dfrac{1}{49}}$

**6.** $\sqrt{x^{64}}$

**7.** $-\sqrt{36}$

**8.** $\sqrt[3]{64}$

**9.** $\sqrt[3]{-a^6b^9}$

**10.** $\sqrt{16a^4b^{12}}$

**11.** $\sqrt[5]{32a^5b^{10}}$

**12.** $\sqrt[5]{-32x^{15}y^{20}}$

**13.** $\sqrt{\dfrac{x^{12}}{36y^2}}$

**14.** $\sqrt[3]{\dfrac{27y^3}{z^{12}}}$

*Simplify. Use absolute value bars when necessary.*

**15.** $\sqrt{(-x)^2}$

**16.** $\sqrt[4]{(x^2 - 4)^4}$

**17.** $\sqrt[3]{(-27)^3}$

**18.** $\sqrt[5]{(-5)^5}$

**19.** $-\sqrt[5]{x^5}$

**20.** $\sqrt[4]{16(2y + z)^{12}}$

**21.** $\sqrt{25(x-y)^{10}}$        **22.** $\sqrt[5]{-y^5}$

**23.** $\sqrt[9]{-x^9}$

*Identify the domain and then graph each function.*

**24.** $f(x) = \sqrt{x} + 3$

**25.** $g(x) = \sqrt[3]{x} - 3$; use the accompanying table.

| $x$ | −5 | 2 | 3 | 4 | 11 |
|---|---|---|---|---|---|
| $g(x)$ | | | ○ | | |

**(10.2)** *Evaluate the following.*

**26.** $\left(\dfrac{1}{81}\right)^{1/4}$        **27.** $\left(-\dfrac{1}{27}\right)^{1/3}$

**28.** $(-27)^{-1/3}$        **29.** $(-64)^{-1/3}$

**30.** $-9^{3/2}$        **31.** $64^{-1/3}$

**32.** $(-25)^{5/2}$        **33.** $\left(\dfrac{25}{49}\right)^{-3/2}$

**34.** $\left(\dfrac{8}{27}\right)^{-2/3}$        **35.** $\left(-\dfrac{1}{36}\right)^{-1/4}$

*Write with rational exponents.*

**36.** $\sqrt[3]{x^2}$        **37.** $\sqrt[5]{5x^2y^3}$

*Write with radical notation.*

**38.** $y^{4/5}$        **39.** $5(xy^2z^5)^{1/3}$

**40.** $(x+2y)^{-1/2}$

*Simplify each expression. Assume that all variables represent positive numbers. Write with only positive exponents.*

**41.** $a^{1/3}a^{4/3}a^{1/2}$        **42.** $\dfrac{b^{1/3}}{b^{4/3}}$

**43.** $(a^{1/2}a^{-2})^3$        **44.** $(x^{-3}y^6)^{1/3}$

**45.** $\left(\dfrac{b^{3/4}}{a^{-1/2}}\right)^8$        **46.** $\dfrac{x^{1/4}x^{-1/2}}{x^{2/3}}$

**47.** $\left(\dfrac{49c^{5/3}}{a^{-1/4}b^{5/6}}\right)^{-1}$        **48.** $a^{-1/4}(a^{5/4} - a^{9/4})$

*Use a calculator and write a three-decimal-place approximation.*

**49.** $\sqrt{20}$        **50.** $\sqrt[3]{-39}$

**51.** $\sqrt[4]{726}$        **52.** $56^{1/3}$

**53.** $-78^{3/4}$        **54.** $105^{-2/3}$

*Use rational exponents to write each radical with the same index. Then multiply.*

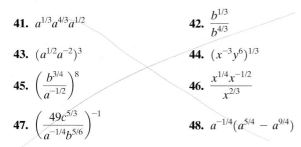

**55.** $\sqrt[3]{2} \cdot \sqrt{7}$        **56.** $\sqrt[3]{3} \cdot \sqrt[4]{x}$

**(10.3)** *Perform the indicated operations and then simplify if possible. For the remainder of this review, assume that variables represent positive numbers only.*

**57.** $\sqrt[3]{3} \cdot \sqrt{8}$        **58.** $\sqrt[3]{7y} \cdot \sqrt[3]{x^2z}$

**59.** $\dfrac{\sqrt{44x^3}}{\sqrt{11x}}$        **60.** $\dfrac{\sqrt[4]{a^6b^{13}}}{\sqrt[4]{a^2b}}$

*Simplify.*

**61.** $\sqrt{60}$        **62.** $-\sqrt{75}$

**63.** $\sqrt[3]{162}$        **64.** $\sqrt[3]{-32}$

**65.** $\sqrt{36x^7}$        **66.** $\sqrt[3]{24a^5b^7}$

**67.** $\sqrt{\dfrac{p^{17}}{121}}$        **68.** $\sqrt[3]{\dfrac{y^5}{27x^6}}$

**69.** $\sqrt[4]{\dfrac{xy^6}{81}}$        **70.** $\sqrt{\dfrac{2x^3}{49y^4}}$

△ **71.** The formula for the radius $r$ of a circle of area $A$ is

$$r = \sqrt{\dfrac{A}{\pi}}$$

    **a.** Find the exact radius of a circle whose area is 25 square meters.

    **b.** Approximate to two decimal places the radius of a circle whose area is 104 square inches.

*Find the distance between each pair of points. Give an exact value and a three-decimal-place approximation.*

**72.** $(-6, 3)$ and $(8, 4)$

**73.** $(-4, -6)$ and $(-1, 5)$

**74.** $(-1, 5)$ and $(2, -3)$

**75.** $(-\sqrt{2}, 0)$ and $(0, -4\sqrt{6})$

**76.** $(-\sqrt{5}, -\sqrt{11})$ and $(-\sqrt{5}, -3\sqrt{11})$

**77.** $(7.4, -8.6)$ and $(-1.2, 5.6)$

*Find the midpoint of each line segment whose endpoints are given.*

**78.** $(2, 6)$; $(-12, 4)$        **79.** $(-6, -5)$; $(-9, 7)$

**80.** $(4, -6)$; $(-15, 2)$        **81.** $\left(0, -\dfrac{3}{8}\right)$; $\left(\dfrac{1}{10}, 0\right)$

**82.** $\left(\dfrac{3}{4}, -\dfrac{1}{7}\right)$; $\left(-\dfrac{1}{4}, -\dfrac{3}{7}\right)$

**83.** $(\sqrt{3}, -2\sqrt{6})$ and $(\sqrt{3}, -4\sqrt{6})$

**(10.4)** *Perform the indicated operation.*

**84.** $2\sqrt{50} - 3\sqrt{125} + \sqrt{98}$

**85.** $x\sqrt{75xy} - \sqrt{27x^3y}$

**86.** $\sqrt[3]{128} + \sqrt[3]{250}$

**87.** $3\sqrt[4]{32a^5} - a\sqrt[4]{162a}$

**88.** $\dfrac{5}{\sqrt{4}} + \dfrac{\sqrt{3}}{3}$

**89.** $\sqrt{\dfrac{8}{x^2}} - \sqrt{\dfrac{50}{16x^2}}$

**90.** $2\sqrt{32x^2y^3} - xy\sqrt{98y}$

**91.** $2a\sqrt[4]{32b^5} - 3b\sqrt[4]{162a^4b} + \sqrt[4]{2a^4b^5}$

*Multiply and then simplify if possible.*

**92.** $\sqrt{3}\left(\sqrt{27} - \sqrt{3}\right)$

**93.** $\left(\sqrt{x} - 3\right)^2$

**94.** $\left(\sqrt{5} - 5\right)\left(2\sqrt{5} + 2\right)$

**95.** $\left(2\sqrt{x} - 3\sqrt{y}\right)\left(2\sqrt{x} + 3\sqrt{y}\right)$

**96.** $\left(\sqrt{a} + 3\right)\left(\sqrt{a} - 3\right)$

**97.** $\left(\sqrt[3]{a} + 2\right)^2$

**98.** $\left(\sqrt[3]{5x} + 9\right)\left(\sqrt[3]{5x} - 9\right)$

**99.** $\left(\sqrt[3]{a} + 4\right)\left(\sqrt[3]{a^2} - 4\sqrt[3]{a} + 16\right)$

***(10.5)*** *Rationalize each denominator.*

**100.** $\dfrac{3}{\sqrt{7}}$

**101.** $\sqrt{\dfrac{x}{12}}$

**102.** $\dfrac{5}{\sqrt[3]{4}}$

**103.** $\sqrt{\dfrac{24x^5}{3y^2}}$

**104.** $\sqrt[3]{\dfrac{15x^6y^7}{z^2}}$

**105.** $\dfrac{5}{2 - \sqrt{7}}$

**106.** $\dfrac{3}{\sqrt{y} - 2}$

**107.** $\dfrac{\sqrt{2} - \sqrt{3}}{\sqrt{2} + \sqrt{3}}$

*Rationalize each numerator.*

**108.** $\dfrac{\sqrt{11}}{3}$

**109.** $\sqrt{\dfrac{18}{y}}$

**110.** $\dfrac{\sqrt[3]{9}}{7}$

**111.** $\sqrt{\dfrac{24x^5}{3y^2}}$

**112.** $\sqrt[3]{\dfrac{xy^2}{10z}}$

**113.** $\dfrac{\sqrt{x} + 5}{-3}$

***(10.6)*** *Solve each equation for the variable.*

**114.** $\sqrt{y - 7} = 5$

**115.** $\sqrt{2x} + 10 = 4$

**116.** $\sqrt[3]{2x - 6} = 4$

**117.** $\sqrt{x + 6} = \sqrt{x + 2}$

**118.** $2x - 5\sqrt{x} = 3$

**119.** $\sqrt{x + 9} = 2 + \sqrt{x - 7}$

*Find each unknown length.*

△ **120.**

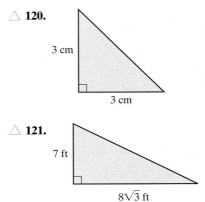

3 cm

3 cm

△ **121.**

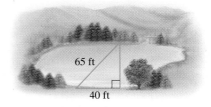

7 ft

$8\sqrt{3}$ ft

**122.** Beverly Hillis wants to determine the distance $x$ across a pond on her property. She is able to measure the distances shown on the following diagram. Find how wide the lake is at the crossing point, indicated by the triangle, to the nearest tenth of a foot.

65 ft

40 ft

△ **123.** A pipe fitter needs to connect two underground pipelines that are offset by 3 feet, as pictured in the diagram. Neglecting the joints needed to join the pipes, find the length of the shortest possible connecting pipe rounded to the nearest hundredth of a foot.

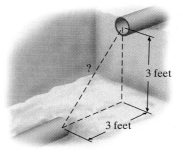

? 

3 feet

3 feet

***(10.7)*** *Perform the indicated operation and simplify. Write the result in the form $a + bi$.*

**124.** $\sqrt{-8}$

**125.** $-\sqrt{-6}$

**126.** $\sqrt{-4} + \sqrt{-16}$

**127.** $\sqrt{-2} \cdot \sqrt{-5}$

**128.** $(12 - 6i) + (3 + 2i)$

**129.** $(-8 - 7i) - (5 - 4i)$

**130.** $(2i)^6$

**131.** $-3i(6 - 4i)$

**132.** $(3 + 2i)(1 + i)$

**133.** $(2 - 3i)^2$

**134.** $\left(\sqrt{6} - 9i\right)\left(\sqrt{6} + 9i\right)$

**135.** $\dfrac{2 + 3i}{2i}$

**136.** $\dfrac{1 + i}{-3i}$

## MIXED REVIEW

*Simplify. Use absolute value bars when necessary.*

**137.** $\sqrt[3]{x^3}$

**138.** $\sqrt{(x + 2)^2}$

*Simplify. Assume that all variables represent positive real numbers. If necessary, write answers with positive exponents only.*

**139.** $-\sqrt{100}$

**140.** $\sqrt[3]{-x^{12}y^3}$

**141.** $\sqrt[4]{\dfrac{y^{20}}{16x^{12}}}$

**142.** $9^{1/2}$

**143.** $64^{-1/2}$

**144.** $\left(\dfrac{27}{64}\right)^{-2/3}$

**145.** $\dfrac{(x^{2/3}x^{-3})^3}{x^{-1/2}}$

**146.** $\sqrt{200x^9}$

**147.** $\sqrt{\dfrac{3n^3}{121m^{10}}}$

**148.** $3\sqrt{20} - 7x\sqrt[3]{40} + 3\sqrt[3]{5x^3}$

**149.** $(2\sqrt{x} - 5)^2$

**150.** Find the distance between $(-3, 5)$ and $(-8, 9)$.

**151.** Find the midpoint of the line segment joining $(-3, 8)$ and $(11, 24)$.

*Rationalize each denominator.*

**152.** $\dfrac{7}{\sqrt{13}}$

**153.** $\dfrac{2}{\sqrt{x} + 3}$

*Solve.*

**154.** $\sqrt{x} + 2 = x$

## CHAPTER 10 TEST TEST PREP VIDEO

Remember to use the Chapter Test Prep Video CD to see the fully worked-out solutions to any of the exercises you want to review.

*Raise to the power or find the root. Assume that all variables represent positive numbers. Write with only positive exponents.*

**1.** $\sqrt{216}$

**2.** $-\sqrt[4]{x^{64}}$

**3.** $\left(\dfrac{1}{125}\right)^{1/3}$

**4.** $\left(\dfrac{1}{125}\right)^{-1/3}$

**5.** $\left(\dfrac{8x^3}{27}\right)^{2/3}$

**6.** $\sqrt[3]{-a^{18}b^9}$

**7.** $\left(\dfrac{64c^{4/3}}{a^{-2/3}b^{5/6}}\right)^{1/2}$

**8.** $a^{-2/3}(a^{5/4} - a^3)$

*Find the root. Use absolute value bars when necessary.*

**9.** $\sqrt[4]{(4xy)^4}$

**10.** $\sqrt[3]{(-27)^3}$

*Rationalize the denominator. Assume that all variables represent positive numbers.*

**11.** $\sqrt{\dfrac{9}{y}}$

**12.** $\dfrac{4 - \sqrt{x}}{4 + 2\sqrt{x}}$

**13.** $\dfrac{\sqrt[3]{ab}}{\sqrt[3]{ab^2}}$

**14.** Rationalize the numerator of $\dfrac{\sqrt{6} + x}{8}$ and simplify.

*Perform the indicated operations. Assume that all variables represent positive numbers.*

**15.** $\sqrt{125x^3} - 3\sqrt{20x^3}$

**16.** $\sqrt{3}(\sqrt{16} - \sqrt{2})$

**17.** $(\sqrt{x} + 1)^2$

**18.** $(\sqrt{2} - 4)(\sqrt{3} + 1)$

**19.** $(\sqrt{5} + 5)(\sqrt{5} - 5)$

*Use a calculator to approximate each to three decimal places.*

**20.** $\sqrt{561}$

**21.** $386^{-2/3}$

*Solve.*

**22.** $x = \sqrt{x - 2} + 2$

**23.** $\sqrt{x^2 - 7} + 3 = 0$

**24.** $\sqrt[3]{x + 5} = \sqrt[3]{2x - 1}$

*Perform the indicated operation and simplify. Write the result in the form $a + bi$.*

**25.** $\sqrt{-2}$

**26.** $-\sqrt{-8}$

**27.** $(12 - 6i) - (12 - 3i)$

**28.** $(6 - 2i)(6 + 2i)$

**29.** $(4 + 3i)^2$

**30.** $\dfrac{1 + 4i}{1 - i}$

△ **31.** Find $x$.

**32.** Identify the domain of $g(x)$. Then complete the accompanying table and graph $g(x)$.

$$g(x) = \sqrt{x} + 2$$

| $x$ | $-2$ | $-1$ | $2$ | $7$ |
|---|---|---|---|---|
| $g(x)$ | | | | |

**33.** Find the distance between the points $(-6, 3)$ and $(-8, -7)$.

**34.** Find the distance between the points $(-2\sqrt{5}, \sqrt{10})$ and $(-\sqrt{5}, 4\sqrt{10})$.

**35.** Find the midpoint of the line segment whose endpoints are $(-2, -5)$ and $(-6, 12)$.

**36.** Find the midpoint of the line segment whose endpoints are $\left(-\dfrac{2}{3}, -\dfrac{1}{5}\right)$ and $\left(-\dfrac{1}{3}, \dfrac{4}{5}\right)$.

*Solve.*

**37.** The function $V(r) = \sqrt{2.5r}$ can be used to estimate the maximum safe velocity $V$ in miles per hour at which a car can travel if it is driven along a curved road with a *radius of curvature r* in feet. To the nearest whole number, find the maximum safe speed if a cloverleaf exit on an expressway has a radius of curvature of 300 feet.

**38.** Use the formula from Exercise 37 to find the radius of curvature if the safe velocity is 30 mph.

# CHAPTER 10 CUMULATIVE REVIEW

**1.** Simplify each expression.

  **a.** $-3 + [(-2 - 5) - 2]$

  **b.** $2^3 - |10| + [-6 - (-5)]$

**2.** Simplify each expression.

  **a.** $2(x - 3) + (5x + 3)$

  **b.** $4(3x + 2) - 3(5x - 1)$

  **c.** $7x + 2(x - 7) - 3x$

**3.** Solve: $\dfrac{x}{2} - 1 = \dfrac{2}{3}x - 3$

**4.** Solve: $\dfrac{a - 1}{2} + a = 2 - \dfrac{2a + 7}{8}$

**5.** A 48-inch Balsa wood stick is to be cut into two pieces so that the longer piece is 3 times the shorter. Find the length of each piece.

**6.** The Smith family owns a lake house 121.5 miles from home. If it takes them $4\frac{1}{2}$ hours round-trip to drive from their house to their lake house, find their average speed.

**7.** Determine the number of solutions of the system.

$$\begin{cases} 3x - y = 4 \\ x + 2y = 8 \end{cases}$$

**8.** Solve: $|3x - 2| + 5 = 5$

**9.** Solve the system:

$$\begin{cases} x + 2y = 7 \\ 2x + 2y = 13 \end{cases}$$

**10.** Solve: $\left|\dfrac{x}{2} - 1\right| \le 0$.

**11.** Solve the system:

$$\begin{cases} 2x - y = 7 \\ 8x - 4y = 1 \end{cases}$$

**12.** Graph $y = |x - 2|$.

**13.** Lynn Pike, a pharmacist, needs 70 liters of a 50% alcohol solution. She has available a 30% alcohol solution and an 80% alcohol solution. How many liters of each solution should she mix to obtain 70 liters of a 50% alcohol solution?

**14.** Find the domain and the range of each relation. Use the vertical line test to determine whether each graph is the graph of a function.

  **a.**

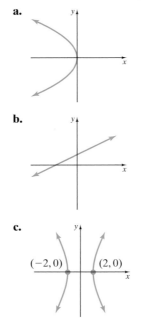

  **b.**

  **c.**

**15.** If $P(x) = 3x^2 - 2x - 5$, find the following.

  **a.** $P(1)$          **b.** $P(-2)$

**16.** Graph $f(x) = -2$.

**17.** Divide $6m^2 + 2m$ by $2m$.

**18.** Find the slope of $y = -3$.

**19.** Use synthetic division to divide $2x^3 - x^2 - 13x + 1$ by $x - 3$.

**20.** Use the substitution method to solve the system.

$$\begin{cases} \dfrac{x}{6} - \dfrac{y}{2} = 1 \\ \dfrac{x}{3} - \dfrac{y}{4} = 2 \end{cases}$$

**21.** Factor $40 - 13t + t^2$.

**22.** At a seasonal clearance sale, Nana Long spent $33.75. She paid $3.50 for tee-shirts and $4.25 for shorts. If she bought 9 items, how many of each item did she buy?

**23.** Simplify each rational expression.

**a.** $\dfrac{x^3 + 8}{2 + x}$  **b.** $\dfrac{2y^2 + 2}{y^3 - 5y^2 + y - 5}$

**24.** Use scientific notation to simplify and write the answer in scientific notation. $\dfrac{0.0000035 \times 4000}{0.28}$

**25.** Solve: $|x - 3| = |5 - x|$

**26.** Subtract $(2x - 5)$ from the sum of $(5x^2 - 3x + 6)$ and $(4x^2 + 5x - 3)$.

**27.** Subtract: $\dfrac{3x^2 + 2x}{x - 1} - \dfrac{10x - 5}{x - 1}$

**28.** Multiply and simplify the product if possible.

**a.** $(y - 2)(3y + 4)$

**b.** $(3y - 1)(2y^2 + 3y - 1)$

**29.** Add: $1 + \dfrac{m}{m + 1}$

**30.** Factor. $x^3 - x^2 + 4x - 4$

**31.** Simply each complex fraction.

**a.** $\dfrac{\dfrac{5x}{x + 2}}{\dfrac{10}{x - 2}}$  **b.** $\dfrac{\dfrac{x}{y^2} + \dfrac{1}{y}}{\dfrac{y}{x^2} + \dfrac{1}{x}}$

**32.** Simplify each rational expression.

**a.** $\dfrac{a^3 - 8}{2 - a}$

**b.** $\dfrac{3a^2 - 3}{a^3 + 5a^2 - a - 5}$

**33.** Solve: $|5x + 1| + 1 \le 10$

**34.** Perform the indicated operations.

**a.** $\dfrac{3}{xy^2} - \dfrac{2}{3x^2y}$

**b.** $\dfrac{5x}{x + 3} - \dfrac{2x}{x - 3}$

**c.** $\dfrac{x}{x - 2} - \dfrac{5}{2 - x}$

**35.** If the following two triangles are similar, find the missing length $x$.

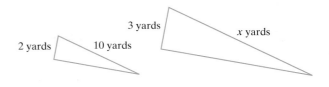

**36.** Simplify each complex fraction.

**a.** $\dfrac{\dfrac{y - 2}{16}}{\dfrac{2y + 3}{12}}$  **b.** $\dfrac{\dfrac{x}{16} - \dfrac{1}{x}}{1 - \dfrac{4}{x}}$

**37.** Find the cube roots.

**a.** $\sqrt[3]{1}$  **b.** $\sqrt[3]{-64}$

**c.** $\sqrt[3]{\dfrac{8}{125}}$  **d.** $\sqrt[3]{x^6}$

**e.** $\sqrt[3]{-27x^9}$

**38.** Divide $x^3 - 2x^2 + 3x - 6$ by $x - 2$.

**39.** Write each expression with a positive exponent, and then simplify.

**a.** $16^{-3/4}$  **b.** $(-27)^{-2/3}$

**40.** Use synthetic division to divide $4y^3 - 12y^2 - y + 12$ by $y - 3$.

**41.** Rational the numerator of $\dfrac{\sqrt{x} + 2}{5}$

**42.** Solve: $\dfrac{28}{9 - a^2} = \dfrac{2a}{a - 3} + \dfrac{6}{a + 3}$

**43.** Suppose that $u$ varies inversely as $w$. If $u$ is 3 when $w$ is 5, find the constant of variation and the inverse variation equation.

**44.** Suppose that $y$ varies directly as $x$. If $y = 0.51$ when $x = 3$, find the constant of variation and the direct variation equation.

An important part of the study of algebra is learning to model and solve problems. Often, the model of a problem is a quadratic equation or a function containing a second-degree polynomial. In this chapter, we continue the work begun in Chapter 6, when we solved polynomial equations in one variable by factoring. Two additional methods of solving quadratic equations are analyzed, as well as methods of solving nonlinear inequalities in one variable.

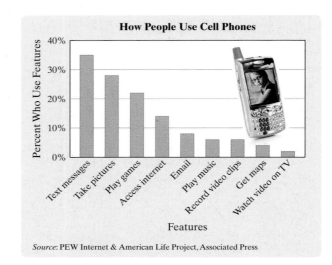

**How People Use Cell Phones**

*Source*: PEW Internet & American Life Project, Associated Press

The growth of cell phones, shown below, can be approximated by a quadratic function. More interesting information is probably given on the graph above. As shown, the cell phone is certainly no longer just a phone.

On page 688, Section 11.5, Exercises 79 and 80, you will have the opportunity to use the quadratic function below and discuss its limitations.

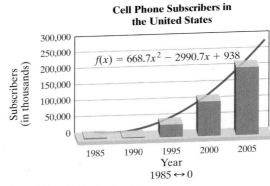

**Cell Phone Subscribers in
the United States**

$$f(x) = 668.7x^2 - 2990.7x + 938$$

Year
1985 ↔ 0

*Source*: CTIA – The Wireless Association

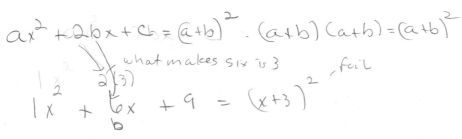

# 11.1 SOLVING QUADRATIC EQUATIONS BY COMPLETING THE SQUARE

## OBJECTIVES

**1** Use the square root property to solve quadratic equations.

**2** Solve quadratic equations by completing the square.

**3** Use quadratic equations to solve problems.

**OBJECTIVE 1 ▶ Using the square root property.** In Chapter 6, we solved quadratic equations by factoring. Recall that a **quadratic**, or **second-degree, equation** is an equation that can be written in the form $ax^2 + bx + c = 0$, where $a$, $b$, and $c$ are real numbers and $a$ is not 0. To solve a quadratic equation such as $x^2 = 9$ by factoring, we use the zero-factor theorem. To use the zero-factor theorem, the equation must first be written in standard form, $ax^2 + bx + c = 0$.

$$x^2 = 9$$
$$x^2 - 9 = 0 \qquad \text{Subtract 9 from both sides.}$$
$$(x + 3)(x - 3) = 0 \qquad \text{Factor.}$$
$$x + 3 = 0 \quad \text{or} \quad x - 3 = 0 \qquad \text{Set each factor equal to 0.}$$
$$x = -3 \qquad\qquad x = 3 \qquad \text{Solve.}$$

The solution set is $\{-3, 3\}$, the positive and negative square roots of 9. Not all quadratic equations can be solved by factoring, so we need to explore other methods. Notice that the solutions of the equation $x^2 = 9$ are two numbers whose square is 9.

$$3^2 = 9 \qquad \text{and} \qquad (-3)^2 = 9$$

Thus, we can solve the equation $x^2 = 9$ by taking the square root of both sides. Be sure to include both $\sqrt{9}$ and $-\sqrt{9}$ as solutions since both $\sqrt{9}$ and $-\sqrt{9}$ are numbers whose square is 9.

$$x^2 = 9$$
$$\sqrt{x^2} = \pm\sqrt{9} \qquad \text{The notation } \pm\sqrt{9} \text{ (read as "plus or minus } \sqrt{9}\text{")}$$
$$x = \pm 3 \qquad \text{indicates the pair of numbers } +\sqrt{9} \text{ and } -\sqrt{9}.$$

This illustrates the square root property.

---

▶ **Helpful Hint**

The notation $\pm 3$, for example, is read as "plus or minus 3." It is a shorthand notation for the pair of numbers $+3$ and $-3$.

---

**Square Root Property**

If $b$ is a real number and if $a^2 = b$, then $a = \pm\sqrt{b}$.

---

**EXAMPLE 1**   Use the square root property to solve $x^2 = 50$.

*Solution*
$$x^2 = 50$$
$$x = \pm\sqrt{50} \qquad \text{Use the square root property.}$$
$$x = \pm 5\sqrt{2} \qquad \text{Simplify the radical.}$$

**Check:**    Let $x = 5\sqrt{2}$. 　　　　　　　 Let $x = -5\sqrt{2}$.
$$x^2 = 50 \qquad\qquad\qquad x^2 = 50$$
$$\left(5\sqrt{2}\right)^2 \overset{?}{=} 50 \qquad\qquad \left(-5\sqrt{2}\right)^2 \overset{?}{=} 50$$
$$25 \cdot 2 \overset{?}{=} 50 \qquad\qquad\quad 25 \cdot 2 \overset{?}{=} 50$$
$$50 = 50 \quad \text{True} \qquad\qquad 50 = 50 \quad \text{True}$$

The solutions are $5\sqrt{2}$ and $-5\sqrt{2}$, or the solution set is $\{-5\sqrt{2}, 5\sqrt{2}\}$. 　　□

**PRACTICE**

**1**　Use the square root property to solve $x^2 = 18$.

**EXAMPLE 2** Use the square root property to solve $2x^2 - 14 = 0$.

_Solution_ First we get the squared variable alone on one side of the equation.

$$2x^2 - 14 = 0$$
$$2x^2 = 14 \qquad \text{Add 14 to both sides.}$$
$$x^2 = 7 \qquad \text{Divide both sides by 2.}$$
$$x = \pm\sqrt{7} \quad \text{Use the square root property.}$$

Check to see that the solutions are $\sqrt{7}$ and $-\sqrt{7}$, or the solution set is $\{-\sqrt{7}, \sqrt{7}\}$. □

**PRACTICE**
**2** Use the square root property to solve $3x^2 - 30 = 0$.

**EXAMPLE 3** Use the square root property to solve $(x + 1)^2 = 12$.

_Solution_
$$(x + 1)^2 = 12$$
$$x + 1 = \pm\sqrt{12} \qquad \text{Use the square root property.}$$
$$x + 1 = \pm 2\sqrt{3} \qquad \text{Simplify the radical.}$$
$$x = -1 \pm 2\sqrt{3} \quad \text{Subtract 1 from both sides.}$$

**Check:** Below is a check for $-1 + 2\sqrt{3}$. The check for $-1 - 2\sqrt{3}$ is almost the same and is left for you to do on your own.

$$(x + 1)^2 = 12$$
$$\left(-1 + 2\sqrt{3} + 1\right)^2 \stackrel{?}{=} 12$$
$$\left(2\sqrt{3}\right)^2 \stackrel{?}{=} 12$$
$$4 \cdot 3 \stackrel{?}{=} 12$$
$$12 = 12 \quad \text{True}$$

The solutions are $-1 + 2\sqrt{3}$ and $-1 - 2\sqrt{3}$. □

**PRACTICE**
**3** Use the square root property to solve $(x + 3)^2 = 20$.

**EXAMPLE 4** Use the square root property to solve $(2x - 5)^2 = -16$.

_Solution_
$$(2x - 5)^2 = -16$$
$$2x - 5 = \pm\sqrt{-16} \quad \text{Use the square root property.}$$
$$2x - 5 = \pm 4i \qquad \text{Simplify the radical.}$$
$$2x = 5 \pm 4i \qquad \text{Add 5 to both sides.}$$
$$x = \frac{5 \pm 4i}{2} \qquad \text{Divide both sides by 2.}$$

The solutions are $\dfrac{5 + 4i}{2}$ and $\dfrac{5 - 4i}{2}$. □

**PRACTICE**
**4** Use the square root property to solve $(5x - 2)^2 = -9$.

**Concept Check** ☑

How do you know just by looking that $(x - 2)^2 = -4$ has complex, but not real solutions?

**Answer to Concept Check:**
answers may vary

**OBJECTIVE 2 ▶ Solving by completing the square.** Notice from Examples 3 and 4 that, if we write a quadratic equation so that one side is the square of a binomial, we can solve by using the square root property. To write the square of a binomial, we write perfect square trinomials. Recall that a perfect square trinomial is a trinomial that can be factored into two identical binomial factors.

| *Perfect Square Trinomials* | *Factored Form* |
|:---:|:---:|
| $x^2 + 8x + 16$ | $(x + 4)^2$ |
| $x^2 - 6x + 9$ | $(x - 3)^2$ |
| $x^2 + 3x + \dfrac{9}{4}$ | $\left(x + \dfrac{3}{2}\right)^2$ |

Notice that for each perfect square trinomial, **the constant term of the trinomial is the square of half the coefficient of the $x$-term.** For example,

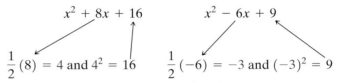

$$\frac{1}{2}(8) = 4 \text{ and } 4^2 = 16 \qquad \frac{1}{2}(-6) = -3 \text{ and } (-3)^2 = 9$$

The process of writing a quadratic equation so that one side is a perfect square trinomial is called **completing the square.**

**EXAMPLE 5**   Solve $p^2 + 2p = 4$ by completing the square.

*Solution*   First, add the square of half the coefficient of $p$ to both sides so that the resulting trinomial will be a perfect square trinomial. The coefficient of $p$ is 2.

$$\frac{1}{2}(2) = 1 \quad \text{and} \quad 1^2 = 1$$

Add 1 to both sides of the original equation.

$$p^2 + 2p = 4$$
$$p^2 + 2p + 1 = 4 + 1 \quad \text{Add 1 to both sides.}$$
$$(p + 1)^2 = 5 \qquad \text{Factor the trinomial; simplify the right side.}$$

We may now use the square root property and solve for $p$.

$$p + 1 = \pm\sqrt{5} \qquad \text{Use the square root property.}$$
$$p = -1 \pm \sqrt{5} \quad \text{Subtract 1 from both sides.}$$

Notice that there are two solutions: $-1 + \sqrt{5}$ and $-1 - \sqrt{5}$.   □

**PRACTICE**
**5**   Solve $b^2 + 4b = 3$ by completing the square.

**EXAMPLE 6**   Solve $m^2 - 7m - 1 = 0$ for $m$ by completing the square.

*Solution*   First, add 1 to both sides of the equation so that the left side has no constant term.

$$m^2 - 7m - 1 = 0$$
$$m^2 - 7m = 1$$

Now find the constant term that makes the left side a perfect square trinomial by squaring half the coefficient of $m$. Add this constant to both sides of the equation.

$$\frac{1}{2}(-7) = -\frac{7}{2} \quad \text{and} \quad \left(-\frac{7}{2}\right)^2 = \frac{49}{4}$$

$$m^2 - 7m + \frac{49}{4} = 1 + \frac{49}{4} \qquad \text{Add } \frac{49}{4} \text{ to both sides of the equation.}$$

$$\left(m - \frac{7}{2}\right)^2 = \frac{53}{4}$$

Factor the perfect square trinomial and simplify the right side.

$$m - \frac{7}{2} = \pm\sqrt{\frac{53}{4}}$$

Apply the square root property.

$$m = \frac{7}{2} \pm \frac{\sqrt{53}}{2}$$

Add $\frac{7}{2}$ to both sides and simplify $\sqrt{\frac{53}{4}}$.

$$m = \frac{7 \pm \sqrt{53}}{2}$$

Simplify.

The solutions are $\dfrac{7 + \sqrt{53}}{2}$ and $\dfrac{7 - \sqrt{53}}{2}$. ☐

**6**   Solve $p^2 - 3p + 1 = 0$ by completing the square.

---

**EXAMPLE 7**   Solve: $2x^2 - 8x + 3 = 0$.

<u>Solution</u>   Our procedure for finding the constant term to complete the square works only if the coefficient of the squared variable term is 1. Therefore, to solve this equation, the first step is to divide both sides by 2, the coefficient of $x^2$.

$$2x^2 - 8x + 3 = 0$$

$$x^2 - 4x + \frac{3}{2} = 0 \qquad \text{Divide both sides by 2.}$$

$$x^2 - 4x = -\frac{3}{2} \qquad \text{Subtract } \frac{3}{2} \text{ from both sides.}$$

Next find the square of half of $-4$.

$$\frac{1}{2}(-4) = -2 \quad \text{and} \quad (-2)^2 = 4$$

Add 4 to both sides of the equation to complete the square.

$$x^2 - 4x + 4 = -\frac{3}{2} + 4$$

$$(x - 2)^2 = \frac{5}{2} \qquad \text{Factor the perfect square and simplify the right side.}$$

$$x - 2 = \pm\sqrt{\frac{5}{2}} \qquad \text{Apply the square root property.}$$

$$x - 2 = \pm\frac{\sqrt{10}}{2} \qquad \text{Rationalize the denominator.}$$

$$x = 2 \pm \frac{\sqrt{10}}{2} \qquad \text{Add 2 to both sides.}$$

$$= \frac{4}{2} \pm \frac{\sqrt{10}}{2} \qquad \text{Find the common denominator.}$$

$$= \frac{4 \pm \sqrt{10}}{2} \qquad \text{Simplify.}$$

The solutions are $\dfrac{4 + \sqrt{10}}{2}$ and $\dfrac{4 - \sqrt{10}}{2}$. ☐

**7**   Solve: $3x^2 - 12x + 1 = 0$.

---

The following steps may be used to solve a quadratic equation such as $ax^2 + bx + c = 0$ by completing the square. This method may be used whether or not the polynomial $ax^2 + bx + c$ is factorable.

> **Solving a Quadratic Equation in x by Completing the Square**
> **STEP 1.** If the coefficient of $x^2$ is 1, go to Step 2. Otherwise, divide both sides of the equation by the coefficient of $x^2$.
>
> **STEP 2.** Isolate all variable terms on one side of the equation.
>
> **STEP 3.** Complete the square for the resulting binomial by adding the square of half of the coefficient of $x$ to both sides of the equation.
>
> **STEP 4.** Factor the resulting perfect square trinomial and write it as the square of a binomial.
>
> **STEP 5.** Use the square root property to solve for $x$.

**EXAMPLE 8**   Solve $3x^2 - 9x + 8 = 0$ by completing the square.

_Solution_   $3x^2 - 9x + 8 = 0$

**STEP 1.** $x^2 - 3x + \dfrac{8}{3} = 0$       Divide both sides of the equation by 3.

**STEP 2.** $x^2 - 3x = -\dfrac{8}{3}$       Subtract $\dfrac{8}{3}$ from both sides.

Since $\dfrac{1}{2}(-3) = -\dfrac{3}{2}$ and $\left(-\dfrac{3}{2}\right)^2 = \dfrac{9}{4}$, we add $\dfrac{9}{4}$ to both sides of the equation.

**STEP 3.** $x^2 - 3x + \dfrac{9}{4} = -\dfrac{8}{3} + \dfrac{9}{4}$

**STEP 4.** $\left(x - \dfrac{3}{2}\right)^2 = -\dfrac{5}{12}$       Factor the perfect square trinomial.

**STEP 5.** $x - \dfrac{3}{2} = \pm\sqrt{-\dfrac{5}{12}}$       Apply the square root property.

$x - \dfrac{3}{2} = \pm\dfrac{i\sqrt{5}}{2\sqrt{3}}$       Simplify the radical.

$x - \dfrac{3}{2} = \pm\dfrac{i\sqrt{15}}{6}$       Rationalize the denominator.

$x = \dfrac{3}{2} \pm \dfrac{i\sqrt{15}}{6}$       Add $\dfrac{3}{2}$ to both sides.

$= \dfrac{9}{6} \pm \dfrac{i\sqrt{15}}{6}$       Find a common denominator.

$= \dfrac{9 \pm i\sqrt{15}}{6}$       Simplify.

The solutions are $\dfrac{9 + i\sqrt{15}}{6}$ and $\dfrac{9 - i\sqrt{15}}{6}$.

**PRACTICE**
**8**   Solve $2x^2 - 5x + 7 = 0$ by completing the square.

**OBJECTIVE 3 ▶ Solving problems modeled by quadratic equations.** Recall the **simple interest** formula $I = Prt$, where $I$ is the interest earned, $P$ is the principal, $r$ is the rate of interest, and $t$ is time in years. If \$100 is invested at a simple interest rate of 5% annually, at the end of 3 years the total interest $I$ earned is

$$I = P \cdot r \cdot t$$

or

$$I = 100 \cdot 0.05 \cdot 3 = \$15$$

and the new principal is

$$\$100 + \$15 = \$115$$

Most of the time, the interest computed on money borrowed or money deposited is **compound interest.** Compound interest, unlike simple interest, is computed on original principal *and* on interest already earned. To see the difference between simple interest and compound interest, suppose that $100 is invested at a rate of 5% compounded annually. To find the total amount of money at the end of 3 years, we calculate as follows.

$$I = P \cdot r \cdot t$$

First year:　Interest $= \$100 \cdot 0.05 \cdot 1 = \$5.00$
　　　　　　New principal $= \$100.00 + \$5.00 = \$105.00$

Second year:　Interest $= \$105.00 \cdot 0.05 \cdot 1 = \$5.25$
　　　　　　New principal $= \$105.00 + \$5.25 = \$110.25$

Third year:　Interest $= \$110.25 \cdot 0.05 \cdot 1 \approx \$5.51$
　　　　　　New principal $= \$110.25 + \$5.51 = \$115.76$

At the end of the third year, the total compound interest earned is $15.76, whereas the total simple interest earned is $15.

It is tedious to calculate compound interest as we did above, so we use a compound interest formula. The formula for calculating the total amount of money when interest is compounded annually is

$$A = P(1 + r)^t$$

where $P$ is the original investment, $r$ is the interest rate per compounding period, and $t$ is the number of periods. For example, the amount of money $A$ at the end of 3 years if $100 is invested at 5% compounded annually is

$$A = \$100(1 + 0.05)^3 \approx \$100(1.1576) = \$115.76$$

as we previously calculated.

### EXAMPLE 9　Finding Interest Rates

Use the formula $A = P(1 + r)^t$ to find the interest rate $r$ if $2000 compounded annually grows to $2420 in 2 years.

#### Solution

1. UNDERSTAND the problem. Since the $2000 is compounded annually, we use the compound interest formula. For this example, make sure that you understand the formula for compounding interest annually.

2. TRANSLATE. We substitute the given values into the formula.

$$A = P(1 + r)^t$$
$$2420 = 2000(1 + r)^2 \qquad \text{Let } A = 2420, P = 2000, \text{ and } t = 2.$$

3. SOLVE. Solve the equation for $r$.

$$2420 = 2000(1 + r)^2$$
$$\frac{2420}{2000} = (1 + r)^2 \qquad \text{Divide both sides by 2000.}$$
$$\frac{121}{100} = (1 + r)^2 \qquad \text{Simplify the fraction.}$$
$$\pm\sqrt{\frac{121}{100}} = 1 + r \qquad \text{Use the square root property.}$$

*Handwritten margin notes:*

Remember
　isolate the one
we are looking
for.

$A = P(1+r)^t$

To find r

$\dfrac{A}{P} = \dfrac{P(1+r)^t}{P}$

$= (1+r)^t$

$\pm\sqrt{\dfrac{A}{P}} = \left(\sqrt{1+r}\right)^t$

$\pm\sqrt{\dfrac{A}{P}} = 1 + r$

$-1 \pm\sqrt{\dfrac{A}{P}} = r$

$$\pm\frac{11}{10} = 1 + r \qquad\qquad \text{Simplify.}$$

$$-1 \pm \frac{11}{10} = r$$

$$-\frac{10}{10} \pm \frac{11}{10} = r$$

$$\frac{1}{10} = r \quad \text{or} \quad -\frac{21}{10} = r$$

**4. INTERPRET.** The rate cannot be negative, so we reject $-\dfrac{21}{10}$.

**Check:**  $\dfrac{1}{10} = 0.10 = 10\%$ per year. If we invest \$2000 at 10% compounded annually, in 2 years the amount in the account would be $2000(1 + 0.10)^2 = 2420$ dollars, the desired amount.

**State:**   The interest rate is 10% compounded annually.  ☐

#### PRACTICE

**9**   Use the formula from Example 9 to find the interest rate $r$ if \$5000 compounded annually grows to \$5618 in 2 years.

---

### Graphing Calculator Explorations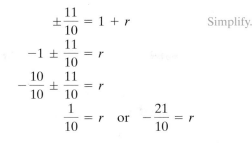

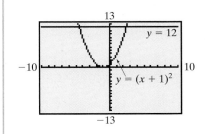

In Section 6.6, we showed how we can use a grapher to approximate real number solutions of a quadratic equation written in standard form. We can also use a grapher to solve a quadratic equation when it is not written in standard form. For example, to solve $(x + 1)^2 = 12$, the quadratic equation in Example 3, we graph the following on the same set of axes. Use Xmin $= -10$, Xmax $= 10$, Ymin $= -13$, and Ymax $= 13$.

$$Y_1 = (x + 1)^2 \quad \text{and} \quad Y_2 = 12$$

Use the Intersect feature or the Zoom and Trace features to locate the points of intersection of the graphs. (See your manuals for specific instructions.) The $x$-values of these points are the solutions of $(x + 1)^2 = 12$. The solutions, rounded to two decimal places, are 2.46 and $-4.46$.

Check to see that these numbers are approximations of the exact solutions $-1 \pm 2\sqrt{3}$.

*Use a graphing calculator to solve each quadratic equation. Round all solutions to the nearest hundredth.*

**1.** $x(x - 5) = 8$                  **2.** $x(x + 2) = 5$

**3.** $x^2 + 0.5x = 0.3x + 1$       **4.** $x^2 - 2.6x = -2.2x + 3$

**5.** Use a graphing calculator and solve $(2x - 5)^2 = -16$, Example 4 in this section, using the window

$$
\begin{aligned}
\text{Xmin} &= -20 \\
\text{Xmax} &= 20 \\
\text{Xscl} &= 1 \\
\text{Ymin} &= -20 \\
\text{Ymax} &= 20 \\
\text{Yscl} &= 1
\end{aligned}
$$

Explain the results. Compare your results with the solution found in Example 4.

**6.** What are the advantages and disadvantages of using a graphing calculator to solve quadratic equations?

# VOCABULARY & READINESS CHECK

*For Exercises 1 through 6, use the choices below to fill in each blank. Not all choices will be used.*

binomial $\quad \sqrt{b} \quad \pm\sqrt{b} \quad b^2 \quad 9 \quad 25 \quad$ completing the square

quadratic $\quad -\sqrt{b} \quad \dfrac{b}{2} \quad \left(\dfrac{b}{2}\right)^2 \quad 3 \quad 5$

**1.** By the square root property, if $b$ is a real number, and $a^2 = b$, then $a =$ _____.

**2.** A _____ equation can be written in the form $ax^2 + bx + c = 0, a \neq 0$.

**3.** The process of writing a quadratic equation so that one side is a perfect square trinomial is called _____.

**4.** A perfect square trinomial is one that can be factored as a _____ squared.

**5.** To solve $x^2 + 6x = 10$ by completing the square, add _____ to both sides.

**6.** To solve $x^2 + bx = c$ by completing the square, add _____ to both sides.

*Fill in the blank with the number needed to make the expression a perfect square trinomial.*

**7.** $m^2 + 2m +$ _____
**8.** $m^2 - 2m +$ _____
**9.** $y^2 - 14y +$ _____
**10.** $z^2 + z +$ ____

## 11.1 | EXERCISE SET

*Use the square root property to solve each equation. These equations have real-number solutions. See Examples 1 through 3.*

**1.** $x^2 = 16$
**2.** $x^2 = 49$
**3.** $x^2 - 7 = 0$
**4.** $x^2 - 11 = 0$
**5.** $x^2 = 18$
**6.** $y^2 = 20$
**7.** $3z^2 - 30 = 0$
**8.** $2x^2 - 4 = 0$
**9.** $(x + 5)^2 = 9$
**10.** $(y - 3)^2 = 4$
**11.** $(z - 6)^2 = 18$
**12.** $(y + 4)^2 = 27$
**13.** $(2x - 3)^2 = 8$
**14.** $(4x + 9)^2 = 6$

*Use the square root property to solve each equation. See Examples 1 through 4.*

**15.** $x^2 + 9 = 0$
**16.** $x^2 + 4 = 0$
**17.** $x^2 - 6 = 0$
**18.** $y^2 - 10 = 0$
**19.** $2z^2 + 16 = 0$
**20.** $3p^2 + 36 = 0$
**21.** $(x - 1)^2 = -16$
**22.** $(y + 2)^2 = -25$
**23.** $(z + 7)^2 = 5$
**24.** $(x + 10)^2 = 11$
**25.** $(x + 3)^2 = -8$
**26.** $(y - 4)^2 = -18$

*Add the proper constant to each binomial so that the resulting trinomial is a perfect square trinomial. Then factor the trinomial.*

**27.** $x^2 + 16x +$ _____
**28.** $y^2 + 2y +$ _____
**29.** $z^2 - 12z +$ _____
**30.** $x^2 - 8x +$ _____
**31.** $p^2 + 9p +$ _____
**32.** $n^2 + 5n +$ _____
**33.** $x^2 + x +$ _____
**34.** $y^2 - y +$ _____

### MIXED PRACTICE

*Solve each equation by completing the square. These equations have real number solutions. See Examples 5 through 7.*

**35.** $x^2 + 8x = -15$
**36.** $y^2 + 6y = -8$
**37.** $x^2 + 6x + 2 = 0$
**38.** $x^2 - 2x - 2 = 0$
**39.** $x^2 + x - 1 = 0$
**40.** $x^2 + 3x - 2 = 0$
**41.** $x^2 + 2x - 5 = 0$
**42.** $y^2 + y - 7 = 0$
**43.** $3p^2 - 12p + 2 = 0$

**44.** $2x^2 + 14x - 1 = 0$

**45.** $4y^2 - 12y - 2 = 0$

**46.** $6x^2 - 3 = 6x$

**47.** $2x^2 + 7x = 4$

**48.** $3x^2 - 4x = 4$

**49.** $x^2 - 4x - 5 = 0$

**50.** $y^2 + 6y - 8 = 0$

**51.** $x^2 + 8x + 1 = 0$

**52.** $x^2 - 10x + 2 = 0$

**53.** $3y^2 + 6y - 4 = 0$

**54.** $2y^2 + 12y + 3 = 0$

**55.** $2x^2 - 3x - 5 = 0$

**56.** $5x^2 + 3x - 2 = 0$

*Solve each equation by completing the square. See Examples 5 through 8.*

**57.** $y^2 + 2y + 2 = 0$

**58.** $x^2 + 4x + 6 = 0$

**59.** $x^2 - 6x + 3 = 0$

**60.** $x^2 - 7x - 1 = 0$

**61.** $2a^2 + 8a = -12$

**62.** $3x^2 + 12x = -14$

**63.** $5x^2 + 15x - 1 = 0$

**64.** $16y^2 + 16y - 1 = 0$

**65.** $2x^2 - x + 6 = 0$

**66.** $4x^2 - 2x + 5 = 0$

**67.** $x^2 + 10x + 28 = 0$

**68.** $y^2 + 8y + 18 = 0$

**69.** $z^2 + 3z - 4 = 0$

**70.** $y^2 + y - 2 = 0$

**71.** $2x^2 - 4x = -3$

**72.** $9x^2 - 36x = -40$

**73.** $3x^2 + 3x = 5$

**74.** $5y^2 - 15y = 1$

*Use the formula $A = P(1 + r)^t$ to solve Exercises 75 through 78. See Example 9.*

**75.** Find the rate $r$ at which $3000 compounded annually grows to $4320 in 2 years.

**76.** Find the rate $r$ at which $800 compounded annually grows to $882 in 2 years.

**77.** Find the rate at which $15,000 compounded annually grows to $16,224 in 2 years.

**78.** Find the rate at which $2000 compounded annually grows to $2880 in 2 years.

**79.** In your own words, what is the difference between simple interest and compound interest?

*Neglecting air resistance, the distance $s(t)$ in feet traveled by a freely falling object is given by the function $s(t) = 16t^2$, where $t$ is time in seconds. Use this formula to solve Exercises 80 through 83. Round answers to two decimal places.*

**80.** The Petronas Towers in Kuala Lumpur, built in 1997, are the tallest buildings in Malaysia. Each tower is 1483 feet tall. How long would it take an object to fall to the ground from the top of one of the towers? (*Source:* Council on Tall Buildings and Urban Habitat, Lehigh University)

**81.** The height of the Chicago Beach Tower Hotel, built in 1998 in Dubai, United Arab Emirates, is 1053 feet. How long would it take an object to fall to the ground from the top of the building? (*Source:* Council on Tall Buildings and Urban Habitat, Lehigh University)

**82.** The height of the Nurek Dam in Tajikistan (part of the former USSR that borders Afghanistan) is 984 feet. How long would it take an object to fall from the top to the base of the dam? (*Source:* U.S. Committee on Large Dams of the International Commission on Large Dams)

**83.** The Hoover Dam, located on the Colorado River on the border of Nevada and Arizona near Las Vegas, is 725 feet tall. How long would it take an object to fall from the top to the base of the dam? (*Source:* U.S. Committee on Large Dams of the International Commission on Large Dams)

**84.** If you are depositing money in an account that pays 4%, would you prefer the interest to be simple or compound? Explain why.

**85.** If you are borrowing money at a rate of 10%, would you prefer the interest to be simple or compound? Explain why.

**REVIEW AND PREVIEW**

*Simplify each expression. See Section 10.1.*

**86.** $\dfrac{3}{4} - \sqrt{\dfrac{25}{16}}$

**87.** $\dfrac{3}{5} + \sqrt{\dfrac{16}{25}}$

**88.** $\dfrac{1}{2} - \sqrt{\dfrac{9}{4}}$

**89.** $\dfrac{9}{10} - \sqrt{\dfrac{49}{100}}$

*Simplify each expression. See Section 10.5.*

**90.** $\dfrac{6 + 4\sqrt{5}}{2}$

**91.** $\dfrac{10 - 20\sqrt{3}}{2}$

**92.** $\dfrac{3 - 9\sqrt{2}}{6}$

**93.** $\dfrac{12 - 8\sqrt{7}}{16}$

*Evaluate $\sqrt{b^2 - 4ac}$ for each set of values. See Section 10.3.*

**94.** $a = 2, b = 4, c = -1$

**95.** $a = 1, b = 6, c = 2$

**96.** $a = 3, b = -1, c = -2$

**97.** $a = 1, b = -3, c = -1$

## CONCEPT EXTENSIONS

*Without solving, determine whether the solutions of each equation are real numbers or complex, but not real numbers. See the Concept Check in this section.*

**98.** $(x + 1)^2 = -1$

**99.** $(y - 5)^2 = -9$

**100.** $3z^2 = 10$

**101.** $4x^2 = 17$

**102.** $(2y - 5)^2 + 7 = 3$

**103.** $(3m + 2)^2 + 4 = 1$

*Find two possible missing terms so that each is a perfect square trinomial.*

**104.** $x^2 + \quad + 16$

**105.** $y^2 + \quad + 9$

**106.** $z^2 + \quad + \dfrac{25}{4}$

**107.** $x^2 + \quad + \dfrac{1}{4}$

*Solve.*

△ **108.** The area of a square room is 225 square feet. Find the dimensions of the room.

△ **109.** The area of a circle is $36\pi$ square inches. Find the radius of the circle.

△ **110.** An isosceles right triangle has legs of equal length. If the hypotenuse is 20 centimeters long, find the length of each leg.

△ **111.** A 27-inch TV is advertised in the *Daily Sentry* newspaper. If 27 inches is the measure of the diagonal of the picture tube, find the measure of each side of the picture tube.

*A common equation used in business is a demand equation. It expresses the relationship between the unit price of some commodity and the quantity demanded. For Exercises 112 and 113, $p$ represents the unit price and $x$ represents the quantity demanded in thousands.*

**112.** A manufacturing company has found that the demand equation for a certain type of scissors is given by the equation $p = -x^2 + 47$. Find the demand for the scissors if the price is $11 per pair.

**113.** Acme, Inc., sells desk lamps and has found that the demand equation for a certain style of desk lamp is given by the equation $p = -x^2 + 15$. Find the demand for the desk lamp if the price is $7 per lamp.

## 11.2 SOLVING QUADRATIC EQUATIONS BY THE QUADRATIC FORMULA

### OBJECTIVES

1 Solve quadratic equations by using the quadratic formula.

2 Determine the number and type of solutions of a quadratic equation by using the discriminant.

3 Solve geometric problems modeled by quadratic equations.

**OBJECTIVE 1 ▶ Solving quadratic equations by using the quadratic formula.** Any quadratic equation can be solved by completing the square. Since the same sequence of steps is repeated each time we complete the square, let's complete the square for a general quadratic equation, $ax^2 + bx + c = 0, a \neq 0$. By doing so, we find a pattern for the solutions of a quadratic equation known as the **quadratic formula.**

Recall that to complete the square for an equation such as $ax^2 + bx + c = 0$, we first divide both sides by the coefficient of $x^2$.

$$ax^2 + bx + c = 0$$

$$x^2 + \dfrac{b}{a}x + \dfrac{c}{a} = 0 \qquad \text{Divide both sides by } a, \text{ the coefficient of } x^2.$$

$$x^2 + \dfrac{b}{a}x = -\dfrac{c}{a} \qquad \text{Subtract the constant } \dfrac{c}{a} \text{ from both sides.}$$

Next, find the square of half $\dfrac{b}{a}$, the coefficient of $x$.

$$\frac{1}{2}\left(\frac{b}{a}\right) = \frac{b}{2a} \quad \text{and} \quad \left(\frac{b}{2a}\right)^2 = \frac{b^2}{4a^2}$$

Add this result to both sides of the equation.

$$x^2 + \frac{b}{a}x + \frac{b^2}{4a^2} = -\frac{c}{a} + \frac{b^2}{4a^2} \qquad \text{Add } \frac{b^2}{4a^2} \text{ to both sides.}$$

$$x^2 + \frac{b}{a}x + \frac{b^2}{4a^2} = \frac{-c \cdot 4a}{a \cdot 4a} + \frac{b^2}{4a^2} \qquad \begin{array}{l}\text{Find a common denominator} \\ \text{on the right side.}\end{array}$$

$$x^2 + \frac{b}{a}x + \frac{b^2}{4a^2} = \frac{b^2 - 4ac}{4a^2} \qquad \text{Simplify the right side.}$$

$$\left(x + \frac{b}{2a}\right)^2 = \frac{b^2 - 4ac}{4a^2} \qquad \begin{array}{l}\text{Factor the perfect square} \\ \text{trinomial on the left side.}\end{array}$$

$$x + \frac{b}{2a} = \pm\sqrt{\frac{b^2 - 4ac}{4a^2}} \qquad \text{Apply the square root property.}$$

$$x + \frac{b}{2a} = \pm\frac{\sqrt{b^2 - 4ac}}{2a} \qquad \text{Simplify the radical.}$$

$$x = -\frac{b}{2a} \pm \frac{\sqrt{b^2 - 4ac}}{2a} \qquad \text{Subtract } \frac{b}{2a} \text{ from both sides.}$$

$$x = \frac{-b \pm \sqrt{b^2 - 4ac}}{2a} \qquad \text{Simplify.}$$

This equation identifies the solutions of the general quadratic equation in standard form and is called the quadratic formula. It can be used to solve any equation written in standard form $ax^2 + bx + c = 0$ as long as $a$ is not 0.

---

**Quadratic Formula**

A quadratic equation written in the form $ax^2 + bx + c = 0$ has the solutions

$$x = \frac{-b \pm \sqrt{b^2 - 4ac}}{2a}$$

---

**EXAMPLE 1**   Solve $3x^2 + 16x + 5 = 0$ for $x$.

**Solution**   This equation is in standard form, so $a = 3$, $b = 16$, and $c = 5$. Substitute these values into the quadratic formula.

$$x = \frac{-b \pm \sqrt{b^2 - 4ac}}{2a} \qquad \text{Quadratic formula}$$

$$= \frac{-16 \pm \sqrt{16^2 - 4(3)(5)}}{2 \cdot 3} \qquad \text{Use } a = 3, b = 16, \text{ and } c = 5.$$

$$= \frac{-16 \pm \sqrt{256 - 60}}{6}$$

$$= \frac{-16 \pm \sqrt{196}}{6} = \frac{-16 \pm 14}{6}$$

$$x = \frac{-16 + 14}{6} = -\frac{1}{3} \quad \text{or} \quad x = \frac{-16 - 14}{6} = -\frac{30}{6} = -5$$

The solutions are $-\dfrac{1}{3}$ and $-5$, or the solution set is $\left\{-\dfrac{1}{3}, -5\right\}$.

**PRACTICE**

**1**   Solve $3x^2 - 5x - 2 = 0$ for $x$.

> ▶ **Helpful Hint**
>
> To replace $a, b,$ and $c$ correctly in the quadratic formula, write the quadratic equation in standard form $ax^2 + bx + c = 0$.

**EXAMPLE 2**  Solve: $2x^2 - 4x = 3$.

*Solution*  First write the equation in standard form by subtracting 3 from both sides.

$$2x^2 - 4x - 3 = 0$$

Now $a = 2, b = -4,$ and $c = -3$. Substitute these values into the quadratic formula.

$$x = \frac{-b \pm \sqrt{b^2 - 4ac}}{2a}$$

$$= \frac{-(-4) \pm \sqrt{(-4)^2 - 4(2)(-3)}}{2 \cdot 2}$$

$$= \frac{4 \pm \sqrt{16 + 24}}{4}$$

$$= \frac{4 \pm \sqrt{40}}{4} = \frac{4 \pm 2\sqrt{10}}{4}$$

$$= \frac{2(2 \pm \sqrt{10})}{2 \cdot 2} = \frac{2 \pm \sqrt{10}}{2}$$

The solutions are $\dfrac{2 + \sqrt{10}}{2}$ and $\dfrac{2 - \sqrt{10}}{2}$, or the solution set is $\left\{ \dfrac{2 - \sqrt{10}}{2}, \dfrac{2 + \sqrt{10}}{2} \right\}$.

**PRACTICE**
**2**  Solve: $3x^2 - 8x = 2$.

---

> ▶ **Helpful Hint**
>
> To simplify the expression $\dfrac{4 \pm 2\sqrt{10}}{4}$ in the preceding example, note that 2 is factored out of both terms of the numerator *before* simplifying.
>
> $$\frac{4 \pm 2\sqrt{10}}{4} = \frac{2(2 \pm \sqrt{10})}{2 \cdot 2} = \frac{2 \pm \sqrt{10}}{2}$$

**Concept Check** ☑

For the quadratic equation $x^2 = 7$, which substitution is correct?

**a.** $a = 1, b = 0,$ and $c = -7$

**b.** $a = 1, b = 0,$ and $c = 7$

**c.** $a = 0, b = 0,$ and $c = 7$

**d.** $a = 1, b = 1,$ and $c = -7$

**EXAMPLE 3**  Solve: $\dfrac{1}{4}m^2 - m + \dfrac{1}{2} = 0$.

*Solution*  We could use the quadratic formula with $a = \dfrac{1}{4}, b = -1,$ and $c = \dfrac{1}{2}$. Instead, we find a simpler, equivalent standard form equation whose coefficients are not fractions. Multiply both sides of the equation by the LCD 4 to clear fractions.

$$4\left( \frac{1}{4}m^2 - m + \frac{1}{2} \right) = 4 \cdot 0$$

$$m^2 - 4m + 2 = 0 \qquad \text{Simplify.}$$

Substitute $a = 1$, $b = -4$, and $c = 2$ into the quadratic formula.

$$m = \frac{-(-4) \pm \sqrt{(-4)^2 - 4(1)(2)}}{2 \cdot 1} = \frac{4 \pm \sqrt{16 - 8}}{2}$$

$$= \frac{4 \pm \sqrt{8}}{2} = \frac{4 \pm 2\sqrt{2}}{2} = \frac{2(2 \pm \sqrt{2})}{2}$$

$$= 2 \pm \sqrt{2}$$

The solutions are $2 + \sqrt{2}$ and $2 - \sqrt{2}$.     □

**PRACTICE**
**3**   Solve: $\dfrac{1}{8}x^2 - \dfrac{1}{4}x - 2 = 0$.

---

**EXAMPLE 4**   Solve: $x = -3x^2 - 3$.

**Solution**   The equation in standard form is $3x^2 + x + 3 = 0$. Thus, let $a = 3$, $b = 1$, and $c = 3$ in the quadratic formula.

$$x = \frac{-1 \pm \sqrt{1^2 - 4(3)(3)}}{2 \cdot 3} = \frac{-1 \pm \sqrt{1 - 36}}{6} = \frac{-1 \pm \sqrt{-35}}{6} = \frac{-1 \pm i\sqrt{35}}{6}$$

The solutions are $\dfrac{-1 + i\sqrt{35}}{6}$ and $\dfrac{-1 - i\sqrt{35}}{6}$.     □

**PRACTICE**
**4**   Solve: $x = -2x^2 - 2$.

---

**Concept Check** ☑

What is the first step in solving $-3x^2 = 5x - 4$ using the quadratic formula?

In Example 1, the equation $3x^2 + 16x + 5 = 0$ had 2 real roots, $-\dfrac{1}{3}$ and $-5$. In Example 4, the equation $3x^2 + x + 3 = 0$ (written in standard form) had no real roots. How do their related graphs compare? Recall that the $x$-intercepts of $f(x) = 3x^2 + 16x + 5$ occur where $f(x) = 0$ or where $3x^2 + 16x + 5 = 0$. Since this equation has 2 real roots, the graph has 2 $x$-intercepts. Similarly, since the equation $3x^2 + x + 3 = 0$ has no real roots, the graph of $f(x) = 3x^2 + x + 3$ has no $x$-intercepts.

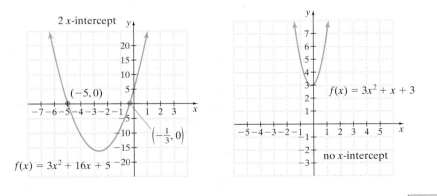

**OBJECTIVE 2 ▶ Using the discriminant.**   In the quadratic formula, $x = \dfrac{-b \pm \sqrt{b^2 - 4ac}}{2a}$,

the radicand $b^2 - 4ac$ is called the **discriminant** because, by knowing its value, we can **discriminate** among the possible number and type of solutions of a quadratic equation. Possible values of the discriminant and their meanings are summarized next.

**Answer to Concept Check:**
Write the equation in standard form.

a and simplify:

**Discriminant**

The following table corresponds the discriminant $b^2 - 4ac$ of a quadratic equation of the form $ax^2 + bx + c = 0$ with the number and type of solutions of the equation.

| $b^2 - 4ac$ | *Number and Type of Solutions* |
|---|---|
| Positive | Two real solutions |
| Zero | One real solution |
| Negative | Two complex but not real solutions |

**EXAMPLE 5**   Use the discriminant to determine the number and type of solutions of each quadratic equation.

**a.** $x^2 + 2x + 1 = 0$      **b.** $3x^2 + 2 = 0$      **c.** $2x^2 - 7x - 4 = 0$

*Solution*

**a.** In $x^2 + 2x + 1 = 0$, $a = 1$, $b = 2$, and $c = 1$. Thus,

$$b^2 - 4ac = 2^2 - 4(1)(1) = 0$$

Since $b^2 - 4ac = 0$, this quadratic equation has one real solution.

**b.** In this equation, $a = 3$, $b = 0$, $c = 2$. Then $b^2 - 4ac = 0 - 4(3)(2) = -24$. Since $b^2 - 4ac$ is negative, the quadratic equation has two complex but not real solutions.

**c.** In this equation, $a = 2$, $b = -7$, and $c = -4$. Then

$$b^2 - 4ac = (-7)^2 - 4(2)(-4) = 81$$

Since $b^2 - 4ac$ is positive, the quadratic equation has two real solutions.   ☐

**PRACTICE**

**5**   Use the discriminant to determine the number and type of solutions of each quadratic equation.

**a.** $x^2 - 6x + 9 = 0$      **b.** $x^2 - 3x - 1 = 0$      **c.** $7x^2 + 11 = 0$

The discriminant helps us determine the number and type of solutions of a quadratic equation, $ax^2 + bx + c = 0$. Recall that the solutions of this equation are the same as the $x$-intercepts of its related graph $f(x) = ax^2 + bx + c$. This means that the discriminant of $ax^2 + bx + c = 0$ also tells us the number of $x$-intercepts for the graph of $f(x) = ax^2 + bx + c$, or equivalently $y = ax^2 + bx + c$.

**Graph of $f(x) = ax^2 + bx + c$ or $y = ax^2 + bx + c$**

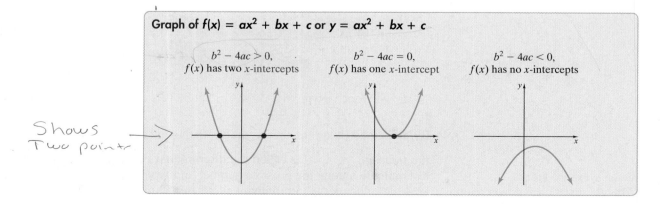

$b^2 - 4ac > 0$, $f(x)$ has two $x$-intercepts   |   $b^2 - 4ac = 0$, $f(x)$ has one $x$-intercept   |   $b^2 - 4ac < 0$, $f(x)$ has no $x$-intercepts

Shows Two points →

$A = p(1+r)^t$

find "r"

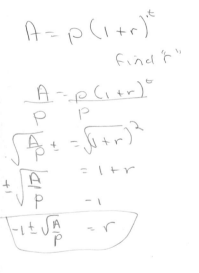

**OBJECTIVE 3 ▶ Solving problems modeled by quadratic equations.** The quadratic formula is useful in solving problems that are modeled by quadratic equations.

⚠ **EXAMPLE 6**   **Calculating Distance Saved**

At a local university, students often leave the sidewalk and cut across the lawn to save walking distance. Given the diagram below of a favorite place to cut across the lawn, approximate how many feet of walking distance a student saves by cutting across the lawn instead of walking on the sidewalk.

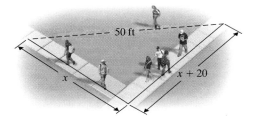

50 ft

$x$

$x + 20$

### Solution

1. **UNDERSTAND.** Read and reread the problem. In the diagram, notice that a triangle is formed. Since the corner of the block forms a right angle, we use the Pythagorean theorem for right triangles. You may want to review this theorem.

2. **TRANSLATE.** By the Pythagorean theorem, we have

$$\text{In words: } (\text{leg})^2 + (\text{leg})^2 = (\text{hypotenuse})^2$$
$$\text{Translate: } x^2 + (x + 20)^2 = 50^2$$

3. **SOLVE.** Use the quadratic formula to solve.

$$x^2 + x^2 + 40x + 400 = 2500 \quad \text{Square } (x + 20) \text{ and 50.}$$
$$2x^2 + 40x - 2100 = 0 \quad \text{Set the equation equal to 0.}$$
$$x^2 + 20x - 1050 = 0 \quad \text{Divide by 2.}$$

Here, $a = 1, b = 20, c = -1050$. By the quadratic formula,

$$x = \frac{-20 \pm \sqrt{20^2 - 4(1)(-1050)}}{2 \cdot 1}$$
$$= \frac{-20 \pm \sqrt{400 + 4200}}{2} = \frac{-20 \pm \sqrt{4600}}{2}$$
$$= \frac{-20 \pm \sqrt{100 \cdot 46}}{2} = \frac{-20 \pm 10\sqrt{46}}{2}$$
$$= -10 \pm 5\sqrt{46} \quad \text{Simplify.}$$

4. **INTERPRET.**

**Check:**  Your calculations in the quadratic formula. The length of a side of a triangle can't be negative, so we reject $-10 - 5\sqrt{46}$. Since $-10 + 5\sqrt{46} \approx 24$ feet, the walking distance along the sidewalk is

$$x + (x + 20) \approx 24 + (24 + 20) = 68 \text{ feet.}$$

**State:**  A student saves about $68 - 50$ or 18 feet of walking distance by cutting across the lawn.  ☐

15 ft

$x$

$x + 3$

**PRACTICE**

**6**   Given the diagram, approximate to the nearest foot how many feet of walking distance a person can save by cutting across the lawn instead of walking on the sidewalk.

**EXAMPLE 7** **Calculating Landing Time**

An object is thrown upward from the top of a 200-foot cliff with a velocity of 12 feet per second. The height $h$ in feet of the object after $t$ seconds is

$$h = -16t^2 + 12t + 200$$

How long after the object is thrown will it strike the ground? Round to the nearest tenth of a second.

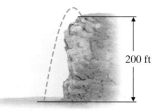

200 ft

## Solution

1. UNDERSTAND. Read and reread the problem.
2. TRANSLATE. Since we want to know when the object strikes the ground, we want to know when the height $h = 0$, or

$$0 = -16t^2 + 12t + 200$$

3. SOLVE. First we divide both sides of the equation by $-4$.

$$0 = 4t^2 - 3t - 50 \quad \text{Divide both sides by } -4.$$

Here, $a = 4$, $b = -3$, and $c = -50$. By the quadratic formula,

$$t = \frac{-(-3) \pm \sqrt{(-3)^2 - 4(4)(-50)}}{2 \cdot 4}$$

$$= \frac{3 \pm \sqrt{9 + 800}}{8}$$

$$= \frac{3 \pm \sqrt{809}}{8}$$

4. INTERPRET.

**Check:** We check our calculations from the quadratic formula. Since the time won't be negative, we reject the proposed solution

$$\frac{3 - \sqrt{809}}{8}.$$

**State:** The time it takes for the object to strike the ground is exactly

$$\frac{3 + \sqrt{809}}{8} \text{ seconds} \approx 3.9 \text{ seconds.} \qquad \square$$

**PRACTICE**

**7** A toy rocket is shot upward at the edge of a building, 45 feet high, with an initial velocity of 20 feet per second. The height $h$ in feet of the rocket after $t$ seconds is

$$h = -16t^2 + 20t + 45$$

How long after the rocket is launched will it strike the ground? Round to the nearest tenth of a second.

## VOCABULARY & READINESS CHECK

*Fill in each blank.*

**1.** The quadratic formula is _____ .

**2.** For $2x^2 + x + 1 = 0$, if $a = 2$, then $b =$ _____ and $c =$ _____ .

**3.** For $5x^2 - 5x - 7 = 0$, if $a = 5$, then $b =$ _____ and $c =$ _____ .

**4.** For $7x^2 - 4 = 0$, if $a = 7$, then $b =$ _____ and $c =$ _____ .

**5.** For $x^2 + 9 = 0$, if $c = 9$, then $a =$ _____ and $b =$ _____ .

**6.** The correct simplified form of $\dfrac{5 \pm 10\sqrt{2}}{5}$ is _____ .

     **a.** $1 \pm 10\sqrt{2}$      **b.** $2\sqrt{2}$      **c.** $1 \pm 2\sqrt{2}$      **d.** $\pm 5\sqrt{2}$

## 11.2 EXERCISE SET

*Use the quadratic formula to solve each equation. These equations have real number solutions only. See Examples 1 through 3.*

**1.** $m^2 + 5m - 6 = 0$

**2.** $p^2 + 11p - 12 = 0$

**3.** $2y = 5y^2 - 3$

**4.** $5x^2 - 3 = 14x$

**5.** $x^2 - 6x + 9 = 0$

**6.** $y^2 + 10y + 25 = 0$

**7.** $x^2 + 7x + 4 = 0$

**8.** $y^2 + 5y + 3 = 0$

**9.** $8m^2 - 2m = 7$

**10.** $11n^2 - 9n = 1$

**11.** $3m^2 - 7m = 3$

**12.** $x^2 - 13 = 5x$

**13.** $\dfrac{1}{2}x^2 - x - 1 = 0$

**14.** $\dfrac{1}{6}x^2 + x + \dfrac{1}{3} = 0$

**15.** $\dfrac{2}{5}y^2 + \dfrac{1}{5}y = \dfrac{3}{5}$

**16.** $\dfrac{1}{8}x^2 + x = \dfrac{5}{2}$

**17.** $\dfrac{1}{3}y^2 = y + \dfrac{1}{6}$

**18.** $\dfrac{1}{2}y^2 = y + \dfrac{1}{2}$

**19.** $x^2 + 5x = -2$

**20.** $y^2 - 8 = 4y$

**21.** $(m + 2)(2m - 6) = 5(m - 1) - 12$

**22.** $7p(p - 2) + 2(p + 4) = 3$

### MIXED PRACTICE

*Use the quadratic formula to solve each equation. These equations have real solutions and complex, but not real, solutions. See Examples 1 through 4.*

**23.** $x^2 + 6x + 13 = 0$

**24.** $x^2 + 2x + 2 = 0$

**25.** $(x + 5)(x - 1) = 2$

**26.** $x(x + 6) = 2$

**27.** $6 = -4x^2 + 3x$

**28.** $2 = -9x^2 - x$

**29.** $\dfrac{x^2}{3} - x = \dfrac{5}{3}$

**30.** $\dfrac{x^2}{2} - 3 = -\dfrac{9}{2}x$

**31.** $10y^2 + 10y + 3 = 0$

**32.** $3y^2 + 6y + 5 = 0$

**33.** $x(6x + 2) = 3$

**34.** $x(7x + 1) = 2$

**35.** $\dfrac{2}{5}y^2 + \dfrac{1}{5}y + \dfrac{3}{5} = 0$

**36.** $\dfrac{1}{8}x^2 + x + \dfrac{5}{2} = 0$

**37.** $\dfrac{1}{2}y^2 = y - \dfrac{1}{2}$

**38.** $\dfrac{2}{3}x^2 - \dfrac{20}{3}x = -\dfrac{100}{6}$

**39.** $(n - 2)^2 = 2n$

**40.** $\left(p - \dfrac{1}{2}\right)^2 = \dfrac{p}{2}$

*Use the discriminant to determine the number and types of solutions of each equation. See Example 5.*

**41.** $x^2 - 5 = 0$

**42.** $x^2 - 7 = 0$

**43.** $4x^2 + 12x = -9$

**44.** $9x^2 + 1 = 6x$

**45.** $3x = -2x^2 + 7$

**46.** $3x^2 = 5 - 7x$

**47.** $6 = 4x - 5x^2$

**48.** $8x = 3 - 9x^2$

**49.** $9x - 2x^2 + 5 = 0$

**50.** $5 - 4x + 12x^2 = 0$

*Solve. See Examples 6 and 7.*

**51.** Nancy, Thelma, and John Varner live on a corner lot. Often, neighborhood children cut across their lot to save walking distance. Given the diagram below, approximate to the nearest foot how many feet of walking distance is saved by cutting across their property instead of walking around the lot.

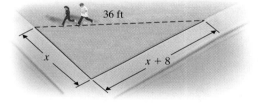

**52.** Given the diagram below, approximate to the nearest foot how many feet of walking distance a person saves by cutting across the lawn instead of walking on the sidewalk.

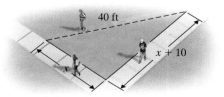

**53.** The hypotenuse of an isosceles right triangle is 2 centimeters longer than either of its legs. Find the exact length of each side. (*Hint:* An isosceles right triangle is a right triangle whose legs are the same length.)

**54.** The hypotenuse of an isosceles right triangle is one meter longer than either of its legs. Find the length of each side.

**55.** Bailey's rectangular dog pen for his Irish setter must have an area of 400 square feet. Also, the length must be 10 feet longer than the width. Find the dimensions of the pen.

**56.** An entry in the Peach Festival Poster Contest must be rectangular and have an area of 1200 square inches. Furthermore, its length must be 20 inches longer than its width. Find the dimensions each entry must have.

**57.** A holding pen for cattle must be square and have a diagonal length of 100 meters.

    **a.** Find the length of a side of the pen.

    **b.** Find the area of the pen.

**58.** A rectangle is three times longer than it is wide. It has a diagonal of length 50 centimeters.

    **a.** Find the dimensions of the rectangle.

    **b.** Find the perimeter of the rectangle.

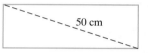

**59.** The heaviest reported door in the world is the 708.6 ton radiation shield door in the National Institute for Fusion Science at Toki, Japan. If the height of the door is 1.1 feet longer than its width, and its front area (neglecting depth) is 1439.9 square feet, find its width and height [Interesting note: the door is 6.6 feet thick.] (*Source: Guiness World Records*)

**60.** Christi and Robbie Wegmann are constructing a rectangular stained glass window whose length is 7.3 inches longer than its width. If the area of the window is 569.9 square inches, find its width and length.

**61.** The base of a triangle is four more than twice its height. If the area of the triangle is 42 square centimeters, find its base and height.

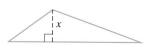

**62.** If a point $B$ divides a line segment such that the smaller portion is to the larger portion as the larger is to the whole, the whole is the length of the *golden ratio.*

The golden ratio was thought by the Greeks to be the most pleasing to the eye, and many of their buildings contained numerous examples of the golden ratio. The value of the golden ratio is the positive solution of

$$\text{(smaller)} \quad \frac{x - 1}{1} = \frac{1}{x} \quad \text{(larger)} \atop \text{(larger)} \qquad\qquad\qquad \text{(whole)}$$

Find this value.

*The Wollomombi Falls in Australia have a height of 1100 feet. A pebble is thrown upward from the top of the falls with an initial velocity of 20 feet per second. The height of the pebble h after t seconds is given by the equation $h = -16t^2 + 20t + 1100$. Use this equation for Exercises 63 and 64.*

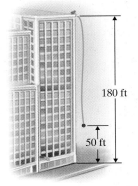

**63.** How long after the pebble is thrown will it hit the ground? Round to the nearest tenth of a second.

**64.** How long after the pebble is thrown will it be 550 feet from the ground? Round to the nearest tenth of a second.

*A ball is thrown downward from the top of a 180-foot building with an initial velocity of 20 feet per second. The height of the ball h after t seconds is given by the equation $h = -16t^2 - 20t + 180$. Use this equation to answer Exercises 65 and 66.*

**65.** How long after the ball is thrown will it strike the ground? Round the result to the nearest tenth of a second.

**66.** How long after the ball is thrown will it be 50 feet from the ground? Round the result to the nearest tenth of a second.

**REVIEW AND PREVIEW**

*Solve each equation. See Sections 7.5 and 10.6.*

**67.** $\sqrt{5x - 2} = 3$

**68.** $\sqrt{y + 2} + 7 = 12$

**69.** $\frac{1}{x} + \frac{2}{5} = \frac{7}{x}$

**70.** $\frac{10}{z} = \frac{5}{z} - \frac{1}{3}$

*Factor. See Sections 6.3 through 6.5.*

**71.** $x^4 + x^2 - 20$

**72.** $2y^4 + 11y^2 - 6$

**73.** $z^4 - 13z^2 + 36$

**74.** $x^4 - 1$

**CONCEPT EXTENSIONS**

*For each quadratic equation, choose the correct substitution for a, b, and c in the standard form $ax^2 + bx + c = 0$.*

**75.** $x^2 = -10$
   **a.** $a = 1, b = 0, c = -10$
   **b.** $a = 1, b = 0, c = 10$
   **c.** $a = 0, b = 1, c = -10$
   **d.** $a = 1, b = 1, c = 10$

**76.** $x^2 + 5 = -x$
   **a.** $a = 1, b = 5, c = -1$
   **b.** $a = 1, b = -1, c = 5$
   **c.** $a = 1, b = 5, c = 1$
   **d.** $a = 1, b = 1, c = 5$

**77.** Solve Exercise 1 by factoring. Explain the result.

**78.** Solve Exercise 2 by factoring. Explain the result.

*Use the quadratic formula and a calculator to approximate each solution to the nearest tenth.*

**79.** $2x^2 - 6x + 3 = 0$

**80.** $3.6x^2 + 1.8x - 4.3 = 0$

*The accompanying graph shows the daily low temperatures for one week in New Orleans, Louisiana.*

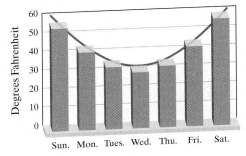

180 ft

50 ft

**81.** Which day of the week shows the greatest decrease in low temperature?

**82.** Which day of the week shows the greatest increase in low temperature?

**83.** Which day of the week had the lowest low temperature?

**84.** Use the graph to estimate the low temperature on Thursday.

*Notice that the shape of the temperature graph is similar to the curve drawn. In fact, this graph can be modeled by the quadratic function $f(x) = 3x^2 - 18x + 56$, where $f(x)$ is the temperature in degrees Fahrenheit and x is the number of days from Sunday. (This graph is shown in blue.) Use this function to answer Exercises 85 and 86.*

**85.** Use the quadratic function given to approximate the temperature on Thursday. Does your answer agree with the graph?

**86.** Use the function given and the quadratic formula to find when the temperature was $35°$ F. [*Hint:* Let $f(x) = 35$ and solve for $x$.] Round your answer to one decimal place and interpret your result. Does your answer agree with the graph?

**87.** The number of Starbucks stores can be modeled by the quadratic function $f(x) = 115x^2 + 711x + 3946$, where $f(x)$ is the number of Starbucks and x is the number of years after 2000. (*Source: Starbuck's Annual Report 2006*)

   **a.** Find the number of Starbucks in 2004.

   **b.** If the trend described by the model continues, predict the years after 2000 in which the number of Starbucks will be 25,000. Round to the nearest whole year.

**88.** The number of visitors to U.S. theme parks can be modeled by the quadratic equation $v(x) = 0.25x^2 + 2.6x + 315.6$, where $v(x)$ is the number of visitors (in millions) and x is the number of years after 2000. (*Source:* Price Waterhouse Coopers)

   **a.** Find the number of visitors to U.S. theme parks in 2005. Round to the nearest million.

   **b.** Find the projected number of visitors to U.S. theme parks in 2010. Round to the nearest million.

*The solutions of the quadratic equation $ax^2 + bx + c = 0$ are $\dfrac{-b + \sqrt{b^2 - 4ac}}{2a}$ and $\dfrac{-b - \sqrt{b^2 - 4ac}}{2a}$.*

**89.** Show that the sum of these solutions is $\dfrac{-b}{a}$.

**90.** Show that the product of these solutions is $\dfrac{c}{a}$.

*Use the quadratic formula to solve each quadratic equation.*

**91.** $3x^2 - \sqrt{12}x + 1 = 0$,
   (*Hint:* $a = 3, b = -\sqrt{12}, c = 1$)

**92.** $5x^2 + \sqrt{20}x + 1 = 0$

**93.** $x^2 + \sqrt{2}x + 1 = 0$

**94.** $x^2 - \sqrt{2}x + 1 = 0$

**95.** $2x^2 - \sqrt{3}x - 1 = 0$

**96.** $7x^2 + \sqrt{7}x - 2 = 0$

**97.** Use a graphing calculator to solve Exercises 63 and 65.

**98.** Use a graphing calculator to solve Exercises 64 and 66.

*Recall that the discriminant also tells us the number of x-intercepts of the related function.*

**99.** Check the results of Exercise 49 by graphing $y = 9x - 2x^2 + 5$.

**100.** Check the results of Exercise 50 by graphing $y = 5 - 4x + 12x^2$.

---

📖 **STUDY SKILLS BUILDER**

**How Well Do You Know Your Textbook?**

Let's check to see whether you are familiar with your textbook yet. For help, see Section 1.1 in this text.

**1.** What does the 🖐 icon mean?

**2.** What does the ＼ icon mean?

**3.** What does the △ icon mean?

**4.** Where can you find a review for each chapter? What answers to this review can be found in the back of your text?

**5.** Each chapter contains an overview of the chapter along with examples. What is this feature called?

**6.** Each chapter contains a review of vocabulary. What is this feature called?

**7.** There are free CDs in your text. What content is contained on these CDs?

**8.** What is the location of the section that is entirely devoted to study skills?

**9.** There are Practice Problems that are contained in the margin of the text. What are they and how can they be used?

# 11.3 SOLVING EQUATIONS BY USING QUADRATIC METHODS

## OBJECTIVES

1 Solve various equations that are quadratic in form.

2 Solve problems that lead to quadratic equations.

**OBJECTIVE 1 ▶ Solving equations that are quadratic in form.** In this section, we discuss various types of equations that can be solved in part by using the methods for solving quadratic equations.

Once each equation is simplified, you may want to use these steps when deciding what method to use to solve the quadratic equation.

---

**Solving a Quadratic Equation**

**STEP 1.** If the equation is in the form $(ax + b)^2 = c$, use the square root property and solve. If not, go to Step 2.

**STEP 2.** Write the equation in standard form: $ax^2 + bx + c = 0$.

**STEP 3.** Try to solve the equation by the factoring method. If not possible, go to Step 4.

**STEP 4.** Solve the equation by the quadratic formula.

---

The first example is a radical equation that becomes a quadratic equation once we square both sides.

**EXAMPLE 1** Solve: $x - \sqrt{x} - 6 = 0$.

*Solution* Recall that to solve a radical equation, first get the radical alone on one side of the equation. Then square both sides.

$$x - 6 = \sqrt{x} \qquad \text{Add } \sqrt{x} \text{ to both sides.}$$
$$(x - 6)^2 = \left(\sqrt{x}\right)^2 \qquad \text{Square both sides.}$$
$$x^2 - 12x + 36 = x$$
$$x^2 - 13x + 36 = 0 \qquad \text{Set the equation equal to 0.}$$
$$(x - 9)(x - 4) = 0$$
$$x - 9 = 0 \quad \text{or} \quad x - 4 = 0$$
$$x = 9 \qquad\qquad x = 4$$

**Check:**

| Let $x = 9$ | Let $x = 4$ |
|---|---|
| $x - \sqrt{x} - 6 = 0$ | $x - \sqrt{x} - 6 = 0$ |
| $9 - \sqrt{9} - 6 \stackrel{?}{=} 0$ | $4 - \sqrt{4} - 6 \stackrel{?}{=} 0$ |
| $9 - 3 - 6 \stackrel{?}{=} 0$ | $4 - 2 - 6 \stackrel{?}{=} 0$ |
| $0 = 0$  True | $-4 = 0$  False |

The solution is 9 or the solution set is $\{9\}$. ☐

**PRACTICE**

**1** Solve: $x - \sqrt{x + 1} - 5 = 0$.

---

**EXAMPLE 2** Solve: $\dfrac{3x}{x - 2} - \dfrac{x + 1}{x} = \dfrac{6}{x(x - 2)}$.

*Solution* In this equation, $x$ cannot be either 2 or 0, because these values cause denominators to equal zero. To solve for $x$, we first multiply both sides of the equation by

$x(x - 2)$ to clear the fractions. By the distributive property, this means that we multiply each term by $x(x - 2)$.

$$x(x - 2)\left(\frac{3x}{x - 2}\right) - x(x - 2)\left(\frac{x + 1}{x}\right) = x(x - 2)\left[\frac{6}{x(x - 2)}\right]$$

$$3x^2 - (x - 2)(x + 1) = 6 \quad \text{Simplify.}$$
$$3x^2 - (x^2 - x - 2) = 6 \quad \text{Multiply.}$$
$$3x^2 - x^2 + x + 2 = 6$$
$$2x^2 + x - 4 = 0 \quad \text{Simplify.}$$

This equation cannot be factored using integers, so we solve by the quadratic formula.

$$x = \frac{-1 \pm \sqrt{1^2 - 4(2)(-4)}}{2 \cdot 2} \quad \text{Use } a = 2, b = 1, \text{ and } c = -4 \text{ in the quadratic formula.}$$

$$= \frac{-1 \pm \sqrt{1 + 32}}{4} \quad \text{Simplify.}$$

$$= \frac{-1 \pm \sqrt{33}}{4}$$

Neither proposed solution will make the denominators 0.

The solutions are $\dfrac{-1 + \sqrt{33}}{4}$ and $\dfrac{-1 - \sqrt{33}}{4}$ or the solution set is $\left\{\dfrac{-1 + \sqrt{33}}{4}, \dfrac{-1 - \sqrt{33}}{4}\right\}$.

**PRACTICE**
**2** Solve: $\dfrac{5x}{x + 1} - \dfrac{x + 4}{x} = \dfrac{3}{x(x + 1)}$.

**EXAMPLE 3** Solve: $p^4 - 3p^2 - 4 = 0$.

*Solution* First we factor the trinomial.

$$p^4 - 3p^2 - 4 = 0$$
$$(p^2 - 4)(p^2 + 1) = 0 \qquad\qquad \text{Factor.}$$
$$(p - 2)(p + 2)(p^2 + 1) = 0 \qquad\qquad \text{Factor further.}$$
$$p - 2 = 0 \quad \text{or} \quad p + 2 = 0 \quad \text{or} \quad p^2 + 1 = 0 \quad \text{Set each factor equal to 0 and solve.}$$
$$p = 2 \qquad\qquad p = -2 \qquad\qquad p^2 = -1$$
$$p = \pm\sqrt{-1} = \pm i$$

The solutions are $2, -2, i$ and $-i$.

**PRACTICE**
**3** Solve: $p^4 - 7p^2 - 144 = 0$.

> ▶ **Helpful Hint**
>
> Example 3 can be solved using substitution also. Think of $p^4 - 3p^2 - 4 = 0$ as
>
> $$(p^2)^2 - 3p^2 - 4 = 0 \quad \text{Then let } x = p^2, \text{ and solve and substitute back. The solutions will be the same.}$$
> $$x^2 - 3x - 4 = 0$$

**Concept Check** ☑

**a.** True or false? The maximum number of solutions that a quadratic equation can have is 2.

**b.** True or false? The maximum number of solutions that an equation in quadratic form can have is 2.

**EXAMPLE 4**  Solve: $(x - 3)^2 - 3(x - 3) - 4 = 0$.

*Solution*  Notice that the quantity $(x - 3)$ is repeated in this equation. Sometimes it is helpful to substitute a variable (in this case other than $x$) for the repeated quantity. We will let $y = x - 3$. Then

$$(x - 3)^2 - 3(x - 3) - 4 = 0$$

becomes

$$y^2 - 3y - 4 = 0 \quad \text{Let } x - 3 = y.$$
$$(y - 4)(y + 1) = 0 \quad \text{Factor.}$$

To solve, we use the zero factor property.

$$y - 4 = 0 \quad \text{or} \quad y + 1 = 0 \quad \text{Set each factor equal to 0.}$$
$$y = 4 \qquad\qquad y = -1 \quad \text{Solve.}$$

> ▶ **Helpful Hint**
> When using substitution, don't forget to substitute back to the original variable.

To find values of $x$, we substitute back. That is, we substitute $x - 3$ for $y$.

$$x - 3 = 4 \quad \text{or} \quad x - 3 = -1$$
$$x = 7 \qquad\qquad x = 2$$

Both 2 and 7 check. The solutions are 2 and 7.  □

**PRACTICE**
**4**  Solve: $(x + 2)^2 - 2(x + 2) - 3 = 0$.

**EXAMPLE 5**  Solve: $x^{2/3} - 5x^{1/3} + 6 = 0$.

*Solution*  The key to solving this equation is recognizing that $x^{2/3} = (x^{1/3})^2$. We replace $x^{1/3}$ with $m$ so that

$$(x^{1/3})^2 - 5x^{1/3} + 6 = 0$$

becomes

$$m^2 - 5m + 6 = 0$$

Now we solve by factoring.

$$m^2 - 5m + 6 = 0$$
$$(m - 3)(m - 2) = 0 \qquad\qquad \text{Factor.}$$
$$m - 3 = 0 \quad \text{or} \quad m - 2 = 0 \quad \text{Set each factor equal to 0.}$$
$$m = 3 \qquad\qquad m = 2$$

Since $m = x^{1/3}$, we have

$$x^{1/3} = 3 \qquad \text{or} \quad x^{1/3} = 2$$
$$x = 3^3 = 27 \quad \text{or} \qquad x = 2^3 = 8$$

Both 8 and 27 check. The solutions are 8 and 27.  □

**PRACTICE**
**5**  Solve: $x^{2/3} - 5x^{1/3} + 4 = 0$.

**OBJECTIVE 2** ▶ **Solving problems that lead to quadratic equations.** The next example is a work problem. This problem is modeled by a rational equation that simplifies to a quadratic equation.

**EXAMPLE 6**   **Finding Work Time**

Together, an experienced word processor and an apprentice word processor can create a word document in 6 hours. Alone, the experienced word processor can create the document 2 hours faster than the apprentice word processor can. Find the time in which each person can create the word document alone.

*Solution*

1. UNDERSTAND. Read and reread the problem. The key idea here is the relationship between the *time* (hours) it takes to complete the job and the *part of the job* completed in one unit of time (hour). For example, because they can complete the job together in 6 hours, the *part of the job* they can complete in 1 hour is $\frac{1}{6}$.

   Let

   $x$ = the *time* in hours it takes the apprentice word processor to complete the job alone

   $x - 2$ = the *time* in hours it takes the experienced word processor to complete the job alone

We can summarize in a chart the information discussed

|  | *Total Hours to Complete Job* | *Part of Job Completed in 1 Hour* |
|---|---|---|
| *Apprentice Word Processor* | $x$ | $\frac{1}{x}$ |
| *Experienced Word Processor* | $x - 2$ | $\frac{1}{x-2}$ |
| *Together* | 6 | $\frac{1}{6}$ |

2. TRANSLATE.

| In words: | part of job completed by apprentice word processor in 1 hour | added to | part of job completed by experienced word processor in 1 hour | is equal to | part of job completed together in 1 hour |
|---|---|---|---|---|---|
| | ↓ | ↓ | ↓ | ↓ | ↓ |
| Translate: | $\frac{1}{x}$ | $+$ | $\frac{1}{x-2}$ | $=$ | $\frac{1}{6}$ |

3. SOLVE.

$$\frac{1}{x} + \frac{1}{x-2} = \frac{1}{6}$$

$$6x(x-2)\left(\frac{1}{x} + \frac{1}{x-2}\right) = 6x(x-2)\cdot\frac{1}{6} \quad \text{Multiply both sides by the LCD } 6x(x-2).$$

$$6x(x-2)\cdot\frac{1}{x} + 6x(x-2)\cdot\frac{1}{x-2} = 6x(x-2)\cdot\frac{1}{6} \quad \text{Use the distributive property.}$$

$$6(x-2) + 6x = x(x-2)$$

$$6x - 12 + 6x = x^2 - 2x$$

$$0 = x^2 - 14x + 12$$

Now we can substitute $a = 1$, $b = -14$, and $c = 12$ into the quadratic formula and simplify.

$$x = \frac{-(-14) \pm \sqrt{(-14)^2 - 4(1)(12)}}{2\cdot 1} = \frac{14 \pm \sqrt{148}}{2}^*$$

(*This expression can be simplified further, but this will suffice as we are approximating.)

Using a calculator or a square root table, we see that $\sqrt{148} \approx 12.2$ rounded to one decimal place. Thus,

$$x \approx \frac{14 \pm 12.2}{2}$$

$$x \approx \frac{14 + 12.2}{2} = 13.1 \quad \text{or} \quad x \approx \frac{14 - 12.2}{2} = 0.9$$

**4. INTERPRET.**

**Check:** If the apprentice word processor completes the job alone in 0.9 hours, the experienced word processor completes the job alone in $x - 2 = 0.9 - 2 = -1.1$ hours. Since this is not possible, we reject the solution of 0.9. The approximate solution thus is 13.1 hours.

**State:** The apprentice word processor can complete the job alone in approximately 13.1 hours, and the experienced word processor can complete the job alone in approximately

$$x - 2 = 13.1 - 2 = 11.1 \text{ hours.} \qquad \square$$

**PRACTICE**

**6** Together, Katy and Steve can groom all the dogs at the Barkin' Doggie Day Care in 4 hours. Alone, Katy can groom the dogs 1 hour faster than Steve can groom the dogs alone. Find the time in which each of them can groom the dogs alone.

---

**EXAMPLE 7** **Finding Driving Speeds**

Beach and Fargo are about 400 miles apart. A salesperson travels from Fargo to Beach one day at a certain speed. She returns to Fargo the next day and drives 10 mph faster. Her total travel time was $14\frac{2}{3}$ hours. Find her speed to Beach and the return speed to Fargo.

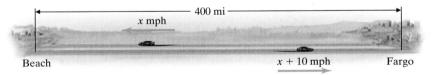

**Solution**

**1. UNDERSTAND.** Read and reread the problem. Let

$$x = \text{the speed to Beach, so}$$
$$x + 10 = \text{the return speed to Fargo.}$$

Then organize the given information in a table.

> **Helpful Hint**
> Since $d = rt$, then $t = \dfrac{d}{r}$. The time column was completed using $\dfrac{d}{r}$.

|  | distance | = | rate | · | time |  |
|---|---|---|---|---|---|---|
| **To Beach** | 400 |  | $x$ |  | $\dfrac{400}{x}$ | ← distance<br>← rate |
| **Return to Fargo** | 400 |  | $x + 10$ |  | $\dfrac{400}{x + 10}$ | ← distance<br>← rate |

**2.** TRANSLATE.

| | time to Beach | + | return time to Fargo | = | $14\frac{2}{3}$ hours |
|---|---|---|---|---|---|

In words: ↓ ↓ ↓

Translate: $\dfrac{400}{x}$ + $\dfrac{400}{x+10}$ = $\dfrac{44}{3}$

**3.** SOLVE.

$$\frac{400}{x} + \frac{400}{x+10} = \frac{44}{3}$$

$$\frac{100}{x} + \frac{100}{x+10} = \frac{11}{3} \qquad \text{Divide both sides by 4.}$$

$$3x(x+10)\left(\frac{100}{x} + \frac{100}{x+10}\right) = 3x(x+10)\cdot\frac{11}{3} \qquad \begin{array}{l}\text{Multiply both sides by}\\ \text{the LCD } 3x(x+10).\end{array}$$

$$3x(x+10)\cdot\frac{100}{x} + 3x(x+10)\cdot\frac{100}{x+10} = 3x(x+10)\cdot\frac{11}{3} \qquad \begin{array}{l}\text{Use the distributive}\\ \text{property.}\end{array}$$

$$3(x+10)\cdot 100 + 3x\cdot 100 = x(x+10)\cdot 11$$

$$300x + 3000 + 300x = 11x^2 + 110x$$

$$0 = 11x^2 - 490x - 3000 \qquad \text{Set equation equal to 0.}$$

$$0 = (11x+60)(x-50) \qquad \text{Factor.}$$

$$11x + 60 = 0 \quad \text{or} \quad x - 50 = 0 \qquad \begin{array}{l}\text{Set each factor equal}\\ \text{to 0.}\end{array}$$

$$x = -\frac{60}{11} \text{ or } -5\frac{5}{11}; \quad x = 50$$

**4.** INTERPRET.

**Check:** The speed is not negative, so it's not $-5\frac{5}{11}$. The number 50 does check.

**State:** The speed to Beach was 50 mph and her return speed to Fargo was 60 mph. □

**PRACTICE**
**7** The 36-km S-shaped Hangzhou Bay Bridge is the longest cross-sea bridge in the world, linking Ningbo and Shanghai, China. A merchant drives over the bridge one morning from Ningbo to Shanghai in very heavy traffic and returns home that night driving 50 km per hour faster. The total travel time was 1.3 hours. Find the speed to Shanghai and the return speed to Ningbo.

---

## 11.3 | EXERCISE SET

**MyMathLab** Powered by CourseCompass™ and MathXL®

Math XL PRACTICE    WATCH    DOWNLOAD    READ    REVIEW

*Solve. See Example 1.*

**1.** $2x = \sqrt{10 + 3x}$

**2.** $3x = \sqrt{8x + 1}$

**3.** $x - 2\sqrt{x} = 8$

**4.** $x - \sqrt{2x} = 4$

**5.** $\sqrt{9x} = x + 2$

**6.** $\sqrt{16x} = x + 3$

*Solve. See Example 2.*

**7.** $\dfrac{2}{x} + \dfrac{3}{x-1} = 1$

**8.** $\dfrac{6}{x^2} = \dfrac{3}{x+1}$

**9.** $\dfrac{3}{x} + \dfrac{4}{x+2} = 2$

**10.** $\dfrac{5}{x-2} + \dfrac{4}{x+2} = 1$

**11.** $\dfrac{7}{x^2 - 5x + 6} = \dfrac{2x}{x-3} - \dfrac{x}{x-2}$

**12.** $\dfrac{11}{2x^2 + x - 15} = \dfrac{5}{2x-5} - \dfrac{x}{x+3}$

*Solve. See Example 3.*

**13.** $p^4 - 16 = 0$

**14.** $x^4 + 2x^2 - 3 = 0$

**15.** $4x^4 + 11x^2 = 3$

**16.** $z^4 = 81$

**17.** $z^4 - 13z^2 + 36 = 0$

**18.** $9x^4 + 5x^2 - 4 = 0$

*Solve. See Examples 4 and 5.*

**19.** $x^{2/3} - 3x^{1/3} - 10 = 0$

**20.** $x^{2/3} + 2x^{1/3} + 1 = 0$

**21.** $(5n + 1)^2 + 2(5n + 1) - 3 = 0$

**22.** $(m - 6)^2 + 5(m - 6) + 4 = 0$

**23.** $2x^{2/3} - 5x^{1/3} = 3$

**24.** $3x^{2/3} + 11x^{1/3} = 4$

**25.** $1 + \dfrac{2}{3t - 2} = \dfrac{8}{(3t - 2)^2}$

**26.** $2 - \dfrac{7}{x + 6} = \dfrac{15}{(x + 6)^2}$

**27.** $20x^{2/3} - 6x^{1/3} - 2 = 0$

**28.** $4x^{2/3} + 16x^{1/3} = -15$

## MIXED PRACTICE

*Solve. See Examples 1 through 5.*

**29.** $a^4 - 5a^2 + 6 = 0$

**30.** $x^4 - 12x^2 + 11 = 0$

**31.** $\dfrac{2x}{x - 2} + \dfrac{x}{x + 3} = -\dfrac{5}{x + 3}$

**32.** $\dfrac{5}{x - 3} + \dfrac{x}{x + 3} = \dfrac{19}{x^2 - 9}$

**33.** $(p + 2)^2 = 9(p + 2) - 20$

**34.** $2(4m - 3)^2 - 9(4m - 3) = 5$

**35.** $2x = \sqrt{11x + 3}$

**36.** $4x = \sqrt{2x + 3}$

**37.** $x^{2/3} - 8x^{1/3} + 15 = 0$

**38.** $x^{2/3} - 2x^{1/3} - 8 = 0$

**39.** $y^3 + 9y - y^2 - 9 = 0$

**40.** $x^3 + x - 3x^2 - 3 = 0$

**41.** $2x^{2/3} + 3x^{1/3} - 2 = 0$

**42.** $6x^{2/3} - 25x^{1/3} - 25 = 0$

**43.** $x^{-2} - x^{-1} - 6 = 0$

**44.** $y^{-2} - 8y^{-1} + 7 = 0$

**45.** $x - \sqrt{x} = 2$

**46.** $x - \sqrt{3x} = 6$

**47.** $\dfrac{x}{x - 1} + \dfrac{1}{x + 1} = \dfrac{2}{x^2 - 1}$

**48.** $\dfrac{x}{x - 5} + \dfrac{5}{x + 5} = -\dfrac{1}{x^2 - 25}$

**49.** $p^4 - p^2 - 20 = 0$

**50.** $x^4 - 10x^2 + 9 = 0$

**51.** $(x + 3)(x^2 - 3x + 9) = 0$

**52.** $(x - 6)(x^2 + 6x + 36) = 0$

**53.** $1 = \dfrac{4}{x - 7} + \dfrac{5}{(x - 7)^2}$

**54.** $3 + \dfrac{1}{2p + 4} = \dfrac{10}{(2p + 4)^2}$

**55.** $27y^4 + 15y^2 = 2$

**56.** $8z^4 + 14z^2 = -5$

*Solve. See Examples 6 and 7.*

**57.** A jogger ran 3 miles, decreased her speed by 1 mile per hour, and then ran another 4 miles. If her total time jogging was $1\frac{3}{5}$ hours, find her speed for each part of her run.

**58.** Mark Keaton's workout consists of jogging for 3 miles, and then riding his bike for 5 miles at a speed 4 miles per hour faster than he jogs. If his total workout time is 1 hour, find his jogging speed and his biking speed.

**59.** A Chinese restaurant in Mandeville, Louisiana, has a large goldfish pond around the restaurant. Suppose that an inlet pipe and a hose together can fill the pond in 8 hours. The inlet pipe alone can complete the job in one hour less time than the hose alone. Find the time that the hose can complete the job alone and the time that the inlet pipe can complete the job alone. Round each to the nearest tenth of an hour.

**60.** A water tank on a farm in Flatonia, Texas, can be filled with a large inlet pipe and a small inlet pipe in 3 hours. The large inlet pipe alone can fill the tank in 2 hours less time than the small inlet pipe alone. Find the time to the nearest tenth of an hour each pipe can fill the tank alone.

**61.** Roma Sherry drove 330 miles from her hometown to Tucson. During her return trip, she was able to increase her speed by 11 mph. If her return trip took 1 hour less time, find her original speed and her speed returning home.

**62.** A salesperson drove to Portland, a distance of 300 miles. During the last 80 miles of his trip, heavy rainfall forced him to decrease his speed by 15 mph. If his total driving time was 6 hours, find his original speed and his speed during the rainfall.

**63.** Bill Shaughnessy and his son Billy can clean the house together in 4 hours. When the son works alone, it takes him an hour longer to clean than it takes his dad alone. Find how long to the nearest tenth of an hour it takes the son to clean alone.

**64.** Together, Noodles and Freckles eat a 50-pound bag of dog food in 30 days. Noodles by himself eats a 50-pound bag in 2 weeks less time than Freckles does by himself. How many days to the nearest whole day would a 50-pound bag of dog food last Freckles?

**65.** The product of a number and 4 less than the number is 96. Find the number.

**66.** A whole number increased by its square is two more than twice itself. Find the number.

△ **67.** Suppose that an open box is to be made from a square sheet of cardboard by cutting out squares from each corner as shown and then folding along the dotted lines. If the box is to have a volume of 300 cubic centimeters, find the original dimensions of the sheet of cardboard.

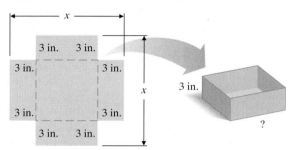

**a.** The ? in the drawing above will be the length (and also the width) of the box as shown. Represent this length in terms of $x$.

**b.** Use the formula for volume of a box, $V = l \cdot w \cdot h$, to write an equation in $x$.

**c.** Solve the equation for $x$ and give the dimensions of the sheet of cardboard. Check your solution.

△ **68.** Suppose that an open box is to be made from a square sheet of cardboard by cutting out squares from each corner as shown and then folding along the dotted lines. If the box is to have a volume of 128 cubic inches, find the original dimensions of the sheet of cardboard.

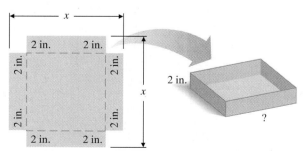

**a.** The ? in the drawing above will be the length (and also the width) of the box as shown. Represent this length in terms of $x$.

**b.** Use the formula for volume of a box, $V = l \cdot w \cdot h$, to write an equation in $x$.

**c.** Solve the equation for $x$ and give the dimensions of the sheet of cardboard. Check your solution.

△ **69.** A sprinkler that sprays water in a circular motion is to be used to water a square garden. If the area of the garden is 920 square feet, find the smallest whole number *radius* that the sprinkler can be adjusted to so that the entire garden is watered.

△ **70.** Suppose that a square field has an area of 6270 square feet. See Exercise 69 and find a new sprinkler radius.

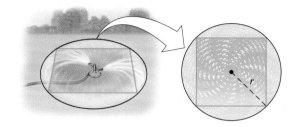

**REVIEW AND PREVIEW**

*Solve each inequality. See Section 2.8.*

**71.** $\dfrac{5x}{3} + 2 \le 7$

**72.** $\dfrac{2x}{3} + \dfrac{1}{6} \ge 2$

**73.** $\dfrac{y-1}{15} > -\dfrac{2}{5}$

**74.** $\dfrac{z-2}{12} < \dfrac{1}{4}$

*Find the domain and range of each graphed relation. Decide which relations are also functions. See Section 3.6.*

**75.**

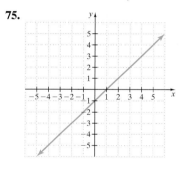

**76.**

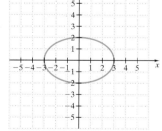

**77.**

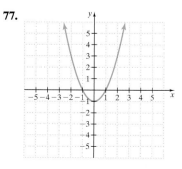

**78.**

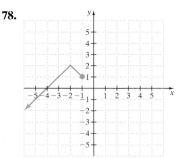

**CONCEPT EXTENSIONS**

*Solve.*

**79.** $y^3 + 9y - y^2 - 9 = 0$

**80.** $x^3 + x - 3x^2 - 3 = 0$

**81.** $x^{-2} - x^{-1} - 6 = 0$

**82.** $y^{-2} - 8y^{-1} + 7 = 0$

**83.** $2x^3 = -54$

**84.** $y^3 - 216 = 0$

**85.** Write a polynomial equation that has three solutions: 2, 5, and $-7$.

**86.** Write a polynomial equation that has three solutions: 0, $2i$, and $-2i$.

**87.** At the 2007 Grand Prix of Long Beach auto race, Simon Pagenaud posted the fastest lap speed, but Sebastian Bourdais won the race. One lap through the streets of Long Beach is 10,391 feet (1.968 miles) long. Pagenaud's fastest lap speed was 0.55 foot per second faster than Bourdais's fastest lap speed. Traveling at these fastest speeds, Bourdais would have taken 0.25 second longer than Pagenaud to complete a lap. (*Source:* Championship Auto Racing Teams, Inc.)

  **a.** Find Sebastian Bourdais's fastest lap speed during the race. Round to two decimal places.

  **b.** Find Simon Pagenaud's fastest last speed during the race. Round to two decimal places.

  **c.** Convert each speed to miles per hour. Round to one decimal place.

**88.** Use a graphing calculator to solve Exercise 29. Compare the solution with the solution from Exercise 29. Explain any differences.

# INTEGRATED REVIEW   SUMMARY ON SOLVING QUADRATIC EQUATIONS

Sections 11.1–11.3

*Use the square root property to solve each equation.*

**1.** $x^2 - 10 = 0$

**2.** $x^2 - 14 = 0$

**3.** $(x - 1)^2 = 8$

**4.** $(x + 5)^2 = 12$

*Solve each equation by completing the square.*

**5.** $x^2 + 2x - 12 = 0$

**6.** $x^2 - 12x + 11 = 0$

**7.** $3x^2 + 3x = 5$

**8.** $16y^2 + 16y = 1$

*Use the quadratic formula to solve each equation*

**9.** $2x^2 - 4x + 1 = 0$

**10.** $\frac{1}{2}x^2 + 3x + 2 = 0$

**11.** $x^2 + 4x = -7$

**12.** $x^2 + x = -3$

*Solve each equation. Use a method of your choice.*

**13.** $x^2 + 3x + 6 = 0$

**14.** $2x^2 + 18 = 0$

**15.** $x^2 + 17x = 0$

**16.** $4x^2 - 2x - 3 = 0$

**17.** $(x - 2)^2 = 27$

**18.** $\frac{1}{2}x^2 - 2x + \frac{1}{2} = 0$

**19.** $3x^2 + 2x = 8$

**20.** $2x^2 = -5x - 1$

**21.** $x(x - 2) = 5$

**22.** $x^2 - 31 = 0$

**23.** $5x^2 - 55 = 0$

**24.** $5x^2 + 55 = 0$

**25.** $x(x + 5) = 66$

**26.** $5x^2 + 6x - 2 = 0$

**27.** $2x^2 + 3x = 1$

△ **28.** The diagonal of a square room measures 20 feet. Find the exact length of a side of the room. Then approximate the length to the nearest tenth of a foot.

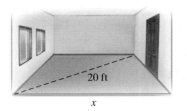

20 ft

*x*

**29.** Together, Jack and Lucy Hoag can prepare a crawfish boil for a large party in 4 hours. Lucy alone can complete the job in 2 hours less time than Jack alone. Find the time that each person can prepare the crawfish boil alone. Round each time to the nearest tenth of an hour.

**30.** Diane Gray exercises at Total Body Gym. On the treadmill, she runs 5 miles, then increases her speed by 1 mile per hour and runs an additional 2 miles. If her total time on the tread mill is $1\frac{1}{3}$ hours, find her speed during each part of her run.

# 11.4 NONLINEAR INEQUALITIES IN ONE VARIABLE

**OBJECTIVES**

**1** Solve polynomial inequalities of degree 2 or greater.

**2** Solve inequalities that contain rational expressions with variables in the denominator.

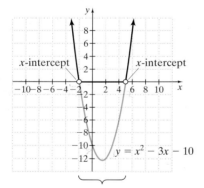

*x*-values corresponding to *negative* *y*-values

**OBJECTIVE 1 ▶ Solving polynomial inequalities.** Just as we can solve linear inequalities in one variable, so can we also solve quadratic inequalities in one variable. A **quadratic inequality** is an inequality that can be written so that one side is a quadratic expression and the other side is 0. Here are examples of quadratic inequalities in one variable. Each is written in **standard form.**

$$x^2 - 10x + 7 \leq 0 \qquad 3x^2 + 2x - 6 > 0$$
$$2x^2 + 9x - 2 < 0 \qquad x^2 - 3x + 11 \geq 0$$

A solution of a quadratic inequality in one variable is a value of the variable that makes the inequality a true statement.

The value of an expression such as $x^2 - 3x - 10$ will sometimes be positive, sometimes negative, and sometimes 0, depending on the value substituted for $x$. To solve the inequality $x^2 - 3x - 10 < 0$, we are looking for all values of $x$ that make the expression $x^2 - 3x - 10$ **less than 0,** or **negative.** To understand how we find these values, we'll study the graph of the quadratic function $y = x^2 - 3x - 10$.

Notice that the $x$-values for which $y$ is positive are separated from the $x$ values for which $y$ is negative by the $x$-intercepts. (Recall that the $x$-intercepts correspond to values of $x$ for which $y = 0$.) Thus, the solution set of $x^2 - 3x - 10 < 0$ consists of all real numbers from $=2$ to 5, or in interval notation, $(-2, 5)$.

It is not necessary to graph $y = x^2 - 3x - 10$ to solve the related inequality $x^2 - 3x - 10 < 0$. Instead, we can draw a number line representing the $x$-axis and keep the following in mind: *A region on the number line for which the value of $x^2 - 3x - 10$ is positive is separated from a region on the number line for which the value of $x^2 - 3x - 10$ is negative by a value for which the expression is* 0.

Let's find these values for which the expression is 0 by solving the related equation:

$$x^2 - 3x - 10 = 0$$
$$(x - 5)(x + 2) = 0 \qquad \text{Factor.}$$
$$x - 5 = 0 \quad \text{or} \quad x + 2 = 0 \qquad \text{Set each factor equal to 0.}$$
$$x = 5 \quad x = -2 \qquad \text{Solve.}$$

These two numbers $-2$ and 5, divide the number line into three regions. We will call the regions $A$, $B$, and $C$. These regions are important because, if the value of $x^2 - 3x - 10$ is negative when a number from a region is substituted for $x$, then $x^2 - 3x - 10$ is negative when any number in that region is substituted for $x$. The same is true if the value of $x^2 - 3x - 10$ is positive for a particular value of $x$ in a region.

To see whether the inequality $x^2 - 3x - 10 < 0$ is true or false in each region, we choose a test point from each region and substitute its value for $x$ in the inequality

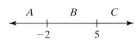

$x^2 - 3x - 10 < 0$. If the resulting inequality is true, the region containing the test point is a solution region.

| Region | Test Point Value | $(x - 5)(x + 2) < 0$ | Result |
|--------|------------------|----------------------|--------|
| $A$ | $-3$ | $(-8)(-1) < 0$ | False |
| $B$ | $0$ | $(-5)(2) < 0$ | True |
| $C$ | $6$ | $(1)(8) < 0$ | False |

The values in region $B$ satisfy the inequality. The numbers $-2$ and 5 are not included in the solution set since the inequality symbol is $<$. The solution set is $(-2, 5)$, and its graph is shown.

$$
\begin{array}{c}
A \qquad\quad B \qquad\quad C \\
\xleftarrow{\qquad\;(\;\;\;\;\;\;\;\;)\;\qquad} \\
\text{F } -2 \quad\; \text{T} \quad\; 5 \;\; \text{F}
\end{array}
$$

**EXAMPLE 1**   Solve: $(x + 3)(x - 3) > 0$.

*Solution*   First we solve the related equation, $(x + 3)(x - 3) = 0$.

$$(x + 3)(x - 3) = 0$$
$$x + 3 = 0 \quad \text{or} \quad x - 3 = 0$$
$$x = -3 \qquad\qquad x = 3$$

The two numbers $-3$ and 3 separate the number line into three regions, $A$, $B$, and $C$.
   Now we substitute the value of a test point from each region. If the test value satisfies the inequality, every value in the region containing the test value is a solution.

| Region | Test Point Value | $(x + 3)(x - 3) > 0$ | Result |
|--------|------------------|----------------------|--------|
| $A$ | $-4$ | $(-1)(-7) > 0$ | True |
| $B$ | $0$ | $(3)(-3) > 0$ | False |
| $C$ | $4$ | $(7)(1) > 0$ | True |

The points in regions $A$ and $C$ satisfy the inequality. The numbers $-3$ and 3 are not included in the solution since the inequality symbol is $>$. The solution set is $(-\infty, -3) \cup (3, \infty)$, and its graph is shown.

$$
\begin{array}{c}
A \qquad\quad B \qquad\quad C \\
\xleftarrow{\;\;)\;\qquad\qquad(\;\;\;} \\
\text{T } -3 \quad\; \text{F} \quad\; 3 \;\; \text{T}
\end{array}
$$

**PRACTICE**
**1**   Solve: $(x - 4)(x + 3) > 0$.

The following steps may be used to solve a polynomial inequality.

> **Solving a Polynomial Inequality**
> **STEP 1.** Write the inequality in standard form and then solve the related equation.
>
> **STEP 2.** Separate the number line into regions with the solutions from Step 1.
>
> **STEP 3.** For each region, choose a test point and determine whether its value satisfies the *original inequality*.
>
> **STEP 4.** The solution set includes the regions whose test point value is a solution. If the inequality symbol is $\leq$ or $\geq$, the values from Step 1 are solutions; if $<$ or $>$, they are not.

**Concept Check** ☑

When choosing a test point in Step 4, why would the solutions from Step 2 not make good choices for test points?

**EXAMPLE 2**  Solve: $x^2 - 4x \leq 0$.

*Solution*  First we solve the related equation, $x^2 - 4x = 0$.

$$x^2 - 4x = 0$$
$$x(x - 4) = 0$$
$$x = 0 \quad \text{or} \quad x = 4$$

The numbers 0 and 4 separate the number line into three regions, $A$, $B$, and $C$.

We check a test value in each region in the original inequality. Values in region $B$ satisfy the inequality. The numbers 0 and 4 are included in the solution since the inequality symbol is $\leq$. The solution set is $[0, 4]$, and its graph is shown.

**PRACTICE**
**2**  Solve: $x^2 - 8x \leq 0$.

**EXAMPLE 3**  Solve: $(x + 2)(x - 1)(x - 5) \leq 0$.

*Solution*  First we solve $(x + 2)(x - 1)(x - 5) = 0$. By inspection, we see that the solutions are $-2, 1$ and $5$. They separate the number line into four regions, $A$, $B$, $C$, and $D$. Next we check test points from each region.

| Region | Test Point Value | $(x + 2)(x - 1)(x - 5) \leq 0$ | Result |
|--------|------------------|-------------------------------|--------|
| $A$ | $-3$ | $(-1)(-4)(-8) \leq 0$ | True |
| $B$ | $0$ | $(2)(-1)(-5) \leq 0$ | False |
| $C$ | $2$ | $(4)(1)(-3) \leq 0$ | True |
| $D$ | $6$ | $(8)(5)(1) \leq 0$ | False |

The solution set is $(-\infty, -2] \cup [1, 5]$, and its graph is shown. We include the numbers $-2, 1,$ and $5$ because the inequality symbol is $\leq$.

**PRACTICE**
**3**  Solve: $(x + 3)(x - 2)(x + 1) \leq 0$.

**OBJECTIVE 2 ▶ Solving rational inequalities.** Inequalities containing rational expressions with variables in the denominator are solved by using a similar procedure.

**Answer to Concept Check:**
The solutions found in Step 2 have a value of 0 in the original inequality.

**EXAMPLE 4**   Solve: $\dfrac{x + 2}{x - 3} \leq 0$.

*Solution*   First we find all values that make the denominator equal to 0. To do this, we solve $x - 3 = 0$ and find that $x = 3$.

Next, we solve the related equation $\dfrac{x + 2}{x - 3} = 0$.

$$\frac{x + 2}{x - 3} = 0$$

$$x + 2 = 0 \qquad \text{Multiply both sides by the LCD, } x - 3.$$

$$x = -2$$

Now we place these numbers on a number line and proceed as before, checking test point values in the original inequality.

| *Choose* $-3$ *from region A.* | *Choose* $0$ *from region B.* | *Choose* $4$ *from region C.* |
|---|---|---|
| $\dfrac{x + 2}{x - 3} \leq 0$ | $\dfrac{x + 2}{x - 3} \leq 0$ | $\dfrac{x + 2}{x - 3} \leq 0$ |
| $\dfrac{-3 + 2}{-3 - 3} \leq 0$ | $\dfrac{0 + 2}{0 - 3} \leq 0$ | $\dfrac{4 + 2}{4 - 3} \leq 0$ |
| $\dfrac{-1}{-6} \leq 0$ | $-\dfrac{2}{3} \leq 0$   True | $6 \leq 0$   False |
| $\dfrac{1}{6} \leq 0$   False | | |

The solution set is $[-2, 3)$. This interval includes $-2$ because $-2$ satisfies the original inequality. This interval does not include 3, because 3 would make the denominator 0.

**PRACTICE**
**4**   Solve: $\dfrac{x - 5}{x + 4} \leq 0$.

The following steps may be used to solve a rational inequality with variables in the denominator.

**Solving a Rational Inequality**
**STEP 1.**  Solve for values that make all denominators 0.

**STEP 2.**  Solve the related equation.

**STEP 3.**  Separate the number line into regions with the solutions from Steps 1 and 2.

**STEP 4.**  For each region, choose a test point and determine whether its value satisfies the *original inequality.*

**STEP 5.**  The solution set includes the regions whose test point value is a solution. Check whether to include values from Step 2. Be sure *not* to include values that make any denominator 0.

**EXAMPLE 5** Solve: $\dfrac{5}{x + 1} < -2$.

_Solution_ First we find values for $x$ that make the denominator equal to 0.

$$x + 1 = 0$$
$$x = -1$$

Next we solve $\dfrac{5}{x + 1} = -2$.

$$(x + 1) \cdot \frac{5}{x + 1} = (x + 1) \cdot -2 \quad \text{Multiply both sides by the LCD, } x + 1.$$
$$5 = -2x - 2 \qquad \text{Simplify.}$$
$$7 = -2x$$
$$-\frac{7}{2} = x$$

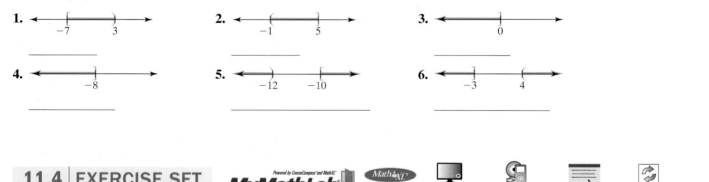

We use these two solutions to divide a number line into three regions and choose test points. Only a test point value from region $B$ satisfies the *original inequality*. The solution set is $\left( -\dfrac{7}{2}, -1 \right)$, and its graph is shown.

**PRACTICE**
**5** Solve: $\dfrac{7}{x + 3} < 5$.

---

# VOCABULARY & READINESS CHECK

*Write the graphed solution set in interval notation.*

**1.** (graph) $-7$ $3$

**2.** (graph) $-1$ $5$

**3.** (graph) $0$

**4.** (graph) $-8$

**5.** (graph) $-12$ $-10$

**6.** (graph) $-3$ $4$

---

## 11.4 EXERCISE SET

**MyMathLab** Powered by CourseCompass™ and MathXL®

MathXL PRACTICE | WATCH | DOWNLOAD | READ | REVIEW

*Solve each quadratic inequality. Write the solution set in interval notation. See Examples 1 through 3.*

**1.** $(x + 1)(x + 5) > 0$

**2.** $(x + 1)(x + 5) \le 0$

**3.** $(x - 3)(x + 4) \le 0$

**4.** $(x + 4)(x - 1) > 0$

**5.** $x^2 - 7x + 10 \le 0$

**6.** $x^2 + 8x + 15 \ge 0$

**7.** $3x^2 + 16x < -5$

**8.** $2x^2 - 5x < 7$

**9.** $(x - 6)(x - 4)(x - 2) > 0$

**10.** $(x - 6)(x - 4)(x - 2) \le 0$

**11.** $x(x - 1)(x + 4) \le 0$

**12.** $x(x - 6)(x + 2) > 0$

**13.** $(x^2 - 9)(x^2 - 4) > 0$

**14.** $(x^2 - 16)(x^2 - 1) \le 0$

*Solve each inequality. Write the solution set in interval notation.
See Example 4.*

**15.** $\dfrac{x + 7}{x - 2} < 0$

**16.** $\dfrac{x - 5}{x - 6} > 0$

**17.** $\dfrac{5}{x + 1} > 0$

**18.** $\dfrac{3}{y - 5} < 0$

**19.** $\dfrac{x + 1}{x - 4} \geq 0$

**20.** $\dfrac{x + 1}{x - 4} \leq 0$

*Solve each inequality. Write the solution set in interval notation.
See Example 5.*

**21.** $\dfrac{3}{x - 2} < 4$

**22.** $\dfrac{-2}{y + 3} > 2$

**23.** $\dfrac{x^2 + 6}{5x} \geq 1$

**24.** $\dfrac{y^2 + 15}{8y} \leq 1$

## MIXED PRACTICE

*Solve each inequality. Write the solution set in interval notation.*

**25.** $(x - 8)(x + 7) > 0$

**26.** $(x - 5)(x + 1) < 0$

**27.** $(2x - 3)(4x + 5) \leq 0$

**28.** $(6x + 7)(7x - 12) > 0$

**29.** $x^2 > x$

**30.** $x^2 < 25$

**31.** $(2x - 8)(x + 4)(x - 6) \leq 0$

**32.** $(3x - 12)(x + 5)(2x - 3) \geq 0$

**33.** $6x^2 - 5x \geq 6$

**34.** $12x^2 + 11x \leq 15$

**35.** $4x^3 + 16x^2 - 9x - 36 > 0$

**36.** $x^3 + 2x^2 - 4x - 8 < 0$

**37.** $x^4 - 26x^2 + 25 \geq 0$

**38.** $16x^4 - 40x^2 + 9 \leq 0$

**39.** $(2x - 7)(3x + 5) > 0$

**40.** $(4x - 9)(2x + 5) < 0$

**41.** $\dfrac{x}{x - 10} < 0$

**42.** $\dfrac{x + 10}{x - 10} > 0$

**43.** $\dfrac{x - 5}{x + 4} \geq 0$

**44.** $\dfrac{x - 3}{x + 2} \leq 0$

**45.** $\dfrac{x(x + 6)}{(x - 7)(x + 1)} \geq 0$

**46.** $\dfrac{(x - 2)(x + 2)}{(x + 1)(x - 4)} \leq 0$

**47.** $\dfrac{-1}{x - 1} > -1$

**48.** $\dfrac{4}{y + 2} < -2$

**49.** $\dfrac{x}{x + 4} \leq 2$

**50.** $\dfrac{4x}{x - 3} \geq 5$

**51.** $\dfrac{z}{z - 5} \geq 2z$

**52.** $\dfrac{p}{p + 4} \leq 3p$

**53.** $\dfrac{(x + 1)^2}{5x} > 0$

**54.** $\dfrac{(2x - 3)^2}{x} < 0$

## REVIEW AND PREVIEW

*Recall that the graph of $f(x) + K$ is the same as the graph of $f(x)$
shifted $K$ units upward if $K > 0$ and $|K|$ units downward if
$K < 0$. Use the graph of $f(x) = |x|$ below to sketch the graph of
each function. See Section 8.3.*

**55.** $g(x) = |x| + 2$

**56.** $H(x) = |x| - 2$

**57.** $F(x) = |x| - 1$

**58.** $h(x) = |x| + 5$

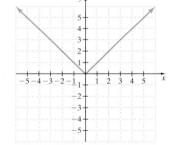

*Use the graph of $f(x) = x^2$ below to sketch the graph of each function.*

**59.** $F(x) = x^2 - 3$

**60.** $h(x) = x^2 - 4$

**61.** $H(x) = x^2 + 1$

**62.** $g(x) = x^2 + 3$

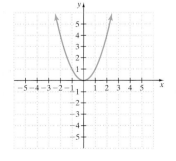

## CONCEPT EXTENSIONS

**63.** Explain why $\dfrac{x+2}{x-3} > 0$ and $(x+2)(x-3) > 0$ have the same solutions.

**64.** Explain why $\dfrac{x+2}{x-3} \geq 0$ and $(x+2)(x-3) \geq 0$ do not have the same solutions.

*Find all numbers that satisfy each of the following.*

**65.** A number minus its reciprocal is less than zero. Find the numbers.

**66.** Twice a number added to its reciprocal is nonnegative. Find the numbers.

**67.** The total profit function $P(x)$ for a company producing $x$ thousand units is given by

$$P(x) = -2x^2 + 26x - 44$$

Find the values of $x$ for which the company makes a profit. [*Hint:* The company makes a profit when $P(x) > 0$.]

**68.** A projectile is fired straight up from the ground with an initial velocity of 80 feet per second. Its height $s(t)$ in feet at any time $t$ is given by the function

$$s(t) = -16t^2 + 80t$$

Find the interval of time for which the height of the projectile is greater than 96 feet.

*Use a graphing calculator to check each exercise.*

**69.** Exercise 25
**70.** Exercise 26
**71.** Exercise 37
**72.** Exercise 38

---

## THE BIGGER PICTURE   SIMPLIFYING EXPRESSIONS AND SOLVING EQUATIONS AND INEQUALITIES

We now continue the outline from Sections 1.7, 2.3, 2.8, 5.6, 6.6, 7.4, 7.6, 9.1, 9.3, 10.5, and 10.6. Although suggestions are given, as usual, this outline should be in your own words. Try to include "how to recognize" and "how to begin to solve" in your outline. Once you complete this new portion, complete the exercises.

   **I.** Simplifying Expressions (See Section 10.5, page 614)

   **II.** Solving Equations

    **A.** Linear Equations (Section 2.3)

    **B. Quadratic and Higher Degree Equations**
    (Sections 6.6, 11.1, 11.2, 11.3)

| Solving by Factoring: | Solving by the Quadratic Formula: |
|---|---|
| $2x^2 - 7x = 9$ | $2x^2 + x - 2 = 0$ |
| $2x^2 - 7x - 9 = 0$ | $a = 2, b = 1, c = -2$ |
| $(2x - 9)(x + 1) = 0$ | $x = \dfrac{-1 \pm \sqrt{(1)^2 - 4(2)(-2)}}{2 \cdot 2}$ |
| $\begin{array}{cc} 2x - 9 = 0 \\ \text{or} \quad x + 1 = 0 \end{array}$ | |
| $x = \dfrac{9}{2} \quad \text{or} \quad x = -1$ | $x = \dfrac{-1 \pm \sqrt{17}}{4}$ |

    **C.** Equations with Rational Expressions (Section 7.5)

    **D.** Proportions (Section 7.6)

    **E.** Absolute Value Equations (Section 9.2)

    **F.** Equations with Radicals (Section 10.6)

   **III.** Solving Inequalities

    **A.** Linear Inequalities (Section 2.8)

    **B.** Compound Inequalities (Section 9.1)

    **C.** Absolute Value Inequalities (Section 9.3)

   **D. Nonlinear Inequalities**

| Polynomial Inequality | Rational Inequality with variable in denominator |
|---|---|
| $x^2 - x < 6$ | |
| $x^2 - x - 6 < 0$ | $\dfrac{x - 5}{x + 1} \geq 0$ |
| $(x - 3)(x + 2) < 0$ | |

| $(-2, 3)$ | $(-\infty, -1) \cup [5, \infty)$ |

*Solve. Write solutions to inequalities in interval notation.*

**1.** $|x - 8| = |2x + 1|$

**2.** $0 < -x + 7 < 3$

**3.** $\sqrt{3x - 11} + 3 = x$

**4.** $x(3x + 1) = 1$

**5.** $\dfrac{x + 2}{x - 7} \leq 0$

**6.** $x(x - 6) + 4 = x^2 - 2(3 - x)$

**7.** $x(5x - 36) = -7$

**8.** $2x^2 - 4 \geq 7x$

**9.** $\left| \dfrac{x - 7}{3} \right| > 5$

**10.** $2(x - 5) + 4 < 1 + 7(x - 5) - x$

# 11.5 QUADRATIC FUNCTIONS AND THEIR GRAPHS

**OBJECTIVES**

**1** Graph quadratic functions of the form $f(x) = x^2 + k$.

**2** Graph quadratic functions of the form $f(x) = (x - h)^2$.

**3** Graph quadratic functions of the form $f(x) = (x - h)^2 + k$.

**4** Graph quadratic functions of the form $f(x) = ax^2$.

**5** Graph quadratic functions of the form $f(x) = a(x - h)^2 + k$.

**OBJECTIVE 1 ▶ Graphing $f(x) = x^2 + k$.** We graphed the quadratic function $f(x) = x^2$ in Section 8.2. In Sections 8.2 and 8.3, we discovered that the graph of a quadratic function is a parabola opening upward or downward. In this section, we continue our study of quadratic functions and their graphs. (Much of the contents of this section is a review of shifting and reflecting techniques from Section 8.3.)

First, let's recall the definition of a quadratic function.

> **Quadratic Function**
>
> A quadratic function is a function that can be written in the form $f(x) = ax^2 + bx + c$, where $a$, $b$, and $c$ are real numbers and $a \neq 0$.

Notice that equations of the form $y = ax^2 + bx + c$, where $a \neq 0$, define quadratic functions, since $y$ is a function of $x$ or $y = f(x)$.

Recall that if $a > 0$, the parabola opens upward and if $a < 0$, the parabola opens downward. Also, the vertex of a parabola is the lowest point if the parabola opens upward and the highest point if the parabola opens downward. The axis of symmetry is the vertical line that passes through the vertex.

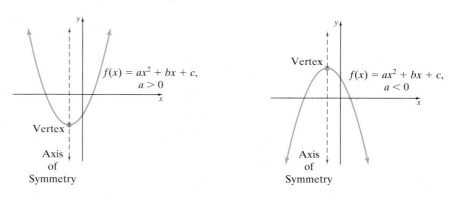

**EXAMPLE 1**    Graph $f(x) = x^2$ and $g(x) = x^2 + 6$ on the same set of axes.

**Solution**    First we construct a table of values for $f(x)$ and plot the points. Notice that for each $x$-value, the corresponding value of $g(x)$ must be 6 more than the corresponding value of $f(x)$ since $f(x) = x^2$ and $g(x) = x^2 + 6$. In other words, the graph of $g(x) = x^2 + 6$ is the same as the graph of $f(x) = x^2$ shifted upward 6 units. The axis of symmetry for both graphs is the $y$-axis.

| $x$ | $f(x) = x^2$ | $g(x) = x^2 + 6$ |
|---|---|---|
| $-2$ | 4 | 10 |
| $-1$ | 1 | 7 |
| 0 | 0 | 6 |
| 1 | 1 | 7 |
| 2 | 4 | 10 |

Each $y$-value is increased by 6.

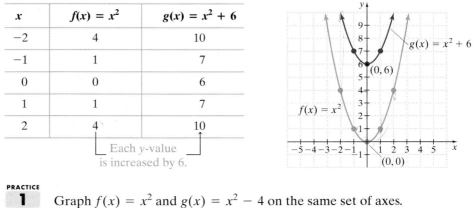

**PRACTICE**

**1**    Graph $f(x) = x^2$ and $g(x) = x^2 - 4$ on the same set of axes.

In general, we have the following properties.

---

**Graphing the Parabola Defined by $f(x) = x^2 + k$**

If $k$ is positive, the graph of $f(x) = x^2 + k$ is the graph of $y = x^2$ shifted upward $k$ units.

If $k$ is negative, the graph of $f(x) = x^2 - k$ is the graph of $y = x^2$ shifted downward $|k|$ units.

The vertex is $(0, k)$, and the axis of symmetry is the $y$-axis.

---

**EXAMPLE 2**   Graph each function.

**a.** $F(x) = x^2 + 2$     **b.** $g(x) = x^2 - 3$

**Solution**

**a.** $F(x) = x^2 + 2$

The graph of $F(x) = x^2 + 2$ is obtained by shifting the graph of $y = x^2$ upward 2 units.

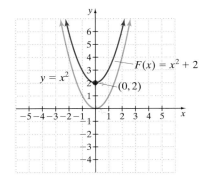

**b.** $g(x) = x^2 - 3$

The graph of $g(x) = x^2 - 3$ is obtained by shifting the graph of $y = x^2$ downward 3 units.

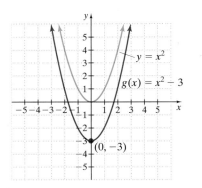

**PRACTICE**

**2**   Graph each function.

**a.** $f(x) = x^2 - 5$          **b.** $g(x) = x^2 + 3$

----

**OBJECTIVE 2 ▶ Graphing $f(x) = (x - h)^2$.** Now we will graph functions of the form $f(x) = (x - h)^2$.

**EXAMPLE 3**   Graph $f(x) = x^2$ and $g(x) = (x - 2)^2$ on the same set of axes.

**Solution**   By plotting points, we see that for each $x$-value, the corresponding value of $g(x)$ is the same as the value of $f(x)$ when the $x$-value is increased by 2. Thus, the graph of $g(x) = (x - 2)^2$ is the graph of $f(x) = x^2$ shifted to the right 2 units. The axis of symmetry for the graph of $g(x) = (x - 2)^2$ is also shifted 2 units to the right and is the line $x = 2$.

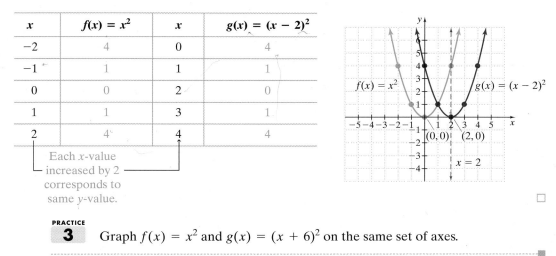

| $x$ | $f(x) = x^2$ | $x$ | $g(x) = (x - 2)^2$ |
|---|---|---|---|
| $-2$ | 4 | 0 | 4 |
| $-1$ | 1 | 1 | 1 |
| 0 | 0 | 2 | 0 |
| 1 | 1 | 3 | 1 |
| 2 | 4 | 4 | 4 |

Each $x$-value increased by 2 corresponds to same $y$-value.

**PRACTICE**
**3**   Graph $f(x) = x^2$ and $g(x) = (x + 6)^2$ on the same set of axes.

In general, we have the following properties.

---

**Graphing the Parabola Defined by $f(x) = (x - h)^2$**

If $h$ is positive, the graph of $f(x) = (x - h)^2$ is the graph of $y = x^2$ shifted to the right $h$ units.
If $h$ is negative, the graph of $f(x) = (x - h)^2$ is the graph of $y = x^2$ shifted to the left $|h|$ units.
The vertex is $(h, 0)$, and the axis of symmetry is the vertical line $x = h$.

---

**EXAMPLE 4**   Graph each function.

**a.** $G(x) = (x - 3)^2$      **b.** $F(x) = (x + 1)^2$

**Solution**

**a.** The graph of $G(x) = (x - 3)^2$ is obtained by shifting the graph of $y = x^2$ to the right 3 units. The graph of $G(x)$ is below on the left.

**b.** The equation $F(x) = (x + 1)^2$ can be written as $F(x) = [x - (-1)]^2$. The graph of $F(x) = [x - (-1)]^2$ is obtained by shifting the graph of $y = x^2$ to the left 1 unit. The graph of $F(x)$ is below on the right.

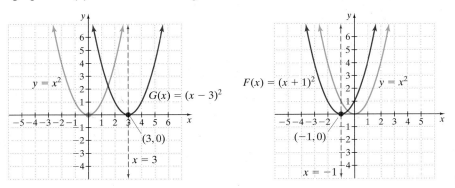

PRACTICE
**4**  Graph each function.

**a.** $G(x) = (x + 4)^2$      **b.** $H(x) = (x - 7)^2$

---

**OBJECTIVE 3 ▶ Graphing $f(x) = (x - h)^2 + k$.** As we will see in graphing functions of the form $f(x) = (x - h)^2 + k$, it is possible to combine vertical and horizontal shifts.

> **Graphing the Parabola Defined by $f(x) = (x - h)^2 + k$**
> The parabola has the same shape as $y = x^2$.
> The vertex is $(h, k)$, and the axis of symmetry is the vertical line $x = h$.

**EXAMPLE 5**  Graph $F(x) = (x - 3)^2 + 1$.

*Solution*   The graph of $F(x) = (x - 3)^2 + 1$ is the graph of $y = x^2$ shifted 3 units to the right and 1 unit up. The vertex is then $(3, 1)$, and the axis of symmetry is $x = 3$. A few ordered pair solutions are plotted to aid in graphing.

| $x$ | $F(x) = (x - 3)^2 + 1$ |
|-----|------------------------|
| 1 | 5 |
| 2 | 2 |
| 4 | 2 |
| 5 | 5 |

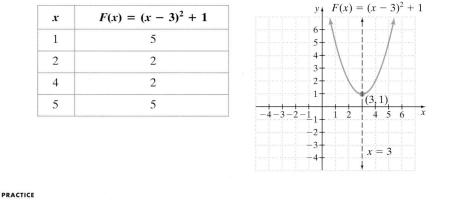

PRACTICE
**5**  Graph $f(x) = (x + 2)^2 + 2$.

---

**OBJECTIVE 4 ▶ Graphing $f(x) = ax^2$.** Next, we discover the change in the shape of the graph when the coefficient of $x^2$ is not 1.

**EXAMPLE 6**  Graph $f(x) = x^2$, $g(x) = 3x^2$, and $h(x) = \dfrac{1}{2}x^2$ on the same set of axes.

*Solution*   Comparing the tables of values, we see that for each $x$-value, the corresponding value of $g(x)$ is triple the corresponding value of $f(x)$. Similarly, the value of $h(x)$ is half the value of $f(x)$.

| $x$ | $f(x) = x^2$ |
|-----|-------------|
| $-2$ | 4 |
| $-1$ | 1 |
| 0 | 0 |
| 1 | 1 |
| 2 | 4 |

| $x$ | $g(x) = 3x^2$ |
|-----|--------------|
| $-2$ | 12 |
| $-1$ | 3 |
| 0 | 0 |
| 1 | 3 |
| 2 | 12 |

| $x$ | $h(x) = \dfrac{1}{2}x^2$ |
|-----|-------------------------|
| $-2$ | 2 |
| $-1$ | $\dfrac{1}{2}$ |
| 0 | 0 |
| 1 | $\dfrac{1}{2}$ |
| 2 | 2 |

The result is that the graph of $g(x) = 3x^2$ is narrower than the graph of $f(x) = x^2$ and the graph of $h(x) = \dfrac{1}{2}x^2$ is wider. The vertex for each graph is $(0, 0)$, and the axis of symmetry is the $y$-axis.

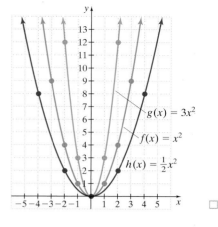

**PRACTICE**

**6**   Graph $f(x) = x^2$, $g(x) = 4x^2$, and $h(x) = \dfrac{1}{4}x^2$ on the same set of axes.

---

**Graphing the Parabola Defined by $f(x) = ax^2$**

If $a$ is positive, the parabola opens upward, and if $a$ is negative, the parabola opens downward.

If $|a| > 1$, the graph of the parabola is narrower than the graph of $y = x^2$.

If $|a| < 1$, the graph of the parabola is wider than the graph of $y = x^2$.

**EXAMPLE 7**   Graph $f(x) = -2x^2$.

**Solution**   Because $a = -2$, a negative value, this parabola opens downward. Since $|-2| = 2$ and $2 > 1$, the parabola is narrower than the graph of $y = x^2$. The vertex is $(0, 0)$, and the axis of symmetry is the $y$-axis. We verify this by plotting a few points.

| $x$ | $f(x) = -2x^2$ |
|-----|----------------|
| $-2$ | $-8$ |
| $-1$ | $-2$ |
| $0$ | $0$ |
| $1$ | $-2$ |
| $2$ | $-8$ |

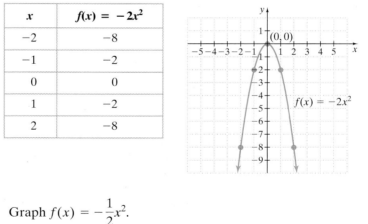

**PRACTICE**

**7**   Graph $f(x) = -\dfrac{1}{2}x^2$.

---

**OBJECTIVE 5 ▶ Graphing $f(x) = a(x - h)^2 + k$.** Now we will see the shape of the graph of a quadratic function of the form $f(x) = a(x - h)^2 + k$.

**EXAMPLE 8**   Graph $g(x) = \frac{1}{2}(x + 2)^2 + 5$. Find the vertex and the axis of symmetry.

*Solution*   The function $g(x) = \frac{1}{2}(x + 2)^2 + 5$ may be written as $g(x) = \frac{1}{2}[x - (-2)]^2 + 5$. Thus, this graph is the same as the graph of $y = x^2$ shifted 2 units to the left and 5 units up, and it is wider because $a$ is $\frac{1}{2}$. The vertex is $(-2, 5)$, and the axis of symmetry is $x = -2$. We plot a few points to verify.

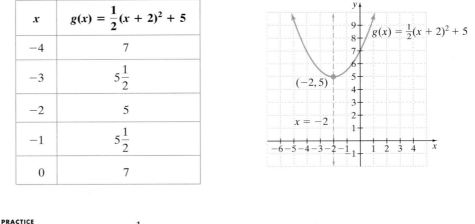

| $x$ | $g(x) = \frac{1}{2}(x + 2)^2 + 5$ |
|---|---|
| $-4$ | $7$ |
| $-3$ | $5\frac{1}{2}$ |
| $-2$ | $5$ |
| $-1$ | $5\frac{1}{2}$ |
| $0$ | $7$ |

**PRACTICE**
**8**   Graph $h(x) = \frac{1}{3}(x - 4)^2 - 3$.

In general, the following holds.

**Graph of a Quadratic Function**
The graph of a quadratic function written in the form $f(x) = a(x - h)^2 + k$ is a parabola with vertex $(h, k)$. If $a > 0$, the parabola opens upward, and if $a < 0$, the parabola opens downward. The axis of symmetry is the line whose equation is $x = h$.

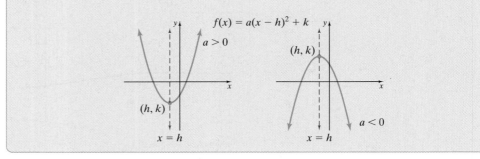

**Concept Check** ☑
Which description of the graph of $f(x) = -0.35(x + 3)^2 - 4$ is correct?

**a.** The graph opens downward and has its vertex at $(-3, 4)$.
**b.** The graph opens upward and has its vertex at $(-3, 4)$.
**c.** The graph opens downward and has its vertex at $(-3, -4)$.
**d.** The graph is narrower than the graph of $y = x^2$.

**Answer to Concept Check:**   c

**Graphing Calculator Explorations**

*Use a graphing calculator to graph the first function of each pair that follows. Then use its graph to predict the graph of the second function. Check your prediction by graphing both on the same set of axes.*

**1.** $F(x) = \sqrt{x}$; $G(x) = \sqrt{x} + 1$

**2.** $g(x) = x^3$; $H(x) = x^3 - 2$

**3.** $H(x) = |x|$; $f(x) = |x - 5|$

**4.** $h(x) = x^3 + 2$; $g(x) = (x - 3)^3 + 2$

**5.** $f(x) = |x + 4|$; $F(x) = |x + 4| + 3$

**6.** $G(x) = \sqrt{x} - 2$; $g(x) = \sqrt{x - 4} - 2$

## VOCABULARY & READINESS CHECK

*Use the choices below to fill in each blank. Some choices will be used more than once.*

   upward     highest     parabola     downward     lowest     quadratic

**1.** A _____ function is one that can be written in the form $f(x) = ax^2 + bx + c, a \neq 0$.

**2.** The graph of a quadratic function is a _____ opening _____ or _____.

**3.** If $a > 0$, the graph of the quadratic function opens _____.

**4.** If $a < 0$, the graph of the quadratic function opens _____.

**5.** The vertex of a parabola is the _____ point if $a > 0$.

**6.** The vertex of a parabola is the _____ point if $a < 0$.

*State the vertex of the graph of each quadratic function.*

**7.** $f(x) = x^2$

**8.** $f(x) = -5x^2$

**9.** $g(x) = (x - 2)^2$

**10.** $g(x) = (x + 5)^2$

**11.** $f(x) = 2x^2 + 3$

**12.** $h(x) = x^2 - 1$

**13.** $g(x) = (x + 1)^2 + 5$

**14.** $h(x) = (x - 10)^2 - 7$

## 11.5 | EXERCISE SET    MyMathLab    *Powered by CourseCompass™ and MathXL®*    PRACTICE    WATCH    DOWNLOAD    READ    REVIEW

**MIXED PRACTICE**

*Sketch the graph of each quadratic function. Label the vertex, and sketch and label the axis of symmetry. See Examples 1 through 5.*

**1.** $f(x) = x^2 - 1$

**2.** $g(x) = x^2 + 3$

**3.** $h(x) = x^2 + 5$

**4.** $h(x) = x^2 - 4$

**5.** $g(x) = x^2 + 7$

**6.** $f(x) = x^2 - 2$

**7.** $f(x) = (x - 5)^2$

**8.** $g(x) = (x + 5)^2$

**9.** $h(x) = (x + 2)^2$

**10.** $H(x) = (x - 1)^2$

**11.** $G(x) = (x + 3)^2$

**12.** $f(x) = (x - 6)^2$

**13.** $f(x) = (x - 2)^2 + 5$

**14.** $g(x) = (x - 6)^2 + 1$

**15.** $h(x) = (x + 1)^2 + 4$

**16.** $G(x) = (x + 3)^2 + 3$

**17.** $g(x) = (x + 2)^2 - 5$

**18.** $h(x) = (x + 4)^2 - 6$

*Sketch the graph of each quadratic function. Label the vertex, and sketch and label the axis of symmetry. See Examples 6 and 7.*

**19.** $g(x) = -x^2$

**20.** $f(x) = 5x^2$

**21.** $h(x) = \dfrac{1}{3}x^2$

**22.** $f(x) = -\dfrac{1}{4}x^2$

**23.** $H(x) = 2x^2$

**24.** $g(x) = -3x^2$

*Sketch the graph of each quadratic function. Label the vertex, and sketch and label the axis of symmetry. See Example 8.*

**25.** $f(x) = 2(x - 1)^2 + 3$  **26.** $g(x) = 4(x - 4)^2 + 2$

**27.** $h(x) = -3(x + 3)^2 + 1$  **28.** $f(x) = -(x - 2)^2 - 6$

**29.** $H(x) = \frac{1}{2}(x - 6)^2 - 3$  **30.** $G(x) = \frac{1}{5}(x + 4)^2 + 3$

## MIXED PRACTICE

*Sketch the graph of each quadratic function. Label the vertex, and sketch and label the axis of symmetry.*

**31.** $f(x) = -(x - 2)^2$  **32.** $g(x) = -(x + 6)^2$

**33.** $F(x) = -x^2 + 4$  **34.** $H(x) = -x^2 + 10$

**35.** $F(x) = 2x^2 - 5$  **36.** $g(x) = \frac{1}{2}x^2 - 2$

**37.** $h(x) = (x - 6)^2 + 4$  **38.** $f(x) = (x - 5)^2 + 2$

**39.** $F(x) = \left(x + \frac{1}{2}\right)^2 - 2$  **40.** $H(x) = \left(x + \frac{1}{2}\right)^2 - 3$

**41.** $F(x) = \frac{3}{2}(x + 7)^2 + 1$  **42.** $g(x) = -\frac{3}{2}(x - 1)^2 - 5$

**43.** $f(x) = \frac{1}{4}x^2 - 9$  **44.** $H(x) = \frac{3}{4}x^2 - 2$

**45.** $G(x) = 5\left(x + \frac{1}{2}\right)^2$  **46.** $F(x) = 3\left(x - \frac{3}{2}\right)^2$

**47.** $h(x) = -(x - 1)^2 - 1$  **48.** $f(x) = -3(x + 2)^2 + 2$

**49.** $g(x) = \sqrt{3}(x + 5)^2 + \frac{3}{4}$  **50.** $G(x) = \sqrt{5}(x - 7)^2 - \frac{1}{2}$

**51.** $h(x) = 10(x + 4)^2 - 6$  **52.** $h(x) = 8(x + 1)^2 + 9$

**53.** $f(x) = -2(x - 4)^2 + 5$  **54.** $G(x) = -4(x + 9)^2 - 1$

## REVIEW AND PREVIEW

*Add the proper constant to each binomial so that the resulting trinomial is a perfect square trinomial. See Section 11.1.*

**55.** $x^2 + 8x$  **56.** $y^2 + 4y$

**57.** $z^2 - 16z$  **58.** $x^2 - 10x$

**59.** $y^2 + y$  **60.** $z^2 - 3z$

*Solve by completing the square. See Section 11.1.*

**61.** $x^2 + 4x = 12$  **62.** $y^2 + 6y = -5$

**63.** $z^2 + 10z - 1 = 0$  **64.** $x^2 + 14x + 20 = 0$

**65.** $z^2 - 8z = 2$  **66.** $y^2 - 10y = 3$

## CONCEPT EXTENSIONS

*Solve. See the Concept Check in this section.*

**67.** Which description of $f(x) = -213(x - 0.1)^2 + 3.6$ is correct?

| Graph Opens | Vertex |
|---|---|
| a. upward | $(0.1, 3.6)$ |
| b. upward | $(-213, 3.6)$ |
| c. downward | $(0.1, 3.6)$ |
| d. downward | $(-0.1, 3.6)$ |

**68.** Which description of $f(x) = 5\left(x + \frac{1}{2}\right)^2 + \frac{1}{2}$ is correct?

| Graph Opens | Vertex |
|---|---|
| a. upward | $\left(\frac{1}{2}, \frac{1}{2}\right)$ |
| b. upward | $\left(-\frac{1}{2}, \frac{1}{2}\right)$ |
| c. downward | $\left(\frac{1}{2}, -\frac{1}{2}\right)$ |
| d. downward | $\left(-\frac{1}{2}, -\frac{1}{2}\right)$ |

*Write the equation of the parabola that has the same shape as $f(x) = 5x^2$ but with the following vertex.*

**69.** $(2, 3)$  **70.** $(1, 6)$

**71.** $(-3, 6)$  **72.** $(4, -1)$

*The shifting properties covered in this section apply to the graphs of all functions. Given the accompanying graph of $y = f(x)$, sketch the graph of each of the following.*

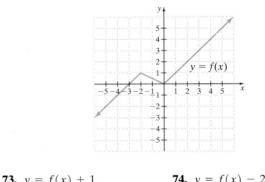

**73.** $y = f(x) + 1$  **74.** $y = f(x) - 2$

**75.** $y = f(x - 3)$  **76.** $y = f(x + 3)$

**77.** $y = f(x + 2) + 2$  **78.** $y = f(x - 1) + 1$

**79.** The quadratic function $f(x) = 668.7x^2 - 2990.7x + 938$ approximates the U.S. growth of cell phone subscribers between 1985 and 2005 where $x$ is the number of years past 1985 and $f(x)$ is the number of subscribers in thousands.

 **a.** Use this function to approximate the number of subscribers in 2004.

 **b.** Use this function to predict the number of subscribers in 2007.

**80.** Use the function in Exercise 79.

 **a.** Predict the number of cell phone subscribers in 2010.

 **b.** Look up the current population of the U.S.

 **c.** Based on your answers for parts **a.** and **b.**, discuss some limitations of using this quadratic function to predict data.

# 11.6 FURTHER GRAPHING OF QUADRATIC FUNCTIONS

**OBJECTIVES**

**1** Write quadratic functions in the form $y = a(x - h)^2 + k$.

**2** Derive a formula for finding the vertex of a parabola.

**3** Find the minimum or maximum value of a quadratic function.

**OBJECTIVE 1 ▶ Writing quadratic functions in the form $y = a(x - h)^2 + k$.** We know that the graph of a quadratic function is a parabola. If a quadratic function is written in the form

$$f(x) = a(x - h)^2 + k$$

we can easily find the vertex $(h, k)$ and graph the parabola. To write a quadratic function in this form, complete the square. (See Section 11.1 for a review of completing the square.)

**EXAMPLE 1**   Graph $f(x) = x^2 - 4x - 12$. Find the vertex and any intercepts.

_Solution_   The graph of this quadratic function is a parabola. To find the vertex of the parabola, we will write the function in the form $y = (x - h)^2 + k$. To do this, we complete the square on the binomial $x^2 - 4x$. To simplify our work, we let $f(x) = y$.

$$y = x^2 - 4x - 12 \quad \text{Let } f(x) = y.$$

$$y + 12 = x^2 - 4x \qquad \text{Add 12 to both sides to get the } x\text{-variable terms alone.}$$

Now we add the square of half of $-4$ to both sides.

$$\frac{1}{2}(-4) = -2 \quad \text{and} \quad (-2)^2 = 4$$

$$y + 12 + 4 = x^2 - 4x + 4 \qquad \text{Add 4 to both sides.}$$

$$y + 16 = (x - 2)^2 \qquad \text{Factor the trinomial.}$$

$$y = (x - 2)^2 - 16 \qquad \text{Subtract 16 from both sides.}$$

$$f(x) = (x - 2)^2 - 16 \qquad \text{Replace } y \text{ with } f(x).$$

From this equation, we can see that the vertex of the parabola is $(2, -16)$, a point in quadrant IV, and the axis of symmetry is the line $x = 2$.

Notice that $a = 1$. Since $a > 0$, the parabola opens upward. This parabola opening upward with vertex $(2, -16)$ will have two $x$-intercepts and one $y$-intercept. (See the Helpful Hint after this example.)

$x$-intercepts: let $y$ or $f(x) = 0$        $y$-intercept: let $x = 0$

$$f(x) = x^2 - 4x - 12 \qquad\qquad f(x) = x^2 - 4x - 12$$

$$0 = x^2 - 4x - 12 \qquad\qquad\quad f(0) = 0^2 - 4 \cdot 0 - 12$$

$$0 = (x - 6)(x + 2) \qquad\qquad\qquad\quad = -12$$

$$0 = x - 6 \quad \text{or} \quad 0 = x + 2$$

$$6 = x \qquad\qquad -2 = x$$

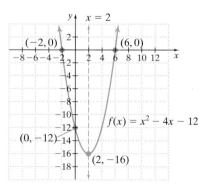

The two $x$-intercepts are $(6, 0)$ and $(-2, 0)$. The $y$-intercept is $(0, -12)$. The sketch of $f(x) = x^2 - 4x - 12$ is shown.

Notice that the axis of symmetry is always halfway between the $x$-intercepts. For the example above, halfway between $-2$ and $6$ is $\dfrac{-2 + 6}{2} = 2$, and the axis of symmetry is $x = 2$.                                    □

**PRACTICE**

**1**   Graph $g(x) = x^2 - 2x - 3$. Find the vertex and any intercepts.

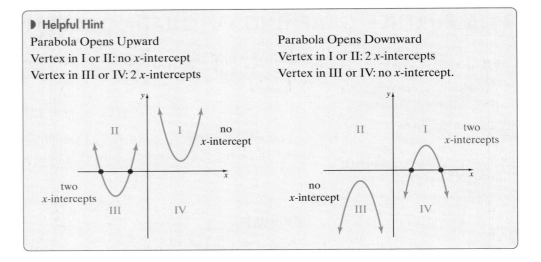

▶ **Helpful Hint**

| Parabola Opens Upward | Parabola Opens Downward |
|---|---|
| Vertex in I or II: no $x$-intercept | Vertex in I or II: 2 $x$-intercepts |
| Vertex in III or IV: 2 $x$-intercepts | Vertex in III or IV: no $x$-intercept. |

**EXAMPLE 2**    Graph $f(x) = 3x^2 + 3x + 1$. Find the vertex and any intercepts.

**Solution**    Replace $f(x)$ with $y$ and complete the square on $x$ to write the equation in the form $y = a(x - h)^2 + k$.

$$y = 3x^2 + 3x + 1 \quad \text{Replace } f(x) \text{ with } y.$$
$$y - 1 = 3x^2 + 3x \quad \text{Isolate } x\text{-variable terms.}$$

Factor 3 from the terms $3x^2 + 3x$ so that the coefficient of $x^2$ is 1.

$$y - 1 = 3(x^2 + x) \quad \text{Factor out 3.}$$

The coefficient of $x$ in the parentheses above is 1. Then $\frac{1}{2}(1) = \frac{1}{2}$ and $\left(\frac{1}{2}\right)^2 = \frac{1}{4}$.

Since we are adding $\frac{1}{4}$ inside the parentheses, we are really adding $3\left(\frac{1}{4}\right)$, so we *must* add $3\left(\frac{1}{4}\right)$ to the left side.

$$y - 1 + 3\left(\frac{1}{4}\right) = 3\left(x^2 + x + \frac{1}{4}\right)$$

$$y - \frac{1}{4} = 3\left(x + \frac{1}{2}\right)^2 \quad \text{Simplify the left side and factor the right side.}$$

$$y = 3\left(x + \frac{1}{2}\right)^2 + \frac{1}{4} \quad \text{Add } \frac{1}{4} \text{ to both sides.}$$

$$f(x) = 3\left(x + \frac{1}{2}\right)^2 + \frac{1}{4} \quad \text{Replace } y \text{ with } f(x).$$

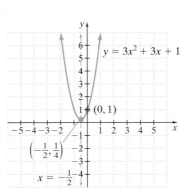

Then $a = 3$, $h = -\frac{1}{2}$, and $k = \frac{1}{4}$. This means that the parabola opens upward with vertex $\left(-\frac{1}{2}, \frac{1}{4}\right)$ and that the axis of symmetry is the line $x = -\frac{1}{2}$.

To find the $y$-intercept, let $x = 0$. Then

$$f(0) = 3(0)^2 + 3(0) + 1 = 1$$

Thus the $y$-intercept is $(0, 1)$.

This parabola has no $x$-intercepts since the vertex is in the second quadrant and opens upward. Use the vertex, axis of symmetry, and $y$-intercept to sketch the parabola.    ☐

**PRACTICE**
**2**    Graph $g(x) = 4x^2 + 4x + 3$. Find the vertex and any intercepts.

**EXAMPLE 3**  Graph $f(x) = -x^2 - 2x + 3$. Find the vertex and any intercepts.

*Solution*  We write $f(x)$ in the form $a(x - h)^2 + k$ by completing the square. First we replace $f(x)$ with $y$.

$$f(x) = -x^2 - 2x + 3$$

$$y = -x^2 - 2x + 3$$

$$y - 3 = -x^2 - 2x \qquad \text{Subtract 3 from both sides to get the } x\text{-variable terms alone.}$$

$$y - 3 = -1(x^2 + 2x) \qquad \text{Factor } -1 \text{ from the terms } -x^2 - 2x.$$

The coefficient of $x$ is 2. Then $\frac{1}{2}(2) = 1$ and $1^2 = 1$. We add 1 to the right side inside the parentheses and add $-1(1)$ to the left side.

$$y - 3 - 1(1) = -1(x^2 + 2x + 1)$$

$$y - 4 = -1(x + 1)^2 \qquad \text{Simplify the left side and factor the right side.}$$

$$y = -1(x + 1)^2 + 4 \qquad \text{Add 4 to both sides.}$$

$$\underline{f(x) = -1(x + 1)^2 + 4} \qquad \text{Replace } y \text{ with } f(x).$$

> ▶ **Helpful Hint**
> This can be written as
> $f(x) = -1[x - (-1)]^2 + 4$.
> Notice that the vertex is
> $(-1, 4)$.

Since $a = -1$, the parabola opens downward with vertex $(-1, 4)$ and axis of symmetry $x = -1$.

To find the $y$-intercept, we let $x = 0$ and solve for $y$. Then

$$f(0) = -0^2 - 2(0) + 3 = 3$$

Thus, $(0, 3)$ is the $y$-intercept.

To find the $x$-intercepts, we let $y$ or $f(x) = 0$ and solve for $x$.

$$f(x) = -x^2 - 2x + 3$$

$$0 = -x^2 - 2x + 3 \qquad \text{Let } f(x) = 0.$$

Now we divide both sides by $-1$ so that the coefficient of $x^2$ is 1.

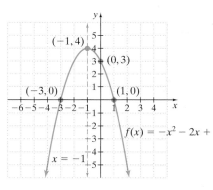

$$\frac{0}{-1} = \frac{-x^2}{-1} - \frac{2x}{-1} + \frac{3}{-1} \qquad \text{Divide both sides by } -1.$$

$$0 = x^2 + 2x - 3 \qquad \text{Simplify.}$$

$$0 = (x + 3)(x - 1) \qquad \text{Factor.}$$

$$x + 3 = 0 \quad \text{or} \quad x - 1 = 0 \qquad \text{Set each factor equal to 0.}$$

$$x = -3 \qquad\qquad x = 1 \qquad \text{Solve.}$$

The $x$-intercepts are $(-3, 0)$ and $(1, 0)$. Use these points to sketch the parabola.  □

**PRACTICE**
**3**  Graph $g(x) = -x^2 + 5x + 6$. Find the vertex and any intercepts.

**OBJECTIVE 2 ▶ Deriving a formula for finding the vertex.**  There is also a formula that may be used to find the vertex of a parabola. Now that we have practiced completing the square, we will show that the $x$-coordinate of the vertex of the graph of $f(x)$ or $y = ax^2 + bx + c$ can be found by the formula $x = \dfrac{-b}{2a}$. To do so, we complete the square on $x$ and write the equation in the form $y = a(x - h)^2 + k$.

First, isolate the $x$-variable terms by subtracting $c$ from both sides.

$$y = ax^2 + bx + c$$

$$y - c = ax^2 + bx$$

Next, factor $a$ from the terms $ax^2 + bx$.

$$y - c = a\left(x^2 + \frac{b}{a}x\right)$$

Next, add the square of half of $\frac{b}{a}$, or $\left(\frac{b}{2a}\right)^2 = \frac{b^2}{4a^2}$, to the right side inside the parentheses. Because of the factor $a$, what we really added was $a\left(\frac{b^2}{4a^2}\right)$ and this must be added to the left side.

$$y - c + a\left(\frac{b^2}{4a^2}\right) = a\left(x^2 + \frac{b}{a}x + \frac{b^2}{4a^2}\right)$$

$$y - c + \frac{b^2}{4a} = a\left(x + \frac{b}{2a}\right)^2 \qquad \text{Simplify the left side and factor the right side.}$$

$$y = a\left(x + \frac{b}{2a}\right)^2 + c - \frac{b^2}{4a} \qquad \text{Add } c \text{ to both sides and subtract } \frac{b^2}{4a} \text{ from both sides.}$$

Compare this form with $f(x)$ or $y = a(x - h)^2 + k$ and see that $h$ is $\frac{-b}{2a}$, which means that the $x$-coordinate of the vertex of the graph of $f(x) = ax^2 + bx + c$ is $\frac{-b}{2a}$.

---

**Vertex Formula**

The graph of $f(x) = ax^2 + bx + c$, when $a \neq 0$, is a parabola with vertex

$$\left(\frac{-b}{2a}, f\left(\frac{-b}{2a}\right)\right)$$

---

Let's use this formula to find the vertex of the parabola we graphed in Example 1.

**EXAMPLE 4**   Find the vertex of the graph of $f(x) = x^2 - 4x - 12$.

**Solution**   In the quadratic function $f(x) = x^2 - 4x - 12$, notice that $a = 1, b = -4$, and $c = -12$. Then

$$\frac{-b}{2a} = \frac{-(-4)}{2(1)} = 2$$

The $x$-value of the vertex is 2. To find the corresponding $f(x)$ or $y$-value, find $f(2)$. Then

$$f(2) = 2^2 - 4(2) - 12 = 4 - 8 - 12 = -16$$

The vertex is $(2, -16)$. These results agree with our findings in Example 1.   □

**PRACTICE**
**4**   Find the vertex of the graph of $g(x) = x^2 - 2x - 3$.

---

**OBJECTIVE 3 ▶ Finding minimum and maximum values.** The vertex of a parabola gives us some important information about its corresponding quadratic function. The quadratic function whose graph is a parabola that opens upward has a minimum value, and the quadratic function whose graph is a parabola that opens downward has a maximum value. The $f(x)$ or $y$-value of the vertex is the minimum or maximum value of the function.

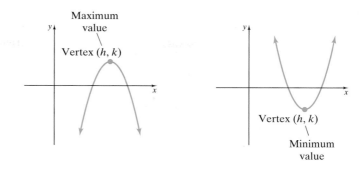

Maximum value

Vertex $(h, k)$

Vertex $(h, k)$

Minimum value

**Concept Check** ☑

Without making any calculations, tell whether the graph of $f(x) = 7 - x - 0.3x^2$ has a maximum value or a minimum value. Explain your reasoning.

### EXAMPLE 5   Finding Maximum Height

A rock is thrown upward from the ground. Its height in feet above ground after $t$ seconds is given by the function $f(t) = -16t^2 + 20t$. Find the maximum height of the rock and the number of seconds it took for the rock to reach its maximum height.

*Solution*

1. UNDERSTAND. The maximum height of the rock is the largest value of $f(t)$. Since the function $f(t) = -16t^2 + 20t$ is a quadratic function, its graph is a parabola. It opens downward since $-16 < 0$. Thus, the maximum value of $f(t)$ is the $f(t)$ or $y$-value of the vertex of its graph.

2. TRANSLATE. To find the vertex $(h, k)$, notice that for $f(t) = -16t^2 + 20t$, $a = -16, b = 20$, and $c = 0$. We will use these values and the vertex formula

$$\left(\frac{-b}{2a}, f\left(\frac{-b}{2a}\right)\right)$$

3. SOLVE.

$$h = \frac{-b}{2a} = \frac{-20}{-32} = \frac{5}{8}$$

$$f\left(\frac{5}{8}\right) = -16\left(\frac{5}{8}\right)^2 + 20\left(\frac{5}{8}\right)$$

$$= -16\left(\frac{25}{64}\right) + \frac{25}{2}$$

$$= -\frac{25}{4} + \frac{50}{4} = \frac{25}{4}$$

4. INTERPRET. The graph of $f(t)$ is a parabola opening downward with vertex $\left(\frac{5}{8}, \frac{25}{4}\right)$. This means that the rock's maximum height is $\frac{25}{4}$ feet, or $6\frac{1}{4}$ feet, which was reached in $\frac{5}{8}$ second. ☐

**PRACTICE**

**5**   A ball is tossed upward from the ground. Its height in feet above ground after $t$ seconds is given by the function $h(t) = -16t^2 + 24t$. Find the maximum height of the ball and the number of seconds it took for the ball to reach the maximum height.

## VOCABULARY & READINESS CHECK

*Fill in each blank.*

**1.** If a quadratic function is in the form $f(x) = a(x - h)^2 + k$, the vertex of its graph is _____.

**2.** The graph of $f(x) = ax^2 + bx + c, a \neq 0$ is a parabola whose vertex has $x$-value of _____.

| | *Parabola Opens* | *Vertex Location* | *Number of x-intercept(s)* | *Number of y-intercept(s)* |
|---|---|---|---|---|
| **3.** | up | Q I | | |
| **4.** | up | Q III | | |
| **5.** | down | Q II | | |
| **6.** | down | Q IV | | |
| **7.** | up | $x$-axis | | |
| **8.** | down | $x$-axis | | |
| **9.** | | Q III | 0 | |
| **10.** | | Q I | 2 | |
| **11.** | | Q IV | 2 | |
| **12.** | | Q II | 0 | |

## 11.6 | EXERCISE SET

**MyMathLab** Powered by CourseCompass™ and MathXL®

MathXL PRACTICE · WATCH · DOWNLOAD · READ · REVIEW

*Find the vertex of the graph of each quadratic function. See Examples 1 through 4.*

**1.** $f(x) = x^2 + 8x + 7$

**2.** $f(x) = x^2 + 6x + 5$

**3.** $f(x) = -x^2 + 10x + 5$

**4.** $f(x) = -x^2 - 8x + 2$

**5.** $f(x) = 5x^2 - 10x + 3$

**6.** $f(x) = -3x^2 + 6x + 4$

**7.** $f(x) = -x^2 + x + 1$

**8.** $f(x) = x^2 - 9x + 8$

*Match each function with its graph. See Examples 1 through 4.*

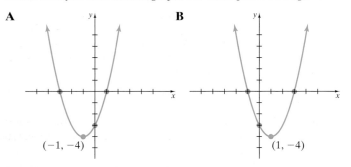

A

B

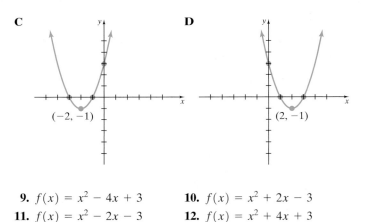

C

D

(−2, −1)          (2, −1)

**9.** $f(x) = x^2 - 4x + 3$      **10.** $f(x) = x^2 + 2x - 3$

**11.** $f(x) = x^2 - 2x - 3$      **12.** $f(x) = x^2 + 4x + 3$

**MIXED PRACTICE**

*Find the vertex of the graph of each quadratic function. Determine whether the graph opens upward or downward, find any intercepts, and sketch the graph. See Examples 1 through 4.*

**13.** $f(x) = x^2 + 4x - 5$      **14.** $f(x) = x^2 + 2x - 3$

**15.** $f(x) = -x^2 + 2x - 1$      **16.** $f(x) = -x^2 + 4x - 4$

**17.** $f(x) = x^2 - 4$      **18.** $f(x) = x^2 - 1$

**19.** $f(x) = 4x^2 + 4x - 3$      **20.** $f(x) = 2x^2 - x - 3$

**21.** $f(x) = x^2 + 8x + 15$      **22.** $f(x) = x^2 + 10x + 9$

**23.** $f(x) = x^2 - 6x + 5$

**24.** $f(x) = x^2 - 4x + 3$

**25.** $f(x) = x^2 - 4x + 5$

**26.** $f(x) = x^2 - 6x + 11$

**27.** $f(x) = 2x^2 + 4x + 5$

**28.** $f(x) = 3x^2 + 12x + 16$

**29.** $f(x) = -2x^2 + 12x$

**30.** $f(x) = -4x^2 + 8x$

**31.** $f(x) = x^2 + 1$

**32.** $f(x) = x^2 + 4$

**33.** $f(x) = x^2 - 2x - 15$

**34.** $f(x) = x^2 - x - 12$

**35.** $f(x) = -5x^2 + 5x$

**36.** $f(x) = 3x^2 - 12x$

**37.** $f(x) = -x^2 + 2x - 12$

**38.** $f(x) = -x^2 + 8x - 17$

**39.** $f(x) = 3x^2 - 12x + 15$

**40.** $f(x) = 2x^2 - 8x + 11$

**41.** $f(x) = x^2 + x - 6$

**42.** $f(x) = x^2 + 3x - 18$

**43.** $f(x) = -2x^2 - 3x + 35$

**44.** $f(x) = 3x^2 - 13x - 10$

*Solve. See Example 5.*

**45.** If a projectile is fired straight upward from the ground with an initial speed of 96 feet per second, then its height $h$ in feet after $t$ seconds is given by the equation

$$h(t) = -16t^2 + 96t$$

Find the maximum height of the projectile.

**46.** If Rheam Gaspar throws a ball upward with an initial speed of 32 feet per second, then its height $h$ in feet after $t$ seconds is given by the equation

$$h(t) = -16t^2 + 32t$$

Find the maximum height of the ball.

**47.** The cost $C$ in dollars of manufacturing $x$ bicycles at Holladay's Production Plant is given by the function

$$C(x) = 2x^2 - 800x + 92,000.$$

**a.** Find the number of bicycles that must be manufactured to minimize the cost.

**b.** Find the minimum cost.

**48.** The Utah Ski Club sells calendars to raise money. The profit $P$, in cents, from selling $x$ calendars is given by the equation $P(x) = 360x - x^2$.

**a.** Find how many calendars must be sold to maximize profit.

**b.** Find the maximum profit.

**49.** Find two numbers whose sum is 60 and whose product is as large as possible. [*Hint:* Let $x$ and $60 - x$ be the two positive numbers. Their product can be described by the function $f(x) = x(60 - x)$.]

**50.** Find two numbers whose sum is 11 and whose product is as large as possible. (Use the hint for Exercise 49.)

**51.** Find two numbers whose difference is 10 and whose product is as small as possible. (Use the hint for Exercise 49.)

**52.** Find two numbers whose difference is 8 and whose product is as small as possible.

△ **53.** The length and width of a rectangle must have a sum of 40. Find the dimensions of the rectangle that will have the maximum area. (Use the hint for Exercise 49.)

△ **54.** The length and width of a rectangle must have a sum of 50. Find the dimensions of the rectangle that will have maximum area.

**REVIEW AND PREVIEW**

*Sketch the graph of each function. See Section 11.5.*

**55.** $f(x) = x^2 + 2$

**56.** $f(x) = (x - 3)^2$

**57.** $g(x) = x + 2$

**58.** $h(x) = x - 3$

**59.** $f(x) = (x + 5)^2 + 2$

**60.** $f(x) = 2(x - 3)^2 + 2$

**61.** $f(x) = 3(x - 4)^2 + 1$

**62.** $f(x) = (x + 1)^2 + 4$

**63.** $f(x) = -(x - 4)^2 + \dfrac{3}{2}$

**64.** $f(x) = -2(x + 7)^2 + \dfrac{1}{2}$

**CONCEPT EXTENSIONS**

*Without calculating, tell whether each graph has a minimum value or a maximum value. See the Concept Check in the section.*

**65.** $f(x) = 2x^2 - 5$

**66.** $g(x) = -7x^2 + x + 1$

**67.** $F(x) = 3 - \dfrac{1}{2}x^2$

**68.** $G(x) = 3 - \dfrac{1}{2}x + 0.8x^2$

*Find the vertex of the graph of each quadratic function. Determine whether the graph opens upward or downward, find the y-intercept, approximate the x-intercepts to one decimal place, and sketch the graph.*

**69.** $f(x) = x^2 + 10x + 15$

**70.** $f(x) = x^2 - 6x + 4$

**71.** $f(x) = 3x^2 - 6x + 7$

**72.** $f(x) = 2x^2 + 4x - 1$

*Find the maximum or minimum value of each function. Approximate to two decimal places.*

**73.** $f(x) = 2.3x^2 - 6.1x + 3.2$

**74.** $f(x) = 7.6x^2 + 9.8x - 2.1$

**75.** $f(x) = -1.9x^2 + 5.6x - 2.7$

**76.** $f(x) = -5.2x^2 - 3.8x + 5.1$

**77.** The number of McDonald's restaurants worldwide can be modeled by the quadratic equation $f(x) = -96x^2 + 1018x + 28{,}824$, where $f(x)$ is the number of McDonald's restaurants and $x$ is the number of years after 2000. (*Source:* Based on data from McDonald's Corporation)

**a.** Will this function have a maximum or minimum? How can you tell?

**b.** According to this model, in what year will the number of McDonald's restaurants be at its maximum/minimum?

**c.** What is the maximum/minimum number of McDonald's restaurants predicted?

**78.** Methane is a gas produced by landfills, natural gas systems, and coal mining that contributes to the greenhouse effect and global warming. Projected methane emissions in the United States can be modeled by the quadratic function

$$f(x) = -0.072x^2 + 1.93x + 173.9$$

where $f(x)$ is the amount of methane produced in million metric tons and $x$ is the number of years after 2000. (*Source:* Based on data from the U.S. Environmental Protection Agency, 2000–2020)

**a.** According to this model, what will U.S. emissions of methane be in 2009? (Round to 2 decimal places.)

**b.** Will this function have a maximum or a minimum? How can you tell?

**c.** In what year will methane emissions in the United States be at their maximum/minimum? Round to the nearest whole year.

**d.** What is the level of methane emissions for that year? (Use your rounded answer from part c.) (Round this answer to 2 decimals places.)

*Use a graphing calculator to check each exercise.*

**79.** Exercise 27          **80.** Exercise 28

**81.** Exercise 37          **82.** Exercise 38

# CHAPTER 11 GROUP ACTIVITY

## Fitting a Quadratic Model to Data

Throughout the twentieth century, the eating habits of Americans changed noticeably. Americans started consuming less whole milk and butter, and started consuming more skim and low-fat milk and margarine. We also started eating more poultry and fish. In this project, you will have the opportunity to investigate trends in per capita consumption of poultry during the twentieth century. This project may be completed by working in groups or individually.

We will start by finding a quadratic model, $y = ax^2 + bx + c$, that has ordered pair solutions that correspond to the data for U.S. per capita consumption of poultry given in the table. To do so, substitute each data pair into the equation. Each time, the result is an equation in three unknowns: $a$, $b$, and $c$. Because there are three pairs of data, we can form a system of three linear equations in three unknowns. Solving for the values of $a$, $b$, and $c$ gives a quadratic model that represents the given data.

| U.S. per Capita Consumption of Poultry (in Pounds) | | |
| --- | --- | --- |
| Year | x | Poultry Consumption, y (in pounds) |
| 1909 | 9 | 11 |
| 1957 | 57 | 22 |
| 2005 | 105 | 66 |

(*Source:* Economic Research Service, U.S. Department of Agriculture)

**1.** Write the system of equations that must be solved to find the values of $a$, $b$, and $c$ needed for a quadratic model of the given data.

**2.** Solve the system of equations for $a$, $b$, and $c$. Recall the various methods of solving linear systems used in Chapter 4. You might consider using matrices, Cramer's rule, or a graphing calculator to do so. Round to the nearest thousandth.

**3.** Write the quadratic model for the data. Note that the variable $x$ represents the number of years after 1900.

**4.** In 1939, the actual U.S. per capita consumption of poultry was 12 pounds per person. Based on this information, how accurate do you think this model is for years other than those given in the table?

**5.** Use your model to estimate the per capita consumption of poultry in 1950.

**6.** According to the model, in what year was per capita consumption of poultry 50 pounds per person?

**7.** In what year was the per capita consumption of poultry at its lowest level? What was that level?

**8.** Who might be interested in a model like this and how would it be helpful?

📖 **STUDY SKILLS BUILDER**

### Are You Preparing for a Test on Chapter 11?

Below I have listed some common trouble areas for students in Chapter 11. After studying for your test—but before taking your test—read these.

- Don't forget that to solve a quadratic equation such as $x^2 + 6x = 1$, by completing the square, add the square of half of 6 to both sides.

$$x^2 + 6x = 1$$
$$x^2 + 6x + 9 = 1 + 9 \qquad \text{Add 9 to both sides, } \left(\frac{1}{2}(6) = 3 \text{ and } 3^2 = 9\right)$$
$$(x + 3)^2 = 10$$
$$x + 3 = \pm\sqrt{10}$$
$$x = -3 \pm \sqrt{10}$$

- Remember to write a quadratic equation in standard form $(ax^2 + bx + c = 0)$ before using the quadratic formula to solve.

$$x(4x - 1) = 1$$
$$4x^2 - x - 1 = 0 \qquad \text{Write in standard form.}$$
$$x = \frac{-(-1) \pm \sqrt{(-1)^2 - 4(4)(-1)}}{2 \cdot 4} \qquad \begin{array}{l}\text{Use the quadratic formula with}\\ a = 4, b = -1, \text{ and } c = -1.\end{array}$$
$$x = \frac{1 \pm \sqrt{17}}{8} \qquad \text{Simplify.}$$

- Review the steps for solving a quadratic equation in general on page 665.
- Don't forget how to graph a quadratic function in the form $f(x) = a(x - h)^2 + k$.

The graph of $f(x) = -2(x - 3)^2 - 1$.

opens downward, narrower    shift 3 units right    shift 1 unit down

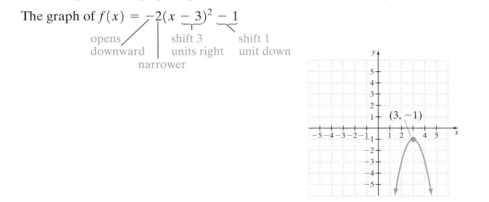

Remember: This is simply a checklist of common trouble areas. For a review of Chapter 11, see the Highlights and Chapter Review at the end of this chapter.

## CHAPTER 11 VOCABULARY CHECK

*Fill in each blank with one of the words or phrases listed below.*

| | | | |
|---|---|---|---|
| quadratic formula | quadratic | discriminant | $\pm\sqrt{b}$ |
| completing the square | quadratic inequality | $(h, k)$ | $(0, k)$ |
| $(h, 0)$ | $\dfrac{-b}{2a}$ | | |

**1.** The _____ helps us find the number and type of solutions of a quadratic equation.

**2.** If $a^2 = b$, then $a =$ _____.

**3.** The graph of $f(x) = ax^2 + bx + c$ where $a$ is not 0 is a parabola whose vertex has $x$-value of _____.

**4.** A(n) _____ is an inequality that can be written so that one side is a quadratic expression and the other side is 0.

**5.** The process of writing a quadratic equation so that one side is a perfect square trinomial is called _____ .

**6.** The graph of $f(x) = x^2 + k$ has vertex _____ .

**7.** The graph of $f(x) = (x - h)^2$ has vertex _____ .

**8.** The graph of $f(x) = (x - h)^2 + k$ has vertex _____ .

**9.** The formula $x = \dfrac{-b \pm \sqrt{b^2 - 4ac}}{2a}$ is called the _____ .

**10.** A _____ equation is one that can be written in the form $ax^2 + bx + c = 0$ where $a, b,$ and $c$ are real numbers and $a$ is not 0.

> ▶ **Helpful Hint**
>
> Are you preparing for your test? Don't forget to take the Chapter 11 Test on page 702. Then check your answers at the back of the text and use the Chapter Test Prep Video CD to see the fully worked-out solutions to any of the exercises you want to review.

# CHAPTER 11 HIGHLIGHTS

| DEFINITIONS AND CONCEPTS | EXAMPLES |
|---|---|

### SECTION 11.1  SOLVING QUADRATIC EQUATIONS BY COMPLETING THE SQUARE

| | |
|---|---|
| ***Square root property*** | Solve: $(x + 3)^2 = 14$. |
| If $b$ is a real number and if $a^2 = b$, then $a = \pm\sqrt{b}$. | $x + 3 = \pm\sqrt{14}$ |
| | $x = -3 \pm \sqrt{14}$ |
| ***To solve a quadratic equation in x by completing the square*** | Solve: $3x^2 - 12x - 18 = 0$. |
| **Step 1.** If the coefficient of $x^2$ is not 1, divide both sides of the equation by the coefficient of $x^2$. | **1.** $x^2 - 4x - 6 = 0$ |
| **Step 2.** Isolate the variable terms. | **2.** $x^2 - 4x = 6$ |
| **Step 3.** Complete the square by adding the square of half of the coefficient of $x$ to both sides. | **3.** $\frac{1}{2}(-4) = -2$ and $(-2)^2 = 4$ <br> $x^2 - 4x + 4 = 6 + 4$ |
| **Step 4.** Write the resulting trinomial as the square of a binomial. | **4.** $(x - 2)^2 = 10$ |
| **Step 5.** Apply the square root property and solve for $x$. | **5.** $x - 2 = \pm\sqrt{10}$ <br> $x = 2 \pm \sqrt{10}$ |

### SECTION 11.2  SOLVING QUADRATIC EQUATIONS BY THE QUADRATIC FORMULA

| | |
|---|---|
| A quadratic equation written in the form $ax^2 + bx + c = 0$ has solutions | Solve: $x^2 - x - 3 = 0$. |
| $$x = \dfrac{-b \pm \sqrt{b^2 - 4ac}}{2a}$$ | $a = 1, b = -1, c = -3$ <br> $x = \dfrac{-(-1) \pm \sqrt{(-1)^2 - 4(1)(-3)}}{2 \cdot 1}$ <br> $x = \dfrac{1 \pm \sqrt{13}}{2}$ |

| **DEFINITIONS AND CONCEPTS** | **EXAMPLES** |
|---|---|

SECTION 11.3   SOLVING EQUATIONS BY USING QUADRATIC METHODS

Substitution is often helpful in solving an equation that contains a repeated variable expression.

Solve: $(2x + 1)^2 - 5(2x + 1) + 6 = 0$.

Let $m = 2x + 1$. Then

$$m^2 - 5m + 6 = 0 \qquad \text{Let } m = 2x + 1.$$

$$(m - 3)(m - 2) = 0$$

$$m = 3 \quad \text{or} \quad m = 2$$

$$2x + 1 = 3 \quad \text{or} \quad 2x + 1 = 2 \quad \text{Substitute back.}$$

$$x = 1 \quad \text{or} \quad x = \frac{1}{2}$$

SECTION 11.4   NONLINEAR INEQUALITIES IN ONE VARIABLE

*To solve a polynomial inequality*

**Step 1.** Write the inequality in standard form.

**Step 2.** Solve the related equation.

**Step 3.** Use solutions from Step 2 to separate the number line into regions.

**Step 4.** Use test points to determine whether values in each region satisfy the original inequality.

**Step 5.** Write the solution set as the union of regions whose test point value is a solution.

Solve: $x^2 \geq 6x$.

**1.** $x^2 - 6x \geq 0$

**2.** $x^2 - 6x = 0$

$x(x - 6) = 0$

$x = 0 \quad \text{or} \quad x = 6$

**3.**

**4.**

| Region | Test Point Value | $x^2 \geq 6x$ | Result |
|---|---|---|---|
| $A$ | $-2$ | $(-2)^2 \geq 6(-2)$ | True |
| $B$ | $1$ | $1^2 \geq 6(1)$ | False |
| $C$ | $7$ | $7^2 \geq 6(7)$ | True |

**5.**

The solution set is $(-\infty, 0] \cup [6, \infty)$.

*To solve a rational inequality*

**Step 1.** Solve for values that make all denominators 0.

**Step 2.** Solve the related equation.

**Step 3.** Use solutions from Steps 1 and 2 to separate the number line into regions.

**Step 4.** Use test points to determine whether values in each region satisfy the original inequality.

**Step 5.** Write the solution set as the union of regions whose test point value is a solution.

Solve: $\dfrac{6}{x - 1} < -2$.

**1.** $x - 1 = 0$   Set denominator equal to 0.

$x = 1$

**2.** $\dfrac{6}{x - 1} = -2$

$6 = -2(x - 1)$   Multiply by $(x - 1)$.

$6 = -2x + 2$

$4 = -2x$

$-2 = x$

**3.**

**4.** Only a test value from region $B$ satisfies the original inequality.

**5.**

The solution set is $(-2, 1)$.

| **DEFINITIONS AND CONCEPTS** | **EXAMPLES** |
|---|---|

*Graph of a quadratic function*

The graph of a quadratic function written in the form $f(x) = a(x - h)^2 + k$ is a parabola with vertex $(h, k)$. If $a > 0$, the parabola opens upward; if $a < 0$, the parabola opens downward. The axis of symmetry is the line whose equation is $x = h$.

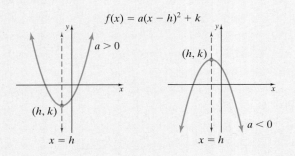

Graph $g(x) = 3(x - 1)^2 + 4$.

The graph is a parabola with vertex $(1, 4)$ and axis of symmetry $x = 1$. Since $a = 3$ is positive, the graph opens upward.

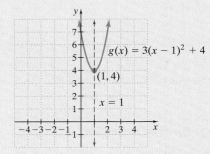

The graph of $f(x) = ax^2 + bx + c$, where $a \neq 0$, is a parabola with vertex

$$\left(\frac{-b}{2a}, f\left(\frac{-b}{2a}\right)\right)$$

Graph $f(x) = x^2 - 2x - 8$. Find the vertex and $x$- and $y$-intercepts.

$$\frac{-b}{2a} = \frac{-(-2)}{2 \cdot 1} = 1$$

$$f(1) = 1^2 - 2(1) - 8 = -9$$

The vertex is $(1, -9)$.

$$0 = x^2 - 2x - 8$$

$$0 = (x - 4)(x + 2)$$

$$x = 4 \quad \text{or} \quad x = -2$$

The $x$-intercepts are $(4, 0)$ and $(-2, 0)$.

$$f(0) = 0^2 - 2 \cdot 0 - 8 = -8$$

The $y$-intercept is $(0, -8)$.

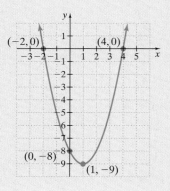

# CHAPTER 11 REVIEW

*(11.1) Solve by factoring.*

**1.** $x^2 - 15x + 14 = 0$      **2.** $7a^2 = 29a + 30$

*Solve by using the square root property.*

**3.** $4m^2 = 196$      **4.** $(5x - 2)^2 = 2$

*Solve by completing the square.*

**5.** $z^2 + 3z + 1 = 0$

**6.** $(2x + 1)^2 = x$

**7.** If $P$ dollars are originally invested, the formula $A = P(1 + r)^2$ gives the amount $A$ in an account paying interest rate $r$ compounded annually after 2 years. Find the interest rate $r$ such that \$2500 increases to \$2717 in 2 years. Round the result to the nearest hundredth of a percent.

△ **8.** Two ships leave a port at the same time and travel at the same speed. One ship is traveling due north and the other due east. In a few hours, the ships are 150 miles apart. How many miles has each ship traveled? Give an exact answer and a one-decimal-place approximation.

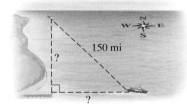

*(11.2) If the discriminant of a quadratic equation has the given value, determine the number and type of solutions of the equation.*

**9.** $-8$      **10.** $48$

**11.** $100$      **12.** $0$

*Solve by using the quadratic formula.*

**13.** $x^2 - 16x + 64 = 0$    **14.** $x^2 + 5x = 0$

**15.** $2x^2 + 3x = 5$      **16.** $9a^2 + 4 = 2a$

**17.** $6x^2 + 7 = 5x$      **18.** $(2x - 3)^2 = x$

**19.** Cadets graduating from military school usually toss their hats high into the air at the end of the ceremony. One cadet threw his hat so that its distance $d(t)$ in feet above the ground $t$ seconds after it was thrown was $d(t) = -16t^2 + 30t + 6$.

   **a.** Find the distance above the ground of the hat 1 second after it was thrown.

   **b.** Find the time it takes the hat to hit the ground. Give an exact time and a one-decimal-place approximation.

△ **20.** The hypotenuse of an isosceles right triangle is 6 centimeters longer than either of the legs. Find the length of the legs.

*(11.3) Solve each equation for the variable.*

**21.** $x^3 = 27$

**22.** $y^3 = -64$

**23.** $\dfrac{5}{x} + \dfrac{6}{x - 2} = 3$

**24.** $x^4 - 21x^2 - 100 = 0$

**25.** $x^{2/3} - 6x^{1/3} + 5 = 0$

**26.** $5(x + 3)^2 - 19(x + 3) = 4$

**27.** $a^6 - a^2 = a^4 - 1$

**28.** $y^{-2} + y^{-1} = 20$

**29.** Two postal workers, Jerome Grant and Tim Bozik, can sort a stack of mail in 5 hours. Working alone, Tim can sort the mail in 1 hour less time than Jerome can. Find the time that each postal worker can sort the mail alone. Round the result to one decimal place.

**30.** A negative number decreased by its reciprocal is $-\dfrac{24}{5}$. Find the number.

*(11.4) Solve each inequality for x. Write each solution set in interval notation.*

**31.** $2x^2 - 50 \le 0$

**32.** $\dfrac{1}{4}x^2 < \dfrac{1}{16}$

**33.** $\dfrac{x - 5}{x - 6} < 0$

**34.** $(x^2 - 16)(x^2 - 1) > 0$

**35.** $\dfrac{(4x + 3)(x - 5)}{x(x + 6)} > 0$

**36.** $(x + 5)(x - 6)(x + 2) \le 0$

**37.** $x^3 + 3x^2 - 25x - 75 > 0$

**38.** $\dfrac{x^2 + 4}{3x} \le 1$

**39.** $\dfrac{(5x + 6)(x - 3)}{x(6x - 5)} < 0$

**40.** $\dfrac{3}{x - 2} > 2$

*(11.5) Sketch the graph of each function. Label the vertex and the axis of symmetry.*

**41.** $f(x) = x^2 - 4$

**42.** $g(x) = x^2 + 7$

**43.** $H(x) = 2x^2$

**44.** $h(x) = -\dfrac{1}{3}x^2$

**45.** $F(x) = (x - 1)^2$

**46.** $G(x) = (x + 5)^2$

**47.** $f(x) = (x - 4)^2 - 2$

**48.** $f(x) = -3(x - 1)^2 + 1$

*(11.6) Sketch the graph of each function. Find the vertex and the intercepts.*

**49.** $f(x) = x^2 + 10x + 25$

**50.** $f(x) = -x^2 + 6x - 9$

**51.** $f(x) = 4x^2 - 1$

**52.** $f(x) = -5x^2 + 5$

**53.** Find the vertex of the graph of $f(x) = -3x^2 - 5x + 4$. Determine whether the graph opens upward or downward, find the $y$-intercept, approximate the $x$-intercepts to one decimal place, and sketch the graph.

**54.** The function $h(t) = -16t^2 + 120t + 300$ gives the height in feet of a projectile fired from the top of a building in $t$ seconds.

   **a.** When will the object reach a height of 350 feet? Round your answer to one decimal place.

   **b.** Explain why part **a** has two answers.

**55.** Find two numbers whose product is as large as possible, given that their sum is 420.

**56.** Write an equation of a quadratic function whose graph is a parabola that has vertex $(-3, 7)$ and that passes through the origin.

**MIXED REVIEW**

*Solve each equation.*

**57.** $x^2 - x - 30 = 0$

**58.** $10x^2 = 3x + 4$

**59.** $9y^2 = 36$

**60.** $(9n + 1)^2 = 9$

**61.** $x^2 + x + 7 = 0$

**62.** $(3x - 4)^2 = 10x$

**63.** $x^2 + 11 = 0$

**64.** $(5a - 2)^2 - a = 0$

**65.** $\dfrac{7}{8} = \dfrac{8}{x^2}$

**66.** $x^{2/3} - 6x^{1/3} = -8$

**67.** $(2x - 3)(4x + 5) \geq 0$

**68.** $\dfrac{x(x + 5)}{4x - 3} \geq 0$

**69.** $\dfrac{3}{x - 2} > 2$

**70.** The total amount of passenger traffic at Phoenix Sky Harbor International Airport in Phoenix, Arizona, during the period 1980 through 2005 can be modeled by the equation $y = 6.46x^2 + 1236.5x + 7289$, where $y$ is the number of passengers enplaned and deplaned in thousands and $x$ is the number of years after 1980. (*Source*: Based on data from The City of Phoenix Aviation Department, 1980–2005)

   **a.** Estimate the passenger traffic at Phoenix Sky Harbor International Airport in 2000.

   **b.** According to this model, in what year will passenger traffic at Phoenix Sky Harbor International Airport reach 60,000,000 passengers?

# CHAPTER 11 TEST  TEST PREP VIDEO

Remember to use the Chapter Test Prep Video CD to see the fully worked-out solutions to any of the exercises you want to review.

*Solve each equation for the variable.*

**1.** $5x^2 - 2x = 7$

**2.** $(x + 1)^2 = 10$

**3.** $m^2 - m + 8 = 0$

**4.** $u^2 - 6u + 2 = 0$

**5.** $7x^2 + 8x + 1 = 0$

**6.** $y^2 - 3y = 5$

**7.** $\dfrac{4}{x + 2} + \dfrac{2x}{x - 2} = \dfrac{6}{x^2 - 4}$

**8.** $x^5 + 3x^4 = x + 3$

**9.** $x^6 + 1 = x^4 + x^2$

**10.** $(x + 1)^2 - 15(x + 1) + 56 = 0$

*Solve the equation for the variable by completing the square.*

**11.** $x^2 - 6x = -2$

**12.** $2a^2 + 5 = 4a$

*Solve each inequality for x. Write the solution set in interval notation.*

**13.** $2x^2 - 7x > 15$

**14.** $(x^2 - 16)(x^2 - 25) \geq 0$

**15.** $\dfrac{5}{x + 3} < 1$

**16.** $\dfrac{7x - 14}{x^2 - 9} \le 0$

*Graph each function. Label the vertex.*

**17.** $f(x) = 3x^2$

**18.** $G(x) = -2(x - 1)^2 + 5$

*Graph each function. Find and label the vertex, y-intercept, and x-intercepts (if any).*

**19.** $h(x) = x^2 - 4x + 4$

**20.** $F(x) = 2x^2 - 8x + 9$

**21.** Dave and Sandy Hartranft can paint a room together in 4 hours. Working alone, Dave can paint the room in 2 hours less time than Sandy can. Find how long it takes Sandy to paint the room alone.

**22.** A stone is thrown upward from a bridge. The stone's height in feet, $s(t)$, above the water $t$ seconds after the stone is thrown is a function given by the equation $s(t) = -16t^2 + 32t + 256$.

**a.** Find the maximum height of the stone.

**b.** Find the time it takes the stone to hit the water. Round the answer to two decimal places.

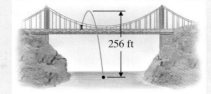

**23.** Given the diagram shown, approximate to the nearest foot how many feet of walking distance a person saves by cutting across the lawn instead of walking on the sidewalk.

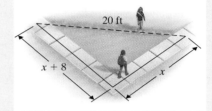

# CHAPTER 11 CUMULATIVE REVIEW

**1.** Find the value of each expression when $x = 2$ and $y = -5$.

**a.** $\dfrac{x - y}{12 + x}$

**b.** $x^2 - 3y$.

**2.** Solve $|3x - 2| = -5$.

**3.** Simplify each expression by combining like terms.

**a.** $2x + 3x + 5 + 2$

**b.** $-5a - 3 + a + 2$

**c.** $4y - 3y^2$

**d.** $2.3x + 5x - 6$

**e.** $-\dfrac{1}{2}b + b$

**4.** Use the addition method to solve the system.
$$\begin{cases} -6x + \ y = 5 \\ \ \ 4x - 2y = 6 \end{cases}$$

**5.** Solve the following system of equations by graphing.
$$\begin{cases} 2x + y = 7 \\ 2y = -4x \end{cases}$$

**6.** Simplify. Use positive exponents to write each answer.

**a.** $(a^{-2}bc^3)^{-3}$

**b.** $\left(\dfrac{a^{-4}b^2}{c^3}\right)^{-2}$

**c.** $\left(\dfrac{3a^8b^2}{12a^5b^5}\right)^{-2}$

**7.** Solve the system:
$$\begin{cases} 7x - 3y = -14 \\ -3x + y = 6 \end{cases}$$

**8.** Multiply.

**a.** $(4a - 3)(7a - 2)$

**b.** $(2a + b)(3a - 5b)$

**9.** Simplify each quotient.

**a.** $\dfrac{x^5}{x^2}$

**b.** $\dfrac{4^7}{4^3}$

**c.** $\dfrac{(-3)^5}{(-3)^2}$

**d.** $\dfrac{s^2}{t^3}$

**e.** $\dfrac{2x^5y^2}{xy}$

**10.** Factor.

**a.** $9x^3 + 27x^2 - 15x$

**b.** $2x(3y - 2) - 5(3y - 2)$

**c.** $2xy + 6x - y - 3$

**11.** If $P(x) = 2x^3 - 4x^2 + 5$

**a.** Find $P(2)$ by substitution.

**b.** Use synthetic division to find the remainder when $P(x)$ is divided by $x - 2$.

**12.** Factor $x^2 - 2x - 48$.

**13.** Solve $(5x - 1)(2x^2 + 15x + 18) = 0$.

**14.** Factor. $2ax^2 - 12axy + 18ay^2$

**15.** Write the rational expression in lowest terms.
$$\dfrac{2x^2}{10x^3 - 2x^2}$$

**16.** Solve $2(a^2 + 2) - 8 = -2a(a - 2) - 5$.

**17.** Simplify.
$$\dfrac{x^{-1} + 2xy^{-1}}{x^{-2} - x^{-2}y^{-1}}$$

18. Find the vertex and any intercepts of $f(x) = x^2 + x - 12$.

19. Factor $4m^4 - 4m^2 + 1$.

20. Simplify. $\dfrac{x^2 - 4x + 4}{2 - x}$

21. The square of a number plus three times the number is 70. Find the number.

22. Subtract. $\dfrac{a + 1}{a^2 - 6a + 8} - \dfrac{3}{16 - a^2}$

23. Use the product rule to simplify.

    a. $\sqrt{25x^3}$         b. $\sqrt[3]{54x^6y^8}$

    c. $\sqrt[4]{81z^{11}}$

24. Simplify. $\dfrac{(2a)^{-1} + b^{-1}}{a^{-1} + (2b)^{-1}}$

25. Rationalize the denominator of each expression.

    a. $\dfrac{2}{\sqrt{5}}$         b. $\dfrac{2\sqrt{16}}{\sqrt{9x}}$

    c. $\sqrt[3]{\dfrac{1}{2}}$

26. Divide $x^3 - 3x^2 - 10x + 24$ by $x + 3$.

27. Solve $\sqrt{2x + 5} + \sqrt{2x} = 3$.

28. If $P(x) = 4x^3 - 2x^2 + 3$,

    a. Find $P(-2)$ by substitution.

    b. Use synthetic division to find the remainder when $P(x)$ is divided by $x + 2$.

29. Solve $\dfrac{x}{2} + \dfrac{8}{3} = \dfrac{1}{6}$.

30. Solve $\dfrac{x + 3}{x^2 + 5x + 6} = \dfrac{3}{2x + 4} - \dfrac{1}{x + 3}$.

31. The quotient of a number and 6, minus $\dfrac{5}{3}$, is the quotient of the number and 2. Find the number.

32. Mr. Briley can roof his house in 24 hours. His son can roof the same house in 40 hours. If they work together, how long will it take to roof the house?

33. Suppose that $y$ varies directly as $x$. If $y$ is 5 when $x$ is 30, find the constant of variation and the direct variation equation.

34. Suppose that $y$ varies inversely as $x$. If $y$ is 8 when $x$ is 24, find the constant of variation and the inverse variation equation.

35. Simplify.

    a. $\sqrt{(-3)^2}$         b. $\sqrt{x^2}$

    c. $\sqrt[4]{(x - 2)^4}$         d. $\sqrt[3]{(-5)^3}$

    e. $\sqrt[5]{(2x - 7)^5}$         f. $\sqrt{25x^2}$

    g. $\sqrt{x^2 + 2x + 1}$

36. Simplify. Assume that the variables represent any real number.

    a. $\sqrt{(-2)^2}$         b. $\sqrt{y^2}$

    c. $\sqrt[4]{(a - 3)^4}$         d. $\sqrt[3]{(-6)^3}$

    e. $\sqrt[5]{(3x - 1)^5}$

37. Use rational exponents to simplify. Assume that variables represent positive numbers.

    a. $\sqrt[8]{x^4}$

    b. $\sqrt[6]{25}$

    c. $\sqrt[4]{r^2s^6}$

38. Use rational exponents to simplify. Assume that variables represent positive numbers.

    a. $\sqrt[4]{5^2}$

    b. $\sqrt[12]{x^3}$

    c. $\sqrt[6]{x^2y^4}$

39. Divide. Write in the form $a + bi$.

    a. $\dfrac{2 + i}{1 - i}$

    b. $\dfrac{7}{3i}$

40. Write each product in the form of $a + bi$.

    a. $3i(5 - 2i)$

    b. $(6 - 5i)^2$

    c. $\left(\sqrt{3} + 2i\right)\left(\sqrt{3} - 2i\right)$

41. Use the square root property to solve $(x + 1)^2 = 12$.

42. Use the square root property to solve $(y - 1)^2 = 24$.

43. Solve $x - \sqrt{x} - 6 = 0$.

44. Use the quadratic formula to solve. $m^2 = 4m + 8$

CHAPTER

# 12

# Exponential and Logarithmic Functions

pH (Potential of Hydrogen) is a measure of the acidity or alkalinity of a solution. Solutions with a pH less than 7 are considered acidic, those with a pH greater than 7 are considered basic (alkaline) and those equal to 7 are defined as "neutral." The pH scale is logarithmic and some examples are in the table below. Since pH is dependent on ionic activity, it can't easily be measured. One of the oldest ways to measure the pH of a solution is litmus paper. Litmus is a water-soluble mixture of different dyes extracted from lichens.

In Section 12.4, Exercise 106, page 737, we will calculate the pH for lemonade.

In this chapter, we discuss two closely related functions: exponential and logarithmic functions. These functions are vital to applications in economics, finance, engineering, the sciences, education, and other fields. Models of tumor growth and learning curves are two examples of the uses of exponential and logarithmic functions.

**Representative pH values**

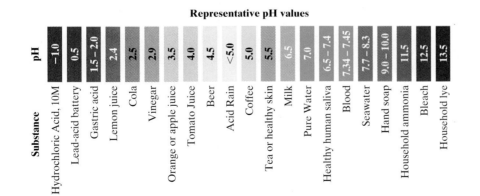

| Substance | pH |
|---|---|
| Hydrochloric Acid, 10M | −1.0 |
| Lead-acid battery | 0.5 |
| Gastric acid | 1.5 − 2.0 |
| Lemon juice | 2.4 |
| Cola | 2.5 |
| Vinegar | 2.9 |
| Orange or apple juice | 3.5 |
| Tomato Juice | 4.0 |
| Beer | 4.5 |
| Acid Rain | <5.0 |
| Coffee | 5.0 |
| Tea or healthy skin | 5.5 |
| Milk | 6.5 |
| Pure Water | 7.0 |
| Healthy human saliva | 6.5 − 7.4 |
| Blood | 7.34 − 7.45 |
| Seawater | 7.7 − 8.3 |
| Hand soap | 9.0 − 10.0 |
| Household ammonia | 11.5 |
| Bleach | 12.5 |
| Household lye | 13.5 |

# 12.1 THE ALGEBRA OF FUNCTIONS; COMPOSITE FUNCTIONS

## OBJECTIVES

1 Add, subtract, multiply, and divide functions.

2 Construct composite functions.

**OBJECTIVE 1 ▶ Adding, subtracting, multiplying, and dividing functions.** As we have seen in earlier chapters, it is possible to add, subtract, multiply, and divide functions. Although we have not stated it as such, the sums, differences, products, and quotients of functions are themselves functions. For example, if $f(x) = 3x$ and $g(x) = x + 1$, their product, $f(x) \cdot g(x) = 3x(x + 1) = 3x^2 + 3x$, is a new function. We can use the notation $(f \cdot g)(x)$ to denote this new function. Finding the sum, difference, product, and quotient of functions to generate new functions is called the **algebra of functions.**

---

**Algebra of Functions**

Let $f$ and $g$ be functions. New functions from $f$ and $g$ are defined as follows.

| | |
|---|---|
| **Sum** | $(f + g)(x) = f(x) + g(x)$ |
| **Difference** | $(f - g)(x) = f(x) - g(x)$ |
| **Product** | $(f \cdot g)(x) = f(x) \cdot g(x)$ |
| **Quotient** | $\left(\dfrac{f}{g}\right)(x) = \dfrac{f(x)}{g(x)}, \quad g(x) \neq 0$ |

---

**EXAMPLE 1** If $f(x) = x - 1$ and $g(x) = 2x - 3$, find

**a.** $(f + g)(x)$     **b.** $(f - g)(x)$     **c.** $(f \cdot g)(x)$     **d.** $\left(\dfrac{f}{g}\right)(x)$

*Solution* Use the algebra of functions and replace $f(x)$ by $x - 1$ and $g(x)$ by $2x - 3$. Then we simplify.

**a.** $(f + g)(x) = f(x) + g(x)$
$\qquad\qquad = (x - 1) + (2x - 3)$
$\qquad\qquad = 3x - 4$

**b.** $(f - g)(x) = f(x) - g(x)$
$\qquad\qquad = (x - 1) - (2x - 3)$
$\qquad\qquad = x - 1 - 2x + 3$
$\qquad\qquad = -x + 2$

**c.** $(f \cdot g)(x) = f(x) \cdot g(x)$
$\qquad\qquad = (x - 1)(2x - 3)$
$\qquad\qquad = 2x^2 - 5x + 3$

**d.** $\left(\dfrac{f}{g}\right)(x) = \dfrac{f(x)}{g(x)} = \dfrac{x - 1}{2x - 3}$, where $x \neq \dfrac{3}{2}$

**PRACTICE**

**1** If $f(x) = x + 2$ and $g(x) = 3x + 5$, find

**a.** $(f + g)(x)$     **b.** $(f - g)(x)$     **c.** $(f \cdot g)(x)$     **d.** $\left(\dfrac{f}{g}\right)(x)$

There is an interesting but not surprising relationship between the graphs of functions and the graphs of their sum, difference, product, and quotient. For example, the graph of $(f + g)(x)$ can be found by adding the graph of $f(x)$ to the graph of $g(x)$. We add two graphs by adding $y$-values of corresponding $x$-values.

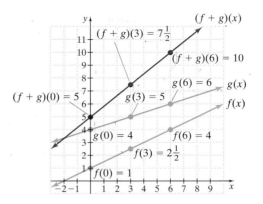

**OBJECTIVE 2 ▶ Constructing composite functions.** Another way to combine functions is called **function composition.** To understand this new way of combining functions, study the diagrams below. The left diagram shows degrees Celsius $f(x)$ as a function of degrees Fahrenheit $x$. The right diagram shows Kelvins $g(x)$ as a function of degrees Celsius $x$. (The Kelvin scale is a temperature scale devised by Lord Kelvin in 1848.) The function represented by the first diagram we will call $f$, and the second function we will call $g$.

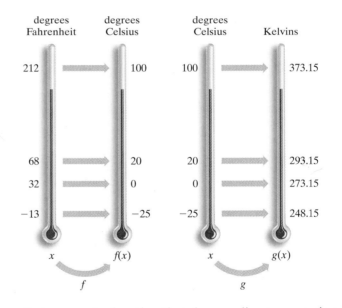

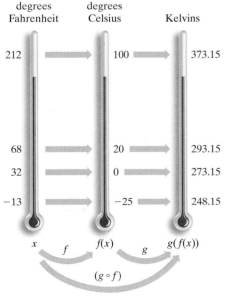

Suppose that we want a function that shows a direct conversion from degrees Fahrenheit to Kelvins. In other words, suppose that a function is needed that shows Kelvins as a function of degrees Fahrenheit. This can easily be done because the output of the first function $f(x)$ is the same as the input of the second function. If we use $f(x)$ to represent this, then we get the left diagram.

For example $g(f(-13)) = 248.15$, and so on.

Since the output of the first function is used as the input of the second function, we write the new function as $g(f(x))$. The new function is formed from the composition of the other two functions. The mathematical symbol for this composition is $(g \circ f)(x)$. Thus, $(g \circ f)(x) = g(f(x))$.

It is possible to find an equation for the composition of the two functions $f$ and $g$. In other words, we can find a function that converts degrees Fahrenheit directly to

Kelvins. The function $f(x) = \dfrac{5}{9}(x - 32)$ converts degrees Fahrenheit to degrees Celsius, and the function $g(x) = x + 273.15$ converts degrees Celsius to Kelvins. Thus,

$$(g \circ f)(x) = g(f(x)) = g\left(\dfrac{5}{9}(x - 32)\right) = \dfrac{5}{9}(x - 32) + 273.15$$

In general, the notation $g(f(x))$ means "g composed with $f$" and can be written as $(g \circ f)(x)$. Also $f(g(x))$, or $(f \circ g)(x)$, means "$f$ composed with $g$."

---

**Composition of Functions**
The composition of functions $f$ and $g$ is
$$(f \circ g)(x) = f(g(x))$$

---

▶ **Helpful Hint**
$(f \circ g)(x)$ does not mean the same as $(f \cdot g)(x)$.

$$(f \circ g)(x) = f(g(x)) \text{ while } (f \cdot g)(x) = f(x) \cdot g(x)$$
      ↑                      ↑
 Composition of functions       Multiplication of functions

---

**EXAMPLE 2** If $f(x) = x^2$ and $g(x) = x + 3$, find each composition.

**a.** $(f \circ g)(2)$ and $(g \circ f)(2)$      **b.** $(f \circ g)(x)$ and $(g \circ f)(x)$

*Solution*

**a.** $(f \circ g)(2) = f(g(2))$
            $= f(5)$         Replace $g(2)$ with 5. [Since $g(x) = x + 3$, then
            $= 5^2 = 25$     $g(2) = 2 + 3 = 5$.]

   $(g \circ f)(2) = g(f(2))$
            $= g(4)$         Since $f(x) = x^2$, then $f(2) = 2^2 = 4$.
            $= 4 + 3 = 7$

**b.** $(f \circ g)(x) = f(g(x))$
            $= f(x + 3)$       Replace $g(x)$ with $x + 3$.
            $= (x + 3)^2$      $f(x + 3) = (x + 3)^2$
            $= x^2 + 6x + 9$   Square $(x + 3)$.

   $(g \circ f)(x) = g(f(x))$
            $= g(x^2)$         Replace $f(x)$ with $x^2$.
            $= x^2 + 3$       $g(x^2) = x^2 + 3$                  □

PRACTICE
   **2**    If $f(x) = x^2 + 1$ and $g(x) = 3x - 5$, find

**a.** $(f \circ g)(4)$            **b.** $(f \circ g)(x)$
   $(g \circ f)(4)$              $(g \circ f)(x)$

**EXAMPLE 3**  If $f(x) = |x|$ and $g(x) = x - 2$, find each composition.

**a.** $(f \circ g)(x)$          **b.** $(g \circ f)(x)$

*Solution*

**a.** $(f \circ g)(x) = f(g(x)) = f(x - 2) = |x - 2|$
**b.** $(g \circ f)(x) = g(f(x)) = g(|x|) = |x| - 2$          □

▶ **Helpful Hint**

In Examples 2 and 3, notice that $(g \circ f)(x) \neq (f \circ g)(x)$. In general, $(g \circ f)(x)$ *may* or *may not* equal $(f \circ g)(x)$.

PRACTICE

**3**  If $f(x) = x^2 + 5$ and $g(x) = x + 3$, find each composition.

**a.** $(f \circ g)(x)$          **b.** $(g \circ f)(x)$

**EXAMPLE 4**  If $f(x) = 5x$, $g(x) = x - 2$, and $h(x) = \sqrt{x}$, write each function as a composition using two of the given functions.

**a.** $F(x) = \sqrt{x - 2}$          **b.** $G(x) = 5x - 2$

*Solution*

**a.** Notice the order in which the function $F$ operates on an input value $x$. First, 2 is subtracted from $x$. This is the function $g(x) = x - 2$. Then the square root *of that result* is taken. The square root function is $h(x) = \sqrt{x}$. This means that $F = h \circ g$. To check, we find $h \circ g$.

$$F(x) = (h \circ g)(x) = h(g(x)) = h(x - 2) = \sqrt{x - 2}$$

**b.** Notice the order in which the function $G$ operates on an input value $x$. First, $x$ is multiplied by 5, and then 2 is subtracted from the result. This means that $G = g \circ f$. To check, we find $g \circ f$.

$$G(x) = (g \circ f)(x) = g(f(x)) = g(5x) = 5x - 2$$          □

PRACTICE

**4**  If $f(x) = 3x$, $g(x) = x - 4$, and $h(x) = |x|$, write each function as a composition using two of the given functions.

**a.** $F(x) = |x - 4|$          **b.** $G(x) = 3x - 4$

---

**Graphing Calculator Explorations**

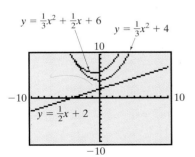

$y = \frac{1}{3}x^2 + \frac{1}{2}x + 6$

$y = \frac{1}{3}x^2 + 4$

$y = \frac{1}{2}x + 2$

If $f(x) = \frac{1}{2}x + 2$ and $g(x) = \frac{1}{3}x^2 + 4$, then

$$(f + g)(x) = f(x) + g(x)$$
$$= \left(\frac{1}{2}x + 2\right) + \left(\frac{1}{3}x^2 + 4\right)$$
$$= \frac{1}{3}x^2 + \frac{1}{2}x + 6.$$

To visualize this addition of functions with a graphing calculator, graph

$$Y_1 = \frac{1}{2}x + 2, \qquad Y_2 = \frac{1}{3}x^2 + 4, \qquad Y_3 = \frac{1}{3}x^2 + \frac{1}{2}x + 6$$

Use a TABLE feature to verify that for a given $x$ value, $Y_1 + Y_2 = Y_3$. For example, verify that when $x = 0$, $Y_1 = 2$, $Y_2 = 4$, and $Y_3 = 2 + 4 = 6$.

## VOCABULARY & READINESS CHECK

*Match each function with its definition.*

**1.** $(f \circ g)(x)$      **4.** $(g \circ f)(x)$      **A.** $g(f(x))$      **D.** $\dfrac{f(x)}{g(x)}, g(x) \neq 0$

**2.** $(f \cdot g)(x)$      **5.** $\left(\dfrac{f}{g}\right)(x)$      **B.** $f(x) + g(x)$      **E.** $f(x) \cdot g(x)$

**3.** $(f - g)(x)$      **6.** $(f + g)(x)$      **C.** $f(g(x))$      **F.** $f(x) - g(x)$

## 12.1 | EXERCISE SET

*For the functions f and g, find **a.** $(f + g)(x)$, **b.** $(f - g)(x)$, **c.** $(f \cdot g)(x)$, and **d.** $\left(\dfrac{f}{g}\right)(x)$. See Example 1.*

**1.** $f(x) = x - 7, g(x) = 2x + 1$

**2.** $f(x) = x + 4, g(x) = 5x - 2$

**3.** $f(x) = x^2 + 1, g(x) = 5x$

**4.** $f(x) = x^2 - 2, g(x) = 3x$

**5.** $f(x) = \sqrt{x}, g(x) = x + 5$

**6.** $f(x) = \sqrt[3]{x}, g(x) = x - 3$

**7.** $f(x) = -3x, g(x) = 5x^2$

**8.** $f(x) = 4x^3, g(x) = -6x$

*If $f(x) = x^2 - 6x + 2$, $g(x) = -2x$, and $h(x) = \sqrt{x}$, find each composition. See Example 2.*

**9.** $(f \circ g)(2)$      **10.** $(h \circ f)(-2)$

**11.** $(g \circ f)(-1)$      **12.** $(f \circ h)(1)$

**13.** $(g \circ h)(0)$      **14.** $(h \circ g)(0)$

*Find $(f \circ g)(x)$ and $(g \circ f)(x)$. See Examples 2 and 3.*

**15.** $f(x) = x^2 + 1, g(x) = 5x$

**16.** $f(x) = x - 3, g(x) = x^2$

**17.** $f(x) = 2x - 3, g(x) = x + 7$

**18.** $f(x) = x + 10, g(x) = 3x + 1$

**19.** $f(x) = x^3 + x - 2, g(x) = -2x$

**20.** $f(x) = -4x, g(x) = x^3 + x^2 - 6$

**21.** $f(x) = |x|; g(x) = 10x - 3$

**22.** $f(x) = |x|; g(x) = 14x - 8$

**23.** $f(x) = \sqrt{x}, g(x) = -5x + 2$

**24.** $f(x) = 7x - 1, g(x) = \sqrt[3]{x}$

*If $f(x) = 3x$, $g(x) = \sqrt{x}$, and $h(x) = x^2 + 2$, write each function as a composition using two of the given functions. See Example 4.*

**25.** $H(x) = \sqrt{x^2 + 2}$

**26.** $G(x) = \sqrt{3x}$

**27.** $F(x) = 9x^2 + 2$

**28.** $H(x) = 3x^2 + 6$

**29.** $G(x) = 3\sqrt{x}$

**30.** $F(x) = x + 2$

*Find $f(x)$ and $g(x)$ so that the given function $h(x) = (f \circ g)(x)$.*

**31.** $h(x) = (x + 2)^2$

**32.** $h(x) = |x - 1|$

**33.** $h(x) = \sqrt{x + 5} + 2$

**34.** $h(x) = (3x + 4)^2 + 3$

**35.** $h(x) = \dfrac{1}{2x - 3}$

**36.** $h(x) = \dfrac{1}{x + 10}$

## REVIEW AND PREVIEW

*Solve each equation for y. See Section 2.3.*

**37.** $x = y + 2$        **38.** $x = y - 5$

**39.** $x = 3y$        **40.** $x = -6y$

**41.** $x = -2y - 7$        **42.** $x = 4y + 7$

## CONCEPT EXTENSIONS

*Given that* $f(-1) = 4$    $g(-1) = -4$

$f(0) = 5$    $g(0) = -3$

$f(2) = 7$    $g(2) = -1$

$f(7) = 1$    $g(7) = 4$

*Find each function value.*

**43.** $(f + g)(2)$        **44.** $(f - g)(7)$

**45.** $(f \circ g)(2)$        **46.** $(g \circ f)(2)$

**47.** $(f \cdot g)(7)$        **48.** $(f \cdot g)(0)$

**49.** $\left(\dfrac{f}{g}\right)(-1)$        **50.** $\left(\dfrac{g}{f}\right)(-1)$

**51.** If you are given $f(x)$ and $g(x)$, explain in your own words how to find $(f \circ g)(x)$, and then how to find $(g \circ f)(x)$.

**52.** Given $f(x)$ and $g(x)$, describe in your own words the difference between $(f \circ g)(x)$ and $(f \cdot g)(x)$.

*Solve.*

**53.** Business people are concerned with cost functions, revenue functions, and profit functions. Recall that the profit $P(x)$ obtained from $x$ units of a product is equal to the revenue $R(x)$ from selling the $x$ units minus the cost $C(x)$ of manufacturing the $x$ units. Write an equation expressing this relationship among $C(x)$, $R(x)$, and $P(x)$.

**54.** Suppose the revenue $R(x)$ for $x$ units of a product can be described by $R(x) = 25x$, and the cost $C(x)$ can be described by $C(x) = 50 + x^2 + 4x$. Find the profit $P(x)$ for $x$ units. (See Exercise 53.)

---

### 📖 STUDY SKILLS BUILDER

#### Tips for Studying for an Exam

To prepare for an exam, try the following study techniques.

- Start the study process days before your exam.
- Make sure that you are up-to-date on your assignments.
- If there is a topic that you are unsure of, use one of the many resources that are available to you. For example,

  See your instructor.

  Visit a learning resource center on campus.

  Read the textbook material and examples on the topic.

  View a video on the topic.

- Reread your notes and carefully review the Chapter Highlights at the end of any chapter.
- Work the review exercises at the end of the chapter. Check your answers and correct any mistakes. If you have trouble, use a resource listed above.
- Find a quiet place to take the Chapter Test found at the end of the chapter. Do not use any resources when taking this sample test. This way, you will have a clear indication of how prepared you are for your exam.

  Check your answers and make sure that you correct any missed exercises.

- Get lots of rest the night before the exam. It's hard to show how well you know the material if your brain is foggy from lack of sleep.

Good luck and keep a positive attitude.

*Let's see how you did on your last exam.*

1. How many days before your last exam did you start studying?
2. Were you up-to-date on your assignments at that time or did you need to catch up on assignments?
3. List the most helpful text supplement (if you used one).
4. List the most helpful campus supplement (if you used one).
5. List your process for preparing for a mathematics test.
6. Was this process helpful? In other words, were you satisfied with your performance on your exam?
7. If not, what changes can you make in your process that will make it more helpful to you?

# 12.2 INVERSE FUNCTIONS

**OBJECTIVE 1 ▶ Determining whether a function is one-to-one.** In the next section, we begin a study of two new functions: exponential and logarithmic functions. As we learn more about these functions, we will discover that they share a special relation to each other: They are inverses of each other.

Before we study these functions, we need to learn about inverses. We begin by defining one-to-one functions.

Study the following diagram.

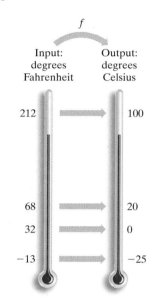

Recall that since each degrees Fahrenheit (input) corresponds to exactly one degrees Celsius (output), this pairing of inputs and outputs does describe a function. Also notice that each output corresponds to exactly one input. This type of function is given a special name—a one-to-one function.

Does the set $f = \{(0, 1), (2, 2), (-3, 5), (7, 6)\}$ describe a one-to-one function? It is a function since each $x$-value corresponds to a unique $y$-value. For this particular function $f$, each $y$-value also corresponds to a unique $x$-value. Thus, this function is also a **one-to-one function.**

---

**One-to-One Function**

For a **one-to-one function,** each $x$-value (input) corresponds to only one $y$-value (output), and each $y$-value (output) corresponds to only one $x$-value (input).

---

**EXAMPLE 1**   Determine whether each function described is one-to-one.

**a.** $f = \{(6, 2), (5, 4), (-1, 0), (7, 3)\}$

**b.** $g = \{(3, 9), (-4, 2), (-3, 9), (0, 0)\}$

**c.** $h = \{(1, 1), (2, 2), (10, 10), (-5, -5)\}$

**d.**

| Mineral (Input) | Talc | Gypsum | Diamond | Topaz | Stibnite |
|---|---|---|---|---|---|
| Hardness on the Mohs Scale (Output) | 1 | 2 | 10 | 8 | 2 |

**e.**

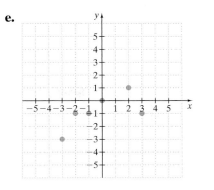

**f.**

| Cities | Percent of Cell Phone Subscribers |
|---|---|
| Atlanta (highest) | → 75 |
| Buffalo | → 53 |
| Austin | → 72 |
| Washington, D.C. | |
| Charleston (lowest) | → 47 |
| Detroit | → 74 |

### Solution

**a.** $f$ is one-to-one since each $y$-value corresponds to only one $x$-value.

**b.** $g$ is not one-to-one because the $y$-value 9 in $(3, 9)$ and $(-3, 9)$ corresponds to two different $x$-values.

**c.** $h$ is a one-to-one function since each $y$-value corresponds to only one $x$-value.

**d.** This table does not describe a one-to-one function since the output 2 corresponds to two different inputs, gypsum and stibnite.

**e.** This graph does not describe a one-to-one function since the $y$-value $-1$ corresponds to three different $x$-values, $-2$, $-1$, and 3.

**f.** The mapping is not one-to-one since 72% corresponds to Austin and Washington, D.C.  □

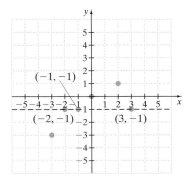

**PRACTICE**

**1** Determine whether each function described is one-to-one.

**a.** $f = \{(4, -3), (3, -4), (2, 7), (5, 0)\}$

**b.** $g = \{(8, 4), (-2, 0), (6, 4), (2, 6)\}$

**c.** $h = \{(2, 4), (1, 3), (4, 6), (-2, 4)\}$

**d.**

| Year | 1950 | 1963 | 1968 | 1975 | 1997 | 2002 |
|---|---|---|---|---|---|---|
| Federal Minimum Wage | $0.75 | $1.25 | $1.60 | $2.10 | $5.15 | $5.15 |

**e.**

**f.**

| State | Mean SAT Math Scores |
|---|---|
| North Dakota (highest) | →617 |
| Connecticut | →516 |
| Indiana | →509 |
| Maryland | →506 |
| Texas | →472 |
| Washington, D.C. (lowest) | |

**OBJECTIVE 2 ▶ Using the horizontal line test.** Recall that we recognize the graph of a function when it passes the vertical line test. Since every $x$-value of the function corresponds to exactly one $y$-value, each vertical line intersects the function's graph at most once. The graph shown (left), for instance, is the graph of a function.

Is this function a *one-to-one* function? The answer is no. To see why not, notice that the $y$-value of the ordered pair $(-3, 3)$, for example, is the same as the $y$-value of the ordered pair $(3, 3)$. In other words, the $y$-value 3 corresponds to two $x$-values, $-3$ and 3. This function is therefore not one-to-one.

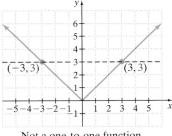

Not a one-to-one function.

To test whether a graph is the graph of a one-to-one function, apply the vertical line test to see if it is a function, and then apply a similar **horizontal line test** to see if it is a one-to-one function.

> **Horizontal Line Test**
>
> If every horizontal line intersects the graph of a function at most once, then the function is a one-to-one function.

**EXAMPLE 2**    Determine whether each graph is the graph of a one-to-one function.

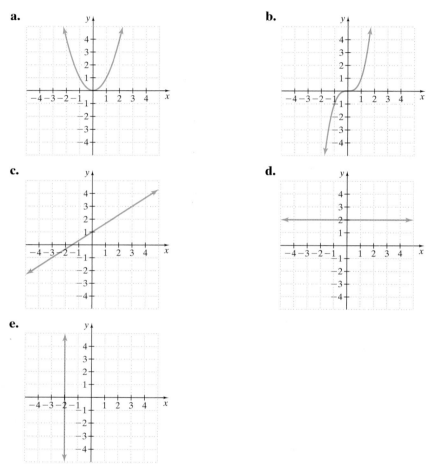

**a.**

**b.**

**c.**

**d.**

**e.**

*Solution*    Graphs **a, b, c,** and **d** all pass the vertical line test, so only these graphs are graphs of functions. But, of these, only **b** and **c** pass the horizontal line test, so only **b** and **c** are graphs of one-to-one functions.                                      □

**PRACTICE**

**2**    Determine whether each graph is the graph of a one-to-one function.

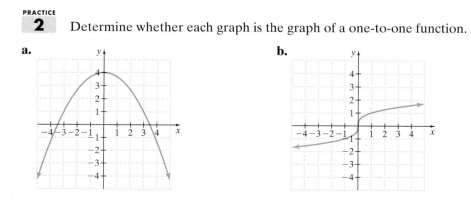

**a.**

**b.**

**c.**

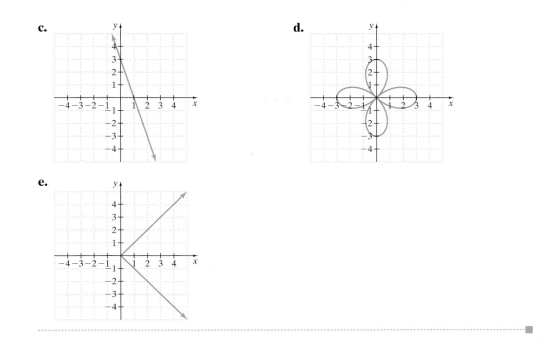

**d.**

**e.**

---

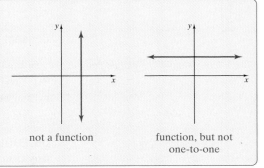

> ▶ **Helpful Hint**
>
> All linear equations are one-to-one functions except those whose graphs are horizontal or vertical lines. A vertical line does not pass the vertical line test and hence is not the graph of a function. A horizontal line is the graph of a function but does not pass the horizontal line test and hence is not the graph of a one-to-one function.
>
> not a function            function, but not one-to-one

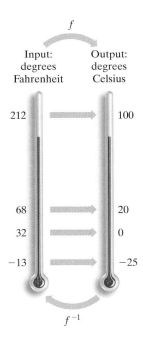

**OBJECTIVE 3 ▶ Finding the inverse of a function.** One-to-one functions are special in that their graphs pass both the vertical and horizontal line tests. They are special, too, in another sense: For each one-to-one function, we can find its **inverse function** by switching the coordinates of the ordered pairs of the function, or the inputs and the outputs. For example,

the inverse of the one-to-one function $f = \{(2, -3), (5, 10), (9, 1)\}$ is $\{(-3, 2), (10, 5), (1, 9)\}$.

For a function $f$, we use the notation $f^{-1}$, read "$f$ inverse," to denote its inverse function. Notice that since the coordinates of each ordered pair have been switched, the domain (set of inputs) of $f$ is the range (set of outputs) of $f^{-1}$, and the range of $f$ is the domain of $f^{-1}$.

The diagram to the left shows the inverse of the one-to-one function $f$ with ordered pairs of the form (degrees Fahrenheit, degrees Celsuis) is the function $f^{-1}$ with ordered pairs of the form (degrees Celsuis, degrees Fahrenheit). Notice that the ordered pair $(-13, -25)$ of the function, for example, becomes the ordered pair $(-25, -13)$ of its inverse.

> **Inverse Function**
>
> The inverse of a one-to-one function $f$ is the one-to-one function $f^{-1}$ that consists of the set of all ordered pairs $(y, x)$ where $(x, y)$ belongs to $f$.

> ▶ **Helpful Hint**
> If a function is not one-to-one, it does not have an inverse function.

**EXAMPLE 3** Find the inverse of the one-to-one function.

$$f = \{(0, 1), (-2, 7), (3, -6), (4, 4)\}$$

*Solution* $f^{-1} = \{(1, 0), (7, -2), (-6, 3), (4, 4)\}$

Switch coordinates of each ordered pair.

**PRACTICE**
**3** Find the inverse of the one-to-one function.

$$f(x) = \{(3, 4), (-2, 0), (2, 8), (6, 6)\}$$

> ▶ **Helpful Hint**
> The symbol $f^{-1}$ is the single symbol used to denote the inverse of the function $f$.
> It is read as "$f$ inverse." This symbol *does not mean* $\dfrac{1}{f}$.

**Concept Check** ☑

Suppose that $f$ is a one-to-one function and that $f(1) = 5$.

**a.** Write the corresponding ordered pair.

**b.** Write one point that we know must belong to the inverse function $f^{-1}$.

**OBJECTIVE 4 ▶ Finding the equation of the inverse of a function.** If a one-to-one function $f$ is defined as a set of ordered pairs, we can find $f^{-1}$ by interchanging the $x$- and $y$-coordinates of the ordered pairs. If a one-to-one function $f$ is given in the form of an equation, we can find $f^{-1}$ by using a similar procedure.

---

**Finding the Inverse of a One-to-One Function f(x)**

**STEP 1.** Replace $f(x)$ with $y$.

**STEP 2.** Interchange $x$ and $y$.

**STEP 3.** Solve the equation for $y$.

**STEP 4.** Replace $y$ with the notation $f^{-1}(x)$.

---

**EXAMPLE 4** Find an equation of the inverse of $f(x) = x + 3$.

*Solution* $f(x) = x + 3$

**STEP 1.** $y = x + 3$     Replace $f(x)$ with $y$.

**STEP 2.** $x = y + 3$     Interchange $x$ and $y$.

**STEP 3.** $x - 3 = y$     Solve for $y$.

**STEP 4.** $f^{-1}(x) = x - 3$     Replace $y$ with $f^{-1}(x)$.

**Answers to Concept Check:**
**a.** $(1, 5)$, **b.** $(5, 1)$

The inverse of $f(x) = x + 3$ is $f^{-1}(x) = x - 3$. Notice that, for example,

$$f(1) = 1 + 3 = 4 \quad \text{and} \quad f^{-1}(4) = 4 - 3 = 1$$

Ordered pair: $(1, 4)$          Ordered pair: $(4, 1)$

The coordinates are
switched, as expected.           ☐

**PRACTICE**
**4**     Find the equation of the inverse of $f(x) = 6 - x$.

---

**EXAMPLE 5**     Find the equation of the inverse of $f(x) = 3x - 5$. Graph $f$ and $f^{-1}$ on the same set of axes.

*Solution*     $f(x) = 3x - 5$

**STEP 1.**   $y = 3x - 5$        Replace $f(x)$ with $y$.

**STEP 2.**   $x = 3y - 5$        Interchange $x$ and $y$.

**STEP 3.**  $3y = x + 5$         Solve for $y$.

$$y = \frac{x + 5}{3}$$

**STEP 4.**  $f^{-1}(x) = \dfrac{x + 5}{3}$   Replace $y$ with $f^{-1}(x)$.

Now we graph $f(x)$ and $f^{-1}(x)$ on the same set of axes. Both $f(x) = 3x - 5$ and $f^{-1}(x) = \dfrac{x + 5}{3}$ are linear functions, so each graph is a line.

| $f(x) = 3x - 5$ | |
|---|---|
| $x$ | $y = f(x)$ |
| 1 | $-2$ |
| 0 | $-5$ |
| $\dfrac{5}{3}$ | 0 |

| $f^{-1}(x) = \dfrac{x + 5}{3}$ | |
|---|---|
| $x$ | $y = f^{-1}(x)$ |
| $-2$ | 1 |
| $-5$ | 0 |
| 0 | $\dfrac{5}{3}$ |

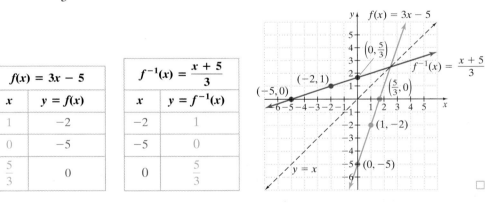

**PRACTICE**
**5**     Find the equation of the inverse of $f(x) = 5x + 2$. Graph $f$ and $f^{-1}$ on the same set of axes.

---

**OBJECTIVE 5 ▶ Graphing inverse functions.** Notice that the graphs of $f$ and $f^{-1}$ in Example 5 are mirror images of each other, and the "mirror" is the dashed line $y = x$. This is true for every function and its inverse. For this reason, we say that *the graphs of $f$ and $f^{-1}$ are symmetric about the line $y = x$.*

To see why this happens, study the graph of a few ordered pairs and their switched coordinates in the diagram to the right.

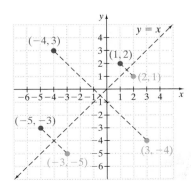

**EXAMPLE 6**  Graph the inverse of each function.

*Solution*  The function is graphed in blue and the inverse is graphed in red.

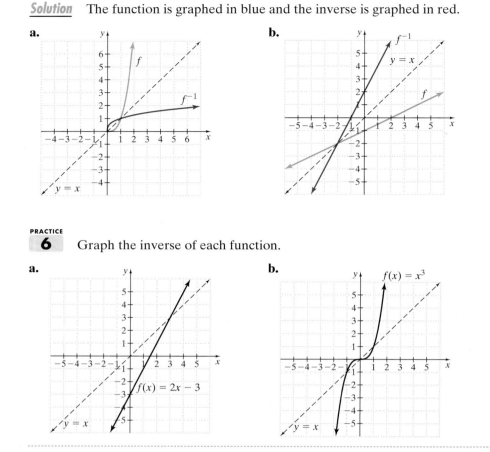

a.

b.

PRACTICE

**6**  Graph the inverse of each function.

a.

$f(x) = 2x - 3$

b.

$f(x) = x^3$

**OBJECTIVE 6 ▶ Determining whether functions are inverses of each other.** Notice in the table of values in Example 5 that $f(0) = -5$ and $f^{-1}(-5) = 0$, as expected. Also, for example, $f(1) = -2$ and $f^{-1}(-2) = 1$. In words, we say that for some input $x$, the function $f^{-1}$ takes the output of $x$, called $f(x)$, back to $x$.

$$x \rightarrow f(x) \quad \text{and} \quad f^{-1}(f(x)) \rightarrow x$$
$$\downarrow \quad \downarrow \qquad\qquad \downarrow \quad\quad \downarrow$$
$$f(0) = -5 \quad \text{and} \quad f^{-1}(-5) = 0$$
$$f(1) = -2 \quad \text{and} \quad f^{-1}(-2) = 1$$

In general,

---

If $f$ is a one-to-one function, then the inverse of $f$ is the function $f^{-1}$ such that

$$(f^{-1} \circ f)(x) = x \quad \text{and} \quad (f \circ f^{-1})(x) = x$$

**EXAMPLE 7**   Show that if $f(x) = 3x + 2$, then $f^{-1}(x) = \dfrac{x - 2}{3}$.

*Solution*   See that $(f^{-1} \circ f)(x) = x$ and $(f \circ f^{-1})(x) = x$.

$$(f^{-1} \circ f)(x) = f^{-1}(f(x))$$

$$= f^{-1}(3x + 2) \qquad \text{Replace } f(x) \text{ with } 3x + 2.$$

$$= \frac{3x + 2 - 2}{3}$$

$$= \frac{3x}{3}$$

$$= x$$

$$(f \circ f^{-1})(x) = f(f^{-1}(x))$$

$$= f\left(\frac{x - 2}{3}\right) \qquad \text{Replace } f^{-1}(x) \text{ with } \frac{x - 2}{3}.$$

$$= 3\left(\frac{x - 2}{3}\right) + 2$$

$$= x - 2 + 2$$

$$= x$$

**PRACTICE**
**7**   Show that if $f(x) = 4x - 1$, then $f^{-1}(x) = \dfrac{x + 1}{4}$.

---

**Graphing Calculator Explorations**

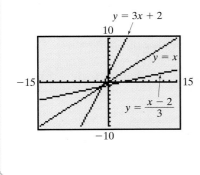

$y = 3x + 2$

$y = x$

$y = \dfrac{x - 2}{3}$

A graphing calculator can be used to visualize the results of Example 7. Recall that the graph of a function $f$ and its inverse $f^{-1}$ are mirror images of each other across the line $y = x$. To see this for the function from Example 7, use a square window and graph

the given function:   $Y_1 = 3x + 2$

its inverse:   $Y_2 = \dfrac{x - 2}{3}$

and the line:   $Y_3 = x$

Exercises will follow in Exercise Set 12.2.

---

## VOCABULARY & READINESS CHECK

*Use the choices below to fill in each blank. Some choices will not be used and some will be used more than once.*

| vertical | (3, 7) | (11, 2) | $y = x$ | $x$ |
|---|---|---|---|---|
| horizontal | (7, 3) | (2, 11) | $\dfrac{1}{f}$ | the inverse of $f$ |

**1.** If $f(2) = 11$, the corresponding ordered pair is _____.

**2.** The symbol $f^{-1}$ means _____.

**3.** If $(7, 3)$ is an ordered pair solution of $f(x)$, and $f(x)$ has an inverse, then an ordered pair solution of $f^{-1}(x)$ is _____.

**4.** To tell whether a graph is the graph of a function, use the _____ line test.

**5.** To tell whether the graph of a function is also a one-to-one function, use the _____ line test.

**6.** The graphs of $f$ and $f^{-1}$ are symmetric about the _____ line.

**7.** Two functions are inverse of each other if $(f \circ f^{-1})(x) = $ _____ and $(f^{-1} \circ f)(x) = $ ____.

## 12.2 | EXERCISE SET

*Powered by CourseCompass™ and MathXL®*
**MyMathLab**

MathⓍL
PRACTICE | WATCH | DOWNLOAD | READ | REVIEW

*Determine whether each function is a one-to-one function. If it is one-to-one, list the inverse function by switching coordinates, or inputs and outputs. See Examples 1 and 3.*

**1.** $f = \{(-1, -1), (1, 1), (0, 2), (2, 0)\}$

**2.** $g = \{(8, 6), (9, 6), (3, 4), (-4, 4)\}$

**3.** $h = \{(10, 10)\}$

**4.** $r = \{(1, 2), (3, 4), (5, 6), (6, 7)\}$

**5.** $f = \{(11, 12), (4, 3), (3, 4), (6, 6)\}$

**6.** $g = \{(0, 3), (3, 7), (6, 7), (-2, -2)\}$

**7.**

| Month of 2007 (Input) | January | February | March | April |
|---|---|---|---|---|
| Unemployment Rate in Percent | 4.6 | 4.6 | 4.4 | 4.4 |

(*Source:* Bureau of Labor Statistics, U.S. Department of Housing and Urban Development)

**8.**

| State (Input) | Wisconsin | Ohio | Georgia | Colorado | California | Arizona |
|---|---|---|---|---|---|---|
| Electoral Votes (Output) | 10 | 20 | 15 | 9 | 55 | 10 |

**9.**

| State (Input) | California | Maryland | Nevada | Florida | North Dakota |
|---|---|---|---|---|---|
| Rank in Population (Output) | 1 | 19 | 35 | 4 | 48 |

(*Source:* U.S. Bureau of the Census)

△ **10.**

| Shape (Input) | Triangle | Pentagon | Quadrilateral | Hexagon | Decagon |
|---|---|---|---|---|---|
| Number of Sides (Output) | 3 | 5 | 4 | 6 | 10 |

(*Source:* U.S. Bureau of the Census)

*Given the one-to-one function $f(x) = x^3 + 2$, find the following.*
[*Hint:* You do not need to find the equation for $f^{-1}(x)$.]

**11. a.** $f(1)$
  **b.** $f^{-1}(3)$
**13. a.** $f(-1)$
  **b.** $f^{-1}(1)$

**12. a.** $f(0)$
  **b.** $f^{-1}(2)$
**14. a.** $f(-2)$
  **b.** $f^{-1}(-6)$

*Determine whether the graph of each function is the graph of a one-to-one function. See Example 2.*

**15.**

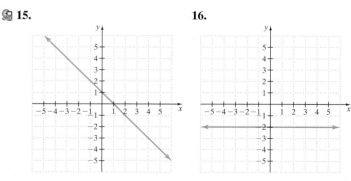

**16.**

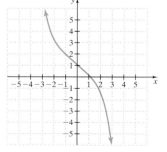

**17.**

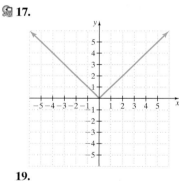

**18.**

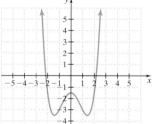

**19.**

**20.**

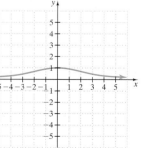

**21.**                                    **22.**

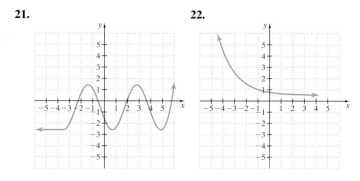

## MIXED PRACTICE

*Each of the following functions is one-to-one. Find the inverse of each function and graph the function and its inverse on the same set of axes. See Examples 4 and 5.*

**23.** $f(x) = x + 4$

**24.** $f(x) = x - 5$

**25.** $f(x) = 2x - 3$

**26.** $f(x) = 4x + 9$

**27.** $f(x) = \dfrac{1}{2}x - 1$

**28.** $f(x) = -\dfrac{1}{2}x + 2$

**29.** $f(x) = x^3$

**30.** $f(x) = x^3 - 1$

*Find the inverse of each one-to-one function. See Examples 4 and 5.*

**31.** $f(x) = 5x + 2$

**32.** $f(x) = 6x - 1$

**33.** $f(x) = \dfrac{x - 2}{5}$

**34.** $f(x) = \dfrac{4x - 3}{2}$

**35.** $f(x) = \sqrt[3]{x}$

**36.** $f(x) = \sqrt[3]{x + 1}$

**37.** $f(x) = \dfrac{5}{3x + 1}$

**38.** $f(x) = \dfrac{7}{2x + 4}$

**39.** $f(x) = (x + 2)^3$

**40.** $f(x) = (x - 5)^3$

*Graph the inverse of each function on the same set of axes. See Example 6.*

**41.**                                    **42.**

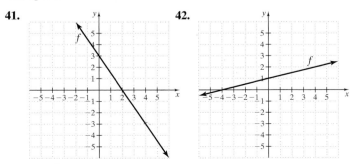

**43.**                  **44.**

**45.**                  **46.**

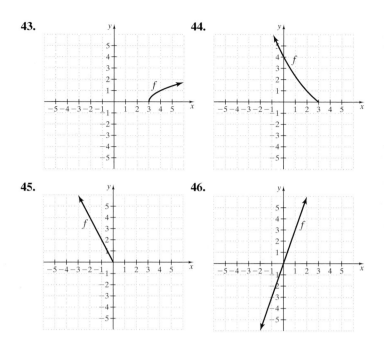

*Solve. See Example 7.*

**47.** If $f(x) = 2x + 1$, show that $f^{-1}(x) = \dfrac{x - 1}{2}$.

**48.** If $f(x) = 3x - 10$, show that $f^{-1}(x) = \dfrac{x + 10}{3}$.

**49.** If $f(x) = x^3 + 6$, show that $f^{-1}(x) = \sqrt[3]{x - 6}$.

**50.** If $f(x) = x^3 - 5$, show that $f^{-1}(x) = \sqrt[3]{x + 5}$.

## REVIEW AND PREVIEW

*Evaluate each of the following. See Section 10.2.*

**51.** $25^{1/2}$                **52.** $49^{1/2}$

**53.** $16^{3/4}$                **54.** $27^{2/3}$

**55.** $9^{-3/2}$                **56.** $81^{-3/4}$

*If $f(x) = 3^x$, find the following. In Exercises 59 and 60, give an exact answer and a two-decimal-place approximation. See Sections 3.6 and 10.2.*

**57.** $f(2)$                    **58.** $f(0)$

**59.** $f\left(\dfrac{1}{2}\right)$          **60.** $f\left(\dfrac{2}{3}\right)$

## CONCEPT EXTENSIONS

*Solve. See the Concept Check in this section.*

**61.** Suppose that $f$ is a one-to-one function and that $f(2) = 9$.

  **a.** Write the corresponding ordered pair.

  **b.** Name one ordered-pair that we know is a solution of the inverse of $f$, or $f^{-1}$.

**62.** Suppose that $F$ is a one-to-one function and that $F\left(\dfrac{1}{2}\right) = -0.7$.

    **a.** Write the corresponding ordered pair.

    **b.** Name one ordered pair that we know is a solution of the inverse of $F$, or $F^{-1}$.

*For Exercises 63 and 64,*

**a.** *Write the ordered pairs for $f(x)$ whose points are highlighted. (Include the points whose coordinates are given.)*

**b.** *Write the corresponding ordered pairs for the inverse of $f$, $f^{-1}$.*

**c.** *Graph the ordered pairs for $f^{-1}$ found in part **b**.*

**d.** *Graph $f^{-1}(x)$ by drawing a smooth curve through the plotted points.*

**63.**

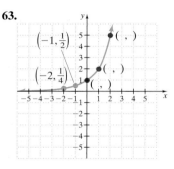

**64.**

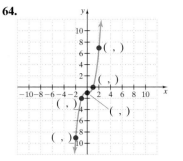

**65.** If you are given the graph of a function, describe how you can tell from the graph whether a function has an inverse.

**66.** Describe the appearance of the graphs of a function and its inverse.

*Find the inverse of each given one-to-one function. Then use a graphing calculator to graph the function and its inverse on a square window.*

**67.** $f(x) = 3x + 1$

**68.** $f(x) = -2x - 6$

**69.** $f(x) = \sqrt[3]{x + 1}$

**70.** $f(x) = x^3 - 3$

---

## 12.3 EXPONENTIAL FUNCTIONS

**OBJECTIVES**

**1** Graph exponential functions.

**2** Solve equations of the form $b^x = b^y$.

**3** Solve problems modeled by exponential equations.

**OBJECTIVE 1 ▶ Graphing exponential functions.** In earlier chapters, we gave meaning to exponential expressions such as $2^x$, where $x$ is a rational number. For example,

$$2^3 = 2 \cdot 2 \cdot 2 \qquad \text{Three factors; each factor is 2}$$
$$2^{3/2} = (2^{1/2})^3 = \sqrt{2} \cdot \sqrt{2} \cdot \sqrt{2} \qquad \text{Three factors; each factor is } \sqrt{2}$$

When $x$ is an irrational number (for example, $\sqrt{3}$), what meaning can we give to $2^{\sqrt{3}}$?

It is beyond the scope of this book to give precise meaning to $2^x$ if $x$ is irrational. We can confirm your intuition and say that $2^{\sqrt{3}}$ is a real number, and since $1 \le \sqrt{3} < 2$, then $2^1 < 2^{\sqrt{3}} < 2^2$. We can also use a calculator and approximate $2^{\sqrt{3}}$: $2^{\sqrt{3}} \approx 3.321997$. In fact, as long as the base $b$ is positive, $b^x$ is a real number for all real numbers $x$. Finally, the rules of exponents apply whether $x$ is rational or irrational, as long as $b$ is positive. In this section, we are interested in functions of the form $f(x) = b^x$, where $b > 0$. A function of this form is called an **exponential function.**

> **Exponential Function**
>
> A function of the form
>
> $$f(x) = b^x$$
>
> is called an **exponential function** if $b > 0$, $b$ is not 1, and $x$ is a real number.

Next, we practice graphing exponential functions.

**EXAMPLE 1** Graph the exponential functions defined by $f(x) = 2^x$ and $g(x) = 3^x$ on the same set of axes.

**Solution** Graph each function by plotting points. Set up a table of values for each of the two functions.

If each set of points is plotted and connected with a smooth curve, the following graphs result.

| $f(x) = 2^x$ | $x$ | 0 | 1 | 2 | 3 | $-1$ | $-2$ |
|---|---|---|---|---|---|---|---|
| | $f(x)$ | 1 | 2 | 4 | 8 | $\dfrac{1}{2}$ | $\dfrac{1}{4}$ |

| $g(x) = 3^x$ | $x$ | 0 | 1 | 2 | 3 | $-1$ | $-2$ |
|---|---|---|---|---|---|---|---|
| | $g(x)$ | 1 | 3 | 9 | 27 | $\dfrac{1}{3}$ | $\dfrac{1}{9}$ |

**PRACTICE**
**1** Graph the exponential functions defined by $f(x) = 2^x$ and $g(x) = 7^x$ on the same set of axes.

A number of things should be noted about the two graphs of exponential functions in Example 1. First, the graphs show that $f(x) = 2^x$ and $g(x) = 3^x$ are one-to-one functions since each graph passes the vertical and horizontal line tests. The $y$-intercept of each graph is $(0, 1)$, but neither graph has an $x$-intercept. From the graph, we can also see that the domain of each function is all real numbers and that the range is $(0, \infty)$. We can also see that as $x$-values are increasing, $y$-values are increasing also.

**EXAMPLE 2** Graph the exponential functions $y = \left(\dfrac{1}{2}\right)^x$ and $y = \left(\dfrac{1}{3}\right)^x$ on the same set of axes.

**Solution** As before, plot points and connect them with a smooth curve.

| $y = \left(\dfrac{1}{2}\right)^x$ | $x$ | 0 | 1 | 2 | 3 | $-1$ | $-2$ |
|---|---|---|---|---|---|---|---|
| | $y$ | 1 | $\dfrac{1}{2}$ | $\dfrac{1}{4}$ | $\dfrac{1}{8}$ | 2 | 4 |

| $y = \left(\dfrac{1}{3}\right)^x$ | $x$ | 0 | 1 | 2 | 3 | $-1$ | $-2$ |
|---|---|---|---|---|---|---|---|
| | $y$ | 1 | $\dfrac{1}{3}$ | $\dfrac{1}{9}$ | $\dfrac{1}{27}$ | 3 | 9 |

**PRACTICE**
**2** Graph the exponential functions $f(x) = \left(\dfrac{1}{3}\right)^x$ and $g(x) = \left(\dfrac{1}{5}\right)^x$ on the same set of axes.

Each function in Example 2 again is a one-to-one function. The $y$-intercept of both is $(0, 1)$. The domain is the set of all real numbers, and the range is $(0, \infty)$.

Notice the difference between the graphs of Example 1 and the graphs of Example 2. An exponential function is always increasing if the base is greater than 1.

When the base is between 0 and 1, the graph is always decreasing. The following figures summarize these characteristics of exponential functions.

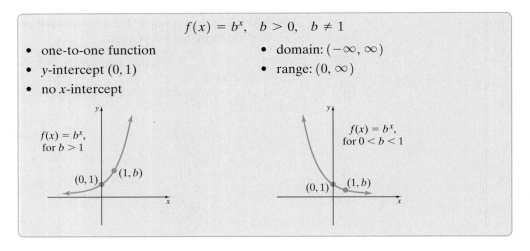

$$f(x) = b^x, \quad b > 0, \quad b \neq 1$$

- one-to-one function
- $y$-intercept $(0, 1)$
- no $x$-intercept
- domain: $(-\infty, \infty)$
- range: $(0, \infty)$

**EXAMPLE 3**   Graph the exponential function $f(x) = 3^{x+2}$.

*Solution*   As before, we find and plot a few ordered pair solutions. Then we connect the points with a smooth curve.

| $y = 3^{x+2}$ | |
|---|---|
| $x$ | $y$ |
| 0 | 9 |
| $-1$ | 3 |
| $-2$ | 1 |
| $-3$ | $\dfrac{1}{3}$ |
| $-4$ | $\dfrac{1}{9}$ |

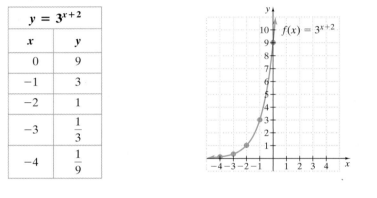

**PRACTICE**
**3**   Graph the exponential function $f(x) = 2^{x+3}$.

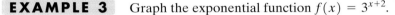

**Concept Check** ☑

Which functions are exponential functions?

**a.** $f(x) = x^3$   **b.** $g(x) = \left(\dfrac{2}{3}\right)^x$   **c.** $h(x) = 5^{x-2}$   **d.** $w(x) = (2x)^2$

**OBJECTIVE 2 ▶ Solving equations of the form $b^x = b^y$.** We have seen that an exponential function $y = b^x$ is a one-to-one function. Another way of stating this fact is a property that we can use to solve exponential equations.

**Uniqueness of $b^x$**
Let $b > 0$ and $b \neq 1$. Then $b^x = b^y$ is equivalent to $x = y$.

**EXAMPLE 4**   Solve each equation for $x$.

**a.** $2^x = 16$   **b.** $9^x = 27$   **c.** $4^{x+3} = 8^x$

*Solution*

**a.** We write 16 as a power of 2 and then use the uniqueness of $b^x$ to solve.

$$2^x = 16$$
$$2^x = 2^4$$

Since the bases are the same and are nonnegative, by the uniqueness of $b^x$, we then have that the exponents are equal. Thus,

$$x = 4$$

The solution is 4, or the solution set is $\{4\}$.

**b.** Notice that both 9 and 27 are powers of 3.

$$9^x = 27$$
$$(3^2)^x = 3^3 \qquad \text{Write 9 and 27 as powers of 3.}$$
$$3^{2x} = 3^3$$
$$2x = 3 \qquad \text{Apply the uniqueness of } b^x.$$
$$x = \frac{3}{2} \qquad \text{Divide by 2.}$$

To check, replace $x$ with $\dfrac{3}{2}$ in the original expression, $9^x = 27$. The solution is $\dfrac{3}{2}$.

**c.** Write both 4 and 8 as powers of 2.

$$4^{x+3} = 8^x$$
$$(2^2)^{x+3} = (2^3)^x$$
$$2^{2x+6} = 2^{3x}$$
$$2x + 6 = 3x \qquad \text{Apply the uniqueness of } b^x.$$
$$6 = x \qquad \text{Subtract } 2x \text{ from both sides.}$$

The solution is 6. ☐

**PRACTICE**
**4**    Solve each equation for $x$.

**a.** $3^x = 9$          **b.** $8^x = 16$          **c.** $125^x = 25^{x-2}$

There is one major problem with the preceding technique. Often the two sides of an equation cannot easily be written as powers of a common base. We explore how to solve an equation such as $4 = 3^x$ with the help of **logarithms** later.

**OBJECTIVE 3 ▶ Solving problems modeled by exponential equations.** The bar graph below shows the increase in the number of cellular phone users. Notice that the graph of the exponential function $y = 110.6(1.132)^x$ approximates the heights of the bars.

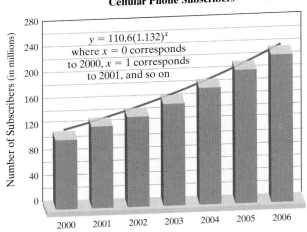

**Cellular Phone Subscribers**

$y = 110.6(1.132)^x$
where $x = 0$ corresponds
to 2000, $x = 1$ corresponds
to 2001, and so on

*Source*: Cellular Telecommunications & Internet Association

**Note:** Compare this Chapter 12 graph with the Chapter 11 opener graph, page 644, on the bottom of the page. Notice how using different data years leads to different models—one quadratic and one exponential.

The graph on the previous page shows just one example of how the world abounds with patterns that can be modeled by exponential functions. To make these applications realistic, we use numbers that warrant a calculator. Another application of an exponential function has to do with interest rates on loans.

The exponential function defined by $A = P\left(1 + \dfrac{r}{n}\right)^{nt}$ models the dollars $A$ accrued (or owed) after $P$ dollars are invested (or loaned) at an annual rate of interest $r$ compounded $n$ times each year for $t$ years. This function is known as the compound interest formula.

**EXAMPLE 5** **Using the Compound Interest Formula**

Find the amount owed at the end of 5 years if $1600 is loaned at a rate of 9% compounded monthly.

*Solution* We use the formula $A = P\left(1 + \dfrac{r}{n}\right)^{nt}$, with the following values.

$P = \$1600$ (the amount of the loan)

$r = 9\% = 0.09$ (the annual rate of interest)

$n = 12$ (the number of times interest is compounded each year)

$t = 5$ (the duration of the loan, in years)

$A = P\left(1 + \dfrac{r}{n}\right)^{nt}$    Compound interest formula

$= 1600\left(1 + \dfrac{0.09}{12}\right)^{12(5)}$    Substitute known values.

$= 1600(1.0075)^{60}$

To approximate $A$, use the $\boxed{y^x}$ or $\boxed{\wedge}$ key on your calculator.

$\boxed{2505.0896}$

Thus, the amount $A$ owed is approximately $2505.09.    ☐

**PRACTICE**

**5** Find the amount owed at the end of 4 years if $3000 is loaned at a rate of 7% compounded semiannually (twice a year).

**EXAMPLE 6** **Estimating Percent of Radioactive Material**

As a result of the Chernobyl nuclear accident, radioactive debris was carried through the atmosphere. One immediate concern was the impact that the debris had on the milk supply. The percent $y$ of radioactive material in raw milk after $t$ days is estimated by $y = 100(2.7)^{-0.1t}$. Estimate the expected percent of radioactive material in the milk after 30 days.

*Solution* Replace $t$ with 30 in the given equation.

$y = 100(2.7)^{-0.1t}$

$= 100(2.7)^{-0.1(30)}$    Let $t = 30$.

$= 100(2.7)^{-3}$

To approximate the percent $y$, the following keystrokes may be used on a scientific calculator.

The display should read

Thus, approximately 5% of the radioactive material still remained in the milk supply after 30 days.

**PRACTICE**
**6** If a single sheet of glass prevents 5% of the incoming light from passing through it, then the percent $p$ of light that passes through $n$ successive sheets of glass is given approximately by the function $p(n) = 100(2.7)^{-0.05n}$. Estimate the expected percent of light that will pass through 10 sheets of glass. Round to the nearest hundredth of a percent.

## Graphing Calculator Explorations

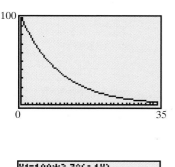

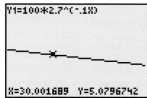

We can use a graphing calculator and its TRACE feature to solve Example 6 graphically.

To estimate the expected percent of radioactive material in the milk after 30 days, enter $Y_1 = 100(2.7)^{-0.1x}$. (The variable $t$ in Example 6 is changed to $x$ here to better accomodate our work on the graphing calculator.) The graph does not appear on a standard viewing window, so we need to determine an appropriate viewing window. Because it doesn't make sense to look at radioactivity *before* the Chernobyl nuclear accident, we use Xmin = 0. We are interested in finding the percent of radioactive material in the milk when $x = 30$, so we choose Xmax = 35 to leave enough space to see the graph at $x = 30$. Because the values of $y$ are percents, it seems appropriate that $0 \leq y \leq 100$. (We also use Xscl = 1 and Yscl = 10.) Now we graph the function.

We can use the TRACE feature to obtain an approximation of the expected percent of radioactive material in the milk when $x = 30$. (A TABLE feature may also be used to approximate the percent.) To obtain a better approximation, let's use the ZOOM feature several times to zoom in near $x = 30$.

The percent of radioactive material in the milk 30 days after the Chernobyl accident was 5.08%, accurate to two decimal places.

*Use a graphing calculator to find each percent. Approximate your solutions so that they are accurate to two decimal places.*

**1.** Estimate the expected percent of radioactive material in the milk 2 days after the Chernobyl nuclear accident.

**2.** Estimate the expected percent of radioactive material in the milk 10 days after the Chernobyl nuclear accident.

**3.** Estimate the expected percent of radioactive material in the milk 15 days after the Chernobyl nuclear accident.

**4.** Estimate the expected percent of radioactive material in the milk 25 days after the Chernobyl nuclear accident.

# VOCABULARY & READINESS CHECK

*Use the choices to fill in each blank.*

1. A function such as $f(x) = 2^x$ is a(n) _____ function.
   **A.** linear      **B.** quadratic      **C.** exponential

2. If $7^x = 7^y$, then _____.
   **A.** $x = 7^y$      **B.** $x = y$      **C.** $y = 7^x$      **D.** $7 = 7^y$

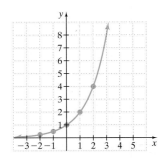

*Answer the questions about the graph of $y = 2^x$, shown to the right.*

3. Is this a one-to-one function? _____

4. Is there an $x$-intercept? _____ If so, name the coordinates. _____

5. Is there a $y$-intercept? _____ If so, name the coordinates. _____

6. The domain of this function, in interval notation, is _____.

7. The range of this function, in interval notation, is _____.

## 12.3 | EXERCISE SET

*Graph each exponential function. See Examples 1 through 3.*

1. $y = 4^x$
2. $y = 5^x$
3. $y = 2^x + 1$
4. $y = 3^x - 1$
5. $y = \left(\dfrac{1}{4}\right)^x$
6. $y = \left(\dfrac{1}{5}\right)^x$
7. $y = \left(\dfrac{1}{2}\right)^x - 2$
8. $y = \left(\dfrac{1}{3}\right)^x + 2$
9. $y = -2^x$
10. $y = -3^x$
11. $y = -\left(\dfrac{1}{4}\right)^x$
12. $y = -\left(\dfrac{1}{5}\right)^x$
13. $f(x) = 2^{x+1}$
14. $f(x) = 3^{x-1}$
15. $f(x) = 4^{x-2}$
16. $f(x) = 2^{x+3}$

*Match each exponential equation with its graph below or in the next column. See Examples 1 through 3.*

17. $f(x) = \left(\dfrac{1}{2}\right)^x$
18. $f(x) = \left(\dfrac{1}{4}\right)^x$
19. $f(x) = 2^x$
20. $f(x) = 3^x$

**A.**

**B.**

**C.**

**D.**

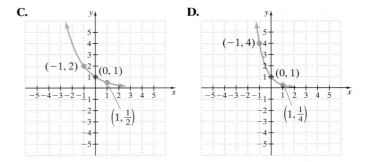

*Solve each equation for x. See Example 4.*

21. $3^x = 27$
22. $6^x = 36$
23. $16^x = 8$
24. $64^x = 16$
25. $32^{2x-3} = 2$
26. $9^{2x+1} = 81$
27. $\dfrac{1}{4} = 2^{3x}$
28. $\dfrac{1}{27} = 3^{2x}$
29. $5^x = 625$
30. $2^x = 64$
31. $4^x = 8$
32. $32^x = 4$
33. $27^{x+1} = 9$
34. $125^{x-2} = 25$
35. $81^{x-1} = 27^{2x}$
36. $4^{3x-7} = 32^{2x}$

*Solve. Unless otherwise indicated, round results to one decimal place. See Example 6.*

37. One type of uranium has a daily radioactive decay rate of 0.4%. If 30 pounds of this uranium is available today, find how much will still remain after 50 days. Use $y = 30(2.7)^{-0.004t}$, and let $t$ be 50.

38. The nuclear waste from an atomic energy plant decays at a rate of 3% each century. If 150 pounds of nuclear waste is

disposed of, find how much of it will still remain after 10 centuries. Use $y = 150(2.7)^{-0.03t}$, and let $t$ be 10.

**39.** National Park Service personnel are trying to increase the size of the bison population of Theodore Roosevelt National Park. If 260 bison currently live in the park, and if the population's rate of growth is 2.5% annually, find how many bison (rounded to the nearest whole) there should be in 10 years. Use $y = 260(2.7)^{0.025t}$.

**40.** The size of the rat population of a wharf area grows at a rate of 8% monthly. If there are 200 rats in January, find how many rats (rounded to the nearest whole) should be expected by next January. Use $y = 200(2.7)^{0.08t}$.

**41.** A rare isotope of a nuclear material is very unstable, decaying at a rate of 15% each second. Find how much isotope remains 10 seconds after 5 grams of the isotope is created. Use $y = 5(2.7)^{-0.15t}$.

**42.** An accidental spill of 75 grams of radioactive material in a local stream has led to the presence of radioactive debris decaying at a rate of 4% each day. Find how much debris still remains after 14 days. Use $y = 75(2.7)^{-0.04t}$.

**43.** The atmospheric pressure $p$, in Pascals, on a weather balloon decreases with increasing height. This pressure, measured in millimeters of mercury, is related to the number of kilometers $h$ above sea level by the function $p(h) = 760(2.7)^{-0.145h}$. Round to the nearest tenth of a Pascal.

  **a.** Find the atmospheric pressure at a height of 1 kilometer.

  **b.** Find the atmospheric pressure at a height of 10 kilometers.

**44.** An unusually wet spring has caused the size of the Cape Cod mosquito population to increase by 8% each day. If an estimated 200,000 mosquitoes are on Cape Cod on May 12, find how many mosquitoes will inhabit the Cape on May 25. Use $y = 200,000(2.7)^{0.08t}$. Round to the nearest thousand.

**45.** The equation $y = 84,949(1.096)^x$ models the number of American college students who study abroad each year from 1995 through 2006. In the equation, $y$ is the number of American students studying abroad and $x$ represents the number of years after 1995. Round answers to the nearest whole. (*Source:* Based on data from Institute of International Education, Open Doors 2006)

  **a.** Estimate the number of American students studying abroad in 2000.

  **b.** Assuming this equation continues to be valid in the future, use this equation to predict the number of American students studying abroad in 2020.

**46.** Carbon dioxide ($CO_2$) is a greenhouse gas that contributes to global warming. Partially due to the combustion of fossil fuels, the amount of $CO_2$ in Earth's atmosphere has been increasing by 0.4% annually over the past century. In 2000, the concentration of $CO_2$ in the atmosphere was 369.4 parts per million by volume. To make the following predictions, use $y = 369.4(1.004)^t$ where $y$ is the concentration of $CO_2$ in parts per million and $t$ is the number of years after 2000. Round answers to the nearest tenth. (*Sources:* Based on data from the United Nations Environment Programme and the Carbon Dioxide Information Analysis Center)

  **a.** Predict the concentration of $CO_2$ in the atmosphere in the year 2006.

  **b.** Predict the concentration of $CO_2$ in the atmosphere in the year 2030.

*Solve. Use* $A = P\left(1 + \dfrac{r}{n}\right)^{nt}$. *Round answers to two decimal places. See Example 5.*

**47.** Find the amount Erica owes at the end of 3 years if $6000 is loaned to her at a rate of 8% compounded monthly.

**48.** Find the amount owed at the end of 5 years if $3000 is loaned at a rate of 10% compounded quarterly.

**49.** Find the total amount Janina has in a college savings account if $2000 was invested and earned 6% compounded semiannually for 12 years.

**50.** Find the amount accrued if $500 is invested and earns 7% compounded monthly for 4 years.

*The formula* $y = 18(1.24)^x$ *gives the number of cellular phone users $y$ (in millions) in the United States for the years 1994 through 2006. In this formula, $x = 0$ corresponds to 1994, $x = 1$ corresponds to 1995, and so on. Use this formula to solve Exercises 51 and 52. Round results to the nearest whole million.*

**51.** Use this model to predict the number of cellular phone users in the year 2010.

**52.** Use this model to predict the number of cellular phone users in the year 2014.

## REVIEW AND PREVIEW

*Solve each equation. See Sections 2.3 and 6.6.*

**53.** $5x - 2 = 18$

**54.** $3x - 7 = 11$

**55.** $3x - 4 = 3(x + 1)$

**56.** $2 - 6x = 6(1 - x)$

**57.** $x^2 + 6 = 5x$

**58.** $18 = 11x - x^2$

*By inspection, find the value for x that makes each statement true. See Sections 5.1 and 5.5.*

**59.** $2^x = 8$

**60.** $3^x = 9$

**61.** $5^x = \dfrac{1}{5}$

**62.** $4^x = 1$

**CONCEPT EXTENSIONS**

63. Explain why the graph of an exponential function $y = b^x$ contains the point $(1, b)$.

64. Explain why an exponential function $y = b^x$ has a $y$-intercept of $(0, 1)$.

*Graph.*

65. $y = |3^x|$

66. $y = \left|\left(\dfrac{1}{3}\right)^x\right|$

67. $y = 3^{|x|}$

68. $y = \left(\dfrac{1}{3}\right)^{|x|}$

69. Graph $y = 2^x$ and $y = \left(\dfrac{1}{2}\right)^{-x}$ on the same set of axes. Describe what you see and why.

70. Graph $y = 2^x$ and $x = 2^y$ on the same set of axes. Describe what you see.

📷 *Use a graphing calculator to solve. Estimate each result to two decimal places.*

71. Verify the results of Exercise 37.

72. From Exercise 37, estimate the number of pounds of uranium that will be available after 100 days.

73. From Exercise 37, estimate the number of pounds of uranium that will be available after 120 days.

74. Verify the results of Exercise 42.

75. From Exercise 42, estimate the amount of debris that remains after 10 days.

76. From Exercise 42, estimate the amount of debris that remains after 20 days.

---

## 12.4 LOGARITHMIC FUNCTIONS

**OBJECTIVES**

1. Write exponential equations with logarithmic notation and write logarithmic equations with exponential notation.

2. Solve logarithmic equations by using exponential notation.

3. Identify and graph logarithmic functions.

**OBJECTIVE 1 ▶ Using logarithmic notation.** Since the exponential function $f(x) = 2^x$ is a one-to-one function, it has an inverse.

We can create a table of values for $f^{-1}$ by switching the coordinates in the accompanying table of values for $f(x) = 2^x$.

$$f(x) = 2^x$$

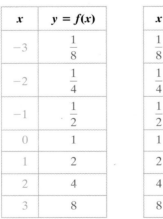

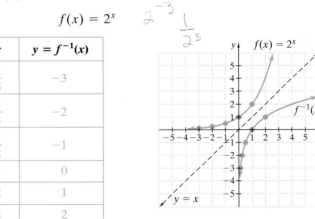

| $x$ | $y = f(x)$ |
|---|---|
| $-3$ | $\dfrac{1}{8}$ |
| $-2$ | $\dfrac{1}{4}$ |
| $-1$ | $\dfrac{1}{2}$ |
| $0$ | $1$ |
| $1$ | $2$ |
| $2$ | $4$ |
| $3$ | $8$ |

| $x$ | $y = f^{-1}(x)$ |
|---|---|
| $\dfrac{1}{8}$ | $-3$ |
| $\dfrac{1}{4}$ | $-2$ |
| $\dfrac{1}{2}$ | $-1$ |
| $1$ | $0$ |
| $2$ | $1$ |
| $4$ | $2$ |
| $8$ | $3$ |

The graphs of $f(x)$ and its inverse are shown above. Notice that the graphs of $f$ and $f^{-1}$ are symmetric about the line $y = x$, as expected.

Now we would like to be able to write an equation for $f^{-1}$. To do so, we follow the steps for finding an inverse.

$$f(x) = 2^x$$

**STEP 1.** Replace $f(x)$ by $y$.      $y = 2^x$

**STEP 2.** Interchange $x$ and $y$.      $x = 2^y$

**STEP 3.** Solve for $y$.

At this point, we are stuck. To solve this equation for $y$, a new notation, the **logarithmic notation,** is needed. The symbol $\log_b x$ means "the power to which $b$ is raised in order to produce a result of $x$."

$$\log_b x = y \quad \text{means} \quad b^y = x$$

We say that $\log_b x$ is "the logarithm of $x$ to the base $b$" or "the log of $x$ to the base $b$."

---

**Logarithmic Definition**

If $b > 0$ and $b \neq 1$, then

$$y = \log_b x \text{ means } x = b^y$$

for every $x > 0$ and every real number $y$.

---

Before returning to the function $x = 2^y$ and solving it for $y$ in terms of $x$, let's practice using the new notation $\log_b x$.

It is important to be able to write exponential equations from logarithmic notation, and vice versa. The following table shows examples of both forms.

> ▶ **Helpful Hint**
>
> Notice that a *logarithm* is an *exponent*. In other words, $\log_3 9$ is the *power* that we raise 3 to in order to get 9.

| *Logarithmic Equation* | *Corresponding Exponential Equation* |
|:---:|:---:|
| $\log_3 9 = 2$ | $3^2 = 9$ |
| $\log_6 1 = 0$ | $6^0 = 1$ |
| $\log_2 8 = 3$ | $2^3 = 8$ |
| $\log_4 \dfrac{1}{16} = -2$ | $4^{-2} = \dfrac{1}{16}$ |
| $\log_8 2 = \dfrac{1}{3}$ | $8^{1/3} = 2$ |

**EXAMPLE 1**   Write as an exponential equation.

**a.** $\log_5 25 = 2$ **b.** $\log_6 \dfrac{1}{6} = -1$ **c.** $\log_2 \sqrt{2} = \dfrac{1}{2}$ **d.** $\log_7 x = 5$

*Solution*

**a.** $\log_5 25 = 2$ means $5^2 = 25$

**b.** $\log_6 \dfrac{1}{6} = -1$ means $6^{-1} = \dfrac{1}{6}$

**c.** $\log_2 \sqrt{2} = \dfrac{1}{2}$ means $2^{1/2} = \sqrt{2}$

**d.** $\log_7 x = 5$ means $7^5 = x$

□

**PRACTICE**
**1**   Write as an exponential equation.

**a.** $\log_3 81 = 4$ **b.** $\log_5 \dfrac{1}{5} = -1$ **c.** $\log_7 \sqrt{7} = \dfrac{1}{2}$ **d.** $\log_{13} y = 4$

**EXAMPLE 2**   Write as a logarithmic equation.

**a.** $9^3 = 729$ **b.** $6^{-2} = \dfrac{1}{36}$ **c.** $5^{1/3} = \sqrt[3]{5}$ **d.** $\pi^4 = x$

*Solution*

**a.** $9^3 = 729$ means $\log_9 729 = 3$

**b.** $6^{-2} = \dfrac{1}{36}$ means $\log_6 \dfrac{1}{36} = -2$

**c.** $5^{1/3} = \sqrt[3]{5}$ means $\log_5 \sqrt[3]{5} = \dfrac{1}{3}$

**d.** $\pi^4 = x$ means $\log_\pi x = 4$ ☐

**PRACTICE**
**2** Write as a logarithmic equation.

**a.** $4^3 = 64$      **b.** $6^{1/3} = \sqrt[3]{6}$      **c.** $5^{-3} = \dfrac{1}{125}$      **d.** $\pi^7 = z$

**EXAMPLE 3**    Find the value of each logarithmic expression.

**a.** $\log_4 16$      **b.** $\log_{10} \dfrac{1}{10}$      **c.** $\log_9 3$

*Solution*

**a.** $\log_4 16 = 2$ because $4^2 = 16$

**b.** $\log_{10} \dfrac{1}{10} = -1$ because $10^{-1} = \dfrac{1}{10}$

**c.** $\log_9 3 = \dfrac{1}{2}$ because $9^{1/2} = \sqrt{9} = 3$ ☐

**PRACTICE**
**3** Find the value of each logarithmic expression.

**a.** $\log_3 9$      **b.** $\log_2 \dfrac{1}{8}$      **c.** $\log_{49} 7$

> ▶ **Helpful Hint**
> Another method for evaluating logarithms such as those in Example 3 is to set the expression equal to $x$ and then write them in exponential form to find $x$. For example:
>
> **a.** $\log_4 16 = x$ means $4^x = 16$. Since $4^2 = 16$, $x = 2$ or $\log_4 16 = 2$.
>
> **b.** $\log_{10} \dfrac{1}{10} = x$ means $10^x = \dfrac{1}{10}$. Since $10^{-1} = \dfrac{1}{10}$, $x = -1$ or $\log_{10} \dfrac{1}{10} = -1$.
>
> **c.** $\log_9 3 = x$ means $9^x = 3$. Since $9^{1/2} = 3$, $x = \dfrac{1}{2}$ or $\log_9 3 = \dfrac{1}{2}$.

**OBJECTIVE 2** ▶ **Solving logarithmic equations.** The ability to interchange the logarithmic and exponential forms of a statement is often the key to solving logarithmic equations.

**EXAMPLE 4**    Solve each equation for $x$.

**a.** $\log_4 \dfrac{1}{4} = x$      **b.** $\log_5 x = 3$      **c.** $\log_x 25 = 2$      **d.** $\log_3 1 = x$      **e.** $\log_b 1 = x$

*Solution*

**a.** $\log_4 \dfrac{1}{4} = x$ means $4^x = \dfrac{1}{4}$. Solve $4^x = \dfrac{1}{4}$ for $x$.

$$4^x = \dfrac{1}{4}$$

$$4^x = 4^{-1}$$

Since the bases are the same, by the uniqueness of $b^x$, we have that

$$x = -1$$

The solution is $-1$ or the solution set is $\{-1\}$. To check, see that $\log_4 \dfrac{1}{4} = -1$, since $4^{-1} = \dfrac{1}{4}$.

**b.** $\log_5 x = 3$

$\qquad 5^3 = x \quad$ Write as an exponential equation.

$\qquad 125 = x$

The solution is 125.

**c.** $\log_x 25 = 2$

$\qquad x^2 = 25 \quad$ Write as an exponential equation. Here $x > 0$, $x \neq 1$.

$\qquad x = 5$

Even though $(-5)^2 = 25$, the base $b$ of a logarithm must be positive. The solution is 5.

**d.** $\log_3 1 = x$

$\qquad 3^x = 1 \quad$ Write as an exponential equation.

$\qquad 3^x = 3^0 \quad$ Write 1 as $3^0$.

$\qquad x = 0 \quad$ Use the uniqueness of $b^x$.

The solution is 0.

**e.** $\log_b 1 = x$

$\qquad b^x = 1 \quad$ Write as an exponential equation. Here, $b > 0$ and $b \neq 1$.

$\qquad b^x = b^0 \quad$ Write 1 as $b^0$.

$\qquad x = 0 \quad$ Apply the uniqueness of $b^x$.

The solution is 0.  $\square$

**PRACTICE**

**4**  Solve each equation for $x$.

**a.** $\log_5 \dfrac{1}{25} = x$ $\qquad$ **b.** $\log_x 8 = 3$ $\qquad$ **c.** $\log_6 x = 2$

**d.** $\log_{13} 1 = x$ $\qquad$ **e.** $\log_h 1 = x$

In Example 4e we proved an important property of logarithms. That is, $\log_b 1$ is always 0. This property as well as two important others are given next.

---

**Properties of Logarithms**

If $b$ is a real number, $b > 0$, and $b \neq 1$, then

**1.** $\log_b 1 = 0$

**2.** $\log_b b^x = x$

**3.** $b^{\log_b x} = x$

---

To see that **2.** $\log_b b^x = x$, change the logarithmic form to exponential form. Then, $\log_b b^x = x$ means $b^x = b^x$. In exponential form, the statement is true, so in logarithmic form, the statement is also true. To understand **3.** $b^{\log_b x} = x$, write this exponential equation as an equivalent logarithm.

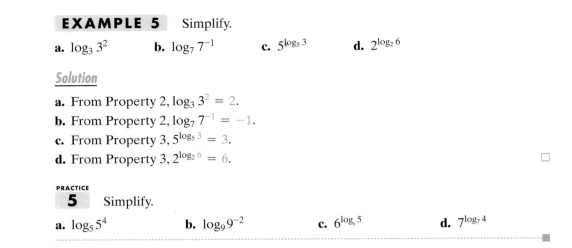

**EXAMPLE 5** Simplify.

**a.** $\log_3 3^2$      **b.** $\log_7 7^{-1}$      **c.** $5^{\log_5 3}$      **d.** $2^{\log_2 6}$

*Solution*

**a.** From Property 2, $\log_3 3^2 = 2$.

**b.** From Property 2, $\log_7 7^{-1} = -1$.

**c.** From Property 3, $5^{\log_5 3} = 3$.

**d.** From Property 3, $2^{\log_2 6} = 6$.

**PRACTICE**
**5** Simplify.

**a.** $\log_5 5^4$      **b.** $\log_9 9^{-2}$      **c.** $6^{\log_6 5}$      **d.** $7^{\log_7 4}$

**OBJECTIVE 3** ▶ **Graphing logarithmic functions.** Let us now return to the function $f(x) = 2^x$ and write an equation for its inverse, $f^{-1}(x)$. Recall our earlier work.

$$f(x) = 2^x$$

**STEP 1.** Replace $f(x)$ by $y$.            $y = 2^x$

**STEP 2.** Interchange $x$ and $y$.            $x = 2^y$

Having gained proficiency with the notation $\log_b x$, we can now complete the steps for writing the inverse equation by writing $x = 2^y$ as an equivalent logarithm.

**STEP 1.** Solve for $y$.            $y = \log_2 x$

**STEP 2.** Replace $y$ with $f^{-1}(x)$.      $f^{-1}(x) = \log_2 x$

Thus, $f^{-1}(x) = \log_2 x$ defines a function that is the inverse function of the function $f(x) = 2^x$. The function $f^{-1}(x)$ or $y = \log_2 x$ is called a **logarithmic function.**

> **Logarithmic Function**
>
> If $x$ is a positive real number, $b$ is a constant positive real number, and $b$ is not 1, then a **logarithmic function** is a function that can be defined by
>
> $$f(x) = \log_b x$$
>
> The domain of $f$ is the set of positive real numbers, and the range of $f$ is the set of real numbers.

**Concept Check** ☑

Let $f(x) = \log_3 x$ and $g(x) = 3^x$. These two functions are inverses of each other. Since $(2, 9)$ is an ordered pair solution of $g(x)$ or $g(2) = 9$, what ordered pair do we know to be a solution of $f(x)$? Also, find $f(9)$. Explain why.

We can explore logarithmic functions by graphing them.

**EXAMPLE 6** Graph the logarithmic function $y = \log_2 x$.

*Solution* First we write the equation with exponential notation as $2^y = x$. Then we find some ordered pair solutions that satisfy this equation. Finally, we plot the points and connect them with a smooth curve. The domain of this function is $(0, \infty)$, and the range is all real numbers.

**Answer to Concept Check:**

$(9, 2); f(9) = 2$; answers may vary

Since $x = 2^y$ is solved for $x$, we choose $y$-values and compute corresponding $x$-values.

If $y = 0, x = 2^0 = 1$

If $y = 1, x = 2^1 = 2$

If $y = 2, x = 2^2 = 4$

If $y = -1, x = 2^{-1} = \dfrac{1}{2}$

| $x = 2^y$ | $y$ |
|---|---|
| 1 | 0 |
| 2 | 1 |
| 4 | 2 |
| $\dfrac{1}{2}$ | $-1$ |

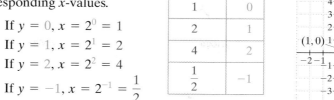

Notice that the $x$-intercept is $(1, 0)$ and there is no $y$-intercept.  □

**PRACTICE**
**6**  Graph the logarithmic function $y = \log_7 x$.

**EXAMPLE 7**  Graph the logarithmic function $f(x) = \log_{1/3} x$.

_Solution_  Replace $f(x)$ with $y$, and write the result with exponential notation.

$$f(x) = \log_{1/3} x$$
$$y = \log_{1/3} x \quad \text{Replace } f(x) \text{ with } y.$$
$$\left(\frac{1}{3}\right)^y = x \quad \text{Write in exponential form.}$$

Now we can find ordered pair solutions that satisfy $\left(\dfrac{1}{3}\right)^y = x$, plot these points, and connect them with a smooth curve.

If $y = 0, x = \left(\dfrac{1}{3}\right)^0 = 1$

If $y = 1, x = \left(\dfrac{1}{3}\right)^1 = \dfrac{1}{3}$

If $y = -1, x = \left(\dfrac{1}{3}\right)^{-1} = 3$

If $y = -2, x = \left(\dfrac{1}{3}\right)^{-2} = 9$

| $x = \left(\dfrac{1}{3}\right)^y$ | $y$ |
|---|---|
| 1 | 0 |
| $\dfrac{1}{3}$ | 1 |
| 3 | $-1$ |
| 9 | $-2$ |

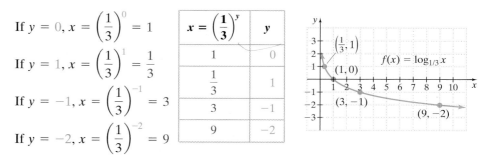

The domain of this function is $(0, \infty)$, and the range is the set of all real numbers. The $x$-intercept is $(1, 0)$ and there is no $y$-intercept.  □

**PRACTICE**
**7**  Graph the logarithmic function $y = \log_{1/4} x$.

The following figures summarize characteristics of logarithmic functions.

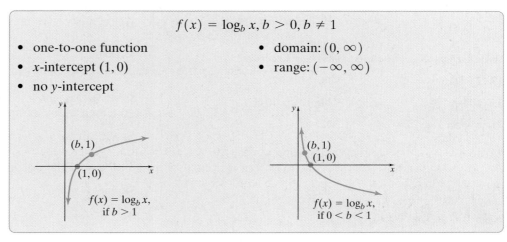

$$f(x) = \log_b x, b > 0, b \neq 1$$

- one-to-one function
- $x$-intercept $(1, 0)$
- no $y$-intercept
- domain: $(0, \infty)$
- range: $(-\infty, \infty)$

## VOCABULARY & READINESS CHECK

*Use the choices to fill in each blank.*

**1.** A function, such as $y = \log_2 x$ is a(n) _____ function.

    **A.** linear      **B.** logarithmic      **C.** quadratic      **D.** exponential

**2.** If $y = \log_2 x$, then _____.

    **A.** $x = y$      **B.** $2^x = y$      **C.** $2^y = x$      **D.** $2y = x$

*Answer the questions about the graph of $y = \log_2 x$, shown to the left.*

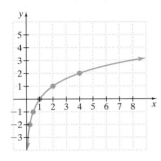

**3.** Is this a one-to-one function? _____

**4.** Is there an $x$-intercept? _____ If so, name the coordinates. _____

**5.** Is there a $y$-intercept? _____ If so, name the coordinates. _____

**6.** The domain of this function, in interval notation, is _____.

**7.** The range of this function, in interval notation, is _____.

## 12.4 EXERCISE SET

*Write each as an exponential equation. See Example 1.*

**1.** $\log_6 36 = 2$

**2.** $\log_2 32 = 5$

**3.** $\log_3 \dfrac{1}{27} = -3$

**4.** $\log_5 \dfrac{1}{25} = -2$

**5.** $\log_{10} 1000 = 3$

**6.** $\log_{10} 10 = 1$

**7.** $\log_9 x = 4$

**8.** $\log_8 y = 7$

**9.** $\log_\pi \dfrac{1}{\pi^2} = -2$

**10.** $\log_e \dfrac{1}{e} = -1$

**11.** $\log_7 \sqrt{7} = \dfrac{1}{2}$

**12.** $\log_{11} \sqrt[4]{11} = \dfrac{1}{4}$

**13.** $\log_{0.7} 0.343 = 3$

**14.** $\log_{1.2} 1.44 = 2$

**15.** $\log_3 \dfrac{1}{81} = -4$

**16.** $\log_{1/4} 16 = -2$

*Write each as a logarithmic equation. See Example 2.*

**17.** $2^4 = 16$

**18.** $5^3 = 125$

**19.** $10^2 = 100$

**20.** $10^4 = 10,000$

**21.** $\pi^3 = x$

**22.** $\pi^5 = y$

**23.** $10^{-1} = \dfrac{1}{10}$

**24.** $10^{-2} = \dfrac{1}{100}$

**25.** $4^{-2} = \dfrac{1}{16}$

**26.** $3^{-4} = \dfrac{1}{81}$

**27.** $5^{1/2} = \sqrt{5}$

**28.** $4^{1/3} = \sqrt[3]{4}$

*Find the value of each logarithmic expression. See Examples 3 and 5.*

**29.** $\log_2 8$

**30.** $\log_3 9$

**31.** $\log_3 \dfrac{1}{9}$

**32.** $\log_2 \dfrac{1}{32}$

**33.** $\log_{25} 5$

**34.** $\log_8 \dfrac{1}{2}$

**35.** $\log_{1/2} 2$

**36.** $\log_{2/3} \dfrac{4}{9}$

**37.** $\log_6 1$

**38.** $\log_9 9$

**39.** $\log_{10} 100$

**40.** $\log_{10} \dfrac{1}{10}$

**41.** $\log_3 81$

**42.** $\log_2 16$

**43.** $\log_4 \dfrac{1}{64}$

**44.** $\log_3 \dfrac{1}{9}$

*Solve. See Example 4.*

**45.** $\log_3 9 = x$

**46.** $\log_2 8 = x$

**47.** $\log_3 x = 4$

**48.** $\log_2 x = 3$

**49.** $\log_x 49 = 2$

**50.** $\log_x 8 = 3$

**51.** $\log_2 \dfrac{1}{8} = x$

**52.** $\log_3 \dfrac{1}{81} = x$

**53.** $\log_3 \dfrac{1}{27} = x$

**54.** $\log_5 \dfrac{1}{125} = x$

**55.** $\log_8 x = \dfrac{1}{3}$

**56.** $\log_9 x = \dfrac{1}{2}$

**57.** $\log_4 16 = x$

**58.** $\log_2 16 = x$

**59.** $\log_{3/4} x = 3$

**60.** $\log_{2/3} x = 2$

**61.** $\log_x 100 = 2$

**62.** $\log_x 27 = 3$

**63.** $\log_2 2^4 = x$

**64.** $\log_6 6^{-2} = x$

**65.** $3^{\log_3 5} = x$

**66.** $5^{\log_5 7} = x$

**67.** $\log_x \dfrac{1}{7} = \dfrac{1}{2}$

**68.** $\log_x 2 = -\dfrac{1}{3}$

*Simplify. See Example 5.*

**69.** $\log_5 5^3$

**70.** $\log_6 6^2$

**71.** $2^{\log_2 3}$

**72.** $7^{\log_7 4}$

**73.** $\log_9 9$

**74.** $\log_8 (8)^{-1}$

*Graph each logarithmic function. Label any intercepts. See Examples 6 and 7.*

**75.** $y = \log_3 x$

**76.** $y = \log_8 x$

**77.** $f(x) = \log_{1/4} x$

**78.** $f(x) = \log_{1/2} x$

**79.** $f(x) = \log_5 x$

**80.** $f(x) = \log_6 x$

**81.** $f(x) = \log_{1/6} x$

**82.** $f(x) = \log_{1/5} x$

### REVIEW AND PREVIEW

*Simplify each rational expression. See Section 7.1.*

**83.** $\dfrac{x + 3}{3 + x}$

**84.** $\dfrac{x - 5}{5 - x}$

**85.** $\dfrac{x^2 - 8x + 16}{2x - 8}$

**86.** $\dfrac{x^2 - 3x - 10}{2 + x}$

*Add or subtract as indicated. See Section 7.4.*

**87.** $\dfrac{2}{x} + \dfrac{3}{x^2}$

**88.** $\dfrac{3x}{x + 3} + \dfrac{9}{x + 3}$

**89.** $\dfrac{m^2}{m + 1} - \dfrac{1}{m + 1}$

**90.** $\dfrac{5}{y + 1} - \dfrac{4}{y - 1}$

### CONCEPT EXTENSIONS

*Solve. See the Concept Check in this section.*

**91.** Let $f(x) = \log_5 x$. Then $g(x) = 5^x$ is the inverse of $f(x)$. The ordered pair $(2, 25)$ is a solution of the function $g(x)$.

  **a.** Write this solution using function notation.

  **b.** Write an ordered pair that we know to be a solution of $f(x)$.

  **c.** Use the answer to part b and write the solution using function notation.

**92.** Let $f(x) = \log_{0.3} x$. Then $g(x) = 0.3^x$ is the inverse of $f(x)$. The ordered pair $(3, 0.027)$ is a solution of the function $g(x)$.

  **a.** Write this solution using function notation.

  **b.** Write an ordered pair that we know to be a solution of $f(x)$.

  **c.** Use the answer to part b and write the solution using function notation.

**93.** Explain why negative numbers are not included as logarithmic bases.

**94.** Explain why 1 is not included as a logarithmic base.

*Solve by first writing as an exponential.*

**95.** $\log_7 (5x - 2) = 1$

**96.** $\log_3 (2x + 4) = 2$

**97.** Simplify: $\log_3(\log_5 125)$

**98.** Simplify: $\log_7(\log_4(\log_2 16))$

*Graph each function and its inverse function on the same set of axes. Label any intercepts.*

**99.** $y = 4^x$; $y = \log_4 x$

**100.** $y = 3^x$; $y = \log_3 x$

**101.** $y = \left(\dfrac{1}{3}\right)^x$; $y = \log_{1/3} x$

**102.** $y = \left(\dfrac{1}{2}\right)^x$; $y = \log_{1/2} x$

**103.** Explain why the graph of the function $y = \log_b x$ contains the point $(1, 0)$ no matter what $b$ is.

**104.** $\log_3 10$ is between which two integers? Explain your answer.

**105.** The formula $\log_{10}(1 - k) = \dfrac{-0.3}{H}$ models the relationship between the half-life $H$ of a radioactive material and its rate of decay $k$. Find the rate of decay of the iodine isotope I-131 if its half-life is 8 days. Round to four decimal places.

**106.** The formula $\text{pH} = -\log_{10}(\text{H}^+)$ provides the pH for a liquid, where $\text{H}^+$ stands for the concentration of hydronium ions. Find the pH of lemonade, whose concentration of hydronium ions is 0.0050 moles/liter.

# 12.5 PROPERTIES OF LOGARITHMS

**OBJECTIVES**

1 Use the product property of logarithms.

2 Use the quotient property of logarithms.

3 Use the power property of logarithms.

4 Use the properties of logarithms together.

In the previous section we explored some basic properties of logarithms. We now introduce and explore additional properties. Because a logarithm is an exponent, logarithmic properties are just restatements of exponential properties.

**OBJECTIVE 1 ▶ Using the product property.** The first of these properties is called the **product property of logarithms,** because it deals with the logarithm of a product.

> **Product Property of Logarithms**
> If $x$, $y$, and $b$ are positive real numbers and $b \neq 1$, then
> $$\log_b xy = \log_b x + \log_b y$$

To prove this, let $\log_b x = M$ and $\log_b y = N$. Now write each logarithm with exponential notation.

$$\log_b x = M \quad \text{is equivalent to} \quad b^M = x$$
$$\log_b y = N \quad \text{is equivalent to} \quad b^N = y$$

Multiply the left sides and the right sides of the exponential equations, and we have that

$$xy = (b^M)(b^N) = b^{M+N}$$

If we write the equation $xy = b^{M+N}$ in equivalent logarithmic form, we have

$$\log_b xy = M + N$$

But since $M = \log_b x$ and $N = \log_b y$, we can write

$$\log_b xy = \log_b x + \log_b y \quad \text{Let } M = \log_b x \text{ and } N = \log_b y.$$

In other words, the logarithm of a product is the sum of the logarithms of the factors. This property is sometimes used to simplify logarithmic expressions.

In the examples that follow, assume that variables represent positive numbers.

**EXAMPLE 1** Write each sum as a single logarithm.

**a.** $\log_{11} 10 + \log_{11} 3$     **b.** $\log_3 \dfrac{1}{2} + \log_3 12$     **c.** $\log_2(x + 2) + \log_2 x$

_Solution_

In each case, both terms have a common logarithmic base.

**a.** $\log_{11} 10 + \log_{11} 3 = \log_{11}(10 \cdot 3)$    Apply the product property.
$$= \log_{11} 30$$

**b.** $\log_3 \dfrac{1}{2} + \log_3 12 = \log_3 \left( \dfrac{1}{2} \cdot 12 \right) = \log_3 6$

**c.** $\log_2(x + 2) + \log_2 x = \log_2[(x + 2) \cdot x] = \log_2(x^2 + 2x)$

▶ **Helpful Hint**

Check your logarithm properties. Make sure you understand that $\log_2(x + 2)$ *is not* $\log_2 x + \log_2 2$.

**PRACTICE**

**1** Write each sum as a single logarithm.

**a.** $\log_8 5 + \log_8 3$

**b.** $\log_2 \dfrac{1}{3} + \log_2 18$

**c.** $\log_5(x - 1) + \log_5(x + 1)$

**OBJECTIVE 2 ▶ Using the quotient property.** The second property is the **quotient property of logarithms.**

> **Quotient Property of Logarithms**
> If $x$, $y$, and $b$ are positive real numbers and $b \neq 1$, then
> $$\log_b \frac{x}{y} = \log_b x - \log_b y$$

The proof of the quotient property of logarithms is similar to the proof of the product property. Notice that the quotient property says that the logarithm of a quotient is the difference of the logarithms of the dividend and divisor.

**Concept Check ☑**

Which of the following is the correct way to rewrite $\log_5 \frac{7}{2}$?

**a.** $\log_5 7 - \log_5 2$   **b.** $\log_5(7 - 2)$   **c.** $\dfrac{\log_5 7}{\log_5 2}$   **d.** $\log_5 14$

---

**EXAMPLE 2**   Write each difference as a single logarithm.

**a.** $\log_{10} 27 - \log_{10} 3$   **b.** $\log_5 8 - \log_5 x$   **c.** $\log_3(x^2 + 5) - \log_3(x^2 + 1)$

*Solution*   All terms have a common logarithmic base.

**a.** $\log_{10} 27 - \log_{10} 3 = \log_{10} \dfrac{27}{3} = \log_{10} 9$

**b.** $\log_5 8 - \log_5 x = \log_5 \dfrac{8}{x}$

**c.** $\log_3(x^2 + 5) - \log_3(x^2 + 1) = \log_3 \dfrac{x^2 + 5}{x^2 + 1}$   Apply the quotient property.   ☐

**PRACTICE**
**2**   Write each difference as a single logarithm.

**a.** $\log_5 18 - \log_5 6$   **b.** $\log_6 x - \log_6 3$   **c.** $\log_4(x^2 + 1) - \log_4(x^2 + 3)$

---

**OBJECTIVE 3 ▶ Using the power property.** The third and final property we introduce is the **power property of logarithms.**

> **Power Property of Logarithms**
> If $x$ and $b$ are positive real numbers, $b \neq 1$, and $r$ is a real number, then
> $$\log_b x^r = r \log_b x$$

**EXAMPLE 3**   Use the power property to rewrite each expression.
**a.** $\log_5 x^3$   **b.** $\log_4 \sqrt{2}$

*Solution*

**a.** $\log_5 x^3 = 3 \log_5 x$   **b.** $\log_4 \sqrt{2} = \log_4 2^{1/2} = \dfrac{1}{2} \log_4 2$   ☐

**PRACTICE**
**3**   Use the power property to rewrite each expression.

**a.** $\log_7 x^8$   **b.** $\log_5 \sqrt[4]{7}$

**OBJECTIVE 4 ▶ Using the properties together.** Many times we must use more than one property of logarithms to simplify a logarithmic expression.

**EXAMPLE 4** Write as a single logarithm.

**a.** $2 \log_5 3 + 3 \log_5 2$    **b.** $3 \log_9 x - \log_9(x + 1)$    **c.** $\log_4 25 + \log_4 3 - \log_4 5$

*Solution* In each case, all terms have a common logarithmic base.

**a.** $2 \log_5 3 + 3 \log_5 2 = \log_5 3^2 + \log_5 2^3$    Apply the power property.

$\qquad\qquad\qquad\qquad = \log_5 9 + \log_5 8$

$\qquad\qquad\qquad\qquad = \log_5(9 \cdot 8)$    Apply the product property.

$\qquad\qquad\qquad\qquad = \log_5 72$

**b.** $3 \log_9 x - \log_9(x + 1) = \log_9 x^3 - \log_9(x + 1)$    Apply the power property.

$\qquad\qquad\qquad\qquad\qquad = \log_9 \dfrac{x^3}{x + 1}$    Apply the quotient property.

**c.** Use both the product and quotient properties.

$\log_4 25 + \log_4 3 - \log_4 5 = \log_4(25 \cdot 3) - \log_4 5$    Apply the product property.

$\qquad\qquad\qquad\qquad\qquad = \log_4 75 - \log_4 5$    Simplify.

$\qquad\qquad\qquad\qquad\qquad = \log_4 \dfrac{75}{5}$    Apply the quotient property.

$\qquad\qquad\qquad\qquad\qquad = \log_4 15$    Simplify.    □

**PRACTICE**
**4** Write as a single logarithm.

**a.** $2 \log_5 4 + 5 \log_5 2$    **b.** $2 \log_8 x - \log_8(x + 3)$    **c.** $\log_7 12 + \log_7 5 - \log_7 4$

---

**EXAMPLE 5** Write each expression as sums or differences of multiples of logarithms.

**a.** $\log_3 \dfrac{5 \cdot 7}{4}$    **b.** $\log_2 \dfrac{x^5}{y^2}$

*Solution*

**a.** $\log_3 \dfrac{5 \cdot 7}{4} = \log_3(5 \cdot 7) - \log_3 4$    Apply the quotient property.

$\qquad\qquad\quad = \log_3 5 + \log_3 7 - \log_3 4$    Apply the product property.

**b.** $\log_2 \dfrac{x^5}{y^2} = \log_2(x^5) - \log_2(y^2)$    Apply the quotient property.

$\qquad\qquad = 5 \log_2 x - 2 \log_2 y$    Apply the power property.    □

**PRACTICE**
**5** Write each expression as sums or differences of multiples of logarithms.

**a.** $\log_5 \dfrac{4 \cdot 3}{7}$    **b.** $\log_4 \dfrac{a^2}{b^5}$

---

▶ **Helpful Hint**

Notice that we are not able to simplify further a logarithmic expression such as $\log_5(2x - 1)$. None of the basic properties gives a way to write the logarithm of a difference in some equivalent form.

**Concept Check** ☑

What is wrong with the following?

$$\log_{10}(x^2 + 5) = \log_{10} x^2 + \log_{10} 5$$
$$= 2 \log_{10} x + \log_{10} 5$$

Use a numerical example to demonstrate that the result is incorrect.

**EXAMPLE 6**   If $\log_b 2 = 0.43$ and $\log_b 3 = 0.68$, use the properties of logarithms to evaluate.

**a.** $\log_b 6$           **b.** $\log_b 9$           **c.** $\log_b \sqrt{2}$

*Solution*

**a.** $\log_b 6 = \log_b(2 \cdot 3)$        Write 6 as $2 \cdot 3$.

$\phantom{\log_b 6} = \log_b 2 + \log_b 3$   Apply the product property.

$\phantom{\log_b 6} = 0.43 + 0.68$        Substitute given values.

$\phantom{\log_b 6} = 1.11$             Simplify.

**b.** $\log_b 9 = \log_b 3^2$         Write 9 as $3^2$.

$\phantom{\log_b 9} = 2 \log_b 3$

$\phantom{\log_b 9} = 2(0.68)$         Substitute 0.68 for $\log_b 3$.

$\phantom{\log_b 9} = 1.36$           Simplify.

**c.** First, recall that $\sqrt{2} = 2^{1/2}$. Then

$\log_b \sqrt{2} = \log_b 2^{1/2}$        Write $\sqrt{2}$ as $2^{1/2}$.

$\phantom{\log_b \sqrt{2}} = \dfrac{1}{2} \log_b 2$         Apply the power property.

$\phantom{\log_b \sqrt{2}} = \dfrac{1}{2}(0.43)$         Substitute the given value.

$\phantom{\log_b \sqrt{2}} = 0.215$          Simplify.           ☐

**PRACTICE**

**6**   If $\log_b 5 = 0.83$ and $\log_b 3 = 0.56$, use the properties of logarithms to evaluate.

**a.** $\log_b 15$           **b.** $\log_b 25$           **c.** $\log_b \sqrt{3}$

A summary of the basic properties of logarithms that we have developed so far is given next.

---

**Properties of Logarithms**

If $x$, $y$, and $b$ are positive real numbers, $b \neq 1$, and $r$ is a real number, then

**1.** $\log_b 1 = 0$  ✓                     ✓ **2.** $\log_b b^x = x$

**3.** $b^{\log_b x} = x$  ∟                     **4.** $\log_b xy = \log_b x + \log_b y$   Product property.

**5.** $\log_b \dfrac{x}{y} = \log_b x - \log_b y$   Quotient property.   **6.** $\log_b x^r = r \log_b x$        Power property.

---

**Answer to Concept Check:**
The properties do not give any way to simplify the logarithm of a sum; answers may vary.

# VOCABULARY & READINESS CHECK

*Select the correct choice.*

**1.** $\log_b 12 + \log_b 3 = \log_b$ ____

   **a.** 36   **b.** 15   **c.** 4   **d.** 9

**2.** $\log_b 12 - \log_b 3 = \log_b$ ____

   **a.** 36   **b.** 15   **c.** 4   **d.** 9

**3.** $7 \log_b 2 =$ _____

   **a.** $\log_b 14$   **b.** $\log_b 2^7$   **c.** $\log_b 7^2$   **d.** $(\log_b 2)^7$

**4.** $\log_b 1 =$ ____

   **a.** $b$   **b.** 1   **c.** 0   **d.** no answer

**5.** $b^{\log_b x} =$ ____

   **a.** $x$   **b.** $b$   **c.** 1   **d.** 0

**6.** $\log_5 5^2 =$ ____

   **a.** 25   **b.** 2   **c.** $5^{5^2}$   **d.** 32

## 12.5 | EXERCISE SET

**MyMathLab**

PRACTICE  WATCH  DOWNLOAD  READ  REVIEW

*Write each sum as a single logarithm. Assume that variables represent positive numbers. See Example 1.*

**1.** $\log_5 2 + \log_5 7$

**2.** $\log_3 8 + \log_3 4$

**3.** $\log_4 9 + \log_4 x$

**4.** $\log_2 x + \log_2 y$

**5.** $\log_6 x + \log_6 (x + 1)$

**6.** $\log_5 y^3 + \log_5 (y - 7)$

**7.** $\log_{10} 5 + \log_{10} 2 + \log_{10} (x^2 + 2)$

**8.** $\log_6 3 + \log_6 (x + 4) + \log_6 5$

*Write each difference as a single logarithm. Assume that variables represent positive numbers. See Examples 2 and 4.*

**9.** $\log_5 12 - \log_5 4$

**10.** $\log_7 20 - \log_7 4$

**11.** $\log_3 8 - \log_3 2$

**12.** $\log_5 12 - \log_5 3$

**13.** $\log_2 x - \log_2 y$

**14.** $\log_3 12 - \log_3 z$

**15.** $\log_2 (x^2 + 6) - \log_2 (x^2 + 1)$

**16.** $\log_7 (x + 9) - \log_7 (x^2 + 10)$

*Use the power property to rewrite each expression. See Example 3.*

**17.** $\log_3 x^2$

**18.** $\log_2 x^5$

**19.** $\log_4 5^{-1}$

**20.** $\log_6 7^{-2}$

**21.** $\log_5 \sqrt{y}$

**22.** $\log_5 \sqrt[3]{x}$

### MIXED PRACTICE

*Write each as a single logarithm. Assume that variables represent positive numbers. See Example 4.*

**23.** $\log_2 5 + \log_2 x^3$

**24.** $\log_5 2 + \log_5 y^2$

**25.** $3 \log_4 2 + \log_4 6$

**26.** $2 \log_3 5 + \log_3 2$

**27.** $3 \log_5 x + 6 \log_5 z$

**28.** $2 \log_7 y + 6 \log_7 z$

**29.** $\log_4 2 + \log_4 10 - \log_4 5$

**30.** $\log_6 18 + \log_6 2 - \log_6 9$

**31.** $\log_7 6 + \log_7 3 - \log_7 4$

**32.** $\log_8 5 + \log_8 15 - \log_8 20$

**33.** $\log_{10} x - \log_{10} (x + 1) + \log_{10} (x^2 - 2)$

**34.** $\log_9 (4x) - \log_9 (x - 3) + \log_9 (x^3 + 1)$

**35.** $3 \log_2 x + \dfrac{1}{2} \log_2 x - 2 \log_2 (x + 1)$

**36.** $2 \log_5 x + \dfrac{1}{3} \log_5 x - 3 \log_5 (x + 5)$

**37.** $2 \log_8 x - \dfrac{2}{3} \log_8 x + 4 \log_8 x$

**38.** $5 \log_6 x - \dfrac{3}{4} \log_6 x + 3 \log_6 x$

### MIXED PRACTICE

*Write each expression as a sum or difference of logarithms. Assume that variables represent positive numbers. See Example 5.*

**39.** $\log_3 \dfrac{4y}{5}$

**40.** $\log_7 \dfrac{5x}{4}$

**41.** $\log_4 \dfrac{2}{9z}$

**42.** $\log_9 \dfrac{7}{8y}$

**43.** $\log_2 \dfrac{x^3}{y}$

**44.** $\log_5 \dfrac{x}{y^4}$

**45.** $\log_b \sqrt{7x}$

**46.** $\log_b \sqrt{\dfrac{3}{y}}$

**47.** $\log_6 x^4 y^5$

**48.** $\log_2 y^3 z$

**49.** $\log_5 x^3 (x + 1)$

**50.** $\log_3 x^2 (x - 9)$

**51.** $\log_6 \dfrac{x^2}{x + 3}$

**52.** $\log_3 \dfrac{(x + 5)^2}{x}$

*If $\log_b 3 = 0.5$ and $\log_b 5 = 0.7$, evaluate each expression. See Example 6.*

**53.** $\log_b 15$

**54.** $\log_b 25$

**55.** $\log_b \dfrac{5}{3}$

**56.** $\log_b \dfrac{3}{5}$

**57.** $\log_b \sqrt{5}$

**58.** $\log_b \sqrt[4]{3}$

*If $\log_b 2 = 0.43$ and $\log_b 3 = 0.68$, evaluate each expression. See Example 6.*

**59.** $\log_b 8$

**60.** $\log_b 81$

**61.** $\log_b \dfrac{3}{9}$

**62.** $\log_b \dfrac{4}{32}$

**63.** $\log_b \sqrt{\dfrac{2}{3}}$

**64.** $\log_b \sqrt{\dfrac{3}{2}}$

### REVIEW AND PREVIEW

**65.** Graph the functions $y = 10^x$ and $y = \log_{10} x$ on the same set of axes. See Section 12.4.

*Evaluate each expression. See Section 12.4.*

**66.** $\log_{10} 100$

**67.** $\log_{10} \dfrac{1}{10}$

**68.** $\log_7 7^2$

**69.** $\log_7 \sqrt{7}$

### CONCEPT EXTENSIONS

*Solve. See the Concept Checks in this section.*

**70.** Which of the following is the correct way to rewrite $\log_3 \dfrac{14}{11}$?

    **a.** $\dfrac{\log_3 14}{\log_3 11}$

    **b.** $\log_3 14 - \log_3 11$

    **c.** $\log_3 (14 - 11)$

    **d.** $\log_3 154$

**71.** Which of the following is the correct way to rewrite $\log_9 \dfrac{21}{3}$?

    **a.** $\log_9 7$

    **b.** $\log_9 (21 - 3)$

    **c.** $\dfrac{\log_9 21}{\log_9 3}$

    **d.** $\log_9 21 - \log_9 3$

*Answer the following true or false. Study your logarithm properties carefully before answering.*

**72.** $\log_2 x^3 = 3 \log_2 x$

**73.** $\log_3(x + y) = \log_3 x + \log_3 y$

**74.** $\dfrac{\log_7 10}{\log_7 5} = \log_7 2$

**75.** $\log_7 \dfrac{14}{8} = \log_7 14 - \log_7 8$

**76.** $\dfrac{\log_7 x}{\log_7 y} = (\log_7 x) - (\log_7 y)$

**77.** $(\log_3 6) \cdot (\log_3 4) = \log_3 24$

**78.** It is true that $\log 8 = \log(8 \cdot 1) = \log 8 + \log 1$. Explain how $\log 8$ can equal $\log 8 + \log 1$.

## INTEGRATED REVIEW FUNCTIONS AND PROPERTIES OF LOGARITHMS

Sections 12.1–12.5

*If $f(x) = x - 6$ and $g(x) = x^2 + 1$, find each value.*

**1.** $(f + g)(x)$

**2.** $(f - g)(x)$

**3.** $(f \cdot g)(x)$

**4.** $\left(\dfrac{f}{g}\right)(x)$

*If $f(x) = \sqrt{x}$ and $g(x) = 3x - 1$, find each function.*

**5.** $(f \circ g)(x)$

**6.** $(g \circ f)(x)$

*Determine whether each is a one-to-one function. If it is, find its inverse.*

**7.** $f = \{(-2, 6), (4, 8), (2, -6), (3, 3)\}$

**8.** $g = \{(4, 2), (-1, 3), (5, 3), (7, 1)\}$

*Determine whether the graph of each function is one-to-one.*

**9.**  **10.**  **11.**

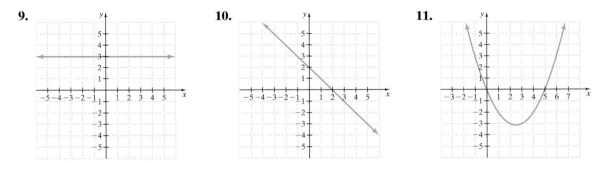

*Each function listed is one-to-one. Find the inverse of each function.*

**12.** $f(x) = 3x$

**13.** $f(x) = x + 4$

**14.** $f(x) = 5x - 1$

**15.** $f(x) = 3x + 2$

*Graph each function.*

**16.** $y = \left(\dfrac{1}{2}\right)^x$

**17.** $y = 2^x + 1$

**18.** $y = \log_3 x$

**19.** $y = \log_{1/3} x$

*Solve.*

**20.** $2^x = 8$

**21.** $9 = 3^{x-5}$

**22.** $4^{x-1} = 8^{x+2}$

**23.** $25^x = 125^{x-1}$

**24.** $\log_4 16 = x$

**25.** $\log_{49} 7 = x$

**26.** $\log_2 x = 5$

**27.** $\log_x 64 = 3$

**28.** $\log_x \dfrac{1}{125} = -3$

**29.** $\log_3 x = -2$

*Write each as a single logarithm.*

**30.** $5 \log_2 x$                      **31.** $x \log_2 5$           **32.** $3 \log_5 x - 5 \log_5 y$      **33.** $9 \log_5 x + 3 \log_5 y$

**34.** $\log_2 x + \log_2(x - 3) - \log_2(x^2 + 4)$                  **35.** $\log_3 y - \log_3(y + 2) + \log_3(y^3 + 11)$

*Write each expression as sums or differences of multiples of logarithms.*

**36.** $\log_7 \dfrac{9x^2}{y}$                                        **37.** $\log_6 \dfrac{5y}{z^2}$

---

# 12.6 COMMON LOGARITHMS, NATURAL LOGARITHMS, AND CHANGE OF BASE

## OBJECTIVES

1 Identify common logarithms and approximate them by calculator.

2 Evaluate common logarithms of powers of 10.

3 Identify natural logarithms and approximate them by calculator.

4 Evaluate natural logarithms of powers of e.

5 Use the change of base formula.

In this section we look closely at two particular logarithmic bases. These two logarithmic bases are used so frequently that logarithms to their bases are given special names. **Common logarithms** are logarithms to base 10. **Natural logarithms** are logarithms to base $e$, which we introduce in this section. The work in this section is based on the use of the calculator, which has both the common "log" | LOG | and the natural "log" | LN | keys.

**OBJECTIVE 1 ▶ Approximating common logarithms.** Logarithms to base 10, common logarithms, are used frequently because our number system is a base 10 decimal system. The notation $\log x$ means the same as $\log_{10} x$.

> **Common Logarithms**
>
> $$\log x \text{ means } \log_{10} x$$

**EXAMPLE 1** Use a calculator to approximate log 7 to four decimal places.

*Solution* Press the following sequence of keys.

| 7 | LOG |    or    | LOG | 7 | ENTER |

To four decimal places,

$$\log 7 \approx 0.8451$$

**PRACTICE**

**1** Use a calculator to approximate log 15 to four decimal places.

**OBJECTIVE 2 ▶ Evaluating common logarithms of powers of 10.** To evaluate the common log of a power of 10, a calculator is not needed. According to the property of logarithms,

$$\log_b b^x = x$$

It follows that if $b$ is replaced with 10, we have

$$\log 10^x = x$$

> ▶ **Helpful Hint**
> Remember that $\log 10^x$ means $\log_{10} 10^x = x$.

**EXAMPLE 2**    Find the exact value of each logarithm.

**a.** $\log 10$        **b.** $\log 1000$        **c.** $\log \dfrac{1}{10}$        **d.** $\log \sqrt{10}$

*Solution*

**a.** $\log 10 = \log 10^1 = 1$        **b.** $\log 1000 = \log 10^3 = 3$

**c.** $\log \dfrac{1}{10} = \log 10^{-1} = -1$        **d.** $\log \sqrt{10} = \log 10^{1/2} = \dfrac{1}{2}$

**PRACTICE**
**2**    Find the exact value of each logarithm.

**a.** $\log \dfrac{1}{100}$        **b.** $\log 100{,}000$        **c.** $\log \sqrt[5]{10}$        **d.** $\log 0.001$

As we will soon see, equations containing common logs are useful models of many natural phenomena.

**EXAMPLE 3**    Solve $\log x = 1.2$ for $x$. Give an exact solution, and then approximate the solution to four decimal places.

*Solution*    Remember that the base of a common log is understood to be 10.

$$\log x = 1.2$$

*help*

> **Helpful Hint**
> The understood base is 10.

$$10^{1.2} = x \qquad \text{Write with exponential notation.}$$

The exact solution is $10^{1.2}$. To four decimal places, $x \approx 15.8489$.

**PRACTICE**
**3**    Solve $\log x = 3.4$ for $x$. Give an exact solution, and then approximate the solution to four decimal places.

The Richter scale measures the intensity, or magnitude, of an earthquake. The formula for the magnitude $R$ of an earthquake is $R = \log\left(\dfrac{a}{T}\right) + B$, where $a$ is the amplitude in micrometers of the vertical motion of the ground at the recording station, $T$ is the number of seconds between successive seismic waves, and $B$ is an adjustment factor that takes into account the weakening of the seismic wave as the distance increases from the epicenter of the earthquake.

**EXAMPLE 4**    **Finding the Magnitude of an Earthquake**

Find an earthquake's magnitude on the Richter scale if a recording station measures an amplitude of 300 micrometers and 2.5 seconds between waves. Assume that $B$ is 4.2. Approximate the solution to the nearest tenth.

*Solution*    Substitute the known values into the formula for earthquake intensity.

$$R = \log\left(\frac{a}{T}\right) + B \qquad \text{Richter scale formula}$$

$$= \log\left(\frac{300}{2.5}\right) + 4.2 \quad \text{Let } a = 300, T = 2.5, \text{ and } B = 4.2.$$

$$= \log(120) + 4.2$$

$$\approx 2.1 + 4.2 \qquad \text{Approximate log 120 by 2.1.}$$

$$= 6.3$$

This earthquake had a magnitude of 6.3 on the Richter scale.

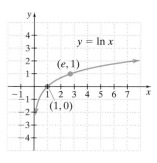

$y = \ln x$

$(e, 1)$

$(1, 0)$

**PRACTICE**

**4**  Find an earthquake's magnitude on the Richter scale if a recording station measures an amplitude of 450 micrometers and 4.2 seconds between waves with $B = 3.6$. Approximate the solution to the nearest tenth.

**OBJECTIVE 3 ▶ Approximating natural logarithms. Natural logarithms** are also frequently used, especially to describe natural events; hence the label "natural logarithm." Natural logarithms are logarithms to the base $e$, which is a constant approximately equal to 2.7183. The number $e$ is an irrational number, as is $\pi$. The notation $\log_e x$ is usually abbreviated to $\ln x$. (The abbreviation ln is read "el en.")

---

**Natural Logarithms**

$$\ln x \text{ means } \log_e x$$

---

The graph of $y = \ln x$ is shown to the left.

**EXAMPLE 5**  Use a calculator to approximate ln 8 to four decimal places.

*Solution*  Press the following sequence of keys.

8  LN    or    LN  8  ENTER

To four decimal places,

$$\ln 8 \approx 2.0794$$

**PRACTICE**

**5**  Use a calculator to approximate ln 13 to four decimal places.

---

help

**OBJECTIVE 4 ▶ Evaluating natural logarithms of powers of e.**  As a result of the property $\log_b b^x = x$, we know that $\log_e e^x = x$, or $\mathbf{\ln e^x = x.}$
     Since $\ln e^x = x$, then $\ln e^5 = 5$, $\ln e^{22} = 22$ and so on. Also,

$$\ln e^1 = 1 \text{ or simply } \ln e = 1.$$

That is why the graph of $y = \ln x$ shown on the previous page passes through $(e, 1)$.
     If $x = e$, then $y = \ln e = 1$, thus the ordered pair is $(e, 1)$.

**EXAMPLE 6**  Find the exact value of each natural logarithm.

**a.** $\ln e^3$             **b.** $\ln \sqrt[5]{e}$

*Solution*

**a.** $\ln e^3 = 3$          **b.** $\ln \sqrt[5]{e} = \ln e^{1/5} = \dfrac{1}{5}$

**PRACTICE**

**6**  Find the exact value of each natural logarithm.

**a.** $\ln e^4$       **b.** $\ln \sqrt[3]{e}$

---

**EXAMPLE 7**  Solve $\ln 3x = 5$. Give an exact solution, and then approximate the solution to four decimal places.

*Solution*  Remember that the base of a natural logarithm is understood to be $e$.

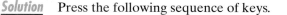

$$\ln 3x = 5$$
$$e^5 = 3x \quad \text{Write with exponential notation.}$$
$$\frac{e^5}{3} = x \quad \text{Solve for } x.$$

▶ **Helpful Hint**
The understood base is $e$.

The exact solution is $\dfrac{e^5}{3}$. To four decimal places,

$$x \approx 49.4711.$$   □

**PRACTICE**

**7**   Solve $\ln 5x = 8$. Give an exact solution, and then approximate the solution to four decimal places.

---

Recall from Section 12.3 the formula $A = P\left(1 + \dfrac{r}{n}\right)^{nt}$ for compound interest, where $n$ represents the number of compoundings per year. When interest is compounded continuously, the formula $A = Pe^{rt}$ is used, where $r$ is the annual interest rate and interest is compounded continuously for $t$ years.

**EXAMPLE 8**   **Finding Final Loan Payment**

Find the amount owed at the end of 5 years if \$1600 is loaned at a rate of 9% compounded continuously.

*Solution*   Use the formula $A = Pe^{rt}$, where

$$P = \$1600 \text{ (the size of the loan)}$$
$$r = 9\% = 0.09 \text{ (the rate of interest)}$$
$$t = 5 \text{ (the 5-year duration of the loan)}$$
$$A = Pe^{rt}$$
$$= 1600e^{0.09(5)} \quad \text{Substitute in known values.}$$
$$= 1600e^{0.45}$$

Now we can use a calculator to approximate the solution.

$$A \approx 2509.30$$

The total amount of money owed is \$2509.30.   □

**PRACTICE**

**8**   Find the amount owed at the end of 4 years if \$2400 is borrowed at a rate of 6% compounded continuously.

---

**OBJECTIVE 5 ▶ Using the change of base formula.** Calculators are handy tools for approximating natural and common logarithms. Unfortunately, some calculators cannot be used to approximate logarithms to bases other than $e$ or 10—at least not directly. In such cases, we use the change of base formula.

> **Change of Base**
> If $a$, $b$, and $c$ are positive real numbers and neither $b$ nor $c$ is 1, then
>
> $$\log_b a = \frac{\log_c a}{\log_c b}$$

**EXAMPLE 9**   Approximate $\log_5 3$ to four decimal places.

*Solution*   Use the change of base property to write $\log_5 3$ as a quotient of logarithms to base 10.

$$\log_5 3 = \frac{\log 3}{\log 5} \qquad \text{Use the change of base property. In the change of base property, we let } a = 3, b = 5, \text{ and } c = 10.$$

$$\approx \frac{0.4771213}{0.69897} \qquad \text{Approximate logarithms by calculator.}$$

$$\approx 0.6826062 \quad \text{Simplify by calculator.}$$

To four decimal places, $\log_5 3 \approx 0.6826$.   □

Approximate $\log_8 5$ to four decimal places.

**Answer to Concept Check:**

$$f(x) = \frac{\log x}{\log 5}$$

**Concept Check** ✓

If a graphing calculator cannot directly evaluate logarithms to base 5, describe how you could use the graphing calculator to graph the function $f(x) = \log_5 x$.

## VOCABULARY & READINESS CHECK

*Use the choices to fill in each blank.*

**1.** The base of $\log 7$ is ___.
   **a.** $e$   **b.** 7   **c.** 10   **d.** no answer

**2.** The base of $\ln 7$ is ___.
   **a.** $e$   **b.** 7   **c.** 10   **d.** no answer

**3.** $\log_{10} 10^7 =$ ___.
   **a.** $e$   **b.** 7   **c.** 10   **d.** no answer

**4.** $\log_7 1 =$ ___.
   **a.** $e$   **b.** 7   **c.** 10   **d.** 0

**5.** $\log_e e^5 =$ ___.
   **a.** $e$   **b.** 5   **c.** 0   **d.** 1

**6.** Study exercise 5 to the left. Then answer: $\ln e^5 =$ ___.
   **a.** $e$   **b.** 5   **c.** 0   **d.** 1

**7.** $\log_2 7 =$ _____   (There may be more than one answer.)

   **a.** $\dfrac{\log 7}{\log 2}$   **b.** $\dfrac{\ln 7}{\ln 2}$   **c.** $\dfrac{\log 2}{\log 7}$   **d.** $\log \dfrac{7}{2}$

## 12.6 EXERCISE SET

*MyMathLab*  PRACTICE  WATCH  DOWNLOAD  READ  REVIEW

### MIXED PRACTICE

*Use a calculator to approximate each logarithm to four decimal places. See Examples 1 and 5.*

**1.** $\log 8$

**2.** $\log 6$

**3.** $\log 2.31$

**4.** $\log 4.86$

**5.** $\ln 2$

**6.** $\ln 3$

**7.** $\ln 0.0716$

**8.** $\ln 0.0032$

**9.** $\log 12.6$

**10.** $\log 25.9$

**11.** $\ln 5$

**12.** $\ln 7$

**13.** $\log 41.5$

**14.** $\ln 41.5$

**15.** Use a calculator and try to approximate $\log 0$. Describe what happens and explain why.

**16.** Use a calculator and try to approximate $\ln 0$. Describe what happens and explain why.

### MIXED PRACTICE

*Find the exact value. See Examples 2 and 6.*

**17.** $\log 100$

**18.** $\log 10{,}000$

**19.** $\log\left(\dfrac{1}{1000}\right)$

**20.** $\log\left(\dfrac{1}{100}\right)$

**21.** $\ln e^2$

**22.** $\ln e^4$

**23.** $\ln \sqrt[4]{e}$

**24.** $\ln \sqrt[5]{e}$

**25.** $\log 10^3$

**26.** $\log 10^7$

**27.** $\ln e^{-7}$

**28.** $\ln e^{-5}$

**29.** $\log 0.0001$

**30.** $\log 0.001$

**31.** $\ln \sqrt{e}$

**32.** $\log \sqrt{10}$

*Solve each equation for x. Give an exact solution and a four-decimal-place approximation. See Examples 3 and 7.*

**33.** $\ln 2x = 7$

**34.** $\ln 5x = 9$

**35.** $\log x = 1.3$

**36.** $\log x = 2.1$

**37.** $\log 2x = 1.1$

**38.** $\log 3x = 1.3$

**39.** $\ln x = 1.4$

**40.** $\ln x = 2.1$

**41.** $\ln(3x - 4) = 2.3$

**42.** $\ln(2x + 5) = 3.4$

**43.** $\log x = 2.3$

**44.** $\log x = 3.1$

**45.** $\ln x = -2.3$

**46.** $\ln x = -3.7$

**47.** $\log(2x + 1) = -0.5$

**48.** $\log(3x - 2) = -0.8$

**49.** $\ln 4x = 0.18$

**50.** $\ln 3x = 0.76$

*Approximate each logarithm to four decimal places. See Example 9.*

**51.** $\log_2 3$                  **52.** $\log_3 2$

**53.** $\log_{1/2} 5$                **54.** $\log_{1/3} 2$

**55.** $\log_4 9$                  **56.** $\log_9 4$

**57.** $\log_3 \dfrac{1}{6}$              **58.** $\log_6 \dfrac{2}{3}$

**59.** $\log_8 6$                  **60.** $\log_6 8$

*Use the formula $R = \log\left(\dfrac{a}{T}\right) + B$ to find the intensity $R$ on the Richter scale of the earthquakes that fit the descriptions given. Round answers to one decimal place. See Example 4.*

**61.** Amplitude $a$ is 200 micrometers, time $T$ between waves is 1.6 seconds, and $B$ is 2.1.

**62.** Amplitude $a$ is 150 micrometers, time $T$ between waves is 3.6 seconds, and $B$ is 1.9.

**63.** Amplitude $a$ is 400 micrometers, time $T$ between waves is 2.6 seconds, and $B$ is 3.1.

**64.** Amplitude $a$ is 450 micrometers, time $T$ between waves is 4.2 seconds, and $B$ is 2.7.

*Use the formula $A = Pe^{rt}$ to solve. See Example 8.*

**65.** Find how much money Dana Jones has after 12 years if $1400 is invested at 8% interest compounded continuously.

**66.** Determine the size of an account in which $3500 earns 6% interest compounded continuously for 1 year.

**67.** Find the amount of money Barbara Mack owes at the end of 4 years if 6% interest is compounded continuously on her $2000 debt.

**68.** Find the amount of money for which a $2500 certificate of deposit is redeemable if it has been paying 10% interest compounded continuously for 3 years.

**REVIEW AND PREVIEW**

*Solve each equation for x. See Sections 2.3 and 6.6.*

**69.** $6x - 3(2 - 5x) = 6$       **70.** $2x + 3 = 5 - 2(3x - 1)$

**71.** $2x + 3y = 6x$            **72.** $4x - 8y = 10x$

**73.** $x^2 + 7x = -6$           **74.** $x^2 + 4x = 12$

*Solve each system of equations. See Sections 4.1 through 4.3.*

**75.** $\begin{cases} x + 2y = -4 \\ 3x - y = 9 \end{cases}$     **76.** $\begin{cases} 5x + y = 5 \\ -3x - 2y = -10 \end{cases}$

**CONCEPT EXTENSIONS**

**77.** Without using a calculator, explain which of $\log 50$ or $\ln 50$ must be larger and why.

**78.** Without using a calculator, explain which of $\log 50^{-1}$ or $\ln 50^{-1}$ must be larger and why.

*Graph each function by finding ordered pair solutions, plotting the solutions, and then drawing a smooth curve through the plotted points.*

**79.** $f(x) = e^x$              **80.** $f(x) = e^{2x}$

**81.** $f(x) = e^{-3x}$           **82.** $f(x) = e^{-x}$

**83.** $f(x) = e^x + 2$          **84.** $f(x) = e^x - 3$

**85.** $f(x) = e^{x-1}$           **86.** $f(x) = e^{x+4}$

**87.** $f(x) = 3e^x$             **88.** $f(x) = -2e^x$

**89.** $f(x) = \ln x$            **90.** $f(x) = \log x$

**91.** $f(x) = -2\log x$          **92.** $f(x) = 3\ln x$

**93.** $f(x) = \log(x + 2)$        **94.** $f(x) = \log(x - 2)$

**95.** $f(x) = \ln x - 3$         **96.** $f(x) = \ln x + 3$

**97.** Graph $f(x) = e^x$ (Exercise 79), $f(x) = e^x + 2$ (Exercise 83), and $f(x) = e^x - 3$ (Exercise 84) on the same screen. Discuss any trends shown on the graphs.

**98.** Graph $f(x) = \ln x$ (Exercise 89), $f(x) = \ln x - 3$ (Exercise 95), and $f(x) = \ln x + 3$ (Exercise 96). Discuss any trends shown on the graphs.

---

### 📖 STUDY SKILLS BUILDER

**What to Do the Day of an Exam**

On the day of an exam, don't forget to try the following:

- Allow yourself plenty of time to arrive.
- Read the directions on the test carefully.
- Read each problem carefully as you take your test. Make sure that you answer the question asked.
- Watch your time and pace yourself so that you may attempt each problem on your test.
- Check your work and answers.
- ***Do not turn your test in early.*** If you have extra time, spend it double-checking your work.

Good luck!

*Answer the following questions based on your most recent mathematics exam, whenever that was.*

**1.** How soon before class did you arrive?

**2.** Did you read the directions on the test carefully?

**3.** Did you make sure you answered the question asked for each problem on the exam?

**4.** Were you able to attempt each problem on your exam?

**5.** If your answer to Question 4 is no, list reasons why.

**6.** Did you have extra time on your exam?

**7.** If your answer to Question 6 is yes, describe how you spent that extra time.

## 12.7 EXPONENTIAL AND LOGARITHMIC EQUATIONS AND APPLICATIONS

**OBJECTIVES**

1 Solve exponential equations.

2 Solve logarithmic equations.

3 Solve problems that can be modeled by exponential and logarithmic equations.

**OBJECTIVE 1 ▶ Solving exponential equations.** In Section 12.3 we solved exponential equations such as $2^x = 16$ by writing 16 as a power of 2 and applying the uniqueness of $b^x$.

$$2^x = 16$$
$$2^x = 2^4 \quad \text{Write 16 as } 2^4.$$
$$x = 4 \quad \text{Use the uniqueness of } b^x.$$

Solving the equation in this manner is possible since 16 is a power of 2. If solving an equation such as $2^x = a$ *number*, where the number is not a power of 2, we use logarithms. For example, to solve an equation such as $3^x = 7$, we use the fact that $f(x) = \log_b x$ is a one-to-one function. Another way of stating this fact is as a property of equality.

---

**Logarithm Property of Equality**

Let $a$, $b$, and $c$ be real numbers such that $\log_b a$ and $\log_b c$ are real numbers and $b$ is not 1. Then

$$\log_b a = \log_b c \text{ is equivalent to } a = c$$

---

**EXAMPLE 1**  Solve: $3^x = 7$.

*Solution*  To solve, we use the logarithm property of equality and take the logarithm of both sides. For this example, we use the common logarithm.

$$3^x = 7$$
$$\log 3^x = \log 7 \quad \text{Take the common log of both sides.}$$
$$x \log 3 = \log 7 \quad \text{Apply the power property of logarithms.}$$
$$x = \frac{\log 7}{\log 3} \quad \text{Divide both sides by } \log 3.$$

The exact solution is $\dfrac{\log 7}{\log 3}$. If a decimal approximation is preferred,

$$\frac{\log 7}{\log 3} \approx \frac{0.845098}{0.4771213} \approx 1.7712 \text{ to four decimal places.}$$

The solution is $\dfrac{\log 7}{\log 3}$, or *approximately* 1.7712.                □

**PRACTICE**
**1**  Solve: $5^x = 9$.

---

**OBJECTIVE 2 ▶ Solving logarithmic equations.** By applying the appropriate properties of logarithms, we can solve a broad variety of logarithmic equations.

**EXAMPLE 2**  Solve: $\log_4(x - 2) = 2$.

*Solution*  Notice that $x - 2$ must be positive, so $x$ must be greater than 2. With this in mind, we first write the equation with exponential notation.

$$\log_4(x - 2) = 2$$
$$4^2 = x - 2$$
$$16 = x - 2$$
$$18 = x \quad \text{Add 2 to both sides.}$$

**Check:** To check, we replace $x$ with 18 in the original equation.

$$\log_4(x - 2) = 2$$
$$\log_4(18 - 2) \overset{?}{=} 2 \qquad \text{Let } x = 18.$$
$$\log_4 16 \overset{?}{=} 2$$
$$4^2 = 16 \qquad \text{True}$$

The solution is 18. □

**PRACTICE**
**2** Solve: $\log_2(x - 1) = 5$.

**EXAMPLE 3** Solve: $\log_2 x + \log_2(x - 1) = 1$.

_Solution_ Notice that $x - 1$ must be positive, so $x$ must be greater than 1. We use the product property on the left side of the equation.

$$\log_2 x + \log_2(x - 1) = 1$$
$$\log_2 x(x - 1) = 1 \qquad \text{Apply the product property.}$$
$$\log_2(x^2 - x) = 1$$

Next we write the equation with exponential notation and solve for $x$.

$$2^1 = x^2 - x$$
$$0 = x^2 - x - 2 \qquad \text{Subtract 2 from both sides.}$$
$$0 = (x - 2)(x + 1) \qquad \text{Factor.}$$
$$0 = x - 2 \quad \text{or} \quad 0 = x + 1 \qquad \text{Set each factor equal to 0.}$$
$$2 = x \qquad \qquad -1 = x$$

Recall that $-1$ cannot be a solution because $x$ must be greater than 1. If we forgot this, we would still reject $-1$ after checking. To see this, we replace $x$ with $-1$ in the original equation.

$$\log_2 x + \log_2(x - 1) = 1$$
$$\log_2(-1) + \log_2(-1 - 1) \overset{?}{=} 1 \qquad \text{Let } x = -1.$$

Because the logarithm of a negative number is undefined, $-1$ is rejected. Check to see that the solution is 2. □

**PRACTICE**
**3** Solve: $\log_5 x + \log_5(x + 4) = 1$.

**EXAMPLE 4** Solve: $\log(x + 2) - \log x = 2$.

We use the quotient property of logarithms on the left side of the equation.

_Solution_ $\log(x + 2) - \log x = 2$

$$\log\frac{x + 2}{x} = 2 \qquad \text{Apply the quotient property.}$$

$$10^2 = \frac{x + 2}{x} \qquad \text{Write using exponential notation.}$$

$$100 = \frac{x + 2}{x} \qquad \text{Simplify.}$$

$$100x = x + 2 \qquad \text{Multiply both sides by } x.$$

$$99x = 2 \qquad \text{Subtract } x \text{ from both sides.}$$

$$x = \frac{2}{99} \qquad \text{Divide both sides by 99.}$$

Verify that the solution is $\frac{2}{99}$. □

**PRACTICE**
**4** Solve: $\log(x + 3) - \log x = 1$.

**OBJECTIVE 3 ▶ Solving problems modeled by exponential and logarithmic equations.**
Logarithmic and exponential functions are used in a variety of scientific, technical, and business settings. A few examples follow.

**EXAMPLE 5** **Estimating Population Size**

The population size $y$ of a community of lemmings varies according to the relationship $y = y_0 e^{0.15t}$. In this formula, $t$ is time in months, and $y_0$ is the initial population at time 0. Estimate the population after 6 months if there were originally 5000 lemmings.

**Solution** We substitute 5000 for $y_0$ and 6 for $t$.

$$
\begin{aligned}
y &= y_0 e^{0.15t} \\
&= 5000 e^{0.15(6)} \quad \text{Let } t = 6 \text{ and } y_0 = 5000. \\
&= 5000 e^{0.9} \quad \text{Multiply.}
\end{aligned}
$$

Using a calculator, we find that $y \approx 12{,}298.016$. In 6 months the population will be approximately 12,300 lemmings. □

**PRACTICE**
**5** The population size $y$ of a group of rabbits varies according to the relationship $y = y_0 e^{0.916t}$. In this formula, $t$ is time in years and $y_0$ is the initial population at time $t = 0$. Estimate the population in three years if there were originally 60 rabbits.

**EXAMPLE 6** **Doubling an Investment**

How long does it take an investment of \$2000 to double if it is invested at 5% interest compounded quarterly? The necessary formula is $A = P\left(1 + \dfrac{r}{n}\right)^{nt}$, where $A$ is the accrued (or owed) amount, $P$ is the principal invested, $r$ is the annual rate of interest, $n$ is the number of compounding periods per year, and $t$ is the number of years.

**Solution** We are given that $P = \$2000$ and $r = 5\% = 0.05$. Compounding quarterly means 4 times a year, so $n = 4$. The investment is to double, so $A$ must be \$4000. Substitute these values and solve for $t$.

$$
\begin{aligned}
A &= P\left(1 + \frac{r}{n}\right)^{nt} \\
4000 &= 2000\left(1 + \frac{0.05}{4}\right)^{4t} \quad \text{Substitute in known values.} \\
4000 &= 2000(1.0125)^{4t} \quad \text{Simplify } 1 + \frac{0.05}{4}. \\
2 &= (1.0125)^{4t} \quad \text{Divide both sides by 2000.} \\
\log 2 &= \log 1.0125^{4t} \quad \text{Take the logarithm of both sides.} \\
\log 2 &= 4t(\log 1.0125) \quad \text{Apply the power property.}
\end{aligned}
$$

$$\frac{\log 2}{4 \log 1.0125} = t$$     Divide both sides by 4 log 1.0125.

$$13.949408 \approx t$$     Approximate by calculator.

Thus, it takes nearly 14 years for the money to double in value.     □

**PRACTICE**
**6**     How long does it take for an investment of $3000 to double if it is invested at 7% interest compounded monthly? Round to the nearest whole year.

---

## Graphing Calculator Explorations

Use a graphing calculator to find how long it takes an investment of $1500 to triple if it is invested at 8% interest compounded monthly.

First, let $P = \$1500$, $r = 0.08$, and $n = 12$ (for 12 months) in the formula

$$A = P\left(1 + \frac{r}{n}\right)^{nt}$$

Notice that when the investment has tripled, the accrued amount $A$ is $4500. Thus,

$$4500 = 1500\left(1 + \frac{0.08}{12}\right)^{12t}$$

Determine an appropriate viewing window and enter and graph the equations

$$Y_1 = 1500\left(1 + \frac{0.08}{12}\right)^{12x}$$

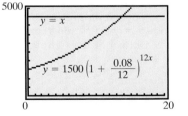

and

$$Y_2 = 4500$$

The point of intersection of the two curves is the solution. The $x$-coordinate tells how long it takes for the investment to triple.

Use a TRACE feature or an INTERSECT feature to approximate the coordinates of the point of intersection of the two curves. It takes approximately 13.78 years, or 13 years and 9 months, for the investment to triple in value to $4500.

*Use this graphical solution method to solve each problem. Round each answer to the nearest hundredth.*

**1.** Find how long it takes an investment of $5000 to grow to $6000 if it is invested at 5% interest compounded quarterly.

**2.** Find how long it takes an investment of $1000 to double if it is invested at 4.5% interest compounded daily. (Use 365 days in a year.)

**3.** Find how long it takes an investment of $10,000 to quadruple if it is invested at 6% interest compounded monthly.

**4.** Find how long it takes $500 to grow to $800 if it is invested at 4% interest compounded semiannually.

**12.7 | EXERCISE SET**

*Solve each equation. Give an exact solution, and also approximate the solution to four decimal places. See Example 1.*

**1.** $3^x = 6$

**2.** $4^x = 7$

**3.** $3^{2x} = 3.8$

**4.** $5^{3x} = 5.6$

**5.** $2^{x-3} = 5$

**6.** $8^{x-2} = 12$

**7.** $9^x = 5$

**8.** $3^x = 11$

**9.** $4^{x+7} = 3$

**10.** $6^{x+3} = 2$

**MIXED PRACTICE**

*Solve each equation. See Examples 1 through 4.*

**11.** $7^{3x-4} = 11$

**12.** $5^{2x-6} = 12$

**13.** $e^{6x} = 5$

**14.** $e^{2x} = 8$

**15.** $\log_2(x + 5) = 4$

**16.** $\log_6(x^2 - x) = 1$

**17.** $\log_3 x^2 = 4$

**18.** $\log_2 x^2 = 6$

**19.** $\log_4 2 + \log_4 x = 0$

**20.** $\log_3 5 + \log_3 x = 1$

**21.** $\log_2 6 - \log_2 x = 3$

**22.** $\log_4 10 - \log_4 x = 2$

**23.** $\log_4 x + \log_4(x + 6) = 2$

**24.** $\log_3 x + \log_3(x + 6) = 3$

**25.** $\log_5(x + 3) - \log_5 x = 2$

**26.** $\log_6(x + 2) - \log_6 x = 2$

**27.** $\log_3(x - 2) = 2$

**28.** $\log_2(x - 5) = 3$

**29.** $\log_4(x^2 - 3x) = 1$

**30.** $\log_8(x^2 - 2x) = 1$

**31.** $\ln 5 + \ln x = 0$

**32.** $\ln 3 + \ln (x - 1) = 0$

**33.** $3 \log x - \log x^2 = 2$

**34.** $2 \log x - \log x = 3$

**35.** $\log_2 x + \log_2(x + 5) = 1$

**36.** $\log_4 x + \log_4(x + 7) = 1$

**37.** $\log_4 x - \log_4(2x - 3) = 3$

**38.** $\log_2 x - \log_2(3x + 5) = 4$

**39.** $\log_2 x + \log_2(3x + 1) = 1$

**40.** $\log_3 x + \log_3(x - 8) = 2$

*Solve. See Example 5.*

**41.** The size of the wolf population at Isle Royale National Park increases at a rate of 4.3% per year. If the size of the current population is 83 wolves, find how many there should be in 5 years. Use $y = y_0 e^{0.043t}$ and round to the nearest whole.

**42.** The number of victims of a flu epidemic is increasing at a rate of 7.5% per week. If 20,000 persons are currently infected, find in how many days we can expect 45,000 to have the flu. Use $y = y_0 e^{0.075t}$ and round to the nearest whole. (*Hint:* Don't forget to convert your answer to days.)

**43.** The size of the population of Belize is increasing at a rate of 2.3% per year. If 294,380 people lived in Belize in 2007, find how many inhabitants there will be by 2015. Round to the nearest thousand. Use $y = y_0 e^{0.023t}$. (*Source: CIA 2007 World Factbook*)

**44.** In 2007, 1730 million people were citizens of India. Find how long it will take India's population to reach a size of 2000 million (that is, 2 billion) if the population size is growing at a rate of 1.6% per year. Use $y = y_0 e^{0.016t}$ and round to the nearest tenth. (*Source:* U.S. Bureau of the Census, International Data Base)

**45.** In 2007, Germany had a population of 82,400 thousand. At that time, Germany's population was declining at a rate of 0.033% per year. If this continues, how long will it take Germany's population to reach 82,000 thousand? Use $y = y_0 e^{-0.00033t}$ and round to the nearest tenth. (*Source: CIA 2007 World Factbook*)

**46.** The population of the United States has been increasing at a rate of 0.894% per year. If there were 301,140,000 people living in the United States in 2007, how many inhabitants will there be by 2020? Use $y = y_0 e^{0.00894t}$ and round to the nearest ten-thousand. (*Source: CIA 2007 World Factbook*)

*Use the formula $A = P\left(1 + \dfrac{r}{n}\right)^{nt}$ to solve these compound interest problems. Round to the nearest tenth. See Example 6.*

**47.** Find how long it takes $600 to double if it is invested at 7% interest compounded monthly.

**48.** Find how long it takes $600 to double if it is invested at 12% interest compounded monthly.

**49.** Find how long it takes a $1200 investment to earn $200 interest if it is invested at 9% interest compounded quarterly.

**50.** Find how long it takes a $1500 investment to earn $200 interest if it is invested at 10% compounded semiannually.

**51.** Find how long it takes $1000 to double if it is invested at 8% interest compounded semiannually.

**52.** Find how long it takes $1000 to double if it is invested at 8% interest compounded monthly.

*The formula* $w = 0.00185h^{2.67}$ *is used to estimate the normal weight w of a boy h inches tall. Use this formula to solve the height-weight problems. Round to the nearest tenth.*

**53.** Find the expected weight of a boy who is 35 inches tall.

**54.** Find the expected weight of a boy who is 43 inches tall.

**55.** Find the expected height of a boy who weighs 85 pounds.

**56.** Find the expected height of a boy who weighs 140 pounds.

*The formula* $P = 14.7e^{-0.21x}$ *gives the average atmospheric pressure P, in pounds per square inch, at an altitude x, in miles above sea level. Use this formula to solve these pressure problems. Round answers to the nearest tenth.*

**57.** Find the average atmospheric pressure of Denver, which is 1 mile above sea level.

**58.** Find the average atmospheric pressure of Pikes Peak, which is 2.7 miles above sea level.

**59.** Find the elevation of a Delta jet if the atmospheric pressure outside the jet is 7.5 lb/in.$^2$.

**60.** Find the elevation of a remote Himalayan peak if the atmospheric pressure atop the peak is 6.5 lb/in.$^2$.

*Psychologists call the graph of the formula* $t = \dfrac{1}{c}\ln\left(\dfrac{A}{A - N}\right)$ *the learning curve, since the formula relates time t passed, in weeks, to a measure N of learning achieved, to a measure A of maximum learning possible, and to a measure c of an individual's learning style. Round to the nearest week.*

**61.** Norman is learning to type. If he wants to type at a rate of 50 words per minute (*N* is 50) and his expected maximum rate is 75 words per minute (*A* is 75), find how many weeks it should take him to achieve his goal. Assume that *c* is 0.09.

**62.** An experiment with teaching chimpanzees sign language shows that a typical chimp can master a maximum of 65 signs. Find how many weeks it should take a chimpanzee to master 30 signs if *c* is 0.03.

**63.** Janine is working on her dictation skills. She wants to take dictation at a rate of 150 words per minute and believes that the maximum rate she can hope for is 210 words per minute. Find how many weeks it should take her to achieve the 150 words per minute level if *c* is 0.07.

**64.** A psychologist is measuring human capability to memorize nonsense syllables. Find how many weeks it should take a subject to learn 15 nonsense syllables if the maximum possible to learn is 24 syllables and *c* is 0.17.

**REVIEW AND PREVIEW**

*If* $x = -2, y = 0,$ *and* $z = 3,$ *find the value of each expression. See Section 1.7.*

**65.** $\dfrac{x^2 - y + 2z}{3x}$

**66.** $\dfrac{x^3 - 2y + z}{2z}$

**67.** $\dfrac{3z - 4x + y}{x + 2z}$

**68.** $\dfrac{4y - 3x + z}{2x + y}$

*Find the inverse function of each one-to-one function. See Section 12.2.*

**69.** $f(x) = 5x + 2$

**70.** $f(x) = \dfrac{x - 3}{4}$

**CONCEPT EXTENSIONS**

*The formula* $y = y_0e^{kt}$ *gives the population size y of a population that experiences an annual rate of population growth k (given as a decimal). In this formula, t is time in years and $y_0$ is the initial population at time 0. Use this formula to solve Exercises 71 and 72.*

**71.** In 2000, the population of Arizona was 5,130,632. By 2006, the population had grown to 6,123,106. Find the annual rate of population growth over this period. Round your answer to the nearest tenth of a percent. (*Source:* State of Arizona)

**72.** In 2000, the population of Nevada was 2,018,456. By 2006, the population had grown to 2,495,529. Find the annual rate of population growth over this period. Round your answer to the nearest tenth of a percent. (*Source:* State of Nevada)

**73.** When solving a logarithmic equation, explain why you must check possible solutions in the original equation.

**74.** Solve $5^x = 9$ by taking the common logarithm of both sides of the equation. Next, solve this equation by taking the natural logarithm of both sides. Compare your solutions. Are they the same? Why or why not?

*Use a graphing calculator to solve each equation. For example, to solve Exercise 75, let* $Y_1 = e^{0.3x}$ *and* $Y_2 = 8,$ *and graph the equations. The x-value of the point of intersection is the solution. Round all solutions to two decimal places.*

**75.** $e^{0.3x} = 8$

**76.** $10^{0.5x} = 7$

**77.** $2\log(-5.6x + 1.3) + x + 1 = 0$

**78.** $\ln(1.3x - 2.1) + 3.5x - 5 = 0$

**79.** Check Exercise 11. Graph $7^{3x-4} - 11 = 0$

**80.** Check Exercise 12. Graph $5^{2x-6} - 12 = 0$

**81.** Check Exercise 31.

**82.** Check Exercise 32.

---

## THE BIGGER PICTURE   SIMPLIFYING EXPRESSIONS AND SOLVING EQUATIONS AND INEQUALITIES

We now complete the outline from Sections 1.7, 2.3, 2.8, 5.6, 6.6, 7.4, 7.6, 9.1, 9.3, 10.5, 10.6 and 11.4. Although suggestions are given, as usual, this outline should be in your own words. Try to include "how to recognize" and "how to begin to solve" in your outline. Once you complete this new portion, complete the exercises.

**I.** Simplifying Expressions (See Section 10.5, page 614)

**II.** Solving Equations

    **A.** Linear Equations (Section 2.3)

    **B.** Quadratic and Higher Degree Equations (Sections 6.6, 11.1, 11.2, 11.3)

    **C.** Equations with Rational Expressions (Section 7.5)

    **D.** Proportions (Section 7.6)

    **E.** Absolute Value Equations (Section 9.2)

    **F.** Equations with Radicals (Section 10.6)

    **G. Exponential Equations—equations with variables in the exponent.**

**1.** If we can write both expressions with the same base, then set the exponents equal to each other and solve

$$9^x = 27^{x+1}$$
$$(3^2)^x = (3^3)^{x+1}$$
$$3^{2x} = 3^{3x+3}$$
$$2x = 3x + 3$$
$$-3 = x$$

**2.** If we can't write both expressions with the same base, then solve using logarithms

$$5^x = 7$$
$$\log 5^x = \log 7$$
$$x \log 5 = \log 7$$
$$x = \frac{\log 7}{\log 5}$$
$$x \approx 1.2091$$

    **H. Logarithmic Equations—equations with logarithms of variable expressions**

$$\log 7 + \log(x + 3) = 2$$   Write equation so that single logarithm on
$$\log 7(x + 3) = 2$$   one side and constant on the other side.
$$10^2 = 7(x + 3)$$   Use definition of logarithm.
$$100 = 7x + 21$$   Multiply.
$$79 = 7x$$
$$\frac{79}{7} = x$$   Solve.

**III.** Solving Inequalities

    **A.** Linear Inequalities (Section 2.8)

    **B.** Compound Inequalities (Section 9.1)

    **C.** Absolute Value Inequalities (Section 9.3)

    **D.** Nonlinear Inequalities (Section 11.4)

        **1.** Polynomial inequalities

        **2.** Rational inequalities

*Solve. Write solutions to inequalities in interval notation.*

**1.** $8^x = 2^{x-3}$

**2.** $11^x = 5$

**3.** $-7x + 3 \le -5x + 13$

**4.** $-7 \le 3x + 6 \le 0$

**5.** $|5y + 3| < 3$

**6.** $(x - 6)(5x + 1) = 0$

**7.** $\log_{13} 8 + \log_{13}(x - 1) = 1$

**8.** $\left| \dfrac{3x - 1}{4} \right| = 2$

**9.** $|7x + 1| > -2$

**10.** $x^2 = 4$

**11.** $(x + 5)^2 = 3$

**12.** $\log_7(4x^2 - 27x) = 1$

---

# CHAPTER 12 GROUP ACTIVITY

## Sound Intensity

The decibel (dB) measures sound intensity, or the relative loudness or strength of a sound. One decibel is the smallest difference in sound levels that is detectable by humans. The decibel is a logarithmic unit. This means that for approximately every 3-decibel increase in sound intensity, the relative loudness of the sound is doubled. For example, a 35 dB sound is twice as loud as a 32 dB sound.

In the modern world, noise pollution has increasingly become a concern. Sustained exposure to high sound intensities can lead to hearing loss. Regular exposure to 90 dB sounds can eventually lead to loss of hearing. Sounds of 130 dB and more can cause permanent loss of hearing instantaneously.

The relative loudness of a sound $D$ in decibels is given by the equation

$$D = 10\log_{10}\frac{I}{10^{-16}}$$

where $I$ is the intensity of a sound given in watts per square centimeter. Some sound intensities of common noises are listed in the table in order of increasing sound intensity.

## Group Activity

1. Work together to create a table of the relative loudness (in decibels) of the sounds listed in the table.

2. Research the loudness of other common noises. Add these sounds and their decibel levels to your table. Be sure to list the sounds in order of increasing sound intensity.

| *Some Sound Intensities of Common Noises* | | |
|---|---|---|
| *Noise* | *Intensity* (watts/cm$^2$) | *Decibels* |
| Whispering | $10^{-15}$ | |
| Rustling leaves | $10^{-14.2}$ | |
| Normal conversation | $10^{-13}$ | |
| Background noise in a quiet residence | $10^{-12.2}$ | |
| Typewriter | $10^{-11}$ | |
| Air conditioning | $10^{-10}$ | |
| Freight train at 50 feet | $10^{-8.5}$ | |
| Vacuum cleaner | $10^{-8}$ | |
| Nearby thunder | $10^{-7}$ | |
| Air hammer | $10^{-6.5}$ | |
| Jet plane at takeoff | $10^{-6}$ | |
| Threshold of pain | $10^{-4}$ | |

---

### 📖 STUDY SKILLS BUILDER

#### Are You Prepared for a Test on Chapter 12?

Below I have listed some common trouble areas for students in Chapter 12. After studying for your test—but before taking your test—read these.

- Don't forget how to find the composition of two functions.

  If $f(x) = x^2 + 5$ and $g(x) = 3x$, then

  $$(f \circ g)(x) = f[g(x)] = f(3x)$$
  $$= (3x)^2 + 5 = 9x^2 + 5$$
  $$(g \circ f)(x) = g[f(x)] = g(x^2 + 5)$$
  $$= 3(x^2 + 5) = 3x^2 + 15$$

- Don't forget that $f^{-1}$ is a special notation used to denote the inverse of a function.

  Let's find the inverse of the one-to-one function $f(x) = 3x - 5$.

$$
\begin{aligned}
f(x) &= 3x - 5 \\
y &= 3x - 5 \quad \text{Replace } f(x) \text{ with } y. \\
x &= 3y - 5 \quad \text{Interchange } x \text{ and } y. \\
x + 5 &= 3y \\
\frac{x + 5}{3} &= y \quad\quad \text{Solve for } y. \\
f^{-1}(x) &= \frac{x + 5}{3} \quad \text{Replace } y \text{ with } f^{-1}(x).
\end{aligned}
$$

- Don't forget that $y = \log_b x$ means $b^y = x$.
  Thus, $3 = \log_5 125$ means $5^3 = 125$.

- Remember rules for logarithms.
  $$\log_b 3x = \log_b 3 + \log_b x$$
  $\log_b(3 + x)$ cannot be simplified in the same manner.

Remember: This is simply a checklist of common trouble areas. For a review of Chapter 12, see the Highlights and Chapter Review at the end of this chapter.

---

## CHAPTER 12 VOCABULARY CHECK

*Fill in each blank with one of the words or phrases listed below.*

| inverse | common | composition | symmetric | exponential |
|---|---|---|---|---|
| vertical | logarithmic | natural | horizontal | |

1. For each one-to-one function, we can find its _____ function by switching the coordinates of the ordered pairs of the function.

2. The _____ of functions $f$ and $g$ is $(f \circ g)(x) = f(g(x))$.

3. A function of the form $f(x) = b^x$ is called an _____ function if $b > 0$, $b$ is not 1, and $x$ is a real number.

4. The graphs of $f$ and $f^{-1}$ are _____ about the line $y = x$.

**5.** _____ logarithms are logarithms to base $e$.

**6.** _____ logarithms are logarithms to base 10.

**7.** To see whether a graph is the graph of a one-to-one function, apply the _____ line test to see if it is a function, and then apply the _____ line test to see if it is a one-to-one function.

**8.** A _____ function is a function that can be defined by $f(x) = \log_b x$ where $x$ is a positive real number, $b$ is a constant positive real number, and $b$ is not 1.

> ▶ **Helpful Hint**
>
> Are you preparing for your test? Don't forget to take the Chapter 12 Test on page 764. Then check your answers at the back of the text and use the Chapter Test Prep Video CD to see the fully worked-out solutions to any of the exercises you want to review.

# CHAPTER 12 HIGHLIGHTS

| DEFINITIONS AND CONCEPTS | EXAMPLES |
| --- | --- |

### SECTION 12.1   THE ALGEBRA OF FUNCTIONS; COMPOSITE FUNCTIONS

***Algebra of Functions***

Sum        $(f + g)(x) = f(x) + g(x)$

Difference   $(f - g)(x) = f(x) - g(x)$

Product     $(f \cdot g)(x) = f(x) \cdot g(x)$

Quotient    $\left(\dfrac{f}{g}\right)(x) = \dfrac{f(x)}{g(x)}, g(x) \neq 0$

***Composite Functions***

The notation $(f \circ g)(x)$ means "$f$ composed with $g$."

$$(f \circ g)(x) = f(g(x))$$
$$(g \circ f)(x) = g(f(x))$$

If $f(x) = 7x$ and $g(x) = x^2 + 1$,

$$(f + g)(x) = f(x) + g(x) = 7x + x^2 + 1$$
$$(f - g)(x) = f(x) - g(x) = 7x - (x^2 + 1)$$
$$= 7x - x^2 - 1$$
$$(f \cdot g)(x) = f(x) \cdot g(x) = 7x(x^2 + 1)$$
$$= 7x^3 + 7x$$
$$\left(\frac{f}{g}\right)(x) = \frac{f(x)}{g(x)} = \frac{7x}{x^2 + 1}$$

If $f(x) = x^2 + 1$ and $g(x) = x - 5$, find $(f \circ g)(x)$.

$$(f \circ g)(x) = f(g(x))$$
$$= f(x - 5)$$
$$= (x - 5)^2 + 1$$
$$= x^2 - 10x + 26$$

### SECTION 12.2   INVERSE FUNCTIONS

If $f$ is a function, then $f$ is a **one-to-one function** only if each $y$-value (output) corresponds to only one $x$-value (input).

***Horizontal Line Test***

If every horizontal line intersects the graph of a function at most once, then the function is a one-to-one function.

Determine whether each graph is a one-to-one function.

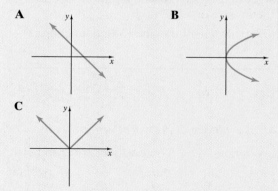

Graphs **A** and **C** pass the vertical line test, so only these are graphs of functions. Of graphs **A** and **C,** only graph **A** passes the horizontal line test, so only graph **A** is the graph of a one-to-one function.

| **DEFINITIONS AND CONCEPTS** | **EXAMPLES** |
|---|---|

The **inverse** of a one-to-one function *f* is the one-to-one function $f^{-1}$ that is the set of all ordered pairs $(b, a)$ such that $(a, b)$ belongs to *f*.

***To Find the Inverse of a One-to-One Function f(x)***

**Step 1.** Replace $f(x)$ with *y*.

**Step 2.** Interchange *x* and *y*.

**Step 3.** Solve for *y*.

**Step 4.** Replace *y* with $f^{-1}(x)$.

Find the inverse of $f(x) = 2x + 7$.

$$y = 2x + 7 \quad \text{Replace } f(x) \text{ with } y.$$

$$x = 2y + 7 \quad \text{Interchange } x \text{ and } y.$$

$$2y = x - 7 \quad \text{Solve for } y.$$

$$y = \frac{x - 7}{2}$$

$$f^{-1}(x) = \frac{x - 7}{2} \quad \text{Replace } y \text{ with } f^{-1}(x).$$

The inverse of $f(x) = 2x + 7$ is $f^{-1}(x) = \dfrac{x - 7}{2}$.

A function of the form $f(x) = b^x$ is an **exponential function,** where $b > 0, b \neq 1$, and *x* is a real number.

Graph the exponential function $y = 4^x$.

| *x* | *y* |
|---|---|
| $-2$ | $\dfrac{1}{16}$ |
| $-1$ | $\dfrac{1}{4}$ |
| $0$ | $1$ |
| $1$ | $4$ |
| $2$ | $16$ |

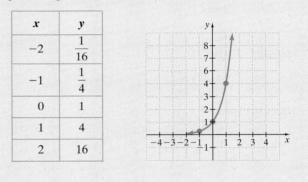

***Uniqueness of $b^x$***

If $b > 0$ and $b \neq 1$, then $b^x = b^y$ is equivalent to $x = y$.

Solve $2^{x+5} = 8$.

$$2^{x+5} = 2^3 \quad \text{Write 8 as } 2^3.$$

$$x + 5 = 3 \quad \text{Use the uniqueness of } b^x.$$

$$x = -2 \quad \text{Subtract 5 from both sides.}$$

***Logarithmic Definition***

If $b > 0$ and $b \neq 1$, then

$$y = \log_b x \quad \text{means} \quad x = b^y$$

for any positive number *x* and real number *y*.

***Properties of Logarithms***

If *b* is a real number, $b > 0$ and $b \neq 1$, then

$$\log_b 1 = 0, \quad \log_b b^x = x, \quad b^{\log_b x} = x$$

| ***Logarithmic Form*** | ***Corresponding Exponential Statement*** |
|---|---|
| $\log_5 25 = 2$ | $5^2 = 25$ |
| $\log_9 3 = \dfrac{1}{2}$ | $9^{1/2} = 3$ |

$$\log_5 1 = 0, \quad \log_7 7^2 = 2, \quad 3^{\log_3 6} = 6$$

*(continued)*

| **DEFINITIONS AND CONCEPTS** | **EXAMPLES** |
|---|---|

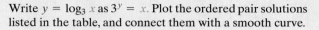

### SECTION 12.4 LOGARITHMIC FUNCTIONS (continued)

**Logarithmic Function**

If $b > 0$ and $b \neq 1$, then a **logarithmic function** is a function that can be defined as

$$f(x) = \log_b x$$

The domain of $f$ is the set of positive real numbers, and the range of $f$ is the set of real numbers.

Graph $y = \log_3 x$.

Write $y = \log_3 x$ as $3^y = x$. Plot the ordered pair solutions listed in the table, and connect them with a smooth curve.

| $x$ | $y$ |
|---|---|
| 3 | 1 |
| 1 | 0 |
| $\dfrac{1}{3}$ | $-1$ |
| $\dfrac{1}{9}$ | $-2$ |

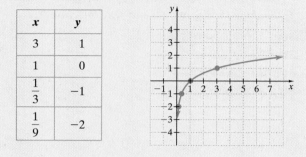

### SECTION 12.5 PROPERTIES OF LOGARITHMS

Let $x$, $y$, and $b$ be positive numbers and $b \neq 1$.

**Product Property**

$$\log_b xy = \log_b x + \log_b y$$

**Quotient Property**

$$\log_b \frac{x}{y} = \log_b x - \log_b y$$

**Power Property**

$$\log_b x^r = r \log_b x$$

Write as a single logarithm.

$2 \log_5 6 + \log_5 x - \log_5(y + 2)$

$= \log_5 6^2 + \log_5 x - \log_5(y + 2)$  Power property

$= \log_5 36 \cdot x - \log_5(y + 2)$  Product property

$= \log_5 \dfrac{36x}{y + 2}$  Quotient property

### SECTION 12.6 COMMON LOGARITHMS, NATURAL LOGARITHMS, AND CHANGE OF BASE

**Common Logarithms**

$\log x$ means $\log_{10} x$

**Natural Logarithms**

$\ln x$ means $\log_e x$

**Continuously Compounded Interest Formula**

$$A = Pe^{rt}$$

where $r$ is the annual interest rate for $P$ dollars invested for $t$ years.

$\log 5 = \log_{10} 5 \approx 0.69897$

$\ln 7 = \log_e 7 \approx 1.94591$

Find the amount in an account at the end of 3 years if $1000 is invested at an interest rate of 4% compounded continuously.

Here, $t = 3$ years, $P = \$1000$, and $r = 0.04$.

$A = Pe^{rt}$

$= 1000e^{0.04(3)}$

$\approx \$1127.50$

### SECTION 12.7 EXPONENTIAL AND LOGARITHMIC EQUATIONS AND APPLICATIONS

**Logarithm Property of Equality**

Let $\log_b a$ and $\log_b c$ be real numbers and $b \neq 1$. Then

$$\log_b a = \log_b c \text{ is equivalent to } a = c$$

Solve $2^x = 5$.

$\log 2^x = \log 5$  Log property of equality

$x \log 2 = \log 5$  Power property

$x = \dfrac{\log 5}{\log 2}$  Divide both sides by $\log 2$.

$x \approx 2.3219$  Use a calculator.

# CHAPTER 12  REVIEW

*(12.1)* If $f(x) = x - 5$ and $g(x) = 2x + 1$, find

**1.** $(f + g)(x)$

**2.** $(f - g)(x)$

**3.** $(f \cdot g)(x)$

**4.** $\left(\dfrac{g}{f}\right)(x)$

If $f(x) = x^2 - 2$, $g(x) = x + 1$, and $h(x) = x^3 - x^2$, find each composition.

**5.** $(f \circ g)(x)$

**6.** $(g \circ f)(x)$

**7.** $(h \circ g)(2)$

**8.** $(f \circ f)(x)$

**9.** $(f \circ g)(-1)$

**10.** $(h \circ h)(2)$

*(12.2)* Determine whether each function is a one-to-one function. If it is one-to-one, list the elements of its inverse.

**11.** $h = \{(-9, 14), (6, 8), (-11, 12), (15, 15)\}$

**12.** $f = \{(-5, 5), (0, 4), (13, 5), (11, -6)\}$

**13.**

| U.S. Region (Input) | West | Midwest | South | Northeast |
|---|---|---|---|---|
| Rank in Automobile Thefts (Output) | 2 | 4 | 1 | 3 |

△ **14.**

| Shape (Input) | Square | Triangle | Parallelogram | Rectangle |
|---|---|---|---|---|
| Number of Sides (Output) | 4 | 3 | 4 | 4 |

Given that $f(x) = \sqrt{x + 2}$ is a one-to-one function, find the following.

**15. a.** $f(7)$

     **b.** $f^{-1}(3)$

**16. a.** $f(-1)$

     **b.** $f^{-1}(1)$

Determine whether each function is a one-to-one function.

**17.**

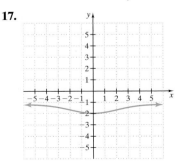

**18.**

**19.**

**20.**

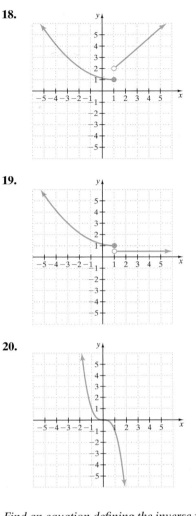

Find an equation defining the inverse function of the given one-to-one function.

**21.** $f(x) = x - 9$

**22.** $f(x) = x + 8$

**23.** $f(x) = 6x + 11$

**24.** $f(x) = 12x$

**25.** $f(x) = x^3 - 5$

**26.** $f(x) = \sqrt[3]{x + 2}$

**27.** $g(x) = \dfrac{12x - 7}{6}$

**28.** $r(x) = \dfrac{13}{2}x - 4$

On the same set of axes, graph the given one-to-one function and its inverse.

**29.** $g(x) = \sqrt{x}$

**30.** $h(x) = 5x - 5$

**31.** Find the inverse of the one-to-one function $f(x) = 2x - 3$. Then graph both $f(x)$ and $f^{-1}(x)$ with a square window.

*(12.3)* Solve each equation for x.

**32.** $4^x = 64$

**33.** $3^x = \dfrac{1}{9}$

**34.** $2^{3x} = \dfrac{1}{16}$

**35.** $5^{2x} = 125$

**36.** $9^{x+1} = 243$

**37.** $8^{3x-2} = 4$

*Graph each exponential function.*

**38.** $y = 3^x$

**39.** $y = \left(\dfrac{1}{3}\right)^x$

**40.** $y = 4 \cdot 2^x$

**41.** $y = 2^x + 4$

*Use the formula* $A = P\left(1 + \dfrac{r}{n}\right)^{nt}$ *to solve the interest problems. In this formula,*

> $A$ = amount accrued (or owed)
> $P$ = principal invested (or loaned)
> $r$ = rate of interest
> $n$ = number of compounding periods per year
> $t$ = time in years

**42.** Find the amount accrued if $1600 is invested at 9% interest compounded semiannually for 7 years.

**43.** A total of $800 is invested in a 7% certificate of deposit for which interest is compounded quarterly. Find the value that this certificate will have at the end of 5 years.

**44.** Use a graphing calculator to verify the results of Exercise 40.

*(12.4) Write each equation with logarithmic notation.*

**45.** $49 = 7^2$

**46.** $2^{-4} = \dfrac{1}{16}$

*Write each logarithmic equation with exponential notation.*

**47.** $\log_{1/2} 16 = -4$

**48.** $\log_{0.4} 0.064 = 3$

*Solve for x.*

**49.** $\log_4 x = -3$

**50.** $\log_3 x = 2$

**51.** $\log_3 1 = x$

**52.** $\log_4 64 = x$

**53.** $\log_x 64 = 2$

**54.** $\log_x 81 = 4$

**55.** $\log_4 4^5 = x$

**56.** $\log_7 7^{-2} = x$

**57.** $5^{\log_5 4} = x$

**58.** $2^{\log_2 9} = x$

**59.** $\log_2(3x - 1) = 4$

**60.** $\log_3(2x + 5) = 2$

**61.** $\log_4(x^2 - 3x) = 1$

**62.** $\log_8(x^2 + 7x) = 1$

*Graph each pair of equations on the same coordinate system.*

**63.** $y = 2^x$ and $y = \log_2 x$

**64.** $y = \left(\dfrac{1}{2}\right)^x$ and $y = \log_{1/2} x$

*(12.5) Write each of the following as single logarithms.*

**65.** $\log_3 8 + \log_3 4$

**66.** $\log_2 6 + \log_2 3$

**67.** $\log_7 15 - \log_7 20$

**68.** $\log 18 - \log 12$

**69.** $\log_{11} 8 + \log_{11} 3 - \log_{11} 6$

**70.** $\log_5 14 + \log_5 3 - \log_5 21$

**71.** $2 \log_5 x - 2 \log_5(x + 1) + \log_5 x$

**72.** $4 \log_3 x - \log_3 x + \log_3(x + 2)$

*Use properties of logarithms to write each expression as a sum or difference of multiples of logarithms.*

**73.** $\log_3 \dfrac{x^3}{x + 2}$

**74.** $\log_4 \dfrac{x + 5}{x^2}$

**75.** $\log_2 \dfrac{3x^2 y}{z}$

**76.** $\log_7 \dfrac{yz^3}{x}$

*If* $\log_b 2 = 0.36$ *and* $\log_b 5 = 0.83$, *find the following.*

**77.** $\log_b 50$

**78.** $\log_b \dfrac{4}{5}$

*(12.6) Use a calculator to approximate the logarithm to four decimal places.*

**79.** $\log 3.6$

**80.** $\log 0.15$

**81.** $\ln 1.25$

**82.** $\ln 4.63$

*Find the exact value.*

**83.** $\log 1000$

**84.** $\log \dfrac{1}{10}$

**85.** $\ln \dfrac{1}{e}$

**86.** $\ln e^4$

*Solve each equation for x.*

**87.** $\ln (2x) = 2$

**88.** $\ln (3x) = 1.6$

**89.** $\ln (2x - 3) = -1$

**90.** $\ln (3x + 1) = 2$

*Use the formula* $\ln \dfrac{I}{I_0} = -kx$ *to solve radiation problems. In this formula,*

> $x$ = depth in millimeters
> $I$ = intensity of radiation
> $I_0$ = initial intensity
> $k$ = a constant measure dependent on the material

*Round answers to two decimal places.*

**91.** Find the depth at which the intensity of the radiation passing through a lead shield is reduced to 3% of the original intensity if the value of $k$ is 2.1.

**92.** If $k$ is 3.2, find the depth at which 2% of the original radiation will penetrate.

*Approximate the logarithm to four decimal places.*

**93.** $\log_5 1.6$

**94.** $\log_3 4$

*Use the formula $A = Pe^{rt}$ to solve the interest problems in which interest is compounded continuously. In this formula,*

$A$ = amount accrued (or owed)
$P$ = principal invested (or loaned)
$r$ = rate of interest
$t$ = time in years

**95.** Bank of New York offers a 5-year, 6% continuously compounded investment option. Find the amount accrued if $1450 is invested.

**96.** Find the amount to which a $940 investment grows if it is invested at 11% compounded continuously for 3 years.

*(12.7) Solve each exponential equation for x. Give an exact solution and also approximate the solution to four decimal places.*

**97.** $3^{2x} = 7$

**98.** $6^{3x} = 5$

**99.** $3^{2x+1} = 6$

**100.** $4^{3x+2} = 9$

**101.** $5^{3x-5} = 4$

**102.** $8^{4x-2} = 3$

**103.** $2 \cdot 5^{x-1} = 1$

**104.** $3 \cdot 4^{x+5} = 2$

*Solve the equation for x.*

**105.** $\log_5 2 + \log_5 x = 2$

**106.** $\log_3 x + \log_3 10 = 2$

**107.** $\log (5x) - \log (x + 1) = 4$

**108.** $\ln (3x) - \ln (x - 3) = 2$

**109.** $\log_2 x + \log_2 2x - 3 = 1$

**110.** $-\log_6(4x + 7) + \log_6 x = 1$

*Use the formula $y = y_0 e^{kt}$ to solve the population growth problems. In this formula,*

$y$ = size of population
$y_0$ = initial count of population
$k$ = rate of growth written as a decimal
$t$ = time

*Round each answer to the nearest whole.*

**111.** The population of mallard ducks in Nova Scotia is expected to grow at a rate of 6% per week during the spring migration. If 155,000 ducks are already in Nova Scotia, find how many are expected by the end of 4 weeks.

**112.** The population of Armenia is declining at a rate of 0.129% per year. If the population in 2007 was 2,971,650, find the expected population by the year 2015. (*Source:* U.S. Bureau of the Census, International Data Base)

**113.** China is experiencing an annual growth rate of 0.606%. In 2007, the population of China was 1,321,851,888. How long will it take for the population to be 1,500,000,000? Round to the nearest tenth. (*Source: CIA 2007 World Factbook*)

**114.** In 2007, Canada had a population of 33,390,141. How long will it take for Canada to double its population if the growth rate is 0.9% annually? Round to the nearest tenth. (*Source: CIA 2007 World Factbook*)

**115.** Malaysia's population is increasing at a rate of 1.8% per year. How long will it take the 2007 population of 24,821,286 to double in size? Round to the nearest tenth. (*Source: CIA 2007 World Factbook*)

*Use the compound interest equation $A = P\left(1 + \dfrac{r}{n}\right)^{nt}$ to solve the following. (See the directions for Exercises 42 and 43 for an explanation of this formula. Round answers to the nearest tenth.)*

**116.** Find how long it will take a $5000 investment to grow to $10,000 if it is invested at 8% interest compounded quarterly.

**117.** An investment of $6000 has grown to $10,000 while the money was invested at 6% interest compounded monthly. Find how long it was invested.

*Use a graphing calculator to solve each equation. Round all solutions to two decimal places.*

**118.** $e^x = 2$          **119.** $10^{0.3x} = 7$

## MIXED REVIEW

*Solve each equation.*

**120.** $3^x = \dfrac{1}{81}$          **121.** $7^{4x} = 49$

**122.** $8^{3x-2} = 32$          **123.** $\log_4 4 = x$

**124.** $\log_3 x = 4$          **125.** $\log_5 (x^2 - 4x) = 1$

**126.** $\log_4 (3x - 1) = 2$          **127.** $\ln x = -3.2$

**128.** $\log_5 x + \log_5 10 = 2$          **129.** $\ln x - \ln 2 = 1$

**130.** $\log_6 x - \log_6 (4x + 7) = 1$

# CHAPTER 12 TEST

TEST PREP VIDEO   Remember to use the Chapter Test Prep Video CD to see the fully worked-out solutions to any of the exercises you want to review.

If $f(x) = x$ and $g(x) = 2x - 3$, find the following.

**1.** $(f \cdot g)(x)$                    **2.** $(f - g)(x)$

If $f(x) = x$, $g(x) = x - 7$, and $h(x) = x^2 - 6x + 5$, find the following.

**3.** $(f \circ h)(0)$                    **4.** $(g \circ f)(x)$

**5.** $(g \circ h)(x)$

On the same set of axes, graph the given one-to-one function and its inverse.

**6.** $f(x) = 7x - 14$

Determine whether the given graph is the graph of a one-to-one function.

**7.**                                        **8.**

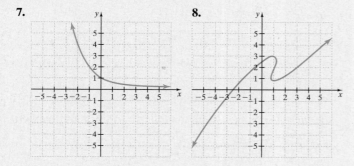

Determine whether each function is one-to-one. If it is one-to-one, find an equation or a set of ordered pairs that defines the inverse function of the given function.

**9.** $f(x) = 6 - 2x$

**10.** $f = \{(0,0), (2,3), (-1,5)\}$

**11.**

| Word (Input) | Dog | Cat | House | Desk | Circle |
|---|---|---|---|---|---|
| First Letter of Word (Output) | d | c | h | d | c |

Use the properties of logarithms to write each expression as a single logarithm.

**12.** $\log_3 6 + \log_3 4$

**13.** $\log_5 x + 3\log_5 x - \log_5(x + 1)$

**14.** Write the expression $\log_6 \dfrac{2x}{y^3}$ as the sum or difference of multiples of logarithms.

**15.** If $\log_b 3 = 0.79$ and $\log_b 5 = 1.16$, find the value of $\log_b \dfrac{3}{25}$.

**16.** Approximate $\log_7 8$ to four decimal places.

**17.** Solve $8^{x-1} = \dfrac{1}{64}$ for $x$. Give an exact solution.

**18.** Solve $3^{2x+5} = 4$ for $x$. Give an exact solution, and also approximate the solution to four decimal places.

Solve each logarithmic equation for x. Give an exact solution.

**19.** $\log_3 x = -2$

**20.** $\ln \sqrt{e} = x$

**21.** $\log_8(3x - 2) = 2$

**22.** $\log_5 x + \log_5 3 = 2$

**23.** $\log_4(x + 1) - \log_4(x - 2) = 3$

**24.** Solve $\ln(3x + 7) = 1.31$ accurate to four decimal places.

**25.** Graph $y = \left(\dfrac{1}{2}\right)^x + 1$.

**26.** Graph the functions $y = 3^x$ and $y = \log_3 x$ on the same coordinate system.

Use the formula $A = P\left(1 + \dfrac{r}{n}\right)^{nt}$ to solve Exercises 27 and 28.

**27.** Find the amount in the account if $4000 is invested for 3 years at 9% interest compounded monthly.

**28.** Find how long it will take $2000 to grow to $3000 if the money is invested at 7% interest compounded semiannually. Round to the nearest whole.

Use the population growth formula $y = y_0 e^{kt}$ to solve Exercises 29 and 30.

**29.** The prairie dog population of the Grand Rapids area now stands at 57,000 animals. If the population is growing at a rate of 2.6% annually, find how many prairie dogs there will be in that area 5 years from now.

**30.** In an attempt to save an endangered species of wood duck, naturalists would like to increase the wood duck population from 400 to 1000 ducks. If the annual population growth rate is 6.2%, find how long it will take the naturalists to reach their goal. Round to the nearest whole year.

**31.** The formula $\log(1 + k) = \dfrac{0.3}{D}$ relates the doubling time $D$, in days, and the growth rate $k$ for a population of mice. Find the rate at which the population is increasing if the doubling time is 56 days. Round to the nearest tenth of a percent.

## CHAPTER 12 CUMULATIVE REVIEW

1. Divide. Write all quotients in lowest terms.

   **a.** $\dfrac{4}{5} \div \dfrac{5}{16}$   **b.** $\dfrac{7}{10} \div 14$   **c.** $\dfrac{3}{8} \div \dfrac{3}{10}$

2. Solve $\dfrac{1}{3}(x - 2) = \dfrac{1}{4}(x + 1)$.

3. Graph $f(x) = x^2$.

4. Find the equation of a line through $(-2, 6)$ and perpendicular to $f(x) = -3x + 4$. Write the equation using function notation.

5. Solve the system.

$$\begin{cases} x - 5y - 2z = 6 \\ -2x + 10y + 4z = -12 \\ \dfrac{1}{2}x - \dfrac{5}{2}y - z = 3 \end{cases}$$

6. Line $l$ and line $m$ are parallel lines cut by transversal $t$. Find the values of $x$ and $y$.

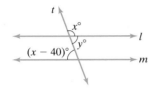

7. Simplify each expression.

   **a.** $-3 + [(-2 - 5) - 2]$   **b.** $2^3 - |10| + [-6 - (-5)]$

8. Use the power rules to simplify the following. Use positive exponents to write all results.

   **a.** $(4a^3)^2$   **b.** $\left(-\dfrac{2}{3}\right)^3$

   **c.** $\left(\dfrac{4a^5}{b^3}\right)^3$   **d.** $\left(\dfrac{3^{-2}}{x}\right)^{-3}$

   **e.** $(a^{-2}b^3c^{-4})^{-2}$

9. For the ICL Production Company, the rational function $C(x) = \dfrac{2.6x + 10,000}{x}$ describes the company's cost per disc of pressing $x$ compact discs. Find the cost per disc for pressing:

   **a.** 100 compact discs   **b.** 1000 compact discs

10. Multiply.

    **a.** $(3x - 1)^2$

    **b.** $\left(\dfrac{1}{2}x + 3\right)\left(\dfrac{1}{2}x - 3\right)$

    **c.** $(2x - 5)(6x + 7)$

11. Solve $12a - 8a = 10 + 2a - 13 - 7$.

12. Perform the indicated operation and simplify if possible.

$$\dfrac{5}{x - 2} + \dfrac{3}{x^2 + 4x + 4} - \dfrac{6}{x + 2}$$

13. Divide $\dfrac{8x^2y^2 - 16xy + 2x}{4xy}$.

14. Simplify each complex fraction.

    **a.** $\dfrac{\dfrac{a}{5}}{\dfrac{a - 1}{10}}$   **b.** $\dfrac{\dfrac{3}{2 + a} + \dfrac{6}{2 - a}}{\dfrac{5}{a + 2} - \dfrac{1}{a - 2}}$   **c.** $\dfrac{x^{-1} + y^{-1}}{xy}$

15. Factor $3m^2 - 24m - 60$.

16. Factor $5x^2 - 85x + 350$.

17. Subtract $\dfrac{3x^2 + 2x}{x - 1} - \dfrac{10x - 5}{x - 1}$.

18. Use synthetic division to divide $(8x^2 - 12x - 7) \div (x - 2)$.

19. Simplify the following expressions.

    **a.** $\sqrt[4]{81}$   **b.** $\sqrt[5]{-243}$   **c.** $-\sqrt{25}$

    **d.** $\sqrt[4]{-81}$   **e.** $\sqrt[3]{64x^3}$

20. Solve $\dfrac{1}{a + 5} = \dfrac{1}{3a + 6} - \dfrac{a + 2}{a^2 + 7x + 10}$.

21. Use rational exponents to write as a single radical.

    **a.** $\sqrt{x} \cdot \sqrt[4]{x}$   **b.** $\dfrac{\sqrt{x}}{\sqrt[3]{x}}$   **c.** $\sqrt[3]{3} \cdot \sqrt{2}$

22. Suppose that $y$ varies directly as $x$. If $y = \dfrac{1}{2}$ when $x = 12$, find the constant of variation and the direct variation equation.

23. Multiply.

    **a.** $\sqrt{3}\left(5 + \sqrt{30}\right)$

    **b.** $\left(\sqrt{5} - \sqrt{6}\right)\left(\sqrt{7} + 1\right)$

    **c.** $\left(7\sqrt{x} + 5\right)\left(3\sqrt{x} - \sqrt{5}\right)$

    **d.** $\left(4\sqrt{3} - 1\right)^2$

    **e.** $\left(\sqrt{2x} - 5\right)\left(\sqrt{2x} + 5\right)$

    **f.** $\left(\sqrt{x - 3} + 5\right)^2$

24. Find each root. Assume that all variables represent nonnegative real numbers.

    **a.** $\sqrt[4]{81}$   **b.** $\sqrt[3]{-27}$   **c.** $\sqrt{\dfrac{9}{64}}$

    **d.** $\sqrt[4]{x^{12}}$   **e.** $\sqrt[3]{-125y^6}$

25. Rationalize the denominator of $\dfrac{\sqrt[4]{x}}{\sqrt[4]{81y^5}}$.

26. Multiply.

    **a.** $a^{1/4}(a^{3/4} - a^8)$

    **b.** $(x^{1/2} - 3)(x^{1/2} + 5)$

27. Solve $\sqrt{4 - x} = x - 2$.

28. Use the quotient rule to divide and simplify if possible.

    **a.** $\dfrac{\sqrt{54}}{\sqrt{6}}$

    **b.** $\dfrac{\sqrt{108a^2}}{3\sqrt{3}}$

    **c.** $\dfrac{3\sqrt[3]{81a^5b^{10}}}{\sqrt[3]{3b^4}}$

**29.** Solve $3x^2 - 9x + 8 = 0$ by completing the square.

**30.** Add or subtract as indicated.

**a.** $\dfrac{\sqrt{20}}{3} + \dfrac{\sqrt{5}}{4}$

**b.** $\sqrt[3]{\dfrac{24x}{27}} - \dfrac{\sqrt[3]{3x}}{2}$

**31.** Solve $\dfrac{3x}{x - 2} - \dfrac{x + 1}{x} = \dfrac{6}{x(x - 2)}$.

**32.** Rationalize the denominator. $\sqrt[3]{\dfrac{27}{m^4 n^8}}$

**33.** Solve $x^2 - 4x \le 0$.

**34.** Find the length of the unknown side of the triangle.

8 in.  4 in.

**35.** Graph $F(x) = (x - 3)^2 + 1$.

**36.** Find the following powers of $i$.

**a.** $i^8$   **b.** $i^{21}$

**c.** $i^{42}$   **d.** $i^{-13}$

**37.** Solve $\dfrac{45}{x} = \dfrac{5}{7}$.

**38.** Solve $4x^2 + 8x - 1 = 0$ by completing the square.

**39.** Find an equation of the inverse of $f(x) = x + 3$.

**40.** Solve by using the quadratic formula.

$$\left(x - \frac{1}{2}\right)^2 = \frac{x}{2}$$

**41.** Find the value of each logarithmic expression.

**a.** $\log_4 16$   **b.** $\log_{10} \dfrac{1}{10}$   **c.** $\log_9 3$

**42.** Graph $f(x) = -(x + 1)^2 + 1$. Find the vertex and the axis of symmetry.

# CHAPTER

# 13 Conic Sections

In Chapter 11, we analyzed some of the important connections between a parabola and its equation. Parabolas are interesting in their own right but are more interesting still because they are part of a collection of curves known as conic sections. This chapter is devoted to quadratic equations in two variables and their conic section graphs: the parabola, circle, ellipse, and hyperbola.

When the sun rises above the horizon on June 21 or 22, thousands of people gather to witness and celebrate summer solstice at Stonehenge. Stonehenge is a megalithic ruin located on the Salisbury Plain in Wiltshire, England. It is a series of earth, timber and stone structures that were constructed, revised, and reconstructed over a period of 1400 years or so. Although no one can say for certain what its true purpose was, there have been multiple theories. The best-known theory is the one from eighteenth century British antiquarian, William Stukeley. He believed that Stonehenge was a temple, possibly an ancient cult center for the Druids. Despite the fact that we don't know its purpose for certain, Stonehenge acts as a prehistoric timepiece, allowing us to theorize what it would have been like during the Neolithic Period, and who could have built this megalithic wonder. In Exercise 72, Section 13.1, on page 775, we will explore the dimensions of the outer stone circle, known as the Sarsen circle, of Stonehenge. (*Source:* The Discovery Channel)

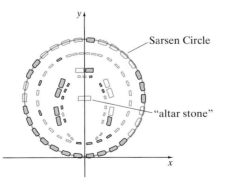

# 13.1 THE PARABOLA AND THE CIRCLE

**Conic sections** derive their name because each conic section is the intersection of a right circular cone and a plane. The circle, parabola, ellipse, and hyperbola are the conic sections.

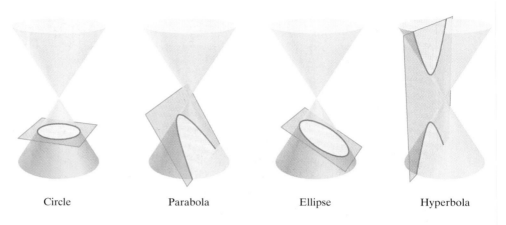

Circle  Parabola  Ellipse  Hyperbola

**OBJECTIVE 1 ▶ Graphing parabolas in standard form.** Thus far, we have seen that $f(x)$ or $y = a(x - h)^2 + k$ is the equation of a parabola that opens upward if $a > 0$ or downward if $a < 0$. Parabolas can also open left or right, or even on a slant. Equations of these parabolas are not functions of $x$, of course, since a parabola opening any way other than upward or downward fails the vertical line test. In this section, we introduce parabolas that open to the left and to the right. Parabolas opening on a slant will not be developed in this book.

Just as $y = a(x - h)^2 + k$ is the equation of a parabola that opens upward or downward, $x = a(y - k)^2 + h$ is the equation of a parabola that opens to the right or to the left. The parabola opens to the right if $a > 0$ and to the left if $a < 0$. The parabola has vertex $(h, k)$, and its axis of symmetry is the line $y = k$.

**Parabolas**

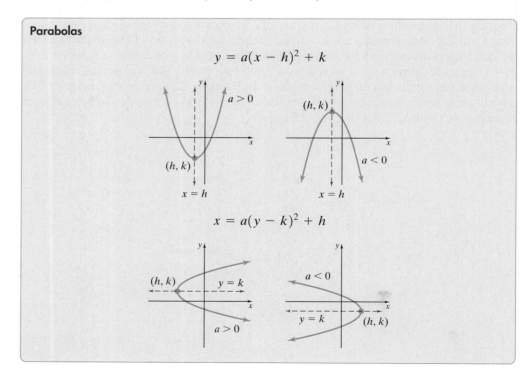

The equations $y = a(x - h)^2 + k$ and $x = a(y - k)^2 + h$ are called **standard forms.**

**Concept Check** ✓

Does the graph of the parabola given by the equation $x = -3y^2$ open to the left, to the right, upward, or downward?

**EXAMPLE 1**  Graph the parabola $x = 2y^2$.

*Solution*  Written in standard form, the equation $x = 2y^2$ is $x = 2(y - 0)^2 + 0$ with $a = 2, h = 0$, and $k = 0$. Its graph is a parabola with vertex $(0, 0)$, and its axis of symmetry is the line $y = 0$. Since $a > 0$, this parabola opens to the right. The table shows a few more ordered pair solutions of $x = 2y^2$. Its graph is also shown.

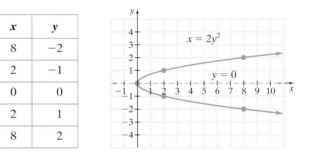

| $x$ | $y$ |
|-----|-----|
| 8 | $-2$ |
| 2 | $-1$ |
| 0 | 0 |
| 2 | 1 |
| 8 | 2 |

**PRACTICE**
**1**  Graph the parabola $x = \dfrac{1}{2}y^2$.

**EXAMPLE 2**  Graph the parabola $x = -3(y - 1)^2 + 2$.

*Solution*  The equation $x = -3(y - 1)^2 + 2$ is in the form $x = a(y - k)^2 + h$ with $a = -3$, $k = 1$, and $h = 2$. Since $a < 0$, the parabola opens to the left. The vertex $(h, k)$ is $(2, 1)$, and the axis of symmetry is the line $y = 1$. When $y = 0$, $x = -1$, so the $x$-intercept is $(-1, 0)$. Again, we obtain a few ordered pair solutions and then graph the parabola.

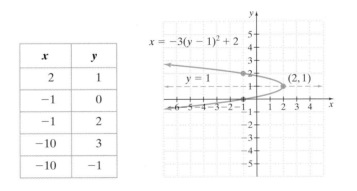

| $x$ | $y$ |
|-----|-----|
| 2 | 1 |
| $-1$ | 0 |
| $-1$ | 2 |
| $-10$ | 3 |
| $-10$ | $-1$ |

**PRACTICE**
**2**  Graph the parabola $x = -2(y + 4)^2 - 1$.

**EXAMPLE 3**  Graph $y = -x^2 - 2x + 15$.

*Solution*  Complete the square on $x$ to write the equation in standard form.

$$y - 15 = -x^2 - 2x \qquad \text{Subtract 15 from both sides.}$$

$$y - 15 = -1(x^2 + 2x) \qquad \text{Factor } -1 \text{ from the terms } -x^2 - 2x.$$

The coefficient of $x$ is 2. Find the square of half of 2.

$$\frac{1}{2}(2) = 1 \quad \text{and} \quad 1^2 = 1$$

$$y - 15 - 1(1) = -1(x^2 + 2x + 1) \quad \text{Add } -1(1) \text{ to both sides.}$$

$$y - 16 = -1(x + 1)^2 \quad \text{Simplify the left side and factor the right side.}$$

$$y = -(x + 1)^2 + 16 \quad \text{Add 16 to both sides.}$$

The equation is now in standard form $y = a(x - h)^2 + k$ with $a = -1, h = -1,$ and $k = 16.$

The vertex is then $(h, k),$ or $(-1, 16).$

A second method for finding the vertex is by using the formula $\frac{-b}{2a}.$

$$x = \frac{-(-2)}{2(-1)} = \frac{2}{-2} = -1$$

$$y = -(-1)^2 - 2(-1) + 15 = -1 + 2 + 15 = 16$$

Again, we see that the vertex is $(-1, 16),$ and the axis of symmetry is the vertical line $x = -1.$ The $y$-intercept is $(0, 15).$ Now we can use a few more ordered pair solutions to graph the parabola.

| $x$ | $y$ |
|-----|-----|
| $-1$ | 16 |
| 0 | 15 |
| $-2$ | 15 |
| 1 | 12 |
| $-3$ | 12 |
| 3 | 0 |
| $-5$ | 0 |

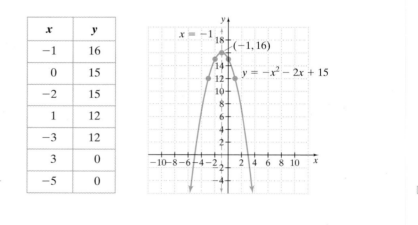

**PRACTICE**
**3** Graph $y = -x^2 + 4x + 6.$

**EXAMPLE 4** Graph $x = 2y^2 + 4y + 5.$

*Solution* Notice that this equation is quadratic in $y,$ so its graph is a parabola that opens to the left or the right. We can complete the square on $y$ or we can use the formula $\frac{-b}{2a}$ to find the vertex.

Since the equation is quadratic in $y,$ the formula gives us the $y$-value of the vertex.

$$y = \frac{-4}{2 \cdot 2} = \frac{-4}{4} = -1$$

$$x = 2(-1)^2 + 4(-1) + 5 = 2 \cdot 1 - 4 + 5 = 3$$

The vertex is $(3, -1)$, and the axis of symmetry is the line $y = -1$. The parabola opens to the right since $a > 0$. The $x$-intercept is $(5, 0)$.

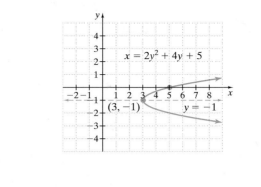

**PRACTICE**
**4**   Graph $x = 3y^2 + 6y + 4$.

**OBJECTIVE 2 ▶ Graphing circles in standard form.** Another conic section is the **circle.** A circle is the set of all points in a plane that are the same distance from a fixed point called the **center.** The distance is called the **radius** of the circle. To find a standard equation for a circle, let $(h, k)$ represent the center of the circle, and let $(x, y)$ represent any point on the circle. The distance between $(h, k)$ and $(x, y)$ is defined to be the circle's radius, $r$ units. We can find this distance $r$ by using the distance formula.

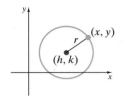

$$r = \sqrt{(x - h)^2 + (y - k)^2}$$
$$r^2 = (x - h)^2 + (y - k)^2 \qquad \text{Square both sides.}$$

**Circle**

The graph of $(x - h)^2 + (y - k)^2 = r^2$ is a circle with center $(h, k)$ and radius $r$.

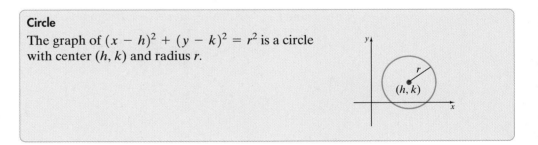

The equation $(x - h)^2 + (y - k)^2 = r^2$ is called **standard form.**
  If an equation can be written in the standard form

$$(x - h)^2 + (y - k)^2 = r^2$$

then its graph is a circle, which we can draw by graphing the center $(h, k)$ and using the radius $r$.

**▶ Helpful Hint**

Notice that the radius is the *distance* from the center of the circle to any point of the circle. Also notice that the *midpoint* of a diameter of a circle is the center of the circle.

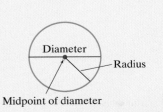

**EXAMPLE 5** Graph $x^2 + y^2 = 4$.

**Solution** The equation can be written in standard form as

$$(x - 0)^2 + (y - 0)^2 = 2^2$$

The center of the circle is $(0, 0)$, and the radius is 2. Its graph is shown.

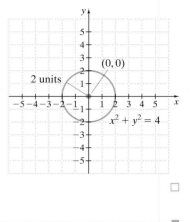

**PRACTICE**
**5** Graph $x^2 + y^2 = 25$.

> ▶ **Helpful Hint**
> Notice the difference between the equation of a circle and the equation of a parabola. The equation of a circle contains both $x^2$ and $y^2$ terms on the same side of the equation with equal coefficients. The equation of a parabola has either an $x^2$ term or a $y^2$ term but not both.

**EXAMPLE 6** Graph $(x + 1)^2 + y^2 = 8$.

**Solution** The equation can be written as $(x + 1)^2 + (y - 0)^2 = 8$ with $h = -1, k = 0$, and $r = \sqrt{8}$. The center is $(-1, 0)$, and the radius is $\sqrt{8} = 2\sqrt{2} \approx 2.8$.

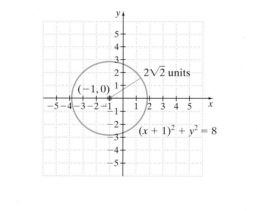

**PRACTICE**
**6** Graph $(x - 3)^2 + (y + 2)^2 = 4$.

**Concept Check** ☑
In the graph of the equation $(x - 3)^2 + (y - 2)^2 = 5$, what is the distance between the center of the circle and any point on the circle?

**Answer to Concept Check:**
$\sqrt{5}$ units

**OBJECTIVE 3** ▶ **Writing equations of circles.** Since a circle is determined entirely by its center and radius, this information is all we need to write the equation of a circle.

**EXAMPLE 7**   Find an equation of the circle with center $(-7, 3)$ and radius 10.

_Solution_   Using the given values $h = -7$, $k = 3$, and $r = 10$, we write the equation

$$(x - h)^2 + (y - k)^2 = r^2$$

or

$$[x - (-7)]^2 + (y - 3)^2 = 10^2 \quad \text{Substitute the given values.}$$

or

$$(x + 7)^2 + (y - 3)^2 = 100 \qquad \square$$

**PRACTICE**
**7**   Find the equation of a circle with center $(-2, -5)$ and radius 9.

**OBJECTIVE 4 ▶ Finding the center and the radius of a circle.** To find the center and the radius of a circle from its equation, write the equation in standard form. To write the equation of a circle in standard form, we complete the square on both $x$ and $y$.

**EXAMPLE 8**   Graph $x^2 + y^2 + 4x - 8y = 16$.

_Solution_   Since this equation contains $x^2$ and $y^2$ terms on the same side of the equation with equal coefficients, its graph is a circle. To write the equation in standard form, group the terms involving $x$ and the terms involving $y$, and then complete the square on each variable.

$$(x^2 + 4x) + (y^2 - 8y) = 16$$

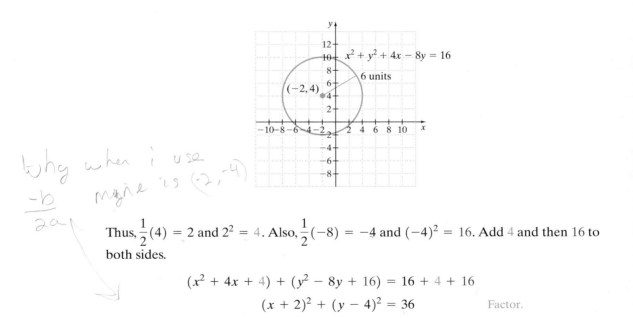

*(handwritten note:)* why when i use $\frac{-b}{2a}$ myne is $(-2, -4)$

Thus, $\frac{1}{2}(4) = 2$ and $2^2 = 4$. Also, $\frac{1}{2}(-8) = -4$ and $(-4)^2 = 16$. Add 4 and then 16 to both sides.

$$(x^2 + 4x + 4) + (y^2 - 8y + 16) = 16 + 4 + 16$$
$$(x + 2)^2 + (y - 4)^2 = 36 \qquad \text{Factor.}$$

This circle has the center $(-2, 4)$ and radius 6, as shown. $\qquad \square$

**PRACTICE**
**8**   Graph $x^2 + y^2 + 6x - 2y = 6$.

## Graphing Calculator Explorations

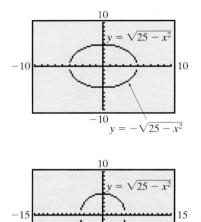

To graph an equation such as $x^2 + y^2 = 25$ with a graphing calculator, we first solve the equation for $y$.

$$x^2 + y^2 = 25$$
$$y^2 = 25 - x^2$$
$$y = \pm\sqrt{25 - x^2}$$

The graph of $y = \sqrt{25 - x^2}$ will be the top half of the circle, and the graph of $y = -\sqrt{25 - x^2}$ will be the bottom half of the circle.

To graph, press $\boxed{Y=}$ and enter $Y_1 = \sqrt{25 - x^2}$ and $Y_2 = -\sqrt{25 - x^2}$. Insert parentheses around $25 - x^2$ so that $\sqrt{25 - x^2}$ and not $\sqrt{25} - x^2$ is graphed.

The top graph to the left does not appear to be a circle because we are currently using a standard window and the screen is rectangular. This causes the tick marks on the $x$-axis to be farther apart than the tick marks on the $y$-axis and, thus, creates the distorted circle. If we want the graph to appear circular, we must define a square window by using a feature of the graphing calculator or by redefining the window to show the $x$-axis from $-15$ to $15$ and the $y$-axis from $-10$ to $10$. Using a square window, the graph appears as shown on the bottom to the left.

*Use a graphing calculator to graph each circle.*

**1.** $x^2 + y^2 = 55$

**2.** $x^2 + y^2 = 20$

**3.** $5x^2 + 5y^2 = 50$

**4.** $6x^2 + 6y^2 = 105$

**5.** $2x^2 + 2y^2 - 34 = 0$

**6.** $4x^2 + 4y^2 - 48 = 0$

**7.** $7x^2 + 7y^2 - 89 = 0$

**8.** $3x^2 + 3y^2 - 35 = 0$

## VOCABULARY & READINESS CHECK

*Use the choices below to fill in each blank. Some choices may be used more than once.*

| | | |
|---|---|---|
| radius | center | vertex |
| diameter | circle | conic sections |

**1.** The circle, parabola, ellipse, and hyperbola are called the _____.

**2.** For a parabola that opens upward the lowest point is the _____.

**3.** A _____ is the set of all points in a plane that are the same distance from a fixed point. The fixed point is called the _____.

**4.** The midpoint of a diameter of a circle is the _____.

**5.** The distance from the center of a circle to any point of the circle is called the _____.

**6.** Twice a circle's radius is its _____.

*The graph of each equation is a parabola. Determine whether the parabola opens upward, downward, to the left, or to the right.*

**7.** $y = x^2 - 7x + 5$

**8.** $y = -x^2 + 16$

**9.** $x = -y^2 - y + 2$

**10.** $x = 3y^2 + 2y - 5$

**11.** $y = -x^2 + 2x + 1$

**12.** $x = -y^2 + 2y - 6$

## 13.1 | EXERCISE SET

*The graph of each equation is a parabola. Find the vertex of the parabola and sketch its graph. See Examples 1 through 4.*

**1.** $x = 3y^2$

**2.** $x = -2y^2$

**3.** $x = (y - 2)^2 + 3$

**4.** $x = (y - 4)^2 - 1$

**5.** $y = 3(x - 1)^2 + 5$

**6.** $x = -4(y - 2)^2 + 2$

**7.** $x = y^2 + 6y + 8$

**8.** $x = y^2 - 6y + 6$

**9.** $y = x^2 + 10x + 20$

**10.** $y = x^2 + 4x - 5$

**11.** $x = -2y^2 + 4y + 6$

**12.** $x = 3y^2 + 6y + 7$

*The graph of each equation is a circle. Find the center and the radius, and then sketch. See Examples 5, 6, and 8.*

**13.** $x^2 + y^2 = 9$

**14.** $x^2 + y^2 = 100$

**15.** $x^2 + (y - 2)^2 = 1$

**16.** $(x - 3)^2 + y^2 = 9$

**17.** $(x - 5)^2 + (y + 2)^2 = 1$

**18.** $(x + 3)^2 + (y + 3)^2 = 4$

**19.** $x^2 + y^2 + 6y = 0$

**20.** $x^2 + 10x + y^2 = 0$

**21.** $x^2 + y^2 + 2x - 4y = 4$

**22.** $x^2 + 6x - 4y + y^2 = 3$

**23.** $x^2 + y^2 - 4x - 8y - 2 = 0$  **24.** $x^2 + y^2 - 2x - 6y - 5 = 0$

*Write an equation of the circle with the given center and radius. See Example 7.*

**25.** $(2, 3); 6$

**26.** $(-7, 6); 2$

**27.** $(0, 0); \sqrt{3}$

**28.** $(0, -6); \sqrt{2}$

**29.** $(-5, 4); 3\sqrt{5}$

**30.** the origin; $4\sqrt{7}$

**31.** Explain the error in the statement: The graph of $x^2 + (y + 3)^2 = 10$ is a circle with center $(0, -3)$ and radius 5.

### MIXED PRACTICE

*Sketch the graph of each equation. If the graph is a parabola, find its vertex. If the graph is a circle, find its center and radius.*

**32.** $x = y^2 + 2$

**33.** $x = y^2 - 3$

**34.** $y = (x + 3)^2 + 3$

**35.** $y = (x - 2)^2 - 2$

**36.** $x^2 + y^2 = 49$

**37.** $x^2 + y^2 = 1$

**38.** $x = (y - 1)^2 + 4$

**39.** $x = (y + 3)^2 - 1$

**40.** $(x + 3)^2 + (y - 1)^2 = 9$  **41.** $(x - 2)^2 + (y - 2)^2 = 16$

**42.** $x = -2(y + 5)^2$

**43.** $x = -(y - 1)^2$

**44.** $x^2 + (y + 5)^2 = 5$

**45.** $(x - 4)^2 + y^2 = 7$

**46.** $y = 3(x - 4)^2 + 2$

**47.** $y = 5(x + 5)^2 + 3$

**48.** $2x^2 + 2y^2 = \dfrac{1}{2}$

**49.** $\dfrac{x^2}{8} + \dfrac{y^2}{8} = 2$

**50.** $y = x^2 - 2x - 15$

**51.** $y = x^2 + 7x + 6$

**52.** $x^2 + y^2 + 6x + 10y - 2 = 0$  **53.** $x^2 + y^2 + 2x + 12y - 12 = 0$

**54.** $x = y^2 + 6y + 2$

**55.** $x = y^2 + 8y - 4$

**56.** $x^2 + y^2 - 8y + 5 = 0$

**57.** $x^2 - 10y + y^2 + 4 = 0$

**58.** $x = -2y^2 - 4y$

**59.** $x = -3y^2 + 30y$

**60.** $\dfrac{x^2}{3} + \dfrac{y^2}{3} = 2$

**61.** $5x^2 + 5y^2 = 25$

**62.** $y = 4x^2 - 40x + 105$

**63.** $y = 5x^2 - 20x + 16$

### REVIEW AND PREVIEW

*Graph each equation. See Sections 3.2 and 3.3.*

**64.** $y = 2x + 5$

**65.** $y = -3x + 3$

**66.** $y = 3$

**67.** $x = -2$

*Rationalize each denominator and simplify, if possible. See Section 10.5.*

**68.** $\dfrac{1}{\sqrt{3}}$

**69.** $\dfrac{\sqrt{5}}{\sqrt{8}}$

**70.** $\dfrac{4\sqrt{7}}{\sqrt{6}}$

**71.** $\dfrac{10}{\sqrt{5}}$

### CONCEPT EXTENSIONS

**72. The Sarsen Circle:** The first image that comes to mind when one thinks of Stonehenge is the very large sandstone blocks with sandstone lintels across the top. The Sarsen Circle of Stonehenge is the outer circle of the sandstone blocks, each of which weighs up to 50 tons. There were originally 30 of these monolithic blocks, but only 17 remain upright to this day. The "altar stone" lies at the center of this circle, which has a diameter of 33 meters.

**a.** What is the radius of the Sarsen circle?

**b.** What is the circumference of the Sarsen circle? Round your result to 2 decimal places.

**c.** Since there were originally 30 Sarsen stones located on the circumference, how far apart would the centers of the stones have been? Round to the nearest tenth of a meter.

**d.** Using the axes in the drawing, what are the coordinates of the center of the circle?

**e.** Use parts **a** and **d** to write the equation of the Sarsen circle.

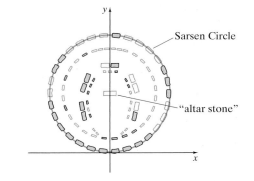

**73.** Opened in 2000 to honor the millennium, the British Airways London Eye is the world's biggest observation wheel. Each of the 32 enclosed capsules, which each hold 25 passengers, completes a full rotation every 30 minutes. Its diameter is

135 meters, and it is constructed on London's South Bank, to allow passengers to enter the Eye at ground level. (*Source: Guinness Book of World Records*)

**a.** What is the radius of the London Eye?

**b.** How close is the wheel to the ground?

**c.** How high is the center of the wheel from the ground?

**d.** Using the axes in the drawing, what are the coordinates of the center of the wheel?

**e.** Use parts **a** and **d** to write the equation of the Eye.

**74.** In 1893, Pittsburgh bridge builder George Ferris designed and built a gigantic revolving steel wheel whose height was 264 feet and diameter was 250 feet. This Ferris wheel opened at the 1893 exposition in Chicago. It had 36 wooden cars, each capable of holding 60 passengers. (*Source: The Handy Science Answer Book*)

**a.** What was the radius of this Ferris wheel?

**b.** How close is the wheel to the ground?

**c.** How high is the center of the wheel from the ground?

**d.** Using the axes in the drawing, what are the coordinates of the center of the wheel?

**e.** Use parts **a** and **d** to write the equation of the wheel.

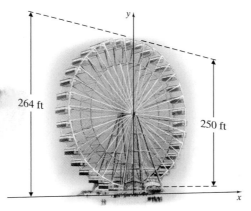

**75.** As of this writing, the world's largest-diameter Ferris wheel currently in operation is the Star of Nanchung in Jiangxi Province, China. It has 60 compartments, each of which carries eight people. It is 160 meters tall, and the diameter of the wheel is 153 meters. (*Source:* China News Agency)

**a.** What is the radius of this Ferris wheel?

**b.** How close is the wheel to the ground?

**c.** How high is the center of the wheel from the ground?

**d.** Using the axes in the drawing, what are the coordinates of the center of the wheel?

**e.** Use parts **a** and **d** to write the equation of the wheel.

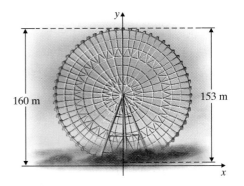

**76.** If you are given a list of equations of circles and parabolas and none are in standard form, explain how you would determine which is an equation of a circle and which is an equation of a parabola. Explain also how you would distinguish the upward or downward parabolas from the left-opening or right-opening parabolas.

*Solve.*

**77.** Cindy Brown, an architect, is drawing plans on grid paper for a circular pool with a fountain in the middle. The paper is marked off in centimeters, and each centimeter represents 1 foot. On the paper, the diameter of the "pool" is 20 centimeters, and "fountain" is the point $(0, 0)$.

**a.** Sketch the architect's drawing. Be sure to label the axes.

**b.** Write an equation that describes the circular pool.

**c.** Cindy plans to place a circle of lights around the fountain such that each light is 5 feet from the fountain. Write an equation for the circle of lights and sketch the circle on your drawing.

**78.** A bridge constructed over a bayou has a supporting arch in the shape of a parabola. Find an equation of the parabolic arch if the length of the road over the arch is 100 meters and the maximum height of the arch is 40 meters.

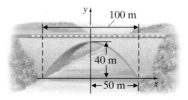

*Use a graphing calculator to verify each exercise. Use a square viewing window.*

**79.** Exercise 61.

**80.** Exercise 60.

**81.** Exercise 63.

**82.** Exercise 62.

# 13.2 THE ELLIPSE AND THE HYPERBOLA

**OBJECTIVES**

1  Define and graph an ellipse.

2  Define and graph a hyperbola.

**OBJECTIVE 1 ▶ Graphing ellipses.** An **ellipse** can be thought of as the set of points in a plane such that the sum of the distances of those points from two fixed points is constant. Each of the two fixed points is called a **focus.** (The plural of focus is **foci.**) The point midway between the foci is called the **center.**

An ellipse may be drawn by hand by using two thumbtacks, a piece of string, and a pencil. Secure the two thumbtacks in a piece of cardboard, for example, and tie each end of the string to a tack. Use your pencil to pull the string tight and draw the ellipse. The two thumbtacks are the foci of the drawn ellipse.

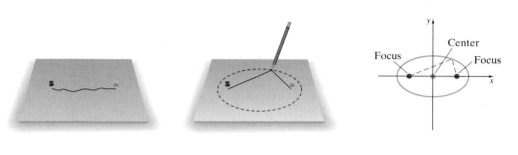

**Ellipse with Center (0, 0)**

The graph of an equation of the form $\dfrac{x^2}{a^2} + \dfrac{y^2}{b^2} = 1$ is an ellipse with center $(0, 0)$.

The $x$-intercepts are $(a, 0)$ and $(-a, 0)$, and the $y$-intercepts are $(0, b)$, and $(0, -b)$.

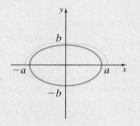

The **standard form** of an ellipse with center $(0, 0)$ is $\dfrac{x^2}{a^2} + \dfrac{y^2}{b^2} = 1$.

**EXAMPLE 1**   Graph $\dfrac{x^2}{9} + \dfrac{y^2}{16} = 1$.

**Solution**   The equation is of the form $\dfrac{x^2}{a^2} + \dfrac{y^2}{b^2} = 1$, with $a = 3$ and $b = 4$, so its graph is an ellipse with center $(0, 0)$, $x$-intercepts $(3, 0)$ and $(-3, 0)$, and $y$-intercepts $(0, 4)$ and $(0, -4)$.

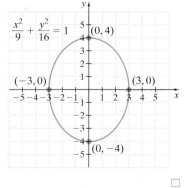

**PRACTICE**
**1**   Graph $\dfrac{x^2}{25} + \dfrac{y^2}{4} = 1$.

**EXAMPLE 2**   Graph $4x^2 + 16y^2 = 64$.

*Solution*   Although this equation contains a sum of squared terms in $x$ and $y$ on the same side of an equation, this is not the equation of a circle since the coefficients of $x^2$ and $y^2$ are not the same. The graph of this equation is an ellipse. Since the standard form of the equation of an ellipse has 1 on one side, divide both sides of this equation by 64.

$$4x^2 + 16y^2 = 64$$

$$\frac{4x^2}{64} + \frac{16y^2}{64} = \frac{64}{64} \quad \text{Divide both sides by 64.}$$

$$\frac{x^2}{16} + \frac{y^2}{4} = 1 \quad \text{Simplify.}$$

We now recognize the equation of an ellipse with $a = 4$ and $b = 2$. This ellipse has center $(0, 0)$, $x$-intercepts $(4, 0)$ and $(-4, 0)$, and $y$-intercepts $(0, 2)$ and $(0, -2)$.

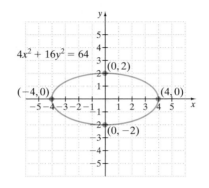

**PRACTICE**
**2**   Graph $9x^2 + 4y^2 = 36$.

The center of an ellipse is not always $(0, 0)$, as shown in the next example.

**EXAMPLE 3**   Graph $\dfrac{(x + 3)^2}{25} + \dfrac{(y - 2)^2}{36} = 1$.

*Solution*   The center of this ellipse is found in a way that is similar to finding the center of a circle. This ellipse has center $(-3, 2)$. Notice that $a = 5$ and $b = 6$. To find four points on the graph of the ellipse, first graph the center, $(-3, 2)$. Since $a = 5$, count 5 units right and then 5 units left of the point with coordinates $(-3, 2)$. Next, since $b = 6$, start at $(-3, 2)$ and count 6 units up and then 6 units down to find two more points on the ellipse.

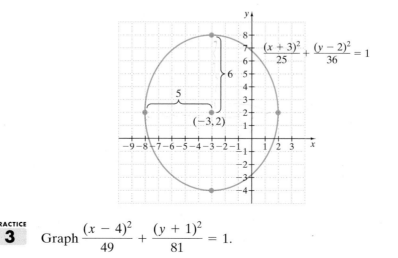

**PRACTICE**
**3**   Graph $\dfrac{(x - 4)^2}{49} + \dfrac{(y + 1)^2}{81} = 1.$

**Concept Check** ☑

In the graph of the equation $\dfrac{x^2}{64} + \dfrac{y^2}{36} = 1$, which distance is longer: the distance between the $x$-intercepts or the distance between the $y$-intercepts? How much longer? Explain.

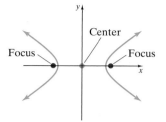

**OBJECTIVE 2 ▶ Graphing hyperbolas.** The final conic section is the **hyperbola.** A hyperbola is the set of points in a plane such that the absolute value of the difference of the distances from two fixed points is constant. Each of the two fixed points is called a **focus.** The point midway between the foci is called the **center.**

Using the distance formula, we can show that the graph of $\dfrac{x^2}{a^2} - \dfrac{y^2}{b^2} = 1$ is a hyperbola with center $(0, 0)$ and $x$-intercepts $(a, 0)$ and $(-a, 0)$. Also, the graph of $\dfrac{y^2}{b^2} - \dfrac{x^2}{a^2} = 1$ is a hyperbola with center $(0,0)$ and $y$-intercepts $(0, b)$ and $(0, -b)$.

---

**Hyperbola with Center (0, 0)**

The graph of an equation of the form $\dfrac{x^2}{a^2} - \dfrac{y^2}{b^2} = 1$ is a hyperbola with center $(0,0)$ and $x$-intercepts $(a, 0)$ and $(-a, 0)$.

The graph of an equation of the form $\dfrac{y^2}{b^2} - \dfrac{x^2}{a^2} = 1$ is a hyperbola with center $(0,0)$ and $y$-intercepts $(0, b)$ and $(0, -b)$.

---

The equations $\dfrac{x^2}{a^2} - \dfrac{y^2}{b^2} = 1$ and $\dfrac{y^2}{b^2} - \dfrac{x^2}{a^2} = 1$ are the **standard forms** for the equation of a hyperbola.

---

**▶ Helpful Hint**

Notice the difference between the equation of an ellipse and a hyperbola. The equation of the ellipse contains $x^2$ and $y^2$ terms on the same side of the equation with same-sign coefficients. For a hyperbola, the coefficients on the same side of the equation have different signs.

**Answer to Concept Check:**
$x$-intercepts, by 4 units

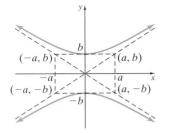

Graphing a hyperbola such as $\dfrac{y^2}{b^2} - \dfrac{x^2}{a^2} = 1$ is made easier by recognizing one of its important characteristics. Examining the figure to the left, notice how the sides of the branches of the hyperbola extend indefinitely and seem to approach the dashed lines in the figure. These dashed lines are called the **asymptotes** of the hyperbola.

To sketch these lines, or asymptotes, draw a rectangle with vertices $(a, b), (-a, b)$, $(a, -b)$, and $(-a, -b)$. The asymptotes of the hyperbola are the extended diagonals of this rectangle.

**EXAMPLE 4**  Graph $\dfrac{x^2}{16} - \dfrac{y^2}{25} = 1$.

*Solution*  This equation has the form $\dfrac{x^2}{a^2} - \dfrac{y^2}{b^2} = 1$, with $a = 4$ and $b = 5$. Thus, its graph is a hyperbola that opens to the left and right. It has center $(0, 0)$ and $x$-intercepts $(4, 0)$ and $(-4, 0)$. To aid in graphing the hyperbola, we first sketch its asymptotes. The extended diagonals of the rectangle with corners $(4, 5), (4, -5), (-4, 5)$, and $(-4, -5)$ are the asymptotes of the hyperbola. Then we use the asymptotes to aid in sketching the hyperbola.

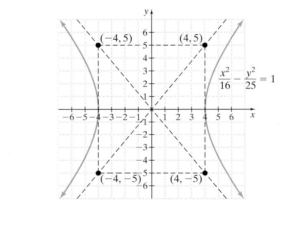

**PRACTICE**
**4**  Graph $\dfrac{x^2}{9} - \dfrac{y^2}{16} = 1$.

**EXAMPLE 5**  Graph $4y^2 - 9x^2 = 36$.

*Solution*  Since this is a difference of squared terms in $x$ and $y$ on the same side of the equation, its graph is a hyperbola, as opposed to an ellipse or a circle. The standard form of the equation of a hyperbola has a 1 on one side, so divide both sides of the equation by 36.

$$4y^2 - 9x^2 = 36$$

$$\frac{4y^2}{36} - \frac{9x^2}{36} = \frac{36}{36} \qquad \text{Divide both sides by 36.}$$

$$\frac{y^2}{9} - \frac{x^2}{4} = 1 \qquad \text{Simplify.}$$

The equation is of the form $\frac{y^2}{b^2} - \frac{x^2}{a^2} = 1$, with $a = 2$ and $b = 3$, so the hyperbola is centered at $(0, 0)$ with $y$-intercepts $(0, 3)$ and $(0, -3)$. The sketch of the hyperbola is shown.

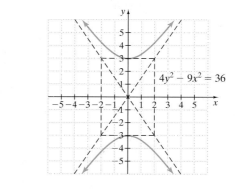

$4y^2 - 9x^2 = 36$

**PRACTICE**

**5**   Graph $9y^2 - 25x^2 = 225$.

## Graphing Calculator Explorations

To find the graph of an ellipse by using a graphing calculator, use the same procedure as for graphing a circle. For example, to graph $x^2 + 3y^2 = 22$, first solve for $y$.

$$3y^2 = 22 - x^2$$

$$y^2 = \frac{22 - x^2}{3}$$

$$y = \pm\sqrt{\frac{22 - x^2}{3}}$$

Next press the $\boxed{\text{Y=}}$ key and enter $Y_1 = \sqrt{\frac{22 - x^2}{3}}$ and $Y_2 = -\sqrt{\frac{22 - x^2}{3}}$.

(Insert two sets of parentheses in the radicand as $\sqrt{((22 - x^2)/3)}$ so that the desired graph is obtained.) The graph appears as follows.

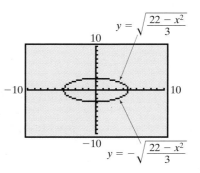

$y = \sqrt{\frac{22 - x^2}{3}}$

$y = -\sqrt{\frac{22 - x^2}{3}}$

*Use a graphing calculator to graph each ellipse.*

**1.** $10x^2 + y^2 = 32$        **2.** $x^2 + 6y^2 = 35$

**3.** $20x^2 + 5y^2 = 100$        **4.** $4y^2 + 12x^2 = 48$

**5.** $7.3x^2 + 15.5y^2 = 95.2$        **6.** $18.8x^2 + 36.1y^2 = 205.8$

# VOCABULARY & READINESS CHECK

*Use the choices below to fill in each blank. Some choices will be used more than once and some not at all.*

| | | | | |
|---|---|---|---|---|
| ellipse | $(0, 0)$ | focus | $(a, 0)$ and $(-a, 0)$ | $(0, a)$ and $(0, -a)$ |
| hyperbola | center | $x$ | $(b, 0)$ and $(-b, 0)$ | $(0, b)$ and $(0, -b)$ |
| | | $y$ | | |

1. A(n) _____ is the set of points in a plane such that the absolute value of the differences of their distances from two fixed points is constant.

2. A(n) _____ is the set of points in a plane such that the sum of their distances from two fixed points is constant.

*For exercises 1 and 2 above,*

3. The two fixed points are each called a _____ .

4. The point midway between the foci is called the _____ .

5. The graph of $\dfrac{x^2}{a^2} - \dfrac{y^2}{b^2} = 1$ is a(n) _____ with center _____ and _____-intercepts of _____ .

6. The graph of $\dfrac{x^2}{b^2} + \dfrac{y^2}{a^2} = 1$ is a(n) _____ with center _____ and $x$-intercepts of _____ .

*Identify the graph of each equation as an ellipse or a hyperbola.*

7. $\dfrac{x^2}{16} + \dfrac{y^2}{4} = 1$

8. $\dfrac{x^2}{16} - \dfrac{y^2}{4} = 1$

9. $x^2 - 5y^2 = 3$

10. $-x^2 + 5y^2 = 3$

11. $-\dfrac{y^2}{25} + \dfrac{x^2}{36} = 1$

12. $\dfrac{y^2}{25} + \dfrac{x^2}{36} = 1$

## 13.2 EXERCISE SET

**MyMathLab®** Powered by CourseCompass™ and MathXL®   MathXL PRACTICE   WATCH   DOWNLOAD   READ   REVIEW

*Sketch the graph of each equation. See Examples 1 and 2.*

1. $\dfrac{x^2}{4} + \dfrac{y^2}{25} = 1$

2. $\dfrac{x^2}{16} + \dfrac{y^2}{9} = 1$

3. $\dfrac{x^2}{9} + y^2 = 1$

4. $x^2 + \dfrac{y^2}{4} = 1$

5. $9x^2 + y^2 = 36$

6. $x^2 + 4y^2 = 16$

7. $4x^2 + 25y^2 = 100$

8. $36x^2 + y^2 = 36$

*Sketch the graph of each equation. See Example 3.*

9. $\dfrac{(x + 1)^2}{36} + \dfrac{(y - 2)^2}{49} = 1$

10. $\dfrac{(x - 3)^2}{9} + \dfrac{(y + 3)^2}{16} = 1$

11. $\dfrac{(x - 1)^2}{4} + \dfrac{(y - 1)^2}{25} = 1$

12. $\dfrac{(x + 3)^2}{16} + \dfrac{(y + 2)^2}{4} = 1$

*Sketch the graph of each equation. See Examples 4 and 5.*

13. $\dfrac{x^2}{4} - \dfrac{y^2}{9} = 1$

14. $\dfrac{x^2}{36} - \dfrac{y^2}{36} = 1$

15. $\dfrac{y^2}{25} - \dfrac{x^2}{16} = 1$

16. $\dfrac{y^2}{25} - \dfrac{x^2}{49} = 1$

17. $x^2 - 4y^2 = 16$

18. $4x^2 - y^2 = 36$

19. $16y^2 - x^2 = 16$

20. $4y^2 - 25x^2 = 100$

21. If you are given a list of equations of circles, parabolas, ellipses, and hyperbolas, explain how you could distinguish the different conic sections from their equations.

### MIXED PRACTICE

*Identify whether each equation, when graphed, will be a parabola, circle, ellipse, or hyperbola. Sketch the graph of each equation.*

22. $(x - 7)^2 + (y - 2)^2 = 4$

23. $y = x^2 + 4$

24. $y = x^2 + 12x + 36$

25. $\dfrac{x^2}{4} + \dfrac{y^2}{9} = 1$

26. $\dfrac{y^2}{9} - \dfrac{x^2}{9} = 1$

27. $\dfrac{x^2}{16} - \dfrac{y^2}{4} = 1$

28. $\dfrac{x^2}{16} + \dfrac{y^2}{4} = 1$

29. $x^2 + y^2 = 16$

30. $x = y^2 + 4y - 1$

31. $x = -y^2 + 6y$

32. $9x^2 - 4y^2 = 36$

33. $9x^2 + 4y^2 = 36$

34. $\dfrac{(x - 1)^2}{49} + \dfrac{(y + 2)^2}{25} = 1$

35. $y^2 = x^2 + 16$

36. $\left(x + \dfrac{1}{2}\right)^2 + \left(y - \dfrac{1}{2}\right)^2 = 1$

37. $y = -2x^2 + 4x - 3$

## REVIEW AND PREVIEW

*Solve each inequality. See Section 2.8.*

**38.** $x < 5$ and $x < 1$          **39.** $x < 5$ or $x < 1$

**40.** $2x - 1 \geq 7$ or $-3x \leq -6$   **41.** $2x - 1 \geq 7$ and $-3x \leq -6$

*Perform the indicated operations. See Sections 5.1 and 5.2.*

**42.** $(2x^3)(-4x^2)$          **43.** $2x^3 - 4x^3$

**44.** $-5x^2 + x^2$            **45.** $(-5x^2)(x^2)$

## CONCEPT EXTENSIONS

*The graph of each equation is an ellipse. Determine which distance is longer. The distance between the x-intercepts or the distance between the y-intercepts. How much longer? See the Concept Check in this section.*

**46.** $\dfrac{x^2}{16} + \dfrac{y^2}{25} = 1$          **47.** $\dfrac{x^2}{100} + \dfrac{y^2}{49} = 1$

**48.** $4x^2 + y^2 = 16$          **49.** $x^2 + 4y^2 = 36$

**50.** We know that $x^2 + y^2 = 25$ is the equation of a circle. Rewrite the equation so that the right side is equal to 1. Which type of conic section does this equation form resemble? In fact, the circle is a special case of this type of conic section. Describe the conditions under which this type of conic section is a circle.

*The orbits of stars, planets, comets, asteroids, and satellites all have the shape of one of the conic sections. Astronomers use a measure called eccentricity to describe the shape and elongation of an orbital path. For the circle and ellipse, eccentricity e is calculated with the formula $e = \dfrac{c}{d}$, where $c^2 = |a^2 - b^2|$ and d is the larger value of a or b. For a hyperbola, eccentricity e is calculated with the formula $e = \dfrac{c}{d}$, where $c^2 = a^2 + b^2$ and the value of d is equal to a if the hyperbola has x-intercepts or equal to b if the hyperbola has y-intercepts. Use equations A–H to answer Exercises 51–60.*

**A.** $\dfrac{x^2}{36} - \dfrac{y^2}{13} = 1$   **B.** $\dfrac{x^2}{4} + \dfrac{y^2}{4} = 1$   **C.** $\dfrac{x^2}{25} + \dfrac{y^2}{16} = 1$

**D.** $\dfrac{y^2}{25} - \dfrac{x^2}{39} = 1$   **E.** $\dfrac{x^2}{17} + \dfrac{y^2}{81} = 1$   **F.** $\dfrac{x^2}{36} + \dfrac{y^2}{36} = 1$

**G.** $\dfrac{x^2}{16} - \dfrac{y^2}{65} = 1$   **H.** $\dfrac{x^2}{144} + \dfrac{y^2}{140} = 1$

**51.** Identify the type of conic section represented by each of the equations A–H.

**52.** For each of the equations A–H, identify the values of $a^2$ and $b^2$.

**53.** For each of the equations A–H, calculate the value of $c^2$ and $c$.

**54.** For each of the equations A–H, find the value of $d$.

**55.** For each of the equations A–H, calculate the eccentricity $e$.

**56.** What do you notice about the values of $e$ for the equations you identified as ellipses?

**57.** What do you notice about the values of $e$ for the equations you identified as circles?

**58.** What do you notice about the values of $e$ for the equations you identified as hyperbolas?

**59.** The eccentricity of a parabola is exactly 1. Use this information and the observations you made in Exercises 31, 32, and 33 to describe a way that could be used to identify the type of conic section based on its eccentricity value.

**60.** Graph each of the conic sections given in equations A–H. What do you notice about the shape of the ellipses for increasing values of eccentricity? Which is the most elliptical? Which is the least elliptical, that is, the most circular?

**61.** A planet's orbit about the Sun can be described as an ellipse. Consider the Sun as the origin of a rectangular coordinate system. Suppose that the x-intercepts of the elliptical path of the planet are ±130,000,000 and that the y-intercepts are ±125,000,000. Write the equation of the elliptical path of the planet.

**62.** Comets orbit the Sun in elongated ellipses. Consider the Sun as the origin of a rectangular coordinate system. Suppose that the equation of the path of the comet is

$$\frac{(x - 1{,}782{,}000{,}000)^2}{3.42 \cdot 10^{23}} + \frac{(y - 356{,}400{,}000)^2}{1.368 \cdot 10^{22}} = 1$$

Find the center of the path of the comet.

**63.** Use a graphing calculator to verify Exercise 33.

**64.** Use a graphing calculator to verify Exercise 6.

*For Exercises 65 through 70, see the example below.*

**Example**

Sketch the graph of $\dfrac{(x - 2)^2}{25} - \dfrac{(y - 1)^2}{9} = 1$.

*Solution*

This hyperbola has center $(2, 1)$. Notice that $a = 5$ and $b = 3$.

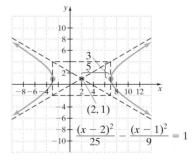

$$\frac{(x - 2)^2}{25} - \frac{(x - 1)^2}{9} = 1$$

*Sketch the graph of each equation.*

**65.** $\dfrac{(x - 1)^2}{4} - \dfrac{(y + 1)^2}{25} = 1$   **66.** $\dfrac{(x + 2)^2}{9} - \dfrac{(y - 1)^2}{4} = 1$

**67.** $\dfrac{y^2}{16} - \dfrac{(x + 3)^2}{9} = 1$   **68.** $\dfrac{(y + 4)^2}{4} - \dfrac{x^2}{25} = 1$

**69.** $\dfrac{(x + 5)^2}{16} - \dfrac{(y + 2)^2}{25} = 1$   **70.** $\dfrac{(x - 3)^2}{9} - \dfrac{(y - 2)^2}{4} = 1$

# INTEGRATED REVIEW GRAPHING CONIC SECTIONS

Sections 13.1, 13.2

Following is a summary of conic sections.

**Conic Sections**

| | Standard Form | Graph |
|---|---|---|
| *Parabola* | $y = a(x - h)^2 + k$ 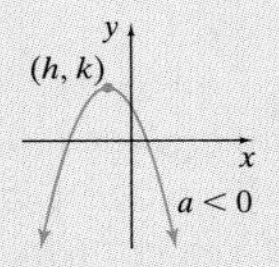 | 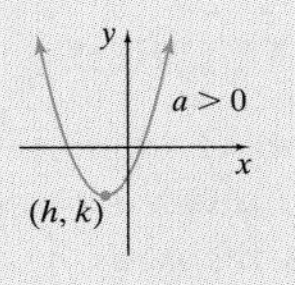 |

| | | |
|---|---|---|
| *Parabola* | $x = a(y - k)^2 + h$ 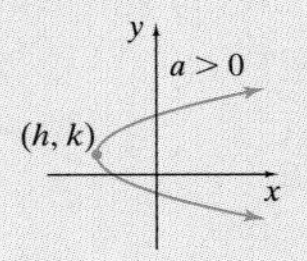 | 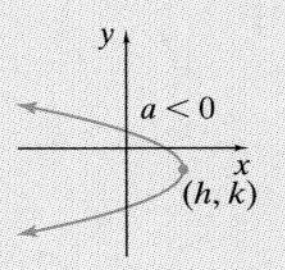 |

| | | |
|---|---|---|
| *Circle* | $(x - h)^2 + (y - k)^2 = r^2$ | 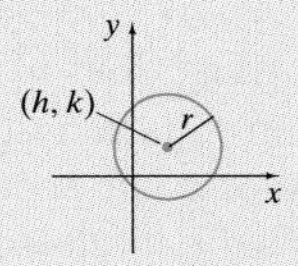 |

| | | |
|---|---|---|
| *Ellipse* center **(0, 0)** | $\dfrac{x^2}{a^2} + \dfrac{y^2}{b^2} = 1$ | 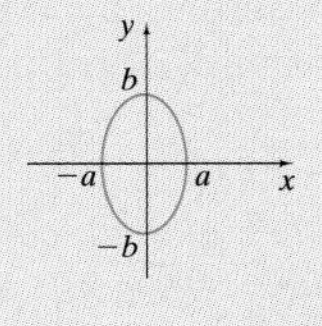 |

| | | |
|---|---|---|
| *Hyperbola* center **(0, 0)** | $\dfrac{x^2}{a^2} - \dfrac{y^2}{b^2} = 1$ | 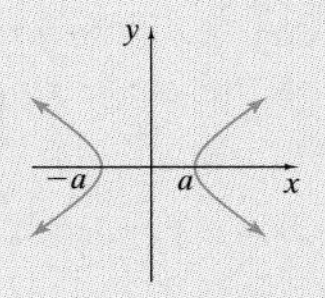 |

| | | |
|---|---|---|
| *Hyperbola* center **(0, 0)** | $\dfrac{y^2}{b^2} - \dfrac{x^2}{a^2} = 1$ | 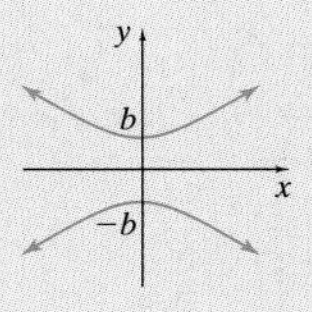 |

*Identify whether each equation, when graphed, will be a parabola, circle, ellipse, or hyperbola. Then graph each equation.*

**1.** $(x - 7)^2 + (y - 2)^2 = 4$

**2.** $y = x^2 + 4$

**3.** $y = x^2 + 12x + 36$

**4.** $\dfrac{x^2}{4} + \dfrac{y^2}{9} = 1$

**5.** $\dfrac{y^2}{9} - \dfrac{x^2}{9} = 1$

**6.** $\dfrac{x^2}{16} - \dfrac{y^2}{4} = 1$

**7.** $\dfrac{x^2}{16} + \dfrac{y^2}{4} = 1$

**8.** $x^2 + y^2 = 16$

**9.** $x = y^2 + 4y - 1$

**10.** $x = -y^2 + 6y$

**11.** $9x^2 - 4y^2 = 36$

**12.** $9x^2 + 4y^2 = 36$

**13.** $\dfrac{(x - 1)^2}{49} + \dfrac{(y + 2)^2}{25} = 1$

**14.** $y^2 = x^2 + 16$

**15.** $\left(x + \dfrac{1}{2}\right)^2 + \left(y - \dfrac{1}{2}\right)^2 = 1$

# 13.3 SOLVING NONLINEAR SYSTEMS OF EQUATIONS

**OBJECTIVES**

**1** Solve a nonlinear system by substitution.

**2** Solve a nonlinear system by elimination.

In Sections 4.1 through 4.3, we used graphing, substitution, and elimination (addition) methods to find solutions of systems of linear equations in two variables. We now apply these same methods to nonlinear systems of equations in two variables. A **nonlinear system of equations** is a system of equations at least one of which is not linear. Since we will be graphing the equations in each system, we are interested in real number solutions only.

**OBJECTIVE 1** ▶ **Solving nonlinear systems by substitution.**   First, nonlinear systems are solved by the substitution method.

**EXAMPLE 1**   Solve the system

$$\begin{cases} x^2 - 3y = 1 \\ x - y = 1 \end{cases}$$

*Solution*   We can solve this system by substitution if we solve one equation for one of the variables. Solving the first equation for $x$ is not the best choice since doing so introduces a radical. Also, solving for $y$ in the first equation introduces a fraction. We solve the second equation for $y$.

$$x - y = 1 \quad \text{Second equation}$$
$$x - 1 = y \quad \text{Solve for } y.$$

Replace $y$ with $x - 1$ in the first equation, and then solve for $x$.

$$x^2 - 3y = 1 \quad \text{First equation}$$
$$x^2 - 3(x - 1) = 1 \quad \text{Replace } y \text{ with } x - 1.$$
$$x^2 - 3x + 3 = 1$$
$$x^2 - 3x + 2 = 0$$
$$(x - 2)(x - 1) = 0$$
$$x = 2 \quad \text{or} \quad x = 1$$

Let $x = 2$ and then let $x = 1$ in the equation $y = x - 1$ to find corresponding $y$-values.

| Let $x = 2$. | Let $x = 1$. |
|---|---|
| $y = x - 1$ | $y = x - 1$ |
| $y = 2 - 1 = 1$ | $y = 1 - 1 = 0$ |

The solutions are $(2, 1)$ and $(1, 0)$ or the solution set is $\{(2, 1), (1, 0)\}$. Check both solutions in both equations. Both solutions satisfy both equations, so both are solutions of the system. The graph of each equation in the system is shown next. Intersections of the graphs are at $(2, 1)$ and $(1, 0)$.

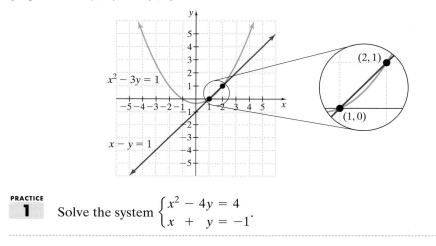

**PRACTICE**
**1**   Solve the system $\begin{cases} x^2 - 4y = 4 \\ x + y = -1 \end{cases}$.

**EXAMPLE 2**  Solve the system

$$\begin{cases} y = \sqrt{x} \\ x^2 + y^2 = 6 \end{cases}$$

*Solution*  This system is ideal for substitution since $y$ is expressed in terms of $x$ in the first equation. Notice that if $y = \sqrt{x}$, then both $x$ and $y$ must be nonnegative if they are real numbers. Substitute $\sqrt{x}$ for $y$ in the second equation, and solve for $x$.

$$x^2 + y^2 = 6$$
$$x^2 + \left(\sqrt{x}\right)^2 = 6 \quad \text{Let } y = \sqrt{x}$$
$$x^2 + x = 6$$
$$x^2 + x - 6 = 0$$
$$(x + 3)(x - 2) = 0$$
$$x = -3 \quad \text{or} \quad x = 2$$

The solution $-3$ is discarded because we have noted that $x$ must be nonnegative. To see this, let $x = -3$ in the first equation. Then let $x = 2$ in the first equation to find a corresponding $y$-value.

| Let $x = -3$. | Let $x = 2$. |
|---|---|
| $y = \sqrt{x}$ | $y = \sqrt{x}$ |
| $y = \sqrt{-3}$  Not a real number | $y = \sqrt{2}$ |

Since we are interested only in real number solutions, the only solution is $\left(2, \sqrt{2}\right)$. Check to see that this solution satisfies both equations. The graph of each equation in the system is shown next.

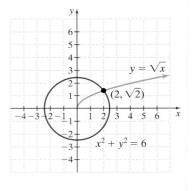

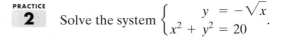

**PRACTICE**
**2**  Solve the system $\begin{cases} y = -\sqrt{x} \\ x^2 + y^2 = 20 \end{cases}$.

**EXAMPLE 3**  Solve the system

$$\begin{cases} x^2 + y^2 = 4 \\ x + y = 3 \end{cases}$$

*Solution*  We use the substitution method and solve the second equation for $x$.

$$x + y = 3 \qquad \text{Second equation}$$
$$x = 3 - y$$

Now we let $x = 3 - y$ in the first equation.

$$x^2 + y^2 = 4 \quad \text{First equation}$$

$$(3 - y)^2 + y^2 = 4 \quad \text{Let } x = 3 - y.$$
$$9 - 6y + y^2 + y^2 = 4$$
$$2y^2 - 6y + 5 = 0$$

By the quadratic formula, where $a = 2, b = -6$, and $c = 5$, we have

$$y = \frac{6 \pm \sqrt{(-6)^2 - 4 \cdot 2 \cdot 5}}{2 \cdot 2} = \frac{6 \pm \sqrt{-4}}{4}$$

Since $\sqrt{-4}$ is not a real number, there is no real solution, or $\varnothing$. Graphically, the circle and the line do not intersect, as shown below.

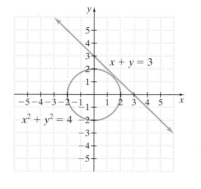

**PRACTICE**
**3**   Solve the system $\begin{cases} x^2 + y^2 = 9 \\ x - y = 5 \end{cases}$.

## Concept Check ☑

Without solving, how can you tell that $x^2 + y^2 = 9$ and $x^2 + y^2 = 16$ do not have any points of intersection?

**OBJECTIVE 2** ▶ **Solving nonlinear systems by elimination.** Some nonlinear systems may be solved by the elimination method.

**EXAMPLE 4**   Solve the system

$$\begin{cases} x^2 + 2y^2 = 10 \\ x^2 - y^2 = 1 \end{cases}$$

*Solution*   We will use the elimination, or addition, method to solve this system. To eliminate $x^2$ when we add the two equations, multiply both sides of the second equation by $-1$. Then

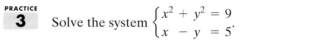

$$\begin{cases} x^2 + 2y^2 = 10 \\ (-1)(x^2 - y^2) = -1 \cdot 1 \end{cases} \text{ is equivalent to } \begin{cases} \underline{\begin{array}{r} x^2 + 2y^2 = 10 \\ -x^2 + y^2 = -1 \end{array}} \\ \begin{array}{r} 3y^2 = 9 \quad \text{Add.} \\ y^2 = 3 \quad \text{Divide both} \\ y = \pm\sqrt{3} \quad \text{sides by 3.} \end{array} \end{cases}$$

To find the corresponding $x$-values, we let $y = \sqrt{3}$ and $y = -\sqrt{3}$ in either original equation. We choose the second equation.

| Let $y = \sqrt{3}$. | Let $y = -\sqrt{3}$. |
|---|---|
| $x^2 - y^2 = 1$ | $x^2 - y^2 = 1$ |
| $x^2 - (\sqrt{3})^2 = 1$ | $x^2 - (-\sqrt{3})^2 = 1$ |
| $x^2 - 3 = 1$ | $x^2 - 3 = 1$ |
| $x^2 = 4$ | $x^2 = 4$ |
| $x = \pm\sqrt{4} = \pm 2$ | $x = \pm\sqrt{4} = \pm 2$ |

**Answer to Concept Check:**
$x^2 + y^2 = 9$ is a circle inside the circle $x^2 + y^2 = 16$, therefore they do not have any points of intersection.

The solutions are $(2, \sqrt{3})$, $(-2, \sqrt{3})$, $(2, -\sqrt{3})$, and $(-2, -\sqrt{3})$. Check all four ordered pairs in both equations of the system. The graph of each equation in this system is shown.

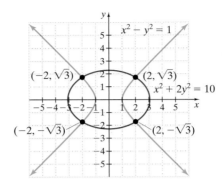

**PRACTICE**

**4**　Solve the system $\begin{cases} x^2 + 4y^2 = 16 \\ x^2 - y^2 = 1 \end{cases}$ .

## 13.3 | EXERCISE SET

*MyMathLab* Powered by CourseCompass™ and MathXL®

Math XP
PRACTICE　WATCH　DOWNLOAD　READ　REVIEW

### MIXED PRACTICE

*Solve each nonlinear system of equations for real solutions. See Examples 1 through 4.*

**1.** $\begin{cases} x^2 + y^2 = 25 \\ 4x + 3y = 0 \end{cases}$

**2.** $\begin{cases} x^2 + y^2 = 25 \\ 3x + 4y = 0 \end{cases}$

**3.** $\begin{cases} x^2 + 4y^2 = 10 \\ y = x \end{cases}$

**4.** $\begin{cases} 4x^2 + y^2 = 10 \\ y = x \end{cases}$

**5.** $\begin{cases} y^2 = 4 - x \\ x - 2y = 4 \end{cases}$

**6.** $\begin{cases} x^2 + y^2 = 4 \\ x + y = -2 \end{cases}$

**7.** $\begin{cases} x^2 + y^2 = 9 \\ 16x^2 - 4y^2 = 64 \end{cases}$

**8.** $\begin{cases} 4x^2 + 3y^2 = 35 \\ 5x^2 + 2y^2 = 42 \end{cases}$

**9.** $\begin{cases} x^2 + 2y^2 = 2 \\ x - y = 2 \end{cases}$

**10.** $\begin{cases} x^2 + 2y^2 = 2 \\ x^2 - 2y^2 = 6 \end{cases}$

**11.** $\begin{cases} y = x^2 - 3 \\ 4x - y = 6 \end{cases}$

**12.** $\begin{cases} y = x + 1 \\ x^2 - y^2 = 1 \end{cases}$

**13.** $\begin{cases} y = x^2 \\ 3x + y = 10 \end{cases}$

**14.** $\begin{cases} 6x - y = 5 \\ xy = 1 \end{cases}$

**15.** $\begin{cases} y = 2x^2 + 1 \\ x + y = -1 \end{cases}$

**16.** $\begin{cases} x^2 + y^2 = 9 \\ x + y = 5 \end{cases}$

**17.** $\begin{cases} y = x^2 - 4 \\ y = x^2 - 4x \end{cases}$

**18.** $\begin{cases} x = y^2 - 3 \\ x = y^2 - 3y \end{cases}$

**19.** $\begin{cases} 2x^2 + 3y^2 = 14 \\ -x^2 + y^2 = 3 \end{cases}$

**20.** $\begin{cases} 4x^2 - 2y^2 = 2 \\ -x^2 + y^2 = 2 \end{cases}$

**21.** $\begin{cases} x^2 + y^2 = 1 \\ x^2 + (y + 3)^2 = 4 \end{cases}$

**22.** $\begin{cases} x^2 + 2y^2 = 4 \\ x^2 - y^2 = 4 \end{cases}$

**23.** $\begin{cases} y = x^2 + 2 \\ y = -x^2 + 4 \end{cases}$

**24.** $\begin{cases} x = -y^2 - 3 \\ x = y^2 - 5 \end{cases}$

**25.** $\begin{cases} 3x^2 + y^2 = 9 \\ 3x^2 - y^2 = 9 \end{cases}$

**26.** $\begin{cases} x^2 + y^2 = 25 \\ \quad x = y^2 - 5 \end{cases}$

**27.** $\begin{cases} x^2 + 3y^2 = 6 \\ x^2 - 3y^2 = 10 \end{cases}$

**28.** $\begin{cases} x^2 + y^2 = 1 \\ \quad y = x^2 - 9 \end{cases}$

**29.** $\begin{cases} x^2 + y^2 = 36 \\ \quad y = \dfrac{1}{6}x^2 - 6 \end{cases}$

**30.** $\begin{cases} x^2 + y^2 = 16 \\ \quad y = -\dfrac{1}{4}x^2 + 4 \end{cases}$

## REVIEW AND PREVIEW

*Graph each inequality in two variables. See Section 9.4.*

**31.** $x > -3$

**32.** $y \le 1$

**33.** $y < 2x - 1$

**34.** $3x - y \le 4$

*Find the perimeter of each geometric figure. See Section 5.2.*

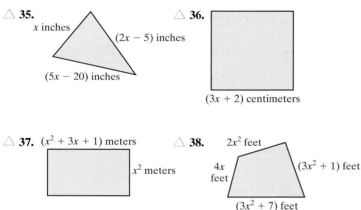

△ **35.** $x$ inches, $(2x - 5)$ inches, $(5x - 20)$ inches

△ **36.** $(3x + 2)$ centimeters

△ **37.** $(x^2 + 3x + 1)$ meters, $x^2$ meters

△ **38.** $2x^2$ feet, $4x$ feet, $(3x^2 + 1)$ feet, $(3x^2 + 7)$ feet

## CONCEPT EXTENSIONS

*For the exercises below, see the Concept Check in this section.*

**39.** Without graphing, how can you tell that the graph of $x^2 + y^2 = 1$ and $x^2 + y^2 = 4$ do not have any points of intersection?

**40.** Without solving, how can you tell that the graphs of $y = 2x + 3$ and $y = 2x + 7$ do not have any points of intersection?

**41.** How many real solutions are possible for a system of equations whose graphs are a circle and a parabola? Draw diagrams to illustrate each possibility.

**42.** How many real solutions are possible for a system of equations whose graphs are an ellipse and a line? Draw diagrams to illustrate each possibility.

*Solve.*

**43.** The sum of the squares of two numbers is 130. The difference of the squares of the two numbers is 32. Find the two numbers.

**44.** The sum of the squares of two numbers is 20. Their product is 8. Find the two numbers.

△ **45.** During the development stage of a new rectangular keypad for a security system, it was decided that the area of the rectangle should be 285 square centimeters and the perimeter should be 68 centimeters. Find the dimensions of the keypad.

△ **46.** A rectangular holding pen for cattle is to be designed so that its perimeter is 92 feet and its area is 525 feet. Find the dimensions of the holding pen.

*Recall that in business, a demand function expresses the quantity of a commodity demanded as a function of the commodity's unit price. A supply function expresses the quantity of a commodity supplied as a function of the commodity's unit price. When the quantity produced and supplied is equal to the quantity demanded, then we have what is called **market equilibrium.***

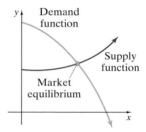

**47.** The demand function for a certain compact disc is given by the function

$$p = -0.01x^2 - 0.2x + 9$$

and the corresponding supply function is given by

$$p = 0.01x^2 - 0.1x + 3$$

where $p$ is in dollars and $x$ is in thousands of units. Find the equilibrium quantity and the corresponding price by solving the system consisting of the two given equations.

**48.** The demand function for a certain style of picture frame is given by the function

$$p = -2x^2 + 90$$

and the corresponding supply function is given by

$$p = 9x + 34$$

where $p$ is in dollars and $x$ is in thousands of units. Find the equilibrium quantity and the corresponding price by solving the system consisting of the two given equations.

*Use a graphing calculator to verify the results of each exercise.*

**49.** Exercise 3.

**50.** Exercise 4.

**51.** Exercise 23.

**52.** Exercise 24.

**Are You Preparing for Your Final Exam?**

To prepare for your final exam, try the following study techniques:

- Review the material that you will be responsible for on your exam. This includes material from your textbook, your notebook, and any handouts from your instructor.
- Review any formulas that you may need to memorize.
- Check to see if your instructor or mathematics department will be conducting a final exam review.
- Check with your instructor to see whether final exams from previous semesters/quarters are available to students for review.

- Use your previously taken exams as a practice final exam. To do so, rewrite the test questions in mixed order on blank sheets of paper. This will help you prepare for exam conditions.
- If you are unsure of a few concepts, see your instructor or visit a learning lab for assistance. Also, view the video segment of any troublesome sections.
- If you need further exercises to work, try the Cumulative Reviews at the end of the chapters.

Once again, good luck! I hope you have enjoyed this textbook and your mathematics course.

# 13.4 NONLINEAR INEQUALITIES AND SYSTEMS OF INEQUALITIES

**OBJECTIVES**

1 Graph a nonlinear inequality.

2 Graph a system of nonlinear inequalities.

**OBJECTIVE 1 ▶ Graphing nonlinear inequalities.** We can graph a nonlinear inequality in two variables such as $\frac{x^2}{9} + \frac{y^2}{16} \leq 1$ in a way similar to the way we graphed a linear inequality in two variables in Section 9.4. First, we graph the related equation $\frac{x^2}{9} + \frac{y^2}{16} = 1$. The graph of the equation is our boundary. Then, using test points, we determine and shade the region whose points satisfy the inequality.

**EXAMPLE 1**   Graph $\frac{x^2}{9} + \frac{y^2}{16} \leq 1$.

*Solution*   First, graph the equation $\frac{x^2}{9} + \frac{y^2}{16} = 1$. Sketch a solid curve since the graph of $\frac{x^2}{9} + \frac{y^2}{16} \leq 1$ includes the graph of $\frac{x^2}{9} + \frac{y^2}{16} = 1$. The graph is an ellipse, and it divides the plane into two regions, the "inside" and the "outside" of the ellipse. To determine which region contains the solutions, select a test point in either region and determine whether the coordinates of the point satisfy the inequality. We choose $(0, 0)$ as the test point.

$$\frac{x^2}{9} + \frac{y^2}{16} \leq 1$$

$$\frac{0^2}{9} + \frac{0^2}{16} \leq 1 \quad \text{Let } x = 0 \text{ and } y = 0.$$

$$0 \leq 1 \quad \text{True}$$

Since this statement is true, the solution set is the region containing $(0, 0)$. The graph of the solution set includes the points on and inside the ellipse, as shaded in the figure.

$$\frac{x^2}{9} + \frac{y^2}{16} \leq 1$$

**PRACTICE 1**   Graph $\frac{x^2}{36} + \frac{y^2}{16} \geq 1$.

**EXAMPLE 2**   Graph $4y^2 > x^2 + 16$.

*Solution*   The related equation is $4y^2 = x^2 + 16$. Subtract $x^2$ from both sides and divide both sides by 16, and we have $\dfrac{y^2}{4} - \dfrac{x^2}{16} = 1$, which is a hyperbola. Graph the hyperbola as a dashed curve since the graph of $4y^2 > x^2 + 16$ does *not* include the graph of $4y^2 = x^2 + 16$. The hyperbola divides the plane into three regions. Select a test point in each region—not on a boundary line—to determine whether that region contains solutions of the inequality.

| *Test Region A with* $(0, 4)$ | *Test Region B with* $(0, 0)$ | *Test Region C with* $(0, -4)$ |
|---|---|---|
| $4y^2 > x^2 + 16$ | $4y^2 > x^2 + 16$ | $4y^2 > x^2 + 16$ |
| $4(4)^2 > 0^2 + 16$ | $4(0)^2 > 0^2 + 16$ | $4(-4)^2 > 0^2 + 16$ |
| $64 > 16$   True | $0 > 16$   False | $64 > 16$   True |

The graph of the solution set includes the shaded regions $A$ and $C$ only, not the boundary.

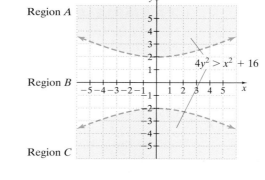

**PRACTICE**
**2**   Graph $16y^2 > 9x^2 + 144$.

**OBJECTIVE 2 ▶ Graphing systems of nonlinear inequalities.**   In Section 3.6 we graphed systems of linear inequalities. Recall that the graph of a system of inequalities is the intersection of the graphs of the inequalities.

**EXAMPLE 3**   Graph the system
$$\begin{cases} x \le 1 - 2y \\ y \le x^2 \end{cases}$$

*Solution*   We graph each inequality on the same set of axes. The intersection is shown in the third graph on the following page. It is the darkest shaded (appears purple) region along with its boundary lines. The coordinates of the points of intersection can be found by solving the related system.

$$\begin{cases} x = 1 - 2y \\ y = x^2 \end{cases}$$

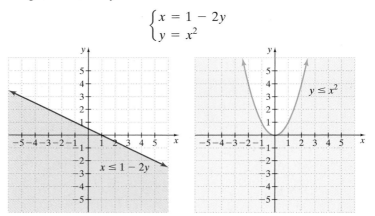

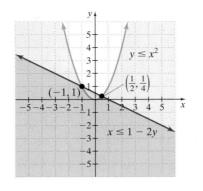

PRACTICE
**3** Graph the system $\begin{cases} y \geq x^2 \\ y \leq -3x + 2 \end{cases}$.

---

**EXAMPLE 4** Graph the system

$$\begin{cases} x^2 + y^2 < 25 \\ \dfrac{x^2}{9} - \dfrac{y^2}{25} < 1 \\ y < x + 3 \end{cases}$$

*Solution* We graph each inequality. The graph of $x^2 + y^2 < 25$ contains points "inside" the circle that has center $(0, 0)$ and radius 5. The graph of $\dfrac{x^2}{9} - \dfrac{y^2}{25} < 1$ is the region between the two branches of the hyperbola with $x$-intercepts $-3$ and $3$ and center $(0, 0)$. The graph of $y < x + 3$ is the region "below" the line with slope 1 and $y$-intercept $(0, 3)$. The graph of the solution set of the system is the intersection of all the graphs, the darkest shaded region shown. The boundary of this region is not part of the solution.

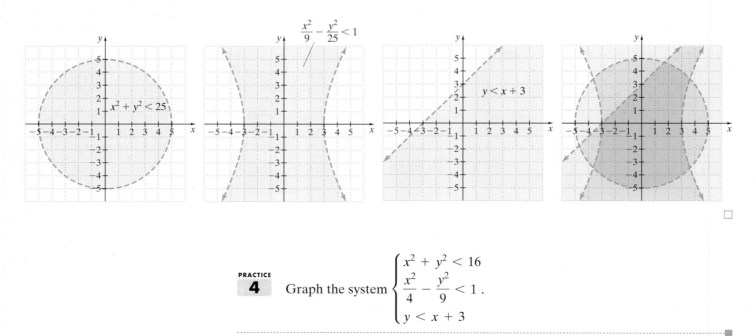

PRACTICE
**4** Graph the system $\begin{cases} x^2 + y^2 < 16 \\ \dfrac{x^2}{4} - \dfrac{y^2}{9} < 1 \\ y < x + 3 \end{cases}$.

## 13.4 EXERCISE SET

MyMathLab®

PRACTICE  WATCH  DOWNLOAD  READ  REVIEW

*Graph each inequality. See Examples 1 and 2.*

**1.** $y < x^2$

**2.** $y < -x^2$

**3.** $x^2 + y^2 \geq 16$

**4.** $x^2 + y^2 < 36$

**5.** $\dfrac{x^2}{4} - y^2 < 1$

**6.** $x^2 - \dfrac{y^2}{9} \geq 1$

**7.** $y > (x - 1)^2 - 3$

**8.** $y > (x + 3)^2 + 2$

**9.** $x^2 + y^2 \leq 9$

**10.** $x^2 + y^2 > 4$

**11.** $y > -x^2 + 5$

**12.** $y < -x^2 + 5$

**13.** $\dfrac{x^2}{4} + \dfrac{y^2}{9} \leq 1$

**14.** $\dfrac{x^2}{25} + \dfrac{y^2}{4} \geq 1$

**15.** $\dfrac{y^2}{4} - x^2 \leq 1$

**16.** $\dfrac{y^2}{16} - \dfrac{x^2}{9} > 1$

**17.** $y < (x - 2)^2 + 1$

**18.** $y > (x - 2)^2 + 1$

**19.** $y \leq x^2 + x - 2$

**20.** $y > x^2 + x - 2$

*Graph each system. See Examples 3 and 4.*

**21.** $\begin{cases} 4x + 3y \geq 12 \\ x^2 + y^2 < 16 \end{cases}$

**22.** $\begin{cases} 3x - 4y \leq 12 \\ x^2 + y^2 < 16 \end{cases}$

**23.** $\begin{cases} x^2 + y^2 \leq 9 \\ x^2 + y^2 \geq 1 \end{cases}$

**24.** $\begin{cases} x^2 + y^2 \geq 9 \\ x^2 + y^2 \geq 16 \end{cases}$

**25.** $\begin{cases} y > x^2 \\ y \geq 2x + 1 \end{cases}$

**26.** $\begin{cases} y \leq -x^2 + 3 \\ y \leq 2x - 1 \end{cases}$

**27.** $\begin{cases} x^2 + y^2 > 9 \\ y > x^2 \end{cases}$

**28.** $\begin{cases} x^2 + y^2 \leq 9 \\ y < x^2 \end{cases}$

**29.** $\begin{cases} \dfrac{x^2}{4} + \dfrac{y^2}{9} \geq 1 \\ x^2 + y^2 \geq 4 \end{cases}$

**30.** $\begin{cases} x^2 + (y - 2)^2 \geq 9 \\ \dfrac{x^2}{4} + \dfrac{y^2}{25} < 1 \end{cases}$

**31.** $\begin{cases} x^2 - y^2 \geq 1 \\ y \geq 0 \end{cases}$

**32.** $\begin{cases} x^2 - y^2 \geq 1 \\ x \geq 0 \end{cases}$

**33.** $\begin{cases} x + y \geq 1 \\ 2x + 3y < 1 \\ x > -3 \end{cases}$

**34.** $\begin{cases} x - y < -1 \\ 4x - 3y > 0 \\ y > 0 \end{cases}$

**35.** $\begin{cases} x^2 - y^2 < 1 \\ \dfrac{x^2}{16} + y^2 \leq 1 \\ x \geq -2 \end{cases}$

**36.** $\begin{cases} x^2 - y^2 \geq 1 \\ \dfrac{x^2}{16} + \dfrac{y^2}{4} \leq 1 \\ y \geq 1 \end{cases}$

### REVIEW AND PREVIEW

*Determine which graph is the graph of a function. See Section 3.6.*

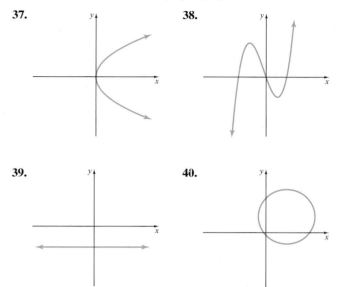

**37.**

**38.**

**39.**

**40.**

*Find each function value if $f(x) = 3x^2 - 2$. See Section 3.6.*

**41.** $f(-1)$

**42.** $f(-3)$

**43.** $f(a)$

**44.** $f(b)$

### CONCEPT EXTENSIONS

**45.** Discuss how graphing a linear inequality such as $x + y < 9$ is similar to graphing a nonlinear inequality such as $x^2 + y^2 < 9$.

**46.** Discuss how graphing a linear inequality such as $x + y < 9$ is different from graphing a nonlinear inequality such as $x^2 + y^2 < 9$.

**47.** Graph the system $\begin{cases} y \leq x^2 \\ y \geq x + 2 \\ x \geq 0 \\ y \geq 0 \end{cases}$.

## CHAPTER 13 GROUP ACTIVITY

### Modeling Conic Sections

In this project, you will have the opportunity to construct and investigate a model of an ellipse. You will need two thumbtacks or nails, graph paper, cardboard, tape, string, a pencil, and a ruler. This project may be completed by working in groups or individually.

Follow these steps, answering any questions as you go.

**1.** Draw an *x*-axis and a *y*-axis on the graph paper as shown in Figure 1.

**2.** Place the graph paper on the cardboard and attach it with tape.

**3.** Locate two points on the *x*-axis, each about $1\frac{1}{2}$ inches from the origin and on opposite sides of the origin (see Figure 1). Insert thumbtacks (or nails) at each of these locations.

**4.** Fasten a 9-inch piece of string to the thumbtacks as shown in Figure 2. Use your pencil to draw and keep the string taut while you carefully move the pencil in a path all around the thumbtacks.

**5.** Using the grid of the graph paper as a guide, find an approximate equation of the ellipse you drew.

**6.** Experiment by moving the tacks closer together or farther apart and drawing new ellipses. What do you observe?

**7.** Write a paragraph explaining why the figure drawn by the pencil is an ellipse. How might you use the same materials to draw a circle?

**8.** (Optional) Choose one of the ellipses you drew with the string and pencil. Use a ruler to draw any six tangent lines to the ellipse. (A line is tangent to the ellipse if it intersects, or just touches, the ellipse at only one point. See Figure 3.) Extend the tangent lines to yield six points of intersection among the tangents. Use a straightedge to draw a line connecting each pair of opposite points of intersection. What do you observe? Repeat with a different ellipse. Can you make a conjecture about the relationship among the lines that connect opposite points of intersection?

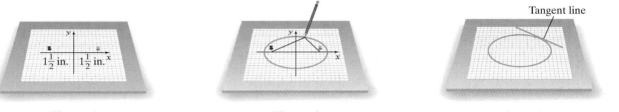

| Figure 1 | Figure 2 | Figure 3 |

---

### 📖 STUDY SKILLS BUILDER

#### Are You Prepared for a Test on Chapter 13?

Below I have listed some common trouble areas for students in Chapter 13. After studying for your test—but before taking your test—read these.

- Don't forget to review all the standard forms for the conic sections.

- Don't forget that both methods, substitution and elimination, are available for solving nonlinear systems of equations.

$$\begin{cases} x^2 + y^2 = 7 \\ 2x^2 - 3y^2 = 4 \end{cases} \text{ is equivalent to}$$

$$\begin{cases} 3x^2 + 3y^2 = 21 \\ \underline{2x^2 - 3y^2 = \phantom{0}4} \\ 5x^2 \phantom{+ 3y^2} = 25 \\ x^2 = 5 \\ x = \pm\sqrt{5} \end{cases}$$

Let $x = \pm\sqrt{5}$ in either original equation, and $y = \pm\sqrt{2}$, the solution set is $\{(\sqrt{5}, \sqrt{2}), (-\sqrt{5}, \sqrt{2}), (\sqrt{5}, -\sqrt{2}), (-\sqrt{5}, -\sqrt{2})\}$.

Remember: This is simply a checklist of common trouble areas. For a review of Chapter 13, see the Highlights and Chapter Review at the end of this chapter.

---

## CHAPTER 13 VOCABULARY CHECK

*Fill in each blank with one of the words or phrases listed below.*

circle          radius          center

ellipse          hyperbola          nonlinear system of equations

**1.** A(n) _____ is the set of all points in a plane that are the same distance from a fixed point, called the _____ .

**2.** A _____ is a system of equations at least one of which is not linear.

**3.** A(n) _____ is the set of points on a plane such that the sum of the distances of those points from two fixed points is a constant.

**4.** In a circle, the distance from the center to a point of the circle is called its _____ .

**5.** A(n) _____ is the set of points in a plane such that the absolute value of the difference of the distance from two fixed points is constant.

▶ **Helpful Hint**

Are you preparing for your test? Don't forget to take the Chapter 13 Test on page 798. Then check your answers at the back of the text and use the Chapter Test Prep Video CD to see the fully worked-out solutions to any of the exercises you want to review.

# CHAPTER 13 HIGHLIGHTS

| **DEFINITIONS AND CONCEPTS** | **EXAMPLES** |
|---|---|

### SECTION 13.1    THE PARABOLA AND THE CIRCLE

**Parabolas**

$$y = a(x - h)^2 + k$$

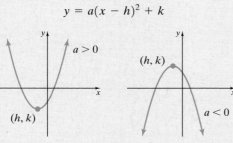

$$x = a(y - k)^2 + h$$

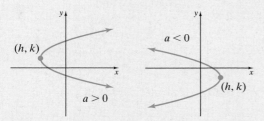

Graph

$$x = 3y^2 - 12y + 13.$$
$$x - 13 = 3y^2 - 12y$$
$$x - 13 + 3(4) = 3(y^2 - 4y + 4) \quad \text{Add } 3(4) \text{ to}$$
$$x = 3(y - 2)^2 + 1 \quad \text{both sides.}$$

Since $a = 3$, this parabola opens to the right with vertex $(1, 2)$. Its axis of symmetry is $y = 2$. The $x$-intercept is $(13, 0)$.

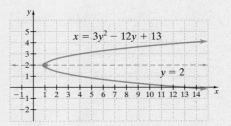

**Circle**

The graph of $(x - h)^2 + (y - k)^2 = r^2$ is a circle with center $(h, k)$ and radius $r$.

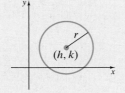

Graph $x^2 + (y + 3)^2 = 5$.

This equation can be written as

$$(x - 0)^2 + (y + 3)^2 = 5 \text{ with } h = 0,$$
$$k = -3, \text{ and } r = \sqrt{5}.$$

The center of this circle is $(0, -3)$, and the radius is $\sqrt{5}$.

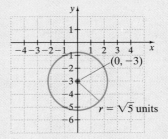

| **DEFINITIONS AND CONCEPTS** | **EXAMPLES** |
|---|---|

*Ellipse with Center* **(0, 0)**

The graph of an equation of the form $\dfrac{x^2}{a^2} + \dfrac{y^2}{b^2} = 1$ is an ellipse with center $(0,0)$. The $x$-intercepts are $(a,0)$ and $(-a,0)$, and the $y$-intercepts are $(0,b)$ and $(0,-b)$.

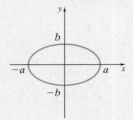

Graph $4x^2 + 9y^2 = 36$.

$$\frac{x^2}{9} + \frac{y^2}{4} = 1 \quad \text{Divide by 36.}$$

$$\frac{x^2}{3^2} + \frac{y^2}{2^2} = 1$$

The ellipse has center $(0,0)$, $x$-intercepts $(3,0)$ and $(-3,0)$, and $y$-intercepts $(0,2)$ and $(0,-2)$.

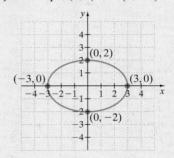

*Hyperbola with Center* **(0, 0)**

The graph of an equation of the form $\dfrac{x^2}{a^2} - \dfrac{y^2}{b^2} = 1$ is a hyperbola with center $(0,0)$ and $x$-intercepts $(a,0)$ and $(-a,0)$.

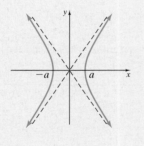

Graph $\dfrac{x^2}{9} - \dfrac{y^2}{4} = 1$. Here $a = 3$ and $b = 2$.

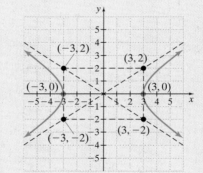

The graph of an equation of the form $\dfrac{y^2}{b^2} - \dfrac{x^2}{a^2} = 1$ is a hyperbola with center $(0,0)$ and $y$-intercepts $(0,b)$ and $(0,-b)$.

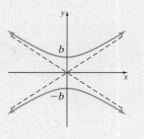

| DEFINITIONS AND CONCEPTS | EXAMPLES |
|---|---|

### SECTION 13.3 SOLVING NONLINEAR SYSTEMS OF EQUATIONS

A **nonlinear system of equations** is a system of equations at least one of which is not linear. Both the substitution method and the elimination method may be used to solve a nonlinear system of equations.

Solve the nonlinear system $\begin{cases} y = x + 2 \\ 2x^2 + y^2 = 3 \end{cases}$.

Substitute $x + 2$ for $y$ in the second equation.

$$2x^2 + y^2 = 3$$
$$2x^2 + (x + 2)^2 = 3$$
$$2x^2 + x^2 + 4x + 4 = 3$$
$$3x^2 + 4x + 1 = 0$$
$$(3x + 1)(x + 1) = 0$$
$$x = -\frac{1}{3}, x = -1$$

If $x = -\frac{1}{3}, y = x + 2 = -\frac{1}{3} + 2 = \frac{5}{3}$.

If $x = -1, y = x + 2 = -1 + 2 = 1$.

The solutions are $\left(-\frac{1}{3}, \frac{5}{3}\right)$ and $(-1, 1)$.

### SECTION 13.4 NONLINEAR INEQUALITIES AND SYSTEMS OF INEQUALITIES

The graph of a system of inequalities is the intersection of the graphs of the inequalities.

Graph the system $\begin{cases} x \geq y^2 \\ x + y \leq 4 \end{cases}$.

The graph of the system is the pink shaded region along with its boundary lines.

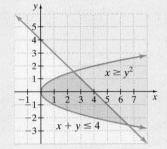

## CHAPTER 13 REVIEW

**(13.1)** *Write an equation of the circle with the given center and radius.*

**1.** center $(-4, 4)$ radius 3

**2.** center $(5, 0)$, radius 5

**3.** center $(-7, -9)$, radius $\sqrt{11}$

**4.** center $(0, 0)$, radius $\frac{7}{2}$

*Sketch the graph of the equation. If the graph is a circle, find its center. If the graph is a parabola, find its vertex.*

**5.** $x^2 + y^2 = 7$

**6.** $x = 2(y - 5)^2 + 4$

**7.** $x = -(y + 2)^2 + 3$

**8.** $(x - 1)^2 + (y - 2)^2 = 4$

**9.** $y = -x^2 + 4x + 10$

**10.** $x = -y^2 - 4y + 6$

**11.** $x = \frac{1}{2}y^2 + 2y + 1$

**12.** $y = -3x^2 + \frac{1}{2}x + 4$

**13.** $x^2 + y^2 + 2x + y = \dfrac{3}{4}$    **14.** $x^2 + y^2 - 3y = \dfrac{7}{4}$

**15.** $4x^2 + 4y^2 + 16x + 8y = 1$

*(13.2) Sketch the graph of each equation.*

**16.** $x^2 + \dfrac{y^2}{4} = 1$    **17.** $x^2 - \dfrac{y^2}{4} = 1$

**18.** $\dfrac{x^2}{5} + \dfrac{y^2}{5} = 1$    **19.** $\dfrac{x^2}{5} - \dfrac{y^2}{5} = 1$

**20.** $-5x^2 + 25y^2 = 125$    **21.** $4y^2 + 9x^2 = 36$

**22.** $x^2 - y^2 = 1$    **23.** $\dfrac{(x+3)^2}{9} + \dfrac{(y-4)^2}{25} = 1$

**24.** $y^2 = x^2 + 9$    **25.** $x^2 = 4y^2 - 16$

**26.** $100 - 25x^2 = 4y^2$

*(13.3) Solve each system of equations.*

**27.** $\begin{cases} y = 2x - 4 \\ y^2 = 4x \end{cases}$

**28.** $\begin{cases} x^2 + y^2 = 4 \\ x - y = 4 \end{cases}$

**29.** $\begin{cases} y = x + 2 \\ y = x^2 \end{cases}$

**30.** $\begin{cases} x^2 + 4y^2 = 16 \\ x^2 + y^2 = 4 \end{cases}$

**31.** $\begin{cases} 4x - y^2 = 0 \\ 2x^2 + y^2 = 16 \end{cases}$

**32.** $\begin{cases} x^2 + 2y = 9 \\ 5x - 2y = 5 \end{cases}$

**33.** $\begin{cases} y = 3x^2 + 5x - 4 \\ y = 3x^2 - x + 2 \end{cases}$

**34.** $\begin{cases} x^2 - 3y^2 = 1 \\ 4x^2 + 5y^2 = 21 \end{cases}$

△ **35.** Find the length and the width of a room whose area is 150 square feet and whose perimeter is 50 feet.

**36.** What is the greatest number of real solutions possible for a system of two equations whose graphs are an ellipse and a hyperbola?

*(13.4) Graph the inequality or system of inequalities.*

**37.** $y \le -x^2 + 3$    **38.** $x^2 + y^2 < 9$

**39.** $\begin{cases} 2x \le 4 \\ x + y \ge 1 \end{cases}$    **40.** $\dfrac{x^2}{4} + \dfrac{y^2}{9} \ge 1$

**41.** $\begin{cases} x^2 + y^2 < 4 \\ x^2 - y^2 \le 1 \end{cases}$    **42.** $\begin{cases} x^2 + y^2 \le 16 \\ x^2 + y^2 \ge 4 \end{cases}$

## MIXED REVIEW

**43.** Write an equation of the circle with center $(-7, 8)$ and radius 5.

*Graph each equation.*

**44.** $3x^2 + 6x + 3y^2 = 9$    **45.** $y = x^2 + 6x + 9$

**46.** $x = y^2 + 6y + 9$    **47.** $\dfrac{y^2}{4} - \dfrac{x^2}{16} = 1$

**48.** $\dfrac{y^2}{4} + \dfrac{x^2}{16} = 1$    **49.** $\dfrac{(x-2)^2}{4} + (y-1)^2 = 1$

**50.** $y^2 = x^2 + 6$    **51.** $y^2 + x^2 = 4x + 6$

**52.** $x^2 + y^2 - 8y = 0$    **53.** $6(x-2)^2 + 9(y+5)^2 = 36$

**54.** $\dfrac{x^2}{16} - \dfrac{y^2}{25} = 1$

*Solve each system of equations.*

**55.** $\begin{cases} y = x^2 - 5x + 1 \\ y = -x + 6 \end{cases}$

**56.** $\begin{cases} x^2 + y^2 = 10 \\ 9x^2 + y^2 = 18 \end{cases}$

*Graph each inequality or system of inequalities.*

**57.** $x^2 - y^2 < 1$    **58.** $\begin{cases} y > x^2 \\ x + y \ge 3 \end{cases}$

## CHAPTER 13 TEST TEST PREP VIDEO

Remember to use the Chapter Test Prep Video CD to see the fully worked-out solutions to any of the exercises you want to review.

*Sketch the graph of each equation.*

**1.** $x^2 + y^2 = 36$    **2.** $x^2 - y^2 = 36$

**3.** $16x^2 + 9y^2 = 144$    **4.** $y = x^2 - 8x + 16$

**5.** $x^2 + y^2 + 6x = 16$    **6.** $x = y^2 + 8y - 3$

**7.** $\dfrac{(x-4)^2}{16} + \dfrac{(y-3)^2}{9} = 1$    **8.** $y^2 - x^2 = 1$

*Solve each system.*

**9.** $\begin{cases} x^2 + y^2 = 26 \\ x^2 - 2y^2 = 23 \end{cases}$    **10.** $\begin{cases} y = x^2 - 5x + 6 \\ y = 2x \end{cases}$

*Graph the solution of each system.*

**11.** $\begin{cases} 2x + 5y \geq 10 \\ \quad\; y \geq x^2 + 1 \end{cases}$

**12.** $\begin{cases} \dfrac{x^2}{4} + y^2 \leq 1 \\ x + y > 1 \end{cases}$

**13.** $\begin{cases} x^2 + y^2 \geq 4 \\ x^2 + y^2 < 16 \\ \quad\; y \geq 0 \end{cases}$

**14.** A bridge has an arch in the shape of a half-ellipse. If the equation of the ellipse, measured in feet, is $100x^2 + 225y^2 = 22{,}500$, find the height of the arch from the road and the width of the arch.

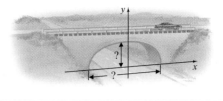

# CHAPTER 13 CUMULATIVE REVIEW

**1.** Solve $2x \geq 0$ and $4x - 1 \leq -9$.

**2.** Solve $3x + 4 > 1$ and $2x - 5 \leq 9$. Write the solution in interval notation.

**3.** Solve $5x - 3 \leq 10$ or $x + 1 \geq 5$.

**4.** Find the slope of the line that goes through $(3, 2)$ and $(1, -4)$.

*Solve.*

**5.** $|5w + 3| = 7$

**6.** Two planes leave Greensboro, one traveling north and the other south. After 2 hours they are 650 miles apart. If one plane is flying 25 mph faster than the other, what is the speed of each?

**7.** $\left| \dfrac{x}{2} - 1 \right| = 11$

**8.** Use the quotient rule to simplify.

    **a.** $\dfrac{4^8}{4^3}$     **b.** $\dfrac{y^{11}}{y^5}$

    **c.** $\dfrac{32x^7}{4x^6}$     **d.** $\dfrac{18a^{12}b^6}{12a^8b^6}$

**9.** Solve $|3x + 2| = |5x - 8|$.

**10.** Factor.

    **a.** $3y^2 + 14y + 15$

    **b.** $20a^5 + 54a^4 + 10a^3$

    **c.** $(y - 3)^2 - 2(y - 3) - 8$

**11.** Solve $|m - 6| < 2$.

**12.** Perform the indicated operation and simplify if possible.

$$\frac{2}{3a - 15} - \frac{a}{25 - a^2}$$

**13.** Simplify $\dfrac{x^{-1} + 2xy^{-1}}{x^{-2} - x^{-2}y^{-1}}$.

**14.** Simplify each complex fraction.

    **a.** $(a^{-1} - b^{-1})^{-1}$

    **b.** $\dfrac{2 - \dfrac{1}{x}}{4x - \dfrac{1}{x}}$

**15.** Solve $|2x + 9| + 5 > 3$.

**16.** Solve $\dfrac{2}{x + 3} = \dfrac{1}{x^2 - 9} - \dfrac{1}{x - 3}$.

**17.** Use the remainder theorem and synthetic division to find $P(4)$ if

$$P(x) = 4x^6 - 25x^5 + 35x^4 + 17x^2.$$

**18.** Suppose that $y$ varies inversely as $x$. If $y = 3$ when $x = \dfrac{2}{3}$, find the constant of variation and the inverse variation equation.

**19.** Find the cube roots.

    **a.** $\sqrt[3]{1}$     **b.** $\sqrt[3]{-64}$

    **c.** $\sqrt[3]{\dfrac{8}{125}}$     **d.** $\sqrt[3]{x^6}$

    **e.** $\sqrt[3]{-27x^9}$

**20.** Multiply and simplify if possible.

    **a.** $\sqrt{5}(2 + \sqrt{15})$

    **b.** $(\sqrt{3} - \sqrt{5})(\sqrt{7} - 1)$

    **c.** $(2\sqrt{5} - 1)^2$

    **d.** $(3\sqrt{2} + 5)(3\sqrt{2} - 5)$

**21.** Multiply.

    **a.** $z^{2/3}(z^{1/3} - z^5)$     **b.** $(x^{1/3} - 5)(x^{1/3} + 2)$

**22.** Rationalize the denominator $\dfrac{-2}{\sqrt{3} + 3}$.

**23.** Use the quotient rule to divide, and simplify if possible.

   **a.** $\dfrac{\sqrt{20}}{\sqrt{5}}$          **b.** $\dfrac{\sqrt{50x}}{2\sqrt{2}}$

   **c.** $\dfrac{7\sqrt[3]{48x^4y^8}}{\sqrt[3]{6y^2}}$     **d.** $\dfrac{2\sqrt[4]{32a^8b^6}}{\sqrt[4]{a^{-1}b^2}}$

**24.** Solve $\sqrt{2x-3} = x-3$.

**25.** Add or subtract as indicated.

   **a.** $\dfrac{\sqrt{45}}{4} - \dfrac{\sqrt{5}}{3}$     **b.** $\sqrt[3]{\dfrac{7x}{8}} + 2\sqrt[3]{7x}$

**26.** Use the discriminant to determine the number and type of solutions for $9x^2 - 6x = -4$.

**27.** Rationalize the denominator of $\sqrt{\dfrac{7x}{3y}}$.

**28.** Solve $\dfrac{4}{x-2} - \dfrac{x}{x+2} = \dfrac{16}{x^2-4}$.

**29.** Solve $\sqrt{2x-3} = 9$.

**30.** Solve $x^3 + 2x^2 - 4x \geq 8$.

**31.** Find the following powers of $i$.

   **a.** $i^7$          **b.** $i^{20}$

   **c.** $i^{46}$        **d.** $i^{-12}$

**32.** Graph $f(x) = (x+2)^2 - 1$

**33.** Solve $p^2 + 2p = 4$ by completing the square.

**34.** Find the maximum value of $f(x) = -x^2 - 6x + 4$.

**35.** Solve $\dfrac{1}{4}m^2 - m + \dfrac{1}{2} = 0$.

**36.** Find the inverse of $f(x) = \dfrac{x+1}{2}$.

**37.** Solve $p^4 - 3p^2 - 4 = 0$.

**38.** Use the quotient rule to simplify.

   **a.** $\dfrac{\sqrt{32}}{\sqrt{4}}$          **b.** $\dfrac{\sqrt[3]{240y^2}}{5\sqrt[3]{3y^{-4}}}$

   **c.** $\dfrac{\sqrt[5]{64x^9y^2}}{\sqrt[5]{2x^2y^{-8}}}$

**39.** Solve $\dfrac{x+2}{x-3} \leq 0$.

**40.** Graph $4x^2 + 9y^2 = 36$.

**41.** Graph $g(x) = \dfrac{1}{2}(x+2)^2 + 5$. Find the vertex and the axis of symmetry.

**42.** Solve each equation for $x$.

   **a.** $64^x = 4$       **b.** $125^{x-3} = 25$

   **c.** $\dfrac{1}{81} = 3^{2x}$

**43.** Find the vertex of the graph of $f(x) = x^2 - 4x - 12$.

**44.** Graph the system: $\begin{cases} x + 2y < 8 \\ \quad\quad y \geq x^2 \end{cases}$

**45.** Find the distance between $(2, -5)$ and $(1, -4)$. Give an exact distance and a three-decimal-place approximation.

**46.** Solve the system $\begin{cases} x^2 + y^2 = 36 \\ \quad\quad y = x + 6 \end{cases}$

# 14 Sequences, Series, and the Binomial Theorem

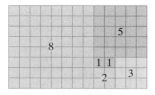

**A tiling with squares whose sides are successive Fibonacci numbers in length**

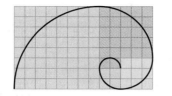

**A Fibonacci spiral, created by drawing arcs connecting the opposite corners of squares in the Fibonacci tiling**

The Fibonacci Sequence is a special sequence in which the first two terms are 1 and each term thereafter is the sum of the two previous terms:

$$1, 1, 2, 3, 5, 8, 13, 21, \ldots$$

The Fibonacci numbers are named after Leonardo of Pisa, known as Fibonacci, although there is some evidence that these numbers had been described earlier in India.

There are numerous interesting facts about this sequence, and some are shown on the diagrams on this page. In Section 14.1, page 805, Exercise 46, you will have the opportunity to check a formula for this sequence.

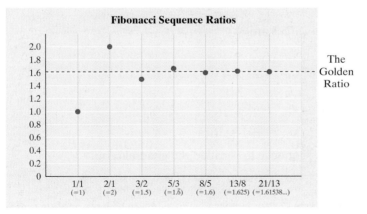

**Fibonacci Sequence Ratios**

The Golden Ratio

Having explored in some depth the concept of function, we turn now in this final chapter to *sequences*. In one sense, a sequence is simply an ordered list of numbers. In another sense, a sequence is itself a function. Phenomena modeled by such functions are everywhere around us. The starting place for all mathematics is the sequence of natural numbers: 1, 2, 3, 4, and so on.

Sequences lead us to *series*, which are a sum of ordered numbers. Through series we gain new insight, for example about the expansion of a binomial $(a + b)^n$, the concluding topic of this book.

The ratio of successive numbers in the Fibonacci Sequence approaches a number called the golden ratio or golden number, which is approximately 1.618034.

# 14.1 SEQUENCES

Suppose that a town's present population of 100,000 is growing by 5% each year. After the first year, the town's population will be

$$100{,}000 + 0.05(100{,}000) = 105{,}000$$

After the second year, the town's population will be

$$105{,}000 + 0.05(105{,}000) = 110{,}250$$

After the third year, the town's population will be

$$110{,}250 + 0.05(110{,}250) \approx 115{,}763$$

If we continue to calculate, the town's yearly population can be written as the **infinite sequence** of numbers

$$105{,}000,\ 110{,}250,\ 115{,}763,\ldots$$

If we decide to stop calculating after a certain year (say, the fourth year), we obtain the **finite sequence**

$$105{,}000,\ 110{,}250,\ 115{,}763,\ 121{,}551$$

---

**Sequences**

An infinite sequence is a function whose domain is the set of natural numbers $\{1, 2, 3, 4, \ldots\}$.

A finite sequence is a function whose domain is the set of natural numbers $\{1, 2, 3, 4, \ldots, n\}$, where $n$ is some natural number.

---

**OBJECTIVE 1 ▸ Writing the terms of a sequence.** Given the sequence $2, 4, 8, 16, \ldots$, we say that each number is a **term** of the sequence. Because a sequence is a function, we could describe it by writing $f(n) = 2^n$, where $n$ is a natural number. Instead, we use the notation

$$a_n = 2^n$$

Some function values are

$$
\begin{array}{ll}
a_1 = 2^1 = 2 & \text{First term of the sequence} \\
a_2 = 2^2 = 4 & \text{Second term} \\
a_3 = 2^3 = 8 & \text{Third term} \\
a_4 = 2^4 = 16 & \text{Fourth term} \\
a_{10} = 2^{10} = 1024 & \text{Tenth term}
\end{array}
$$

The $n$th term of the sequence $a_n$ is called the **general term.**

---

▸ **Helpful Hint**

If it helps, think of a sequence as simply a list of values in which a position is assigned. For the sequence directly above,

Value:    2,   4,   8,   16, $\ldots$, 1024
          ↑   ↑   ↑   ↑      ↑
Position   1ˢᵗ 2ⁿᵈ 3ʳᵈ 4ᵗʰ     10ᵗʰ

---

**EXAMPLE 1**    Write the first five terms of the sequence whose general term is given by

$$a_n = n^2 - 1$$

*Solution*   Evaluate $a_n$, where $n$ is $1, 2, 3, 4,$ and $5$.

$$a_n = n^2 - 1$$
$$a_1 = 1^2 - 1 = 0 \qquad \text{Replace } n \text{ with 1.}$$

$$a_2 = 2^2 - 1 = 3 \qquad \text{Replace } n \text{ with 2.}$$
$$a_3 = 3^2 - 1 = 8 \qquad \text{Replace } n \text{ with 3.}$$
$$a_4 = 4^2 - 1 = 15 \qquad \text{Replace } n \text{ with 4.}$$
$$a_5 = 5^2 - 1 = 24 \qquad \text{Replace } n \text{ with 5.}$$

Thus, the first five terms of the sequence $a_n = n^2 - 1$ are $0, 3, 8, 15,$ and $24$. □

**PRACTICE**

**1**  Write the first five terms of the sequence whose general term is given by $a_n = 5 + n^2$.

**EXAMPLE 2**  If the general term of a sequence is given by $a_n = \dfrac{(-1)^n}{3n}$, find

**a.** the first term of the sequence          **b.** $a_8$

**c.** the one-hundredth term of the sequence          **d.** $a_{15}$

*Solution*

**a.** $a_1 = \dfrac{(-1)^1}{3(1)} = -\dfrac{1}{3}$          Replace $n$ with 1.

**b.** $a_8 = \dfrac{(-1)^8}{3(8)} = \dfrac{1}{24}$          Replace $n$ with 8.

**c.** $a_{100} = \dfrac{(-1)^{100}}{3(100)} = \dfrac{1}{300}$          Replace $n$ with 100.

**d.** $a_{15} = \dfrac{(-1)^{15}}{3(15)} = -\dfrac{1}{45}$          Replace $n$ with 15.          □

**PRACTICE**

**2**  If the general term of a sequence is given by $a_n = \dfrac{(-1)^n}{5n}$, find

**a.** the first term of the sequence          **b.** $a_4$

**c.** The thirtieth term of the sequence          **d.** $a_{19}$

**OBJECTIVE 2 ▶ Finding the general term of a sequence.** Suppose we know the first few terms of a sequence and want to find a general term that fits the pattern of the first few terms.

**EXAMPLE 3**  Find a general term $a_n$ of the sequence whose first few terms are given.

**a.** $1, 4, 9, 16, \ldots$          **b.** $\dfrac{1}{1}, \dfrac{1}{2}, \dfrac{1}{3}, \dfrac{1}{4}, \dfrac{1}{5}, \ldots$

**c.** $-3, -6, -9, -12, \ldots$          **d.** $\dfrac{1}{2}, \dfrac{1}{4}, \dfrac{1}{8}, \dfrac{1}{16}, \ldots$

*Solution*

**a.** These numbers are the squares of the first four natural numbers, so a general term might be $a_n = n^2$.

**b.** These numbers are the reciprocals of the first five natural numbers, so a general term might be $a_n = \dfrac{1}{n}$.

**c.** These numbers are the product of $-3$ and the first four natural numbers, so a general term might be $a_n = -3n$.

**d.** Notice that the denominators double each time.

$$\frac{1}{2}, \quad \frac{1}{2\cdot 2}, \quad \frac{1}{2(2\cdot 2)}, \quad \frac{1}{2(2\cdot 2\cdot 2)}$$

or

$$\frac{1}{2^1}, \quad \frac{1}{2^2}, \quad \frac{1}{2^3}, \quad \frac{1}{2^4}$$

We might then suppose that the general term is $a_n = \dfrac{1}{2^n}$. □

**PRACTICE**
**3**  Find the general term $a_n$ of the sequence whose first few terms are given.

**a.** $1, 3, 5, 7, \ldots$                     **b.** $3, 9, 27, 81, \ldots$

**c.** $\dfrac{1}{2}, \dfrac{2}{3}, \dfrac{3}{4}, \dfrac{4}{5}, \ldots$          **d.** $-\dfrac{1}{2}, -\dfrac{1}{3}, -\dfrac{1}{4}, -\dfrac{1}{5}, \ldots$

**OBJECTIVE 3** ▶ **Solving applications modeled by sequences.** Sequences model many phenomena of the physical world, as illustrated by the following example.

**EXAMPLE 4**  **Finding a Puppy's Weight Gain**

The amount of weight, in pounds, a puppy gains in each month of its first year is modeled by a sequence whose general term is $a_n = n + 4$, where $n$ is the number of the month. Write the first five terms of the sequence, and find how much weight the puppy should gain in its fifth month.

**Solution**  Evaluate $a_n = n + 4$ when $n$ is $1, 2, 3, 4,$ and $5$.

$$a_1 = 1 + 4 = 5$$
$$a_2 = 2 + 4 = 6$$
$$a_3 = 3 + 4 = 7$$
$$a_4 = 4 + 4 = 8$$
$$a_5 = 5 + 4 = 9$$

The puppy should gain 9 pounds in its fifth month. □

**PRACTICE**
**4**  The value $v$, in dollars, of an office copier depreciates according to the sequence $v_n = 3950(0.8)^n$, where $n$ is the time in years. Find the value of the copier after three years.

## VOCABULARY & READINESS CHECK

*Use the choices below to fill in each blank.*

infinite          finite          general

**1.** The $n$th term of the sequence $a_n$ is called the _____ term.
**2.** A(n) _____ sequence is a function whose domain is $\{1, 2, 3, 4, \ldots, n\}$ where $n$ is some natural number.
**3.** A(n) _____ sequence is a function whose domain is $\{1, 2, 3, 4, \ldots\}$.

*Write the first term of each sequence.*

**4.** $a_n = 7^n; a_1 =$ _____.          **5.** $a_n = \dfrac{(-1)^n}{n}; a_1 =$ _____.          **6.** $a_n = (-1)^n \cdot n^4; a_1 =$ _____.

## 14.1 | EXERCISE SET

*Write the first five terms of each sequence whose general term is given. See Example 1.*

**1.** $a_n = n + 4$

**2.** $a_n = 5 - n$

**3.** $a_n = (-1)^n$

**4.** $a_n = (-2)^n$

**5.** $a_n = \dfrac{1}{n + 3}$

**6.** $a_n = \dfrac{1}{7 - n}$

**7.** $a_n = 2n$

**8.** $a_n = -6n$

**9.** $a_n = -n^2$

**10.** $a_n = n^2 + 2$

**11.** $a_n = 2^n$

**12.** $a_n = 3^{n-2}$

**13.** $a_n = 2n + 5$

**14.** $a_n = 1 - 3n$

**15.** $a_n = (-1)^n n^2$

**16.** $a_n = (-1)^{n+1}(n - 1)$

*Find the indicated term for each sequence whose general term is given. See Example 2.*

**17.** $a_n = 3n^2; a_5$

**18.** $a_n = -n^2; a_{15}$

**19.** $a_n = 6n - 2; a_{20}$

**20.** $a_n = 100 - 7n; a_{50}$

**21.** $a_n = \dfrac{n + 3}{n}; a_{15}$

**22.** $a_n = \dfrac{n}{n + 4}; a_{24}$

**23.** $a_n = (-3)^n; a_6$

**24.** $a_n = 5^{n+1}; a_3$

**25.** $a_n = \dfrac{n - 2}{n + 1}; a_6$

**26.** $a_n = \dfrac{n + 3}{n + 4}; a_8$

**27.** $a_n = \dfrac{(-1)^n}{n}; a_8$

**28.** $a_n = \dfrac{(-1)^n}{2n}; a_{100}$

**29.** $a_n = -n^2 + 5; a_{10}$

**30.** $a_n = 8 - n^2; a_{20}$

**31.** $a_n = \dfrac{(-1)^n}{n + 6}; a_{19}$

**32.** $a_n = \dfrac{n - 4}{(-2)^n}; a_6$

*Find a general term $a_n$ for each sequence whose first four terms are given. See Example 3.*

**33.** $3, 7, 11, 15$

**34.** $2, 7, 12, 17$

**35.** $-2, -4, -8, -16$

**36.** $-4, 16, -64, 256$

**37.** $\dfrac{1}{3}, \dfrac{1}{9}, \dfrac{1}{27}, \dfrac{1}{81}$

**38.** $\dfrac{2}{5}, \dfrac{2}{25}, \dfrac{2}{125}, \dfrac{2}{625}$

*Solve. See Example 4.*

**39.** The distance, in feet, that a Thermos dropped from a cliff falls in each consecutive second is modeled by a sequence whose general term is $a_n = 32n - 16$, where $n$ is the number of seconds. Find the distance the Thermos falls in the second, third, and fourth seconds.

**40.** The population size of a culture of bacteria triples every hour such that its size is modeled by the sequence $a_n = 50(3)^{n-1}$, where $n$ is the number of the hour just beginning. Find the size of the culture at the beginning of the fourth hour and the size of the culture at the beginning of the first hour.

**41.** Mrs. Laser agrees to give her son Mark an allowance of $0.10 on the first day of his 14-day vacation, $0.20 on the second day, $0.40 on the third day, and so on. Write an equation of a sequence whose terms correspond to Mark's allowance. Find the allowance Mark will receive on the last day of his vacation.

**42.** A small theater has 10 rows with 12 seats in the first row, 15 seats in the second row, 18 seats in the third row, and so on. Write an equation of a sequence whose terms correspond to the seats in each row. Find the number of seats in the eighth row.

**43.** The number of cases of a new infectious disease is doubling every year such that the number of cases is modeled by a sequence whose general term is $a_n = 75(2)^{n-1}$, where $n$ is the number of the year just beginning. Find how many cases there will be at the beginning of the sixth year. Find how many cases there were at the beginning of the first year.

**44.** A new college had an initial enrollment of 2700 students in 2000, and each year the enrollment increases by 150 students. Find the enrollment for each of 5 years, beginning with 2000.

**45.** An endangered species of sparrow had an estimated population of 800 in 2000, and scientists predict that its population will decrease by half each year. Estimate the population in 2004. Estimate the year the sparrow will be extinct.

**46.** A **Fibonacci sequence** is a special type of sequence in which the first two terms are 1, and each term thereafter is the sum of the two previous terms: 1, 1, 2, 3, 5, 8, etc. The formula for the $n$th Fibonacci term is $a_n = \dfrac{1}{\sqrt{5}}\left[\left(\dfrac{1 + \sqrt{5}}{2}\right)^n - \left(\dfrac{1 - \sqrt{5}}{2}\right)^n\right]$.

Verify that the first two terms of the Fibonacci sequence are each 1.

### REVIEW AND PREVIEW

*Sketch the graph of each quadratic function. See Section 11.5.*

**47.** $f(x) = (x - 1)^2 + 3$

**48.** $f(x) = (x - 2)^2 + 1$

**49.** $f(x) = 2(x + 4)^2 + 2$

**50.** $f(x) = 3(x - 3)^2 + 4$

*Find the distance between each pair of points. See Section 10.3.*

**51.** $(-4, -1)$ and $(-7, -3)$

**52.** $(-2, -1)$ and $(-1, 5)$

**53.** $(2, -7)$ and $(-3, -3)$

**54.** $(10, -14)$ and $(5, -11)$

### CONCEPT EXTENSIONS

*Find the first five terms of each sequence. Round each term after the first to four decimal places.*

**55.** $a_n = \dfrac{1}{\sqrt{n}}$

**56.** $\dfrac{\sqrt{n}}{\sqrt{n} + 1}$

**57.** $a_n = \left(1 + \dfrac{1}{n}\right)^n$

**58.** $a_n = \left(1 + \dfrac{0.05}{n}\right)^n$

# 14.2 ARITHMETIC AND GEOMETRIC SEQUENCES

**OBJECTIVES**

1 Identify arithmetic sequences and their common differences.

2 Identify geometric sequences and their common ratios.

**OBJECTIVE 1 ▶ Identifying arithmetic sequences.** Find the first four terms of the sequence whose general term is $a_n = 5 + (n-1)3$.

$$a_1 = 5 + (1-1)3 = 5 \quad \text{Replace } n \text{ with 1.}$$
$$a_2 = 5 + (2-1)3 = 8 \quad \text{Replace } n \text{ with 2.}$$
$$a_3 = 5 + (3-1)3 = 11 \quad \text{Replace } n \text{ with 3.}$$
$$a_4 = 5 + (4-1)3 = 14 \quad \text{Replace } n \text{ with 4.}$$

The first four terms are $5, 8, 11,$ and $14$. Notice that the difference of any two successive terms is 3.

$$8 - 5 = 3$$
$$11 - 8 = 3$$
$$14 - 11 = 3$$
$$\vdots$$
$$a_n - a_{n-1} = 3$$

$n$th    previous
term    term

Because the difference of any two successive terms is a constant, we call the sequence an **arithmetic sequence,** or an **arithmetic progression.** The constant difference $d$ in successive terms is called the **common difference.** In this example, $d$ is 3.

---

**Arithmetic Sequence and Common Difference**

An **arithmetic sequence** is a sequence in which each term (after the first) differs from the preceding term by a constant amount $d$. The constant $d$ is called the **common difference** of the sequence.

---

The sequence $2, 6, 10, 14, 18, \ldots$ is an arithmetic sequence. Its common difference is 4. Given the first term $a_1$ and the common difference $d$ of an arithmetic sequence, we can find any term of the sequence.

**EXAMPLE 1**   Write the first five terms of the arithmetic sequence whose first term is 7 and whose common difference is 2.

*Solution*

$$a_1 = 7$$
$$a_2 = 7 + 2 = 9$$
$$a_3 = 9 + 2 = 11$$
$$a_4 = 11 + 2 = 13$$
$$a_5 = 13 + 2 = 15$$

The first five terms are $7, 9, 11, 13, 15$.      ☐

**PRACTICE**

**1**   Write the first five terms of the arithmetic sequence whose first term is 4 and whose common difference is 5.

Notice the general pattern of the terms in Example 1.

$$a_1 = 7$$
$$a_2 = 7 + 2 = 9 \quad \text{or} \quad a_2 = a_1 + d$$
$$a_3 = 9 + 2 = 11 \quad \text{or} \quad a_3 = a_2 + d = (a_1 + d) + d = a_1 + 2d$$
$$a_4 = 11 + 2 = 13 \quad \text{or} \quad a_4 = a_3 + d = (a_1 + 2d) + d = a_1 + 3d$$
$$a_5 = 13 + 2 = 15 \quad \text{or} \quad a_5 = a_4 + d = (a_1 + 3d) + d = a_1 + 4d$$

$\longrightarrow$ (subscript $-1$) is multiplier $\longrightarrow$

The pattern on the right suggests that the general term $a_n$ of an arithmetic sequence is given by

$$a_n = a_1 + (n - 1)d$$

> **General Term of an Arithmetic Sequence**
> The general term $a_n$ of an arithmetic sequence is given by
> $$a_n = a_1 + (n - 1)d$$
> where $a_1$ is the first term and $d$ is the common difference.

**EXAMPLE 2**   Consider the arithmetic sequence whose first term is 3 and common difference is $-5$.

**a.** Write an expression for the general term $a_n$.
**b.** Find the twentieth term of this sequence.

*Solution*

**a.** Since this is an arithmetic sequence, the general term $a_n$ is given by $a_n = a_1 + (n - 1)d$. Here, $a_1 = 3$ and $d = -5$, so

$$
\begin{aligned}
a_n &= 3 + (n - 1)(-5) &&\text{Let } a_1 = 3 \text{ and } d = -5. \\
&= 3 - 5n + 5 &&\text{Multiply.} \\
&= 8 - 5n &&\text{Simplify.}
\end{aligned}
$$

**b.** $a_n = 8 - 5n$
$$
\begin{aligned}
a_{20} &= 8 - 5 \cdot 20 &&\text{Let } n = 20. \\
&= 8 - 100 = -92
\end{aligned}
$$
□

PRACTICE
**2**   Consider the arithmetic sequence whose first term is 2 and whose common difference is $-3$.

**a.** Write an expression for the general term $a_n$.
**b.** Find the twelfth term of the sequence.

- - - - - - - - - - - - - - - - - - - - - - - - - - - - - - - - - - - - - - - - - ■

**EXAMPLE 3**   Find the eleventh term of the arithmetic sequence whose first three terms are 2, 9, and 16.

*Solution*   Since the sequence is arithmetic, the eleventh term is

$$a_{11} = a_1 + (11 - 1)d = a_1 + 10d$$

We know $a_1$ is the first term of the sequence, so $a_1 = 2$. Also, $d$ is the constant difference of terms, so $d = a_2 - a_1 = 9 - 2 = 7$. Thus,

$$
\begin{aligned}
a_{11} &= a_1 + 10d \\
&= 2 + 10 \cdot 7 &&\text{Let } a_1 = 2 \text{ and } d = 7. \\
&= 72
\end{aligned}
$$
□

PRACTICE
**3**   Find the ninth term of the arithmetic sequence whose first three terms are 3, 9, and 15.

- - - - - - - - - - - - - - - - - - - - - - - - - - - - - - - - - - - - - - - - - ■

**EXAMPLE 4** If the third term of an arithmetic sequence is 12 and the eighth term is 27, find the fifth term.

_Solution_ We need to find $a_1$ and $d$ to write the general term, which then enables us to find $a_5$, the fifth term. The given facts about terms $a_3$ and $a_8$ lead to a system of linear equations.

$$\begin{cases} a_3 = a_1 + (3-1)d \\ a_8 = a_1 + (8-1)d \end{cases} \text{ or } \begin{cases} 12 = a_1 + 2d \\ 27 = a_1 + 7d \end{cases}$$

Next, we solve the system $\begin{cases} 12 = a_1 + 2d \\ 27 = a_1 + 7d \end{cases}$ by elimination. Multiply both sides of the second equation by $-1$ so that

$$\begin{cases} 12 = a_1 + 2d \\ -1(27) = -1(a_1 + 7d) \end{cases} \begin{array}{c} \text{simplifies} \\ \text{to} \end{array} \begin{cases} 12 = a_1 + 2d \\ \underline{-27 = -a_1 - 7d} \\ -15 = -5d \quad \text{Add the equations.} \\ 3 = d \quad \text{Divide both sides by } -5. \end{cases}$$

To find $a_1$, let $d = 3$ in $12 = a_1 + 2d$. Then

$$12 = a_1 + 2(3)$$
$$12 = a_1 + 6$$
$$6 = a_1$$

Thus, $a_1 = 6$ and $d = 3$, so

$$a_n = 6 + (n-1)(3)$$
$$= 6 + 3n - 3$$
$$= 3 + 3n$$

and

$$a_5 = 3 + 3 \cdot 5 = 18 \qquad \square$$

**PRACTICE**
**4** If the third term of an arithmetic sequence is 23 and the eighth term is 63, find the sixth term.

---

**EXAMPLE 5** Finding Salary

Donna Theime has an offer for a job starting at \$40,000 per year and guaranteeing her a raise of \$1600 per year for the next 5 years. Write the general term for the arithmetic sequence that models Donna's potential annual salaries, and find her salary for the fourth year.

_Solution_ The first term, $a_1$, is 40,000, and $d$ is 1600. So

$$a_n = 40{,}000 + (n-1)(1600) = 38{,}400 + 1600n$$
$$a_4 = 38{,}400 + 1600 \cdot 4 = 44{,}800$$

Her salary for the fourth year will be \$44,800. $\qquad \square$

**PRACTICE**
**5** A starting salary for a consulting company is \$57,000 per year, with guaranteed annual increases of \$2200 for the next 4 years. Write the general term for the arithmetic sequence that models the potential annual salaries, and find the salary for the third year.

---

**OBJECTIVE 2 ▶ Identifying geometric sequences.** We now investigate a **geometric sequence**, also called a **geometric progression**. In the sequence $5, 15, 45, 135, \ldots$, each term after the first is the *product* of 3 and the preceding term. This pattern of multiplying by a constant to get the next term defines a geometric sequence. The constant is called the **common ratio** because it is the ratio of any term (after the first) to its preceding term.

$$\frac{15}{5} = 3$$

$$\frac{45}{15} = 3$$

$$\frac{135}{45} = 3$$

$$\vdots$$

$n$th term $\longrightarrow$    $\dfrac{a_n}{a_{n-1}} = 3$

previous term $\longrightarrow$

---

**Geometric Sequence and Common Ratio**

A **geometric sequence** is a sequence in which each term (after the first) is obtained by multiplying the preceding term by a constant $r$. The constant $r$ is called the **common ratio** of the sequence.

---

The sequence $12, 6, 3, \dfrac{3}{2}, \ldots$ is geometric since each term after the first is the product of the previous term and $\dfrac{1}{2}$.

**EXAMPLE 6**    Write the first five terms of a geometric sequence whose first term is 7 and whose common ratio is 2.

*Solution*
$$a_1 = 7$$
$$a_2 = 7(2) = 14$$
$$a_3 = 14(2) = 28$$
$$a_4 = 28(2) = 56$$
$$a_5 = 56(2) = 112$$

The first five terms are $7, 14, 28, 56,$ and $112$.    □

**PRACTICE**

**6**    Write the first four terms of a geometric sequence whose first term is 8 and whose common ratio is $-3$.

Notice the general pattern of the terms in Example 6.

$$a_1 = 7$$
$$a_2 = 7(2) = 14 \quad \text{or} \quad a_2 = a_1(r)$$
$$a_3 = 14(2) = 28 \quad \text{or} \quad a_3 = a_2(r) = (a_1 \cdot r) \cdot r = a_1 r^2$$
$$a_4 = 28(2) = 56 \quad \text{or} \quad a_4 = a_3(r) = (a_1 \cdot r^2) \cdot r = a_1 r^3$$
$$a_5 = 56(2) = 112 \quad \text{or} \quad a_5 = a_4(r) = (a_1 \cdot r^3) \cdot r = a_1 r^4$$

$\longrightarrow$ (subscript $-1$) is power

The pattern on the right above suggests that the general term of a geometric sequence is given by $a_n = a_1 r^{n-1}$.

---

**General Term of a Geometric Sequence**

The general term $a_n$ of a geometric sequence is given by

$$a_n = a_1 r^{n-1}$$

where $a_1$ is the first term and $r$ is the common ratio.

---

**EXAMPLE 7**   Find the eighth term of the geometric sequence whose first term is 12 and whose common ratio is $\frac{1}{2}$.

*Solution*   Since this is a geometric sequence, the general term $a_n$ is given by

$$a_n = a_1 r^{n-1}$$

Here $a_1 = 12$ and $r = \frac{1}{2}$, so $a_n = 12\left(\frac{1}{2}\right)^{n-1}$. Evaluate $a_n$ for $n = 8$.

$$a_8 = 12\left(\frac{1}{2}\right)^{8-1} = 12\left(\frac{1}{2}\right)^{7} = 12\left(\frac{1}{128}\right) = \frac{3}{32}$$

**PRACTICE**
**7**   Find the seventh term of the geometric sequence whose first term is 64 and whose common ratio is $\frac{1}{4}$.

---

**EXAMPLE 8**   Find the fifth term of the geometric sequence whose first three terms are 2, −6, and 18.

*Solution*   Since the sequence is geometric and $a_1 = 2$, the fifth term must be $a_1 r^{5-1}$, or $2r^4$. We know that $r$ is the common ratio of terms, so $r$ must be $\frac{-6}{2}$, or −3. Thus,

$$a_5 = 2r^4$$
$$a_5 = 2(-3)^4 = 162$$

**PRACTICE**
**8**   Find the seventh term of the geometric sequence whose first three terms are −3, 6, and −12.

---

**EXAMPLE 9**   If the second term of a geometric sequence is $\frac{5}{4}$ and the third term is $\frac{5}{16}$, find the first term and the common ratio.

*Solution*   Notice that $\frac{5}{16} \div \frac{5}{4} = \frac{1}{4}$, so $r = \frac{1}{4}$. Then

$$a_2 = a_1\left(\frac{1}{4}\right)^{2-1}$$
$$\frac{5}{4} = a_1\left(\frac{1}{4}\right)^{1}, \quad \text{or} \quad a_1 = 5 \quad \text{Replace } a_2 \text{ with } \frac{5}{4}.$$

The first term is 5.

**PRACTICE**
**9**   If the second term of a geometric sequence is $\frac{9}{2}$ and the third term is $\frac{27}{4}$, find the first term and the common ratio.

---

**EXAMPLE 10**   **Predicting Population of a Bacterial Culture**

The population size of a bacterial culture growing under controlled conditions is doubling each day. Predict how large the culture will be at the beginning of day 7 if it measures 10 units at the beginning of day 1.

*Solution*   Since the culture doubles in size each day, the population sizes are modeled by a geometric sequence. Here $a_1 = 10$ and $r = 2$. Thus,

$$a_n = a_1 r^{n-1} = 10(2)^{n-1} \quad \text{and} \quad a_7 = 10(2)^{7-1} = 640$$

The bacterial culture should measure 640 units at the beginning of day 7.

**PRACTICE**
**10**   After applying a test antibiotic, the population of a bacterial culture is reduced by one-half every day. Predict how large the culture will be at the start of day 7 if it measures 4800 units at the beginning of day 1.

## VOCABULARY & READINESS CHECK

*Use the choices below to fill in each blank. Some choices may be used more than once and some not at all.*

| first | arithmetic | difference |
|-------|-----------|-----------|
| last | geometric | ratio |

1. A(n) _____ sequence is one in which each term (after the first) is obtained by multiplying the preceding term by a constant $r$. The constant $r$ is called the common _____.

2. A(n) _____ sequence is one in which each term (after the first) differs from the preceding term by a constant amount $d$. The constant $d$ is called the common _____.

3. The general term of an arithmetic sequence is $a_n = a_1 + (n-1)d$ where $a_1$ is the _____ term and $d$ is the common _____.

4. The general term of a geometric sequence is $a_n = a_1 r^{n-1}$ where $a_1$ is the _____ term and $r$ is the common _____.

## 14.2 EXERCISE SET

*Write the first five terms of the arithmetic or geometric sequence whose first term, $a_1$, and common difference, $d$, or common ratio, $r$, are given. See Examples 1 and 6.*

1. $a_1 = 4; d = 2$

2. $a_1 = 3; d = 10$

3. $a_1 = 6; d = -2$

4. $a_1 = -20; d = 3$

5. $a_1 = 1; r = 3$

6. $a_1 = -2; r = 2$

7. $a_1 = 48; r = \dfrac{1}{2}$

8. $a_1 = 1; r = \dfrac{1}{3}$

*Find the indicated term of each sequence. See Examples 2 and 7.*

9. The eighth term of the arithmetic sequence whose first term is 12 and whose common difference is 3

10. The twelfth term of the arithmetic sequence whose first term is 32 and whose common difference is $-4$

11. The fourth term of the geometric sequence whose first term is 7 and whose common ratio is $-5$

12. The fifth term of the geometric sequence whose first term is 3 and whose common ratio is 3

13. The fifteenth term of the arithmetic sequence whose first term is $-4$ and whose common difference is $-4$

14. The sixth term of the geometric sequence whose first term is 5 and whose common ratio is $-4$

*Find the indicated term of each sequence. See Examples 3 and 8.*

15. The ninth term of the arithmetic sequence $0, 12, 24, \ldots$

16. The thirteenth term of the arithmetic sequence $-3, 0, 3, \ldots$

17. The twenty-fifth term of the arithmetic sequence $20, 18, 16, \ldots$

18. The ninth term of the geometric sequence $5, 10, 20, \ldots$

19. The fifth term of the geometric sequence $2, -10, 50, \ldots$

20. The sixth term of the geometric sequence $\dfrac{1}{2}, \dfrac{3}{2}, \dfrac{9}{2}, \ldots$

*Find the indicated term of each sequence. See Examples 4 and 9.*

21. The eighth term of the arithmetic sequence whose fourth term is 19 and whose fifteenth term is 52

22. If the second term of an arithmetic sequence is 6 and the tenth term is 30, find the twenty-fifth term.

23. If the second term of an arithmetic progression is $-1$ and the fourth term is 5, find the ninth term.

24. If the second term of a geometric progression is 15 and the third term is 3, find $a_1$ and $r$.

25. If the second term of a geometric progression is $-\dfrac{4}{3}$ and the third term is $\dfrac{8}{3}$, find $a_1$ and $r$.

26. If the third term of a geometric sequence is 4 and the fourth term is $-12$, find $a_1$ and $r$.

27. Explain why 14, 10, and 6 may be the first three terms of an arithmetic sequence when it appears we are subtracting instead of adding to get the next term.

**28.** Explain why 80, 20, and 5 may be the first three terms of a geometric sequence when it appears we are dividing instead of multiplying to get the next term.

## MIXED PRACTICE

*Given are the first three terms of a sequence that is either arithmetic or geometric. If the sequence is arithmetic, find $a_1$ and d. If a sequence is geometric, find $a_1$ and r.*

**29.** 2, 4, 6

**30.** 8, 16, 24

**31.** 5, 10, 20

**32.** 2, 6, 18

**33.** $\frac{1}{2}, \frac{1}{10}, \frac{1}{50}$

**34.** $\frac{2}{3}, \frac{4}{3}, 2$

**35.** $x, 5x, 25x$

**36.** $y, -3y, 9y$

**37.** $p, p + 4, p + 8$

**38.** $t, t - 1, t - 2$

*Find the indicated term of each sequence.*

**39.** The twenty-first term of the arithmetic sequence whose first term is 14 and whose common difference is $\frac{1}{4}$

**40.** The fifth term of the geometric sequence whose first term is 8 and whose common ratio is $-3$

**41.** The fourth term of the geometric sequence whose first term is 3 and whose common ratio is $-\frac{2}{3}$

**42.** The fourth term of the arithmetic sequence whose first term is 9 and whose common difference is 5

**43.** The fifteenth term of the arithmetic sequence $\frac{3}{2}, 2, \frac{5}{2}, \ldots$

**44.** The eleventh term of the arithmetic sequence $2, \frac{5}{3}, \frac{4}{3}, \ldots$

**45.** The sixth term of the geometric sequence $24, 8, \frac{8}{3}, \ldots$

**46.** The eighteenth term of the arithmetic sequence $5, 2, -1, \ldots$

**47.** If the third term of an arithmetic sequence is 2 and the seventeenth term is $-40$, find the tenth term.

**48.** If the third term of a geometric sequence is $-28$ and the fourth term is $-56$, find $a_1$ and r.

*Solve. See Examples 5 and 10.*

**49.** An auditorium has 54 seats in the first row, 58 seats in the second row, 62 seats in the third row, and so on. Find the general term of this arithmetic sequence and the number of seats in the twentieth row.

**50.** A triangular display of cans in a grocery store has 20 cans in the first row, 17 cans in the next row, and so on, in an arithmetic sequence. Find the general term and the number of cans in the fifth row. Find how many rows there are in the display and how many cans are in the top row.

**51.** The initial size of a virus culture is 6 units, and it triples its size every day. Find the general term of the geometric sequence that models the culture's size.

**52.** A real estate investment broker predicts that a certain property will increase in value 15% each year. Thus, the yearly property values can be modeled by a geometric sequence whose common ratio r is 1.15. If the initial property value was $500,000, write the first four terms of the sequence and predict the value at the end of the third year.

**53.** A rubber ball is dropped from a height of 486 feet, and it continues to bounce one-third the height from which it last fell. Write out the first five terms of this geometric sequence and find the general term. Find how many bounces it takes for the ball to rebound less than 1 foot.

**54.** On the first swing, the length of the arc through which a pendulum swings is 50 inches. The length of each successive swing is 80% of the preceding swing. Determine whether this sequence is arithmetic or geometric. Find the length of the fourth swing.

**55.** Jose takes a job that offers a monthly starting salary of $4000 and guarantees him a monthly raise of $125 during his first year of training. Find the general term of this arithmetic sequence and his monthly salary at the end of his training.

**56.** At the beginning of Claudia Schaffer's exercise program, she rides 15 minutes on the Lifecycle. Each week she increases her riding time by 5 minutes. Write the general term of this arithmetic sequence, and find her riding time after 7 weeks. Find how many weeks it takes her to reach a riding time of 1 hour.

**57.** If a radioactive element has a half-life of 3 hours, then $x$ grams of the element dwindles to $\frac{x}{2}$ grams after 3 hours. If a nuclear reactor has 400 grams of that radioactive element, find the amount of radioactive material after 12 hours.

## REVIEW AND PREVIEW

*Evaluate. See Section 1.7.*

**58.** $5(1) + 5(2) + 5(3) + 5(4)$

**59.** $\frac{1}{3(1)} + \frac{1}{3(2)} + \frac{1}{3(3)}$

**60.** $2(2 - 4) + 3(3 - 4) + 4(4 - 4)$

**61.** $3^0 + 3^1 + 3^2 + 3^3$

**62.** $\frac{1}{4(1)} + \frac{1}{4(2)} + \frac{1}{4(3)}$

**63.** $\frac{8 - 1}{8 + 1} + \frac{8 - 2}{8 + 2} + \frac{8 - 3}{8 + 3}$

**CONCEPT EXTENSIONS**

*Write the first four terms of the arithmetic or geometric sequence whose first term, $a_1$, and common difference, d, or common ratio, r, are given.*

**64.** $a_1 = \$3720, d = -\$268.50$

**65.** $a_1 = \$11{,}782.40, r = 0.5$

**66.** $a_1 = 26.8, r = 2.5$

**67.** $a_1 = 19.652; d = -0.034$

**68.** Describe a situation in your life that can be modeled by a geometric sequence. Write an equation for the sequence.

**69.** Describe a situation in your life that can be modeled by an arithmetic sequence. Write an equation for the sequence.

---

## 14.3 SERIES

**OBJECTIVES**

**1** Identify finite and infinite series and use summation notation.

**2** Find partial sums.

**OBJECTIVE 1 ▶ Identifying finite and infinite series and using summation notation.** A person who conscientiously saves money by saving first $100 and then saving $10 more each month than he saved the preceding month is saving money according to the arithmetic sequence

$$a_n = 100 + 10(n - 1)$$

Following this sequence, he can predict how much money he should save for any particular month. But if he also wants to know how much money *in total* he has saved, say, by the fifth month, he must find the *sum* of an infinite first five terms of the sequence

$$\underbrace{100}_{a_1} + \underbrace{100 + 10}_{a_2} + \underbrace{100 + 20}_{a_3} + \underbrace{100 + 30}_{a_4} + \underbrace{100 + 40}_{a_5}$$

A sum of the terms of a sequence is called a **series** (the plural is also "series"). As our example here suggests, series are frequently used to model financial and natural phenomena.

A series is a **finite series** if it is the sum of a finite number of terms. A series is an **infinite series** if it is the sum of all the terms of an infinite sequence. For example,

| *Sequence* | *Series* | |
|---|---|---|
| $5, 9, 13$ | $5 + 9 + 13$ | Finite; sum of 3 terms |
| $5, 9, 13, \ldots$ | $5 + 9 + 13 + \cdots$ | Infinite |
| $4, -2, 1, -\dfrac{1}{2}, \dfrac{1}{4}$ | $4 + (-2) + 1 + \left(-\dfrac{1}{2}\right) + \left(\dfrac{1}{4}\right)$ | Finite; sum of 5 terms |
| $4, -2, 1, \ldots$ | $4 + (-2) + 1 + \cdots$ | Infinite |
| $3, 6, \ldots, 99$ | $3 + 6 + \cdots + 99$ | Finite; sum of 33 terms |

A shorthand notation for denoting a series when the general term of the sequence is known is called **summation notation.** The Greek uppercase letter **sigma, $\Sigma$,** is used to mean "sum." The expression $\displaystyle\sum_{n=1}^{5}(3n + 1)$ is read "the sum of $3n + 1$ as $n$ goes from 1 to 5"; this expression means the sum of the first five terms of the sequence whose general term is $a_n = 3n + 1$. Often, the variable $i$ is used instead of $n$ in summation notation: $\displaystyle\sum_{i=1}^{5}(3i + 1)$. Whether we use $n, i, k$, or some other variable, the variable is called the **index of summation.** The notation $i = 1$ below the symbol $\Sigma$ indicates the beginning value of $i$, and the number 5 above the symbol $\Sigma$ indicates the ending value of $i$. Thus, the terms of the sequence are found by successively replacing $i$ with the natural numbers $1, 2, 3, 4, 5$. To find the sum, we write out the terms and then add.

$$\sum_{i=1}^{5}(3i + 1) = (3 \cdot 1 + 1) + (3 \cdot 2 + 1) + (3 \cdot 3 + 1)$$
$$+ (3 \cdot 4 + 1) + (3 \cdot 5 + 1)$$
$$= 4 + 7 + 10 + 13 + 16 = 50$$

**EXAMPLE 1** Evaluate.

**a.** $\displaystyle\sum_{i=0}^{6}\dfrac{i - 2}{2}$

**b.** $\displaystyle\sum_{i=3}^{5}2^i$

*Solution*

**a.** $\displaystyle\sum_{i=0}^{6} \frac{i-2}{2} = \frac{0-2}{2} + \frac{1-2}{2} + \frac{2-2}{2} + \frac{3-2}{2} + \frac{4-2}{2} + \frac{5-2}{2} + \frac{6-2}{2}$

$\qquad = (-1) + \left(-\dfrac{1}{2}\right) + 0 + \dfrac{1}{2} + 1 + \dfrac{3}{2} + 2$

$\qquad = \dfrac{7}{2}, \text{ or } 3\dfrac{1}{2}$

**b.** $\displaystyle\sum_{i=3}^{5} 2^i = 2^3 + 2^4 + 2^5$

$\qquad = 8 + 16 + 32$

$\qquad = 56$ □

**PRACTICE**
**1** Evaluate.

**a.** $\displaystyle\sum_{i=0}^{4} \frac{i-3}{4}$                  **b.** $\displaystyle\sum_{i=2}^{5} 3^i$

---

**EXAMPLE 2** Write each series with summation notation.

**a.** $3 + 6 + 9 + 12 + 15$        **b.** $\dfrac{1}{2} + \dfrac{1}{4} + \dfrac{1}{8} + \dfrac{1}{16}$

*Solution*

**a.** Since the *difference* of each term and the preceding term is 3, the terms correspond to the first five terms of the arithmetic sequence $a_n = a_1 + (n-1)d$ with $a_1 = 3$ and $d = 3$. So $a_n = 3 + (n-1)3 = 3n$, when simplified. Thus, in summation notation,

$$3 + 6 + 9 + 12 + 15 = \sum_{i=1}^{5} 3i.$$

**b.** Since each term is the *product* of the preceding term and $\dfrac{1}{2}$, these terms correspond to the first four terms of the geometric sequence $a_n = a_1 r^{n-1}$. Here $a_1 = \dfrac{1}{2}$ and $r = \dfrac{1}{2}$, so $a_n = \left(\dfrac{1}{2}\right)\left(\dfrac{1}{2}\right)^{n-1} = \left(\dfrac{1}{2}\right)^{1+(n-1)} = \left(\dfrac{1}{2}\right)^{n}$. In summation notation,

$$\frac{1}{2} + \frac{1}{4} + \frac{1}{8} + \frac{1}{16} = \sum_{i=1}^{4} \left(\frac{1}{2}\right)^{i}$$ □

**PRACTICE**
**2** Write each series with summation notation.

**a.** $5 + 10 + 15 + 20 + 25 + 30$        **b.** $\dfrac{1}{5} + \dfrac{1}{25} + \dfrac{1}{125} + \dfrac{1}{625}$

---

**OBJECTIVE 2 ▶ Finding partial sums.** The sum of the first $n$ terms of a sequence is a finite series known as a **partial sum,** $S_n$. Thus, for the sequence $a_1, a_2, \ldots, a_n$, the first three partial sums are

$$S_1 = a_1$$
$$S_2 = a_1 + a_2$$
$$S_3 = a_1 + a_2 + a_3$$

In general, $S_n$ is the sum of the first $n$ terms of a sequence.

$$S_n = \sum_{i=1}^{n} a_n$$

**EXAMPLE 3**   Find the sum of the first three terms of the sequence whose general term is $a_n = \dfrac{n + 3}{2n}$.

*Solution*

$$S_3 = \sum_{i=1}^{3} \frac{i + 3}{2i} = \frac{1 + 3}{2 \cdot 1} + \frac{2 + 3}{2 \cdot 2} + \frac{3 + 3}{2 \cdot 3}$$

$$= 2 + \frac{5}{4} + 1 = 4\frac{1}{4}$$

□

**PRACTICE**
**3**   Find the sum of the first four terms of the sequence whose general term is $a_n = \dfrac{2 + 3n}{n^2}$.

The next example illustrates how these sums model real-life phenomena.

**EXAMPLE 4**   **Number of Baby Gorillas Born**

The number of baby gorillas born at the San Diego Zoo is a sequence defined by $a_n = n(n - 1)$, where $n$ is the number of years the zoo has owned gorillas. Find the *total* number of baby gorillas born in the *first 4 years*.

*Solution*   To solve, find the sum

$$S_4 = \sum_{i=1}^{4} i(i - 1)$$

$$= 1(1 - 1) + 2(2 - 1) + 3(3 - 1) + 4(4 - 1)$$

$$= 0 + 2 + 6 + 12 = 20$$

There were 20 gorillas born in the first 4 years.

□

**PRACTICE**
**4**   The number of strawberry plants growing in a garden is a sequence defined by $a_n = n(2n - 1)$, where $n$ is the number of years after planting a strawberry plant. Find the total number of strawberry plants after 5 years.

# VOCABULARY & READINESS CHECK

*Use the choices below to fill in each blank. Not all choices may be used.*

| | | | | |
|---|---|---|---|---|
| index of summation | infinite | sigma | 1 | 7 |
| partial sum | finite | summation | 5 | |

1. A series is a(n) _____ series if it is the sum of all the terms of the sequence.

2. A series is a(n) _____ series if it is the sum of a finite number of terms.

3. A shorthand notation for denoting a series when the general term of the sequence is known is called _____ notation.

4. In the notation $\displaystyle\sum_{i=1}^{7}(5i - 2)$, the $\Sigma$ is the Greek uppercase letter _____ and the $i$ is called the _____.

5. The sum of the first $n$ terms of a sequence is a finite series known as a _____.

6. For the notation in Exercise 4 above, the beginning value of $i$ is _____ and the ending value of $i$ is _____.

## 14.3 | EXERCISE SET

*Evaluate. See Example 1.*

**1.** $\displaystyle\sum_{i=1}^{4}(i-3)$

**2.** $\displaystyle\sum_{i=1}^{5}(i+6)$

**3.** $\displaystyle\sum_{i=4}^{7}(2i+4)$

**4.** $\displaystyle\sum_{i=2}^{3}(5i-1)$

**5.** $\displaystyle\sum_{i=2}^{4}(i^2-3)$

**6.** $\displaystyle\sum_{i=3}^{5}i^3$

**7.** $\displaystyle\sum_{i=1}^{3}\left(\frac{1}{i+5}\right)$

**8.** $\displaystyle\sum_{i=2}^{4}\left(\frac{2}{i+3}\right)$

**9.** $\displaystyle\sum_{i=1}^{3}\frac{1}{6i}$

**10.** $\displaystyle\sum_{i=1}^{3}\frac{1}{3i}$

**11.** $\displaystyle\sum_{i=2}^{6}3i$

**12.** $\displaystyle\sum_{i=3}^{6}-4i$

**13.** $\displaystyle\sum_{i=3}^{5}i(i+2)$

**14.** $\displaystyle\sum_{i=2}^{4}i(i-3)$

**15.** $\displaystyle\sum_{i=1}^{5}2^i$

**16.** $\displaystyle\sum_{i=1}^{4}3^{i-1}$

**17.** $\displaystyle\sum_{i=1}^{4}\frac{4i}{i+3}$

**18.** $\displaystyle\sum_{i=2}^{5}\frac{6-i}{6+i}$

*Write each series with summation notation. See Example 2.*

**19.** $1 + 3 + 5 + 7 + 9$

**20.** $4 + 7 + 10 + 13$

**21.** $4 + 12 + 36 + 108$

**22.** $5 + 10 + 20 + 40 + 80 + 160$

**23.** $12 + 9 + 6 + 3 + 0 + (-3)$

**24.** $5 + 1 + (-3) + (-7)$

**25.** $12 + 4 + \dfrac{4}{3} + \dfrac{4}{9}$

**26.** $80 + 20 + 5 + \dfrac{5}{4} + \dfrac{5}{16}$

**27.** $1 + 4 + 9 + 16 + 25 + 36 + 49$

**28.** $1 + (-4) + 9 + (-16)$

*Find each partial sum. See Example 3.*

**29.** Find the sum of the first two terms of the sequence whose general term is $a_n = (n+2)(n-5)$.

**30.** Find the sum of the first two terms of the sequence whose general term is $a_n = n(n-6)$.

**31.** Find the sum of the first six terms of the sequence whose general term is $a_n = (-1)^n$.

**32.** Find the sum of the first seven terms of the sequence whose general term is $a_n = (-1)^{n-1}$.

**33.** Find the sum of the first four terms of the sequence whose general term is $a_n = (n+3)(n+1)$.

**34.** Find the sum of the first five terms of the sequence whose general term is $a_n = \dfrac{(-1)^n}{2n}$.

**35.** Find the sum of the first four terms of the sequence whose general term is $a_n = -2n$.

**36.** Find the sum of the first five terms of the sequence whose general term is $a_n = (n-1)^2$.

**37.** Find the sum of the first three terms of the sequence whose general term is $a_n = -\dfrac{n}{3}$.

**38.** Find the sum of the first three terms of the sequence whose general term is $a_n = (n+4)^2$.

*Solve. See Example 4.*

**39.** A gardener is making a triangular planting with 1 tree in the first row, 2 trees in the second row, 3 trees in the third row, and so on for 10 rows. Write the sequence that describes the number of trees in each row. Find the total number of trees planted.

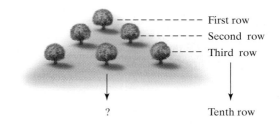

First row

Second row

Third row

?                    Tenth row

**40.** Some surfers at the beach form a human pyramid with 2 surfers in the top row, 3 surfers in the second row, 4 surfers in the third row, and so on. If there are 6 rows in the pyramid, write the sequence that describes the number of surfers in each row of the pyramid. Find the total number of surfers.

Top row

Second row

Third row

Sixth row          ?

**41.** A culture of fungus starts with 6 units and doubles every day. Write the general term of the sequence that describes the growth of this fungus. Find the number of fungus units there will be at the beginning of the fifth day.

**42.** A bacterial colony begins with 100 bacteria and doubles every 6 hours. Write the general term of the sequence describing the growth of the bacteria. Find the number of bacteria there will be after 24 hours.

**43.** A bacterial colony begins with 50 bacteria and doubles every 12 hours. Write the sequence that describes the growth of the bacteria. Find the number of bacteria there will be after 48 hours.

**44.** The number of otters born each year in a new aquarium forms a sequence whose general term is $a_n = (n - 1)(n + 3)$. Find the number of otters born in the third year, and find the total number of otters born in the first three years.

**45.** The number of opossums killed each month on a new highway forms the sequence whose general term is $a_n = (n + 1)(n + 2)$, where $n$ is the number of the months. Find the number of opossums killed in the fourth month, and find the total number killed in the first four months.

**46.** In 2007, the population of the Northern Spotted Owl continued to decline, and the owl remained on the endangered species list, as old-growth Northwest forests were logged. The size of the decrease in the population in a given year can be estimated by $200 - 6n$ pairs of birds. Find the decrease in population in 2010 if year 1 is 2007. Find the estimated total decrease in the spotted owl population for the years 2007 through 2010. (*Source:* United States Forest Service)

**47.** The amount of decay in pounds of a radioactive isotope each year is given by the sequence whose general term is $a_n = 100(0.5)^n$, where $n$ is the number of the year. Find the amount of decay in the fourth year, and find the total amount of decay in the first four years.

**48.** Susan has a choice between two job offers. Job $A$ has an annual starting salary of $20,000 with guaranteed annual raises of $1200 for the next four years, whereas job $B$ has an annual starting salary of $18,000 with guaranteed annual raises of $2500 for the next four years. Compare the fifth partial sums for each sequence to determine which job would pay Susan more money over the next 5 years.

**49.** A pendulum swings a length of 40 inches on its first swing. Each successive swing is $\dfrac{4}{5}$ of the preceding swing. Find the length of the fifth swing and the total length swung during the first five swings. (Round to the nearest tenth of an inch.)

**50.** Explain the difference between a sequence and a series.

REVIEW AND PREVIEW

*Evaluate. See Sections 1.7 and 7.7.*

**51.** $\dfrac{5}{1 - \dfrac{1}{2}}$

**52.** $\dfrac{-3}{1 - \dfrac{1}{7}}$

**53.** $\dfrac{\dfrac{1}{3}}{1 - \dfrac{1}{10}}$

**54.** $\dfrac{\dfrac{6}{11}}{1 - \dfrac{1}{10}}$

**55.** $\dfrac{3(1 - 2^4)}{1 - 2}$

**56.** $\dfrac{2(1 - 5^3)}{1 - 5}$

**57.** $\dfrac{10}{2}(3 + 15)$

**58.** $\dfrac{12}{2}(2 + 19)$

**CONCEPT EXTENSIONS**

**59. a.** Write the sum $\displaystyle\sum_{i=1}^{7}(i + i^2)$ without summation notation.

**b.** Write the sum $\displaystyle\sum_{i=1}^{7}i + \sum_{i=1}^{7}i^2$ without summation notation.

**c.** Compare the results of parts **a** and **b**.

**d.** Do you think the following is true or false? Explain your answer.

$$\sum_{i=1}^{n}(a_n + b_n) = \sum_{i=1}^{n}a_n + \sum_{i=1}^{n}b_n$$

**60. a.** Write the sum $\displaystyle\sum_{i=1}^{6}5i^3$ without summation notation.

**b.** Write the expression $5 \cdot \displaystyle\sum_{i=1}^{6}i^3$ without summation notation.

**c.** Compare the results of parts **a** and **b**.

**d.** Do you think the following is true or false? Explain your answer.

$$\sum_{i=1}^{n}c \cdot a_n = c \cdot \sum_{i=1}^{n}a_n, \text{ where } c \text{ is a constant}$$

# INTEGRATED REVIEW SEQUENCES AND SERIES

Sections 14.1–14.3

*Write the first five terms of each sequence whose general term is given.*

**1.** $a_n = n - 3$

**2.** $a_n = \dfrac{7}{1 + n}$

**3.** $a_n = 3^{n-1}$

**4.** $a_n = n^2 - 5$

*Find the indicated term for each sequence.*

**5.** $(-2)^n$; $a_6$

**6.** $-n^2 + 2$; $a_4$

**7.** $\dfrac{(-1)^n}{n}$; $a_{40}$

**8.** $\dfrac{(-1)^n}{2n}$; $a_{41}$

*Write the first five terms of the arithmetic or geometric sequence whose first term is $a_1$, and common difference, d, or common ratio, r, are given.*

**9.** $a_1 = 7; d = -3$

**10.** $a_1 = -3; r = 5$

**11.** $a_1 = 45; r = \dfrac{1}{3}$

**12.** $a_1 = -12; d = 10$

*Find the indicated term of each sequence.*

**13.** The tenth term of the arithmetic sequence whose first term is 20 and whose common difference is 9.

**14.** The sixth term of the geometric sequence whose first term is 64 and whose common ratio is $\dfrac{3}{4}$.

**15.** The seventh term of the geometric sequence $6, -12, 24, \ldots$

**16.** The twentieth term of the arithmetic sequence $-100, -85, -70, \ldots$

**17.** The fifth term of the arithmetic sequence whose fourth term is $-5$ and whose tenth term is $-35$.

**18.** The fifth term of the geometric sequence whose fourth term is 1 and whose seventh term is $\dfrac{1}{125}$.

*Evaluate.*

**19.** $\displaystyle\sum_{i=1}^{4} 5i$

**20.** $\displaystyle\sum_{i=1}^{7} (3i + 2)$

**21.** $\displaystyle\sum_{i=3}^{7} 2^{i-4}$

**22.** $\displaystyle\sum_{i=2}^{5} \dfrac{i}{i + 1}$

*Find each partial sum.*

**23.** Find the sum of the first three terms of the sequence whose general term is $a_n = n(n - 4)$.

**24.** Find the sum of the first ten terms of the sequence whose general term is $a_n = (-1)^n(n + 1)$.

## 14.4 PARTIAL SUMS OF ARITHMETIC AND GEOMETRIC SEQUENCES

**OBJECTIVES**

**1** Find the partial sum of an arithmetic sequence.

**2** Find the partial sum of a geometric sequence.

**3** Find the sum of the terms of an infinite geometric sequence.

**OBJECTIVE 1 ▶ Finding partial sums of arithmetic sequences.** Partial sums $S_n$ are relatively easy to find when $n$ is small—that is, when the number of terms to add is small. But when $n$ is large, finding $S_n$ can be tedious. For a large $n$, $S_n$ is still relatively easy to find if the addends are terms of an arithmetic sequence or a geometric sequence.

For an arithmetic sequence, $a_n = a_1 + (n - 1)d$ for some first term $a_1$ and some common difference $d$. So $S_n$, the sum of the first $n$ terms, is

$$S_n = a_1 + (a_1 + d) + (a_1 + 2d) + \cdots + (a_1 + (n - 1)d)$$

We might also find $S_n$ by "working backward" from the $n$th term $a_n$, finding the preceding term $a_{n-1}$, by subtracting $d$ each time.

$$S_n = a_n + (a_n - d) + (a_n - 2d) + \cdots + (a_n - (n - 1)d)$$

Now add the left sides of these two equations and add the right sides.

$$2S_n = (a_1 + a_n) + (a_1 + a_n) + (a_1 + a_n) + \cdots + (a_1 + a_n)$$

The $d$ terms subtract out, leaving $n$ sums of the first term, $a_1$, and last term, $a_n$. Thus, we write

$$2S_n = n(a_1 + a_n)$$

or

$$S_n = \dfrac{n}{2}(a_1 + a_n)$$

> **Partial Sum $S_n$ of an Arithmetic Sequence**
> The partial sum $S_n$ of the first $n$ terms of an arithmetic sequence is given by
> $$S_n = \frac{n}{2}(a_1 + a_n)$$
> where $a_1$ is the first term of the sequence and $a_n$ is the $n$th term.

**EXAMPLE 1**   Use the partial sum formula to find the sum of the first six terms of the arithmetic sequence $2, 5, 8, 11, 14, 17, \ldots$.

*Solution*   Use the formula for $S_n$ of an arithmetic sequence, replacing $n$ with 6, $a_1$ with 2, and $a_n$ with 17.

$$S_n = \frac{n}{2}(a_1 + a_n)$$
$$S_6 = \frac{6}{2}(2 + 17) = 3(19) = 57$$

**PRACTICE**
**1**   Use the partial sum formula to find the sum of the first five terms of the arithmetic sequence $2, 9, 16, 23, 30$.

**EXAMPLE 2**   Find the sum of the first 30 positive integers.

*Solution*   Because $1, 2, 3, \ldots, 30$ is an arithmetic sequence, use the formula for $S_n$ with $n = 30$, $a_1 = 1$, and $a_n = 30$. Thus,

$$S_n = \frac{n}{2}(a_1 + a_n)$$
$$S_{30} = \frac{30}{2}(1 + 30) = 15(31) = 465$$

**PRACTICE**
**2**   Find the sum of the first 50 positive integers.

**EXAMPLE 3**   **Stacking Rolls of Carpet**

Rolls of carpet are stacked in 20 rows with 3 rolls in the top row, 4 rolls in the next row, and so on, forming an arithmetic sequence. Find the total number of carpet rolls if there are 22 rolls in the bottom row.

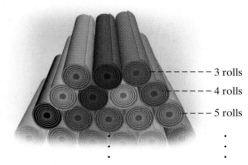

3 rolls
4 rolls
5 rolls

*Solution*   The list $3, 4, 5, \ldots, 22$ is the first 20 terms of an arithmetic sequence. Use the formula for $S_n$ with $a_1 = 3$, $a_n = 22$, and $n = 20$ terms. Thus,

$$S_{20} = \frac{20}{2}(3 + 22) = 10(25) = 250$$

There are a total of 250 rolls of carpet.

**PRACTICE**
**3**   An ice sculptor is creating a gigantic castle-facade ice sculpture for First Night festivities in Boston. To get the volume of ice necessary, large blocks of ice were stacked atop each other. The topmost row was comprised of 6 blocks of ice, the next row of 7 blocks of ice, and so on, forming an arithmetic sequence. Find the total number of ice blocks needed if there were 15 blocks in the bottom row.

**OBJECTIVE 2 ▶ Finding partial sums of geometric sequences.** We can also derive a formula for the partial sum $S_n$ of the first $n$ terms of a geometric series. If $a_n = a_1 r^{n-1}$, then

$$S_n = a_1 + a_1 r + a_1 r^2 + \cdots + a_1 r^{n-1}$$

↑   ↑   ↑        ↑

1st  2nd  3rd      $n$th

term term term    term

Multiply each side of the equation by $-r$.

$$-rS_n = -a_1 r - a_1 r^2 - a_1 r^3 - \cdots - a_1 r^n$$

Add the two equations.

$$S_n - rS_n = a_1 + (a_1 r - a_1 r) + (a_1 r^2 - a_1 r^2) + (a_1 r^3 - a_1 r^3) + \cdots - a_1 r^n$$
$$S_n - rS_n = a_1 - a_1 r^n$$

Now factor each side.

$$S_n(1 - r) = a_1(1 - r^n)$$

Solve for $S_n$ by dividing both sides by $1 - r$. Thus,

$$S_n = \frac{a_1(1 - r^n)}{1 - r}$$

as long as $r$ is not 1.

---

**Partial Sum $S_n$ of a Geometric Sequence**

The partial sum $S_n$ of the first $n$ terms of a geometric sequence is given by

$$S_n = \frac{a_1(1 - r^n)}{1 - r}$$

where $a_1$ is the first term of the sequence, $r$ is the common ratio, and $r \neq 1$.

---

**EXAMPLE 4**   Find the sum of the first six terms of the geometric sequence 5, 10, 20, 40, 80, 160.

*Solution*   Use the formula for the partial sum $S_n$ of the terms of a geometric sequence. Here, $n = 6$, the first term $a_1 = 5$, and the common ratio $r = 2$.

$$S_n = \frac{a_1(1 - r^n)}{1 - r}$$

$$S_6 = \frac{5(1 - 2^6)}{1 - 2} = \frac{5(-63)}{-1} = 315 \qquad \square$$

**PRACTICE**
**4**   Find the sum of the first five terms of the geometric sequence $32, 8, 2, \frac{1}{2}, \frac{1}{8}$.

---

**EXAMPLE 5**   **Finding Amount of Donation**

A grant from an alumnus to a university specified that the university was to receive $800,000 during the first year and 75% of the preceding year's donation during each of the following 5 years. Find the total amount donated during the 6 years.

**Solution**   The donations are modeled by the first six terms of a geometric sequence. Evaluate $S_n$ when $n = 6$, $a_1 = 800{,}000$, and $r = 0.75$.

$$S_6 = \frac{800{,}000[1 - (0.75)^6]}{1 - 0.75}$$

$$= \$2{,}630{,}468.75$$

The total amount donated during the 6 years is $\$2{,}630{,}468.75$.

**PRACTICE**

**5**   A new youth center is being established in a downtown urban area. A philanthropic charity has agreed to help it get off the ground. The charity has pledged to donate $\$250{,}000$ in the first year, with 80% of the preceding year's donation for each of the following 6 years. Find the total amount donated during the 7 years.

**OBJECTIVE 3 ▶ Finding sums of terms of infinite geometric sequences.**   Is it possible to find the sum of all the terms of an infinite sequence? Examine the partial sums of the geometric sequence $\dfrac{1}{2}, \dfrac{1}{4}, \dfrac{1}{8}, \dots$ .

$$S_1 = \frac{1}{2}$$

$$S_2 = \frac{1}{2} + \frac{1}{4} = \frac{3}{4}$$

$$S_3 = \frac{1}{2} + \frac{1}{4} + \frac{1}{8} = \frac{7}{8}$$

$$S_4 = \frac{1}{2} + \frac{1}{4} + \frac{1}{8} + \frac{1}{16} = \frac{15}{16}$$

$$S_5 = \frac{1}{2} + \frac{1}{4} + \frac{1}{8} + \frac{1}{16} + \frac{1}{32} = \frac{31}{32}$$

$$\vdots$$

$$S_{10} = \frac{1}{2} + \frac{1}{4} + \frac{1}{8} + \cdots + \frac{1}{2^{10}} = \frac{1023}{1024}$$

Even though each partial sum is larger than the preceding partial sum, we see that each partial sum is closer to 1 than the preceding partial sum. If $n$ gets larger and larger, then $S_n$ gets closer and closer to 1. We say that 1 is the **limit** of $S_n$ and also that 1 is the sum of the terms of this infinite sequence. In general, if $|r| < 1$, the following formula gives the sum of the terms of an infinite geometric sequence.

**Sum of the Terms of an Infinite Geometric Sequence**
The sum $S_\infty$ of the terms of an infinite geometric sequence is given by

$$S_\infty = \frac{a_1}{1 - r}$$

where $a_1$ is the first term of the sequence, $r$ is the common ratio, and $|r| < 1$. If $|r| \geq 1$, $S_\infty$ does not exist.

What happens for other values of $r$? For example, in the following geometric sequence, $r = 3$.

$$6, 18, 54, 162, \dots$$

Here, as $n$ increases, the sum $S_n$ increases also. This time, though, $S_n$ does not get closer and closer to a fixed number but instead increases without bound.

**EXAMPLE 6**  Find the sum of the terms of the geometric sequence $2, \dfrac{2}{3}, \dfrac{2}{9}, \dfrac{2}{27}, \dots$.

**Solution**  For this geometric sequence, $r = \dfrac{1}{3}$. Since $|r| < 1$, we may use the formula for $S_\infty$ of a geometric sequence with $a_1 = 2$ and $r = \dfrac{1}{3}$.

$$S_\infty = \frac{a_1}{1-r} = \frac{2}{1 - \dfrac{1}{3}} = \frac{2}{\dfrac{2}{3}} = 3 \qquad \square$$

The formula for the sum of the terms of an infinite geometric sequence can be used to write a repeating decimal as a fraction. For example,

$$0.33\overline{3} = \frac{3}{10} + \frac{3}{100} + \frac{3}{1000} + \cdots$$

This sum is the sum of the terms of an infinite geometric sequence whose first term $a_1$ is $\dfrac{3}{10}$ and whose common ratio $r$ is $\dfrac{1}{10}$. Using the formula for $S_\infty$,

$$S_\infty = \frac{a_1}{1-r} = \frac{\dfrac{3}{10}}{1 - \dfrac{1}{10}} = \frac{1}{3}$$

So, $0.33\overline{3} = \dfrac{1}{3}$.

**PRACTICE**
**6**  Find the sum of the terms of the geometric sequence $7, \dfrac{7}{4}, \dfrac{7}{16}, \dfrac{7}{64}, \dots$.

---

**EXAMPLE 7**  **Distance Traveled by a Pendulum**

On its first pass, a pendulum swings through an arc whose length is 24 inches. On each pass thereafter, the arc length is 75% of the arc length on the preceding pass. Find the total distance the pendulum travels before it comes to rest.

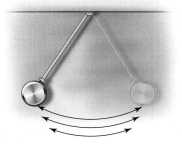

**Solution**  We must find the sum of the terms of an infinite geometric sequence whose first term, $a_1$, is 24 and whose common ratio, $r$, is 0.75. Since $|r| < 1$, we may use the formula for $S_\infty$.

$$S_\infty = \frac{a_1}{1-r} = \frac{24}{1 - 0.75} = \frac{24}{0.25} = 96$$

The pendulum travels a total distance of 96 inches before it comes to rest. $\qquad \square$

**PRACTICE**
**7**  The manufacturers of the "perpetual bouncing ball" claim that the ball rises to 96% of its dropped height on each bounce of the ball. Find the total distance the ball travels before it comes to rest, if it is dropped from a height of 36 inches.

# VOCABULARY & READINESS CHECK

*Decide whether each sequence is geometric or arithmetic.*

**1.** $5, 10, 15, 20, 25, \ldots$; _____

**2.** $5, 10, 20, 40, 80, \ldots$; _____

**3.** $-1, 3, -9, 27, -81 \ldots$; _____

**4.** $-1, 1, 3, 5, 7, \ldots$; _____

**5.** $-7, 0, 7, 14, 21, \ldots$; _____

**6.** $-7, 7, -7, 7, -7, \ldots$; _____

## 14.4 EXERCISE SET

*Use the partial sum formula to find the partial sum of the given arithmetic or geometric sequence. See Examples 1 and 4.*

**1.** Find the sum of the first six terms of the arithmetic sequence $1, 3, 5, 7, \ldots$.

**2.** Find the sum of the first seven terms of the arithmetic sequence $-7, -11, -15, \ldots$.

**3.** Find the sum of the first five terms of the geometric sequence $4, 12, 36, \ldots$.

**4.** Find the sum of the first eight terms of the geometric sequence $-1, 2, -4, \ldots$.

**5.** Find the sum of the first six terms of the arithmetic sequence $3, 6, 9, \ldots$.

**6.** Find the sum of the first four terms of the arithmetic sequence $-4, -8, -12, \ldots$.

**7.** Find the sum of the first four terms of the geometric sequence $2, \dfrac{2}{5}, \dfrac{2}{25}, \ldots$.

**8.** Find the sum of the first five terms of the geometric sequence $\dfrac{1}{3}, -\dfrac{2}{3}, \dfrac{4}{3}, \ldots$.

*Solve. See Example 2.*

**9.** Find the sum of the first ten positive integers.

**10.** Find the sum of the first eight negative integers.

**11.** Find the sum of the first four positive odd integers.

**12.** Find the sum of the first five negative odd integers.

*Find the sum of the terms of each infinite geometric sequence. See Example 6.*

**13.** $12, 6, 3, \ldots$

**14.** $45, 15, 5, \ldots$

**15.** $\dfrac{1}{10}, \dfrac{1}{100}, \dfrac{1}{1000}, \ldots$

**16.** $\dfrac{3}{5}, \dfrac{3}{20}, \dfrac{3}{80}, \ldots$

**17.** $-10, -5, -\dfrac{5}{2}, \ldots$

**18.** $-16, -4, -1, \ldots$

**19.** $2, -\dfrac{1}{4}, \dfrac{1}{32}, \ldots$

**20.** $-3, \dfrac{3}{5}, -\dfrac{3}{25}, \ldots$

**21.** $\dfrac{2}{3}, -\dfrac{1}{3}, \dfrac{1}{6}, \ldots$

**22.** $6, -4, \dfrac{8}{3}, \ldots$

## MIXED PRACTICE

*Solve.*

**23.** Find the sum of the first ten terms of the sequence $-4, 1, 6, \ldots, 41$ where $41$ is the tenth term.

**24.** Find the sum of the first twelve terms of the sequence $-3, -13, -23, \ldots, -113$ where $-113$ is the twelfth term.

**25.** Find the sum of the first seven terms of the sequence $3, \dfrac{3}{2}, \dfrac{3}{4}, \ldots$.

**26.** Find the sum of the first five terms of the sequence $-2, -6, -18, \ldots$.

**27.** Find the sum of the first five terms of the sequence $-12, 6, -3, \ldots$.

**28.** Find the sum of the first four terms of the sequence $-\dfrac{1}{4}, -\dfrac{3}{4}, -\dfrac{9}{4}, \ldots$.

**29.** Find the sum of the first twenty terms of the sequence $\dfrac{1}{2}, \dfrac{1}{4}, 0, \ldots, -\dfrac{17}{4}$ where $-\dfrac{17}{4}$ is the twentieth term.

**30.** Find the sum of the first fifteen terms of the sequence $-5, -9, -13, \ldots, -61$ where $-61$ is the fifteenth term.

**31.** If $a_1$ is $8$ and $r$ is $-\dfrac{2}{3}$, find $S_3$.

**32.** If $a_1$ is $10$, $a_{18}$ is $\dfrac{3}{2}$, and $d$ is $-\dfrac{1}{2}$, find $S_{18}$.

*Solve. See Example 3.*

**33.** Modern Car Company has come out with a new car model. Market analysts predict that 4000 cars will be sold in the first month and that sales will drop by 50 cars per month after that during the first year. Write out the first five terms of the sequence, and find the number of sold cars predicted for the twelfth month. Find the total predicted number of sold cars for the first year.

**34.** A company that sends faxes charges $3 for the first page sent and $0.10 less than the preceding page for each additional page sent. The cost per page forms an arithmetic sequence. Write the first five terms of this sequence, and use a partial sum to find the cost of sending a nine-page document.

**35.** Sal has two job offers: Firm A starts at $22,000 per year and guarantees raises of $1000 per year, whereas Firm B starts at $20,000 and guarantees raises of $1200 per year. Over a 10-year period, determine the more profitable offer.

**36.** The game of pool uses 15 balls numbered 1 to 15. In the variety called rotation, a player who sinks a ball receives as many points as the number on the ball. Use an arithmetic series to find the score of a player who sinks all 15 balls.

*Solve. See Example 5.*

**37.** A woman made $30,000 during the first year she owned her business and made an additional 10% over the previous year in each subsequent year. Find how much she made during her fourth year of business. Find her total earnings during the first four years.

**38.** In free fall, a parachutist falls 16 feet during the first second, 48 feet during the second second, 80 feet during the third second, and so on. Find how far she falls during the eighth second. Find the total distance she falls during the first 8 seconds.

**39.** A trainee in a computer company takes 0.9 times as long to assemble each computer as he took to assemble the preceding computer. If it took him 30 minutes to assemble the first computer, find how long it takes him to assemble the fifth computer. Find the total time he takes to assemble the first five computers (round to the nearest minute).

**40.** On a gambling trip to Reno, Carol doubled her bet each time she lost. If her first losing bet was $5 and she lost six consecutive bets, find how much she lost on the sixth bet. Find the total amount lost on these six bets.

*Solve. See Example 7.*

**41.** A ball is dropped from a height of 20 feet and repeatedly rebounds to a height that is $\frac{4}{5}$ of its previous height. Find the total distance the ball covers before it comes to rest.

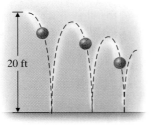

20 ft

**42.** A rotating flywheel coming to rest makes 300 revolutions in the first minute and in each minute thereafter makes $\frac{2}{5}$ as many revolutions as in the preceding minute. Find how many revolutions the wheel makes before it comes to rest.

**MIXED PRACTICE**

*Solve.*

**43.** In the pool game of rotation, player *A* sinks balls numbered 1 to 9, and player *B* sinks the rest of the balls. Use arithmetic series to find each player's score (see Exercise 36).

**44.** A godfather deposited $250 in a savings account on the day his godchild was born. On each subsequent birthday he deposited $50 more than he deposited the previous year. Find how much money he deposited on his godchild's twenty-first birthday. Find the total amount deposited over the 21 years.

**45.** During the holiday rush a business can rent a computer system for $200 the first day, with the rental fee decreasing $5 for each additional day. Find the fee paid for 20 days during the holiday rush.

**46.** The spraying of a field with insecticide killed 6400 weevils the first day, 1600 the second day, 400 the third day, and so on. Find the total number of weevils killed during the first 5 days.

**47.** A college student humorously asks his parents to charge him room and board according to this geometric sequence: $0.01 for the first day of the month, $0.02 for the second day, $0.04 for the third day, and so on. Find the total room and board he would pay for 30 days.

**48.** Following its television advertising campaign, a bank attracted 80 new customers the first day, 120 the second day, 160 the third day, and so on, in an arithmetic sequence. Find how many new customers were attracted during the first 5 days following its television campaign.

**REVIEW AND PREVIEW**

*Evaluate. See Section 1.7.*

**49.** $6 \cdot 5 \cdot 4 \cdot 3 \cdot 2 \cdot 1$

**50.** $8 \cdot 7 \cdot 6 \cdot 5 \cdot 4 \cdot 3 \cdot 2 \cdot 1$

**51.** $\dfrac{3 \cdot 2 \cdot 1}{2 \cdot 1}$

**52.** $\dfrac{5 \cdot 4 \cdot 3 \cdot 2 \cdot 1}{3 \cdot 2 \cdot 1}$

*Multiply. See Section 5.4.*

**53.** $(x + 5)^2$

**54.** $(x - 2)^2$

**55.** $(2x - 1)^3$

**56.** $(3x + 2)^3$

**CONCEPT EXTENSIONS**

**57.** Write $0.88\overline{8}$ as an infinite geometric series and use the formula for $S_\infty$ to write it as a rational number.

**58.** Write $0.54\overline{54}$ as an infinite geometric series and use the formula $S_\infty$ to write it as a rational number.

**59.** Explain whether the sequence $5, 5, 5, \ldots$ is arithmetic, geometric, neither, or both.

**60.** Describe a situation in everyday life that can be modeled by an infinite geometric series.

# 14.5 THE BINOMIAL THEOREM

**OBJECTIVES**

1  Use Pascal's triangle to expand binomials.

2  Evaluate factorials.

3  Use the binomial theorem to expand binomials.

4  Find the *n*th term in the expansion of a binomial raised to a positive power.

In this section, we learn how to **expand** binomials of the form $(a + b)^n$ easily. Expanding a binomial such as $(a + b)^n$ means to write the factored form as a sum. First, we review the patterns in the expansions of $(a + b)^n$.

$$(a + b)^0 = 1 \qquad \text{1 term}$$
$$(a + b)^1 = a + b \qquad \text{2 terms}$$
$$(a + b)^2 = a^2 + 2ab + b^2 \qquad \text{3 terms}$$
$$(a + b)^3 = a^3 + 3a^2b + 3ab^2 + b^3 \qquad \text{4 terms}$$
$$(a + b)^4 = a^4 + 4a^3b + 6a^2b^2 + 4ab^3 + b^4 \qquad \text{5 terms}$$
$$(a + b)^5 = a^5 + 5a^4b + 10a^3b^2 + 10a^2b^3 + 5ab^4 + b^5 \qquad \text{6 terms}$$

Notice the following patterns.

1. The expansion of $(a + b)^n$ contains $n + 1$ terms. For example, for $(a + b)^3$, $n = 3$, and the expansion contains $3 + 1$ terms, or 4 terms.

2. The first term of the expansion of $(a + b)^n$ is $a^n$, and the last term is $b^n$.

3. The powers of $a$ decrease by 1 for each term, whereas the powers of $b$ increase by 1 for each term.

4. For each term of the expansion of $(a + b)^n$, the sum of the exponents of $a$ and $b$ is $n$. (For example, the sum of the exponents of $5a^4b$ is $4 + 1$, or 5, and the sum of the exponents of $10a^3b^2$ is $3 + 2$, or 5.)

**OBJECTIVE 1 ▶ Using Pascal's triangle.** There are patterns in the coefficients of the terms as well. Written in a triangular array, the coefficients are called **Pascal's triangle.**

$$
\begin{array}{llccccccc}
(a + b)^0: & & & & & 1 & & & & & n = 0\\
(a + b)^1: & & & & 1 & & 1 & & & & n = 1\\
(a + b)^2: & & & 1 & & 2 & & 1 & & & n = 2\\
(a + b)^3: & & 1 & & 3 & & 3 & & 1 & & n = 3\\
(a + b)^4: & 1 & & 4 & & 6 & & 4 & & 1 & n = 4\\
(a + b)^5: & 1 & 5 & & 10 & & 10 & & 5 & 1 & n = 5
\end{array}
$$

Each row in Pascal's triangle begins and ends with 1. Any other number in a row is the sum of the two closest numbers above it. Using this pattern, we can write the next row, for $n = 6$, by first writing the number 1. Then we can add the consecutive numbers in the row for $n = 5$ and write each sum "between and below" the pair. We complete the row by writing a 1.

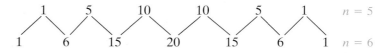

We can use Pascal's triangle and the patterns noted to expand $(a + b)^n$ without actually multiplying any terms.

### EXAMPLE 1   Expand $(a + b)^6$.

*Solution*   Using the $n = 6$ row of Pascal's triangle as the coefficients and following the patterns noted, $(a + b)^6$ can be expanded as

$$a^6 + 6a^5b + 15a^4b^2 + 20a^3b^3 + 15a^2b^4 + 6ab^5 + b^6 \qquad \square$$

**PRACTICE**
**1**   Expand $(p + r)^7$.

**OBJECTIVE 2 ▶ Evaluating factorials.** For a large $n$, the use of Pascal's triangle to find coefficients for $(a + b)^n$ can be tedious. An alternative method for determining these coefficients is based on the concept of a **factorial.**

The **factorial of $n$,** written $n!$ (read "$n$ factorial"), is the product of the first $n$ consecutive natural numbers.

---

**Factorial of $n$: $n!$**

If $n$ is a natural number, then $n! = n(n - 1)(n - 2)(n - 3) \cdots 3 \cdot 2 \cdot 1$. The factorial of 0, written $0!$, is defined to be 1.

---

For example, $3! = 3 \cdot 2 \cdot 1 = 6$,   $5! = 5 \cdot 4 \cdot 3 \cdot 2 \cdot 1 = 120$, and $0! = 1$.

**EXAMPLE 2**   Evaluate each expression.

**a.** $\dfrac{5!}{6!}$     **b.** $\dfrac{10!}{7!3!}$     **c.** $\dfrac{3!}{2!1!}$     **d.** $\dfrac{7!}{7!0!}$

*Solution*

**a.** $\dfrac{5!}{6!} = \dfrac{5 \cdot 4 \cdot 3 \cdot 2 \cdot 1}{6 \cdot 5 \cdot 4 \cdot 3 \cdot 2 \cdot 1} = \dfrac{1}{6}$

**b.** $\dfrac{10!}{7!3!} = \dfrac{10 \cdot 9 \cdot 8 \cdot 7!}{7! \cdot 3 \cdot 2 \cdot 1} = \dfrac{10 \cdot 9 \cdot 8}{3 \cdot 2 \cdot 1} = 10 \cdot 3 \cdot 4 = 120$

**c.** $\dfrac{3!}{2!1!} = \dfrac{3 \cdot 2 \cdot 1}{2 \cdot 1 \cdot 1} = 3$

**d.** $\dfrac{7!}{7!0!} = \dfrac{7!}{7! \cdot 1} = 1$

**PRACTICE**
**2**   Evaluate each expression.

**a.** $\dfrac{6!}{7!}$     **b.** $\dfrac{8!}{4!2!}$     **c.** $\dfrac{5!}{4!1!}$     **d.** $\dfrac{9!}{9!0!}$

---

**▶ Helpful Hint**

We can use a calculator with a factorial key to evaluate a factorial. A calculator uses scientific notation for large results.

---

**OBJECTIVE 3 ▶ Using the binomial theorem.** It can be proved, although we won't do so here, that the coefficients of terms in the expansion of $(a + b)^n$ can be expressed in terms of factorials. Following patterns 1 through 4 given earlier and using the factorial expressions of the coefficients, we have what is known as the **binomial theorem.**

---

**Binomial Theorem**
If $n$ is a positive integer, then

$$(a + b)^n = a^n + \frac{n}{1!}a^{n-1}b^1 + \frac{n(n - 1)}{2!}a^{n-2}b^2$$

$$+ \frac{n(n - 1)(n - 2)}{3!}a^{n-3}b^3 + \cdots + b^n$$

---

We call the formula for $(a + b)^n$ given by the binomial theorem the **binomial formula.**

**EXAMPLE 3**   Use the binomial theorem to expand $(x + y)^{10}$.

*Solution*   Let $a = x, b = y$, and $n = 10$ in the binomial formula.

$$(x + y)^{10} = x^{10} + \frac{10}{1!}x^9y + \frac{10 \cdot 9}{2!}x^8y^2 + \frac{10 \cdot 9 \cdot 8}{3!}x^7y^3 + \frac{10 \cdot 9 \cdot 8 \cdot 7}{4!}x^6y^4$$

$$+ \frac{10 \cdot 9 \cdot 8 \cdot 7 \cdot 6}{5!}x^5y^5 + \frac{10 \cdot 9 \cdot 8 \cdot 7 \cdot 6 \cdot 5}{6!}x^4y^6$$

$$+ \frac{10 \cdot 9 \cdot 8 \cdot 7 \cdot 6 \cdot 5 \cdot 4}{7!}x^3y^7$$

$$+ \frac{10 \cdot 9 \cdot 8 \cdot 7 \cdot 6 \cdot 5 \cdot 4 \cdot 3}{8!}x^2y^8$$

$$+ \frac{10 \cdot 9 \cdot 8 \cdot 7 \cdot 6 \cdot 5 \cdot 4 \cdot 3 \cdot 2}{9!}xy^9 + y^{10}$$

$$= x^{10} + 10x^9y + 45x^8y^2 + 120x^7y^3 + 210x^6y^4 + 252x^5y^5 + 210x^4y^6$$

$$+ 120x^3y^7 + 45x^2y^8 + 10xy^9 + y^{10}$$

**PRACTICE**
**3**   Use the binomial theorem to expand $(a + b)^9$.

**EXAMPLE 4**   Use the binomial theorem to expand $(x + 2y)^5$.

*Solution*   Let $a = x$ and $b = 2y$ in the binomial formula.

$$(x + 2y)^5 = x^5 + \frac{5}{1!}x^4(2y) + \frac{5 \cdot 4}{2!}x^3(2y)^2 + \frac{5 \cdot 4 \cdot 3}{3!}x^2(2y)^3$$

$$+ \frac{5 \cdot 4 \cdot 3 \cdot 2}{4!}x(2y)^4 + (2y)^5$$

$$= x^5 + 10x^4y + 40x^3y^2 + 80x^2y^3 + 80xy^4 + 32y^5$$

**PRACTICE**
**4**   Use the binomial theorem to expand $(a + 5b)^3$.

**EXAMPLE 5**   Use the binomial theorem to expand $(3m - n)^4$.

*Solution*   Let $a = 3m$ and $b = -n$ in the binomial formula.

$$(3m - n)^4 = (3m)^4 + \frac{4}{1!}(3m)^3(-n) + \frac{4 \cdot 3}{2!}(3m)^2(-n)^2$$

$$+ \frac{4 \cdot 3 \cdot 2}{3!}(3m)(-n)^3 + (-n)^4$$

$$= 81m^4 - 108m^3n + 54m^2n^2 - 12mn^3 + n^4$$

**PRACTICE**
**5**   Use the binomial theorem to expand $(3x - 2y)^3$.

**OBJECTIVE 4 ▶ Finding the *n*th term of a binomial expansion.** Sometimes it is convenient to find a specific term of a binomial expansion without writing out the entire expansion. By studying the expansion of binomials, a pattern forms for each term. This pattern is most easily stated for the $(r + 1)$st term.

**(r + 1)st Term in a Binomial Expansion**

The $(r + 1)$st term of the expansion of $(a + b)^n$ is $\dfrac{n!}{r!(n - r)!}a^{n-r}b^r$.

**EXAMPLE 6** Find the eighth term in the expansion of $(2x - y)^{10}$.

*Solution* Use the formula, with $n = 10$, $a = 2x$, $b = -y$, and $r + 1 = 8$. Notice that, since $r + 1 = 8$, $r = 7$.

$$\frac{n!}{r!(n-r)!}a^{n-r}b^r = \frac{10!}{7!3!}(2x)^3(-y)^7$$

$$= 120(8x^3)(-y^7)$$

$$= -960x^3y^7$$

**PRACTICE**
**6** Find the seventh term in the expansion of $(x - 4y)^{11}$.

## VOCABULARY & READINESS CHECK

*Fill in each blank.*

**1.** $0! =$ _____  **2.** $1! =$ _____  **3.** $4! =$ _____  **4.** $2! =$ _____  **5.** $3!0! =$ _____  **6.** $0!2! =$ _____

## 14.5 | EXERCISE SET

*MyMathLab* Powered by CourseCompass and MathXL

Math XL PRACTICE   WATCH   DOWNLOAD   READ   REVIEW

*Use Pascal's triangle to expand the binomial. See Example 1.*

**1.** $(m + n)^3$

**2.** $(x + y)^4$

**3.** $(c + d)^5$

**4.** $(a + b)^6$

**5.** $(y - x)^5$

**6.** $(q - r)^7$

**7.** Explain how to generate a row of Pascal's triangle.

**8.** Write the $n = 8$ row of Pascal's triangle.

*Evaluate each expression. See Example 2.*

**9.** $\dfrac{8!}{7!}$  **10.** $\dfrac{6!}{0!}$

**11.** $\dfrac{7!}{5!}$  **12.** $\dfrac{8!}{5!}$

**13.** $\dfrac{10!}{7!2!}$  **14.** $\dfrac{9!}{5!3!}$

**15.** $\dfrac{8!}{6!0!}$  **16.** $\dfrac{10!}{4!6!}$

### MIXED PRACTICE

*Use the binomial formula to expand each binomial. See Examples 3 through 5.*

**17.** $(a + b)^7$

**18.** $(x + y)^8$

**19.** $(a + 2b)^5$

**20.** $(x + 3y)^6$

**21.** $(q + r)^9$

**22.** $(b + c)^6$

**23.** $(4a + b)^5$

**24.** $(3m + n)^4$

**25.** $(5a - 2b)^4$

**26.** $(m - 4)^6$

**27.** $(2a + 3b)^3$

**28.** $(4 - 3x)^5$

**29.** $(x + 2)^5$

**30.** $(3 + 2a)^4$

*Find the indicated term. See Example 6.*

**31.** The fifth term of the expansion of $(c - d)^5$

**32.** The fourth term of the expansion of $(x - y)^6$

**33.** The eighth term of the expansion of $(2c + d)^7$

**34.** The tenth term of the expansion of $(5x - y)^9$

**35.** The fourth term of the expansion of $(2r - s)^5$

**36.** The first term of the expansion of $(3q - 7r)^6$

**37.** The third term of the expansion of $(x + y)^4$

**38.** The fourth term of the expansion of $(a + b)^8$

**39.** The second term of the expansion of $(a + 3b)^{10}$

**40.** The third term of the expansion of $(m + 5n)^7$

## REVIEW AND PREVIEW

*Sketch the graph of each function. Decide whether each function is one-to-one. See Sections 3.6 and 12.2.*

**41.** $f(x) = |x|$

**42.** $g(x) = 3(x-1)^2$

**43.** $H(x) = 2x + 3$

**44.** $F(x) = -2$

**45.** $f(x) = x^2 + 3$

**46.** $h(x) = -(x+1)^2 - 4$

## CONCEPT EXTENSIONS

**47.** Expand the expression $\left(\sqrt{x} + \sqrt{3}\right)^5$.

**48.** Find the term containing $x^2$ in the expansion of $\left(\sqrt{x} - \sqrt{5}\right)^6$.

*Evaluate the following.*

The notation $\binom{n}{r}$ means $\dfrac{n!}{r!(n-r)!}$. For example,

$$\binom{5}{3} = \frac{5!}{3!(5-3)!} = \frac{5!}{3!2!} = \frac{5\cdot4\cdot3\cdot2\cdot1}{(3\cdot2\cdot1)\cdot(2\cdot1)} = 10.$$

**49.** $\binom{9}{5}$

**50.** $\binom{4}{3}$

**51.** $\binom{8}{2}$

**52.** $\binom{12}{11}$

**53.** Show that $\binom{n}{n} = 1$ for any whole number $n$.

---

### 📖 STUDY SKILLS BUILDER

**Are You Preparing for Your Final Exam?**

To prepare for your final exam, try the following study techniques:

- Review the material that you will be responsible for on your exam. This includes material from your textbook, your notebook, and any handouts from your instructor.
- Review any formulas that you may need to memorize.
- Check to see if your instructor or mathematics department will be conducting a final exam review.
- Check with your instructor to see whether final exams from previous semesters/quarters are available to students for review.

- Use your previously taken exams as a practice final exam. To do so, rewrite the test questions in mixed order on blank sheets of paper. This will help you prepare for exam conditions.
- If you are unsure of a few concepts, see your instructor or visit a learning lab for assistance. Also, view the video segment of any troublesome sections.
- If you need further exercises to work, try the Cumulative Reviews at the end of the chapters.

Once again, good luck! I hope you have enjoyed this textbook and your mathematics course.

---

# CHAPTER 14 GROUP ACTIVITY

## Modeling College Tuition

Annual college tuition has steadily increased since 1970. According to the College Board, by the 2004–2005 academic year, the average annual tuition at a public 4-year university had increased to $5132. Similarly, average annual tuition at private 4-year universities had grown to $20,082.

Over the past few years, annual tuition at 4-year public universities has been increasing at an average rate of 7.9% per year. Over the same time period, annual tuition at 4-year private universities has been increasing at an average rate of $1068 per year. In this project, you will have the opportunity to model and investigate the trend in increasing tuition at public and private universities. This project may be completed by working in groups or individually.

1. Using the information given in the introductory paragraphs, decide whether the sequence of public university tuitions is arithmetic or geometric.

2. Using the information given in the introductory paragraphs, decide whether the sequence of private university tuitions is arithmetic or geometric.

3. Find the general term of the sequence that describes the pattern of average annual tuition for 4-year public universities. Let $n = 1$ represent the 2004–2005 academic year.

| Academic Year | n |
|---|---|
| 2004–2005 | 1 |
| 2005–2006 | 2 |
| 2006–2007 | 3 |
| 2007–2008 | 4 |
| 2008–2009 | 5 |
| 2009–2010 | 6 |
| 2010–2011 | 7 |

4. Find the general term of the sequence that describes the pattern of average annual tuition for private 4-year universities. Let $n = 1$ represent the 2004–2005 academic year.

5. Assuming that the rate of tuition increase remains the same, use the general term equation from Question 3 to find the average annual tuition at a 4-year public university for the 2007–2008 academic year.

6. Assuming that the rate of tuition increase remains the same, use the general term equation from Question 4 to find the average annual tuition at a 4-year private university for the 2007–2008 academic year.

7. Use partial sums to find the average cost of a 4-year college education at a public university for a student who started college in the 2004–2005 academic year.

8. Use partial sums to find the average cost of a 4-year college education at a private university for a student who starts college in the 2007–2008 academic year. (*Hint:* One way to so this is to find $S_7$ and subtract $S_3$ from it. If you use this method, explain why this gives the desired sum.)

9. (Optional) Use newspapers or news magazines to find a situation that can be modeled by a sequence. Briefly describe the situation, and decide whether it is an arithmetic or geometric sequence. Find an equation of the general term of the sequence.

## CHAPTER 14 VOCABULARY CHECK

*Fill in each blank with one of the words or phrases listed below.*

| | | | | |
|---|---|---|---|---|
| general term | common difference | finite sequence | common ratio | Pascal's triangle |
| infinite sequence | factorial of $n$ | arithmetic sequence | geometric sequence | series |

1. A(n) _____ is a function whose domain is the set of natural numbers $\{1, 2, 3, \ldots, n\}$, where $n$ is some natural number.

2. The _____, written $n!$, is the product of the first $n$ consecutive natural numbers.

3. A(n) _____ is a function whose domain is the set of natural numbers.

4. A(n) _____ is a sequence in which each term (after the first) is obtained by multiplying the preceding term by a constant amount $r$. The constant $r$ is called the _____ of the sequence.

5. The sum of the terms of a sequence is called a _____.

6. The $n$th term of the sequence $a_n$ is called the _____.

7. A(n) _____ is a sequence in which each term (after the first) differs from the preceding term by a constant amount $d$. The constant $d$ is called the _____ of the sequence.

8. A triangle array of the coefficients of the terms of the expansions of $(a + b)^n$ is called _____.

> ▶ **Helpful Hint**
>
> Are you preparing for your test? Don't forget to take the Chapter 14 Test on page 834. Then check your answers at the back of the text and use the Chapter Test Prep Video CD to see the fully worked-out solutions to any of the exercises you want to review.

## CHAPTER 14 HIGHLIGHTS

| **DEFINITIONS AND CONCEPTS** | **EXAMPLES** |
|---|---|

### SECTION 14.1   SEQUENCES

An **infinite sequence** is a function whose domain is the set of natural numbers $\{1, 2, 3, 4, \ldots\}$.

A **finite sequence** is a function whose domain is the set of natural numbers $\{1, 2, 3, 4, \ldots, n\}$, where $n$ is some natural number.

The notation $a_n$, where $n$ is a natural number, is used to denote a sequence.

*Infinite Sequence*

$$2, 4, 6, 8, 10, \ldots$$

*Finite Sequence*

$$1, -2, 3, -4, 5, -6$$

Write the first four terms of the sequence whose general term is $a_n = n^2 + 1$.

$$a_1 = 1^2 + 1 = 2$$
$$a_2 = 2^2 + 1 = 5$$
$$a_3 = 3^2 + 1 = 10$$
$$a_4 = 4^2 + 1 = 17$$

**DEFINITIONS AND CONCEPTS**                                    **EXAMPLES**

An **arithmetic sequence** is a sequence in which each term differs from the preceding term by a constant amount $d$, called the **common difference**.

The **general term** $a_n$ of an arithmetic sequence is given by

$$a_n = a_1 + (n - 1)d$$

where $a_1$ is the first term and $d$ is the common difference.

A **geometric sequence** is a sequence in which each term is obtained by multiplying the preceding term by a constant $r$, called the **common ratio**.

The **general term** $a_n$ of a geometric sequence is given by

$$a_n = a_1 r^{n-1}$$

where $a_1$ is the first term and $r$ is the common ratio.

*Arithmetic Sequence*

$$5, 8, 11, 14, 17, 20, \ldots$$

Here, $a_1 = 5$ and $d = 3$.
The general term is

$$a_n = a_1 + (n - 1)d \text{ or}$$
$$a_n = 5 + (n - 1)3$$

*Geometric Sequence*

$$12, -6, 3, -\frac{3}{2}, \ldots$$

Here $a_1 = 12$ and $r = -\frac{1}{2}$.

The general term is

$$a_n = a_1 r^{n-1} \text{ or}$$
$$a_n = 12\left(-\frac{1}{2}\right)^{n-1}$$

A sum of the terms of a sequence is called a **series**.

A shorthand notation for denoting a series is called **summation notation**:

$$\text{index of summation} \rightarrow \sum_{i=1}^{4} \leftarrow \begin{array}{l}\text{Greek letter sigma}\\ \text{used to mean sum}\end{array}$$

| Sequence | Series | |
|---|---|---|
| $3, 7, 11, 15$ | $3 + 7 + 11 + 15$ | finite |
| $3, 7, 11, 15, \ldots$ | $3 + 7 + 11 + 15 + \cdots$ | infinite |

$$\sum_{i=1}^{4} 3^i = 3^1 + 3^2 + 3^3 + 3^4$$
$$= 3 + 9 + 27 + 81$$
$$= 120$$

**Partial sum,** $S_n$, of the first $n$ terms of an arithmetic sequence:

$$s_n = \frac{n}{2}(a_1 + a_n)$$

where $a_1$ is the first term and $a_n$ is the $n$th term.

**Partial sum,** $S_n$, of the first $n$ terms of a geometric sequence:

$$S_n = \frac{a_1(1 - r^n)}{1 - r}$$

where $a_1$ is the first term, $r$ is the common ratio, and $r \neq 1$.

Sum of the terms of an infinite geometric sequence:

$$S_\infty = \frac{a_1}{1 - r}$$

where $a_1$ is the first term, $r$ is the common ratio, and $|r| < 1$. (If $|r| \geq 1$, $S_\infty$ does not exist.)

The sum of the first five terms of the arithmetic sequence

$$12, 24, 36, 48, 60, \ldots \text{ is}$$

$$S_5 = \frac{5}{2}(12 + 60) = 180$$

The sum of the first five terms of the geometric sequence

$$15, 30, 60, 120, 240, \ldots \text{ is}$$

$$S_5 = \frac{15(1 - 2^5)}{1 - 2} = 465$$

The sum of the terms of the infinite geometric sequence

$$1, \frac{1}{3}, \frac{1}{9}, \frac{1}{27}, \ldots \text{ is}$$

$$S_\infty = \frac{1}{1 - \frac{1}{3}} = \frac{3}{2}$$

| DEFINITIONS AND CONCEPTS | EXAMPLES |
|---|---|

SECTION 14.5   THE BINOMIAL THEOREM

The **factorial of $n$,** written $n!$, is the product of the first $n$ consecutive natural numbers.

**Binomial Theorem**

If $n$ is a positive integer, then

$$(a + b)^n = a^n + \frac{n}{1!}a^{n-1}b^1 + \frac{n(n-1)}{2!}a^{n-2}b^2$$
$$+ \frac{n(n-1)(n-2)}{3!}a^{n-3}b^3 + \cdots + b^n$$

$5! = 5 \cdot 4 \cdot 3 \cdot 2 \cdot 1 = 120$

Expand $(3x + y)^4$.

$$(3x + y)^4 = (3x)^4 + \frac{4}{1!}(3x)^3(y)^1$$
$$+ \frac{4 \cdot 3}{2!}(3x)^2(y)^2 + \frac{4 \cdot 3 \cdot 2}{3!}(3x)^1y^3 + y^4$$
$$= 81x^4 + 108x^3y + 54x^2y^2 + 12xy^3 + y^4$$

# CHAPTER 14 REVIEW

*(14.1) Find the indicated term(s) of the given sequence.*

**1.** The first five terms of the sequence $a_n = -3n^2$

**2.** The first five terms of the sequence $a_n = n^2 + 2n$

**3.** The one-hundredth term of the sequence $a_n = \dfrac{(-1)^n}{100}$

**4.** The fiftieth term of the sequence $a_n = \dfrac{2n}{(-1)^2}$

**5.** The general term $a_n$ of the sequence $\dfrac{1}{6}, \dfrac{1}{12}, \dfrac{1}{18}, \dots.$

**6.** The general term $a_n$ of the sequence $-1, 4, -9, 16, \dots.$

*Solve the following applications.*

**7.** The distance in feet that an olive falling from rest in a vacuum will travel during each second is given by an arithmetic sequence whose general term is $a_n = 32n - 16$, where $n$ is the number of the second. Find the distance the olive will fall during the fifth, sixth, and seventh seconds.

**8.** A culture of yeast doubles every day in a geometric progression whose general term is $a_n = 100(2)^{n-1}$, where $n$ is the number of the day just ending. Find how many days it takes the yeast culture to measure at least 10,000. Find the original measure of the yeast culture.

**9.** The Colorado Forest Service reported that western pine beetle infestation, which kills trees, affected approximately 660,000 acres of lodgepole forests in Colorado in 2006. The forest service predicts that during the next 5 years, the beetles will infest twice the number of acres per year as the year before. Write out the first 5 terms of this geometric sequence, and predict the number of acres of infested trees there will be in 2010.

**10.** The first row of an amphitheater contains 50 seats, and each row thereafter contains 8 additional seats. Write the first ten terms of this arithmetic progression, and find the number of seats in the tenth row.

**(14.2)**

**11.** Find the first five terms of the geometric sequence whose first term is $-2$ and whose common ratio is $\dfrac{2}{3}$.

**12.** Find the first five terms of the arithmetic sequence whose first term is 12 and whose common difference is $-1.5$.

**13.** Find the thirtieth term of the arithmetic sequence whose first term is $-5$ and whose common difference is 4.

**14.** Find the eleventh term of the arithmetic sequence whose first term is 2 and whose common difference is $\dfrac{3}{4}$.

**15.** Find the twentieth term of the arithmetic sequence whose first three terms are 12, 7, and 2.

**16.** Find the sixth term of the geometric sequence whose first three terms are 4, 6, and 9.

**17.** If the fourth term of an arithmetic sequence is 18 and the twentieth term is 98, find the first term and the common difference.

**18.** If the third term of a geometric sequence is $-48$ and the fourth term is 192, find the first term and the common ratio.

**19.** Find the general term of the sequence $\dfrac{3}{10}, \dfrac{3}{100}, \dfrac{3}{1000}, \dots.$

**20.** Find a general term that satisfies the terms shown for the sequence $50, 58, 66, \dots.$

*Determine whether each of the following sequences is arithmetic, geometric, or neither. If a sequence is arithmetic, find $a_1$ and d. If a sequence is geometric, find $a_1$ and r.*

**21.** $\dfrac{8}{3}, 4, 6, \ldots$

**22.** $-10.5, -6.1, -1.7$

**23.** $7x, -14x, 28x$

**24.** $3x^2, 9x^4, 81x^8, \ldots$

*Solve the following applications.*

**25.** To test the bounce of a racquetball, the ball is dropped from a height of 8 feet. The ball is judged "good" if it rebounds at least 75% of its previous height with each bounce. Write out the first six terms of this geometric sequence (round to the nearest tenth). Determine if a ball is "good" that rebounds to a height of 2.5 feet after the fifth bounce.

**26.** A display of oil cans in an auto parts store has 25 cans in the bottom row, 21 cans in the next row, and so on, in an arithmetic progression. Find the general term and the number of cans in the top row.

**27.** Suppose that you save $1 the first day of a month, $2 the second day, $4 the third day, continuing to double your savings each day. Write the general term of this geometric sequence and find the amount you will save on the tenth day. Estimate the amount you will save on the thirtieth day of the month, and check your estimate with a calculator.

**28.** On the first swing, the length of an arc through which a pendulum swings is 30 inches. The length of the arc for each successive swing is 70% of the preceding swing. Find the length of the arc for the fifth swing.

**29.** Rosa takes a job that has a monthly starting salary of $900 and guarantees her a monthly raise of $150 during her 6-month training period. Find the general term of this sequence and her salary at the end of her training.

**30.** A sheet of paper is $\dfrac{1}{512}$-inch thick. By folding the sheet in half, the total thickness will be $\dfrac{1}{256}$ inch. A second fold produces a total thickness of $\dfrac{1}{128}$ inch. Estimate the thickness of the stack after 15 folds, and then check your estimate with a calculator.

*(14.3) Write out the terms and find the sum for each of the following.*

**31.** $\displaystyle\sum_{i=1}^{5}(2i-1)$

**32.** $\displaystyle\sum_{i=1}^{5}i(i+2)$

**33.** $\displaystyle\sum_{i=2}^{4}\dfrac{(-1)^i}{2i}$

**34.** $\displaystyle\sum_{i=3}^{5}5(-1)^{i-1}$

*Find the partial sum of the given sequence.*

**35.** $S_4$ of the sequence $a_n = (n-3)(n+2)$

**36.** $S_6$ of the sequence $a_n = n^2$

**37.** $S_5$ of the sequence $a_n = -8 + (n-1)3$

**38.** $S_3$ of the sequence $a_n = 5(4)^{n-1}$

*Write the sum with $\Sigma$ notation.*

**39.** $1 + 3 + 9 + 27 + 81 + 243$

**40.** $6 + 2 + (-2) + (-6) + (-10) + (-14) + (-18)$

**41.** $\dfrac{1}{4} + \dfrac{1}{16} + \dfrac{1}{64} + \dfrac{1}{256}$

**42.** $1 + \left(-\dfrac{3}{2}\right) + \dfrac{9}{4}$

*Solve.*

**43.** A yeast colony begins with 20 yeast and doubles every 8 hours. Write the sequence that describes the growth of the yeast, and find the total yeast after 48 hours.

**44.** The number of cranes born each year in a new aviary forms a sequence whose general term is $a_n = n^2 + 2n - 1$. Find the number of cranes born in the fourth year and the total number of cranes born in the first four years.

**45.** Harold has a choice between two job offers. Job $A$ has an annual starting salary of $39,500 with guaranteed annual raises of $2200 for the next four years, whereas job $B$ has an annual starting salary of $41,000 with guaranteed annual raises of $1400 for the next four years. Compare the salaries for the fifth year under each job offer.

**46.** A sample of radioactive waste is decaying such that the amount decaying in kilograms during year $n$ is $a_n = 200(0.5)^n$. Find the amount of decay in the third year, and the total amount of decay in the first three years.

*(14.4) Find the partial sum of the given sequence.*

**47.** The sixth partial sum of the sequence $15, 19, 23, \ldots$

**48.** The ninth partial sum of the sequence $5, -10, 20, \ldots$

**49.** The sum of the first 30 odd positive integers

**50.** The sum of the first 20 positive multiples of 7

**51.** The sum of the first 20 terms of the sequence $8, 5, 2, \ldots$

**52.** The sum of the first eight terms of the sequence $\dfrac{3}{4}, \dfrac{9}{4}, \dfrac{27}{4}, \ldots$

**53.** $S_4$ if $a_1 = 6$ and $r = 5$

**54.** $S_{100}$ if $a_1 = -3$ and $d = -6$

*Find the sum of each infinite geometric sequence.*

**55.** $5, \dfrac{5}{2}, \dfrac{5}{4}, \ldots$

**56.** $18, -2, \dfrac{2}{9}, \ldots$

**57.** $-20, -4, -\dfrac{4}{5}, \ldots$

**58.** $0.2, 0.02, 0.002, \ldots$

*Solve.*

**59.** A frozen yogurt store owner cleared $20,000 the first year he owned his business and made an additional 15% over the previous year in each subsequent year. Find how much he made during his fourth year of business. Find his total earnings during the first 4 years (round to the nearest dollar).

**60.** On his first morning in a television assembly factory, a trainee takes 0.8 times as long to assemble each television as he took to assemble the one before. If it took him 40 minutes to assemble the first television, find how long it takes him to assemble the fourth television. Find the total time he takes to assemble the first four televisions (round to the nearest minute).

**61.** During the harvest season a farmer can rent a combine machine for $100 the first day, with the rental fee decreasing $7 for each additional day. Find how much the farmer pays for the rental on the seventh day. Find how much total rent the farmer pays for 7 days.

**62.** A rubber ball is dropped from a height of 15 feet and rebounds 80% of its previous height after each bounce. Find the total distance the ball travels before it comes to rest.

**63.** After a pond was sprayed once with insecticide, 1800 mosquitoes were killed the first day, 600 the second day, 200 the third day, and so on. Find the total number of mosquitoes killed during the first 6 days after the spraying (round to the nearest unit).

**64.** See Exercise 63. Find the day on which the insecticide is no longer effective, and find the total number of mosquitoes killed (round to the nearest mosquito).

**65.** Use the formula $S_\infty$ to write $0.55\overline{5}$ as a fraction.

**66.** A movie theater has 27 seats in the first row, 30 seats in the second row, 33 seats in the third row, and so on. Find the total number of seats in the theater if there are 20 rows.

*(14.5) Use Pascal's triangle to expand each binomial.*

**67.** $(x + z)^5$

**68.** $(y - r)^6$

**69.** $(2x + y)^4$

**70.** $(3y - z)^4$

*Use the binomial formula to expand the following.*

**71.** $(b + c)^8$

**72.** $(x - w)^7$

**73.** $(4m - n)^4$

**74.** $(p - 2r)^5$

*Find the indicated term.*

**75.** The fourth term of the expansion of $(a + b)^7$

**76.** The eleventh term of the expansion of $(y + 2z)^{10}$

**MIXED REVIEW**

**77.** Evaluate: $\sum_{i=1}^{4} i^2(i + 1)$

**78.** Find the fifteenth term of the arithmetic sequence whose first three terms are 14, 8, and 2.

**79.** Find the sum of the infinite geometric sequence $27, 9, 3, 1, \ldots$

**80.** Expand: $(2x - 3)^4$

---

# CHAPTER 14 TEST  TEST PREP VIDEO

Remember to use the Chapter Test Prep Video CD to see the fully worked-out solutions to any of the exercises you want to review.

*Find the indicated term(s) of the given sequence.*

**1.** The first five terms of the sequence $a_n = \dfrac{(-1)^n}{n + 4}$

**2.** The eightieth term of the sequence $a_n = 10 + 3(n - 1)$

**3.** The general term of the sequence $\dfrac{2}{5}, \dfrac{2}{25}, \dfrac{2}{125}, \ldots$

**4.** The general term of the sequence $-9, 18, -27, 36, \ldots$

*Find the partial sum of the given sequence.*

**5.** $S_5$ of the sequence $a_n = 5(2)^{n-1}$

**6.** $S_{30}$ of the sequence $a_n = 18 + (n - 1)(-2)$

**7.** $S_\infty$ of the sequence $a_1 = 24$ and $r = \dfrac{1}{6}$

**8.** $S_\infty$ of the sequence $\dfrac{3}{2}, -\dfrac{3}{4}, \dfrac{3}{8}, \ldots$

**9.** $\sum_{i=1}^{4} i(i - 2)$

**10.** $\sum_{i=2}^{4} 5(2)^i(-1)^{i-1}$

*Expand each binomial.*

**11.** $(a - b)^6$

**12.** $(2x + y)^5$

*Solve the following applications.*

**13.** The population of a small town is growing yearly according to the sequence defined by $a_n = 250 + 75(n - 1)$, where $n$ is the number of the year just beginning. Predict the population at the beginning of the tenth year. Find the town's initial population.

**14.** A gardener is making a triangular planting with one shrub in the first row, three shrubs in the second row, five shrubs in the third row, and so on, for eight rows. Write the finite series of this sequence, and find the total number of shrubs planted.

**15.** A pendulum swings through an arc of length 80 centimeters on its first swing. On each successive swing, the length of the arc is $\frac{3}{4}$ the length of the arc on the preceding swing.

Find the length of the arc on the fourth swing, and find the total arc length for the first four swings.

**16.** See Exercise 15. Find the total arc length before the pendulum comes to rest.

**17.** A parachutist in free-fall falls 16 feet during the first second, 48 feet during the second second, 80 feet during the third second, and so on. Find how far he falls during the tenth second. Find the total distance he falls during the first 10 seconds.

**18.** Use the formula $S_\infty$ to write $0.42\overline{42}$ as a fraction.

# CHAPTER 14 CUMULATIVE REVIEW

**1.** Evaluate.

  **a.** $(-2)^3$      **b.** $-2^3$

  **c.** $(-3)^2$      **d.** $-3^2$

**2.** Simplify each expression.

  **a.** $3a - (4a + 3)$

  **b.** $(5x - 3) + (2x + 6)$

  **c.** $4(2x - 5) - 3(5x + 1)$

**3.** Subtract $4x - 2$ from $2x - 3$. Simplify, if possible.

**4.** Sara bought a digital camera for $344.50 including tax. If the tax rate is 6%, what was the price of the camera before taxes?

**5.** Write an equation of the line with $y$-intercept $(0, -3)$ and slope of $\frac{1}{4}$.

**6.** Find an equation of a line through $(3, -2)$ and parallel to $3x - 2y = 6$. Write the equation using function notation.

**7.** Find an equation of the line through $(2, 5)$ and $(-3, 4)$. Write the equation in standard form.

**8.** Solve $y^3 + 5y^2 - y = 5$

**9.** Use synthetic division to divide $(x^4 - 2x^3 - 11x^2 + 5x + 34)$ by $(x + 2)$.

**10.** Perform the indicated operation and simplify if possible. $\frac{5}{3a - 6} - \frac{a}{a - 2} + \frac{3 + 2a}{5a - 10}$

**11.** Simplify the following.

  **a.** $\sqrt{50}$      **b.** $\sqrt[3]{24}$

  **c.** $\sqrt{26}$      **d.** $\sqrt[4]{32}$

**12.** Solve $\sqrt{3x + 6} - \sqrt{7x - 6} = 0$.

**13.** Find the interest rate $r$ if $2000 compounded annually grows to $2420 in 2 years.

**14.** Rationalize each denominator.

  **a.** $\sqrt[3]{\frac{4}{3x}}$      **b.** $\frac{\sqrt{2} + 1}{\sqrt{2} - 1}$

**15.** Solve $(x - 3)^2 - 3(x - 3) - 4 = 0$.

**16.** Solve $\frac{10}{(2x + 4)^2} - \frac{1}{2x + 4} = 3$.

**17.** Solve $\frac{5}{x + 1} < -2$.

**18.** Graph $f(x) = (x + 2)^2 - 6$. Find the vertex and axis of symmetry.

**19.** A rock is thrown upward from the ground. Its height in feet above ground after $t$ seconds is given by the function $f(t) = -16t^2 + 20t$. Find the maximum height of the rock and the number of seconds it took for the rock to reach its maximum height.

**20.** Find the vertex of $f(x) = x^2 + 3x - 18$.

**21.** If $f(x) = x^2$ and $g(x) = x + 3$, find each composition.

  **a.** $(f \circ g)(2)$ and $(g \circ f)(2)$

  **b.** $(f \circ g)(x)$ and $(g \circ f)(x)$

**22.** Find the inverse of $f(x) = -2x + 3$.

**23.** Find the inverse of the one-to-one function. $f = \{(0, 1), (-2, 7), (3, -6), (4, 4)\}$

**24.** If $f(x) = x^2 - 2$ and $g(x) = x + 1$, find each composition.

  **a.** $(f \circ g)(2)$ and $(g \circ f)(2)$

  **b.** $(f \circ g)(x)$ and $(g \circ f)(x)$

**25.** Solve each equation for $x$.

  **a.** $2^x = 16$      **b.** $9^x = 27$

  **c.** $4^{x+3} = 8^x$

**26.** Solve each equation.

    **a.** $\log_2 32 = x$                  **b.** $\log_4 \dfrac{1}{64} = x$

    **c.** $\log_{\frac{1}{2}} x = 5$

**27.** Simplify.

    **a.** $\log_3 3^2$                     **b.** $\log_7 7^{-1}$

    **c.** $5^{\log_5 3}$                     **d.** $2^{\log_2 6}$

**28.** Solve each equation for $x$.

    **a.** $4^x = 64$                 **b.** $8^x = 32$

    **c.** $9^{x+4} = 243^x$

**29.** Write each sum as a single logarithm.

    **a.** $\log_{11} 10 + \log_{11} 3$        **b.** $\log_3 \dfrac{1}{2} + \log_3 12$

    **c.** $\log_2(x + 2) + \log_2 x$

**30.** Find the exact value.

    **a.** $\log 100{,}000$

    **b.** $\log 10^{-3}$

    **c.** $\ln \sqrt[5]{e}$

    **d.** $\ln e^4$

**31.** Find the amount owed at the end of 5 years if $1600 is loaned at a rate of 9% compounded continuously.

**32.** Write each expression as a single logarithm.

    **a.** $\log_6 5 + \log_6 4$

    **b.** $\log_8 12 - \log_8 4$

    **c.** $2\log_2 x + 3\log_2 x - 2\log_2(x - 1)$

**33.** Solve $3^x = 7$.

**34.** Using $A = P\left(1 + \dfrac{r}{n}\right)^{nt}$, find how long it takes $5000 to double if it is invested at 2% interest compounded quarterly. Round to the nearest tenth.

**35.** Solve $\log_4(x - 2) = 2$.

**36.** Solve $\log_4 10 - \log_4 x = 2$.

**37.** Graph $\dfrac{x^2}{16} - \dfrac{y^2}{25} = 1$.

**38.** Find the distance between $(8, 5)$ and $(-2, 4)$.

**39.** Solve the system: $\begin{cases} y = \sqrt{x} \\ x^2 + y^2 = 6 \end{cases}$

**40.** Solve the system: $\begin{cases} x^2 + y^2 = 36 \\ x - y = 6 \end{cases}$

**41.** Graph $\dfrac{x^2}{9} + \dfrac{y^2}{16} \leq 1$.

**42.** Graph $\begin{cases} y \geq x^2 \\ y \leq 4 \end{cases}$.

**43.** Write the first five terms of the sequence whose general term is given by $a_n = n^2 - 1$.

**44.** If the general term of a sequence is $a_n = \dfrac{n}{n + 4}$, find $a_8$.

**45.** Find the eleventh term of the arithmetic sequence whose first three terms are $2, 9$, and $16$.

**46.** Find the sixth term of the geometric sequence $2, 10, 50, \ldots$

**47.** Evaluate.

    **a.** $\displaystyle\sum_{i=0}^{6} \dfrac{i - 2}{2}$           **b.** $\displaystyle\sum_{i=3}^{5} 2^i$

**48.** Evaluate.

    **a.** $\displaystyle\sum_{i=0}^{4} i(i + 1)$        **b.** $\displaystyle\sum_{i=0}^{3} 2^i$

**49.** Find the sum of the first 30 positive integers.

**50.** Find the third term of the expansion of $(x - y)^6$.

# Appendix A

## The Bigger Picture/Practice Final Exam

### A.1  THE BIGGER PICTURE: SIMPLIFYING EXPRESSIONS AND SOLVING EQUATIONS AND INEQUALITIES

**I. Simplifying Expressions**

  **A. Real Numbers**

    **1. Add:** (Sec. 1.5)

$$-1.7 + (-0.21) = -1.91 \quad \text{Adding like signs. Add absolute values. Attach common sign.}$$

$$-7 + 3 = -4 \quad \text{Adding different signs. Subtract absolute values. Attach the sign of the number with the larger absolute value.}$$

    **2. Subtract:** Add the first number to the opposite of the second number. (Sec. 1.6)

$$17 - 25 = 17 + (-25) = -8$$

    **3. Multiply or divide:** Multiply or divide the two numbers as usual. If the signs are the same, the answer is positive. If the signs are different, the answer is negative. (Sec. 1.7)

$$-10 \cdot 3 = -30, \qquad -81 \div (-3) = 27$$

  **B. Exponents** (Sec. 5.1 and 5.5)

$$x^7 \cdot x^5 = x^{12}; \ (x^7)^5 = x^{35}; \ \frac{x^7}{x^5} = x^2; \ x^0 = 1; \ 8^{-2} = \frac{1}{8^2} = \frac{1}{64}$$

  **C. Polynomials**

    **1. Add:** Combine like terms. (Sec. 5.2)

$$(3y^2 + 6y + 7) + (9y^2 - 11y - 15) = 3y^2 + 6y + 7 + 9y^2 - 11y - 15$$
$$= 12y^2 - 5y - 8$$

    **2. Subtract:** Change the sign of the terms of the polynomial being subtracted, then add. (Sec. 5.2)

$$(3y^2 + 6y + 7) - (9y^2 - 11y - 15) = 3y^2 + 6y + 7 - 9y^2 + 11y + 15$$
$$= -6y^2 + 17y + 22$$

    **3. Multiply:** Multiply each term of one polynomial by each term of the other polynomial. (Sec. 5.3 and 5.4)

$$(x + 5)(2x^2 - 3x + 4) = x(2x^2 - 3x + 4) + 5(2x^2 - 3x + 4)$$
$$= 2x^3 - 3x^2 + 4x + 10x^2 - 15x + 20$$
$$= 2x^3 + 7x^2 - 11x + 20$$

4. **Divide:** (Sec. 5.6)
  a. To divide by a monomial, divide each term of the polynomial by the monomial.

$$\frac{8x^2 + 2x - 6}{2x} = \frac{8x^2}{2x} + \frac{2x}{2x} - \frac{6}{2x} = 4x + 1 - \frac{3}{x}$$

  b. To divide by a polynomial other than a monomial, use long division.

$$
\begin{array}{r}
x - 6 + \dfrac{40}{2x + 5} \\[4pt]
2x + 5 \overline{)\, 2x^2 - 7x + 10\,} \\
\underline{2x^2 + 5x} \\
-12x + 10 \\
\underline{-12x - 30} \\
40
\end{array}
$$

D. **Factoring Polynomials**
  See the Chapter 6 Integrated Review for steps.

$$3x^4 - 78x^2 + 75 = 3(x^4 - 26x^2 + 25) \quad \text{Factor out GCF—always first step.}$$

$$= 3(x^2 - 25)(x^2 - 1) \quad \text{Factor trinomial.}$$

$$= 3(x + 5)(x - 5)(x + 1)(x - 1) \quad \text{Factor further—each difference of squares.}$$

E. **Rational Expressions**
  1. **Simplify:** Factor the numerator and denominator. Then divide out factors of 1 by dividing out common factors in the numerator and denominator. (Sec. 7.1)

$$\frac{x^2 - 9}{7x^2 - 21x} = \frac{(x + 3)(x - 3)}{7x(x - 3)} = \frac{x + 3}{7x}$$

  2. **Multiply:** Multiply numerators, then multiply denominators. (Sec. 7.2)

$$\frac{5z}{2z^2 - 9z - 18} \cdot \frac{22z + 33}{10z} = \frac{5 \cdot z}{(2z + 3)(z - 6)} \cdot \frac{11(2z + 3)}{2 \cdot 5 \cdot z} = \frac{11}{2(z - 6)}$$

  3. **Divide:** First fraction times the reciprocal of the second fraction. (Sec. 7.2)

$$\frac{14}{x + 5} \div \frac{x + 1}{2} = \frac{14}{x + 5} \cdot \frac{2}{x + 1} = \frac{28}{(x + 5)(x + 1)}$$

  4. **Add or subtract:** Must have same denominator. If not, find the LCD and write each fraction as an equivalent fraction with the LCD as denominator. (Sec. 7.4)

$$\frac{9}{10} - \frac{x + 1}{x + 5} = \frac{9(x + 5)}{10(x + 5)} - \frac{10(x + 1)}{10(x + 5)}$$

$$= \frac{9x + 45 - 10x - 10}{10(x + 5)} = \frac{-x + 35}{10(x + 5)}$$

F. **Radicals**
  1. **Simplify square roots:** If possible, factor the radicand so that one factor is a perfect square. Then use the product rule and simplify. (Sec. 10.3)

$$\sqrt{75} = \sqrt{25 \cdot 3} = \sqrt{25} \cdot \sqrt{3} = 5\sqrt{3}$$

  2. **Add or subtract:** Only like radicals (same index and radicand) can be added or subtracted. (Sec. 10.4)

$$8\sqrt{10} - \sqrt{40} + \sqrt{5} = 8\sqrt{10} - 2\sqrt{10} + \sqrt{5} = 6\sqrt{10} + \sqrt{5}$$

3. **Multiply or divide:** $\sqrt{a} \cdot \sqrt{b} = \sqrt{ab}$; $\dfrac{\sqrt{a}}{\sqrt{b}} = \sqrt{\dfrac{a}{b}}$. (Sec. 10.4)

$$\sqrt{11} \cdot \sqrt{3} = \sqrt{33}; \quad \dfrac{\sqrt{140}}{\sqrt{7}} = \sqrt{\dfrac{140}{7}} = \sqrt{20} = \sqrt{4 \cdot 5} = 2\sqrt{5}$$

4. **Rationalizing the denominator:** (Sec. 10.5)
   **a.** If denominator is one term,

$$\dfrac{5}{\sqrt{11}} = \dfrac{5 \cdot \sqrt{11}}{\sqrt{11} \cdot \sqrt{11}} = \dfrac{5\sqrt{11}}{11}$$

   **b.** If denominator is two terms, multiply by 1 in the form of $\dfrac{\text{conjugate of denominator}}{\text{conjugate of denominator}}$.

$$\dfrac{13}{3 + \sqrt{2}} = \dfrac{13}{3 + \sqrt{2}} \cdot \dfrac{3 - \sqrt{2}}{3 - \sqrt{2}} = \dfrac{13(3 - \sqrt{2})}{9 - 2} = \dfrac{13(3 - \sqrt{2})}{7}$$

## II. Solving Equations

**A. Linear Equations** (Sec. 2.3)

$$5(x - 2) = \dfrac{4(2x + 1)}{3}$$
$$3 \cdot 5(x - 2) = \cancel{3} \cdot \dfrac{4(2x + 1)}{\cancel{3}}$$
$$15x - 30 = 8x + 4$$
$$7x = 34$$
$$x = \dfrac{34}{7}$$

**B. Quadratic and Higher-Degree Equations** (Secs. 6.6, 11.1, 11.2, 11.3)

$$2x^2 - 7x = 9$$
$$2x^2 - 7x - 9 = 0$$
$$(2x - 9)(x + 1) = 0$$
$$2x - 9 = 0 \quad \text{or} \quad x + 1 = 0$$
$$x = \dfrac{9}{2} \quad \text{or} \quad x = -1$$

$$2x^2 + x - 2 = 0$$
$$a = 2, \; b = 1, \; c = -2$$
$$x = \dfrac{-1 \pm \sqrt{1^2 - 4(2)(-2)}}{2 \cdot 2}$$
$$x = \dfrac{-1 \pm \sqrt{17}}{4}$$

**C. Equations with Rational Expressions** (Sec. 7.5)

$$\dfrac{7}{x - 1} + \dfrac{3}{x + 1} = \dfrac{x + 3}{x^2 - 1}$$
$$\cancel{(x - 1)}(x + 1) \cdot \dfrac{7}{\cancel{x - 1}} + (x - 1)\cancel{(x + 1)} \cdot \dfrac{3}{\cancel{x + 1}}$$
$$= \cancel{(x - 1)}\cancel{(x + 1)} \cdot \dfrac{x + 3}{\cancel{(x - 1)}\cancel{(x + 1)}}$$
$$7(x + 1) + 3(x - 1) = x + 3$$
$$7x + 7 + 3x - 3 = x + 3$$
$$9x = -1$$
$$x = -\dfrac{1}{9}$$

**D. Proportions:** An equation with two ratios equal. Set cross products equal, then solve. Make sure the proposed solution does not make the denominator 0. (Sec. 7.6)

$$\frac{5}{x} \nearrow \frac{9}{2x-3}$$

$$5(2x-3) = 9\cdot x \quad \text{Set cross products equal.}$$
$$10x - 15 = 9x \quad \text{Multiply.}$$
$$x = 15 \quad \text{Write equation with variable terms on one side and constants on the other.}$$

**E. Absolute Value Equations** (Sec. 9.2)

$$|3x - 1| = 8$$
$$3x - 1 = 8 \quad \text{or} \quad 3x - 1 = -8$$
$$3x = 9 \quad \text{or} \quad 3x = -7$$
$$x = 3 \quad \text{or} \quad x = -\frac{7}{3}$$

$$|x - 5| = |x + 1|$$
$$x - 5 = x + 1 \quad \text{or} \quad x - 5 = -(x + 1)$$
$$-5 = 1 \qquad\qquad \text{or} \quad x - 5 = -x - 1$$
$$\text{No solution} \qquad \text{or} \quad 2x = 4$$
$$x = 2$$

**F. Equations with Radicals** (Sec. 10.6)

$$\sqrt{5x + 10} - 2 = x$$
$$\sqrt{5x + 10} = x + 2$$
$$(\sqrt{5x + 10})^2 = (x + 2)^2$$
$$5x + 10 = x^2 + 4x + 4$$
$$0 = x^2 - x - 6$$
$$0 = (x - 3)(x + 2)$$
$$x - 3 = 0 \quad \text{or} \quad x + 2 = 0$$
$$x = 3 \quad \text{or} \quad x = -2$$

Both solutions check.

**G. Exponential Equations** (Secs. 12.3, 12.7)

$$9^x = 27^{x+1} \qquad\qquad 5^x = 7$$
$$(3^2)^x = (3^3)^{x+1} \qquad \log 5^x = \log 7$$
$$3^{2x} = 3^{3x+3} \qquad\qquad x \log 5 = \log 7$$
$$2x = 3x + 3 \qquad\qquad x = \frac{\log 7}{\log 5}$$
$$-3 = x$$

**H. Logarithmic Equations** (Sec. 12.7)

$$\log 7 + \log(x + 3) = \log 5$$
$$\log 7(x + 3) = \log 5$$
$$7(x + 3) = 5$$
$$7x + 21 = 5$$
$$7x = -16$$
$$x = \frac{-16}{7}$$

**III. Solving Inequalities**

**A. Linear Inequalities** (Sec. 2.8)

$$-3(x + 2) \geq 6$$
$$-3x - 6 \geq 6$$
$$-3x \geq 12$$
$$\frac{-3x}{-3} \leq \frac{12}{-3}$$
$$x \leq -4 \quad \text{or} \quad (-\infty, -4]$$

**B. Compound Inequalities** (Sec. 9.1)

$$x \leq 3 \quad \text{and} \quad x < -7 \qquad\qquad x \leq 3 \quad \text{or} \quad x < -7$$

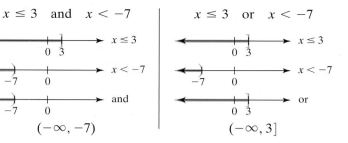

**C. Absolute Value Inequalities** (Sec. 9.3)

$$|x - 5| - 8 < -2 \qquad\qquad |2x + 1| \geq 17$$
$$|x - 5| < 6 \qquad\qquad\qquad 2x + 1 \geq 17 \quad \text{or} \quad 2x + 1 \leq -17$$
$$-6 < x - 5 < 6 \qquad\qquad\quad 2x \geq 16 \quad \text{or} \qquad 2x \leq -18$$
$$-1 < x < 11 \qquad\qquad\qquad x \geq 8 \quad \text{or} \qquad x \leq -9$$
$$(-1, 11) \qquad\qquad\qquad (-\infty, -9] \cup [8, \infty)$$

**D. Nonlinear Inequalities** (Sec. 11.4)

$$x^2 - x < 6 \qquad\qquad\qquad \frac{x - 5}{x + 1} \geq 0$$
$$x^2 - x - 6 < 0$$
$$(x - 3)(x + 2) < 0$$

| | ✗ | ✓ | ✗ | |
|---|---|---|---|---|
| | | -2 | 3 | |

$$(-2, 3)$$

| | ✓ | ✗ | ✓ | |
|---|---|---|---|---|
| | | -1 | 5 | |

$$(-\infty, -1) \cup [5, \infty)$$

> ▶ **Helpful Hint**
> - If your course in this text ended with Chapter 6, work Exercises 1–30.
> - If your course in this text ended with Chapter 7, work Exercises 1–39.
> - If your course in this text started with Chapter 7, start with Exercise 31.
> - If your course in this text started with Chapter 8, start with Exercise 40.

## A.2  PRACTICE FINAL EXAM  TEST PREP VIDEO

Preparing for your Final Exam? Take this Practice Final and watch the full video solutions to any of the exercises you want to review. You will find the Practice Final video in the Video Lecture Series. The video also provides you with an overview to help you approach different problem types just as you will need to do on a Final Exam. To build your own study guide, use the Bigger Picture feature in the text. See Appendix A.1 for an example.

*Evaluate.*

**1.** $6[5 + 2(3 - 8) - 3]$

**2.** $-3^4$

**3.** $4^{-3}$

**4.** $\dfrac{1}{2} - \dfrac{5}{6}$

*Perform the indicated operations and simplify if possible.*

**5.** $(5x^3 + x^2 + 5x - 2) - (8x^3 - 4x^2 + x - 7)$

**6.** $(4x - 2)^2$

**7.** $(3x + 7)(x^2 + 5x + 2)$

*Factor.*

**8.** $y^2 - 8y - 48$

**9.** $9x^3 + 39x^2 + 12x$

**10.** $180 - 5x^2$

**11.** $3a^2 + 3ab - 7a - 7b$

**12.** $8y^3 - 64$

*Simplify. Write answers with positive exponents only.*

**13.** $\left(\dfrac{x^2 y^3}{x^3 y^{-4}}\right)^2$

*Solve each equation or inequality. Write inequality answers using interval notation.*

**14.** $-4(a + 1) - 3a = -7(2a - 3)$

**15.** $3x - 5 \geq 7x + 3$

**16.** $x(x + 6) = 7$

*Graph the following.*

**17.** $5x - 7y = 10$

**18.** $x - 3 = 0$

*Find the slope of each line.*

**19.** through $(6, -5)$ and $(-1, 2)$

**20.** $-3x + y = 5$

*Write equations of the following lines. Write each equation in standard form.*

**21.** through $(2, -5)$ and $(1, 3)$

**22.** through $(-5, -1)$ and parallel to $x = 7$

*Solve each system of equations.*

**23.** $\begin{cases} \dfrac{1}{2}x + 2y = -\dfrac{15}{4} \\ 4x = -y \end{cases}$

**24.** $\begin{cases} 4x - 6y = 7 \\ -2x + 3y = 0 \end{cases}$

**25.** Divide by long division: $\dfrac{27x^3 - 8}{3x + 2}$

*Answer the questions about functions.*

**26.** If $h(x) = x^3 - x$, find

 **a.** $h(-1)$   **b.** $h(0)$   **c.** $h(4)$

**27.** Find the domain and range of the function graphed.

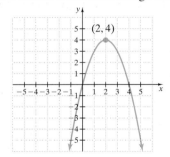

*Solve each application.*

**28.** Some states have a single area code for the entire state. Two such states have area codes where one is double the other. If the sum of these integers is 1203, find the two area codes.

**29.** Two trains leave Los Angeles simultaneously traveling on the same track in opposite directions at speeds of 50 and 64 mph. How long will it take before they are 285 miles apart?

**30.** Find the amount of a 12% saline solution a lab assistant should add to 80 cc (cubic centimeters) of a 22% saline solution in order to have a 16% solution.

**31.** Find the domain of the rational function

$$g(x) = \frac{9x^2 - 9}{x^2 + 4x + 3}.$$

*Perform the indicated operations and simplify if possible.*

**32.** $\dfrac{15x}{2x + 5} - \dfrac{6 - 4x}{2x + 5}$

**33.** $\dfrac{x^2 - 9}{x^2 - 3x} \div \dfrac{xy + 5x + 3y + 15}{2x + 10}$

**34.** $\dfrac{5a}{a^2 - a - 6} - \dfrac{2}{a - 3}$

**35.** $\dfrac{5 - \dfrac{1}{y^2}}{\dfrac{1}{y} + \dfrac{2}{y^2}}$

*Solve each equation.*

**36.** $\dfrac{4}{y} - \dfrac{5}{3} = -\dfrac{1}{5}$

**37.** $\dfrac{5}{y + 1} = \dfrac{4}{y + 2}$

**38.** $\dfrac{a}{a - 3} = \dfrac{3}{a - 3} - \dfrac{3}{2}$

*Solve.*

**39.** One number plus five times its reciprocal is equal to six. Find the number.

*Simplify. If needed, write answers with positive exponents only.*

**40.** $\sqrt{216}$

**41.** $\left(\dfrac{1}{125}\right)^{-1/3}$

**42.** $\left(\dfrac{64c^{4/3}}{a^{-2/3}b^{5/6}}\right)^{1/2}$

*Perform the indicated operations and simplify if possible.*

**43.** $\sqrt{125x^3} - 3\sqrt{20x^3}$

**44.** $(\sqrt{5} + 5)(\sqrt{5} - 5)$

*Solve each equation or inequality. Write inequality solutions using interval notation.*

**45.** $|6x - 5| - 3 = -2$

**46.** $-3 < 2(x - 3) \le 4$

**47.** $|3x + 1| > 5$

**48.** $y^2 - 3y = 5$

**49.** $x = \sqrt{x - 2} + 2$

**50.** $2x^2 - 7x > 15$

*Graph the following.*

**51.** $y > -4x$

**52.** $g(x) = -|x + 2| - 1$. Also, find the domain and range of this function.

**53.** $h(x) = x^2 - 4x + 4$. Label the vertex and any intercepts.

**54.** $f(x) = \begin{cases} -\dfrac{1}{2}x & \text{if } x \le 0 \\ 2x - 3 & \text{if } x > 0 \end{cases}$. Also, find the domain and range of this function.

*Write equations of the following lines. Write each equation using function notation.*

**55.** through $(4, -2)$ and $(6, -3)$

**56.** through $(-1, 2)$ and perpendicular to $3x - y = 4$

*Find the distance or midpoint.*

**57.** Find the distance between the points $(-6, 3)$ and $(-8, -7)$.

**58.** Find the midpoint of the line segment whose endpoints are $(-2, -5)$ and $(-6, 12)$.

*Rationalize each denominator. Assume that variables represent positive numbers.*

**59.** $\sqrt{\dfrac{9}{y}}$

**60.** $\dfrac{4 - \sqrt{x}}{4 + 2\sqrt{x}}$

*Solve.*

**61.** Suppose that $W$ is inversely proportional to $V$. If $W = 20$ when $V = 12$, find $W$ when $V = 15$.

**62.** Given the diagram shown, approximate to the nearest foot, how many feet of walking distance a person saves by cutting across the lawn instead of walking on the sidewalk.

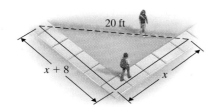

**63.** A stone is thrown upward from a bridge. The stone's height in feet, $s(t)$, above the water $t$ seconds after the stone is thrown is a function given by the equation

$$s(t) = -16t^2 + 32t + 256$$

**a.** Find the maximum height of the stone.

**b.** Find the time it takes the stone to hit the water. Round the answer to two decimal places.

**Complex Numbers: Chapter 10**

*Perform the indicated operation and simplify. Write the result in the form $a + bi$.*

**64.** $-\sqrt{-8}$

**65.** $(12 - 6i) - (12 - 3i)$

**66.** $(4 + 3i)^2$

**67.** $\dfrac{1 + 4i}{1 - i}$

**Inverse, Exponential, and Logarithmic Functions: Chapter 12**

**68.** If $g(x) = x - 7$ and $h(x) = x^2 - 6x + 5$, find $(g \circ h)(x)$.

**69.** Decide whether $f(x) = 6 - 2x$ is a one-to-one function. If it is, find its inverse.

**70.** Use properties of logarithms to write the expression as a single logarithm.

$$\log_5 x + 3\log_5 x - \log_5(x + 1)$$

*Solve. Give exact solutions.*

**71.** $8^{x-1} = \dfrac{1}{64}$

**72.** $3^{2x+5} = 4$ Give an exact solution and a 4-decimal place approximation.

**73.** $\log_8(3x - 2) = 2$

**74.** $\log_4(x + 1) - \log_4(x - 2) = 3$

**75.** $\ln\sqrt{e} = x$

**76.** Graph $y = \left(\dfrac{1}{2}\right)^x + 1$

**77.** The prairie dog population of the Grand Rapids area now stands at 57,000 animals. If the population is growing at a rate of 2.6% annually, use the formula $y = y_0 e^{kt}$ to find how many prairie dogs there will be in that area 5 years from now.

**Conic Sections: Chapter 13**

*Sketch the graph of each equation.*

**78.** $x^2 - y^2 = 36$

**79.** $16x^2 + 9y^2 = 144$

**80.** $x^2 + y^2 + 6x = 16$

**81.** Solve the system:

$$\begin{cases} x^2 + y^2 = 26 \\ x^2 - 2y^2 = 23 \end{cases}$$

**Sequences, Series and the Binomial Theorem: Chapter 14**

**82.** Find the first five terms of the sequence $a_n = \dfrac{(-1)^n}{n + 4}$.

**83.** Find the partial sum, $S_5$; of the sequence $a_n = 5(2)^{n-1}$.

**84.** Find $S_\infty$ of the sequence $\dfrac{3}{2}, -\dfrac{3}{4}, \dfrac{3}{8}, \ldots$

**85.** Find $\displaystyle\sum_{i=1}^{4} i(i - 2)$

**86.** Expand: $(2x + y)^5$

# Appendix B

## Operations on Decimals/Percent, Decimal, and Fraction Table

### B.1 OPERATIONS ON DECIMALS

To **add** or **subtract** decimals, write the numbers vertically with decimal points lined up. Add or subtract as with whole numbers and place the decimal point in the answer directly below the decimal points in the problem.

**EXAMPLE 1**   Add $5.87 + 23.279 + 0.003$.

*Solution*

$$
\begin{array}{r}
5.87 \\
23.279 \\
+\,0.003 \\
\hline
29.152
\end{array}
$$

**EXAMPLE 2**   Subtract $32.15 - 11.237$.

*Solution*

$$
\begin{array}{r}
3\;\overset{1}{\cancel{2}}\,.\,\overset{11}{\cancel{1}}\;\overset{4}{\cancel{5}}\;\overset{10}{\cancel{0}} \\
-\;1\;\;1\,.\,2\;\;3\;\;7 \\
\hline
2\;\;0\,.\,9\;\;1\;\;3
\end{array}
$$

To **multiply** decimals, multiply the numbers as if they were whole numbers. The decimal point in the product is placed so that the number of decimal places in the product is the same as the sum of the number of decimal places in the factors.

**EXAMPLE 3**   Multiply $0.072 \times 3.5$.

*Solution*

$$
\begin{array}{r}
0.072 \quad \text{3 decimal places}\\
\times\quad 3.5 \quad \text{1 decimal place}\\
\hline
360\phantom{00}\\
216\phantom{0}\\
\hline
0.2520 \quad \text{4 decimal places}
\end{array}
$$

To **divide** decimals, move the decimal point in the divisor to the right of the last digit. Move the decimal point in the dividend the same number of places that the decimal point in the divisor was moved. The decimal point in the quotient lies directly above the decimal point in the dividend.

**EXAMPLE 4**   Divide $9.46 \div 0.04$.

*Solution*

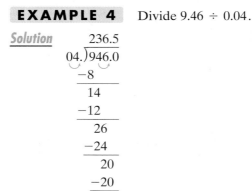

$$
\begin{array}{r}
236.5\phantom{0} \\
04.\overline{)946.0} \\
-8\phantom{46.0} \\
\hline
14\phantom{6.0} \\
-12\phantom{6.0} \\
\hline
26\phantom{.0} \\
-24\phantom{.0} \\
\hline
20\phantom{0} \\
-20\phantom{0}
\end{array}
$$

# APPENDIX B | EXERCISE SET

*Perform the indicated operations.*

**1.** $9.076 + 8.004$

**2.**
$$\begin{array}{r} 6.3 \\ \times\ 0.05 \\ \hline \end{array}$$

**3.**
$$\begin{array}{r} 27.004 \\ -14.2 \\ \hline \end{array}$$

**4.**
$$\begin{array}{r} 0.0036 \\ 7.12 \\ 32.502 \\ +0.05 \\ \hline \end{array}$$

**5.**
$$\begin{array}{r} 107.92 \\ +3.04 \\ \hline \end{array}$$

**6.** $7.2 \div 4$

**7.** $10 - 7.6$

**8.** $40 \div 0.25$

**9.** $126.32 - 97.89$

**10.**
$$\begin{array}{r} 3.62 \\ 7.11 \\ 12.36 \\ 4.15 \\ +2.29 \\ \hline \end{array}$$

**11.**
$$\begin{array}{r} 3.25 \\ \times\ 70 \\ \hline \end{array}$$

**12.**
$$\begin{array}{r} 26.014 \\ -\ 7.8 \\ \hline \end{array}$$

**13.** $8.1 \div 3$

**14.**
$$\begin{array}{r} 1.2366 \\ 0.005 \\ 15.17 \\ +\ 0.97 \\ \hline \end{array}$$

**15.** $55.405 - 6.1711$

**16.** $8.09 + 0.22$

**17.** $60 \div 0.75$

**18.** $20 - 12.29$

**19.** $7.612 \div 100$

**20.**
$$\begin{array}{r} 8.72 \\ 1.12 \\ 14.86 \\ 3.98 \\ +\ 1.99 \\ \hline \end{array}$$

**21.** $12.312 \div 2.7$

**22.** $0.443 \div 100$

**23.**
$$\begin{array}{r} 569.2 \\ 71.25 \\ +\ 8.01 \\ \hline \end{array}$$

**24.** $3.706 - 2.91$

**25.** $768 - 0.17$

**26.** $63 \div 0.28$

**27.** $12 + 0.062$

**28.** $0.42 + 18$

**29.** $76 - 14.52$

**30.** $1.1092 \div 0.47$

**31.** $3.311 \div 0.43$

**32.** $7.61 + 0.0004$

**33.**
$$\begin{array}{r} 762.12 \\ 89.7 \\ +\ 11.55 \\ \hline \end{array}$$

**34.** $444 \div 0.6$

**35.** $23.4 - 0.821$

**36.** $3.7 + 5.6$

**37.** $476.12 - 112.97$

**38.** $19.872 \div 0.54$

**39.** $0.007 + 7$

**40.**
$$\begin{array}{r} 51.77 \\ +\ 3.6 \\ \hline \end{array}$$

## B.2 TABLE OF PERCENT, DECIMAL, AND FRACTION EQUIVALENTS

| Percent, Decimal, and Fraction Equivalents | | |
|---|---|---|
| **Percent** | **Decimal** | **Fraction** |
| 1% | 0.01 | $\frac{1}{100}$ |
| 5% | 0.05 | $\frac{1}{20}$ |
| 10% | 0.1 | $\frac{1}{10}$ |
| 12.5% or $12\frac{1}{2}\%$ | 0.125 | $\frac{1}{8}$ |
| $16.\overline{6}\%$ or $16\frac{2}{3}\%$ | $0.1\overline{6}$ | $\frac{1}{6}$ |
| 20% | 0.2 | $\frac{1}{5}$ |
| 25% | 0.25 | $\frac{1}{4}$ |
| 30% | 0.3 | $\frac{3}{10}$ |
| $33.\overline{3}\%$ or $33\frac{1}{3}\%$ | $0.\overline{3}$ | $\frac{1}{3}$ |
| 37.5% or $37\frac{1}{2}\%$ | 0.375 | $\frac{3}{8}$ |
| 40% | 0.4 | $\frac{2}{5}$ |
| 50% | 0.5 | $\frac{1}{2}$ |
| 60% | 0.6 | $\frac{3}{5}$ |
| 62.5% or $62\frac{1}{2}\%$ | 0.625 | $\frac{5}{8}$ |
| $66.\overline{6}\%$ or $66\frac{2}{3}\%$ | $0.\overline{6}$ | $\frac{2}{3}$ |
| 70% | 0.7 | $\frac{7}{10}$ |
| 75% | 0.75 | $\frac{3}{4}$ |
| 80% | 0.8 | $\frac{4}{5}$ |
| $83.\overline{3}\%$ or $83\frac{1}{3}\%$ | $08.\overline{3}$ | $\frac{5}{6}$ |
| 87.5% or $87\frac{1}{2}\%$ | 0.875 | $\frac{7}{8}$ |
| 90% | 0.9 | $\frac{9}{10}$ |
| 100% | 1.0 | 1 |
| 110% | 1.1 | $1\frac{1}{10}$ |
| 125% | 1.25 | $1\frac{1}{4}$ |
| $133.\overline{3}\%$ or $133\frac{1}{3}\%$ | $1.\overline{3}$ | $1\frac{1}{3}$ |
| 150% | 1.5 | $1\frac{1}{2}$ |
| $166.\overline{6}\%$ or $166\frac{2}{3}\%$ | $1.\overline{6}$ | $1\frac{2}{3}$ |
| 175% | 1.75 | $1\frac{3}{4}$ |
| 200% | 2.0 | 2 |

# Appendix C

## Review of Algebra Topics

Recall that equations model many real-life problems. For example, we can use a linear equation to calculate the increase in the percent of households with digital cameras.

With the help of your computer, digital cameras allow you to see your pictures and make copies immediately, send them in e-mail, or use them on a Web page. Current projected percentage of households with these cameras is shown in the graph below.

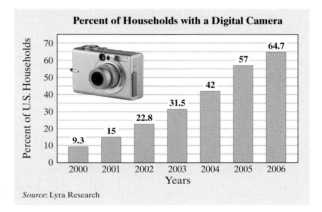

**Percent of Households with a Digital Camera**

*Source*: Lyra Research

To find the increase in percent of households from 2005 to 2006, for example, we can use the equation below:

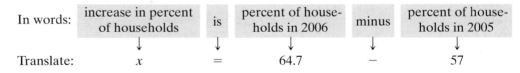

| In words: | increase in percent of households | is | percent of households in 2006 | minus | percent of households in 2005 |
|---|---|---|---|---|---|
| | ↓ | ↓ | ↓ | ↓ | ↓ |
| Translate: | $x$ | = | 64.7 | − | 57 |

Since our variable $x$ (increase in percent of households) is by itself on one side of the equation, we can find the value of $x$ by simplifying the right side.

$$x = 7.7$$

The increase in the households with digital cameras from 2005 to 2006 is 7.7%.

The **equation** $x = 64.7 - 57$ is a linear equation in one variable. In this section, we review solving linear equations and quadratic equations that can be solved by factoring. We will learn other methods for solving quadratic equations in Chapter 11.

### C.1 EQUATIONS (LINEAR AND QUADRATIC SOLVED BY FACTORING)

**OBJECTIVE 1 ▶ Solving linear and quadratic equations.**

**EXAMPLE 1**    Solve: $2(x - 3) = 5x - 9$.

*Solution*    First, use the distributive property.

$$2(x - 3) = 5x - 9$$
$$2x - 6 = 5x - 9 \quad \text{Use the distributive property.}$$

**847**

Next, get variable terms on the same side of the equation by subtracting $5x$ from both sides.

$$2x - 6 - 5x = 5x - 9 - 5x \quad \text{Subtract } 5x \text{ from both sides.}$$
$$-3x - 6 = -9 \quad \text{Simplify.}$$
$$-3x - 6 + 6 = -9 + 6 \quad \text{Add 6 to both sides.}$$
$$-3x = -3 \quad \text{Simplify.}$$
$$\frac{-3x}{-3} = \frac{-3}{-3} \quad \text{Divide both sides by } -3.$$
$$x = 1$$

Let $x = 1$ in the original equation to see that 1 is the solution. □

**PRACTICE**
**1** Solve: $3(x - 5) = 6x - 3$.

Don't forget, if an equation contains fractions, you may want to first clear the equation of fractions by multiplying both sides of the equation by the *least common denominator* (LCD) of all fractions in the equation.

**EXAMPLE 2** Solve for $y$: $\frac{y}{3} - \frac{y}{4} = \frac{1}{6}$.

*Solution* First, clear the equation of fractions by multiplying both sides of the equation by 12, the LCD of denominators 3, 4, and 6.

$$\frac{y}{3} - \frac{y}{4} = \frac{1}{6}$$
$$12\left(\frac{y}{3} - \frac{y}{4}\right) = 12\left(\frac{1}{6}\right) \quad \text{Multiply both sides by the LCD 12.}$$
$$12\left(\frac{y}{3}\right) - 12\left(\frac{y}{4}\right) = 2 \quad \text{Apply the distributive property.}$$
$$4y - 3y = 2 \quad \text{Simplify.}$$
$$y = 2 \quad \text{Simplify.}$$

**Check:** To check, let $y = 2$ in the original equation.

$$\frac{y}{3} - \frac{y}{4} = \frac{1}{6} \quad \text{Original equation}$$
$$\frac{2}{3} - \frac{2}{4} \stackrel{?}{=} \frac{1}{6} \quad \text{Let } y = 2.$$
$$\frac{8}{12} - \frac{6}{12} \stackrel{?}{=} \frac{1}{6} \quad \text{Write fractions with the LCD.}$$
$$\frac{2}{12} \stackrel{?}{=} \frac{1}{6} \quad \text{Subtract.}$$
$$\frac{1}{6} = \frac{1}{6} \quad \text{Simplify.}$$

This is a true statement, so the solution is 2. □

**PRACTICE**
**2** Solve for $y$: $\frac{y}{2} - \frac{y}{5} = \frac{1}{4}$.

**EXAMPLE 3**   Solve: $3(x^2 + 4) + 5 = -6(x^2 + 2x) + 13$.

*Solution*   Rewrite the equation so that one side is 0.

$$3(x^2 + 4) + 5 = -6(x^2 + 2x) + 13.$$

| | |
|---|---|
| $3x^2 + 12 + 5 = -6x^2 - 12x + 13$ | Apply the distributive property. |
| $9x^2 + 12x + 4 = 0$ | Rewrite the equation so that one side is 0. |
| $(3x + 2)(3x + 2) = 0$ | Factor. |
| $3x + 2 = 0 \quad \text{or} \quad 3x + 2 = 0$ | Set each factor equal to 0. |
| $3x = -2 \quad \text{or} \quad 3x = -2$ | |
| $x = -\dfrac{2}{3} \quad \text{or} \quad x = -\dfrac{2}{3}$ | Solve each equation. |

The solution is $-\dfrac{2}{3}$. Check by substituting $-\dfrac{2}{3}$ into the original equation.   ☐

**PRACTICE**
**3**   Solve: $8(x^2 + 3) + 4 = -8x(x + 3) + 19$.

---

**EXAMPLE 4**   Solve for $x$: $\dfrac{x + 5}{2} + \dfrac{1}{2} = 2x - \dfrac{x - 3}{8}$.

*Solution*   Multiply both sides of the equation by 8, the LCD of 2 and 8.

> ▶ **Helpful Hint**
> When we multiply both sides of an equation by a number, the distributive property tells us that each term of the equation is multiplied by the number.

| | |
|---|---|
| $8\left(\dfrac{x + 5}{2} + \dfrac{1}{2}\right) = 8\left(2x - \dfrac{x - 3}{8}\right)$ | Multiply both sides by 8. |
| $8\left(\dfrac{x + 5}{2}\right) + 8 \cdot \dfrac{1}{2} = 8 \cdot 2x - 8\left(\dfrac{x - 3}{8}\right)$ | Apply the distributive property. |
| $4(x + 5) + 4 = 16x - (x - 3)$ | Simplify. |
| $4x + 20 + 4 = 16x - x + 3$ | Use the distributive property to remove parentheses. |
| $4x + 24 = 15x + 3$ | Combine like terms. |
| $-11x + 24 = 3$ | Subtract $15x$ from both sides. |
| $-11x = -21$ | Subtract 24 from both sides. |
| $\dfrac{-11x}{-11} = \dfrac{-21}{-11}$ | Divide both sides by $-11$. |
| $x = \dfrac{21}{11}$ | Simplify. |

**Check:**   To check, verify that replacing $x$ with $\dfrac{21}{11}$ makes the original equation true. The solution is $\dfrac{21}{11}$.   ☐

**PRACTICE**
**4**   Solve for $x$: $x - \dfrac{x - 2}{12} = \dfrac{x + 3}{4} + \dfrac{1}{4}$.

---

**EXAMPLE 5**   Solve: $2x^2 = \dfrac{17}{3}x + 1$.

*Solution*

| | |
|---|---|
| $2x^2 = \dfrac{17}{3}x + 1$ | |
| $3(2x^2) = 3\left(\dfrac{17}{3}x + 1\right)$ | Clear the equation of fractions. |
| $6x^2 = 17x + 3$ | Apply the distributive property. |
| $6x^2 - 17x - 3 = 0$ | Rewrite the equation in standard form. |
| $(6x + 1)(x - 3) = 0$ | Factor. |

$$6x + 1 = 0 \quad \text{or} \quad x - 3 = 0 \quad \text{Set each factor equal to zero.}$$

$$6x = -1$$

$$x = -\frac{1}{6} \quad \text{or} \quad x = 3 \quad \text{Solve each equation.}$$

The solutions are $-\dfrac{1}{6}$ and 3. ☐

**PRACTICE**
**5**  Solve: $4x^2 = \dfrac{15}{2}x + 1$.

---

## C.1 EXERCISE SET

MyMathLab  Math XL  PRACTICE  WATCH  DOWNLOAD  READ  REVIEW

**MIXED PRACTICE**

*Solve each equation. See Examples 1 through 5.*

**1.** $x^2 + 11x + 24 = 0$

**2.** $y^2 - 10y + 24 = 0$

**3.** $3x - 4 - 5x = x + 4 + x$

**4.** $13x - 15x + 8 = 4x + 2 - 24$

**5.** $12x^2 + 5x - 2 = 0$

**6.** $3y^2 - y - 14 = 0$

**7.** $z^2 + 9 = 10z$

**8.** $n^2 + n = 72$

**9.** $5(y + 4) = 4(y + 5)$

**10.** $6(y - 4) = 3(y - 8)$

**11.** $0.6x - 10 = 1.4x - 14$

**12.** $0.3x + 2.4 = 0.1x + 4$

**13.** $x(5x + 2) = 3$

**14.** $n(2n - 3) = 2$

**15.** $6x - 2(x - 3) = 4(x + 1) + 4$

**16.** $10x - 2(x + 4) = 8(x - 2) + 6$

**17.** $\dfrac{3}{8} + \dfrac{b}{3} = \dfrac{5}{12}$

**18.** $\dfrac{a}{2} + \dfrac{7}{4} = 5$

**19.** $x^2 - 6x = x(8 + x)$

**20.** $n(3 + n) = n^2 + 4n$

**21.** $\dfrac{z^2}{6} - \dfrac{z}{2} - 3 = 0$

**22.** $\dfrac{c^2}{20} - \dfrac{c}{4} + \dfrac{1}{5} = 0$

**23.** $z + 3(2 + 4z) = 6(z + 1) + 5z$

**24.** $4(m - 6) - m = 8(m - 3) - 5m$

**25.** $\dfrac{x^2}{2} + \dfrac{x}{20} = \dfrac{1}{10}$

**26.** $\dfrac{y^2}{30} = \dfrac{y}{15} + \dfrac{1}{2}$

**27.** $\dfrac{4t^2}{5} = \dfrac{t}{5} + \dfrac{3}{10}$

**28.** $\dfrac{5x^2}{6} - \dfrac{7x}{2} + \dfrac{2}{3} = 0$

**29.** $\dfrac{3t + 1}{8} = \dfrac{5 + 2t}{7} + 2$

**30.** $4 - \dfrac{2z + 7}{9} = \dfrac{7 - z}{12}$

**31.** $\dfrac{m - 4}{3} - \dfrac{3m - 1}{5} = 1$

**32.** $\dfrac{n + 1}{8} - \dfrac{2 - n}{3} = \dfrac{5}{6}$

**33.** $3x^2 = -x$

**34.** $y^2 = -5y$

**35.** $x(x - 3) = x^2 + 5x + 7$

**36.** $z^2 - 4z + 10 = z(z - 5)$

**37.** $3(t - 8) + 2t = 7 + t$

**38.** $7c - 2(3c + 1) = 5(4 - 2c)$

**39.** $-3(x - 4) + x = 5(3 - x)$

**40.** $-4(a + 1) - 3a = -7(2a - 3)$

**41.** $(x - 1)(x + 4) = 24$

**42.** $(2x - 1)(x + 2) = -3$

**43.** $\dfrac{x^2}{4} - \dfrac{5}{2}x + 6 = 0$

**44.** $\dfrac{x^2}{18} + \dfrac{x}{2} + 1 = 0$

**45.** $y^2 + \dfrac{1}{4} = -y$

**46.** $\dfrac{x^2}{10} + \dfrac{5}{2} = x$

**47.** Which solution strategies are incorrect? Why?

　　**a.** Solve $(y - 2)(y + 2) = 4$ by setting each factor equal to 4.

**b.** Solve $(x + 1)(x + 3) = 0$ by setting each factor equal to 0.

**c.** Solve $z^2 + 5z + 6 = 0$ by factoring $z^2 + 5z + 6$ and setting each factor equal to 0.

**d.** Solve $x^2 + 6x + 8 = 10$ by factoring $x^2 + 6x + 8$ and setting each factor equal to 0.

**48.** Describe two ways a linear equation differs from a quadratic equation.

*Find the value of K such that the equations are equivalent.*

**49.** $3.2x + 4 = 5.4x - 7$
$3.2x = 5.4x + K$

**50.** $-7.6y - 10 = -1.1y + 12$
$-7.6y = -1.1y + K$

**51.** $\dfrac{x}{6} + 4 = \dfrac{x}{3}$
$x + K = 2x$

**52.** $\dfrac{5x}{4} + \dfrac{1}{2} = \dfrac{x}{2}$
$5x + K = 2x$

*Solve and check.*

**53.** $2.569x = -12.48534$

**54.** $-9.112y = -47.537304$

**55.** $2.86z - 8.1258 = -3.75$

**56.** $1.25x - 20.175 = -8.15$

**REVIEW AND PREVIEW**

*Translate each phrase into an expression. Use the variable x to represent each unknown number. See Section 1.4.*

**57.** the quotient of 8 and a number

**58.** the sum of 8 and a number

**59.** the product of 8 and a number

**60.** the difference of 8 and a number

**61.** 2 more than three times a number

**62.** 5 subtracted from twice a number

---

**OBJECTIVES**

**1** Write algebraic expressions that can be simplified.

**2** Apply the steps for problem solving.

## C.2 PROBLEM SOLVING

**OBJECTIVE 1 ▶ Writing and simplifying algebraic expressions.** In order to prepare for problem solving, we practice writing algebraic expressions that can be simplified.

Our first example involves consecutive integers and perimeter. Recall that *consecutive integers* are integers that follow one another in order. Study the examples of consecutive, even, and odd integers and their representations.

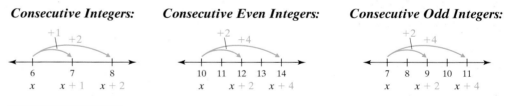

**Consecutive Integers:**   **Consecutive Even Integers:**   **Consecutive Odd Integers:**

**EXAMPLE 1** Write the following as algebraic expressions. Then simplify.

**a.** The sum of three consecutive integers, if $x$ is the first consecutive integer.

△ **b.** The perimeter of the triangle with sides of length $x$, $5x$, and $6x - 3$.

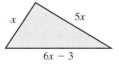

*Solution*

**a.** Recall that if $x$ is the first integer, then the next consecutive integer is 1 more, or $x + 1$ and the next consecutive integer is 1 more than $x + 1$, or $x + 2$.

| In words: | first integer | plus | next consecutive integer | plus | next consecutive integer |
|---|---|---|---|---|---|
| | ↓ | ↓ | ↓ | ↓ | ↓ |
| Translate: | $x$ | $+$ | $(x + 1)$ | $+$ | $(x + 2)$ |

Then　　　　　$x + (x + 1) + (x + 2) = x + x + 1 + x + 2$

$= 3x + 3$　Simplify by combining like terms.

**b.** The perimeter of a triangle is the sum of the lengths of the sides.

In words:   side   +   side   +   side

Translate: $\qquad x \qquad + \qquad 5x \qquad + \quad (6x - 3)$

Then $\qquad x + 5x + (6x - 3) = x + 5x + 6x - 3$

$\qquad\qquad\qquad\qquad\qquad\qquad\quad = 12x - 3 \qquad$ Simplify. $\qquad\square$

**PRACTICE**

**1**   Write the following algebraic expressions. Then simplify.

**a.** The sum of three consecutive odd integers, if $x$ is the first consecutive odd integer

**b.** The perimeter of a trapezoid with bases $x$ and $2x$, and sides of $x + 2$ and $2x - 3$

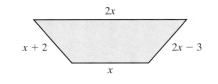

$2x$

$x + 2 \qquad\qquad\qquad\qquad 2x - 3$

$x$

**EXAMPLE 2**   The three busiest airports in the United States are in Chicago, Atlanta, and Los Angeles. The airport in Atlanta has 13.3 million more arrivals and departures than the Los Angeles airport. The Chicago airport has 1.3 million more arrivals and departures than the Los Angeles airport. Write the sum of the arrivals and departures from these three cities as a simplified algebraic expression. Let $x$ be the number of arrivals and departures at the Los Angeles airport. (*Source:* U.S. Department of Transportation)

**Solution**   If $x$ = millions of arrivals and departures at the Los Angeles airport, then

$\qquad x + 13.3$ = millions of arrivals and departures at the Atlanta airport and

$\qquad x + 1.3$ = millions of arrivals and departures at the Chicago airport.

Since we want their sum, we have

In words:

| arrivals and departures at Los Angeles | + | arrivals and departures at Atlanta | + | arrivals and departures at Chicago |
|---|---|---|---|---|

Translate: $\qquad x \qquad\qquad + \qquad (x + 13.3) \qquad + \qquad (x + 1.3)$

Then $\qquad x + (x + 13.3) + (x + 1.3) = x + x + 13.3 + x + 1.3$

$\qquad\qquad\qquad\qquad\qquad\qquad\qquad\qquad = 3x + 14.6 \qquad$ Combine like terms.

In Exercise 57, we will find the actual number of arrivals and departures at these airports. $\qquad\square$

**PRACTICE**

**2**   The three busiest airports in Europe are in London, England; Paris, France; and Frankfurt, Germany. The airport in London has 15.7 million more arrivals and departures than the Frankfurt airport. The Paris airport has 1.6 million more arrivals and departures than the Frankfurt airport. Write the sum of the arrivals and departures from these three cities as a simplified algebraic expression. Let $x$ be the number of arrivals and departures at the Frankfurt airport. (*Source:* Association of European Airlines)

**OBJECTIVE 2** ▶ **Applying steps for problem solving.** Our main purpose for studying algebra is to solve problems. The following problem-solving strategy will be used throughout this text and may also be used to solve real-life problems that occur outside the mathematics classroom.

**General Strategy for Problem Solving**

**1.** UNDERSTAND the problem. During this step, become comfortable with the problem. Some ways of doing this are:

Read and reread the problem.

Propose a solution and check. Pay careful attention to how you check your proposed solution. This will help when writing an equation to model the problem.

Construct a drawing.

**Choose a variable to represent the unknown.** (Very important part)

**2.** TRANSLATE the problem into an equation.

**3.** SOLVE the equation.

**4.** INTERPRET the results: *Check* the proposed solution in the stated problem and *state* your conclusion.

Let's review this strategy by solving a problem involving unknown numbers.

**EXAMPLE 3**   **Finding Unknown Numbers**

Find three numbers such that the second number is 3 more than twice the first number, and the third number is four times the first number. The sum of the three numbers is 164.

*Solution*

**1.** UNDERSTAND the problem. First let's read and reread the problem and then propose a solution. For example, if the first number is 25, then the second number is 3 more than twice 25, or 53. The third number is four times 25, or 100. The sum of 25, 53, and 100 is 178, not the required sum, but we have gained some valuable information about the problem. First, we know that the first number is less than 25 since our guess led to a sum greater than the required sum. Also, we have gained some information as to how to model the problem.

Next let's assign a variable and use this variable to represent any other unknown quantities. If we let

$$x = \text{the first number, then}$$

$$2x + 3 = \text{the second number}$$

3 more than
twice the second number

$$4x = \text{the third number}$$

**2.** TRANSLATE the problem into an equation. To do so, we use the fact that the sum of the numbers is 164. First let's write this relationship in words and then translate to an equation.

| In words: | first number | added to | second number | added to | third number | is | 164 |
|---|---|---|---|---|---|---|---|
| Translate: | $x$ | $+$ | $(2x + 3)$ | $+$ | $4x$ | $=$ | $164$ |

**3.** SOLVE the equation.

$$x + (2x + 3) + 4x = 164$$

$$x + 2x + 4x + 3 = 164 \quad \text{Remove parentheses.}$$

$$7x + 3 = 164 \quad \text{Combine like terms.}$$

$$7x = 161 \quad \text{Subtract 3 from both sides.}$$

$$x = 23 \quad \text{Divide both sides by 7.}$$

**4.** INTERPRET. Here, we *check* our work and *state* the solution. Recall that if the first number $x = 23$, then the second number $2x + 3 = 2 \cdot 23 + 3 = 49$ and the third number $4x = 4 \cdot 23 = 92$.

**Check:** Is the second number 3 more than twice the first number? Yes, since 3 more than twice 23 is $46 + 3$, or 49. Also, their sum, $23 + 49 + 92 = 164$, is the required sum.

**State:** The three numbers are 23, 49, and 92.  □

**PRACTICE**

**3** Find three numbers such that the second number is 8 less than triple the first number, the third number is five times the first number, and the sum of the three numbers is 118.

----

Many of today's rates and statistics are given as percents. Interest rates, tax rates, nutrition labeling, and percent of households in a given category are just a few examples. Before we practice solving problems containing percents, let's briefly take a moment and review the meaning of percent and how to find a percent of a number.

The word *percent* means "per hundred," and the symbol % is used to denote percent. This means that 23% is 23 per hundred, or $\frac{23}{100}$. Also,

$$41\% = \frac{41}{100} = 0.41$$

To find a percent of a number, we multiply.

$$16\% \text{ of } 25 = 16\% \cdot 25 = 0.16 \cdot 25 = 4$$

Thus, 16% of 25 is 4.

Study the table below. It will help you become more familiar with finding percents.

| Percent | Meaning/Shortcut | Example |
|---------|------------------|---------|
| 50% | $\frac{1}{2}$ or half of a number | 50% of 60 is 30. |
| 25% | $\frac{1}{4}$ or a quarter of a number | 25% of 60 is 15. |
| 10% | 0.1 or $\frac{1}{10}$ of a number (move the decimal point 1 place to the left) | 10% of 60 is 6.0 or 6. |
| 1% | 0.01 or $\frac{1}{100}$ of a number (move the decimal point 2 places to the left) | 1% of 60 is 0.60 or 0.6. |
| 100% | 1 or all of a number | 100% of 60 is 60. |
| 200% | 2 or double a number | 200% of 60 is 120. |

**Concept Check** ☑️

Suppose you are finding 112% of a number $x$. Which of the following is a correct description of the result? Explain.

**a.** The result is less than $x$.          **b.** The result is equal to $x$.          **c.** The result is greater than $x$.

Next, we solve a problem containing a percent.

**EXAMPLE 4**    **Finding the Original Price of a Computer**

Suppose that a computer store just announced an 8% decrease in the price of a particular computer model. If this computer sells for $2162 after the decrease, find the original price of this computer.

*Solution*

1. **UNDERSTAND.** Read and reread the problem. Recall that a percent decrease means a percent of the original price. Let's guess that the original price of the computer is $2500. The amount of decrease is then 8% of $2500, or $(0.08)(\$2500) = \$200$. This means that the new price of the computer is the original price minus the decrease, or $\$2500 - \$200 = \$2300$. Our guess is incorrect, but we now have an idea of how to model this problem. In our model, we will let $x$ = the original price of the computer.

2. **TRANSLATE.**

| In words: | original price of computer | minus | 8% of original price | is | new price |
|---|---|---|---|---|---|
| | ↓ | ↓ | ↓ | ↓ | ↓ |
| Translate: | $x$ | $-$ | $0.08x$ | $=$ | $2162$ |

3. **SOLVE** the equation.

$$x - 0.08x = 2162$$
$$0.92x = 2162 \quad \text{Combine like terms.}$$
$$x = \frac{2162}{0.92} = 2350 \quad \text{Divide both sides by 0.92.}$$

4. **INTERPRET.**

**Check:**   If the original price of the computer was $2350, the new price is

$$\$2350 - (0.08)(\$2350) = \$2350 - \$188$$
$$= \$2162 \quad \text{The given new price}$$

**State:**   The original price of the computer was $2350.                    □

**PRACTICE**

**4**    At the end of the season, the cost of a snowboard was reduced by 40%. If the snowboard sells for $270 after the decrease, find the original price of the board.

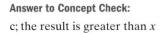

**Answer to Concept Check:**

c; the result is greater than $x$

# VOCABULARY & READINESS CHECK

*Fill in each blank with* $<$, $>$, *or* $=$. *(Assume that the unknown number is a positive number.)*

**1.** 130% of a number _____ the number.

**2.** 70% of a number _____ the number.

**3.** 100% of a number _____ the number.

**4.** 200% of a number _____ the number.

*Complete the table. The first row has been completed for you.*

| | First Integer | All Described Integers |
|---|---|---|
| Three consecutive integers | 18 | 18, 19, 20 |
| **5.** Four consecutive integers | 31 | |
| **6.** Three consecutive odd integers | 31 | |
| **7.** Three consecutive even integers | 18 | |
| **8.** Four consecutive even integers | 92 | |
| **9.** Three consecutive integers | $y$ | |
| **10.** Three consecutive even integers | $z$ ($z$ is even) | |
| **11.** Four consecutive integers | $p$ | |
| **12.** Three consecutive odd integers | $s$ ($s$ is odd) | |

## C.2 EXERCISE SET

MyMathLab    Math XP PRACTICE    WATCH    DOWNLOAD    READ    REVIEW

*Write the following as algebraic expressions. Then simplify. See Examples 1 and 2.*

△ **1.** The perimeter of the square with side length $y$.

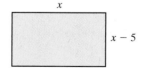

△ **2.** The perimeter of the rectangle with length $x$ and width $x - 5$.

**3.** The sum of three consecutive integers if the first is $z$.

**4.** The sum of three consecutive odd integers if the first integer is $x$.

**5.** The total amount of money (in cents) in $x$ nickels, $(x + 3)$ dimes, and $2x$ quarters. (*Hint:* the value of a nickel is 5 cents, the value of a dime is 10 cents, and the value of a quarter is 25 cents)

**6.** The total amount of money (in cents) in $y$ quarters, $7y$ dimes, and $(2y - 1)$ nickels. (Use the hint for Exercise 5.)

△ **7.** A piece of land along Bayou Liberty is to be fenced and subdivided as shown so that each rectangle has the same dimensions. Express the total amount of fencing needed as an algebraic expression in $x$.

**8.** A flooded piece of land near the Mississippi River in New Orleans is to be surveyed and divided into 4 rectangles of equal dimension. Express the total amount of fencing needed as an algebraic expression in $x$.

△ **9.** Write the perimeter of the floor plan shown as an algebraic expression in $x$.

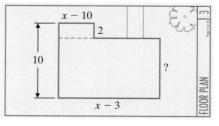

**10.** Write the perimeter of the floor plan shown as an algebraic expression in *x*.

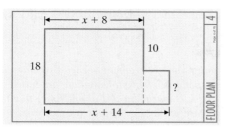

*Solve. See Example 3.*

**11.** Four times the difference of a number and 2 is the same as 2 increased by four times the number plus twice the number. Find the number.

**12.** Twice the sum of a number and 3 is the same as five times the number minus 1 minus four times the number. Find the number.

**13.** A second number is five times a first number. A third number is 100 more than the first number. If the sum of the three numbers is 415, find the numbers.

**14.** A second number is 6 less than a first number. A third number is twice the first number. If the sum of the three numbers is 306, find the numbers.

*Solve. See Example 4.*

**15.** The United States consists of 2271 million acres of land. Approximately 29% of this land is federally owned. Find the number of acres that are not federally owned. (*Source:* U.S. General Services Administration)

**16.** The state of Nevada contains the most federally owned acres of land in the United States. If 90% of the state's 70 million acres of land is federally owned, find the number of acres that are not federally owned. (*Source:* U.S. General Services Administration)

**17.** In 2006, a total of 2748 earthquakes occurred in the United States. Of these, 85.3% were minor tremors with magnitudes of 3.9 or less on the Richter scale. How many minor earthquakes occurred in the United States in 2006? Round to the nearest whole number. (*Source:* U.S. Geological Survey National Earthquake Information Center)

**18.** Of the 958 tornadoes that occurred in the United States during 2006, 25.5% occurred during the month of April. How many tornadoes occurred during April 2006? Round to the nearest whole. (*Source:* Storm Prediction Center)

**19.** In a recent survey, 15% of online shoppers in the United States say that they prefer to do business only with large, well-known retailers. In a group of 1500 online shoppers, how many are willing to do business with any size retailers? (*Source:* Inc.com)

**20.** In 2006, the restaurant and food service industry employed 9.1% of the total employees in the state of California. In 2006, California had approximately 17,029,300 employed workers. How many people worked in the restaurant and food service industry in California? Round to the nearest whole number. (*Source:* California Employment Development Center.)

*The following graph is called a circle graph or a pie chart. The circle represents a whole, or in this case, 100%. This particular graph shows the number of minutes per day that people use e-mail at work. Use this graph to answer Exercises 21 through 24.*

**Time Spent on E-Mail at Work**

*Source*: Pew Internet & American Life Project

**21.** What percent of e-mail users at work spend less than 15 minutes on e-mail per day?

**22.** Among e-mail users at work, what is the most common time spent on e-mail per day?

**23.** If it were estimated that a large company has 5957 employees, how many of these would you expect to be using e-mail more than 3 hours per day? Round to the nearest whole employee.

**24.** If it were estimated that a medium size company has 278 employees, how many of these would you expect to be using e-mail between 2 and 3 hours per day? Round to the nearest whole employee.

**25.** In 2006, the population of Canada was 31.6 million. This represented an increase in population of 5.4% since 2001. What was the population of Canada in 2001? (Round to the nearest hundredth of a million.) (*Source:* Statistics Canada)

**26.** In 2006, the cost of an average hotel room per night was $96.73. This was an increase of 6.8% over the average cost in 2005. Find the average hotel room cost in 2005. (*Source:* Smith Travel Research)

## MIXED PRACTICE

*Use the diagrams to find the unknown measures of angles or lengths of sides. Recall that the sum of the angle measures of a triangle is 180°.*

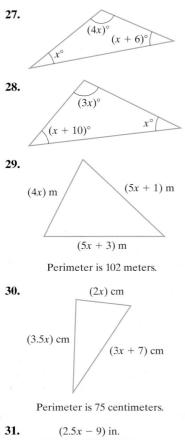

**27.**

**28.**

**29.**

$(4x)$ m  $(5x + 1)$ m

$(5x + 3)$ m

Perimeter is 102 meters.

**30.** $(2x)$ cm

$(3.5x)$ cm

$(3x + 7)$ cm

Perimeter is 75 centimeters.

**31.** $(2.5x - 9)$ in.

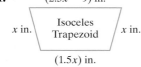

$x$ in.  Isoceles Trapezoid  $x$ in.

$(1.5x)$ in.

Perimeter is 99 inches.

**32.** $(9.2x - 3)$ ft

$(7.3x)$ ft  Parallelogram  $(7.3x)$ ft

$(9.2x - 3)$ ft

Perimeter is 324 feet.

*Solve.*

**33.** The sum of three consecutive integers is 228. Find the integers.

**34.** The sum of three consecutive odd integers is 327. Find the integers.

**35.** The zip codes of three Nevada locations—Fallon, Fernley, and Gardnerville Ranchos—are three consecutive even integers. If twice the first integer added to the third is 268,222, find each zip code.

**36.** During a recent year, the average SAT scores in math for the states of Alabama, Louisiana, and Michigan were 3 consecutive integers. If the sum of the first integer, second integer, and three times the third integer is 2637, find each score.

*Many companies and government agencies predict the growth or decline of occupations. The following data are based on information from the U.S. Department of Labor. Notice that the first table shows the increase in number of jobs (in thousands) and the second table shows the percent increase in number of jobs.*

**37.** Use the table to find the actual number of jobs for each occupation.

| Occupation | Increase in Number of Jobs (in thousands) from 2000 to 2012 |
|---|---|
| Security guards | $2x - 51$ |
| Home health aides | $\frac{3}{2}x + 3$ |
| Computer systems analysts | $x$ |
| Total | 780 thousand |

**38.** Use the table to find the actual percent increase in number of jobs for each occupation.

| Occupation | Precent Increase in Number of Jobs from 2000 to 2012 |
|---|---|
| Computer software engineers | $\frac{3}{2}x + 1$ |
| Management analysts | $x$ |
| Receptionist and information clerks | $x - 1$ |
| Total | 105% |

**39.** The occupations of postsecondary teachers, registered nurses, and medical assistants are among the ten with the largest growth from 2000 to 2012. (See the Chapter 9 opener.) The number of postsecondary teacher jobs will grow 173 thousand more than twice the number of medical assistant jobs. The number of registered nurse jobs will grow 22 thousand less than three times the number of medical assistant jobs. If the total growth of these three jobs is predicted to be 1441 thousand, find the predicted growth of each job.

**40.** The occupations of telephone operators, fishers, and sewing machine operators are among the ten with the largest job decline from 2000 to 2012, according to the U.S. Department of Labor. The number of telephone operators will decline 8 thousand more than twice the number of fishers. The number of sewing machine operators will decline 1 thousand less than

10 times the number of fishers. If the total decline of these three jobs is predicted to be 137 thousand, find the predicted decline of each job.

**41.** America West Airlines has 3 models of Boeing aircraft in their fleet. The 737-300 contains 21 more seats than the 737-200. The 757-200 contains 36 less seats than twice the number of seats in the 737-200. Find the number of seats for each aircraft if the total number of seats for the 3 models is 437. (*Source:* America West Airlines)

**42.** The new governor of California makes $27,500 more than the governor of New York, and $120,724 more than the governor of Alaska. If the sum of these 3 salaries is $471,276, find the salary of each governor. (*Source:* State Web sites)

**43.** A new fax machine was recently purchased for an office in Hopedale for $464.40 including tax. If the tax rate in Hopedale is 8%, find the price of the fax machine before tax.

**44.** A premedical student at a local university was complaining that she had just paid $158.60 for her human anatomy book, including tax. Find the price of the book before taxes if the tax rate at this university is 9%.

**45.** In 2006, the population of South Africa was 44.2 million people. From 2006 to 2050, South Africa's population is expected to decrease by 5.6%. Find the expected population of South Africa in 2050. Round to the nearest tenth of a million. (*Source:* Population Reference Bureau)

**46.** In 2006, the population of Morocco was 33.2 million. This represented an increase in population of 1.5% from a year earlier. What was the population of Morocco in 2005? Round to the nearest tenth of a million. (*Source:* Population Reference Bureau)

*Recall that two angles are complements of each other if their sum is 90°. Two angles are supplements of each other if their sum is 180°. Find the measure of each angle.*

**47.** One angle is three times its supplement increased by 20°. Find the measures of the two supplementary angles.

**48.** One angle is twice its complement increased by 30°. Find the measure of the two complementary angles.

*Recall that the sum of the angle measures of a triangle is 180°.*

△ **49.** Find the measures of the angles of a triangle if the measure of one angle is twice the measure of a second angle and the third angle measures 3 times the second angle decreased by 12.

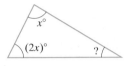

△ **50.** Find the angles of an isoceles triangle whose two base angles are equal and whose third angle is 10° less than three times a base angle.

**51.** Two frames are needed with the same perimeter: one frame in the shape of a square and one in the shape of an equilateral triangle. Each side of the triangle is 6 centimeters longer than each side of the square. Find the side length of each frame. (An equilateral triangle has sides that are the same length.)

**52.** Two frames are needed with the same perimeter: one frame in the shape of a square and one in the shape of a regular pentagon. Each side of the square is 7 inches longer than each side of the pentagon. Find the side length of each frame. (A regular polygon has sides that are the same length.)

**53.** The sum of the first and third of three consecutive even integers is 156. Find the three even integers.

**54.** The sum of the second and fourth of four consecutive integers is 110. Find the four integers.

△ **55.** The perimeter of the triangle in Example 1b in this section is 483 feet. Find the length of each side.

△ **56.** The perimeter of the trapezoid in Practice 1b in this section is 110 meters. Find the lengths of its sides and bases.

**57.** The airports in Chicago, Atlanta, and Los Angeles have a total of 197.6 million annual arrivals and departures. Use this information and Example 2 in this section to find the number from each individual airport.

**58.** The airports in London, Paris, and Frankfurt have a total of 173.9 million annual arrivals and departures. Use this information and Practice 2 in this section to find the number from each airport.

**59.** Incandescent, fluorescent, and halogen bulbs are lasting longer today than ever before. On average, the number of bulb hours for a fluorescent bulb is 25 times the number of bulb hours for a halogen bulb. The number of bulb hours for an incandescent bulb is 2,500 less than the halogen bulb. If the total number of bulb hours for the three types of bulbs is 105,500, find the number of bulb hours for each type. (*Source: Popular Science Magazine*)

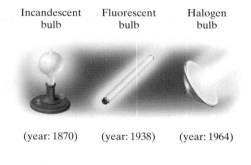

| Incandescent bulb | Fluorescent bulb | Halogen bulb |
|---|---|---|
| (year: 1870) | (year: 1938) | (year: 1964) |

**60.** The three tallest hospitals in the world are Guy's Tower in London, Queen Mary Hospital in Hong Kong, and Galter Pavilion in Chicago. These buildings have a total height of 1320 feet. Guy's Tower is 67 feet taller than Galter Pavilion and the Queen Mary Hospital is 47 feet taller than Galter Pavilion. Find the heights of the three hospitals.

△ **61.** The official manual for traffic signs is the *Manual on Uniform Traffic Control Devices* published by the Government Printing Office. The rectangular sign below has a length 12 inches more than twice its height. If the perimeter of the sign is 312 inches, find its dimensions.

**62.** INVESCO Field at Mile High, home to the Denver Broncos, has 11,675 more seats than Heinz Field, home to the Pittsburgh Steelers. Together, these two stadiums can seat a total of 140,575 NFL fans. How many seats does each stadium have? (*Sources:* Denver Broncos, Pittsburgh Steelers)

**OBJECTIVES**

1 Plot ordered pairs.

2 Graph linear equations.

## C.3 GRAPHING

**OBJECTIVE 1 ▶ Plotting ordered pairs.** Graphs are widely used today in newspapers, magazines, and all forms of newsletters. A few examples of graphs are shown here.

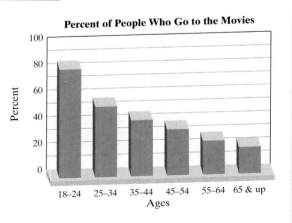

**Percent of People Who Go to the Movies**

*Source*: TELENATION/Market Facts, Inc.

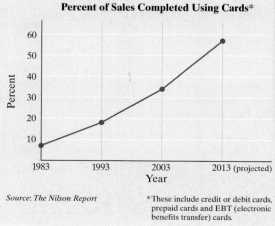

**Percent of Sales Completed Using Cards\***

*Source: The Nilson Report*

\*These include credit or debit cards, prepaid cards and EBT (electronic benefits transfer) cards.

To review how to read these graphs, we review their origin—the rectangular coordinate system. One way to locate points on a plane is by using a **rectangular coordinate system,** which is also called a **Cartesian coordinate system** after its inventor, René Descartes (1596–1650). The next diagram to the left shows the rectangular coordinate system. For further review of this system, see Section 3.1.

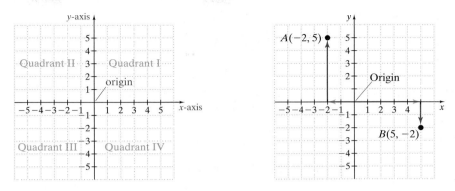

Recall that, the location of point $A$ in the figure above is described as 2 units to the left of the origin along the $x$-axis and 5 units upward parallel to the $y$-axis. Thus, we identify point $A$ with the ordered pair $(-2, 5)$. Notice that the order of these numbers is *critical*. The $x$-value $-2$ is called the **$x$-coordinate** and is associated with the $x$-axis. The $y$-value 5 is called the **$y$-coordinate** and is associated with the $y$-axis. Compare the location of point $A$ with the location of point $B$, which corresponds to the ordered pair $(5, -2)$.

Keep in mind that **each ordered pair corresponds to exactly one point in the real plane and that each point in the plane corresponds to exactly one ordered pair.** Thus, we may refer to the ordered pair $(x, y)$ as the point $(x, y)$.

**EXAMPLE 1**   Plot each ordered pair on a Cartesian coordinate system and name the quadrant or axis in which the point is located.

**a.** $(2, -1)$   **b.** $(0, 5)$   **c.** $(-3, 5)$   **d.** $(-2, 0)$   **e.** $\left(-\frac{1}{2}, -4\right)$   **f.** $(1.5, 1.5)$

*Solution*   The six points are graphed as shown.

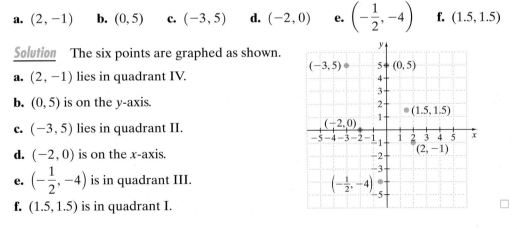

**a.** $(2, -1)$ lies in quadrant IV.

**b.** $(0, 5)$ is on the $y$-axis.

**c.** $(-3, 5)$ lies in quadrant II.

**d.** $(-2, 0)$ is on the $x$-axis.

**e.** $\left(-\frac{1}{2}, -4\right)$ is in quadrant III.

**f.** $(1.5, 1.5)$ is in quadrant I.

**PRACTICE**

**1**   Plot each ordered pair on a Cartesian coordinate system and name the quadrant or axis in which the point is located.

**a.** $(3, -4)$   **b.** $(0, -2)$   **c.** $(-2, 4)$   **d.** $(4, 0)$   **e.** $\left(-1\frac{1}{2}, -2\right)$   **f.** $(2.5, 3.5)$

Notice that the $y$-coordinate of any point on the $x$-axis is 0. For example, the point with coordinates $(-2, 0)$ lies on the $x$-axis. Also, the $x$-coordinate of any point on the $y$-axis is 0. For example, the point with coordinates $(0, 5)$ lies on the $y$-axis. These points that lie on the axes do not lie in any quadrants.

**Concept Check** ☑

Which of the following correctly describes the location of the point $(3, -6)$ in a rectangular coordinate system?

**a.** 3 units to the left of the $y$-axis and 6 units above the $x$-axis

**b.** 3 units above the $x$-axis and 6 units to the left of the $y$-axis

**c.** 3 units to the right of the $y$-axis and 6 units below the $x$-axis

**d.** 3 units below the $x$-axis and 6 units to the right of the $y$-axis

**OBJECTIVE 2** ▶ **Graphing linear equations.** Recall that an equation such as $3x - y = 12$ is called a linear equation in two variables, and **the graph of every linear equation in two variables is a line.**

---

**Linear Equation in Two Variables**

A linear equation in two variables is an equation that can be written in the form

$$Ax + By = C$$

where $A$ and $B$ are not both 0. This form is called **standard form.**

---

Some examples of equations in standard form:

$$3x - y = 12$$
$$-2.1x + 5.6y = 0$$

---

▶ **Helpful Hint**

Remember: A linear equation is written in standard form when all of the variable terms are on one side of the equation and the constant is on the other side.

---

**EXAMPLE 2**   Graph the equation $y = -2x + 3$.

*Solution*   This is a linear equation. (In standard form it is $2x + y = 3$.) Find three ordered pair solutions, and plot the ordered pairs. The line through the plotted points is the graph. Since the equation is solved for $y$, let's choose three $x$-values. We'll choose 0, 2, and then $-1$ for $x$ to find our three ordered pair solutions.

| Let $x = 0$ | Let $x = 2$ | Let $x = -1$ |
|---|---|---|
| $y = -2x + 3$ | $y = -2x + 3$ | $y = -2x + 3$ |
| $y = -2 \cdot 0 + 3$ | $y = -2 \cdot 2 + 3$ | $y = -2(-1) + 3$ |
| $y = 3$  Simplify. | $y = -1$  Simplify. | $y = 5$  Simplify. |

The three ordered pairs $(0, 3)$, $(2, -1)$ and $(-1, 5)$ are listed in the table and the graph is shown.

| $x$ | $y$ |
|---|---|
| 0 | 3 |
| 2 | -1 |
| -1 | 5 |

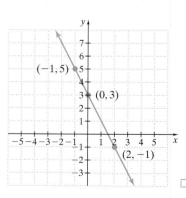

**PRACTICE**
**2**   Graph the equation $y = -3x - 2$.

Notice that the graph crosses the $y$-axis at the point $(0, 3)$. This point is called the **$y$-intercept.** (You may sometimes see just the number 3 called the $y$-intercept.) This graph also crosses the $x$-axis at the point $\left(\frac{3}{2}, 0\right)$. This point is called the **$x$-intercept.** (You may also see just the number $\frac{3}{2}$ called the $x$-intercept.)

Since every point on the $y$-axis has an $x$-value of 0, we can find the $y$-intercept of a graph by letting $x = 0$ and solving for $y$. Also, every point on the $x$-axis has a $y$-value of 0. To find the $x$-intercept, we let $y = 0$ and solve for $x$.

> **Finding x- and y-Intercepts**
> To find an $x$-intercept, let $y = 0$ and solve for $x$.
> To find a $y$-intercept, let $x = 0$ and solve for $y$.

**EXAMPLE 3**   Graph the linear equation $y = \frac{1}{3}x$.

_Solution_   To graph, we find ordered pair solutions, plot the ordered pairs, and draw a line through the plotted points. We will choose $x$-values and substitute in the equation. To avoid fractions, we choose $x$-values that are multiples of 3. To find the $y$-intercept, we let $x = 0$.

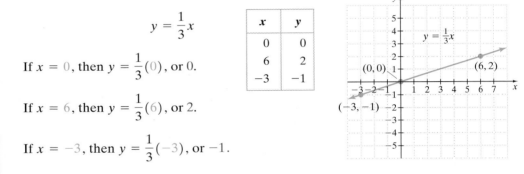

> ▶ **Helpful Hint**
> Notice that by using multiples of 3 for $x$, we avoid fractions.

$$y = \frac{1}{3}x$$

If $x = 0$, then $y = \frac{1}{3}(0)$, or 0.

If $x = 6$, then $y = \frac{1}{3}(6)$, or 2.

| $x$ | $y$ |
|---|---|
| 0 | 0 |
| 6 | 2 |
| -3 | -1 |

> ▶ **Helpful Hint**
> Since the equation $y = \frac{1}{3}x$ is solved for $y$, we choose $x$-values for finding points. This way, we simply need to evaluate an expression to find the $y$-value, as shown.

If $x = -3$, then $y = \frac{1}{3}(-3)$, or $-1$.

This graph crosses the $x$-axis at $(0, 0)$ and the $y$-axis at $(0, 0)$. This means that the $x$-intercept is $(0, 0)$ and that the $y$-intercept is $(0, 0)$.   □

**PRACTICE**
**3**   Graph the linear equation $y = -\frac{1}{2}x$.

---

**C.3 EXERCISE SET**

_Determine the coordinates of each point on the graph._

**1.** Point $A$
**2.** Point $B$
**3.** Point $C$
**4.** Point $D$
**5.** Point $E$
**6.** Point $F$
**7.** Point $G$
**8.** Point $H$

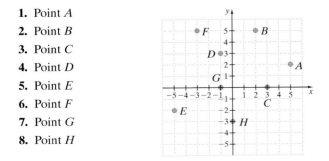

_Without graphing, visualize the location of each point. Then give its location by quadrant or $x$- or $y$-axis._

**9.** $(2, 3)$
**10.** $(0, 5)$
**11.** $(-2, 7)$
**12.** $(-3, 0)$
**13.** $(-1, -4)$
**14.** $(4, -2)$
**15.** $(0, -100)$
**16.** $(10, 30)$
**17.** $(-10, -30)$
**18.** $(0, 0)$
**19.** $(-87, 0)$
**20.** $(-42, 17)$

*Given that x is a positive number and that y is a positive number, determine the quadrant or axis in which each point lies.*

**21.** $(x, -y)$

**22.** $(-x, y)$

**23.** $(x, 0)$

**24.** $(0, -y)$

**25.** $(-x, -y)$

**26.** $(0, 0)$

*Graph each linear equation. See Examples 2 and 3.*

**27.** $y = -x - 2$

**28.** $y = -2x + 1$

**29.** $3x - 4y = 8$

**30.** $x - 9y = 3$

**31.** $y = \dfrac{1}{3}x$

**32.** $y = \dfrac{3}{2}x$

**33.** $y + 4 = 0$

**34.** $x = -1.5$

*Recall that if $f(2) = 7$, for example, this corresponds to the ordered pair (2, 7), for the graph of f. Use this information and the graphs of f and g below to answer Exercises 35 through 42.*

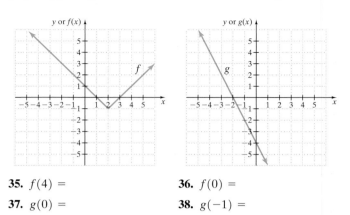

**35.** $f(4) =$

**36.** $f(0) =$

**37.** $g(0) =$

**38.** $g(-1) =$

**39.** Find all values for $x$ such that $f(x) = 0$.

**40.** Find all values for $x$ such that $g(x) = 0$.

**41.** If $(-1, -2)$ is a point on the graph of $g$, write this using function notation.

**42.** If $(-1, 2)$ is a point on the graph of $f$, write this using function notation.

---

**OBJECTIVES**

1 Review operations on polynomials.

2 Review factoring polynomials.

## C.4 POLYNOMIALS AND FACTORING

**OBJECTIVE 1 ▶ Operations on polynomials**

*Perform each indicated operation.*

**1.** $(-y^2 + 6y - 1) + (3y^2 - 4y - 10)$

**2.** $(5z^4 - 6z^2 + z + 1) - (7z^4 - 2z + 1)$

**3.** Subtract $(x - 5)$ from $(x^2 - 6x + 2)$.

**4.** $(2x^2 + 6x - 5) + (5x^2 - 10x)$

**5.** $(5x - 3)^2$

**6.** $(5x^2 - 14x - 3) \div (5x + 1)$

**7.** $(2x^4 - 3x^2 + 5x - 2) \div (x + 2)$

**8.** $(4x - 1)(x^2 - 3x - 2)$

**OBJECTIVE 2 ▶ Factoring strategies.** The key to proficiency in factoring polynomials is to practice until you are comfortable with each technique. A strategy for factoring polynomials completely is given next.

**Factoring a Polynomial**

**STEP 1.** Are there any common factors? If so, factor out the greatest common factor.

**STEP 2.** How many terms are in the polynomial?

    **a.** If there are *two* terms, decide if one of the following formulas may be applied:

      **i.** Difference of two squares: $a^2 - b^2 = (a - b)(a + b)$

      **ii.** Difference of two cubes: $a^3 - b^3 = (a - b)(a^2 + ab + b^2)$

      **iii.** Sum of two cubes: $a^3 + b^3 = (a + b)(a^2 - ab + b^2)$

    **b.** If there are *three* terms, try one of the following:

      **i.** Perfect square trinomial: $a^2 + 2ab + b^2 = (a + b)^2$
$$a^2 - 2ab + b^2 = (a - b)^2$$

      **ii.** If not a perfect square trinomial, factor by using the methods presented in Section 6.2.

    **c.** If there are *four* or more terms, try factoring by grouping.

**STEP 3.** See whether any factors in the factored polynomial can be factored further.

A few examples are worked for you below.

**EXAMPLE 1**    Factor each polynomial completely.

**a.** $8a^2b - 4ab$      **b.** $36x^2 - 9$      **c.** $2x^2 - 5x - 7$

**d.** $5p^2 + 5 + qp^2 + q$      **e.** $9x^2 + 24x + 16$      **f.** $y^2 + 25$

*Solution*

**a. STEP 1.** The terms have a common factor of $4ab$, which we factor out.
$$8a^2b - 4ab = 4ab(2a - 1)$$

  **STEP 2.** There are two terms, but the binomial $2a - 1$ is not the difference of two squares or the sum or difference of two cubes.

  **STEP 3.** The factor $2a - 1$ cannot be factored further.

**b. STEP 1.** Factor out a common factor of 9.
$$36x^2 - 9 = 9(4x^2 - 1)$$

  **STEP 2.** The factor $4x^2 - 1$ has two terms, and it is the difference of two squares.
$$9(4x^2 - 1) = 9(2x + 1)(2x - 1)$$

  **STEP 3.** No factor with more than one term can be factored further.

**c. STEP 1.** The terms of $2x^2 - 5x - 7$ contain no common factor other than 1 or $-1$.

  **STEP 2.** There are three terms. The trinomial is not a perfect square, so we factor by methods from Section 6.3.
$$2x^2 - 5x - 7 = (2x - 7)(x + 1)$$

  **STEP 3.** No factor with more than one term can be factored further.

**d. STEP 1.** There is no common factor of all terms of $5p^2 + 5 + qp^2 + q$.

    **STEP 2.** The polynomial has four terms, so try factoring by grouping.

$$5p^2 + 5 + qp^2 + q = (5p^2 + 5) + (qp^2 + q) \quad \text{Group the terms.}$$
$$= 5(p^2 + 1) + q(p^2 + 1)$$
$$= (p^2 + 1)(5 + q)$$

    **STEP 3.** No factor can be factored further.

**e. STEP 1.** The terms of $9x^2 + 24x + 16$ contain no common factor other than 1 or $-1$.

    **STEP 2.** The trinomial $9x^2 + 24x + 16$ is a perfect square trinomial, and $9x^2 + 24x + 16 = (3x + 4)^2$.

    **STEP 3.** No factor can be factored further.

**f. STEP 1.** There is no common factor of $y^2 + 25$ other than 1.

    **STEP 2.** This binomial is the sum of two squares and is prime.

    **STEP 3.** The binomial $y^2 + 25$ cannot be factored further.     ☐

**PRACTICE**

**1**    Factor each polynomial completely.

**a.** $12x^2y - 3xy$                            **b.** $49x^2 - 4$

**c.** $5x^2 + 2x - 3$                       **d.** $3x^2 + 6 + x^3 + 2x$

**e.** $4x^2 + 20x + 25$                  **f.** $b^2 + 100$

**EXAMPLE 2**    Factor each completely.

**a.** $27a^3 - b^3$       **b.** $3n^2m^4 - 48m^6$      **c.** $2x^2 - 12x + 18 - 2z^2$

**d.** $8x^4y^2 + 125xy^2$     **e.** $(x - 5)^2 - 49y^2$

*Solution*

**a.** This binomial is the difference of two cubes.

$$27a^3 - b^3 = (3a)^3 - b^3$$
$$= (3a - b)[(3a)^2 + (3a)(b) + b^2]$$
$$= (3a - b)(9a^2 + 3ab + b^2)$$

**b.** $3n^2m^4 - 48m^6 = 3m^4(n^2 - 16m^2)$      Factor out the GCF, $3m^4$.

$$= 3m^4(n + 4m)(n - 4m) \quad \text{Factor the difference of squares.}$$

**c.** $2x^2 - 12x + 18 - 2z^2 = 2(x^2 - 6x + 9 - z^2)$      The GCF is 2.

$$= 2[(x^2 - 6x + 9) - z^2] \quad \text{Group the first three terms together.}$$
$$= 2[(x - 3)^2 - z^2] \quad \text{Factor the perfect square trinomial.}$$
$$= 2[(x - 3) + z][(x - 3) - z] \quad \text{Factor the difference of squares.}$$
$$= 2(x - 3 + z)(x - 3 - z)$$

**d.** $8x^4y^2 + 125xy^2 = xy^2(8x^3 + 125)$         The GCF is $xy^2$.

$$= xy^2[(2x)^3 + 5^3]$$

$$= xy^2(2x + 5)[(2x)^2 - (2x)(5) + 5^2]$$    Factor the sum of cubes.

$$= xy^2(2x + 5)(4x^2 - 10x + 25)$$

**e.** This binomial is the difference of squares.

$$(x - 5)^2 - 49y^2 = (x - 5)^2 - (7y)^2$$

$$= [(x - 5) + 7y][(x - 5) - 7y]$$

$$= (x - 5 + 7y)(x - 5 - 7y)$$      □

**PRACTICE**
**2**    Factor each polynomial completely.

**a.** $64x^3 + y^3$

**b.** $7x^2y^2 - 63y^4$

**c.** $3x^2 + 12x + 12 - 3b^2$

**d.** $x^5y^4 + 27x^2y$

**e.** $(x + 7)^2 - 81y^2$

---

*Factor completely.*

**9.** $x^2 - 8x + 16 - y^2$

**10.** $12x^2 - 22x - 20$

**11.** $x^4 - x$

**12.** $(2x + 1)^2 - 3(2x + 1) + 2$

**13.** $14x^2y - 2xy$

**14.** $24ab^2 - 6ab$

**15.** $4x^2 - 16$

**16.** $9x^2 - 81$

**17.** $3x^2 - 8x - 11$

**18.** $5x^2 - 2x - 3$

**19.** $4x^2 + 8x - 12$

**20.** $6x^2 - 6x - 12$

**21.** $4x^2 + 36x + 81$

**22.** $25x^2 + 40x + 16$

**23.** $8x^3 + 125y^3$

**24.** $27x^3 - 64y^3$

**25.** $64x^2y^3 - 8x^2$

**26.** $27x^5y^4 - 216x^2y$

**27.** $(x + 5)^3 + y^3$

**28.** $(y - 1)^3 + 27x^3$

**29.** $(5a - 3)^2 - 6(5a - 3) + 9$

**30.** $(4r + 1)^2 + 8(4r + 1) + 16$

**31.** $7x^2 - 63x$

**32.** $20x^2 + 23x + 6$

**33.** $ab - 6a + 7b - 42$

**34.** $20x^2 - 220x + 600$

**35.** $x^4 - 1$

**36.** $15x^2 - 20x$

**37.** $10x^2 - 7x - 33$

**38.** $45m^3n^3 - 27m^2n^2$

**39.** $5a^3b^3 - 50a^3b$

**40.** $x^4 + x$

**41.** $16x^2 + 25$

**42.** $20x^3 + 20y^3$

**43.** $10x^3 - 210x^2 + 1100x$

**44.** $9y^2 - 42y + 49$

**45.** $64a^3b^4 - 27a^3b$

**46.** $y^4 - 16$

**47.** $2x^3 - 54$

**48.** $2sr + 10s - r - 5$

**49.** $3y^5 - 5y^4 + 6y - 10$

**50.** $64a^2 + b^2$

**51.** $100z^3 + 100$

**52.** $250x^4 - 16x$

**53.** $4b^2 - 36b + 81$

**54.** $2a^5 - a^4 + 6a - 3$

**55.** $(y - 6)^2 + 3(y - 6) + 2$

**56.** $(c + 2)^2 - 6(c + 2) + 5$

△ **57.** Express the area of the shaded region as a polynomial. Factor the polynomial completely.

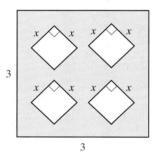

### OBJECTIVE

1 Perform operations on rational expressions and solve equations containing rational expressions.

## C.5 RATIONAL EXPRESSIONS

**OBJECTIVE 1 ▶ Performing operations on rational expressions and solving equations containing rational expressions.** It is very important that you understand the difference between an expression and an equation containing rational expressions. An equation contains an equal sign; an expression does not.

| **Expression to be Simplified** | **Equation to be Solved** |
|---|---|
| $$\frac{x}{2} + \frac{x}{6}$$ | $$\frac{x}{2} + \frac{x}{6} = \frac{2}{3}$$ |

Write both rational expressions with the LCD, 6, as the denominator.

$$\frac{x}{2} + \frac{x}{6} = \frac{x \cdot 3}{2 \cdot 3} + \frac{x}{6}$$

$$= \frac{3x}{6} + \frac{x}{6}$$

$$= \frac{4x}{6} = \frac{2x}{3}$$

Multiply both sides by the LCD, 6.

$$6\left(\frac{x}{2} + \frac{x}{6}\right) = 6\left(\frac{2}{3}\right)$$

$$3x + x = 4$$

$$4x = 4$$

$$x = 1$$

Check to see that the solution set is 1.

▶ **Helpful Hint**

Remember: Equations can be cleared of fractions; expressions cannot.

**EXAMPLE 1**  Multiply. $\dfrac{x^3 - 1}{-3x + 3} \cdot \dfrac{15x^2}{x^2 + x + 1}$

*Solution*

$$\dfrac{x^3 - 1}{-3x + 3} \cdot \dfrac{15x^2}{x^2 + x + 1} = \dfrac{(x - 1)(x^2 + x + 1)}{-3(x - 1)} \cdot \dfrac{15x^2}{x^2 + x + 1} \qquad \text{Factor.}$$

$$= \dfrac{(x - 1)(x^2 + x + 1) \cdot 3 \cdot 5x^2}{-1 \cdot 3(x - 1)(x^2 + x + 1)} \qquad \text{Factor.}$$

$$= \dfrac{5x^2}{-1} = -5x^2 \qquad \text{Simplest form} \qquad \square$$

**PRACTICE**
**1**  Multiply.

**a.** $\dfrac{2 + 5n}{3n} \cdot \dfrac{6n + 3}{5n^2 - 3n - 2}$ 
**b.** $\dfrac{x^3 - 8}{-6x + 12} \cdot \dfrac{6x^2}{x^2 + 2x + 4}$

---

**EXAMPLE 2**  Divide. $\dfrac{8m^2}{3m^2 - 12} \div \dfrac{40}{2 - m}$

*Solution*

$$\dfrac{8m^2}{3m^2 - 12} \div \dfrac{40}{2 - m} = \dfrac{8m^2}{3m^2 - 12} \cdot \dfrac{2 - m}{40} \qquad \begin{array}{l}\text{Multiply by the reciprocal of}\\ \text{the divisor.}\end{array}$$

$$= \dfrac{8m^2(2 - m)}{3(m + 2)(m - 2) \cdot 40} \qquad \text{Factor and multiply.}$$

$$= \dfrac{8\,m^2 \cdot -1\,(m - 2)}{3(m + 2)\,(m - 2) \cdot 8 \cdot 5} \qquad \text{Write } (2 - m) \text{ as } -1(m - 2).$$

$$= -\dfrac{m^2}{15(m + 2)} \qquad \text{Simplify.} \qquad \square$$

**PRACTICE**
**2**  Divide.

**a.** $\dfrac{6y^3}{3y^2 - 27} \div \dfrac{42}{3 - y}$ 
**b.** $\dfrac{10x^2 + 23x - 5}{5x^2 - 51x + 10} \div \dfrac{2x^2 + 9x + 10}{7x^2 - 68x - 20}$

---

**EXAMPLE 3**  Perform the indicated operation.

$$\dfrac{3}{x + 2} + \dfrac{2x}{x - 2}$$

*Solution*  The LCD is the product of the two denominators: $(x + 2)(x - 2)$.

$$\dfrac{3}{x + 2} + \dfrac{2x}{x - 2} = \dfrac{3 \cdot (x - 2)}{(x + 2) \cdot (x - 2)} + \dfrac{2x \cdot (x + 2)}{(x - 2) \cdot (x + 2)} \qquad \begin{array}{l}\text{Write equivalent rational}\\ \text{expressions.}\end{array}$$

$$= \dfrac{3x - 6}{(x + 2)(x - 2)} + \dfrac{2x^2 + 4x}{(x + 2)(x - 2)} \qquad \text{Multiply in the numerators.}$$

$$= \dfrac{3x - 6 + 2x^2 + 4x}{(x + 2)(x - 2)} \qquad \text{Add the numerators.}$$

$$= \dfrac{2x^2 + 7x - 6}{(x + 2)(x - 2)} \qquad \text{Simplify the numerator.} \qquad \square$$

PRACTICE
**3** Perform the indicated operation.

**a.** $\dfrac{4}{p^3q} + \dfrac{3}{5p^4q}$  **b.** $\dfrac{4}{y+3} + \dfrac{5y}{y-3}$  **c.** $\dfrac{3z-18}{z-5} - \dfrac{3}{5-z}$

**EXAMPLE 4**  Solve: $\dfrac{2x}{x-3} + \dfrac{6-2x}{x^2-9} = \dfrac{x}{x+3}$.

_Solution_  We factor the second denominator to find that the LCD is $(x+3)(x-3)$. We multiply both sides of the equation by $(x+3)(x-3)$. By the distributive property, this is the same as multiplying each term by $(x+3)(x-3)$.

$$\dfrac{2x}{x-3} + \dfrac{6-2x}{x^2-9} = \dfrac{x}{x+3}$$

$$(x+3)(x-3) \cdot \dfrac{2x}{x-3} + (x+3)(x-3) \cdot \dfrac{6-2x}{(x+3)(x-3)}$$

$$= (x+3)(x-3)\left(\dfrac{x}{x+3}\right)$$

$$2x(x+3) + (6-2x) = x(x-3) \quad \text{Simplify.}$$

$$2x^2 + 6x + 6 - 2x = x^2 - 3x \quad \text{Use the distributive property.}$$

Next we solve this quadratic equation by the factoring method. To do so, we first write the equation so that one side is 0.

$$x^2 + 7x + 6 = 0$$

$$(x+6)(x+1) = 0 \quad \text{Factor.}$$

$$x = -6 \text{ or } x = -1 \quad \text{Set each factor equal to 0.}$$

Neither $-6$ nor $-1$ makes any denominator 0 so they are both solutions. The solutions are $-6$ and $-1$. $\qquad\square$

PRACTICE
**4**  Solve: $\dfrac{2}{x-2} - \dfrac{5+2x}{x^2-4} = \dfrac{x}{x+2}$.

## C.5 | EXERCISE SET

**MyMathLab** Powered by CourseCompass™ and MathXL®

MathXL PRACTICE  WATCH  DOWNLOAD  READ  REVIEW

_Perform each indicated operation and simplify, or solve the equation for the variable._

**1.** $\dfrac{x}{2} = \dfrac{1}{8} + \dfrac{x}{4}$

**2.** $\dfrac{x}{4} = \dfrac{3}{2} + \dfrac{x}{10}$

**3.** $\dfrac{1}{8} + \dfrac{x}{4}$

**4.** $\dfrac{3}{2} + \dfrac{x}{10}$

**5.** $\dfrac{4}{x+2} - \dfrac{2}{x-1}$

**6.** $\dfrac{5}{x-2} - \dfrac{10}{x+4}$

**7.** $\dfrac{4}{x+2} = \dfrac{2}{x-1}$

**8.** $\dfrac{5}{x-2} = \dfrac{10}{x+4}$

**9.** $\dfrac{2}{x^2-4} = \dfrac{1}{x+2} - \dfrac{3}{x-2}$

> ▶ **Helpful Hint**
> Remember: Equations can be cleared of fractions; expressions cannot.

10. $\dfrac{3}{x^2 - 25} = \dfrac{1}{x + 5} + \dfrac{2}{x - 5}$

11. $\dfrac{5}{x^2 - 3x} + \dfrac{4}{2x - 6}$

12. $\dfrac{5}{x^2 - 3x} \div \dfrac{4}{2x - 6}$

13. $\dfrac{x - 1}{x + 1} + \dfrac{x + 7}{x - 1} = \dfrac{4}{x^2 - 1}$

14. $\left(1 - \dfrac{y}{x}\right) \div \left(1 - \dfrac{x}{y}\right)$

15. $\dfrac{a^2 - 9}{a - 6} \cdot \dfrac{a^2 - 5a - 6}{a^2 - a - 6}$

16. $\dfrac{2}{a - 6} + \dfrac{3a}{a^2 - 5a - 6} - \dfrac{a}{5a + 5}$

17. $\dfrac{2x + 3}{3x - 2} = \dfrac{4x + 1}{6x + 1}$

18. $\dfrac{5x - 3}{2x} = \dfrac{10x + 3}{4x + 1}$

19. $\dfrac{a}{9a^2 - 1} + \dfrac{2}{6a - 2}$

20. $\dfrac{3}{4a - 8} - \dfrac{a + 2}{a^2 - 2a}$

21. $-\dfrac{3}{x^2} - \dfrac{1}{x} + 2 = 0$

22. $\dfrac{x}{2x + 6} + \dfrac{5}{x^2 - 9}$

23. $\dfrac{x - 8}{x^2 - x - 2} + \dfrac{2}{x - 2}$

24. $\dfrac{x - 8}{x^2 - x - 2} + \dfrac{2}{x - 2} = \dfrac{3}{x + 1}$

25. $\dfrac{3}{a} - 5 = \dfrac{7}{a} - 1$

26. $\dfrac{7}{3z - 9} + \dfrac{5}{z}$

*Use* $\dfrac{x}{5} - \dfrac{x}{4} = \dfrac{1}{10}$ *and* $\dfrac{x}{5} - \dfrac{x}{4} + \dfrac{1}{10}$ *for Exercises 27 and 28.*

27. **a.** Which one above is an expression?
   **b.** Describe the first step to simplify this expression.
   **c.** Simplify the expression.

28. **a.** Which one above is an equation?
   **b.** Describe the first step to solve this equation.
   **c.** Solve the equation.

*For each exercise, choose the correct statement.\* Each figure represents a real number and no denominators are 0.*

29. **a.** $\dfrac{\triangle + \square}{\triangle} = \square$    **b.** $\dfrac{\triangle + \square}{\triangle} = 1 + \dfrac{\square}{\triangle}$
   **c.** $\dfrac{\triangle + \square}{\triangle} = \dfrac{\square}{\triangle}$    **d.** $\dfrac{\triangle + \square}{\triangle} = 1 + \square$
   **e.** $\dfrac{\triangle + \square}{\triangle - \square} = -1$

30. **a.** $\dfrac{\triangle}{\square} + \dfrac{\square}{\triangle} = \dfrac{\triangle + \square}{\square + \triangle} = 1$    **b.** $\dfrac{\triangle}{\square} + \dfrac{\square}{\triangle} = \dfrac{\triangle + \square}{\triangle\square}$
   **c.** $\dfrac{\triangle}{\square} + \dfrac{\square}{\triangle} = \triangle\triangle + \square\square$    **d.** $\dfrac{\triangle}{\square} + \dfrac{\square}{\triangle} = \dfrac{\triangle\triangle + \square\square}{\square\triangle}$
   **e.** $\dfrac{\triangle}{\square} + \dfrac{\square}{\triangle} = \dfrac{\triangle\square}{\square\triangle} = 1$

31. **a.** $\dfrac{\triangle}{\square} \cdot \dfrac{\bigcirc}{\square} = \dfrac{\triangle\bigcirc}{\square}$    **b.** $\dfrac{\triangle}{\square} \cdot \dfrac{\bigcirc}{\square} = \triangle\bigcirc$
   **c.** $\dfrac{\triangle}{\square} \cdot \dfrac{\bigcirc}{\square} = \dfrac{\triangle + \bigcirc}{\square + \square}$    **d.** $\dfrac{\triangle}{\square} \cdot \dfrac{\bigcirc}{\square} = \dfrac{\triangle\bigcirc}{\square\square}$

32. **a.** $\dfrac{\triangle}{\square} \div \dfrac{\bigcirc}{\triangle} = \dfrac{\triangle\triangle}{\square\bigcirc}$    **b.** $\dfrac{\triangle}{\square} \div \dfrac{\bigcirc}{\triangle} = \dfrac{\bigcirc\square}{\triangle\triangle}$
   **c.** $\dfrac{\triangle}{\square} \div \dfrac{\bigcirc}{\triangle} = \dfrac{\bigcirc}{\square}$    **d.** $\dfrac{\triangle}{\square} \div \dfrac{\bigcirc}{\triangle} = \dfrac{\triangle + \triangle}{\square + \bigcirc}$

33. **a.** $\dfrac{\dfrac{\triangle + \square}{\bigcirc}}{\dfrac{\triangle}{\bigcirc}} = \square$    **b.** $\dfrac{\dfrac{\triangle + \square}{\bigcirc}}{\dfrac{\triangle}{\bigcirc}} = \dfrac{\triangle\triangle + \triangle\square}{\bigcirc\bigcirc}$
   **c.** $\dfrac{\dfrac{\triangle + \square}{\bigcirc}}{\dfrac{\triangle}{\bigcirc}} = 1 + \square$    **d.** $\dfrac{\dfrac{\triangle + \square}{\bigcirc}}{\dfrac{\triangle}{\bigcirc}} = \dfrac{\triangle + \square}{\triangle}$

*\*My thanks to Kelly Champagne for permission to use her Exercises for 29 through 33.*

# Appendix D

# An Introduction to
# Using a Graphing Utility

### THE VIEWING WINDOW AND INTERPRETING WINDOW SETTINGS

In this appendix, we will use the term **graphing utility** to mean a graphing calculator or a computer software graphing package. All graphing utilities graph equations by plotting points on a screen. While plotting several points can be slow and sometimes tedious for us, a graphing utility can quickly and accurately plot hundreds of points. How does a graphing utility show plotted points? A computer or calculator screen is made up of a grid of small rectangular areas called **pixels.** If a pixel contains a point to be plotted, the pixel is turned "on"; otherwise, the pixel remains "off." The graph of an equation is then a collection of pixels turned "on." The graph of $y = 3x + 1$ from a graphing calculator is shown in Figure A-1. Notice the irregular shape of the line caused by the rectangular pixels.

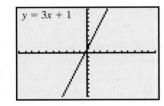

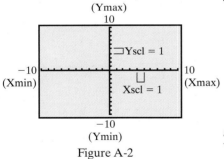

Figure A-1

The portion of the coordinate plane shown on the screen in Figure A-1 is called the **viewing window** or the **viewing rectangle.** Notice the x-axis and the y-axis on the graph. While tick marks are shown on the axes, they are not labeled. This means that from this screen alone, we do not know how many units each tick mark represents. To see what each tick mark represents and the minimum and maximum values on the axes, check the window setting of the graphing utility. It defines the viewing window. The window of the graph of $y = 3x + 1$ shown in Figure A-1 has the following setting (Figure A-2):

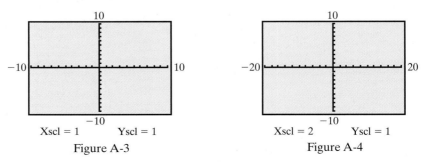

Figure A-2

| | |
|---|---|
| $\text{Xmin} = -10$ | The minimum x-value is $-10$. |
| $\text{Xmax} = 10$ | The maximum x-value is 10. |
| $\text{Xscl} = 1$ | The x-axis scale is 1 unit per tick mark. |
| $\text{Ymin} = -10$ | The minimum y-value is $-10$. |
| $\text{Ymax} = 10$ | The maximum y-value is 10. |
| $\text{Yscl} = 1$ | The y-axis scale is 1 unit per tick mark. |

By knowing the scale, we can find the minimum and the maximum values on the axes simply by counting tick marks. For example, if both the Xscl (x-axis scale) and the Yscl are 1 unit per tick mark on the graph in Figure A-3, we can count the tick marks and find that the minimum x-value is $-10$ and the maximum x-value is 10. Also, the minimum y-value is $-10$ and the maximum y-value is 10. If the Xscl (x-axis scale) changes to 2 units per tick mark (shown in Figure A-4), by counting tick marks, we see that the minimum x-value is now $-20$ and the maximum x-value is now 20.

Figure A-3

Xscl = 1    Yscl = 1

Figure A-4

Xscl = 2    Yscl = 1

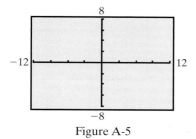

Figure A-5

It is also true that if we know the Xmin and the Xmax values, we can calculate the Xscl by the displayed axes. For example, the Xscl of the graph in Figure A-5 must be 3 units per tick mark for the maximum and minimum $x$-values to be as shown. Also, the Yscl of that graph must be 2 units per tick mark for the maximum and minimum $y$-values to be as shown.

We will call the viewing window in Figure A-3 a *standard* viewing window or rectangle. Although a standard viewing window is sufficient for much of this text, special care must be taken to ensure that all key features of a graph are shown. Figures A-6, A-7, and A-8 show the graph of $y = x^2 + 11x - 1$ on three different viewing windows. Note that certain viewing windows for this equation are misleading.

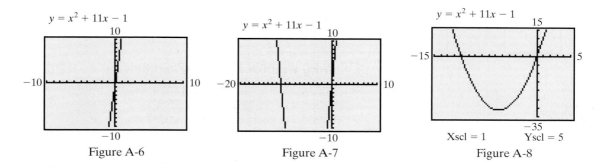

Figure A-6         Figure A-7         Figure A-8

How do we ensure that all distinguishing features of the graph of an equation are shown? It helps to know about the equation that is being graphed. For example, the equation $y = x^2 + 11x - 1$ is not a linear equation and its graph is not a line. This equation is a quadratic equation and, therefore, its graph is a parabola. By knowing this information, we know that the graph shown in Figure A-6, although correct, is misleading. Of the three viewing rectangles shown, the graph in Figure A-8 is best because it shows more of the distinguishing features of the parabola. Properties of equations needed for graphing will be studied in this text.

## VIEWING WINDOW AND INTERPRETING WINDOW SETTINGS EXERCISE SET

*In Exercises 1–4, determine whether all ordered pairs listed will lie within a standard viewing rectangle.*

**1.** $(-9, 0), (5, 8), (1, -8)$
**2.** $(4, 7), (0, 0), (-8, 9)$
**3.** $(-11, 0), (2, 2), (7, -5)$
**4.** $(3, 5), (-3, -5), (15, 0)$

*In Exercises 5–10, choose an Xmin, Xmax, Ymin, and Ymax so that all ordered pairs listed will lie within the viewing rectangle.*

**5.** $(-90, 0), (55, 80), (0, -80)$
**6.** $(4, 70), (20, 20), (-18, 90)$
**7.** $(-11, 0), (2, 2), (7, -5)$
**8.** $(3, 5), (-3, -5), (15, 0)$
**9.** $(200, 200), (50, -50), (70, -50)$
**10.** $(40, 800), (-30, 500), (15, 0)$

*Write the window setting for each viewing window shown. Use the following format:*

| | |
|---|---|
| Xmin = | Ymin = |
| Xmax = | Ymax = |
| Xscl = | Yscl = |

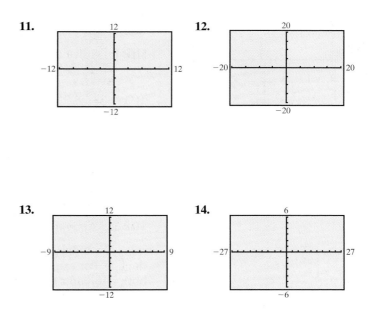

**15.**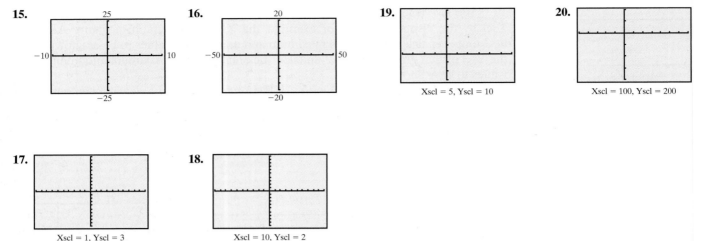

**16.**

**19.**

Xscl = 5, Yscl = 10

**20.**

Xscl = 100, Yscl = 200

**17.**

Xscl = 1, Yscl = 3

**18.**

Xscl = 10, Yscl = 2

## GRAPHING EQUATIONS AND SQUARE VIEWING WINDOW

In general, the following steps may be used to graph an equation on a standard viewing window.

---

**Graphing an Equation in *X* and *Y* with a Graphing Utility on a Standard Viewing Window**

**Step 1:** Solve the equation for *y*.

**Step 2:** Use your graphing utility and enter the equation in the form
Y = *expression involving x*.

**Step 3:** Activate the graphing utility.

---

Special care must be taken when entering the *expression involving x* in Step 2. You must be sure that the graphing utility you are using interprets the expression as you want it to. For example, let's graph $3y = 4x$. To do so,

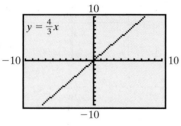

$y = \frac{4}{3}x$

Figure A-9

**STEP 1:** Solve the equation for *y*.

$$3y = 4x$$

$$\frac{3y}{3} = \frac{4x}{3}$$

$$y = \frac{4}{3}x$$

**STEP 2:** Using your graphing utility, enter the expression $\frac{4}{3}x$ after the Y = prompt.
In order for your graphing utility to correctly interpret the expression, you may need to enter $(4/3)x$ or $(4 \div 3)x$.

**STEP 3:** Activate the graphing utility. The graph should appear as in Figure A-9.

Distinguishing features of the graph of a line include showing all the intercepts of the line. For example, the window of the graph of the line in Figure A-10 does not show both intercepts of the line, but the window of the graph of the same line in Figure A-11 does show both intercepts. Notice the notation below each graph. This is a shorthand notation of the range setting of the graph. This notation means [Xmin, Xmax] by [Ymin, Ymax].

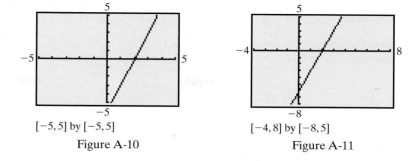

[-5, 5] by [-5, 5]
Figure A-10

[-4, 8] by [-8, 5]
Figure A-11

On a standard viewing window, the tick marks on the *y*-axis are closer together than the tick marks on the *x*-axis. This happens because the viewing window is a rectangle, and so 10 equally spaced tick marks on the positive *y*-axis will be closer together than 10 equally spaced tick marks on the positive *x*-axis. This causes the appearance of graphs to be distorted.

For example, notice the different appearances of the same line graphed using different viewing windows. The line in Figure A-12 is distorted because the tick marks along the *x*-axis are farther apart than the tick marks along the *y*-axis. The graph of the same line in Figure A-13 is not distorted because the viewing rectangle has been selected so that there is equal spacing between tick marks on both axes.

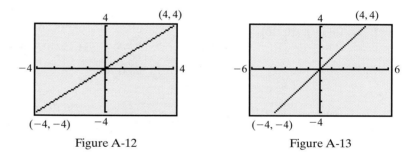

Figure A-12

Figure A-13

We say that the line in Figure A-13 is graphed on a *square* setting. Some graphing utilities have a built-in program that, if activated, will automatically provide a square setting. A square setting is especially helpful when we are graphing perpendicular lines, circles, or when a true geometric perspective is desired. Some examples of square screens are shown in Figures A-14 and A-15.

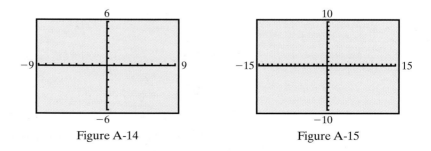

Figure A-14

Figure A-15

Other features of a graphing utility such as Trace, Zoom, Intersect, and Table are discussed in appropriate Graphing Calculator Explorations in this text.

# GRAPHING EQUATIONS AND SQUARE VIEWING WINDOW EXERCISE SET

*Graph each linear equation in two variables, using the two different range settings given. Determine which setting shows all intercepts of a line.*

**1.** $y = 2x + 12$
Setting A: $[-10, 10]$ by $[-10, 10]$
Setting B: $[-10, 10]$ by $[-10, 15]$

**2.** $y = -3x + 25$
Setting A: $[-5, 5]$ by $[-30, 10]$
Setting B: $[-10, 10]$ by $[-10, 30]$

**3.** $y = -x - 41$
Setting A: $[-50, 10]$ by $[-10, 10]$
Setting B: $[-50, 10]$ by $[-50, 15]$

**4.** $y = 6x - 18$
Setting A: $[-10, 10]$ by $[-20, 10]$
Setting B: $[-10, 10]$ by $[-10, 10]$

**5.** $y = \dfrac{1}{2}x - 15$
Setting B: $[-10, 10]$ by $[-20, 10]$
Setting B: $[-10, 35]$ by $[-20, 15]$

**6.** $y = -\dfrac{2}{3}x - \dfrac{29}{3}$
Setting A: $[-10, 10]$ by $[-10, 10]$
Setting B: $[-15, 5]$ by $[-15, 5]$

*The graph of each equation is a line. Use a graphing utility and a standard viewing window to graph each equation.*

**7.** $3x = 5y$     **8.** $7y = -3x$     **9.** $9x - 5y = 30$
**10.** $4x + 6y = 20$     **11.** $y = -7$     **12.** $y = 2$
**13.** $x + 10y = -5$     **14.** $x - 5y = 9$

*Graph the following equations using the square setting given. Some keystrokes that may be helpful are given.*

**15.** $y = \sqrt{x}$    $[-12, 12]$ by $[-8, 8]$
Suggested keystrokes: $\sqrt{\phantom{x}} x$
**16.** $y = \sqrt{2x}$    $[-12, 12]$ by $[-8, 8]$
Suggested keystrokes: $\sqrt{\phantom{x}}(2x)$
**17.** $y = x^2 + 2x + 1$    $[-15, 15]$ by $[-10, 10]$
Suggested keystrokes: $x^\wedge 2 + 2x + 1$
**18.** $y = x^2 - 5$    $[-15, 15]$ by $[-10, 10]$
Suggested keystrokes: $x^\wedge 2 - 5$
**19.** $y = |x|$    $[-9, 9]$ by $[-6, 6]$
Suggested keystrokes: $ABS(x)$
**20.** $y = |x - 2|$    $[-9, 9]$ by $[-6, 6]$
Suggested keystrokes: $ABS(x - 2)$

*Graph each line. Use a standard viewing window; then, if necessary, change the viewing window so that all intercepts of each line show.*

**21.** $x + 2y = 30$         **22.** $1.5x - 3.7y = 40.3$

# Appendix E

# Solving Systems of Equations by Matrices

By now, you may have noticed that the solution of a system of equations depends on the coefficients of the equations in the system and not on the variables. In this section, we introduce solving a system of equations by a **matrix.**

**OBJECTIVE 1 ▶ Using matrices to solve a system of two equations.**  A matrix (plural: **matrices**) is a rectangular array of numbers. The following are examples of matrices.

$$\begin{bmatrix} 1 & 0 \\ 0 & 1 \end{bmatrix} \quad \begin{bmatrix} 2 & 1 & 3 & -1 \\ 0 & -1 & 4 & 5 \\ -6 & 2 & 1 & 0 \end{bmatrix} \quad \begin{bmatrix} a & b & c \\ d & e & f \end{bmatrix}$$

The numbers aligned horizontally in a matrix are in the same **row.** The numbers aligned vertically are in the same **column.**

$$\begin{array}{l} \text{row 1} \rightarrow \\ \text{row 2} \rightarrow \end{array} \begin{bmatrix} 2 & 1 & 0 \\ -1 & 6 & 2 \end{bmatrix}$$

This matrix has 2 rows and 3 columns. It is called a 2 × 3 (read "two by three") matrix.

column 1
column 2
column 3

To see the relationship between systems of equations and matrices, study the example below.

*System of Equations (in standard form)*

$$\begin{cases} 2x - 3y = 6 & \text{Equation 1} \\ x + y = 0 & \text{Equation 2} \end{cases}$$

*Corresponding Matrix*

$$\begin{bmatrix} 2 & -3 & \vdots & 6 \\ 1 & 1 & \vdots & 0 \end{bmatrix} \begin{array}{l} \text{Row 1} \\ \text{Row 2} \end{array}$$

Notice that the rows of the matrix correspond to the equations in the system. The coefficients of each variable are placed to the left of a vertical dashed line. The constants are placed to the right. Each of these numbers in the matrix is called an **element.**

The method of solving systems by matrices is to write this matrix as an equivalent matrix from which we easily identify the solution. Two matrices are equivalent if they represent systems that have the same solution set. The following **row operations** can be performed on matrices, and the result is an equivalent matrix.

---

**Elementary Row Operations**

1.  Any two rows in a matrix may be interchanged.

2.  The elements of any row may be multiplied (or divided) by the same nonzero number.

3.  The elements of any row may be multiplied (or divided) by a nonzero number and added to their corresponding elements in any other row.

---

▶ **Helpful Hint**

Notice that these *row* operations are the same operations that we can perform on *equations* in a system.

To solve a system of two equations in $x$ and $y$ by matrices, write the corresponding matrix associated with the system. Then use elementary row operations to write equivalent matrices until you have a matrix of the form

$$\begin{bmatrix} 1 & a & b \\ 0 & 1 & c \end{bmatrix},$$

where $a$, $b$, and $c$ are constants. Why? If a matrix associated with a system of equations is in this form, we can easily solve for $x$ and $y$. For example,

| *Matrix* | | *System of Equations* | |
|---|---|---|---|
| $\begin{bmatrix} 1 & 2 & -3 \\ 0 & 1 & 5 \end{bmatrix}$ | corresponds to | $\begin{cases} 1x + 2y = -3 \\ 0x + 1y = 5 \end{cases}$ or | $\begin{cases} x + 2y = -3 \\ y = 5 \end{cases}$ |

In the second equation, we have $y = 5$. Substituting this in the first equation, we have $x + 2(5) = -3$ or $x = -13$. The solution of the system is the ordered pair $(-13, 5)$.

**EXAMPLE 1** Use matrices to solve the system.

$$\begin{cases} x + 3y = 5 \\ 2x - y = -4 \end{cases}$$

*Solution* The corresponding matrix is $\begin{bmatrix} 1 & 3 & 5 \\ 2 & -1 & -4 \end{bmatrix}$. We use elementary row operations to write an equivalent matrix that looks like $\begin{bmatrix} 1 & a & b \\ 0 & 1 & c \end{bmatrix}$.

For the matrix given, the element in the first row, first column is already 1, as desired. Next we write an equivalent matrix with a 0 below the 1. To do this, we multiply row 1 by $-2$ and add to row 2. *We will change only row 2.*

$$\begin{bmatrix} 1 & 3 & 5 \\ -2(1) + 2 & -2(3) + (-1) & -2(5) + (-4) \end{bmatrix} \text{ simplifies to } \begin{bmatrix} 1 & 3 & 5 \\ 0 & -7 & -14 \end{bmatrix}$$

↑   ↑    ↑    ↑    ↑    ↑

row 1   row 2   row 1   row 2   row 1   row 2
element element element element element element

Now we change the $-7$ to a 1 by use of an elementary row operation. We divide row 2 by $-7$, then

$$\begin{bmatrix} 1 & 3 & 5 \\ \dfrac{0}{-7} & \dfrac{-7}{-7} & \dfrac{-14}{-7} \end{bmatrix} \text{ simplifies to } \begin{bmatrix} 1 & 3 & 5 \\ 0 & 1 & 2 \end{bmatrix}$$

This last matrix corresponds to the system

$$\begin{cases} x + 3y = 5 \\ y = 2 \end{cases}$$

To find $x$, we let $y = 2$ in the first equation, $x + 3y = 5$.

$$x + 3y = 5 \quad \text{First equation}$$
$$x + 3(2) = 5 \quad \text{Let } y = 2.$$
$$x = -1$$

The ordered pair solution is $(-1, 2)$. Check to see that this ordered pair satisfies both equations. □

**PRACTICE**
**1** Use matrices to solve the system.

$$\begin{cases} x + 4y = -2 \\ 3x - y = 7 \end{cases}$$

**EXAMPLE 2** Use matrices to solve the system.

$$\begin{cases} 2x - y = 3 \\ 4x - 2y = 5 \end{cases}$$

**Solution** The corresponding matrix is $\begin{bmatrix} 2 & -1 & 3 \\ 4 & -2 & 5 \end{bmatrix}$. To get 1 in the row 1, column 1 position, we divide the elements of row 1 by 2.

$$\begin{bmatrix} \frac{2}{2} & -\frac{1}{2} & \frac{3}{2} \\ 4 & -2 & 5 \end{bmatrix} \quad \text{simplifies to} \quad \begin{bmatrix} 1 & -\frac{1}{2} & \frac{3}{2} \\ 4 & -2 & 5 \end{bmatrix}$$

To get 0 under the 1, we multiply the elements of row 1 by $-4$ and add the new elements to the elements of row 2.

$$\begin{bmatrix} 1 & -\frac{1}{2} & \frac{3}{2} \\ -4(1) + 4 & -4\left(-\frac{1}{2}\right) - 2 & -4\left(\frac{3}{2}\right) + 5 \end{bmatrix} \quad \text{simplifies to} \quad \begin{bmatrix} 1 & -\frac{1}{2} & \frac{3}{2} \\ 0 & 0 & -1 \end{bmatrix}$$

The corresponding system is $\begin{cases} x - \frac{1}{2}y = \frac{3}{2} \\ 0 = -1 \end{cases}$. The equation $0 = -1$ is false for all $y$ or $x$ values; hence the system is inconsistent and has no solution. □

**PRACTICE**
**2** Use matrices to solve the system.

$$\begin{cases} x - 3y = 3 \\ -2x + 6y = 4 \end{cases}$$

**Concept Check** ✓
Consider the system

$$\begin{cases} 2x - 3y = 8 \\ x + 5y = -3 \end{cases}$$

What is wrong with its corresponding matrix shown below?

$$\begin{bmatrix} 2 & -3 & 8 \\ 0 & 5 & -3 \end{bmatrix}$$

**OBJECTIVE 2 ▶ Using matrices to solve a system of three equations.** To solve a system of three equations in three variables using matrices, we will write the corresponding matrix in the form

$$\begin{bmatrix} 1 & a & b & d \\ 0 & 1 & c & e \\ 0 & 0 & 1 & f \end{bmatrix}$$

**Answer to Concept Check:**

matrix should be $\begin{bmatrix} 2 & -3 & 8 \\ 1 & 5 & -3 \end{bmatrix}$

**EXAMPLE 3**  Use matrices to solve the system.

$$\begin{cases} x + 2y + z = 2 \\ -2x - y + 2z = 5 \\ x + 3y - 2z = -8 \end{cases}$$

**Solution**  The corresponding matrix is $\begin{bmatrix} 1 & 2 & 1 & \vdots & 2 \\ -2 & -1 & 2 & \vdots & 5 \\ 1 & 3 & -2 & \vdots & -8 \end{bmatrix}$. Our goal is to write an equivalent matrix with 1's along the diagonal (see the numbers in red) and 0's below the 1's. The element in row 1, column 1 is already 1. Next we get 0's for each element in the rest of column 1. To do this, first we multiply the elements of row 1 by 2 and add the new elements to row 2. Also, we multiply the elements of row 1 by −1 and add the new elements to the elements of row 3. We *do not change row 1*. Then

$$\begin{bmatrix} 1 & 2 & 1 & \vdots & 2 \\ 2(1) - 2 & 2(2) - 1 & 2(1) + 2 & \vdots & 2(2) + 5 \\ -1(1) + 1 & -1(2) + 3 & -1(1) - 2 & \vdots & -1(2) - 8 \end{bmatrix} \text{ simplifies to } \begin{bmatrix} 1 & 2 & 1 & \vdots & 2 \\ 0 & 3 & 4 & \vdots & 9 \\ 0 & 1 & -3 & \vdots & -10 \end{bmatrix}$$

We continue down the diagonal and use elementary row operations to get 1 where the element 3 is now. To do this, we interchange rows 2 and 3.

$$\begin{bmatrix} 1 & 2 & 1 & \vdots & 2 \\ 0 & 3 & 4 & \vdots & 9 \\ 0 & 1 & -3 & \vdots & -10 \end{bmatrix} \text{ is equivalent to } \begin{bmatrix} 1 & 2 & 1 & \vdots & 2 \\ 0 & 1 & -3 & \vdots & -10 \\ 0 & 3 & 4 & \vdots & 9 \end{bmatrix}$$

Next we want the new row 3, column 2 element to be 0. We multiply the elements of row 2 by −3 and add the result to the elements of row 3.

$$\begin{bmatrix} 1 & 2 & 1 & \vdots & 2 \\ 0 & 1 & -3 & \vdots & -10 \\ -3(0) + 0 & -3(1) + 3 & -3(-3) + 4 & \vdots & -3(-10) + 9 \end{bmatrix} \text{ simplifies to }$$

$$\begin{bmatrix} 1 & 2 & 1 & \vdots & 2 \\ 0 & 1 & -3 & \vdots & -10 \\ 0 & 0 & 13 & \vdots & 39 \end{bmatrix}$$

Finally, we divide the elements of row 3 by 13 so that the final diagonal element is 1.

$$\begin{bmatrix} 1 & 2 & 1 & \vdots & 2 \\ 0 & 1 & -3 & \vdots & -10 \\ \frac{0}{13} & \frac{0}{13} & \frac{13}{13} & \vdots & \frac{39}{13} \end{bmatrix} \text{ simplifies to } \begin{bmatrix} 1 & 2 & 1 & \vdots & 2 \\ 0 & 1 & -3 & \vdots & -10 \\ 0 & 0 & 1 & \vdots & 3 \end{bmatrix}$$

This matrix corresponds to the system

$$\begin{cases} x + 2y + z = 2 \\ y - 3z = -10 \\ z = 3 \end{cases}$$

We identify the $z$-coordinate of the solution as 3. Next we replace $z$ with 3 in the second equation and solve for $y$.

$$y - 3z = -10 \quad \text{Second equation}$$
$$y - 3(3) = -10 \quad \text{Let } z = 3.$$
$$y = -1$$

To find $x$, we let $z = 3$ and $y = -1$ in the first equation.

$$x + 2y + z = 2 \quad \text{First equation}$$
$$x + 2(-1) + 3 = 2 \quad \text{Let } z = 3 \text{ and } y = -1.$$
$$x = 1$$

The ordered triple solution is $(1, -1, 3)$. Check to see that it satisfies all three equations in the original system.  $\square$

**PRACTICE**
**3**  Use matrices to solve the system.

$$\begin{cases} x + 3y - z = 0 \\ 2x + y + 3z = 5 \\ -x - 2y + 4z = 7 \end{cases}$$

## APPENDIX E | EXERCISE SET

*Powered by CourseCompass™ and MathXL®*  **MyMathLab**

Math XL  PRACTICE   WATCH   DOWNLOAD   READ   REVIEW

*Solve each system of linear equations using matrices. See Example 1.*

**1.** $\begin{cases} x + y = 1 \\ x - 2y = 4 \end{cases}$

**2.** $\begin{cases} 2x - y = 8 \\ x + 3y = 11 \end{cases}$

**3.** $\begin{cases} x + 3y = 2 \\ x + 2y = 0 \end{cases}$

**4.** $\begin{cases} 4x - y = 5 \\ 3x + 3y = 0 \end{cases}$

*Solve each system of linear equations using matrices. See Example 2.*

**5.** $\begin{cases} x - 2y = 4 \\ 2x - 4y = 4 \end{cases}$

**6.** $\begin{cases} -x + 3y = 6 \\ 3x - 9y = 9 \end{cases}$

**7.** $\begin{cases} 3x - 3y = 9 \\ 2x - 2y = 6 \end{cases}$

**8.** $\begin{cases} 9x - 3y = 6 \\ -18x + 6y = -12 \end{cases}$

*Solve each system of linear equations using matrices. See Example 3.*

**9.** $\begin{cases} x + y = 3 \\ 2y = 10 \\ 3x + 2y - 4z = 12 \end{cases}$

**10.** $\begin{cases} 5x = 5 \\ 2x + y = 4 \\ 3x + y - 5z = -15 \end{cases}$

**11.** $\begin{cases} 2y - z = -7 \\ x + 4y + z = -4 \\ 5x - y + 2z = 13 \end{cases}$

**12.** $\begin{cases} 4y + 3z = -2 \\ 5x - 4y = 1 \\ -5x + 4y + z = -3 \end{cases}$

## MIXED PRACTICE

*Solve each system of linear equations using matrices. See Examples 1 through 3.*

**13.** $\begin{cases} x - 4 = 0 \\ x + y = 1 \end{cases}$

**14.** $\begin{cases} 3y = 6 \\ x + y = 7 \end{cases}$

**15.** $\begin{cases} x + y + z = 2 \\ 2x - z = 5 \\ 3y + z = 2 \end{cases}$

**16.** $\begin{cases} x + 2y + z = 5 \\ x - y - z = 3 \\ y + z = 2 \end{cases}$

**17.** $\begin{cases} 5x - 2y = 27 \\ -3x + 5y = 18 \end{cases}$

**18.** $\begin{cases} 4x - y = 9 \\ 2x + 3y = -27 \end{cases}$

**19.** $\begin{cases} 4x - 7y = 7 \\ 12x - 21y = 24 \end{cases}$

**20.** $\begin{cases} 2x - 5y = 12 \\ -4x + 10y = 20 \end{cases}$

**21.** $\begin{cases} 4x - y + 2z = 5 \\ 2y + z = 4 \\ 4x + y + 3z = 10 \end{cases}$

**22.** $\begin{cases} 5y - 7z = 14 \\ 2x + y + 4z = 10 \\ 2x + 6y - 3z = 30 \end{cases}$

**23.** $\begin{cases} 4x + y + z = 3 \\ -x + y - 2z = -11 \\ x + 2y + 2z = -1 \end{cases}$

**24.** $\begin{cases} x + y + z = 9 \\ 3x - y + z = -1 \\ -2x + 2y - 3z = -2 \end{cases}$

## CONCEPT EXTENSIONS

*Solve. See the Concept Check in the section.*

**25.** For the system $\begin{cases} x + z = 7 \\ y + 2z = -6, \\ 3x - y = 0 \end{cases}$ which is the correct corresponding matrix?

**a.** $\begin{bmatrix} 1 & 1 & | & 7 \\ 1 & 2 & | & -6 \\ 3 & -1 & | & 0 \end{bmatrix}$

**b.** $\begin{bmatrix} 1 & 0 & 1 & | & 7 \\ 1 & 2 & 0 & | & -6 \\ 3 & -1 & 0 & | & 0 \end{bmatrix}$

**c.** $\begin{bmatrix} 1 & 0 & 1 & | & 7 \\ 0 & 1 & 2 & | & -6 \\ 3 & -1 & 0 & | & 0 \end{bmatrix}$

# Appendix F

## Solving Systems of Equations by Determinants

We have solved systems of two linear equations in two variables in four different ways: graphically, by substitution, by elimination, and by matrices. Now we analyze another method called **Cramer's rule.**

**OBJECTIVE 1 ▶ Evaluating 2 × 2 determinants.** Recall that a matrix is a rectangular array of numbers. If a matrix has the same number of rows and columns, it is called a **square matrix.** Examples of square matrices are

$$\begin{bmatrix} 1 & 6 \\ 5 & 2 \end{bmatrix} \qquad \begin{bmatrix} 2 & 4 & 1 \\ 0 & 5 & 2 \\ 3 & 6 & 9 \end{bmatrix}$$

A **determinant** is a real number associated with a square matrix. The determinant of a square matrix is denoted by placing vertical bars about the array of numbers. Thus,

The determinant of the square matrix $\begin{bmatrix} 1 & 6 \\ 5 & 2 \end{bmatrix}$ is $\begin{vmatrix} 1 & 6 \\ 5 & 2 \end{vmatrix}$.

The determinant of the square matrix $\begin{bmatrix} 2 & 4 & 1 \\ 0 & 5 & 2 \\ 3 & 6 & 9 \end{bmatrix}$ is $\begin{vmatrix} 2 & 4 & 1 \\ 0 & 5 & 2 \\ 3 & 6 & 9 \end{vmatrix}$.

We define the determinant of a 2 × 2 matrix first. (Recall that 2 × 2 is read "two by two." It means that the matrix has 2 rows and 2 columns.)

---

**Determinant of a 2 × 2 Matrix**

$$\begin{vmatrix} a & b \\ c & d \end{vmatrix} = ad - bc$$

---

**EXAMPLE 1**    Evaluate each determinant

**a.** $\begin{vmatrix} -1 & 2 \\ 3 & -4 \end{vmatrix}$       **b.** $\begin{vmatrix} 2 & 0 \\ 7 & -5 \end{vmatrix}$

_Solution_    First we identify the values of $a, b, c,$ and $d$. Then we perform the evaluation.

**a.** Here $a = -1, b = 2, c = 3,$ and $d = -4$.

$$\begin{vmatrix} -1 & 2 \\ 3 & -4 \end{vmatrix} = ad - bc = (-1)(-4) - (2)(3) = -2$$

**b.** In this example, $a = 2, b = 0, c = 7,$ and $d = -5$.

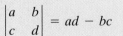

$$\begin{vmatrix} 2 & 0 \\ 7 & -5 \end{vmatrix} = ad - bc = 2(-5) - (0)(7) = -10$$

☐

**OBJECTIVE 2 ▶ Using Cramer's rule to solve a system of two linear equations.** To develop Cramer's rule, we solve the system $\begin{cases} ax + by = h \\ cx + dy = k \end{cases}$ using elimination. First, we eliminate $y$ by multiplying both sides of the first equation by $d$ and both sides of the second equation by $-b$ so that the coefficients of $y$ are opposites. The result is that

$$\begin{cases} d(ax + by) = d \cdot h \\ -b(cx + dy) = -b \cdot k \end{cases} \quad \text{simplifies to} \quad \begin{cases} adx + bdy = hd \\ -bcx - bdy = -kb \end{cases}$$

We now add the two equations and solve for $x$.

$$
\begin{aligned}
adx + bdy &= hd \\
\underline{-bcx - bdy} &= \underline{-kb} \\
adx - bcx &= hd - kb \quad \text{Add the equations.} \\
(ad - bc)x &= hd - kb \\
x &= \frac{hd - kb}{ad - bc} \quad \text{Solve for } x.
\end{aligned}
$$

When we replace $x$ with $\dfrac{hd - kb}{ad - bc}$ in the equation $ax + by = h$ and solve for $y$, we find that $y = \dfrac{ak - ch}{ad - bc}$.

Notice that the numerator of the value of $x$ is the determinant of

$$\begin{vmatrix} h & b \\ k & d \end{vmatrix} = hd - kb$$

Also, the numerator of the value of $y$ is the determinant of

$$\begin{vmatrix} a & h \\ c & k \end{vmatrix} = ak - hc$$

Finally, the denominators of the values of $x$ and $y$ are the same and are the determinant of

$$\begin{vmatrix} a & b \\ c & d \end{vmatrix} = ad - bc$$

This means that the values of $x$ and $y$ can be written in determinant notation:

$$x = \frac{\begin{vmatrix} h & b \\ k & d \end{vmatrix}}{\begin{vmatrix} a & b \\ c & d \end{vmatrix}} \quad \text{and} \quad y = \frac{\begin{vmatrix} a & h \\ c & k \end{vmatrix}}{\begin{vmatrix} a & b \\ c & d \end{vmatrix}}$$

For convenience, we label the determinants $D$, $D_x$, and $D_y$.

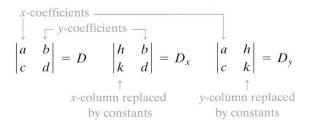

These determinant formulas for the coordinates of the solution of a system are known as **Cramer's rule.**

---

**Cramer's Rule for Two Linear Equations in Two Variables**

The solution of the system $\begin{cases} ax + by = h \\ cx + dy = k \end{cases}$ is given by

$$x = \frac{\begin{vmatrix} h & b \\ k & d \end{vmatrix}}{\begin{vmatrix} a & b \\ c & d \end{vmatrix}} = \frac{D_x}{D} \qquad y = \frac{\begin{vmatrix} a & h \\ c & k \end{vmatrix}}{\begin{vmatrix} a & b \\ c & d \end{vmatrix}} = \frac{D_y}{D}$$

as long as $D = ad - bc$ is not 0.

---

When $D = 0$, the system is either inconsistent or the equations are dependent. When this happens, we need to use another method to see which is the case.

**EXAMPLE 2**   Use Cramer's rule to solve the system

$$\begin{cases} 3x + 4y = -7 \\ x - 2y = -9 \end{cases}$$

*Solution*   First we find $D$, $D_x$, and $D_y$.

$$\begin{array}{ccc} a & b & h \\ \downarrow & \downarrow & \downarrow \end{array}$$

$$\begin{cases} 3x + 4y = -7 \\ x - 2y = -9 \end{cases}$$

$$\begin{array}{ccc} \uparrow & \uparrow & \uparrow \\ c & d & k \end{array}$$

$$D = \begin{vmatrix} a & b \\ c & d \end{vmatrix} = \begin{vmatrix} 3 & 4 \\ 1 & -2 \end{vmatrix} = 3(-2) - 4(1) = -10$$

$$D_x = \begin{vmatrix} h & b \\ k & d \end{vmatrix} = \begin{vmatrix} -7 & 4 \\ -9 & -2 \end{vmatrix} = (-7)(-2) - 4(-9) = 50$$

$$D_y = \begin{vmatrix} a & h \\ c & k \end{vmatrix} = \begin{vmatrix} 3 & -7 \\ 1 & -9 \end{vmatrix} = 3(-9) - (-7)(1) = -20$$

Then $x = \dfrac{D_x}{D} = \dfrac{50}{-10} = -5$ and $y = \dfrac{D_y}{D} = \dfrac{-20}{-10} = 2$.

The ordered pair solution is $(-5, 2)$.

As always, check the solution in both original equations.   □

**EXAMPLE 3**   Use Cramer's rule to solve the system

$$\begin{cases} 5x + y = 5 \\ -7x - 2y = -7 \end{cases}$$

*Solution*   First we find $D$, $D_x$, and $D_y$.

$$D = \begin{vmatrix} 5 & 1 \\ -7 & -2 \end{vmatrix} = 5(-2) - (-7)(1) = -3$$

$$D_x = \begin{vmatrix} 5 & 1 \\ -7 & -2 \end{vmatrix} = 5(-2) - (-7)(1) = -3$$

$$D_y = \begin{vmatrix} 5 & 5 \\ -7 & -7 \end{vmatrix} = 5(-7) - 5(-7) = 0$$

Then

$$x = \frac{D_x}{D} = \frac{-3}{-3} = 1 \qquad y = \frac{D_y}{D} = \frac{0}{-3} = 0$$

The ordered pair solution is $(1, 0)$.

**OBJECTIVE 3 ▶ Evaluating 3 × 3 determinants.** A 3 × 3 determinant can be used to solve a system of three equations in three variables. The determinant of a 3 × 3 matrix, however, is considerably more complex than a 2 × 2 one.

---

**Determinant of a 3 × 3 Matrix**

$$\begin{vmatrix} a_1 & b_1 & c_1 \\ a_2 & b_2 & c_2 \\ a_3 & b_3 & c_3 \end{vmatrix} = a_1 \cdot \begin{vmatrix} b_2 & c_2 \\ b_3 & c_3 \end{vmatrix} - a_2 \cdot \begin{vmatrix} b_1 & c_1 \\ b_3 & c_3 \end{vmatrix} + a_3 \cdot \begin{vmatrix} b_1 & c_1 \\ b_2 & c_2 \end{vmatrix}$$

---

Notice that the determinant of a 3 × 3 matrix is related to the determinants of three 2 × 2 matrices. Each determinant of these 2 × 2 matrices is called a **minor,** and every element of a 3 × 3 matrix has a minor associated with it. For example, the minor of $c_2$ is the determinant of the 2 × 2 matrix found by deleting the row and column containing $c_2$.

$$\begin{matrix} a_1 & b_1 & c_1 \\ a_2 & b_2 & c_2 \\ a_3 & b_3 & c_3 \end{matrix} \qquad \text{The minor of } c_2 \text{ is} \qquad \begin{vmatrix} a_1 & b_1 \\ a_3 & b_3 \end{vmatrix}$$

Also, the minor of element $a_1$ is the determinant of the 2 × 2 matrix that has no row or column containing $a_1$.

$$\begin{matrix} a_1 & b_1 & c_1 \\ a_2 & b_2 & c_2 \\ a_3 & b_3 & c_3 \end{matrix} \qquad \text{The minor of } a_1 \text{ is} \qquad \begin{vmatrix} b_2 & c_2 \\ b_3 & c_3 \end{vmatrix}$$

So the determinant of a 3 × 3 matrix can be written as

$$a_1 \cdot (\text{minor of } a_1) - a_2 \cdot (\text{minor of } a_2) + a_3 \cdot (\text{minor of } a_3)$$

Finding the determinant by using minors of elements in the first column is called **expanding** by the minors of the first column. *The value of a determinant can be found by expanding by the minors of any row or column.* The following **array of signs** is helpful in determining whether to add or subtract the product of an element and its minor.

$$\begin{matrix} + & - & + \\ - & + & - \\ + & - & + \end{matrix}$$

If an element is in a position marked $+$, we add. If marked $-$, we subtract.

**EXAMPLE 4** Evaluate by expanding by the minors of the given row or column.

$$\begin{vmatrix} 0 & 5 & 1 \\ 1 & 3 & -1 \\ -2 & 2 & 4 \end{vmatrix}$$

**a.** First column       **b.** Second row

*Solution*

**a.** The elements of the first column are 0, 1, and $-2$. The first column of the array of signs is $+, -, +$.

$$\begin{vmatrix} 0 & 5 & 1 \\ 1 & 3 & -1 \\ -2 & 2 & 4 \end{vmatrix} = 0 \cdot \begin{vmatrix} 3 & -1 \\ 2 & 4 \end{vmatrix} - 1 \cdot \begin{vmatrix} 5 & 1 \\ 2 & 4 \end{vmatrix} + (-2) \cdot \begin{vmatrix} 5 & 1 \\ 3 & -1 \end{vmatrix}$$

$$= 0(12 - (-2)) - 1(20 - 2) + (-2)(-5 - 3)$$
$$= 0 - 18 + 16 = -2$$

**b.** The elements of the second row are 1, 3, and $-1$. This time, the signs begin with $-$ and again alternate.

$$\begin{vmatrix} 0 & 5 & 1 \\ 1 & 3 & -1 \\ -2 & 2 & 4 \end{vmatrix} = -1 \cdot \begin{vmatrix} 5 & 1 \\ 2 & 4 \end{vmatrix} + 3 \cdot \begin{vmatrix} 0 & 1 \\ -2 & 4 \end{vmatrix} - (-1) \cdot \begin{vmatrix} 0 & 5 \\ -2 & 2 \end{vmatrix}$$

$$= -1(20 - 2) + 3(0 - (-2)) - (-1)(0 - (-10))$$
$$= -18 + 6 + 10 = -2$$

Notice that the determinant of the $3 \times 3$ matrix is the same regardless of the row or column you select to expand by. □

**Concept Check** ✓

Why would expanding by minors of the second row be a good choice for the determinant

$$\begin{vmatrix} 3 & 4 & -2 \\ 5 & 0 & 0 \\ 6 & -3 & 7 \end{vmatrix}?$$

**OBJECTIVE 4 ▶ Using Cramer's rule to solve a system of three linear equations.** A system of three equations in three variables may be solved with Cramer's rule also. Using the elimination process to solve a system with unknown constants as coefficients leads to the following.

**Cramer's Rule for Three Equations in Three Variables**

The solution of the system $\begin{cases} a_1x + b_1y + c_1z = k_1 \\ a_2x + b_2y + c_2z = k_2 \\ a_3x + b_3y + c_3z = k_3 \end{cases}$ is given by

$$x = \frac{D_x}{D} \qquad y = \frac{D_y}{D} \qquad \text{and} \qquad z = \frac{D_z}{D}$$

where

$$D = \begin{vmatrix} a_1 & b_1 & c_1 \\ a_2 & b_2 & c_2 \\ a_3 & b_3 & c_3 \end{vmatrix} \qquad D_x = \begin{vmatrix} k_1 & b_1 & c_1 \\ k_2 & b_2 & c_2 \\ k_3 & b_3 & c_3 \end{vmatrix}$$

$$D_y = \begin{vmatrix} a_1 & k_1 & c_1 \\ a_2 & k_2 & c_2 \\ a_3 & k_3 & c_3 \end{vmatrix} \qquad D_z = \begin{vmatrix} a_1 & b_1 & k_1 \\ a_2 & b_2 & k_2 \\ a_3 & b_3 & k_3 \end{vmatrix}$$

as long as $D$ is not 0.

**Answer to Concept Check:**
Two elements of the second row are 0, which makes calculations easier.

**EXAMPLE 5**   Use Cramer's rule to solve the system

$$\begin{cases} x - 2y + z = 4 \\ 3x + y - 2z = 3 \\ 5x + 5y + 3z = -8 \end{cases}$$

*Solution*   First we find $D$, $D_x$, $D_y$, and $D_z$. Beginning with $D$, we expand by the minors of the first column.

$$D = \begin{vmatrix} 1 & -2 & 1 \\ 3 & 1 & -2 \\ 5 & 5 & 3 \end{vmatrix} = 1 \cdot \begin{vmatrix} 1 & -2 \\ 5 & 3 \end{vmatrix} - 3 \cdot \begin{vmatrix} -2 & 1 \\ 5 & 3 \end{vmatrix} + 5 \cdot \begin{vmatrix} -2 & 1 \\ 1 & -2 \end{vmatrix}$$

$$= 1(3 - (-10)) - 3(-6 - 5) + 5(4 - 1)$$

$$= 13 + 33 + 15 = 61$$

$$D_x = \begin{vmatrix} 4 & -2 & 1 \\ 3 & 1 & -2 \\ -8 & 5 & 3 \end{vmatrix} = 4 \cdot \begin{vmatrix} 1 & -2 \\ 5 & 3 \end{vmatrix} - 3 \cdot \begin{vmatrix} -2 & 1 \\ 5 & 3 \end{vmatrix} + (-8) \cdot \begin{vmatrix} -2 & 1 \\ 1 & -2 \end{vmatrix}$$

$$= 4(3 - (-10)) - 3(-6 - 5) + (-8)(4 - 1)$$

$$= 52 + 33 - 24 = 61$$

$$D_y = \begin{vmatrix} 1 & 4 & 1 \\ 3 & 3 & -2 \\ 5 & -8 & 3 \end{vmatrix} = 1 \cdot \begin{vmatrix} 3 & -2 \\ -8 & 3 \end{vmatrix} - 3 \cdot \begin{vmatrix} 4 & 1 \\ -8 & 3 \end{vmatrix} + 5 \cdot \begin{vmatrix} 4 & 1 \\ 3 & -2 \end{vmatrix}$$

$$= 1(9 - 16) - 3(12 - (-8)) + 5(-8 - 3)$$

$$= -7 - 60 - 55 = -122$$

$$D_z = \begin{vmatrix} 1 & -2 & 4 \\ 3 & 1 & 3 \\ 5 & 5 & -8 \end{vmatrix} = 1 \cdot \begin{vmatrix} 1 & 3 \\ 5 & -8 \end{vmatrix} - 3 \cdot \begin{vmatrix} -2 & 4 \\ 5 & -8 \end{vmatrix} + 5 \cdot \begin{vmatrix} -2 & 4 \\ 1 & 3 \end{vmatrix}$$

$$= 1(-8 - 15) - 3(16 - 20) + 5(-6 - 4)$$

$$= -23 + 12 - 50 = -61$$

From these determinants, we calculate the solution:

$$x = \frac{D_x}{D} = \frac{61}{61} = 1 \quad y = \frac{D_y}{D} = \frac{-122}{61} = -2 \quad z = \frac{D_z}{D} = \frac{-61}{61} = -1$$

The ordered triple solution is $(1, -2, -1)$. Check this solution by verifying that it satisfies each equation of the system.   $\square$

# APPENDIX F | EXERCISE SET MyMathLab

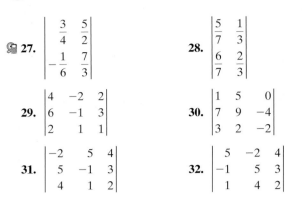

*Evaluate. See Example 1.*

**1.** $\begin{vmatrix} 3 & 5 \\ -1 & 7 \end{vmatrix}$

**2.** $\begin{vmatrix} -5 & 1 \\ 0 & -4 \end{vmatrix}$

**3.** $\begin{vmatrix} 9 & -2 \\ 4 & -3 \end{vmatrix}$

**4.** $\begin{vmatrix} 4 & 0 \\ 9 & 8 \end{vmatrix}$

**5.** $\begin{vmatrix} -2 & 9 \\ 4 & -18 \end{vmatrix}$

**6.** $\begin{vmatrix} -40 & 8 \\ 70 & -14 \end{vmatrix}$

*Use Cramer's rule, if possible, to solve each system of linear equations. See Examples 2 and 3.*

**7.** $\begin{cases} 2y - 4 = 0 \\ x + 2y = 5 \end{cases}$

**8.** $\begin{cases} 4x - y = 5 \\ 3x - 3 = 0 \end{cases}$

**9.** $\begin{cases} 3x + y = 1 \\ 2y = 2 - 6x \end{cases}$

**10.** $\begin{cases} y = 2x - 5 \\ 8x - 4y = 20 \end{cases}$

**11.** $\begin{cases} 5x - 2y = 27 \\ -3x + 5y = 18 \end{cases}$

**12.** $\begin{cases} 4x - y = 9 \\ 2x + 3y = -27 \end{cases}$

*Evaluate. See Example 4.*

**13.** $\begin{vmatrix} 2 & 1 & 0 \\ 0 & 5 & -3 \\ 4 & 0 & 2 \end{vmatrix}$

**14.** $\begin{vmatrix} -6 & 4 & 2 \\ 1 & 0 & 5 \\ 0 & 3 & 1 \end{vmatrix}$

**15.** $\begin{vmatrix} 4 & -6 & 0 \\ -2 & 3 & 0 \\ 4 & -6 & 1 \end{vmatrix}$

**16.** $\begin{vmatrix} 5 & 2 & 1 \\ 3 & -6 & 0 \\ -2 & 8 & 0 \end{vmatrix}$

**17.** $\begin{vmatrix} 3 & 6 & -3 \\ -1 & -2 & 3 \\ 4 & -1 & 6 \end{vmatrix}$

**18.** $\begin{vmatrix} 2 & -2 & 1 \\ 4 & 1 & 3 \\ 3 & 1 & 2 \end{vmatrix}$

*Use Cramer's rule, if possible, to solve each system of linear equations. See Example 5.*

**19.** $\begin{cases} 3x + z = -1 \\ -x - 3y + z = 7 \\ 3y + z = 5 \end{cases}$

**20.** $\begin{cases} 4y - 3z = -2 \\ 8x - 4y = 4 \\ -8x + 4y + z = -2 \end{cases}$

**21.** $\begin{cases} x + y + z = 8 \\ 2x - y - z = 10 \\ x - 2y + 3z = 22 \end{cases}$

**22.** $\begin{cases} 5x + y + 3z = 1 \\ x - y - 3z = -7 \\ -x + y = 1 \end{cases}$

*Evaluate.*

**23.** $\begin{vmatrix} 10 & -1 \\ -4 & 2 \end{vmatrix}$

**24.** $\begin{vmatrix} -6 & 2 \\ 5 & -1 \end{vmatrix}$

**25.** $\begin{vmatrix} 1 & 0 & 4 \\ 1 & -1 & 2 \\ 3 & 2 & 1 \end{vmatrix}$

**26.** $\begin{vmatrix} 0 & 1 & 2 \\ 3 & -1 & 2 \\ 3 & 2 & -2 \end{vmatrix}$

**27.** $\begin{vmatrix} \frac{3}{4} & \frac{5}{2} \\ -\frac{1}{6} & \frac{7}{3} \end{vmatrix}$

**28.** $\begin{vmatrix} \frac{5}{7} & \frac{1}{3} \\ \frac{6}{7} & \frac{2}{3} \end{vmatrix}$

**29.** $\begin{vmatrix} 4 & -2 & 2 \\ 6 & -1 & 3 \\ 2 & 1 & 1 \end{vmatrix}$

**30.** $\begin{vmatrix} 1 & 5 & 0 \\ 7 & 9 & -4 \\ 3 & 2 & -2 \end{vmatrix}$

**31.** $\begin{vmatrix} -2 & 5 & 4 \\ 5 & -1 & 3 \\ 4 & 1 & 2 \end{vmatrix}$

**32.** $\begin{vmatrix} 5 & -2 & 4 \\ -1 & 5 & 3 \\ 1 & 4 & 2 \end{vmatrix}$

## MIXED PRACTICE

*Use Cramer's rule, if possible, to solve each system of linear equations.*

**33.** $\begin{cases} 2x - 5y = 4 \\ x + 2y = -7 \end{cases}$

**34.** $\begin{cases} 3x - y = 2 \\ -5x + 2y = 0 \end{cases}$

**35.** $\begin{cases} 4x + 2y = 5 \\ 2x + y = -1 \end{cases}$

**36.** $\begin{cases} 3x + 6y = 15 \\ 2x + 4y = 3 \end{cases}$

**37.** $\begin{cases} 2x + 2y + z = 1 \\ -x + y + 2z = 3 \\ x + 2y + 4z = 0 \end{cases}$

**38.** $\begin{cases} 2x - 3y + z = 5 \\ x + y + z = 0 \\ 4x + 2y + 4z = 4 \end{cases}$

**39.** $\begin{cases} \frac{2}{3}x - \frac{3}{4}y = -1 \\ -\frac{1}{6}x + \frac{3}{4}y = \frac{5}{2} \end{cases}$

**40.** $\begin{cases} \frac{1}{2}x - \frac{1}{3}y = -3 \\ \frac{1}{8}x + \frac{1}{6}y = 0 \end{cases}$

**41.** $\begin{cases} 0.7x - 0.2y = -1.6 \\ 0.2x - y = -1.4 \end{cases}$

**42.** $\begin{cases} -0.7x + 0.6y = 1.3 \\ 0.5x - 0.3y = -0.8 \end{cases}$

**43.** $\begin{cases} -2x + 4y - 2z = 6 \\ x - 2y + z = -3 \\ 3x - 6y + 3z = -9 \end{cases}$

**44.** $\begin{cases} -x - y + 3z = 2 \\ 4x + 4y - 12z = -8 \\ -3x - 3y + 9z = 6 \end{cases}$

**45.** $\begin{cases} x - 2y + z = -5 \\ 3y + 2z = 4 \\ 3x - y = -2 \end{cases}$

**46.** $\begin{cases} 4x + 5y = 10 \\ 3y + 2z = -6 \\ x + y + z = 3 \end{cases}$

## CONCEPT EXTENSIONS

*Find the value of x such that each is a true statement.*

**47.** $\begin{vmatrix} 1 & x \\ 2 & 7 \end{vmatrix} = -3$

**48.** $\begin{vmatrix} 6 & 1 \\ -2 & x \end{vmatrix} = 26$

**49.** If all the elements in a single row of a square matrix are zero, to what does the determinant evaluate? Explain your answer.

**50.** If all the elements in a single column of a square matrix are 0, to what does the determinant evaluate? Explain your answer.

**51.** Suppose you are interested in finding the determinant of a $4 \times 4$ matrix. Study the pattern shown in the array of signs for a $3 \times 3$ matrix. Use the pattern to expand the array of signs for use with a $4 \times 4$ matrix.

**52.** Why would expanding by minors of the second row be a good choice for the determinant $\begin{vmatrix} 3 & 4 & -2 \\ 5 & 0 & 0 \\ 6 & -3 & 7 \end{vmatrix}$?

*Find the value of each determinant. To evaluate a $4 \times 4$ determinant, select any row or column and expand by the minors. The array of signs for a $4 \times 4$ determinant is the same as for a $3 \times 3$ determinant except expanded.*

**53.** $\begin{vmatrix} 5 & 0 & 0 & 0 \\ 0 & 4 & 2 & -1 \\ 1 & 3 & -2 & 0 \\ 0 & -3 & 1 & 2 \end{vmatrix}$

**54.** $\begin{vmatrix} 1 & 7 & 0 & -1 \\ 1 & 3 & -2 & 0 \\ 1 & 0 & -1 & 2 \\ 0 & -6 & 2 & 4 \end{vmatrix}$

**55.** $\begin{vmatrix} 4 & 0 & 2 & 5 \\ 0 & 3 & -1 & 1 \\ 0 & 0 & 2 & 0 \\ 0 & 0 & 0 & 1 \end{vmatrix}$

**56.** $\begin{vmatrix} 2 & 0 & -1 & 4 \\ 6 & 0 & 4 & 1 \\ 2 & 4 & 3 & -1 \\ 4 & 0 & 5 & -4 \end{vmatrix}$

# Appendix G

## Mean, Median, and Mode

It is sometimes desirable to be able to describe a set of data, or a set of numbers, by a single "middle" number. Three such **measures of central tendency** are the mean, the median, and the mode.

The most common measure of central tendency is the mean (sometimes called the arithmetic mean or the average). The **mean** of a set of data items, denoted by $\overline{x}$, is the sum of the items divided by the number of items.

**EXAMPLE 1**  Seven students in a psychology class conducted an experiment on mazes. Each student was given a pencil and asked to successfully complete the same maze. The timed results are below.

| Student | Ann | Thanh | Carlos | Jesse | Melinda | Ramzi | Dayni |
|---|---|---|---|---|---|---|---|
| Time (Seconds) | 13.2 | 11.8 | 10.7 | 16.2 | 15.9 | 13.8 | 18.5 |

a. Who completed the maze in the shortest time? Who completed the maze in the longest time?
b. Find the mean.
c. How many students took longer than the mean time? How many students took shorter than the mean time?

### Solution

a. Carlos completed the maze in 10.7 seconds, the shortest time. Dayni completed the maze in 18.5 seconds, the longest time.
b. To find the mean, $\overline{x}$, find the sum of the data items and divide by 7, the number of items.

$$\overline{x} = \frac{13.2 + 11.8 + 10.7 + 16.2 + 15.9 + 13.8 + 18.5}{7} = \frac{100.1}{7} = 14.3$$

c. Three students, Jesse, Melinda, and Dayni, had times longer than the mean time. Four students, Ann, Thanh, Carlos, and Ramzi, had times shorter than the mean time.

Two other measures of central tendency are the median and the mode.

The **median** of an ordered set of numbers is the middle number. If the number of items is even, the median is the mean of the two middle numbers. The **mode** of a set of numbers is the number that occurs most often. It is possible for a data set to have no mode or more than one mode.

**EXAMPLE 2**  Find the median and the mode of the following list of numbers. These numbers were high temperatures for fourteen consecutive days in a city in Montana.

76, 80, 85, 86, 89, 87, 82, 77, 76, 79, 82, 89, 89, 92

*Solution*

First, write the numbers in order.

76, 76, 77, 79, 80, 82, 82, 85, 86, 87, 89, 89, 89, 92

two middle numbers      mode

Since there are an even number of items, the median is the mean of the two middle numbers.

$$\text{median} = \frac{82 + 85}{2} = 83.5$$

The mode is 89, since 89 occurs most often.

**APPENDIX G | EXERCISE SET**  *MyMathLab*  Powered by CourseCompass™ and MathXL®   MathXL PRACTICE   WATCH   DOWNLOAD   READ   REVIEW

*For each of the following data sets, find the mean, the median, and the mode. If necessary, round the mean to one decimal place.*

**1.** 21, 28, 16, 42, 38

**2.** 42, 35, 36, 40, 50

**3.** 7.6, 8.2, 8.2, 9.6, 5.7, 9.1

**4.** 4.9, 7.1, 6.8, 6.8, 5.3, 4.9

**5.** 0.2, 0.3, 0.5, 0.6, 0.6, 0.9, 0.2, 0.7, 1.1

**6.** 0.6, 0.6, 0.8, 0.4, 0.5, 0.3, 0.7, 0.8, 0.1

**7.** 231, 543, 601, 293, 588, 109, 334, 268

**8.** 451, 356, 478, 776, 892, 500, 467, 780

*The eight tallest buildings in the United States are listed below. Use this table for Exercises 9 through 12.*

| Building | Height (feet) |
|---|---|
| Sears Tower, Chicago, IL | 1454 |
| Empire State, New York, NY | 1250 |
| Amoco, Chicago, IL | 1136 |
| John Hancock Center, Chicago, IL | 1127 |
| First Interstate World Center, Los Angeles, CA | 1107 |
| Chrysler, New York, NY | 1046 |
| NationsBank Tower, Atlanta, GA | 1023 |
| Texas Commerce Tower, Houston, TX | 1002 |

**9.** Find the mean height for the five tallest buildings.
**10.** Find the median height for the five tallest buildings.
**11.** Find the median height for the eight tallest buildings.
**12.** Find the mean height for the eight tallest buildings.

*During an experiment, the following times (in seconds) were recorded: 7.8, 6.9, 7.5, 4.7, 6.9, 7.0.*

**13.** Find the mean. Round to the nearest tenth.
**14.** Find the median.
**15.** Find the mode.

*In a mathematics class, the following test scores were recorded for a student: 86, 95, 91, 74, 77, 85.*

**16.** Find the mean. Round to the nearest hundredth.
**17.** Find the median.
**18.** Find the mode.

*The following pulse rates were recorded for a group of fifteen students: 78, 80, 66, 68, 71, 64, 82, 71, 70, 65, 70, 75, 77, 86, 72.*

**19.** Find the mean.
**20.** Find the median.
**21.** Find the mode.
**22.** How many rates were higher than the mean?
**23.** How many rates were lower than the mean?
**24.** Have each student in your algebra class take his/her pulse rate. Record the data and find the mean, the median, and the mode.

*Find the missing numbers in each list of numbers. (These numbers are not necessarily in numerical order.)*

**25.** __, __, 16, 18, __
The mode is 21.      The mean is 20.

**26.** __, __, __, __, 40
The mode is 35.      The median is 37.      The mean is 38.

# Appendix H

## Review of Angles, Lines, and Special Triangles

The word **geometry** is formed from the Greek words, **geo,** meaning earth, and **metron,** meaning measure. Geometry literally means to measure the earth.

   This section contains a review of some basic geometric ideas. It will be assumed that fundamental ideas of geometry such as point, line, ray, and angle are known. In this appendix, the notation $\angle 1$ is read "angle 1" and the notation $m\angle 1$ is read "the measure of angle 1."

   We first review types of angles.

---

**Angles**

A **right angle** is an angle whose measure is 90°. A right angle can be indicated by a square drawn at the vertex of the angle, as shown below.

An angle whose measure is more than 0° but less than 90° is called an **acute angle.**

An angle whose measure is greater than 90° but less than 180° is called an **obtuse angle.**

An angle whose measure is 180° is called a **straight angle.**

Two angles are said to be **complementary** if the sum of their measures is 90°. Each angle is called the **complement** of the other.

Two angles are said to be **supplementary** if the sum of their measures is 180°. Each angle is called the **supplement** of the other.

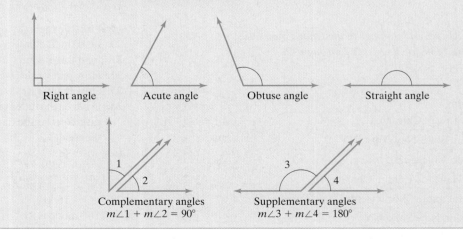

Right angle    Acute angle    Obtuse angle    Straight angle

Complementary angles
$m\angle 1 + m\angle 2 = 90°$

Supplementary angles
$m\angle 3 + m\angle 4 = 180°$

---

**EXAMPLE 1**    If an angle measures 28°, find its complement.

*Solution*    Two angles are complementary if the sum of their measures is 90°. The complement of a 28° angle is an angle whose measure is $90° - 28° = 62°$. To check, notice that $28° + 62° = 90°$.  □

**Plane** is an undefined term that we will describe. A plane can be thought of as a flat surface with infinite length and width, but no thickness. A plane is two dimensional. The arrows in the following diagram indicate that a plane extends indefinitely and has no boundaries.

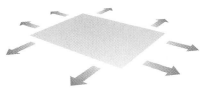

Figures that lie on a plane are called **plane figures.** Lines that lie in the same plane are called **coplanar.**

---

**Lines**

Two lines are **parallel** if they lie in the same plane but never meet.

**Intersecting lines** meet or cross in one point.

Two lines that form right angles when they intersect are said to be **perpendicular.**

| Parallel lines | Intersecting lines | Intersecting lines that are perpendicular |

---

Two intersecting lines form **vertical angles.** Angles 1 and 3 are vertical angles. Also angles 2 and 4 are vertical angles. It can be shown that **vertical angles have equal measures.**

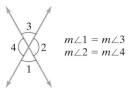

$$m\angle 1 = m\angle 3$$
$$m\angle 2 = m\angle 4$$

**Adjacent angles** have the same vertex and share a side. Angles 1 and 2 are adjacent angles. Other pairs of adjacent angles are angles 2 and 3, angles 3 and 4, and angles 4 and 1.

A **transversal** is a line that intersects two or more lines in the same plane. Line *l* is a transversal that intersects lines *m* and *n*. The eight angles formed are numbered and certain pairs of these angles are given special names.

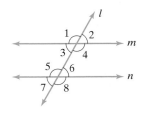

**Corresponding angles:** ∠1 and ∠5, ∠3 and ∠7, ∠2 and ∠6, and ∠4 and ∠8.

**Exterior angles:** ∠1, ∠2, ∠7, and ∠8.

**Interior angles:** ∠3, ∠4, ∠5, and ∠6.

**Alternate interior angles:** ∠3 and ∠6, ∠4 and ∠5.

These angles and parallel lines are related in the following manner.

---

**Parallel Lines Cut by a Transversal**

1. If two parallel lines are cut by a transversal, then
   a. corresponding angles are equal and
   b. alternate interior angles are equal.

2. If corresponding angles formed by two lines and a transversal are equal, then the lines are parallel.

3. If alternate interior angles formed by two lines and a transversal are equal, then the lines are parallel.

---

**EXAMPLE 2**  Given that lines $m$ and $n$ are parallel and that the measure of angle 1 is 100°, find the measures of angles 2, 3, and 4.

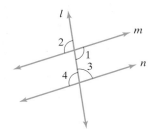

*Solution*  $m\angle 2 = 100°$, since angles 1 and 2 are vertical angles.

$m\angle 4 = 100°$, since angles 1 and 4 are alternate interior angles.

$m\angle 3 = 180° - 100° = 80°$, since angles 4 and 3 are supplementary angles.

A **polygon** is the union of three or more coplanar line segments that intersect each other only at each end point, with each end point shared by exactly two segments.

A **triangle** is a polygon with three sides. The sum of the measures of the three angles of a triangle is 180°. In the following figure, $m\angle 1 + m\angle 2 + m\angle 3 = 180°$.

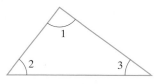

**EXAMPLE 3**  Find the measure of the third angle of the triangle shown.

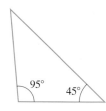

*Solution*  The sum of the measures of the angles of a triangle is 180°. Since one angle measures 45° and the other angle measures 95°, the third angle measures $180° - 45° - 95° = 40°$.

Two triangles are **congruent** if they have the same size and the same shape. In congruent triangles, the measures of corresponding angles are equal and the lengths of corresponding sides are equal. The following triangles are congruent.

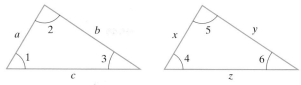

Corresponding angles are equal: $m\angle 1 = m\angle 4$, $m\angle 2 = m\angle 5$, and $m\angle 3 = m\angle 6$. Also, lengths of corresponding sides are equal: $a = x$, $b = y$, and $c = z$.

Any one of the following may be used to determine whether two triangles are congruent.

**Congruent Triangles**

1. If the measures of two angles of a triangle equal the measures of two angles of another triangle and the lengths of the sides between each pair of angles are equal, the triangles are congruent.

$$m\angle 1 = m\angle 3$$
$$m\angle 2 = m\angle 4$$
and
$$a = x$$

2. If the lengths of the three sides of a triangle equal the lengths of corresponding sides of another triangle, the triangles are congruent.

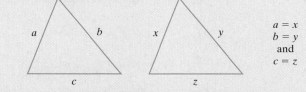

$$a = x$$
$$b = y$$
and
$$c = z$$

3. If the lengths of two sides of a triangle equal the lengths of corresponding sides of another triangle, and the measures of the angles between each pair of sides are equal, the triangles are congruent.

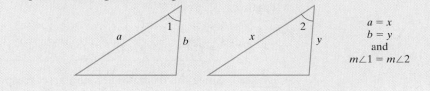

$$a = x$$
$$b = y$$
and
$$m\angle 1 = m\angle 2$$

Two triangles are **similar** if they have the same shape. In similar triangles, the measures of corresponding angles are equal and corresponding sides are in proportion. The following triangles are similar. (All similar triangles drawn in this appendix will be oriented the same.)

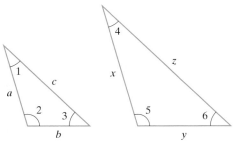

Corresponding angles are equal: $m\angle 1 = m\angle 4$, $m\angle 2 = m\angle 5$, and $m\angle 3 = m\angle 6$. Also, corresponding sides are proportional: $\dfrac{a}{x} = \dfrac{b}{y} = \dfrac{c}{z}$.

Any one of the following may be used to determine whether two triangles are similar.

> ### Similar Triangles
>
> **1.** If the measures of two angles of a triangle equal the measures of two angles of another triangle, the triangles are similar.
>
>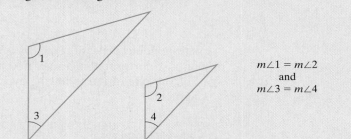
>
> $$m\angle 1 = m\angle 2$$
> and
> $$m\angle 3 = m\angle 4$$
>
> **2.** If three sides of one triangle are proportional to three sides of another triangle, the triangles are similar.
>
>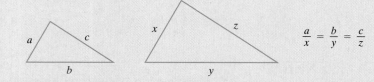
>
> $$\frac{a}{x} = \frac{b}{y} = \frac{c}{z}$$
>
> **3.** If two sides of a triangle are proportional to two sides of another triangle and the measures of the included angles are equal, the triangles are similar.
>
>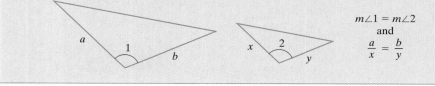
>
> $$m\angle 1 = m\angle 2$$
> and
> $$\frac{a}{x} = \frac{b}{y}$$

**EXAMPLE 4**   Given that the following triangles are similar, find the missing length $x$.

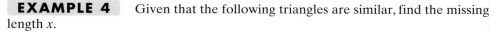

**Solution**   Since the triangles are similar, corresponding sides are in proportion. Thus, $\dfrac{2}{3} = \dfrac{10}{x}$. To solve this equation for $x$, we multiply both sides by the LCD, $3x$.

$$3x\left(\frac{2}{3}\right) = 3x\left(\frac{10}{x}\right)$$
$$2x = 30$$
$$x = 15$$

The missing length is 15 units.     □

A **right triangle** contains a right angle. The side opposite the right angle is called the **hypotenuse,** and the other two sides are called the **legs.** The **Pythagorean theorem** gives a formula that relates the lengths of the three sides of a right triangle.

> ### The Pythagorean Theorem
>
> If $a$ and $b$ are the lengths of the legs of a right triangle, and $c$ is the length of the hypotenuse, then $a^2 + b^2 = c^2$.
>
>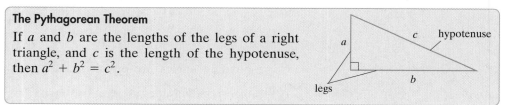

**EXAMPLE 5**   Find the length of the hypotenuse of a right triangle whose legs have lengths of 3 centimeters and 4 centimeters.

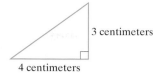

3 centimeters

4 centimeters

_Solution_   Because we have a right triangle, we use the Pythagorean theorem. The legs are 3 centimeters and 4 centimeters, so let $a = 3$ and $b = 4$ in the formula.

$$a^2 + b^2 = c^2$$
$$3^2 + 4^2 = c^2$$
$$9 + 16 = c^2$$
$$25 = c^2$$

Since $c$ represents a length, we assume that $c$ is positive. Thus, if $c^2$ is 25, $c$ must be 5. The hypotenuse has a length of 5 centimeters.  □

## APPENDIX H | EXERCISE SET

_Find the complement of each angle. See Example 1._

**1.** $19°$

**2.** $65°$

**3.** $70.8°$

**4.** $45\frac{2}{3}°$

**5.** $11\frac{1}{4}°$

**6.** $19.6°$

_Find the supplement of each angle._

**7.** $150°$

**8.** $90°$

**9.** $30.2°$

**10.** $81.9°$

**11.** $79\frac{1}{2}°$

**12.** $165\frac{8}{9}°$

**13.** If lines $m$ and $n$ are parallel, find the measures of angles 1 through 7. See Example 2.

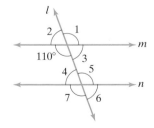

**14.** If lines $m$ and $n$ are parallel, find the measures of angles 1 through 5. See Example 2.

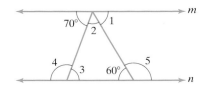

_In each of the following, the measures of two angles of a triangle are given. Find the measure of the third angle. See Example 3._

**15.** $11°, 79°$

**16.** $8°, 102°$

**17.** $25°, 65°$

**18.** $44°, 19°$

**19.** $30°, 60°$

**20.** $67°, 23°$

_In each of the following, the measure of one angle of a right triangle is given. Find the measures of the other two angles._

**21.** $45°$

**22.** $60°$

**23.** $17°$

**24.** $30°$

**25.** $39\frac{3}{4}°$

**26.** $72.6°$

_Given that each of the following pairs of triangles is similar, find the missing lengths. See Example 4._

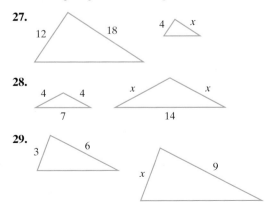

**27.**

**28.**

**29.**

**30.**

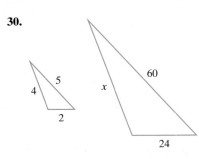

*Use the Pythagorean theorem to find the missing lengths in the right triangles. See Example 5.*

**31.**

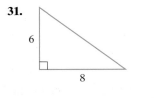

**32.**

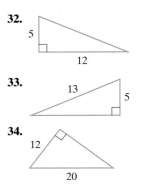

**33.**

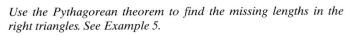

**34.**

# Answers to Selected Exercises

## CHAPTER 1 REVIEW OF REAL NUMBERS

### Section 1.2
#### Practice Exercises
**1. a.** $<$ **b.** $>$ **c.** $<$ **2. a.** True **b.** False **c.** True **d.** True **3. a.** $3 < 8$ **b.** $15 \geq 9$ **c.** $6 \neq 7$ **4.** $-52$ **5. a.** 25 **b.** 25
**c.** 25, $-15$, $-99$ **d.** 25, $\frac{7}{3}$, $-15$, $-\frac{3}{4}$, $-3.7$, 8.8, $-99$ **e.** $\sqrt{5}$ **f.** 25, $\frac{7}{3}$, $-15$, $-\frac{3}{4}$, $\sqrt{5}$, $-3.7$, 8.8, $-99$ **6. a.** $<$ **b.** $>$ **c.** $=$ **7. a.** 8
**b.** 9 **c.** 2.5 **d.** $\frac{5}{11}$ **e.** $\sqrt{3}$ **8. a.** $=$ **b.** $>$ **c.** $<$ **d.** $>$ **e.** $<$

#### Vocabulary and Readiness Check 1.2
**1.** whole **3.** inequality **5.** real **7.** irrational

#### Exercise Set 1.2
**1.** $>$ **3.** $=$ **5.** $<$ **7.** $<$ **9.** $32 < 212$ **11.** $2631 > 2456$ **13.** true **15.** false **17.** false **19.** true **21.** $30 \leq 45$ **23.** $8 < 12$
**25.** $5 \geq 4$ **27.** $15 \neq -2$ **29.** $535; -8$ **31.** $-21,350$ **33.** $350; -126$ **35.** 1998, 1999 **37.** 1998, 1999, 2000 **39.** 279 million $>$ 273 million
**41.** whole, integers, rational, real **43.** integers, rational, real **45.** natural, whole, integers, rational, real **47.** rational, real **49.** irrational, real
**51.** false **53.** true **55.** true **57.** true **59.** false **61.** $>$ **63.** $>$ **65.** $<$ **67.** $<$ **69.** $>$ **71.** $=$ **73.** $<$ **75.** $<$
**77.** $-0.04 > -26.7$ **79.** sun **81.** sun **83.** $20 \leq 25$ **85.** $6 > 0$ **87.** $-12 < -10$ **89.** answers may vary

### Section 1.3
#### Practice Exercises
**1. a.** $2 \cdot 2 \cdot 3 \cdot 3$ **b.** $3 \cdot 5 \cdot 5$ **2. a.** $\frac{7}{8}$ **b.** $\frac{16}{3}$ **c.** $\frac{7}{25}$ **3.** $\frac{7}{24}$ **4. a.** $\frac{27}{16}$ **b.** $\frac{1}{36}$ **c.** $\frac{5}{2}$ **5. a.** 1 **b.** $\frac{6}{5}$ **c.** $\frac{4}{5}$ **d.** $\frac{1}{2}$ **6.** $\frac{14}{21}$
**7. a.** $\frac{46}{77}$ **b.** $3\frac{15}{26}$ **c.** $\frac{1}{2}$

#### Vocabulary and Readiness Check 1.3
**1.** fraction **3.** product **5.** factors, product **7.** equivalent **9.** $\frac{1}{4}$ **11.** $\frac{2}{5}$

#### Exercise Set 1.3
**1.** $3 \cdot 11$ **3.** $2 \cdot 7 \cdot 7$ **5.** $2 \cdot 2 \cdot 5$ **7.** $3 \cdot 5 \cdot 5$ **9.** $3 \cdot 3 \cdot 5$ **11.** $\frac{1}{2}$ **13.** $\frac{2}{3}$ **15.** $\frac{3}{7}$ **17.** $\frac{3}{5}$ **19.** $\frac{3}{8}$ **21.** $\frac{1}{2}$ **23.** $\frac{6}{7}$ **25.** 15 **27.** $\frac{1}{6}$
**29.** $\frac{25}{27}$ **31.** $\frac{11}{20}$ sq mi **33.** $\frac{3}{5}$ **35.** 1 **37.** $\frac{1}{3}$ **39.** $\frac{9}{35}$ **41.** $\frac{21}{30}$ **43.** $\frac{4}{18}$ **45.** $\frac{16}{20}$ **47.** $\frac{23}{21}$ **49.** $1\frac{2}{3}$ **51.** $\frac{5}{66}$ **53.** $\frac{7}{5}$ **55.** $\frac{1}{5}$
**57.** $\frac{3}{8}$ **59.** $\frac{1}{9}$ **61.** $\frac{5}{7}$ **63.** $\frac{65}{21}$ **65.** $\frac{2}{5}$ **67.** $\frac{9}{7}$ **69.** $\frac{3}{4}$ **71.** $\frac{17}{3}$ **73.** $\frac{7}{26}$ **75.** 1 **77.** $\frac{1}{5}$ **79.** $5\frac{1}{6}$ **81.** $\frac{17}{18}$ **83.** $55\frac{1}{4}$ ft
**85.** $6\frac{7}{50}$ m **87.** answers may vary **89.** $3\frac{3}{8}$ mi **91.** $\frac{7}{50}$ **93.** $\frac{1}{4}$ **95.** $\frac{160}{509}$ **97.** $\frac{7}{36}$ sq ft

### Section 1.4
#### Practice Exercises
**1. a.** 1 **b.** 25 **c.** $\frac{1}{100}$ **d.** 9 **e.** $\frac{8}{125}$ **2. a.** 33 **b.** 11 **c.** $\frac{32}{9}$ or $3\frac{5}{9}$ **d.** 36 **e.** $\frac{3}{16}$ **3.** $\frac{31}{11}$ **4.** 4 **5.** $\frac{9}{22}$ **6. a.** 9 **b.** $\frac{8}{15}$
**c.** $\frac{19}{10}$ **d.** 33 **7.** No **8. a.** $6x$ **b.** $x - 8$ **c.** $x \cdot 9$ or $9x$ **d.** $2x + 3$ **e.** $7 + x$ **9. a.** $x + 7 = 13$ **b.** $x - 2 = 11$ **c.** $2x + 9 \neq 25$
**d.** $5(11) \geq x$

#### Calculator Explorations 1.4
**1.** 625 **3.** 59,049 **5.** 30 **7.** 9857 **9.** 2376

#### Vocabulary and Readiness Check 1.4
**1.** base, exponent **3.** variable **5.** equation **7.** solving **9.** add **11.** divide

#### Exercise Set 1.4
**1.** 243 **3.** 27 **5.** 1 **7.** 5 **9.** $\frac{1}{125}$ **11.** $\frac{16}{81}$ **13.** 49 **15.** 16 **17.** 1.44 **19.** 17 **21.** 20 **23.** 10 **25.** 21 **27.** 45 **29.** 0
**31.** $\frac{2}{7}$ **33.** 30 **35.** 2 **37.** $\frac{7}{18}$ **39.** $\frac{27}{10}$ **41.** $\frac{7}{5}$ **43.** no **45. a.** 64 **b.** 43 **c.** 19 **d.** 22 **47.** 9 **49.** 1 **51.** 1 **53.** 11

**55.** 45    **57.** 27    **59.** 132    **61.** $\dfrac{37}{18}$    **63.** 16, 64, 144, 256    **65.** yes    **67.** no    **69.** no    **71.** yes    **73.** no    **75.** $x + 15$    **77.** $x - 5$

**79.** $3x + 22$    **81.** $1 + 2 = 9 \div 3$    **83.** $3 \neq 4 \div 2$    **85.** $5 + x = 20$    **87.** $13 - 3x = 13$    **89.** $\dfrac{12}{x} = \dfrac{1}{2}$    **91.** answers may vary

**93.** $(20 - 4) \cdot 4 \div 2$    **95.** 28 m    **97.** 12,000 sq ft    **99.** 6.5%    **101.** \$13.08

## Section 1.5
### Practice Exercises

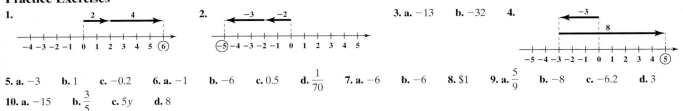

**1.**    **2.**    **3. a.** $-13$   **b.** $-32$   **4.**

**5. a.** $-3$   **b.** 1   **c.** $-0.2$   **6. a.** $-1$   **b.** $-6$   **c.** 0.5   **d.** $\dfrac{1}{70}$   **7. a.** $-6$   **b.** $-6$   **8.** \$1   **9. a.** $\dfrac{5}{9}$   **b.** $-8$   **c.** $-6.2$   **d.** 3

**10. a.** $-15$   **b.** $\dfrac{3}{5}$   **c.** $5y$   **d.** 8

### Vocabulary and Readiness Check 1.5

**1.** opposites    **3.** $n$    **5.** positive number    **7.** negative number    **9.** 0

### Exercise Set 1.5

**1.** 9    **3.** $-14$    **5.** 1    **7.** $-12$    **9.** $-5$    **11.** $-12$    **13.** $-4$    **15.** 7    **17.** $-2$    **19.** 0    **21.** $-19$    **23.** 31    **25.** $-47$    **27.** $-2.1$
**29.** $-8$    **31.** 38    **33.** $-13.1$    **35.** $\dfrac{2}{8} = \dfrac{1}{4}$    **37.** $-\dfrac{3}{16}$    **39.** $-\dfrac{13}{10}$    **41.** $-8$    **43.** $-59$    **45.** $-9$    **47.** 5    **49.** 11    **51.** $-18$    **53.** 19
**55.** $-0.7$    **57.** $-6°$    **59.** $-16,427$ ft    **61.** $-\$9250$ million    **63.** $-9$    **65.** $-6$    **67.** 2    **69.** 0    **71.** $-6$    **73.** answers may vary    **75.** $-2$
**77.** 0    **79.** $-\dfrac{2}{3}$    **81.** answers may vary    **83.** yes    **85.** no    **87.** July    **89.** October    **91.** 4.7°F    **93.** negative    **95.** positive

## Section 1.6
### Practice Exercises

**1. a.** $-13$   **b.** $-7$   **c.** 12   **d.** $-2$   **2. a.** 10.9   **b.** $-\dfrac{1}{2}$   **c.** $-\dfrac{19}{20}$   **3.** $-7$   **4. a.** $-6$   **b.** 6.1   **5. a.** $-20$   **b.** 13   **6. a.** 2   **b.** 13

**7.** \$357   **8. a.** 28°   **b.** 137°

### Vocabulary and Readiness Check 1.6

**1.** $7 - x$    **3.** $x - 7$    **5.** $7 - x$

### Exercise Set 1.6

**1.** $-10$    **3.** $-5$    **5.** 19    **7.** $\dfrac{1}{6}$    **9.** 2    **11.** $-11$    **13.** 11    **15.** 5    **17.** 37    **19.** $-6.4$    **21.** $-71$    **23.** 0    **25.** 4.1    **27.** $\dfrac{2}{11}$
**29.** $-\dfrac{11}{12}$    **31.** 8.92    **33.** 13    **35.** $-5$    **37.** $-1$    **39.** $-23$    **41.** answers may vary    **43.** $-26$    **45.** $-24$    **47.** 3    **49.** $-45$    **51.** $-4$
**53.** 13    **55.** 6    **57.** 9    **59.** $-9$    **61.** $-7$    **63.** $\dfrac{7}{5}$    **65.** 21    **67.** $\dfrac{1}{4}$    **69.** 100°    **71.** $-23$ yd or 23 yd loss    **73.** $-569$ or 569 B.C.
**75.** $-308$ ft    **77.** 19,852 ft    **79.** 130°    **81.** 30°    **83.** no    **85.** no    **87.** yes    **89.** $-4.4°$; $2.6°$; $12°$; $23.5°$; $15.3°$; $3.9°$; $-0.3°$; $-6.3°$; $-18.2°$; $-15.7°$; $-10.3°$    **91.** October    **93.** true    **95.** true    **97.** negative, $-2.6466$

### Integrated Review

**1.** negative    **2.** negative    **3.** positive    **4.** 0    **5.** positive    **6.** 0    **7.** positive    **8.** positive    **9.** $-\dfrac{1}{7}; \dfrac{1}{7}$    **10.** $\dfrac{12}{5}; \dfrac{12}{5}$    **11.** 3; 3
**12.** $-\dfrac{9}{11}; \dfrac{9}{11}$    **13.** $-42$    **14.** 10    **15.** 2    **16.** $-18$    **17.** $-7$    **18.** $-39$    **19.** $-2$    **20.** $-9$    **21.** $-3.4$    **22.** $-9.8$    **23.** $-\dfrac{25}{28}$    **24.** $-\dfrac{5}{24}$
**25.** $-4$    **26.** $-24$    **27.** 6    **28.** 20    **29.** 6    **30.** 61    **31.** $-6$    **32.** $-16$    **33.** $-19$    **34.** $-13$    **35.** $-4$    **36.** $-1$    **37.** $\dfrac{13}{20}$    **38.** $-\dfrac{29}{40}$
**39.** 4    **40.** 9    **41.** $-1$    **42.** $-3$    **43.** 8    **44.** 10    **45.** 47    **46.** $\dfrac{2}{3}$

## Section 1.7
### Practice Exercises

**1. a.** $-40$   **b.** 12   **c.** $-54$   **2. a.** $-30$   **b.** 24   **c.** 0   **d.** 26   **3. a.** $-0.046$   **b.** $-\dfrac{4}{15}$   **c.** 14   **4. a.** 36   **b.** $-36$   **c.** $-64$   **d.** $-64$

**5. a.** $\dfrac{3}{8}$   **b.** $\dfrac{1}{15}$   **c.** $-\dfrac{7}{2}$   **d.** $-\dfrac{1}{5}$   **6. a.** $-8$   **b.** $-4$   **c.** 5   **7. a.** 3   **b.** $-16$   **c.** $-\dfrac{6}{5}$   **d.** $-\dfrac{1}{18}$   **8. a.** 0   **b.** undefined

**c.** undefined   **9. a.** $\dfrac{-84}{5}$   **b.** 11   **10. a.** $-9$   **b.** 33   **c.** $\dfrac{5}{3}$

## Calculator Explorations 1.7
**1.** 38    **3.** $-441$    **5.** $163.\overline{3}$    **7.** 54,499    **9.** 15,625

## Vocabulary and Readiness Check 1.7
**1.** 0, 0    **3.** positive    **5.** negative    **7.** positive

## Exercise Set 1.7
**1.** $-24$    **3.** $-2$    **5.** 50    **7.** $-12$    **9.** 0    **11.** $-18$    **13.** $\frac{3}{10}$    **15.** $\frac{2}{3}$    **17.** $-7$    **19.** 0.14    **21.** $-800$    **23.** $-28$    **25.** 25    **27.** $-\frac{8}{27}$

**29.** $-121$    **31.** $-\frac{1}{4}$    **33.** $-30$    **35.** 23    **37.** $-7$    **39.** true    **41.** false    **43.** 16    **45.** $-1$    **47.** 25    **49.** $-49$    **51.** $\frac{1}{9}$    **53.** $\frac{3}{2}$

**55.** $-\frac{1}{14}$    **57.** $-\frac{11}{3}$    **59.** $\frac{1}{0.2}$    **61.** $-6.3$    **63.** $-9$    **65.** 4    **67.** $-4$    **69.** 0    **71.** $-5$    **73.** undefined    **75.** 3    **77.** $-15$    **79.** $-\frac{18}{7}$

**81.** $\frac{20}{27}$    **83.** $-1$    **85.** $-\frac{9}{2}$    **87.** $-4$    **89.** 16    **91.** $-3$    **93.** $-\frac{16}{7}$    **95.** 2    **97.** $\frac{6}{5}$    **99.** $-5$    **101.** $\frac{3}{2}$    **103.** $-21$    **105.** 41

**107.** $-134$    **109.** 3    **111.** 0    **113.** $-\$24{,}812$ million    **115.** yes    **117.** no    **119.** yes    **121.** answers may vary    **123.** 1, $-1$

**125.** positive    **127.** not possible    **129.** negative    **131.** $-2 + \frac{-15}{3}; -7$    **133.** $2[-5 + (-3)]; -16$

## The Bigger Picture
**1.** $-5$    **2.** $-14$    **3.** $-\frac{26}{35}$    **4.** 8    **5.** 49    **6.** $-49$    **7.** $-21$    **8.** undefined    **9.** 0    **10.** $-12$    **11.** $-16.6$    **12.** $\frac{1}{6}$    **13.** $-79$

**14.** $\frac{10}{13}$    **15.** 50    **16.** $-12$

## Section 1.8
### Practice Exercises
**1. a.** $8 \cdot x$    **b.** $17 + x$    **2. a.** $2 + (9 + 7)$    **b.** $(-4 \cdot 2) \cdot 7$    **3. a.** $x + 14$    **b.** $-30x$    **4. a.** $5x - 5y$    **b.** $-24 - 12t$    **c.** $6x - 8y - 2z$
**d.** $-3 + y$    **e.** $-x + 7 - 2s$    **f.** $14x + 14$    **5. a.** $5(w + 3)$    **b.** $9(w + z)$    **6. a.** commutative property of multiplication    **b.** associative
property of addition    **c.** identity element for addition    **d.** multiplicative inverse property    **e.** commutative property of addition    **f.** additive
inverse property    **g.** commutative and associative properties of multiplication

### Vocabulary and Readiness Check 1.8
**1.** commutative property of addition    **3.** distributive property    **5.** associative property of addition    **7.** opposites or additive inverses

### Exercise Set 1.8
**1.** $16 + x$    **3.** $y \cdot (-4)$    **5.** $yx$    **7.** $13 + 2x$    **9.** $x \cdot (yz)$    **11.** $(2 + a) + b$    **13.** $(4a) \cdot b$    **15.** $a + (b + c)$    **17.** $17 + b$    **19.** $24y$
**21.** $y$    **23.** $26 + a$    **25.** $-72x$    **27.** $s$    **29.** answers may vary    **31.** $4x + 4y$    **33.** $9x - 54$    **35.** $6x + 10$    **37.** $28x - 21$    **39.** $18 + 3x$
**41.** $-2y + 2z$    **43.** $-21y - 35$    **45.** $5x + 20m + 10$    **47.** $-4 + 8m - 4n$    **49.** $-5x - 2$    **51.** $-r + 3 + 7p$    **53.** $3x + 4$    **55.** $-x + 3y$
**57.** $6r + 8$    **59.** $-36x - 70$    **61.** $-16x - 25$    **63.** $4(1 + y)$    **65.** $11(x + y)$    **67.** $-1(5 + x)$    **69.** $30(a + b)$    **71.** commutative property
of multiplication    **73.** associative property of addition    **75.** distributive property    **77.** associative property of multiplication    **79.** identity
element of addition    **81.** distributive property    **83.** commutative and associative properties of multiplication    **85.** $-8; \frac{1}{8}$    **87.** $-x; \frac{1}{x}$

**89.** $2x; -2x$    **91.** no    **93.** yes    **95.** answers may vary

## Chapter 1 Vocabulary Check
**1.** inequality symbols    **2.** equation    **3.** absolute value    **4.** variable    **5.** opposites    **6.** numerator    **7.** solution    **8.** reciprocals
**9.** base; exponent    **10.** denominator    **11.** grouping symbols    **12.** set

## Chapter 1 Review
**1.** $<$    **3.** $>$    **5.** $<$    **7.** $=$    **9.** $>$    **11.** $4 \geq -3$    **13.** $0.03 < 0.3$    **15. a.** 1, 3    **b.** 0, 1, 3    **c.** $-6, 0, 1, 3$    **d.** $-6, 0, 1, 1\frac{1}{2}, 3, 9.62$

**e.** $\pi$    **f.** $-6, 0, 1, 1\frac{1}{2}, 3, \pi, 9.62$    **17.** Friday    **19.** $2 \cdot 2 \cdot 3 \cdot 3$    **21.** $\frac{12}{25}$    **23.** $\frac{13}{10}$    **25.** $9\frac{3}{8}$    **27.** 15    **29.** $\frac{7}{12}$    **31.** $A = \frac{34}{121}$ sq in.; $P = 2\frac{4}{11}$ in.

**33.** $2\frac{15}{16}$ lb    **35.** $11\frac{5}{16}$ lb    **37.** Odera    **39.** $3\frac{7}{8}$ lb    **41.** 16    **43.** $\frac{4}{49}$    **45.** 70    **47.** 37    **49.** $\frac{18}{7}$    **51.** $20 - 12 = 2 \cdot 4$    **53.** 18    **55.** 5

**57.** $63°$    **59.** no    **61.** $-\frac{2}{3}$    **63.** 7    **65.** $-17$    **67.** $-5$    **69.** 3.9    **71.** $-14$    **73.** 5    **75.** $-19$    **77.** 15    **79.** $-\frac{1}{6}$    **81.** $-48$    **83.** 3

**85.** undefined    **87.** undefined    **89.** $-12$    **91.** 9    **93.** $-5$    **95.** commutative property of addition    **97.** distributive property
**99.** associative property of addition    **101.** distributive property    **103.** multiplicative inverse    **105.** $5y - 10$    **107.** $-7 + x - 4z$

**109.** $-12x - 27$    **111.** $<$    **113.** $-15.3$    **115.** $-80$    **117.** $-\frac{1}{4}$    **119.** 16    **121.** $-5$    **123.** $-\frac{5}{6}$

## Chapter 1 Test

**1.** $|-7| > 5$   **2.** $9 + 5 \geq 4$   **3.** $-5$   **4.** $-11$   **5.** $-3$   **6.** $-39$   **7.** $12$   **8.** $-2$   **9.** undefined   **10.** $-8$   **11.** $-\frac{1}{3}$   **12.** $4\frac{5}{8}$

**13.** $-\frac{5}{2}$ or $-2\frac{1}{2}$   **14.** $-32$   **15.** $-48$   **16.** $3$   **17.** $0$   **18.** $>$   **19.** $>$   **20.** $<$   **21.** $=$   **22.** $2221 < 10{,}993$ or $10{,}993 > 2221$

**23. a.** $1, 7$   **b.** $0, 1, 7$   **c.** $-5, -1, 0, 1, 7$   **d.** $-5, -1, 0, \frac{1}{4}, 1, 7, 11.6$   **e.** $\sqrt{7}, 3\pi$   **f.** $-5, -1, 0, \frac{1}{4}, 1, 7, 11.6, \sqrt{7}, 3\pi$   **24.** $40$   **25.** $12$

**26.** $22$   **27.** $-1$   **28.** associative property of addition   **29.** commutative property of multiplication   **30.** distributive property
**31.** multiplicative inverse property   **32.** $9$   **33.** $-3$   **34.** second down   **35.** yes   **36.** $17°$   **37.** $\$650$ million   **38.** $\$420$

# CHAPTER 2  EQUATIONS, INEQUALITIES, AND PROBLEM SOLVING

## Section 2.1

### Practice Exercises

**1. a.** $1$   **b.** $-7$   **c.** $-\frac{1}{5}$   **d.** $43$   **e.** $-1$   **2. a.** like terms   **b.** unlike terms   **c.** like terms   **d.** like terms   **3. a.** $7x^2$   **b.** $-2y$

**c.** $5x + 5x^2$   **4. a.** $11y - 5$   **b.** $5x - 6$   **c.** $-\frac{1}{4}t$   **d.** $12.2y + 13$   **e.** $5z - 3z^4$   **5. a.** $6x - 21$   **b.** $-15x + 20z + 25$

**c.** $-2x + y - z + 2$   **6. a.** $36x + 10$   **b.** $-11x + 1$   **c.** $-30x - 17$   **7.** $-5x + 4$   **8. a.** $3 + 2x$   **b.** $x - 1$   **c.** $2x + 10$   **d.** $\frac{13}{2}x$

### Vocabulary and Readiness Check 2.1

**1.** expression; term   **3.** numerical coefficient   **5.** numerical coefficient   **7.** $-7$   **9.** $1$   **11.** $-\frac{5}{3}$   **13.** like   **15.** unlike

### Exercise Set 2.1

**1.** $15y$   **3.** $13w$   **5.** $-7b - 9$   **7.** $-m - 6$   **9.** $-8$   **11.** $7.2x - 5.2$   **13.** $4x - 3$   **15.** $5x^2$   **17.** $1.3x + 3.5$   **19.** answers may vary
**21.** $5y - 20$   **23.** $-2x - 4$   **25.** $7d - 11$   **27.** $-10x + 15y - 30$   **29.** $-3x + 2y - 1$   **31.** $2x + 14$   **33.** $10x - 3$   **35.** $-4x - 9$

**37.** $-4m - 3$   **39.** $k - 6$   **41.** $-15x + 18$   **43.** $16$   **45.** $x + 5$   **47.** $x + 2$   **49.** $2k + 10$   **51.** $-3x + 5$   **53.** $-11$   **55.** $3y + \frac{5}{6}$

**57.** $-22 + 24x$   **59.** $0.9m + 1$   **61.** $10 - 6x - 9y$   **63.** $-x - 38$   **65.** $5x - 7$   **67.** $2x - 4$   **69.** $2x + 7$   **71.** $\frac{3}{4}x + 12$

**73.** $-2 + 12x$   **75.** $8(x + 6)$ or $8x + 48$   **77.** $x - 10$   **79.** $\frac{7x}{6}$   **81.** $7x - 7$   **83.** $2$   **85.** $-23$   **87.** $-25$   **89.** $(18x - 2)$ ft

**91.** balanced   **93.** balanced   **95.** answers may vary   **97.** $(15x + 23)$ in.   **99.** $5b^2c^3 + b^3c^2$   **101.** $5x^2 + 9x$   **103.** $-7x^2y$

## Section 2.2

### Practice Exercises

**1.** $-8$   **2.** $-1.8$   **3.** $-10$   **4.** $18$   **5.** $20$   **6.** $-12$   **7.** $65$   **8.** $-3$   **9. a.** $7$   **b.** $9 - x$   **c.** $(9 - x)$ ft   **10.** $3x + 6$

### Vocabulary and Readiness Check 2.2

**1.** equation; expression   **3.** solution   **5.** addition   **7.** multiplication   **9.** false   **11.** $9$   **13.** $2$

### Exercise Set 2.2

**1.** $3$   **3.** $-2$   **5.** $-14$   **7.** $0.5$   **9.** $-3$   **11.** $-0.7$   **13.** $3$   **15.** $11$   **17.** $0$   **19.** $-3$   **21.** $16$   **23.** $-4$   **25.** $0$

**27.** $12$   **29.** $10$   **31.** $-12$   **33.** $3$   **35.** $-2$   **37.** $0$   **39.** answers may vary   **41.** $10$   **43.** $-20$   **45.** $0$   **47.** $-5$   **49.** $0$   **51.** $-\frac{3}{2}$

**53.** $-21$   **55.** $\frac{11}{2}$   **57.** $1$   **59.** $-\frac{1}{4}$   **61.** $12$   **63.** $-30$   **65.** $\frac{9}{10}$   **67.** $-30$   **69.** $2$   **71.** $-2$   **73.** $23$   **75.** $20 - p$   **77.** $(10 - x)$ ft

**79.** $(180 - x)°$   **81.** $(n + 284)$ votes   **83.** $(m - 60)$ ft   **85.** $(n + 47{,}628)$ votes   **87.** $7x$ sq mi   **89.** $2x + 2$   **91.** $2x + 2$

**93.** $5x + 20$   **95.** $7x - 12$   **97.** $1$   **99.** $>$   **101.** $=$   **103.** $(173 - 3x)°$   **105.** answers may vary   **107.** $4$

**109.** answers may vary   **111.** answers may vary   **113.** $-48$   **115.** $\frac{700}{3}$ mg   **117.** solution   **119.** $-2.95$   **121.** $0.02$

## Section 2.3
### Practice Exercises

**1.** 3   **2.** $\frac{21}{13}$   **3.** $-15$   **4.** 3   **5.** 0   **6.** no solution   **7.** all real numbers

### Calculator Explorations 2.3

**1.** solution   **3.** not a solution   **5.** solution

### Vocabulary and Readiness Check 2.3

**1.** equation   **3.** expression   **5.** expression   **7.** equation

### Exercise Set 2.3

**1.** $-6$   **3.** 3   **5.** 1   **7.** $\frac{3}{2}$   **9.** 0   **11.** $-1$   **13.** 4   **15.** $-4$   **17.** $-3$   **19.** 2   **21.** 50   **23.** 1   **25.** $\frac{7}{3}$   **27.** 0.2

**29.** all real numbers   **31.** no solution   **33.** no solution   **35.** all real numbers   **37.** 18   **39.** $\frac{19}{9}$   **41.** $\frac{14}{3}$   **43.** 13   **45.** 4

**47.** all real numbers   **49.** $-\frac{3}{5}$   **51.** $-5$   **53.** 10   **55.** no solution   **57.** 3   **59.** $-17$   **61.** $-4$   **63.** 3   **65.** all real numbers

**67.** $-8 - x$   **69.** $-3 + 2x$   **71.** $9(x + 20)$   **73.** $(6x - 8)$ m   **75. a.** all real numbers   **b.** answers may vary   **c.** answers may vary   **77.** a
**79.** b   **81.** c   **83.** answers may vary   **85. a.** $x + x + x + 2x + 2x = 28$   **b.** $x = 4$   **c.** $x = 4$ cm; $2x = 8$ cm   **87.** answers may vary

**89.** 15.3   **91.** $-0.2$   **93.** $-\frac{7}{8}$   **95.** no solution

### The Bigger Picture

**1.** $-\frac{8}{3}$   **2.** 0   **3.** $-1$   **4.** 3   **5.** 6   **6.** no solution   **7.** $\frac{3}{2}$   **8.** all real numbers

### Integrated Review

**1.** 6   **2.** $-17$   **3.** 12   **4.** $-26$   **5.** $-3$   **6.** $-1$   **7.** $\frac{27}{2}$   **8.** $\frac{25}{2}$   **9.** 8   **10.** $-64$   **11.** 2   **12.** $-3$   **13.** no solution

**14.** no solution   **15.** $-2$   **16.** $-2$   **17.** $-\frac{5}{6}$   **18.** $\frac{1}{6}$   **19.** 1   **20.** 6   **21.** 4   **22.** 1   **23.** $\frac{9}{5}$   **24.** $-\frac{6}{5}$   **25.** all real numbers

**26.** all real numbers   **27.** 0   **28.** $-1.6$   **29.** $\frac{4}{19}$   **30.** $-\frac{5}{19}$   **31.** $\frac{7}{2}$   **32.** $-\frac{1}{4}$   **33.** no solution   **34.** no solution   **35.** $\frac{7}{6}$   **36.** $\frac{1}{15}$

## Section 2.4
### Practice Exercises

**1.** 9   **2.** 2   **3.** 9 in. and 36 in.   **4.** 22 Republican and 28 Democratic Governors   **5.** $25°, 75°, 80°$   **6.** 46, 48, 50

### Vocabulary and Readiness Check 2.4

**1.** $2x; 2x - 31$   **3.** $x + 5; 2(x + 5)$   **5.** $20 - y; \frac{20 - y}{3}$ or $(20 - y) \div 3$

### Exercise Set 2.4

**1.** $2x + 7 = x + 6; -1$   **3.** $3x - 6 = 2x + 8; 14$   **5.** $2(x - 8) = 3(x + 3); -25$   **7.** $4(-2 + x) = 5x + \frac{1}{2}; -\frac{17}{2}$   **9.** 5 ft, 12 ft

**11.** Armanty: 22 tons; Hoba West: 66 tons   **13.** China: 42,400; U.S.: 36,594   **15.** 1st angle: $37.5°$; 2nd angle: $37.5°$; 3rd angle: $105°$

**17.** $3x + 3$   **19.** $x + 2, x + 4, 2x + 4$   **21.** $x + 1; x + 2; x + 3; 4x + 6$   **23.** $x + 2; x + 4; 2x + 6$   **25.** 234, 235

**27.** Belgium: 32; France: 33; Spain: 34   **29.** 3 in.; 6 in.; 16 in.   **31.** $\frac{5}{4}$   **33.** Botswana: 32,000,000 carats; Angola: 8,000,000 carats   **35.** $58°, 60°, 62°$

**37.** Russia: 22; Austria: 23; Canada: 24; United States: 25   **39.** $-16$   **41.** Weller: 108,375; Pavich: 88,179   **43.** $43°, 137°$   **45.** 1

**47.** 1st angle: $65°$; 2nd angle: $115°$   **49.** Maglev: 361 mph; TGV: 357.2 mph   **51.** $\frac{5}{2}$   **53.** California: 58; Montana: 56   **55.** Colts: 29; Bears: 17

**57.** $34.5°, 34.5°, 111°$   **59.** 1st piece: 5 in.; 2nd piece: 10 in.; 3rd piece: 25 in.   **61.** Hawaii   **63.** Texas: $31.1 million; Florida: $29.4 million
**65.** answers may vary   **67.** 34   **69.** $225\pi$   **71.** answers may vary

## Section 2.5
### Practice Exercises
**1.** 116 sec or 1 min 56 sec **2.** width: 9 ft **3.** 46.4°F **4.** length: 28 in.; width: 5 in. **5.** $r = \dfrac{I}{Pt}$ **6.** $s = \dfrac{H - 10a}{5a}$ **7.** $d = \dfrac{N - F}{n - 1}$
**8.** $B = \dfrac{2A - ab}{a}$

### Exercise Set 2.5
**1.** $h = 3$ **3.** $h = 3$ **5.** $h = 20$ **7.** $c = 12$ **9.** $r \approx 2.5$ **11.** $T = 3$ **13.** $h \approx 15$ **15.** $h = \dfrac{f}{5g}$ **17.** $w = \dfrac{V}{lh}$ **19.** $y = 7 - 3x$
**21.** $R = \dfrac{A - P}{PT}$ **23.** $A = \dfrac{3V}{h}$ **25.** $a = P - b - c$ **27.** $h = \dfrac{S - 2\pi r^2}{2\pi r}$ **29. a.** area: 103.5 sq ft; perimeter: 41 ft

**b.** baseboard: perimeter; carpet: area **31. a.** area: 480 sq in.; perimeter: 120 in. **b.** frame: perimeter; glass: area **33.** 70 ft **35.** −10°C
**37.** 6.25 hr **39.** length: 78 ft; width: 52 ft **41.** 18 ft, 36 ft, 48 ft **43.** 55.2 mph **45.** 96 piranhas **47.** 2 bags **49.** one 16-in. pizza
**51.** $x = 6$ m, $2.5x = 15$ m **53.** 0.2 hr or 12 min **55.** 13 in. **57.** 2.25 hr **59.** 12,090 ft **61.** 50°C **63.** 515,509.5 cu in. **65.** 449 cu in.

**67.** 332.6°F **69.** $\dfrac{9}{x + 5}$ **71.** $3(x + 4)$ **73.** $3(x - 12)$ **75.** 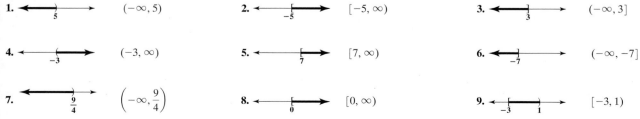 **77.** −109.3°F **79.** 500 sec or $8\frac{1}{3}$ min

**81.** 608.33 ft **83.** 565.5 cu in. **85.** It multiplies the area by 4.

## Section 2.6
### Practice Exercises
**1.** 62.5% **2.** 360 **3. a.** 4% **b.** 87% **c.** 13 people **4.** discount: $408; new price: $72 **5.** 50.7% **6.** 520 new films
**7.** 2 liters of 5% eyewash; 4 liters of 2% eyewash

### Vocabulary and Readiness Check 2.6
**1.** no **3.** yes

### Exercise Set 2.6
**1.** 11.2 **3.** 55% **5.** 180 **7.** 69% **9.** 1896.3 million bushels or 1,896,300,000 bushels **11.** discount: $1480; new price: $17,020
**13.** $46.58 **15.** 35% **17.** 30% **19.** $104 **21.** $42,500 **23.** 2 gal **25.** 7 lb **27.** 4.6 **29.** 50 **31.** 30% **33.** 71%
**35.** 178,778 **37.** 46%; 28%; 9%; 6%; total: 100% **39.** decrease: $64; sale price: $192 **41.** 115% increase **43.** 230 million **45.** 400 oz
**47.** markup: $18.90; adult ticket price: $45.90 **49.** 300% **51.** 120 employees **53.** 5 lb **55.** 4.6% **57.** 335 decisions
**59.** 854 thousand Scoville units **61.** 361 college students **63.** > **65.** = **67.** > **69.** no; answers may vary **71.** no; answers may vary
**73.** 9.6% **75.** 26.9%; yes **77.** 17.1%

## Section 2.7
### Practice Exercises
**1.** 2.2 hr **2.** eastbound: 62 mph; westbound: 52 mph **3.** 106 $5 bills; 59 $20 bills **4.** $18,000 at 11.5%; $12,000 at 6%

### Exercise Set 2.7
**1.** $666\frac{2}{3}$ mi **3.** 55 mph **5.** $0.10\,y$ **7.** $0.05(x + 7)$ **9.** $20(4y)$ or $80y$ **11.** $50(35 - x)$ **13.** 12 $10 bills; 32 $5 bills
**15.** $11,500 at 8%; $13,500 at 9% **17.** $7000 at 11% profit; $3000 at 4% loss **19.** 187 adult tickets; 313 child tickets
**21.** $30,000 at 8%; $24,000 at 10% **23.** 2 hr $37\frac{1}{2}$ min **25.** 483 dimes; 161 nickels **27.** $4500 **29.** 2.2 mph; 3.3 mph **31.** 27.5 mi **33.** −4
**35.** $\dfrac{9}{16}$ **37.** −4 **39.** 25 $100 bills; 71 $50 bills; 175 $20 bills **41.** 25 skateboards **43.** 800 books **45.** answers may vary

## Section 2.8
### Practice Exercises
**1.** $(-\infty, 5)$ **2.** $[-5, \infty)$ **3.** $(-\infty, 3]$

**4.** $(-3, \infty)$ **5.** $[7, \infty)$ **6.** $(-\infty, -7]$

**7.** $\left(-\infty, \dfrac{9}{4}\right)$ **8.** $[0, \infty)$ **9.** $[-3, 1)$

**10.**    $(-2, 2]$    **11.** $-\dfrac{16}{3}$ ... $\dfrac{4}{3}$       $\left(-\dfrac{16}{3}, \dfrac{4}{3}\right)$    **12.** $x \le 3.2$; Kasonga can afford at most 3 classes.

## Vocabulary and Readiness Check 2.8

**1.** expression    **3.** inequality    **5.** equation    **7.** $-5$    **9.** 4.1

## Exercise Set 2.8

**1.**    $x \ge 2$    **3.**    $x < -5$    **5.**    $(-\infty, -1]$

**7.**    $\left(-\infty, \dfrac{1}{2}\right)$    **9.**    $[5, \infty)$

**11.**    $(-\infty, -3)$    **13.**    $[-5, \infty)$

**15.**    $[-2, \infty)$    **17.**    $(-3, \infty)$

**19.**    $(-\infty, 1]$    **21.**    $(-5, \infty)$

**23.**    $(-\infty, -2]$    **25.**    $(-\infty, -8]$

**27.**    $(4, \infty)$    **29.**    $[20, \infty)$

**31.**    $(16, \infty)$    **33.**    $(-3, \infty)$

**35.**    $\left(-\infty, -\dfrac{2}{3}\right]$    **37.**    $\left(\dfrac{8}{3}, \infty\right)$

**39.**    $(-13, \infty)$    **41.**    $(-\infty, 0)$

**43.**    $(-\infty, 0]$    **45.**    $(3, \infty)$

**47.**    $(-\infty, 0]$    **49.** answers may vary    **51.**    $(-1, 3)$

**53.**    $[0, 2)$    **55.**    $(-1, 2)$

**57.**    $[4, 5]$    **59.**    $(1, 5]$

**61.**    $(1, 4)$    **63.**    $\left(0, \dfrac{14}{3}\right]$

**65.** answers may vary    **67.** $x > -10$    **69.** 86 people    **71.** $x \le 35$    **73.** at least 10%    **75.** at least 193
**77.** $x < 200$ recommended; $200 \le x \le 240$ borderline; $x > 240$ high    **79.** $-3 < x < 3$    **81.** $-38.2° \le F \le 113°$    **83.** 8

**85.** 1    **87.** $\dfrac{16}{49}$    **89.** \$52.70    **91.** 2005    **93.** $0.924 \le d \le 0.987$    **95.**    $(1, \infty)$

**97.**    $\left(-\infty, \dfrac{5}{8}\right)$

## The Bigger Picture

**1.** $-3$    **2.** $(-\infty, -3)$    **3.** $\dfrac{2}{9}$    **4.** $-\dfrac{1}{4}$    **5.** $[-15, \infty)$    **6.** no solution    **7.** 7    **8.** $(-\infty, 37)$    **9.** all real numbers    **10.** $\dfrac{41}{29}$

## Chapter 2 Vocabulary Check

**1.** like terms    **2.** linear equation in one variable    **3.** equivalent equations    **4.** compound inequalities    **5.** formula
**6.** linear inequality in one variable    **7.** numerical coefficient

## Chapter 2 Review

**1.** $6x$    **3.** $4x - 2$    **5.** $3n - 18$    **7.** $-6x + 7$    **9.** $3x - 7$    **11.** 4    **13.** 6    **15.** 0    **17.** $-23$    **19.** 5; 5    **21.** b    **23.** b    **25.** $-12$

**27.** 0    **29.** 0.75    **31.** $-6$    **33.** $-1$    **35.** $-\dfrac{1}{5}$    **37.** $3x + 3$    **39.** $-4$    **41.** 2    **43.** no solution    **45.** $\dfrac{3}{4}$    **47.** 20    **49.** $\dfrac{23}{7}$    **51.** 102

**53.** 6665.5 in.    **55.** Kellogg: 35 plants; Keebler: 18 plants    **57.** 3    **59.** $w = 9$    **61.** $m = \dfrac{y - b}{x}$    **63.** $x = \dfrac{2y - 7}{5}$    **65.** $\pi = \dfrac{C}{D}$    **67.** 15 m

**69.** 1 hr 20 min    **71.** 20%    **73.** 110    **75.** mark-up: $209; new price: $2109    **77.** 40% solution: 10 gal; 10% solution: 20 gal    **79.** 18%
**81.** 966 customers    **83.** 50 km    **85.** 80 nickels    **87.** $(0, \infty)$

**89.** $[0.5, 1.5)$

**91.** $(-\infty, -4)$

**93.** $(-\infty, 4]$

**95.** $\left(-\dfrac{1}{2}, \dfrac{3}{4}\right)$

**97.** $\left(-\infty, \dfrac{19}{3}\right]$    **99.** $2500    **101.** $x = 4$    **103.** $a = -\dfrac{3}{2}$    **105.** all real numbers

**107.** $-13$    **109.** $h = \dfrac{3V}{A}$    **111.** 160    **113.** $(9, \infty)$    **115.** $(-\infty, 0]$

## Chapter 2 Test

**1.** $y - 10$    **2.** $5.9x + 1.2$    **3.** $-2x + 10$    **4.** $10y + 1$    **5.** $-5$    **6.** 8    **7.** $\dfrac{7}{10}$    **8.** 0    **9.** 27    **10.** 3    **11.** 0.25    **12.** $\dfrac{25}{7}$

**13.** no solution    **14.** 21    **15.** 7 gal    **16.** 401, 802    **17.** $8500 @ 10%; $17,000 @ 12%    **18.** $2\dfrac{1}{2}$ hr    **19.** $x = 6$    **20.** $h = \dfrac{V}{\pi r^2}$

**21.** $y = \dfrac{3x - 10}{4}$    **22.** $(-\infty, -2]$    **23.** $(-\infty, 4)$

**24.** $\left(-1, \dfrac{7}{3}\right)$    **25.** $\left(\dfrac{2}{5}, \infty\right)$

## Chapter 2 Cumulative Review

**1. a.** 11, 112    **b.** 0, 11, 112    **c.** $-3, -2, 0, 11, 112$    **d.** $-3, -2, -1.5, 0, \dfrac{1}{4}, 11, 112$    **e.** $\sqrt{2}$    **f.** all numbers in the given set; Sec. 1.2, Ex. 5

**3. a.** 4    **b.** 5    **c.** 0    **d.** $\dfrac{1}{2}$    **e.** 5.6; Sec. 1.2, Ex. 7    **5. a.** $2 \cdot 2 \cdot 2 \cdot 5$    **b.** $3 \cdot 3 \cdot 7$; Sec. 1.3, Ex. 1    **7.** $\dfrac{8}{20}$; Sec. 1.3, Ex. 6    **9.** 66; Sec. 1.4, Ex. 4

**11.** 2 is a solution; Sec. 1.4, Ex. 7    **13.** $-3$; Sec. 1.5, Ex. 2    **15.** 2; Sec. 1.5, Ex. 4    **17. a.** 10    **b.** $\dfrac{1}{2}$    **c.** $2x$    **d.** $-6$; Sec. 1.5, Ex. 10

**19. a.** 9.9    **b.** $-\dfrac{4}{5}$    **c.** $\dfrac{2}{15}$; Sec. 1.6, Ex. 2    **21. a.** $52°$    **b.** $118°$; Sec. 1.6, Ex. 8    **23. a.** $-0.06$    **b.** $-\dfrac{7}{15}$    **c.** 16; Sec. 1.7, Ex. 3    **25. a.** 6

**b.** $-12$    **c.** $-\dfrac{8}{15}$    **d.** $-\dfrac{1}{6}$; Sec. 1.7, Ex. 7    **27. a.** $5 + x$    **b.** $x \cdot 3$; Sec. 1.8, Ex. 1    **29. a.** $8(2 + x)$    **b.** $7(s + t)$; Sec. 1.8, Ex. 5

**31.** $-2x - 1$; Sec. 2.1, Ex. 7    **33.** $-1.6$; Sec. 2.2, Ex. 2    **35.** 8; Sec. 2.2, Ex. 4    **37.** 140; Sec. 2.2, Ex. 7    **39.** 2; Sec. 2.3, Ex. 1    **41.** 10; Sec. 2.4, Ex. 2

**43.** $\dfrac{V}{wh} = l$; Sec. 2.5, Ex. 5    **45.** $(-\infty, -10]$ ; Sec. 2.8, Ex. 2

# CHAPTER 3  GRAPHS AND INTRODUCTION TO FUNCTIONS
## Section 3.1
### Practice Exercises

**1. a.** Germany, 45 million Internet users    **b.** approximately 5 million Internet users    **2. a.** 70 beats per minute    **b.** 60 beats per minute
**c.** 5 minutes after lighting    **3.**    **4. a.** $(2000, 92), (2001, 84), (2002, 73), (2003, 64), (2004, 65), (2005, 67), (2006, 96)$

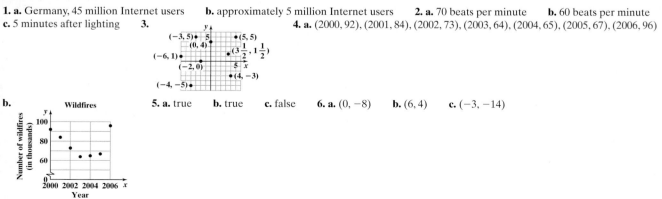

**5. a.** true    **b.** true    **c.** false    **6. a.** $(0, -8)$    **b.** $(6, 4)$    **c.** $(-3, -14)$

**7.**

| | x | y |
|---|---|---|
| **a.** | −2 | 8 |
| **b.** | 3 | −12 |
| **c.** | 0 | 0 |

**8.**

| | x | y |
|---|---|---|
| **a.** | −10 | −4 |
| **b.** | 0 | −2 |
| **c.** | 10 | 0 |

**9.**

| x | 0 | 1 | 2 | 3 | 4 |
|---|---|---|---|---|---|
| y | 12,000 | 10,200 | 8400 | 6600 | 4800 |

## Vocabulary and Readiness Check 3.1

**1.** x-axis    **3.** origin    **5.** x-coordinate; y-coordinate    **7.** solution

## Exercise Set 3.1

**1.** France    **3.** France, U.S., Spain, and China    **5.** 30 million    **7.** 72,600    **9.** 2007; 104,000    **11.** 15.9    **13.** from 1996 to 1998    **15.** 2002

**17.** 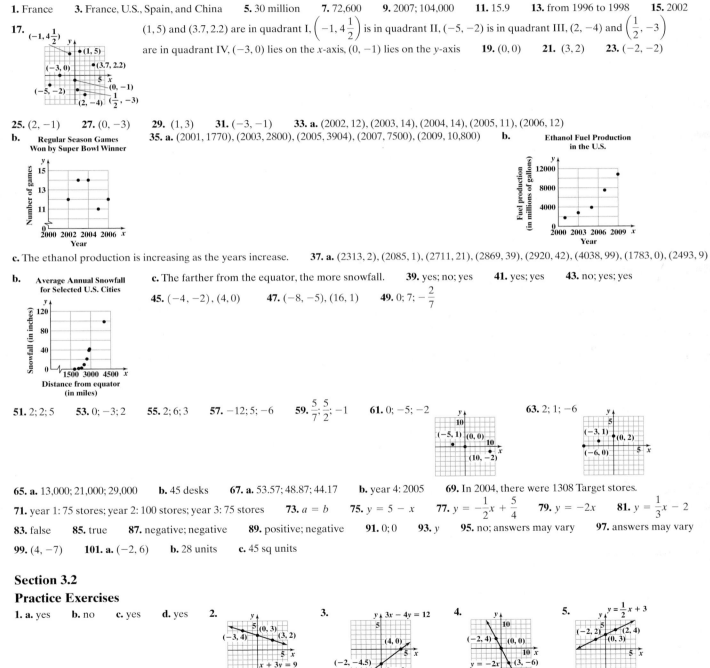 $(1, 5)$ and $(3.7, 2.2)$ are in quadrant I, $\left(-1, 4\frac{1}{2}\right)$ is in quadrant II, $(-5, -2)$ is in quadrant III, $(2, -4)$ and $\left(\frac{1}{2}, -3\right)$ are in quadrant IV, $(-3, 0)$ lies on the x-axis, $(0, -1)$ lies on the y-axis    **19.** $(0, 0)$    **21.** $(3, 2)$    **23.** $(-2, -2)$

**25.** $(2, -1)$    **27.** $(0, -3)$    **29.** $(1, 3)$    **31.** $(-3, -1)$    **33. a.** $(2002, 12), (2003, 14), (2004, 14), (2005, 11), (2006, 12)$

**b.** **Regular Season Games Won by Super Bowl Winner**

**35. a.** $(2001, 1770), (2003, 2800), (2005, 3904), (2007, 7500), (2009, 10,800)$    **b.** **Ethanol Fuel Production in the U.S.**

**c.** The ethanol production is increasing as the years increase.    **37. a.** $(2313, 2), (2085, 1), (2711, 21), (2869, 39), (2920, 42), (4038, 99), (1783, 0), (2493, 9)$

**b.** **Average Annual Snowfall for Selected U.S. Cities**    **c.** The farther from the equator, the more snowfall.    **39.** yes; no; yes    **41.** yes; yes    **43.** no; yes; yes

**45.** $(-4, -2), (4, 0)$    **47.** $(-8, -5), (16, 1)$    **49.** $0; 7; -\frac{2}{7}$

**51.** $2; 2; 5$    **53.** $0; -3; 2$    **55.** $2; 6; 3$    **57.** $-12; 5; -6$    **59.** $\frac{5}{7}; \frac{5}{2}; -1$    **61.** $0; -5; -2$    **63.** $2; 1; -6$

**65. a.** $13,000; 21,000; 29,000$    **b.** 45 desks    **67. a.** $53.57; 48.87; 44.17$    **b.** year 4: 2005    **69.** In 2004, there were 1308 Target stores.

**71.** year 1: 75 stores; year 2: 100 stores; year 3: 75 stores    **73.** $a = b$    **75.** $y = 5 - x$    **77.** $y = -\frac{1}{2}x + \frac{5}{4}$    **79.** $y = -2x$    **81.** $y = \frac{1}{3}x - 2$

**83.** false    **85.** true    **87.** negative; negative    **89.** positive; negative    **91.** $0; 0$    **93.** $y$    **95.** no; answers may vary    **97.** answers may vary

**99.** $(4, -7)$    **101. a.** $(-2, 6)$    **b.** 28 units    **c.** 45 sq units

## Section 3.2
## Practice Exercises

**1. a.** yes    **b.** no    **c.** yes    **d.** yes    **2.**    **3.**    **4.**    **5.**

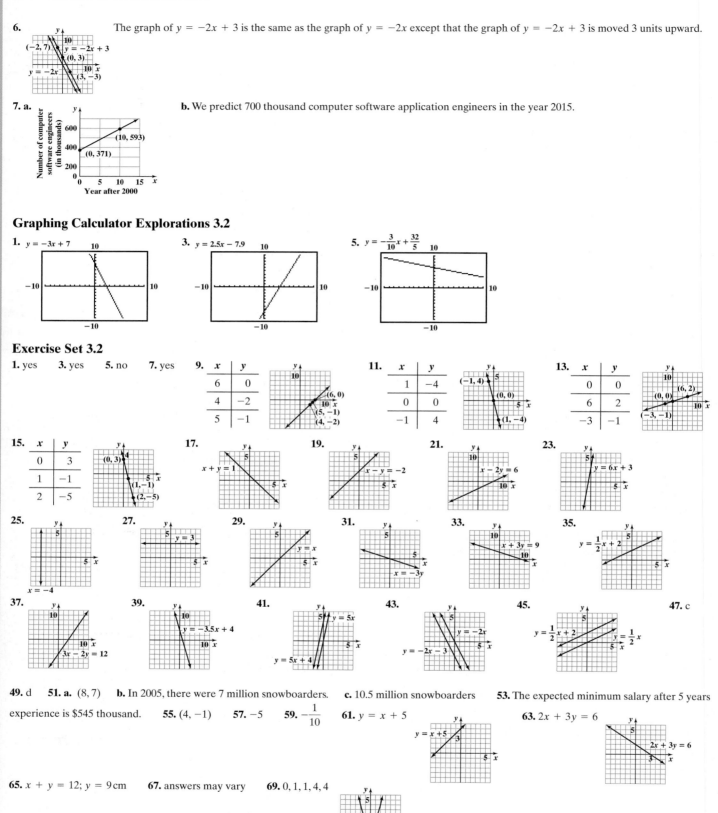

**6.** The graph of $y = -2x + 3$ is the same as the graph of $y = -2x$ except that the graph of $y = -2x + 3$ is moved 3 units upward.

**7. a.**  **b.** We predict 700 thousand computer software application engineers in the year 2015.

## Graphing Calculator Explorations 3.2

**1.** $y = -3x + 7$    **3.** $y = 2.5x - 7.9$    **5.** $y = -\dfrac{3}{10}x + \dfrac{32}{5}$

## Exercise Set 3.2

**1.** yes    **3.** yes    **5.** no    **7.** yes

**9.**

| x | y |
|---|---|
| 6 | 0 |
| 4 | -2 |
| 5 | -1 |

**11.**

| x | y |
|---|---|
| 1 | -4 |
| 0 | 0 |
| -1 | 4 |

**13.**

| x | y |
|---|---|
| 0 | 0 |
| 6 | 2 |
| -3 | -1 |

**15.**

| x | y |
|---|---|
| 0 | 3 |
| 1 | -1 |
| 2 | -5 |

**17.**  **19.**  **21.**  **23.**

**25.**  **27.**  **29.**  **31.**  **33.**  **35.**

**37.**  **39.**  **41.**  **43.**  **45.**  **47.** c

**49.** d    **51. a.** $(8, 7)$    **b.** In 2005, there were 7 million snowboarders.    **c.** 10.5 million snowboarders    **53.** The expected minimum salary after 5 years experience is $545 thousand.    **55.** $(4, -1)$    **57.** $-5$    **59.** $-\dfrac{1}{10}$    **61.** $y = x + 5$    **63.** $2x + 3y = 6$

**65.** $x + y = 12; y = 9$ cm    **67.** answers may vary    **69.** $0, 1, 1, 4, 4$

## Section 3.3

### Practice Exercises

**1.** $x$-intercept: $(-4, 0)$   **2.** $x$-intercepts: $(-1, 0), (-0.5, 0)$   **3.** $x$-intercept: $(0, 0)$   **4.** $x$-intercept: none   **5.** $x$-intercepts: $(-1, 0), (5, 0)$
   $y$-intercept: $(0, -6)$      $y$-intercept: $(0, 1)$      $y$-intercept: $(0, 0)$      $y$-intercept: $(0, 3)$      $y$-intercepts: $(0, 2), (0, -2)$

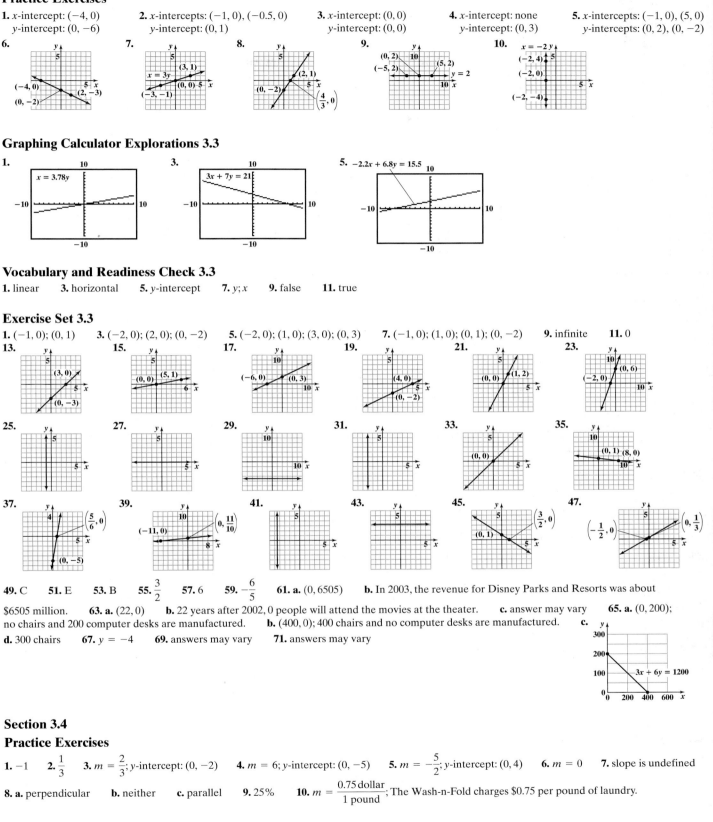

### Graphing Calculator Explorations 3.3

### Vocabulary and Readiness Check 3.3

**1.** linear   **3.** horizontal   **5.** $y$-intercept   **7.** $y; x$   **9.** false   **11.** true

### Exercise Set 3.3

**1.** $(-1, 0); (0, 1)$   **3.** $(-2, 0); (2, 0); (0, -2)$   **5.** $(-2, 0); (1, 0); (3, 0); (0, 3)$   **7.** $(-1, 0); (1, 0); (0, 1); (0, -2)$   **9.** infinite   **11.** 0

**49.** C   **51.** E   **53.** B   **55.** $\dfrac{3}{2}$   **57.** 6   **59.** $-\dfrac{6}{5}$   **61. a.** $(0, 6505)$   **b.** In 2003, the revenue for Disney Parks and Resorts was about $6505 million.   **63. a.** $(22, 0)$   **b.** 22 years after 2002, 0 people will attend the movies at the theater.   **c.** answer may vary   **65. a.** $(0, 200)$; no chairs and 200 computer desks are manufactured.   **b.** $(400, 0)$; 400 chairs and no computer desks are manufactured.   **c.**
**d.** 300 chairs   **67.** $y = -4$   **69.** answers may vary   **71.** answers may vary

## Section 3.4

### Practice Exercises

**1.** $-1$   **2.** $\dfrac{1}{3}$   **3.** $m = \dfrac{2}{3}$; $y$-intercept: $(0, -2)$   **4.** $m = 6$; $y$-intercept: $(0, -5)$   **5.** $m = -\dfrac{5}{2}$; $y$-intercept: $(0, 4)$   **6.** $m = 0$   **7.** slope is undefined

**8. a.** perpendicular   **b.** neither   **c.** parallel   **9.** 25%   **10.** $m = \dfrac{0.75\,\text{dollar}}{1\,\text{pound}}$; The Wash-n-Fold charges $0.75 per pound of laundry.

## Graphing Calculator Explorations 3.4

**1.** 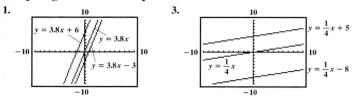 **3.**

## Vocabulary and Readiness Check 3.4

**1.** slope **3.** 0 **5.** positive **7.** $y; x$ **9.** positive **11.** 0 **13.** downward **15.** vertical

## Exercise Set 3.4

**1.** $-1$ **3.** undefined **5.** $-\dfrac{2}{3}$ **7.** 0 **9.** $m = -\dfrac{4}{3}$ **11.** undefined slope **13.** $m = \dfrac{5}{2}$ **15.** line 1 **17.** line 2 **19.** D **21.** B **23.** E

**25.** undefined slope **27.** $m = 0$ **29.** undefined slope **31.** $m = 0$ **33.** $m = 5$ **35.** $m = -0.3$ **37.** $m = -2$ **39.** $m = \dfrac{2}{3}$

**41.** undefined slope **43.** $m = \dfrac{1}{2}$ **45.** $m = 0$ **47.** $m = -\dfrac{3}{4}$ **49.** $m = 4$ **51.** neither **53.** neither **55.** parallel **57.** perpendicular

**59.** $\dfrac{3}{5}$ **61.** 12.5% **63.** 40% **65.** 79% **67.** $m = 3$; Every 1 year, there are/should be 3 million more U.S. households with personal computers.

**69.** $m = 0.42$; It costs \$0.42 per 1 mile to own and operate a compact car. **71.** $y = 2x - 14$ **73.** $y = -6x - 11$ **75. a.** 1 **b.** $-1$ **77. a.** $\dfrac{9}{11}$
**b.** $-\dfrac{11}{9}$ **79.** $m = \dfrac{1}{2}$ **81.** answers may vary **83.** 28.5 mi per gal **85.** 2000; 28.1 mi per gal **87.** from 2000 to 2001 **89.** $x = 6$

**91. a.** $(2001, 1132), (2006, 1657)$ **b.** 105 **c.** For the years 2001 through 2006, the price per acre of U.S. farmland rose approximately \$105 per year.
**93.** The slope through $(-3, 0)$ and $(1, 1)$ is $\dfrac{1}{4}$. The slope through $(-3, 0)$ and $(-4, 4)$ is $-4$. The product of the slopes is $-1$, so the sides are perpendicular.
**95.** $-0.25$ **97.** 0.875 **99.** The line becomes steeper.

## Integrated Review

**1.** $m = 2$ **2.** $m = 0$ **3.** $m = -\dfrac{2}{3}$ **4.** undefined slope **5.** **6.** $x + y = 3$ **7.**

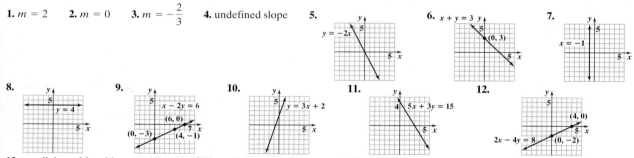

**8.** **9.** **10.** **11.** **12.**

**13.** parallel **14.** neither **15. a.** $(0, 1650)$ **b.** In 2002, there were 1650 million admissions to movie theaters in the U.S. **c.** $-75$
**d.** For the years 2002 through 2005, the number of movie theater admissions decreased at a rate of 75 million per year. **16. a.** $(9, 26.6)$
**b.** In 2009, the predicted revenue for online advertising is \$26.6 billion.

## Section 3.5

## Practice Exercises

**1.** $y = \dfrac{1}{2}x + 7$ **2.** **3.** **4.** $4x - y = 5$ **5.** $5x + 4y = 19$ **6.** $x = 3$ **7.** $y = 3$

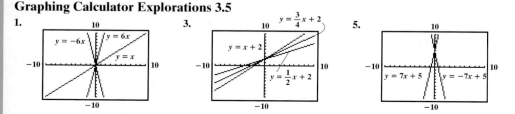

**8. a.** $y = -1500x + 195{,}000$ **b.** \$105,000

## Graphing Calculator Explorations 3.5

**1.** **3.** **5.**

## Vocabulary and Readiness Check 3.5
**1.** slope-intercept; $m$; $b$     **3.** point-slope     **5.** horizontal     **7.** slope-intercept

## Exercise Set 3.5
**1.** $y = 5x + 3$     **3.** $y = -4x - \dfrac{1}{6}$     **5.** $y = \dfrac{2}{3}x$     **7.** $y = -8$     **9.** $y = -\dfrac{1}{5}x + \dfrac{1}{9}$     **11.**     **13.**

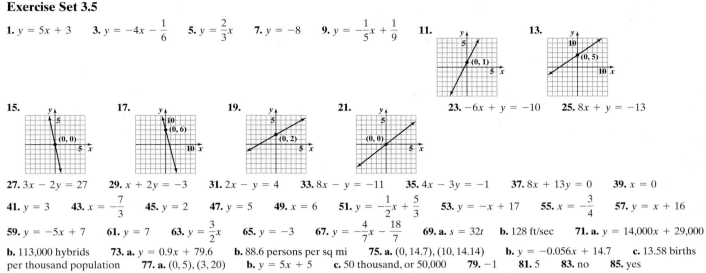

**23.** $-6x + y = -10$     **25.** $8x + y = -13$

**27.** $3x - 2y = 27$     **29.** $x + 2y = -3$     **31.** $2x - y = 4$     **33.** $8x - y = -11$     **35.** $4x - 3y = -1$     **37.** $8x + 13y = 0$     **39.** $x = 0$
**41.** $y = 3$     **43.** $x = -\dfrac{7}{3}$     **45.** $y = 2$     **47.** $y = 5$     **49.** $x = 6$     **51.** $y = -\dfrac{1}{2}x + \dfrac{5}{3}$     **53.** $y = -x + 17$     **55.** $x = -\dfrac{3}{4}$     **57.** $y = x + 16$
**59.** $y = -5x + 7$     **61.** $y = 7$     **63.** $y = \dfrac{3}{2}x$     **65.** $y = -3$     **67.** $y = -\dfrac{4}{7}x - \dfrac{18}{7}$     **69. a.** $s = 32t$     **b.** 128 ft/sec     **71. a.** $y = 14{,}000x + 29{,}000$
**b.** 113,000 hybrids     **73. a.** $y = 0.9x + 79.6$     **b.** 88.6 persons per sq mi     **75. a.** $(0, 14.7), (10, 14.14)$     **b.** $y = -0.056x + 14.7$     **c.** 13.58 births
per thousand population     **77. a.** $(0, 5), (3, 20)$     **b.** $y = 5x + 5$     **c.** 50 thousand, or 50,000     **79.** $-1$     **81.** 5     **83.** no     **85.** yes
**87.** answers may vary     **89. a.** $3x - y = -5$     **b.** $x + 3y = 5$     **91. a.** $3x + 2y = -1$     **b.** $2x - 3y = 21$

## Section 3.6
### Practice Exercises
**1.** Domain: $\{0, 1, 5\}$; Range: $\{-2, 0, 3, 4\}$     **2. a.** function     **b.** not a function     **3. a.** not a function     **b.** function     **4. a.** function
**b.** function     **c.** function     **d.** not a function     **5. a.** function     **b.** function     **c.** function     **d.** not a function     **6. a.** 69°F     **b.** November
**c.** yes     **7. a.** $h(2) = 9; (2, 9)$     **b.** $h(-5) = 30; (-5, 30)$     **c.** $h(0) = 5; (0, 5)$     **8. a.** domain: $(-\infty, \infty)$     **b.** domain: $(-\infty, 0) \cup (0, \infty)$
**9. a.** domain: $[-4, 6]$; range: $[-2, 3]$     **b.** domain: $(-\infty, \infty)$; range: $(-\infty, 3]$

## Vocabulary and Readiness Check 3.6
**1.** relation     **3.** range     **5.** vertical     **7.** $(3, 7)$

## Exercise Set 3.6
**1.** $\{-7, 0, 2, 10\}$; $\{-7, 0, 4, 10\}$     **3.** $\{0, 1, 5\}$; $\{-2\}$     **5.** yes     **7.** no     **9.** no     **11.** yes     **13.** yes     **15.** no     **17.** yes     **19.** yes     **21.** yes
**23.** no     **25.** no     **27.** 9:30 p.m.     **29.** January 1 and December 1     **31.** yes; it passes the vertical line test     **33.** \$4.25 per hour     **35.** 2009
**37.** yes; answers may vary     **39.** $-9, -5, 1$     **41.** $6, 2, 11$     **43.** $-6, 0, 9$     **45.** $2, 0, 3$     **47.** $5, 0, -20$     **49.** $5, 3, 35$     **51.** $(3, 6)$
**53.** $\left(0, -\dfrac{1}{2}\right)$     **55.** $(-2, 9)$     **57.** $(-\infty, \infty)$     **59.** all real number except $-5$ or $(-\infty, -5) \cup (-5, \infty)$     **61.** $(-\infty, \infty)$     **63.** domain: $(-\infty, \infty)$;
range: $[-4, \infty)$     **65.** domain: $(-\infty, \infty)$; range: $(-\infty, \infty)$     **67.** domain: $(-\infty, \infty)$; range: $\{2\}$     **69.** $(-2, 1)$     **71.** $(-3, -1)$     **73.** $f(-5) = 12$
**75.** $(3, -4)$     **77.** $f(5) = 0$     **79. a.** 166.38 cm     **b.** 148.25 cm     **81.** answers may vary     **83.** $f(x) = x + 7$     **85. a.** $-3s + 12$     **b.** $-3r + 12$
**87. a.** 132     **b.** $a^2 - 12$

## Chapter 3 Vocabulary Check
**1.** solution     **2.** $y$-axis     **3.** linear     **4.** $x$-intercept     **5.** standard     **6.** $y$-intercept     **7.** slope-intercept     **8.** point-slope     **9.** $y$     **10.** $x$-axis
**11.** $x$     **12.** slope     **13.** function     **14.** domain     **15.** range     **16.** relation

## Chapter 3 Review
**1.**     **3.**     **5.**     **7. a.** $(8.00, 1); (7.50, 10); (6.50, 25); (5.00, 50); (2.00, 100)$     **b.**

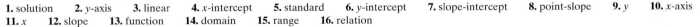

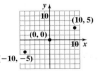

**9.** no; yes     **11.** yes; yes     **13.** $(7, 44)$     **15.** $(-3, 0); (1, 3); (9, 9)$     **17.** $(0, 0); (10, 5); (-10, -5)$

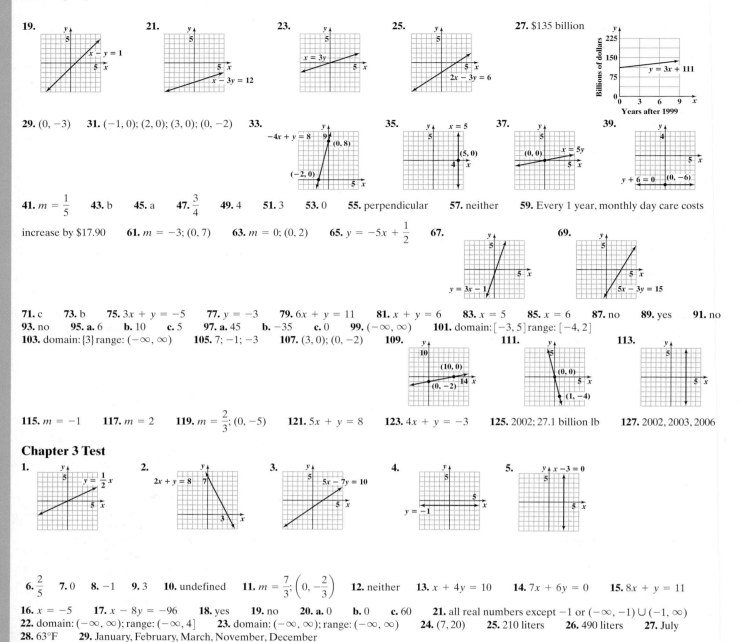

**19.** **21.** **23.** **25.** **27.** $135 billion

**29.** $(0, -3)$ **31.** $(-1, 0); (2, 0); (3, 0); (0, -2)$ **33.** **35.** **37.** **39.**

**41.** $m = \dfrac{1}{5}$ **43.** b **45.** a **47.** $\dfrac{3}{4}$ **49.** 4 **51.** 3 **53.** 0 **55.** perpendicular **57.** neither **59.** Every 1 year, monthly day care costs

increase by $17.90 **61.** $m = -3; (0, 7)$ **63.** $m = 0; (0, 2)$ **65.** $y = -5x + \dfrac{1}{2}$ **67.** **69.**

**71.** c **73.** b **75.** $3x + y = -5$ **77.** $y = -3$ **79.** $6x + y = 11$ **81.** $x + y = 6$ **83.** $x = 5$ **85.** $x = 6$ **87.** no **89.** yes **91.** no
**93.** no **95. a.** 6 **b.** 10 **c.** 5 **97. a.** 45 **b.** $-35$ **c.** 0 **99.** $(-\infty, \infty)$ **101.** domain: $[-3, 5]$ range: $[-4, 2]$
**103.** domain: $\{3\}$ range: $(-\infty, \infty)$ **105.** 7; $-1$; $-3$ **107.** $(3, 0); (0, -2)$ **109.** **111.** **113.**

**115.** $m = -1$ **117.** $m = 2$ **119.** $m = \dfrac{2}{3}; (0, -5)$ **121.** $5x + y = 8$ **123.** $4x + y = -3$ **125.** 2002; 27.1 billion lb **127.** 2002, 2003, 2006

## Chapter 3 Test

**1.** **2.** **3.** **4.** **5.**

**6.** $\dfrac{2}{5}$ **7.** 0 **8.** $-1$ **9.** 3 **10.** undefined **11.** $m = \dfrac{7}{3}; \left(0, -\dfrac{2}{3}\right)$ **12.** neither **13.** $x + 4y = 10$ **14.** $7x + 6y = 0$ **15.** $8x + y = 11$
**16.** $x = -5$ **17.** $x - 8y = -96$ **18.** yes **19.** no **20. a.** 0 **b.** 0 **c.** 60 **21.** all real numbers except $-1$ or $(-\infty, -1) \cup (-1, \infty)$
**22.** domain: $(-\infty, \infty)$; range: $(-\infty, 4]$ **23.** domain: $(-\infty, \infty)$; range: $(-\infty, \infty)$ **24.** $(7, 20)$ **25.** 210 liters **26.** 490 liters **27.** July
**28.** $63°F$ **29.** January, February, March, November, December

## Chapter 3 Cumulative Review

**1. a.** $<$ **b.** $>$ **c.** $>$; Sec. 1.2, Ex. 1 **3.** $\dfrac{2}{39}$; Sec. 1.3, Ex. 3 **5.** $\dfrac{8}{3}$; Sec. 1.4, Ex. 3 **7. a.** $-19$ **b.** 30 **c.** $-0.5$ **d.** $-\dfrac{4}{5}$ **e.** 6.7

**f.** $\dfrac{1}{40}$; Sec. 1.5, Ex. 6 **9. a.** $-6$; **b.** 6.3; Sec. 1.6, Ex. 4 **11. a.** $-6$ **b.** 0 **c.** $\dfrac{3}{4}$; Sec. 1.7, Ex. 10 **13. a.** $22 + x$ **b.** $-21x$; Sec. 1.8, Ex. 3

**15. a.** $-3$ **b.** 22 **c.** 1 **d.** $-1$ **e.** $\dfrac{1}{7}$; Sec. 2.1, Ex. 1 **17.** 17; Sec. 2.2, Ex. 1 **19.** $-5$; Sec. 2.2, Ex. 8 **21.** $3x + 3$; Sec. 2.2, Ex. 10

**23.** 0; Sec. 2.3, Ex. 4 **25.** 202 Republicans, 235 Democrats; Sec. 2.4, Ex. 4 **27.** 40 ft; Sec. 2.5, Ex. 2 **29.** $\dfrac{y - b}{m} = x$; Sec. 2.5, Ex. 6

**31.** 40% solution: 8 liters; 70% solution: 4 liters; Sec. 2.6, Ex. 7 **33.** ; Sec. 2.8, Ex. 1 **35.** $[1, 4)$; Sec. 2.8, Ex. 10

**37. a.** solution **b.** not a solution **c.** solution; Sec. 3.1, Ex. 5 **39. a.** yes **b.** yes **c.** no **d.** yes; Sec. 3.2, Ex. 1 **41.** 0; Sec. 3.4, Ex. 6

**43.** $y = \dfrac{1}{4}x - 3$; Sec. 3.5, Ex. 1

# CHAPTER 4  SOLVING SYSTEMS OF LINEAR EQUATIONS

## Section 4.1

### Practice Exercises

**1.** no    **2.** yes    **3.** $(8, 5)$    **4.** $(-3, -5)$    **5.** no solution; inconsistent, independent

**6.** infinite number of solutions; consistent, dependent    **7.** one solution    **8.** no solution

## Graphing Calculator Explorations 4.1

**1.** $(0.37, 0.23)$    **3.** $(0.03, -1.89)$

## Vocabulary and Readiness Check 4.1

**1.** dependent    **3.** consistent    **5.** inconsistent    **7.** one solution, $(-1, 3)$    **9.** infinite number of solutions

## Exercise Set 4.1

**1. a.** no    **b.** yes    **3. a.** yes    **b.** no    **5. a.** yes    **b.** yes    **7. a.** no    **b.** no    **9.**    **11.**    **13.**

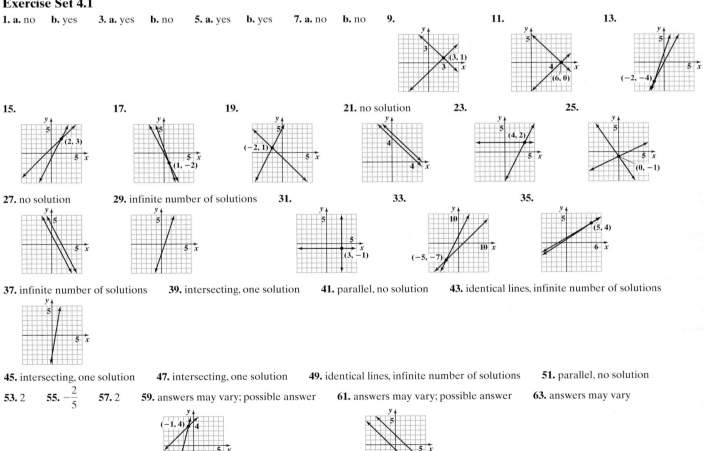

**15.**    **17.**    **19.**    **21.** no solution    **23.**    **25.**

**27.** no solution    **29.** infinite number of solutions    **31.**    **33.**    **35.**

**37.** infinite number of solutions    **39.** intersecting, one solution    **41.** parallel, no solution    **43.** identical lines, infinite number of solutions

**45.** intersecting, one solution    **47.** intersecting, one solution    **49.** identical lines, infinite number of solutions    **51.** parallel, no solution

**53.** 2    **55.** $-\dfrac{2}{5}$    **57.** 2    **59.** answers may vary; possible answer    **61.** answers may vary; possible answer    **63.** answers may vary

**65.** 2000, 2001, 2002    **67.** 2001, 2002, 2003    **69.** answers may vary    **71.** answers may vary    **73. a.** $(4, 9)$    **b.**     **c.** yes
**75.** answers may vary

## Section 4.2
### Practice Exercises

**1.** $(8, 7)$    **2.** $(-3, -6)$    **3.** $\left(4, \dfrac{2}{3}\right)$    **4.** $(-3, 2)$    **5.** infinite number of solutions    **6.** no solution

### Vocabulary and Readiness Check 4.2
**1.** $(1, 4)$    **3.** infinite number of solutions    **5.** $(0, 0)$

### Exercise Set 4.2

**1.** $(2, 1)$    **3.** $(-3, 9)$    **5.** $(2, 7)$    **7.** $\left(-\dfrac{1}{5}, \dfrac{43}{5}\right)$    **9.** $(2, -1)$    **11.** $(-2, 4)$    **13.** $(4, 2)$    **15.** $(-2, -1)$    **17.** no solution    **19.** $(3, -1)$

**21.** $(3, 5)$    **23.** $\left(\dfrac{2}{3}, -\dfrac{1}{3}\right)$    **25.** $(-1, -4)$    **27.** $(-6, 2)$    **29.** $(2, 1)$    **31.** no solution    **33.** infinite number of solutions    **35.** $\left(\dfrac{1}{2}, 2\right)$

**37.** $(1, -3)$    **39.** $-6x - 4y = -12$    **41.** $-12x + 3y = 9$    **43.** $5n$    **45.** $-15b$    **47.** answers may vary    **49.** no; answers may vary
**51.** c; answers may vary    **53. a.** $(13, 492)$    **b.** In $1970 + 13 = 1983$, the number of men and women receiving bachelor's degrees was the same.
**c.** answers may vary;    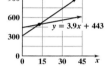    **55.** $(-2.6, 1.3)$    **57.** $(3.28, 2.1)$

## Section 4.3
### Practice Exercises

**1.** $(5, 3)$    **2.** $(3, -4)$    **3.** no solution    **4.** infinite number of solutions    **5.** $(2, 2)$    **6.** $\left(-\dfrac{8}{5}, \dfrac{6}{5}\right)$

### Exercise Set 4.3

**1.** $(1, 2)$    **3.** $(2, -3)$    **5.** $(-2, -5)$    **7.** $(5, -2)$    **9.** $(-7, 5)$    **11.** $(6, 0)$    **13.** no solution    **15.** infinite number of solutions    **17.** $\left(2, -\dfrac{1}{2}\right)$

**19.** $(-2, 0)$    **21.** $(1, -1)$    **23.** infinite number of solutions    **25.** $\left(\dfrac{12}{11}, -\dfrac{4}{11}\right)$    **27.** $\left(\dfrac{3}{2}, 3\right)$    **29.** infinite number of solutions    **31.** $(1, 6)$

**33.** $\left(-\dfrac{1}{2}, -2\right)$    **35.** infinite number of solutions    **37.** $\left(-\dfrac{2}{3}, \dfrac{2}{5}\right)$    **39.** $(2, 4)$    **41.** $(-0.5, 2.5)$    **43.** $(2, 5)$    **45.** $(-3, 2)$    **47.** $(0, 3)$    **49.** $(5, 7)$

**51.** $\left(\dfrac{1}{3}, 1\right)$    **53.** infinite number of solutions    **55.** $(-8.9, 10.6)$    **57.** $2x + 6 = x - 3$    **59.** $20 - 3x = 2$    **61.** $4(n + 6) = 2n$

**63.** 2; $6x - 2y = -24$    **65.** b; answers may vary    **67.** answers may vary    **69. a.** $b = 15$    **b.** any real number except 15    **71.** $(-4.2, 9.6)$
**73. a.** $(5, 294)$ or $(5, 295)$ or $(5, 296)$    **b.** In 2009 $(2004 + 5)$, the number of pharmacy technician jobs equals the number of network and data analyst jobs.
**c.** 294–296 thousand

## Integrated Review

**1.** $(2, 5)$    **2.** $(4, 2)$    **3.** $(5, -2)$    **4.** $(6, -14)$    **5.** $(-3, 2)$    **6.** $(-4, 3)$    **7.** $(0, 3)$    **8.** $(-2, 4)$    **9.** $(5, 7)$    **10.** $(-3, -23)$    **11.** $\left(\dfrac{1}{3}, 1\right)$

**12.** $\left(-\dfrac{1}{4}, 2\right)$    **13.** no solution    **14.** infinite number of solutions    **15.** $(0.5, 3.5)$    **16.** $(-0.75, 1.25)$    **17.** infinite number of solutions
**18.** no solution    **19.** $(7, -3)$    **20.** $(-1, -3)$    **21.** answers may vary    **22.** answers may vary

## Section 4.4
### Practice Exercises
**1.** $(-1, 2, 1)$    **2.** $\{ \}$ or $\varnothing$    **3.** $\left(\dfrac{2}{3}, -\dfrac{1}{2}, 0\right)$    **4.** $\{(x, y, z) | 2x + y - 3z = 6\}$    **5.** $(6, 15, -5)$

### Exercise Set 4.4

**1.** a, b, d    **3.** yes; answers may vary    **5.** $(-1, 5, 2)$    **7.** $(-2, 5, 1)$    **9.** $(-2, 3, -1)$    **11.** $\{(x, y, z) | x - 2y + z = -5\}$    **13.** $\varnothing$
**15.** $(0, 0, 0)$    **17.** $(-3, -35, -7)$    **19.** $(6, 22, -20)$    **21.** $\varnothing$    **23.** $(3, 2, 2)$    **25.** $\{(x, y, z) | x + 2y - 3z = 4\}$    **27.** $(-3, -4, -5)$
**29.** $\left(0, \dfrac{1}{2}, -4\right)$    **31.** $(12, 6, 4)$    **33.** 15 and 30    **35.** 5    **37.** $-\dfrac{5}{3}$    **39.** answers may vary    **41.** answers may vary    **43.** $(1, 1, -1)$
**45.** $(1, 1, 0, 2)$    **47.** $(1, -1, 2, 3)$    **49.** answers may vary

## Section 4.5

### Practice Exercises

**1. a.** 2037    **b.** yes; answers may vary    **2.** 12 and 17    **3. a.** Adult: $19    **b.** Child: $6    **c.** No, the regular rates are less than the group rate.
**4.** Atlantique: 500 kph; V150: 575 kph    **5.** 0.95 liter of water; 0.05 liter of 99% HCL    **6.** 1500 packages    **7.** $40°, 60°, 80°$

### Exercise Set 4.5

**1.** c    **3.** b    **5.** a    **7.** $\begin{cases} x + y = 15 \\ x - y = 7 \end{cases}$    **9.** $\begin{cases} x + y = 6500 \\ x = y + 800 \end{cases}$    **11.** 33 and 50    **13.** 14 and $-3$    **15.** Taurasi: 860 points; Augustus: 744 points

**17.** child's ticket: $18; adult's ticket: $29    **19.** quarters: 53; nickels: 27    **21.** Apple: $87.97; Microsoft: $27.29    **23.** daily fee: $32; mileage

charge: $0.25 per mi    **25.** distance downstream = distance upstream = 18 mi; time downstream: 2 hr; time upstream: $4\frac{1}{2}$ hr; still water: 6.5 mph;

current: 2.5 mph    **27.** still air: 455 mph; wind: 65 mph    **29.** $4\frac{1}{2}$ hr    **31.** 12% solution: $7\frac{1}{2}$ oz; 4% solution: $4\frac{1}{2}$ oz    **33.** $4.95 beans: 113 lb;

$2.65 beans: 87 lb    **35.** $60°, 30°$    **37.** $20°, 70°$    **39.** number sold at $9.50: 23; number sold at $7.50: 67    **41.** $2\frac{1}{4}$ mph and $2\frac{3}{4}$ mph

**43.** 30%: 50 gal; 60%: 100 gal    **45.** length: 42 in.; width: 30 in.    **47.** 2007    **49. a.** answers vary, but depend on slope    **b.** $-3.89$ years, or during
1991    **51.** $x = 75; y = 105$    **53.** 625 units    **55.** 3000 units    **57.** 1280 units    **59. a.** $R(x) = 450x$    **b.** $C(x) = 200x + 6000$    **c.** 24 desks
**61.** 2 units of mix A; 3 units of mix B; 1 unit of mix C    **63.** 5 in.; 7 in.; 10 in.    **65.** 18, 13, and 9    **67.** free throws: 143; two-point field goals;

177; three-point field goals: 121    **69.** $x = 60; y = 55; z = 65$    **71.** $(3, \infty)$    **73.** $\left[\frac{1}{2}, \infty\right)$    **75.** a    **77.** width: 9 ft; length: 15 ft

**79. a.** $(27.1, 39.5)$    **b.** For viewers 27.1 years over 18 (or 45.1 years old) the percent who watch cable news and network news is the same, or 39.5%.
**c.** answers may vary    **81a.** $(112, 137)$    **b.** June 2019

## Chapter 4 Vocabulary Check

**1.** dependent    **2.** system of linear equations    **3.** consistent    **4.** solution    **5.** addition; substitution    **6.** inconsistent    **7.** independent

## Chapter 4 Review

**1. a.** no    **b.** yes    **c.** no    **3. a.** no    **b.** no    **c.** yes    **5.**    **7.**    **9.**

**11.** no solution    **13.** $(-1, 4)$    **15.** $(3, -2)$    **17.** infinite number of solutions    **19.** no solution    **21.** $(-6, 2)$    **23.** $(3, 7)$

**25.** infinite number of solutions    **27.** $(8, -6)$    **29.** $(2, 0, 2)$    **31.** $\left(-\frac{1}{2}, \frac{3}{4}, 1\right)$    **33.** $\varnothing$    **35.** $(1, 1, -2)$    **37.** $-6$ and 22    **39.** current of river: 3.2 mph;
speed in still water: 21.1 mph    **41.** egg: $0.40; strip of bacon: $0.65    **43.** 17 pennies; 20 nickels; 16 dimes    **45.** two sides: 22 cm each; third side: 29 cm

**47.**    **49.** $(3, 2)$    **51.** $\left(1\frac{1}{2}, -3\right)$    **53.** infinite number of solutions    **55.** $(-5, 2)$    **57.** $(-1, 3, 5)$    **59.** 4 and 8
**61.** 24 nickels and 41 dimes    **63.** 28 units, 42 units, 56 units

## Chapter 4 Test

**1.** false    **2.** false    **3.** true    **4.** false    **5.** no    **6.** yes

**7.** $(-4, 2)$    **8.** $(-4, 1)$    **9.** $\left(\frac{1}{2}, -2\right)$    **10.** $(4, -2)$    **11.** no solution

**12.** $(4, -5)$    **13.** $(7, 2)$    **14.** $(5, -2)$    **15.** 78, 46    **16.** 120 cc

**17.** Texas: 226 thousand; Missouri: 110 thousand    **18.** $(-1, -2, 4)$    **19.** $\varnothing$    **20.** $23°, 45°, 112°$

## Chapter 4 Cumulative Review

**1. a.** $<$　**b.** $=$　**c.** $>$; Sec. 1.2, Ex. 6　**3. a.** commutative property of multiplication　**b.** associative property of addition
**c.** identity element for addition　**d.** commutative property of multiplication　**e.** multiplicative inverse property　**f.** additive inverse property
**g.** commutative and associative properties of multiplication; Sec. 1.8, Ex. 6　**5.** $-2x - 1$; Sec. 2.1, Ex. 7　**7.** 8; Sec. 2.2, Ex. 4
**9.** 6; Sec. 2.2, Ex. 5

**11.** 12; Sec. 2.3, Ex. 3　**13.** 10; Sec. 2.4, Ex. 2　**15.** $x = \dfrac{y - b}{m}$; Sec. 2.5, Ex. 6　**17.** $[2, \infty)$; Sec. 2.8, Ex. 3

**19.**

; Sec. 3.3, Ex. 7　**21.** $-\dfrac{8}{3}$; Sec. 3.4, Ex. 1　**23.** $\dfrac{3}{4}$; Sec. 3.4, Ex. 3　**25.** slope: $\dfrac{3}{4}$; $y$-intercept: $(0, -1)$; Sec. 3.4, Ex. 5

**27.** $2x + y = 3$; Sec. 3.5, Ex. 4　**29.** $x = -1$; Sec. 3.5, Ex. 6　**31.** domain: $\{-1, 0, 3\}$; range: $\{-2, 0, 2, 3\}$; Sec. 3.6, Ex. 1

**33. a.** function　**b.** not a function; Sec. 3.6, Ex. 2　**35.** one solution; Sec. 4.1, Ex. 8　**37.** $\left(6, \dfrac{1}{2}\right)$; Sec. 4.2, Ex. 3

**39.** $(6, 1)$; Sec. 4.3, Ex. 1　**41.** $(-4, 2, -1)$; Sec. 4.4, Ex. 1　**43.** 7 and 11; Sec. 4.5, Ex. 2

# CHAPTER 5　EXPONENTS AND POLYNOMIALS

## Section 5.1

### Practice Exercises

**1. a.** 27　**b.** 4　**c.** 64　**d.** $-64$　**e.** $\dfrac{27}{64}$　**f.** 0.0081　**g.** 75　**2. a.** 243　**b.** $\dfrac{3}{8}$　**3. a.** $3^{10}$　**b.** $y^5$　**c.** $z^5$　**d.** $x^{11}$　**e.** $(-2)^8$　**f.** $b^3 \cdot t^5$

**4.** $15y^7$　**5. a.** $y^{12}z^4$　**b.** $-7m^5n^{14}$　**6. a.** $x^{12}$　**b.** $z^{21}$　**c.** $(-2)^{15}$　**7. a.** $p^5r^5$　**b.** $36b^2$　**c.** $\dfrac{1}{64}x^6y^3$　**d.** $81a^{12}b^{16}c^4$　**8. a.** $\dfrac{x^5}{y^{10}}$　**b.** $\dfrac{32a^{20}}{b^{15}}$

**9. a.** $z^4$　**b.** 25　**c.** 64　**d.** $\dfrac{q^5}{t^2}$　**e.** $6x^2y^2$　**10. a.** $-1$　**b.** 1　**c.** 1　**d.** 1　**e.** 1　**11. a.** $\dfrac{125}{x^3z^3}$　**b.** $16z^{32}x^{20}$　**c.** $\dfrac{-27x^9}{y^{12}}$

### Vocabulary and Readiness Check 5.1

**1.** exponent　**3.** add　**5.** 1　**7.** base: 3; exponent: 2　**9.** base: 4; exponent: 2　**11.** base: 5; exponent: 1; base: $x$; exponent: 2

### Exercise Set 5.1

**1.** 49　**3.** $-5$　**5.** $-16$　**7.** 16　**9.** 0.00001　**11.** $\dfrac{1}{81}$　**13.** 224　**15.** $-250$　**17.** answers may vary　**19.** 4　**21.** 135　**23.** 150

**25.** $\dfrac{32}{5}$　**27.** $x^7$　**29.** $(-3)^{12}$　**31.** $15y^5$　**33.** $x^{19}y^6$　**35.** $-72m^3n^8$　**37.** $-24z^{20}$　**39.** $20x^5$ sq ft　**41.** $x^{36}$　**43.** $p^8q^8$　**45.** $8a^{15}$

**47.** $x^{10}y^{15}$　**49.** $49a^4b^{10}c^2$　**51.** $\dfrac{r^9}{s^9}$　**53.** $\dfrac{m^5p^5}{n^5}$　**55.** $\dfrac{4x^2z^2}{y^{10}}$　**57.** $64z^{10}$ sq dm　**59.** $27y^{12}$ cu ft　**61.** $x^2$　**63.** $-64$　**65.** $p^6q^5$　**67.** $\dfrac{y^3}{2}$

**69.** 1　**71.** 1　**73.** $-7$　**75.** 2　**77.** $-81$　**79.** $\dfrac{1}{64}$　**81.** $\dfrac{81}{q^2r^2}$　**83.** $a^6$　**85.** $-16x^7$　**87.** $a^{11}b^{20}$　**89.** $26m^9n^7$　**91.** $z^{40}$　**93.** $36x^2y^2z^6$

**95.** $3x$　**97.** $81x^2y^2$　**99.** 33　**101.** $\dfrac{y^{15}}{8x^{12}}$　**103.** $2x^2y$　**105.** $2y - 10$　**107.** $-x - 4$　**109.** $-x + 5$　**111.** c　**113.** e

**115.** answers may vary　**117.** answers may vary　**119.** 343 cu m　**121.** volume　**123.** answers may vary　**125.** $x^{9a}$　**127.** $a^{5b}$
**129.** $x^{5a}$　**131.** $1045.85

## Section 5.2

### Practice Exercises

**1. a.** degree 3　**b.** degree 2　**c.** degree 1　**d.** degree 8　**e.** degree 0　**2. a.** trinomial, degree 2　**b.** binomial, degree 1　**c.** none of these, degree 3

**3.**

| Term | Numerical Coefficient | Degree of Term |
|------|----------------------|----------------|
| $-3x^3y^2$ | $-3$ | 5 |
| $4xy^2$ | 4 | 3 |
| $-y^2$ | $-1$ | 2 |
| $3x$ | 3 | 1 |
| $-2$ | $-2$ | 0 |

**4. a.** 4　**b.** $-21$　**5.** 114 ft; 66 ft　**6. a.** $-2y$　**b.** $z + 5z^3$　**c.** $4a^2 - 12$
**d.** $\dfrac{1}{3}x^4 + \dfrac{11}{24}x^3 - x^2$　**7.** $-3x^2 + 5xy + 5y^2$　**8.** $4x^2 + 7x + 4$
**9. a.** $5y^2 - 3y + 2x - 9$　**b.** $-8a^2b - 3ab^2$　**10.** $7x^3 - 6x + 7$
**11.** $2x^3 - 4x^2 + 4x - 6$　**12.** $7x - 11$　**13. a.** $-5a^2 - ab + 6b^2$
**b.** $3x^2y^2 - 10xy - 4xy^2 - 6y^2 + 5$

## Graphing Calculator Explorations 5.2

**1.** $x^3 - 4x^2 + 7x - 8$　**3.** $-2.1x^2 - 3.2x - 1.7$　**5.** $7.69x^2 - 1.26x + 5.3$

## Vocabulary and Readiness Check 5.2

**1.** binomial     **3.** trinomial     **5.** constant     **7.** $-14y$     **9.** $7x$     **11.** $5m^2 + 2m$

## Exercise Set 5.2

**1.** 1; binomial     **3.** 3; none of these     **5.** 6; trinomial     **7.** 2; binomial     **9.** 3     **11.** 2     **13.** 57     **15.** 499     **17.** 1     **19.** $-\dfrac{11}{16}$     **21.** 1134 ft
**23.** 1006 ft     **25.** $23x^2$     **27.** $12x^2 - y$     **29.** $7s$     **31.** $-1.1y^2 + 4.8$     **33.** $-\dfrac{7}{12}x^3 + \dfrac{7}{5}x^2 + 6$     **35.** $5a^2 - 9ab + 16b^2$     **37.** $-3x^2 + 10$
**39.** $-x^2 + 14$     **41.** $-2x + 9$     **43.** $2x^2 + 7x - 16$     **45.** $8t^2 - 4$     **47.** $-2z^2 - 16z + 6$     **49.** $2x^3 - 2x^2 + 7x + 2$     **51.** $62x^2 + 5$
**53.** $12x + 2$     **55.** $-y^2 - 3y - 1$     **57.** $2x^2 + 11x$     **59.** $-16x^4 + 8x + 9$     **61.** $7x^2 + 14x + 18$     **63.** $3x - 3$     **65.** $7x^2 - 2x + 2$
**67.** $4y^2 + 12y + 19$     **69.** $6x^2 - 5x + 21$     **71.** $4x^2 + 7x + x^2 + 5x$; $5x^2 + 12x$     **73.** $18x + 44$     **75.** $(x^2 + 7x + 4)$ ft     **77.** $(3y^2 + 4y + 11)$ m
**79.** $-2a - b + 1$     **81.** $3x^2 + 5$     **83.** $6x^2 - 2xy + 19y^2$     **85.** $8r^2s + 16rs - 8 + 7r^2s^2$     **87.** $-5.42x^2 + 7.75x - 19.61$
**89.** $3.7y^4 - 0.7y^3 + 2.2y - 4$     **91.** $6x^2$     **93.** $-12x^8$     **95.** $200x^3y^2$     **97., 99.** answers may vary     **101.** b     **103.** e     **105. a.** $4z$     **b.** $3z^2$
**c.** $-4z$     **d.** $3z^2$; answers may vary     **107.** $3x^{2a} + 2x^a + 0.7$     **109.** $4x^{2y} + 2x^y - 11$     **111.** $4x^2 - 3x + 6$     **113.** $-x^2 - 6x + 10$
**115.** $3x^2 - 12x + 13$     **117. a.** $2a - 3$     **b.** $-2x - 3$     **c.** $2x + 2h - 3$     **119. a.** $4a$     **b.** $-4x$     **c.** $4x + 4h$     **121.** \$8169
**123.** $10.84x^2 + 20.43x + 3285$

## Section 5.3
## Practice Exercises

**1.** $10y^2$     **2.** $-2z^8$     **3.** $\dfrac{7}{72}b^9$     **4. a.** $15x^6 + 15x$     **b.** $-10x^5 + 45x^4 - 10x^3$     **5.** $10x^2 + 11x - 6$     **6.** $25x^2 - 30xy + 9y^2$
**7.** $2y^3 + 5y^2 - 7y + 20$     **8.** $s^3 + 6s^2t + 12st^2 + 8t^3$     **9.** $5x^3 - 23x^2 + 17x - 20$     **10.** $x^5 - 2x^4 + 2x^3 - 3x^2 + 2$
**11.** $5x^4 - 3x^3 + 11x^2 + 8x - 6$

## Vocabulary and Readiness Check 5.3

**1.** distributive     **3.** $(5y - 1)(5y - 1)$     **5.** $x^8$     **7.** cannot simplify     **9.** $x^{14}$     **11.** $2x^7$

## Exercise Set 5.3

**1.** $-28n^{10}$     **3.** $-12.4x^{12}$     **5.** $-\dfrac{2}{15}y^3$     **7.** $-24x^8$     **9.** $6x^2 + 15x$     **11.** $-2a^2 - 8a$     **13.** $6x^3 - 9x^2 + 12x$     **15.** $-6a^4 + 4a^3 - 6a^2$

**17.** $-4x^3y + 7x^2y^2 - xy^3 - 3y^4$     **19.** $4x^4 - 3x^3 + \dfrac{1}{2}x^2$     **21.** $x^2 + 7x + 12$     **23.** $a^2 + 5a - 14$     **25.** $x^2 + \dfrac{1}{3}x - \dfrac{2}{9}$     **27.** $12x^4 + 25x^2 + 7$

**29.** $4y^2 - 16y + 16$     **31.** $12x^2 - 29x + 15$     **33.** $9x^4 + 6x^2 + 1$     **35. a.** $6x + 12$     **b.** $9x^2 + 36x + 35$     **c.** answers may vary
**37.** $x^3 - 5x^2 + 13x - 14$     **39.** $x^4 + 5x^3 - 3x^2 - 11x + 20$     **41.** $10a^3 - 27a^2 + 26a - 12$     **43.** $x^3 + 6x^2 + 12x + 8$
**45.** $8y^3 - 36y^2 + 54y - 27$     **47.** $12x^2 - 64x - 11$     **49.** $10x^3 + 22x^2 - x - 1$     **51.** $2x^4 + 3x^3 - 58x^2 + 4x + 63$     **53.** $8.4y^7$

**55.** $-3x^3 - 6x^2 + 24x$     **57.** $2x^2 + 39x + 19$     **59.** $x^2 - \dfrac{2}{7}x - \dfrac{3}{49}$     **61.** $9y^2 + 30y + 25$     **63.** $a^3 - 2a^2 - 18a + 24$

**65.** $8x^3 - 60x^2 + 150x - 125$     **67.** $32x^3 + 48x^2 - 6x - 20$     **69.** $6x^4 - 8x^3 - 7x^2 + 22x - 12$     **71.** $(4x^2 - 25)$ sq yd
**73.** $(6x^2 - 4x)$ sq in.     **75.** $25x^2$     **77.** $9y^6$     **79.** $x^2 + 3x$     **81.** $x^2 + 5x + 6$     **83.** $11a$     **85.** $25x^2 + 4y^2$     **87.** $13x - 7$
**89.** $30x^2 - 28x + 6$     **91.** $-7x + 5$     **93. a.** $a^2 - b^2$     **b.** $4x^2 - 9y^2$     **c.** $16x^2 - 49$     **d.** answers may vary     **95.** $(x^2 + 6x + 5)$ sq units

## Section 5.4
## Practice Exercises

**1.** $x^2 - 3x - 10$     **2.** $4x^2 - 13x + 9$     **3.** $9x^2 + 42x - 15$     **4.** $16x^2 - 8x + 1$     **5. a.** $b^2 + 6b + 9$     **b.** $x^2 - 2xy + y^2$     **c.** $9y^2 + 12y + 4$
**d.** $a^4 - 10a^2b + 25b^2$     **6. a.** $3x^2 - 75$     **b.** $16b^2 - 9$     **c.** $x^2 - \dfrac{4}{9}$     **d.** $25s^2 - t^2$     **e.** $4y^2 - 9z^4$     **7. a.** $4x^2 - 21x - 18$     **b.** $49b^2 - 28b + 4$
**c.** $x^2 - 0.16$     **d.** $3x^6 - 9x^4 + 2x^2 - 6$     **e.** $x^3 + 6x^2 + 3x - 2$

## Vocabulary and Readiness Check 5.4

**1.** false     **3.** false

## Exercise Set 5.4

**1.** $x^2 + 7x + 12$     **3.** $x^2 + 5x - 50$     **5.** $5x^2 + 4x - 12$     **7.** $4y^2 - 25y + 6$     **9.** $6x^2 + 13x - 5$     **11.** $x^2 - 4x + 4$     **13.** $4x^2 - 4x + 1$

**15.** $9a^2 - 30a + 25$     **17.** $25x^2 + 90x + 81$     **19.** answers may vary     **21.** $a^2 - 49$     **23.** $9x^2 - 1$     **25.** $9x^2 - \dfrac{1}{4}$     **27.** $81x^2 - y^2$

**29.** $4x^2 - 0.01$     **31.** $a^2 + 9a + 20$     **33.** $a^2 + 14a + 49$     **35.** $12a^2 - a - 1$     **37.** $x^2 - 4$     **39.** $9a^2 + 6a + 1$     **41.** $4x^3 - x^2y^4 + 4xy - y^5$

**43.** $x^3 - 3x^2 - 17x + 3$     **45.** $4a^2 - 12a + 9$     **47.** $25x^2 - 36z^2$     **49.** $x^{10} - 8x^5 + 15$     **51.** $x^2 - \dfrac{1}{9}$     **53.** $a^7 - 3a^3 + 11a^4 - 33$

**55.** $3x^2 - 12x + 12$     **57.** $6b^2 - b - 35$     **59.** $49p^2 - 64$     **61.** $\dfrac{1}{9}a^4 - 49$     **63.** $15x^4 - 5x^3 + 10x^2$     **65.** $4r^2 - 9s^2$     **67.** $9x^2 - 42xy + 49y^2$

**69.** $16x^2 - 25$     **71.** $64x^2 + 64x + 16$     **73.** $a^2 - \dfrac{1}{4}y^2$     **75.** $\dfrac{1}{25}x^2 - y^2$     **77.** $3a^3 + 2a^2 + 1$     **79.** $(2x + 1)(2x + 1)$ sq ft or $(4x^2 + 4x + 1)$ sq ft

**81.** $\dfrac{5b^5}{7}$     **83.** $-2a^{10}b^5$     **85.** $\dfrac{2y^8}{3}$     **87.** $\dfrac{1}{3}$     **89.** 1     **91.** c     **93.** d     **95.** 2     **97.** $\left(\dfrac{25}{2}a^2 - \dfrac{1}{2}b^2\right)$ sq units     **99.** $(24x^2 - 32x + 8)$ sq m

**101.** $(x^2 + 10x + 25)$ sq units     **103.** answers may vary     **105.** $x^2 + 2xy + y^2 - 9$     **107.** $a^2 - 6a + 9 - b^2$

## Integrated Review

**1.** $35x^5$    **2.** $32y^9$    **3.** $-16$    **4.** $16$    **5.** $2x^2 - 9x - 5$    **6.** $3x^2 + 13x - 10$    **7.** $3x - 4$    **8.** $4x + 3$    **9.** $7x^6y^2$    **10.** $\dfrac{10b^6}{7}$

**11.** $144m^{14}n^{12}$    **12.** $64y^{27}z^{30}$    **13.** $48y^2 - 27$    **14.** $98x^2 - 2$    **15.** $x^{63}y^{45}$    **16.** $27x^{27}$    **17.** $2x^2 - 2x - 6$    **18.** $6x^2 + 13x - 11$
**19.** $2.5y^2 - 6y - 0.2$    **20.** $8.4x^2 - 6.8x - 5.7$    **21.** $x^2 + 8xy + 16y^2$    **22.** $y^2 - 18yz + 81z^2$    **23.** $2x + 8y$    **24.** $2y - 18z$
**25.** $7x^2 - 10xy + 4y^2$    **26.** $-a^2 - 3ab + 6b^2$    **27.** $x^3 + 2x^2 - 16x + 3$    **28.** $x^3 - 2x^2 - 5x - 2$    **29.** $6x^5 + 20x^3 - 21x^2 - 70$
**30.** $20x^7 + 25x^3 - 4x^4 - 5$    **31.** $2x^3 - 19x^2 + 44x - 7$    **32.** $5x^3 + 9x^2 - 17x + 3$    **33.** cannot simplify    **34.** $25x^3y^3$    **35.** $125x^9$

**36.** $\dfrac{x^3}{y^3}$    **37.** $2x$    **38.** $x^2$

## Section 5.5

### Practice Exercises

**1. a.** $\dfrac{1}{125}$    **b.** $\dfrac{3}{y^4}$    **c.** $\dfrac{5}{6}$    **d.** $\dfrac{1}{25}$    **e.** $x^5$    **f.** $64$    **2. a.** $s^5$    **b.** $8$    **c.** $\dfrac{y^5}{x^7}$    **d.** $\dfrac{9}{64}$    **3. a.** $\dfrac{1}{x^5}$    **b.** $5y^7$    **c.** $z^5$    **4. a.** $\dfrac{16}{9}$    **b.** $x^{10}$

**c.** $\dfrac{q^2}{25p^{16}}$    **d.** $\dfrac{6y^2}{x^7}$    **e.** $\dfrac{b^{15}}{a^{20}}$    **f.** $-27x^6y^9$    **5. a.** $7 \times 10^{-6}$    **b.** $2.07 \times 10^7$    **c.** $4.3 \times 10^{-3}$    **d.** $8.12 \times 10^8$    **6. a.** $0.000367$    **b.** $8,954,000$

**c.** $0.00002009$    **d.** $4054$    **7. a.** $4000$    **b.** $20,000,000,000$

### Graphing Calculator Explorations 5.5

**1.** 5.31 EE 3    **3.** 6.6 EE $-9$    **5.** $1.5 \times 10^{13}$    **7.** $8.15 \times 10^{19}$

### Vocabulary and Readiness Check 5.5

**1.** $\dfrac{1}{x^3}$    **3.** scientific notation    **5.** $\dfrac{5}{x^2}$    **7.** $y^6$    **9.** $4y^3$

### Exercise Set 5.5

**1.** $\dfrac{1}{64}$    **3.** $\dfrac{1}{16}$    **5.** $\dfrac{7}{x^3}$    **7.** $32$    **9.** $-64$    **11.** $\dfrac{5}{6}$    **13.** $p^3$    **15.** $\dfrac{q^4}{p^5}$    **17.** $\dfrac{1}{x^3}$    **19.** $z^3$    **21.** $\dfrac{4}{9}$    **23.** $-p^4$    **25.** $-2$    **27.** $x^4$    **29.** $p^4$

**31.** $m^{11}$    **33.** $r^6$    **35.** $\dfrac{1}{x^{15}y^9}$    **37.** $\dfrac{1}{x^4}$    **39.** $\dfrac{1}{a^2}$    **41.** $4k^3$    **43.** $3m$    **45.** $-\dfrac{4a^5}{b}$    **47.** $\dfrac{6x^2}{y^3}$    **49.** $\dfrac{a^{30}}{b^{12}}$    **51.** $\dfrac{1}{x^{10}y^6}$    **53.** $\dfrac{z^2}{4}$    **55.** $\dfrac{1}{32x^5}$

**57.** $\dfrac{49a^4}{b^6}$    **59.** $a^{24}b^8$    **61.** $x^9y^{19}$    **63.** $-\dfrac{y^8}{8x^2}$    **65.** $-\dfrac{6x}{7y^2}$    **67.** $\dfrac{25b^{33}}{a^{16}}$    **69.** $7.8 \times 10^4$    **71.** $1.67 \times 10^{-6}$    **73.** $6.35 \times 10^{-3}$    **75.** $1.16 \times 10^6$

**77.** $2 \times 10^9$    **79.** $1.212 \times 10^9$    **81.** $0.0000000008673$    **83.** $0.033$    **85.** $20,320$    **87.** $700,000,000$    **89.** $9,460,000,000,000$
**91.** Yahoo! sites, 130,000,000, $1.3 \times 10^8$    **93.** $1 \times 10^9$    **95.** $57,000,000$    **97.** $0.000036$    **99.** $0.0000000000000000028$    **101.** $0.0000005$

**103.** $200,000$    **105.** $\dfrac{5x^3}{3}$    **107.** $\dfrac{5z^3y^2}{7}$    **109.** $5y - 6 + \dfrac{5}{y}$    **111.** $\dfrac{27}{x^6z^3}$ cu in.    **113.** $9a^{13}$    **115.** $-5$    **117.** answers may vary

**119. a.** $1.3 \times 10^1$    **b.** $4.4 \times 10^7$    **c.** $6.1 \times 10^{-2}$    **121. a.** false    **b.** true    **c.** false    **123.** $\dfrac{1}{x^{9s}}$    **125.** $a^{4m+5}$    **127.** $31,753,800$    **129.** $500$ sec

## Section 5.6

### Practice Exercises

**1.** $2t + 1$    **2.** $4x^4 + 5x - \dfrac{3}{x}$    **3.** $3x^3y^3 - 2 + \dfrac{1}{5x}$    **4.** $x + 3$    **5.** $2x + 3 + \dfrac{-10}{2x + 1}$    **6.** $3x^2 - 2x + 5 + \dfrac{-13}{3x + 2}$    **7.** $3x^2 - 2x - 9 + \dfrac{5x + 22}{x^2 + 2}$

### Vocabulary and Readiness Check 5.6

**1.** dividend, quotient, divisor    **3.** $a^2$    **5.** $y$

### Exercise Set 5.6

**1.** $12x^3 + 3x$    **3.** $4x^3 - 6x^2 + x + 1$    **5.** $5p^2 + 6p$    **7.** $-\dfrac{3}{2x} + 3$    **9.** $-3x^2 + x - \dfrac{4}{x^3}$    **11.** $-1 + \dfrac{3}{2x} - \dfrac{7}{4x^4}$    **13.** $x + 1$    **15.** $2x + 3$

**17.** $2x + 1 + \dfrac{7}{x - 4}$    **19.** $3a^2 - 3a + 1 + \dfrac{2}{3a + 2}$    **21.** $4x + 3 - \dfrac{2}{2x + 1}$    **23.** $2x^2 + 6x - 5 - \dfrac{2}{x - 2}$    **25.** $x + 6$    **27.** $x^2 + 3x + 9$

**29.** $-3x + 6 - \dfrac{11}{x + 2}$    **31.** $2b - 1 - \dfrac{6}{2b - 1}$    **33.** $ab - b^2$    **35.** $4x + 9$    **37.** $x + 4xy - \dfrac{y}{2}$    **39.** $2b^2 + b + 2 - \dfrac{12}{b + 4}$

**41.** $5x - 2 + \dfrac{2}{x + 6}$    **43.** $x^2 - \dfrac{12x}{5} - 1$    **45.** $6x - 1 - \dfrac{1}{x + 3}$    **47.** $6x - 1$    **49.** $-x^3 + 3x^2 - \dfrac{4}{x}$    **51.** $x^2 + 3x + 9$

**53.** $y^2 + 5y + 10 + \dfrac{24}{y - 2}$    **55.** $-6x - 12 - \dfrac{19}{x - 2}$    **57.** $x^3 - x^2 + x$    **59.** $2a^3 + 2a$    **61.** $2x^3 + 14x^2 - 10x$    **63.** $-3x^2y^3 - 21x^3y^2 - 24xy$

**65.** $9a^2b^3c + 36ab^2c - 72ab$    **67.** The Rolling Stones (2005)    **69.** \$139 million    **71.** $(3x^3 + x - 4)$ ft    **73.** c    **75.** answers may vary
**77.** $(7x - 10)$ in.    **79.** $5y^{10b} + y^{5b} - 4y^{2b} + 20$

## The Bigger Picture

**1.** $-5.93$ **2.** $-\dfrac{2}{5}$ **3.** $5x^9y^4$ **4.** $\dfrac{1}{8a^4}$ **5.** $6y^3 - 2y^2 - 6y$ **6.** $8y^2 - 3y - 7$ **7.** $4x^3 - 13x^2 + 10x - 21$ **8.** $36m^2 - 60m + 25$
**9.** $4n - 1 + \dfrac{2}{n}$ **10.** $2x - 6 + \dfrac{14}{3x - 1}$ **11.** $-0.6$ **12.** $(-0.6, \infty)$ **13.** $(-\infty, 2]$ **14.** $\dfrac{2}{3}$

## Section 5.7
### Practice Exercises

**1.** $4x^2 + x + 7 + \dfrac{12}{x - 1}$ **2.** $x^3 - 5x + 21 - \dfrac{51}{x + 3}$ **3. a.** $-4$ **b.** $-4$ **4.** $15$

### Exercise Set 5.7

**1.** $x + 8$ **3.** $x - 1$ **5.** $x^2 - 5x - 23 - \dfrac{41}{x - 2}$ **7.** $4x + 8 + \dfrac{7}{x - 2}$ **9.** $3$ **11.** $73$ **13.** $-8$ **15.** $x^2 + \dfrac{2}{x - 3}$ **17.** $6x + 7 + \dfrac{1}{x + 1}$
**19.** $2x^3 - 3x^2 + x - 4$ **21.** $3x - 9 + \dfrac{12}{x + 3}$ **23.** $3x^2 - \dfrac{9}{2}x + \dfrac{7}{4} + \dfrac{47}{8(x - \frac{1}{2})}$ **25.** $3x^2 + 3x - 3$ **27.** $3x^2 + 4x - 8 + \dfrac{20}{x + 1}$
**29.** $x^2 + x + 1$ **31.** $x - 6$ **33.** $1$ **35.** $-133$ **37.** $3$ **39.** $-\dfrac{187}{81}$ **41.** $\dfrac{95}{32}$ **43.** answers may vary **45.** $-\dfrac{5}{6}$ **47.** $54$ **49.** $8$
**51.** $-32$ **53.** $48$ **55.** $25$ **57.** $-2$ **59.** yes **61.** no **63.** $(x^3 - 5x^2 + 2x - 1)$ cm **65.** $x^3 + \dfrac{5}{3}x^2 + \dfrac{5}{3}x + \dfrac{8}{3} + \dfrac{8}{3(x - 1)}$
**67.** $(x + 3)(x^2 + 4) = x^3 + 3x^2 + 4x + 12$ **69.** $0$ **71.** $x^3 + 2x^2 + 7x + 28$ **73.** answers may vary

## Chapter 5 Vocabulary Check

**1.** term **2.** FOIL **3.** trinomial **4.** degree of a polynomial **5.** binomial **6.** coefficient **7.** degree of a term **8.** monomial
**9.** polynomials

## Chapter 5 Review

**1.** base: 7; exponent: 9 **3.** base: 5; exponent: 4 **5.** $512$ **7.** $-36$ **9.** $1$ **11.** $y^9$ **13.** $-6x^{11}$ **15.** $x^8$ **17.** $81y^{24}$ **19.** $x^5$ **21.** $a^4b^3$
**23.** $\dfrac{4}{x^3y^4}$ **25.** $40a^{19}$ **27.** $3$ **29.** b **31.** $7$ **33.** $8$ **35.** $5$ **37.** $5$ **39. a.** $1, 4; 5, 2; -7, 2; 11, 2; -1, 0$ **b.** $4$ **41.** $15a^2 + 4a$
**43.** $-6a^2b - 3b^2 - q^2$ **45.** $8x^2 + 3x + 6$ **47.** $-7y^2 - 1$ **49.** $4x - 13y$ **51.** $290$ **53.** $(6x^2y - 12x + 12)$ cm **55.** $8a + 28$
**57.** $-7x^3 - 35x$ **59.** $-6a^4 + 8a^2 - 2a$ **61.** $2x^2 - 12x - 14$ **63.** $x^2 - 18x + 81$ **65.** $4a^2 + 27a - 7$ **67.** $25x^2 + 20x + 4$
**69.** $x^4 + 7x^3 + 4x^2 + 23x - 35$ **71.** $x^4 + 4x^3 + 4x^2 - 16$ **73.** $x^3 + 21x^2 + 147x + 343$ **75.** $x^2 + 14x + 49$ **77.** $9x^2 - 42x + 49$
**79.** $25x^2 - 90x + 81$ **81.** $49x^2 - 16$ **83.** $4x^2 - 36$ **85.** $(9x^2 - 6x + 1)$ sq m **87.** $\dfrac{1}{49}$ **89.** $\dfrac{2}{x^4}$ **91.** $125$ **93.** $\dfrac{17}{16}$ **95.** $x^8$ **97.** $r$
**99.** $c^4$ **101.** $\dfrac{1}{x^6y^{13}}$ **103.** $a^{11m}$ **105.** $27x^3y^{6z}$ **107.** $2.7 \times 10^{-4}$ **109.** $8.08 \times 10^7$ **111.** $9.1 \times 10^7$ **113.** $867,000$ **115.** $0.00086$
**117.** $1,431,280,000,000,000$ **119.** $0.016$ **121.** $\dfrac{1}{7} + \dfrac{3}{x} + \dfrac{7}{x^2}$ **123.** $a + 1 + \dfrac{6}{a - 2}$ **125.** $a^2 + 3a + 8 + \dfrac{22}{a - 2}$
**127.** $2x^3 - x^2 + 2 - \dfrac{1}{2x - 1}$ **129.** $\left(5x - 1 + \dfrac{20}{x^2}\right)$ ft **131.** $3x^2 + 6x + 24 + \dfrac{44}{x - 2}$ **133.** $x^4 - x^3 + x^2 - x + 1 - \dfrac{2}{x + 1}$
**135.** $3x^3 + 13x^2 + 51x + 204 + \dfrac{814}{x - 4}$ **137.** $3043$ **139.** $-\dfrac{1}{8}$ **141.** $\dfrac{2x^6}{3}$ **143.** $\dfrac{x^{16}}{16y^{12}}$ **145.** $11x - 5$ **147.** $5y^2 - 3y - 1$
**149.** $28x^3 + 12x$ **151.** $x^3 + x^2 - 18x + 18$ **153.** $25x^2 + 40x + 16$ **155.** $4a - 1 + \dfrac{2}{a^2} - \dfrac{5}{2a^3}$ **157.** $2x^2 + 7x + 5 + \dfrac{19}{2x - 3}$

## Chapter 5 Test

**1.** $32$ **2.** $81$ **3.** $-81$ **4.** $\dfrac{1}{64}$ **5.** $-15x^{11}$ **6.** $y^5$ **7.** $\dfrac{1}{r^5}$ **8.** $\dfrac{y^{14}}{x^2}$ **9.** $\dfrac{1}{6xy^8}$ **10.** $5.63 \times 10^5$ **11.** $8.63 \times 10^{-5}$ **12.** $0.0015$
**13.** $62,300$ **14.** $0.036$ **15. a.** $4, 3; 7, 3; 1, 4; -2, 0$ **b.** $4$ **16.** $-2x^2 + 12xy + 11$ **17.** $16x^3 + 7x^2 - 3x - 13$ **18.** $-3x^3 + 5x^2 + 4x + 5$
**19.** $x^3 + 8x^2 + 3x - 5$ **20.** $3x^3 + 22x^2 + 41x + 14$ **21.** $6x^4 - 9x^3 + 21x^2$ **22.** $3x^2 + 16x - 35$ **23.** $9x^2 - \dfrac{1}{25}$ **24.** $16x^2 - 16x + 4$
**25.** $x^4 - 81b^2$ **26.** $1001$ ft; $985$ ft; $857$ ft; $601$ ft **27.** $(4x^2 - 9)$ sq in. **28.** $\dfrac{x}{2y} + 3 - \dfrac{7}{8y}$ **29.** $x + 2$ **30.** $9x^2 - 6x + 4 - \dfrac{16}{3x + 2}$
**31. a.** $960$ ft **b.** $953.44$ ft **c.** $11$ sec. **32.** $4x^3 - 15x^2 + 45x - 136 + \dfrac{407}{x + 3}$ **33.** $91$

## Chapter 5 Cumulative Review

**1. a.** true **b.** true **c.** false **d.** true; Sec. 1.2, Ex. 2 **3. a.** $\dfrac{64}{25}$ **b.** $\dfrac{1}{20}$ **c.** $\dfrac{5}{4}$; Sec. 1.3, Ex. 4 **5. a.** $9$ **b.** $125$ **c.** $16$ **d.** $7$
**e.** $\dfrac{9}{49}$; Sec. 1.4, Ex. 1 **7. a.** $-10$ **b.** $-21$ **c.** $-12$; Sec. 1.5, Ex. 3 **9.** $-12$; Sec. 1.6, Ex. 3 **11. a.** $\dfrac{1}{22}$ **b.** $\dfrac{16}{3}$ **c.** $-\dfrac{1}{10}$
**d.** $-\dfrac{13}{9}$; Sec. 1.7, Ex. 5 **13. a.** $(5 + 4) + 6$ **b.** $-1 \cdot (2 \cdot 5)$; Sec. 1.8, Ex. 2 **15. a.** $22 + x$ **b.** $-21x$; Sec. 1.8, Ex. 3 **17. a.** $5x + 10$
**b.** $-2y - 0.6z + 2$ **c.** $-x - y + 2z - 6$; Sec. 2.1, Ex. 5 **19.** $17$; Sec. 2.2, Ex. 1 **21.** $6$; Sec. 2.2, Ex. 5 **23.** $-10$; Sec. 2.4, Ex. 1

**25.** 10; Sec. 2.4, Ex. 2    **27.** width: 4 ft; length: 10 ft; Sec. 2.5, Ex. 4    **29.** $\dfrac{5F - 160}{9} = C$; Sec. 2.5, Ex. 8

**31.** 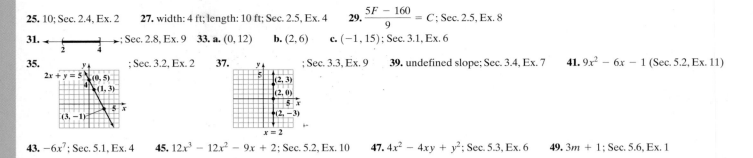; Sec. 2.8, Ex. 9    **33. a.** $(0, 12)$    **b.** $(2, 6)$    **c.** $(-1, 15)$; Sec. 3.1, Ex. 6

**35.** ; Sec. 3.2, Ex. 2    **37.** ; Sec. 3.3, Ex. 9    **39.** undefined slope; Sec. 3.4, Ex. 7    **41.** $9x^2 - 6x - 1$ (Sec. 5.2, Ex. 11)

**43.** $-6x^7$; Sec. 5.1, Ex. 4    **45.** $12x^3 - 12x^2 - 9x + 2$; Sec. 5.2, Ex. 10    **47.** $4x^2 - 4xy + y^2$; Sec. 5.3, Ex. 6    **49.** $3m + 1$; Sec. 5.6, Ex. 1

# CHAPTER 6  FACTORING POLYNOMIALS

## Section 6.1

### Practice Exercises

**1. a.** 6    **b.** 1    **c.** 4    **2. a.** $y^4$    **b.** $y$    **3. a.** $5y^2$    **b.** $x^2$    **c.** $a^2b^2$    **4. a.** $4(t + 3)$    **b.** $y^4(y^4 + 1)$

**5.** $8b^2(-b^4 + 2b^2 - 1)$ or $-8b^2(b^4 - 2b^2 + 1)$    **6.** $5x(x^3 - 4)$    **7.** $\dfrac{1}{9}z^3(5z^2 + z - 2)$    **8.** $4ab^3(2ab - 5a^2 + 3)$    **9.** $(y - 2)(8 + x)$

**10.** $(p + q)(7xy^3 - 1)$    **11.** $(x + 3)(y + 4)$    **12.** $(2x + 3y)(y - 1)$    **13.** $(7a + 5)(a^2 + 1)$    **14.** $(y - 3)(4x - 5)$

**15.** cannot be factored by grouping    **16.** $3(x - a)(y - 2a)$

### Vocabulary and Readiness Check 6.1

**1.** factors    **3.** least    **5.** false    **7.** $2 \cdot 7$    **9.** 3    **11.** 5

### Exercise Set 6.1

**1.** 4    **3.** 6    **5.** 1    **7.** $y^2$    **9.** $z^7$    **11.** $xy^2$    **13.** 7    **15.** $4y^3$    **17.** $5x^2$    **19.** $3x^3$    **21.** $9x^2y$    **23.** $10a^6b$    **25.** $3(a + 2)$    **27.** $15(2x - 1)$
**29.** $x^2(x + 5)$    **31.** $2y^3(3y + 1)$    **33.** $4(x - 2y + 1)$    **35.** $3x(2x^2 - 3x + 4)$    **37.** $a^2b^2(a^5b^4 - a + b^3 - 1)$    **39.** $4(2x^5 + 4x^4 - 5x^3 + 3)$
**41.** $\dfrac{1}{3}x(x^3 + 2x^2 - 4x^4 + 1)$    **43.** $(x^2 + 2)(y + 3)$    **45.** $(y + 4)(z - 3)$    **47.** $(z^2 - 6)(r + 1)$    **49.** $-2(x + 7)$    **51.** $-x^5(2 - x^2)$
**53.** $-3a^2(2a^2 - 3a + 1)$    **55.** $(x + 2)(x^2 + 5)$    **57.** $(x + 3)(5 + y)$    **59.** $(3x - 2)(2x^2 + 5)$    **61.** $(5m^2 + 6n)(m + 1)$    **63.** $(y - 4)(2 + x)$
**65.** $(2x - 1)(x^2 + 4)$    **67.** $(x - 2y)(4x - 3)$    **69.** $(5q - 4p)(q - 1)$    **71.** $x(x^2 + 1)(2x + 5)$    **73.** $2(2y - 7)(3x^2 - 1)$    **75.** $2x(16y - 9x)$
**77.** $(x + 2)(y - 3)$    **79.** $7xy(2x^2 + x - 1)$    **81.** $(4x - 1)(7x^2 + 3)$    **83.** $-8x^8y^5(5y + 2x)$    **85.** $3(2a + 3b^2)(a + b)$    **87.** $x^2 + 7x + 10$
**89.** $b^2 - 3b - 4$    **91.** 2, 6    **93.** $-1, -8$    **95.** $-2, 5$    **97.** b    **99.** factored    **101.** not factored    **103.** answers may vary
**105.** answers may vary    **107. a.** 3088 thousand or 3,088,000    **b.** 3092 thousand or 3,092,000    **c.** $-2(4x^2 - 25x - 1510)$    **109.** $4x^2 - \pi x^2$; $x^2(4 - \pi)$
**111.** $(x^3 - 1)$ units    **113.** $(x^n + 6)(x^n + 10)$    **115.** $(2x^n - 5)(6x^n - 5)$

## Section 6.2

### Practice Exercises

**1.** $(x + 2)(x + 3)$    **2.** $(x - 10)(x - 7)$    **3.** $(x + 7)(x - 2)$    **4.** $(p - 9)(p + 7)$    **5.** prime polynomial    **6.** $(x + 3y)(x + 4y)$
**7.** $(x^2 + 12)(x^2 + 1)$    **8.** $(x - 6)(x - 8)$    **9.** $4(x - 3)(x - 3)$    **10.** $3y^2(y - 7)(y + 1)$

### Vocabulary and Readiness Check 6.2

**1.** true    **3.** false    **5.** $+5$    **7.** $-3$    **9.** $+2$

### Exercise Set 6.2

**1.** $(x + 6)(x + 1)$    **3.** $(y - 9)(y - 1)$    **5.** $(x - 3)(x - 3)$ or $(x - 3)^2$    **7.** $(x - 6)(x + 3)$    **9.** $(x + 10)(x - 7)$    **11.** prime
**13.** $(x + 5y)(x + 3y)$    **15.** $(a^2 - 5)(a^2 + 3)$    **17.** $(m + 13)(m + 1)$    **19.** $(t - 2)(t + 12)$    **21.** $(a - 2b)(a - 8b)$    **23.** $2(z + 8)(z + 2)$
**25.** $2x(x - 5)(x - 4)$    **27.** $(x - 4y)(x + y)$    **29.** $(x + 12)(x + 3)$    **31.** $(x - 2)(x + 1)$    **33.** $(r - 12)(r - 4)$    **35.** $(x + 2y)(x - y)$
**37.** $3(x + 5)(x - 2)$    **39.** $3(x - 18)(x - 2)$    **41.** $(x - 24)(x + 6)$    **43.** prime    **45.** $(x - 5)(x - 3)$    **47.** $6x(x + 4)(x + 5)$
**49.** $4y(x^2 + x - 3)$    **51.** $(x - 7)(x + 3)$    **53.** $(x + 5y)(x + 2y)$    **55.** $2(t + 8)(t + 4)$    **57.** $x(x - 6)(x + 4)$    **59.** $2t^3(t - 4)(t - 3)$

**61.** $5xy(x - 8y)(x + 3y)$    **63.** $3(m - 9)(m - 6)$    **65.** $-1(x - 11)(x - 1)$    **67.** $\dfrac{1}{2}(y - 11)(y + 2)$    **69.** $x(xy - 4)(xy + 5)$

**71.** $2x^2 + 11x + 5$    **73.** $15y^2 - 17y + 4$    **75.** $9a^2 + 23ab - 12b^2$    **77.** $x^2 + 5x - 24$    **79.** answers may vary

**81.** $2x^2 + 28x + 66$; $2(x + 3)(x + 11)$    **83.** $-16(t - 5)(t + 1)$    **85.** $\left(x + \dfrac{1}{4}\right)\left(x + \dfrac{1}{4}\right)$ or $\left(x + \dfrac{1}{4}\right)^2$    **87.** $(x + 1)(z - 10)(z + 7)$

**89.** $(x^n + 10)(x^n - 2)$    **91.** 7; 12; 15; 16    **93.** 15; 28; 39; 48; 55; 60; 63; 64    **95.** 9; 12; 21    **97.** 5; 13

## Section 6.3

### Practice Exercises

**1.** $(2x + 5)(x + 3)$    **2.** $(5x - 4)(3x - 2)$    **3.** $(4x - 1)(x + 3)$    **4.** $(7x - y)(3x + 2y)$    **5.** $(2x^2 - 7)(x^2 + 1)$    **6.** $x(3x + 2)(x + 5)$
**7.** $-1(4x - 3)(2x + 1)$    **8.** $(x + 7)^2$    **9.** $(2x + 9y)(2x + y)$    **10.** $(6n^2 - 1)^2$    **11.** $3x(2x - 7)^2$

## Vocabulary and Readiness Check 6.3

**1.** perfect square trinomial     **3.** perfect square trinomial     **5.** no     **7.** $8^2$     **9.** $(11a)^2$     **11.** $(6p^2)^2$

## Exercise Set 6.3

**1.** $x + 4$     **3.** $10x - 1$     **5.** $5x - 2$     **7.** $(2x + 3)(x + 5)$     **9.** $(y - 1)(8y - 9)$     **11.** $(2x + 1)(x - 5)$     **13.** $(4r - 1)(5r + 8)$
**15.** $(10x + 1)(x + 3)$     **17.** prime     **19.** $(3x - 5y)(2x - y)$     **21.** $(3m - 5)(5m + 3)$     **23.** $x(3x + 2)(4x + 1)$     **25.** $3(7b + 5)(b - 3)$
**27.** $(3z + 4)(4z - 3)$     **29.** $2y^2(3x - 10)(x + 3)$     **31.** $(2x - 7)(2x + 3)$     **33.** $-1(x - 6)(x + 4)$     **35.** $x(4x + 3)(x - 3)$     **37.** $(4x - 9)(6x - 1)$
**39.** $(x + 11)^2$     **41.** $(x - 8)^2$     **43.** $(4a - 3)^2$     **45.** $(x^2 + 2)^2$     **47.** $2(n - 7)^2$     **49.** $(4y + 5)^2$     **51.** $(2x + 11)(x - 9)$     **53.** $(8x + 3)(3x + 4)$
**55.** $(3a + b)(a + 3b)$     **57.** $(x - 4)(x - 5)$     **59.** $(p + 6q)^2$     **61.** $(xy - 5)^2$     **63.** $b(8a - 3)(5a + 3)$     **65.** $2x(3x + 2)(5x + 3)$
**67.** $2y(3y + 5)(y - 3)$     **69.** $5x^2(2x - y)(x + 3y)$     **71.** $-1(2x - 5)(7x - 2)$     **73.** $p^2(4p - 5)(4p - 5)$ or $p^2(4p - 5)^2$     **75.** $(3x - 2)(x + 1)$
**77.** $(4x + 9y)(2x - 3y)$     **79.** prime     **81.** $(3x - 4y)^2$     **83.** $(6x - 7)(3x + 2)$     **85.** $(7t + 1)(t - 4)$     **87.** $(7p + 1)(7p - 2)$     **89.** $m(m + 9)^2$
**91.** prime     **93.** $a(6a^2 + b^2)(a^2 + 6b^2)$     **95.** $x^2 - 4$     **97.** $a^3 + 27$     **99.** \$75,000 and above     **101.** answers may vary     **103.** no

**105.** answers may vary     **107.** $4x^2 + 21x + 5; (4x + 1)(x + 5)$     **109.** $\left(2x + \dfrac{1}{2}\right)\left(2x + \dfrac{1}{2}\right)$ or $\left(2x + \dfrac{1}{2}\right)^2$     **111.** $(y - 1)^2(4x^2 + 10x + 25)$     **113.** 8

**115.** $a^2 + 2ab + b^2$     **117.** 2; 14     **119.** 2     **121.** $-3xy^2(4x - 5)(x + 1)$     **123.** $(y - 1)^2(2x + 5)^2$     **125.** $(3x^n + 2)(x^n + 5)$     **127.** answers may vary

## Section 6.4

### Practice Exercises

**1.** $(5x + 1)(x + 12)$     **2.** $(4x - 5)(3x - 1)$     **3.** $2(5x + 1)(3x - 2)$     **4.** $5m^2(8m - 7)(m + 1)$     **5.** $(4x + 3)^2$

## Exercise Set 6.4

**1.** $(x + 3)(x + 2)$     **3.** $(y + 8)(y - 2)$     **5.** $(8x - 5)(x - 3)$     **7.** $(5x^2 - 3)(x^2 + 5)$     **9. a.** $9, 2$     **b.** $9x + 2x$     **c.** $(2x + 3)(3x + 1)$
**11. a.** $-20, -3$     **b.** $-20x - 3x$     **c.** $(3x - 4)(5x - 1)$     **13.** $(3y + 2)(7y + 1)$     **15.** $(7x - 11)(x + 1)$     **17.** $(5x - 2)(2x - 1)$
**19.** $(2x - 5)(x - 1)$     **21.** $(2x + 3)(2x + 3)$ or $(2x + 3)^2$     **23.** $(2x + 3)(2x - 7)$     **25.** $(5x - 4)(2x - 3)$     **27.** $x(2x + 3)(x + 5)$
**29.** $2(8y - 9)(y - 1)$     **31.** $(2x - 3)(3x - 2)$     **33.** $3(3a + 2)(6a - 5)$     **35.** $a(4a + 1)(5a + 8)$     **37.** $3x(4x + 3)(x - 3)$     **39.** $y(3x + y)(x + y)$
**41.** prime     **43.** $5(x + 5y)^2$     **45.** $6(a + b)(4a - 5b)$     **47.** $p^2(15p + q)(p + 2q)$     **49.** $2(9a^2 - 2)^2$     **51.** $(7 + x)(5 + x)$ or $(x + 7)(x + 5)$
**53.** $(6 - 5x)(1 - x)$ or $(5x - 6)(x - 1)$     **55.** $x^2 - 4$     **57.** $y^2 + 8y + 16$     **59.** $81z^2 - 25$     **61.** $x^3 - 27$
**63.** $10x^2 + 45x + 45; 5(2x + 3)(x + 3)$     **65.** $(x^n + 2)(x^n + 3)$     **67.** $(3x^n - 5)(x^n + 7)$     **69.** answers may vary

## Section 6.5

### Practice Exercises

**1.** $(x + 9)(x - 9)$     **2. a.** $(3x - 1)(3x + 1)$     **b.** $(6a - 7b)(6a + 7b)$     **c.** $\left(p + \dfrac{5}{6}\right)\left(p - \dfrac{5}{6}\right)$     **3.** $(p^2 - q^5)(p^2 + q^5)$

**4. a.** $(z^2 + 9)(z + 3)(z - 3)$     **b.** prime polynomial     **5.** $y(6y + 5)(6y - 5)$     **6.** $5(4y^2 + 1)(2y + 1)(2y - 1)$
**7.** $-1(3x + 10)(3x - 10)$ or $(10 + 3x)(10 - 3x)$     **8.** $(x + 4)(x^2 - 4x + 16)$     **9.** $(x - 5)(x^2 + 5x + 25)$     **10.** $(3y + 1)(9y^2 - 3y + 1)$
**11.** $4(2x - 5y)(4x^2 + 10xy + 25y^2)$

## Graphing Calculator Explorations 6.5

|  | $x^2 - 2x + 1$ | $x^2 - 2x - 1$ | $(x - 1)^2$ |
|---|---|---|---|
| $x = 5$ | 16 | 14 | 16 |
| $x = -3$ | 16 | 14 | 16 |
| $x = 2.7$ | 2.89 | 0.89 | 2.89 |
| $x = -12.1$ | 171.61 | 169.61 | 171.61 |
| $x = 0$ | 1 | $-1$ | 1 |

## Vocabulary and Readiness Check 6.5

**1.** difference of two cubes     **3.** sum of two cubes     **5.** $8^2$     **7.** $(7x)^2$     **9.** $4^3$     **11.** $(2y)^3$

## Exercise Set 6.5

**1.** $(x + 2)(x - 2)$     **3.** $(9p + 1)(9p - 1)$     **5.** $(5y - 3)(5y + 3)$     **7.** $(11m + 10n)(11m - 10n)$     **9.** $(xy - 1)(xy + 1)$     **11.** $\left(x - \dfrac{1}{2}\right)\left(x + \dfrac{1}{2}\right)$

**13.** $-1(2r + 1)(2r - 1)$     **15.** prime     **17.** $(-6 + x)(6 + x)$ or $-1(6 + x)(6 - x)$     **19.** $(m^2 + 1)(m + 1)(m - 1)$     **21.** $(m^2 + n^9)(m^2 - n^9)$
**23.** $(x + 5)(x^2 - 5x + 25)$     **25.** $(2a - 1)(4a^2 + 2a + 1)$     **27.** $(m + 3n)(m^2 - 3mn + 9n^2)$     **29.** $5(k + 2)(k^2 - 2k + 4)$
**31.** $(xy - 4)(x^2y^2 + 4xy + 16)$     **33.** $2(5r - 4t)(25r^2 + 20rt + 16t^2)$     **35.** $(r + 8)(r - 8)$     **37.** $(x + 13y)(x - 13y)$     **39.** $(3 - t)(9 + 3t + t^2)$
**41.** $2(3r + 2)(3r - 2)$     **43.** $x(3y + 2)(3y - 2)$     **45.** $8(m + 2)(m^2 - 2m + 4)$     **47.** $xy(y - 3z)(y + 3z)$     **49.** $4(3x - 4y)(3x + 4y)$

**51.** $9(4 - 3x)(4 + 3x)$     **53.** $(xy - z^2)(x^2y^2 + xyz^2 + z^4)$     **55.** $\left(7 - \dfrac{3}{5}m\right)\left(7 + \dfrac{3}{5}m\right)$     **57.** $(t + 7)(t^2 - 7t + 49)$     **59.** $n(n^2 + 49)$

**61.** $x^2(x^2 + 9)(x + 3)(x - 3)$     **63.** $pq(8p + 9q)(8p - 9q)$     **65.** $xy^2(27xy + 1)$     **67.** $a(5a - 4b)(25a^2 + 20ab + 16b^2)$     **69.** $16x^2(x + 2)(x - 2)$

**71.** 6    **73.** −2    **75.** $\dfrac{1}{5}$    **77.** 81.2%    **79.** −4(0.3$x^2$ − $x$ − 20)    **81.** ($x$ + 2 + $y$)($x$ + 2 − $y$)    **83.** ($a$ + 4)($a$ − 4)($b$ − 4)
**85.** ($x$ + 3 + 2$y$)($x$ + 3 − 2$y$)    **87.** ($x^n$ + 10)($x^n$ − 10)    **89.** ($x$ + 6)    **91.** answers may vary    **93. a.** 777 ft    **b.** 441 ft    **c.** 7 sec
**d.** (29 + 4$t$)(29 − 4$t$)    **95. a.** 1456 ft    **b.** 816 ft    **c.** 10 sec    **d.** 16(10 + $t$)(10 − $t$)

## Integrated Review
### Practice Exercises 6.1–6.5
**1.** (3$x$ − 1)(2$x$ − 3)    **2.** (3$x$ + 1)($x$ − 2)($x$ + 2)    **3.** 3(3$x$ − $y$)(3$x$ + $y$)    **4.** (2$a$ + $b$)(4$a^2$ − 2$ab$ + $b^2$)    **5.** 6$xy^2$(5$x$ + 2)(2$x$ − 3)

## Exercise Set 6.1–6.5
**1.** ($x$ + $y$)$^2$    **2.** ($x$ − $y$)$^2$    **3.** ($a$ + 12)($a$ − 1)    **4.** ($a$ − 10)($a$ − 1)    **5.** ($a$ + 2)($a$ − 3)    **6.** ($a$ − 1)$^2$    **7.** ($x$ + 1)$^2$    **8.** ($x$ + 2)($x$ − 1)
**9.** ($x$ + 1)($x$ + 3)    **10.** ($x$ + 3)($x$ − 2)    **11.** ($x$ + 3)($x$ + 4)    **12.** ($x$ + 4)($x$ − 3)    **13.** ($x$ + 4)($x$ − 1)    **14.** ($x$ − 5)($x$ − 2)
**15.** ($x$ + 5)($x$ − 3)    **16.** ($x$ + 6)($x$ + 5)    **17.** ($x$ − 6)($x$ + 5)    **18.** ($x$ + 8)($x$ + 3)    **19.** 2($x$ + 7)($x$ − 7)    **20.** 3($x$ + 5)($x$ − 5)
**21.** ($x$ + 3)($x$ + $y$)    **22.** ($y$ − 7)(3 + $x$)    **23.** ($x$ + 8)($x$ − 2)    **24.** ($x$ − 7)($x$ + 4)    **25.** 4$x$($x$ + 7)($x$ − 2)    **26.** 6$x$($x$ − 5)($x$ + 4)
**27.** 2(3$x$ + 4)(2$x$ + 3)    **28.** (2$a$ − $b$)(4$a$ + 5$b$)    **29.** (2$a$ + $b$)(2$a$ − $b$)    **30.** (4 − 3$x$)(7 + 2$x$)    **31.** (5 − 2$x$)(4 + $x$)    **32.** prime    **33.** prime
**34.** (3$y$ + 5)(2$y$ − 3)    **35.** (4$x$ − 5)($x$ + 1)    **36.** $y$($x$ + $y$)($x$ − $y$)    **37.** 4($t^2$ + 9)    **38.** ($x$ + 1)($x$ + $y$)    **39.** ($x$ + 1)($a$ + 2)
**40.** 9$x$(2$x^2$ − 7$x$ + 1)    **41.** 4$a$(3$a^2$ − 6$a$ + 1)    **42.** ($x$ + 16)($x$ − 2)    **43.** prime    **44.** (4$a$ − 7$b$)$^2$    **45.** (5$p$ − 7$q$)$^2$    **46.** (7$x$ + 3$y$)($x$ + 3$y$)
**47.** (5 − 2$y$)(25 + 10$y$ + 4$y^2$)    **48.** (4$x$ + 3)(16$x^2$ − 12$x$ + 9)    **49.** −($x$ − 5)($x$ + 6)    **50.** −($x$ − 2)($x$ − 4)    **51.** (7 − $x$)(2 + $x$)
**52.** (3 + $x$)(1 − $x$)    **53.** 3$x^2y$($x$ + 6)($x$ − 4)    **54.** 2$xy$($x$ + 5$y$)($x$ − $y$)    **55.** 5$xy^2$($x$ − 7$y$)($x$ − $y$)    **56.** 4$x^2y$($x$ − 5)($x$ + 3)
**57.** 3$xy$(4$x^2$ + 81)    **58.** 2$xy^2$(3$x^2$ + 4)    **59.** (2 + $x$)(2 − $x$)    **60.** (3 + $y$)(3 − $y$)    **61.** ($s$ + 4)(3$r$ − 1)    **62.** ($x$ − 2)($x^2$ + 3)
**63.** (4$x$ − 3)($x$ − 2$y$)    **64.** (2$x$ − $y$)(2$x$ + 7$z$)    **65.** 6($x$ + 2$y$)($x$ + $y$)    **66.** 2($x$ + 4$y$)(6$x$ − $y$)    **67.** ($x$ + 3)($y$ + 2)($y$ − 2)
**68.** ($y$ + 3)($y$ − 3)($x^2$ + 3)    **69.** (5 + $x$)($x$ + $y$)    **70.** ($x$ − $y$)(7 + $y$)    **71.** (7$t$ − 1)(2$t$ − 1)    **72.** prime    **73.** (3$x$ + 5)($x$ − 1)
**74.** (7$x$ − 2)($x$ + 3)    **75.** ($x$ + 12$y$)($x$ − 3$y$)    **76.** (3$x$ − 2$y$)($x$ + 4$y$)    **77.** (1 − 10$ab$)(1 + 2$ab$)    **78.** (1 + 5$ab$)(1 − 12$ab$)
**79.** (3 + $x$)(3 − $x$)(1 + $x$)(1 − $x$)    **80.** (3 + $x$)(3 − $x$)(2 + $x$)(2 − $x$)    **81.** ($x$ + 4)($x$ − 4)($x^2$ + 2)    **82.** ($x$ + 5)($x$ − 5)($x^2$ + 3)
**83.** ($x$ − 15)($x$ − 8)    **84.** ($y$ + 16)($y$ + 6)    **85.** 2$x$(3$x$ − 2)($x$ − 4)    **86.** 2$y$(3$y$ + 5)($y$ − 3)    **87.** (3$x$ − 5$y$)(9$x^2$ + 15$xy$ + 25$y^2$)
**88.** (6$y$ − $z$)(36$y^2$ + 6$yz$ + $z^2$)    **89.** ($xy$ + 2$z$)($x^2y^2$ − 2$xyz$ + 4$z^2$)    **90.** (3$ab$ + 2)(9$a^2b^2$ − 6$ab$ + 4)    **91.** 2$xy$(1 + 6$x$)(1 − 6$x$)
**92.** 2$x$($x$ + 3)($x$ − 3)    **93.** ($x$ + 2)($x$ − 2)($x$ + 6)    **94.** ($x$ − 2)($x$ + 6)($x$ − 6)    **95.** 2$a^2$(3$a$ + 5)    **96.** 2$n$(2$n$ − 3)    **97.** ($a^2$ + 2)($a$ + 2)
**98.** ($a$ − $b$)(1 + $x$)    **99.** ($x$ + 2)($x$ − 2)($x$ + 7)    **100.** ($a$ + 3)($a$ − 3)($a$ + 5)    **101.** ($x$ − $y$ + $z$)($x$ − $y$ − $z$)
**102.** ($x$ + 2$y$ + 3)($x$ + 2$y$ − 3)    **103.** (9 + 5$x$ + 1)(9 − 5$x$ − 1)    **104.** ($b$ + 4$a$ + $c$)($b$ − 4$a$ − $c$)    **105.** answers may vary
**106.** yes; 9($x^2$ + 9$y^2$)    **107.** a, c

## Section 6.6
### Practice Exercises
**1.** −4, 5    **2.** 0, $\dfrac{6}{7}$    **3.** −4, 12    **4.** $\dfrac{4}{3}$    **5.** −3, $\dfrac{2}{3}$    **6.** −6, 4    **7.** −3, 0, 3    **8.** $\dfrac{2}{3}$, $\dfrac{3}{2}$, 5    **9.** −3, 0, 2    **10.** The $x$-intercepts are (2, 0) and (4, 0).

### Graphing Calculator Explorations 6.6
**1.** −0.9, 2.2    **3.** no real solution    **5.** −1.8, 2.8

### Vocabulary and Readiness Check 6.6
**1.** quadratic    **3.** 3, −5    **5.** 3, 7    **7.** −8, −6    **9.** −1, 3

## Exercise Set 6.6
**1.** 2, −1    **3.** −9, −17    **5.** 0, −6    **7.** 0, 8    **9.** −$\dfrac{3}{2}$, $\dfrac{5}{4}$    **11.** $\dfrac{7}{2}$, −$\dfrac{2}{7}$    **13.** $\dfrac{1}{2}$, −$\dfrac{1}{3}$    **15.** −0.2, −1.5

**17.** answers may vary; for example, ($x$ − 6)($x$ + 1) = 0    **19.** 9, 4    **21.** −4, 2    **23.** 0, 7    **25.** 8, −4    **27.** 4, −4    **29.** −3, 12    **31.** $\dfrac{7}{3}$, −2
**33.** −5, 5    **35.** −2, $\dfrac{1}{6}$    **37.** 0, 4, 8    **39.** $\dfrac{3}{4}$    **41.** −$\dfrac{1}{2}$, 0, $\dfrac{1}{2}$    **43.** −$\dfrac{3}{8}$, 0, $\dfrac{1}{2}$    **45.** −3, 2    **47.** −20, 0    **49.** $\dfrac{17}{2}$    **51.** −$\dfrac{1}{2}$, $\dfrac{1}{2}$
**53.** −$\dfrac{3}{2}$, −$\dfrac{1}{2}$, 3    **55.** −5, 3    **57.** −$\dfrac{5}{6}$, $\dfrac{6}{5}$    **59.** 2, −$\dfrac{4}{5}$    **61.** −$\dfrac{4}{3}$, 5    **63.** −4, 3    **65.** $\dfrac{8}{3}$, −9, 0    **67.** −7    **69.** 0, $\dfrac{3}{2}$    **71.** 0, 1, −1
**73.** −6, $\dfrac{4}{3}$    **75.** $\dfrac{6}{7}$, 1    **77.** $\left(-\dfrac{4}{3}, 0\right)$, (1, 0)    **79.** (−2, 0), (5, 0)    **81.** (−6, 0), $\left(\dfrac{1}{2}, 0\right)$    **83.** e    **85.** b    **87.** c    **89.** $\dfrac{47}{45}$    **91.** $\dfrac{17}{60}$    **93.** $\dfrac{15}{8}$
**95.** $\dfrac{7}{10}$    **97.** didn't write equation in standard form; should be $x$ = 4 or $x$ = −2    **99.** answers may vary; for example, $x^2$ − 12$x$ + 35 = 0

**101. a.** 300; 304; 276; 216; 124; 0; −156    **b.** 5 sec    **c.** 304 ft    **d.**    **103.** 0, $\dfrac{1}{2}$
**105.** 0, −15

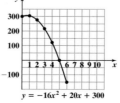

$y = -16x^2 + 20x + 300$

## The Bigger Picture

**1.** $-34$    **2.** $x^{22}$    **3.** $-4x^3 - 6x^2 + 8$    **4.** $y - 1 + \dfrac{3}{y^2}$    **5.** $10x(x + 5)(x - 5)$    **6.** $(x - 1)(x - 35)$    **7.** $3(2y + 5)(x - 1)$

**8.** $x(5y - 7)(y + 1)$    **9.** $5, -\dfrac{1}{2}$    **10.** $1$    **11.** $-2, 14$    **12.** $\dfrac{33}{17}$

## Section 6.7

### Practice Exercises

**1.** 2 sec    **2.** There are 2 numbers. They are $-4$ and 12.    **3.** base: 35 ft; height: 12 ft    **4.** 7 and 8 or $-6$ and $-5$
**5.** leg: 8 units; leg: 15 units; hypotenuse: 17 units

### Exercise Set 6.7

**1.** width $= x$; length $= x + 4$    **3.** $x$ and $x + 2$ if $x$ is an odd integer    **5.** base $= x$; height $= 4x + 1$    **7.** 11 units
**9.** 15 cm, 13 cm, 70 cm, 22 cm    **11.** base $= 16$ mi; height $= 6$ mi    **13.** 5 sec    **15.** length $= 5$ cm; width $= 6$ cm    **17.** 54 diagonals
**19.** 10 sides    **21.** $-12$ or 11    **23.** 14, 15    **25.** 13 feet    **27.** 5 in.    **29.** 12 mm, 16 mm, 20 mm    **31.** 10 km    **33.** 36 ft    **35.** 9.5 sec

**37.** 20%    **39.** length: 15 mi; width: 8 mi    **41.** 105 units    **43.** 175 acres    **45.** 6.25 million    **47.** 1966    **49.** answers may vary    **51.** $\dfrac{3}{4}$

**53.** $\dfrac{5}{9}$    **55.** $\dfrac{9}{10}$    **57.** 8 m    **59.** 10 and 15    **61.** width: 29 m; length: 35 m    **63.** answers may vary

## Chapter 6 Vocabulary Check

**1.** quadratic equation    **2.** factoring    **3.** greatest common factor    **4.** perfect square trinomial    **5.** difference of two squares
**6.** difference of two cubes    **7.** sum of two cubes    **8.** 0

## Chapter 6 Review

**1.** $2x - 5$    **3.** $4x(5x + 3)$    **5.** $-2x^2y(4x - 3y)$    **7.** $(x + 1)(5x - 1)$    **9.** $(2x - 1)(3x + 5)$    **11.** $(x + 4)(x + 2)$    **13.** prime
**15.** $(x + 4)(x - 2)$    **17.** $(x + 5y)(x + 3y)$    **19.** $2(3 - x)(12 + x)$    **21.** $(2x - 1)(x + 6)$    **23.** $(2x + 3)(2x - 1)$    **25.** $(6x - y)(x - 4y)$
**27.** $(2x + 3y)(x - 13y)$    **29.** $(6x + 5y)(3x - 4y)$    **31.** $(2x + 3)(2x - 3)$    **33.** prime    **35.** $(2x + 3)(4x^2 - 6x + 9)$

**37.** $2(3 - xy)(9 + 3xy + x^2y^2)$    **39.** $(4x^2 + 1)(2x + 1)(2x - 1)$    **41.** $-6, 2$    **43.** $-\dfrac{1}{5}, -3$    **45.** $-4, 6$    **47.** 2, 8    **49.** $-\dfrac{2}{7}, \dfrac{3}{8}$    **51.** $-\dfrac{2}{5}$

**53.** 3    **55.** $0, -\dfrac{7}{4}, 3$    **57.** c    **59.** 9 units    **61.** width: 20 in.; length: 25 in.    **63.** 19 and 20    **65.** 6.25 sec    **67.** $7(x - 9)$

**69.** $\left( m + \dfrac{2}{5} \right)\left( m - \dfrac{2}{5} \right)$    **71.** $(y + 2)(x - 1)$    **73.** $3x(x - 9)(x - 1)$    **75.** $2(x + 3)(x - 3)$    **77.** $5(x + 2)^2$    **79.** $2xy(2x - 3y)$

**81.** $(5x + 3)(25x^2 - 15x + 9)$    **83.** $(x + 7 + y)(x + 7 - y)$    **85.** $2b(3a - 1)(9a^2 + 3a + 1)$    **87.** factor out the GCF, 3
**89.** $16x^2 - 28x + 6; 2(4x - 1)(2x - 3)$    **91.** $-3, 5$    **93.** 3, 2    **95.** 19 in., 8 in., 21 in.    **97.** 6.75 sec

## Chapter 6 Test

**1.** $(x + 7)(x + 4)$    **2.** $(7 - m)(7 + m)$    **3.** $(y + 11)^2$    **4.** $(a + 3)(4 - y)$    **5.** prime    **6.** $(y - 12)(y + 4)$    **7.** prime
**8.** $3x(3x + 1)(x + 4)$    **9.** $(3a - 7)(a + b)$    **10.** $(3x - 2)(x - 1)$    **11.** $(x + 12y)(x + 2y)$    **12.** $5(6 + x)(6 - x)$    **13.** $(6t + 5)(t - 1)$
**14.** $(y + 2)(y - 2)(x - 7)$    **15.** $x(1 + x^2)(1 + x)(1 - x)$    **16.** $-xy(y^2 + x^2)$    **17.** $(4x - 1)(16x^2 + 4x + 1)$    **18.** $8(y - 2)(y^2 + 2y + 4)$

**19.** $-9, 3$    **20.** $-7, 2$    **21.** $-7, 1$    **22.** $0, \dfrac{3}{2}, -\dfrac{4}{3}$    **23.** $0, 3, -3$    **24.** $-3, 5$    **25.** $0, \dfrac{5}{2}$    **26.** 17 ft    **27.** 8 and 9    **28.** 7 sec

**29.** hypotenuse: 25 cm; legs: 15 cm, 20 cm

## Chapter 6 Cumulative Review

**1. a.** $9 \le 11$    **b.** $8 > 1$    **c.** $3 \ne 4$; Sec. 1.2, Ex. 3    **3. a.** $\dfrac{6}{7}$    **b.** $\dfrac{11}{27}$    **c.** $\dfrac{22}{5}$; Sec. 1.3, Ex. 2    **5.** $\dfrac{14}{3}$; Sec. 1.4, Ex. 5

**7. a.** $-12$    **b.** $-1$; Sec. 1.5, Ex. 7    **9. a.** $-32$    **b.** $-14$    **c.** 90; Sec. 1.7, Ex. 1    **11. a.** $4x$    **b.** $11y^2$    **c.** $8x^2 - x$; Sec. 2.1, Ex. 3

**13.** $140$; Sec. 2.2, Ex. 7    **15.** $-11$; Sec. 2.2, Ex. 6    **17.** $\dfrac{16}{3}$; Sec. 2.3, Ex. 2    **19.** shorter: 12 in.; longer: 36 in.; Sec. 2.4, Ex. 3

**21.** 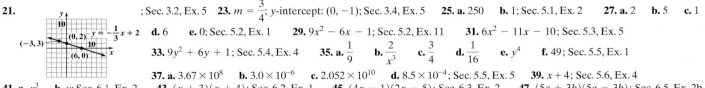    ; Sec. 3.2, Ex. 5    **23.** $m = \dfrac{3}{4}$; $y$-intercept: $(0, -1)$; Sec. 3.4, Ex. 5    **25. a.** 250    **b.** 1; Sec. 5.1, Ex. 2    **27. a.** 2    **b.** 5    **c.** 1

**d.** 6    **e.** 0; Sec. 5.2, Ex. 1    **29.** $9x^2 - 6x - 1$; Sec. 5.2, Ex. 11    **31.** $6x^2 - 11x - 10$; Sec. 5.3, Ex. 5

**33.** $9y^2 + 6y + 1$; Sec. 5.4, Ex. 4    **35. a.** $\dfrac{1}{9}$    **b.** $\dfrac{2}{x^3}$    **c.** $\dfrac{3}{4}$    **d.** $\dfrac{1}{16}$    **e.** $y^4$    **f.** 49; Sec. 5.5, Ex. 1

**37. a.** $3.67 \times 10^8$    **b.** $3.0 \times 10^{-6}$    **c.** $2.052 \times 10^{10}$    **d.** $8.5 \times 10^{-4}$; Sec. 5.5, Ex. 5    **39.** $x + 4$; Sec. 5.6, Ex. 4
**41. a.** $x^3$    **b.** $y$; Sec. 6.1, Ex. 2    **43.** $(x + 3)(x + 4)$; Sec. 6.2, Ex. 1    **45.** $(4x - 1)(2x - 5)$; Sec. 6.3, Ex. 2    **47.** $(5a + 3b)(5a - 3b)$; Sec. 6.5, Ex. 2b
**49.** $3, -1$; Sec. 6.6, Ex. 1

# CHAPTER 7  RATIONAL EXPRESSIONS

## Section 7.1
### Practice Exercises

**1. a.** $\{x|x \text{ is a real number}\}$  **b.** $\{x|x \text{ is a real number and } x \neq -3\}$  **c.** $\{x|x \text{ is a real number and } x \neq 2, x \neq 3\}$  **2. a.** $\dfrac{1}{2z-1}$  **b.** $\dfrac{5x+3}{6x-5}$

**3. a.** 1  **b.** $-1$  **4.** $-\dfrac{5(2+x)}{x+3}$  **5. a.** $x^2-4x+16$  **b.** $\dfrac{5}{z-3}$  **6.** $\dfrac{-(x+3)}{6x-11}; \dfrac{-x-3}{6x-11}; \dfrac{x+3}{-(6x-11)}; \dfrac{x+3}{-6x+11}; \dfrac{x+3}{11-6x}$

**7. a.** \$7.20  **b.** \$3.60

### Graphing Calculator Explorations 7.1

**1.** $\{x|x \text{ is a real number and } x \neq -2, x \neq 2\}$ 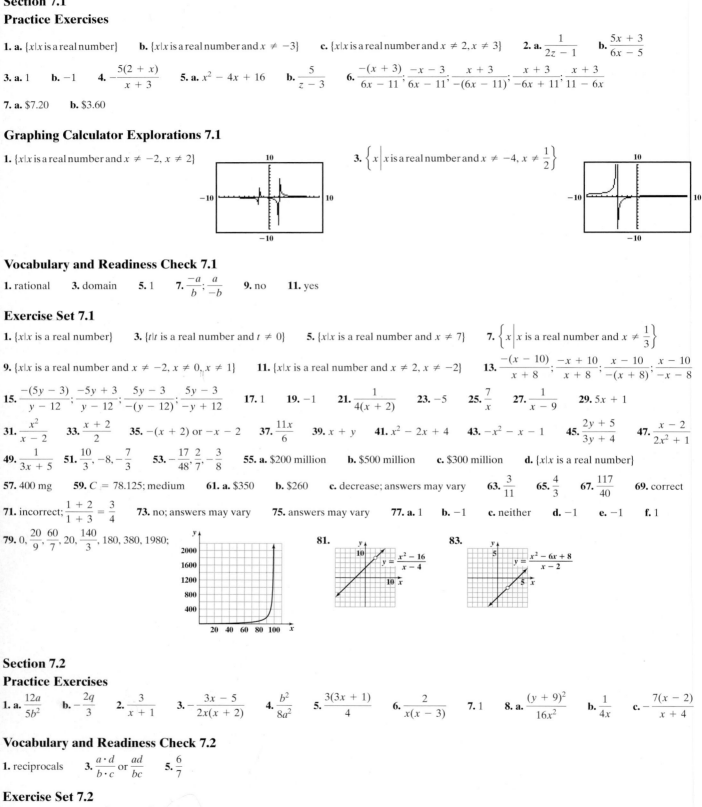  **3.** $\left\{x \middle| x \text{ is a real number and } x \neq -4, x \neq \dfrac{1}{2}\right\}$

### Vocabulary and Readiness Check 7.1

**1.** rational  **3.** domain  **5.** 1  **7.** $\dfrac{-a}{b}; \dfrac{a}{-b}$  **9.** no  **11.** yes

### Exercise Set 7.1

**1.** $\{x|x \text{ is a real number}\}$  **3.** $\{t|t \text{ is a real number and } t \neq 0\}$  **5.** $\{x|x \text{ is a real number and } x \neq 7\}$  **7.** $\left\{x \middle| x \text{ is a real number and } x \neq \dfrac{1}{3}\right\}$

**9.** $\{x|x \text{ is a real number and } x \neq -2, x \neq 0, x \neq 1\}$  **11.** $\{x|x \text{ is a real number and } x \neq 2, x \neq -2\}$  **13.** $\dfrac{-(x-10)}{x+8}; \dfrac{-x+10}{x+8}; \dfrac{x-10}{-(x+8)}; \dfrac{x-10}{-x-8}$

**15.** $\dfrac{-(5y-3)}{y-12}; \dfrac{-5y+3}{y-12}; \dfrac{5y-3}{-(y-12)}; \dfrac{5y-3}{-y+12}$  **17.** 1  **19.** $-1$  **21.** $\dfrac{1}{4(x+2)}$  **23.** $-5$  **25.** $\dfrac{7}{x}$  **27.** $\dfrac{1}{x-9}$  **29.** $5x+1$

**31.** $\dfrac{x^2}{x-2}$  **33.** $\dfrac{x+2}{2}$  **35.** $-(x+2)$ or $-x-2$  **37.** $\dfrac{11x}{6}$  **39.** $x+y$  **41.** $x^2-2x+4$  **43.** $-x^2-x-1$  **45.** $\dfrac{2y+5}{3y+4}$  **47.** $\dfrac{x-2}{2x^2+1}$

**49.** $\dfrac{1}{3x+5}$  **51.** $\dfrac{10}{3}, -8, -\dfrac{7}{3}$  **53.** $-\dfrac{17}{48}, \dfrac{2}{7}, -\dfrac{3}{8}$  **55. a.** \$200 million  **b.** \$500 million  **c.** \$300 million  **d.** $\{x|x \text{ is a real number}\}$

**57.** 400 mg  **59.** $C = 78.125$; medium  **61. a.** \$350  **b.** \$260  **c.** decrease; answers may vary  **63.** $\dfrac{3}{11}$  **65.** $\dfrac{4}{3}$  **67.** $\dfrac{117}{40}$  **69.** correct

**71.** incorrect; $\dfrac{1+2}{1+3} = \dfrac{3}{4}$  **73.** no; answers may vary  **75.** answers may vary  **77. a.** 1  **b.** $-1$  **c.** neither  **d.** $-1$  **e.** $-1$  **f.** 1

**79.** $0, \dfrac{20}{9}, \dfrac{60}{7}, 20, \dfrac{140}{3}, 180, 380, 1980$;  **81.** $y = \dfrac{x^2-16}{x-4}$  **83.** $y = \dfrac{x^2-6x+8}{x-2}$

## Section 7.2
### Practice Exercises

**1. a.** $\dfrac{12a}{5b^2}$  **b.** $-\dfrac{2q}{3}$  **2.** $\dfrac{3}{x+1}$  **3.** $-\dfrac{3x-5}{2x(x+2)}$  **4.** $\dfrac{b^2}{8a^2}$  **5.** $\dfrac{3(3x+1)}{4}$  **6.** $\dfrac{2}{x(x-3)}$  **7.** 1  **8. a.** $\dfrac{(y+9)^2}{16x^2}$  **b.** $\dfrac{1}{4x}$  **c.** $-\dfrac{7(x-2)}{x+4}$

### Vocabulary and Readiness Check 7.2

**1.** reciprocals  **3.** $\dfrac{a \cdot d}{b \cdot c}$ or $\dfrac{ad}{bc}$  **5.** $\dfrac{6}{7}$

### Exercise Set 7.2

**1.** $\dfrac{21}{4y}$  **3.** $x^4$  **5.** $-\dfrac{b^2}{6}$  **7.** $\dfrac{x^2}{10}$  **9.** $\dfrac{1}{3}$  **11.** $\dfrac{m+n}{m-n}$  **13.** $\dfrac{x+5}{x}$  **15.** $\dfrac{(x+2)(x-3)}{(x-4)(x+4)}$  **17.** $\dfrac{2x^4}{3}$  **19.** $\dfrac{12}{y^6}$  **21.** $x(x+4)$  **23.** $\dfrac{3(x+1)}{x^3(x-1)}$

**25.** $m^2-n^2$  **27.** $-\dfrac{x+2}{x-3}$  **29.** $\dfrac{x+2}{x-3}$  **31.** $\dfrac{5}{6}$  **33.** $\dfrac{3x}{8}$  **35.** $\dfrac{3}{2}$  **37.** $\dfrac{3x+4y}{2(x+2y)}$  **39.** $\dfrac{2(x+2)}{x-2}$  **41.** $-\dfrac{y(x+2)}{4}$  **43.** $\dfrac{(a+5)(a+3)}{(a+2)(a+1)}$

**45.** $\dfrac{5}{x}$   **47.** $\dfrac{2(n-8)}{3n-1}$   **49.** $4x^3(x-3)$   **51.** $\dfrac{(a+b)^2}{a-b}$   **53.** $\dfrac{3x+5}{x^2+4}$   **55.** $\dfrac{4}{x-2}$   **57.** $\dfrac{a-b}{6(a^2+ab+b^2)}$   **59.** 1   **61.** $-\dfrac{10}{9}$   **63.** $-\dfrac{1}{5}$

**65.** 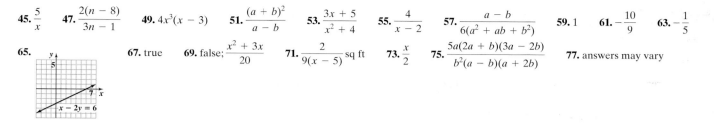   **67.** true   **69.** false; $\dfrac{x^2+3x}{20}$   **71.** $\dfrac{2}{9(x-5)}$ sq ft   **73.** $\dfrac{x}{2}$   **75.** $\dfrac{5a(2a+b)(3a-2b)}{b^2(a-b)(a+2b)}$   **77.** answers may vary

## Section 7.3
### Practice Exercises

**1.** $\dfrac{2a}{b}$   **2.** 1   **3.** $4x-5$   **4. a.** 42   **b.** $45y^3$   **5. a.** $(y-5)(y-4)$   **b.** $a(a+2)$   **6.** $3(2x-1)^2$   **7.** $(x+4)(x-4)(x+1)$

**8.** $(3-x)$ or $(x-3)$   **9. a.** $\dfrac{21x^2y}{35xy^2}$   **b.** $\dfrac{18x}{8x+14}$   **10.** $\dfrac{3x-6}{(x-2)(x+3)(x-5)}$

## Vocabulary and Readiness Check 7.3

**1.** $\dfrac{9}{11}$   **3.** $\dfrac{a+c}{b}$   **5.** $\dfrac{5-(6+x)}{x}$

## Exercise Set 7.3

**1.** $\dfrac{a+9}{13}$   **3.** $\dfrac{3m}{n}$   **5.** 4   **7.** $\dfrac{y+10}{3+y}$   **9.** $5x+3$   **11.** $\dfrac{4}{a+5}$   **13.** $\dfrac{1}{x-6}$   **15.** $\dfrac{5x+7}{x-3}$   **17.** $x+5$   **19.** 3   **21.** $4x^3$

**23.** $8x(x+2)$   **25.** $(x+3)(x-2)$   **27.** $3(x+6)$   **29.** $5(x-6)^2$   **31.** $6(x+1)^2$   **33.** $x-8$ or $8-x$   **35.** $(x-1)(x+4)(x+3)$

**37.** $(3x+1)(x+1)(x-1)(2x+1)$   **39.** $2x^2(x+4)(x-4)$   **41.** $\dfrac{6x}{4x^2}$   **43.** $\dfrac{24b^2}{12ab^2}$   **45.** $\dfrac{9y}{2y(x+3)}$   **47.** $\dfrac{9ab+2b}{5b(a+2)}$

**49.** $\dfrac{x^2+x}{x(x+4)(x+2)(x+1)}$   **51.** $\dfrac{18y-2}{30x^2-60}$   **53.** $2x$   **55.** $\dfrac{x+3}{2x-1}$   **57.** $x+1$   **59.** $\dfrac{1}{x^2-8}$   **61.** $\dfrac{6(4x+1)}{x(2x+1)}$   **63.** $\dfrac{29}{21}$   **65.** $-\dfrac{5}{12}$

**67.** $\dfrac{7}{30}$   **69.** d   **71.** c   **73.** b   **75.** $-\dfrac{5}{x-2}$   **77.** $\dfrac{7+x}{x-2}$   **79.** $\dfrac{20}{x-2}$ m   **81.** answers may vary   **83.** 95,304 Earth days
**85.** answers may vary   **87.** answers may vary

## Section 7.4
### Practice Exercises

**1. a.** 0   **b.** $\dfrac{21a+10}{24a^2}$   **2.** $\dfrac{6}{x-5}$   **3.** $\dfrac{13y+3}{5y(y+1)}$   **4.** $\dfrac{13}{x-5}$   **5.** $\dfrac{3b+6}{b+3}$ or $\dfrac{3(b+2)}{b+3}$   **6.** $\dfrac{10-3x^2}{2x(2x+3)}$   **7.** $\dfrac{x(5x+6)}{(x+4)(x+3)(x-3)}$

## Vocabulary and Readiness Check 7.4
**1.** d   **3.** a

## Exercise Set 7.4

**1.** $\dfrac{5}{x}$   **3.** $\dfrac{75a-6b^2}{5b}$   **5.** $\dfrac{6x+5}{2x^2}$   **7.** $\dfrac{11}{x+1}$   **9.** $\dfrac{x-6}{(x-2)(x+2)}$   **11.** $\dfrac{35x-6}{4x(x-2)}$   **13.** $-\dfrac{2}{x-3}$   **15.** 0   **17.** $-\dfrac{1}{x^2-1}$   **19.** $\dfrac{5+2x}{x}$

**21.** $\dfrac{6x-7}{x-2}$   **23.** $-\dfrac{y+4}{y+3}$   **25.** $\dfrac{-5x+14}{4x}$ or $-\dfrac{5x-14}{4x}$   **27.** 2   **29.** $\dfrac{9x^4-4x^2}{21}$   **31.** $\dfrac{x+2}{(x+3)^2}$   **33.** $\dfrac{9b-4}{5b(b-1)}$   **35.** $\dfrac{2+m}{m}$

**37.** $\dfrac{x^2+3x}{(x-7)(x-2)}$ or $\dfrac{x(x+3)}{(x-7)(x-2)}$   **39.** $\dfrac{10}{1-2x}$   **41.** $\dfrac{15x-1}{(x+1)^2(x-1)}$   **43.** $\dfrac{x^2-3x-2}{(x-1)^2(x+1)}$   **45.** $\dfrac{a+2}{2(a+3)}$   **47.** $\dfrac{y(2y+1)}{(2y+3)^2}$

**49.** $\dfrac{x-10}{2(x-2)}$   **51.** $\dfrac{2x+21}{(x+3)^2}$   **53.** $\dfrac{-5x+23}{(x-2)(x-3)}$   **55.** $\dfrac{7}{2(m-10)}$   **57.** $\dfrac{2x^2-2x-46}{(x+1)(x-6)(x-5)}$ or $\dfrac{2(x^2-x-23)}{(x+1)(x-6)(x-5)}$

**59.** $\dfrac{n+4}{4n(n-1)(n-2)}$   **61.** 10   **63.** 2   **65.** $\dfrac{25a}{9(a-2)}$   **67.** $\dfrac{x+4}{(x-2)(x-1)}$   **69.** $x=\dfrac{2}{3}$   **71.** $x=-\dfrac{1}{2}$, $x=1$   **73.** $x=-\dfrac{15}{2}$

**75.** $\dfrac{6x^2-5x-3}{x(x+1)(x-1)}$   **77.** $\dfrac{4x^2-15x+6}{(x-2)^2(x+2)(x-3)}$   **79.** $\dfrac{-2x^2+14x+55}{(x+2)(x+7)(x+3)}$   **81.** $\dfrac{2x-16}{(x+4)(x-4)}$ in.   **83.** $\dfrac{P-G}{P}$   **85.** answers may vary

**87.** $\left(\dfrac{90x-40}{x}\right)^\circ$   **89.** answers may vary

## The Bigger Picture

**1.** $-17.7$   **2.** $78.26$   **3.** $28$   **4.** $4x^4 - x^2 - 17$   **5.** $\dfrac{x-1}{5}$   **6.** $\dfrac{14}{x+1}$   **7.** $-\dfrac{11}{18}$   **8.** $\dfrac{-4x-27}{45}$ or $-\dfrac{4x+27}{45}$   **9.** $x(9x-11)(x+1)$

**10.** $(4y-7)(3x+1)$   **11.** $12$   **12.** $\left(\frac{1}{2}, \infty\right)$   **13.** $-\dfrac{8}{3}$   **14.** $-4, 6$

## Section 7.5
## Practice Exercises

**1.** $-2$   **2.** $13$   **3.** $-1, 7$   **4.** $-\dfrac{19}{2}$   **5.** $3$   **6.** $-8$   **7.** $b = \dfrac{ax}{a-x}$

## Graphing Calculator Explorations 7.5

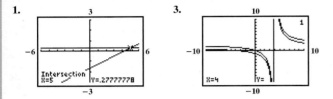

**1.**   **3.**

## Exercise Set 7.5

**1.** $30$   **3.** $0$   **5.** $-2$   **7.** $-5, 2$   **9.** $5$   **11.** $3$   **13.** $1$   **15.** $5$   **17.** no solution   **19.** $4$   **21.** $-8$   **23.** $6, -4$   **25.** $1$   **27.** $3, -4$

**29.** $-3$   **31.** $0$   **33.** $-2$   **35.** $8, -2$   **37.** no solution   **39.** $3$   **41.** $-11, 1$   **43.** $I = \dfrac{E}{R}$   **45.** $B = \dfrac{2U - TE}{T}$   **47.** $W = \dfrac{Bh^2}{705}$

**49.** $G = \dfrac{V}{N - R}$   **51.** $r = \dfrac{C}{2\pi}$   **53.** $x = \dfrac{3y}{3+y}$   **55.** $\dfrac{1}{x}$   **57.** $\dfrac{1}{x} + \dfrac{1}{2}$   **59.** $\dfrac{1}{3}$   **61.** $(2,0), (0,-2)$   **63.** $(-4,0), (-2,0), (3,0), (0,4)$

**65.** answers may vary   **67.** $\dfrac{5x+9}{9x}$   **69.** no solution   **71.** $100°, 80°$   **73.** $22.5°, 67.5°$   **75.** $\dfrac{17}{4}$

## Integrated Review

**1.** expression; $\dfrac{3+2x}{3x}$   **2.** expression; $\dfrac{18+5a}{6a}$   **3.** equation; $3$   **4.** equation; $18$   **5.** expression; $\dfrac{x+1}{x(x-1)}$   **6.** expression; $\dfrac{3(x+1)}{x(x-3)}$

**7.** equation; no solution   **8.** equation; $1$   **9.** expression; $10$   **10.** expression; $\dfrac{z}{3(9z-5)}$   **11.** expression; $\dfrac{5x+7}{x-3}$   **12.** expression; $\dfrac{7p+5}{2p+7}$

**13.** equation; $23$   **14.** equation; $5$   **15.** expression; $\dfrac{25a}{9(a-2)}$   **16.** expression; $\dfrac{4x+5}{(x+1)(x-1)}$   **17.** expression; $\dfrac{3x^2+5x+3}{(3x-1)^2}$

**18.** expression; $\dfrac{2x^2-3x-1}{(2x-5)^2}$   **19.** expression; $\dfrac{4x-37}{5x}$   **20.** equation; $-\dfrac{7}{3}$   **21.** equation; $\dfrac{8}{5}$   **22.** expression; $\dfrac{29x-23}{3x}$

## Section 7.6
## Practice Exercises

**1.** $99$   **2.** $\dfrac{13}{3}$   **3.** $\$9.03$   **4.** $6$   **5.** $15$   **6.** $1\dfrac{5}{7}$ hr   **7.** bus: 45 mph; car: 60 mph

## Vocabulary and Readiness Check 7.6
**1.** c

## Exercise Set 7.6

**1.** $4$   **3.** $\dfrac{50}{9}$   **5.** $-3$   **7.** $\dfrac{14}{9}$   **9.** $123$ lb   **11.** $165$ cal   **13.** $y = 21.25$   **15.** $y = 5\dfrac{5}{7}$ ft   **17.** $2$   **19.** $-3$   **21.** $2\dfrac{2}{9}$ hr   **23.** $1\dfrac{1}{2}$ min

**25.** trip to park rate: $r$; to park time: $\dfrac{12}{r}$; return trip rate: $r$; return time: $\dfrac{18}{r} = \dfrac{12}{r} + 1$; $r = 6$ mph   **27.** 1st portion: 10 mph; cooldown: 8 mph

**29.** $360$ sq ft   **31.** $2$   **33.** $\$108.00$   **35.** $20$ mph   **37.** $y = 37\dfrac{1}{2}$ ft   **39.** $337$ yd/game   **41.** $5$   **43.** $217$ mph   **45.** $9$ gal   **47.** $8$ mph

**49.** 35 mph; 75 mph   **51.** 3 hr   **53.** $26\dfrac{2}{3}$ ft   **55.** 216 nuts   **57.** 510 mph   **59.** 20 hr   **61.** car: 70 mph; motorcycle: 60 mph   **63.** $5\dfrac{1}{4}$ hr

**65.** first pump: 28 min; second pump: 84 min   **67.** $x = 5$   **69.** $x = 13.5$   **71.** $-\dfrac{4}{3}$; downward   **73.** $\dfrac{1}{2}$   **75.** $\dfrac{3}{7}$   **77.** 2000 megawatts

**79.** 8,190,000 people   **81.** answers may vary   **83.** none; answers may vary   **85.** 84 yr   **87.** 30 yr   **89.** 3.75 min

## The Bigger Picture

**1.** $12x^3 - 11x^2 - 13x + 10$   **2.** $4x^2 - 4xy + y^2$   **3.** $4y^3(2 - 5y^2)$   **4.** $(9m - 2n)(m - n)$   **5.** $-35$   **6.** $\dfrac{16x - 70}{x(x - 10)}$ or $\dfrac{2(8x - 35)}{x(x - 10)}$

**7.** $-12x^{13}$   **8.** 2   **9.** $-\dfrac{1}{2}$   **10.** $[4, \infty)$   **11.** $-\dfrac{27}{23}$   **12.** $\dfrac{a^{14}}{b^{14}}$

## Section 7.7

### Practice Exercises

**1. a.** $\dfrac{1}{12m}$   **b.** $\dfrac{8x(x + 4)}{3(x - 4)}$   **c.** $\dfrac{b^2}{a^2}$   **2. a.** $\dfrac{8x(x + 4)}{3(x - 4)}$   **b.** $\dfrac{b^2}{a^2}$   **3.** $\dfrac{y(3xy + 1)}{x^2(1 + xy)}$   **4.** $\dfrac{1 - 6x}{15 + 6x}$

### Vocabulary and Readiness Check 7.7

**1.** $\dfrac{7}{1 + z}$   **3.** $\dfrac{1}{x^2}$   **5.** $\dfrac{2}{x}$   **7.** $\dfrac{1}{9y}$

### Exercise Set 7.7

**1.** 4   **3.** $\dfrac{7}{13}$   **5.** $\dfrac{4}{x}$   **7.** $\dfrac{9(x - 2)}{9x^2 + 4}$   **9.** $2x + y$   **11.** $\dfrac{2(x + 1)}{2x - 1}$   **13.** $\dfrac{2x + 3}{4 - 9x}$   **15.** $\dfrac{1}{x^2 - 2x + 4}$   **17.** $\dfrac{x}{5(x - 2)}$

**19.** $\dfrac{x - 2}{2x - 1}$   **21.** $\dfrac{x}{2 - 3x}$   **23.** $-\dfrac{y}{x + y}$   **25.** $-\dfrac{2x^3}{y(x - y)}$   **27.** $\dfrac{2x + 1}{y}$   **29.** $\dfrac{x - 3}{9}$   **31.** $\dfrac{1}{x + 2}$   **33.** 2

**35.** $\dfrac{xy^2}{x^2 + y^2}$   **37.** $\dfrac{2b^2 + 3a}{b(b - a)}$   **39.** $\dfrac{x}{(x + 1)(x - 1)}$   **41.** $\dfrac{1 + a}{1 - a}$   **43.** $\dfrac{x(x + 6y)}{2y}$   **45.** $\dfrac{5a}{2(a + 2)}$   **47.** $xy(5y + 2x)$

**49.** $\dfrac{xy}{2x + 5y}$   **51.** $\dfrac{x^2y^2}{4}$   **53.** $-9x^3y^4$   **55.** $-9$   **57.** a and c   **59.** $\dfrac{770a}{770 - s}$   **61.** a, b   **63.** $\dfrac{1 + x}{2 + x}$   **65.** $x(x + 1)$

**67.** $\dfrac{x - 3y}{x + 3y}$   **69.** $3a^2 + 4a + 4$   **71. a.** $\dfrac{1}{a + h}$   **b.** $\dfrac{1}{a}$   **c.** $\dfrac{\dfrac{1}{a + h} - \dfrac{1}{a}}{h}$   **d.** $\dfrac{-1}{a(a + h)}$   **73. a.** $\dfrac{3}{a + h + 1}$

**b.** $\dfrac{3}{a + 1}$   **c.** $\dfrac{\dfrac{3}{a + h + 1} - \dfrac{3}{a + 1}}{h}$   **d.** $\dfrac{-3}{(a + h + 1)(a + 1)}$

## Chapter 7 Vocabulary Check

**1.** ratio   **2.** proportion   **3.** cross products   **4.** rational expression   **5.** complex fraction   **6.** domain   **7.** reciprocal

## Chapter 7 Review

**1.** $\{x \mid x \text{ is a real number}\}$   **3.** $\{x \mid x \text{ is a real number and } x \neq 5\}$   **5.** $\{x \mid x \text{ is a real number and } x \neq 0, x \neq -8\}$   **7.** $-1$   **9.** $\dfrac{1}{x - 1}$   **11.** $\dfrac{2(x - 3)}{x - 4}$

**13.** $\dfrac{x + a}{x - c}$   **15.** $-\dfrac{1}{x^2 + 4x + 16}$   **17.** $\dfrac{3x^2}{y}$   **19.** $\dfrac{x - 3}{x + 2}$   **21.** $\dfrac{x + 3}{x - 4}$   **23.** $(x - 6)(x - 3)$   **25.** $\dfrac{1}{2}$   **27.** $-\dfrac{2(2x + 3)}{y - 2}$   **29.** $\dfrac{1}{x + 2}$

**31.** $\dfrac{2x - 10}{3x^2}$   **33.** $14x$   **35.** $\dfrac{10x^2y}{14x^3y}$   **37.** $\dfrac{x^2 - 3x - 10}{(x + 2)(x - 5)(x + 9)}$   **39.** $\dfrac{4y - 30x^2}{5x^2y}$   **41.** $\dfrac{-2x - 2}{x + 3}$   **43.** $\dfrac{x - 4}{3x}$   **45.** $\dfrac{x^2 + 2x + 4}{4x}; \dfrac{x + 2}{32}$

**47.** 30   **49.** no solution   **51.** $\dfrac{9}{7}$   **53.** $b = \dfrac{4A}{5x^2}$   **55.** $x = 6$   **57.** $x = 9$   **59.** 675 parts   **61.** 3   **63.** fast car speed: 30 mph; slow car speed: 20 mph

**65.** $17\dfrac{1}{2}$ hr   **67.** $x = 15$   **69.** $-\dfrac{7}{18y}$   **71.** $\dfrac{3y - 1}{2y - 1}$   **73.** $-\dfrac{x^2 + 9}{6x}$   **75.** $\dfrac{xy + 1}{x}$   **77.** $\dfrac{1}{2x}$   **79.** $\dfrac{x - 4}{x + 4}$   **81.** $\dfrac{1}{x - 6}$   **83.** $\dfrac{2}{(x + 3)(x - 2)}$

**85.** $\dfrac{1}{2}$   **87.** 1   **89.** $x = 6$   **91.** $\dfrac{3}{10}$   **93.** $\dfrac{1}{y^2 - 1}$

## Chapter 7 Test

**1.** $\{x \mid x \text{ is a real number}, x \neq -1, x \neq -3\}$   **2. a.** $\$115$   **b.** $\$103$   **3.** $\dfrac{3}{5}$   **4.** $\dfrac{1}{x + 6}$   **5.** $\dfrac{1}{x^2 - 3x + 9}$   **6.** $\dfrac{2m(m + 2)}{m - 2}$   **7.** $\dfrac{a + 2}{a + 5}$

**8.** $-\dfrac{1}{x + y}$   **9.** 15   **10.** $\dfrac{y - 2}{4}$   **11.** $\dfrac{19x - 6}{2x + 5}$   **12.** $\dfrac{3a - 4}{(a - 3)(a + 2)}$   **13.** $\dfrac{3}{x - 1}$   **14.** $\dfrac{2(x + 5)}{x(y + 5)}$   **15.** $\dfrac{x^2 + 2x + 35}{(x + 9)(x + 2)(x - 5)}$

**16.** $\dfrac{30}{11}$   **17.** $-6$   **18.** no solution   **19.** $-2, 5$   **20.** no solution   **21.** $\dfrac{xz}{2y}$   **22.** $\dfrac{5y^2 - 1}{y + 2}$   **23.** 18 bulbs   **24.** 5 or 1   **25.** 30 mph

**26.** $6\dfrac{2}{3}$ hr   **27.** $x = 12$

## Chapter 7 Cumulative Review

**1. a.** $\dfrac{15}{x} = 4$   **b.** $12 - 3 = x$   **c.** $4x + 17 \neq 21$   **d.** $3x < 48$; Sec. 1.4, Ex. 9   **3.** amount at 7%: $12,500; amount at 9%: $7500; Sec. 2.7, Ex. 4

**5.**    ; Sec. 3.3, Ex. 6   **7. a.** $4^7$   **b.** $x^{10}$   **c.** $y^4$   **d.** $y^{12}$   **e.** $(-5)^{15}$;   **f.** $a^2b^2$; Sec. 5.1, Ex. 3   **9.** $12z + 16$; Sec. 5.2, Ex. 12

**11.** $27a^3 + 27a^2b + 9ab^2 + b^3$; Sec. 5.3, Ex. 8   **13. a.** $t^2 + 4t + 4$   **b.** $p^2 - 2pq + q^2$   **c.** $4x^2 + 20x + 25$   **d.** $x^4 - 14x^2y + 49y^2$; Sec. 5.4, Ex. 5

**15. a.** $x^3$   **b.** 81   **c.** $\dfrac{q^9}{p^4}$   **d.** $\dfrac{32}{125}$; Sec. 5.5, Ex. 2   **17.** $4x^2 - 4x + 6 + \dfrac{-11}{2x + 3}$; Sec. 5.6, Ex. 6   **19. a.** 4   **b.** 1   **c.** 3; Sec. 6.1, Ex. 1

**21.** $-3a(3a^4 - 6a + 1)$; Sec. 6.1, Ex. 5   **23.** $3(m + 2)(m - 10)$; Sec. 6.2, Ex. 9   **25.** $(3x + 2)(x + 3)$; Sec. 6.3, Ex. 1   **27.** $(x + 6)^2$; Sec. 6.3, Ex. 8
**29.** prime polynomial; Sec. 6.5, Ex. 4b   **31.** $(x + 2)(x^2 - 2x + 4)$; Sec. 6.5, Ex. 8   **33.** $(2x + 3)(x + 1)(x - 1)$; Ch. 6 Int. Rev., Ex. 2

**35.** $3(2m + n)(2m - n)$; Ch. 6 Int. Rev., Ex. 3   **37.** $-\dfrac{1}{2}$, 4; Sec. 6.6, Ex. 5   **39.** $(1, 0), (4, 0)$; Sec. 6.6, Ex. 10   **41.** base: 6 m; height: 10 m; Sec. 6.7, Ex. 3

**43.** $-\dfrac{2(3 + x)}{x + 1}$; Sec. 7.1, Ex. 4   **45.** $\dfrac{2}{x(x + 1)}$; Sec. 7.2, Ex. 6   **47.** $\dfrac{1 + 2x}{2(2 - x)}$; Sec. 7.7, Ex. 4

# CHAPTER 8  MORE ON FUNCTIONS AND GRAPHS

## Section 8.1
## Practice Exercises

**1.**    **2.**    **3.** $f(x) = -4x - 3$   **4.** $f(x) = -\dfrac{2}{3}x + \dfrac{4}{3}$   **5.** $f(x) = -2$   **6.** $3x + 4y = 12$

**7.** $f(x) = \dfrac{4}{3}x - \dfrac{41}{3}$

## Graphing Calculator Explorations 8.1

**1.** $y = \dfrac{x}{3.5}$    **3.** $y = -\dfrac{5.78}{2.31}x + \dfrac{10.98}{2.31}$

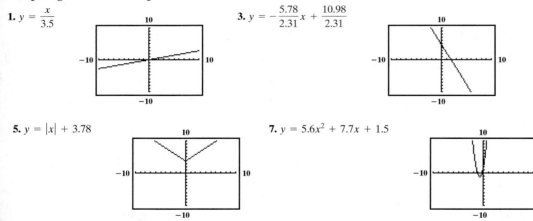

**5.** $y = |x| + 3.78$    **7.** $y = 5.6x^2 + 7.7x + 1.5$

## Vocabulary and Readiness Check 8.1

**1.** linear   **3.** $m = -4$, $y$-intercept: $(0, 12)$   **5.** $m = 5$, $y$-intercept: $(0, 0)$   **7.** parallel   **9.** neither

## Exercise Set 8.1

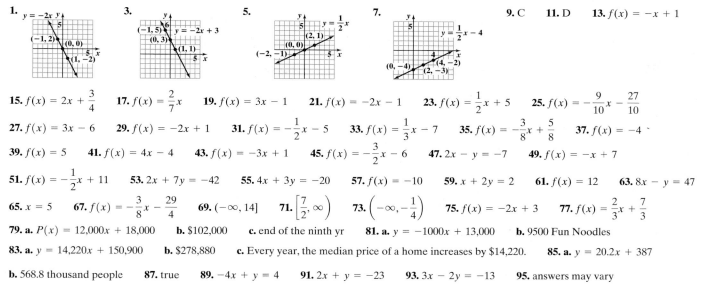

**1.** $y = -2x$    **3.**    **5.** $y = \frac{1}{2}x$    **7.**    **9.** C    **11.** D    **13.** $f(x) = -x + 1$

**15.** $f(x) = 2x + \frac{3}{4}$    **17.** $f(x) = \frac{2}{7}x$    **19.** $f(x) = 3x - 1$    **21.** $f(x) = -2x - 1$    **23.** $f(x) = \frac{1}{2}x + 5$    **25.** $f(x) = -\frac{9}{10}x - \frac{27}{10}$

**27.** $f(x) = 3x - 6$    **29.** $f(x) = -2x + 1$    **31.** $f(x) = -\frac{1}{2}x - 5$    **33.** $f(x) = \frac{1}{3}x - 7$    **35.** $f(x) = -\frac{3}{8}x + \frac{5}{8}$    **37.** $f(x) = -4$

**39.** $f(x) = 5$    **41.** $f(x) = 4x - 4$    **43.** $f(x) = -3x + 1$    **45.** $f(x) = -\frac{3}{2}x - 6$    **47.** $2x - y = -7$    **49.** $f(x) = -x + 7$

**51.** $f(x) = -\frac{1}{2}x + 11$    **53.** $2x + 7y = -42$    **55.** $4x + 3y = -20$    **57.** $f(x) = -10$    **59.** $x + 2y = 2$    **61.** $f(x) = 12$    **63.** $8x - y = 47$

**65.** $x = 5$    **67.** $f(x) = -\frac{3}{8}x - \frac{29}{4}$    **69.** $(-\infty, 14]$    **71.** $\left[\frac{7}{2}, \infty\right)$    **73.** $\left(-\infty, -\frac{1}{4}\right)$    **75.** $f(x) = -2x + 3$    **77.** $f(x) = \frac{2}{3}x + \frac{7}{3}$

**79. a.** $P(x) = 12,000x + 18,000$    **b.** \$102,000    **c.** end of the ninth yr    **81. a.** $y = -1000x + 13,000$    **b.** 9500 Fun Noodles

**83. a.** $y = 14,220x + 150,900$    **b.** \$278,880    **c.** Every year, the median price of a home increases by \$14,220.    **85. a.** $y = 20.2x + 387$

**b.** 568.8 thousand people    **87.** true    **89.** $-4x + y = 4$    **91.** $2x + y = -23$    **93.** $3x - 2y = -13$    **95.** answers may vary

## Section 8.2

## Practice Exercises

**1. a.** $-3$    **b.** $-2$    **c.** 3    **d.** 1    **e.** $-1$ and 3    **f.** $-3$    **2.** \$35 billion    **3.** \$57.224 billion    **4. a.** 11    **b.** $\frac{1}{4}$    **c.** $-8$

**d.** not a real number    **e.** 10    **5.**    **6.**    **7.**

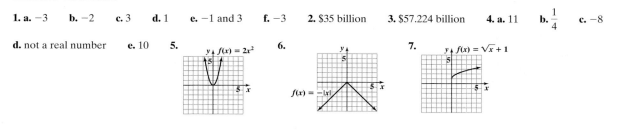

$f(x) = -|x|$

## Graphing Calculator Explorations 8.2

**1.**    **3.**    **5.**

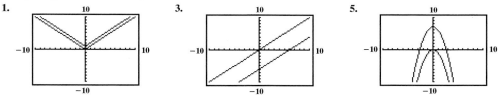

## Vocabulary and Readiness Check 8.2

**1.** V-shaped    **3.** $(-2, 1.7)$

## Exercise Set 8.2

**1.** 0    **3.** $-4$    **5.** 3    **7.** $(1, -10)$    **9.** $(4, 56)$    **11.** $f(-1) = -2$    **13.** $g(2) = 0$    **15.** $-4, 0$    **17.** 3    **19.** 7    **21.** $-\frac{2}{3}$    **23.** 8

**25.** 9    **27.** not a real number    **29.**    **31.**    **33.**    **35.**

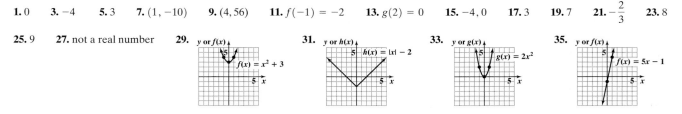

**37.** 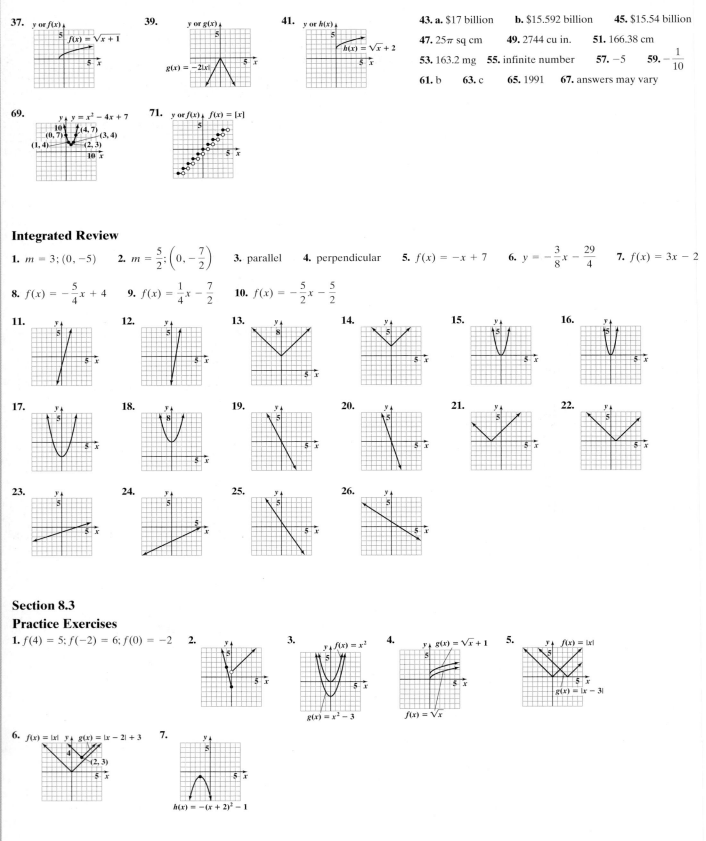 **39.** **41.**

**43. a.** $17 billion    **b.** $15.592 billion    **45.** $15.54 billion

**47.** $25\pi$ sq cm    **49.** 2744 cu in.    **51.** 166.38 cm

**53.** 163.2 mg    **55.** infinite number    **57.** $-5$    **59.** $-\dfrac{1}{10}$

**61.** b    **63.** c    **65.** 1991    **67.** answers may vary

**69.** **71.**

## Integrated Review

**1.** $m = 3; (0, -5)$    **2.** $m = \dfrac{5}{2}; \left(0, -\dfrac{7}{2}\right)$    **3.** parallel    **4.** perpendicular    **5.** $f(x) = -x + 7$    **6.** $y = -\dfrac{3}{8}x - \dfrac{29}{4}$    **7.** $f(x) = 3x - 2$

**8.** $f(x) = -\dfrac{5}{4}x + 4$    **9.** $f(x) = \dfrac{1}{4}x - \dfrac{7}{2}$    **10.** $f(x) = -\dfrac{5}{2}x - \dfrac{5}{2}$

**11.** **12.** **13.** **14.** **15.** **16.**

**17.** **18.** **19.** **20.** **21.** **22.**

**23.** **24.** **25.** **26.**

## Section 8.3
## Practice Exercises

**1.** $f(4) = 5; f(-2) = 6; f(0) = -2$    **2.** **3.** **4.** **5.**

**6.** **7.**

## Vocabulary and Readiness Check 8.3
**1.** C    **3.** D

## Exercise Set 8.3

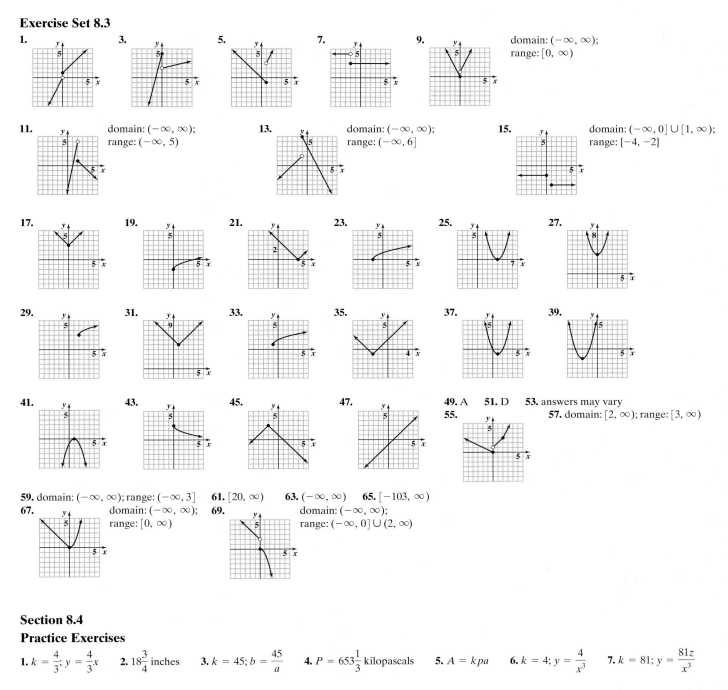

**1.**    **3.**    **5.**    **7.**    **9.** domain: $(-\infty, \infty)$; range: $[0, \infty)$

**11.** domain: $(-\infty, \infty)$; range: $(-\infty, 5)$    **13.** domain: $(-\infty, \infty)$; range: $(-\infty, 6]$    **15.** domain: $(-\infty, 0] \cup [1, \infty)$; range: $\{-4, -2\}$

**17.**    **19.**    **21.**    **23.**    **25.**    **27.**

**29.**    **31.**    **33.**    **35.**    **37.**    **39.**

**41.**    **43.**    **45.**    **47.**    **49.** A    **51.** D    **53.** answers may vary    **55.**    **57.** domain: $[2, \infty)$; range: $[3, \infty)$

**59.** domain: $(-\infty, \infty)$; range: $(-\infty, 3]$    **61.** $[20, \infty)$    **63.** $(-\infty, \infty)$    **65.** $[-103, \infty)$    **67.** domain: $(-\infty, \infty)$; range: $[0, \infty)$    **69.** domain: $(-\infty, \infty)$; range: $(-\infty, 0] \cup (2, \infty)$

## Section 8.4
## Practice Exercises
**1.** $k = \dfrac{4}{3}$; $y = \dfrac{4}{3}x$    **2.** $18\dfrac{3}{4}$ inches    **3.** $k = 45$; $b = \dfrac{45}{a}$    **4.** $P = 653\dfrac{1}{3}$ kilopascals    **5.** $A = kpa$    **6.** $k = 4$; $y = \dfrac{4}{x^3}$    **7.** $k = 81$; $y = \dfrac{81z}{x^3}$

## Vocabulary and Readiness Check 8.4
**1.** direct    **3.** joint    **5.** inverse    **7.** direct

## Exercise Set 8.4

**1.** $k = \frac{1}{5}$; $y = \frac{1}{5}x$ **3.** $k = \frac{3}{2}$; $y = \frac{3}{2}x$ **5.** $k = 14$; $y = 14x$ **7.** $k = 0.25$; $y = 0.25x$ **9.** 4.05 lb **11.** 204,706 tons **13.** $k = 30$; $y = \frac{30}{x}$

**15.** $k = 700$; $y = \frac{700}{x}$ **17.** $k = 2$; $y = \frac{2}{x}$ **19.** $k = 0.14$; $y = \frac{0.14}{x}$ **21.** 54 mph **23.** 72 amps **25.** divided by 4 **27.** $x = kyz$

**29.** $r = kst^3$ **31.** $k = \frac{1}{3}$; $y = \frac{1}{3}x^3$ **33.** $k = 0.2$; $y = 0.2\sqrt{x}$ **35.** $k = 1.3$; $y = \frac{1.3}{x^2}$ **37.** $k = 3$; $y = 3xz^3$ **39.** 22.5 tons

**41.** $15\pi$ cu in. **43.** 8 ft **45.** $y = kx$ **47.** $a = \frac{k}{b}$ **49.** $y = kxz$ **51.** $y = \frac{k}{x^3}$ **53.** $y = \frac{kx}{p^2}$ **55.** $C = 8\pi$ in.; $A = 16\pi$ sq in.

**57.** $C = 18\pi$ cm; $A = 81\pi$ sq cm **59.** 1.2 **61.** $-7$ **63.** $-\frac{1}{2}$ **65.** $\frac{8}{27}$ **67.** a **69.** c **71.** multiplied by 8 **73.** multiplied by 2

**75.** **77.**

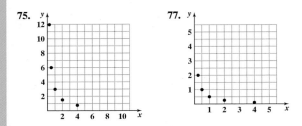

## Chapter 8 Vocabulary Check

**1.** parallel **2.** slope-intercept **3.** function **4.** slope **5.** perpendicular **6.** linear function **7.** directly **8.** inversely **9.** jointly

## Chapter 8 Review

**1.** **3.** **5.** C **7.** B **9.** $m = \frac{2}{5}$; y-intercept $\left(0, -\frac{4}{3}\right)$ **11.** $2x - y = 12$ **13.** $11x + y = -52$

**15.** $y = -5$ **17.** $f(x) = -1$ **19.** $f(x) = -x - 2$ **21.** $f(x) = -\frac{3}{2}x - 8$ **23.** $f(x) = -\frac{3}{2}x - 1$ **25. a.** $y = \frac{17}{22}x + 43$ **b.** 52 million

**27.** 0 **29.** $-2, 4$ **31.** linear **33.** nonlinear **35.** linear

**37.** linear $y = -1.36x$ **39.** nonlinear **41.** **43.** **45.**

**47.** 9 **49.** 3.125 cu ft **51.** $f(x) = \frac{9}{2}$ **53.** $f(x) = -5x - 7$ **55.** $f(x) = -\frac{4}{5}x + 3$ **57.** **59.**

## Chapter 8 Test

**1.** 3    **2.** −5    **3.** 2, −2    **4.** 0    **5.**    **6.**    **7.** $y = -8$    **8.** $3x + y = 11$    **9.** $5x - y = 2$

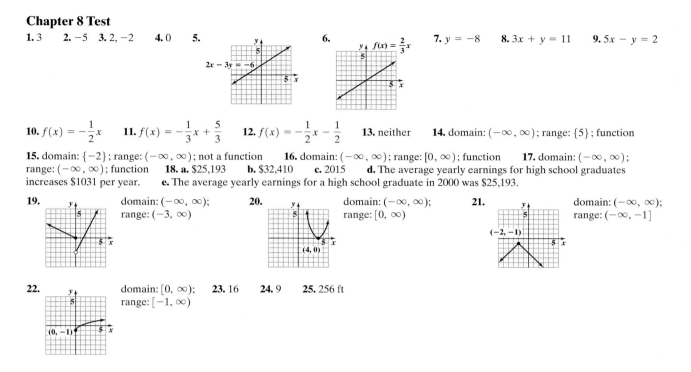

**10.** $f(x) = -\dfrac{1}{2}x$    **11.** $f(x) = -\dfrac{1}{3}x + \dfrac{5}{3}$    **12.** $f(x) = -\dfrac{1}{2}x - \dfrac{1}{2}$    **13.** neither    **14.** domain: $(-\infty, \infty)$; range: $\{5\}$; function

**15.** domain: $\{-2\}$; range: $(-\infty, \infty)$; not a function    **16.** domain: $(-\infty, \infty)$; range: $[0, \infty)$; function    **17.** domain: $(-\infty, \infty)$; range: $(-\infty, \infty)$; function    **18. a.** $25,193    **b.** $32,410    **c.** 2015    **d.** The average yearly earnings for high school graduates increases $1031 per year.    **e.** The average yearly earnings for a high school graduate in 2000 was $25,193.

**19.** domain: $(-\infty, \infty)$; range: $(-3, \infty)$    **20.** domain: $(-\infty, \infty)$; range: $[0, \infty)$    **21.** domain: $(-\infty, \infty)$; range: $(-\infty, -1]$

**22.** domain: $[0, \infty)$; range: $[-1, \infty)$    **23.** 16    **24.** 9    **25.** 256 ft

## Chapter 8 Cumulative Review

**1.** 66; Sec. 1.4, Ex. 4    **3. a.** $\dfrac{1}{2}$    **b.** 19; Sec. 1.6, Ex. 6    **5.** −11; Sec. 2.2, Ex. 6    **7.** $\dfrac{16}{3}$; Sec. 2.3, Ex. 2    **9.** $x = \dfrac{y - b}{m}$; Sec. 2.5, Ex. 6

**11.** $(-\infty, 4]$; Sec. 2.8, Ex. 5    **13.** $\dfrac{3}{5}$; Sec. 3.4, Ex. 6    **15. a.** 1    **b.** 1    **c.** −3; Sec. 3.6, Ex. 7    **17.** $(4, 2)$; Sec. 4.2, Ex. 1

**19.** $\left(-\dfrac{5}{4}, -\dfrac{5}{2}\right)$; Sec. 4.3, Ex. 6    **21.** $x + 4$; Sec. 5.6, Ex. 4    **23. a.** $6(t + 3)$    **b.** $y^5(1 - y^2)$; Sec. 6.1, Ex. 4

**25.** $(x - 2)(x + 6)$; Sec. 6.2, Ex. 3    **27.** $(2y - 3y)(5x + y)$; Sec. 6.3, Ex. 4    **29.** $(x + 2)(x^2 - 2x + 4)$; Sec. 6.5, Ex. 8

**31.** 11, −2; Sec. 6.6, Ex. 3    **33. a.** $\dfrac{1}{5x - 1}$    **b.** $\dfrac{9x + 4}{8x - 7}$; Sec. 7.1, Ex. 2    **35.** $-\dfrac{3(x + 1)}{5x(2x - 3)}$; Sec. 7.2, Ex. 3    **37.** $3x - 5$; Sec. 7.3, Ex. 3

**39.** −3, −2; Sec. 7.5, Ex. 3    **41.** $f(x) = \dfrac{5}{8}x - \dfrac{5}{2}$; Sec. 8.1, Ex. 4

# CHAPTER 9    INEQUALITIES AND ABSOLUTE VALUE

## Section 9.1

### Practice Exercises

**1.** $\{1, 3\}$    **2.** $(-\infty, 2)$    **3.** $\{\ \}$ or $\varnothing$    **4.** $(-4, 2)$    **5.** $[-6, 8]$    **6.** $\{1, 2, 3, 4, 5, 6, 7, 9\}$    **7.** $\left(-\infty, \dfrac{3}{8}\right] \cup [3, \infty)$    **8.** $(-\infty, \infty)$

### Vocabulary and Readiness Check 9.1

**1.** compound    **3.** or    **5.** $\cup$    **7.** and

### Exercise Set 9.1

**1.** $\{2, 3, 4, 5, 6, 7\}$    **3.** $\{4, 6\}$    **5.** $\{\ldots, -2, -1, 0, 1, \ldots\}$    **7.** $\{5, 7\}$    **9.** $\{x \mid x \text{ is an odd integer or } x = 2 \text{ or } x = 4\}$    **11.** $\{2, 4\}$

**13.** $(-3, 1)$    **15.** $\varnothing$    **17.** $(-\infty, -1)$    **19.** $[6, \infty)$    **21.** $(-\infty, -3]$    **23.** $(4, 10)$

**25.** $(11, 17)$    **27.** $[1, 4]$    **29.** $\left[-3, \dfrac{3}{2}\right]$    **31.** $\left[-\dfrac{7}{3}, 7\right]$    **33.** $(-\infty, 5)$    **35.** $(-\infty, -4] \cup [1, \infty)$

**37.** $(-\infty, \infty)$    **39.** $[2, \infty)$    **41.** $(-\infty, -4) \cup (-2, \infty)$    **43.** $(-\infty, \infty)$    **45.** $\left(-\dfrac{1}{2}, \dfrac{2}{3}\right)$    **47.** $(-\infty, \infty)$    **49.** $\left[\dfrac{3}{2}, 6\right]$

**51.** $\left(\dfrac{5}{4}, \dfrac{11}{4}\right)$    **53.** $\varnothing$    **55.** $\left(-\infty, -\dfrac{56}{5}\right) \cup \left(\dfrac{5}{3}, \infty\right)$    **57.** $\left(-5, \dfrac{5}{2}\right)$    **59.** $\left(0, \dfrac{14}{3}\right]$    .**61.** $(-\infty, -3]$    **63.** $(-\infty, 1] \cup \left(\dfrac{29}{7}, \infty\right)$    **65.** $\varnothing$

**67.** $\left[-\dfrac{1}{2}, \dfrac{3}{2}\right)$    **69.** $\left(-\dfrac{4}{3}, \dfrac{7}{3}\right)$    **71.** $(6, 12)$    **73.** $-12$    **75.** $-4$    **77.** $-7, 7$    **79.** $0$    **81.** 2003, 2004, 2005    **83.** registered nurses

**85.** $-20.2° \le F \le 95°$    **87.** $67 \le$ final score $\le 94$    **89.** $(6, \infty)$    **91.** $[3, 7]$    **93.** $(-\infty, -1)$

## The Bigger Picture

**1.** $-\dfrac{1}{8}$    **2.** $4xy - 6y^2$    **3.** $\dfrac{1}{xy - z^2}$    **4.** $(x + 2)(x^2 + 2x + 4)$    **5.** $[-1, 3]$    **6.** $(-1, 6)$    **7.** $5.1$    **8.** $(-\infty, -4)$    **9.** $(-\infty, -3]$
**10.** $(-\infty, -2) \cup (2, \infty)$    **11.** $4$    **12.** $[19, \infty)$

## Section 9.2
## Practice Exercises

**1.** $-7, 7$    **2.** $-1, 4$    **3.** $-80, 70$    **4.** $-2, 2$    **5.** $0$    **6.** $\{\ \}$ or $\varnothing$    **7.** $\{\ \}$ or $\varnothing$    **8.** $-\dfrac{3}{5}, 5$    **9.** $5$

## Vocabulary and Readiness Check 9.2
**1.** C    **3.** B    **5.** D

## Exercise Set 9.2

**1.** $7, -7$    **3.** $4.2, -4.2$    **5.** $7, -2$    **7.** $8, 4$    **9.** $5, -5$    **11.** $3, -3$    **13.** $0$    **15.** $\varnothing$    **17.** $\dfrac{1}{5}$    **19.** $|x| = 5$    **21.** $9, -\dfrac{1}{2}$    **23.** $-\dfrac{5}{2}$

**25.** answers may vary    **27.** $4, -4$    **29.** $0$    **31.** $\varnothing$    **33.** $0, \dfrac{14}{3}$    **35.** $2, -2$    **37.** $\varnothing$    **39.** $7, -1$    **41.** $\varnothing$    **43.** $\varnothing$    **45.** $-\dfrac{1}{8}$    **47.** $\dfrac{1}{2}, -\dfrac{5}{6}$

**49.** $2, -\dfrac{12}{5}$    **51.** $3, -2$    **53.** $-8, \dfrac{2}{3}$    **55.** $\varnothing$    **57.** $4$    **59.** $13, -8$    **61.** $3, -3$    **63.** $8, -7$    **65.** $2, 3$    **67.** $2, -\dfrac{10}{3}$    **69.** $\dfrac{3}{2}$    **71.** $\varnothing$

**73.** answers may vary    **75.** $34\%$    **77.** $39.6$ lb    **79.** answers may vary    **81.** no solution    **83.** $|x - 7| = 2$    **85.** $|2x - 1| = 4$
**87. a.** if $c = 0$    **b.** if $c$ is a negative number    **c.** if $c$ is a positive number

## Section 9.3
## Practice Exercises

**1.** $(-2, 2)$    **2.** $(-4, 2)$    **3.** $\left[-\dfrac{2}{3}, 2\right]$    **4.** $\{\ \}$ or $\varnothing$    **5.** $(-\infty, -10] \cup [2, \infty)$
**6.** $(-\infty, \infty)$    **7.** $(-\infty, 0) \cup (12, \infty)$    **8.** $\{2\}$

## Vocabulary and Readiness Check 9.3
**1.** D    **3.** C    **5.** A

## Exercise Set 9.3

**1.** ; $[-4, 4]$    **3.** ; $(1, 5)$    **5.** ; $(-5, -1)$    **7.** ; $[-10, 3]$    **9.** ; $[-5, 5]$

**11.** ; $\varnothing$    **13.** ; $[0, 12]$    **15.** ; $(-\infty, -3) \cup (3, \infty)$    **17.** ; $(-\infty, -24] \cup [4, \infty)$

**19.** 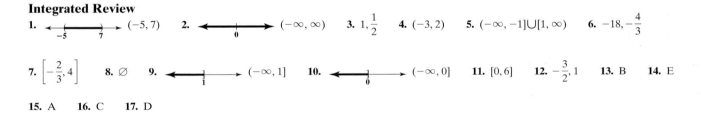 ; $(-\infty, -4) \cup (4, \infty)$  **21.** ; $(-\infty, \infty)$  **23.** ; $\left(-\infty, \frac{2}{3}\right) \cup (2, \infty)$  **25.** ; $\{0\}$

**27.** ; $\left(-\infty, -\frac{3}{8}\right) \cup \left(-\frac{3}{8}, \infty\right)$  **29.** ; $[-2, 2]$  **31.** ; $(-\infty, -1) \cup (1, \infty)$

**33.** ; $(-5, 11)$  **35.** ; $(-\infty, 4) \cup (6, \infty)$  **37.** ; $\varnothing$  **39.** ; $(-\infty, \infty)$

**41.** ; $[-2, 9]$  **43.** ; $(-\infty, -11] \cup [1, \infty)$  **45.** ; $(-\infty, 0) \cup (0, \infty)$  **47.** ; $(-\infty, \infty)$

**49.** ; $\left[-\frac{1}{2}, 1\right]$  **51.** ; $(-\infty, -3) \cup (0, \infty)$  **53.** ; $\varnothing$  **55.** ; $\left\{\frac{3}{8}\right\}$

**57.** ; $\left(-\frac{2}{3}, 0\right)$  **59.** ; $(-\infty, -12) \cup (0, \infty)$  **61.** ; $[-1, 8]$  **63.** ; $\left[-\frac{23}{8}, \frac{17}{8}\right]$

**65.** $(-2, 5)$  **67.** $5, -2$  **69.** $(-\infty, -7] \cup [17, \infty)$  **71.** $-\frac{9}{4}$  **73.** $(-2, 1)$  **75.** $2, \frac{4}{3}$  **77.** $\varnothing$  **79.** $\frac{19}{2}, -\frac{17}{2}$  **81.** $\left(-\infty, -\frac{25}{3}\right) \cup \left(\frac{35}{3}, \infty\right)$

**83.** $-1.5$  **85.** $0$  **87.** $|x| < 7$  **89.** $|x| \le 5$  **91.** answers may vary  **93.** $3.45 < x < 3.55$

## The Bigger Picture

**1.** $-8$  **2.** $-13, 21$  **3.** $(-\infty, 6]$  **4.** $(-7, 5]$  **5.** $-\frac{13}{2}$  **6.** $\varnothing$  **7.** $-50$  **8.** $-5$  **9.** $\left[-\frac{9}{5}, 1\right]$  **10.** $(-\infty, -13) \cup (-9, \infty)$  **11.** $-3, \frac{23}{9}$

**12.** $-11, -\frac{9}{7}$

## Integrated Review

**1.** $(-5, 7)$  **2.** $(-\infty, \infty)$  **3.** $1, \frac{1}{2}$  **4.** $(-3, 2)$  **5.** $(-\infty, -1] \cup [1, \infty)$  **6.** $-18, -\frac{4}{3}$

**7.** $\left[-\frac{2}{3}, 4\right]$  **8.** $\varnothing$  **9.** $(-\infty, 1]$  **10.** $(-\infty, 0]$  **11.** $[0, 6]$  **12.** $-\frac{3}{2}, 1$  **13.** B  **14.** E

**15.** A  **16.** C  **17.** D

## Section 9.4
## Practice Exercises

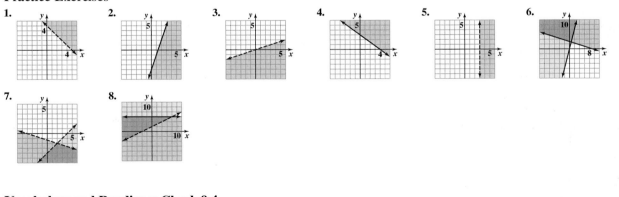

## Vocabulary and Readiness Check 9.4

**1.** linear inequality in two variables  **3.** false  **5.** true  **7.** yes  **9.** yes  **11.** yes  **13.** no

## Exercise Set 9.4

**1.** no; yes   **3.** no; no   **5.** no; yes   **7.**   **9.**   **11.**   **13.**

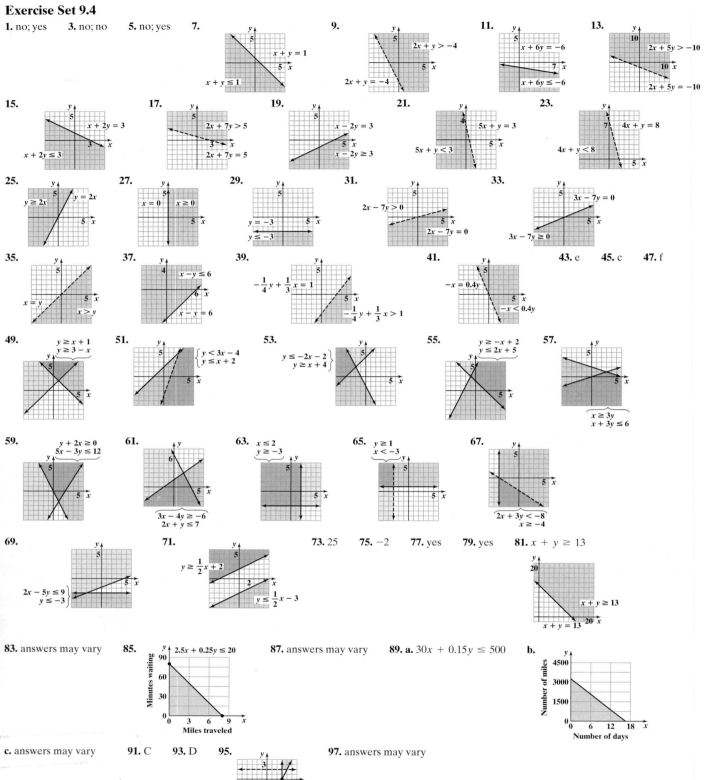

**43.** e   **45.** c   **47.** f

**73.** 25   **75.** −2   **77.** yes   **79.** yes   **81.** $x + y \geq 13$

**83.** answers may vary   **85.**   **87.** answers may vary   **89. a.** $30x + 0.15y \leq 500$   **b.**

**c.** answers may vary   **91.** C   **93.** D   **95.**   **97.** answers may vary

## Chapter 9 Vocabulary Check

**1.** compound inequality　　**2.** intersection　　**3.** union　　**4.** absolute value　　**5.** solution　　**6.** system of linear inequalities

## Chapter 9 Review

**1.** $\left(\dfrac{1}{8}, 2\right)$　　**3.** $\left(\dfrac{7}{8}, \dfrac{27}{20}\right]$　　**5.** $\left(\dfrac{11}{3}, \infty\right)$　　**7.** $5, 11$　　**9.** $-1, \dfrac{11}{3}$　　**11.** $-\dfrac{1}{6}$　　**13.** $\varnothing$　　**15.** $5, -\dfrac{1}{3}$　　**17.** $7, -\dfrac{8}{5}$　　**19.** $\left(-\dfrac{8}{5}, 2\right)$

**21.** $(-\infty, -3) \cup (3, \infty)$　　**23.** $\varnothing$　　**25.** $\left(-\infty, -\dfrac{22}{15}\right] \cup \left[\dfrac{6}{5}, \infty\right)$

**27.** $(-\infty, -27) \cup (-9, \infty)$　　**29.**　　**31.**　　**33.**

**35.** $y \geq 2x - 3$, $y \leq -2x + 1$　　**37.** $x + 2y > 0$, $x - y \leq 6$　　**39.** $3x - 2y \leq 4$, $2x + y \geq 5$　　**41.** $\left[-\dfrac{4}{3}, \dfrac{7}{6}\right]$　　**43.** $(5, \infty)$　　**45.** $\varnothing$　　**47.** $(-\infty, \infty)$

**49.**　　**51.**

## Chapter 9 Test

**1.** $1, \dfrac{2}{3}$　　**2.** $\varnothing$　　**3.** $\dfrac{3}{2}$　　**4.** $\left(\dfrac{3}{2}, 5\right]$　　**5.** $(-\infty, -2) \cup \left(\dfrac{4}{3}, \infty\right)$　　**6.** $(3, 7)$　　**7.** $(-\infty, -5]$　　**8.** $(-\infty, -2]$　　**9.** $[-3, -1)$

**10.** $(-\infty, \infty)$　　**11.** $3, -\dfrac{1}{5}$　　**12.** $(-\infty, \infty)$　　**13.** $\left[1, \dfrac{11}{2}\right)$　　**14.**　　**15.**　　**16.** $y + 2x \leq 4$, $y \geq 2$

**17.** $2y - x \geq 1$, $x + y \geq -4$

## Chapter 9 Cumulative Review

**1. a.** $\dfrac{1}{2}$　　**b.** $19$; Sec. 1.6, Ex. 6　　**3. a.** $-\dfrac{39}{5}$　　**b.** $2$; Sec. 1.7, Ex. 9　　**5. a.** $5x + 7$　　**b.** $-4a - 1$　　**c.** $4y - 3y^2$　　**d.** $7.3x - 6$

**e.** $\dfrac{1}{2}b$; Sec. 2.1, Ex. 4　　**7.** $-3$; Sec. 2.2, Ex. 3　　**9.**

| $x$ | $y$ |
|---|---|
| $-1$ | $-3$ |
| $0$ | $0$ |
| $-3$ | $-9$ |

; Sec. 3.1, Ex. 7　　**11. a.** $x$-int: $(-3, 0)$; $y$-int: $(0, 2)$　　**b.** $x$-int: $(-4, 0)$, $(-1, 0)$; $y$-int: $(0, 1)$

**c.** $x$-int and $y$-int: $(0, 0)$　　**d.** $x$-int: $(2, 0)$; $y$-int: none　　**e.** $x$-int: $(-1, 0)$, $(3, 0)$; $y$-int: $(0, -1)$, $(0, 2)$; Sec. 3.3, Ex. 1–5　　**13.** parallel; Sec. 3.4, Ex. 8a

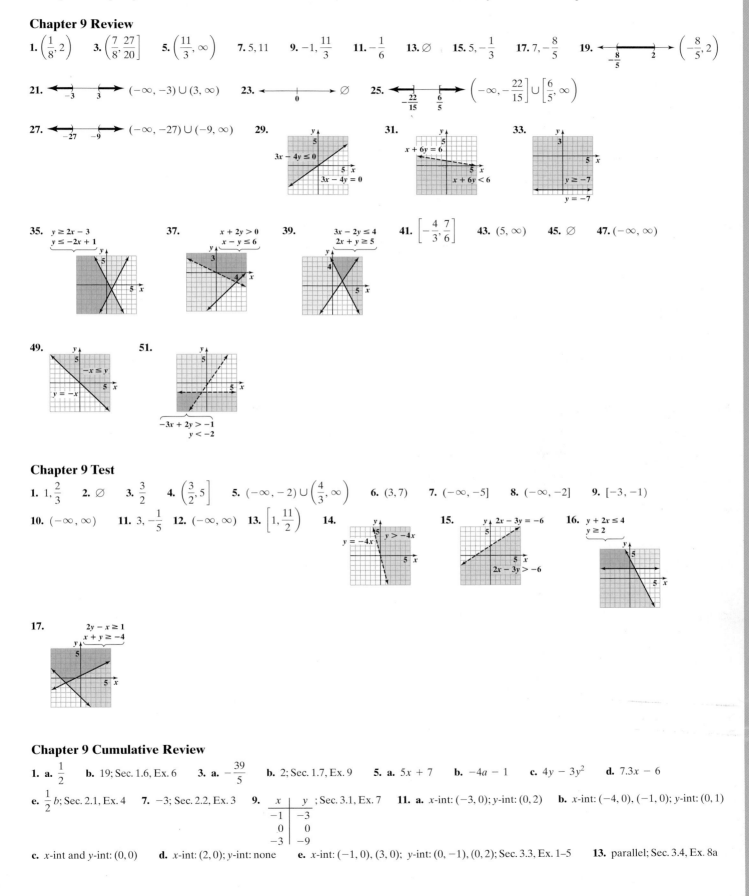

**15.** $y = \frac{1}{4}x - 3$; Sec. 3.5, Ex. 1     **17.** $y = -3$; Sec. 3.5, Ex. 7     **19.** a, b, c; Sec. 3.6, Ex. 5     **21.** solution; Sec. 4.1, Ex. 1

**23.** $12x^3 - 12x^2 - 9x + 2$; Sec. 5.2, Ex. 10     **25.** $(x + 6y)(x + y)$; Sec. 6.2, Ex. 6     **27.** $\frac{3y^9}{160}$; Sec. 7.2, Ex. 4     **29.** 1; Sec. 7.3, Ex. 2

**31.** $\frac{x(3x - 1)}{(x + 1)^2(x - 1)}$; Sec. 7.4, Ex. 7     **33.** $-5$; Sec. 7.5, Ex. 1     **35.** no solution; Sec. 4.1, Ex. 5     **37.** $(-2, 0)$; Sec. 4.2, Ex. 4

**39.** infinite number of solutions; Sec. 4.3, Ex. 4     **41.** ; Sec. 9.4, Ex. 8     **43. a.** $\frac{s^4 t^4}{16}$     **b.** $81y^{10}z^{14}$     **c.** $\frac{25x^4}{y^6}$; Sec. 5.1, Ex. 11

**45.** $-6, -\frac{3}{2}, \frac{1}{5}$; Sec. 6.6, Ex. 8     **47.** 63; Sec. 7.6, Ex. 1

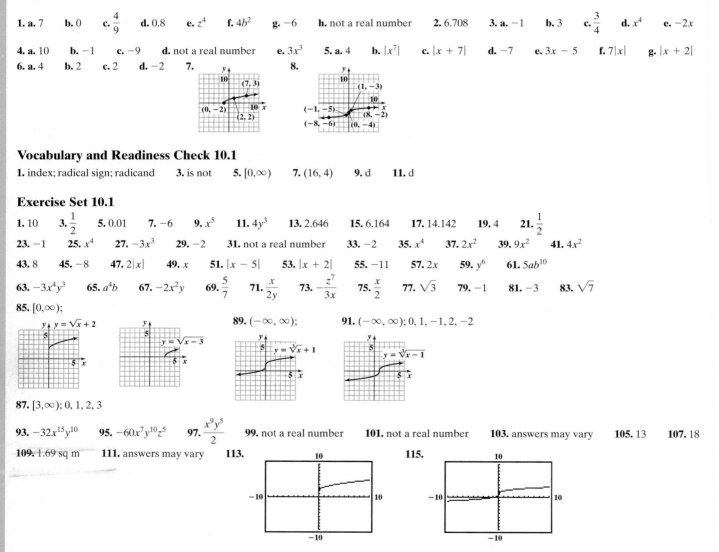

# CHAPTER 10 RATIONAL EXPONENTS, RADICALS, AND COMPLEX NUMBERS

## Section 10.1
### Practice Exercises

**1. a.** 7     **b.** 0     **c.** $\frac{4}{9}$     **d.** 0.8     **e.** $z^4$     **f.** $4b^2$     **g.** $-6$     **h.** not a real number     **2.** 6.708     **3. a.** $-1$     **b.** 3     **c.** $\frac{3}{4}$     **d.** $x^4$     **e.** $-2x$

**4. a.** 10     **b.** $-1$     **c.** $-9$     **d.** not a real number     **e.** $3x^3$     **5. a.** 4     **b.** $|x^7|$     **c.** $|x + 7|$     **d.** $-7$     **e.** $3x - 5$     **f.** $7|x|$     **g.** $|x + 2|$

**6. a.** 4     **b.** 2     **c.** 2     **d.** $-2$     **7.**     **8.**

### Vocabulary and Readiness Check 10.1

**1.** index; radical sign; radicand     **3.** is not     **5.** $[0,\infty)$     **7.** $(16, 4)$     **9.** d     **11.** d

### Exercise Set 10.1

**1.** 10     **3.** $\frac{1}{2}$     **5.** 0.01     **7.** $-6$     **9.** $x^5$     **11.** $4y^3$     **13.** 2.646     **15.** 6.164     **17.** 14.142     **19.** 4     **21.** $\frac{1}{2}$

**23.** $-1$     **25.** $x^4$     **27.** $-3x^3$     **29.** $-2$     **31.** not a real number     **33.** $-2$     **35.** $x^4$     **37.** $2x^2$     **39.** $9x^2$     **41.** $4x^2$

**43.** 8     **45.** $-8$     **47.** $2|x|$     **49.** $x$     **51.** $|x - 5|$     **53.** $|x + 2|$     **55.** $-11$     **57.** $2x$     **59.** $y^6$     **61.** $5ab^{10}$

**63.** $-3x^4y^3$     **65.** $a^4b$     **67.** $-2x^2y$     **69.** $\frac{5}{7}$     **71.** $\frac{x}{2y}$     **73.** $-\frac{z^7}{3x}$     **75.** $\frac{x}{2}$     **77.** $\sqrt{3}$     **79.** $-1$     **81.** $-3$     **83.** $\sqrt{7}$

**85.** $[0,\infty)$;

**89.** $(-\infty, \infty)$;     **91.** $(-\infty, \infty)$; 0, 1, $-1$, 2, $-2$

**87.** $[3,\infty)$; 0, 1, 2, 3

**93.** $-32x^{15}y^{10}$     **95.** $-60x^7y^{10}z^5$     **97.** $\frac{x^9y^5}{2}$     **99.** not a real number     **101.** not a real number     **103.** answers may vary     **105.** 13     **107.** 18

**109.** 1.69 sq m     **111.** answers may vary     **113.**     **115.**

## Section 10.2
### Practice Exercises
**1. a.** 6   **b.** 10   **c.** $\sqrt[5]{x}$   **d.** 1   **e.** $-8$   **f.** $5x^3$   **g.** $\sqrt[4]{3x}$   **2. a.** 64   **b.** $-1$   **c.** $-27$   **d.** $\dfrac{1}{125}$   **e.** $\sqrt[9]{(3x+2)^5}$

**3. a.** $\dfrac{1}{27}$   **b.** $\dfrac{1}{16}$   **4. a.** $y^{10/3}$   **b.** $x^{17/20}$   **c.** $\dfrac{1}{9}$   **d.** $b^{2/9}$   **e.** $\dfrac{81}{x^3 y^{11/3}}$   **5. a.** $x^{14/15} - x^{13/5}$   **b.** $x + 4x^{1/2} - 12$   **6.** $x^{-1/5}(2 - 7x)$

**7. a.** $\sqrt[3]{x}$   **b.** $\sqrt{6}$   **c.** $\sqrt[4]{a^2 b}$   **8. a.** $\sqrt[12]{x^7}$   **b.** $\sqrt[15]{y^2}$   **c.** $\sqrt[6]{675}$

## Vocabulary and Readiness Check 10.2
**1.** true   **3.** true   **5.** multiply   **7.** A   **9.** C   **11.** B

## Exercise Set 10.2
**1.** 7   **3.** 3   **5.** $\dfrac{1}{2}$   **7.** 13   **9.** $2\sqrt[3]{m}$   **11.** $3x^2$   **13.** $-3$   **15.** $-2$   **17.** 8   **19.** 16   **21.** not a real number   **23.** $\sqrt[5]{(2x)^3}$

**25.** $\sqrt[3]{(7x+2)^2}$   **27.** $\dfrac{64}{27}$   **29.** $\dfrac{1}{16}$   **31.** $\dfrac{1}{16}$   **33.** not a real number   **35.** $\dfrac{1}{x^{1/4}}$   **37.** $a^{2/3}$   **39.** $\dfrac{5x^{3/4}}{7}$   **41.** $a^{7/3}$   **43.** $x$   **45.** $3^{5/8}$

**47.** $y^{1/6}$   **49.** $8u^3$   **51.** $-b$   **53.** $\dfrac{1}{x^2}$   **55.** $27x^{2/3}$   **57.** $\dfrac{y}{z^{1/6}}$   **59.** $\dfrac{1}{x^{7/4}}$   **61.** $y - y^{7/6}$   **63.** $x^{5/3} - 2x^{2/3}$   **65.** $4x^{2/3} - 9$   **67.** $x^{8/3}(1 + x^{2/3})$

**69.** $x^{1/5}(x^{1/5} - 3)$   **71.** $x^{-1/3}(5 + x)$   **73.** $\sqrt{x}$   **75.** $\sqrt[3]{2}$   **77.** $2\sqrt{x}$   **79.** $\sqrt{xy}$   **81.** $\sqrt[3]{a^2 b}$   **83.** $\sqrt{x+3}$   **85.** $\sqrt[15]{y^{11}}$

**87.** $\sqrt[12]{b^5}$   **89.** $\sqrt[24]{x^{23}}$   **91.** $\sqrt{a}$   **93.** $\sqrt[6]{432}$   **95.** $\sqrt[15]{343y^5}$   **97.** $\sqrt[6]{125r^3 s^2}$   **99.** $25 \cdot 3$   **101.** $16 \cdot 3$ or $4 \cdot 12$   **103.** $8 \cdot 2$   **105.** $27 \cdot 2$

**107.** 1509 calories   **109.** 210.1 million   **111.** $a^{1/3}$   **113.** $x^{1/5}$   **115.** 1.6818   **117.** 5.6645   **119.** $\dfrac{t^{1/2}}{u^{1/2}}$

## Section 10.3
### Practice Exercises
**1. a.** $\sqrt{35}$   **b.** $\sqrt{13z}$   **c.** 5   **d.** $\sqrt[3]{15x^2 y}$   **e.** $\sqrt{\dfrac{5t}{2m}}$   **2. a.** $\dfrac{6}{7}$   **b.** $\dfrac{\sqrt{z}}{4}$   **c.** $\dfrac{5}{2}$   **d.** $\dfrac{\sqrt[4]{5}}{3x^2}$   **3. a.** $7\sqrt{2}$   **b.** $3\sqrt[3]{2}$   **c.** $\sqrt{35}$   **d.** $3\sqrt[4]{3}$

**4. a.** $6z^3 \sqrt{z}$   **b.** $2pq^2 \sqrt[3]{4pq}$   **c.** $2x^3 \sqrt[4]{x^3}$   **5. a.** 4   **b.** $\dfrac{7}{3}\sqrt{z}$   **c.** $10xy^2 \sqrt[3]{x^2}$   **d.** $6x^2 y \sqrt[5]{2y}$   **6.** $\sqrt{17} \approx 4.123$   **7.** $\left(\dfrac{13}{2}, -4\right)$

## Vocabulary and Readiness Check 10.3
**1.** midpoint; point   **3.** midpoint   **5.** false   **7.** true   **9.** false

## Exercise Set 10.3
**1.** $\sqrt{14}$   **3.** 2   **5.** $\sqrt[3]{36}$   **7.** $\sqrt{6x}$   **9.** $\sqrt{\dfrac{14}{xy}}$   **11.** $\sqrt[4]{20x^3}$   **13.** $\dfrac{\sqrt{6}}{7}$   **15.** $\dfrac{\sqrt{2}}{7}$   **17.** $\dfrac{\sqrt[4]{x^3}}{2}$   **19.** $\dfrac{\sqrt[3]{4}}{3}$   **21.** $\dfrac{\sqrt[4]{8}}{x^2}$   **23.** $\dfrac{\sqrt[3]{2x}}{3y^4 \sqrt[3]{3}}$

**25.** $\dfrac{x\sqrt{y}}{10}$   **27.** $\dfrac{x\sqrt{5}}{2y}$   **29.** $-\dfrac{z^2 \sqrt[3]{z}}{3x}$   **31.** $4\sqrt{2}$   **33.** $4\sqrt[3]{3}$   **35.** $25\sqrt{3}$   **37.** $2\sqrt{6}$   **39.** $10x^2 \sqrt{x}$   **41.** $2y^2 \sqrt[3]{2y}$   **43.** $a^2 b \sqrt[4]{b^3}$

**45.** $y^2 \sqrt{y}$   **47.** $5ab\sqrt{b}$   **49.** $-2x^2 \sqrt[5]{y}$   **51.** $x^4 \sqrt[3]{50x^2}$   **53.** $-4a^4 b^3 \sqrt{2b}$   **55.** $3x^3 y^4 \sqrt{xy}$   **57.** $5r^3 s^4$   **59.** $\sqrt{2}$   **61.** 2   **63.** 10

**65.** $x^2 y$   **67.** $24m^2$   **69.** $\dfrac{15x\sqrt{2x}}{2}$ or $\dfrac{15x}{2}\sqrt{2x}$   **71.** $2a^2 \sqrt[4]{2}$   **73.** 5 units   **75.** $\sqrt{41}$ units $\approx 6.403$   **77.** $\sqrt{10}$ units $\approx 3.162$

**79.** $\sqrt{5}$ units $\approx 2.236$   **81.** $\sqrt{192.58}$ units $\approx 13.877$   **83.** $(4, -2)$   **85.** $\left(-5, \dfrac{5}{2}\right)$   **87.** $(3, 0)$   **89.** $\left(-\dfrac{1}{2}, \dfrac{1}{2}\right)$   **91.** $\left(\sqrt{2}, \dfrac{\sqrt{5}}{2}\right)$

**93.** $(6.2, -6.65)$   **95.** $14x$   **97.** $2x^2 - 7x - 15$   **99.** $y^2$   **101.** $-3x - 15$   **103.** $x^2 - 8x + 16$   **105.** $\dfrac{\sqrt[3]{64}}{\sqrt{64}} = \dfrac{4}{8} = \dfrac{1}{2}$   **107.** $x^7$   **109.** $a^3 bc^5$

**111.** $z^{10} \sqrt[3]{z^2}$   **113.** $q^2 r^5 s \sqrt[4]{q^3 r^5}$   **115.** $r = 1.6\,\text{m}$   **117. a.** 3.8 times   **b.** 2.9 times   **c.** answers may vary

## Section 10.4
### Practice Exercises
**1. a.** $8\sqrt{17}$   **b.** $-5\sqrt[3]{5z}$   **c.** $3\sqrt{2} + 5\sqrt[3]{2}$   **2. a.** $11\sqrt{6}$   **b.** $-9\sqrt[3]{3}$   **c.** $-2\sqrt{3x}$   **d.** $2\sqrt{10} + 2\sqrt[3]{5}$   **e.** $4x\sqrt[3]{3x}$   **3. a.** $\dfrac{5\sqrt{7}}{12}$

**b.** $\dfrac{13\sqrt[3]{6y}}{4}$   **4. a.** $2\sqrt{5} + 5\sqrt{3}$   **b.** $2\sqrt{3} + 2\sqrt{2} - \sqrt{30} - 2\sqrt{5}$   **c.** $6z + \sqrt{z} - 12$   **d.** $-6\sqrt{6} + 15$   **e.** $5x - 9$   **f.** $6\sqrt{x+2} + x + 11$

## Vocabulary and Readiness Check 10.4

**1.** Unlike    **3.** Like    **5.** $6\sqrt{3}$    **7.** $7\sqrt{x}$    **9.** $8\sqrt[3]{x}$    **11.** $\sqrt{11} + \sqrt[3]{11}$    **13.** $10\sqrt[3]{2x}$

## Exercise Set 10.4

**1.** $-2\sqrt{2}$   **3.** $10x\sqrt{2x}$   **5.** $17\sqrt{2} - 15\sqrt{5}$   **7.** $-\sqrt[3]{2x}$   **9.** $5b\sqrt{b}$   **11.** $\dfrac{31\sqrt{2}}{15}$   **13.** $\dfrac{\sqrt[3]{11}}{3}$   **15.** $\dfrac{5\sqrt{5x}}{9}$   **17.** $14 + \sqrt{3}$   **19.** $7 - 3y$

**21.** $6\sqrt{3} - 6\sqrt{2}$   **23.** $-23\sqrt[3]{5}$   **25.** $2b\sqrt{b}$   **27.** $20y\sqrt{2y}$   **29.** $2y\sqrt[3]{2x}$   **31.** $6\sqrt[3]{11} - 4\sqrt{11}$   **33.** $4x\sqrt[4]{x^3}$   **35.** $\dfrac{2\sqrt{3}}{3}$   **37.** $\dfrac{5x\sqrt[3]{x}}{7}$

**39.** $\dfrac{5\sqrt{7}}{2x}$   **41.** $\dfrac{\sqrt[3]{2}}{6}$   **43.** $\dfrac{14x\sqrt[3]{2x}}{9}$   **45.** $15\sqrt{3}$ in.   **47.** $\sqrt{35} + \sqrt{21}$   **49.** $7 - 2\sqrt{10}$   **51.** $3\sqrt{x} - x\sqrt{3}$   **53.** $6x - 13\sqrt{x} - 5$

**55.** $\sqrt[3]{a^2} + \sqrt[3]{a} - 20$   **57.** $6\sqrt{2} - 12$   **59.** $2 + 2x\sqrt{3}$   **61.** $-16 - \sqrt{35}$   **63.** $x - y^2$   **65.** $3 + 2x\sqrt{3} + x^2$   **67.** $23x - 5x\sqrt{15}$

**69.** $2\sqrt[3]{2} - \sqrt[3]{4}$   **71.** $x + 1$   **73.** $x + 24 + 10\sqrt{x-1}$   **75.** $2x + 6 - 2\sqrt{2x+5}$   **77.** $x - 7$   **79.** $\dfrac{7}{x+y}$   **81.** $2a - 3$   **83.** $\dfrac{-2 + \sqrt{3}}{3}$

**85.** $22\sqrt{5}$ ft; $150$ sq ft   **87. a.** $2\sqrt{3}$   **b.** $3$   **c.** answers may vary   **89.** answers may vary

## Section 10.5
### Practice Exercises

**1. a.** $\dfrac{5\sqrt{3}}{3}$   **b.** $\dfrac{15\sqrt{x}}{2x}$   **c.** $\dfrac{\sqrt[3]{6}}{3}$   **2.** $\dfrac{\sqrt{15yz}}{5y}$   **3.** $\dfrac{\sqrt[3]{z^2x^2}}{3x^2}$   **4. a.** $\dfrac{5(3\sqrt{5} - 2)}{41}$   **b.** $\dfrac{\sqrt{6} + 5\sqrt{3} + \sqrt{10} + 5\sqrt{5}}{-2}$   **c.** $\dfrac{6x - 3\sqrt{xy}}{4x - y}$

**5.** $\dfrac{2}{\sqrt{10}}$   **6.** $\dfrac{5b}{\sqrt[3]{50ab^2}}$   **7.** $\dfrac{x - 9}{4(\sqrt{x} + 3)}$

## Vocabulary and Readiness Check 10.5

**1.** conjugate   **3.** rationalizing the numerator   **5.** $\sqrt{2} - x$   **7.** $5 + \sqrt{a}$   **9.** $-7\sqrt{5} - 8\sqrt{x}$

## Exercise Set 10.5

**1.** $\dfrac{\sqrt{14}}{7}$   **3.** $\dfrac{\sqrt{5}}{5}$   **5.** $\dfrac{2\sqrt{x}}{x}$   **7.** $\dfrac{4\sqrt[3]{9}}{3}$   **9.** $\dfrac{3\sqrt{2x}}{4x}$   **11.** $\dfrac{3\sqrt[3]{2x}}{2x}$   **13.** $\dfrac{3\sqrt{3a}}{a}$   **15.** $\dfrac{3\sqrt[3]{4}}{2}$   **17.** $\dfrac{2\sqrt{21}}{7}$   **19.** $\dfrac{\sqrt{10xy}}{5y}$   **21.** $\dfrac{\sqrt[3]{75}}{5}$

**23.** $\dfrac{\sqrt{6x}}{10}$   **25.** $\dfrac{\sqrt{3z}}{6z}$   **27.** $\dfrac{\sqrt[3]{6xy^2}}{3x}$   **29.** $\dfrac{3\sqrt[4]{2}}{2}$   **31.** $\dfrac{2\sqrt[4]{9x}}{3x^2}$   **33.** $\dfrac{5a\sqrt[5]{4ab^4}}{2a^2b^3}$   **35.** $-2(2 + \sqrt{7})$   **37.** $\dfrac{7(3 + \sqrt{x})}{9 - x}$   **39.** $-5 + 2\sqrt{6}$

**41.** $\dfrac{2a + 2\sqrt{a} + \sqrt{ab} + \sqrt{b}}{4a - b}$   **43.** $-\dfrac{8(1 - \sqrt{10})}{9}$   **45.** $\dfrac{x - \sqrt{xy}}{x - y}$   **47.** $\dfrac{5 + 3\sqrt{2}}{7}$   **49.** $\dfrac{5}{\sqrt{15}}$   **51.** $\dfrac{6}{\sqrt{10}}$   **53.** $\dfrac{2x}{7\sqrt{x}}$   **55.** $\dfrac{5y}{\sqrt[3]{100xy}}$

**57.** $\dfrac{2}{\sqrt{10}}$   **59.** $\dfrac{2x}{11\sqrt{2x}}$   **61.** $\dfrac{7}{2\sqrt[3]{49}}$   **63.** $\dfrac{3x^2}{10\sqrt[3]{9x}}$   **65.** $\dfrac{6x^2y^3}{\sqrt{6z}}$   **67.** answers may vary   **69.** $\dfrac{-7}{12 + 6\sqrt{11}}$   **71.** $\dfrac{3}{10 + 5\sqrt{7}}$

**73.** $\dfrac{x - 9}{x - 3\sqrt{x}}$   **75.** $\dfrac{1}{3 + 2\sqrt{2}}$   **77.** $\dfrac{x - 1}{x - 2\sqrt{x} + 1}$   **79.** $5$   **81.** $-\dfrac{1}{2}, 6$   **83.** $2, 6$   **85.** $\sqrt[3]{25}$   **87.** $r = \dfrac{\sqrt{A\pi}}{2\pi}$   **89.** answers may vary

## The Bigger Picture

**1.** $2\sqrt{14}$   **2.** $\dfrac{2x^2\sqrt{5x}}{7}$   **3.** $-15x^5y^{11}$   **4.** $\dfrac{\sqrt{110}}{11}$   **5.** $2(\sqrt{5} + 1)$ or $2\sqrt{5} + 2$   **6.** $5x^2 - 19$   **7.** $\dfrac{\sqrt{26x}}{2x^3}$   **8.** $\dfrac{-y + 6}{y^2 + 1}$   **9.** $\dfrac{x(x + y)}{4}$  
**10.** $2y^6\sqrt[3]{2y^2}$

## Integrated Review

**1.** $9$   **2.** $-2$   **3.** $\dfrac{1}{2}$   **4.** $x^3$   **5.** $y^3$   **6.** $2y^5$   **7.** $-2y$   **8.** $3b^3$   **9.** $6$   **10.** $\sqrt[4]{3y}$   **11.** $\dfrac{1}{16}$   **12.** $\sqrt[5]{(x + 1)^3}$   **13.** $y$   **14.** $16x^{1/2}$

**15.** $x^{5/4}$   **16.** $4^{11/15}$   **17.** $2x^2$   **18.** $\sqrt[4]{a^3b^2}$   **19.** $\sqrt[4]{x^3}$   **20.** $\sqrt[6]{500}$   **21.** $2\sqrt{10}$   **22.** $2xy^2\sqrt[4]{x^3y^2}$   **23.** $3x\sqrt[3]{2x}$   **24.** $-2b^2\sqrt[5]{2}$

**25.** $\sqrt{5x}$   **26.** $4x$   **27.** $7y^2\sqrt{y}$   **28.** $2a^2\sqrt[4]{3}$   **29.** $2\sqrt{5} - 5\sqrt{3} + 5\sqrt{7}$   **30.** $y\sqrt[3]{2y}$   **31.** $\sqrt{15} - \sqrt{6}$   **32.** $10 + 2\sqrt{21}$

**33.** $4x^2 - 5$   **34.** $x + 2 - 2\sqrt{x + 1}$   **35.** $\dfrac{\sqrt{21}}{3}$   **36.** $\dfrac{5\sqrt[3]{4x}}{2x}$   **37.** $\dfrac{13 - 3\sqrt{21}}{5}$   **38.** $\dfrac{7}{\sqrt{21}}$   **39.** $\dfrac{3y}{\sqrt[3]{33y^2}}$   **40.** $\dfrac{x - 4}{x + 2\sqrt{x}}$

## Section 10.6
### Practice Exercises

**1.** $18$   **2.** $\dfrac{3}{8}$   **3.** $10$   **4.** $9$   **5.** $\dfrac{3}{25}$   **6.** $6\sqrt{3}$ m   **7.** $\sqrt{193}$ in. $\approx 13.89$ in.

## Graphing Calculator Explorations 10.6

**1.** 3.19    **3.** $\varnothing$    **5.** 3.23

## Vocabulary and Readiness Check 10.6

**1.** extraneous solution    **3.** $x^2 - 10x + 25$

## Exercise Set 10.6

**1.** 8    **3.** 7    **5.** $\varnothing$    **7.** 7    **9.** 6    **11.** $-\dfrac{9}{2}$    **13.** 29    **15.** 4    **17.** $-4$    **19.** $\varnothing$    **21.** 7    **23.** 9    **25.** 50    **27.** $\varnothing$    **29.** $\dfrac{15}{4}$

**31.** 13    **33.** 5    **35.** $-12$    **37.** 9    **39.** $-3$    **41.** 1    **43.** 1    **45.** $\dfrac{1}{2}$    **47.** 0, 4    **49.** $\dfrac{37}{4}$    **51.** $3\sqrt{5}$ ft    **53.** $2\sqrt{10}$ m

**55.** $2\sqrt{131}$ m $\approx 22.9$ m    **57.** $\sqrt{100.84}$ mm $\approx 10.0$ mm    **59.** 17 ft    **61.** 13 ft    **63.** 14,657, 415 sq mi    **65.** 100 ft    **67.** 100

**69.** $\dfrac{\pi}{2}$ sec $\approx 1.57$ sec    **71.** 12.97 ft    **73.** answers may vary    **75.** $15\sqrt{3}$ sq mi $\approx 25.98$ sq mi    **77.** answers may vary    **79.** 0.51 km

**81.** function    **83.** function    **85.** not a function    **87.** $\dfrac{x}{4x+3}$    **89.** $-\dfrac{4z+2}{3z}$

**91.** $\sqrt{5x-1}+4=7$
$\sqrt{5x-1}=3$
$(\sqrt{5x-1})^2=3^2$
$5x-1=9$
$5x=10$
$x=2$

**93.** 1    **95. a.–b.** answers may vary    **97.** $-1, 2$    **99.** $-8, -6, 0, 2$

## The Bigger Picture

**1.** $-19$    **2.** $-\dfrac{5}{3}, 5$    **3.** $-\dfrac{9}{2}, 5$    **4.** $\left[\dfrac{-11}{5}, 1\right]$    **5.** $\left(-\dfrac{7}{5}, \infty\right)$    **6.** 25    **7.** $(-5, \infty)$    **8.** $\varnothing$    **9.** $(-\infty, -13) \cup (17, \infty)$    **10.** $\dfrac{17}{25}$

## Section 10.7

### Practice Exercises

**1. a.** $2i$   **b.** $i\sqrt{7}$   **c.** $-3i\sqrt{2}$    **2. a.** $-\sqrt{30}$   **b.** $-3$   **c.** $25i$   **d.** $3i$    **3. a.** $-1 - 4i$   **b.** $-3 + 5i$   **c.** $3 - 2i$    **4. a.** 20   **b.** $-5 + 10i$

**c.** $15 + 16i$   **d.** $8 - 6i$   **e.** 85    **5. a.** $\dfrac{11}{10} - \dfrac{7i}{10}$   **b.** $0 - \dfrac{5i}{2}$    **6. a.** $i$   **b.** 1   **c.** $-1$   **d.** 1

## Vocabulary and Readiness Check 10.7

**1.** complex    **3.** $-1$    **5.** real    **7.** $9i$    **9.** $i\sqrt{7}$    **11.** $-4$    **13.** $8i$

## Exercise Set 10.7

**1.** $2i\sqrt{6}$    **3.** $-6i$    **5.** $24i\sqrt{7}$    **7.** $-3\sqrt{6}$    **9.** $-\sqrt{14}$    **11.** $-5\sqrt{2}$    **13.** $4i$    **15.** $i\sqrt{3}$    **17.** $2\sqrt{2}$    **19.** $6 - 4i$    **21.** $-2 + 6i$

**23.** $-2 - 4i$    **25.** $-40$    **27.** $18 + 12i$    **29.** 7    **31.** $12 - 16i$    **33.** $-4i$    **35.** $\dfrac{28}{25} - \dfrac{21}{25}i$    **37.** $4 + i$    **39.** $\dfrac{17}{13} + \dfrac{7}{13}i$    **41.** 63

**43.** $2 - i$    **45.** $27 + 3i$    **47.** $-\dfrac{5}{2} - 2i$    **49.** $18 + 13i$    **51.** 20    **53.** 10    **55.** 2    **57.** $-5 + \dfrac{16}{3}i$    **59.** $17 + 144i$    **61.** $\dfrac{3}{5} - \dfrac{1}{5}i$

**63.** $5 - 10i$    **65.** $\dfrac{1}{5} - \dfrac{8}{5}i$    **67.** $8 - i$    **69.** 7    **71.** $12 - 16i$    **73.** 1    **75.** $i$    **77.** $-i$    **79.** $-1$    **81.** $-64$    **83.** $-243i$    **85.** $40°$

**87.** $x^2 - 5x - 2 - \dfrac{6}{x-1}$    **89.** 5 people    **91.** 14 people    **93.** 16.7%    **95.** $-1 - i$    **97.** 0    **99.** $2 + 3i$    **101.** $2 + i\sqrt{2}$    **103.** $\dfrac{1}{2} - \dfrac{\sqrt{3}}{2}i$

**105.** answers may vary    **107.** $6 - 6i$    **109.** yes

## Chapter 10 Vocabulary Check

**1.** conjugate    **2.** principal square root    **3.** rationalizing    **4.** imaginary unit    **5.** cube root    **6.** index, radicand    **7.** like radicals
**8.** complex number    **9.** distance    **10.** midpoint

## Chapter 10 Review

**1.** 9    **3.** $-2$    **5.** $-\dfrac{1}{7}$    **7.** $-6$    **9.** $-a^2b^3$    **11.** $2ab^2$    **13.** $\dfrac{x^6}{6y}$    **15.** $|x|$    **17.** $-27$    **19.** $-x$    **21.** $5|(x-y)^5|$    **23.** $-x$

**25.** $(-\infty, \infty)$; $-2, -1, 0, 1, 2$    **27.** $-\dfrac{1}{3}$    **29.** $-\dfrac{1}{4}$    **31.** $\dfrac{1}{4}$    **33.** $\dfrac{343}{125}$    **35.** not a real number    **37.** $5^{1/5}x^{2/5}y^{3/5}$    **39.** $5\sqrt[3]{xy^2z^5}$

**41.** $a^{13/6}$    **43.** $\dfrac{1}{a^{9/2}}$    **45.** $a^4b^6$    **47.** $\dfrac{b^{5/6}}{49a^{1/4}c^{5/3}}$    **49.** 4.472    **51.** 5.191    **53.** $-26.246$    **55.** $\sqrt[6]{1372}$

**57.** $2\sqrt{6}$    **59.** $2x$    **61.** $2\sqrt{15}$    **63.** $3\sqrt[3]{6}$    **65.** $6x^3\sqrt{x}$    **67.** $\dfrac{p^8\sqrt{p}}{11}$    **69.** $\dfrac{y\sqrt[4]{xy^2}}{3}$    **71. a.** $\dfrac{5}{\sqrt{\pi}}$ m or $\dfrac{5\sqrt{\pi}}{\pi}$ m

**b.** 5.75 in.    **73.** $\sqrt{130}$ units $\approx 11.402$    **75.** $7\sqrt{2}$ units $\approx 9.899$    **77.** $\sqrt{275.6}$ units $\approx 16.601$    **79.** $\left(-\dfrac{15}{2}, 1\right)$    **81.** $\left(\dfrac{1}{20}, -\dfrac{3}{16}\right)$

**83.** $\left(\sqrt{3}, -3\sqrt{6}\right)$   **85.** $2x\sqrt{3xy}$   **87.** $3a\sqrt[4]{2a}$   **89.** $\dfrac{3\sqrt{2}}{4x}$   **91.** $-4ab\sqrt[4]{2b}$   **93.** $x - 6\sqrt{x} + 9$   **95.** $4x - 9y$   **97.** $\sqrt[3]{a^2} + 4\sqrt[3]{a} + 4$

**99.** $a + 64$   **101.** $\dfrac{\sqrt{3x}}{6}$   **103.** $\dfrac{2x^2\sqrt{2x}}{y}$   **105.** $-\dfrac{10 + 5\sqrt{7}}{3}$   **107.** $-5 + 2\sqrt{6}$   **109.** $\dfrac{6}{\sqrt{2y}}$   **111.** $\dfrac{4x^3}{y\sqrt{2x}}$   **113.** $\dfrac{x - 25}{-3\sqrt{x} + 15}$

**115.** $\varnothing$   **117.** $\varnothing$   **119.** 16   **121.** $\sqrt{241}$   **123.** 4.24 ft   **125.** $-i\sqrt{6}$   **127.** $-\sqrt{10}$   **129.** $-13 - 3i$   **131.** $-12 - 18i$   **133.** $-5 - 12i$

**135.** $\dfrac{3}{2} - i$   **137.** $x$   **139.** $-10$   **141.** $\dfrac{y^5}{2x^3}$   **143.** $\dfrac{1}{8}$   **145.** $\dfrac{1}{x^{13/2}}$   **147.** $\dfrac{n\sqrt{3n}}{11m^5}$   **149.** $4x - 20\sqrt{x} + 25$   **151.** $(4, 16)$   **153.** $\dfrac{2\sqrt{x} - 6}{x - 9}$

## Chapter 10 Test

**1.** $6\sqrt{6}$   **2.** $-x^{16}$   **3.** $\dfrac{1}{5}$   **4.** 5   **5.** $\dfrac{4x^2}{9}$   **6.** $-a^6b^3$   **7.** $\dfrac{8a^{1/3}c^{2/3}}{b^{5/12}}$   **8.** $a^{7/12} - a^{7/3}$   **9.** $|4xy|$ or $4|xy|$   **10.** $-27$   **11.** $\dfrac{3\sqrt{y}}{y}$

**12.** $\dfrac{8 - 6\sqrt{x} + x}{8 - 2x}$   **13.** $\dfrac{\sqrt[3]{b^2}}{b}$   **14.** $\dfrac{6 - x^2}{8(\sqrt{6} - x)}$   **15.** $-x\sqrt{5x}$   **16.** $4\sqrt{3} - \sqrt{6}$   **17.** $x + 2\sqrt{x} + 1$   **18.** $\sqrt{6} - 4\sqrt{3} + \sqrt{2} - 4$

**19.** $-20$   **20.** 23.685   **21.** 0.019   **22.** 2, 3   **23.** $\varnothing$   **24.** 6   **25.** $i\sqrt{2}$   **26.** $-2i\sqrt{2}$   **27.** $-3i$   **28.** 40   **29.** $7 + 24i$   **30.** $-\dfrac{3}{2} + \dfrac{5}{2}i$

**31.** $\dfrac{5\sqrt{2}}{2}$   **32.** $[-2, \infty)$; 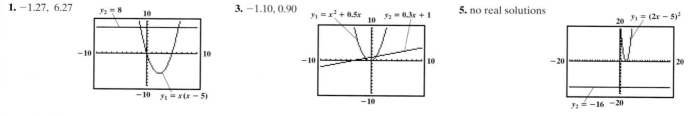 ; 0, 1, 2, 3   **33.** $2\sqrt{26}$ units   **34.** $\sqrt{95}$ units   **35.** $\left(-4, \dfrac{7}{2}\right)$   **36.** $\left(-\dfrac{1}{2}, \dfrac{3}{10}\right)$

**37.** 27 mph   **38.** 360 ft

## Chapter 10 Cumulative Review

**1. a.** $-12$   **b.** $-3$; Sec. 1.6, Ex. 5   **3.** 12; Sec. 2.3, Ex. 3   **5.** 12 in., 36 in.; Sec. 2.4, Ex. 3   **7.** one; Sec. 4.1, Ex. 8   **9.** $\left(6, \dfrac{1}{2}\right)$; Sec. 4.2, Ex. 3

**11.** no solution; Sec. 4.3, Ex. 3   **13.** 30% solution: 42 L; 80% solution: 28 L; Sec. 4.5, Ex. 4   **15. a.** $-4$   **b.** 11; Sec. 5.2, Ex. 4

**17.** $3m + 1$; Sec. 5.6, Ex. 1   **19.** $2x^2 + 5x + 2 + \dfrac{7}{x - 3}$; Sec. 5.7, Ex. 1   **21.** $(t - 8)(t - 5)$; Sec. 6.2, Ex. 8   **23. a.** $x^2 - 2x + 4$

**b.** $\dfrac{2}{y - 5}$; Sec. 7.1, Ex. 5   **25.** 4; Sec. 9.2, Ex. 9   **27.** $3x - 5$; Sec. 7.3, Ex. 3   **29.** $\dfrac{2m + 1}{m + 1}$; Sec. 7.4, Ex. 5   **31. a.** $\dfrac{x(x - 2)}{2(x + 2)}$

**b.** $\dfrac{x^2}{y^2}$; Sec. 7.7, Ex. 2   **33.** $\left[-2, \dfrac{8}{5}\right]$; Sec. 9.3, Ex. 3   **35.** 15 yd; Sec. 7.6, Ex. 4   **37. a.** 1   **b.** $-4$   **c.** $\dfrac{2}{5}$   **d.** $x^2$   **e.** $-3x^3$; Sec. 10.1, Ex. 3

**39. a.** $\dfrac{1}{8}$   **b.** $\dfrac{1}{9}$; Sec. 10.2, Ex. 3   **41.** $\dfrac{x - 4}{5(\sqrt{x} - 2)}$; Sec. 10.5, Ex. 7   **43.** constant of variation: 15, $u = \dfrac{15}{w}$; Sec. 8.4, Ex. 3

# CHAPTER 11   QUADRATIC EQUATIONS AND FUNCTIONS

## Section 11.1

### Practice Exercises

**1.** $-3\sqrt{2}, 3\sqrt{2}$   **2.** $\pm\sqrt{10}$   **3.** $-3 \pm 2\sqrt{5}$   **4.** $\dfrac{2 + 3i}{5}, \dfrac{2 - 3i}{5}$   **5.** $-2 \pm \sqrt{7}$   **6.** $\dfrac{3 \pm \sqrt{5}}{2}$   **7.** $\dfrac{6 \pm \sqrt{33}}{3}$   **8.** $\dfrac{5 \pm i\sqrt{31}}{4}$   **9.** 6%

### Graphing Calculator Explorations 11.1

**1.** $-1.27, 6.27$   **3.** $-1.10, 0.90$   **5.** no real solutions

### Vocabulary and Readiness Check 11.1

**1.** $\pm\sqrt{b}$   **3.** completing the square   **5.** 9   **7.** 1   **9.** 49

## Exercise Set 11.1

**1.** $-4, 4$   **3.** $-\sqrt{7}, \sqrt{7}$   **5.** $-3\sqrt{2}, 3\sqrt{2}$   **7.** $-\sqrt{10}, \sqrt{10}$   **9.** $-8, -2$   **11.** $6 - 3\sqrt{2}, 6 + 3\sqrt{2}$   **13.** $\dfrac{3 - 2\sqrt{2}}{2}, \dfrac{3 + 2\sqrt{2}}{2}$

**15.** $-3i, 3i$   **17.** $-\sqrt{6}, \sqrt{6}$   **19.** $-2i\sqrt{2}, 2i\sqrt{2}$   **21.** $1 - 4i, 1 + 4i$   **23.** $-7 - \sqrt{5}, -7 + \sqrt{5}$   **25.** $-3 - 2i\sqrt{2}, -3 + 2i\sqrt{2}$

**27.** $x^2 + 16x + 64 = (x + 8)^2$   **29.** $z^2 - 12z + 36 = (z - 6)^2$   **31.** $p^2 + 9p + \dfrac{81}{4} = \left(p + \dfrac{9}{2}\right)^2$   **33.** $x^2 + x + \dfrac{1}{4} = \left(x + \dfrac{1}{2}\right)^2$   **35.** $-5, -3$

**37.** $-3 - \sqrt{7}, -3 + \sqrt{7}$   **39.** $\dfrac{-1 - \sqrt{5}}{2}, \dfrac{-1 + \sqrt{5}}{2}$   **41.** $-1 - \sqrt{6}, -1 + \sqrt{6}$   **43.** $\dfrac{6 - \sqrt{30}}{3}, \dfrac{6 + \sqrt{30}}{3}$   **45.** $\dfrac{3 - \sqrt{11}}{2}, \dfrac{3 + \sqrt{11}}{2}$

**47.** $-4, \dfrac{1}{2}$   **49.** $-1, 5$   **51.** $-4 - \sqrt{15}, -4 + \sqrt{15}$   **53.** $\dfrac{-3 - \sqrt{21}}{3}, \dfrac{-3 + \sqrt{21}}{3}$   **55.** $-1, \dfrac{5}{2}$   **57.** $-1 - i, -1 + i$   **59.** $3 - \sqrt{6}, 3 + \sqrt{6}$

**61.** $-2 - i\sqrt{2}, -2 + i\sqrt{2}$   **63.** $\dfrac{-15 - 7\sqrt{5}}{10}, \dfrac{-15 + 7\sqrt{5}}{10}$   **65.** $\dfrac{1 - i\sqrt{47}}{4}, \dfrac{1 + i\sqrt{47}}{4}$   **67.** $-5 - i\sqrt{3}, -5 + i\sqrt{3}$   **69.** $-4, 1$

**71.** $\dfrac{2 - i\sqrt{2}}{2}, \dfrac{2 + i\sqrt{2}}{2}$   **73.** $\dfrac{-3 - \sqrt{69}}{6}, \dfrac{-3 + \sqrt{69}}{6}$   **75.** $20\%$   **77.** $4\%$   **79.** answers may vary   **81.** $8.11$ sec   **83.** $6.73$ sec

**85.** simple; answers may vary   **87.** $\dfrac{7}{5}$   **89.** $\dfrac{1}{5}$   **91.** $5 - 10\sqrt{3}$   **93.** $\dfrac{3 - 2\sqrt{7}}{4}$   **95.** $2\sqrt{7}$   **97.** $\sqrt{13}$   **99.** complex, but not real numbers

**101.** real solutions   **103.** complex, but not real numbers   **105.** $-6y, 6y$   **107.** $-x, x$   **109.** 6 in.   **111.** 16.2 in. by 21.6 in.
**113.** 2.828 thousand units or 2828 units

## Section 11.2
### Practice Exercises

**1.** $2, -\dfrac{1}{3}$   **2.** $\dfrac{4 \pm \sqrt{22}}{3}$   **3.** $1 \pm \sqrt{17}$   **4.** $\dfrac{-1 \pm i\sqrt{15}}{4}$   **5. a.** one real solution   **b.** two real solutions   **c.** two complex but not real

solutions   **6.** 6 ft   **7.** 2.4 sec

## Vocabulary and Readiness Check 11.2

**1.** $x = \dfrac{-b \pm \sqrt{b^2 - 4ac}}{2a}$   **3.** $-5; -7$   **5.** $1; 0$

## Exercise Set 11.2

**1.** $-6, 1$   **3.** $-\dfrac{3}{5}, 1$   **5.** $3$   **7.** $\dfrac{-7 - \sqrt{33}}{2}, \dfrac{-7 + \sqrt{33}}{2}$   **9.** $\dfrac{1 - \sqrt{57}}{8}, \dfrac{1 + \sqrt{57}}{8}$   **11.** $\dfrac{7 - \sqrt{85}}{6}, \dfrac{7 + \sqrt{85}}{6}$   **13.** $1 - \sqrt{3}, 1 + \sqrt{3}$

**15.** $-\dfrac{3}{2}, 1$   **17.** $\dfrac{3 - \sqrt{11}}{2}, \dfrac{3 + \sqrt{11}}{2}$   **19.** $\dfrac{-5 - \sqrt{17}}{2}, \dfrac{-5 + \sqrt{17}}{2}$   **21.** $\dfrac{5}{2}, 1$   **23.** $-3 - 2i, -3 + 2i$   **25.** $-2 - \sqrt{11}, -2 + \sqrt{11}$

**27.** $\dfrac{3 - i\sqrt{87}}{8}, \dfrac{3 + i\sqrt{87}}{8}$   **29.** $\dfrac{3 - \sqrt{29}}{2}, \dfrac{3 + \sqrt{29}}{2}$   **31.** $\dfrac{-5 - i\sqrt{5}}{10}, \dfrac{-5 + i\sqrt{5}}{10}$   **33.** $\dfrac{-1 - \sqrt{19}}{6}, \dfrac{-1 + \sqrt{19}}{6}$

**35.** $\dfrac{-1 - i\sqrt{23}}{4}, \dfrac{-1 + i\sqrt{23}}{4}$   **37.** $1$   **39.** $3 + \sqrt{5}, 3 - \sqrt{5}$   **41.** two real solutions   **43.** one real solution   **45.** two real solutions

**47.** two complex but not real solutions   **49.** two real solutions   **51.** 14 ft   **53.** $2 + 2\sqrt{2}$ cm, $2 + 2\sqrt{2}$ cm, $4 + 2\sqrt{2}$ cm

**55.** width: $-5 + 5\sqrt{17}$ ft; length: $5 + 5\sqrt{17}$ ft   **57. a.** $50\sqrt{2}$ m   **b.** 5000 sq m   **59.** 37.4 ft by 38.5 ft   **61.** base, $2 + 2\sqrt{43}$ cm;

height, $-1 + \sqrt{43}$ cm   **63.** 8.9 sec   **65.** 2.8 sec   **67.** $\dfrac{11}{5}$   **69.** 15   **71.** $(x^2 + 5)(x + 2)(x - 2)$   **73.** $(z + 3)(z - 3)(z + 2)(z - 2)$

**75.** b   **77.** answers may vary   **79.** 0.6, 2.4   **81.** Sunday to Monday   **83.** Wednesday   **85.** $f(4) = 32$; yes   **87. a.** 8630 stores   **b.** 2011

**89.** answers may vary   **91.** $\dfrac{\sqrt{3}}{3}$   **93.** $\dfrac{-\sqrt{2} - i\sqrt{2}}{2}, \dfrac{-\sqrt{2} + i\sqrt{2}}{2}$   **95.** $\dfrac{\sqrt{3} - \sqrt{11}}{4}, \dfrac{\sqrt{3} + \sqrt{11}}{4}$

**97.** 8.9 sec; ![graph, 1200, Zero X=8.9400842 Y=0, 10, −100]   2.8 sec; ![graph, 200, Zero X=2.786836 Y=0, 4, −10]   **99.** two real solutions

## Section 11.3
### Practice Exercises

**1.** 8   **2.** $\dfrac{5 \pm \sqrt{137}}{8}$   **3.** $4, -4, 3i, -3i$   **4.** $1, -3$   **5.** $1, 64$   **6.** Katy: $\dfrac{7 + \sqrt{65}}{2} \approx 7.5$ hr; Steve: $\dfrac{9 + \sqrt{65}}{2} \approx 8.5$ hr

**7.** to Shanghai: 40 km/hr;  to Ningbo: 90 km/hr

## Exercise Set 11.3

**1.** 2    **3.** 16    **5.** 1, 4    **7.** $3 - \sqrt{7}, 3 + \sqrt{7}$    **9.** $\dfrac{3 - \sqrt{57}}{4}, \dfrac{3 + \sqrt{57}}{4}$    **11.** $\dfrac{1 - \sqrt{29}}{2}, \dfrac{1 + \sqrt{29}}{2}$    **13.** $-2, 2, -2i, 2i$

**15.** $-\dfrac{1}{2}, \dfrac{1}{2}, -i\sqrt{3}, i\sqrt{3}$    **17.** $-3, 3, -2, 2$    **19.** $125, -8$    **21.** $-\dfrac{4}{5}, 0$    **23.** $-\dfrac{1}{8}, 27$    **25.** $-\dfrac{2}{3}, \dfrac{4}{3}$    **27.** $-\dfrac{1}{125}, \dfrac{1}{8}$    **29.** $-\sqrt{2}, \sqrt{2}, -\sqrt{3}, \sqrt{3}$

**31.** $\dfrac{-9 - \sqrt{201}}{6}, \dfrac{-9 + \sqrt{201}}{6}$    **33.** 2, 3    **35.** 3    **37.** 27, 125    **39.** $1, -3i, 3i$    **41.** $\dfrac{1}{8}, -8$    **43.** $-\dfrac{1}{2}, \dfrac{1}{3}$    **45.** 4

**47.** $-3$    **49.** $-\sqrt{5}, \sqrt{5}, -2i, 2i$    **51.** $-3, \dfrac{3 - 3i\sqrt{3}}{2}, \dfrac{3 + 3i\sqrt{3}}{2}$    **53.** 6, 12    **55.** $-\dfrac{1}{3}, \dfrac{1}{3}, -\dfrac{i\sqrt{6}}{3}, \dfrac{i\sqrt{6}}{3}$    **57.** 5 mph, then 4 mph

**59.** inlet pipe: 15.5 hr; hose: 16.5 hr    **61.** 55 mph, 66 mph    **63.** 8.5 hr    **65.** 12 or $-8$    **67. a.** $(x - 6)$ in.    **b.** $300 = (x - 6) \cdot (x - 6) \cdot 3$

**c.** 16 cm by 16 cm    **69.** 22 feet    **71.** $(-\infty, 3]$    **73.** $(-5, \infty)$    **75.** domain: $(-\infty, \infty)$; range: $(-\infty, \infty)$; function

**77.** domain: $(-\infty, \infty)$; range: $[-1, \infty)$; function    **79.** $1, -3i, 3i$    **81.** $-\dfrac{1}{2}, \dfrac{1}{3}$    **83.** $-3, \dfrac{3 - 3i\sqrt{3}}{2}, \dfrac{3 + 3i\sqrt{3}}{2}$    **85.** answers may vary

**87. a.** 150.94 ft/sec    **b.** 151.49 ft/sec    **c.** Bourdais: 102.9 mph; Pagenaud: 103.3 mph

## Integrated Review

**1.** $-\sqrt{10}, \sqrt{10}$    **2.** $-\sqrt{14}, \sqrt{14}$    **3.** $1 - 2\sqrt{2}, 1 + 2\sqrt{2}$    **4.** $-5 - 2\sqrt{3}, -5 + 2\sqrt{3}$    **5.** $-1 - \sqrt{13}, -1 + \sqrt{13}$

**6.** 1, 11    **7.** $\dfrac{-3 - \sqrt{69}}{6}, \dfrac{-3 + \sqrt{69}}{6}$    **8.** $\dfrac{-2 - \sqrt{5}}{4}, \dfrac{-2 + \sqrt{5}}{4}$    **9.** $\dfrac{2 - \sqrt{2}}{2}, \dfrac{2 + \sqrt{2}}{2}$    **10.** $-3 - \sqrt{5}, -3 + \sqrt{5}$

**11.** $-2 + i\sqrt{3}, -2 - i\sqrt{3}$    **12.** $\dfrac{-1 - i\sqrt{11}}{2}, \dfrac{-1 + i\sqrt{11}}{2}$    **13.** $\dfrac{-3 + i\sqrt{15}}{2}, \dfrac{-3 - i\sqrt{15}}{2}$    **14.** $3i, -3i$    **15.** $0, -17$

**16.** $\dfrac{1 + \sqrt{13}}{4}, \dfrac{1 - \sqrt{13}}{4}$    **17.** $2 + 3\sqrt{3}, 2 - 3\sqrt{3}$    **18.** $2 + \sqrt{3}, 2 - \sqrt{3}$    **19.** $-2, \dfrac{4}{3}$    **20.** $\dfrac{-5 + \sqrt{17}}{4}, \dfrac{-5 - \sqrt{17}}{4}$    **21.** $1 - \sqrt{6}, 1 + \sqrt{6}$

**22.** $-\sqrt{31}, \sqrt{31}$    **23.** $-\sqrt{11}, \sqrt{11}$    **24.** $-i\sqrt{11}, i\sqrt{11}$    **25.** $-11, 6$    **26.** $\dfrac{-3 + \sqrt{19}}{5}, \dfrac{-3 - \sqrt{19}}{5}$    **27.** $\dfrac{-3 + \sqrt{17}}{4}, \dfrac{-3 - \sqrt{17}}{4}$

**28.** $10\sqrt{2}\,\text{ft} \approx 14.1\,\text{ft}$    **29.** Jack: 9.1 hr; Lucy: 7.1 hr    **30.** 5 mph during the first part, then 6 mph

## Section 11.4
### Practice Exercises

**1.** $(-\infty, -3) \cup (4, \infty)$    **2.** $[0, 8]$    **3.** $(-\infty, -3] \cup [-1, 2]$    **4.** $(-4, 5]$    **5.** $(-\infty, -3) \cup \left(-\dfrac{8}{5}, \infty\right)$

## Vocabulary and Readiness Check 11.4

**1.** $[-7, 3)$    **3.** $(-\infty, 0]$    **5.** $(-\infty, -12) \cup [-10, \infty)$

## Exercise Set 11.4

**1.** $(-\infty, -5) \cup (-1, \infty)$    **3.** $[-4, 3]$    **5.** $[2, 5]$    **7.** $\left(-5, -\dfrac{1}{3}\right)$    **9.** $(2, 4) \cup (6, \infty)$    **11.** $(-\infty, -4] \cup [0, 1]$

**13.** $(-\infty, -3) \cup (-2, 2) \cup (3, \infty)$    **15.** $(-7, 2)$    **17.** $(-1, \infty)$    **19.** $(-\infty, -1] \cup (4, \infty)$    **21.** $(-\infty, 2) \cup \left(\dfrac{11}{4}, \infty\right)$    **23.** $(0, 2] \cup [3, \infty)$

**25.** $(-\infty, -7) \cup (8, \infty)$    **27.** $\left[-\dfrac{5}{4}, \dfrac{3}{2}\right]$    **29.** $(-\infty, 0) \cup (1, \infty)$    **31.** $(-\infty, -4] \cup [4, 6]$    **33.** $\left(-\infty, -\dfrac{2}{3}\right] \cup \left[\dfrac{3}{2}, \infty\right)$

**35.** $\left(-4, -\dfrac{3}{2}\right) \cup \left(\dfrac{3}{2}, \infty\right)$    **37.** $(-\infty, -5] \cup [-1, 1] \cup [5, \infty)$    **39.** $\left(-\infty, -\dfrac{5}{3}\right) \cup \left(\dfrac{7}{2}, \infty\right)$    **41.** $(0, 10)$    **43.** $(-\infty, -4) \cup [5, \infty)$

**45.** $(-\infty, -6] \cup (-1, 0] \cup (7, \infty)$    **47.** $(-\infty, 1) \cup (2, \infty)$    **49.** $(-\infty, -8] \cup (-4, \infty)$    **51.** $(-\infty, 0] \cup \left(5, \dfrac{11}{2}\right]$    **53.** $(0, \infty)$

**55.**    **57.**    **59.**    **61.**    **63.** answers may vary

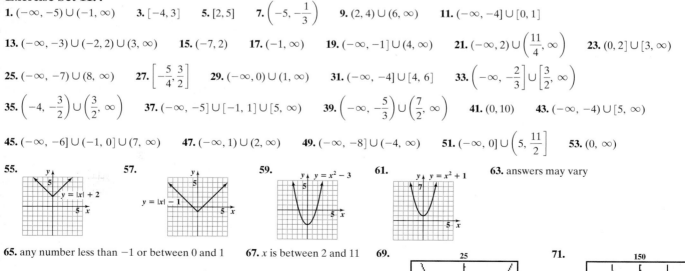

**65.** any number less than $-1$ or between 0 and 1    **67.** $x$ is between 2 and 11    **69.**    **71.**

## The Bigger Picture

**1.** $-9, \dfrac{7}{3}$    **2.** $(4, 7)$    **3.** $4, 5$    **4.** $\dfrac{-1 - \sqrt{13}}{6}, \dfrac{-1 + \sqrt{13}}{6}$    **5.** $[-2, 7)$    **6.** $\dfrac{5}{4}$    **7.** $\dfrac{1}{5}, 7$    **8.** $\left(-\infty, -\dfrac{1}{2}\right] \cup [4, \infty)$

**9.** $(-\infty, -8) \cup (22, \infty)$    **10.** $(7, \infty)$

## Section 11.5
### Practice Exercises

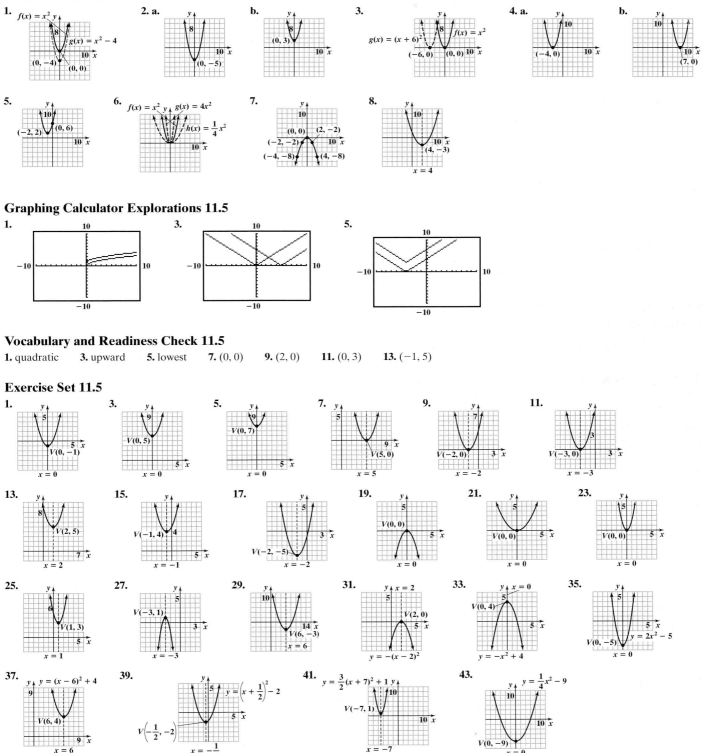

## Graphing Calculator Explorations 11.5

## Vocabulary and Readiness Check 11.5

**1.** quadratic    **3.** upward    **5.** lowest    **7.** $(0, 0)$    **9.** $(2, 0)$    **11.** $(0, 3)$    **13.** $(-1, 5)$

## Exercise Set 11.5

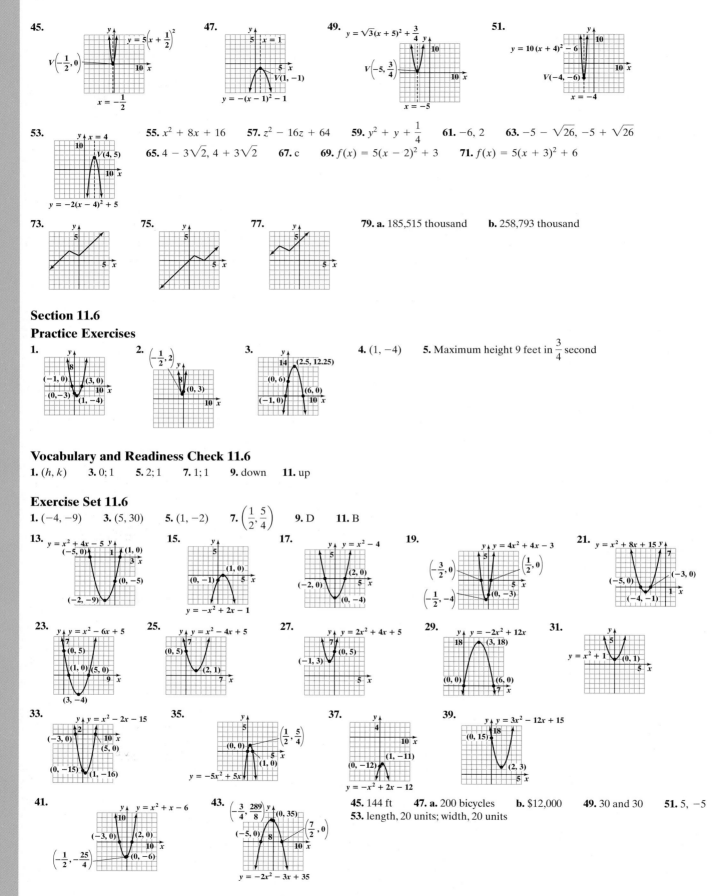

**45.** $V\left(-\frac{1}{2}, 0\right)$, $y = 5\left(x + \frac{1}{2}\right)^2$, $x = -\frac{1}{2}$

**47.** $x = 1$, $V(1, -1)$, $y = -(x - 1)^2 - 1$

**49.** $y = \sqrt{3}(x + 5)^2 + \frac{3}{4}$, $V\left(-5, \frac{3}{4}\right)$, $x = -5$

**51.** $y = 10(x + 4)^2 - 6$, $V(-4, -6)$, $x = -4$

**53.** $x = 4$, $V(4, 5)$, $y = -2(x - 4)^2 + 5$

**55.** $x^2 + 8x + 16$   **57.** $z^2 - 16z + 64$   **59.** $y^2 + y + \frac{1}{4}$   **61.** $-6, 2$   **63.** $-5 - \sqrt{26}, -5 + \sqrt{26}$

**65.** $4 - 3\sqrt{2}, 4 + 3\sqrt{2}$   **67.** c   **69.** $f(x) = 5(x - 2)^2 + 3$   **71.** $f(x) = 5(x + 3)^2 + 6$

**73.** **75.** **77.**   **79. a.** 185,515 thousand   **b.** 258,793 thousand

## Section 11.6
### Practice Exercises

**1.** $(-1, 0)$, $(3, 0)$, $(0, -3)$, $(1, -4)$

**2.** $\left(-\frac{1}{2}, 2\right)$, $(0, 3)$

**3.** $(2.5, 12.25)$, $(0, 6)$, $(-1, 0)$, $(6, 0)$

**4.** $(1, -4)$   **5.** Maximum height 9 feet in $\frac{3}{4}$ second

## Vocabulary and Readiness Check 11.6

**1.** $(h, k)$   **3.** 0; 1   **5.** 2; 1   **7.** 1; 1   **9.** down   **11.** up

## Exercise Set 11.6

**1.** $(-4, -9)$   **3.** $(5, 30)$   **5.** $(1, -2)$   **7.** $\left(\frac{1}{2}, \frac{5}{4}\right)$   **9.** D   **11.** B

**13.** $y = x^2 + 4x - 5$, $(-5, 0)$, $(1, 0)$, $(0, -5)$, $(-2, -9)$

**15.** $(1, 0)$, $(0, -1)$, $y = -x^2 + 2x - 1$

**17.** $y = x^2 - 4$, $(2, 0)$, $(-2, 0)$, $(0, -4)$

**19.** $y = 4x^2 + 4x - 3$, $\left(-\frac{3}{2}, 0\right)$, $\left(\frac{1}{2}, 0\right)$, $\left(-\frac{1}{2}, -4\right)$, $(0, -3)$

**21.** $y = x^2 + 8x + 15$, $(-5, 0)$, $(-3, 0)$, $(-4, -1)$

**23.** $y = x^2 - 6x + 5$, $(0, 5)$, $(1, 0)$, $(5, 0)$, $(3, -4)$

**25.** $y = x^2 - 4x + 5$, $(0, 5)$, $(2, 1)$

**27.** $y = 2x^2 + 4x + 5$, $(-1, 3)$, $(0, 5)$

**29.** $y = -2x^2 + 12x$, $(3, 18)$, $(0, 0)$, $(6, 0)$

**31.** $y = x^2 + 1$, $(0, 1)$

**33.** $y = x^2 - 2x - 15$, $(-3, 0)$, $(5, 0)$, $(0, -15)$, $(1, -16)$

**35.** $(0, 0)$, $\left(\frac{1}{2}, \frac{5}{4}\right)$, $(1, 0)$, $y = -5x^2 + 5x$

**37.** $(1, -11)$, $(0, -12)$, $y = -x^2 + 2x - 12$

**39.** $y = 3x^2 - 12x + 15$, $(0, 15)$, $(2, 3)$

**41.** $y = x^2 + x - 6$, $(-3, 0)$, $(2, 0)$, $\left(-\frac{1}{2}, -\frac{25}{4}\right)$, $(0, -6)$

**43.** $\left(-\frac{3}{4}, \frac{289}{8}\right)$, $(0, 35)$, $\left(\frac{7}{2}, 0\right)$, $(-5, 0)$, $y = -2x^2 - 3x + 35$

**45.** 144 ft   **47. a.** 200 bicycles   **b.** \$12,000   **49.** 30 and 30   **51.** 5, $-5$

**53.** length, 20 units; width, 20 units

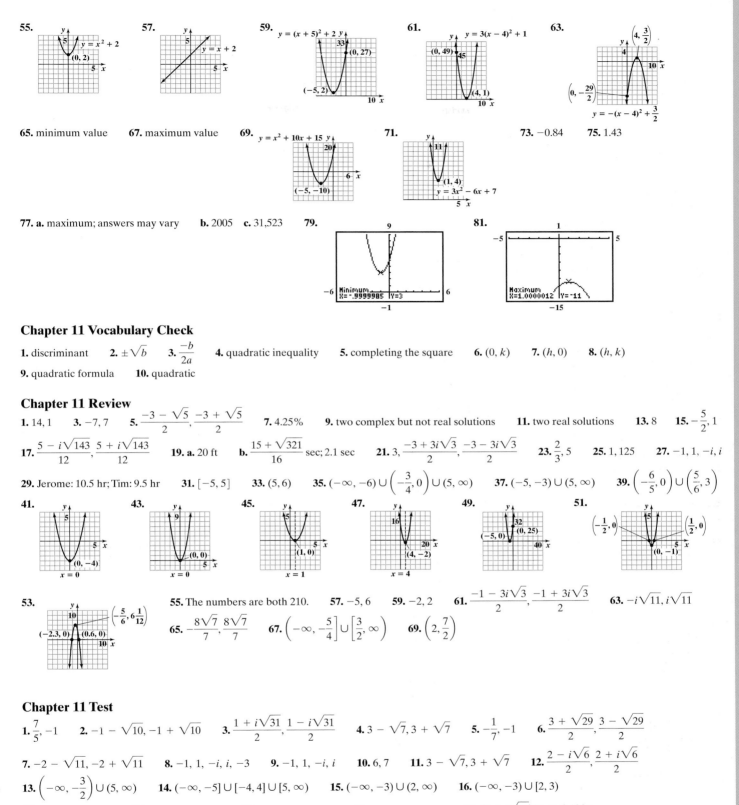

**55.** $y = x^2 + 2$  (0, 2)

**57.** $y = x + 2$

**59.** $y = (x + 5)^2 + 2$  (0, 27)  (−5, 2)

**61.** $y = 3(x − 4)^2 + 1$  (0, 49)  (4, 1)

**63.** $\left(4, \frac{3}{2}\right)$  $\left(0, -\frac{29}{2}\right)$  $y = -(x − 4)^2 + \frac{3}{2}$

**65.** minimum value   **67.** maximum value   **69.** $y = x^2 + 10x + 15$  (−5, −10)   **71.** (1, 4)  $y = 3x^2 − 6x + 7$   **73.** −0.84   **75.** 1.43

**77. a.** maximum; answers may vary   **b.** 2005   **c.** 31,523   **79.** Minimum X=−.9999985 Y=3   **81.** Maximum X=1.0000012 Y=−11

## Chapter 11 Vocabulary Check

**1.** discriminant   **2.** $\pm\sqrt{b}$   **3.** $\dfrac{-b}{2a}$   **4.** quadratic inequality   **5.** completing the square   **6.** $(0, k)$   **7.** $(h, 0)$   **8.** $(h, k)$
**9.** quadratic formula   **10.** quadratic

## Chapter 11 Review

**1.** 14, 1   **3.** −7, 7   **5.** $\dfrac{-3 − \sqrt{5}}{2}, \dfrac{-3 + \sqrt{5}}{2}$   **7.** 4.25%   **9.** two complex but not real solutions   **11.** two real solutions   **13.** 8   **15.** $-\dfrac{5}{2}, 1$

**17.** $\dfrac{5 − i\sqrt{143}}{12}, \dfrac{5 + i\sqrt{143}}{12}$   **19. a.** 20 ft   **b.** $\dfrac{15 + \sqrt{321}}{16}$ sec; 2.1 sec   **21.** 3, $\dfrac{-3 + 3i\sqrt{3}}{2}, \dfrac{-3 − 3i\sqrt{3}}{2}$   **23.** $\dfrac{2}{3}, 5$   **25.** 1, 125   **27.** −1, 1, −i, i

**29.** Jerome: 10.5 hr; Tim: 9.5 hr   **31.** $[−5, 5]$   **33.** (5, 6)   **35.** $(−\infty, −6) \cup \left(-\dfrac{3}{4}, 0\right) \cup (5, \infty)$   **37.** $(−5, −3) \cup (5, \infty)$   **39.** $\left(-\dfrac{6}{5}, 0\right) \cup \left(\dfrac{5}{6}, 3\right)$

**41.** (0, −4)  x = 0
**43.** (0, 0)  x = 0
**45.** (1, 0)  x = 1
**47.** (4, −2)  x = 4
**49.** (−5, 0)  (0, 25)
**51.** $\left(-\dfrac{1}{2}, 0\right)$  $\left(\dfrac{1}{2}, 0\right)$  (0, −1)

**53.** (−2.3, 0)  (0.6, 0)  $\left(-\dfrac{5}{6}, 6\dfrac{1}{12}\right)$   **55.** The numbers are both 210.   **57.** −5, 6   **59.** −2, 2   **61.** $\dfrac{-1 − 3i\sqrt{3}}{2}, \dfrac{-1 + 3i\sqrt{3}}{2}$   **63.** $-i\sqrt{11}, i\sqrt{11}$

**65.** $-\dfrac{8\sqrt{7}}{7}, \dfrac{8\sqrt{7}}{7}$   **67.** $\left(-\infty, -\dfrac{5}{4}\right] \cup \left[\dfrac{3}{2}, \infty\right)$   **69.** $\left(2, \dfrac{7}{2}\right)$

## Chapter 11 Test

**1.** $\dfrac{7}{5}, -1$   **2.** $-1 − \sqrt{10}, -1 + \sqrt{10}$   **3.** $\dfrac{1 + i\sqrt{31}}{2}, \dfrac{1 − i\sqrt{31}}{2}$   **4.** $3 − \sqrt{7}, 3 + \sqrt{7}$   **5.** $-\dfrac{1}{7}, -1$   **6.** $\dfrac{3 + \sqrt{29}}{2}, \dfrac{3 − \sqrt{29}}{2}$

**7.** $-2 − \sqrt{11}, -2 + \sqrt{11}$   **8.** −1, 1, −i, i, −3   **9.** −1, 1, −i, i   **10.** 6, 7   **11.** $3 − \sqrt{7}, 3 + \sqrt{7}$   **12.** $\dfrac{2 − i\sqrt{6}}{2}, \dfrac{2 + i\sqrt{6}}{2}$

**13.** $\left(-\infty, -\dfrac{3}{2}\right) \cup (5, \infty)$   **14.** $(−\infty, −5] \cup [−4, 4] \cup [5, \infty)$   **15.** $(−\infty, −3) \cup (2, \infty)$   **16.** $(−\infty, −3) \cup [2, 3)$

**17.** (0, 0)   **18.** (1, 5)   **19.** (0, 4)  (2, 0)   **20.** (0, 9)  (2, 1)   **21.** $(5 + \sqrt{17})$ hr $\approx 9.12$ hr
**22. a.** 272 ft   **b.** 5.12 sec   **23.** 7 ft

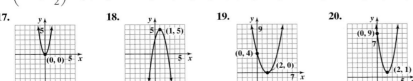

## Chapter 11 Cumulative Review

**1. a.** $\dfrac{1}{2}$ **b.** 19; Sec. 1.6, Ex. 6 **3. a.** $5x + 7$ **b.** $-4a - 1$ **c.** $4y - 3y^2$ **d.** $7.3x - 6$ **e.** $\dfrac{1}{2}b$; Sec. 2.1, Ex. 4 **5.** no solution; Sec. 4.1, Ex. 5

**7.** $(-2, 0)$; Sec. 4.2, Ex. 4 **9. a.** $x^3$ **b.** 256 **c.** $-27$ **d.** cannot be simplified **e.** $2x^4y$; Sec. 5.1, Ex. 9 **11. a.** 5 **b.** 5; Sec. 5.7, Ex. 3

**13.** $-6, -\dfrac{3}{2}, \dfrac{1}{5}$; Sec. 6.6, Ex. 8 **15.** $\dfrac{1}{5x - 1}$; Sec. 7.1, Ex. 2a **17.** $\dfrac{xy + 2x^3}{y - 1}$; Sec. 7.7, Ex. 3 **19.** $(2m^2 - 1)^2$; Sec. 6.3, Ex. 10

**21.** $-10$ and 7; Sec. 6.7, Ex. 2 **23. a.** $5x\sqrt{x}$ **b.** $3x^2y^2\sqrt[3]{2y^2}$ **c.** $3z^2\sqrt[4]{z^3}$; Sec. 10.3, Ex. 4 **25. a.** $\dfrac{2\sqrt{5}}{5}$ **b.** $\dfrac{8\sqrt{x}}{3x}$ **c.** $\dfrac{\sqrt[3]{4}}{2}$; Sec. 10.5, Ex. 1

**27.** $\dfrac{2}{9}$; Sec. 10.6, Ex. 5 **29.** $-5$; Sec. 7.5, Ex. 1 **31.** $-5$; Sec. 7.6, Ex. 5 **33.** $\dfrac{1}{6}$; $y = \dfrac{1}{6}x$; Sec. 8.4, Ex. 1 **35. a.** 3 **b.** $|x|$ **c.** $|x - 2|$

**d.** $-5$ **e.** $2x - 7$ **f.** $5|x|$ **g.** $|x + 1|$; Sec. 10.1, Ex. 5 **37. a.** $\sqrt{x}$ **b.** $\sqrt[3]{5}$ **c.** $\sqrt{rs^3}$; Sec. 10.2, Ex. 7 **39. a.** $\dfrac{1}{2} + \dfrac{3}{2}i$

**b.** $-\dfrac{7}{3}i$; Sec. 10.7, Ex. 5 **41.** $-1 + 2\sqrt{3}, -1 - 2\sqrt{3}$; Sec. 11.1, Ex. 3 **43.** 9; Sec. 11.3, Ex. 1

# CHAPTER 12 EXPONENTIAL AND LOGARITHMIC FUNCTIONS

## Section 12.1

### Practice Exercises

**1. a.** $4x + 7$ **b.** $-2x - 3$ **c.** $3x^2 + 11x + 10$ **d.** $\dfrac{x + 2}{3x + 5}$, where $x \neq -\dfrac{5}{3}$ **2. a.** 50; 46 **b.** $9x^2 - 30x + 26$; $3x^2 - 2$
**3. a.** $x^2 + 6x + 14$ **b.** $x^2 + 8$ **4. a.** $(h \circ g)(x)$ **b.** $(g \circ f)(x)$

### Vocabulary and Readiness Check 12.1

**1.** C **3.** F **5.** D

### Exercise Set 12.1

**1. a.** $3x - 6$ **b.** $-x - 8$ **c.** $2x^2 - 13x - 7$ **d.** $\dfrac{x - 7}{2x + 1}$, where $x \neq -\dfrac{1}{2}$ **3. a.** $x^2 + 5x + 1$ **b.** $x^2 - 5x + 1$ **c.** $5x^3 + 5x$

**d.** $\dfrac{x^2 + 1}{5x}$, where $x \neq 0$ **5. a.** $\sqrt{x} + x + 5$ **b.** $\sqrt{x} - x - 5$ **c.** $x\sqrt{x} + 5\sqrt{x}$ **d.** $\dfrac{\sqrt{x}}{x + 5}$, where $x \neq -5$

**7. a.** $5x^2 - 3x$ **b.** $-5x^2 - 3x$ **c.** $-15x^3$ **d.** $-\dfrac{3}{5x}$, where $x \neq 0$ **9.** 42 **11.** $-18$ **13.** 0

**15.** $(f \circ g)(x) = 25x^2 + 1$; $(g \circ f)(x) = 5x^2 + 5$ **17.** $(f \circ g)(x) = 2x + 11$; $(g \circ f)(x) = 2x + 4$

**19.** $(f \circ g)(x) = -8x^3 - 2x - 2$; $(g \circ f)(x) = -2x^3 - 2x + 4$ **21.** $(f \circ g)(x) = |10x - 3|$; $(g \circ f)(x) = 10|x| - 3$

**23.** $(f \circ g)(x) = \sqrt{-5x + 2}$; $(g \circ f)(x) = -5\sqrt{x} + 2$ **25.** $H(x) = (g \circ h)(x)$ **27.** $F(x) = (h \circ f)(x)$ **29.** $G(x) = (f \circ g)(x)$

**31.** answers may vary; for example $g(x) = x + 2$ and $f(x) = x^2$ **33.** answers may vary; for example, $g(x) = x + 5$ and $f(x) = \sqrt{x} + 2$

**35.** answers may vary; for example, $g(x) = 2x - 3$ and $f(x) = \dfrac{1}{x}$ **37.** $y = x - 2$ **39.** $y = \dfrac{x}{3}$ **41.** $y = -\dfrac{x + 7}{2}$ **43.** 6 **45.** 4 **47.** 4

**49.** $-1$ **51.** answers may vary **53.** $P(x) = R(x) - C(x)$

## Section 12.2

### Practice Exercises

**1. a.** one-to-one **b.** not one-to-one **c.** not one-to-one **d.** not one-to-one **e.** not one-to-one **f.** not one-to-one
**2. a.** no, not one-to-one **b.** yes **c.** yes **d.** no, not a function **e.** no, not a function
**3.** $f^{-1}(x) = \{(4, 3), (0, -2), (8, 2), (6, 6)\}$ **4.** $f^{-1}(x) = 6 - x$

**5.** **6. a.** **b.** **7.** $f(f^{-1}(x)) = f\left(\dfrac{x + 1}{4}\right) = 4\left(\dfrac{x + 1}{4}\right) - 1 = x + 1 - 1 = x$

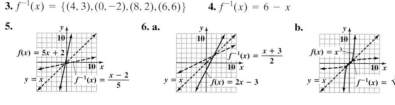

$f^{-1}(f(x)) = f^{-1}(4x - 1) = \dfrac{(4x - 1) + 1}{4} = \dfrac{4x}{4} = x$

### Vocabulary and Readiness Check 12.2

**1.** $(2, 11)$ **3.** $(3, 7)$ **5.** horizontal **7.** $x; x$

### Exercise Set 12.2

**1.** one-to-one; $f^{-1} = \{(-1, -1), (1, 1), (2, 0), (0, 2)\}$ **3.** one-to-one; $h^{-1} = \{(10, 10)\}$ **5.** one-to-one; $f^{-1} = \{(12, 11), (3, 4), (4, 3), (6, 6)\}$
**7.** not one-to-one **9.** one-to-one;

| Rank in Population (Input) | 1 | 19 | 35 | 4 | 48 |
|---|---|---|---|---|---|
| State (Output) | CA | MD | NV | FL | ND |

**11. a.** 3 **b.** 1 **13. a.** 1 **b.** $-1$
**15.** one-to-one **17.** not one-to-one
**19.** one-to-one **21.** not one-to-one

**23.** $f^{-1}(x) = x - 4$   **25.** $f^{-1}(x) = \dfrac{x+3}{2}$   **27.** $f^{-1}(x) = 2x + 2$   **29.** $f^{-1}(x) = \sqrt[3]{x}$

**31.** $f^{-1}(x) = \dfrac{x-2}{5}$   **33.** $f^{-1}(x) = 5x + 2$   **35.** $f^{-1}(x) = x^3$   **37.** $f^{-1}(x) = \dfrac{5-x}{3x}$   **39.** $f^{-1}(x) = \sqrt[3]{x} - 2$

**41.**   **43.**   **45.**   **47.** $(f \circ f^{-1})(x) = x; (f^{-1} \circ f)(x) = x$

**49.** $(f \circ f^{-1})(x) = x; (f^{-1} \circ f)(x) = x$   **51.** 5   **53.** 8   **55.** $\dfrac{1}{27}$   **57.** 9   **59.** $3^{1/2} \approx 1.73$   **61. a.** $(2, 9)$   **b.** $(9, 2)$

**63. a.** $\left(-2, \dfrac{1}{4}\right), \left(-1, \dfrac{1}{2}\right), (0, 1), (1, 2), (2, 5)$   **b.** $\left(\dfrac{1}{4}, -2\right), \left(\dfrac{1}{2}, -1\right), (1, 0), (2, 1), (5, 2)$

**c.**   **d.**   **65.** answers may vary   **67.** $f^{-1}(x) = \dfrac{x-1}{3}$;   **69.** $f^{-1}(x) = x^3 - 1$;

## Section 12.3
### Practice Exercises

**1.**   **2.**   **3.**   **4. a.** 2;   **b.** $\dfrac{4}{3}$;   **c.** $-4$   **5.** \$3950.43   **6.** 60.86%

### Graphing Calculator Explorations 12.3

**1.** 81.98%;   **3.** 22.54%;

### Vocabulary and Readiness Check 12.3

**1.** exponential   **3.** yes   **5.** yes; $(0, 1)$   **7.** $(0, \infty)$

### Exercise Set 12.3

**1.**   **3.**   **5.**   **7.**   **9.**

**11.**   **13.**   **15.**   **17.** C   **19.** B   **21.** 3   **23.** $\dfrac{3}{4}$   **25.** $\dfrac{8}{5}$   **27.** $-\dfrac{2}{3}$   **29.** 4   **31.** $\dfrac{3}{2}$

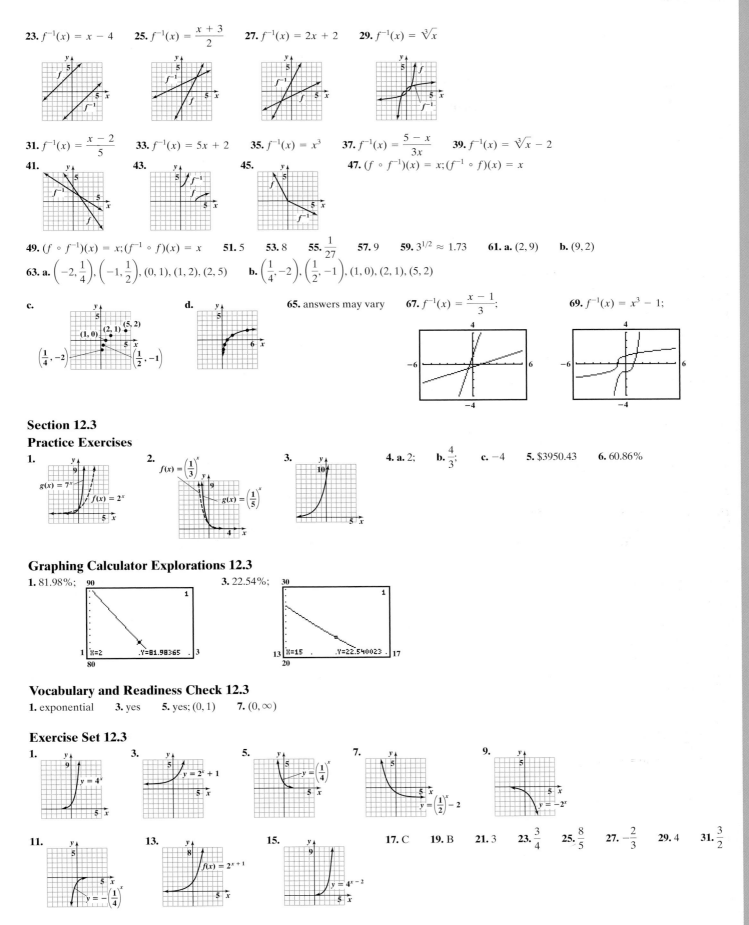

**33.** $-\dfrac{1}{3}$   **35.** $-2$   **37.** 24.6 lb   **39.** 333 bison   **41.** 1.1 g   **43. a.** 658.1 Pascals   **b.** 180.0 Pascals   **45. a.** 134,342 students   **b.** 840,276 students

**47.** \$7621.42   **49.** \$4065.59   **51.** 562 million cell phone users   **53.** 4   **55.** $\varnothing$   **57.** 2, 3   **59.** 3   **61.** $-1$   **63.** answers may vary

**65.**   **67.**   **69.** The graphs are the same since $\left(\dfrac{1}{2}\right)^{-x} = 2x.$   **71.** 24.60 lb;

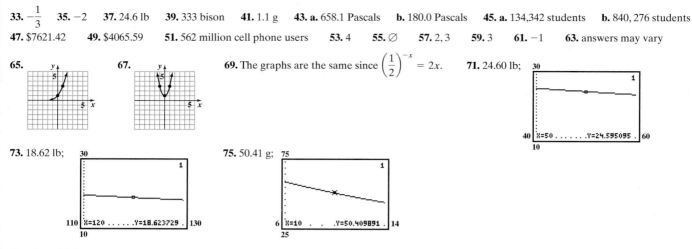

**73.** 18.62 lb;   **75.** 50.41 g;

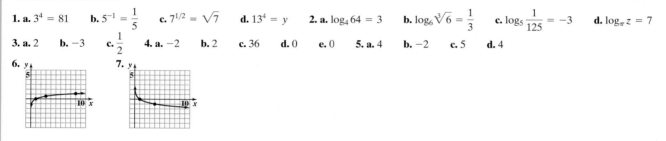

## Section 12.4

### Practice Exercises

**1. a.** $3^4 = 81$   **b.** $5^{-1} = \dfrac{1}{5}$   **c.** $7^{1/2} = \sqrt{7}$   **d.** $13^4 = y$   **2. a.** $\log_4 64 = 3$   **b.** $\log_6 \sqrt[3]{6} = \dfrac{1}{3}$   **c.** $\log_5 \dfrac{1}{125} = -3$   **d.** $\log_\pi z = 7$

**3. a.** 2   **b.** $-3$   **c.** $\dfrac{1}{2}$   **4. a.** $-2$   **b.** 2   **c.** 36   **d.** 0   **e.** 0   **5. a.** 4   **b.** $-2$   **c.** 5   **d.** 4

**6.**   **7.**

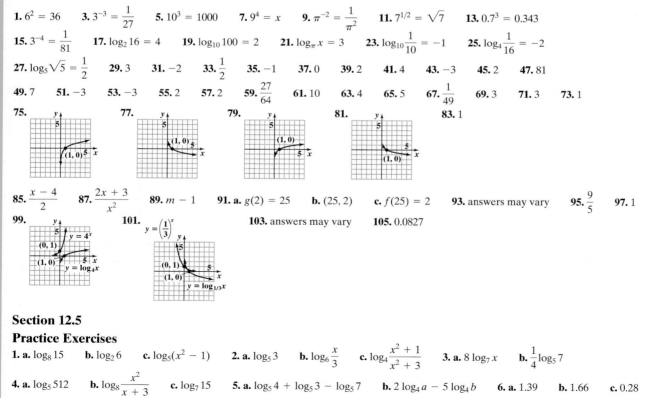

## Vocabulary and Readiness Check 12.4

**1.** logarithmic   **3.** yes   **5.** no; none   **7.** $(-\infty, \infty)$

## Exercise Set 12.4

**1.** $6^2 = 36$   **3.** $3^{-3} = \dfrac{1}{27}$   **5.** $10^3 = 1000$   **7.** $9^4 = x$   **9.** $\pi^{-2} = \dfrac{1}{\pi^2}$   **11.** $7^{1/2} = \sqrt{7}$   **13.** $0.7^3 = 0.343$

**15.** $3^{-4} = \dfrac{1}{81}$   **17.** $\log_2 16 = 4$   **19.** $\log_{10} 100 = 2$   **21.** $\log_\pi x = 3$   **23.** $\log_{10}\dfrac{1}{10} = -1$   **25.** $\log_4 \dfrac{1}{16} = -2$

**27.** $\log_5 \sqrt{5} = \dfrac{1}{2}$   **29.** 3   **31.** $-2$   **33.** $\dfrac{1}{2}$   **35.** $-1$   **37.** 0   **39.** 2   **41.** 4   **43.** $-3$   **45.** 2   **47.** 81

**49.** 7   **51.** $-3$   **53.** $-3$   **55.** 2   **57.** 2   **59.** $\dfrac{27}{64}$   **61.** 10   **63.** 4   **65.** 5   **67.** $\dfrac{1}{49}$   **69.** 3   **71.** 3   **73.** 1

**75.**   **77.**   **79.**   **81.**   **83.** 1

**85.** $\dfrac{x-4}{2}$   **87.** $\dfrac{2x+3}{x^2}$   **89.** $m - 1$   **91. a.** $g(2) = 25$   **b.** $(25, 2)$   **c.** $f(25) = 2$   **93.** answers may vary   **95.** $\dfrac{9}{5}$   **97.** 1

**99.**   **101.** $y = \left(\dfrac{1}{3}\right)^x$   **103.** answers may vary   **105.** 0.0827

## Section 12.5

### Practice Exercises

**1. a.** $\log_8 15$   **b.** $\log_2 6$   **c.** $\log_5(x^2 - 1)$   **2. a.** $\log_5 3$   **b.** $\log_6 \dfrac{x}{3}$   **c.** $\log_4 \dfrac{x^2 + 1}{x^2 + 3}$   **3. a.** $8 \log_7 x$   **b.** $\dfrac{1}{4} \log_5 7$

**4. a.** $\log_5 512$   **b.** $\log_8 \dfrac{x^2}{x+3}$   **c.** $\log_7 15$   **5. a.** $\log_5 4 + \log_5 3 - \log_5 7$   **b.** $2 \log_4 a - 5 \log_4 b$   **6. a.** 1.39   **b.** 1.66   **c.** 0.28

## Vocabulary and Readiness Check 12.5

**1.** 36    **3.** $\log_b 2^7$    **5.** $x$

## Exercise Set 12.5

**1.** $\log_5 14$    **3.** $\log_4 9x$    **5.** $\log_6(x^2 + x)$    **7.** $\log_{10}(10x^2 + 20)$    **9.** $\log_5 3$    **11.** $\log_3 4$    **13.** $\log_2 \dfrac{x}{y}$    **15.** $\log_2 \dfrac{x^2 + 6}{x^2 + 1}$    **17.** $2\log_3 x$

**19.** $-1 \log_4 5 = -\log_4 5$    **21.** $\dfrac{1}{2}\log_5 y$    **23.** $\log_2 5x^3$    **25.** $\log_4 48$    **27.** $\log_5 x^3 z^6$    **29.** $\log_4 4$, or 1    **31.** $\log_7 \dfrac{9}{2}$    **33.** $\log_{10} \dfrac{x^3 - 2x}{x + 1}$

**35.** $\log_2 \dfrac{x^{7/2}}{(x + 1)^2}$    **37.** $\log_8 x^{16/3}$    **39.** $\log_3 4 + \log_3 y - \log_3 5$    **41.** $\log_4 2 - \log_4 9 - \log_4 z$    **43.** $3\log_2 x - \log_2 y$    **45.** $\dfrac{1}{2}\log_b 7 + \dfrac{1}{2}\log_b x$

**47.** $4\log_6 x + 5\log_6 y$    **49.** $3\log_5 x + \log_5(x + 1)$    **51.** $2\log_6 x - \log_6(x + 3)$    **53.** 1.2    **55.** 0.2    **57.** 0.35    **59.** 1.29    **61.** $-0.68$

**63.** $-0.125$    **65.**    **67.** $-1$    **69.** $\dfrac{1}{2}$    **71.** a and d    **73.** false    **75.** true    **77.** false

## Integrated Review

**1.** $x^2 + x - 5$    **2.** $-x^2 + x - 7$    **3.** $x^3 - 6x^2 + x - 6$    **4.** $\dfrac{x - 6}{x^2 + 1}$    **5.** $\sqrt{3x - 1}$    **6.** $3\sqrt{x} - 1$

**7.** one-to-one; $\{(6, -2), (8, 4), (-6, 2), (3, 3)\}$    **8.** not one-to-one    **9.** not one-to-one    **10.** one-to-one

**11.** not one-to-one    **12.** $f^{-1}(x) = \dfrac{x}{3}$    **13.** $f^{-1}(x) = x - 4$    **14.** $f^{-1}(x) = \dfrac{x + 1}{5}$    **15.** $f^{-1}(x) = \dfrac{x - 2}{3}$

**16.**    **17.**    **18.**    **19.**    **20.** 3

**21.** 7    **22.** $-8$    **23.** 3    **24.** 2    **25.** $\dfrac{1}{2}$    **26.** 32    **27.** 4    **28.** 5    **29.** $\dfrac{1}{9}$    **30.** $\log_2 x^5$    **31.** $\log_2 5^x$    **32.** $\log_5 \dfrac{x^3}{y^5}$

**33.** $\log_5 x^9 y^3$    **34.** $\log_2 \dfrac{x^2 - 3x}{x^2 + 4}$    **35.** $\log_3 \dfrac{y^4 + 11y}{y + 2}$    **36.** $\log_7 9 + 2\log_7 x - \log_7 y$    **37.** $\log_6 5 + \log_6 y - 2\log_6 z$

## Section 12.6

### Practice Exercises

**1.** 1.1761    **2. a.** $-2$    **b.** 5    **c.** $\dfrac{1}{5}$    **d.** $-3$    **3.** $10^{3.4} \approx 2511.8864$    **4.** 5.6    **5.** 2.5649    **6. a.** 4    **b.** $\dfrac{1}{3}$

**7.** $\dfrac{e^8}{5} \approx 596.1916$    **8.** \$3051    **9.** 0.7740

## Vocabulary and Readiness Check 12.6

**1.** 10    **3.** 7    **5.** 5    **7.** $\dfrac{\log 7}{\log 2}$ or $\dfrac{\ln 7}{\ln 2}$

## Exercise Set 12.6

**1.** 0.9031    **3.** 0.3636    **5.** 0.6931    **7.** $-2.6367$    **9.** 1.1004    **11.** 1.6094    **13.** 1.6180    **15.** answers may vary    **17.** 2    **19.** $-3$    **21.** 2

**23.** $\dfrac{1}{4}$    **25.** 3    **27.** $-7$    **29.** $-4$    **31.** $\dfrac{1}{2}$    **33.** $\dfrac{e^7}{2} \approx 548.3166$    **35.** $10^{1.3} \approx 19.9526$    **37.** $\dfrac{10^{1.1}}{2} \approx 6.2946$    **39.** $e^{1.4} \approx 4.0552$

**41.** $\dfrac{4 + e^{2.3}}{3} \approx 4.6581$    **43.** $10^{2.3} \approx 199.5262$    **45.** $e^{-2.3} \approx 0.1003$    **47.** $\dfrac{10^{-0.5} - 1}{2} \approx -0.3419$    **49.** $\dfrac{e^{0.18}}{4} \approx 0.2993$    **51.** 1.5850    **53.** $-2.3219$

**55.** 1.5850    **57.** $-1.6309$    **59.** 0.8617    **61.** 4.2    **63.** 5.3    **65.** \$3656.38    **67.** \$2542.50    **69.** $\dfrac{4}{7}$    **71.** $x = \dfrac{3y}{4}$    **73.** $-6, -1$    **75.** $(2, -3)$

**77.** $\ln 50$; answers may vary    **79.**    **81.**    **83.**    **85.**    **87.**

**89.** **91.** **93.** **95.** **97.** answers may vary

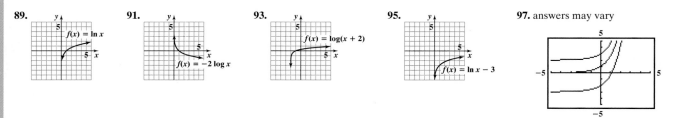

## Section 12.7
### Practice Exercises

**1.** $\dfrac{\log 9}{\log 5} \approx 1.3652$    **2.** 33    **3.** 1    **4.** $\dfrac{1}{3}$    **5.** 937 rabbits    **6.** 10 years

### Graphing Calculator Explorations 12.7

**1.** 3.67 years, or 3 years and 8 months    **3.** 23.16 years, or 23 years and 2 months

### Exercise Set 12.7

**1.** $\dfrac{\log 6}{\log 3}$; 1.6309    **3.** $\dfrac{\log 3.8}{2\log 3}$; 0.6076    **5.** $3 + \dfrac{\log 5}{\log 2}$; 5.3219    **7.** $\dfrac{\log 5}{\log 9}$; 0.7325    **9.** $\dfrac{\log 3}{\log 4} - 7$; $-6.2075$    **11.** $\dfrac{1}{3}\left(4 + \dfrac{\log 11}{\log 7}\right)$; 1.7441

**13.** $\dfrac{\ln 5}{6}$; 0.2682    **15.** 11    **17.** $9, -9$    **19.** $\dfrac{1}{2}$    **21.** $\dfrac{3}{4}$    **23.** 2    **25.** $\dfrac{1}{8}$    **27.** 11    **29.** $4, -1$    **31.** $\dfrac{1}{5}$    **33.** 100    **35.** $\dfrac{-5 + \sqrt{33}}{2}$

**37.** $\dfrac{192}{127}$    **39.** $\dfrac{2}{3}$    **41.** 103 wolves    **43.** 354,000 inhabitants    **45.** 14.7 yr    **47.** 9.9 yr    **49.** 1.7 yr    **51.** 8.8 yr    **53.** 24.5 lb    **55.** 55.7 in.

**57.** 11.9 lb/sq in.    **59.** 3.2 mi    **61.** 12 weeks    **63.** 18 weeks    **65.** $-\dfrac{5}{3}$    **67.** $\dfrac{17}{4}$    **69.** $f^{-1}(x) = \dfrac{x-2}{5}$    **71.** 2.9%    **73.** answers may vary

**75.** 6.93    **77.** $-3.68$    **79.** 1.74    **81.** 0.2

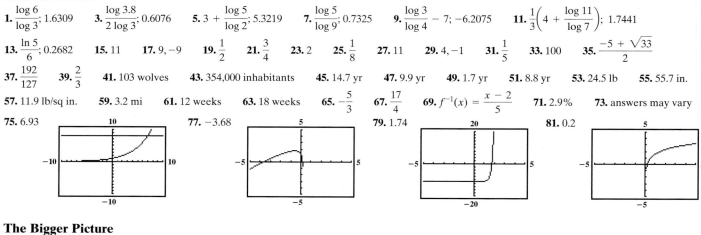

### The Bigger Picture

**1.** $-\dfrac{3}{2}$    **2.** $\dfrac{\log 5}{\log 11} \approx 0.6712$    **3.** $[-5, \infty)$    **4.** $\left[-\dfrac{13}{3}, -2\right]$    **5.** $\left(-\dfrac{6}{5}, 0\right)$    **6.** $6, -\dfrac{1}{5}$    **7.** $\dfrac{21}{8}$    **8.** $-\dfrac{7}{3}, 3$    **9.** $(-\infty, \infty)$    **10.** $-2, 2$

**11.** $-5 + \sqrt{3}, -5 - \sqrt{3}$    **12.** $-\dfrac{1}{4}, 7$

### Chapter 12 Vocabulary Check

**1.** inverse    **2.** composition    **3.** exponential    **4.** symmetric    **5.** Natural    **6.** Common    **7.** vertical; horizontal    **8.** logarithmic

### Chapter 12 Review

**1.** $3x - 4$    **3.** $2x^2 - 9x - 5$    **5.** $x^2 + 2x - 1$    **7.** 18    **9.** $-2$    **11.** one-to-one; $h^{-1} = \{(14, -9), (8, 6), (12, -11), (15, 15)\}$

**13.** one-to-one;

| Rank in Automobile Thefts (Input) | 2 | 4 | 1 | 3 |
|---|---|---|---|---|
| US Region (Output) | W | Midwest | S | NE |

**15. a.** 3    **b.** 7    **17.** not one-to-one    **19.** not one-to-one

**21.** $f^{-1}(x) = x + 9$    **23.** $f^{-1}(x) = \dfrac{x - 11}{6}$

**25.** $f^{-1}(x) = \sqrt[3]{x + 5}$    **27.** $g^{-1}(x) = \dfrac{6x + 7}{12}$

**29.**    **31.** $f^{-1}(x) = \dfrac{x + 3}{2}$;    **33.** $-2$    **35.** $\dfrac{3}{2}$    **37.** $\dfrac{8}{9}$    **39.**    **41.**

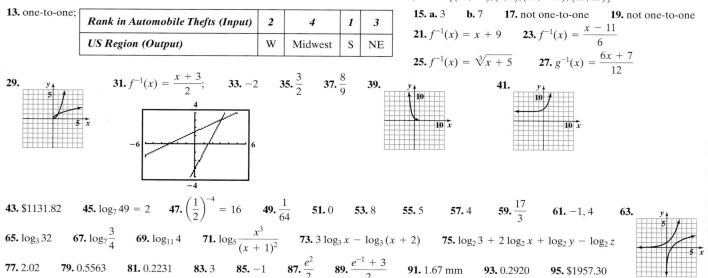

**43.** \$1131.82    **45.** $\log_7 49 = 2$    **47.** $\left(\dfrac{1}{2}\right)^{-4} = 16$    **49.** $\dfrac{1}{64}$    **51.** 0    **53.** 8    **55.** 5    **57.** 4    **59.** $\dfrac{17}{3}$    **61.** $-1, 4$    **63.**

**65.** $\log_3 32$    **67.** $\log_7 \dfrac{3}{4}$    **69.** $\log_{11} 4$    **71.** $\log_5 \dfrac{x^3}{(x + 1)^2}$    **73.** $3 \log_3 x - \log_3 (x + 2)$    **75.** $\log_2 3 + 2 \log_2 x + \log_2 y - \log_2 z$

**77.** 2.02    **79.** 0.5563    **81.** 0.2231    **83.** 3    **85.** $-1$    **87.** $\dfrac{e^2}{2}$    **89.** $\dfrac{e^{-1} + 3}{2}$    **91.** 1.67 mm    **93.** 0.2920    **95.** \$1957.30

**97.** $\dfrac{\log 7}{2 \log 3}$; 0.8856  **99.** $\dfrac{1}{2}\left(\dfrac{\log 6}{\log 3} - 1\right)$; 0.3155  **101.** $\dfrac{1}{3}\left(\dfrac{\log 4}{\log 5} + 5\right)$; 1.9538  **103.** $-\dfrac{\log 2}{\log 5} + 1$; 0.5693  **105.** $\dfrac{25}{2}$  **107.** $\varnothing$  **109.** $2\sqrt{2}$

**111.** 197,044 ducks  **113.** 20.9 yr  **115.** 38.5 yr  **117.** 8.5 yr  **119.** 2.82  **121.** $\dfrac{1}{2}$  **123.** 1  **125.** $-1, 5$  **127.** $e^{-3.2}$  **129.** $2e$

## Chapter 12 Test

**1.** $2x^2 - 3x$  **2.** $3 - x$  **3.** 5  **4.** $x - 7$  **5.** $x^2 - 6x - 2$

**6.** [graph]  **7.** one-to-one  **8.** not one-to-one  **9.** one-to-one; $f^{-1}(x) = \dfrac{-x + 6}{2}$  **10.** one-to-one; $f^{-1} = \{(0, 0), (3, 2)(5, -1)\}$

**11.** not one-to-one  **12.** $\log_3 24$  **13.** $\log_5 \dfrac{x^4}{x + 1}$  **14.** $\log_6 2 + \log_6 x - 3 \log_6 y$  **15.** $-1.53$  **16.** 1.0686  **17.** $-1$

**18.** $\dfrac{1}{2}\left(\dfrac{\log 4}{\log 3} - 5\right)$; $-1.8691$  **19.** $\dfrac{1}{9}$  **20.** $\dfrac{1}{2}$  **21.** 22  **22.** $\dfrac{25}{3}$  **23.** $\dfrac{43}{21}$  **24.** $-1.0979$

**25.** [graph]  **26.** [graph]  **27.** \$5234.58  **28.** 6 yr  **29.** 64,913 prairie dogs  **30.** 15 yr  **31.** 1.2%

## Chapter 12 Cumulative Review

**1. a.** $\dfrac{64}{25}$  **b.** $\dfrac{1}{20}$  **c.** $\dfrac{5}{4}$; Sec. 1.3, Ex. 4  **3.** [graph $y = x^2$]; Sec. 8.2, Ex. 5  **5.** $\{(x, y, z) | x - 5y - 2z = 6\}$; Sec. 4.4, Ex. 4

**7. a.** $-12$  **b.** $-3$; Sec. 1.6, Ex. 5  **9. a.** \$102.60  **b.** \$12.60; Sec. 7.1, Ex. 7

**11.** $-5$; Sec. 2.2, Ex. 8  **13.** $2xy - 4 + \dfrac{1}{2y}$; Sec. 5.6, Ex. 3

**15.** $3(m + 2)(m - 10)$; Sec. 6.2, Ex. 9  **17.** $3x - 5$; Sec. 7.3, Ex. 3  **19. a.** 3  **b.** $-3$  **c.** $-5$  **d.** not a real number  **e.** $4x$; Sec. 10.1, Ex. 4

**21. a.** $\sqrt[4]{x^3}$  **b.** $\sqrt[6]{x}$  **c.** $\sqrt[6]{72}$; Sec. 10.2, Ex. 8  **23. a.** $5\sqrt{3} + 3\sqrt{10}$  **b.** $\sqrt{35} + \sqrt{5} - \sqrt{42} - \sqrt{6}$  **c.** $21x - 7\sqrt{5x} + 15\sqrt{x} - 5\sqrt{5}$

**d.** $49 - 8\sqrt{3}$  **e.** $2x - 25$  **f.** $x + 22 + 10\sqrt{x - 3}$; Sec. 10.4, Ex. 4  **25.** $\dfrac{\sqrt[4]{xy^3}}{3y^2}$; Sec. 10.5, Ex. 3  **27.** 3; Sec. 10.6, Ex. 4

**29.** $\dfrac{9 + i\sqrt{15}}{6}, \dfrac{9 - i\sqrt{15}}{6}$; Sec. 11.1, Ex. 8  **31.** $\dfrac{-1 + \sqrt{33}}{4}, \dfrac{-1 - \sqrt{33}}{4}$; Sec. 11.3, Ex. 2  **33.** $[0, 4]$; Sec. 11.4, Ex. 2

**35.** [graph]; Sec. 11.5, Ex. 5  **37.** 63; Sec. 7.6, Ex. 1  **39.** $f^{-1}(x) = x - 3$; Sec. 12.2, Ex. 4  **41. a.** 2  **b.** $-1$  **c.** $\dfrac{1}{2}$; Sec. 12.4, Ex. 3

# CHAPTER 13  CONIC SECTIONS

## Section 13.1
## Practice Exercises

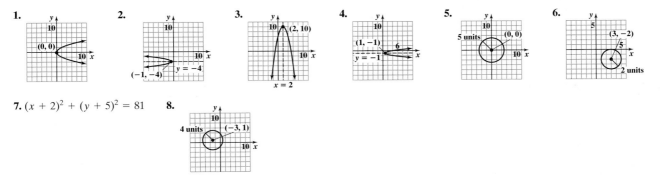

**1.** [graph, (0, 0)]  **2.** [graph, $(-1, -4)$, $y = -4$]  **3.** [graph, (2, 10), $x = 2$]  **4.** [graph, $(1, -1)$, 6, $y = -1$]  **5.** [graph, (0, 0), 5 units]  **6.** [graph, $(3, -2)$, 5, 2 units]

**7.** $(x + 2)^2 + (y + 5)^2 = 81$  **8.** [graph, $(-3, 1)$, 4 units]

## Graphing Calculator Explorations 13.1

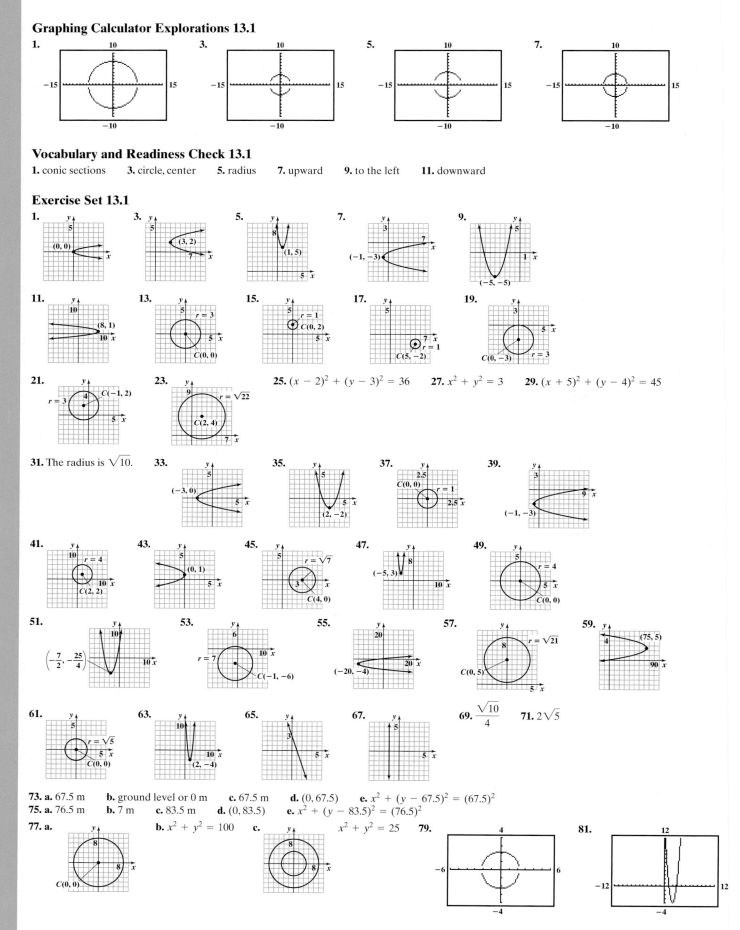

## Vocabulary and Readiness Check 13.1

**1.** conic sections **3.** circle, center **5.** radius **7.** upward **9.** to the left **11.** downward

## Exercise Set 13.1

**25.** $(x - 2)^2 + (y - 3)^2 = 36$ **27.** $x^2 + y^2 = 3$ **29.** $(x + 5)^2 + (y - 4)^2 = 45$

**31.** The radius is $\sqrt{10}$.

**69.** $\dfrac{\sqrt{10}}{4}$ **71.** $2\sqrt{5}$

**73. a.** 67.5 m **b.** ground level or 0 m **c.** 67.5 m **d.** $(0, 67.5)$ **e.** $x^2 + (y - 67.5)^2 = (67.5)^2$
**75. a.** 76.5 m **b.** 7 m **c.** 83.5 m **d.** $(0, 83.5)$ **e.** $x^2 + (y - 83.5)^2 = (76.5)^2$

**77. a.** **b.** $x^2 + y^2 = 100$ **c.** $x^2 + y^2 = 25$ **79.** **81.**

## Section 13.2
## Practice Exercises

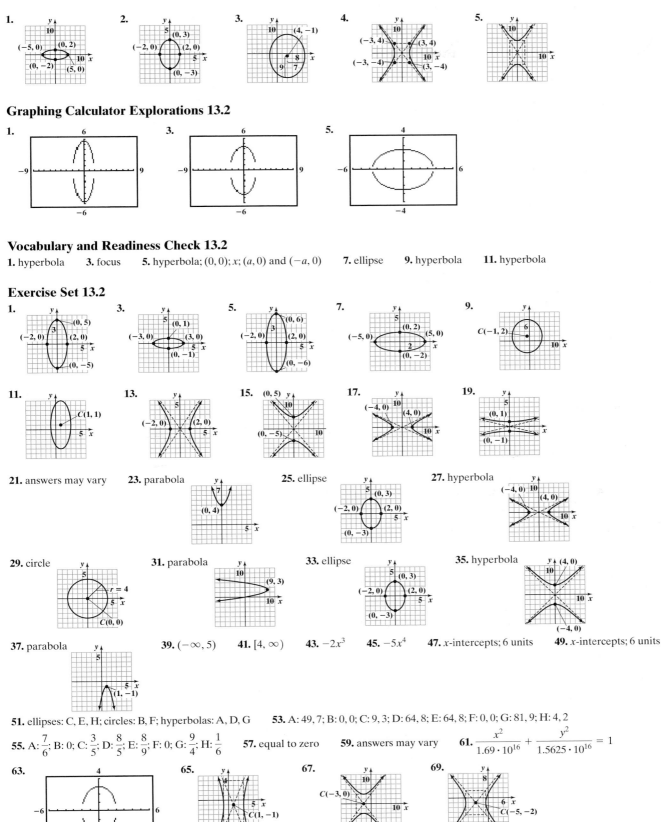

**1.**   **2.**   **3.**   **4.**   **5.**

## Graphing Calculator Explorations 13.2

**1.**   **3.**   **5.**

## Vocabulary and Readiness Check 13.2

**1.** hyperbola   **3.** focus   **5.** hyperbola; $(0,0)$; $x$; $(a, 0)$ and $(-a, 0)$   **7.** ellipse   **9.** hyperbola   **11.** hyperbola

## Exercise Set 13.2

**1.**   **3.**   **5.**   **7.**   **9.**

**11.**   **13.**   **15.**   **17.**   **19.**

**21.** answers may vary   **23.** parabola   **25.** ellipse   **27.** hyperbola

**29.** circle   **31.** parabola   **33.** ellipse   **35.** hyperbola

**37.** parabola   **39.** $(-\infty, 5)$   **41.** $[4, \infty)$   **43.** $-2x^3$   **45.** $-5x^4$   **47.** $x$-intercepts; 6 units   **49.** $x$-intercepts; 6 units

**51.** ellipses: C, E, H; circles: B, F; hyperbolas: A, D, G   **53.** A: 49, 7; B: 0, 0; C: 9, 3; D: 64, 8; E: 64, 8; F: 0, 0; G: 81, 9; H: 4, 2

**55.** A: $\frac{7}{6}$; B: 0; C: $\frac{3}{5}$; D: $\frac{8}{5}$; E: $\frac{8}{9}$; F: 0; G: $\frac{9}{4}$; H: $\frac{1}{6}$   **57.** equal to zero   **59.** answers may vary   **61.** $\dfrac{x^2}{1.69 \cdot 10^{16}} + \dfrac{y^2}{1.5625 \cdot 10^{16}} = 1$

**63.**   **65.**   **67.**   **69.**

## Integrated Review

**1.** circle

**2.** parabola

**3.** parabola

**4.** ellipse

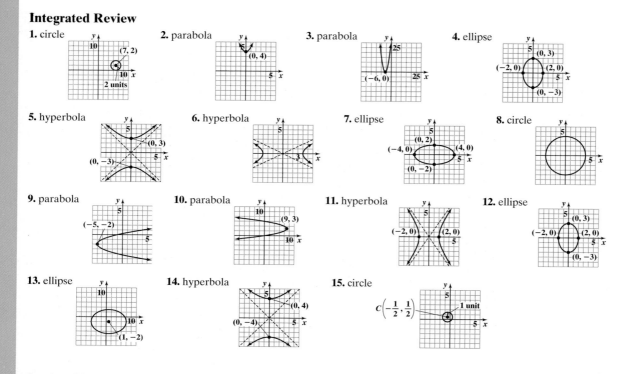

**5.** hyperbola  **6.** hyperbola  **7.** ellipse  **8.** circle

**9.** parabola  **10.** parabola  **11.** hyperbola  **12.** ellipse

**13.** ellipse  **14.** hyperbola  **15.** circle

## Section 13.3

### Practice Exercises

**1.** $(-4, 3)(0, -1)$  **2.** $(4, -2)$  **3.** $\varnothing$  **4.** $(2, \sqrt{3}); (2, -\sqrt{3}); (-2, \sqrt{3}); (-2, -\sqrt{3})$

### Exercise Set 13.3

**1.** $(3, -4), (-3, 4)$  **3.** $(\sqrt{2}, \sqrt{2}), (-\sqrt{2}, -\sqrt{2})$  **5.** $(4, 0), (0, -2)$  **7.** $(-\sqrt{5}, -2), (-\sqrt{5}, 2), (\sqrt{5}, -2), (\sqrt{5}, 2)$  **9.** $\varnothing$  **11.** $(1, -2), (3, 6)$
**13.** $(2, 4), (-5, 25)$  **15.** $\varnothing$  **17.** $(1, -3)$  **19.** $(-1, -2), (-1, 2), (1, -2), (1, 2)$  **21.** $(0, -1)$  **23.** $(-1, 3), (1, 3)$  **25.** $(\sqrt{3}, 0), (-\sqrt{3}, 0)$
**27.** $\varnothing$  **29.** $(-6, 0), (6, 0), (0, -6)$  **31.**  **33.**  **35.** $(8x - 25)$ in.  **37.** $(4x^2 + 6x + 2)$ m

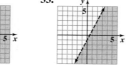

**39.** answers may vary  **41.** 0, 1, 2, 3, or 4; answers may vary  **43.** 9 and 7; 9 and $-7$; $-9$ and 7; $-9$ and $-7$  **45.** 15 cm by 19 cm
**47.** 15 thousand compact discs; price: \$3.75  **49.**  **51.**

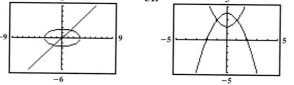

## Section 13.4

### Practice Exercises

**1.**  **2.**  **3.**  **4.**

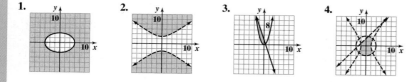

## Exercise Set 13.4

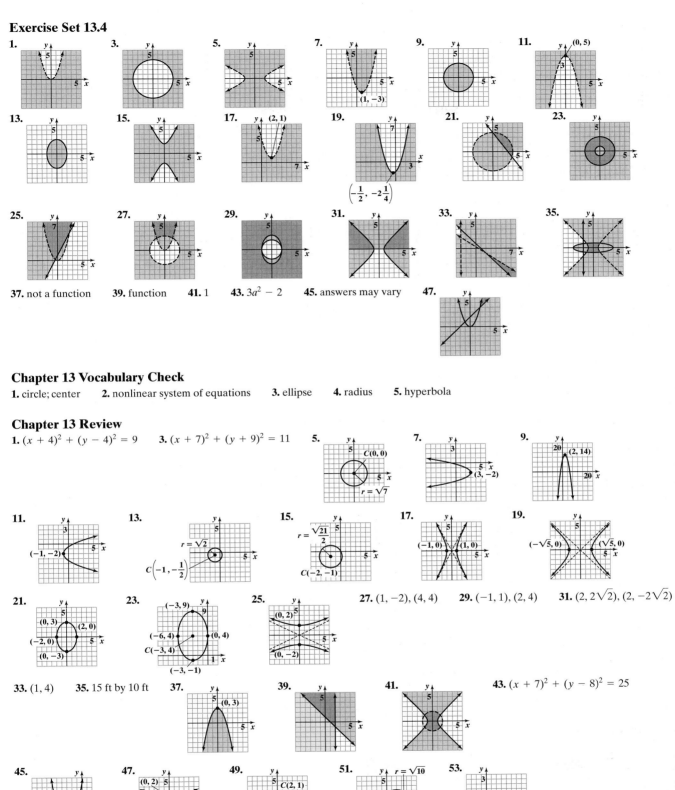

**37.** not a function    **39.** function    **41.** 1    **43.** $3a^2 - 2$    **45.** answers may vary

## Chapter 13 Vocabulary Check

**1.** circle; center    **2.** nonlinear system of equations    **3.** ellipse    **4.** radius    **5.** hyperbola

## Chapter 13 Review

**1.** $(x + 4)^2 + (y - 4)^2 = 9$    **3.** $(x + 7)^2 + (y + 9)^2 = 11$

**27.** $(1, -2), (4, 4)$    **29.** $(-1, 1), (2, 4)$    **31.** $(2, 2\sqrt{2}), (2, -2\sqrt{2})$

**33.** $(1, 4)$    **35.** 15 ft by 10 ft

**43.** $(x + 7)^2 + (y - 8)^2 = 25$

**55.** $(5, 1), (-1, 7)$

## Chapter 13 Test

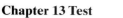

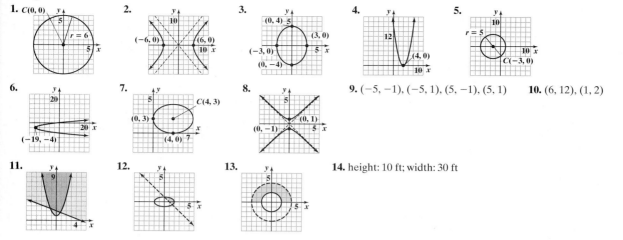

**9.** $(-5, -1), (-5, 1), (5, -1), (5, 1)$ **10.** $(6, 12), (1, 2)$

**14.** height: 10 ft; width: 30 ft

## Chapter 13 Cumulative Review

**1.** $\varnothing$; Sec. 9.1, Ex. 3 **3.** $\left(-\infty, \frac{13}{5}\right] \cup [4, \infty)$; Sec. 9.1, Ex. 7 **5.** $-2, \frac{4}{5}$; Sec. 9.2, Ex. 2 **7.** $24, -20$; Sec. 9.2, Ex. 3 **9.** $\frac{3}{4}, 5$; Sec. 9.2, Ex. 8

**11.** $(4, 8)$; Sec. 9.3, Ex. 2 **13.** $\frac{xy + 2x^3}{y - 1}$; Sec. 7.7, Ex. 3 **15.** $(-\infty, -\infty)$; Sec. 9.3, Ex. 6 **17.** 16; Sec. 5.7, Ex. 4 **19. a.** 1 **b.** $-4$ **c.** $\frac{2}{5}$ **d.** $x^2$

**e.** $-3x^3$; Sec. 10.1, Ex. 3 **21. a.** $z - z^{17/3}$ **b.** $x^{2/3} - 3x^{1/3} - 10$; Sec. 10.2, Ex. 5 **23. a.** 2 **b.** $\frac{5}{2}\sqrt{x}$ **c.** $14xy^2\sqrt[3]{x}$ **d.** $4a^2b\sqrt[4]{2a}$;

Sec. 10.3, Ex. 5 **25. a.** $\frac{5\sqrt{5}}{12}$ **b.** $\frac{5\sqrt[3]{7x}}{2}$; Sec. 10.4, Ex. 3 **27.** $\frac{\sqrt{21xy}}{3y}$; Sec. 10.5, Ex. 2 **29.** 42; Sec. 10.6, Ex. 1 **31. a.** $-i$ **b.** 1 **c.** $-1$

**d.** 1; Sec. 10.7, Ex. 6 **33.** $-1 + \sqrt{5}, -1 - \sqrt{5}$; Sec. 11.1, Ex. 5 **35.** $2 + \sqrt{2}, 2 - \sqrt{2}$; Sec. 11.2, Ex. 3 **37.** $2, -2, i, -i$; Sec. 11.3, Ex. 3

**39.** $[-2, 3)$; Sec. 11.4, Ex. 4 **41.** ; Sec. 11.5, Ex. 8 **43.** $(2, -16)$; Sec. 11.6, Ex. 4 **45.** $\sqrt{2} \approx 1.414$; Sec. 10.3, Ex. 6

# CHAPTER 14 SEQUENCES, SERIES, AND THE BINOMIAL THEOREM

## Section 14.1

### Practice Exercises

**1.** 6, 9, 14, 21, 30 **2. a.** $-\frac{1}{5}$ **b.** $\frac{1}{20}$ **c.** $\frac{1}{150}$ **d.** $-\frac{1}{95}$ **3. a.** $a_n = (2n - 1)$ **b.** $a_n = 3^n$ **c.** $a_n = \frac{n}{n + 1}$ **d.** $a_n = -\frac{1}{n + 1}$ **4.** $2022.40

### Vocabulary and Readiness Check 14.1

**1.** general **3.** infinite **5.** $-1$

### Exercise Set 14.1

**1.** 5, 6, 7, 8, 9 **3.** $-1, 1, -1, 1, -1$ **5.** $\frac{1}{4}, \frac{1}{5}, \frac{1}{6}, \frac{1}{7}, \frac{1}{8}$ **7.** 2, 4, 6, 8, 10 **9.** $-1, -4, -9, -16, -25$ **11.** 2, 4, 8, 16, 32 **13.** 7, 9, 11, 13, 15

**15.** $-1, 4, -9, 16, -25$ **17.** 75 **19.** 118 **21.** $\frac{6}{5}$ **23.** 729 **25.** $\frac{4}{7}$ **27.** $\frac{1}{8}$ **29.** $-95$ **31.** $-\frac{1}{25}$ **33.** $a_n = 4n - 1$ **35.** $a_n = -2^n$

**37.** $a_n = \frac{1}{3^n}$ **39.** 48 ft, 80 ft, and 112 ft **41.** $a_n = 0.10(2)^{n-1}$; $819.20 **43.** 2400 cases; 75 cases **45.** 50 sparrows in 2004: extinct in 2010

**47.** **49.** **51.** $\sqrt{13}$ units **53.** $\sqrt{41}$ units **55.** 1, 0.7071, 0.5774, 0.5, 0.4472

**57.** 2, 2.25, 2.3704, 2.4414, 2.4883

## Section 14.2

### Practice Exercises

**1.** 4, 9, 14, 19, 24 **2. a.** $a_n = 5 - 3n$ **b.** $-31$ **3.** 51 **4.** 47 **5.** $a_n = 54{,}800 + 2200n$; $61{,}400 **6.** 8, $-24$, 72, $-216$ **7.** $\frac{1}{64}$ **8.** $-192$

**9.** $a_1 = 3$; $r = \frac{3}{2}$ **10.** 75 units

## Vocabulary and Readiness Check 14.2

**1.** geometric; ratio     **3.** first; difference

## Exercise Set 14.2

**1.** 4, 6, 8, 10, 12     **3.** 6, 4, 2, 0, −2     **5.** 1, 3, 9, 27, 81     **7.** 48, 24, 12, 6, 3     **9.** 33     **11.** −875     **13.** −60     **15.** 96     **17.** −28     **19.** 1250

**21.** 31     **23.** 20     **25.** $a_1 = \frac{2}{3}; r = -2$     **27.** answers may vary     **29.** $a_1 = 2; d = 2$     **31.** $a_1 = 5; r = 2$     **33.** $a_1 = \frac{1}{2}; r = \frac{1}{5}$     **35.** $a_1 = x; r = 5$

**37.** $a_1 = p; d = 4$     **39.** 19     **41.** $-\frac{8}{9}$     **43.** $\frac{17}{2}$     **45.** $\frac{8}{81}$     **47.** −19     **49.** $a_n = 4n + 50$; 130 seats     **51.** $a_n = 6(3)^{n-1}$

**53.** 486, 162, 54, 18, 6; $a_n = \frac{486}{3^{n-1}}$; 6 bounces     **55.** $a_n = 4000 + 125(n - 1)$ or $a_n = 3875 + 125n$; $5375     **57.** 25 g     **59.** $\frac{11}{18}$     **61.** 40     **63.** $\frac{907}{495}$

**65.** $11,782.40, $5891.20, $2945.60, $1472.80     **67.** 19.652, 19.618, 19.584, 19.55     **69.** answers may vary

## Section 14.3

### Practice Exercises

**1. a.** $-\frac{5}{4}$     **b.** 360     **2. a.** $\sum_{i=1}^{6} 5i$     **b.** $\sum_{i=1}^{4} \left(\frac{1}{5}\right)^i$     **3.** $\frac{655}{72}$     **4.** 95 plants

## Vocabulary and Readiness Check 14.3

**1.** infinite     **3.** summation     **5.** partial sum

## Exercise Set 14.3

**1.** −2     **3.** 60     **5.** 20     **7.** $\frac{73}{168}$     **9.** $\frac{11}{36}$     **11.** 60     **13.** 74     **15.** 62     **17.** $\frac{241}{35}$     **19.** $\sum_{i=1}^{5}(2i - 1)$     **21.** $\sum_{i=1}^{4} 4(3)^{i-1}$

**23.** $\sum_{i=1}^{6}(-3i + 15)$     **25.** $\sum_{i=1}^{4} \frac{4}{3^{i-2}}$     **27.** $\sum_{i=1}^{7} i^2$     **29.** −24     **31.** 0     **33.** 82     **35.** −20     **37.** −2     **39.** 1, 2, 3, . . . , 10; 55 trees

**41.** $a_n = 6(2)^{n-1}$; 96 units     **43.** $a_n = 50(2)^n$; $n$ represents the number of 12-hr periods; 800 bacteria     **45.** 30 opossums; 68 opossums

**47.** 6.25 lb; 93.75 lb     **49.** 16.4 in.; 134.5 in.     **51.** 10     **53.** $\frac{10}{27}$     **55.** 45     **57.** 90     **59. a.** 2 + 6 + 12 + 20 + 30 + 42 + 56

**b.** 1 + 2 + 3 + 4 + 5 + 6 + 7 + 1 + 4 + 9 + 16 + 25 + 36 + 49     **c.** answers may vary     **d.** true; answers may vary

### Integrated Review

**1.** −2, −1, 0, 1, 2     **2.** $\frac{7}{2}, \frac{7}{3}, \frac{7}{4}, \frac{7}{5}, \frac{7}{6}$     **3.** 1, 3, 9, 27, 81     **4.** −4, −1, 4, 11, 20     **5.** 64     **6.** −14     **7.** $\frac{1}{40}$     **8.** $-\frac{1}{82}$     **9.** 7, 4, 1, −2, −5

**10.** −3, −15, −75, −375, −1875     **11.** 45, 15, 5, $\frac{5}{3}, \frac{5}{9}$     **12.** −12, −2, 8, 18, 28     **13.** 101     **14.** $\frac{243}{16}$     **15.** 384     **16.** 185     **17.** −10     **18.** $\frac{1}{5}$

**19.** 50     **20.** 98     **21.** $\frac{31}{2}$     **22.** $\frac{61}{20}$     **23.** −10     **24.** 5

## Section 14.4

### Practice Exercises

**1.** 80     **2.** 1275     **3.** 105 blocks of ice     **4.** $42\frac{5}{8}$     **5.** $987,856     **6.** $9\frac{1}{3}$     **7.** 900 in.

## Vocabulary and Readiness Check 14.4

**1.** arithmetic     **3.** geometric     **5.** arithmetic

## Exercise Set 14.4

**1.** 36     **3.** 484     **5.** 63     **7.** 2.496     **9.** 55     **11.** 16     **13.** 24     **15.** $\frac{1}{9}$     **17.** −20     **19.** $\frac{16}{9}$     **21.** $\frac{4}{9}$     **23.** 185     **25.** $\frac{381}{64}$

**27.** $-\frac{33}{4}$, or −8.25     **29.** $-\frac{75}{2}$     **31.** $\frac{56}{9}$     **33.** 4000, 3950, 3900, 3850, 3800; 3450 cars; 44,700 cars     **35.** Firm $A$ (Firm $A$, $265,000; Firm $B$, $254,000)

**37.** $39,930; $139,230     **39.** 20 min; 123 min     **41.** 180 ft     **43.** Player $A$, 45 points; Player $B$, 75 points     **45.** $3050     **47.** $10,737,418.23

**49.** 720     **51.** 3     **53.** $x^2 + 10x + 25$     **55.** $8x^3 - 12x^2 + 6x - 1$     **57.** $\frac{8}{10} + \frac{8}{100} + \frac{8}{1000} + \cdots; \frac{8}{9}$     **59.** answers may vary

## Section 14.5

### Practice Exercises

**1.** $p^7 + 7p^6r + 21p^5r^2 + 35p^4r^3 + 35p^3r^4 + 21p^2r^5 + 7pr^6 + r^7$     **2. a.** $\frac{1}{7}$     **b.** 840     **c.** 5     **d.** 1

**3.** $a^9 + 9a^8b + 36a^7b^2 + 84a^6b^3 + 126a^5b^4 + 126a^4b^5 + 84a^3b^6 + 36a^2b^7 + 9ab^8 + b^9$     **4.** $a^3 + 15a^2b + 75ab^2 + 125b^3$

**5.** $27x^3 - 54x^2y + 36xy^2 - 8y^3$     **6.** $1,892,352x^5y^6$

## Vocabulary and Readiness Check 14.5

**1.** 1     **3.** 24     **5.** 6

## Exercise Set 14.5

**1.** $m^3 + 3m^2n + 3mn^2 + n^3$  **3.** $c^5 + 5c^4d + 10c^3d^2 + 10c^2d^3 + 5cd^4 + d^5$  **5.** $y^5 - 5y^4x + 10y^3x^2 - 10y^2x^3 + 5yx^4 - x^5$
**7.** answers may vary  **9.** 8  **11.** 42  **13.** 360  **15.** 56  **17.** $a^7 + 7a^6b + 21a^5b^2 + 35a^4b^3 + 35a^3b^4 + 21a^2b^5 + 7ab^6 + b^7$
**19.** $a^5 + 10a^4b^2 + 40a^3b^2 + 80a^2b^3 + 80ab^4 + 32b^5$  **21.** $q^9 + 9q^8r + 36q^7r^2 + 84q^6r^3 + 126q^5r^4 + 126q^4r^5 + 84q^3r^6 + 36q^2r^7 + 9qr^8 + r^9$
**23.** $1024a^5 + 1280a^4b + 640a^3b^2 + 160a^2b^3 + 20ab^4 + b^5$  **25.** $625a^4 - 1000a^3b + 600a^2b^2 - 160ab^3 + 16b^4$  **27.** $8a^3 + 36a^2b + 54ab^2 + 27b^3$
**29.** $x^5 + 10x^4 + 40x^3 + 80x^2 + 80x + 32$  **31.** $5cd^4$  **33.** $d^7$  **35.** $-40r^2s^3$  **37.** $6x^2y^2$  **39.** $30a^9b$

**41.**  **43.**  **45.**  **47.** $x^2\sqrt{x} + 5\sqrt{3}x^2 + 30x\sqrt{x} + 30\sqrt{3}x + 45\sqrt{x} + 9\sqrt{3}$
**49.** 126  **51.** 28  **53.** answers may vary

## Chapter 14 Vocabulary Check

**1.** finite sequence  **2.** factorial of $n$  **3.** infinite sequence  **4.** geometric sequence; common ratio  **5.** series  **6.** general term
**7.** arithmetic sequence; common difference  **8.** Pascal's triangle

## Chapter 14 Review

**1.** $-3, -12, -27, -48, -75$  **3.** $\dfrac{1}{100}$  **5.** $a_n = \dfrac{1}{6n}$  **7.** 144 ft, 176 ft, 208 ft  **9.** 660,000; 1,320,000; 2,640,000; 5,280,000; 10,560,000; 2010:
10,560,000 infested acres  **11.** $-2, -\dfrac{4}{3}, -\dfrac{8}{9}, -\dfrac{16}{27}, -\dfrac{32}{81}$  **13.** 111  **15.** $-83$  **17.** $a_1 = 3; d = 5$  **19.** $a_n = \dfrac{3}{10^n}$  **21.** $a_1 = \dfrac{8}{3}, r = \dfrac{3}{2}$
**23.** $a_1 = 7x, r = -2$  **25.** $8, 6, 4.5, 3.4, 2.5, 1.9$; good  **27.** $a_n = 2^{n-1}$, $\$512, \$536,870,912$  **29.** $a_n = 900 + (n-1)150$ or $a_n = 150n + 750$;
$\$1650$/month  **31.** $1 + 3 + 5 + 7 + 9 = 25$  **33.** $\dfrac{1}{4} - \dfrac{1}{6} + \dfrac{1}{8} = \dfrac{5}{24}$  **35.** $-4$  **37.** $-10$  **39.** $\sum_{i=1}^{6} 3^{i-1}$  **41.** $\sum_{i=1}^{4} \dfrac{1}{4^i}$  **43.** $a_n = 20(2)^n$;
$n$ represents the number of 8-hour periods; 1280 yeast  **45.** Job $A$, $\$48,300$; Job $B$, $\$46,600$  **47.** 150  **49.** 900  **51.** $-410$  **53.** 936  **55.** 10
**57.** $-25$  **59.** $\$30,418; \$99,868$  **61.** $\$58; \$553$  **63.** 2696 mosquitoes  **65.** $\dfrac{5}{9}$  **67.** $x^5 + 5x^4z + 10x^3z^2 + 10x^2z^3 + 5xz^4 + z^5$
**69.** $16x^4 + 32x^3y + 24x^2y^2 + 8xy^3 + y^4$  **71.** $b^8 + 8b^7c + 28b^6c^2 + 56b^5c^3 + 70b^4c^4 + 56b^3c^5 + 28b^2c^6 + 8bc^7 + c^8$
**73.** $256m^4 - 256m^3n + 96m^2n^2 - 16mn^3 + n^4$  **75.** $35a^4b^3$  **77.** 130  **79.** 40.5

## Chapter 14 Test

**1.** $-\dfrac{1}{5}, \dfrac{1}{6}, -\dfrac{1}{7}, \dfrac{1}{8}, -\dfrac{1}{9}$  **2.** 247  **3.** $a_n = \dfrac{2}{5}\left(\dfrac{1}{5}\right)^{n-1}$  **4.** $a_n = (-1)^n 9n$  **5.** 155  **6.** $-330$  **7.** $\dfrac{144}{5}$  **8.** 1  **9.** 10  **10.** $-60$
**11.** $a^6 - 6a^5b + 15a^4b^2 - 20a^3b^3 + 15a^2b^4 - 6ab^5 + b^6$  **12.** $32x^5 + 80x^4y + 80x^3y^2 + 40x^2y^3 + 10xy^4 + y^5$  **13.** 925 people; 250 people initially
**14.** $1 + 3 + 5 + 7 + 9 + 11 + 13 + 15$; 64 shrubs  **15.** 33.75 cm, 218.75 cm  **16.** 320 cm  **17.** 304 ft; 1600 ft  **18.** $\dfrac{14}{33}$

## Chapter 14 Cumulative Review

**1. a.** $-8$  **b.** $-8$  **c.** 9  **d.** $-9$; Sec. 1.7, Ex. 4  **3.** $-2x - 1$; Sec. 2.1, Ex. 7  **5.** $y = \dfrac{1}{4}x - 3$; Sec. 3.5, Ex. 1  **7.** $-x + 5y = 23$ or
$x - 5y = -23$; Sec. 3.5, Ex. 5  **9.** $x^3 - 4x^2 - 3x + 11 + \dfrac{12}{x+2}$; Sec. 5.7, Ex. 2  **11. a.** $5\sqrt{2}$  **b.** $2\sqrt[3]{3}$  **c.** $\sqrt{26}$  **d.** $2\sqrt[4]{2}$; Sec. 10.3, Ex. 3
**13.** 10%; Sec. 11.1, Ex. 9  **15.** 2, 7; Sec. 11.3, Ex. 4  **17.** $\left(-\dfrac{7}{2}, -1\right)$; Sec. 11.4, Ex. 5  **19.** $\dfrac{25}{4}$ ft; $\dfrac{5}{8}$ sec; Sec. 11.6, Ex. 5  **21. a.** 25; 7
**b.** $x^2 + 6x + 9; x^2 + 3$; Sec. 12.1, Ex. 2  **23.** $f^{-1} = \{(1, 0), (7, -2), (-6, 3), (4, 4)\}$; Sec. 12.2, Ex. 3  **25. a.** 4  **b.** $\dfrac{3}{2}$  **c.** 6; Sec. 12.3, Ex. 4
**27. a.** 2  **b.** $-1$  **c.** 3  **d.** 6; Sec. 12.4, Ex. 5  **29. a.** $\log_{11}30$  **b.** $\log_3 6$  **c.** $\log_2(x^2 + 2x)$; Sec. 12.5, Ex. 1  **31.** $\$2509.30$; Sec. 12.6, Ex. 8
**33.** $\dfrac{\log 7}{\log 3} \approx 1.7712$; Sec. 12.7, Ex. 1  **35.** 18; Sec. 12.7, Ex. 2  **37.** ; Sec. 13.2, Ex. 4  **39.** $(2, \sqrt{2})$; Sec. 13.3, Ex. 2
**41.** ; Sec. 13.4, Ex. 1

**43.** 0, 3, 8, 15, 24; Sec. 14.1, Ex. 1  **45.** 72; Sec. 14.2, Ex. 3  **47. a.** $\dfrac{7}{2}$  **b.** 56; Sec. 14.3, Ex. 1  **49.** 465; Sec. 14.4, Ex. 2

# APPENDIX A  THE BIGGER PICTURE/PRACTICE FINAL EXAM

**1.** $-48$    **2.** $-81$    **3.** $\dfrac{1}{64}$    **4.** $-\dfrac{1}{3}$    **5.** $-3x^3 + 5x^2 + 4x + 5$    **6.** $16x^2 - 16x + 4$    **7.** $3x^3 + 22x^2 + 41x + 14$    **8.** $(y - 12)(y + 4)$

**9.** $3x(3x + 1)(x + 4)$    **10.** $5(6 + x)(6 - x)$    **11.** $(3a - 7)(a + b)$    **12.** $8(y - 2)(y^2 + 2y + 4)$    **13.** $\dfrac{y^{14}}{x^2}$    **14.** $\dfrac{25}{7}$    **15.** $(-\infty, -2]$

**16.** $-7, 1$    **17.**    **18.**    **19.** $m = -1$    **20.** $m = 3$    **21.** $8x + y = 11$    **22.** $x = -5$    **23.** $\left(\dfrac{1}{2}, -2\right)$

**24.** no solution    **25.** $9x^2 - 6x + 4 - \dfrac{16}{3x + 2}$    **26. a.** $0$    **b.** $0$    **c.** $60$    **27.** domain: $(-\infty, \infty)$; range: $(-\infty, 4]$    **28.** $401, 802$

**29.** $2\dfrac{1}{2}$ hr    **30.** $120$ cc    **31.** $\{x \mid x \text{ is a real number}, x \neq -1, x \neq -3\}$    **32.** $\dfrac{19x - 6}{2x + 5}$    **33.** $\dfrac{2(x + 5)}{x(y + 5)}$    **34.** $\dfrac{3a - 4}{(a - 3)(a + 2)}$    **35.** $\dfrac{5y^2 - 1}{y + 2}$

**36.** $\dfrac{30}{11}$    **37.** $-6$    **38.** no solution    **39.** $5$ or $1$    **40.** $6\sqrt{6}$    **41.** $5$    **42.** $\dfrac{8a^{1/3}c^{2/3}}{b^{5/12}}$    **43.** $-x\sqrt{5x}$    **44.** $-20$    **45.** $1, \dfrac{2}{3}$    **46.** $\left(\dfrac{3}{2}, 5\right]$

**47.** $(-\infty, -2) \cup \left(\dfrac{4}{3}, \infty\right)$    **48.** $\dfrac{3 \pm \sqrt{29}}{2}$    **49.** $2, 3$    **50.** $\left(-\infty, -\dfrac{3}{2}\right) \cup (5, \infty)$    **51.**    **52.**    domain: $(-\infty, \infty)$; range: $(-\infty, -1]$

**53.**    **54.**    domain: $(-\infty, \infty)$; range: $(-3, \infty)$    **55.** $f(x) = -\dfrac{1}{2}x$    **56.** $f(x) = -\dfrac{1}{3}x + \dfrac{5}{3}$    **57.** $2\sqrt{26}$ units

**58.** $\left(-4, \dfrac{7}{2}\right)$    **59.** $\dfrac{3\sqrt{y}}{y}$    **60.** $\dfrac{8 - 6\sqrt{x} + x}{8 - 2x}$    **61.** $16$    **62.** $7$ ft    **63. a.** $272$ ft    **b.** $5.12$ sec    **64.** $-2i\sqrt{2}$    **65.** $-3i$    **66.** $7 + 24i$

**67.** $-\dfrac{3}{2} + \dfrac{5}{2}i$    **68.** $(g \circ h)(x) = x^2 - 6x - 2$    **69.** $f^{-1}(x) = \dfrac{-x + 6}{2}$    **70.** $\log_5 \dfrac{x^4}{x + 1}$    **71.** $-1$    **72.** $\dfrac{1}{2}\left(\dfrac{\log 4}{\log 3} - 5\right); -1.8691$    **73.** $22$

**74.** $\dfrac{43}{21}$    **75.** $\dfrac{1}{2}$    **76.**    **77.** $64{,}913$ prairie dogs    **78.**    **79.**    **80.**

**81.** $(-5, -1)(-5, 1), (5, -1), (5, 1)$    **82.** $-\dfrac{1}{5}, \dfrac{1}{6}, -\dfrac{1}{7}, \dfrac{1}{8}, -\dfrac{1}{9}$    **83.** $155$    **84.** $1$    **85.** $10$    **86.** $32x^5 + 80x^4y + 80x^3y^2 + 40x^2y^3 + 10xy^4 + y^5$

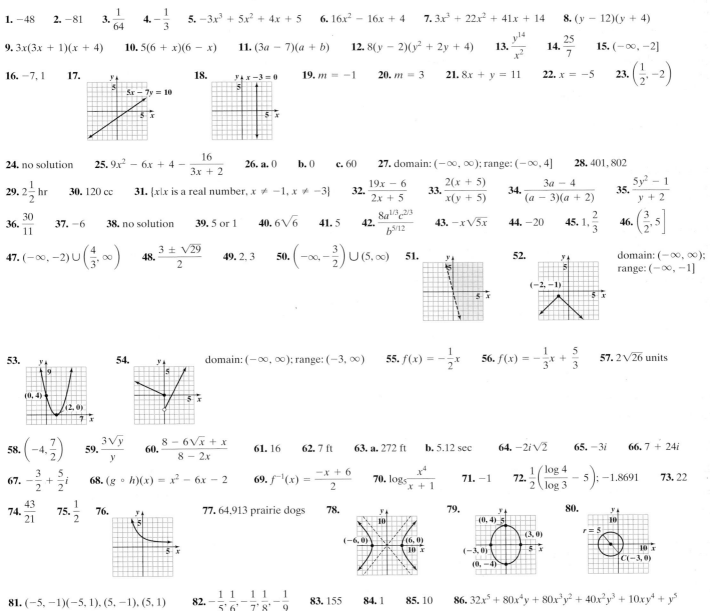

# APPENDIX B  OPERATIONS ON DECIMALS

**1.** $17.08$    **3.** $12.804$    **5.** $110.96$    **7.** $2.4$    **9.** $28.43$    **11.** $227.5$    **13.** $2.7$    **15.** $49.2339$    **17.** $80$    **19.** $0.07612$    **21.** $4.56$    **23.** $648.46$
**25.** $767.83$    **27.** $12.062$    **29.** $61.48$    **31.** $7.7$    **33.** $863.37$    **35.** $22.579$    **37.** $363.15$    **39.** $7.007$

# APPENDIX C  REVIEW OF ALGEBRA TOPICS

## C.1  Practice Exercises

**1.** $-4$    **2.** $\dfrac{5}{6}$    **3.** $-\dfrac{3}{4}$    **4.** $\dfrac{5}{4}$    **5.** $-\dfrac{1}{8}, 2$

## C.1  Exercise Set

**1.** $-3, -8$    **3.** $-2$    **5.** $\dfrac{1}{4}, -\dfrac{2}{3}$    **7.** $1, 9$    **9.** $0$    **11.** $5$    **13.** $\dfrac{3}{5}, -1$    **15.** no solution    **17.** $\dfrac{1}{8}$    **19.** $0$

**21.** $6, -3$    **23.** $0$    **25.** $\dfrac{2}{5}, -\dfrac{1}{2}$    **27.** $\dfrac{3}{4}, -\dfrac{1}{2}$    **29.** $29$    **31.** $-8$    **33.** $-\dfrac{1}{3}, 0$    **35.** $-\dfrac{7}{8}$    **37.** $\dfrac{31}{4}$    **39.** $1$    **41.** $-7, 4$    **43.** $4, 6$

**45.** $-\dfrac{1}{2}$   **47. a.** incorrect   **b.** correct   **c.** correct   **d.** incorrect   **49.** $K = -11$   **51.** $K = 24$   **53.** $-4.86$   **55.** $1.53$   **57.** $\dfrac{8}{x}$

**59.** $8x$   **61.** $3x + 2$

## Appendix C.2 Problem Solving
### C.2 Practice Exercises
**1. a.** $3x + 6$   **b.** $6x - 1$   **2.** $3x + 17.3$   **3.** $14, 34, 70$   **4.** \$450

### C.2 Vocabulary and Readiness Check
**1.** $>$   **3.** $=$   **5.** $31, 32, 33, 34$   **7.** $18, 20, 22$   **9.** $y, y + 1, y + 2$   **11.** $p, p + 1, p + 2, p + 3$

### C.2 Exercise Set
**1.** $4y$   **3.** $3z + 3$   **5.** $(65x + 30)$ cents   **7.** $10x + 3$   **9.** $2x + 14$   **11.** $-5$   **13.** $45, 145, 225$   **15.** approximately 1612.41 million acres
**17.** 2344 earthquakes   **19.** 1275 shoppers   **21.** 22%   **23.** 417 employees   **25.** 29.98 million   **27.** $29°, 35°, 116°$   **29.** 28 m, 36 m, 38 m
**31.** 18 in., 18 in., 27 in., 36 in.   **33.** $75, 76, 77$   **35.** Fallon's zip code is 89406; Fernley's zip code is 89408; Gardnerville Ranchos' zip code is 89410
**37.** 317 thousand; 279 thousand; 184 thousand   **39.** medical assistant: 215 thousand; postsecondary teacher jobs: 603 thousand; registered nurses: 623
thousand   **41.** 757-200: 190 seats; 737-200: 113 seats; 737-300: 134 seats   **43.** \$430.00   **45.** 41.7 million   **47.** $40°, 140°$   **49.** $64°, 32°, 84°$
**51.** square: 18 cm; triangle: 24 cm   **53.** $76, 78, 80$   **55.** 40.5 ft; 202.5 ft; 240 ft   **57.** Los Angeles: 61.0 million, Atlanta: 74.3 million, Chicago: 62.3 million
**59.** incandescent: 1500 bulb hours; fluorescent: 100,000 bulb hours; halogen: 4000 bulb hours   **61.** height: 48 in.; length: 108 in.

### C.3 Practice Exercises
**1.**

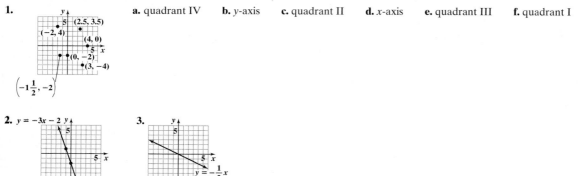

**a.** quadrant IV   **b.** $y$-axis   **c.** quadrant II   **d.** $x$-axis   **e.** quadrant III   **f.** quadrant I

**2.** $y = -3x - 2$   **3.**

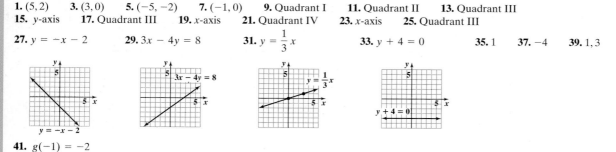

### C.3 Exercise Set
**1.** $(5, 2)$   **3.** $(3, 0)$   **5.** $(-5, -2)$   **7.** $(-1, 0)$   **9.** Quadrant I   **11.** Quadrant II   **13.** Quadrant III
**15.** $y$-axis   **17.** Quadrant III   **19.** $x$-axis   **21.** Quadrant IV   **23.** $x$-axis   **25.** Quadrant III
**27.** $y = -x - 2$   **29.** $3x - 4y = 8$   **31.** $y = \dfrac{1}{3}x$   **33.** $y + 4 = 0$   **35.** 1   **37.** $-4$   **39.** $1, 3$

**41.** $g(-1) = -2$

### C.4 Practice Exercises
**1. a.** $3xy(4x - 1)$   **b.** $(7x + 2)(7x - 2)$   **c.** $(5x - 3)(x + 1)$   **d.** $(3 + x)(x^2 + 2)$   **e.** $(2x + 5)^2$   **f.** cannot be factored
**2. a.** $(4x + y)(16x^2 - 4xy + y^2)$   **b.** $7y^2(x - 3y)(x + 3y)$   **c.** $3(x + 2 + b)(x + 2 - b)$   **d.** $x^2y(xy + 3)(x^2y^2 - 3xy + 9)$
**e.** $(x + 7 + 9y)(x + 7 - 9y)$

## C.4 Exercise Set

**1.** $2y^2 + 2y - 11$    **3.** $x^2 - 7x + 7$    **5.** $25x^2 - 30x + 9$    **7.** $2x^3 - 4x^2 + 5x - 5 + \dfrac{8}{x+2}$    **9.** $(x - 4 + y)(x - 4 - y)$

**11.** $x(x - 1)(x^2 + x + 1)$    **13.** $2xy(7x - 1)$    **15.** $4(x + 2)(x - 2)$    **17.** $(3x - 11)(x + 1)$    **19.** $4(x + 3)(x - 1)$

**21.** $(2x + 9)^2$    **23.** $(2x + 5y)(4x^2 - 10xy + 25y^2)$    **25.** $8x^2(2y - 1)(4y^2 + 2y + 1)$    **27.** $(x + 5 + y)(x^2 + 10x - xy - 5y + y^2 + 25)$

**29.** $(5a - 6)^2$    **31.** $7x(x - 9)$    **33.** $(a + 7)(b - 6)$    **35.** $(x^2 + 1)(x - 1)(x + 1)$    **37.** $(5x - 11)(2x + 3)$    **39.** $5a^3b(b^2 - 10)$

**41.** prime    **43.** $10x(x - 10)(x - 11)$    **45.** $a^3b(4b - 3)(16b^2 + 12b + 9)$    **47.** $2(x - 3)(x^2 + 3x + 9)$    **49.** $(y^4 + 2)(3y - 5)$

**51.** $100(z + 1)(z^2 - z + 1)$    **53.** $(2b - 9)^2$    **55.** $(y - 4)(y - 5)$    **57.** $A = 9 - 4x^2 = (3 + 2x)(3 - 2x)$

## C.5 Practice Exercises

**1. a.** $\dfrac{2n + 1}{n(n - 1)}$    **b.** $-x^2$    **2. a.** $\dfrac{-y^3}{21(y + 3)}$    **b.** $\dfrac{7x + 2}{x + 2}$    **3. a.** $\dfrac{20p + 3}{5p^4q}$    **b.** $\dfrac{5y^2 + 19y - 12}{(y + 3)(y - 3)}$    **c.** 3    **4.** 1

## C.5 Exercise Set

**1.** $\dfrac{1}{2}$    **3.** $\dfrac{1 + 2x}{8}$    **5.** $\dfrac{2(x - 4)}{(x + 2)(x - 1)}$    **7.** 4    **9.** $-5$    **11.** $\dfrac{2x + 5}{x(x - 3)}$    **13.** $-2$    **15.** $\dfrac{(a + 3)(a + 1)}{a + 2}$    **17.** $-\dfrac{1}{5}$

**19.** $\dfrac{4a + 1}{(3a + 1)(3a - 1)}$    **21.** $-1, \dfrac{3}{2}$    **23.** $\dfrac{3}{x + 1}$    **25.** $-1$    **27. a.** $\dfrac{x}{5} - \dfrac{x}{4} + \dfrac{1}{10}$    **b.** Write each rational expression term so that the denominator

is the LCD, 20.    **c.** $\dfrac{-x + 2}{20}$    **29.** b    **31.** d    **33.** d

# APPENDIX D  AN INTRODUCTION TO USING A GRAPHING UTILITY

## Viewing Window and Interpreting Window Settings Exercise Set

**1.** yes    **3.** no    **5.** answers may vary    **7.** answers may vary    **9.** answers may vary

**11.** Xmin $= -12$   Ymin $= -12$    **13.** Xmin $= -9$   Ymin $= -12$    **15.** Xmin $= -10$   Ymin $= -25$    **17.** Xmin $= -10$   Ymin $= -30$
     Xmax $= 12$   Ymax $= 12$       Xmax $= 9$   Ymax $= 12$       Xmax $= 10$   Ymax $= 25$       Xmax $= 10$   Ymax $= 30$
     Xscl $= 3$   Yscl $= 3$        Xscl $= 1$   Yscl $= 2$        Xscl $= 2$   Yscl $= 5$         Xscl $= 1$   Yscl $= 3$

**19.** Xmin $= -20$   Ymin $= -30$
     Xmax $= 30$   Ymax $= 50$
     Xscl $= 5$   Yscl $= 10$

## Graphing Equations and Square Viewing Window Exercise Set

**1.** Setting B    **3.** Setting B    **5.** Setting B

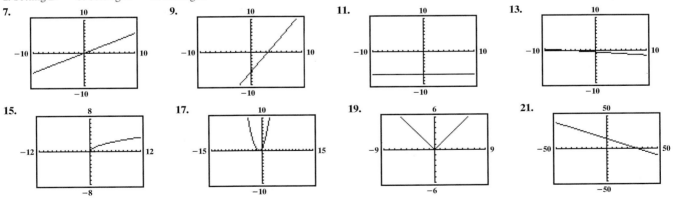

**7.**    **9.**    **11.**    **13.**

**15.**    **17.**    **19.**    **21.**

# APPENDIX E  SOLVING SYSTEMS OF EQUATIONS BY MATRICES

## Practice Exercises

**1.** $(2, -1)$    **2.** $\varnothing$    **3.** $(-1, 1, 2)$

## Exercise Set

**1.** $(2, -1)$    **3.** $(-4, 2)$    **5.** $\varnothing$    **7.** $\{(x, y) \mid 3x - 3y = 9\}$    **9.** $(-2, 5, -2)$    **11.** $(1, -2, 3)$    **13.** $(4, -3)$    **15.** $(2, 1, -1)$    **17.** $(9, 9)$

**19.** $\varnothing$    **21.** $\varnothing$    **23.** $(1, -4, 3)$    **25.** c

# APPENDIX F SOLVING SYSTEMS OF EQUATIONS BY DETERMINANTS

**1.** 26　**3.** $-19$　**5.** 0　**7.** $(1, 2)$　**9.** $\{(x, y) \mid 3x + y = 1\}$　**11.** $(9, 9)$　**13.** 8　**15.** 0　**17.** 54　**19.** $(-2, 0, 5)$　**21.** $(6, -2, 4)$　**23.** 16

**25.** 15　**27.** $\dfrac{13}{6}$　**29.** 0　**31.** 56　**33.** $(-3, -2)$　**35.** $\varnothing$　**37.** $(-2, 3, -1)$　**39.** $(3, 4)$　**41.** $(-2, 1)$　**43.** $\{(x, y, z) \mid x - 2y + z = -3\}$

**45.** $(0, 2, -1)$　**47.** 5　**49.** 0　**51.** $\begin{array}{cccc} + & - & + & - \\ - & + & - & + \\ + & - & + & - \\ - & + & - & + \end{array}$　**53.** $-125$　**55.** 24

# APPENDIX G MEAN, MEDIAN, AND MODE

**1.** mean: 29, median: 28, no mode　**3.** mean: 8.1, median: 8.2, mode: 8.2　**5.** mean: 0.6, median: 0.6, mode: 0.2 and 0.6
**7.** mean: 370.9, median: 313.5, no mode　**9.** 1214.8 ft　**11.** 1117 ft　**13.** 6.8　**15.** 6.9　**17.** 85.5　**19.** 73　**21.** 70 and 71
**23.** 9　**25.** 21, 21, 24

# APPENDIX H REVIEW OF ANGLES, LINES, AND SPECIAL TRIANGLES

**1.** $71°$　**3.** $19.2°$　**5.** $78\dfrac{3}{4}°$　**7.** $30°$　**9.** $149.8°$　**11.** $100\dfrac{1}{2}°$　**13.** $m\angle 1 = m\angle 5 = m\angle 7 = 110°; m\angle 2 = m\angle 3 = m\angle 4 = m\angle 6 = 70°$

**15.** $90°$　**17.** $90°$　**19.** $90°$　**21.** $45°, 90°$　**23.** $73°, 90°$　**25.** $50\dfrac{1}{4}°, 90°$　**27.** $x = 6$　**29.** $x = 4.5$　**31.** 10　**33.** 12

# Index

# Photo Credits

**Page 104** U.S. Congress, Office of Technology Assessment

**Page 105** AP Wide World Photos

**Page 106** © Rolf Bruderer/Masterfile

**Page 107** © Forestier Yves/CORBIS SYGMA

**Page 110** © Liu Jin/Agence France Presse/Getty Images

**Page 110** © G. Bowater/CORBIS All Rights Reserved

**Page 112** Sean Reid/Alaska Stock

**Page 119** © Buddy Mays/CORBIS All Rights Reserved

**Page 121** Colin Braley/Corbis/Reuters America LLC

**Page 121** © 2005 Norbert Wu/www.norbertwu.com

**Page 122** John Elk III/Stock Boston

**Page 126** Blend Images/Alamy Images

**Page 127** First Light/ImageState/International Stock Photography Ltd.

**Page 132** © vario images GmbH & Co. KG/Alamy

**Page 132** Neal Ulevich/Bloomberg News/Landov LLC

**Page 132** Laurance B. Aiuppy/The Stock Connection

**Page 149** Jeff Greenberg/The Image Works

**Page 160** Catherine Karnow/Woodfin Camp & Associates, Inc.

**Chapter 3**

**Page 165** © Editorial Image, LLC/Alamy

**Page 166** Getty Images/Stockbyte

**Page 174** David Young-Wolff/Stone/Getty Images

**Page 186** © Michael Newman/PhotoEdit

**Page 188** Tom Stillo/Omni-Photo Communications, Inc.

**Page 189** Amy C. Etra/PhotoEdit Inc.

**Page 197** © Tony Freeman/PhotoEdit

**Page 198** © Bill Aron/PhotoEdit

**Page 207** © John Neubaur/PhotoEdit

**Page 213** © Editorial Image, LLC/Alamy

**Page 218** © Ian Shaw/Alamy

**Page 221** © Jim West/The Image Works

**Page 221** Geri Engberg Photography

**Page 222** © Editorial Image, LLC/Alamy

**Chapter 4**

**Page 245** © Jose Luis Pelaez, Inc./CORBIS All Rights Reserved

**Page 260** © Peter Hvizdak/The Image Works

**Page 267** © Image100/CORBIS All Rights Reserved